Unit Conversions (Equivalents)

Length

1 in. = 2.54 cm
1 cm = 0.394 in.
1 ft = 30.5 cm
1 m = 39.37 in. = 3.28 ft
1 mi = 5280 ft = 1.61 km
1 km = 0.621 mi
1 nautical mile (U.S.) = 1.15 mi = 6076 ft = 1.852 km
1 fermi = 1 femtometer (fm) = 10^{-15} m
1 angstrom (Å) = 10^{-10} m
1 light-year (ly) = 9.46×10^{15} m
1 parsec = 3.26 ly = 3.09×10^{16} m

Volume

1 liter (L) = 1000 mL = 1000 cm^3 = $1.0 \times 10^{-3}\,m^3$ = 1.057 quart (U.S.) = 61.02 $in.^3$
1 gallon (U.S.) = 4 qt (U.S.) = 231 $in.^3$ = 3.79 L = 0.83 gal (Imperial)
1 m^3 = 35.31 ft^3

Speed

1 mi/h = 1.47 ft/s = 1.609 km/h = 0.447 m/s
1 km/h = 0.278 m/s = 0.621 mi/h
1 ft/s = 0.305 m/s = 0.682 mi/h
1 m/s = 3.28 ft/s = 3.60 km/h
1 knot = 1.151 mi/h = 0.5144 m/s

Angle

1 radian (rad) = 57.30° = 57°18′
1° = 0.01745 rad
1 rev/min (rpm) = 0.1047 rad/s

Time

1 day = 8.64×10^4 s
1 year = 3.156×10^7 s

Mass

1 atomic mass unit (u) = 1.6605×10^{-27} kg
1 kg = 0.0685 slug
[1 kg has a weight of 2.20 lb where g = 9.81 m/s^2.]

Force

1 lb = 4.45 N
1 N = 10^5 dyne = 0.225 lb

Energy and Work

1 J = 10^7 ergs = 0.738 ft·lb
1 ft·lb = 1.36 J = 1.29×10^{-3} Btu = 3.24×10^{-4} kcal
1 kcal = 4.186×10^3 J = 3.97 Btu
1 eV = 1.602×10^{-19} J
1 kWh = 3.60×10^6 J = 860 kcal

Power

1 W = 1 J/s = 0.738 ft·lb/s = 3.42 Btu/h
1 hp = 550 ft·lb/s = 746 W

Pressure

1 atm = 1.013 bar = 1.013×10^5 N/m^2
= 14.7 $lb/in.^2$ = 760 torr
1 $lb/in.^2$ = 6.90×10^3 N/m^2
1 Pa = 1 N/m^2 = 1.45×10^{-4} $lb/in.^2$

SI Derived Units and Their Abbreviations

Quantity	Unit	Abbreviation	In Terms of Base Units[†]
Force	newton	N	$kg \cdot m/s^2$
Energy and work	joule	J	$kg \cdot m^2/s^2$
Power	watt	W	$kg \cdot m^2/s^3$
Pressure	pascal	Pa	$kg/(m \cdot s^2)$
Frequency	hertz	Hz	s^{-1}
Electric charge	coulomb	C	$A \cdot s$
Electric potential	volt	V	$kg \cdot m^2/(A \cdot s^3)$
Electric resistance	ohm	Ω	$kg \cdot m^2/(A^2 \cdot s^3)$
Capacitance	farad	F	$A^2 \cdot s^4/(kg \cdot m^2)$
Magnetic field	tesla	T	$kg/(A \cdot s^2)$
Magnetic flux	weber	Wb	$kg \cdot m^2/(A \cdot s^2)$
Inductance	henry	H	$kg \cdot m^2/(s^2 \cdot A^2)$

[†]kg = kilogram (mass), m = meter (length), s = second (time), A = ampere (electric current).

Metric (SI) Multipliers

Prefix	Abbreviation	Value
exa	E	10^{18}
peta	P	10^{15}
tera	T	10^{12}
giga	G	10^{9}
mega	M	10^{6}
kilo	k	10^{3}
hecto	h	10^{2}
deka	da	10^{1}
deci	d	10^{-1}
centi	c	10^{-2}
milli	m	10^{-3}
micro	μ	10^{-6}
nano	n	10^{-9}
pico	p	10^{-12}
femto	f	10^{-15}
atto	a	10^{-18}

P404

Condition	Name of Pupil	Grade	T'chr
	Kevin Cooper		
	Yaakov Malamet	11	
	Sophia Shoulson	11	
	Danielle Clanaman		
	Joey Ruddel		
	Zach Lipstein	11	

PHYSICS

PRINCIPLES WITH APPLICATIONS

Fifth Edition

Douglas C. Giancoli

PRENTICE HALL, Upper Saddle River, New Jersey 07458

Library of Congress Cataloging-in-Publication Data

Giancoli, Douglas C.
    Physics: principles with applications / Douglas C. Giancoli. —
5th ed.

        p.   cm.
    Includes index.
    ISBN 0-13-611971-9
    1. Physics. I. Title.
QC23.G399 1998
530 — dc21   97-28516
                                                       CIP

Editor-in-Chief: Paul F. Corey
Development Editors: Ray Mullaney, David Chelton
Production Editor: Susan Fisher
Executive Editor: Alison Reeves
Editorial Director: Tim Bozik
Director of Marketing: Kelly McDonald
Senior Marketing Manager: Leslie Cavaliere
Marketing Liaison: Ramona Sherman
Assistant Editor: Wendy Rivers
Assistant Vice President of Production and Manufacturing: David W. Riccardi
Executive Managing Editor: Kathleen Schiaparelli
Art Manager: Gus Vibal
Page Layout: Richard Foster, Karen Stephens
Illustrators: Patrice Van Acker, Tamara Newnam Cavallo
Photo Editor: Warren Ramezzana
Creative Director: Paula Maylahn
Art Director: Amy Rosen
Manager, Art and Formatting: John Jordan
Manufacturing Manager: Trudy Pisciotti
Advertising Design Manager: Meghan Dacey
Advertising and Promotions Manager: Elise Schneider
Additional Formatting: Jeff Henn
Production Assistants: Paula Williams, Nancy Gross, Adam Velthaus
Art Assistants: Pat Gutierrez, Charlie Pelletreau
Photo Research Coordinator: Lorinda Morris-Nantz
Photo Research Administrator: Melinda Reo
Photo Research: Tobi Zausner, Mary Teresa Giancoli, Rhoda Sidney
Editorial Assistants: Marilyn Coco, Jo Marie Jacobs
Cover photo: Malcom Hanes Pressenbild/Adventure Photo & Film

Photo credits appear on page 1095 which constitutes
a continuation of the copyright page.

Printed in the United States of America

15   14   13   12   11

ISBN 0-13-611971-9

Prentice-Hall International (UK) Limited, *London*
Prentice-Hall of Australia Pty. Limited, *Sydney*
Prentice-Hall Canada Inc., *Toronto*
Prentice-Hall Hispanoamericana, S.A., *Mexico City*
Prentice-Hall of India Private Limited, *New Delhi*
Prentice-Hall of Japan, Inc., *Tokyo*
Pearson Education Asia Pte. Ltd., *Singapore*
Editora Prentice-Hall do Brasil, Ltda., *Rio de Janeiro*

C O N T E N T S

PREFACE

A Fifth Edition?

By the fifth edition of an algebra-based physics text, one might expect that the author has at last gotten it right.

I hope the earlier editions weren't all that wrong. The idea of a new edition is to improve, to bring in material, and perhaps to delete material that makes the book longer but isn't all that useful. All these things have been done:

Physics. Physics itself may not change all that rapidly, but over the span of a few years, there may be some new discoveries to include, such as

- planets revolving around distant stars
- information gathered by the Hubble Space Telescope
- updates in particle physics and cosmology (e.g., age of universe).

Pedagogy. One aspect of physics that is changing fairly rapidly is research on how students learn. As a result, this new edition contains some new elements:

Conceptual Examples, an average of 2 or 3 per chapter, are each a sort of brief Socratic question and answer. It is intended that readers will be stimulated by the questions to think, or reflect, and come up with a response—before reading the Response given. Here are a few:

- Velocity vs. acceleration (Chapter 2)
- What exerts the force on a car? (Chapter 4)
- Apple and the wagon (reference frames and projectile motion, Chapter 3)
- Which object rolls down a plane faster? (Chapter 8)
- Finger on a full straw (Chapter 10)
- Suction (Chapter 10)
- Boiling pasta (Chapter 14)
- Electric shielding/safety from lightning (Chapter 16)
- Which part of the photo is the reflection? (Chapter 23).

Estimating Examples, also a new feature of this edition, are intended to show how to make order-of-magnitude estimates even when the data are scarce, even when you might never have guessed that any result was possible at all.

Problem Solving has not been slighted in the least. There are many new worked-out Examples and here are some highlights:

- Air bags (2-10)
- Bungee jumper (6-14)
- Computer hard drive (8-4)
- Loudspeaker (11-7)
- Photocopier (16-5)
- Age of archeological bone (30-10).

Some of the new Examples have replaced older less useful ones. Many other Examples have been improved by more detailed reasoning, by displaying more mathematical steps, and by improving the ambience to make them more real-world and so more inviting and interesting.

New problems have been added and many of the old ones have undergone change.

<u>Example Titles.</u> Examples of all three types (including Conceptual and Estimating) now have titles (for fun and for easy reference).

<u>Emphasized Equations.</u> The great laws of physics are emphasized not only by setting them off, but also by giving them a marginal note in capitals and in a box. The equations that express the great laws, as well as the major equations that one just can't do without, are emphasized with a tan screen.

New Topics in this fifth edition include (these are only a few)

- Rolling motion (Chapter 8)
- Work in rotational motion ($W = \tau\theta$) (Chapter 8)
- v and a for simple harmonic motion (Chapter 11)
- Highway mirages (Chapter 24)
- Hubble Space Telescope (several places)
- Higgs Boson, Symmetry (explained), Supersymmetry (Chapter 32).

Diagrams. There are many more diagrams (over 200 new ones, for an increase of 20 percent), a lot of them to go with Examples and with Problems. Many of the old diagrams have been improved with more realistic backgrounds and figures, and more detail, and the use of photorealistic art.

Photographs. Many of the chapter opening photos now have vectors or other analysis superimposed on them—to give students a richer feeling for the physics. These are visual images of physics that will be fixed in the students' minds.

Many new and interesting photos have been added in the text to bring home the usefulness of physics, a few of which are: diffusion, images in spherical mirrors, depth of field with a camera lens, Hubble Space Telescope, and DNA X-ray diffraction.

Applications. Relevant applications of physics to everyday life and to biology, medicine, architecture, geology, and other fields has always been a strong feature of this book, and continues to be. Among other things, they answer the students' question, "Why study physics?" New applications have been added (and a few older ones dropped), some of which are

- Elevator and counterweight (Chapter 4)
- Shielding (Chapters 16 and 20)
- Dry cell (Chapter 18)
- Aurora borealis (Chapter 20)
- Induction stove (Chapter 21)
- TV and radio antennas (Chapter 22)
- CD player, laser and disk (Chapter 28)
- Smoke detectors (Chapter 29)

and some already mentioned earlier such as airbags, bungee jumping, photocopiers, highway mirages, and computer hard drives.

Revised Physics. No topic, no paragraph in this book was overlooked in the search to improve the clarity of the presentation. Many changes and clarifications have been made, both small and not so small. Here are just a few of the more important ones:

- New tables of typical lengths, times, masses (Chapter 1) and voltages (Chapter 17)
- Chapter 2: rearranged presentation of displacement, velocity, and reference frames
- New diagrams to aid understanding of velocity and acceleration (Chapter 2)
- Unit conversion moved to Chapter 1
- Relative velocity moved to end of Chapter 3
- Simplified introduction to Newton's second law (Chapter 4)
- Simple machines: pulley (Chapter 4), lever (Chapter 9), hydraulic lift (Chapter 10)
- New Section: "Car rounding a curve" (Chapter 5)
- Period and frequency introduced earlier (Chapter 5)
- Work and energy reworked in general, and potential energy especially dealt with in more detail (Chapter 6)
- Angular momentum simplified a bit, especially vector aspects (Chapter 8)
- More formulas for moment of inertia (Fig. 8-20)
- Rotating reference frames, inertial forces, and Coriolis removed from Chapter 8 to an Appendix
- Greatly simplified vertical spring derivation (Chapter 11)
- Energy transported by waves simplified, with more difficult parts in optional Sections (Chapters 11 and 12)
- Speed of light measurement moved from Chapter 23 on Optics to Chapter 22 on EM waves
- Magnifying glass reworked (Chapter 23)
- Relativistic momentum reworked and in more detail (Chapter 26)
- New energy state diagrams for complex atoms (Chapter 28)
- New results in elementary particle physics and cosmology (Chapters 32 and 33).

Page Layout: Complete Derivations. Serious attention has been paid to how each page was formatted, especially for page turns. Great effort has been made to keep important derivations and arguments on facing pages. Thus readers don't have to turn back to check. More importantly, throughout the book, readers see before them, on two facing pages, an important slice of physics. On rare occasions when an argument related to a particular figure requires a page turn, that figure is repeated after the page turn so readers won't have to look back.

Deletions. With all of these additions, something had to go to keep the book from getting too long. Some new Examples simply replaced less interesting old ones. The treatment of quite a few topics was shortened and some were simply dropped. Here are some of the deletions: derived vs. base units; operational definitions (general details); section on "Laws or Definitions" dropped (kept a tiny bit earlier in Chapter 4); vector nature of angular quantities (greatly shortened); Reynolds number, sedimentation, Stokes' equation; flow in tubes (halved); Olber's paradox.

Scope of this Book

This book is written for students. The two motivating factors are to give students a thorough understanding of the basic concepts of physics and, by means of interesting applications, to prepare them to use physics in their own lives and professions. In particular, this book is written for students who are taking a one-year introductory course in physics that uses algebra and trigonometry but not calculus. Many of these students have as their main interest biology, (pre)medicine, architecture, technology, or the earth and environmental sciences. This book contains a wide range of applications to these and other fields, as well as to everyday life. These applications answer that common student query, "Why must I study physics?" The answer, of course, is that physics comes into play in all these fields very importantly, and here they can see how. Physics is all about us. Indeed, it is the goal of this book to help students see the world through eyes that know physics.

Before the applications must come the physics. And this new edition, even more than previous editions, aims to explain physics in a readable and interesting manner that is accessible and clear. It aims to teach students by anticipating their needs and difficulties, but without oversimplifying.

General Approach

This book offers an in-depth presentation of physics, and retains the basic approach of the earlier editions. Rather than using the common, dry, dogmatic approach of treating topics formally and abstractly first, and only later relating the material to the students' own experience, my approach is to recognize that physics is a description of reality and thus to start each topic with concrete observations and experiences that students can directly relate to. Then we move on to the generalizations. Not only does this make the material more interesting and easier to understand, but it is closer to the way physics is actually practiced.

I have sought, where possible, to present the basic concepts of physics in their historical and philosophic context.

As mentioned above, this book includes a wide range of examples and applications from other fields: biology, medicine, architecture, technology, earth sciences, the environment, and daily life. Some applications serve only as examples of physical principles. Others are treated in depth, with whole Sections devoted to them (among these are the study of medical imaging systems, constructing arches and domes, and the effects of radiation). But applications do not dominate the text—this is, after all, a physics book. They have been carefully chosen and integrated into the text so as not to interfere with the development of the physics but rather illuminate it. You won't find essay sidebars here. The applications are integrated right into the physics. Even when an application gets a separate Section all to itself, it is directly tied to the physics just studied. To make it easy to spot these applications, a new *Physics Applied* marginal note has been added.

Mathematics can be an obstacle to student understanding. To avoid frightening students with an initial chapter on mathematics, I have instead incorporated many important mathematical tools, such as addition of vectors and trigonometry, directly in the text where first needed. In addition, the appendices contain a review of many mathematical topics such as algebra and geometry, as well as dimensional analysis. A few advanced topics are also given an Appendix: Rotating frames of reference, Inertial forces, Coriolis effect; Gauss's law; Galilean and Lorentz transformations.

It is necessary, I feel, to pay careful attention to detail, especially when deriving an important result. I have aimed at including all steps in a derivation, and have tried to make clear which equations are general, and which are not, by explicitly stating the limitations of important equations in brackets next to the equation, such as

$$x = x_0 + v_0 t + \frac{1}{2} at^2. \qquad \text{[constant acceleration]}$$

Difficult language, too, can hinder understanding: and I have tried to write in a relaxed style, avoiding jargon, and often talking directly to the students. New or unusual terms are carefully defined when first used.

Color is used pedagogically to bring out the physics. Different types of vectors are given different colors (see the chart on page xxiii). There are many new diagrams to illustrate new Examples (and old ones too) and to enrich the text and problems. The fifth edition features new and revised art—including new photorealistic art, more illustrations to accompany the in-text Examples and end-of-chapter problems, and dozens of new photos.

Problem Solving

Strong attention is given to problem solving. Learning how to approach and solve problems is a basic part of a physics course, and is a highly useful skill in itself. Solving problems is also important because the process brings understanding of the physics. Scattered throughout the book are special Sections and special Boxes devoted to how to approach the solving of problems. Many are found in the early chapters, where students first begin wrestling with problem solving; but many are also found later in the book, throughout mechanics, and in electricity, for example, where problem solving is an emphasized issue, as well as in thermodynamics and in optics. These Problem Solving Boxes provide a summary of how to approach problem solving. They do *not* provide a prescription to be followed. Hence they are often placed *after* a few Examples have been done, as a sort of summary of how we have been approaching Problems.

Over 400 Examples are fully worked out in the text. In this new edition, there are three types of Examples: regular worked Examples, Estimating Examples, and Conceptual Examples. The regular Examples are fully worked out in the text, and most are accompanied by analytical drawings. These Examples are designed to help students develop problem-solving skills and range from simple to fairly complicated. Estimating Examples encourage student analysis and understanding by using "back of the envelope" estimations as a problem-solving technique; they increase awareness of the power of analytical thinking. Conceptual Examples, in contrast to numerical problem solving and the application of formulas, challenge students to explore the basic concepts that are fundamental to understanding physics. Many Examples are taken from everyday life and aim at being realistic applications of physics principles.

There are over 3100 end-of-chapter exercises, including more than 700 questions that require verbal answers based on an understanding of the concepts, and about 2400 problems involving mathematical calculation.

Each chapter contains a large group of problems arranged by Section and graded according to difficulty: level I problems are simple, usually plug-in types, designed to give students confidence; level II are normal problems, requiring more thought and often the combination of two different concepts; level III are the most difficult and serve as a challenge to superior students. The arrangement by Section number means only that those problems depend on material up to and including that Section: earlier material may also be relied upon. The ranking of problems by difficulty (I, II, III) is intended only as a guide.

I suggest that instructors assign a significant number of the level I and level II problems, and reserve level III problems to stimulate the best students. Although most level I problems may seem easy, they help to build self-confidence—an important part of learning, especially in physics.

Each chapter also contains a group of "General Problems" which are unranked and not arranged by Section number.

Answers to odd-numbered problems are given at the back of the book. Throughout the text, *Système International* (SI) units are used. Other metric and British units are defined for informational purposes.

Organization

The general outline of this new edition retains a traditional order of topics: mechanics (Chapters 1 to 12), including vibrations, waves, and sound, followed by kinetic theory and thermodynamics (Chapters 13 to 15), electricity and magnetism (Chapters 16 to 22), light (Chapters 23 to 25), and modern physics (Chapters 26 to 33). Nearly all topics customarily taught in introductory physics courses are included here.

The tradition of beginning with mechanics is sensible, I believe, because it was developed first, historically, and because so much else in physics depends on it. Within mechanics, there are various ways to order topics, and this book allows for considerable flexibility. I prefer, for example, to cover statics after dynamics, partly because many students have trouble with the concept of force without motion. Besides, statics is a special case of dynamics—we study statics so that we can prevent structures from becoming dynamic (falling down)— and that sense of being at the limit of dynamics is intuitively helpful. Nonetheless statics (Chapter 9) can be covered earlier, if desired, before dynamics, after a brief introduction to vectors. Another option is light, which I have placed after electricity and magnetism and EM waves. But light could be treated immediately after the chapters on waves (Chapter 11 and 12). Special relativity (Chapter 26), which is located along with the other chapters on modern physics, could instead be treated along with mechanics—say, after Chapter 7.

Not every chapter need be given equal weight. Whereas Chapter 4 or Chapter 21 might require $1\frac{1}{2}$ to 2 weeks of coverage, Chapter 12 or 22 may need only $\frac{1}{2}$ week.

The book contains more material than can be covered in most one-year courses, so instructors have flexibility in choice of topics. Sections marked with a star (asterisk) are considered optional (if not covered in class, they can be a resource for later study). These Sections contain slightly more advanced physics material, often material not usually covered in typical courses, and/or interesting applications. They contain no material needed in later chapters (except perhaps in later optional Sections). This does not imply that all nonstarred sections must be covered: there still remains considerable flexibility in the choice of material. For a brief course, all optional material could be dropped as well as major parts of Chapters 10, 12, 19, 22, 28, 29, 32, and 33, as well as selected parts of Chapters 7, 8, 9, 15, 21, 24, 25, and 31.

Thanks

More than 60 physics professors provided direct feedback on every aspect of the text: organization, content, figures, and suggestions for new Examples and Problems. These reviewers for this fifth edition are listed below. I owe each of them a debt of gratitude:

David B. Aaron (South Dakota State University)
Zaven Altounian (McGill University)
Atam P. Arya (West Virginia University)
David E. Bannon (Chemeketa Community College)
Jacob Becher (Old Dominion University)
Michael S. Berger (Indiana University, Bloomington)
Donald E. Bowen (Stephen F. Austin University)
Neal M. Cason (University of Notre Dame)
H. R. Chandrasekhar (University of Missouri)
Ram D. Chaudhari (SUNY—Oswego)
K. Kelvin Cheng (Texas Tech University)
Lowell O. Christensen (American River College)
Mark W. Plano Clark (Doane College)
Irvine G. Clator (UNC, Wilmington)
Scott Cohen (Portland State University)
Lattie Collins (East Tennessee State University)
Sally Daniels (Oakland University)
Jack E. Denson (Mississippi State University)
Eric Dietz (California State University, Chico)
Paul Draper (University of Texas, Arlington)
Miles J. Dresser (Washington State University)
F. Eugene Dunnam (University of Florida)
Gregory E. Francis (Montana State University)
Philip Gash (California State University, Chico)
J. David Gavenda (University of Texas, Austin)
Grant W. Hart (Brigham Young University)
Melissa Hill (Marquette University)
Mark Hillery (Hunter College)
Hans Hochheimer (Colorado State University)
Alex Holloway (University of Nebraska, Omaha)

James P. Jacobs (University of Montana)
Larry D. Johnson (Northeast Louisiana University)
David Lamp (Texas Tech University)
Paul Lee (University of California, Northridge)
Daniel J. McLaughlin (University of Hartford)
Victor Montemeyer (Middle Tennessee State Univ.)
Dennis Nemeschansky (USC)
Robert Oakley (University of Southern Maine)
Robert Pelcovits (Brown University)
Brian L. Pickering (Laney College)
T.A.K. Pillai (University of Wisconsin, La Crosse)
Michael Ram (University of Buffalo)
David Reid (Eastern Michigan University)
Charles Richardson (University of Arkansas)
Lawrence Rowan (UNC, Chapel Hill)
Roy S. Rubins (University of Texas, Arlington)
Thomas Sayetta (East Carolina University)
Neil Schiller (Ocean County College)
Juergen Schroeer (Illinois State University)
Marc Sher (College of William and Mary)
James P. Sheerin (Eastern Michigan University)
Donald Sparks (Los Angeles Pierce College)
Michael G. Strauss (University of Oklahoma)
Harold E. Taylor (Stockton State University)
Michael Thoennessen (Michigan State University)
Linn D. Van Woerkom (Ohio State University)
S. L. Varghese (University of South Alabama)
Robert A. Walking (University of Southern Maine)
Lowell Wood (University of Houston)
David Wright (Tidewater Community College)

I am grateful also to all those physicist-reviewers of the earlier editions:

Narahari Achar (Memphis State University)
William T. Achor (Western Maryland College)
Arthur Alt (College of Great Falls)
Zaven Altounian (McGill University)
John Anderson (University of Pittsburgh)
Subhash Antani (Edgewood College)
Sirus Arvainejad (Eastern Illinois University)
Charles R. Bacon (Ferris State University)
Arthur Ballato (Brookhaven National Laboratory)
Gene Barnes (California State U., Sacramento)
Isaac Bass
Paul A. Bender (Washington State University)
Joseph Boyle (Miami–Dade Community College)
Peter Brancazio (Brooklyn College, CUNY)
Michael E. Browne (University of Idaho)
Michael Broyles (Collin County Community College)

Anthony Buffa (Cal Poly S.L.O.)
David Bushnell (Northern Illinois University)
Albert C. Claus (Loyola University of Chicago)
Lawrence Coleman (Univ. of California, Davis)
Waren Deshotels (Marquette University)
Frank Drake (Univ. of California, Santa Cruz)
Miles J. Dresser (Washington State University)
Ryan Droste (The College of Charleston)
Frank A. Ferrone (Drexel University)
Len Feuerhelm (Oklahoma Christian University)
Donald Foster (Wichita State University)
Philip Gash (California State University, Chico)
Simon George (California State Univ., Long Beach)
James Gerhart (University of Washington)
Bernard Gerstman (Florida International Univ.)
Charles Glashausser (Rutgers University)

Hershel J. Hausman (Ohio State University)
Laurent Hodges (Iowa State University)
Joseph M. Hoffman (Frostburg State University)
Peter Hoffmann-Pinther (U. of Houston–Downtown)
Fred W. Inman (Mankato State University)
M. Azad Islan (State Univ. of New York—Potsdam)
James P. Jacobs (Seattle University)
Gordon Jones (Mississippi State University)
Rex Joyner (Indiana Institute of Technology)
Sina David Kaviani (El Camino College)
Joseph A. Keane (St. Thomas Aquinas College)
Kirby W. Kemper (Florida State University)
Sanford Kern (Colorado State University)
James E. Kettler (Ohio University–Eastern Campus)
James R. Kirk (Edinboro University)
Alok Kuman (State Univ. of New York—Oswego)
Sung Kyu Kim (Macalester College)
Amer Lahamer (Berea College)
Clement Y. Lam (North Harris College)
Peter Landry (McGill University, Montreal)
Michael Lieber (University of Arkansas)
Bryan H. Long (Columbia State College)
Michael C. LoPresto (Henry Ford Com. College)
James Madsen (University of Wisconsin, River Falls)
Ponn Mahes (Winthrop University)
Robert H. March (University of Wisconsin–Madison)
David Markowitz (University of Connecticut)
E. R. Menzel (Texas Tech University)
Robert Messina
David Mills (College of the Redwoods)
George K. Miner (University of Dayton)
Marina Morrow (Lansing Community College)
Ed Nelson (University of Iowa)
Gregor Novak (Indiana Univ./Purdue Univ.)
Roy J. Peterson (University of Colorado–Boulder)
Frederick M. Phelps (Central Michigan University)

T. A. K. Pillai (University of Wisconsin–La Crosse)
John Polo (Edinboro University of Pennsylvania)
W. Steve Quon (Ventura College)
John Reading (Texas A&M)
William Riley (Ohio State University)
Larry Rowan (University of North Carolina)
R. S. Rubins (University of Texas, Arlington)
D. Lee Rutledge (Oklahoma State University)
Hajime Sakai (Univ. of Massachusetts at Amherst)
Ann Schmiedekamp (Penn State U., Ogontz Campus)
Mark Semon (Bates College)
Eric Sheldon (University of Massachusetts–Lowell)
K. Y. Shen (California State University, Long Beach)
Joseph Shinar (Iowa State University)
Thomas W. Sills (Wilbur Wright College)
Anthony A. Siluidi (Kent State University)
Michael A. Simon (Housatonic Community College)
Upindranath Singh (Embry-Riddle)
Michael I. Sobel (Brooklyn College)
Thor F. Stromberg (New Mexico State University)
James F. Sullivan (University of Cincinnati)
Kenneth Swinney (Bevill State Community College)
John E. Teggins (Auburn Univ. at Montgomery)
Colin Terry (Ventura College)
Jagdish K. Tuli (Brookhaven National Laboratory)
Kwok Yeung Tsang (Georgia Institute of Technology)
Paul Urone (CSU, Sacramento)
Jearl Walker (Cleveland State University)
Jai-Ching Wang (Alabama A&M University)
John C. Wells (Tennessee Technological)
Gareth Williams (San Jose State University)
Thomas A. Weber (Iowa State University)
Wendall S. Williams (Case Western Reserve Univ.)
Jerry Wilson (Metropolitan State College at Denver)
Peter Zimmerman (Louisiana State University)

I owe special thanks to Irv Miller for working out all the problems and for managing the team that also worked out the problems, each checking the others, and finally, producing the answers at the back of the book as well as producing the Solutions Manual.

I am grateful to Paul Draper, Robert Pelcovits, Gregory E. Francis, and James P. Jacobs, who inspired many of the Conceptual Examples, as well as suggestions for Questions and Problems. I wish also to thank Professors Howard Shugart, John Heilbron, Joe Cerny, and Roger Falcone for helpful discussions, and for hospitality at the University of California, Berkeley, Physics Department. Thanks also to Prof. Tito Arecchi at the Istituto Nazionale di Ottica, Florence, Italy, and to the staff of the Institute for the History of Science, Florence, for their kind hospitality.

Finally, I am most grateful to the many people at Prentice Hall with whom I worked on this project, especially Susan Fisher, Marilyn Coco, David Chelton, Tim Bozik, Gary June, Kathleen Schiaparelli, Richard Foster, Patrice Van Acker and Dave Riccardi. And special thanks to Paul Corey for guiding this project at every stage with clarity and that rare gift of "getting things done," and to Ray Mullaney, whose high level of dedication through edition after edition has helped make this a clear and accurate book. The final responsibility for all errors lies with me, of course. I welcome comments and corrections.

Douglas C. Giancoli

Supplements

For the Instructor

Instructor's Solutions Manual by Irvin A. Miller *Print version*
(0-13-627985-6); *Electronic Versions:* Windows (0-13-627993-7);
Macintosh (0-13-628009-9)
Contains detailed, worked solutions to every problem in the text by Irv
Miller of Drexel University.

Answers to Questions
Prepared by Michelle Rallis and Kurt Reibel of The Ohio State University,
Columbus and Gordon Aubrecht of The Ohio State University, Marion,
this supplement contains answers to all end-of-chapter questions.

Transparency Pack (0-13-628041-2)
Includes 400 four-color transparencies—nearly twice the number of images
as the previous edition.

Test Item File by Bo Lou (0-13-628017-X)
Over 2,400 multiple-choice test questions—30% new! Many new conceptual problems have been added for the Fifth Edition.

Prentice Hall Custom Test
Windows (0-13-628025-0); Macintosh (0-13-628033-1)
Based on the powerful testing technology developed by Engineering Soft-
ware Associates, Inc. (ESA); Prentice Hall Custom Test allows instructors to
create and tailor exams to their own needs. With the Online Testing Program,
exams can also be administered online and data can then be automatically
transferred for evaluation. A comprehensive desk reference guide is included along with online assistance.

For the Student

Student Study Guide by Joseph Boyle (0-13-627944-9)
Complements the strong pedagogy in Giancoli's text with overviews, topic
summaries and exercises, key phrases and terms, self-study exams, and questions for review of each chapter.

MCAT Study Guide by Joseph Boone (0-13-627951-1)
A thoroughly revised study resource that references all of the physics topics
on the MCAT to the appropriate sections in the text. Additional review,
review questions, and problems are provided.

Physics on the Internet: A Student's Guide
by Andrew Stull and Carl Adler (0-13-890153-8)
The perfect tool to help students take advantage of the *Physics: Principles
and Applications, Fifth Edition* Web page. This useful resource gives clear
steps to access Prentice Hall's regularly updated physics resources, along with
an overview of general navigation strategies. Available FREE for students
when purchased in a special package with Giancoli's text.

**Prentice Hall/*The New York Times*
Themes of theTimes—Physics**
This unique newspaper supplement brings together a collection of the latest physics-related articles from the pages of *The New York Times.*

Media Supplements

Interactive Journey through Physics by Cindy Schwarz, Vassar College, CD-ROM for Windows and Macintosh, ©1997 (0-13-254103-3)
Whether your students are interested in exploring fascinating physics concepts, improving their grades, or reviewing for the MCAT, *Interactive Journey through Physics* will augment the traditional learning experiences of lecture, lab, and text.

Interactive Physics Player Workbook by Cindy Schwarz,
Windows book/disk (0-13-667312-0); Macintosh book/disk (0-13-477670-4)
An easy way to use *Interactive Physics* in your courses, this highly interactive workbook/disk package contains 40 simulation projects of varying degrees of difficulty. Each contains a physics review, simulation details, hints, explanation of results, math help, and a self test.

Physics Explorer Runtime Version by LOGAL,
Windows (0-13-627969-4); Macintosh (0-13-627977-5)
Tailored for use with Giancoli's text, *Physics Explorer Runtime Version* contains simulations of over 100 problems and examples directly from Giancoli's text. Students can conduct experiments, interactively record results on a spreadsheet, and generate graphs using each of ten independent learning models:
Particle Mechanics—One Body, Two Body, Gravity, and Harmonic Motion
Wave Mechanics—Waves, Ripple Tank, Diffraction
Electricity and Magnetism—One Body Electrodynamics, AC/DC Circuits, Electrostatics.

Presentation Manager CD-Rom
This CD-ROM contains all the text art and videos from the Physics You Can See video tape as well as additional lab and demonstration videos and animations from the *Interactive Journey through Physics* CD-ROM.

Physics: Principles and Applications Web Site
http://www.prenhall.com/giancoli
Features include practice tests with on-line feedback/grading keyed to the text.

Physics You Can See *Videos* (0-205-12393-7)
Each two- to five-minute segment demonstrates a classical physics experiment. Includes 11 segments such as "Coin and Feather" (acceleration due to gravity); "Monkey & Gun" (projectile motion); "Swivel Hips" (force pairs); and "Collapse a Can" (atmospheric pressure).

NOTES TO STUDENTS AND INSTRUCTORS ON THE FORMAT

1. Sections marked with a star (*) are considered optional. They can be omitted without interrupting the main flow of topics. No later material depends on them except possibly later starred sections. They may be fun to read though.

2. The customary conventions are used: symbols for quantities (such as m for mass) are italicized, whereas units (such as m for meter) are not italicized. Boldface (**F**) is used for vectors.

3. Few equations are valid in all situations. Where practical, the limitations of important equations are stated in square brackets next to the equation. The equations that represent the great laws of physics are displayed with a tan background, as are a few other equations that are so useful that they are indispensable.

4. The number of significant figures (see Section 1–4) should not be assumed to be greater than given: if a number is stated as (say) 6, with its units, it is meant to be 6 and not 6.0 or 6.00.

5. At the end of each chapter is a set of questions that students should attempt to answer (to themselves at least). These are followed by problems which are ranked as level I, II, or III, according to estimated difficulty, with level I problems being easiest. These problems are arranged by Section, but problems for a given Section may depend on earlier material as well. There follows a group of General Problems, which are not arranged by Section nor ranked as to difficulty. Questions and problems that relate to optional Sections are starred.

6. Being able to solve problems is a crucial part of learning physics, and provides a powerful means for understanding the concepts and principles of physics. This book contains many aids to problem solving: (a) worked-out Examples and their solutions in the text, which are set off with a vertical blue line in the margin, and should be studied as an integral part of the text; (b) special "Problem-solving boxes" placed throughout the text to suggest ways to approach problem solving for a particular topic—but don't get the idea that every topic has its own "techniques," because the basics remain the same; (c) special problem-solving Sections (marked in blue in the Table of Contents); (d) marginal notes (see below), many of which refer to hints for solving problems, in which case they are so indicated; (e) problems themselves at the end of each chapter (see point 5 above); (f) some of the worked-out Examples are Estimation Examples, which show how rough or approximate results can be obtained even if the given data are sparse (see Section 1–7).

7. Conceptual Examples look like ordinary Examples but are conceptual rather than numerical. Each proposes a question or two, which hopefully starts you to think and come up with a response. Give yourself a little time to come up with your own response before reading the Response given.

8. Marginal notes: brief notes in the margin of almost every page are printed in blue and are of four types: (a) ordinary notes (the majority) that serve as a sort of outline of the text and can help you later locate important concepts and equations; (b) notes that refer to the great laws and principles of physics, and these are in capital letters and in a box for emphasis; (c) notes that refer to a problem-solving hint or technique treated in the text, and these say "Problem Solving"; (d) notes that refer to a physics application in the text or an Example, and these say "Physics Applied."

9. This book is printed in full color. But not simply to make it more attractive. The color is used above all in the figures, to give them greater clarity for our analysis, and to provide easier learning of the physical principles involved. The table on the next page is a summary of how color is used in the figures, and shows which colors are used for the different kinds of vectors, for field lines, and for other symbols and objects. These colors are used consistently throughout the book.

NOTES ON USE OF COLOR

Vectors

A general vector

    resultant vector (sum) is slightly thicker

    components of any vector are dashed

Displacement (**D, r**)

Velocity (**v**)

Acceleration (**a**)

Force (**F**)

    Force on second or

    third object in same figure

Momentum (**p** or m**v**)

Angular momentum (**L**)

Angular velocity (ω)

Torque (τ)

Electric field (**E**)

Magnetic field (**B**)

Electricity and magnetism

Electric field lines

Equipotential lines

Magnetic field lines

Electric charge (+)   + or +

Electric charge (–)   – or –

Electric circuit symbols

Wire

Resistor

Capacitor

Inductor

Battery

Optics

Light rays

Object

Real image
(dashed)

Virtual image
(dashed and paler)

Other

Energy level
(atom, etc.)

Measurement lines   ⊢—1.0 m—⊣

Path of a moving
object

Direction of motion
or current

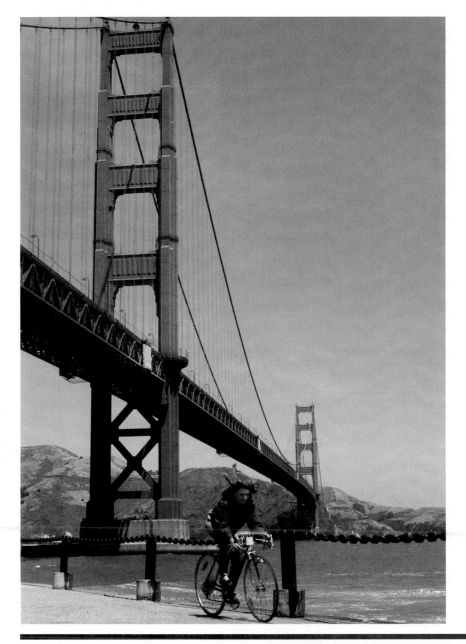

How many aspects of Physics do you see in this photograph? (Partial answer given upside down at bottom of photograph.)

Motion, speed, momentum, work, energy, structures and forces within them (why do they stay up?), rotational motion, torque, angular momentum, fluids, drag forces, friction, wave motion. . . .

CHAPTER 1

INTRODUCTION

Physics is the most basic of the sciences. It deals with the behavior and structure of matter. The field of physics is usually divided into the areas of motion, fluids, heat, sound, light, electricity and magnetism, and the modern topics of relativity, atomic structure, condensed-matter physics, nuclear physics, elementary particles, and astrophysics. We will cover all these topics in this book, beginning with motion (or mechanics, as it is often called). But before we begin on the physics itself, let us take a brief look at how this overall activity called "science," including physics, is actually practiced.

1

FIGURE 1–1 Aristotle is the central figure (dressed in blue) at the top of the stairs (the figure next to him is Plato) in this famous Renaissance portrayal of *The School of Athens*, painted by Raphael around 1510. Also in this painting, considered one of the great masterpieces in art, are Euclid (drawing a circle at the lower right), Ptolemy (extreme right with globe), Pythagoras, Sophocles, and Diogenes.

1–1 Science and Creativity

The principal aim of all sciences, including physics, is generally considered to be the search for order in our observations of the world around us. Many people think that science is a mechanical process of collecting facts and devising theories. This is not the case. Science is a creative activity that in many respects resembles other creative activities of the human mind.

Observation   Let's take some examples to see why this is true. One important aspect of science is *observation* of events. But observation requires imagination, for scientists can never include everything in a description of what they observe. Hence, scientists must make judgments about what is relevant in their observations. As an example, let us consider how two great minds, Aristotle (384–322 B.C.; Fig. 1–1) and Galileo (1564–1642; Fig. 2–15), interpreted motion along a horizontal surface. Aristotle noted that objects given an initial push along the ground (or on a tabletop) always slow down and stop. Consequently, Aristotle believed that the natural state of an object is to be at rest. Galileo, in his reexamination of horizontal motion in the early 1600s, imagined that if friction could be eliminated, an object given an initial push along a horizontal surface would continue to move indefinitely without stopping. He concluded that for an object to be in motion was just as natural as for it to be at rest. By inventing a new approach, Galileo founded our modern view of motion (more details in Chapters 2, 3, and 4), and he did so with a leap of the imagination. Galileo made this leap conceptually, without actually eliminating friction.

Observation and careful experimentation and measurement are one side of the scientific process. The other side is the invention or creation of *theories* to explain and order the observations. Theories are never derived *Theories are creations* directly from observations. They are inspirations that come from the minds of human beings. For example, the idea that matter is made up of atoms (the atomic theory) was certainly not arrived at because someone observed atoms. Rather, the idea sprang from creative minds. The theory of relativity, the electromagnetic theory of light, and Newton's law of universal gravitation were likewise the result of human imagination.

The great theories of science may be compared, as creative achievements, with great works of art or literature. But how does science differ from these other creative activities? One important difference is that science requires *testing* of its ideas or theories to see if their predictions are borne out by experiment.

Although the testing of theories can be considered to distinguish science from other creative fields, it should not be assumed that a theory is "proved" by testing. First of all, no measuring instrument is perfect, so exact confirmation cannot be possible. Furthermore, it is not possible to test a theory for every possible set of circumstances. Hence a theory can never be absolutely "proved." In fact, theories themselves are generally not perfect—a theory rarely agrees with experiment exactly, within experimental error, in every single case in which it is tested. Indeed, the history of science tells us that long-held theories are sometimes replaced by new ones. The process of one theory replacing another is an important subject in the philosophy of science; we can discuss it here only briefly.

Testing a theory

A new theory is accepted by scientists in some cases because its predictions are quantitatively in much better agreement with experiment than those of the older theory. But in many cases, a new theory is accepted only if it explains a greater *range* of phenomena than does the older one. Copernicus's Sun-centered theory of the universe (Fig. 1–2b), for example, was originally no more accurate than Ptolemy's Earth-centered theory (Fig. 1–2a) for predicting the motion of heavenly bodies (Sun, Moon, planets). But Copernicus's theory had consequences that Ptolemy's did not, such as predicting the moonlike phases of Venus. A simpler (or no more complex) and richer theory, one which unifies and explains a greater variety of phenomena, is more useful and beautiful to a scientist. And this aspect,

Theory acceptance

FIGURE 1–2 (a) Ptolemy's geocentric view of the universe. Note at the center the four elements of the ancients: Earth, water, air (clouds around the Earth), and fire; then the circles, with symbols, for the Moon, Mercury, Venus, Sun, Mars, Jupiter, Saturn, the fixed stars, and the signs of the zodiac. (b) An early representation of Copernicus's heliocentric view of the universe with the Sun at the center. (See Chapter 5.)

(a)

(b)

as well as quantitative agreement, plays a major role in the acceptance of a theory.

An important aspect of any theory is how well it can quantitatively predict phenomena, and from this point of view a new theory may often seem to be only a minor advance over the old one. For example, Einstein's theory of relativity gives predictions that differ very little from the older theories of Galileo and Newton in nearly all everyday situations. Its predictions are better mainly in the extreme case of very high speeds close to the speed of light. In this respect, the theory of relativity might be considered as mere "fine-tuning" of the older theory. But quantitative prediction is not the only important outcome of a theory. Our view of the world is affected as well. As a result of Einstein's theory of relativity, for example, our concepts of space and time have been completely altered, and we have come to see mass and energy as a single entity (via the famous equation $E = mc^2$). Indeed, our view of the world underwent a major change when relativity theory came to be accepted.

1–2 Physics and Its Relation to Other Fields

For a long time science was more or less a united whole known as natural philosophy. Not until the last century or two did the distinctions between physics and chemistry and even the life sciences become prominent. Indeed, the sharp distinction we now see between the arts and the sciences is itself but a few centuries old. It is no wonder then that the development of physics has both influenced and been influenced by other fields. For example, the notebooks (Fig. 1–3) of Leonardo da Vinci, the great Renaissance artist, researcher, and engineer, contain the first references to the forces acting within a structure, a subject we consider as physics today; but then, as now, it has great relevance to architecture and building. Early work in electricity that led to the discovery of the electric battery and electric current was done by an eighteenth-century physiologist, Luigi Galvani (1737–1798). He noticed the twitching of frogs' legs in response to an electric spark and later that the muscles twitched when in contact with two dissimilar metals (Chapter 18). At first this phenomenon was known as "animal electricity," but it shortly became clear that electric current itself could exist in the absence of an animal. Later, in the 1930s and 1940s, a number of scientists trained as physicists became interested in applying the ideas and techniques of physics to problems in microbiology. Among the most prominent were Max Delbrück (1906–1981) and Erwin Schrödinger (1887–1961). They hoped, among other things, that studying biological organisms might lead to the discovery of new unsuspected laws of physics. Alas, this hope has not been realized; but their efforts helped give rise to the field we now call molecular biology, which has resulted in a dramatic increase in our understanding of the genetics and structure of living beings.

You do not have to be a research scientist in, say, medicine or molecular biology to be able to use physics in your work. A zoologist, for example, may find physics useful in understanding how prairie dogs and other animals can live underground without suffocating. A physical therapist will do a more effective job if aware of the principles of center of gravity and the action of forces within the human body. A knowledge of the operating principles of optical and electronic equipment is helpful in a variety of fields. Life scientists and architects alike will be interested in the nature of heat loss and gain in human beings and the resulting comfort or discomfort. Architects them-

FIGURE 1–3 Studies on the forces in structures by Leonardo da Vinci (1452–1519).

(a)

(b)

FIGURE 1–4 (a) This Greek temple in Paestum, Italy, was built 2500 years ago, and still stands. (b) Collapse of the Hartford Civic Center in 1978, just two years after it was built.

selves may never have to calculate, for example, the dimensions of the pipes in a heating system or the forces involved in a given structure to determine if it will remain standing (Fig. 1–4). But architects must know the principles behind these analyses in order to make realistic designs and to communicate effectively with engineering consultants and other specialists. From the aesthetic or psychological point of view, too, architects must be aware of the forces involved in a structure—for instability, even if only illusory, can be discomforting to those who must live or work in the structure. Indeed, many of the features we admire in the architecture of the past three millennia were introduced not for their decorative effect but rather for practical purposes. The development of the arch as a means to span a space and at the same time support a heavy load will be discussed in Chapter 9, where we will see that the pointed, or Gothic, arch was not originally a decorative device but a technological development of considerable importance.

The list of ways in which physics relates to other fields is extensive. In the chapters that follow we will discuss many such applications as we carry out our principal aim of explaining basic physics.

1–3 Models, Theories, and Laws

When scientists are trying to understand a particular set of phenomena, they often make use of a **model**. A model, in the scientific sense, is a kind *Models* of analogy or mental image of the phenomena in terms of something we are familiar with. One example is the wave model of light. We cannot see waves of light as we can water waves. But it is valuable to think of light as if it were made up of waves, because experiments indicate that light behaves in many respects as water waves do.

The purpose of a model is to give us an approximate mental or visual picture—something to hold onto—when we cannot see what actually is happening. Models often give us a deeper understanding: the analogy to a known system (for instance, water waves in the above example) can suggest new experiments to perform and can provide ideas about what other related phenomena might occur.

You may wonder what the difference is between a theory and a model. Sometimes the words are used interchangeably. Usually, however, a model is relatively simple and provides a structural similarity to the phenomena being studied, whereas a **theory** is broader, more detailed, and can give quantitatively testable predictions, often with great precision. Sometimes, as a model is developed and modified and corresponds more closely to experiment over a wide range of phenomena, it may come to be referred to as a theory. The atomic theory is an example, as is the wave theory of light.

Theories (vs. models)

Models can be very helpful, and they often lead to important theories. But it is important not to confuse a model, or a theory, with the real system or the phenomena themselves.

Laws

Scientists give the title **law** to certain concise but general statements about how nature behaves (that energy is conserved, for example). Sometimes the statement takes the form of a relationship or equation between quantities (such as Newton's second law, $F = ma$).

and

To be called a law, a statement must be found experimentally valid over a wide range of observed phenomena. In a sense, the law brings a unity to many observations. For less general statements, the term **principle** is often used (such as Archimedes' principle). Where to draw the line between laws and principles is, of course, arbitrary, and there is not always complete consistency.

principles

Scientific laws are different from political laws in that the latter are *prescriptive*: they tell us how we ought to behave. Scientific laws are *descriptive*: they do not say how nature *should* behave, but rather are meant to describe how nature *does* behave. As with theories, laws cannot be tested in the infinite variety of cases possible. So we cannot be sure that any law is absolutely true. We use the term "law" when its validity has been tested over a wide range of cases, and when any limitations and the range of validity are clearly understood. Even then, as new information comes in, certain laws may have to be modified or discarded.

Scientists normally do their work as if the accepted laws and theories were true. But they are obliged to keep an open mind in case new information should alter the validity of any given law or theory.

1–4 Measurement and Uncertainty

In the quest to understand the world around us, scientists seek to find relationships among the various physical quantities they observe and measure.

We may ask, for example, in what way does the magnitude of a force on an object affect the object's speed or acceleration? Or by how much does the pressure of gas in a closed container (such as a tire) change if the temperature is raised or lowered? Scientists normally try to express such relationships quantitatively, in terms of equations whose symbols represent the quantities involved. To determine (or confirm) the form of a relationship, careful experimental measurements are required, although creative thinking also plays a role.

Accurate measurements are an important part of physics. But no measurement is absolutely precise. There is an uncertainty associated with every measurement. Uncertainty arises from different sources. Among the most important, other than blunders, are the limited accuracy of every measuring instrument and the inability to read an instrument beyond some fraction of the smallest division shown. For example, if you were to use a centimeter ruler to measure the width of a board (Fig. 1–5), the result

FIGURE 1–5 Measuring the width of a board with a centimeter ruler. Accuracy is about ±1 mm.

Every measurement has an uncertainty

could be claimed to be accurate to about 0.1 cm, the smallest division on the ruler. The reason is that it is difficult for the observer to interpolate between the smallest divisions, and the ruler itself may not have been manufactured or calibrated to an accuracy very much better than this.

When giving the result of a measurement, it is also important to state the precision, or **estimated uncertainty**, in the measurement. For example, the width of a board might be written as 5.2 ± 0.1 cm. The ± 0.1 cm ("plus or minus 0.1 cm") represents the estimated uncertainty in the measurement, so that the actual width most likely lies between 5.1 and 5.3 cm. The **percent uncertainty** is simply the ratio of the uncertainty to the measured value, multiplied by 100. For example, if the measurement is 5.2 and the uncertainty about 0.1 cm, the percent uncertainty is

Stating the uncertainty

$$\frac{0.1}{5.2} \times 100 = 2\%.$$

Often the uncertainty in a measured value is not specified explicitly. In such cases, the uncertainty is generally assumed to be one or two units (or even three) in the last digit specified. For example, if a length is given as 5.2 cm, the uncertainty is assumed to be about 0.1 cm (or perhaps 0.2 cm). It is important in this case that you do not write 5.20 cm, for this implies an uncertainty on the order of 0.01 cm; it assumes that the length is probably between 5.19 cm and 5.21 cm, when actually you believe it is between 5.1 and 5.3 cm.

Assumed uncertainty

The number of reliably known digits in a number is called the number of **significant figures**. Thus there are four significant figures in the number 23.21 and two in the number 0.062 cm (the zeros in the latter are merely "place holders" that show where the decimal point goes). The number of significant figures may not always be clear. Take, for example, the number 80. Are there one or two significant figures? If we say it is *about* 80 km between two cities, there is only one significant figure (the 8) since the zero is merely a place holder. If it is 80 km within an accuracy of 1 or 2 km, then the 80 has two significant figures. If it is precisely 80 km measured to within ± 0.1 km, then we write 80.0 km.

Which digits are significant?

When making measurements, or when doing calculations, you should avoid the temptation to keep more digits in the final answer than is justified. For example, to calculate the area of a rectangle 11.3 cm by 6.8 cm, the result of multiplication would be 76.84 cm^2. But this answer is clearly not accurate to 0.01 cm^2, since (using the outer limits of the assumed uncertainty for each measurement) the result could be between $11.2 \times 6.7 = 75.04$ cm^2 and $11.4 \times 6.9 = 78.66$ cm^2. At best, we can quote the answer as 77 cm^2, which implies an uncertainty of about 1 or 2 cm^2. The other two digits (in the number 76.84 cm^2) must be dropped since they are not significant. As a general rule, *the final result of a multiplication or division should have only as many digits as the number with the least number of significant figures used in the calculation.* In our example, 6.8 cm has the least number of significant figures, namely two. Thus the result 76.84 cm^2 needs to be rounded off to 77 cm^2.

➡ **PROBLEM SOLVING**

Report only the proper number of significant figures in the final result. An extra digit or two can be kept during the calculation.

Similarly, when adding or subtracting numbers, the final result is no more accurate than the least accurate number used. For example, the result of subtracting 0.57 from 3.6 is 3.0 (and not 3.03). Keep in mind when you use a calculator that all the digits it produces may not be significant. When you divide 2.0 by 3.0, the proper answer is 0.67, and not 0.66666666. Digits should not be quoted (or written down) in a result, unless they are truly significant figures. However, to obtain the most accurate result, it is good

Careful: Electronic calculators err with significant figures

practice to keep an extra significant figure or two throughout a calculation, and round off only in the final result. Note also that calculators sometimes give too few significant figures. For example, when you multiply 2.5×3.2, a calculator may give the answer as simply 8. But the answer is good to two significant figures, so the proper answer is 8.0.

Powers of ten (scientific notation)

It is common in science to write numbers in "powers of ten," or "exponential" notation—for instance 36,900 as 3.69×10^4, or 0.0021 as 2.1×10^{-3}. (For more details, see Appendices A–2 and A–3.) One advantage of exponential notation is that it allows the number of significant figures to be clearly expressed. For example, it is not clear whether 36,900 has three, four, or five significant figures. With exponential notation the ambiguity can be avoided: if the number is known to an accuracy of three significant figures, we write 3.69×10^4, but if it is known to four, we write 3.690×10^4.

CONCEPTUAL EXAMPLE 1–1 **Is the diamond yours?** A friend asks to borrow your precious diamond for a day to show her family. You are a bit worried, so you carefully have your diamond weighed on a scale which reads 8.17 grams. The scale's accuracy is claimed to be ± 0.05 grams. The next day you weigh the returned diamond again, getting 8.09 grams. Is this your diamond?

RESPONSE The scale readings are measurements and do not give the actual value of the mass. Each measurement could have been high or low by up to 0.05 gram or so. The actual mass of your diamond lies most likely between 8.12 grams and 8.22 grams. The actual mass of the returned diamond is most likely between 8.04 grams and 8.14 grams. These two ranges overlap, so there is no reason to doubt that the returned diamond is yours, at least based on the scale readings. (But check the color!)

1–5 Units, Standards, and the SI System

The measurement of any quantity is made relative to a particular standard or **unit**, and this unit must be specified along with the numerical value of the quantity. For example, we can measure length in units such as inches, feet, or miles, or in the metric system in centimeters, meters, or kilometers. To specify that the length of a particular object is 18.6 is meaningless. The unit *must* be given; for clearly, 18.6 meters is very different from 18.6 inches or 18.6 millimeters.

Standard of length (meter)

The first real international standard was the **meter** (abbreviated m) established as the standard of **length** by the French Academy of Sciences in the 1790s. In a spirit of rationality, the standard meter was originally chosen to be one ten-millionth of the distance from the Earth's equator to either pole,[†] and a platinum rod to represent this length was made. (This turns out to be, very roughly, the distance from the tip of your nose to the tip of your longest finger, with arm and hand stretched out horizontally.) In 1889, the meter was defined more precisely as the distance between two finely engraved marks on a particular bar of platinum–iridium alloy. In 1960, to provide greater precision and reproducibility, the meter was redefined as 1,650,763.73 wavelengths of a particular orange light emitted by

---

[†]Modern measurements of the Earth's circumference reveal that the intended length is off by about one-fiftieth of 1 percent. Not bad!

the gas krypton 86. In 1983 the meter was again redefined, this time in terms of the speed of light (whose best measured value in terms of the older definition of the meter was 299,792,458 m/s, with an uncertainty of 1 m/s). The new definition reads: "The meter is the length of path traveled by light in vacuum during a time interval of 1/299,792,458 of a second."[†]

British units of length (inch, foot, mile) are now defined in terms of the meter. The inch (in.) is defined as precisely 2.54 centimeters (cm; 1 cm = 0.01 m). Other conversion factors are given in the table on the inside of the front cover of this book.

Table 1–1 presents some characteristic lengths, from very small to very large.

The standard unit of **time** is the **second** (s). For many years, the second was defined as 1/86,400 of a mean solar day. The standard second is now defined more precisely in terms of the frequency of radiation emit-

TABLE 1–1
Some typical Lengths or Distances (order of magnitude)

Length (or distance)	Meters (approximate)
Neutron or proton (radius)	10^{-15} m
Atom	10^{-10} m
Virus [see Fig. 1–6]	10^{-7} m
Sheet of paper (thickness)	10^{-4} m
Finger width	10^{-2} m
Football field length	10^{2} m
Mt. Everest height [see Fig. 1–6]	10^{4} m
Earth diameter	10^{7} m
Earth to Sun	10^{11} m
Nearest star, distance	10^{16} m
Nearest galaxy	10^{22} m
Farthest galaxy visible	10^{26} m

(a)

(b)

FIGURE 1–6 (a) Some viruses (about 10^{-7} m long) attacking a cell. (b) Mt. Everest's height is on the order of 10^{4} m (8848 m to be precise).

[†]The new definition of the meter has the effect of giving the speed of light the exact value of 299,792,458 m/s.

TABLE 1–2 Some typical Time Intervals

Time interval	Seconds (approximate)
Lifetime of very unstable particle	10^{-23} s
Lifetime of radioactive elements	10^{-22} s to 10^{28} s
Lifetime of muon	10^{-6} s
Time between human heartbeats	10^0 s ($=1$ s)
One day	10^5 s
One year	3×10^7 s
Human life span	2×10^9 s
Length of recorded history	10^{11} s
Humans on Earth	10^{14} s
Life on Earth	10^{17} s
Age of Universe	10^{18} s

TABLE 1–3 Some Masses

Object	Kilograms (approx.)
Electron	10^{-30} kg
Proton, neutron	10^{-27} kg
DNA molecule	10^{-17} kg
Bacterium	10^{-15} kg
Mosquito	10^{-5} kg
Plum	10^{-1} kg
Person	10^2 kg
Ship	10^8 kg
Earth	6×10^{24} kg
Sun	2×10^{30} kg
Galaxy	10^{41} kg

ted by cesium atoms when they pass between two particular states. Specifically, one second is defined as the time required for 9,192,631,770 periods of this radiation. There are, of course, precisely 60 s in one minute (min) and 60 minutes in one hour (h). Note that these two factors of 60 (as well as the 2.54 cm per inch) are definitions and hence have an indefinite number of significant figures. Table 1–2 presents a range of measured time intervals.

Standard of mass (kilogram)

The standard unit of **mass** is the **kilogram** (kg). The standard mass is a particular platinum–iridium cylinder, kept at the International Bureau of Weights and Measures near Paris, France, whose mass is defined as exactly 1 kg. A range of masses is presented in Table 1–3. [For practical purposes, 1 kg weighs about 2.2 pounds.]

When dealing with atoms and molecules, the **unified atomic mass unit** (u) is usually used. In terms of the kilogram

$$1 \text{ u} = 1.6605 \times 10^{-27} \text{ kg}.$$

The definitions of other standard units for other quantities will be given as we encounter them in later chapters.

In the metric system, the larger and smaller units are defined in multiples of 10 from the standard unit, and this makes calculation particularly easy. Thus 1 kilometer (km) is 1000 m, 1 centimeter is $\frac{1}{100}$ m, 1 millimeter (mm) is $\frac{1}{1000}$ m or $\frac{1}{10}$ cm, and so on. The prefixes "centi-," "kilo-," and others are listed in Table 1–4 and can be applied not only to units of length, but to units of volume, mass, or any other metric unit. For example, a centiliter (cL) is $\frac{1}{100}$ liter (L) and a kilogram (kg) is 1000 grams (g).

➡ **PROBLEM SOLVING**

Always use a consistent set of units

SI units

When dealing with the laws and equations of physics it is very important to use a consistent set of units. Several systems of units have been in use over the years. Today the most important is the **Système International** (French for International System), which is abbreviated SI. In SI units, the standard of length is the meter, the standard for time is the second, and the standard for mass is the kilogram. This system used to be called the MKS (meter-kilogram-second) system.

A second metric system is the **cgs system**, in which the centimeter, gram, and second are the standard units of length, mass, and time, as abbreviated in the title. The **British engineering system** takes as its standards the foot for length, the pound for force, and the second for time.

SI units are the principal ones used today in scientific work. We will therefore use SI units almost exclusively in this book, although we will give the cgs and British units for various quantities when introduced.

TABLE 1–4
Metric (SI) Prefixes

Prefix	Abbreviation	Value
exa	E	10^{18}
peta	P	10^{15}
tera	T	10^{12}
giga	G	10^{9}
mega	M	10^{6}
kilo	k	10^{3}
hecto	h	10^{2}
deka	da	10^{1}
deci	d	10^{-1}
centi	c	10^{-2}
milli	m	10^{-3}
micro[†]	μ	10^{-6}
nano	n	10^{-9}
pico	p	10^{-12}
femto	f	10^{-15}
atto	a	10^{-18}

[†]μ is the Greek letter "mu."

1–6 Converting Units

Any quantity we measure, such as a length, a speed, or an electric current, consists of a number *and* a unit. Often we are given a quantity in one set of units, but we want it expressed in another set of units. For example, suppose we measure that a table is 21.5 inches wide, and we want to express this in centimeters. We must use a **conversion factor** which in this case is

$$1 \text{ in.} = 2.54 \text{ cm}$$

or, written another way,

$$1 = 2.54 \text{ cm/in.}$$

Since multiplying by one does not change anything, the width of our table, in cm, is

$$21.5 \text{ inches} = (21.5 \text{ in.}) \times \left(2.54 \frac{\text{cm}}{\text{in.}}\right) = 54.6 \text{ cm}$$

Note how the units (inches in this case) cancelled out. A table containing many unit conversions is found inside the front cover of this book. Let's take some Examples.

EXAMPLE 1–2 **The 100-m dash.** What is the length of the 100-m dash expressed in yards?

SOLUTION Let us assume the distance is accurately known to four significant figures, 100.0 m. One yard (yd) is precisely 3 feet (36 inches), so we can write

$$1 \text{ yd} = 3 \text{ ft} = 36 \text{ in.} = (36 \text{ in.})\left(2.540 \frac{\text{cm}}{\text{in.}}\right) = 91.44 \text{ cm}$$

or,

$$1 \text{ yd} = 0.9144 \text{ m},$$

since 1 m = 100 cm. We can rewrite this result as

$$1 \text{ m} = \frac{1 \text{ yd}}{0.9144} = 1.094 \text{ yd}.$$

Then

$$100 \text{ m} = (100 \text{ m})\left(1.094 \frac{\text{yd}}{\text{m}}\right) = 109.4 \text{ yd},$$

so a 100-m dash is 9.4 yards longer than a 100-yard dash.

EXAMPLE 1–3 **Area of a cell membrane.** A round membrane has an area of 1.25 square inches. Express this in square centimeters.

SOLUTION Because 1 in. = 2.54 cm, then

$$1 \text{ in.}^2 = (2.54 \text{ cm})^2 = 6.45 \text{ cm}^2.$$

So

$$1.25 \text{ in.}^2 = (1.25 \text{ in.}^2)\left(2.54 \frac{\text{cm}}{\text{in.}}\right)^2 = (1.25 \text{ in.}^2)\left(6.45 \frac{\text{cm}^2}{\text{in.}^2}\right) = 8.06 \text{ cm.}^2$$

EXAMPLE 1–4 **Speeds.** Where the posted speed limit is 55 miles per hour (mi/h or mph), what is this speed (*a*) in meters per second (m/s) and (*b*) in kilometers per hour (km/h)?

SOLUTION (*a*) We can write 1 mile as

$$1 \text{ mi} = (5280 \text{ ft})\left(12 \frac{\text{in.}}{\text{ft}}\right)\left(2.54 \frac{\text{cm}}{\text{in.}}\right)\left(\frac{1 \text{ m}}{100 \text{ cm}}\right) = 1609 \text{ m.}$$

Conversion factors = 1

Note that each conversion factor is equal to one. We also know that 1 hour equals (60 min/h) × (60 s/min) = 3600 s/h, so

$$55 \frac{\text{mi}}{\text{h}} = \left(55 \frac{\text{mi}}{\text{h}}\right)\left(1609 \frac{\text{m}}{\text{mi}}\right)\left(\frac{1 \text{ h}}{3600 \text{ s}}\right) = 25 \text{ m/s.}$$

(*b*) Now we use 1 mi = 1609 m = 1.609 km; then

$$55 \frac{\text{mi}}{\text{h}} = \left(55 \frac{\text{mi}}{\text{h}}\right)\left(1.609 \frac{\text{km}}{\text{mi}}\right) = 88 \frac{\text{km}}{\text{h}}.$$

➡ PROBLEM SOLVING

Unit conversion is wrong if units do not cancel

When changing units, you can avoid making an error in the use of conversion factors by checking that units cancel out properly. For example, in our conversion of 1 mi to 1609 m in Example 1–4(*a*), if we had incorrectly used the factor $\left(\frac{100 \text{ cm}}{1 \text{ m}}\right)$ instead of $\left(\frac{1 \text{ m}}{100 \text{ cm}}\right)$, the meter units would not have cancelled out; we would not have ended up with meters.

1–7 Order of Magnitude: Rapid Estimating

We are sometimes interested only in an approximate value for a quantity. This might be because an accurate calculation would take more time than it is worth or would require additional data that are not available. In other cases, we may want to make a rough estimate in order to check an accurate calculation made on a calculator, to make sure that no blunders were made when entering the numbers.

➡ PROBLEM SOLVING

How to make a rough estimate

A rough estimate is made by rounding off all numbers to one significant figure and its power of 10, and after the calculation is made, again only one significant figure is kept. Such an estimate is called an **order-of-magnitude estimate** and can be accurate within a factor of 10, and often better. In fact, the phrase "order of magnitude" is sometimes used to refer simply to the power of 10.

To give you some idea of how useful and powerful rough estimates can be, let us do a few "worked-out Examples."

(a)

(b)

FIGURE 1–7 Example 1–5. (a) How much water is in this lake? (Photo is of one of the Rae Lakes in the Sierra Nevada of California.) (b) Model of the lake as a cylinder. [We could go one step further and estimate the mass or weight of this lake. We will see later that water has a density of 1000 kg/m^3, so this lake has a mass of about $(10^3 \text{ kg/m}^3)(10^7 \text{ m}^3) \approx 10^{10} \text{ kg}$, which is about 10 billion kg or 10 million metric tons. (A metric ton is 1000 kg, about 2200 lbs, slightly larger than a British ton, 2000 lbs.)]

EXAMPLE 1–5 **ESTIMATE** **Volume of a lake.** Estimate how much water there is in a particular lake, Fig. 1–7, which is roughly circular, about 1 km across, and you guess it to have an average depth of about 10 m.

SOLUTION No lake is a perfect circle, nor can lakes be expected to have a perfectly flat bottom. We are only estimating here. To estimate the volume, we use a simple model of the lake as a cylinder: we multiply the average depth of the lake times its roughly circular surface area, as if the lake were a cylinder (Fig. 1–7b). The volume V of a cylinder is the product of its height h times the area of its base: $V = h\pi r^2$, where r is the radius of the circular base. The radius r is $\frac{1}{2}$ km = 500 m, so the volume is approximately

$$V = h\pi r^2 \approx (10 \text{ m}) \times (3) \times (5 \times 10^2 \text{ m})^2 \approx 8 \times 10^6 \text{ m}^3 \approx 10^7 \text{ m}^3,$$

where π was rounded off to 3; the symbol $\approx$ means "approximately equal to." So the volume is on the order of 10^7 m^3, ten million cubic meters. Because of all the estimates that went into this calculation, the order-of-magnitude estimate (10^7 m^3) is probably better to quote than the $8 \times 10^6 \text{ m}^3$ figure.

Here's another Example:

EXAMPLE 1–6 **ESTIMATE** **Thickness of a page.** Estimate the thickness of a page of this book.

SOLUTION At first you might think that a special measuring device, a micrometer (Fig. 1–8), is needed to measure the thickness of one page since an ordinary ruler clearly won't do. But we can use a trick or, to put

FIGURE 1–8 A micrometer, which is used for measuring small thicknesses.

it in physics terms, make use of a *symmetry*: we can make the reasonable assumption that all the pages of this book are equal in thickness. Thus, we can use a ruler to measure hundreds of pages at once. This book is about 1000 pages long, when you count both sides of the page (front and back), so it contains about 500 separate pieces of paper. It is about 4 cm thick (don't include the cover, of course). So if 500 pages are 4 cm thick, one page must be about

$$\frac{4\ \text{cm}}{500\ \text{pages}} \approx 8 \times 10^{-3}\ \text{cm} = 8 \times 10^{-2}\ \text{mm}$$

or, rounding off even more, about a tenth of a millimeter (0.1 mm), or 10^{-4} m.

Now let's take a simple Example of how a diagram can be useful for making an estimate. It cannot be emphasized enough how important it is to draw a diagram when trying to solve a physics problem.

EXAMPLE 1–7 **ESTIMATE** **Height by triangulation.** Estimate the height of the building shown in Fig. 1–9a, by "triangulation," with the help of a bus-stop pole and a friend.

SOLUTION By standing your friend next to the pole, you estimate the height of the pole to be 3 m. You next step away from the pole until the top of the pole is in line with the top of the building, Fig. 1–9a. You are 5 ft 6 in. tall, so your eyes are about 1.5 m above the ground. Your friend is taller, and when she stretches out her arms, one hand touches you, and the other touches the pole, so you estimate that distance as 2 m (Fig. 1–9a). You then pace off the distance from the pole to the base of the building with big, 1-m-long, steps, and you get a total of 16 steps or 16 m. Now you draw, to scale, the diagram shown in Fig. 1–9b using these measurements. You can measure, right on the diagram, the last side of the triangle to be about $x = 13$ m. Alternatively, you can use similar triangles to obtain the height x:

$$\frac{1.5\ \text{m}}{2\ \text{m}} = \frac{x}{18\ \text{m}}, \qquad \text{so} \quad x \approx 13\tfrac{1}{2}\ \text{m}.$$

Finally you add in your eye height of 1.5 m above the ground to get your final result: the building is about 15 m tall.

FIGURE 1–9 Example 1–7. Diagrams are really useful!

(a)

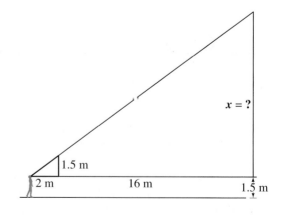

(b)

Another example, this one made famous by the physicist Enrico Fermi, is to estimate the number of piano tuners in a city, say, Chicago or San Francisco. To get a rough order-of-magnitude estimate of the number of piano tuners today in San Francisco, a city of about 700,000 inhabitants, we can proceed by estimating the number of functioning pianos, how often each piano is tuned, and how many pianos each tuner can tune. To estimate the number of pianos in San Francisco, we note that certainly not everyone has a piano. A guess of 1 family in 5 or 6 having a piano would correspond to 1 piano per 20 persons, assuming an average family of 3 or 4 persons. As an order of magnitude, 1 piano per 20 people is certainly more reasonable than 1 per 100 people, or 1 per every person, so let's proceed with the estimate that 1 person in 20 has a piano, or about 35,000 pianos in San Francisco. Now a piano tuner needs an hour or two to tune a piano. So let's estimate that a tuner can tune about 3 pianos a day. A piano ought to be tuned every 6 months or a year—let's say once each year. A piano tuner tuning 3 pianos a day, 5 days a week, 50 weeks a year can tune about 700 pianos a year. So San Francisco, with its (very) roughly 35,000 pianos needs about 50 piano tuners. This is, of course, only a rough estimate. It tells us that there must be many more than 5 piano tuners, and surely not as many as 500. A check of the San Francisco Yellow Pages (done after this calculation) reveals about 50 listings. Each of these listings may employ more than one tuner, but on the other hand, each may also do repairs as well as tuning. In any case, our estimate is reasonable.

➡ **PROBLEM SOLVING**

Estimating how many piano tuners there are in a city

1–8 Mathematics in Physics

Physics is sometimes thought of as being a difficult subject. However, sometimes it is the mathematics used that is the source of difficulties rather than the physics itself. The appendices at the end of this book contain a brief summary of simple mathematical techniques, including algebra, geometry, and trigonometry, that will be used in this book. You may find it useful to examine those appendices now to review old topics or learn any new ones. You may also want to reread them later when you need those concepts. Some mathematical techniques, such as vectors and trigonometric functions, are treated in the text itself, when we first need them.

Check the mathematical Appendices (end of book)

▮ SUMMARY

[The Summary that appears at the end of each chapter in this book gives a brief overview of the main ideas of the chapter. The Summary *cannot* serve to give an understanding of the material, which can be accomplished only by a detailed reading of the chapter.]

Physics, like other sciences, is a creative endeavor. It is not simply a collection of facts. Important theories are created with the idea of explaining observations. To be accepted, theories are "tested" by comparing their predictions with the results of actual experiments. Note that, in general, a theory cannot be "proved" in an absolute sense.

Scientists often devise models of physical phenomena. A **model** is a kind of picture or analogy that seems to explain the phenomena. A **theory**, often developed from a model, is usually deeper and more complex than a simple model.

A scientific **law** is a concise statement, often expressed in the form of an equation, which quantitatively describes a particular range of phenomena over a wide range of cases.

Measurements play a crucial role in physics, but can never be perfectly precise. It is important to specify the **uncertainty** of a measurement either by stating it directly using the ± notation, and/or by keeping only the correct number of **significant figures**.

Physical quantities are always specified relative to a particular standard or **unit**, and the unit used should always be stated. The commonly accepted set of units today is the **Système International** (SI), in which the standard units of length, mass, and time are the **meter**, **kilogram**, and **second**.

When converting units, check all **conversion factors** for correct cancellation of units.

Making rough, **order-of-magnitude estimates** is a very useful technique in science as well as in everyday life.

QUESTIONS

1. It is advantageous that fundamental standards, such as those for length and time, be accessible (easy to compare to), invariable (do not change), indestructible, and reproducible. Discuss why these are advantages and whether any of these criteria can be incompatible with others.

2. What are the merits and drawbacks of using a person's foot as a standard? Discuss in terms of the criteria mentioned in Question 1. Consider both (a) a particular person's foot, and (b) any person's foot.

3. When traveling a highway in the mountains, you may see elevation signs that read "914 m (3000 ft)." Critics of the metric system claim that such numbers show the metric system is more complicated. How would you alter such signs to be more consistent with a switch to the metric system?

4. Suggest a way to measure the distance from Earth to the Sun.

5. What is wrong with this road sign:
   Boston 7 mi (11.263 km)?

6. List assumptions useful to estimate the number of car mechanics in (a) San Francisco, (b) your hometown, and then make the estimates.

7. Discuss how you would estimate the number of hours you have spent in school thus far in your life. Then make the estimate.

8. Discuss how the notion of symmetry could be used to estimate the number of marbles in a one-liter jar.

9. You measure the radius of a wheel to be 4.16 cm. If you multiply by 2 to get the diameter, should you write the result as 8 cm or as 8.32 cm? Justify your answer.

PROBLEMS

[The problems at the end of each chapter are ranked I, II, or III according to estimated difficulty, with I problems being easiest. The problems are arranged by Sections, meaning that the reader should have read up to and including that Section, but not only that Section—problems often depend on earlier material. Each chapter also has a group of General Problems that are not arranged by Section and not ranked.]

SECTION 1–4

1. (I) The age of the universe is thought to be somewhere around 10 billion years. Assuming one significant figure, write this in powers of ten in (a) years, (b) seconds.

2. (I) Write out the following numbers in full with a decimal point and correct number of zeros: (a) 8.69×10^4, (b) 7.1×10^3, (c) 6.6×10^{-1}, (d) 8.76×10^2, and (e) 8.62×10^{-5}.

3. (I) Write the following numbers in powers of ten notation: (a) 1,156,000, (b) 218, (c) 0.0068, (d) 27.635, (e) 0.21, and (f) 22.

4. (I) How many significant figures do each of the following numbers have: (a) 142, (b) 81.60, (c) 7.63, (d) 0.03, (e) 0.0086, (f) 3236, and (g) 8700?

5. (I) What is the percent uncertainty in the measurement 2.26 ± 0.25 m?

6. (I) What, approximately, is the percent uncertainty for the measurement 1.67?

7. (I) Time intervals measured with a stopwatch typically have an uncertainty of about a half second, due to human reaction time at the start and stop moments. What is the percent uncertainty of a hand-timed measurement of (a) 5 s, (b) 50 s, (c) 5 min?

8. (II) Multiply 2.079×10^2 m by 0.072×10^{-1}, taking into account significant figures.

9. (II) Add 7.2×10^3 s $+$ 8.3×10^4 s $+$ 0.09×10^6 s.

10. (II) What is the area, and its approximate uncertainty, of a circle of radius 2.8×10^4 cm?

11. (II) What is the percent uncertainty in the volume of a spherical beach ball whose radius is $r = 3.86 \pm 0.08$ m?

SECTIONS 1–5 AND 1–6

12. (I) Express the following using the prefixes of Table 1–4: (a) 10^6 volts, (b) 10^{-6} meters, (c) 5×10^3 days, (d) 8×10^2 bucks, and (e) 8×10^{-9} pieces.

13. (I) Write the following as full (decimal) numbers with standard units: (a) 86.6 mm, (b) 35 μV, (c) 860 mg, (d) 600 picoseconds, (e) 12.5 femtometers, (f) 250 gigavolts.

14. (I) How many kisses is 50 hectokisses? What would you be if you earned a megabuck a year?

15. (I) Determine your height in meters.

16. (I) The Sun, on average, is 93 million miles from the Earth. How many meters is this? Express (a) using powers of ten, and (b) using a metric prefix.

17. (II) A typical atom has a diameter of about 1.0×10^{-10} m. (a) What is this in inches? (b) How many atoms are there along a 1.0-cm line?

18. (II) Express the following sum with the correct number of significant figures:

$$1.00 \text{ m} + 142.5 \text{ cm} + 1.24 \times 10^5 \ \mu\text{m}.$$

19. (II) Determine the conversion factor between (a) km/h and mi/h, (b) m/s and ft/s, and (c) km/h and m/s.

20. (II) How much longer (percentage) is a one-mile race than a 1500-m race ("the metric mile")?

21. (II) A *light-year* is the distance light (speed $= 2.998 \times 10^8$ m/s) travels in 1.00 year. (a) How many meters are there in 1.00 light-year? (b) An astronomical unit (AU) is the average distance from the Sun to Earth, 1.50×10^8 km. How many AU are there in 1.00 light-year? (c) What is the speed of light in AU/h?

22. (III) The diameter of the moon is 3480 km. What is the surface area, and how does it compare to the land surface area of the Earth?

SECTION 1–7

(*Note*: Remember that for rough estimates, only round numbers are needed both as input to calculations and as final results.)

23. (I) Estimate the order of magnitude (power of ten) of: (a) 7800, (b) 9.630×10^2, (c) 0.00076, and (d) 150×10^8.

24. (II) Estimate how long it would take a good runner to run across the United States from New York to California.

25. (II) Make a rough estimate of the percentage of a house's outside wall area that consists of window area.

26. (II) Estimate the number of times a human heart beats in a lifetime.

27. (II) Make a rough estimate of the volume of your body (in cm^3).

28. (II) Estimate the time to drive from Beijing (Peking) to Paris (a) today, and (b) in 1906 when a great car race was run between those two cities.

29. (II) Estimate the number of dentists (a) in San Francisco and (b) in your town or city.

30. (II) Estimate how long it would take one person to mow a football field using an ordinary home lawn mower.

31. (III) The rubber worn from tires mostly enters the atmosphere as particulate pollution. Estimate how much rubber (in kg) is put into the air in the United States every year. To get you started, a good estimate for a tire tread's depth is 1 cm when new, and the density of rubber is about 1200 kg/m^3.

GENERAL PROBLEMS

32. An angstrom (symbol Å) is an older unit of length, defined as 10^{-10} m. (a) How many nanometers are in 1.0 angstrom? (b) How many femtometers (the common unit of length in nuclear physics) are in 1.0 angstrom? (c) How many angstroms are in 1.0 meter? (d) How many angstroms are in 1.0 light-year (see Problem 21)?

33. (a) How many seconds are there in 1.00 year? (b) How many nanoseconds are there in 1.00 year? (c) How many years are there in 1.00 second?

34. Estimate the number of bus drivers (a) in Washington, D.C., and (b) in your town.

FIGURE 1–10 Problem 35. The wafer held by hand (above) is shown below, enlarged and illuminated by colored light. Visible are rows of integrated circuits (chips).

35. Computer chips (Fig. 1–10) are etched on circular silicon wafers of thickness 0.60 mm that are sliced from a solid cylindrical silicon crystal of length 30 cm. If each wafer can hold 100 chips, what is the maximum number of chips that can be produced from one entire cylinder?

36. Estimate the number of gallons of gasoline consumed by automobile drivers in the United States, per year.

37. Estimate the number of gumballs in the machine shown in Fig. 1–11.

FIGURE 1–11 Problem 37. Estimate the number of gumballs in the machine.

38. An average family of four uses roughly 1200 liters (about 300 gallons) of water per day. (One liter = 1000 cm³.) How much depth would a lake lose per year if it uniformly covered an area of 50 square kilometers and supplied a local town with a population of 40,000 people? Consider only population uses, and neglect evaporation and so on.

39. How big is a ton? That is, what is the volume of something that weighs a ton? To be specific, estimate the diameter of a 1-ton rock, but first make a wild guess: will it be 1 ft across, 3 ft, or the size of a car? [*Hint:* Rock has mass per volume about 3 times that of water, which is 1 kg per liter (10^3 cm³) or 62 lbs per cubic foot.]

40. A violent rainstorm dumps 1.0 cm of rain on a city 5 km wide and 8 km long in a 2-h period. How many metric tons (1 ton = 10^3 kg) of water fell on the city? [1 cm³ of water has a mass of 1 gram = 10^{-3} kg.]

41. Hold a pencil in front of your eye at a position where its end just blocks out the Moon (Fig. 1–12). Make appropriate measurements to estimate the diameter of the Moon, given that the Earth–Moon distance is 3.8×10^5 km.

42. The volume of an object is 1000 m³. Express this volume in (*a*) cm³, (*b*) ft³, (*c*) in.³.

43. Estimate how long it would take to walk around the world.

44. Noah's ark was ordered to be 300 cubits long, 50 cubits wide, and 30 cubits high. The cubit was a unit of measure equal to the length of a human forearm, elbow to the tip of the longest finger. Express the dimensions of Noah's ark in meters.

FIGURE 1–12 Problem 41. How big is the Moon?

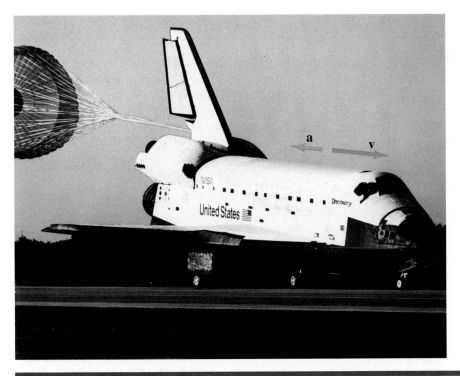

Space shuttle Discovery landing on Earth. The parachute helps it to reduce its speed quickly. The directions of Discovery's velocity and acceleration are shown by the green (**v**) and gold (**a**) arrows. Note that they (**v** and **a**) point in opposite directions.

2

DESCRIBING MOTION: KINEMATICS IN ONE DIMENSION

The motion of objects—baseballs, automobiles, joggers, and even the Sun and Moon—is an obvious part of everyday life. Although the ancients acquired significant insight into motion, it was not until comparatively recently, in the sixteenth and seventeenth centuries, that our modern understanding of motion was established. Many contributed to this understanding, but, as we shall soon see, two individuals stand out above the rest: Galileo Galilei (1564–1642) and Isaac Newton (1642–1727).

The study of the motion of objects, and the related concepts of force and energy, form the field called **mechanics**. Mechanics is customarily divided into two parts: **kinematics**, which is the description of how objects move, and **dynamics**, which deals with force and why objects move as they do. This chapter and the next deal with kinematics.

We start by discussing objects that move without rotating (Fig. 2–1a). Such motion is called **translational motion**. In the present chapter we will be concerned with describing an object that moves along a straight-line path, which is one-dimensional motion. In Chapter 3 we will study how to describe translational motion in two (or three) dimensions.

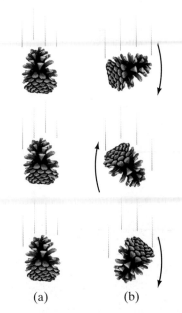

(a)   (b)

FIGURE 2–1 The pinecone in (a) undergoes pure translation as it falls, whereas in (b) it is rotating as well as translating.

2–1 Reference Frames and Displacement

All measurements are made relative to a frame of reference

Any measurement of position, distance, or speed must be made with respect to a **frame of reference**. For example, while you are on a train traveling at 80 km/h, you might notice a person who walks past you toward the front of the train at a speed of, say, 5 km/h (Fig. 2–2). Of course this is the person's speed with respect to the train as frame of reference. With respect to the ground that person is moving at a speed of 80 km/h + 5 km/h = 85 km/h. It is always important to specify the frame of reference when stating a speed. In everyday life, we usually mean "with respect to the Earth" without even thinking about it, but the reference frame should be specified whenever there might be confusion.

FIGURE 2–2 A person walks toward the front of a train at 5 km/h. The train is moving 80 km/h with respect to the ground, so the walking person's speed, relative to the ground, is 85 km/h.

When specifying the motion of an object, it is important to specify not only the speed but also the direction of motion. Often we can specify a direction by using the cardinal points, north, east, south, and west, and by "up" and "down." In physics, we often draw a set of **coordinate axes**, as shown in Fig. 2–3, to represent a frame of reference. We can always place the origin 0, and the directions of the x and y axes, as we like for convenience. Objects positioned to the right of the origin of coordinates (0) on the x axis have an x coordinate which we usually choose to be positive; then points to the left of 0 have a negative x coordinate. The position along the y axis is usually considered positive when above 0, and negative when below 0, although the reverse convention can be used if convenient. Any point on the plane can be specified by giving its x and y coordinates. In three dimensions, a z axis perpendicular to the x and y axes is also used.

For one-dimensional motion, we often choose the x axis as the line along which the motion takes place. Thus the position of an object at any moment is given by its x coordinate.

Displacement

We need to make a distinction between the distance an object has traveled, and its **displacement**, which is defined as the *change in position* of the object. That is, displacement is how far the object is from its starting point. To see the distinction between total distance and displacement, imagine a person walking 70 m to the east and then turning around and walking back (west) a distance of 30 m (see Fig. 2–4). The total *distance*

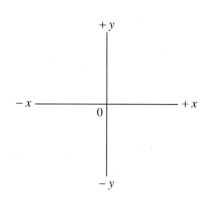

FIGURE 2–3 Standard set of xy coordinate axes.

traveled is 100 m, but the *displacement* is only 40 m since the person is now only 40 m from the starting point.

Displacement is a quantity that has both magnitude and direction. Such quantities are called **vectors**, and are represented in diagrams by arrows. For example, in Fig. 2–4, the blue arrow represents the displacement whose magnitude is 40 m and whose direction is to the right.

We will deal with vectors more fully in Chapter 3. For now, we deal only with motion in one dimension, along a line, and in this case, vectors which point in one direction will have a positive sign, whereas vectors that point in the opposite direction will have a negative sign.

Let's see how this works. Consider the motion of an object over a particular time interval. Suppose that at some initial moment in time, call it t_1, the object is on the x axis at the point x_1 in the coordinate system shown in Fig. 2–5. At some later time, t_2, suppose the object is at point x_2. The displacement of our object is $x_2 - x_1$, and is represented by the arrow pointing to the right in Fig. 2–5. It is convenient to write

$$\Delta x = x_2 - x_1$$

where the symbol Δ (Greek letter delta) means "change in." Then Δx means "the change in x," which is the displacement. Note that the "change in" any quantity means the final value of that quantity minus the initial value.

To be concrete, suppose $x_1 = 10.0$ m and $x_2 = 30.0$ m. Then

$$\Delta x = x_2 - x_1 = 30.0 \text{ m} - 10.0 \text{ m} = 20.0 \text{ m}.$$

See Fig. 2–5.

Now consider a different situation, that of an object moving to the left as shown in Fig. 2–6. Here an object, say a person, starts at $x_1 = 30.0$ m and walks to the left to the point $x_2 = 10.0$ m. In this case

$$\Delta x = x_2 - x_1 = 10.0 \text{ m} - 30.0 \text{ m} = -20.0 \text{ m}$$

and the blue arrow representing the vector displacement points to the left. This example illustrates that when dealing with one-dimensional motion, a vector pointing to the right has a positive value, whereas one pointing to the left has a negative value.

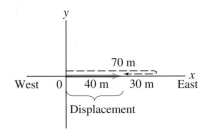

FIGURE 2–4 A person walks 70 m east, then 30 m west. The total distance traveled is 100 m (path is shown in black); but the displacement, shown as a blue arrow, is 40 m to the east.

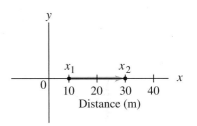

FIGURE 2–5 The arrow represents the displacement $x_2 - x_1$. Distances are in meters.

FIGURE 2–6 For the displacement $\Delta x = x_2 - x_1 = 10.0 \text{ m} - 30.0 \text{ m}$, the displacement vector points to the left.

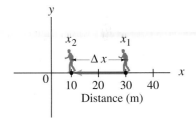

2–2 Average Velocity

The most obvious aspect of the motion of a moving object is how fast it is moving—its speed or velocity.

The term "speed" refers to how far an object travels in a given time interval. If a car travels 240 kilometers (km) in 3 hours, we say its average speed was 80 km/h. In general, the **average speed** of an object is defined as *the distance traveled along its path divided by the time it takes to travel this distance:*

$$\text{average speed} = \frac{\text{distance traveled}}{\text{time elapsed}}. \qquad \textbf{(2–1)} \qquad \textit{Average speed}$$

The terms velocity and speed are often used interchangeably in ordinary language. But in physics we make a distinction between the two. Speed is simply a positive number, with units. **Velocity**, on the other hand, is used to signify both the *magnitude* (numerical value) of how fast an object is moving and the *direction* in which it is moving. (Velocity is therefore a vector.) There

Velocity

is a second difference between speed and velocity: namely, the **average velocity** is defined in terms of *displacement*, rather than total distance traveled:

$$\text{average velocity} = \frac{\text{displacement}}{\text{time elapsed}}.$$

Average speed and average velocity often have the same magnitude, but sometimes they don't. As an example, recall the walk we described earlier, in Fig. 2–4, where a person walked 70 m east and then 30 m west. The total distance traveled was 70 m + 30 m = 100 m, but the displacement was 40 m. Suppose this walk took 70 s to complete. Then the average speed was:

$$\frac{\text{distance}}{\text{time}} = \frac{100 \text{ m}}{70 \text{ s}} = 1.4 \text{ m/s}.$$

The magnitude of the average velocity, on the other hand, was:

$$\frac{\text{displacement}}{\text{time}} = \frac{40 \text{ m}}{70 \text{ s}} = 0.57 \text{ m/s}.$$

This discrepancy between the speed and the magnitude of the velocity occurs in some cases, but only for the *average* values, and we rarely need be concerned with it.

To discuss one-dimensional motion of an object in general, suppose that at some moment in time, call it t_1, the object is on the x axis at point x_1 in a coordinate system, and at some later time, t_2, suppose it is at point x_2. The elapsed time is $t_2 - t_1$, and during this time interval the displacement of our object was $\Delta x = x_2 - x_1$. Then the average velocity, defined as *the displacement divided by the elapsed time*, can be written

Average velocity

$$\bar{v} = \frac{x_2 - x_1}{t_2 - t_1} = \frac{\Delta x}{\Delta t}, \tag{2–2}$$

where v stands for velocity and the bar ($^-$) over the v is a standard symbol meaning "average."

Notice that if x_2 is less than x_1, the object is moving to the left, and then $\Delta x = x_2 - x_1$ is less than zero. The sign of the displacement, and thus of the velocity, indicates the direction: the average velocity is positive for an object moving to the right along the x axis and negative when the object moves to the left. The direction of the velocity is always the same as the direction of the displacement.

➡ **PROBLEM SOLVING**

+ or − sign can signify the direction for linear motion

EXAMPLE 2–1 **Runner's average velocity.** The position of a runner as a function of time is plotted as moving along the x axis of a coordinate system. During a 3.00-s time interval, the runner's position changes from $x_1 = 50.0$ m to $x_2 = 30.5$ m, as shown in Fig. 2–7. What was the runner's average velocity?

SOLUTION Average velocity is the displacement divided by the elapsed time. The displacement is $\Delta x = x_2 - x_1 = 30.5 \text{ m} - 50.0 \text{ m} = -19.5 \text{ m}$. The time interval is $\Delta t = 3.00$ s. Therefore the average velocity is

$$\bar{v} = \frac{\Delta x}{\Delta t} = \frac{-19.5 \text{ m}}{3.00 \text{ s}} = -6.50 \text{ m/s}.$$

FIGURE 2–7 Example 2–1. A person runs from $x_1 = 50.0$ m to $x_2 = 30.5$ m. The displacement is −19.5 m.

The displacement and average velocity are negative, which tells us (if we didn't already know it) that the runner is moving to the left along the x axis, as indicated by the arrow in Fig. 2–7. Thus we can say that the runner's average velocity was 6.50 m/s to the left.

> **EXAMPLE 2–2** **Distance a cyclist travels.** How far can a cyclist travel in 2.5 h along a straight road if her average speed is 18 km/h?
>
> **SOLUTION** We want to find the distance traveled, so we use Eq. 2–2 letting Δx be the distance and $\bar{v}$ be the average speed, and then rewrite it as
>
> $$\Delta x = \bar{v} \, \Delta t = (18 \text{ km/h})(2.5 \text{ h}) = 45 \text{ km}.$$

2–3 Instantaneous Velocity

If you drive a car along a straight road for 150 km in 2.0 h, the magnitude of your average velocity is 75 km/h. It is unlikely, though, that you were moving at precisely 75 km/h at every instant. To deal with this situation we need the concept of *instantaneous velocity*, which is the velocity at any instant of time. (This is the magnitude that a speedometer is supposed to indicate.) More precisely, the **instantaneous velocity** at any moment is defined as *the average velocity over an infinitesimally short time interval.* That is, starting with Eq. 2–2

$$\bar{v} = \frac{\Delta x}{\Delta t},$$

we define instantaneous velocity as the average velocity in the limit of Δt becoming extremely small, approaching zero. We can write the definition of instantaneous velocity, v, for one-dimensional motion as

$$v = \lim_{\Delta t \to 0} \frac{\Delta x}{\Delta t}. \tag{2–3}$$

Instantaneous velocity

The notation $\lim_{\Delta t \to 0}$ means the ratio $\Delta x / \Delta t$ is to be evaluated in the limit of Δt approaching zero. But we do not simply set $\Delta t = 0$ in this definition, for then Δx would also be zero, and we would have an undefined number. Rather, we are considering the *ratio* $\Delta x / \Delta t$, as a whole. As we let Δt approach zero, Δx approaches zero as well. But the ratio $\Delta x / \Delta t$ approaches some definite value, which is the instantaneous velocity at a given instant.[†]

For instantaneous velocity we use the symbol v, whereas for average velocity we use $\bar{v}$, with a bar. In the rest of this book, when we use the term "velocity" it will refer to instantaneous velocity. When we want to speak of the average velocity, we will make this clear by including the word "average." Note that the *instantaneous* speed always equals the magnitude of the instantaneous velocity. Why? Because distance and displacement become the same when they become infinitesimally small.

If an object moves at a uniform (that is, constant) velocity over a particular time interval, then its instantaneous velocity at any instant is the same as its average velocity (see Fig. 2–8a). But in many situations this is not the case. For example, a car may start from rest, speed up to 50 km/h, remain at that velocity for a time, then slow down to 20 km/h in a traffic jam, and finally stop at its destination after traveling a total of 15 km in 30 min. This trip is plotted on the graph of Fig. 2–8b. Also shown on the graph is the average velocity (dashed line), which is $\bar{v} = \Delta x / \Delta t = 15 \text{ km}/0.50 \text{ h} = 30 \text{ km/h}.$

(a)

Average velocity

(b)

FIGURE 2–8 Velocity of a car as a function of time: (a) at constant velocity; (b) with varying velocity.

[†]More on this in Section 2–8.

2–4 Acceleration

An object whose velocity is changing is said to be accelerating. A car whose velocity increases in magnitude from zero to 80 km/h is accelerating. If one car can accomplish this change in velocity in less time than another, it is said to undergo a greater acceleration. That is, acceleration specifies how rapidly the velocity of an object is changing. **Average acceleration** is defined as the change in velocity divided by the time taken to make this change:

$$\text{average acceleration} = \frac{\text{change of velocity}}{\text{time elapsed}}.$$

In symbols, the average acceleration, $\bar{a}$, over a time interval $\Delta t = t_2 - t_1$ during which the velocity changes by $\Delta v = v_2 - v_1$, is defined as

Average acceleration

$$\bar{a} = \frac{v_2 - v_1}{t_2 - t_1} = \frac{\Delta v}{\Delta t}. \tag{2–4}$$

Acceleration is also a vector, but for one-dimensional motion, we need only use a plus or minus sign to indicate direction relative to a chosen coordinate system.

The **instantaneous acceleration**, a, can be defined in analogy to instantaneous velocity, for any specific instant:

Instantaneous acceleration

$$a = \lim_{\Delta t \to 0} \frac{\Delta v}{\Delta t}. \tag{2–5}$$

Here Δv represents the very small change in velocity during the very short time interval Δt.

EXAMPLE 2–3 **Average acceleration.** A car accelerates along a straight road from rest to 75 km/h in 5.0 s, Fig. 2–9. What is the magnitude of its average acceleration?

SOLUTION The car starts from rest, so $v_1 = 0$. The final velocity is $v_2 = 75$ km/h. Then from Eq. 2–4, the average acceleration is

$$\bar{a} = \frac{75 \text{ km/h} - 0 \text{ km/h}}{5.0 \text{ s}} = 15 \frac{\text{km/h}}{\text{s}}.$$

This is read as "fifteen kilometers per hour per second" and means that, on average, the velocity changed by 15 km/h during each second. That is, assuming the acceleration was constant, during the first second the car's velocity increased from zero to 15 km/h. During the next second its velocity increased by another 15 km/h up to 30 km/h, and so on, Fig. 2–9. (Of course, if the instantaneous acceleration was not constant, these numbers could be different.)

Careful:
Do not confuse
velocity with acceleration

Note carefully that *acceleration tells us how fast the velocity changes,* whereas *velocity tells us how fast the position changes.* In this last Example, the calculated acceleration contained two different time units: hours and seconds. We

$t_1 = 0$
$v_1 = 0$

Acceleration
$= 15 \dfrac{km/h}{s}$

at $t = 1.0$ s
$v = 15$ km/h

at $t = 2.0$ s
$v = 30$ km/h

at $t = t_2 = 5.0$ s
$v = v_2 = 75$ km/h

FIGURE 2–9 Example 2–3. The car is shown at the start with $v_1 = 0$ at $t_1 = 0$. It is shown three more times, at $t = 1.0$ s, $t = 2.0$ s, and $t_2 = 5.0$ s. We assume the acceleration is constant and equals 15 km/h/s. The green arrows represent the velocity vectors; the length of each represents the magnitude of the velocity at that moment. The acceleration vector is the orange arrow.

usually prefer to use only seconds. To do so we can change km/h to m/s (see Section 1–6, and Example 1–4):

$$75 \ km/h = \left(75 \ \frac{km}{h}\right)\left(\frac{1000 \ m}{1 \ km}\right)\left(\frac{1 \ h}{3600 \ s}\right) = 21 \ m/s.$$

Then we get

$$\bar{a} = \frac{21 \ m/s - 0.0 \ m/s}{5.0 \ s} = 4.2 \ \frac{m/s}{s} = 4.2 \ \frac{m}{s^2}.$$

We almost always write these units as m/s^2 (meters per second squared), as done here, instead of m/s/s. This is possible because:

$$\frac{m/s}{s} = \frac{m}{s \cdot s} = \frac{m}{s^2}.$$

According to the above calculation, the velocity in Example 2–3 (Fig. 2–9) changed on the average by 4.2 m/s during each second, for a total change of 21 m/s over the 5.0 s.

CONCEPTUAL EXAMPLE 2–4 **Velocity and acceleration.** (*a*) If the velocity of an object is zero, does it mean that the acceleration is zero? (*b*) If the acceleration is zero, does it mean that the velocity is zero? Think of some examples.

RESPONSE A zero velocity does not necessarily mean that the acceleration is zero, nor does a zero acceleration mean that the velocity is zero. (*a*) For example, when you put your foot on the gas pedal of your car which is at rest, the velocity starts from zero but the acceleration is not zero since the velocity of the car changes. (How else could your car start forward if its velocity weren't changing—that is, if the acceleration were zero?) (*b*) As you cruise along a straight highway at a constant velocity of 100 km/h, your acceleration is zero.

at $t_1 = 0$
$v_1 = 15.0$ m/s

Acceleration
$= -2.0$ m/s^2

at $t_2 = 5.0$ s
$v_2 = 5.0$ m/s

FIGURE 2–10 Example 2–5, showing the position of the car at times t_1 and t_2, as well as the car's velocity represented by the green arrows. The acceleration vector (orange) points to the left.

FIGURE 2–11 The same car as in Example 2–5, but now moving to the left and decelerating. The acceleration is

$$a = \frac{v_2 - v_1}{\Delta t} = \frac{-5.0 \text{ m/s} - (-15.0 \text{ m/s})}{5.0 \text{ s}}$$

$$= \frac{-5.0 \text{ m/s} + 15.0 \text{ m/s}}{5.0 \text{ s}} = +2.0 \text{ m/s}^2$$

$v_2 = -5.0$ m/s $v_1 = -15.0$ m/s

a

EXAMPLE 2–5 **Car slowing down.** An automobile is moving to the right along a straight highway, which we choose to be the positive x axis (Fig. 2–10), and the driver puts on the brakes. If the initial velocity is $v_1 = 15.0$ m/s and it takes 5.0 s to slow down to $v_2 = 5.0$ m/s, what was the car's average acceleration?

SOLUTION The average acceleration is equal to the change in velocity divided by the elapsed time, Eq. 2–4. Let us call the initial time $t_1 = 0$; then $t_2 = 5.0$ s. (Note that our choice of $t_1 = 0$ doesn't affect the calculation of $\bar{a}$ because only $\Delta t = t_2 - t_1$ appears in Eq. 2–4.) Then

$$\bar{a} = \frac{5.0 \text{ m/s} - 15.0 \text{ m/s}}{5.0 \text{ s}} = -2.0 \text{ m/s}^2.$$

The negative sign appears because the final velocity is less than the initial velocity. In this case the direction of the acceleration is to the left (in the negative x direction)—even though the velocity is always pointing to the right. We say that the acceleration is 2.0 m/s^2 to the left, and it is shown in Fig. 2–10 as an orange arrow.

When an object is slowing down, we sometimes say it is decelerating. But be careful: deceleration does *not* mean that the acceleration is necessarily negative. For an object moving to the right along the positive x axis and slowing down (as in Fig. 2–10), the acceleration *is* negative. But the same car moving to the left (decreasing x) and slowing down has positive acceleration that points to the right, as shown in Fig. 2–11. We have a deceleration whenever the velocity and acceleration point in opposite directions.

2–5 Motion at Constant Acceleration

Many practical situations occur in which the acceleration is constant or close enough that we can assume it is constant. That is, the acceleration doesn't change over time. We now treat this situation when the magnitude of the acceleration is constant and the motion is in a straight line (sometimes called **uniformly accelerated motion**). In this case, the instantaneous and average accelerations are equal.

Let a = constant

To simplify our notation, let us take the initial time in any discussion to be zero: $t_1 = 0$. We can then let $t_2 = t$ be the elapsed time. The initial position (x_1) and initial velocity (v_1) of an object will now be represented by x_0 and v_0; and at time t the position and velocity will be called x and v (rather than x_2 and v_2). The average velocity during the time t will be (from Eq. 2–2)

$t_1 = 0, t_2 = t$
$x_1 = x_0, x_2 = x$
$v_1 = v_0, v_2 = v$

$$\bar{v} = \frac{x - x_0}{t - t_0} = \frac{x - x_0}{t}$$

since $t_0 = 0$. And the acceleration, which is assumed constant in time, will be (from Eq. 2–4)

$$a = \frac{v - v_0}{t}.$$

A common problem is to determine the velocity of an object after a certain time, given its acceleration. We can solve such problems by solving for v in the last equation: we multiply both sides by t and get

$$at = v - v_0$$

and then add v_0 to both sides to obtain:

$$v = v_0 + at. \qquad \text{[constant acceleration]} \quad \textbf{(2–6)}$$

v related to a and t
(a = constant)

For example, it may be known that the acceleration of a particular motorcycle is $4.0 \, \text{m/s}^2$, and we wish to determine how fast it will be going after, say, 6.0 s. Assuming it starts from rest ($v_0 = 0$), after 6.0 s the velocity will be $v = at = (4.0 \, \text{m/s}^2)(6.0 \, \text{s}) = 24 \, \text{m/s}$.

Next, let us see how to calculate the position of an object after a time t when it is undergoing constant acceleration. The definition of average velocity (Eq. 2–2) is

$$\bar{v} = \frac{x - x_0}{t},$$

which we can rewrite (solving for x) as

$$x = x_0 + \bar{v}t.$$

Because the velocity increases at a uniform rate, the average velocity, $\bar{v}$, will be midway between the initial and final velocities:

$$\bar{v} = \frac{v_0 + v}{2}. \qquad \text{[constant acceleration]} \quad \textbf{(2–7)}$$

Average velocity
(when acceleration is constant)

(Careful: this is not usually valid if the acceleration is not constant.) We combine the last two equations with Eq. 2–6 and find

$$x = x_0 + \bar{v}t = x_0 + \left(\frac{v_0 + v}{2}\right)t$$

$$= x_0 + \left(\frac{v_0 + v_0 + at}{2}\right)t$$

or

$$x = x_0 + v_0 t + \tfrac{1}{2} at^2. \qquad \text{[constant acceleration]} \quad \textbf{(2–8)}$$

x related to a and t
(a = constant)

Equations 2–6, 2–7, and 2–8 are three of the four most useful equations for motion at constant acceleration. We now derive the fourth equation, which is useful in situations where the time t is not known. We begin as above, with Eq. 2–7 and the equation just before it:

$$x = x_0 + \bar{v}t = x_0 + \left(\frac{v + v_0}{2}\right)t$$

Next we solve Eq. 2–6 for t, obtaining

$$t = \frac{v - v_0}{a},$$

and substituting this into the equation above we have

$$x = x_0 + \left(\frac{v + v_0}{2}\right)\left(\frac{v - v_0}{a}\right) = x_0 + \frac{v^2 - v_0^2}{2a}.$$

We solve this for v^2 and obtain

$$v^2 = v_0^2 + 2a(x - x_0), \qquad \text{[constant acceleration]} \quad \textbf{(2–9)}$$

v related to a and x
(a = constant)

which is the useful equation we sought.

SECTION 2–5   Motion at Constant Acceleration   **27**

We now have four equations relating position, velocity, acceleration, and time, when the acceleration a is constant. We collect them here in one place for further reference (the tan background screen is to emphasize their usefulness):

Kinematic equations

for constant acceleration

(we'll use them a lot)

$$v = v_0 + at \qquad\qquad [a = \text{constant}] \quad \textbf{(2–10a)}$$

$$x = x_0 + v_0 t + \tfrac{1}{2}at^2 \qquad [a = \text{constant}] \quad \textbf{(2–10b)}$$

$$v^2 = v_0^2 + 2a(x - x_0) \qquad [a = \text{constant}] \quad \textbf{(2–10c)}$$

$$\bar{v} = \frac{v + v_0}{2}. \qquad\qquad [a = \text{constant}] \quad \textbf{(2–10d)}$$

These useful equations are not valid unless a is a constant. In many cases we can set $x_0 = 0$, and this simplifies the above equations a bit. Note that x represents position, not distance, and $x - x_0$ is the displacement.

EXAMPLE 2–6 **Runway design.** You are designing an airport for small planes. One kind of airplane that might use this airfield must reach a speed before takeoff of at least 27.8 m/s (100 km/h), and can accelerate at 2.00 m/s². (*a*) If the runway is 150 m long, can this airplane reach the proper speed to take off? (*b*) If not, what minimum length must the runway have?

SOLUTION (*a*) We are given the airplane's acceleration ($a = 2.00$ m/s²), and we know the plane can travel a distance of 150 m. We want to find its velocity, to determine if it will be at least 27.8 m/s. We want to find v when we are given:

Known	Wanted
$x_0 = 0$	v
$v_0 = 0$	
$x = 150$ m	
$a = 2.00$ m/s²	

➡ **PROBLEM SOLVING**

Equations 2–10 are valid only when the acceleration is constant, which we assume in this Example

Of the above four equations, Eq. 2–10c will give us v, when we know v_0, a, x, and x_0:

$$v^2 = v_0^2 + 2a(x - x_0)$$

$$= 0 + 2(2.0 \text{ m/s}^2)(150 \text{ m}) = 600 \text{ m}^2/\text{s}^2$$

$$v = \sqrt{600 \text{ m}^2/\text{s}^2} = 24.5 \text{ m/s}.$$

This runway length is *not* sufficient.
(*b*) Now we want $(x - x_0)$ given $v = 27.8$ m/s and $a = 2.0$ m/s². So we use Eq. 2–10c, rewritten as

$$(x - x_0) = \frac{v^2 - v_0^2}{2a} = \frac{(27.8 \text{ m/s})^2 - 0}{2(2.0 \text{ m/s}^2)} = 193 \text{ m}.$$

2–6 Solving Problems

The solving of problems, such as the Examples we have already given, serves two purposes. First, solving problems is useful and practical in itself. Second, solving problems makes you think about the ideas and concepts, and applying the concepts helps you to understand them. But knowing how to do a problem—even to begin it—may not always seem easy. First, it is most important to read the problem through carefully, and more than once. Spend a moment thinking and trying to understand what physics principles, ideas, laws, and definitions might be involved. Up to this point in the book, we have been concerned mainly with the definitions of velocity and acceleration, and the "kinematic equations for constant acceleration," Eqs. 2–10, that we derived from those definitions. For now it is important to note that physics is *not* a collection of equations to be memorized. (In fact, rather than memorizing the very useful Eqs. 2–10, it is better to understand how to derive them from the definitions of velocity and acceleration as we did above.) Simply searching for an equation that might work can be disastrous and can lead you to a wrong result (and will surely not help you understand physics). A better approach is to use the following (rough) procedure, which we put in a special "box" (other such Problem Solving boxes, as an aid, will be found throughout the book):

➡ PROBLEM SOLVING

1. **Read** *and reread* the whole problem carefully before trying to solve it.

2. **Draw** a **diagram** or picture of the situation, with coordinate axes wherever applicable. [You can choose to place the origin of coordinates and the axes wherever you like, so as to make your calculations easier. You also choose which direction is positive and which is negative. Usually we choose the x axis to the right as positive, but you could choose positive to the left.]

3. **Write down** what quantities are "known" or "given," and then what you *want* to know.

4. Think about which principles of physics apply in this problem.

5. Consider which equations (and/or definitions) relate the quantities involved. Before using them, be sure their **range of validity** includes your problem (for example, Eqs. 2–10 are valid only when the acceleration is constant). If you find an applicable equation that involves only known quantities and one desired unknown, **solve** the equation algebraically for the unknown. In many instances several sequential calculations, or a combination of equations, may be

needed. It is often preferable to solve algebraically for the desired unknown before putting in numerical values.

6. Carry out the **calculation** if it is a numerical problem. Keep one or two extra digits during the calculations, but round off the final answers to the correct number of significant figures (Section 1–4).

7. Think carefully about the result you obtain: Is it **reasonable**? Does it make sense according to your own intuition and experience? A good check is to do a rough **estimate** using only powers of ten, as discussed in Section 1–7. Often it is preferable to do a rough estimate at the *start* of a numerical problem because it can help you focus your attention on finding a path toward a solution.

8. A very important aspect of doing problems is keeping track of **units**. Note that an equals sign implies the units on each side must be the same, just as the numbers must. If the units do not balance, a mistake has no doubt been made. This can serve as a **check** on your solution (but it only tells you if you're wrong, not if you're right). And: always use a consistent set of units.

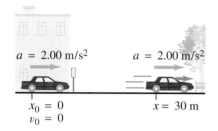

FIGURE 2–12 Example 2–7.

➡ **PROBLEM SOLVING**

Check your answer

➡ **PHYSICS APPLIED**

Braking distances

FIGURE 2–13
Example 2–8: stopping
distance for a braking car.

EXAMPLE 2–7 **Acceleration of car.** How long does it take a car to cross a 30.0-m-wide intersection after the light turns green, if it accelerates from rest at a constant 2.00 m/s²?

SOLUTION First we make a sketch, Fig. 2–12. Next we make a table, choosing $x_0 = 0$ and assume the car moves to the right along the positive x axis, and noting that "starting from rest" means $v = 0$ at $t = 0$ [i.e., $v_0 = 0$].

Known	Wanted
$x_0 = 0$	t
$x = 30.0$ m	
$a = 2.00$ m/s²	
$v_0 = 0$	

Since a is constant, we can use Eqs. 2–10. Equation 2–10b is perfect since the only unknown quantity is t. Setting $v_0 = 0$ and $x_0 = 0$, we can solve Eq. 2–10b for t:

$$x = \tfrac{1}{2} at^2$$

$$t^2 = \frac{2x}{a}$$

$$t = \sqrt{\frac{2x}{a}} = \sqrt{\frac{2(30.0\text{ m})}{2.00\text{ m/s}^2}} = 5.48\text{ s}.$$

This is our answer. Note that the units come out correctly. We can check the reasonableness of the answer by calculating the final velocity $v = at = (2.00\text{ m/s}^2)(5.48\text{ s}) = 10.96\text{ m/s}$, and then finding $x = x_0 + \bar{v}t = 0 + \tfrac{1}{2}(10.96\text{ m/s} + 0)(5.48\text{ s}) = 30.0$ m, which is our given distance.

EXAMPLE 2–8 **ESTIMATE** **Braking distances.** Estimate the minimum stopping distances for a car, which are important for traffic safety and traffic design. The problem is best dealt with in two parts: (1) the time between the decision to apply the brakes and their actual application (the "reaction time"), during which we assume $a = 0$; and (2) the actual braking period when the vehicle slows down ($a \neq 0$). The stopping distance depends on the reaction time of the driver, the initial speed of the car (the final speed is zero), and the acceleration of the car. For a dry road and good tires, good brakes can decelerate a car at a rate of about 5 m/s² to 8 m/s². Calculate the total stopping distance for an initial velocity of 100 km/h (28 m/s ≈ 62 mph) and assume the acceleration of the car is

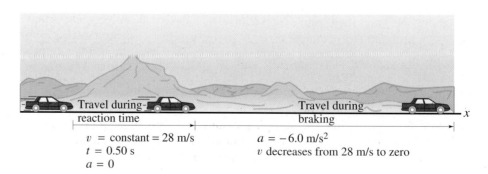

-6.0 m/s^2 (the minus sign appears because the velocity is taken to be in the positive x direction and its magnitude is decreasing). Reaction time for normal drivers varies from perhaps 0.3 s to about 1.0 s; take it to be 0.50 s.

SOLUTION   The car is moving to the right in the positive x direction. We take $x_0 = 0$ for the first part of the problem, in which the car travels at a constant speed of 28 m/s during the time the driver is reacting (0.50 s). See Fig. 2–13. Thus:

Part 1: Reaction time

Part 1:

Known	Wanted
$t\ =0.50 \text{ s}$	x
$v_0 =28 \text{ m/s}$	
$v\ =28 \text{ m/s}$	
$a\ =0$	
$x_0 =0$	

To find x we can use Eq. 2–10b (note that Eq. 2–10c isn't useful because x is multiplied by a, which is zero):

$$x = v_0 t + 0 = (28 \text{ m/s})(0.50 \text{ s}) = 14 \text{ m}.$$

The car travels 14 m during the driver's reaction time, until the moment the brakes are applied. Now for the second part, during which the brakes are applied and the car is brought to rest. We now take $x_0 = 14$ m (result of the first part):

Part 2: Braking

Part 2:

Known	Wanted
$x_0 =14 \text{ m}$	x
$v_0 =28 \text{ m/s}$	
$v\ =0$	
$a\ =-6.0 \text{ m/s}^2$	

Equation 2–10a doesn't contain x; Eq. 2–10b contains x but also the unknown t. Equation 2–10c is what we want; we solve for x (after setting $x_0 = 14$m):

$$v^2 - v_0^2 = 2a(x - x_0)$$

$$x = x_0 + \frac{v^2 - v_0^2}{2a}$$

$$= 14 \text{ m} + \frac{0 - (28 \text{ m/s})^2}{2(-6.0 \text{ m/s}^2)} = 14 \text{ m} + \frac{-784 \text{ m}^2/\text{s}^2}{-12 \text{ m/s}^2}$$

$$= 14 \text{ m} + 65 \text{ m} = 79 \text{ m}.$$

The car traveled 14 m while the driver was reacting and another 65 m during the braking period before coming to a stop. The total distance traveled was then 79 m. Under wet or icy conditions, the value of a may be only one third the value for a dry road since the brakes cannot be applied as hard without skidding, and hence stopping distances are much greater. Note also that the stopping distance during braking increases with the *square* of the speed, not just linearly with speed. If you are traveling twice as fast, it takes four times the distance to stop.

FIGURE 2–14 A baseball pitcher accelerates the ball through a displacement of about 3.5 m.

EXAMPLE 2–9 **ESTIMATE** **The fastball.** A baseball pitcher throws a fastball with a speed of 44 m/s. Estimate the average acceleration of the ball during the throwing motion. It is observed that in throwing the baseball, the pitcher accelerates the ball through a displacement of about 3.5 m from behind the body to the point where it is released (Fig. 2–14).

SOLUTION We want to find the acceleration a given that $x = 3.5$ m, $v_0 = 0$, and $v = 44$ m/s. We use Eq. 2–10c and solve for a:

$$a = \frac{v^2 - v_0^2}{2x}$$

$$= \frac{(44 \text{ m/s})^2 - (0 \text{ m/s})^2}{2(3.5 \text{ m})} = 280 \text{ m/s}^2.$$

This is a very large acceleration!

⇒ **PHYSICS APPLIED**

Car safety—air bags

EXAMPLE 2–10 **ESTIMATE** **Air bags.** Suppose you want to design an air-bag system that can protect the driver in a head-on collision at a speed of 100 km/h (60 mph). Estimate how fast the air bag must inflate to effectively protect the driver. Assume the car crumples upon impact over a distance of about 1 m. How does the use of a seat belt help the driver?

SOLUTION The car decelerates from 100 km/h to zero in a very short time and a very short distance (1 m). Noting that 100 km/h = 100×10^3 m/3600 s = 28 m/s, we can get the acceleration from Eq. 2–10c:

$$a = -\frac{v_0^2}{2x} = -\frac{(28 \text{ m/s})^2}{2.0 \text{ m}} = -390 \text{ m/s}^2.$$

This enormous acceleration takes place in a time given by (Eq. 2–10a):

$$t = \frac{v - v_0}{a} = \frac{0 - 28 \text{ m/s}}{-390 \text{ m/s}^2} = 0.07 \text{ s}.$$

To be effective, the air bag would need to inflate faster than this.

What does the air bag do? First, it spreads the force over a larger area of the chest. This is better than being punctured by the steering column. Also, the pressure in the bag is controlled to minimize the head's maximum deceleration. The seat belt keeps the person in the correct position against the expanding air bag.

FIGURE 2–15
Galileo Galilei (1564–1642).

The analysis of motion we have been discussing in this chapter is basically algebraic. It is sometimes helpful to use a graphical interpretation as well, and this is discussed in the optional Section 2–8.

2–7 Falling Objects

One of the most common examples of uniformly accelerated motion is that of an object allowed to fall freely near the Earth's surface. That a falling object is accelerating may not be obvious at first. And beware of thinking, as was widely believed until the time of Galileo (Fig. 2–15), that heavier objects fall faster than lighter objects and that the speed of fall is proportional to how heavy the object is.

Galileo's analysis made use of his new and creative technique of imagining what would happen in idealized (simplified) cases. For free fall, he postulated that all objects would fall with the *same constant acceleration* in the absence of air or other resistance. He showed that this postulate predicts that for an object falling from rest, the distance traveled will be proportional to the square of the time (Fig. 2–16); that is, $d \propto t^2$. We can see this from Eq. 2–10b, but Galileo was the first to derive this mathematical relation. In fact, among Galileo's great contributions to science was to establish such mathematical relations, and to insist on their importance. Another great contribution of Galileo was to propose theories with specific experimental consequences that could be quantitatively checked (such as $d \propto t^2$).

To support his claim that the speed of falling objects increases as they fall, Galileo made use of a clever argument: a heavy stone dropped from a height of 2 m will drive a stake into the ground much further than will the same stone dropped from a height of only 0.2 m. Clearly, the stone must be moving faster in the former case.

As we saw, Galileo also claimed that *all* objects, light or heavy, fall with the *same* acceleration, at least in the absence of air. If you hold a piece of paper horizontally in one hand and a heavier object—say, a baseball—in the other, and release them at the same time as in Fig. 2–17a, the heavier object will reach the ground first. But if you repeat the experiment, this time crumpling the paper into a small wad (see Fig. 2–17b), you will find that the two objects reach the floor at nearly the same time.

Galileo was sure that air acts as a resistance to very light objects that have a large surface area. But in many ordinary circumstances this air resistance is negligible. In a chamber from which the air has been removed, even light objects like a feather or a horizontally held piece of paper will fall with the same acceleration as any other object (see Fig. 2–18). Such a demonstration in vacuum was of course not possible in Galileo's time, which makes Galileo's achievement all the greater. Galileo is often called the "father of modern science," not only for the content of his science (astronomical discoveries, inertia, free fall), but also for his style or approach to science (idealization and simplification, mathematization of theory, theories that have testable consequences, experiments to test theoretical predictions).

FIGURE 2–16 Multiflash photograph of a falling apple, photographed at equal time intervals. Note that the apple falls farther during each successive time interval, which means it is accelerating.

(a)   (b)

FIGURE 2–17 (a) A ball and a light piece of paper are dropped at the same time. (b) Repeated, with the paper wadded up.

Air-filled tube   Evacuated tube

(a)   (b)

FIGURE 2–18 A rock and a feather are dropped simultaneously (a) in air, (b) in a vacuum.

Galileo's specific contribution to our understanding of the motion of falling objects can be summarized as follows:

Galileo's hypothesis: free fall is at constant acceleration g

at a given location on the Earth and in the absence of air resistance, all objects fall with the same constant acceleration.

We call this acceleration the **acceleration due to gravity** on the Earth, and we give it the symbol g. Its magnitude is approximately

$$g = 9.80 \text{ m/s}^2.$$

In British units g is about 32 ft/s^2. Actually, g varies slightly according to latitude and elevation (see Table 2–1), but these variations are so small that we will ignore them for most purposes. The effects of air resistance are often small, and we will neglect them for the most part. However, air resistance will be noticeable even on a reasonably heavy object if the velocity becomes large.[†]

When dealing with freely falling objects we can make use of Eqs. 2–10, where for a we use the value of g given above. Also, since the motion is vertical we will substitute y in place of x, and y_0 in place of x_0. We take $y_0 = 0$ unless otherwise specified. *It is arbitrary whether we choose y to be positive in the upward direction or in the downward direction; but we must be consistent about it throughout a problem's solution.*

TABLE 2–1
Acceleration Due to Gravity at Various Locations on Earth

Location	Elevation (m)	g (m/s²)
New York	0	9.803
San Francisco	100	9.800
Denver	1650	9.796
Pikes Peak	4300	9.789
Equator	0	9.780
North Pole (calculated)	0	9.832

FIGURE 2–19 Example 2–11. When an object is dropped from a tower, it falls with progressively greater speed and covers greater distance with each successive second. (See also Fig. 2–16.)

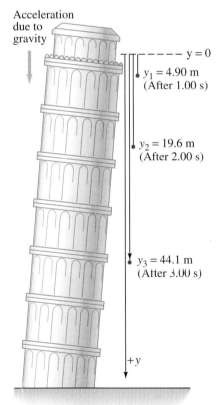

Acceleration due to gravity

y = 0

$y_1 = 4.90$ m (After 1.00 s)

$y_2 = 19.6$ m (After 2.00 s)

$y_3 = 44.1$ m (After 3.00 s)

+y

EXAMPLE 2–11 **Falling from a tower.** Suppose that a ball is dropped from a tower 70.0 m high. How far will it have fallen after 1.00 s, 2.00 s, and 3.00 s? Assume y is positive downward. Neglect air resistance.

SOLUTION We are given the acceleration, $a = g = +9.80 \text{ m/s}^2$, which is positive because we have chosen downward as positive. Since we want to find the distance fallen given the time, t, Eq. 2–10b is the appropriate one, with $v_0 = 0$ and $y_0 = 0$. Then, after 1.00 s, the position of the ball is

$$y_1 = \tfrac{1}{2}at^2 = \tfrac{1}{2}(9.80 \text{ m/s}^2)(1.00 \text{ s})^2 = 4.90 \text{ m},$$

so the ball has fallen a distance of 4.90 m after 1.00 s. Similarly, after 2.00 s,

$$y_2 = \tfrac{1}{2}at^2 = \tfrac{1}{2}(9.80 \text{ m/s}^2)(2.00 \text{ s})^2 = 19.6 \text{ m},$$

and after 3.00 s,

$$y_3 = \tfrac{1}{2}at^2 = \tfrac{1}{2}(9.80 \text{ m/s}^2)(3.00 \text{ s})^2 = 44.1 \text{ m}.$$

See Fig. 2–19.

EXAMPLE 2–12 **Thrown down from a tower.** Suppose the ball in Example 2–11 is *thrown downward* with an initial velocity of 3.00 m/s, instead of being dropped. (*a*) What then would be its position after 1.00 s and 2.00 s? (*b*) What would its speed be after 1.00 s and 2.00 s? Compare to the speeds of a dropped ball.

[†]The speed of an object falling in air (or other fluid) does not increase indefinitely. If the object falls far enough, it will reach a maximum velocity called the **terminal velocity**. Acceleration due to gravity is a vector (as is any acceleration), and its direction is downward, toward the center of the Earth.

SOLUTION (a) We can approach this in the same way as Example 2–11, using Eq. 2–10b, but this time v_0 is not zero but is $v_0 = 3.0$ m/s. Thus, at $t = 1.00$ s, the position of the ball is

$$y = v_0 t + \tfrac{1}{2} a t^2 = (3.00 \text{ m/s})(1.00 \text{ s}) + \tfrac{1}{2}(9.80 \text{ m/s}^2)(1.00 \text{ s})^2 = 7.90 \text{ m,}$$

and at $t = 2.00$ s

$$y = v_0 t + \tfrac{1}{2} a t^2 = (3.00 \text{ m/s})(2.00 \text{ s}) + \tfrac{1}{2}(9.80 \text{ m/s}^2)(2.00 \text{ s})^2 = 25.6 \text{ m.}$$

As expected, the ball falls farther each second than if it were dropped with $v_0 = 0$.

(b) The velocity is readily obtained from Eq. 2–10a:

$$v = v_0 + at$$

$$= 3.00 \text{ m/s} + (9.80 \text{ m/s}^2)(1.00 \text{ s}) = 12.8 \text{ m/s} \quad [\text{at } t = 1.00 \text{ s}]$$

$$= 3.00 \text{ m/s} + (9.80 \text{ m/s}^2)(2.00 \text{ s}) = 22.6 \text{ m/s.} \quad [\text{at } t = 2.00 \text{ s}]$$

When the ball is dropped ($v_0 = 0$), the first term in the above equations is zero, so

$$v = 0 + at$$

$$= (9.80 \text{ m/s}^2)(1.00 \text{ s}) = 9.80 \text{ m/s} \quad [\text{at } t = 1.00 \text{ s}]$$

$$= (9.80 \text{ m/s}^2)(2.00 \text{ s}) = 19.6 \text{ m/s.} \quad [\text{at } t = 2.00 \text{ s}]$$

We see that the speed of a dropped ball increases linearly in time. (In Example 2–11 we saw that the distance fallen increases as the *square* of the time.) The downwardly thrown ball also increases linearly in speed ($\Delta v = 9.80$ m/s each second), but its speed at any moment is always 3.0 m/s (its initial speed) higher than a falling ball.

FIGURE 2–20 An object thrown into the air leaves the thrower's hand at A, reaches its maximum height at B, and returns to the original height at C. Examples 2–13, 2–14, and 2–15.

EXAMPLE 2–13 **Ball thrown upward.** A person throws a ball *upward* into the air with an initial velocity of 15.0 m/s. Calculate (a) how high it goes, and (b) how long the ball is in the air before it comes back to his hand. We are not concerned here with the throwing action, but only with the motion of the ball *after* it leaves the thrower's hand (Fig. 2–20).

SOLUTION Let us choose y to be positive in the upward direction and negative in the downward direction. (Note: This is a different convention from that used in Examples 2–11 and 2–12.) Then the acceleration due to gravity will have a negative sign, $a = -9.80 \text{ m/s}^2$. Note that as the ball rises, its speed decreases until it reaches the highest point (B in Fig. 2–20), where its speed is zero for an instant; then it descends with increasing speed.

(a) To determine the maximum height, we calculate the position of the ball when its velocity equals zero ($v = 0$ at the highest point). At $t = 0$ (point A in Fig. 2–20) we have $y_0 = 0$, $v_0 = 15.0$ m/s, and $a = -9.80 \text{ m/s}^2$. At time t (maximum height), $v = 0$, $a = -9.80 \text{ m/s}^2$, and we wish to find y. We use Eq. 2–10c (replacing x with y) and solve for y:

$$v^2 = v_0^2 + 2ay$$

$$y = \frac{v^2 - v_0^2}{2a} = \frac{0 - (15.0 \text{ m/s})^2}{2(-9.80 \text{ m/s}^2)} = 11.5 \text{ m.}$$

The ball reaches a height of 11.5 m above the hand.

$B(v = 0)$

g g

v v

A C

FIGURE 2–21 Examples 2–13, 2–14, and 2–15.

Careful:
Velocity and acceleration are not always in the same direction

Careful:
$a \neq 0$ *even at the highest point of a trajectory*

(b) Now we need to calculate how long the ball is in the air before it returns to his hand. We could do this calculation in two parts by first determining the time required for the ball to reach its highest point, and then determining the time it takes to fall back down. However, it is simpler to consider the motion from A to B to C (Fig. 2–21) in one step and use Eq. 2–10b. We can do this because y (or x) represents position or displacement, and not the total distance traveled. Thus, at both points A and C, $y = 0$. We use Eq. 2–10b with $a = -9.80 \text{ m/s}^2$ and find

$$y = v_0 t + \tfrac{1}{2} a t^2$$

$$0 = (15.0 \text{ m/s})t + \tfrac{1}{2}(-9.80 \text{ m/s}^2)t^2.$$

This equation is readily factored (we factor out one t):

$$(15.0 \text{ m/s} - 4.90 \text{ m/s}^2 t) t = 0.$$

There are two solutions:

$$t = 0,$$

and

$$t = \frac{15.0 \text{ m/s}}{4.90 \text{ m/s}^2} = 3.06 \text{ s}.$$

The first solution ($t = 0$) corresponds to the initial point (A) in Fig. 2–21, when the ball was first thrown and was also at $y = 0$. The second solution, $t = 3.06 \text{ s}$, corresponds to point C, when the ball has returned to $y = 0$. Thus the ball is in the air 3.06 s.

CONCEPTUAL EXAMPLE 2–14 **Two common misconceptions.** Explain the error in these two common misconceptions: (1) that acceleration and velocity are always in the same direction, and (2) that an object thrown upward has zero acceleration at the highest point (B in Fig. 2–21).

RESPONSE Both are wrong. (1) Velocity and acceleration are *not* necessarily in the same direction. When the ball in Example 2–13 is moving upward, its velocity is positive (upward), whereas the acceleration is negative (downward). (2) At the highest point (B in Fig. 2–21), the ball has zero velocity for an instant. Is the acceleration also zero at this point? No. Gravity does not stop acting, so $a = -g = -9.80 \text{ m/s}^2$ even there. Thinking that $a = 0$ at point B would lead to the conclusion that upon reaching point B, the ball would hover there. For if the acceleration (= rate of change of velocity) were zero, the velocity would remain zero, and the ball could stay up there without falling.

EXAMPLE 2–15 **Ball thrown upward, II.** Let us consider again the ball thrown upward of Example 2–13, and make three more calculations. Calculate (a) how much time it takes for the ball to reach the maximum height (point B in Fig. 2–21), (b) the velocity of the ball when it returns to the thrower's hand (point C), and (c) at what time t the ball passes a point 8.00 m above the person's hand.

SOLUTION Again we take y as positive upward. (a) Both Eqs. 2–10a and 2–10b contain the time t with other quantities known. Let us use Eq. 2–10a with $a = -9.80 \, \text{m/s}^2$, $v_0 = 15.0 \, \text{m/s}$, and $v = 0$:

$$v = v_0 + at,$$

so

$$t = -\frac{v_0}{a} = -\frac{15.0 \, \text{m/s}}{-9.80 \, \text{m/s}^2} = 1.53 \, \text{s}.$$

This is just half the time it takes the ball to go up and fall back to its original position [3.06 s, calculated in part (b) of Example 2–13]. Thus it takes the same time to reach the maximum height as to fall back to the starting point. (b) We use Eq. 2–10a with $v_0 = 15.0 \, \text{m/s}$ and $t = 3.06 \, \text{s}$ (the time calculated in Example 2–13 for the ball to come back to the hand):

$$v = v_0 + at = 15.0 \, \text{m/s} - (9.80 \, \text{m/s}^2)(3.06 \, \text{s}) = -15.0 \, \text{m/s}.$$

The ball has the same magnitude of velocity when it returns to the starting point as it did initially, but in the opposite direction (this is the meaning of the negative sign). Thus, as we gathered from part (a), we see that the motion is symmetrical about the maximum height.

Note the symmetry: the speed at any height is the same when going up as coming down (but the direction is opposite)

(c) We want t, given that $y = 8.00 \, \text{m}$, $y_0 = 0$, $v_0 = 15.0 \, \text{m/s}$, and $a = -9.80 \, \text{m/s}^2$. We use Eq. 2–10b:

$$y = y_0 + v_0 t + \tfrac{1}{2} a t^2$$

$$8.00 \, \text{m} = 0 + (15.0 \, \text{m/s})t + \tfrac{1}{2}(-9.80 \, \text{m/s}^2)t^2.$$

To solve any quadratic equation of the form $at^2 + bt + c = 0$, where a, b, and c are constants, we can use the **quadratic formula** (see Appendix A–4):

$$t = \frac{-b \pm \sqrt{b^2 - 4ac}}{2a}.$$

➡ **PROBLEM SOLVING**

Using the quadratic formula

We rewrite our equation in standard form:

$$(4.90 \, \text{m/s}^2)t^2 - (15.0 \, \text{m/s})t + (8.00 \, \text{m}) = 0.$$

So the coefficient a is $4.90 \, \text{m/s}^2$, b is $-15.0 \, \text{m/s}$, and c is $8.00 \, \text{m}$. Putting these into the quadratic formula, we obtain

$$t = \frac{15.0 \, \text{m/s} \pm \sqrt{(15.0 \, \text{m/s})^2 - 4(4.90 \, \text{m/s}^2)(8.00 \, \text{m})}}{2(4.90 \, \text{m/s}^2)},$$

so $t = 0.69 \, \text{s}$ and $t = 2.37 \, \text{s}$. Why are there two solutions? Are they both valid? Yes, because the ball passes $y = 8.00 \, \text{m}$ when it goes up ($t = 0.69 \, \text{s}$) and again when it comes down ($t = 2.37 \, \text{s}$).

Acceleration expressed in g's

The acceleration of an object, particularly rockets and fast airplanes, is often given as a multiple of $g = 9.80 \text{ m/s}^2$. For example, a plane pulling out of a dive and undergoing $3.00\,g$'s would have an acceleration of $(3.00)(9.80 \text{ m/s}^2) = 29.4 \text{ m/s}^2$.

* 2–8 Graphical Analysis of Linear Motion[†]

Figure 2–8 showed the graph of the velocity of a car versus time for two cases of linear motion: (a) constant velocity, and (b) a particular case in which the magnitude of the velocity varied. It is also useful to graph (or "plot") the position x as a function of time. The time t is considered the independent variable and is measured along the horizontal axis. The position, x, the dependent variable, is measured along the vertical axis.

We make a graph of x vs. t, and we make the choice that at $t = 0$, the position is $x_0 = 0$. First we consider a car moving at a constant velocity of 40 km/h, which is equivalent to 11 m/s. From Eq. 2–10b, $x = vt$, and we see that x increases by 11 m every second. Thus, the position increases linearly in time, so the graph of x vs. t is a straight line, as shown in Fig. 2–22. Each point on this straight line tells us the car's position at a particular time. For example, at $t = 3.0$ s, the position is 33 m, and at $t = 4.0$ s, $x = 44$ m, as indicated by the dashed lines. The small triangle on the graph indicates the **slope** of the straight line, which is defined as the change in the dependent variable (Δx) divided by the corresponding change in the independent variable (Δt):

$$\text{slope} = \frac{\Delta x}{\Delta t}.$$

We see, using the definition of velocity (Eq. 2–2), that the slope of the x vs. t graph is equal to the velocity. And, as can be seen from the small tri-

[†]Some sections of this book, such as this one, may be considered *optional* at the discretion of the instructor. See the Preface for more details.

FIGURE 2–22 Graph of position vs. time for an object moving at a uniform velocity of 11 m/s.

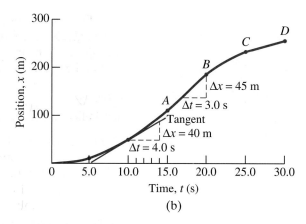

FIGURE 2–23 (a) Velocity vs. time and (b) displacement vs. time for an object with variable velocity. (See text.)

angle on the graph, $\Delta x / \Delta t = (11\ \text{m})/(1.0\ \text{s}) = 11\ \text{m/s}$, which is the given velocity.

The slope of the x vs. t graph is everywhere the same if the velocity is constant, as in Fig. 2–22. But if the velocity changes, as in Fig. 2–23, the slope of the x vs. t graph also varies. Consider, for example, a car that accelerates uniformly from rest to 15 m/s in 15 s, after which it remains at a constant velocity of 15 m/s for the next 5.0 s, then slows down uniformly to 5.0 m/s during the following 5.0 s, and then remains at this constant velocity. The velocity as a function of time is shown in the graph of Fig. 2–23a. Now, to construct the x vs. t graph, we can use Eq. 2–10b with constant acceleration for the intervals $t = 0$ to $t = 15\ \text{s}$ and $t = 20\ \text{s}$ to $t = 25\ \text{s}$, and with constant velocity for the periods $t = 15\ \text{s}$ to $t = 20\ \text{s}$ and after $t = 25\ \text{s}$. The result is the x vs. t graph of Fig. 2–23b.

From the origin to point A on the plot, the x vs. t graph is not a straight line, but a curve. The **slope** of the curve at any point is defined as the *slope of the tangent to the curve at that point*. (The *tangent* is a straight line drawn so it touches the curve only at that one point, but does not pass across or through the curve.) For example, the tangent to the curve at the time $t = 10.0\ \text{s}$ is drawn on the graph (it is labeled "tangent"). A triangle is drawn with Δt chosen to be 4.0 s; Δx can be measured off the graph for this chosen Δt and is found to be 40 m. Thus, the slope of the curve at $t = 10.0\ \text{s}$, which equals the instantaneous velocity at that instant, is $v = \Delta x / \Delta t = 40\ \text{m}/4.0\ \text{s} = 10\ \text{m/s}$. In the region between A and B (Fig. 2–23) the x vs. t graph is a straight line and the slope can be measured using the triangle shown between $t = 17\ \text{s}$ and $t = 20\ \text{s}$, where the increase in x is 45 m; $\Delta x / \Delta t = 45\ \text{m}/3.0\ \text{s} = 15\ \text{m/s}$.

Slope of a curve

Suppose we were given the x vs. t graph of Fig. 2–23b. We could measure the slopes at a number of points and plot these slopes as a function of time. Since the slope equals the velocity, we could thus reconstruct the v vs. t graph! In other words, given the graph of x vs. t, we can determine the velocity as a function of time using graphical methods, instead of using equations. This technique is particularly useful when the acceleration is not constant, for then Eqs. 2–10 cannot be used.

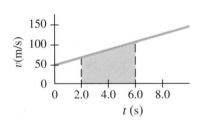

FIGURE 2–24 Determining the displacement from the graph of v vs. t is done by calculating areas.

Displacement = area under v vs. t graph

FIGURE 2–25 Example 2–16: the rose-shaded area represents the displacement during the time interval $t = 2.0$ s to $t = 6.0$ s.

The reverse process is also possible. If we are given the v vs. t graph, we can determine the position, x, as a function of time. To do so, we use the following procedure, which we apply to the v vs. t graph of Fig. 2–24a (which is the same as Fig. 2–23a). We first divide the time axis into many subintervals (in Fig. 2–24a, only six for simplicity), which are indicated in the figure by the dashed vertical lines. In each interval, a *horizontal* dashed line is drawn to indicate the average velocity during that time interval. For example, in the first interval, the velocity increases at a constant rate from zero to 5.0 m/s, so $\bar{v} = 2.5$ m/s; and in the fourth interval the velocity is a constant 15 m/s, so $\bar{v} = 15$ m/s (no horizontal dashed line is shown since it coincides with the curve itself). The displacement (change in position) during any subinterval is $\Delta x = \bar{v}\Delta t$. Thus the displacement during each subinterval equals the product of $\bar{v}$ and Δt, and this is just the *area of the rectangle* (base × height = $\Delta t × v$), shown shaded in rose, for that interval. The total displacement after 25 s, say, will be the sum of the first five rectangles.

If the velocity varies a great deal, it may be difficult to estimate $\bar{v}$ from the graph. To reduce this difficulty, more—but narrower—subintervals are used. That is, we make each Δt smaller, as in Fig. 2–24b. The more intervals give a better approximation. Ideally, we could let Δt approach zero; this leads to the techniques of integral calculus, which we don't discuss here. The result, in any case, is that *the total displacement between any two times is equal to the area under the v vs. t graph between these two times.*

EXAMPLE 2–16 **Displacement from graph.** A space probe accelerates uniformly from 50 m/s at $t = 0$ to 150 m/s at $t = 10$ s. How far did it move between $t = 2.0$ s and $t = 6.0$ s?

SOLUTION A graph of v vs. t can be drawn as shown in Fig. 2–25. We simply need to calculate the area of the shaded region shown in rose, which is a trapezoid. The area will be the average of the heights (in units of velocity) times the width (which is 4.0 s). At $t = 2.0$ s, $v = 70$ m/s; and at $t = 6.0$ s, $v = 110$ m/s. Thus the area, which equals Δx, is

$$\Delta x = \left(\frac{70 \text{ m/s} + 110 \text{ m/s}}{2}\right)(4.0 \text{ s}) = 360 \text{ m}.$$

For this case of constant acceleration, we could use Eqs. 2–10 and we would get the same result: $a = \Delta v/\Delta t = (150 \text{ m/s} - 50 \text{ m/s})/10 \text{ s} = 10 \text{ m/s}$; at $t = 2.0$ s, $v = v_0 + at = 50 \text{ m/s} + (10 \text{ m/s}^2)(2.0 \text{ s}) = 70 \text{ m/s}$, and at $t = 6.0$ s, $v = 50 \text{ m/s} + (10 \text{ m/s}^2)(6.0 \text{ s}) = 110 \text{ m/s}$; then, using Eq. 2–10c,

$$\Delta x = \frac{(v^2 - v_0^2)}{2a}$$

$$= \frac{(110 \text{ m/s})^2 - (70 \text{ m/s})^2}{2(10 \text{ m/s}^2)} = 360 \text{ m}.$$

In cases where the acceleration is not constant, the area can be obtained by counting squares on graph paper.

SUMMARY

[The Summary that appears at the end of each chapter in this book gives a brief overview of the main ideas of the chapter. The Summary *cannot* serve to give an understanding of the material, which can be accomplished only by a detailed reading of the chapter.]

Kinematics deals with the description of how objects move. The description of the motion of any object must always be given relative to some particular **reference frame**.

The **displacement** of an object is the change in position of the object.

Average speed is the distance traveled divided by the elapsed time. An object's **average velocity** over a particular time interval Δt is the displacement Δx divided by Δt:

$$\bar{v} = \frac{\Delta x}{\Delta t}.$$

The **instantaneous velocity**, whose magnitude is the same as the *instantaneous speed*, is the average velocity taken over an infinitesimally short time interval.

Acceleration is the change of velocity per unit time. An object's **average acceleration** over a time interval Δt is

$$\bar{a} = \frac{\Delta v}{\Delta t},$$

where Δv is the change of velocity during the time interval Δt. **Instantaneous acceleration** is the average acceleration taken over an infinitesimally short time interval.

If an object moves in a straight line with constant acceleration, the velocity v and position x are related to the acceleration a, the elapsed time t, and the initial position x_0 and initial velocity v_0, by Eqs. 2–10:

$$v = v_0 + at, \qquad x = x_0 + v_0 t + \tfrac{1}{2} a t^2,$$

$$v^2 = v_0^2 + 2a(x - x_0), \qquad \bar{v} = \frac{v + v_0}{2}.$$

Objects that move vertically near the surface of the Earth, either falling or having been projected vertically up or down, move with the constant downward **acceleration due to gravity** with magnitude of about $g = 9.80\ \text{m/s}^2$, if air resistance can be ignored.

QUESTIONS

1. Does a car speedometer measure speed, velocity, or both?

2. Can an object have a varying velocity if its speed is constant? If yes, give examples.

3. Can an object have a varying speed if its velocity is constant? If yes, give examples.

4. When an object moves with constant velocity, does its average velocity during any time interval differ from its instantaneous velocity at any instant?

5. In drag racing, is it possible for the car with the greatest speed crossing the finish line to lose the race? Explain.

6. If one object has a greater speed than a second object, does the first necessarily have a greater acceleration? Explain, using examples.

7. Compare the acceleration of a motorcycle that accelerates from 80 km/h to 90 km/h with the acceleration of a bicycle that accelerates from rest to 10 km/h in the same time.

8. How is speed represented on a speedometer? How is acceleration represented?

9. Can an object have a northward velocity and a southward acceleration? Explain.

10. Can the velocity of an object be negative when its acceleration is positive? What about vice versa?

11. Give an example where both the velocity and acceleration are negative.

12. Is it possible for an object to have a negative acceleration while increasing in speed? If so, provide an example.

13. Two cars emerge side by side from a tunnel. Car A is traveling with a speed of 60 km/h and has an acceleration of 40 km/h/min. Car B has a speed of 40 km/h and has an acceleration of 60 km/h/min. Which car is passing the other as they come out of the tunnel? Explain your reasoning.

14. Can an object be increasing in speed as its acceleration decreases? If so, give an example. If not, explain.

15. As a freely falling object speeds up, what is happening to its acceleration due to gravity? Does it increase, decrease, or stay the same?

16. How would you estimate the maximum height you could throw a ball vertically upward? How would you estimate the maximum speed you could give it?

17. An object that is thrown vertically upward will return to its original position with the same speed as it had initially if air resistance is negligible. If air resistance is appreciable, will this result be altered, and if so, how? [*Hint:* The acceleration due to air resistance is always in a direction opposite to the motion.]

* 18. Describe in words the motion plotted in Fig. 2–26 in terms of v, a, etc. [*Hint:* First try to duplicate the motion plotted by walking or moving your hand.]

* 19. Describe in words the motion of the object graphed in Fig. 2–27.

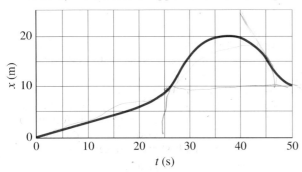

FIGURE 2–26   Question 18, Problems 51, 52, and 57.

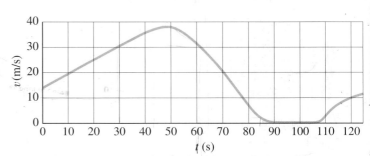

FIGURE 2–27   Question 19, Problems 53, 56, and 58.

PROBLEMS

[The problems at the end of each chapter are ranked I, II, or III according to estimated difficulty, with I problems being easiest. The problems are arranged by Section, meaning that the reader should have read up to and including that Section, but not only that Section—problems often depend on earlier material. Finally, there is a set of unranked "General Problems" not arranged by Section number.]

SECTIONS 2–1 TO 2–3

1. (I) What must be your average speed in order to travel 230 km in 3.25 h?

2. (I) A bird can fly 25 km/h. How long does it take to fly 15 km?

3. (I) If you are driving 110 km/h along a straight road and you look to the side for 2.0 s, how far do you travel during this inattentive period?

4. (I) 65 mph is how many (*a*) km/h, (*b*) m/s, and (*c*) ft/s?

5. (II) You are driving home from school steadily at 65 mph for 130 miles. It then begins to rain and you slow to 55 mph. You arrive home after driving 3 hours and 20 minutes. (*a*) How far is your hometown from school? (*b*) What was your average speed?

6. (II) According to a rule-of-thumb, every five seconds between a lightning flash and the following thunder gives the distance of the storm in miles. Assuming that the flash of light arrives in essentially no time at all, estimate the speed of sound in m/s from this rule.

7. (II) A person jogs eight complete laps around a quarter-mile track in a total time of 12.5 min. Calculate (*a*) the average speed and (*b*) the average velocity, in m/s.

8. (II) A horse canters away from its trainer in a straight line, moving 130 m away in 14.0 s. It then turns abruptly and gallops halfway back in 4.8 s. Calculate (*a*) its average speed and (*b*) its average velocity for the entire trip, using "away from the trainer" as the positive direction.

9. (II) Two locomotives approach each other on parallel tracks. Each has a speed of 95 km/h with respect to the ground. If they are initially 8.5 km apart, how long will it be before they reach each other? (See Fig. 2–28.)

FIGURE 2–28   Problem 9.

10. (II) An airplane travels 2100 km at a speed of 800 km/h, and then encounters a tailwind that boosts its speed to 1000 km/h for the next 1800 km. What was the total time for the trip? What was the average speed of the plane for this trip? [*Hint*: Think carefully before using Eq. 2–10d.]

11. (II) Calculate the average speed and average velocity of a complete round-trip in which the outgoing 200 km is covered at 90 km/h, followed by a one-hour lunch break, and the return 200 km is covered at 50 km/h.

12. (III) A bowling ball traveling with constant speed hits the pins at the end of a bowling lane 16.5 m long. The bowler hears the sound of the ball hitting the pins 2.50 s after the ball is released from his hands. What is the speed of the ball? The speed of sound is 340 m/s.

SECTION 2–4

13. (I) A sports car accelerates from rest to 95 km/h in 6.2 s. What is its average acceleration in m/s^2?

14. (I) At highway speeds, a particular automobile is capable of an acceleration of about $1.6 \, m/s^2$. At this rate, how long does it take to accelerate from 80 km/h to 110 km/h?

15. (I) A sprinter accelerates from rest to 10.0 m/s in 1.35 s. What is her acceleration (a) in m/s^2, and (b) in km/h^2?

16. (II) A sports car is advertised to be able to stop in a distance of 50 m from a speed of 90 km/h. What is its acceleration in m/s^2? How many g's is this ($g = 9.80 \, m/s^2$)?

17. (III) The position of a racing car, which starts from rest at $t = 0$ and moves in a straight line, has been measured as a function of time, as given in the following table. Estimate (a) its velocity and (b) its acceleration as a function of time. Display each in a table and on a graph.

t(s)	0	0.25	0.50	0.75	1.00	1.50	2.00	2.50
x(m)	0	0.11	0.46	1.06	1.94	4.62	8.55	13.79

t(s)	3.00	3.50	4.00	4.50	5.00	5.50	6.00
x(m)	20.36	28.31	37.65	48.37	60.30	73.26	87.16

SECTIONS 2–5 AND 2–6

18. (I) The principal kinematic equations, Eqs. 2–10a through 2–10d, become particularly simple if the initial speed is zero. Write down the equations for this special case. (Also put $x_0 = 0$.)

19. (I) A car accelerates from 12 m/s to 25 m/s in 6.0 s. What was its acceleration? How far did it travel in this time? Assume constant acceleration.

20. (I) A car slows down from 20 m/s to rest in a distance of 85 m. What was its acceleration, assumed constant?

21. (I) A light plane must reach a speed of 30 m/s for takeoff. How long a runway is needed if the (constant) acceleration is $3.0 \, m/s^2$?

22. (II) A world-class sprinter can burst out of the blocks to essentially top speed (of about 11.5 m/s) in the first 15.0 m of the race. What is the average acceleration of this sprinter and how long does it take her to reach that speed?

23. (II) A car slows down from a speed of 25.0 m/s to rest in 5.00 s. How far did it travel in that time?

24. (II) In coming to a stop, a car leaves skid marks 80 m long on the highway. Assuming a deceleration of $7.00 \, m/s^2$, estimate the speed of the car just before braking.

25. (II) A car traveling 45 km/h slows down at a constant $0.50 \, m/s^2$ just by "letting up on the gas." Calculate (a) the distance the car coasts before it stops, (b) the time it takes to stop, and (c) the distance it travels during the first and fifth seconds.

26. (II) A car traveling at 90 km/h strikes a tree. The front end of the car compresses and the driver comes to rest after traveling 0.80 m. What was the average acceleration of the driver during the collision? Express the answer in terms of "g's," where $1.00 \, g = 9.80 \, m/s^2$.

27. (II) Determine the stopping distances for an automobile with an initial speed of 90 km/h and human reaction time of 1.0 s: (a) for an acceleration $a = -4.0 \, m/s^2$; (b) for $a = -8.0 \, m/s^2$.

28. (III) Show that the equation for the stopping distance of a car is $d_S = v_0 t_R - v_0^2/(2a)$, where v_0 is the initial speed of the car, t_R is the driver's reaction time, and a is the constant acceleration (and is negative).

29. (III) A speeding motorist traveling 120 km/h passes a stationary police officer. The officer immediately begins pursuit at a constant acceleration of 10.0 km/h/s (note the mixed units). How much time will it take for the police officer to reach the speeder, assuming that the speeder maintains a constant speed? How fast will the police officer be traveling at this time?

30. (III) A person driving her car at 50 km/h approaches an intersection just as the traffic light turns yellow. She knows that the yellow light lasts only 2.0 s before turning to red, and she is 30 m away from the near side of the intersection (Fig. 2–29). Should she try to stop, or should she make a run for it? The intersection is 15 m wide. Her car's maximum deceleration is $-6.0 \, m/s^2$, whereas it can accelerate from 50 km/h to 70 km/h in 6.0 s. Ignore the length of her car and her reaction time.

FIGURE 2–29 Problem 30.

31. (III) A runner hopes to complete the 10,000-m run in less than 30.0 min. After exactly 27.0 min, there are still 1100 m to go. The runner must then accelerate at 0.20 m/s^2 for how many seconds in order to achieve the desired time?

SECTION 2–7 [neglect air resistance]

32. (I) Calculate the acceleration of the baseball in Example 2–9 in "g's."

33. (I) If a car rolls gently ($v_0 = 0$) off a vertical cliff, how long does it take it to reach 90 km/h?

34. (I) A stone is dropped from the top of a cliff. It is seen to hit the ground below after 3.50 s. How high is the cliff?

35. (I) Calculate (a) how long it took King Kong to fall straight down from the top of the Empire State Building (380 m high), and (b) his velocity just before "landing"?

36. (II) A foul ball is hit straight up into the air with a speed of about 25 m/s. (a) How high does it go? (b) How long is it in the air?

37. (II) A kangaroo jumps to a vertical height of 2.7 m. How long was it in the air before returning to Earth?

38. (II) A ballplayer catches a ball 3.3 s after throwing it vertically upward. With what speed did he throw it, and what height did it reach?

39. (II) Draw graphs of (a) the speed and (b) the distance fallen, as a function of time, for an object falling under the influence of gravity from $t = 0$ to $t = 5.00$ s. Ignore air resistance and assume $v_0 = 0$.

40. (II) The best rebounders in basketball have a vertical leap (that is, the vertical movement of a fixed point on their body) of about 120 cm. (a) What is their initial "launch" speed off the ground? (b) How long are they in the air?

41. (II) A helicopter is ascending vertically with a speed of 5.50 m/s. At a height of 105 m above the Earth, a package is dropped from a window. How much time does it take for the package to reach the ground?

42. (II) For an object falling freely from rest, show that the distance traveled during each successive second increases in the ratio of successive odd integers (1, 3, 5, etc.). (This was first shown by Galileo.) See Figs. 2–16 and 2–19.

43. (II) If air resistance is neglected, show (algebraically) that a ball thrown vertically upward with a speed v_0 will have the same speed, v_0, when it comes back down to the starting point.

44. (II) A stone is thrown vertically upward with a speed of 20.0 m/s. (a) How fast is it moving when it reaches a height of 12.0 m? (b) How long is required to reach this height? (c) Why are there two answers to (b)?

45. (II) Estimate the time between each photoflash of the apple in Fig. 2–16 (or number of photoflashes per second). Assume the apple is about 10 cm in diameter.

FIGURE 2–30   Problem 46.

46. (III) A falling stone takes 0.30 s to travel past a window 2.2 m tall (Fig. 2–30). From what height above the top of the window did the stone fall?

47. (III) A rock is dropped from a sea cliff and the sound of it striking the ocean is heard 3.4 s later. If the speed of sound is 340 m/s, how high is the cliff?

48. (III) Suppose you adjust your garden hose nozzle for a hard stream of water. You point the nozzle vertically upward at a height of 1.5 m above the ground (Fig. 2–31). When you quickly move the nozzle away from the vertical, you hear the water striking the ground next to you for another 2.0 s. What is the water speed as it leaves the nozzle?

FIGURE 2–31   Problem 48.

FIGURE 2–32
Problem 49.

49. (III) A stone is thrown vertically upward with a speed of 12.0 m/s from the edge of a cliff 75.0 m high (Fig. 2–32). (a) How much later does it reach the bottom of the cliff? (b) What is its speed just before hitting? (c) What total distance did it travel?

50. (III) A baseball is seen to pass upward by a window 25 m above the street with a vertical speed of 12 m/s. If the ball was thrown from the street, (a) what was its initial speed, (b) what altitude does it reach, (c) when was it thrown, and (d) when does it reach the street again?

* **SECTION 2–8**

* 51. (I) The position of a rabbit along a straight tunnel as a function of time is plotted in Fig. 2–26. What is its instantaneous velocity (a) at $t = 10.0$ s and (b) at $t = 30.0$ s? What is its average velocity (c) between $t = 0$ and $t = 5.0$ s, (d) between $t = 25.0$ s and $t = 30.0$ s, and (e) between $t = 40.0$ s and $t = 50.0$ s?

* 52. (I) In Fig. 2–26, (a) during what time periods, if any, is the object's velocity constant? (b) At what time is its velocity the greatest? (c) At what time, if any, is the velocity zero? (d) Does the object run in one direction or in both along its tunnel during the time shown?

* 53. (I) Figure 2–27 shows the velocity of a train as a function of time. (a) At what time was its velocity greatest? (b) During what periods, if any, was the velocity constant? (c) During what periods, if any, was the acceleration constant? (d) When was the magnitude of the acceleration greatest?

* 54. (II) A high-performance automobile can accelerate approximately as shown in the velocity–time graph of Fig. 2–33. (The short flat spots in the curve represent shifting of the gears.) (a) Estimate the average acceleration of the car in second gear and in fourth gear. (b) Estimate how far the car traveled while in fourth gear.

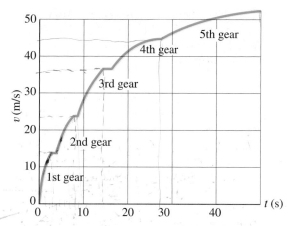

FIGURE 2–33 The velocity of a high-performance automobile as a function of time, starting from a dead stop. The jumps in the curve represent gear shifts. (Problems 54 and 55.)

* 55. (II) Estimate the average acceleration of the car in the previous problem (Fig. 2–33) when it is in (a) first, (b) third, and (c) fifth gear. (d) What is its average acceleration through the first four gears?

* 56. (II) In Fig. 2–27, estimate the distance the object traveled during (a) the first minute and (b) the second minute.

* 57. (II) Construct the v vs. t graph for the object whose displacement as a function of time is given by Fig. 2–26.

* 58. (II) Construct an x vs. t graph for the object whose velocity as a function of time is given by Fig. 2–27.

* 59. (II) Figure 2–34 is a position versus time graph for the motion of an object along the x axis. As the object moves from A to B: (a) Is the object moving in the positive or negative direction? (b) Is the object speeding up or slowing down? (c) Is the acceleration of the object positive or negative? Next, for the time interval from D to E: (d) Is the object moving in the positive or negative direction? (e) Is the object speeding up or slowing down? (f) Is the acceleration of the object positive or negative? (g) Finally, answer these same three questions for the time interval from C to D.

FIGURE 2–34 Problem 59.

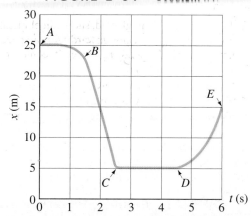

60. A person jumps from a fourth-story window 15.0 m above a firefighter's safety net. The survivor stretches the net 1.0 m before coming to rest, Fig. 2–35. (a) What was the average deceleration experienced by the survivor when slowed to rest by the net? (b) What would you do to make it "safer" (that is, generate a smaller deceleration): would you stiffen or loosen the net? Explain.

15.0 m

1.0 m

FIGURE 2–35 Problem 60.

61. The acceleration due to gravity on the Moon is about one sixth what it is on Earth. If an object is thrown vertically upward on the Moon, how many times higher will it go than it would on Earth, assuming the same initial velocity?

62. A person who is properly constrained by an over-the-shoulder seat belt has a good chance of surviving a car collision if the deceleration does not exceed 30 "g's" (1.00 g = 9.80 m/s^2). Assuming uniform deceleration of this value, calculate the distance over which the front end of the car must be designed to collapse if a crash brings the car to rest from 100 km/h.

63. A race car driver must average 200.0 km/h over the course of a time trial lasting ten laps. If the first nine laps were done at 199.0 km/h, what average speed must be maintained for the last lap?

64. A car manufacturer tests its cars for front-end collisions by hauling them up on a crane and dropping them from a certain height. (a) Show that the speed just before a car hits the ground, after falling from rest a vertical distance H, is given by $\sqrt{2gH}$. What height corresponds to a collision at (b) 50 km/h? (c) 100 km/h?

65. A first stone is dropped from the roof of a building. 2.00 s after that, a second stone is thrown straight down with an initial speed of 30.0 m/s, and it is observed that the two stones land at the same time. (a) How long did it take the first stone to reach the ground? (b) How high is the building? (c) What are the speeds of the two stones just before they hit the ground?

66. A 90-m-long train begins uniform acceleration from rest. The front of the train has a speed of 20 m/s when it passes a railway worker who is standing 180 m from where the front of the train started. What will be the speed of the last car as it passes the worker? (See Fig. 2–36.)

90 m

v = 20 m/s

FIGURE 2–36 Problem 66.

67. A police car at rest, passed by a speeder traveling at a constant 110 km/h, takes off in hot pursuit. The police officer catches up to the speeder in 700 m, maintaining a constant acceleration. (a) Qualitatively plot the position versus time graph for both cars from the police car's start to the catch-up point. (b) Calculate how long it took the police officer to overtake the speeder, (c) calculate the required police car acceleration, and (d) calculate the speed of the police car at the overtaking point.

68. In the design of a rapid transit system, it is necessary to balance out the average speed of a train against the distance between stops. The more stops there are, the slower the train's average speed. To get an idea of this problem, calculate the time it takes a train to make a 36-km trip in two situations: (a) the stations at which the trains must stop are 0.80 km apart; and (b) the stations are 3.0 km apart. Assume that at each station the train accelerates at a rate of 1.1 m/s^2 until it reaches 90 km/h, then stays at this speed until its brakes are applied for arrival at the next station, at which time it decelerates at -2.0 m/s^2. Assume it stops at each intermediate station for 20 s.

69. Pelicans tuck their wings and free fall straight down when diving for fish. Suppose a pelican starts its dive from a height of 16.0 m and cannot change its path once committed. If it takes a fish 0.20 s to perform evasive action, at what minimum height must it spot the pelican to escape? Assume the fish is at the surface of the water.

70. In putting, the force with which a golfer strikes a ball is planned so that the ball will stop within some small distance of the cup, say 1.0 m long or short, in case the putt is missed. Accomplishing this from an uphill lie (that is, putting downhill, see Fig. 2–37) is more difficult than from a downhill lie. To see why, assume that on a particular green the ball decelerates constantly at 2.0 m/s^2 going downhill, and constantly at 3.0 m/s^2 going uphill. Suppose we have an uphill lie 7.0 m from the cup. Calculate the allowable range of initial velocities we may impart to the ball so that it stops in the range 1.0 m short to 1.0 m long of the cup. Do the same for a downhill lie 7.0 m from the cup. What in your results suggests that the downhill putt is more difficult?

71. A car is behind a truck going 25 m/s on the highway. The driver looks for an opportunity to pass, guessing that his car can accelerate at 1.0 m/s^2, and he gauges that he has to cover the 20-m length of the truck, plus 10 m clear room at the rear of the truck and 10 m more at the front of it. In the oncoming lane, he sees a car approaching, probably also traveling at 25 m/s. He estimates that the car is about 400 m away. Should he attempt the pass? Give details.

72. A stone is dropped from the roof of a high building. A second stone is dropped 1.50 s later. How far apart are the stones when the second one has reached a speed of 12.0 m/s?

73. Bond is standing on a bridge, 10 m above the road below, and his pursuers are getting too close for comfort. He spots a flatbed truck loaded with mattresses approaching at 30 m/s, which he measures by knowing that the telephone poles the truck is passing are 20 m apart in this country. The bed of the truck is 1.5 m above the road, and Bond quickly calculates how many poles away the truck should be when he jumps down from the bridge onto the truck, making his getaway. How many poles is it?

FIGURE 2–37 Problem 70. Golf on Wednesday morning.

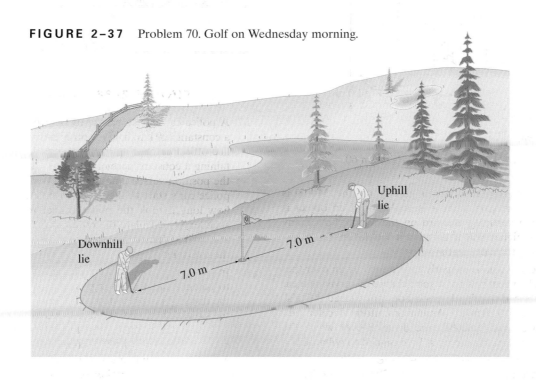

This multiflash photograph of a ping pong ball shows examples of motion in two dimensions. The arcs of the ping pong ball are parabolas that represent "projectile motion." Galileo analyzed projectile motion into its horizontal and vertical components, under the action of gravity (the gold arrow represents the downward acceleration of gravity, **g**).

3 KINEMATICS IN TWO DIMENSIONS; VECTORS

FIGURE 3–1 Car traveling on a road. The green arrows represent the velocity vector at each position.

Scale for velocity:
1 cm = 90 km/h

In Chapter 2 we dealt with motion along a straight line. We now consider the description of the motion of objects that move in paths in two (or three) dimensions. To do so we first need to discuss vectors and how they are added.

3–1 Vectors and Scalars

We mentioned in Chapter 2 that the term *velocity* refers not only to how fast something is moving but also to its direction. A quantity such as velocity, which has *direction* as well as *magnitude*, is a **vector** quantity. Other quantities that are also vectors are displacement, force, and momentum. However, many quantities such as mass, time, and temperature, have no direction associated with them. They are specified completely by giving a number and units. Such quantities are called **scalars**.

Drawing a diagram of a particular physical situation is always helpful in physics, and this is especially true when dealing with vectors. On a diagram, each vector is represented by an arrow. The arrow is always drawn so that it points in the direction of the vector it represents. The length of the arrow is drawn proportional to the magnitude of the vector. For example, in Fig. 3–1, arrows have been drawn representing the velocity of a car at various places as it rounds a curve. The magnitude of the velocity at each point can be read off this figure by measuring the length of the corresponding arrow and using the scale shown (1 cm = 90 km/h).

When we write the symbol for a vector, we will always use boldface type. Thus for velocity we write **v**. (In handwritten work, the symbol for a vector can be indicated by putting an arrow over it, a $\vec{v}$ for velocity.) If we are concerned only with the magnitude of the vector, we will write simply v, in italics.

3–2 Addition of Vectors—Graphical Methods

Because vectors are quantities that have direction as well as magnitude, they must be added in a special way. In this chapter, we will deal mainly with displacement vectors (for which we now use the symbol **D**) and velocity vectors (**v**). But the results will apply for other vectors we encounter later.

We use simple arithmetic for adding scalars. Simple arithmetic can also be used for adding vectors if they are in the same direction. For example, if a person walks 8 km east one day, and 6 km east the next day, the person will be 8 km + 6 km = 14 km east of the point of origin. We say that the *net* or *resultant* displacement is 14 km to the east (Fig. 3–2a). If, on the other hand, the person walks 8 km east on the first day, and 6 km west (in the reverse direction) on the second day, then the person will end up 2 km from the origin (Fig. 3–2b), so the resultant displacement is 2 km to the east. In this case, the resultant displacement is obtained by subtraction: 8 km − 6 km = 2 km.

(a)

(b)

FIGURE 3–2 Combining vectors in one dimension.

But simple arithmetic cannot be used if the two vectors are not along the same line. For example, suppose a person walks 10.0 km east and then walks 5.0 km north. These displacements can be represented on a graph in which the positive y axis points north and the positive x axis points east, Fig. 3–3. On this graph, we draw an arrow, labeled **D**$_1$, to represent the displacement vector of the 10.0-km displacement to the east. Then we draw a second arrow, **D**$_2$, to represent the 5.0-km displacement to the north. Both vectors are drawn to scale, as in Fig. 3–3.

After taking this walk, the person is now 10.0 km east and 5.0 km north of the point of origin. The **resultant displacement** is represented by the arrow labeled **D**$_R$ in Fig. 3–3. Using a ruler and a protractor, you can measure on this diagram that the person is 11.2 km from the origin at an angle of 27° north of east. In other words, the resultant displacement vector has a magnitude of 11.2 km and makes an angle $\theta = 27°$ with the positive x axis. The magnitude (length) of **D**$_R$ can also be obtained using the theorem of Pythagoras in this case, since D_1, D_2, and D_R form a right triangle with D_R as the hypotenuse. Thus

$$D_R = \sqrt{D_1^2 + D_2^2} = \sqrt{(10.0 \text{ km})^2 + (5.0 \text{ km})^2} = \sqrt{125 \text{ km}^2} = 11.2 \text{ km}.$$

You can use the Pythagorean theorem, of course, only when the vectors are *perpendicular* to each other.

FIGURE 3–3 A person walks 10.0 km east and then 5.0 km north. These two displacements are represented by the vectors **D**$_1$ and **D**$_2$, which are shown as arrows. The resultant displacement vector, **D**$_R$, which is the vector sum of **D**$_1$ and **D**$_2$, is also shown. Measurement on the graph with ruler and protractor shows that **D**$_R$ has a magnitude of 11.2 km and points at an angle $\theta = 27°$ north of east.

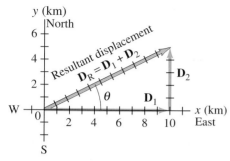

FIGURE 3–3 (Repeated from previous page.) A person walks 10.0 km east and then 5.0 km north. The resultant vector has magnitude $D_R = 11.2$ km at an angle $\theta = 27°$ north of east.

The resultant displacement vector, $\mathbf{D}_R$, is the sum of the vectors $\mathbf{D}_1$ and $\mathbf{D}_2$. That is,

$$\mathbf{D}_R = \mathbf{D}_1 + \mathbf{D}_2.$$

This is a *vector* equation. An important feature of adding two vectors that are not along the same line is that the magnitude of the resultant vector is not equal to the sum of the magnitudes of the two separate vectors, but is smaller than their sum:

$$D_R < D_1 + D_2. \qquad \text{[Vectors not along the same line]}$$

In our example (Fig. 3–3), $D_R = 11.2$ km, whereas $D_1 + D_2$ equals 15 km. We generally are not interested in $D_1 + D_2$; rather we are interested in the *vector* sum of the two vectors and its magnitude, D_R. Note also that we cannot set $\mathbf{D}_R$ equal to 11.2 km, because we have a vector equation and 11.2 km is only a part of the resultant vector, its magnitude. We could write something like this, though: $\mathbf{D}_R = \mathbf{D}_1 + \mathbf{D}_2 = (11.2 \text{ km}, 27° \text{ N of E})$.

Figure 3–3 illustrates the general rules for graphically adding two vectors together, no matter what angles they make, to get their sum. The rules are as follows:

Tail-to-tip method

of

adding vectors

1. On a diagram, draw one of the vectors—call it $\mathbf{V}_1$—to scale.
2. Next draw the second vector, $\mathbf{V}_2$, to scale, placing its tail at the tip of the first vector and being sure its direction is correct.
3. The arrow drawn from the tail of the first vector to the tip of the second represents the *sum*, or **resultant**, of the two vectors.

Note that vectors can be translated parallel to themselves to accomplish these manipulations. The length of the resultant can be measured with a ruler and compared to the scale. Angles can be measured with a protractor. This method is known as the **tail-to-tip method of adding vectors**.

Note that it is not important in which order the vectors are added. For example, a displacement of 5.0 km north, to which is added a displacement of 10.0 km east, yields a resultant of 11.2 km and angle $\theta = 27°$ (see Fig. 3–4), the same as when they were added in reverse order (Fig. 3–3). That is,

$$\mathbf{V}_1 + \mathbf{V}_2 = \mathbf{V}_2 + \mathbf{V}_1.$$

The tail-to-tip method of adding vectors can be extended to three or more vectors. The resultant is drawn from the tail of the first vector to the tip of the last one added. An example is shown in Fig. 3–5; the three vectors could represent displacements (northeast, south, west) or perhaps three forces. Check for yourself that you get the same resultant no matter in which order you add the three vectors.

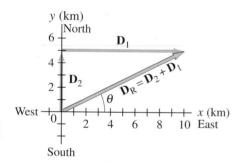

FIGURE 3–4 If the vectors are added in reverse order, the resultant is the same. (Compare Fig. 3–3.)

FIGURE 3–5
The resultant of three vectors, $\mathbf{V}_R = \mathbf{V}_1 + \mathbf{V}_2 + \mathbf{V}_3$.

A second way to add two vectors is the **parallelogram method**. It is fully equivalent to the tail-to-tip method. In this method, the two vectors are drawn starting from a common origin, and a parallelogram is constructed using these two vectors as adjacent sides as shown in Fig. 3–6b. The resultant is the diagonal drawn from the common origin. In Fig. 3–6a, the tail-to-tip method is shown, and it is clear that both methods yield the same result. It is a common error to draw the sum vector as the diagonal running between the tips of the two vectors, as in Fig. 3–6c. *This is incorrect*: it does not represent the sum of the two vectors. (In fact, it represents their difference, $\mathbf{V}_2 - \mathbf{V}_1$, as we will see in the next Section.)

Parallelogram method of adding vectors

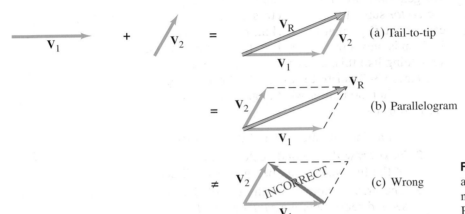

(a) Tail-to-tip

(b) Parallelogram

(c) Wrong

FIGURE 3–6 Vector addition by two different methods, (a) and (b). Part (c) is incorrect.

3–3 Subtraction of Vectors, and Multiplication of a Vector by a Scalar

Given a vector $\mathbf{V}$, we define the *negative* of this vector $(-\mathbf{V})$ to be a vector with the same magnitude as $\mathbf{V}$ but opposite in direction, Fig. 3–7. Note, however, that no vector is ever negative in the sense of its magnitude: the magnitude of every vector is positive. A minus sign tells us about its direction.

We can now define the subtraction of one vector from another: the difference between two vectors, $\mathbf{V}_2 - \mathbf{V}_1$ is defined as

$$\mathbf{V}_2 - \mathbf{V}_1 = \mathbf{V}_2 + (-\mathbf{V}_1).$$

That is, the difference between two vectors is equal to the sum of the first plus the negative of the second. Thus our rules for addition of vectors can be applied as shown in Fig. 3–8 using the tail-to-tip method.

FIGURE 3–7 The negative of a vector is a vector having the same length but opposite direction.

FIGURE 3–8 Subtracting two vectors: $\mathbf{V}_2 - \mathbf{V}_1$.

$V_2 = 1.5 V$

V

$V_3 = -2.0 V$

FIGURE 3–9 Multiplying a vector **V** by a scalar c gives a vector whose magnitude is c times greater and in the same direction as **V** (or opposite direction if c is negative).

A vector **V** can be multiplied by a scalar c. We define this product so that $c\mathbf{V}$ has the same direction as **V** and has magnitude cV. That is, multiplication of a vector by a positive scalar c changes the magnitude of the vector by a factor c but doesn't alter the direction. If c is a negative scalar, the magnitude of the product $c\mathbf{V}$ is still cV (without the minus sign), but the direction is precisely opposite to that of **V**. See Fig. 3–9.

3–4 Adding Vectors by Components

Adding vectors graphically using a ruler and protractor is often not sufficiently accurate and is not useful for vectors in three dimensions. We discuss now a more powerful and precise method for adding vectors.

Resolving a vector into components

Consider first a vector **V** that lies in a particular plane. It can be expressed as the sum of two other vectors, called the **components** of the original vector. The components are usually chosen to be along two perpendicular directions. The process of finding the components is known as **resolving the vector into its components**. An example is shown in Fig. 3–10; the vector **V** could be a displacement vector that points at an angle $\theta = 30°$ north of east, where we have chosen the positive x axis to be to the east and the positive y axis north. This vector **V** is resolved into its x and y components by drawing dashed lines from the tip (A) of the vector and drawing these lines perpendicular to the x and y axes (lines AB and AC). Then the lines OB and OC represent the x and y components of **V**, respectively, as shown in Fig. 3–10b. These *vector components* are written $\mathbf{V}_x$ and $\mathbf{V}_y$. We generally show vector components as arrows, like vectors, but dashed. The *scalar components*, V_x and V_y, are numbers, with units, that are given a positive or negative sign depending on whether they point along the positive or negative x or y axis. As can be seen in Fig. 3–10, $\mathbf{V}_x + \mathbf{V}_y = \mathbf{V}$ by the parallelogram method of adding vectors.

Vector components

FIGURE 3–10 Resolving a vector **V** into its components along an arbitrarily chosen set of x and y axes. Note that the components, once found, themselves represent the vector. That is, the components contain as much information as the vector itself.

(a)

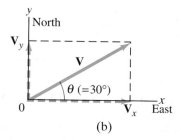

(b)

Space is made up of three dimensions, and sometimes it is necessary to resolve a vector into components along three mutually perpendicular directions. In rectangular coordinates the components are $\mathbf{V}_x$, $\mathbf{V}_y$, and $\mathbf{V}_z$. Resolution of a vector in three dimensions is merely an extension of the above technique. We will mainly be concerned with situations in which the vectors are in a plane and two components are all that are necessary.

In order to add vectors using the method of components, we need to use the trigonometric functions sine, cosine, and tangent, which we now review.

Given any angle, θ, as in Fig. 3–11a, a right triangle can be constructed by drawing a line perpendicular to either of its sides, as in Fig. 3–11b. The longest side of a right triangle, opposite the right angle, is called the hypotenuse, which we label h. The side opposite the angle θ is labeled o, and the side adjacent is labeled a. We let h, o, and a represent the lengths of these sides, respectively. We now define the three trigonometric functions, sine, cosine, and

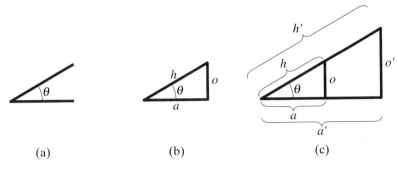

(a)   (b)   (c)

FIGURE 3–11 Starting with an angle θ as in (a), we can construct right triangles of different sizes, (b) and (c), but the ratio of the lengths of the sides does not depend on the size of the triangle.

tangent (abbreviated sin, cos, tan), in terms of the right triangle, as follows:

$$\sin \theta = \frac{\text{side opposite}}{\text{hypotenuse}} = \frac{o}{h}$$

$$\cos \theta = \frac{\text{side adjacent}}{\text{hypotenuse}} = \frac{a}{h} \qquad \textbf{(3–1)}$$

$$\tan \theta = \frac{\text{side opposite}}{\text{side adjacent}} = \frac{o}{a}.$$

Trig. functions defined

Now it is an interesting fact that if we make the triangle bigger, but keep the same angles, then the ratio of the length of one side to the other, or of one side to the hypotenuse, remains the same. That is, in Fig. 3–11c we have: $a/h = a'/h'$; $o/h = o'/h'$; and $o/a = o'/a'$. Thus the values of sine, cosine, and tangent do not depend on how big the triangle is. They depend only on the size of the angle. The values of sine, cosine, and tangent for different angles can be found using a scientific calculator, or from Tables (see inside rear cover).

A useful trigonometric identity is

$$\sin^2 \theta + \cos^2 \theta = 1 \qquad \textbf{(3–2)}$$

which follows from the Pythagorean theorem ($o^2 + a^2 = h^2$ in Fig. 3–11). That is:

$$\sin^2 \theta + \cos^2 \theta = \frac{o^2}{h^2} + \frac{a^2}{h^2} = \frac{o^2 + a^2}{h^2} = \frac{h^2}{h^2} = 1.$$

(See also Appendix A for other details on trigonometric functions and identities.)

The use of trigonometric functions for finding the components of a vector is illustrated in Fig. 3–12, where it is seen that a vector and its two components can be thought of as making up a right triangle. We then see that the sine, cosine, and tangent are as given in the figure. If we multiply the definition of $\sin \theta = V_y/V$ by V on both sides, we get

$$V_y = V \sin \theta. \qquad \textbf{(3–3a)}$$

Similarly, from the definition of $\cos \theta$, we obtain

$$V_x = V \cos \theta. \qquad \textbf{(3–3b)}$$

Note that θ is chosen (by convention) to be the angle that the vector makes with the positive x axis.[†]

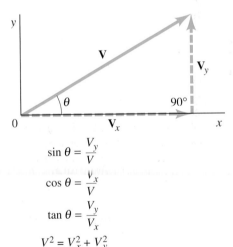

$$\sin \theta = \frac{V_y}{V}$$

$$\cos \theta = \frac{V_x}{V}$$

$$\tan \theta = \frac{V_y}{V_x}$$

$$V^2 = V_x^2 + V_y^2$$

FIGURE 3–12 Finding the components of a vector using trigonometric functions.

Components of a vector

[†]Whatever convention is used, the vector component opposite the angle is proportional to the sine, whether we call that component x or y. Most often we use the convention that it is the y component (Eq. 3–3a).

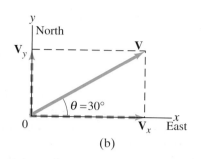

$V_y = V \sin \theta = 250 \, \text{m}$

$V_x = V \cos \theta = 433 \, \text{m}$

$V = \sqrt{V_x^2 + V_y^2} = 500 \, \text{m}$

FIGURE 3–13 (a) Vector **V** represents a displacement of 500 m at a 30° angle north of east. (b) The components of **V** are V_x and V_y whose magnitudes are given on the right.

Using Eqs. 3–3, we can calculate V_x and V_y for any vector, such as that illustrated in Fig. 3–10 or Fig. 3–12. Suppose **V** represents a displacement of 500 m in a direction 30° north of east, as shown in Fig. 3–13. Then $V = 500 \, \text{m}$. From the trigonometric tables, $\sin 30° = 0.500$ and $\cos 30° = 0.866$. Then

$$V_x = V \cos \theta = (500 \, \text{m})(0.866) = 433 \, \text{m (east)},$$

$$V_y = V \sin \theta = (500 \, \text{m})(0.500) = 250 \, \text{m (north)}.$$

Note that there are two ways to specify a vector in a given coordinate system:

Two ways to specify a vector

1. We can give its components, V_x and V_y.

2. We can give its magnitude V and the angle θ it makes with the positive x axis.

We can shift from one description to the other using Eqs. 3–3, and, for the reverse, by using the theorem of Pythagoras† and the definition of tangent:

Components related to magnitude and direction

$$V = \sqrt{V_x^2 + V_y^2} \tag{3–4a}$$

$$\tan \theta = \frac{V_y}{V_x} \tag{3–4b}$$

as can be seen in Fig. 3–12.

We can now discuss how to add vectors using components. The first step is to resolve each vector into its components. Next we can see, using Fig. 3–14,

†In three dimensions, the theorem of Pythagoras becomes $V = \sqrt{V_x^2 + V_y^2 + V_z^2}$, where V_z is the component along the third, or z, axis.

FIGURE 3–14
The components of
$\mathbf{V} = \mathbf{V}_1 + \mathbf{V}_2$ are
$V_x = V_{1x} + V_{2x}$ and
$V_y = V_{1y} + V_{2y}$.

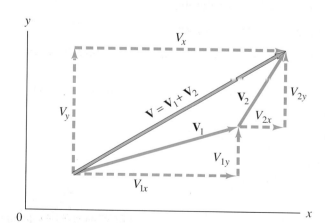

that the addition of any two vectors $\mathbf{V}_1$ and $\mathbf{V}_2$ to give a resultant, $\mathbf{V} = \mathbf{V}_1 + \mathbf{V}_2$, implies that

$$V_x = V_{1x} + V_{2x}$$
$$V_y = V_{1y} + V_{2y}.$$

(3–5)

Adding vectors analytically (by components)

That is, the sum of the x components equals the x component of the resultant, and similarly for y. That this is valid can be verified by a careful examination of Fig. 3–14. But note carefully that we add all the x components together to get the x component of the resultant; and we add all the y components together to get the y component of the resultant. We do *not* add x components to y components.

If the magnitude and direction of the resultant vector are desired, they can be obtained using Eqs. 3–4.

The choice of coordinate axes is, of course, always arbitrary. You can often reduce the work involved in adding vectors by a good choice of axes—for example, by choosing one of the axes to be in the same direction as one of the vectors. Then that vector will have only one nonzero component.

Choice of axes can simplify effort needed

EXAMPLE 3–1 **Mail carrier's displacement.** A rural mail carrier leaves the post office and drives 22.0 km in a northerly direction to the next town. She then drives in a direction 60.0° south of east for 47.0 km (Fig. 3–15a) to another town. What is her displacement from the post office?

SOLUTION We want to find her resultant displacement from the origin. We choose the positive x axis to be east and the positive y axis north, and resolve each displacement vector into its components (Fig. 3–15b). Since $\mathbf{D}_1$ has magnitude 22.0 km and points north, it has only a y component:

$$D_{1x} = 0, \quad D_{1y} = 22.0 \text{ km}$$

whereas $\mathbf{D}_2$ has both x and y components:

$$D_{2x} = +(47.0 \text{ km})(\cos 60°) = +(47.0 \text{ km})(0.500) = +23.5 \text{ km}$$
$$D_{2y} = -(47.0 \text{ km})(\sin 60°) = -(47.0 \text{ km})(0.866) = -40.7 \text{ km}.$$

Notice that D_{2y} is negative because this vector component points along the negative y axis. The resultant vector, $\mathbf{D}$, has components:

$$D_x = D_{1x} + D_{2x} = \quad 0 \text{ km} + \quad 23.5 \text{ km} = +23.5 \text{ km}$$
$$D_y = D_{1y} + D_{2y} = 22.0 \text{ km} + (-40.7 \text{ km}) = \quad 18.7 \text{ km}.$$

This specifies the resultant vector completely:

$$D_x = 23.5 \text{ km}, \quad D_y = -18.7 \text{ km}.$$

We can also specify the resultant vector by giving its magnitude and angle using Eqs. 3–4:

$$D = \sqrt{D_x^2 + D_y^2} = \sqrt{(23.5 \text{ km})^2 + (-18.7 \text{ km})^2} = 30.0 \text{ km}$$

$$\tan \theta = \frac{D_y}{D_x} = \frac{-18.7 \text{ km}}{23.5 \text{ km}} = -0.796.$$

A calculator with an INV TAN or TAN^{-1} key gives $\theta = \tan^{-1}(-0.796) = -38.5°$. The negative sign means $\theta = 38.5°$ below the x axis, Fig. 3–15c.

(a)

(b)

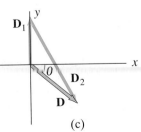

(c)

FIGURE 3–15 Example 3–1.

The signs of trigonometric functions depend on which "quadrant" the angle falls in: for example, the tangent is positive in the first and third quadrants (from 0° to 90°, and 180° to 270°), but negative in the second and fourth quadrants; see Appendix A–8. The best way to keep track of angles, and to check any vector result, is always to draw a vector diagram. A vector diagram gives you something tangible to look at when analyzing a problem, and provides a check on the results.

➡ PROBLEM SOLVING   Adding Vectors

Here is a brief summary of how to add two or more vectors using components:

1. Draw a diagram, adding the vectors graphically.

2. Choose x and y axes. Choose them in a way, if possible, that will make your work easier. (For example, choose one axis along the direction of one of the vectors so that vector will have only one component.)

3. Resolve each vector into its x and y components, showing each component along its appropriate (x or y) axis as a (dashed) arrow.

4. Calculate each component (when not given) using sines and cosines. If θ_1 is the angle vector $\mathbf{V}_1$ makes with the x axis, then:

$$V_{1x} = V_1 \cos \theta_1, \qquad V_{1y} = V_1 \sin \theta_1.$$

Pay careful attention to signs: any component that points along the negative x or y axis gets a negative sign.

5. Add the x components together to get the x component of the resultant. Ditto for y:

$$V_x = V_{1x} + V_{2x} + \text{any others}$$
$$V_y = V_{1y} + V_{2y} + \text{any others}.$$

This is the answer: the components of the resultant vector.

6. If you want to know the magnitude and direction of the resultant vector, use Eqs. 3–4:

$$V = \sqrt{V_x^2 + V_y^2}, \qquad \tan \theta = \frac{V_y}{V_x}.$$

The vector diagram you already drew helps to obtain the correct position (quadrant) of the angle θ.

EXAMPLE 3–2 **Three short trips.** An airplane trip involves three legs, with two stopovers, as shown in Fig. 3–16a. The first leg is due east for 620 km; the second leg is southeast (45°) for 440 km; and the third leg is at 53° south of west, for 550 km, as shown. What is the plane's total displacement?

FIGURE 3–16
Example 3–2.

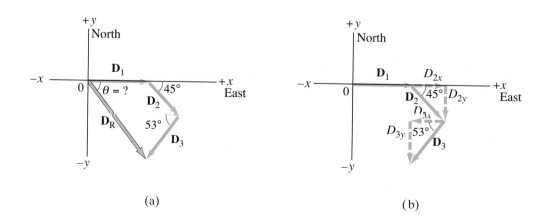

(a)   (b)

SOLUTION We follow the steps in the above Problem Solving box:
(1) and (2): Already shown in Fig. 3–16a, where we have taken the x axis as east (then $\mathbf{D}_1$ has only an x component).
(3): It is imperative to draw a good figure. The components are shown in Fig. 3–16b. Notice that instead of drawing all the vectors starting from a common origin, as we did in Fig. 3–15b, here we have drawn them "tail-to-tip" style, which is just as valid and may make it easier to see.
(4) Now we calculate the components:

$$\mathbf{D}_1:\ D_{1x} = +D_1 \cos 0° \ = D_1 = 620 \text{ km}$$

$$D_{1y} = +D_1 \sin 0° \ = 0 \text{ km}$$

$$\mathbf{D}_2:\ D_{2x} = +D_2 \cos 45° = +(440 \text{ km})(0.707) = +311 \text{ km}$$

$$D_{2y} = -D_2 \sin 45° = -(440 \text{ km})(0.707) = -311 \text{ km}$$

$$\mathbf{D}_3:\ D_{3x} = -D_3 \cos 53° = -(550 \text{ km})(0.602) = -331 \text{ km}$$

$$D_{3y} = -D_3 \sin 53° = -(550 \text{ km})(0.799) = -439 \text{ km}.$$

Note carefully that we have given a minus sign to each component that in Fig. 3–16b points in the negative x or negative y direction. We see why a good drawing is so important. We summarize the components in the table in the margin.
(5): This is easy:

Vector	Components	
	x (km)	y (km)
$\mathbf{D}_1$	620	0
$\mathbf{D}_2$	311	-311
$\mathbf{D}_3$	-331	-439
$\mathbf{D}_R$	600	-750

$$D_x = D_{1x} + D_{2x} + D_{3x} = 620 \text{ km} + 311 \text{ km} - 331 \text{ km} = \ 600 \text{ km}$$

$$D_y = D_{1y} + D_{2y} + D_{3y} = \ 0 \text{ km} - 311 \text{ km} - 439 \text{ km} = -750 \text{ km}.$$

The x and y components are 600 km and -750 km, and point respectively to the east and south. This is one way to give the answer.
(6): We can also give the answer as

$$D_R = \sqrt{D_x^2 + D_y^2} = \sqrt{(600)^2 + (-750)^2} \text{ km} = 960 \text{ km}$$

$$\tan \theta = \frac{D_y}{D_x} = \frac{-750 \text{ km}}{600 \text{ km}} = -1.25, \qquad \text{so } \theta = -51°,$$

where we assume only two significant figures. Thus, the total displacement has magnitude 960 km and points 51° below the x axis (south of east), as was shown in our original sketch, Fig. 3–16a.

3–5 Projectile Motion

In Chapter 2, we studied the motion of objects in one dimension in terms of displacement, velocity, and acceleration, including purely vertical motion of falling bodies undergoing acceleration due to gravity. Now we examine the more general motion of objects moving through the air in two dimensions near the Earth's surface, such as a golf ball, a thrown or batted baseball, kicked footballs, speeding bullets, and athletes doing the long jump or high

FIGURE 3–17 This strobe photograph of a soccer ball in the air shows the characteristic "parabolic" path of projectile motion.

jump. These are all examples of **projectile motion** (see Fig. 3–17), which we can describe as taking place in two dimensions. Although air resistance is often important, in many cases its effect can be ignored, and we will ignore it in the following analysis. We will not be concerned now with the process by which the object is thrown or projected. We consider only its motion *after* it has been projected and is moving freely through the air under the action of gravity alone. Thus the acceleration of the object is that due to gravity, which acts downward with magnitude $g = 9.80 \, \text{m/s}^2$, and we assume it is constant.[†]

Horizontal and vertical motion analyzed separately

Galileo first accurately described projectile motion. He showed that it could be understood by analyzing the horizontal and vertical components of the motion separately. This was an innovative analysis, not done in this way

[†]This restricts us to objects whose distance traveled and maximum height above the Earth are small compared to the Earth's radius (6400 km).

FIGURE 3–18 Projectile motion. (A vertically falling object is shown at the left for comparison.)

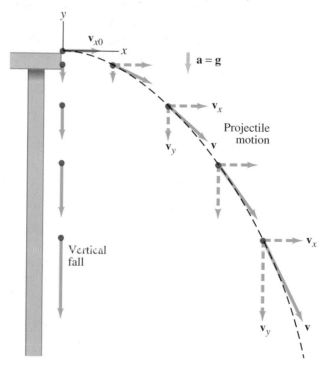

by anyone prior to Galileo. (It was also idealized in that it did not take into account air resistance.) For convenience, we assume that the motion begins at time $t = 0$ at the origin of an xy coordinate system (so $x_0 = y_0 = 0$).

Let us look at a (tiny) ball rolling off the end of a table with an initial velocity v_{x0} in the horizontal (x) direction. See Fig. 3–18 (also shown is an object falling vertically, for comparison). The velocity vector $\mathbf{v}$ at each instant points in the direction of the ball's motion at that instant and is always tangent to the path. Following Galileo's ideas, we treat the horizontal and vertical components of the velocity, v_x and v_y, separately, and we can apply the kinematic equations (Eqs. 2–10a through 2–10d) to each.

First we examine the vertical (y) component of the motion. Once the ball leaves the table (at $t = 0$), it experiences a vertically downward acceleration, g, the acceleration due to gravity. Thus v_y is initially zero but increases continually in the downward direction (until the ball hits the ground). Let us take y to be positive upwards. Then $a_y = -g$, and from Eq. 2–10a we can write $v_y = -gt$ since the initial velocity in the vertical direction (v_{y0}) is zero. The vertical displacement, y, is given by $y = -\frac{1}{2}gt^2$, if we set $y_0 = 0$.

In the horizontal direction, on the other hand, there is no acceleration. So the horizontal component of velocity, v_x, remains constant, equal to its initial value, v_{x0}, and thus has the same magnitude at each point on the path. The two vector components, $\mathbf{v}_x$ and $\mathbf{v}_y$, can be added vectorially to obtain the velocity $\mathbf{v}$ for each point on the path, as shown in Fig. 3–18.

One result of this analysis, which Galileo himself predicted, is that *an object projected horizontally will reach the ground in the same time as an object dropped vertically*. This is because the vertical motions are the same in both cases, as shown in Fig. 3–18 where on the left a falling object is shown. Figure 3–19 is a multiple-exposure photograph of an experiment that confirms this.

If an object is projected at an upward angle, as in Fig. 3–20, the analysis is similar, except that now there is an initial vertical component of velocity, v_{y0}. Because of the downward acceleration of gravity, v_y continually decreases until the object reaches the highest point on its path in Fig. 3–20, at which point $v_y = 0$. Then v_y starts to increase in the downward direction, as shown (that is, becoming negative). As before, v_x remains constant.

FIGURE 3–19 Multiple-exposure photograph showing positions of two balls at equal time intervals. One ball was dropped from rest at the same time the other was projected horizontally outward. The vertical position of each ball is seen to be the same.

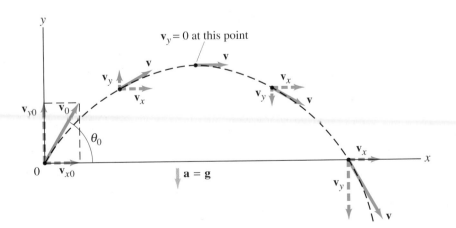

FIGURE 3–20 Path of a projectile fired with initial velocity $\mathbf{v}_0$ at angle θ to the horizontal. Path is shown in black, the velocity vectors are green arrows, and velocity components are dashed.

3–6 Solving Problems Involving Projectile Motion

We now work through several Examples of projectile motion quantitatively. We use the kinematic equations (2–10a through 2–10c) separately for the vertical and horizontal components of the motion. These equations are shown separately for the x and y components of the motion in Table 3–1, for the general case of two-dimensional motion. Note that x and y are the respective displacements, that v_x and v_y are the components of the velocity, and that a_x and a_y are the components of the acceleration. The subscript $_0$ means "at $t = 0$."

TABLE 3–1 General Kinematic Equations for Constant Acceleration in Two Dimensions

x Component (horizontal)		y Component (vertical)
$v_x = v_{x0} + a_x t$	(Eq. 2–10a)	$v_y = v_{y0} + a_y t$
$x = x_0 + v_{x0}t + \frac{1}{2}a_x t^2$	(Eq. 2–10b)	$y = y_0 + v_{y0}t + \frac{1}{2}a_y t^2$
$v_x^2 = v_{x0}^2 + 2a_x (x - x_0)$	(Eq. 2–10c)	$v_y^2 = v_{y0}^2 + 2a_y (y - y_0)$

We can simplify these equations for the case of projectile motion because we can set $a_x = 0$. See Table 3–2, which assumes y is positive upwards, so $a_y = -g = -9.80 \text{ m/s}^2$. Note that if θ is chosen relative to the $+x$ axis, as in Fig. 3–20, then $v_{x0} = v_0 \cos \theta$, and $v_{y0} = v_0 \sin \theta$.

TABLE 3–2 Kinematic Equations for Projectile Motion (y positive upward; $a_x = 0$, $a_y = -g = -9.80 \text{ m/s}^2$)

Horizontal Motion ($a_x = 0, v_x = $ constant)		Vertical Motion[†] ($a_y = -g = $ constant)
$v_x = v_{x0}$	(Eq. 2–10a)	$v_y = v_{y0} - gt$
$x = x_0 + v_{x0}t$	(Eq. 2–10b)	$y = y_0 + v_{y0}t - \frac{1}{2}gt^2$
	(Eq. 2–10c)	$v_y^2 = v_{y0}^2 - 2gy$

[†]If y is taken positive downward, the minus $(-)$ signs become $+$ signs.

The approach to solving problems that we discussed in Section 2–6 also applies here. However, solving problems involving projectile motion can require a bit of creativity, and cannot be done just by simply following some rules. Certainly you must avoid just plugging numbers into equations that seem to "work."

As always, **read** carefully and **draw** a careful diagram.

1. Choose an origin and an xy coordinate system.

2. Analyze the horizontal (x) motion and the vertical (y) motion separately. If you are given the initial velocity, you may want to resolve it into its x and y components.

3. List the known and unknown quantities, choosing $a_x = 0$ and $a_y = -g$ or $+g$, where $g = 9.80 \text{ m/s}^2$, depending on whether you choose y positive up or down. Remember that v_x never changes throughout the trajectory, and that $v_y = 0$ at the highest point of any trajectory that returns downward. The velocity just before hitting the ground is generally not zero.

4. Think for a minute before jumping into the equations. A little planning goes a long way. Apply the relevant equations (Table 3–2), combining equations if necessary. You may need to combine components of a vector to get magnitude and direction (Eqs. 3–4).

EXAMPLE 3–3 **Driving off a cliff.** A movie stunt driver on a motorcycle speeds horizontally off a 50.0-m-high cliff. How fast must the motorcycle leave the cliff-top if it is to land on level ground below, 90.0 m from the base of the cliff (Fig. 3–21) where the cameras are?

SOLUTION We take the y direction to be positive upward, with the top of the cliff as $y_0 = 0$, so the bottom is at $y = -50.0$ m. First, we find how long it takes the motorcycle to reach the ground below. We use Eq. 2–10b for the vertical (y) direction (Table 3–2) with $y_0 = 0$, and $v_{y0} = 0$:

$$y = -\tfrac{1}{2}gt^2$$

We solve for t and set $y = -50.0$ m:

$$t = \sqrt{\frac{2y}{g}} = \sqrt{\frac{2(-50.0 \text{ m})}{9.80 \text{ m/s}^2}} = 3.19 \text{ s}.$$

To calculate the initial velocity, v_{x0}, we again use Eq. 2–10b, but this time for the horizontal (x) direction, with $a_x = 0$ and $x_0 = 0$:

$$x = v_{x0}t$$

$$v_{x0} = \frac{x}{t} = \frac{90.0 \text{ m}}{3.19 \text{ s}} = 28.2 \text{ m/s},$$

which is 101 km/h.

FIGURE 3–21
Example 3–3.

50.0 m

$y = -50.0$ m

90.0 m

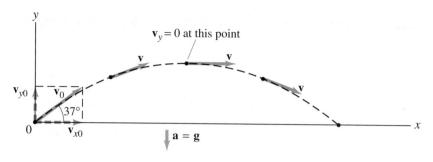

FIGURE 3–22   Example 3–4.

EXAMPLE 3–4 **A kicked football.** A football is kicked at an angle $\theta_0 = 37.0°$ with a velocity of 20.0 m/s, as shown in Fig. 3–22. Calculate (*a*) the maximum height, (*b*) the time of travel before the football hits the ground, (*c*) how far away it hits the ground, (*d*) the velocity vector at the maximum height, and (*e*) the acceleration vector at maximum height. Assume the ball leaves the foot at ground level.

SOLUTION This may seem difficult because there are so many questions. But we can deal with them one at a time. We take the *y* direction as positive upward. The components of the initial velocity are (Fig. 3–22):

$$v_{x0} = v_0 \cos 37.0° = (20.0 \text{ m/s})(0.799) = 16.0 \text{ m/s}$$

$$v_{y0} = v_0 \sin 37.0° = (20.0 \text{ m/s})(0.602) = 12.0 \text{ m/s}.$$

(*a*) At the maximum height, the velocity is horizontal (Fig. 3–22), so $v_y = 0$; and this occurs (see Eq. 2–10a in Table 3–2) at time

$$t = v_{y0}/g = (12.0 \text{ m/s})/(9.80 \text{ m/s}^2) = 1.22 \text{ s}.$$

From Eq. 2–10b, with $y_0 = 0$, we have

$$y = v_{y0}t - \tfrac{1}{2}gt^2$$

$$= (12.0 \text{ m/s})(1.22 \text{ s}) - \tfrac{1}{2}(9.80 \text{ m/s}^2)(1.22 \text{ s})^2 = 7.35 \text{ m}.$$

Alternatively, we could have used Eq. 2–10c, solved for *y*, and found

$$y = \frac{v_{y0}^2 - v_y^2}{2g} = \frac{(12.0 \text{ m/s})^2 - (0 \text{ m/s})^2}{2(9.80 \text{ m/s}^2)} = 7.35 \text{ m}.$$

(*b*) To find the time it takes for the ball to return to the ground, we use Eq. 2–10b with $y_0 = 0$ and also set $y = 0$ (ground level):

$$y = y_0 + v_{y0}t - \tfrac{1}{2}gt^2$$

$$0 = 0 + (12.0 \text{ m/s})t - \tfrac{1}{2}(9.80 \text{ m/s}^2)t^2$$

which is an equation that can be easily factored:

$$[\tfrac{1}{2}(9.80 \text{ m/s}^2)t - 12.0 \text{ m/s}]t = 0.$$

There are two solutions, $t = 0$ (which corresponds to the initial point, y_0), and

$$t = \frac{2(12.0 \text{ m/s})}{(9.80 \text{ m/s}^2)} = 2.45 \text{ s},$$

which is the result we sought.

(c) The total distance traveled in the x direction is found by applying Eq. 2–10b with $x_0 = 0$, $a_x = 0$, $v_{x0} = 16.0$ m/s:

$$x = v_{x0}t = (16.0 \text{ m/s})(2.45 \text{ s}) = 39.2 \text{ m}.$$

(d) At the highest point, there is no vertical component to the velocity. There is only the horizontal component (which remains constant throughout the flight), so $v = v_{x0} = v_0 \cos 37.0° = 16.0$ m/s.

(e) The acceleration vector is the same at the highest point as it is throughout the flight, which is 9.80 m/s^2 downward.

CONCEPTUAL EXAMPLE 3–5 **Where does the apple land?** A child sits upright in a wagon which is moving to the right at constant speed as shown in Fig. 3–23. The child extends her hand and throws an apple straight upward (from her own point of view, Fig. 3–23a), while the wagon continues to travel forward at constant speed. If air resistance is neglected, will the apple land (a) behind the wagon, (b) in the wagon, or (c) in front of the wagon?

RESPONSE The child throws the apple straight up from her own point of view with initial velocity $\mathbf{v}_{0y}$ (Fig. 3–23a). But when viewed by someone on the ground, the apple also has an initial horizontal component of velocity equal to the speed of the wagon, $\mathbf{v}_{0x}$. Thus, to a person on the ground, the apple will follow the path of a projectile as shown in Fig. 3–23b. The apple experiences no horizontal acceleration, so $\mathbf{v}_{0x}$ will stay constant and equal to the speed of the wagon. As the apple follows its arc, the wagon will be directly under the apple at all times because they have the same horizontal velocity. When the apple comes down, it will drop right into the wagon, and into the outstretched hand of the child. The answer is (b).

(a) Wagon reference frame

(b) Ground reference frame

FIGURE 3–23
Conceptual Example 3–5.

CONCEPTUAL EXAMPLE 3–6 **The wrong strategy.** A boy on a small hill aims his water-balloon slingshot horizontally, straight at a second boy hanging from a tree branch a distance d away, Fig. 3–24. At the instant the water balloon is released, the second boy lets go and falls from the tree, hoping to avoid being hit. Show that he made the wrong move. (He hadn't studied physics yet.)

RESPONSE Both the water balloon and the boy in the tree start falling at the same instant, and in a time t they each fall the same vertical distance $y = \frac{1}{2}gt^2$. In the time it takes the water balloon to travel the horizontal distance d, the balloon will have the same y position as the falling boy. Splat. If the boy had stayed in the tree, he'd have saved himself the humiliation.

FIGURE 3–24
Conceptual Example 3–6.

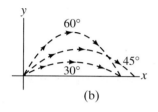

FIGURE 3–25 Example 3–7. (a) The range R of a projectile; (b) shows how generally there are two angles θ_0 that will give the same range. Can you show that if one angle is θ_{01}, the other is $\theta_{02} = 90° - \theta_{01}$?

Level range formula
[y (final) = y_0]

EXAMPLE 3–7 **Level horizontal range.** (a) Derive a formula for the horizontal range R of a projectile in terms of its initial velocity v_0 and angle θ_0. The horizontal range is defined as the horizontal distance the projectile travels before returning to its original height (which is typically the ground); that is, y (final) $= y_0$. See Fig. 3–25. (b) Suppose one of Napoleon's cannons had a muzzle velocity, v_0, of 60.0 m/s. At what angle should it have been aimed (ignore air resistance) to strike a target 320 m away?

SOLUTION (a) We set $x_0 = 0$ and $y_0 = 0$ at $t = 0$. After the projectile travels a horizontal distance R, it returns to the same level, $y = 0$, the final point. So to find a general expression for R, we set both $y = 0$ and $y_0 = 0$ in Eq. 2–10b for the vertical motion, and obtain

$$v_{y0}t - \tfrac{1}{2}gt^2 = 0.$$

We solve for t, which gives two solutions: $t = 0$ and $t = 2v_{y0}/g$. The first solution corresponds to the initial instant of projection and the second is the time when the projectile returns to $y = 0$. Then the range, R, will be equal to x at the moment t has this value, which we put into Eq. 2–10b for the horizontal motion ($x = v_{x0}t$, with $x_0 = 0$). Thus we have:

$$R = x = v_{x0}t = v_{x0}\left(\frac{2v_{y0}}{g}\right) = \frac{2v_{x0}v_{y0}}{g} = \frac{2v_0^2 \sin\theta_0 \cos\theta_0}{g} \qquad [y = y_0]$$

where we have written $v_{x0} = v_0 \cos\theta_0$ and $v_{y0} = v_0 \sin\theta_0$. This is the result we sought, and it can be rewritten, using the trigonometric identity $2\sin\theta\cos\theta = \sin 2\theta$ (Appendix A), as

$$R = \frac{v_0^2 \sin 2\theta_0}{g}. \qquad [y = y_0]$$

We see that the maximum range, for a given initial velocity, v_0, is obtained when the sine takes on its maximum value of 1.0, which occurs for $2\theta_0 = 90°$; so

$$\theta_0 = 45° \text{ for maximum range, and } R_{max} = v_0^2/g.$$

[When air resistance is important, the range is less for a given v_0, and the maximum range is obtained at an angle smaller than 45°.] Note that the maximum range increases by the square of v_0, so doubling the muzzle velocity of a cannon increases its maximum range by a factor of 4.

(b) From the equation we just derived, Napoleon's cannon should be aimed (assuming, unrealistically, no air resistance) at an angle θ_0 given by

$$\sin 2\theta_0 = \frac{Rg}{v_0^2} = \frac{(320 \text{ m})(9.80 \text{ m/s}^2)}{(60.0 \text{ m/s})^2} = 0.871.$$

We want to solve for an angle θ_0 that is between 0° and 90°, which means $2\theta_0$ in this equation can be as large as 180°. Thus, $2\theta_0 = 60.6°$ is a solution, but $2\theta_0 = 180° - 60.6° = 119.4°$ is also a solution (see Appendix A–8). In general we will have two solutions, which in Napoleon's case are given by

$$\theta_0 = 30.3° \text{ or } 59.7°.$$

Either angle gives the same range. Only when $\sin 2\theta_0 = 1$ (so $\theta_0 = 45°$) is there a single solution (that is, both solutions are the same).

FIGURE 3–26
Example 3–8: the football leaves
the punter's foot at $y = 0$, and
reaches the ground where
$y = -1.00$ m.

EXAMPLE 3–8 **A punt.** Suppose the football in Example 3–4 was a punt and left the punter's foot at a height of 1.00 m above the ground. How far did the football travel before hitting the ground? Set $x_0 = 0$, $y_0 = 0$.

SOLUTION We cannot use the range formula from Example 3–7 because it is valid only if y (final) $= y_0$, which is not the case here. Now we have $y_0 = 0$, and the football hits the ground where $y = -1.00$ m (see Fig. 3–26). We can get x from Eq. 2–10b, $x = v_{x0}t$, since we know that $v_{x0} = 16.0$ m/s. But first we must find t, the time at which the ball hits the ground. With $y = -1.00$ m and $v_{y0} = 12.0$ m/s (see Example 3–4), we use the equation

➡ PROBLEM SOLVING

Do not use any formula unless you are sure its range of validity fits the problem. The range formula does not apply here because $y \neq y_0$

$$y = y_0 + v_{y0}t - \tfrac{1}{2}gt^2,$$

and obtain

$$-1.00 \text{ m} = 0 + (12.0 \text{ m/s})t - (4.90 \text{ m/s}^2)t^2.$$

We rearrange this equation into standard form so we can use the quadratic formula (Appendix A–4):

$$(4.90 \text{ m/s}^2)t^2 - (12.0 \text{ m/s})t - (1.00 \text{ m}) = 0.$$

Using the quadratic formula gives

$$t = \frac{12.0 \text{ m/s} \pm \sqrt{(12.0 \text{ m/s})^2 - 4(4.90 \text{ m/s}^2)(-1.00 \text{ m})}}{2(4.90 \text{ m/s}^2)}$$

$$= 2.53 \text{ s} \quad \text{or} \quad -0.081 \text{ s}.$$

The second solution would correspond to a time previous to the kick, so it doesn't apply here. With $t = 2.53$ s for the time at which the ball touches the ground, the distance the ball traveled is (putting $v_{x0} = 16.0$ m/s, from Example 3–4):

$$x = v_{x0}t = (16.0 \text{ m/s})(2.53 \text{ s}) = 40.5 \text{ m}.$$

Note that our assumption in Example 3–4 that the ball leaves the foot at ground level results in an underestimate of about 1.3 m in the distance traveled.

(a)

(b)

FIGURE 3–27 Examples of projectile motion—sparks (small hot glowing pieces of metal) and fireworks. Both exhibit the parabolic path characteristic of projectile motion, although the effects of air resistance can be seen to alter the path of some trajectories.

* 3–7 Projectile Motion Is Parabolic

We now show that the path followed by any projectile is a parabola, if we can ignore air resistance and can assume that **g** is constant. To do so, we need to find y as a function of x by eliminating t between the two equations for horizontal and vertical motion (Eq. 2–10b), and we set $x_0 = y_0 = 0$:

$$x = v_{x0}t$$

$$y = v_{y0}t - \tfrac{1}{2}gt^2.$$

From the first equation, we have $t = x/v_{x0}$, and we substitute this into the second one to obtain

$$y = \left(\frac{v_{y0}}{v_{x0}}\right)x - \left(\frac{g}{2v_{x0}^2}\right)x^2.$$

If we write $v_{x0} = v_0 \cos\theta_0$ and $v_{y0} = v_0 \sin\theta_0$, we can also write

$$y = (\tan\theta_0)x - \left(\frac{g}{2v_0^2 \cos^2\theta_0}\right)x^2.$$

In either case, we see that y as a function of x has the form

$$y = ax - bx^2,$$

where a and b are constants for any specific projectile motion. This is the well-known equation for a parabola. See Figs. 3–17 and 3–27.

The idea that projectile motion is parabolic was, in Galileo's day, at the forefront of physics research. Today we discuss it in Chapter 3 of introductory physics!

* 3–8 Relative Velocity

We now consider how observations made in different reference frames are related to each other. For example, consider two trains approaching one another, each with a speed of 80 km/h with respect to the Earth. Observers on the Earth beside the tracks will measure 80 km/h for the speed of each of the trains. Observers on either of the trains (a different reference frame) will measure a speed of 160 km/h for the other train approaching them. Similarly, when one car traveling 90 km/h passes a second car traveling in the same direction at 75 km/h, the first car has a speed relative to the second car of 90 km/h − 75 km/h = 15 km/h.

When the velocities are along the same line, simple addition or subtraction is sufficient to obtain the relative velocity. But if they are not along the same line, we must make use of vector addition. We emphasize, as mentioned in Section 2–1, that when specifying a velocity, it is important to specify what the reference frame is.

When determining relative velocity, it is easy to make a mistake by adding or subtracting the wrong velocities. It is important, therefore, to draw a diagram and use a careful labeling process that makes things clear. Each velocity is labeled by *two subscripts: the first refers to the object, the second to the reference frame in which it has this velocity*. For example, suppose a boat is to cross a river to the opposite side, as shown in Fig. 3–28. We let $\mathbf{v}_{BW}$ be the velocity of the **B**oat with respect to the **W**ater. (This is also what the boat's velocity would be relative to the shore if the water were still.) Similarly, $\mathbf{v}_{BS}$ is the velocity of the **B**oat with respect to the **S**hore, and $\mathbf{v}_{WS}$ is the velocity of the **W**ater with respect to the **S**hore (this is the river current). Note that $\mathbf{v}_{BW}$ is what the boat's motor produces (against the water), whereas $\mathbf{v}_{BS}$ is equal to $\mathbf{v}_{BW}$ plus the effect of the current. Therefore, the velocity of the boat relative to the shore is (see vector diagram, Fig. 3–28)

➡ **PROBLEM SOLVING**

Subscripts for adding velocities: first subscript for the object; second for the reference frame

$$\mathbf{v}_{BS} = \mathbf{v}_{BW} + \mathbf{v}_{WS}. \tag{3–6}$$

Follow the subscripts

By writing the subscripts via the convention above, we see that the inner subscripts (the two W's) on the right-hand side of Eq. 3–6 are the same, whereas the outer subscripts on the right of Eq. 3–6 (the B and the S) are the same as the two subscripts for the sum vector on the left, $\mathbf{v}_{BS}$. By following this convention (first subscript for the object, second for the reference frame), one can write down the correct equation relating velocities in different reference frames.[†] Equation 3–6 is valid in general and can be extended to three or more velocities. For example, if a fisherman on the boat walks with a velocity $\mathbf{v}_{FB}$ relative to the boat, his velocity relative to the shore is $\mathbf{v}_{FS} = \mathbf{v}_{FB} + \mathbf{v}_{BW} + \mathbf{v}_{WS}$. The equations involving relative velocity will be correct when adjacent inner subscripts are identical and when the outermost ones correspond exactly to the two on the velocity on the left of the equation. But this works only with plus signs (on the right), not minus signs.

It is often useful to remember that for any two objects or reference frames, A and B, the velocity of A relative to B has the same magnitude, but opposite direction, as the velocity of B relative to A:

$$\mathbf{v}_{BA} = -\mathbf{v}_{AB}. \tag{3–7}$$

For example, if a train is traveling 100 km/h relative to the Earth in a certain direction, objects on the Earth (such as trees) appear to an observer on the train to be traveling 100 km/h in the opposite direction.

FIGURE 3–28 The boat must head upstream at an angle θ if it is to move directly across the river. Velocity vectors are shown as green arrows:

$\mathbf{v}_{BS}$ = velocity of **B**oat with respect to the **S**hore,

$\mathbf{v}_{BW}$ = velocity of **B**oat with respect to the **W**ater,

$\mathbf{v}_{WS}$ = velocity of the **W**ater with respect to the **S**hore (river current).

CONCEPTUAL EXAMPLE 3–9 | **Crossing a river.** A man in a rowboat is trying to cross a river that flows due west with a strong current. The man starts on the south bank and is trying to reach the north bank directly north from his starting point. He should:

1. head due north.
2. head due west.
3. head in a northwesterly direction.
4. head in a northeasterly direction.

RESPONSE The current will drag the boat westward, so to counteract that motion the boat must head in a northeasterly direction (see Fig. 3–28). The actual angle depends on the strength of the current and how fast the boat moves relative to the water. If the current is weak and the rower is strong, then the boat can head almost, but not quite, due north.

[†]We thus would know by inspection that (for example) the equation $\mathbf{v}_{BW} = \mathbf{v}_{BS} + \mathbf{v}_{WS}$ is wrong.

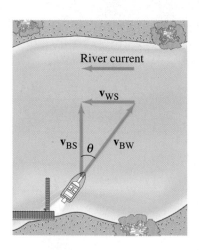

FIGURE 3–29
Example 3–10.

FIGURE 3–30 Example 3–11: a boat heading directly across a river whose current moves at 1.20 m/s.

EXAMPLE 3–10 **Heading upstream.** A boat's speed in still water is $v_{BW} = 1.85$ m/s. If the boat is to travel directly across a river whose current has speed $v_{WS} = 1.20$ m/s, at what upstream angle must the boat head? (See Fig. 3–29.)

SOLUTION Figure 3–29 has been drawn with $\mathbf{v}_{BS}$, the velocity of the **B**oat relative to the **S**hore, pointing directly across the river since this is how the boat is supposed to move. (Note that $\mathbf{v}_{BS} = \mathbf{v}_{BW} + \mathbf{v}_{WS}$.) To accomplish this, the boat needs to head upstream to offset the current pulling it downstream. Thus, $\mathbf{v}_{BW}$ points upstream at an angle θ as shown. From the diagram,

$$\sin\theta = \frac{v_{WS}}{v_{BW}} = \frac{1.20 \text{ m/s}}{1.85 \text{ m/s}} = 0.6486.$$

Thus $\theta = 40.4°$, so the boat must head upstream at a 40.4° angle.

EXAMPLE 3–11 **Heading across the stream.** The same boat ($v_{BW} = 1.85$ m/s) now heads directly across the stream whose current is still 1.20 m/s. (a) What is the velocity (magnitude and direction) of the boat relative to the shore? (b) If the river is 110 m wide, how long will it take to cross and how far downstream will the boat be then?

SOLUTION (a) As shown in Fig. 3–30, the boat is pulled downstream by the current. The boat's velocity with respect to the shore, $\mathbf{v}_{BS}$, is the sum of its velocity with respect to the water, $\mathbf{v}_{BW}$, plus the velocity of the water with respect to the shore, $\mathbf{v}_{WS}$:

$$\mathbf{v}_{BS} = \mathbf{v}_{BW} + \mathbf{v}_{WS},$$

just as before. Since $\mathbf{v}_{BW}$ is perpendicular to $\mathbf{v}_{WS}$, we can get v_{BS} using the theorem of Pythagoras:

$$v_{BS} = \sqrt{v_{BW}^2 + v_{WS}^2} = \sqrt{(1.85 \text{ m/s})^2 + (1.20 \text{ m/s})^2} = 2.21 \text{ m/s}.$$

We can obtain the angle (note how θ is defined in diagram) from:

$$\tan\theta = v_{WS}/v_{BW} = (1.20 \text{ m/s})/(1.85 \text{ m/s}) = 0.6486.$$

A calculator with an INV TAN or TAN^{-1} key gives $\theta = \tan^{-1}(0.6486) = 33.0°$. Note that this angle is not equal to the angle calculated in Example 3–10. (b) Given the river's width $D = 120$ m and using the definition of velocity, we solve for $t = D/v_{BW}$, where we use the velocity component in the direction of D; so $t = 110$ m/1.85 m/s $= 60$ s. The boat will have been carried downstream, in this time, a distance

$$d = v_{WS}t = (1.20 \text{ m/s})(60 \text{ s}) = 72 \text{ m}.$$

EXAMPLE 3–12 **Airplane with crosswind.** A plane whose airspeed is 200 km/h heads due north. But a 100-km/h northeast wind (that is, coming from the northeast) suddenly begins to blow. What is the resulting velocity of the plane with respect to the ground?

SOLUTION The two velocity vectors, and their components, are shown in Fig. 3–31a. They are drawn with a common origin for convenience; $\mathbf{v}_{PA}$ represents the velocity of the plane with respect to the air; and the wind velocity is $\mathbf{v}_{AG}$, the velocity of the air with respect to the ground. The resultant velocity, $\mathbf{v}_{PG}$, the velocity of the plane with respect to the ground, is given by:

$$\mathbf{v}_{PG} = \mathbf{v}_{PA} + \mathbf{v}_{AG}.$$

Note the use of our subscript rule, Eq. 3–6 above. Since $\mathbf{v}_{PA}$ is along the y axis, it has only a y component:

$$v_{PAx} = 0 \text{ km/h}$$

$$v_{PAy} = v_{PA} = 200 \text{ km/h}.$$

The components of v_{AG} are (note unconventional choice of 45° angle, below negative x axis):

$$v_{AGx} = -v_{AG} \cos 45° = -(100 \text{ km/h})(0.707) = -70.7 \text{ km/h}$$

$$v_{AGy} = -v_{AG} \sin 45° = -(100 \text{ km/h})(0.707) = -70.7 \text{ km/h}.$$

Both v_{AGx} and v_{AGy} are negative because their directions are, respectively, along the negative x and negative y axes. The components of the resultant velocity are

$$v_{PGx} = \quad 0 \text{ km/h} - 70.7 \text{ km/h} = -70.7 \text{ km/h},$$

$$v_{PGy} = 200 \text{ km/h} - 70.7 \text{ km/h} = +129 \text{ km/h}.$$

We find the magnitude of the resultant velocity using the Pythagorean theorem:

$$v_{PG} = \sqrt{v_{PGx}^2 + v_{PGy}^2} = 147 \text{ km/h}.$$

To find the angle θ that $\mathbf{v}_{PG}$ makes with the x axis (Fig. 3–31b), we use

$$\tan \theta = \frac{v_{PGy}}{v_{PGx}} = \frac{129 \text{ km/h}}{-70.7 \text{ km/h}} = -1.825.$$

(The negative sign results because θ is with respect to the negative x axis, which we already know from the diagram.) Then

$$\theta = \tan^{-1}(-1.825) = -61.3°,$$

(a)

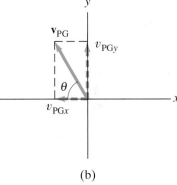

(b)

FIGURE 3–31
Example 3–12.

SUMMARY

A quantity that has both a magnitude and a direction is called a **vector**. A quantity that has only a magnitude is called a **scalar**.

Addition of vectors can be done graphically by placing the tail of each successive arrow at the tip of the previous one. The sum, or **resultant vector**, is the arrow drawn from the tail of the first to the tip of the last. Two vectors can also be added using the parallelogram method.

Vectors can be added more accurately by adding their **components** along chosen axes with the aid of trigonometric functions. A vector of magnitude V making an angle θ with the x axis has components

$$V_x = V \cos \theta \qquad V_y = V \sin \theta.$$

Given the components, we can find the magnitude and direction of a vector from

$$V = \sqrt{V_x^2 + V_y^2}, \qquad \tan \theta = \frac{V_y}{V_x}.$$

Projectile motion of an object moving in an arc near the Earth's surface can be analyzed as two separate motions if air resistance can be ignored. The horizontal component of motion is at constant velocity, whereas the vertical component is at constant acceleration, $\mathbf{g}$, just as for a body falling vertically under the action of gravity.

The velocity of an object relative to one frame of reference can be found by vector addition if its velocity relative to a second frame of reference, and the *relative velocity* of the two reference frames, are known.

QUESTIONS

1. Does the odometer of a car measure a scalar or a vector quantity? What about the speedometer?

2. Two vectors are added together, and their vector sum is zero. What can you say about the magnitude and direction of the two initial vectors?

3. Can the displacement vector for an object moving in two dimensions ever be longer than the length of path traveled by the object over the same time interval? Can it ever be less? Discuss.

4. During baseball practice, a batter hits a very high fly ball, and then runs in a straight line and catches it. Which had the greater displacement, the player or the ball?

5. If $\mathbf{V} = \mathbf{V}_1 + \mathbf{V}_2$, is V necessarily greater than V_1 and/or V_2? Discuss.

6. Two vectors have magnitudes $V_1 = 3.5$ km and $V_2 = 4.0$ km. What are the maximum and minimum magnitudes of their vector sum?

7. Can two vectors, of unequal magnitude, ever add up to give the zero vector? Can *three* unequal vectors? Under what conditions?

8. Can the magnitude of a vector ever (*a*) equal, or (*b*) be less than, one of its components?

9. Can a vector of magnitude zero have a nonzero component?

10. One car travels due east at 50 km/h, and a second car travels north at 50 km/h. Are their velocities equal? Explain.

11. A projectile has the least speed at what point in its path?

12. What physical factors are important for an athlete doing the long jump? What about the high jump?

13. A child wishes to determine the speed a slingshot imparts to a rock. How can this be done using only a meter stick, a rock, and the slingshot?

14. If you are riding on a train that speeds past another train moving in the same direction on an adjacent track, it appears that the slower moving train is moving backwards. Why?

15. Two rowers, who can row at the same speed in still water, set off across a river at the same time. One heads straight across and is pulled downstream somewhat by the current. The other one heads upstream at an angle so as to arrive at a point opposite the starting point. Which rower reaches the opposite side first?

PROBLEMS

SECTIONS 3–2 TO 3–4

1. (I) A car is driven 125 km west and then 65 km southwest. What is the displacement of the car from the point of origin (magnitude and direction)? Draw a diagram.

2. (I) A delivery truck travels 14 blocks north, 16 blocks east, and 26 blocks south. What is its final displacement from the origin? Assume the blocks are equal length.

3. (I) The three vectors in Fig. 3–32 can be added in six different orders ($\mathbf{V}_1 + \mathbf{V}_2 + \mathbf{V}_3$, $\mathbf{V}_1 + \mathbf{V}_3 + \mathbf{V}_2$, etc.). Show on a diagram that the same resultant is obtained no matter what the order.

4. (I) If $V_x = 18.80$ units and $V_y = -16.40$ units, determine the magnitude and direction of $\mathbf{V}$.

5. (II) Graphically determine the resultant of the following three vector displacements: (1) 24 m, 30° north of east; (2) 28 m, 37° east of north; and (3) 20 m, 50° west of south.

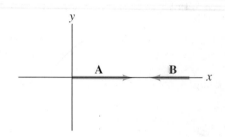

FIGURE 3–33   Problem 7.

6. (II) $\mathbf{V}$ is a vector 24.3 units in magnitude and points at an angle of 54.8° above the negative x axis. (*a*) Sketch this vector. (*b*) Find V_x and V_y. (*c*) Use V_x and V_y to obtain (again) the magnitude and direction of $\mathbf{V}$. [*Note:* Part (*c*) is a good way to check if you've resolved your vector correctly.]

7. (II) Figure 3–33 shows two vectors, $\mathbf{A}$ and $\mathbf{B}$, whose magnitudes are $A = 8.31$ units and $B = 5.55$ units. Determine $\mathbf{C}$ if (*a*) $\mathbf{C} = \mathbf{A} + \mathbf{B}$, (*b*) $\mathbf{C} = \mathbf{A} - \mathbf{B}$, (*c*) $\mathbf{C} = \mathbf{B} - \mathbf{A}$. Give the magnitude and direction for each.

FIGURE 3–32
Problem 3.

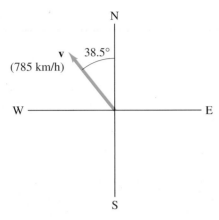

FIGURE 3-34   Problem 9.

8. (II) Vector $\mathbf{V}_1$ is 8.08 units long and points along the negative x axis. Vector $\mathbf{V}_2$ is 4.51 units long and points at $+45.0°$ to the positive x axis. (a) What are the x and y components of each vector? (b) Determine the sum of the two vectors (magnitude and angle).

9. (II) An airplane is traveling 785 km/h in a direction 38.5° west of north (Fig. 3–34). (a) Find the components of the velocity vector in the northerly and westerly directions. (b) How far north and how far west has the plane traveled after 3.00 h?

10. (II) The components of a vector $\mathbf{V}$ are often written ten (V_x, V_y, V_z). What are the components and length of a vector which is the sum of the two vectors, $\mathbf{V}_1$ and $\mathbf{V}_2$, whose components are $(3.0, 2.7, 0.0)$ and $(2.9, -4.1, -1.4)$?

11. (II) Three vectors are shown in Fig. 3–35. Their magnitudes are given in arbitrary units. Determine the sum of the three vectors. Give the resultant in terms of (a) components, (b) magnitude and angle with x axis.

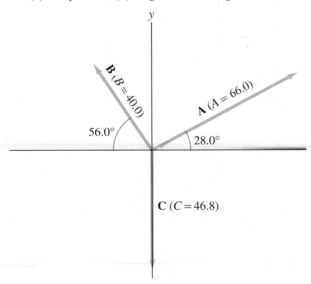

FIGURE 3-35   Problems 11, 12, 13, 14, and 15.
Vector magnitudes are given in arbitrary units.

12. (II) Determine the vector $\mathbf{A} - \mathbf{C}$, given the vectors $\mathbf{A}$ and $\mathbf{C}$ in Fig. 3–35.

13. (II) (a) Given the vectors $\mathbf{A}$ and $\mathbf{B}$ shown in Fig. 3–35, determine $\mathbf{B} - \mathbf{A}$. (b) Determine $\mathbf{A} - \mathbf{B}$ without using your answer in (a). Then compare your results and see if they are opposite.

14. (II) For the vectors given in Fig. 3–35, determine (a) $\mathbf{A} - \mathbf{B} + \mathbf{C}$, (b) $\mathbf{A} + \mathbf{B} - \mathbf{C}$, and (c) $\mathbf{B} - 2\mathbf{A}$.

15. (II) For the vectors shown in Fig. 3–35, determine (a) $\mathbf{C} - \mathbf{A} - \mathbf{B}$, (b) $2\mathbf{A} - 3\mathbf{B} + 2\mathbf{C}$.

16. (II) (a) A skier is accelerating down a 30.0° hill at 3.80 m/s² (Fig. 3–36). What is the vertical component of her acceleration? (b) How long will it take her to reach the bottom of the hill, assuming she starts from rest and accelerates uniformly, if the elevation change is 335 m?

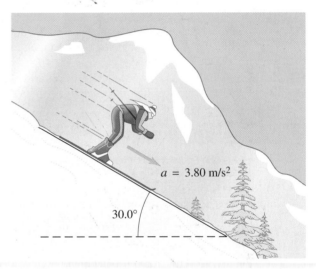

FIGURE 3-36   Problem 16.

17. (II) The summit of a mountain, 2085 m above base camp, is measured on a map to be 4580 m horizontally from the camp in a direction 32.4° west of north. What are the x, y, and z components of the displacement vector from camp to summit? What is its length? Choose the x axis east, y axis north, and z axis up.

18. (III) You are given a vector in the xy plane that has a magnitude of 90.0 units and a y component of -55.0 units. (a) What are the two possibilities for its x component? (b) Assuming the x component is known to be positive, specify the vector which, if you add it to the original one, would give a resultant vector that is 80.0 units long and points entirely in the $-x$ direction.

(neglect air resistance)

19. (I) A tiger leaps horizontally from a 7.5-m-high rock with a speed of 4.5 m/s. How far from the base of the rock will she land?

20. (I) A diver running 1.6 m/s dives out horizontally from the edge of a vertical cliff and reaches the water below 3.0 s later. How high was the cliff and how far from its base did the diver hit the water?

21. (II) A fire hose held near the ground shoots water at a speed of 6.5 m/s. At what angle(s) should the nozzle point in order that the water land 2.0 m away (Fig. 3–37)? Why are there two different angles?

θ_0

|← 2.0 m →|

FIGURE 3–37 Problem 21.

22. (II) Romeo is chucking pebbles gently up to Juliet's window, and he wants the pebbles to hit the window with only a horizontal component of velocity. He is standing at the edge of a rose garden 8.0 m below her window and 9.0 m from the base of the wall (Fig. 3–38). How fast are the pebbles going when they hit her window?

8.0 m

|← 9.0 m →|

FIGURE 3–38 Problem 22.

23. (II) Suppose the kick in Example 3–4 is attempted 36.0 m from the goalposts, whose crossbar is 3.00 m above the ground. If the football is directed correctly between the goalposts, will it pass over the bar and be a field goal? Show why or why not. If not, from what horizontal distance must this kick be made if it is to score?

24. (II) A ball is thrown horizontally from the roof of a building 56 m tall and lands 45 m from the base. What was the ball's initial speed?

25. (II) Show that the speed with which a projectile leaves the ground is equal to its speed just before it strikes the ground at the end of its journey, assuming the firing level equals the landing level.

26. (II) A football is kicked at ground level with a speed of 20.0 m/s at an angle of 37.0° to the horizontal. How much later does it hit the ground?

27. (II) A ball thrown horizontally at 22.2 m/s from the roof of a building lands 36.0 m from the base of the building. How high is the building?

28. (II) A shot-putter throws the shot with an initial speed of 14 m/s at a 40° angle to the horizontal. Calculate the horizontal distance traveled by the shot if it leaves the athlete's hand at a height of 2.2 m above the ground.

29. (II) Determine how much farther a person can jump on the Moon as compared to the Earth if the takeoff speed and angle are the same. The acceleration due to gravity on the Moon is one-sixth what it is on Earth.

30. (II) An athlete executing a long jump leaves the ground at a 30° angle and travels 7.80 m. (a) What was the takeoff speed? (b) If this speed were increased by just 5.0 percent, how much longer would the jump be?

31. (II) The pilot of an airplane traveling 160 km/h wants to drop supplies to flood victims isolated on a patch of land 160 m below. The supplies should be dropped how many seconds before the plane is directly overhead?

32. (II) A hunter aims directly at a target (on the same level) 120 m away. (a) If the bullet leaves the gun at a speed of 250 m/s, by how much will it miss the target? (b) At what angle should the gun be aimed so the target will be hit?

33. (II) Show that the time required for a projectile to reach its highest point is equal to the time for it to return from this highest point to its original height.

34. (II) A projectile is fired with an initial speed of 40.0 m/s. Plot on graph paper its trajectory for initial projection angles of θ = 15°, 30°, 45°, 60°, 75°, and 90°. Plot at least 10 points for each curve.

35. (II) A projectile is fired with an initial speed of 75.2 m/s at an angle of 34.5° above the horizontal on a long flat firing range. Determine (a) the maximum height reached by the projectile, (b) the total time in the air, (c) the total horizontal distance covered (that is, the range), and (d) the velocity of the projectile 1.50 s after firing.

FIGURE 3-39 Problem 36.

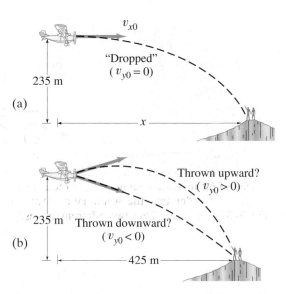

FIGURE 3-41 Problem 38.

36. (II) A projectile is shot from the edge of a cliff 125 m above ground level with an initial speed of 105 m/s at an angle of 37.0° with the horizontal, as shown in Fig. 3–39. (*a*) Determine the time taken by the projectile to hit point P at ground level. (*b*) Determine the range X of the projectile as measured from the base of the cliff. At the instant just before the projectile hits point P, find (*c*) the horizontal and the vertical components of its velocity, (*d*) the magnitude of the velocity, and (*e*) the angle made by the velocity vector with the horizontal.

37. (III) Revisit Conceptual Example 3–6, and assume that the boy with the slingshot is *below* the boy in the tree (Fig. 3–40), and so aims *upward*, directly at the boy in the tree. Show that again the boy in the tree makes the wrong move by letting go at the moment the water balloon is shot.

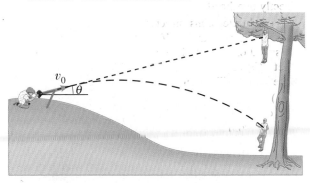

FIGURE 3-40 Problem 37.

38. (III) A rescue plane wants to drop supplies to isolated mountain climbers on a rocky ridge 235 m below. If the plane is traveling horizontally with a speed of 250 km/h (69.4 m/s), (*a*) how far in advance of the recipients (horizontal distance) must the goods be

dropped (Fig. 3–41a)? (*b*) Suppose, instead, that the plane releases the supplies a horizontal distance of 425 m in advance of the mountain climbers. What vertical velocity (up or down) should the supplies be given so that they arrive precisely at the climbers' position (Fig. 3–41b)? (*c*) With what speed do the supplies land in the latter case?

39. (III) A ball is thrown horizontally from the top of a cliff with initial speed v_0 (at $t = 0$). At any moment, its direction of motion makes an angle θ to the horizontal (Fig. 3–42). Derive a formula for θ as a function of time, t, as the ball follows a projectile's path.

FIGURE 3-42 Problem 39.

* **40.** (I) A person going for a morning jog on the deck of a cruise ship is running toward the bow (front) of the ship at 2.0 m/s while the ship is moving ahead at 8.5 m/s. What is the velocity of the jogger relative to the water? Later, the jogger is moving toward the stern (rear) of the ship. What is the jogger's velocity relative to the water now?

FIGURE 3–43   Problem 41.

FIGURE 3–44   Problem 48.

* 41. (II) Huck Finn walks at a speed of 1.0 m/s across his raft (that is, he walks perpendicular to the raft's motion relative to the shore). The raft is traveling down the Mississippi River at a speed of 2.7 m/s relative to the river bank (Fig. 3–43). What is the velocity (speed and direction) of Huck relative to the river bank?

* 42. (II) You are driving south on a highway at 25 m/s (approximately 55 mph) in a snowstorm. When you last stopped, you noticed that the snow was coming down vertically, but it is passing the windows of the moving car at an angle of 30° to the horizontal. Estimate the speed of the snowflakes relative to the car and relative to the ground.

* 43. (II) A boat can travel 2.30 m/s in still water. (a) If the boat points its prow directly across a stream whose current is 1.20 m/s, what is the velocity (magnitude and direction) of the boat relative to the shore? (b) What will be the position of the boat, relative to its point of origin, after 3.00 s? (See Fig. 3–30.)

* 44. (II) Two planes approach each other head-on. Each has a speed of 835 km/h, and they spot each other when they are initially 10.0 km apart. How much time do the pilots have to take evasive action?

* 45. (II) An airplane is heading due south at a speed of 500 km/h. If a wind begins blowing from the southwest at a speed of 100 km/h (average), calculate: (a) the velocity (magnitude and direction) of the plane relative to the ground, and (b) how far off course it will be after 10 min if the pilot takes no corrective action. [*Hint:* First draw a diagram.]

* 46. (II) In what direction should the pilot aim the plane in Problem 45 so that it will fly due south?

* 47. (II) Determine the speed of the boat with respect to the shore in Example 3–10.

* 48. (II) A passenger on a boat moving at 1.50 m/s on a still lake walks up a flight of stairs at a speed of 0.50 m/s, Fig. 3–44. The stairs are angled at 45° pointing in the direction of motion as shown. What is the velocity of the passenger relative to the water?

* 49. (II) A motorboat whose speed in still water is 3.60 m/s must aim upstream at an angle of 27.5° (with respect to a line perpendicular to the shore) in order to travel directly across the stream. (a) What is the speed of the current? (b) What is the resultant speed of the boat with respect to the shore? (See Fig. 3–28.)

* 50. (II) A boat, whose speed in still water is 2.20 m/s, must cross a 260-m-wide river and arrive at a point 110 m upstream from where it starts (Fig. 3–45). To do so, the pilot must head the boat at a 45° upstream angle. What is the speed of the river's current?

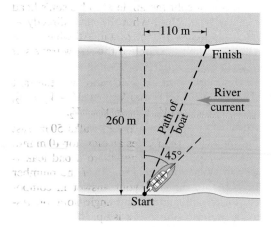

FIGURE 3–45   Problem 50.

* 51. (II) A swimmer is capable of swimming 1.00 m/s in still water. (a) If she aims her body directly across a 150-m-wide river whose current is 0.80 m/s, how far downstream (from a point opposite her starting point) will she land? (b) How long will it take her to reach the other side?

* 52. (II) At what upstream angle must the swimmer in Problem 51 aim, if she is to arrive at a point directly across the stream?

* 53. (III) An airplane, whose air speed is 600 km/h, is supposed to fly in a straight path 35.0° north of east. But a steady 100 km/h wind is blowing from the north. In what direction should the plane head?

* 54. (III) A motorcycle traveling 90.0 km/h approaches a car traveling in the same direction at 75.0 km/h. When the motorcycle is 60.0 m behind the car, the rider accelerates uniformly and passes the car 10.0 s later. What was the acceleration of the motorcycle?

* 55. (III) Two cars approach a street corner at right angles to each other (Fig. 3–46). Car 1 travels at 30 km/h and car 2 at 50 km/h. What is the relative velocity of car 1 as seen by car 2? What is the velocity of car 2 relative to car 1?

* 56. (III) An unmarked police car, traveling a constant 90 km/h, is passed by a speeder traveling 140 km/h. Precisely 1.00 s after the speeder passes, the policeman steps on the accelerator. If the police car's acceleration is 2.00 m/s², how much time elapses after the police car is passed until it overtakes the speeder (assumed moving at constant speed)?

* 57. (III) Assume in the previous problem that the speeder's speed is not known. If the police car accelerates uniformly as given above, and overtakes the speeder after 7.00 s, what was the speeder's speed?

FIGURE 3–46   Problem 55.

GENERAL PROBLEMS

58. William Tell must split the apple atop his son's head from a distance of 27 m. When he aims directly at the apple, the arrow is horizontal. At what angle must he aim it to hit the apple if the arrow travels at a speed of 35 m/s?

59. Two vectors, $\mathbf{V}_1$ and $\mathbf{V}_2$, add to a resultant $\mathbf{V} = \mathbf{V}_1 + \mathbf{V}_2$. Describe $\mathbf{V}_1$ and $\mathbf{V}_2$ if (a) $V = V_1 + V_2$, (b) $V^2 = V_1^2 + V_2^2$, (c) $V_1 + V_2 = V_1 - V_2$.

60. A plumber steps out of his truck, walks 50 m east and 25 m south, and then takes an elevator 10 m into the subbasement of a building where a bad leak is occurring. What is the displacement of the plumber relative to his truck? Give your answer in components and also in magnitude and angle notation. Assume x is east, y is north, and z is up.

61. On mountainous downhill roads, escape routes are sometimes placed to the side of the road for trucks whose brakes might fail. Assuming a constant upward slope of 30°, calculate the horizontal and vertical components of the acceleration of a truck that slowed from 120 km/h to rest in 12 s. See Fig. 3–47.

FIGURE 3–47   Problem 61.

62. What is the y component of a vector in the xy plane whose magnitude is 88.5 and whose x component is 75.4? What is the direction of this vector (angle it makes with the x axis)?

FIGURE 3–48   Problem 63.

63. Raindrops make an angle θ with the vertical when viewed through a moving train window (Fig. 3–48). If the speed of the train is v_T, what is the speed of the raindrops in the reference frame of the Earth in which they are assumed to fall vertically?

64. A light plane is headed due south with a speed relative to still air of 155 km/h. After 1.00 hour, the pilot notices that they have covered only 125 km and their direction is not south but southeast. What is the wind velocity?

65. An automobile traveling 95 km/h overtakes a 1.00-km-long train traveling in the same direction on a track parallel to the road. If the train's speed is 75 km/h, how long does it take the car to pass it and how far will the car have traveled in this time? What are the results if the car and train are traveling in opposite directions?

66. An Olympic long jumper is capable of jumping 8.0 m. Assuming his horizontal speed is 9.1 m/s as he leaves the ground, how long is he in the air and how high does he go? Assume that he lands standing upright—that is, the same way he left the ground.

FIGURE 3-49   Problem 69.

FIGURE 3-51   Problem 71.

67. Apollo astronauts took a "nine iron" to the Moon and hit a golf ball about 180 m! Assuming that the swing, launch angle, and so on, were the same as on Earth where the same astronaut could hit it only 30 m, estimate the acceleration due to gravity on the surface of the Moon. (Neglect air resistance in both cases, but on the Moon there is none!)

68. When Babe Ruth hit a homer over the 12-m-high right-field fence 95 m from home plate, roughly what was the minimum speed of the ball when it left the bat? Assume the ball was hit 1.0 m above the ground and its path initially made a 40° angle with the ground.

69. The cliff divers of Acapulco push off horizontally from rock platforms about 35 m above the water, but they must clear rocky outcrops at water level that extend out into the water 5.0 m from the base of the cliff directly under their launch point. See Fig. 3-49. What minimum pushoff speed is necessary to do this? How long are they in the air?

70. At serve, a tennis player aims to hit the ball horizontally. What minimum speed is required for the ball to clear the 0.90-m-high net about 15.0 m from the server if the ball is "launched" from a height of 2.50 m? Where will the ball land if it just clears the net (and will it be "good" in the sense that it lands within 7.0 m of the net)? How long will it be in the air? See Fig. 3-50.

71. Agent Tim, flying a constant 185 km/h horizontally in a low-flying helicopter, wants to drop a small explosive onto a master criminal's automobile traveling 145 km/h on a level highway 88.0 m below. At what angle (with the horizontal) should the car be in his sights when the bomb is released (Fig. 3-51)?

72. The speed of a boat in still water is v. The boat is to make a round-trip in a river whose current travels at speed u. Derive a formula for the time needed to make a round trip of total distance D if the boat makes the round-trip by moving (a) upstream and back downstream, (b) directly across the river and back. We must assume $u < v$; why?

FIGURE 3-52   Problem 73.

73. A projectile is launched from ground level to the top of a cliff which is 195 m away and 155 m high (see Fig. 3-52). If the projectile lands on top of the cliff 7.6 s after it is fired, find the initial velocity of the projectile (magnitude and direction). Neglect air resistance.

FIGURE 3-50   Problem 70.

Acceleration at the start of a race. Can you describe the force that causes an athlete's acceleration? (Hint: note the force vector **F**.)

MOTION AND FORCE: 4
DYNAMICS

We have discussed how motion is described in terms of velocity and acceleration. Now we deal with the question of *why* objects move as they do: What makes an object at rest begin to move? What causes a body to accelerate or decelerate? What is involved when an object moves in a circle? We can answer in each case that a force is required. In this chapter, we will investigate the connection between force and motion. Before we delve into this subject of *dynamics*, we first discuss the concept of force in a qualitative way.

FIGURE 4-1 Exerting a force on a stalled car.

4-1 Force

Intuitively, we experience **force** as any kind of a push or a pull on an object. When you push a grocery cart or a stalled car (Fig. 4–1), you are exerting a force on it. When a motor lifts an elevator, or a hammer hits a nail, or the wind blows the leaves of a tree, a force is being exerted. We say that an object falls because of the *force of gravity*. Forces do not always give rise to motion. For example, you may push very hard on a heavy desk and it may not move.

FIGURE 4–2 A spring scale used to measure a force.

Measuring force

One way to measure the magnitude (or strength) of a force is to make use of a spring scale (Fig. 4–2). Normally, such a spring scale is used to find the weight of an object; by weight we mean the force of gravity acting on the body (Section 4–6). The spring scale, once calibrated, can be used to measure other kinds of forces as well, such as the pulling force shown in Fig. 4–2.

A force has direction as well as magnitude, and is indeed a vector that follows the rules of vector addition discussed in Chapter 3. We can represent any force on a diagram by an arrow, just as we did with velocity. The direction of the arrow is the direction of the push or pull, and its length is drawn proportional to the magnitude of the force.

4–2 Newton's First Law of Motion

What is the exact connection between force and motion? Aristotle (384–322 B.C.) believed that a force was required to keep an object moving along a horizontal plane. He would argue that to make a book move across the table, you would have to exert a force on it continuously. To Aristotle, the natural state of a body was at rest, and a force was believed necessary to keep a body in motion. Furthermore, Aristotle argued, the greater the force on the body, the greater its speed.

Galileo vs. Aristotle

Some 2000 years later, Galileo questioned these Aristotelian views and came to a radically different conclusion. Galileo maintained that it is just as natural for an object to be in horizontal motion with a constant velocity as it is for it to be at rest!

FIGURE 4–3 Photo of an air table. Air issuing from many tiny holes forms a thin air layer between the table and a puck, which, when given an initial shove, will travel at constant speed in a straight line (until it hits a wall or another puck).

To understand Galileo's idea, consider the following observations involving motion along a horizontal plane. To push an object with a rough surface along a tabletop at constant speed requires a certain amount of force. To push an equally heavy object with a very smooth surface across the table at the same speed will require less force. If a layer of oil or other lubricant is placed between the surface of the object and the table, then almost no force is required to move the object. Notice that in each successive step, less force is required. As the next step, we can imagine a situation in which the object does not rub against the table at all—or there is a perfect lubricant between the object and the table—and theorize that once started, the object would move across the table at constant speed with *no* force applied. A steel ball bearing rolling on a hard horizontal surface approaches this situation. So does a puck on an air table (Fig. 4–3), in which a thin layer of air reduces friction almost to zero.

It was Galileo's genius to imagine such an idealized world—in this case, one where there is no friction—and to see that it could produce a more useful view of the real world. It was this idealization that led him to his remarkable conclusion that if no force is applied to a moving object, it will continue to move with constant speed in a straight line. An object slows down only if a force is exerted on it. Galileo thus interpreted friction as a force akin to ordinary pushes and pulls.

To push an object across a table at constant speed requires a force from your hand only to balance out the force of friction (Fig. 4–4). When the object moves at constant speed, your pushing force is equal in magnitude to the friction force, but these two forces are in opposite directions, so the *net* force on the object (the vector sum of the two forces) is zero. This is consistent with Galileo's viewpoint, for the object moves with constant speed when no net force is exerted on it.

The difference between Aristotle's view and Galileo's is not simply one of right or wrong. Aristotle's view was not really wrong, for our everyday experience indicates that moving objects do tend to come to a stop if not continually pushed. The real difference lies in the fact that Aristotle's view about the "natural state" of a body was essentially a final statement—no further development was possible. Galileo's analysis, on the other hand, could be extended to explain a great many more phenomena, and it provided a quantitative theory allowing verifiable predictions. By making the creative leap of imagining the experimentally unattainable situation of no friction, and by interpreting friction as a force, Galileo was able to reach his conclusion that an object will continue moving with constant velocity if no force acts to change this motion.

FIGURE 4–4 **F** represents the force applied by the person and $\mathbf{F}_{fr}$ represents the force of friction.

Upon this foundation, Isaac Newton (Fig. 4–5) built his great theory of motion. Newton's analysis of motion is summarized in his famous "three laws of motion." In his great work, the *Principia* (published in 1687), Newton readily acknowledged his debt to Galileo. In fact, **Newton's first law of motion** is very close to Galileo's conclusions. It states that

> **Every body continues in its state of rest or of uniform speed in a straight line unless acted on by a nonzero net force.**

NEWTON'S FIRST LAW OF MOTION

The tendency of a body to maintain its state of rest or of uniform motion in a straight line is called **inertia**. As a result, Newton's first law is often called the **law of inertia**.

[Newton's first law does not hold in every reference frame. For example, if your reference frame is fixed in an accelerating car, an object such as a cup resting on the dashboard may begin to move toward you (it stayed at rest as long as the car's velocity remained constant). The cup accelerated toward you but neither you nor anything else exerted a force on it in that direction. In such an accelerating reference frame, Newton's first law does not hold. Reference frames in which Newton's first law does hold are called **inertial reference frames** (the law of inertia is valid in them). For most purposes, we can usually assume that reference frames fixed on the Earth are inertial frames. (This is not precisely true, due to the Earth's rotation, but usually it is close enough.) Any reference frame that moves with constant velocity (say, a car or an airplane) relative to an inertial frame is also an inertial reference frame. Reference frames where the law of inertia does *not*

FIGURE 4–5 Isaac Newton (1642–1727).

hold, such as the accelerating reference frame discussed above, are called **noninertial** reference frames. How can we be sure a reference frame is inertial or not? By checking to see if Newton's first law holds. Thus Newton's first law serves as the definition of inertial reference frames.]

4–3 | Mass

Newton's second law, which we come to in the next section, makes use of the concept of mass. Newton used the term *mass* as a synonym for *quantity of matter*. This intuitive notion of the mass of a body is not very precise because the concept "quantity of matter" is not very well defined. More precisely, we can say that **mass** is a *measure of the inertia of a body*. The more mass a body has, the harder it is to change its state of motion. It is harder to start it moving from rest, or to stop it when it is moving, or to change its motion sideways out of a straight-line path. A truck has much more inertia than a baseball, and it is much harder to speed it up or slow it down. It therefore has much more mass.

Mass as inertia

To quantify the concept of mass, we must define a standard. In SI units, the unit of mass is the **kilogram** (kg) as we discussed in Chapter 1, Section 1–5.

The terms *mass* and *weight* are often confused with one another, but it is important to distinguish between them. Mass is a property of a body itself (it is a measure of a body's inertia, or its "quantity of matter"). Weight, on the other hand, is a force, the force of gravity acting on a body. To see the difference, suppose we take an object to the Moon. The object will weigh only about one sixth as much as it did on Earth, since the force of gravity is weaker, but its mass will be the same. It will have the same amount of matter and it will have just as much inertia—for in the absence of friction, it will be just as hard to start it moving or to stop it once it is moving. (More on weight in Section 4–6.)

Mass vs. weight

4–4 | Newton's Second Law of Motion

FIGURE 4–6 The bobsled accelerates because the team exerts a force.

Newton's first law states that if no net force is acting on a body, it remains at rest, or if moving, it continues moving with constant speed in a straight line. But what happens if a net force is exerted on a body? Newton perceived that the velocity will change (Fig. 4–6). A net force exerted on an object may make its speed increase. Or, if the net force is in a direction opposite to the motion, it will reduce the speed. If the net force acts sideways on a moving object, the *direction* of the velocity changes (and the magnitude may as well). Since a change in speed or velocity is an acceleration (Chapter 2, Section 2–4), we can say that *a net force gives rise to acceleration.*

What precisely is the relationship between acceleration and force? Everyday experience can answer this question. Consider the force required to push a cart whose friction is minimal. (If there is friction, consider the *net* force, which is the force you exert minus the force of friction.) Now if you push with a gentle but constant force for a certain period of time, you will make the cart accelerate from rest up to some speed, say

3 km/h. If you push with twice the force, you will find that the cart will reach 3 km/h in half the time. That is, the acceleration will be twice as great. If you double the force, the acceleration doubles. If you triple the force, the acceleration is tripled, and so on. Thus, the acceleration of a body is directly proportional to the net applied force. But the acceleration depends on the mass of the object as well. If you push an empty grocery cart with the same force as you push one that is filled with groceries, you will find that the latter accelerates more slowly. The greater the mass, the less the acceleration for the same net force. The mathematical relation, as Newton argued, is that the acceleration of a body is inversely proportional to its mass. These relationships are found to hold in general and can be summarized as follows:

> **The acceleration of an object is directly proportional to the net force acting on it and is inversely proportional to its mass. The direction of the acceleration is in the direction of the net force acting on the object.**

NEWTON'S SECOND LAW OF MOTION

This is **Newton's second law of motion**. As an equation, it can be written

$$\mathbf{a} = \frac{\Sigma \mathbf{F}}{m},$$

where **a** stands for acceleration, m for the mass, and $\Sigma \mathbf{F}$ for the *net force*. The symbol Σ (Greek "sigma") stands for "sum of"; **F** stands for force, so $\Sigma\mathbf{F}$ means the *vector sum of all forces* acting on the body, which we define as the **net force**.

Net force

We rearrange this equation to obtain the familiar statement of Newton's second law:

$$\Sigma \mathbf{F} = m\mathbf{a}. \tag{4-1}$$

NEWTON'S SECOND LAW OF MOTION

Newton's second law relates the description of motion to the cause of motion, force. It is one of the most fundamental relationships in physics. From Newton's second law we can make a more precise definition of **force** as *an action capable of accelerating an object.*

Force defined

Every force **F** is a vector, with magnitude and direction. Equation 4–1 is a vector equation valid in any inertial reference frame. It can be written in component form in rectangular coordinates as

$$\Sigma F_x = ma_x, \qquad \Sigma F_y = ma_y, \qquad \Sigma F_z = ma_z.$$

If the motion is all along a line (one dimensional), we can leave out the subscripts and simply write $\Sigma F = ma$.

In SI units, with the mass in kilograms, the unit of force is called the **newton** (N). One newton, then, is the force required to impart an acceleration of 1 m/s^2 to a mass of 1 kg. Thus $1 \text{ N} = 1 \text{ kg} \cdot \text{m/s}^2$.

Unit of force; the newton

In cgs units, the unit of mass is the gram (g) as mentioned earlier.[†] The unit of force is the *dyne*, which is defined as the net force needed to impart an acceleration of 1 cm/s^2 to a mass of 1 g. Thus $1 \text{ dyne} = 1 \text{ g} \cdot \text{cm/s}^2$. It is easy to show that $1 \text{ dyne} = 10^{-5} \text{ N}$.

In the British system, the unit of force is the *pound* (abbreviated lb), where $1 \text{ lb} \approx 4.45 \text{ N}$. The unit of mass is the *slug*, which is defined as that

---

[†]Be careful not to confuse g for gram with g for the acceleration due to gravity. The latter is always italicized (or bold face as a vector).

mass which will undergo an acceleration of 1 ft/s^2 when a force of 1 lb is applied to it. Thus $1 \text{ lb} = 1 \text{ slug·ft/s}^2$. Table 4–1 summarizes the units in the different systems.

It is very important that only one set of units be used in a given calculation or problem, with the SI being preferred. If the force is given in, say, newtons, and the mass in grams, then before attempting to solve for the acceleration in SI units, the mass must be changed to kilograms. For example, if the force is given as 2.0 N along the x axis and the mass is 500 g, we change the latter to 0.50 kg, and the acceleration will then automatically come out in m/s^2 when Newton's second law is used:

$$a_x = \frac{\Sigma F_x}{m} = \frac{2.0 \text{ N}}{0.50 \text{ kg}} = \frac{2.0}{0.50} \frac{\text{kg·m}}{\text{kg·s}^2} = 4.0 \text{ m/s}^2.$$

➡ **PROBLEM SOLVING**

Use a consistent set of units

TABLE 4–1
Units for Mass and Force

System	Mass	Force (including weight)
SI	kilogram (kg)	newton (N) ($= \text{kg·m/s}^2$)
cgs	gram (g)	dyne ($= \text{g·cm/s}^2$)
British	slug	pound (lb)

EXAMPLE 4–1 **ESTIMATE** **Force to accelerate a fast car.** Estimate the net force needed to accelerate a 1000-kg car at $\frac{1}{2}g$.

SOLUTION The car's acceleration is $a = \frac{1}{2}g = \frac{1}{2}(9.8 \text{ m/s}^2) \approx 5 \text{ m/s}^2$. We use Newton's second law to get the net force needed to achieve this acceleration:

$$\Sigma F = ma = (1000 \text{ kg})(5 \text{ m/s}^2) = 5000 \text{ N}.$$

(If you are used to the British units, to get an idea what a 5000 N force is, you can divide by 4.45 N/lb and get a force of about 1000 lb.)

EXAMPLE 4–2 **Force to stop a car.** What net force is required to bring a 1500-kg car to rest from a speed of 100 km/h within a distance of 55 m?

SOLUTION We use Newton's second law, $\Sigma F = ma$, but first we must determine the acceleration a, which we assume is constant. We assume the motion is along the $+x$ axis (Fig. 4–7). We are given the initial velocity $v_0 = 100 \text{ km/h} = 28 \text{ m/s}$, the final velocity $v = 0$, and the distance traveled $x - x_0 = 55 \text{ m}$. From Eq. 2–10c, we have

$$v^2 = v_0^2 + 2a(x - x_0)$$

so

$$a = \frac{v^2 - v_0^2}{2(x - x_0)} = \frac{0 - (28 \text{ m/s})^2}{2(55 \text{ m})} = -7.1 \text{ m/s}^2.$$

The net force required is then

$$\Sigma F = ma = (1500 \text{ kg})(-7.1 \text{ m/s}^2) = -1.1 \times 10^4 \text{ N}.$$

The force must be exerted in the direction *opposite* to the initial velocity, which is what the negative sign tells us.

FIGURE 4–7 Example 4–2.

$v_0 = 100 \text{ km/h}$ ⟶ $v = 0$

$x = 0$ ⟶ $x(m)$ ⟶ $x = 55 \text{ m}$

4–5 Newton's Third Law of Motion

Newton's second law of motion describes quantitatively how forces affect motion. But where, we may ask, do forces come from? Observations suggest that a force applied to any object is always applied *by another object*. A horse pulls a wagon, a person pushes a grocery cart, a hammer pushes on a nail, a magnet attracts a paper clip. In each of these examples, a force is exerted *on* one body, and that force is exerted *by* another body. For example, the force exerted *on* the nail is exerted *by* the hammer.

*A force is exerted **on** an object and is exerted **by** another object*

But Newton realized that things are not so one-sided. True, the hammer exerts a force on the nail (Fig. 4–8). But the nail evidently exerts a force back on the hammer as well, for the hammer's speed is rapidly reduced to zero upon contact. Only a strong force could cause such a rapid change of velocity of the hammer. Thus, said Newton, the two bodies must be treated on an equal basis. The hammer exerts a force on the nail, and the nail exerts a force back on the hammer. This is the essence of **Newton's third law of motion**:

> **Whenever one object exerts a force on a second object, the second exerts an equal and opposite force on the first.**

NEWTON'S THIRD LAW OF MOTION

This law is sometimes paraphrased as "to every action there is an equal and opposite reaction." This is perfectly valid. But to avoid confusion, it is very important to remember that the "action" force and the "reaction" force are acting on *different* objects.

Action and reaction act on different objects

As evidence for the validity of Newton's third law, look at your hand when you push against a grocery cart or against the edge of a desk, Fig. 4–9. Your hand's shape is distorted, clear evidence that a force is being exerted on it. You can *see* the edge of the desk pressing into your hand. You can even *feel* the desk exerting a force on your hand; it hurts! The harder you push against the desk, the harder the desk pushes back on your hand. (Note that you only feel forces exerted *on* you, not forces you exert on something else.)

FIGURE 4–8 Multiflash photo of a hammer striking a nail. In accordance with Newton's third law, the hammer exerts a force on the nail, and the nail exerts a force back on the hammer. The latter force decelerates the hammer and brings it to rest.

Force exerted on hand by desk

Force exerted on desk by hand

FIGURE 4–9 If your hand pushes against the edge of a desk (the force vector is shown in red), the desk pushes back against your hand (this force vector is shown in a different color, purple, to remind us that this force acts on a different object).

FIGURE 4–10 When an ice-skater pushes against the railing, the railing pushes back and this force causes her to move away.

FIGURE 4–11 The launch of a rocket. Newton's third law.

As another demonstration of Newton's third law, consider the ice-skater in Fig. 4–10. Since there is very little friction between her skates and the ice, she will move freely if a force is exerted on her. She pushes against the railing; and then *she* starts moving backward. Clearly, there had to be a force exerted on her to make her move. The force she exerts on the railing cannot make *her* move, for that force acts on the railing. Something had to exert a force on her to make her start moving, and that force could only have been exerted by the railing. The force with which the railing pushes on her is, by Newton's third law, equal and opposite to the force she exerts on the railing.

When a person throws a package out of a boat (initially at rest), the boat starts moving in the opposite direction. The person exerts a force on the package. The package exerts an equal and opposite force back on the person, and this force propels the person (and the boat) backward slightly. Rocket propulsion also is explained using Newton's third law (Fig. 4–11). A common misconception is that rockets accelerate because the gases rushing out the back of the engine push against the ground or the atmosphere. Not true. What happens, instead, is that a rocket exerts a strong force on the gases, expelling them; and the gases exert an equal and opposite force *on the rocket*. It is this latter force that propels the rocket forward. Thus, a space vehicle is maneuvered in empty space by firing its rockets in the direction opposite to that in which it needs to accelerate.

↪ **PHYSICS APPLIED**

How does a rocket accelerate?

How we can walk

Consider how we walk. A person begins walking by pushing with the foot against the ground. The ground then exerts an equal and opposite force back on the person (Fig. 4–12) and it is this force, *on* the person, that moves the person forward. (If you doubt this, try walking on very smooth slippery ice.) In a similar way, a bird flies forward by exerting a force on the air, but it is the air pushing back on the bird's wings that propels the bird forward.

| CONCEPTUAL EXAMPLE 4–3 | **What exerts the force on a car?** What |

makes a car go forward?

RESPONSE A common answer is that the engine makes the car move forward. But it is not so simple. The engine makes the wheels go around. But what good is that if they are on slick ice or mud? They just spin. A car moves forward due to the friction force exerted by the ground on the tires, and this force is the reaction to the force exerted on the ground by the tires.

We tend to associate forces with active bodies such as humans, animals, engines, or a moving object like a hammer. It is often difficult to see how an inanimate object at rest, such as a wall or a desk, can exert a force. The explanation lies in the fact that every material, no matter how hard, is elastic, at least to some degree. No one can deny that a stretched rubber band can exert a force on a wad of paper and send it flying across the room. Other materials may not stretch as easily as rubber, but they do stretch when a force is applied to them. And just as a stretched rubber band exerts a force, so does a stretched (or compressed) wall, desk, or car fender.

From the examples discussed above, it is clear that it is quite important to remember *on* what object a given force is exerted and *by* what object that force is exerted. The point is that a force influences the motion of an object only when it is applied *on* that object. A force exerted *by* a body does not influence that body; it only influences the other body *on* which it is exerted. Thus, to avoid confusion, the two prepositions *on* and *by* must always be used—and used with care.

One way to keep clear which force acts on which object is to use double subscripts. For example the force exerted on the **P**erson by the **G**round in Fig. 4–13 can be labeled F_{PG}. And the force exerted on the ground by the person is F_{GP}, as shown in Fig. 4–13. Note that we have used different colors for the force vectors when they act on different objects. By Newton's third law

$$\mathbf{F}_{GP} = -\mathbf{F}_{PG}. \tag{4–2}$$

$\mathbf{F}_{GP}$ and $\mathbf{F}_{PG}$ have the same magnitude, and the minus sign reminds us that these two forces are in opposite directions.

Inanimate objects can exert a force

➡ **PROBLEM SOLVING**

*For each force, be clear **on** which object it acts, and **by** which object it is exerted. $\Sigma\mathbf{F} = m\mathbf{a}$ applies only to forces acting **on** a body.*

NEWTON'S THIRD LAW OF MOTION

Horizontal force exerted on the ground by person's foot ⟵

⟶ Horizontal force exerted on the person's foot by the ground

$\mathbf{F}_{GP}$ $\mathbf{F}_{PG}$

FIGURE 4–12 We can walk forward because, when one foot pushes backward against the ground, the ground pushes forward on that foot.

FIGURE 4–13 Newton's third law. Subscripts on forces remind us *on* which body a force acts and *by* which body it is exerted.

FIGURE 4–14 Seventy-year-old Michelangelo has selected a fine block of marble for his next sculpture. Shown here is his assistant pulling it on a sled away from the quarry. Forces on the assistant are shown as red (magenta) arrows. Forces on the sled are purple arrows. Forces acting on the ground are orange arrows. Action–reaction forces that are equal and opposite are labeled by the same subscripts but reversed (such as $\mathbf{F}_{GA}$ and $\mathbf{F}_{AG}$) and are of different colors because they act on different objects.

Force on sled exerted by assistant

Force on assistant exerted by sled

$\mathbf{F}_{SA}$ $\quad$ $\mathbf{F}_{AS}$
$(=-\mathbf{F}_{AS})$

$\mathbf{F}_{SG}$ $\quad$ $\mathbf{F}_{GS} (=-\mathbf{F}_{SG})$ $\qquad$ $\mathbf{F}_{GA} (=-\mathbf{F}_{AG})$ $\qquad$ $\mathbf{F}_{AG}$

Friction force on sled exerted by ground

Force on ground exerted by sled

Force on ground exerted by assistant

Force on assistant exerted by ground

➡ **PROBLEM SOLVING**

A study of Newton's second and third laws

CONCEPTUAL EXAMPLE 4–4 **Third law clarification.** Michelangelo's assistant has been assigned the task of moving a block of marble using a sled (Fig. 4–14). He says to his boss, "When I exert a forward force on the sled, the sled exerts an equal and opposite force backward. So how can I ever start it moving? No matter how hard I pull, the backward re-action force always equals my forward force, so the net force must be zero. I'll never be able to move this load." Is this a case of a little knowledge being dangerous? Explain.

RESPONSE Yes. Although it is true that the action and reaction forces are equal in magnitude, the assistant has forgotten that they are exerted on different objects. The forward ("action") force is exerted by the assistant on the sled (Fig. 4–14), whereas the backward "reaction" force is exerted by the sled on the assistant. To determine if the *assistant* moves or not, we must consider only the forces *on the assistant* and then apply $\Sigma \mathbf{F} = m\mathbf{a}$, where $\Sigma \mathbf{F}$ is the net force *on the assistant,* $\mathbf{a}$ is the acceleration of the assistant, and m is the assistant's mass. There are two forces on the assistant that affect his forward motion and he seems to have forgotten one of them. The two forces on the assistant are shown as bright red (magenta) arrows in Fig. 4–14; they are (1) the horizontal force $\mathbf{F}_{AG}$ exerted on the assistant by the ground (the harder he pushes backward against the ground, the harder the ground pushes forward on him—Newton's third law), and (2) the force $\mathbf{F}_{AS}$ exerted on the assistant by the sled, pulling backward on him. When the ground pushes forward on the assistant harder than the sled pulls backward, the assistant accelerates forward (Newton's second law). The sled, on the other hand, accelerates forward when the force on it exerted by the assistant is greater than the frictional force acting backward (that is, when $\mathbf{F}_{SA}$ has greater magnitude than $\mathbf{F}_{SG}$ in Fig. 4–14).

Using double subscripts to clarify Newton's third law can become cumbersome, and we won't usually use them in this way. Nevertheless, if there is any confusion in your mind about a given force, go ahead and use them to identify *on* what object and *by* what object the force is exerted.

4-6 Weight—the Force of Gravity; and the Normal Force

Galileo claimed that objects dropped near the surface of the Earth will all fall with the same acceleration, **g**, if air resistance can be neglected. The force that gives rise to this acceleration is called the force of gravity. We now apply Newton's second law to the gravitational force; and for the acceleration, **a**, we use the downward acceleration due to gravity, **g**. Thus, the force of gravity on an object, $\mathbf{F}_G$, whose magnitude is commonly called its **weight**, can be written as

$$\mathbf{F}_G = m\mathbf{g}. \tag{4–3}$$

Weight = force of gravity

The direction of this force is down toward the center of the Earth.

In SI units, $g = 9.80 \text{ m/s}^2 = 9.80 \text{ N/kg}$,[†] so the weight of a 1.00-kg mass on Earth is $1.00 \text{ kg} \times 9.80 \text{ m/s}^2 = 9.80 \text{ N}$. We will mainly be concerned with the weight of objects on Earth, but we note that on the Moon, on other planets, or in space, the weight of a given mass will be different. For example, on the Moon g is about one sixth what it is on Earth, and a 1.0 kg mass weighs only 1.7 N. Although we will not have occasion to use British units, we note that for practical purposes on the Earth, a mass of 1 kg weighs about 2.2 lb. (On the Moon, 1 kg weighs only about 0.4 lb.)

The force of gravity acts on an object when it is falling. When an object is at rest on the Earth, the gravitational force on it does not disappear, as we know if we weigh it on a spring scale. The same force, given by Eq. 4–3, continues to act. Why, then, doesn't the object move? From Newton's second law, the net force on an object that remains at rest is zero. There must be another force on the object to balance the gravitational force. For an object resting on a table, the table exerts this upward force; see Fig. 4–15a. The table is compressed slightly beneath the object, and due to its elasticity, it pushes up on the object as shown. The force exerted by the table is often called a **contact force**, since it occurs when two objects are in contact. (The force of your hand pushing on a cart is also a contact force.) When a contact force acts *perpendicular* to the common surface of contact, it is usually referred to as the **normal force** ("normal" means perpendicular); hence it is labeled $\mathbf{F}_N$ in the diagram.

The two forces shown in Fig. 4–15a are both acting on the statue, which remains at rest, so the vector sum of these two forces must be zero (Newton's second law). Hence $\mathbf{F}_G$ and $\mathbf{F}_N$ must be of equal magnitude and in opposite directions. But they are *not* the equal and opposite forces spoken of in Newton's third law. The action and reaction forces of Newton's third law act on *different objects*, whereas the two forces shown in Fig. 4–15a act on the *same* object. For each of the forces shown in Fig. 4–15a, we can ask, "What is the reaction force?" The upward force, $\mathbf{F}_N$, on the statue is exerted by the table. The reaction to this force is a force exerted by the statue on the table. It is shown in Fig. 4–15b, where it is labeled $\mathbf{F}'_N$. This force, $\mathbf{F}'_N$, exerted on the table by the statue, is the reaction force to $\mathbf{F}_N$ in accord with Newton's third law. (We could equally well say the reverse: the force $\mathbf{F}_N$ on the statue exerted by the table is the reaction to the force $\mathbf{F}'_N$ exerted on the table by the statue.) Now, what about the other force on the statue, the force of gravity $\mathbf{F}_G$? Can you guess what the reaction is to this force? [We will see in Chapter 5 that the reaction force is also a gravitational force exerted on the Earth by the statue, and can be considered to act at the Earth's center.]

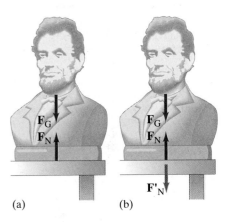

FIGURE 4–15 (a) The net force on an object at rest is zero according to Newton's second law. Therefore the downward force of gravity ($\mathbf{F}_G$) on an object must be balanced by an upward force (the normal force $\mathbf{F}_N$) exerted by the table in this case. (b) $\mathbf{F}'_N$ is the force exerted on the table by the statue and is the reaction force to $\mathbf{F}_N$ as per Newton's third law. ($\mathbf{F}'_N$ is shown in a different color to remind us it acts on a different body.) The reaction to $\mathbf{F}_G$ is not shown.

Careful:
Weight and normal force are **not** *action–reaction pairs*

[†]Since $1 \text{ N} = 1 \text{ kg·m/s}^2$ (Section 4–4), $1 \text{ m/s}^2 = 1 \text{ N/kg}$.

FIGURE 4–16 Example 4–5.
(a) A 10-kg gift box is at rest on a table. (b) A person pushes down on the box with a force of 40.0 N. (c) A person pulls upward on the box with a force of 40.0 N. The forces are all assumed to act along a line; they are shown slightly displaced in order to be distinguishable on the diagram. Only forces acting on the box are shown.

EXAMPLE 4–5 **Weight, normal force, and a box.** A friend has given you a special gift, a box of mass 10.0 kg with a mystery surprise inside. It's a reward for your fine showing on the physics final. The box is resting on the smooth (frictionless) horizontal surface of a table (Fig. 4–16a). (*a*) Determine the weight of the box and the normal force acting on it. (*b*) Now your friend pushes down on the box with a force of 40.0 N, as in Fig. 4–16b. Again determine the normal force acting on the box. (*c*) If your friend pulls upward on the box with a force of 40.0 N (Fig. 4–16c), what now is the normal force on the box?

SOLUTION (*a*) The box is resting on the table. The weight of the box is $mg = (10.0\,\text{kg})(9.80\,\text{m/s}^2) = 98.0\,\text{N}$, and this force acts downward. The only other force on the box is the normal force exerted upward on it by the table, as shown in Fig. 4–16a. We chose the upward direction as the positive y direction, and then the net force ΣF_y on the box is $\Sigma F_y = F_N - mg$. Since the box is at rest, the net force on it must be zero ($\Sigma F_y = ma_y$, and $a_y = 0$). Thus

$$\Sigma F_y = F_N - mg = 0,$$

so we have in this case

$$F_N = mg.$$

The normal force on the box, exerted by the table, is 98.0 N upward, and has magnitude equal to the box's weight.
(*b*) Your friend is pushing down on the box with a force of 40.0 N. So now there are three forces acting on the box, as shown in Fig. 4–16b. The weight of the box is still $mg = 98.0\,\text{N}$. The net force is $\Sigma F_y = F_N - mg - 40.0\,\text{N}$, and is equal to zero since the box remains at rest. Thus, since $a = 0$, Newton's second law gives

$$\Sigma F_y = F_N - mg - 40.0\,\text{N} = 0,$$

so the normal force is now

$$F_N = mg + 40.0\,\text{N} = 98.0\,\text{N} + 40.0\,\text{N} = 138.0\,\text{N},$$

which is greater than in (*a*). The table pushes back with more force.
(*c*) The box's weight is still 98.0 N and acts downward. The force exerted by your friend and the normal force both act upward (positive direction), as shown in Fig. 4–16c. The box doesn't move since your friend's upward force is less than the weight. The net force, again set to zero, is

$$\Sigma F_y = F_N - mg + 40.0\,\text{N} = 0,$$

so

$$F_N = mg - 40.0\,\text{N} = 98.0\,\text{N} - 40.0\,\text{N} = 58.0\,\text{N}.$$

The table does not push against the full weight of the box because of the upward pull exerted by your friend.

Notice that the normal force is elastic in origin (the table in Fig. 4–16 sags slightly under the weight of the box).

EXAMPLE 4–6 **Accelerating the box.** What happens when a person pulls upward on the box in Example 4–5(c) with a force equal to, or greater than, the box's weight, say $F_P = 100.0$ N rather than the 40.0 N shown in Fig. 4–16c?

SOLUTION The net force is now

$$\Sigma F_y = F_N - mg + F_P = F_N - 98.0 \text{ N} + 100.0 \text{ N},$$

and if we set this equal to zero, we would get $F_N = -2.0$ N. This is nonsense, since the negative sign implies F_N points downward, and the table surely cannot pull down on the box (unless there's glue on the table). The least F_N can be is zero, which it will be in this case. What really happens here is clear: the box accelerates upward since the net force is not zero; it is

$$\Sigma F_y = F_P - mg = 100.0 \text{ N} - 98.0 \text{ N} = 2.0 \text{ N}$$

upward. See Fig. 4–17. So the box moves upward with an acceleration of magnitude

$$a_y = \Sigma F_y/m = 2.0 \text{ N}/10.0 \text{ kg} = 0.20 \text{ m/s}^2.$$

FIGURE 4–17 Example 4–6. The box accelerates upwards because $F_P > mg$.

4–7 Solving Problems with Newton's Laws: Vector Forces and Free-Body Diagrams

Newton's second law tells us that the acceleration of an object is proportional to the *net force* acting on the object. The **net force**, as mentioned earlier, is the *vector sum* of all forces acting on the object. Indeed, extensive experiments have shown that forces do add together as vectors precisely according to the rules we developed in Chapter 3. For example, in Fig. 4–18, two forces of equal magnitude (100 N each) are shown acting on an object at right angles to each other. Intuitively, we can see that the object will move at a 45° angle and thus the net force acts at a 45° angle. This is just what the rules of vector addition give. The Pythagorean theorem tells us that the magnitude of the resultant force is $F_R = \sqrt{(100 \text{ N})^2 + (100 \text{ N})^2} = 141$ N.

FIGURE 4–18 (a) Two forces, $\mathbf{F}_1$ and $\mathbf{F}_2$, act on an object. (b) The sum, or resultant, of $\mathbf{F}_1$ and $\mathbf{F}_2$ is $\mathbf{F}_R$.

(a)

(b)

(a)

(b)

(c)

FIGURE 4–19 Two force vectors act on a boat (Example 4–7).

➡ PROBLEM SOLVING

Free-body diagram

EXAMPLE 4–7 **Adding force vectors.** Calculate the sum of the two forces acting on the boat shown in Fig. 4–19a.

SOLUTION These two forces are shown resolved in Fig. 4–19b. We add the forces using the method of components. The components of $\mathbf{F}_1$ are

$$F_{1x} = F_1 \cos 45.0° = (40.0 \text{ N})(0.707) = 28.3 \text{ N},$$

$$F_{1y} = F_1 \sin 45.0° = (40.0 \text{ N})(0.707) = 28.3 \text{ N}.$$

The components of $\mathbf{F}_2$ are

$$F_{2x} = +F_2 \cos 37.0° = +(30.0 \text{ N})(0.799) = +24.0 \text{ N},$$

$$F_{2y} = -F_2 \sin 37.0° = -(30.0 \text{ N})(0.602) = -18.1 \text{ N}.$$

F_{2y} is negative because it points along the negative y axis. The components of the resultant force are (see Fig. 4–19c)

$$F_{Rx} = F_{1x} + F_{2x} = 28.3 \text{ N} + 24.0 \text{ N} = 52.3 \text{ N},$$

$$F_{Ry} = F_{1y} + F_{2y} = 28.3 \text{ N} - 18.1 \text{ N} = 10.2 \text{ N}.$$

To find the magnitude of the resultant force, we use the Pythagorean theorem:

$$F_R = \sqrt{F_{Rx}^2 + F_{Ry}^2} = \sqrt{(52.3)^2 + (10.2)^2} = 53.3 \text{ N}.$$

The only remaining question is the angle θ that the net force $\mathbf{F}_R$ makes with the x axis. We use:

$$\tan \theta = \frac{F_{Ry}}{F_{Rx}} = \frac{10.2 \text{ N}}{52.3 \text{ N}} = 0.195,$$

and $\tan^{-1}(0.195) = 11.0°$.

When solving problems involving Newton's laws and force, it is very important to draw a diagram showing all the forces acting *on* each object involved. Such a diagram is called a **free-body diagram**, or **force diagram**: we draw an arrow to represent each force acting on a given body, being sure to include *every* force acting on that body.

[When concerned only about translational motion, we can draw all the forces on a given body as acting at the center of the object, thus treating the object as a point. However, when doing problems involving rotation or statics, *where* each force acts is also important, as we shall see.]

CONCEPTUAL EXAMPLE 4–8 **The hockey puck.** A hockey puck is sliding at constant velocity across a flat horizontal ice surface that is assumed to be frictionless. Which of the sketches in Fig. 4–20 is the correct free-body diagram for this puck? What would your answer be if the puck slowed down?

RESPONSE Did you choose (a)? If so, can you answer the question: what exerts the horizontal force labeled $\mathbf{F}$? If you say that it is the force needed to maintain the motion (as the ancient Greeks said), ask yourself: what exerts this force? Remember that another object must exert any force—and there simply isn't any possibility here. Therefore, (a) is wrong. Besides, the force $\mathbf{F}$ in Fig. 4–20a would give rise to an accelera-

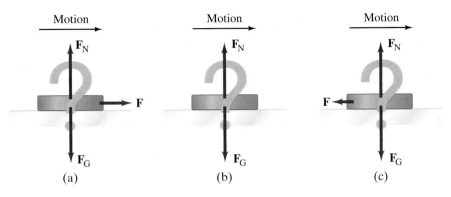

FIGURE 4-20 Which is the correct free-body diagram for a hockey puck sliding across frictionless ice (Example 4–8)?

tion by Newton's second law. It is (b) that is correct, as long as there is no friction. No net force acts on the puck, and the puck slides at constant velocity across the ice. But if someone insists that we come down from the ivory tower of idealized frictionless surfaces, down to the real world where even smooth ice exerts at least a tiny friction force, then (c) is the correct answer. The tiny friction force is in the direction opposite to the motion (it ought to be labeled F_{fr}, not simply F), and the puck's velocity decreases, even if very slowly.

Here now is a brief summary of how to approach solving problems involving Newton's laws:

PROBLEM SOLVING   Newton's Laws; Free-Body Diagrams

1. Draw a sketch of the situation.
2. Consider only one object (at a time), and draw a **free-body diagram** for that body, showing *all* the forces acting *on* that body, including any unknown forces that you have to solve for. Do not show any forces that the body exerts on other bodies. Draw the arrow for each force vector reasonably accurately for direction and magnitude. Label each force, including forces you must solve for, as to its source (gravity, person, friction, and so on). If several bodies are involved, draw a free-body diagram for each body *separately*, showing all the forces acting *on that body* (and *only* forces acting on that body). For each (and every) force, you must be clear about: *on* what

object that force acts; and *by* what object that force is exerted. Only forces acting *on* a given body can be included in $\Sigma F = ma$ for that body.
3. Newton's second law involves vectors, and it is usually important to resolve vectors into components. Choose an x and a y axis in a way that simplifies the calculation.
4. For each body, Newton's second law can be applied to the x and y components separately. That is, the x component of the net force on that body will be related to the x component of that body's acceleration: $\Sigma F_x = ma_x$, and similarly for the y direction.
5. Solve the equation or equations for the unknown(s).

This problem-solving box should not be considered a prescription. Rather it is a summary of things to do that will start your mind thinking and getting involved in the problem at hand.

In the Examples that follow, we assume that all surfaces are very smooth so that friction can be ignored. (Friction, and Examples using it, are discussed in the next Section.) In some of the following Examples, we will deal again with the large gift box on the table, which we first encountered in Example 4–5 (Fig. 4–16). In each successive Example, we add an additional complication so that, step by step, you can see how to approach solving problems.

$F_P = 40.0 \text{ N}$

$30.0°$

(a)

y

F_N

F_P

$30.0°$

x

$m\mathbf{g}$

(b)

FIGURE 4–21 Example 4–9: (b) is the free-body diagram.

EXAMPLE 4–9 Pulling the mystery box. Suppose a friend asks to examine the 10.0-kg box you were given (Example 4–5, Fig. 4–16), hoping to guess what is inside; and you respond, "Sure, pull the box over to you." She then pulls the box by the attached ribbon (or string), as shown in Fig. 4–21a, along the smooth surface of the table. The magnitude of the force exerted by the person is $F_P = 40.0 \text{ N}$, and it is exerted at a 30.0° angle as shown. Calculate (a) the acceleration of the box, and (b) the magnitude of the upward force F_N exerted by the table on the box. Assume that friction can be neglected.

SOLUTION Figure 4–21b shows the free-body diagram of the box, which means we show *all* the forces acting on the box and *only* the forces acting on the box. They are: the force of gravity $m\mathbf{g}$; the normal force exerted by the table $\mathbf{F}_N$; and the force exerted by the person $\mathbf{F}_P$. With the y axis vertical and the x axis horizontal, the pull of 40.0 N has components

$$F_{Px} = (40.0 \text{ N})(\cos 30.0°) = (40.0 \text{ N})(0.866) = 34.6 \text{ N},$$

$$F_{Py} = (40.0 \text{ N})(\sin 30.0°) = (40.0 \text{ N})(0.500) = 20.0 \text{ N}.$$

(a) In the horizontal (x) direction, $\mathbf{F}_N$ and $m\mathbf{g}$ have zero components. Thus the horizontal component of the net force is F_{Px}. From Newton's second law, $\Sigma F_x = ma_x$, we have

$$F_{Px} = ma_x,$$

so

$$a_x = \frac{F_{Px}}{m} = \frac{(34.6 \text{ N})}{(10.0 \text{ kg})} = 3.46 \text{ m/s}^2.$$

The acceleration of the box is thus 3.46 m/s² to the right.
(b) In the vertical (y) direction, with upward as positive, again using Newton's second law we have

$$\Sigma F_y = ma_y$$

$$F_N - mg + F_{Py} = ma_y.$$

Now $mg = (10.0 \text{ kg})(9.80 \text{ m/s}^2) = 98.0 \text{ N}$ and $F_{Py} = 20.0 \text{ N}$ as we calculated above. Furthermore, we know $a_y = 0$ since the box does not even move vertically. Thus

$$F_N - 98.0 \text{ N} + 20.0 \text{ N} = 0$$

which tells us that the normal force is

$$F_N = 78.0 \text{ N}.$$

Notice that F_N is less than mg. The table does not push against the full weight of the box since part of the pull exerted by the person is in the upward direction. Compare this to Example 4–5, part *c*.

Tension in a cord.

When a flexible cord pulls on an object, the cord is said to be under **tension**, and the force it exerts on the object is the tension $\mathbf{F}_T$. If the cord has negligible mass, the force exerted at one end is transmitted undiminished to each adjacent piece of cord along the entire length to the other end. Note that flexible ropes and cords can only pull. They can't push because they bend.

FIGURE 4–22 Example 4–10. (a) Two boxes are connected by a cord. A person pulls horizontally on box 1 with force F_P = 40.0 N. (b) Free-body diagram for box 1. (c) Free-body diagram for box 2.

(a)

(b)

(c)

EXAMPLE 4–10 **Two boxes connected by a cord.** Two boxes are connected by a lightweight cord and are resting on a table. The boxes have masses of 12.0 kg and 10.0 kg. A horizontal force F_P of 40.0 N is applied by a person to the 10.0-kg box, as shown in Fig. 4–22a. Find (a) the acceleration of each box, and (b) the tension in the cord.

SOLUTION (a) The free-body diagram for each of the boxes is shown in Figs. 4–22b and c. We consider each box by itself so that Newton's second law can be applied to each. The cord is light, so we neglect its mass relative to the mass of the boxes. The force F_P acts on box 1. Box 1 exerts a force F_T on the connecting cord, and the cord exerts an opposite but equal magnitude force F_T back on box 1 (Newton's third law). These forces on box 1 are shown in Fig. 4–22b. Because the cord is considered to be massless, the tension at each end is the same.[†] Hence the cord exerts a force F_T on the second box; Fig. 4–22c shows the forces on box 2. There will be only horizontal motion. We take the positive x axis to the right, and we use subscripts 1 and 2 to refer to the two boxes. Applying $\Sigma F_x = ma_x$ to box 1, we have:

$$\Sigma F_x = F_P - F_T = m_1 a_1.$$

For box 2, the only horizontal force is F_T, so

$$\Sigma F_x = F_T = m_2 a_2.$$

The boxes are connected, and if the cord remains taut and doesn't stretch, then the two boxes will have the same acceleration a. Thus $a_1 = a_2 = a$, and we are given $m_1 = 10.0$ kg and $m_2 = 12.0$ kg. We add the two equations above and obtain

$$(m_1 + m_2)a = F_P - F_T + F_T = F_P$$

or

$$a = \frac{F_P}{m_1 + m_2} = \frac{40.0 \text{ N}}{22.0 \text{ kg}} = 1.82 \text{ m/s}^2.$$

This is what we sought. Notice that we would have obtained the same result had we considered a single system, of mass $m_1 + m_2$, acted on by a net horizontal force equal to F_P. (The tension forces F_T would then be considered internal to the system as a whole, and summed together would make zero contribution to the net force on the *whole* system.)
(b) From the equation above for box 2 ($F_T = m_2 a_2$), the tension in the cord is

$$F_T = m_2 a = (12.0 \text{ kg})(1.82 \text{ m/s}^2) = 21.8 \text{ N}.$$

➡ **PROBLEM SOLVING**

An alternate analysis

[†]Since the mass m of the cord is zero, the net force on the cord is $\Sigma F = ma = 0$ no matter what a is. Hence the forces pulling on the cord at its two ends must add up to zero.

FIGURE 4–23 Example 4–11. (a) Atwood's machine in the form of an elevator-counterweight system. (b) and (c) Free-body diagrams for the two masses.

EXAMPLE 4–11 **Elevator and counterweight (Atwood's machine).** Two masses suspended over a pulley by a cable, as shown in Fig. 4–23a, is sometimes referred to generically as an *Atwood's machine*. Consider the real-life application of an elevator (m_1) and its counterweight (m_2). To minimize the work done by the motor to raise and lower the elevator safely, m_1 and m_2 are similar in mass. We leave the motor out of the system for this calculation, and assume the cable's mass is negligible and the pulley is frictionless and massless,† which assures that the tension $\mathbf{F}_T$ in the cord has the same magnitude on both sides of the pulley. Let the mass of the counterweight be $m_2 = 1000$ kg. Assume the mass of the empty elevator is 850 kg, and its mass when carrying four passengers is $m_1 = 1150$ kg. For the latter case ($m_1 = 1150$ kg), calculate (*a*) the acceleration of the elevator and (*b*) the tension in the cable.

SOLUTION (*a*) Figures 4–23b and c show the free-body diagrams for the two masses. It is clear that m_1, being the heavier, will accelerate downward, and m_2 will accelerate upward. The magnitudes of their accelerations will be equal (we assume the cable doesn't stretch). For the counterweight, $m_2 g = (1000\,\text{kg})(9.80\,\text{m/s}^2) = 9800$ N, so F_T must be greater than 9800 N (in order that m_2 will accelerate upward). For the elevator, $m_1 g = (1150\,\text{kg})(9.80\,\text{m/s}^2) = 11,300$ N, which must have greater magnitude than F_T so that m_1 accelerates downward. Thus our calculation must give F_T between 9800 N and 11,300 N. To find F_T as well as the acceleration a, we apply $\Sigma F = ma$ to each mass, where we take upward as the positive y direction for both masses. With this choice of axes, $a_2 = a$, and $a_1 = -a$. Thus

$$F_T - m_1 g = m_1 a_1 = -m_1 a$$
$$F_T - m_2 g = m_2 a_2 = +m_2 a.$$

We subtract the first equation from the second to get

$$(m_1 - m_2)g = (m_1 + m_2)a.$$

We solve this for a:

$$a = \frac{m_1 - m_2}{m_1 + m_2}g = \frac{1150\,\text{kg} - 1000\,\text{kg}}{1150\,\text{kg} + 1000\,\text{kg}}g = 0.070g = 0.68\,\text{m/s}^2.$$

The elevator (m_1) accelerates downward (and the counterweight m_2 upward) at $a = 0.070\,g = 0.68\,\text{m/s}^2$.

(*b*) The tension in the cord, F_T, can be obtained from either of the two $\Sigma F = ma$ equations, setting $a = 0.070\,g$:

$$F_T = m_1 g - m_1 a = m_1 (g - a)$$
$$= 1150\,\text{kg}\,(9.80\,\text{m/s}^2 - 0.68\,\text{m/s}^2) = 10,500\,\text{N},$$
$$F_T = m_2 g + m_2 a = m_2 (g + a)$$
$$= 1000\,\text{kg}\,(9.80\,\text{m/s}^2 + 0.68\,\text{m/s}^2) = 10,500\,\text{N},$$

which are consistent.

We can check our equation for the acceleration a in this Example by noting that if the masses were equal ($m_1 = m_2$), then our equation above for a would give $a = 0$, as we should expect. Also, if one of the masses is zero (say, $m_1 = 0$), then the other mass ($m_2 \neq 0$) would be predicted by our equation to accelerate at $a = g$, again as expected.

†We'll see how to deal with a rotating pulley with mass in Chapter 8.

CONCEPTUAL EXAMPLE 4–12 **The advantage of a pulley.** Muscleman is trying to lift a piano (slowly) up to a second-story apartment (Fig. 4–24). He is using a rope looped over two pulleys as shown. How much of the piano's 2000 N weight does he have to pull on the rope?

RESPONSE Look at the forces acting on the lower pulley at the piano. The weight of the piano is pulling down. The tension in the rope, looped through this pulley, pulls up *twice*, once on each side of the pulley. Thus, Newton's second law gives

$$2F_T - mg = ma.$$

To move the piano with constant speed ($a = 0$) requires a tension in the cord, and hence a pull on the cord, of $F_T = mg/2$. Muscleman exerts a force equal to half the piano's weight. We say the pulley has given a **mechanical advantage** of 2, since without the pulley Muscleman would have to exert twice the force.

FIGURE 4–24 Conceptual Example 4–12.

EXAMPLE 4–13 **Getting the car out of the mud.** Finding her car stuck in the mud, a bright graduate of a good physics course ties a strong rope to the back bumper of the car, and the other end to a tree, as shown in Fig. 4–25a. She pushes at the midpoint of the rope with her maximum effort, which she estimates to be a force $F_P \approx 300\,\text{N}$. The car just begins to budge with the rope at an angle θ (see the figure) which she estimates to be 5°. With what force is the rope pulling on the car? Neglect the mass of the rope.

How to get out of the mud

SOLUTION First, note that the tension in a rope is always along the rope. Any component perpendicular to the rope would cause the rope to bend or buckle (as it does here where $\mathbf{F}_P$ acts)—in other words, a rope can support a tension force only along its length. Let $\mathbf{F}_{T1}$ and $\mathbf{F}_{T2}$ be the forces the rope exerts on the tree and on the car, as shown in Fig. 4–25a. As our "free body," we choose the tiny section of rope where she pushes. The free-body diagram is shown in Fig. 4–25b, which shows $\mathbf{F}_P$ as well as the tensions in the rope (note that we have used Newton's third law). At the moment the car budges, the acceleration is still essentially zero, so $\mathbf{a} = 0$. For the x component of $\Sigma \mathbf{F} = m\mathbf{a} = 0$ on that small section of rope, we have

$$\Sigma F_x = F_{T2x} - F_{T1x} = 0, \quad \text{or} \quad F_{T1}\cos\theta - F_{T2}\cos\theta = 0.$$

Hence $F_{T1} = F_{T2}$, and we can write $F_T = F_{T1} = F_{T2}$. In the y direction, the forces acting are F_P, and the components of $\mathbf{F}_{T1}$ and $\mathbf{F}_{T2}$ that point in the negative y direction (each equal to $F_T \sin\theta$). So for the y component of $\Sigma\mathbf{F} = m\mathbf{a}$, we have

$$\Sigma F_y = F_P - 2F_T \sin\theta = 0.$$

We solve this for F_T and insert $F_P \approx 300\,\text{N}$, which was given:

$$F_T = \frac{F_P}{2\sin\theta} \approx \frac{300\,\text{N}}{2\sin 5°} \approx 1700\,\text{N}.$$

She was able to magnify her effort almost six times using this technique! Notice the symmetry of the problem, which ensures that $F_{T1} = F_{T2}$.

(a)

(b)

FIGURE 4–25 Example 4–13. Getting a car out of the mud.

➡ **PROBLEM SOLVING**

Use any symmetry present to simplify a problem

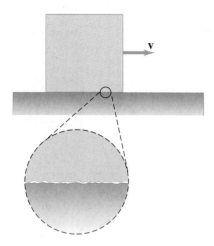

FIGURE 4–26 An object moving to the right on a table or floor. The two surfaces in contact are rough, at least on a microscopic scale.

4-8 Applications Involving Friction, Inclines

Until now we have ignored friction, but it must be taken into account in most practical situations. Friction exists between two solid surfaces because even the smoothest looking surface is quite rough on a microscopic scale, Fig. 4–26. When we try to slide an object across another surface, these microscopic bumps impede the motion. In addition, at the atomic level, the atoms on a bump of one surface come so close to the atoms of the other surface that electric forces between the atoms can form chemical bonds, as a tiny weld between the two surfaces. Sliding an object across a surface is often jerky due to the making and breaking of these bonds. Even when a body rolls across a surface, there is still some friction, called *rolling friction*, although it is generally much less than when a body slides across a surface. We will be concerned mainly with sliding friction in this section, and it is usually called **kinetic friction** (*kinetic* is from the Greek for "moving").

When a body is in motion along a rough surface, the force of kinetic friction acts opposite to the direction of the body's velocity. The magnitude of the force of kinetic friction depends on the nature of the two sliding surfaces. For given surfaces, experiment shows that the friction force is approximately proportional to the *normal force* between the two surfaces, which is the force that either object exerts on the other, perpendicular to their common surface of contact (see Fig. 4–27). The force of friction between hard surfaces depends very little on the total surface area of contact; that is, the friction force on this book is roughly the same whether it is being slid on its wide face or on its spine, assuming the surfaces have the same smoothness. We can write the proportionality as an equation by inserting a constant of proportionality, μ_k:

Kinetic friction

$$F_{fr} = \mu_k F_N.$$

This relation is not a fundamental law; it is an experimental relation between the magnitude of the friction force $\mathbf{F}_{fr}$ which acts parallel to the two surfaces, and the magnitude of the normal force $\mathbf{F}_N$ which acts perpendicular to the surfaces. It is *not* a vector equation since the two forces are perpendicular to one another. The term μ_k is called the *coefficient of kinetic friction*, and its value depends on the nature of the two surfaces. Measured values for a variety of surfaces are given in Table 4–2. These are only approximate, however, since μ depends on whether the surfaces are wet or dry, on how much they have been sanded or rubbed, if any burrs remain, and other such factors. But μ_k is roughly independent of the sliding speed.

$\mathbf{F}_{fr} \perp \mathbf{F}_N$

FIGURE 4–27 When an object is pulled by an applied force ($\mathbf{F}_A$) along a surface, the force of friction $\mathbf{F}_{fr}$ opposes the motion. The magnitude of $\mathbf{F}_{fr}$ is proportional to the magnitude of the normal force (F_N).

What we have been discussing up to now is *kinetic friction*, when one object slides over another. There is also **static friction**, which refers to a force parallel to the two surfaces that can arise even when they are not sliding. Suppose an object such as a desk is resting on a horizontal floor. If no horizontal force is exerted on the desk, there also is no friction force. But now, suppose you try to push the desk, but it doesn't move. You are exerting a horizontal force, but the desk isn't moving, so there must be another force on the desk keeping it from moving (the net force is zero on an object that doesn't move). This is the force of *static friction* exerted by the floor on the desk. If you push with a greater force without moving the desk, the force of static friction also has increased. If you push hard enough, the desk will

TABLE 4–2 Coefficients of Friction[†]

Surfaces	Coefficient of Static Friction, μ_s	Coefficient of Kinetic Friction, μ_k
Wood on wood	0.4	0.2
Ice on ice	0.1	0.03
Metal on metal (lubricated)	0.15	0.07
Steel on steel (unlubricated)	0.7	0.6
Rubber on dry concrete	1.0	0.8
Rubber on wet concrete	0.7	0.5
Rubber on other solid surfaces	1–4	1
Teflon® on Teflon in air	0.04	0.04
Teflon on steel in air	0.04	0.04
Lubricated ball bearings	<0.01	<0.01
Synovial joints (in human limbs)	0.01	0.01

[†]Values are approximate and are intended only as a guide.

eventually start to move, and kinetic friction takes over. At this point, you have exceeded the maximum force of static friction, which is given by $F_{max} = \mu_s F_N$, where μ_s is the *coefficient of static friction* (Table 4–2). Since the force of static friction can vary from zero to this maximum value, we write

$$F_{fr} \leq \mu_s F_N.$$

Static friction

You may have noticed that it is often easier to keep a heavy object moving, such as pushing a table, than it is to start it moving in the first place. This is consistent with the fact (see Table 4–2) that μ_s is generally greater than μ_k. (It can never be less. Why?)

EXAMPLE 4–14 **Friction: static and kinetic.** Our 10.0-kg mystery box rests on a horizontal floor. The coefficient of static friction is $\mu_s = 0.40$ and the coefficient of kinetic friction is $\mu_k = 0.30$. Determine the force of friction, F_{fr}, acting on the box if a horizontal external applied force F_A is exerted on it of magnitude: (a) 0, (b) 10 N, (c) 20 N, (d) 38 N, and (e) 40 N.

SOLUTION The free-body diagram of the box is shown in Fig. 4–27. Examine it carefully. In the vertical direction there is no motion, so $\Sigma F_y = ma_y = 0$ yields $F_N - mg = 0$. Hence the normal force for all cases is

$$F_N = mg = (10.0 \text{ kg})(9.8 \text{ m/s}^2) = 98 \text{ N}.$$

(a) Since no force is applied in this first case, the box doesn't move, and $F_{fr} = 0$.
(b) The force of static friction will oppose any applied force up to a maximum of

$$\mu_s F_N = (0.40)(98 \text{ N}) = 39 \text{ N}.$$

The applied force is $F_A = 10 \text{ N}$. Thus the box will not move; since $\Sigma F_x = F_A - F_{fr} = 0$ then $F_{fr} = 10 \text{ N}$.
(c) An applied force of 20 N is also not sufficient to move the box. Thus $F_{fr} = 20 \text{ N}$ to balance the applied force.
(d) The applied force of 38 N is still not quite large enough to move the box; so the friction force has now increased to 38 N to keep the box at rest.

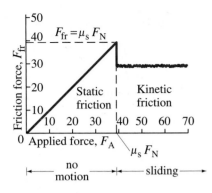

FIGURE 4–28 Magnitude of the force of friction as a function of the external force applied to a body initially at rest. As the applied force is increased in magnitude, the force of static friction increases linearly to just match it, until the applied force equals $\mu_s F_N$. If the applied force increases further, the body will begin to move, and the friction force drops to a roughly constant value characteristic of kinetic friction.

(*e*) A force of 40 N will start the box moving since it exceeds the maximum force of static friction, $\mu_s F_N = (0.40)(98\text{ N}) = 39\text{ N}$. Instead of static friction, we now have kinetic friction, and its magnitude is

$$F_{fr} = \mu_k F_N = (0.30)(98\text{ N}) = 29\text{ N}.$$

There is now a net (horizontal) force on the box of magnitude $F = 40\text{ N} - 29\text{ N} = 11\text{ N}$, so the box will accelerate at a rate

$$a_x = \Sigma F/m = 11\text{ N}/10\text{ kg} = 1.1\text{ m/s}^2$$

as long as the applied force is 40 N. Figure 4–28 shows a graph that summarizes this Example.

Now we look at some Examples involving kinetic friction in a variety of situations. Note that both the normal force and the friction force are forces exerted by one surface on another; one is perpendicular to the contact surfaces (the normal force), and the other is parallel (the friction force).

CONCEPTUAL EXAMPLE 4–15   To push or to pull a sled? Your little sister wants a ride on her sled. If you are on flat ground, will you exert less force if you push her or pull her? See Figs. 4–29a and b. Assume the same angle θ in each case.

RESPONSE Free-body diagrams are shown in Figs. 4-29c and d. If you push her, and $\theta > 0$, there is a vertically downward component to your force. Hence the normal force exerted upward by the ground will be larger than mg (where m is the mass of sister plus sled). If you pull her, your force has a vertically upward component, so the normal force F_N can be less than mg. Because the friction force is proportional to the normal force, it will be less if you pull her. So you exert less force if you pull her.

FIGURE 4–29
Conceptual Example 4–15.

EXAMPLE 4–16 **Two boxes and a pulley.** In Fig. 4–30a, two boxes are connected by a cord running over a pulley. The coefficient of kinetic friction between box I and the table is 0.20. We ignore the mass of the cord and pulley and any friction in the pulley, which means we can assume that a force applied to one end of the cord will have the same magnitude at the other end. We wish to find the acceleration, a, of the system, which will have the same magnitude for both boxes assuming the cord doesn't stretch. As box II moves down, box I moves to the right.

SOLUTION Free-body diagrams are shown for each box in Fig. 4–30b and c. Box I does not move vertically, so the normal force just balances the weight,

$$F_N = m_I g = (5.0\ \text{kg})(9.8\ \text{m/s}^2) = 49\ \text{N}.$$

In the horizontal direction, there are two forces on box I (Fig. 4–30b): F_T, the tension in the cord (whose value we don't know), and the force of friction

$$F_{fr} = \mu_k F_N = (0.20)(49\ \text{N}) = 9.8\ \text{N}.$$

The horizontal acceleration is what we wish to find; we use Newton's second law in the x direction, $\Sigma F_{Ix} = m_I a_x$, which becomes (taking the positive direction to the right and setting $a_{Ix} = a$):

$$\Sigma F_{Ix} = F_T - F_{fr} = m_I a. \qquad \text{[box I]}$$

Next consider box II. The force of gravity $F_G = m_{II} g = 19.6\ \text{N}$ pulls downward; and the cord pulls upward with a force F_T. So we can write Newton's second law for box II (taking the downward direction as positive):

$$\Sigma F_{IIy} = m_{II} g - F_T = m_{II} a. \qquad \text{[box II]}$$

[Note here that if $a \neq 0$, then F_T is not equal to $m_{II} g$.] We have two unknowns, a and F_T, and we also have two equations. We solve the box I equation for F_T:

$$F_T = F_{fr} + m_I a,$$

and substitute this into the box II equation:

$$m_{II} g - F_{fr} - m_I a = m_{II} a.$$

Now we solve for a and put in numerical values:

$$a = \frac{m_{II} g - F_{fr}}{m_I + m_{II}} = \frac{19.6\ \text{N} - 9.8\ \text{N}}{5.0\ \text{kg} + 2.0\ \text{kg}} = 1.4\ \text{m/s}^2,$$

which is the acceleration of box I to the right, and of box II down. If we wish, we can calculate F_T using the first equation:

$$F_T = F_{fr} + m_I a = 9.8\ \text{N} + (5.0\ \text{kg})(1.4\ \text{m/s}^2) = 17\ \text{N}.$$

We now discuss some examples of objects moving on an incline such as a hill or a ramp. Solving problems is usually easier if we choose the xy coordinate system so that the x axis points along the incline (either up the incline, or down the incline), and the y axis perpendicular to the incline, as shown in Fig. 4–31. This helps because then **a** has only one component, and if friction is present, two of the forces will have only one component: $\mathbf{F}_{fr}$ along the plane, opposite to the object's velocity, and $\mathbf{F}_N$ which is not vertical but is perpendicular to the plane.

(a)

(b)

(c)

FIGURE 4–30
Example 4–16.

FIGURE 4–31 Forces on an object sliding down an incline.

(a)

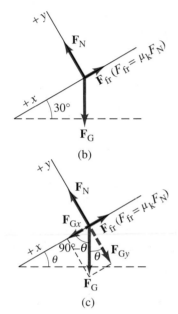

(b)

(c)

FIGURE 4–32 Example 4–17. A skier descending a slope.

EXAMPLE 4–17 **The skier.** The skier in Fig. 4–32a has just begun descending the 30° slope. Assuming the coefficient of kinetic friction is 0.10, (a) first draw the free-body diagram, then calculate (b) her acceleration and (c) the speed she will reach after 4.0 s.

SOLUTION (a) The free-body diagram in Fig. 4–32b shows all the forces acting on the skier: her weight ($\mathbf{F_G} = m\mathbf{g}$) downward, and the two forces exerted on her skis by the snow—the normal force perpendicular to the snow's surface, and the friction force parallel to the surface. These three forces are shown acting at one point in Fig. 4–32b, for convenience. Also for convenience, we choose the x axis parallel to the snow surface with positive direction downhill, and the y axis perpendicular to the surface. With this choice, we only have to resolve one vector into components, the weight. The components of the weight are shown as dashed lines in Fig. 4–32c. They are given by

$$F_{Gx} = mg \sin \theta,$$

$$F_{Gy} = -mg \cos \theta,$$

where we have stayed general, using θ rather than 30° for now.

(b) To calculate her acceleration down the hill, a_x, we apply Newton's second law to the x direction:

$$\Sigma F_x = ma_x$$

$$mg \sin \theta - \mu_k F_N = ma_x$$

where the two forces are the component of the gravity force ($+x$ direction) and the friction force ($-x$ direction). We want to find the value of a_x, but we don't yet know F_N in the last equation. Let's see if we can get F_N from the y component of Newton's second law:

$$\Sigma F_y = ma_y$$

$$F_N - mg \cos \theta = ma_y = 0$$

where we set $a_y = 0$ because there is no motion in the y direction (perpendicular to the slope). Thus we can solve for F_N:

$$F_N = mg \cos \theta$$

and we can substitute this into our equation above for ma_x:

$$mg \sin \theta - \mu_k(mg \cos \theta) = ma_x.$$

There is an m in each term which can be canceled out. Thus (setting $\theta = 30°$ and $\mu_k = 0.10$):

$$a_x = g \sin 30° - \mu_k g \cos 30° = 0.50\,g - (0.10)(0.866)\,g = 0.41\,g.$$

The skier's acceleration is 0.41 times the acceleration of gravity, which in numbers is $a = (0.41)(9.8 \text{ m/s}^2) = 4.0 \text{ m/s}^2$. It is interesting that the mass canceled out and so we have the useful conclusion that *the acceleration doesn't depend on the mass.* That such a cancellation sometimes occurs, and thus may give a useful conclusion as well as saving calculation, is a big advantage of working with the algebraic equations and putting in the numbers only at the end.

(*c*) The speed after 4.0 s is found by using Eq. 2–10a:

$$v = v_0 + at = 0 + (4.0 \text{ m/s}^2)(4.0 \text{ s}) = 16 \text{ m/s},$$

where we assumed a start from rest.

▶ **PROBLEM SOLVING**

It is often helpful to put in numbers only at the end

EXAMPLE 4–18 **Measuring** μ_k. Suppose in Example 4–17 that the snow is slushy and the skier moves down the 30° slope at constant speed. What can you say about the coefficient of friction, μ_k?

SOLUTION Now the skier moves down the slope at constant speed, and we want to find μ_k. The free-body diagram and the $\Sigma F = ma$ equations for the *x* and *y* components will be the same as above, except that now we are given $a_x = 0$. Thus

$$\Sigma F_y = F_N - mg \cos \theta = ma_y = 0$$
$$\Sigma F_x = mg \sin \theta - \mu_k F_N = ma_x = 0.$$

From the first equation, we have $F_N = mg \cos \theta$; we substitute this into the second equation:

$$mg \sin \theta - \mu_k(mg \cos \theta) = 0.$$

Now we solve for μ_k:

$$\mu_k = \frac{mg \sin \theta}{mg \cos \theta} = \frac{\sin \theta}{\cos \theta} = \tan \theta$$

which for $\theta = 30°$ is

$$\mu_k = \tan \theta = \tan 30° = 0.58.$$

Notice that we could use the equation

$$\mu_k = \tan \theta$$

A method for determining μ_k

to determine μ_k under a variety of conditions. All we need to do is observe at what slope angle the skier descends at constant speed. Here is another reason why it is often useful to plug in numbers only at the end: we obtained a general result useful for other situations as well.

4-9 Problem Solving—A General Approach

A basic part of a physics course is solving problems effectively. Problem solving is important not only because it is valuable in itself, but also because it makes you think about the ideas and concepts, and by applying the concepts, you come to understand them better. We have already studied a range of Examples in these first few chapters. It is so important to learn problem solving that we will spend a little time now summarizing how to approach problems in general, even though much of this has been discussed before. The approach discussed here, though emphasizing Newton's laws, can be applied generally for other topics discussed throughout this book.

➡ **PROBLEM SOLVING** In General

1. **Read** and reread written problems carefully. A common error is to leave out a word or two when reading, which can completely change the meaning of a problem.

2. **Draw** an accurate picture or diagram of the situation. (This is probably the most overlooked, yet most crucial, part of solving a problem.) Use arrows to represent vectors such as velocity or force, and label the vectors with appropriate symbols. When dealing with forces and applying Newton's laws, make sure to include all forces on a given body, including unknown ones, and make clear what forces act on what body (otherwise you may make an error in determining the *net force* on a particular body). A separate *free-body diagram* needs to be drawn for each body involved, and it must show *all* the forces acting on a given body (and only on that body). Do not show forces that the body exerts on other bodies.

3. Choose a convenient xy **coordinate system** (choose one that makes your calculations easier). Vectors are to be resolved into components along these axes. When using Newton's second law, apply $\Sigma \mathbf{F} = m\mathbf{a}$ separately to x and y components, remembering that x direction forces are related to a_x, and similarly for y.

4. Note what the unknowns are—that is, what you are trying to determine—and decide what you need in order to find the unknowns. For problems in the present chapter, we use Newton's laws. More generally, it may help to see if

there are one or more **relationships** (or **equations**) that relate the unknowns to the knowns. But be sure each relationship is applicable in the given case. It is very important to know the limitations of each formula or relationship—when it is valid and when not. In this book, the more general equations have been given numbers, but even these can have a limited range of validity (often stated briefly, in brackets, to the right of the equation).

5. Try to solve the problem approximately, to see if it is doable (to check if enough information has been given) and reasonable. Use your intuition, and make **rough calculations**—see "Order of Magnitude Estimating" in Section 1–7. A rough calculation, or a reasonable guess about what the range of final answers might be, is very useful. And a rough calculation can be checked against the final answer to catch errors in calculation (such as in a decimal point or the powers of 10).

6. **Solve** the problem, which may include algebraic manipulation of equations and/or numerical calculations. Substituting numbers into the equations only at the end can give you greater insight into the problem and to related ones.

7. Be sure to keep track of **units**, for they can serve as a check (they must balance on both sides of any equation).

8. Again consider if your answer is **reasonable**. The use of dimensional analysis, described in Appendix B, can also serve as a check for many problems.

Newton's three laws of motion are the basic classical laws describing motion.

Newton's first law (the law of inertia) states that if the net force on an object is zero, an object originally at rest remains at rest, and an object in motion remains in motion in a straight line with constant velocity.

Newton's second law states that the acceleration of a body is directly proportional to the net force acting on it, and inversely proportional to its mass:

$$\Sigma \mathbf{F} = m\mathbf{a}.$$

Newton's second law is one of the most important and fundamental laws in classical physics.

Newton's third law states that whenever one body exerts a force on a second body, the second body always exerts a force on the first body which is equal in magnitude but opposite in direction:

$$\mathbf{F}_{12} = -\mathbf{F}_{21}$$

The tendency of a body to resist a change in its motion is called **inertia**. **Mass** is a measure of the inertia of a body.

Weight refers to the force of gravity on a body, and is equal to the product of the body's mass m and the acceleration of gravity $\mathbf{g}$:

$$\mathbf{F}_G = m\mathbf{g}.$$

Force, which is a vector, can be considered as a push or pull; or, from Newton's second law, force can be defined as an action capable of giving rise to acceleration. The **net force** on an object is the vector sum of all forces acting on it.

When two bodies slide over one another, the force of friction that each body exerts on the other can be written approximately as $F_{fr} = \mu_k F_N$, where F_N is the **normal force** (the force each body exerts on the other perpendicular to their contact surfaces), and μ_k is the coefficient of **kinetic friction**. If the bodies are at rest relative to each other, then F_{fr} is just large enough to hold them at rest and satisfies the inequality $F_{fr} < \mu_s F_N$, where μ_s is the coefficient of **static friction**.

For solving problems involving the forces on one or more bodies, it is essential to draw a **free-body diagram** for each body, showing all the forces acting on only that body. Newton's second law can be applied to the vector components for each body.

1. Why does a child in a wagon seem to fall backward when you give the wagon a sharp pull?
2. What, roughly, does an apple weigh in newtons?
3. If the acceleration of a body is zero, are no forces acting on it?
4. Why do you push harder on the pedals of a bicycle when first starting out than when moving at constant speed?
5. Only one force acts on an object. Can the object have zero acceleration? Can it have zero velocity?
6. When a golf ball is dropped to the pavement, it bounces back up. (a) Is a force needed to make it bounce back up? (b) If so, what exerts the force?
7. Examine, in the light of Newton's first and second laws, the motion of your leg during one stride while walking.
8. Why might your foot hurt if you kick a heavy desk or a wall?
9. When you are running and want to stop quickly, you must decelerate quickly. (a) What is the origin of the force that causes you to stop? (b) Estimate (using your own experience) the maximum rate of deceleration of a person running at top speed to come to rest.

FIGURE 4–33 Question 10.

10. A stone hangs by a fine thread from the ceiling, and a section of the same thread dangles from the bottom of the stone (Fig. 4–33). If a person gives a sharp pull on the dangling thread, where is the thread likely to break: below the stone or above it? What if the person gives a slow and steady pull? Explain your answers.
11. The force of gravity on a 2-kg rock is twice as great as that on a 1-kg rock. Why then doesn't the heavier rock fall faster?

12. Compare the effort (or force) needed to lift a 10-kg object when you are on the Moon as compared to lifting it on Earth. Compare the force needed to throw a 2-kg object horizontally with a given speed when on the Moon as compared to on Earth.

13. Whiplash sometimes results from an automobile accident when the victim's car is struck violently from the rear. Explain why the head of the victim seems to be thrown backward in this situation. Is it really?

14. A person exerts an upward force of 40 N to hold a bag of groceries. Describe the "reaction" force (Newton's third law) by stating (*a*) its magnitude, (*b*) its direction, (*c*) *on* what body it is exerted, and (*d*) *by* what body it is exerted.

15. When you stand still on the ground, how large a force does the ground exert on you? Why doesn't this force make you rise up into the air?

16. According to Newton's third law, each team in a tug of war (Fig. 4–34) pulls with equal force on the other team. What, then, determines which team will win?

17. When driving on slick roads, why is it advisable to apply the brakes slowly?

18. Why is the stopping distance of a truck much shorter than for a train going the same speed?

19. Can a coefficient of friction exceed 1.0?

FIGURE 4–34 A tug of war. Describe the forces on each of the teams and on the rope. Question 16.

20. A block is given a push so that it slides up a ramp. After the block reaches its highest point, it slides back down. Why is the magnitude of its acceleration less on the descent than on the ascent?

21. A heavy crate rests on the bed of a flatbed truck. When the truck accelerates, the crate remains where it is on the truck, so it, too, accelerates. What force causes the crate to accelerate?

22. You can hold a heavy box against a rough wall and prevent it from slipping down by pressing only horizontally. How can the application of a *horizontal* force keep an object from moving vertically?

PROBLEMS

SECTIONS 4–4 TO 4–6

1. (I) What force is needed to accelerate a child on a sled (total mass = 60.0 kg) at 1.15 m/s^2?

2. (I) A net force of 255 N accelerates a bike and rider at 2.20 m/s^2. What is the mass of the bike and rider?

3. (I) How much force is required to accelerate a 9.0-g object at 10,000 "*g*'s" (say, in a centrifuge)?

4. (I) How much tension must a rope withstand if it is used to accelerate a 1050-kg car horizontally at 1.20 m/s^2? Ignore friction.

5. (I) What is the weight of a 66-kg astronaut (*a*) on Earth, (*b*) on the Moon ($g = 1.7 \text{ m/s}^2$), (*c*) on Mars ($g = 3.7 \text{ m/s}^2$), (*d*) in outer space traveling with constant velocity?

6. (II) A 20.0-kg box rests on a table. (*a*) What is the weight of the box and the normal force acting on it? (*b*) A 10.0-kg box is placed on top of the 20.0-kg box, as shown in Fig. 4–35. Determine the normal force that the table exerts on the 20.0-kg box and the normal force that the 20.0-kg box exerts on the 10.0-kg box.

7. (II) What average force is required to stop an 1100-kg car in 8.0 s if it is traveling at 90 km/h?

8. (II) What average force is needed to accelerate a 7.00-gram pellet from rest to 175 m/s over a distance of 0.700 m along the barrel of a rifle?

FIGURE 4–35 Problem 6.

9. (II) A fisherman yanks a fish out of the water with an acceleration of 4.5 m/s^2 using very light fishing line that has a "test" value of 22 N. The fisherman unfortunately loses the fish as the line snaps. What can you say about the mass of the fish?

10. (II) A 0.140-kg baseball traveling 45.0 m/s strikes the catcher's mitt, which, in bringing the ball to rest, recoils backward 11.0 cm. What was the average force applied by the ball on the glove?

11. (II) What is the average force exerted by a shot-putter on a 7.0-kg shot if the shot is moved through a distance of 2.8 m and is released with a speed of 13 m/s?

12. (II) How much tension must a rope withstand if it is used to accelerate a 1200-kg car vertically upward at 0.80 m/s^2? Ignore friction.

13. (II) A 10-kg bucket is lowered by a rope in which there is 63 N of tension. What is the acceleration of the bucket? Is it up or down?

14. (II) An elevator (mass 4850 kg) is to be designed so that the maximum acceleration is 0.0600 g. What are the maximum and minimum forces the motor should exert on the supporting cable?

15. (II) A 75-kg petty thief wants to escape from a third-story jail window. Unfortunately, a makeshift rope made of sheets tied together can support a mass of only 58 kg. How might the thief use this "rope" to escape? Give quantitative answer.

16. (II) A person stands on a bathroom scale in a motionless elevator. When the elevator begins to move, the scale briefly reads only 0.75 of the person's regular weight. Calculate the acceleration of the elevator, and find the direction of acceleration.

17. (II) The cable supporting a 2100-kg elevator has a maximum strength of 21,750 N. What maximum upward acceleration can it give the elevator without breaking?

18. (II) (a) What is the acceleration of two falling sky divers (mass 120.0 kg including parachute) when the upward force of air resistance is equal to one fourth of their weight? (b) After popping open the parachute, the divers descend leisurely to the ground at constant speed. What now is the force of air resistance on the sky divers and their parachute? See Fig. 4–36.

FIGURE 4–36   Problem 18.

19. (II) A Saturn V rocket has a mass of 2.75×10^6 kg and exerts a force of 33×10^6 N on the gases it expels. Determine (a) the initial vertical acceleration of the rocket, (b) its velocity after 8.0 s, and (c) how long it takes to reach an altitude of 9500 m. Ignore mass of gas expelled and assume g remains constant.

20. (III) An exceptional standing jump would raise a person 0.80 m off the ground. To do this, what force must a 66-kg person exert against the ground? Assume the person crouches a distance of 0.20 m prior to jumping, and thus the upward force has this distance to act over before he leaves the ground.

21. (III) A person jumps from the roof of a house 4.5-m high. When he strikes the ground below, he bends his knees so that his torso decelerates over an approximate distance of 0.70 m. If the mass of his torso (excluding legs) is 45 kg, find (a) his velocity just before his feet strike the ground, and (b) the average force exerted on his torso by his legs during deceleration.

22. (III) The 100-m dash can be run by the best sprinters in 10.0 s. A 65-kg sprinter accelerates uniformly for the first 50 m to reach top speed, which he maintains for the remaining 50 m. (a) What is the average horizontal component of force exerted on his feet by the ground during acceleration? (b) What is the speed of the sprinter over the last 50 m of the race (i.e., his top speed)?

SECTION 4–7

23. (I) A box weighing 70 N rests on a table. A rope tied to the box runs vertically upward over a pulley and a weight is hung from the other end (Fig. 4–37). Determine the force that the table exerts on the box if the weight hanging on the other side of the pulley weighs (a) 30 N, (b) 60 N, and (c) 90 N.

24. (I) A 650-N force acts in a northwesterly direction. A second 650-N force must be exerted in what direction so that the resultant of the two forces points westward?

FIGURE 4–37   Problem 23.

25. (I) Draw the free-body diagram for a basketball player (a) just before leaving the ground on a jump, and (b) while in the air. (Fig. 4–38)

FIGURE 4–38
Problem 25.

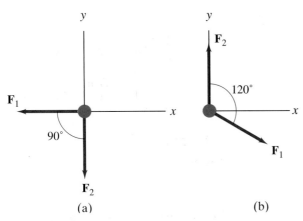

FIGURE 4-39 Problem 27.

26. (I) Sketch the free-body diagram of a baseball (*a*) at the moment it is hit by the bat, and again (*b*) after it has left the bat and is flying toward the outfield.

27. (II) The two forces F_1 and F_2 shown in Fig. 4–39a and b (looking down) act on a 27.0-kg object on a frictionless tabletop. If $F_1 = 10.2$ N and $F_2 = 16.0$ N, find the net force on the object and its acceleration for each situation, (*a*) and (*b*).

28. (II) A person pushes a 14.5-kg lawn mower at constant speed with a force of 88.0 N directed along the handle, which is at an angle of 45.0° to the horizontal (Fig. 4–40). (*a*) Draw the free-body diagram showing all forces acting on the mower. Calculate (*b*) the horizontal retarding force on the mower, then (*c*) the normal force exerted vertically upward on the mower by the ground, and (*d*) the force the person must exert on the lawn mower to accelerate it from rest to 1.5 m/s in 2.5 seconds (assuming the same retarding force).

FIGURE 4-40 Problem 28.

29. (II) At the instant a race began, a 65-kg sprinter was found to exert a force of 800 N on the starting block at a 22° angle with respect to the ground (see chapter opening photo). (*a*) What was the horizontal acceleration of the sprinter? (*b*) If the force was exerted for 0.38 s, with what speed did the sprinter leave the starting block?

FIGURE 4-41 Problem 30.

30. (II) One 3.0-kg paint bucket is hanging by a massless cord from another 3.0-kg paint bucket, also hanging by a massless cord, as shown in Fig. 4–41. (*a*) If the buckets are at rest, what is the tension in each cord? (*b*) If the two buckets are pulled upward with an acceleration of 1.60 m/s² by the upper cord, calculate the tension in each cord.

31. (II) A 6500-kg helicopter accelerates upward at 0.60 m/s² while lifting a 1200-kg car. (*a*) What is the lift force exerted by the air on the rotors? (*b*) What is the tension in the cable (ignore its mass) that connects the car to the helicopter?

32. (II) A window washer pulls herself upward using the bucket-pulley apparatus shown in Fig. 4–42. (*a*) How hard must she pull downward to raise herself slowly at constant speed? (*b*) If she increases this force by 10 percent, what will her acceleration be? The mass of the person plus the bucket is 65 kg.

FIGURE 4-42
Problem 32.

FIGURE 4–43 Problem 34.

33. (II) A train locomotive is pulling two cars of the same mass behind it. Show that the tension in the coupling between the locomotive and the first car is twice that between the first car and the second car, for any nonzero acceleration of the train.

34. (II) A pair of fuzzy dice is hanging by a string from your rearview mirror. While you are accelerating from a stoplight to 20 m/s (in 5.0 seconds), what angle θ does the string make with the vertical? See Fig. 4–43.

FIGURE 4–44 Problem 35.

35. (III) Three blocks on a frictionless horizontal surface are in contact with each other, as shown in Fig. 4–44. A force **F** is applied to block 1 (mass m_1). (a) Draw a free-body diagram for each block. Determine (b) the acceleration of the system (in terms of m_1, m_2, and m_3), (c) the net force on each block, and (d) the force of contact that each block exerts on its neighbor. (e) If $m_1 = m_2 = m_3 = 12.0$ kg and $F = 96.0$ N, give numerical answers to (b), (c), and (d). Do your answers make sense intuitively?

FIGURE 4–45 Problem 36.

36. (III) The two masses shown in Fig. 4–45 are each initially 1.80 m above the ground, and the massless frictionless pulley is 4.8 m above the ground. What maximum height does the lighter object reach after the system is released? [*Hint:* First determine the acceleration of the lighter mass and then its velocity at the moment the heavier one hits the ground. This is its "launch" speed.]

37. (III) Suppose the cord in Example 4–10 and Fig. 4–22 is a heavy rope of mass 1.0 kg. Calculate the acceleration of each box and the tension at each end of the cord, using the free-body diagrams shown in Fig. 4–46. Ignore sagging of the cord.

SECTION 4–8

38. (I) If the coefficient of kinetic friction between a 35-kg crate and the floor is 0.30, what horizontal force is required to move the crate at a steady speed across the floor? What horizontal force is required if μ_k is zero?

39. (I) A force of 40.0 N is required to start a 5.0-kg box moving across a horizontal concrete floor. (a) What is the coefficient of static friction between the box and the floor? (b) If the 40.0-N force continues, the box accelerates at 0.70 m/s². What is the coefficient of kinetic friction?

FIGURE 4–46 Problem 37. Free-body diagrams for each of the objects of the system shown in Fig. 4–22a. Vertical forces, $\mathbf{F}_N$ and $\mathbf{F}_G$, are not shown.

40. (I) (*a*) A box sits at rest on a rough 30° inclined plane. Draw the free-body diagram, showing all the forces acting on the box. (*b*) How would the diagram change if the box were sliding down the plane? (*c*) How would it change if the box were sliding up the plane after an initial shove?

41. (I) A 2.0-kg silverware drawer does not slide readily. The owner gradually pulls with more and more force. When the applied force reaches 8.0 N, the drawer suddenly opens, throwing all the utensils to the floor. Find the coefficient of static friction between the drawer and the cabinet.

42. (II) Drag race tires in contact with an asphalt surface probably have one of the highest coefficients of static friction in the everyday world. Assuming a constant acceleration and no slipping of tires, estimate the coefficient of static friction for a drag racer that covers the quarter mile in 6.0 s.

43. (II) For the system of Fig. 4–30 (Example 4–16), how large a mass would body I have to have to prevent any motion from occurring? Assume $\mu_s = 0.30$.

44. (II) A box is given a push so that it slides across the floor. How far will it go, given that the coefficient of kinetic friction is 0.20 and the push imparts an initial speed of 4.0 m/s?

45. (II) Two crates, of mass 75 kg and 110 kg, are in contact and at rest on a horizontal surface (Fig. 4–47). A 730-N force is exerted on the 75-kg crate. If the coefficient of kinetic friction is 0.15, calculate (*a*) the acceleration of the system, and (*b*) the force that each crate exerts on the other.

FIGURE 4–47 Problem 45.

46. (II) (*a*) Show that the minimum stopping distance for an automobile traveling at speed *v* is equal to $v^2/2\mu_s g$, where μ_s is the coefficient of static friction between the tires and the road, and *g* is the acceleration of gravity. (*b*) What is this distance for a 1200-kg car traveling 95 km/h if $\mu_s = 0.75$? (*c*) What would it be if the car were on the Moon but all else stayed the same?

47. (II) A flatbed truck is carrying a heavy crate. The coefficient of static friction between the crate and the bed of the truck is 0.75. What is the maximum rate at which the driver can decelerate and still avoid having the crate slide against the cab?

48. (II) On an icy day, you worry about parking your car in your driveway, which has an incline of 12°. Your neighbor Ralph's driveway has an incline of 9.0°, and Bonnie's driveway across the street has one of 6.0°. The coefficient of static friction between tire rubber and ice is 0.15. Which driveway(s) will be safe to park in?

49. (II) A child slides down a slide with a 28° incline, and at the bottom her speed is precisely half what it would have been if the slide had been frictionless. Calculate the coefficient of kinetic friction between the slide and the child.

50. (II) A coffee cup on the dashboard of a car slides forward on the dash when the driver decelerates from 40 km/h to rest in 3.5 s or less, but not if he decelerates in a longer time. What is the coefficient of static friction between the cup and the dash?

51. (II) A wet bar of soap (mass = 150 grams) slides without friction down a ramp 2.0 m long inclined at 7.3°. How long does it take to reach the bottom? Neglect friction. How would this change if the soap's mass were 250 grams?

52. (II) The block shown in Fig. 4–48 lies on a smooth plane tilted at an angle $\theta = 22.0°$ to the horizontal. (*a*) Determine the acceleration of the block as it slides down the plane. (*b*) If the block starts from rest 9.10 m up the plane from its base, what will be the block's speed when it reaches the bottom of the incline? Ignore friction.

FIGURE 4–48 Block on inclined plane. Problems 52, 53, and 54.

53. (II) A block is given an initial speed of 3.0 m/s up the 22.0° plane shown in Fig. 4–48. (*a*) How far up the plane will it go? (*b*) How much time elapses before it returns to its starting point? Ignore friction.

54. (II) Repeat (a) Problem 52 and (b) Problem 53, assuming $\mu_k = 0.20$ between the block and plane.

55. (II) A roller coaster reaches the top of the steepest hill with a speed of 6.0 km/h. It then descends the hill, which is at an average angle of 45° and is 45.0 m long. What will its speed be when it reaches the bottom? Assume $\mu_k = 0.12$.

56. (II) An 18.0-kg box is released on a 37.0° incline and accelerates down the incline at 0.270 m/s². Find the friction force impeding its motion. How large is the coefficient of friction?

57. (II) In the design of a supermarket, there are to be several ramps connecting different parts of the store. Customers will have to push grocery carts up the ramps, Fig. 4–49, and it is obviously desired that this not be too difficult. An engineer has done a survey and found that almost no one complains if the force required is no more than 50 N. Will a slope $\theta = 5°$ be too steep, assuming a 30-kg grocery cart (full of groceries)? Assume friction (wheels against ground, wheel on the axles, and so on), can be accounted for by a coefficient $\mu_k = 0.10$.

FIGURE 4–49 Problem 57.

58. (II) Figaro the cat (5.0 kg) is hanging on the tablecloth, pulling Cleo's fishbowl (11 kg) toward the edge of the table (Fig. 4–50). The coefficient of kinetic friction between the tablecloth (ignore its mass) under the fishbowl and the table is 0.44. (a) What is the acceleration of Figaro and the fishbowl? (b) If the fishbowl is 0.90 m from the edge of the table, how much time does it take for Figaro to pull Cleo off the table?

FIGURE 4–50 Problem 58.

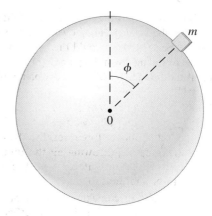

FIGURE 4–51 Problem 59.

59. (III) A small mass m is set on the surface of a sphere, Fig. 4–51. If the coefficient of static friction is $\mu_s = 0.60$, at what angle ϕ would the mass start sliding? [Hint: compare to Fig. 4–48; how are θ and ϕ related?]

60. (III) The 70-kg climber in Fig. 4–52 is supported in the "chimney" by the friction forces exerted on his shoes and back. The static coefficients of friction between his shoes and the wall, and between his back and the wall, are 0.80 and 0.60, respectively. What is the minimum normal force he must exert? Assume the walls are vertical and that friction forces are both at a maximum.

FIGURE 4–52 Problem 60.

FIGURE 4–53   Problems 61, 62, and 63.

61. (III) A block (mass m_1) lying on a frictionless inclined plane is connected to a mass m_2 by a massless cord passing over a pulley, as shown in Fig. 4–53.

(a) Determine a formula for the acceleration of the system in terms of m_1, m_2, θ, and g. (b) What conditions apply to masses m_1 and m_2 for the acceleration to be in one direction (say, m_1 down the plane), or in the opposite direction?

62. (III) Suppose the coefficient of kinetic friction between m_1 and the plane in Fig. 4–53 is $\mu_k = 0.15$, and that $m_1 = m_2 = 2.7$ kg. As m_2 moves down, determine the magnitude and direction of the acceleration of m_1 and m_2, given $\theta = 25°$.

63. (III) What smallest value of μ_k will keep the system of Problem 62 from accelerating?

64. (III) If a bicyclist of mass 65 kg (including the bicycle) can coast down a 6.0° hill at a steady speed of 6.0 km/h because of air resistance, how much force must be applied to climb the hill at the same speed (and the same air resistance)?

GENERAL PROBLEMS

65. According to a simplified model of a mammalian heart, at each pulse, approximately 20 g of blood is accelerated from 0.25 m/s to 0.35 m/s during a period of 0.10 s. What is the magnitude of the force exerted by the heart muscle?

66. A person has a reasonable chance of surviving an automobile crash if the deceleration is no more than 30 "g's." Calculate the force on a 70-kg person accelerating at this rate. What distance is traveled if brought to rest at this rate from 90 km/h?

67. (a) If the horizontal acceleration produced by an earthquake is a, and if an object is going to "hold its place" on the ground, show that the coefficient of static friction with the ground must be at least $\mu_s = a/g$. (b) The famous Loma Prieta earthquake that stopped the 1989 World Series produced ground accelerations of up to $4.0 \, \text{m/s}^2$ in the San Francisco Bay Area. Would a chair have started to slide on a linoleum floor if the coefficient of static friction were 0.25?

68. A 1000-kg car pulls a 450-kg trailer. The car exerts a horizontal force of 3.5×10^3 N against the ground in order to accelerate. What force does the car exert on the trailer? Assume an effective friction coefficient of 0.15 for the trailer.

69. Police lieutenants, examining the scene of an accident involving two cars, measure the skid marks of one of the cars, which nearly came to a stop before colliding, to be 80 m long. The coefficient of kinetic friction between rubber and the pavement is about 0.80. Estimate the initial speed of that car assuming a level road.

70. A car starts rolling down a 1-in-4 hill (1-in-4 means that for each 4 m traveled along the road, the elevation change is 1 m). How fast is it going when it reaches the bottom after traveling 50 m? (a) Ignore friction. (b) Assume an effective coefficient of friction equal to 0.10.

71. A fisherman in a boat is using a "10-lb test" fishing line. This means that the line can exert a force of 45 N without breaking (1 lb = 4.45 N). (a) How heavy a fish can the fisherman land if he pulls the fish up vertically at constant speed? (b) If he accelerates the fish upwards at $2.0 \, \text{m/s}^2$, what maximum weight fish can he land?

72. An elevator in a tall building is allowed to reach a maximum speed of 3.5 m/s going down. What must the tension be in the cable to stop this elevator over a distance of 3.0 m if the elevator has a mass of 1300 kg including occupants?

73. Two boxes, $m_1 = 1.0$ kg with a coefficient of kinetic friction of 0.10, and $m_2 = 2.0$ kg with a coefficient of 0.20, are placed on a plane inclined at $\theta = 30°$. (a) What acceleration does each block experience? (b) If a taut string is connected to the blocks (Fig. 4–54), with m_2 initially farther down the slope, what is the acceleration of each block? (c) If the initial configuration is reversed with m_1 starting lower with a taut string, what is the acceleration of each block?

FIGURE 4–54   Problem 73.

74. A 75.0-kg person stands on a scale in an elevator. What does the scale read (in kg) when (a) the elevator is at rest, (b) the elevator is climbing at a constant speed of 3.0 m/s, (c) the elevator is falling at 3.0 m/s, (d) the elevator is accelerating upward at 3.0 m/s², (e) the elevator is accelerating downward at 3.0 m/s²?

75. A small block of mass m is given an initial speed v_0 up a ramp inclined at angle θ to the horizontal. It travels a distance d up the ramp and comes to rest. (a) Determine a formula for the coefficient of kinetic friction between the block and ramp. (b) What can you say about the value of the coefficient of static friction?

76. A motorcyclist is coasting with the engine off at a steady speed of 17 m/s but enters a sandy stretch where the coefficient of friction is 0.80. Will the cyclist emerge from the sandy stretch without having to start the engine if the sand lasts for 15 m? If so, what will be the speed upon emerging?

77. A city planner is working on the redesign of a hilly portion of a city. An important consideration is how steep the roads can be so that even low-powered cars can get up the hills without slowing down. It is given that a particular small car, with a mass of 1100 kg, can accelerate on a level road from rest to 21 m/s (75 km/h) in 14.0 s. Using this data, calculate the maximum steepness of a hill.

78. A bicyclist can coast down a 5.0° hill at a constant 6.0 km/h. If the force of friction (air resistance) is proportional to the speed v so that $F_{fr} = cv$, calculate (a) the value of the constant c, and (b) the average force that must be applied in order to descend the hill at 20 km/h. The mass of the cyclist plus bicycle is 80 kg.

79. Jean, who likes physics experiments, dangles her watch from a thin piece of string while the jetliner she is in takes off from Dulles Airport (Fig. 4–55). She notices that the string makes an angle of 25° with respect to the vertical while the aircraft accelerates for takeoff, which takes about 18 seconds. Estimate the takeoff speed of the aircraft.

FIGURE 4–55   Problem 79.

80. (a) What minimum force F is needed to lift the piano (mass M) using the pulley apparatus shown in Fig. 4–56? (b) Determine the tension in each section of rope: F_{T1}, F_{T2}, F_{T3}, and F_{T4}.

FIGURE 4–56   Problem 80.

81. A 28.0-kg block is connected to an empty 1.00-kg bucket by a cord running over a frictionless pulley (Fig. 4–57). The coefficient of static friction between the table and the block is 0.450 and the coefficient of kinetic friction between the table and the block is 0.320. Sand is gradually added to the bucket until the system just begins to move. (a) Calculate the mass of sand added to the bucket. (b) Calculate the acceleration of the system.

FIGURE 4–57   Problem 81.

The planet Jupiter showing two of its moons, Io (visible in front of the planet) and Europa at far right. The moons, held by the force of gravity, revolve around Jupiter. Galileo was the first to observe four of Jupiter's moons, a momentous discovery that he used to argue in favor of the Copernican system.

CHAPTER

5 CIRCULAR MOTION; GRAVITATION

An object moves in a straight line if the net force on it acts in the direction of motion, or is zero. If the net force acts at an angle to the direction of motion at any moment, then the object moves in a curved path. An example of the latter is projectile motion, which we have already discussed in Chapter 3. Another important case is that of an object moving in a circle, such as a ball at the end of a string revolving around one's head, or the nearly circular motion of the Moon about the Earth.

In this chapter, we study the circular motion of an object, and how Newton's laws of motion apply to it. We will also discuss how Newton conceived of another great law by applying the concepts of circular motion to the motion of the Moon and the planets. This is the law of universal gravitation, which was the capstone of Newton's analysis of the physical world.

5–1 Kinematics of Uniform Circular Motion

An object that moves in a circle at constant speed v is said to experience **uniform circular motion**. The *magnitude* of the velocity remains constant in this case, but the *direction* of the velocity is continuously changing as the object moves around the circle (Fig. 5–1). Since acceleration is defined as the rate of change of velocity, a change in direction of velocity constitutes an acceleration just as does a change in magnitude of velocity. Thus, an object revolving in a circle is continuously accelerating, even when the speed remains constant ($v_1 = v_2 = v$). We now investigate this acceleration quantitatively.

Acceleration is defined as

$$\mathbf{a} = \frac{\mathbf{v}_2 - \mathbf{v}_1}{\Delta t} = \frac{\Delta \mathbf{v}}{\Delta t},$$

where $\Delta \mathbf{v}$ is the change in velocity during the short time interval Δt. We will eventually consider the situation when Δt approaches zero and thus obtain the instantaneous acceleration. But for purposes of making a clear drawing, Fig. 5–2, we consider a nonzero time interval. During the time Δt, the particle in Fig. 5–2a moves from point A to point B, covering a distance Δl *along the arc* which subtends an angle $\Delta \theta$. The change in the velocity vector is $\mathbf{v}_2 - \mathbf{v}_1 = \Delta \mathbf{v}$, and is shown in Fig. 5–2b. If we let Δt be very small (approaching zero), then Δl and $\Delta \theta$ are also very small and $\mathbf{v}_2$ will be almost parallel to $\mathbf{v}_1$ and $\Delta \mathbf{v}$ will be essentially perpendicular to them. Thus $\Delta \mathbf{v}$ points toward the center of the circle. Since $\mathbf{a}$, by its definition above, is in the same direction as $\Delta \mathbf{v}$, it too must point toward the center of the circle. Therefore, this acceleration is called **centripetal acceleration** ("center-seeking" acceleration) or **radial acceleration** (since it is directed along the radius, toward the center of the circle), and we denote it by $\mathbf{a}_R$.

We next determine the magnitude of the centripetal (radial) acceleration, a_R. Because CA is perpendicular to $\mathbf{v}_1$, and CB is perpendicular to $\mathbf{v}_2$, it follows that the angle $\Delta \theta$, defined as the angle between CA and CB, is also the angle between $\mathbf{v}_1$ and $\mathbf{v}_2$. Hence the vectors $\mathbf{v}_2$, $\mathbf{v}_1$, and $\Delta \mathbf{v}$ in Fig. 5–2b form a triangle that is geometrically similar[†] to triangle ABC in Fig. 5–2a. Taking $\Delta \theta$ small (letting Δt be very small), we can write

$$\frac{\Delta v}{v} \approx \frac{\Delta l}{r}$$

where we have set $v = v_1 = v_2$ because the magnitude of the velocity is assumed not to change. This is an exact equality when Δt approaches zero, for then the arc length Δl equals the cord length AB. Since we want to find the instantaneous acceleration, for which Δt approaches zero, we write the above expression as an equality and solve for Δv:

$$\Delta v = \frac{v}{r} \Delta l.$$

To get the centripetal acceleration, a_R, we divide Δv by Δt:

$$a_R = \frac{\Delta v}{\Delta t} = \frac{v}{r} \frac{\Delta l}{\Delta t};$$

and since $\Delta l / \Delta t$ is the linear speed, v, of the object,

$$a_R = \frac{v^2}{r}. \qquad \textbf{(5–1)}$$

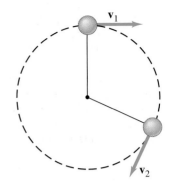

FIGURE 5–1 A small object moving in a circle, showing how the velocity changes. Note that at each point, the instantaneous velocity is in a direction tangent to the circular path.

FIGURE 5–2 Determining the change in velocity, $\Delta \mathbf{v}$, for a particle moving in a circle. The length Δl is the distance along the arc, from A to B.

CENTRIPETAL ACCELERATION

[†]Appendix A contains a review of geometry.

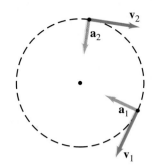

FIGURE 5–3 For uniform circular motion, **a** is always perpendicular to **v**.

Period and frequency

To summarize, *an object moving in a circle of radius r with constant speed v has an acceleration whose direction is toward the center of the circle and whose magnitude is* $a_R = v^2/r$. It is not surprising that this acceleration depends on v and r. For the greater the speed v, the faster the velocity changes direction; and the larger the radius, the less rapidly the velocity changes direction.

The acceleration vector points toward the center of the circle. But the velocity vector always points in the direction of motion, which is tangential to the circle. Thus the velocity and acceleration vectors are perpendicular to each other at every point in the path for uniform circular motion (see Fig. 5–3). This is another example that illustrates the error in thinking that acceleration and velocity are always in the same direction. For an object falling vertically, **a** and **v** are indeed parallel. But in circular motion, **a** and **v** are not parallel—nor are they in projectile motion (Section 3–5), where the acceleration **a** = **g** is always downward but the velocity vector can have various directions (Figs. 3–18 and 3–20).

Circular motion is often described in terms of the **frequency** f as so many revolutions per second. The **period** T of an object revolving in a circle is the time required for one complete revolution. Period and frequency are related by

$$T = \frac{1}{f}. \tag{5–2}$$

For example, if an object revolves at a frequency of 3 rev/s, then each revolution takes $\frac{1}{3}$ s. For an object revolving in a circle at constant speed v, we can write

$$v = \frac{2\pi r}{T}$$

since in one revolution the object travels one circumference ($= 2\pi r$).

EXAMPLE 5–1 **Acceleration of a revolving ball.** A 150-g ball at the end of a string is revolving uniformly in a horizontal circle of radius 0.600 m, as in Fig. 5–1. The ball makes 2.00 revolutions in a second. What is its centripetal acceleration?

SOLUTION The centripetal acceleration is $a_R = v^2/r$. First, we determine the speed of the ball v. If the ball makes two complete revolutions per second, then the ball travels in a complete circle in 0.500 s, which is its period T. The distance traveled in this time is the circumference of the circle, $2\pi r$, where r is the radius of the circle. Therefore, the ball has speed

$$v = \frac{2\pi r}{T} = \frac{2(3.14)(0.600 \text{ m})}{(0.500 \text{ s})} = 7.54 \text{ m/s}.$$

The centripetal acceleration is

$$a_R = \frac{v^2}{r} = \frac{(7.54 \text{ m/s})^2}{(0.600 \text{ m})} = 94.8 \text{ m/s}^2.$$

EXAMPLE 5–2 **Moon's centripetal acceleration.** The Moon's nearly circular orbit about the Earth has a radius of about 384,000 km and a period T of 27.3 days. Determine the acceleration of the Moon toward the Earth.

SOLUTION In orbit around the Earth, the Moon travels a distance $2\pi r$, where $r = 3.84 \times 10^8$ m is the radius of its circular path. The speed of the Moon in its orbit about the Earth is $v = 2\pi r/T$. The period T in seconds is $T = (27.3\,\text{d})(24.0\,\text{h/d})(3600\,\text{s/h}) = 2.36 \times 10^6\,\text{s}$. Therefore,

$$a_R = \frac{v^2}{r} = \frac{(2\pi r)^2}{T^2 r} = \frac{[2(3.14)(3.84 \times 10^8\,\text{m})]^2}{(2.36 \times 10^6\,\text{s})^2 (3.84 \times 10^8\,\text{m})}$$

$$= 0.00272\,\text{m/s}^2 = 2.72 \times 10^{-3}\,\text{m/s}^2.$$

We can write this in terms of $g = 9.80\,\text{m/s}^2$ (the acceleration of gravity at the Earth's surface) as[†]

$$a = 2.72 \times 10^{-3}\,\text{m/s}^2 \left(\frac{g}{9.80\,\text{m/s}^2}\right)$$

$$= 2.78 \times 10^{-4}\,g.$$

5–2 Dynamics of Uniform Circular Motion

According to Newton's second law ($\Sigma\mathbf{F} = m\mathbf{a}$), an object that is accelerating must have a net force acting on it. An object moving in a circle, such as a ball on the end of a string, must therefore have a force applied to it to keep it moving in that circle. That is, a net force is necessary to give it centripetal acceleration. The magnitude of the required force can be calculated using Newton's second law for the radial component, $\Sigma F_R = ma_R$, where a_R is the centripetal acceleration, $a_R = v^2/r$, and ΣF_R is the total (or net) force in the radial direction:

$$\Sigma F_R = ma_R = m\frac{v^2}{r}. \qquad \text{[circular motion]} \quad \textbf{(5–3)}$$

Since a_R is directed toward the center of the circle at any moment, the *net force too must be directed toward the center of the circle.* A net force is clearly necessary because otherwise, if no net force were exerted on the object, it would not move in a circle but in a straight line, as Newton's first law tells us. To pull an object out of its "natural" straight-line path, a net force to the side is necessary. For uniform circular motion, this sideways force must act toward the circle's center (see Fig. 5–4). The direction of the net force is thus continually changing so that it is always directed toward the center of the circle. This force is sometimes called a centripetal ("aiming toward the center") force. But be aware that "centripetal force" does not indicate some new kind of force. The term merely describes the direction of the net force: that the net force is directed toward the circle's center. The force *must be applied by other objects.* For example, when a

Force is needed to provide centripetal acceleration

FIGURE 5–4 A force is required to keep an object moving in a circle. If the speed is constant, the force is directed toward the center of the circle.

Careful:
Centripetal force is not a new kind of force

[†]Note: this acceleration $a = 2.78 \times 10^{-4}\,g$ is *not* the acceleration of gravity for objects at the Moon's surface due to the Moon's gravity. Rather it is the acceleration due to the *Earth's* gravity for any object (such as the Moon) that is 384,000 km from the Earth.

person swings a ball in a circle on the end of a string, the person pulls on the string and the string exerts the force on the ball.

There is a common misconception that an object moving in a circle has an outward force acting on it, a so-called centrifugal ("center-fleeing") force. This is incorrect: *there is no outward force*. Consider, for example, a person swinging a ball on the end of a string around her head (Fig. 5–5). If you have ever done this yourself, you know that you feel a force pulling outward on your hand. The misconception arises when this pull is interpreted as an outward "centrifugal" force pulling on the ball that is transmitted along the string to your hand. This is not what is happening at all. To keep the ball moving in a circle, you pull *inwardly* on the string, which in turn exerts the force on the ball. The ball exerts an equal and opposite force (Newton's third law), and *this* is the force your hand feels (see Fig. 5–5). The force *on the ball* is the one exerted *inwardly* on it by the string. To see even more convincing evidence that a "centrifugal force" does not act on the ball, consider what happens when you let go of the string. If a centrifugal force were acting, the ball would fly outward, as shown in Fig. 5–6a. But it doesn't; the ball flies off tangentially (Fig. 5–6b), in the direction of the velocity it had at the moment it was released, because the inward force no longer acts. Try it and see!

There is no centrifugal force

FIGURE 5–5 Swinging a ball on the end of a string.

Force on ball exerted by string

Force on hand exerted by string

FIGURE 5–6 If centrifugal force existed, the ball would fly off as in (a) when released. In fact, it flies off tangentially as in (b). For example, in (c) sparks fly in straight lines tangentially from the edge of a rotating grinding wheel.

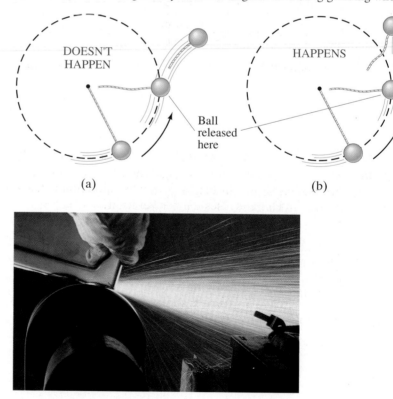

DOESN'T HAPPEN

HAPPENS

Ball released here

(a)

(b)

(c)

mg

FIGURE 5–7   Example 5–3.

EXAMPLE 5–3 **ESTIMATE**   **Force on revolving ball (horizontal).**   Estimate the force a person must exert on a string attached to a 0.150-kg ball to make the ball revolve in a horizontal circle of radius 0.600 m, as in Example 5–1. The ball makes 2.00 revolutions per second.

SOLUTION   First we draw the free-body diagram for the ball, Fig. 5–7, which shows the two forces acting on the ball: the force of gravity, $m\mathbf{g}$; and the tension force $\mathbf{F}_T$ that the string exerts (which occurs because the person exerts that same force on the string). The ball's weight complicates matters and makes it impossible to revolve a ball with the cord horizontal. But if the weight is small enough, we can ignore it. Then $\mathbf{F}_T$ will act nearly horizontally ($\theta \approx 0$ in Fig. 5–7) and provide the force necessary to give the ball its centripetal acceleration. We apply Newton's second law to the radial direction, which now is horizontal, so we call it x:

$$\Sigma F_x = ma_x$$

or (recall from Example 5–1 that $v = 7.54\,\text{m/s}$),

Tension in cord acts to provide the centripetal acceleration

$$F_{Tx} = m\,\frac{v^2}{r} = (0.150\ \text{kg})\,\frac{(7.54\ \text{m/s})^2}{(0.600\ \text{m})} \approx 14\ \text{N},$$

where we have rounded off because our estimate ignores the ball's mass.

CONCEPTUAL EXAMPLE 5–4   **Tetherball.**   The game of tetherball is played with a ball tied to a pole with a string. When the ball is struck, it whirls around the pole as shown in Fig. 5–8. In what direction is the acceleration of the ball, and what causes the acceleration?

RESPONSE   The acceleration points horizontally toward the center of the ball's circular path (not toward the top of the pole). The force responsible for the acceleration may not be obvious at first, since there seems to be no force pointing directly horizontally. But it is the *net* force (the sum of $m\mathbf{g}$ and $\mathbf{F}_T$ here) that must point in the direction of the acceleration. The vertical component of the string tension balances the ball's weight, $m\mathbf{g}$. The horizontal component of the string tension, F_{Tx}, is the force that produces the centripetal acceleration toward the center.

FIGURE 5–8
Conceptual Example 5–4.

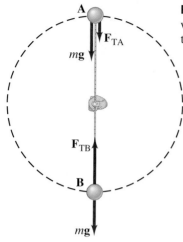

FIGURE 5–9 Example 5–5, with free-body diagrams at the two positions.

Gravity and cord tension together provide centripetal acceleration

Gravity provides centripetal acceleration

String tension and gravity acting in opposite directions provide centripetal acceleration

EXAMPLE 5–5 **Revolving ball (vertical circle).** A 0.150-kg ball on the end of a 1.10-m-long cord (negligible mass) is swung in a *vertical* circle. (*a*) Determine the minimum speed the ball must have at the top of its arc so that it continues moving in a circle. (*b*) Calculate the tension in the cord at the bottom of the arc assuming the ball is moving at twice the speed of part (*a*).

SOLUTION The free-body diagram is shown in Fig. 5–9 for both situations. (*a*) At the top (point A), two forces can act on the ball: $m\mathbf{g}$, its weight; and $\mathbf{F}_{TA}$, the tension force the cord exerts at point A. Both act downward, and their vector sum acts to give the ball its centripetal acceleration a_R. We apply Newton's second law, for the vertical direction, choosing downward as positive (toward the center):

$$\Sigma F_R = ma_R$$

$$F_{TA} + mg = m\frac{v_A^2}{r}.$$

From this equation we can see that the tension force F_{TA} at A will get larger if v_A (ball's speed at top of circle) is made larger, as expected. But we are asked for the *minimum* speed to keep the ball moving in a circle. The cord will remain taut as long as there is tension in it; but if the tension disappears (because v_A is too small) the cord can go limp, and the ball will fall out of its circular path. Thus, the minimum speed will occur if $F_{TA} = 0$, for which we have

$$mg = m\frac{v_A^2}{r}.$$

We solve for v_A:

$$v_A = \sqrt{gr} = \sqrt{(9.80 \text{ m/s}^2)(1.10 \text{ m})} = 3.28 \text{ m/s}.$$

This is the minimum speed at the top of the circle if the ball is to continue moving in a circular path.

(*b*) At the bottom of the circle (see Fig. 5–9) the cord exerts its tension force F_{TB} upward whereas the force of gravity, $m\mathbf{g}$, acts downward. So Newton's second law, this time choosing *upward* as positive (toward the center), gives

$$\Sigma F_R = ma_R$$

$$F_{TB} - mg = m\frac{v_B^2}{r}.$$

The speed v_B is given as twice what we found in (*a*), namely 6.56 m/s. [Note that the speed changes here because gravity acts on the ball at all points along the path, but Eq. 5–3 still remains valid, $\Sigma F_R = mv^2/r$.] We solve for F_{TB} in the last equation:

$$F_{TB} = m\frac{v_B^2}{r} + mg$$

$$= (0.150 \text{ kg})\frac{(6.56 \text{ m/s})^2}{(1.10 \text{ m})} + (0.150 \text{ kg})(9.80 \text{ m/s}^2) = 7.34 \text{ N}.$$

Note that we could not simply set F_{TB} equal to mv_B^2/r; the latter is equal to the *net* force on the ball in the radial direction and so also includes gravity. Indeed, the cord's tension not only provides the centripetal acceleration, but must be even larger than ma_R to compensate for the downward force of gravity.

CONCEPTUAL EXAMPLE 5–6   **Ferris wheel.**   A rider on a Ferris wheel moves in a vertical circle of radius r at constant speed v (Fig. 5–10). Is the normal force that the seat exerts on the rider at the top of the circle (*a*) less than, (*b*) more than, or (*c*) the same as, the force the seat exerts at the bottom of the circle?

RESPONSE   The free-body diagram is shown in Fig. 5–10 and is similar to that for Example 5–5, with $\mathbf{F}_N$ replacing $\mathbf{F}_T$. Because the acceleration points radially *toward* the center, Newton's second law tells us that $F_N < mg$ at the top but $F_N > mg$ at the bottom. So the correct answer is (*a*).

FIGURE 5–10   Conceptual Example 5–6.

5–3   A Car Rounding a Curve

An example of centripetal acceleration occurs when an automobile rounds a curve. In such a situation, you may feel that you are thrust outward. But there is not some mysterious centrifugal force pulling on you. What is happening is that you tend to move in a straight line, whereas the car has begun to follow a curved path. To make you go in the curved path, the seat (friction) or the door of the car (direct contact) exerts a force on you (Fig. 5–11). The car itself must have an inward force exerted on it if it is to move in a curve. On a flat road, this force is supplied by friction between the tires and the pavement. (It is static friction as long

➡ **PHYSICS APPLIED**
Driving on a curve

FIGURE 5–11   The road exerts an inward force (friction against the tires) on a car to make it move in a circle; and the car exerts an inward force on the passenger.

Force on car
(sum of friction forces
acting on each tire)

Force on
passenger

Tendency for
passenger to
go straight

FIGURE 5–12 Race car heading down into a curve. From the tire marks we can see that most cars experienced a sufficient friction force to give them the needed centripetal acceleration for rounding the curve safely. But, we can also see a few tire tracks of cars on which there was not sufficient force—and which followed more nearly straight-line paths.

FIGURE 5–13 Forces on a car rounding a curve on a flat road, Example 5–7. (a) Front view, (b) top view.

$F_G = m\mathbf{g}$

(a)

(b)

as the tires are not slipping.) If the friction force is not great enough, as under icy conditions, sufficient force cannot be applied and the car will skid out of a circular path into a more nearly straight path. See Fig. 5–12.

EXAMPLE 5–7 **Skidding on a curve.** A 1000-kg car rounds a curve on a flat road of radius 50 m at a speed of 50 km/h (14 m/s). Will the car make the turn, or will it skid, if: (*a*) the pavement is dry and the coefficient of static friction is $\mu_s = 0.60$; (*b*) the pavement is icy and $\mu_s = 0.25$?

SOLUTION Figure 5–13 shows the free-body diagram for the car. The normal force, F_N, on the car is equal to the weight since the road is flat and there is no vertical acceleration:

$$F_N = mg = (1000 \text{ kg})(9.8 \text{ m/s}^2) = 9800 \text{ N}.$$

In the horizontal direction the only force is friction, and we must compare it to the force needed to produce the centripetal acceleration to see if it is sufficient. The net horizontal force required to keep the car moving in a circle around the curve is

$$\Sigma F_R = ma_R = m\frac{v^2}{r} = (1000 \text{ kg})\frac{(14 \text{ m/s})^2}{(50 \text{ m})} = 3900 \text{ N}.$$

Naturally we hope the maximum total friction force (the sum of the friction forces acting on each of the four tires) will be at least this large. For (*a*), $\mu_s = 0.60$, and the maximum friction force attainable (recall from Section 4–8 that $F_{fr} \le \mu_s F_N$) is

$$(F_{fr})_{max} = \mu_s F_N = (0.60)(9800 \text{ N}) = 5900 \text{ N}.$$

Since a force of only 3900 N is needed, and that is, in fact, how much will be exerted by the road as a static friction force, the car can make the turn fine. But in (*b*) the maximum friction force possible is

$$(F_{fr})_{max} = \mu_s F_N = (0.25)(9800 \text{ N}) = 2500 \text{ N}.$$

The car will skid because the ground cannot exert sufficient force (3900 N is needed) to keep it moving in a curve of radius 50 m.

The situation is worse if the wheels lock (stop rotating) when the brakes are applied too hard. When the tires are rolling, the bottom of the tire is at rest against the road at each instant, so static friction exists. But if the wheels lock, the tires slide and the friction force, which is now kinetic friction, is less. Moreover, when the road is wet or icy, locking of the wheels occurs with less force on the brake pedal since there is less road friction to keep the wheels turning rather than sliding. Antilock brakes (ABS) are designed to limit brake pressure just before the point where sliding would occur, by means of delicate sensors and a fast computer.

The banking of curves can reduce the chance of skidding because the normal force of the road (acting perpendicular to the road) will have a component toward the center of the circle (Fig. 5–14), thus reducing the reliance on friction. For a given banking angle, θ, there will be one speed for which no friction at all is required. This will be the case when the horizontal component of the normal force toward the center of the curve, $F_N \sin\theta$ (see Fig. 5–14), is just equal to the force required to give a vehicle its centripetal acceleration—that is, when

$$F_N \sin\theta = m\frac{v^2}{r}.$$

The banking angle of a road, θ, is chosen so that this condition holds for a particular speed, called the "design speed."

EXAMPLE 5–8 **Banking angle.** (a) For a car traveling with speed v around a curve of radius r, determine a formula for the angle at which a road should be banked so that no friction is required. (b) What is this angle for an expressway off-ramp curve of radius 50 m at a design speed of 50 km/h?

SOLUTION We choose our x and y axes as horizontal and vertical so that a_R, which is horizontal, is along the x axis. The components of F_N are as shown in Fig. 5–14. (a) For the horizontal direction, $\Sigma F_R = ma_R$ gives

$$F_N \sin\theta = \frac{mv^2}{r}.$$

In the vertical direction, the forces are $F_N \cos\theta$ upward (Fig. 5–14) and the weight of the car (mg) downward. Since there is no vertical motion, the y component of the acceleration is zero, so $\Sigma F_y = ma_y$ gives us

$$F_N \cos\theta - mg = 0.$$

Thus,

$$F_N = \frac{mg}{\cos\theta}.$$

[Note in this case that $F_N \geq mg$ since $\cos\theta \leq 1$.] We substitute this relation for F_N into the equation for the horizontal motion,

$$F_N \sin\theta = m\frac{v^2}{r},$$

and obtain

$$\frac{mg}{\cos\theta}\sin\theta = m\frac{v^2}{r}$$

or

$$mg\tan\theta = m\frac{v^2}{r}$$

so

$$\tan\theta = \frac{v^2}{rg}.$$

This is the formula for the banking angle θ.
(b) For $r = 50$ m and $v = 50$ km/h (or 14 m/s),

$$\tan\theta = \frac{(14 \text{ m/s})^2}{(50 \text{ m})(9.8 \text{ m/s}^2)} = 0.40,$$

so $\theta = 22°$.

Banked curves

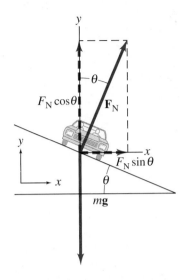

FIGURE 5–14 Normal force on a car rounding a banked curve, resolved into its horizontal and vertical components. Note that the centripetal acceleration is horizontal (and not parallel to the sloping road).

Horizontal component of normal force alone acts to provide centripetal acceleration (friction is desired to be zero—otherwise it too would contribute)

(a)

(b)

FIGURE 5–15 The speed of an object moving in a circle changes if the force on it has a tangential component, F_{tan}. Part (a) shows the force **F** and its vector components; part (b) shows the acceleration vector and its vector components.

Circular motion at constant speed occurs when the net force on an object is exerted toward the center of the circle. If the net force is not directed toward the center but is at an angle, as shown in Fig. 5–15a, the force has two components. The component directed toward the center of the circle, F_R, gives rise to the centripetal acceleration, a_R, and keeps the object moving in a circle. The component tangent to the circle, F_{tan}, acts to increase (or decrease) the speed, and thus gives rise to a component of the acceleration tangent to the circle, a_{tan}. When the speed of the object is changing, a tangential component of force is acting.

When you first start revolving a ball on the end of a string around your head, you must give it tangential acceleration. You do this by pulling on the string with your hand displaced from the center of the circle. In athletics, a hammer thrower accelerates the hammer tangentially in a similar way so that it reaches a high speed before release.

The tangential component of the acceleration, a_{tan}, is equal to the rate of change of the *magnitude* of the velocity of the object:

$$a_{tan} = \frac{\Delta v}{\Delta t}.$$

The radial (centripetal) acceleration arises from the change in *direction* of the velocity and, as we have seen (Eq. 5–1), is given by

$$a_R = \frac{v^2}{r}.$$

The tangential acceleration always points in a direction tangent to the circle, and is in the direction of motion (parallel to **v**) if the speed is increasing, as shown in Fig. 5–15b. If the speed is decreasing, $\mathbf{a}_{tan}$ points antiparallel to **v**. In either case, $\mathbf{a}_{tan}$ and $\mathbf{a}_R$ are always perpendicular to each other; and *their directions change* continually as the object moves along its circular path. The total vector acceleration, **a**, is the sum of these two:

$$\mathbf{a} = \mathbf{a}_{tan} + \mathbf{a}_R.$$

Since $\mathbf{a}_R$ and $\mathbf{a}_{tan}$ are always perpendicular to each other, the magnitude of **a** at any moment is

$$a = \sqrt{a_{tan}^2 + a_R^2}.$$

EXAMPLE 5–9 **Two components of acceleration.** A racing car starts from rest in the pit area and accelerates at a uniform rate to a speed of 35 m/s in 11 s, moving on a circular track of radius 500 m. Assuming constant tangential acceleration, find (a) the tangential acceleration, and (b) the centripetal acceleration when the speed is 30 m/s.

SOLUTION (a) a_{tan} is constant, of magnitude

$$a_{tan} = \frac{\Delta v}{\Delta t} = \frac{(35 \text{ m/s} - 0 \text{ m/s})}{11 \text{ s}} = 3.2 \text{ m/s}^2.$$

(b) $$a_R = \frac{v^2}{r} = \frac{(30 \text{ m/s})^2}{500 \text{ m}} = 1.8 \text{ m/s}^2.$$

A useful device that very nicely illuminates the dynamic aspects of circular motion is the centrifuge, or the very high speed ultracentrifuge. These devices are used to sediment materials quickly or to separate materials with slightly different characteristics. Test tubes or other containers are held in the centrifuge rotor, which is accelerated to very high rotational speeds: see Fig. 5–16, where one test tube is shown in two different positions as the rotor turns. The small green dot represents a small particle, perhaps a macromolecule, in a fluid-filled test tube. When the tube is at position A and the rotor is turning, the particle has a tendency to move in a straight line in the direction of the dashed arrow. But the fluid, resisting the motion of the particles, exerts a centripetal force that keeps the particles moving nearly in a circle. Usually, the resistance of the fluid (which could be a liquid, a gas, or a gel, depending on the application) does not quite equal mv^2/r, and the particles eventually reach the bottom of the tube. If the particles are sedimenting in a semi-rigid medium like a gel, and the rotation is stopped before the particles reach the bottom of the tube, the particles will be separated according to their size or other factors that influence their mobility. If the particles reach the bottom of the tube, then the bottom of the tube exerts a force that keeps the particles moving in a circle. In fact, the bottom of the tube must exert a force on the whole tube of fluid, keeping it moving in a circle. If the tube is not strong enough to exert this force, it will break.

The kinds of materials placed in a centrifuge are those that do not sediment or separate quickly under the action of gravity. The purpose of a centrifuge is to provide an "effective gravity" much larger than normal gravity because of the high rotational speeds, so that the particles move down the tube more rapidly.

➡ **PHYSICS APPLIED**
Centrifuge

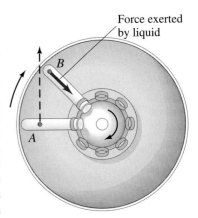

FIGURE 5–16 Rotating test tube in a centrifuge (top view). Tube is shown at two positions. At A, the green dot represents a macromolecule or other particle being sedimented. It would tend to go along the dashed line heading toward the bottom of the tube but the fluid resists this motion by exerting a force on the particle as shown at point B.

EXAMPLE 5–10 Ultracentrifuge. An ultracentrifuge rotor rotates at 50,000 rpm (revolutions per minute). The top of a 4.00-cm long test tube (Fig. 5–16) is 6.00 cm from the rotation axis and is perpendicular to it. The bottom of the tube is 10.00 cm, from the axis of rotation. (*a*) Calculate the centripetal acceleration, in "g's," at the top and the bottom of the tube. (*b*) If the contents of the tube have a total mass of 12.0 g, what force must the bottom of the tube withstand?

SOLUTION We can calculate the centripetal acceleration from $a_R = v^2/r$. (*a*) At the top of the tube, a particle revolves in a circle of circumference $2\pi r$, which is a distance

$$2\pi r = 2(3.14)(0.0600 \text{ m}) = 0.377 \text{ m per revolution.}$$

It makes 5.00×10^4 such revolutions each minute, or, dividing by 60 s/min, 833 rev/s; so the time to make one revolution, the period T, is

$$T = 1/(833 \text{ rev/s}) = 1.20 \times 10^{-3} \text{ s/rev.}$$

The speed of the particle is then

$$v = \frac{2\pi r}{T} = \left(\frac{0.377 \text{ m/rev}}{1.20 \times 10^{-3} \text{ s/rev}} \right) = 3.14 \times 10^2 \text{ m/s.}$$

The centripetal acceleration is

$$a_R = \frac{v^2}{r} = \frac{(3.14 \times 10^2 \text{ m/s})^2}{0.0600 \text{ m}} = 1.64 \times 10^6 \text{ m/s}^2,$$

which, dividing by $g = 9.80 \text{ m/s}^2$, is 1.67×10^5 g's.

At the bottom of the tube ($r = 0.1000$ m), the speed is

$$v = \frac{2\pi r}{T} = \frac{2\pi(0.1000 \text{ m})}{1.20 \times 10^{-3} \text{ s/rev}} = 5.23 \times 10^2 \text{ m/s}.$$

Then

$$a_R = v^2/r = (5.23 \times 10^2 \text{ m/s})^2/(0.1000 \text{ m}) = 2.74 \times 10^6 \text{ m/s}^2$$

$$= 2.79 \times 10^5 \text{ } g\text{'s}.$$

(b) Since the acceleration varies with distance from the axis, we estimate the force using the average acceleration

$$\bar{a} = \frac{(1.64 \times 10^6 \text{ m/s}^2 + 2.74 \times 10^6 \text{ m/s}^2)}{2}$$

$$= 2.19 \times 10^6 \text{ m/s}^2.$$

Then

$$F = m\bar{a} = (0.0120 \text{ kg})(2.19 \times 10^6 \text{ m/s}^2) = 2.63 \times 10^4 \text{ N}.$$

which is equivalent to the weight of a 2680-kg mass [since $m = F/g = (2.63 \times 10^4 \text{ N})/9.80 \text{ m/s}^2 = 2.68 \times 10^3$ kg], or almost 3 tons!

5–6 Newton's Law of Universal Gravitation

Besides developing the three laws of motion, Sir Isaac Newton also examined the motion of the planets and the Moon. In particular, he wondered about the nature of the force that must act to keep the Moon in its nearly circular orbit around the Earth.

Newton was also thinking about the problem of gravity. Since falling bodies accelerate, Newton had concluded that they must have a force exerted on them, a force we call the force of gravity. Whenever a body has a force exerted *on* it, that force is exerted *by* some other body. But what *exerts* the force of gravity? Every object on the surface of the Earth feels the force of gravity, and no matter where the object is, the force is directed toward the center of the Earth, Fig. 5–17. Newton concluded that it must be the Earth itself that exerts the gravitational force on objects at its surface.

According to an early account, Newton was sitting in his garden and noticed an apple drop from a tree. He is said to have been struck with a sudden inspiration: if gravity acts at the tops of trees, and even at the tops of mountains, then perhaps it acts all the way to the Moon! Whether this story is true or not, it does seem to capture something of Newton's reasoning and inspiration. With this idea that it is terrestrial gravity that holds the Moon in its orbit, Newton developed his great theory of gravitation. [But there was controversy at the time. Many thinkers had trouble accept-

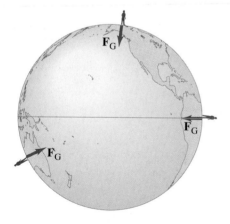

FIGURE 5–17 Anywhere on Earth, whether in Alaska, Peru, or Australia, the force of gravity acts downward toward the center.

Newton's apple

ing the idea of a force "acting at a distance." Typical forces act through contact—your hand pushes a cart and pulls a wagon, a bat hits a ball, and so on. But gravity acts without contact, said Newton: the Earth exerts a force on a falling apple and on the Moon, even though there is no contact, and the two objects may even be very far apart.]

Newton set about determining the magnitude of the gravitational force that the Earth exerts on the Moon as compared to the gravitational force on objects at the Earth's surface. At the surface of the Earth, the force of gravity accelerates objects at 9.80 m/s^2. But what is the centripetal acceleration of the Moon? Since the Moon moves with nearly uniform circular motion, the acceleration can be calculated from $a_R = v^2/r$. We already performed this calculation in Example 5–2 and found that $a_R = 0.00272 \text{ m/s}^2$. In terms of the acceleration of gravity at the Earth's surface, g, this is equivalent to

$$a_R \approx \frac{1}{3600} g.$$

The Moon's acceleration toward Earth

That is, the acceleration of the Moon toward the Earth is about $\frac{1}{3600}$ as great as the acceleration of objects at the Earth's surface. The Moon is 384,000 km from the Earth, which is about 60 times the Earth's radius of 6380 km. That is, the Moon is 60 times farther from the Earth's center than are objects at the Earth's surface. But $60 \times 60 = 60^2 = 3600$. Again that number 3600! Newton concluded that the gravitational force exerted by the Earth on any object decreases with the square of its distance, r, from the Earth's center:

$$\text{force of gravity} \propto \frac{1}{r^2}.$$

The Moon, being 60 Earth radii away, feels a gravitational force only $\frac{1}{60^2} = \frac{1}{3600}$ times as strong as it would if it were at the Earth's surface. Any object placed 384,000 km from the Earth would experience the same acceleration due to the Earth's gravity as the Moon experiences: 0.00272 m/s^2.

Newton realized that the force of gravity on an object depends not only on distance but also on the object's mass. In fact, it is directly proportional to its mass, as we have seen. According to Newton's third law, when the Earth exerts its gravitational force on any body, such as the Moon, that other body exerts an equal and opposite force on the Earth (Fig. 5–18). Because of this symmetry, Newton reasoned, the magnitude of the force of gravity must be proportional to *both* the masses. Thus

$$F \propto \frac{m_E m_B}{r^2}$$

where m_E is the mass of the Earth, m_B the mass of the other body, and r the distance from the Earth's center to the center of the other body.

Newton went a step further in his analysis of gravity. In his examination of the orbits of the planets, he concluded that the force required to hold the different planets in their orbits around the Sun seems to diminish as the inverse square of their distance from the Sun. This led him to believe that it is also the gravitational force that acts between the Sun and each of the planets to keep them in their orbits. And if gravity acts between these objects, why not between all objects? Thus he proposed his

Moon

Gravitational force exerted on Moon by Earth

Earth

Gravitational force exerted on Earth by the Moon

FIGURE 5–18 The gravitational force one body exerts on a second body is directed toward the first body, and is equal and opposite to the force exerted by the second body on the first.

famous **law of universal gravitation**, which we can state as follows:

> Every particle in the universe attracts every other particle with a force that is proportional to the product of their masses and inversely proportional to the square of the distance between them. This force acts along the line joining the two particles.

The magnitude of the gravitational force can be written as

$$F = G \frac{m_1 m_2}{r^2}, \tag{5-4}$$

where m_1 and m_2 are the masses of the two particles, r is the distance between them, and G is a universal constant which must be measured experimentally and has the same numerical value for all objects.

The value of G must be very small, since we are not aware of any force of attraction between ordinary-sized objects, such as between two baseballs. The force between two ordinary objects was first measured, over 100 years after Newton's publication of his law, by Henry Cavendish in 1798. To detect and measure the incredibly small force, he used an apparatus like that shown in Fig. 5–19. Cavendish confirmed Newton's hypothesis that two bodies attract one another, and that Eq. 5–4 accurately describes this force. In addition, because he could measure F, m_1, m_2, and r accurately, he was able to determine the value of the constant G as well. The accepted value today is

$$G = 6.67 \times 10^{-11}\,\text{N·m}^2/\text{kg}^2.$$

[Strictly speaking, Eq. 5–4 gives the magnitude of the gravitational force that one particle exerts on a second particle that is a distance r away. For an extended object (that is, not a point), we must consider how to measure the distance r. You might think that r would be the distance between the centers of the objects. This is often true and often a good approximation even when not quite true, but to do a calculation correctly, each extended body must be considered as a collection of tiny particles. The total force is the sum of the forces due to all the particles. The sum over all these particles is often best done using integral calculus, which Newton himself invented. Newton showed that for two uniform spheres, Eq. 5–4 gives the correct force where r is the distance between their centers. Also, when extended bodies are small compared to the distance between them (as for the Earth–Sun system), little inaccuracy results from considering them as point particles.]

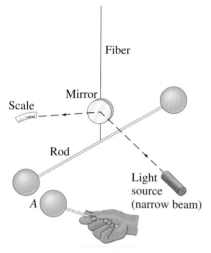

FIGURE 5–19 Schematic diagram of Cavendish's apparatus. Two spheres are attached to a light horizontal rod, which is suspended at its center by a thin fiber. When a third sphere labeled A is brought close to one of the suspended spheres, the gravitational force causes the latter to move, and this twists the fiber slightly. The tiny movement is magnified by the use of a narrow light beam directed at a mirror mounted on the fiber. The beam reflects onto a scale. Previous determination of how large a force will twist the fiber a given amount then allows one to determine the magnitude of the gravitational force between two objects.

EXAMPLE 5–11 **ESTIMATE** **Can you attract another person gravitationally?** A 50-kg person and a 75-kg person are sitting on a bench so that their centers are about 50 cm apart. Estimate the magnitude of the gravitational force each exerts on the other.

SOLUTION We use Eq. 5–4, which gives

$$F = \frac{(6.67 \times 10^{-11}\,\text{N·m}^2/\text{kg}^2)(50\,\text{kg})(75\,\text{kg})}{(0.50\,\text{m})^2} = 1.0 \times 10^{-6}\,\text{N},$$

which is unnoticeably small unless very delicate instruments are used.

EXAMPLE 5–12 **Spacecraft at $2r_E$.** What is the force of gravity acting on a 2000-kg spacecraft when it orbits two Earth radii from the Earth's center (that is, a distance r_E = 6380 km above the Earth's surface, (Fig. 5–20)? The mass of the Earth is $M_E = 5.98 \times 10^{24}$ kg.

SOLUTION We could plug all the numbers into Eq. 5–4, but there is a simpler approach. The spacecraft is twice as far from the Earth's center as when at the surface of the Earth. Therefore, since the force of gravity decreases as the square of the distance (and $\frac{1}{2^2} = \frac{1}{4}$), the force of gravity on it will be only one fourth its weight at the Earth's surface:

$$F_G = \tfrac{1}{4} mg = \tfrac{1}{4}(2000 \text{ kg})(9.80 \text{ m/s}^2) = 4900 \text{ N}.$$

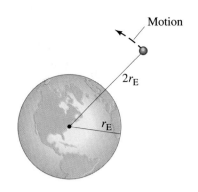

FIGURE 5–20 Example 5–12.

EXAMPLE 5–13 **Force on the moon.** Find the net force on the Moon ($m_M = 7.35 \times 10^{22}$ kg) due to the gravitational attraction of both the Earth ($m_E = 5.98 \times 10^{24}$ kg) and the Sun ($m_S = 1.99 \times 10^{30}$ kg), assuming they are at right angles to each other (Fig. 5–21).

SOLUTION We must add the two forces vectorially. First we calculate their magnitudes. The Earth is 3.84×10^5 km $= 3.84 \times 10^8$ m from the Moon, so F_{ME} (the force on the Moon due to the Earth) is

$$F_{ME} = \frac{(6.67 \times 10^{-11} \text{ N·m}^2/\text{kg}^2)(7.35 \times 10^{22} \text{ kg})(5.98 \times 10^{24} \text{ kg})}{(3.84 \times 10^8 \text{ m})^2}$$

$$= 1.99 \times 10^{20} \text{ N}.$$

The Sun is 1.50×10^8 km from the Earth and the Moon, so F_{MS} (the force on the Moon due to the Sun) is

$$F_{MS} = \frac{(6.67 \times 10^{-11} \text{ N·m}^2/\text{kg}^2)(7.35 \times 10^{22} \text{ kg})(1.99 \times 10^{30} \text{ kg})}{(1.50 \times 10^{11} \text{ m})^2}$$

$$= 4.34 \times 10^{20} \text{ N}.$$

Since the two forces act at right angles in the case we are considering (Fig. 5–21), the total force is

$$F = \sqrt{(1.99)^2 + (4.34)^2} \times 10^{20} \text{ N} = 4.77 \times 10^{20} \text{ N}$$

which acts at an angle $\theta = \tan^{-1}(1.99/4.34) = 24.6°$.

FIGURE 5–21 Orientation of Sun (S), Earth (E), and Moon (M) for Example 5–13 (not to scale).

The law of universal gravitation should not be confused with Newton's second law of motion, $\Sigma F = ma$. The former describes a particular force, gravity, and how its strength varies with the distance and masses involved. Newton's second law, on the other hand, relates the net force on a body (i.e., the vector sum of all the different forces acting on the body whatever their sources) to the mass and acceleration of that body.

Careful:
Distinction between Newton's second law and the law of gravity

5–7 Gravity Near the Earth's Surface; Geophysical Applications

When Eq. 5–4 is applied to the gravitational force between the Earth and an object at its surface, m_1 becomes the mass of the Earth m_E, m_2 becomes the mass of the object, m, and r becomes the distance of the object from the Earth's center,[†] which is the radius of the Earth r_E. This force of gravity due to the Earth is the weight of the object, which we have been writing as mg. Thus,

$$mg = G\frac{mm_E}{r_E^2}.$$

Hence

g in terms of G

$$g = G\frac{m_E}{r_E^2}.$$

(5–5)

Thus, the acceleration of gravity at the surface of the Earth, g, is determined by m_E and r_E. (Be careful not to confuse G with g; they are very different quantities, but are related by Eq. 5–5.)

Until G was measured, the mass of the Earth was not known. But once G was measured, Eq. 5–5 could be used to calculate the Earth's mass, and Cavendish was the first to do so. Since $g = 9.80\ \text{m/s}^2$ and the radius of the Earth is $r_E = 6.38 \times 10^6\ \text{m}$, then, from Eq. 5–5, we obtain

$$m_E = \frac{gr_E^2}{G} = \frac{(9.80\ \text{m/s}^2)(6.38 \times 10^6\ \text{m})^2}{6.67 \times 10^{-11}\ \text{N·m}^2/\text{kg}^2} = 5.98 \times 10^{24}\ \text{kg}$$

for the mass of the Earth.

When dealing with the weight of objects at the surface of the Earth, we can continue to use simply mg. If we wish to calculate the force of gravity on an object some distance from the Earth, or the force due to some other heavenly body, such as that exerted by the Moon or a planet, we can calculate the effective value of g from Eq. 5–5, replacing r_E (and m_E) by the appropriate distance (and mass), or we can use Eq. 5–4 directly.

EXAMPLE 5–14 **ESTIMATE** **Gravity on Everest.** Estimate the effective value of g on the top of Mt. Everest, 8848 m (29,028 ft) above the Earth's surface. That is, what is the acceleration due to gravity of objects allowed to fall freely at this altitude?

SOLUTION Let us call the acceleration of gravity at the given point g'. We use Eq. 5–5, with r_E replaced by $r = 6380\ \text{km} + 8.8\ \text{km} = 6389\ \text{km} = 6.389 \times 10^6\ \text{m}$:

$$g' = G\frac{m_E}{r^2} = \frac{(6.67 \times 10^{-11}\ \text{N·m}^2/\text{kg}^2)(5.98 \times 10^{24}\ \text{kg})}{(6.389 \times 10^6\ \text{m})^2} = 9.77\ \text{m/s}^2$$

which is a reduction of about 3 parts in a thousand (0.3%). Note that we ignored the mass accumulated under the mountain top, and we have used $1\ \text{N/kg} = 1\ \text{m/s}^2$.

Note that Eq. 5–5 does not give precise values for g at different locations because the earth is not a perfect sphere. The Earth not only has mountains and valleys, and bulges at the equator, but its mass is not distributed

---

[†]That the distance is measured from the Earth's center does not imply that the force of gravity somehow emanates from that one point. Rather, all parts of the Earth attract gravitationally, but the net effect is a force acting toward the Earth's center.

128 CHAPTER 5 Circular Motion; Gravitation

precisely uniformly (see Table 2–1). The Earth's rotation also has an effect on the value of g.

The value of g can vary locally on the Earth's surface because of the presence of irregularities and rocks of different densities. Such variations in g, known as "gravity anomalies," are very small—on the order of 1 part per 10^6 or 10^7 in the value of g. But they can be measured ("gravimeters" today can detect variations in g to 1 part in 10^9). Geophysicists use such measurements as part of their investigations into the structure of the Earth's crust, and in mineral and oil exploration. Mineral deposits, for example, often have a greater density than surrounding material. Because of the greater mass in a given volume, g can have a slightly greater value on top of such a deposit than at its flanks. "Salt domes," under which petroleum is often found, have a lower than average density and searches for a slight reduction in the value of g in certain locales have led to the discovery of oil.

Some geophysics

➡ **PHYSICS APPLIED**

Geology—mineral and oil exploration

5–8 Satellites and "Weightlessness"

Artificial satellites circling the Earth are now commonplace (Fig. 5–22). A satellite is put into orbit by accelerating it to a sufficiently high tangential speed with the use of rockets, as shown in Fig. 5–23. If the speed is too high, the spacecraft will not be confined by the Earth's gravity and will escape, never to return. If the speed is too low, it will return to Earth. Satellites are usually put into circular (or nearly circular) orbits, because they require the least takeoff speed. It is sometimes asked: "What keeps a satellite up?" The answer is: its high speed. If a satellite stopped moving, it would, of course, fall directly to Earth. But at the very high speed a satellite has, it would quickly fly out into space (Fig. 5–24) if it weren't for the gravitational force of the Earth pulling it into orbit. In fact, a satellite *is* falling (accelerating toward Earth), but its high tangential speed keeps it from hitting Earth.

For satellites that move in a circle (at least approximately), the acceleration is v^2/r. The force that gives a satellite this acceleration is the force of gravity, and since a satellite may be at a considerable distance from the Earth, we must use Eq. 5–4 for the force acting on it. When we apply

➡ **PHYSICS APPLIED**

Artificial Earth satellites

FIGURE 5–22 A satellite circling the Earth.

FIGURE 5–23 Artificial satellites launched at different speeds.

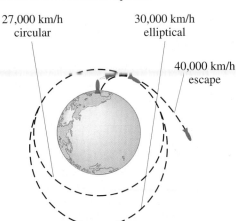

27,000 km/h circular

30,000 km/h elliptical

40,000 km/h escape

FIGURE 5–24 A moving satellite "falls" out of a straight-line path toward the Earth.

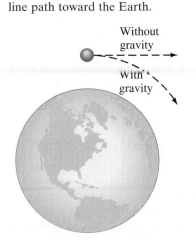

Without gravity

With gravity

Newton's second law, $\Sigma F_R = ma_R$, we find

$$G\frac{mm_E}{r^2} = m\frac{v^2}{r}, \tag{5-6}$$

where m is the mass of the satellite. This equation relates the distance of the satellite from the Earth's center, r, to its speed, v. Note that only one force—gravity—is acting on the satellite, and that r is the sum of the Earth's radius r_E plus the satellite's height h above the Earth: $r = r_E + h$.

➡ PHYSICS APPLIED

*Geosynchronous
satellites*

EXAMPLE 5–15 **Geosynchronous satellite.** A *geosynchronous* satellite is one that stays above the same point on the equator of the Earth. Such satellites are used for such purposes as cable TV transmission, for weather forecasting, and as communication relays. Determine (a) the height above the Earth's surface such a satellite must orbit and (b) such a satellite's speed.

SOLUTION (a) The only force on the satellite is gravity, so we apply Eq. 5–6 assuming the satellite moves in a circle:

$$G\frac{m_{Sat}m_E}{r^2} = m_{Sat}\frac{v^2}{r}.$$

This equation seems to have two unknowns, r and v. But we know that v must be such that the satellite revolves around the Earth with the same period that the Earth rotates on its axis, namely once in 24 hours. Thus the speed of the satellite must be

$$v = \frac{2\pi r}{T}$$

where $T = 1$ day $= (24\,h)(3600\,s/h) = 86,400\,s$. We put this into the first equation above and obtain (after canceling m_{Sat} on both sides):

$$G\frac{m_E}{r^2} = \frac{(2\pi r)^2}{rT^2}.$$

We solve for r:

$$r^3 = \frac{Gm_E T^2}{4\pi^2} = \frac{(6.67 \times 10^{-11}\,N\cdot m^2/kg^2)(5.98 \times 10^{24}\,kg)(86,400\,s)^2}{4\pi^2}$$

$$= 7.54 \times 10^{22}\,m^3,$$

and, taking the cube root, $r = 4.23 \times 10^7$ m, or 42,300 km from the Earth's center. We subtract the Earth's radius of 6380 km to find that the satellite must orbit about 36,000 km (about 6 r_E) above the Earth's surface.
(b) We solve Eq. 5–6 for v:

$$v = \sqrt{\frac{Gm_E}{r}} = \sqrt{\frac{(6.67 \times 10^{-11}\,N\cdot m^2/kg^2)(5.98 \times 10^{24}\,kg)}{(4.23 \times 10^7\,m)}} = 3070\,m/s.$$

We get the same result if we use $v = 2\pi r/T$.

*Apparent weight
and weightlessness*

People and other objects in a satellite circling the Earth are said to experience apparent weightlessness. Before tackling the case of a satellite, however, let us first look at the simpler case of a falling elevator. In Fig. 5–25a, we see an elevator at rest with a bag hanging from a spring scale. The scale reading

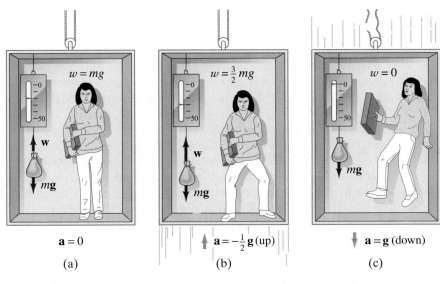

FIGURE 5–25 (a) An object in an elevator at rest exerts a force on a spring scale equal to its weight. (b) In an elevator accelerating upward at $\frac{1}{2}g$, the object's apparent weight is $1\frac{1}{2}$ times larger. (c) In a freely falling elevator, the object experiences "weightlessness."

indicates the downward force exerted on it by the bag. This force, exerted *on* the scale, is equal and opposite to the force exerted *by* the scale upward on the bag. We call this force **w**. (Similarly, if you were standing on a scale in an elevator, the normal force the scale exerts on you would be its reading.) Since the mass, m, is not accelerating, we apply $\Sigma F = ma$ to the bag and obtain

$$w - mg = 0,$$

where mg is the weight of the bag. Thus, $w = mg$, and since the scale indicates the force w exerted on it by the bag, it registers a force equal to the weight of the bag as we expect. If, now, the elevator has an acceleration, a, then applying $\Sigma F = ma$ to the bag, we have

$$w - mg = ma.$$

Solving for w, we have

$$w = mg + ma.$$

We have chosen the positive direction up. Thus, if the acceleration a is up, a is positive; and the scale, which measures w, will read more than mg. We call w the *apparent weight* of the bag, which here is greater than its actual weight (mg). If the elevator accelerates downward, a will be negative and w, the apparent weight, will be less than mg.

For example, if the elevator's acceleration is $\frac{1}{2}g$ upward, then we find $w = mg + m(\frac{1}{2}g) = \frac{3}{2}mg$. That is, the scale reads $1\frac{1}{2}$ times the actual weight (Fig. 5–25b). The apparent weight of the bag is $1\frac{1}{2}$ times its real weight. The same is true of the person: her apparent weight (equal to the normal force exerted on her by the elevator floor) is $1\frac{1}{2}$ times her real weight. We can say that she is experiencing $1\frac{1}{2}$ g's, just as astronauts experience so many g's at a rocket's launch.

If, instead, the elevator's acceleration is $-\frac{1}{2}g$ (downward), then $w = mg - \frac{1}{2}mg = \frac{1}{2}mg$. That is, the scale reads one half the actual weight. If the elevator is in *free fall* (for example, if the cables break), then $a = -g$ and $w = mg - mg = 0$. The scale reads zero! (See Fig. 5–25c.) The bag seems weightless. If the person in the elevator let go of a pencil, say, it would not fall to the floor. True, the pencil would be falling with acceleration g. But so would the floor of the elevator and the person. The pencil would hover right

FIGURE 5–26 The astronaut seems not to know which end is up. Note edge of Earth on the right. (Turn photo upside down for another view.)

in front of the person. This phenomenon is called *apparent weightlessness* because, in fact, gravity is still acting on the object and its weight is still *mg*. The objects seem weightless only because the elevator is in free fall.

The "weightlessness" experienced by people in a satellite orbit close to the Earth (Fig. 5–26) is the same apparent weightlessness experienced in a freely falling elevator. It may seem strange, at first, to think of a satellite as freely falling. But a satellite is indeed falling toward the Earth, as was shown in Fig. 5–24. The force of gravity causes it to "fall" out of its natural straight-line path. The acceleration of the satellite must be the acceleration due to gravity at that point, since the only force acting on it is gravity. (We used this to obtain Eq. 5–6.) Thus, although the force of gravity acts on objects within the satellite, the objects experience an apparent weightlessness because they, and the satellite, are accelerating as in free fall.

Figure 5–27 shows some examples of "free fall," or apparent weightlessness, experienced by people on Earth for brief moments.

A completely different situation occurs when a spacecraft is out in space far from the Earth, the Moon, and other attracting bodies. The force of gravity due to the Earth and other heavenly bodies will then be quite small because of the distances involved, and persons in such a spacecraft will experience real weightlessness.

The effects on human beings of weightlessness (whether real or apparent makes no difference) are interesting. In ordinary circumstances, for example, people can become quite tired holding out their arms horizontally. But for a person experiencing weightlessness, no effort is needed. The arms will just "float" there, since there is no sensation of weight. This effect has many applications in athletics (Fig. 5–27). During a jump or a dive, while on a trampoline, and even between strides while running, a person is experiencing apparent weightlessness or free fall, although only for a short time. During these brief periods, limbs can be moved much more easily, since only inertia needs to be overcome. The loss of control because of lack of contact with the ground is compensated for by the increased mobility. Prolonged weightlessness in space, however, can have harmful effects on health. Red blood cells diminish in number, blood collects in the thorax, bones lose calcium and become brittle, and muscles lose their tone. These effects are being carefully studied.

FIGURE 5–27 Experiencing weightlessness on Earth.

* 5-9 Kepler's Laws and Newton's Synthesis

More than a half century before Newton proposed his three laws of motion and his law of universal gravitation, the German astronomer Johannes Kepler (1571–1630) had written a number of astronomical works in which we can find a detailed description of the motion of the planets about the Sun. Kepler's work resulted in part from the many years he spent examining data collected by Tycho Brahe (1546–1601) on the positions of the planets in their motion through the heavens. Among Kepler's writings were three findings that we now refer to as **Kepler's laws of planetary motion**. These are summarized as follows, with additional explanation in Fig. 5–28:

Kepler's first law: The path of each planet about the Sun is an ellipse (Fig. 5–28a) with the Sun at one focus.

Kepler's second law: Each planet moves so that an imaginary line drawn from the Sun to the planet sweeps out equal areas in equal periods of time (Fig. 5–28b).

Kepler's third law: The ratio of the squares of the periods (the time needed for one revolution about the Sun) of any two planets revolving about the Sun is equal to the ratio of the cubes of their mean distances from the Sun. That is, if T_1 and T_2 represent the periods for any two planets, and r_1 and r_2 represent their mean distances from the Sun, then

$$\left(\frac{T_1}{T_2}\right)^2 = \left(\frac{r_1}{r_2}\right)^3.$$

We can rewrite this as

$$\frac{r_1^3}{T_1^2} = \frac{r_2^3}{T_2^2}$$

meaning that r^3/T^2 should be the same for each planet. (Present-day data are given in Table 5–1; see the last column.)

Newton was able to show that Kepler's laws could be derived mathematically from the law of universal gravitation and the laws of motion. He also showed that among the reasonable possibilities for the gravitational force law, only one that depends on the inverse square of the distance is fully

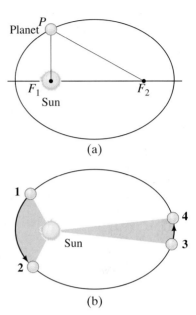

(a)

(b)

FIGURE 5–28 (a) Kepler's first law. An ellipse is a closed curve such that the sum of the distances from any point P on the curve to two fixed points (called the foci, F_1 and F_2) remains constant. That is, the sum of the distances $F_1P + F_2P$ is the same for all points on the curve. A circle is a special case of an ellipse in which the two foci coincide, at the center of the circle. (b) Kepler's second law. The two shaded regions have equal areas. The planet moves from point 1 to point 2 in the same time as it takes to move from point 3 to point 4. Planets move fastest in that part of their orbit where they are closest to the Sun.

TABLE 5–1	Planetary Data Applied to Kepler's Third Law		
Planet	**Mean distance from Sun, r (10^6 km)**	**Period, T (Earth years)**	**r^3/T^2 (10^{24} km³/yr²)**
Mercury	57.9	0.241	3.34
Venus	108.2	0.615	3.35
Earth	149.6	1.0	3.35
Mars	227.9	1.88	3.35
Jupiter	778.3	11.86	3.35
Saturn	1427	29.5	3.34
Uranus	2870	84.0	3.35
Neptune	4497	165	3.34
Pluto	5900	248	3.33

consistent with all three of Kepler's laws. He thus used Kepler's laws as evidence for his law of universal gravitation, Eq. 5–4.

Derivation of
Kepler's third law

Kepler's third law is the easiest to derive, and we do it here for the special case of a circular orbit. (Most of the planetary orbits are fairly close to a circle, which is a special case of an ellipse.) First, we write down Newton's second law of motion, $\Sigma F = ma$. Then for ΣF we substitute the law of universal gravitation, Eq. 5–4, and for a the centripetal acceleration, v^2/r:

$$\Sigma F = ma$$

$$G\frac{m_1 M_S}{r_1^2} = m_1 \frac{v_1^2}{r_1}.$$

Here m_1 is the mass of a particular planet, r_1 its mean distance from the Sun, and v_1 its average speed in orbit. Then M_2 in Eq. 5–4 is taken to be M_S, the mass of the Sun, since it is the gravitational attraction of the Sun that keeps each planet in its orbit. Now the period T_1 of the planet is the time required for one complete orbit, a distance equal to $2\pi r_1$, the circumference of a circle. Thus,

$$v_1 = \frac{2\pi r_1}{T_1}.$$

We substitute this formula for v_1 into the equation above:

$$G\frac{m_1 M_S}{r_1^2} = m_1 \frac{4\pi^2 r_1}{T_1^2}.$$

We rearrange to get

$$\frac{T_1^2}{r_1^3} = \frac{4\pi^2}{GM_S}. \tag{5–7a}$$

We derived this for planet 1 (say, Mars). The same derivation would apply for a second planet (say, Saturn):

$$\frac{T_2^2}{r_2^3} = \frac{4\pi^2}{GM_S},$$

where T_2 and r_2 are the period and orbit radius, respectively, for the second planet. Since the right sides of the two previous equations are equal, we have $T_1^2/r_1^3 = T_2^2/r_2^3$ or, rearranging,

$$\left(\frac{T_1}{T_2}\right)^2 = \left(\frac{r_1}{r_2}\right)^3, \tag{5–7b}$$

which is Kepler's third law.

The derivations of Eqs. 5–7a and b (Kepler's third law) are general enough to be applied to other systems. For example, we could determine the mass of the Earth from Eq. 5–7a using the period of the Moon about the Earth and the Moon's distance from the Earth, or the mass of Jupiter from the period and distance of one of its moons (this is indeed how masses are determined; see the Problems). We can also use Eqs. 5–7a and b to compare objects that orbit other attracting centers, such as the Moon and a weather satellite orbiting Earth. But be careful not to use Eq. 5–7 to compare, say, the Moon's orbit to the orbit of Mars because they depend on different attracting centers.

Careful:
A common error using
Kepler's third law

In the following examples, we assume the orbits are circles, although it is not quite true in general.

EXAMPLE 5–16 **Where is Mars?** Mars' period (its "year") was first noted by Kepler to be about 687 days (Earth-days), which is $(687\,\text{d}/365\,\text{d}) = 1.88\,\text{yr}$. Determine the distance of Mars from the Sun using the Earth as a reference.

SOLUTION The period of the Earth is $T_E = 1\,\text{yr}$, and the distance of Earth from the Sun is $r_{ES} = 1.50 \times 10^{11}\,\text{m}$. From Kepler's third law (Eq. 5–7b):

$$\frac{r_{MS}}{r_{ES}} = \left(\frac{T_M}{T_E}\right)^{2/3} = \left(\frac{1.88\,\text{yr}}{1\,\text{yr}}\right)^{2/3} = 1.52.$$

So Mars is 1.52 the Earth's distance from the Sun, or $2.28 \times 10^{11}\,\text{m}$.

EXAMPLE 5–17 **The Sun's mass determined.** Determine the mass of the Sun given the Earth's distance from the Sun as $r_{ES} = 1.5 \times 10^{11}\,\text{m}$.

➡ **PHYSICS APPLIED**
Determining the Sun's mass

SOLUTION We can use Eq. 5–7a and solve for M_S:

$$M_S = \frac{4\pi^2 r_{ES}^3}{GT_E^2} = \frac{4\pi^2 (1.5 \times 10^{11}\,\text{m})^3}{(6.67 \times 10^{-11}\,\text{N·m}^2/\text{kg}^2)(3.16 \times 10^7\,\text{s})^2}$$

$$= 2.0 \times 10^{30}\,\text{kg}$$

where we used the fact that

$$T_E = 1\,\text{yr} = (365\tfrac{1}{4}\,\text{d})(24\,\text{h/d})(3600\,\text{s/h}) = 3.16 \times 10^7\,\text{s}.$$

EXAMPLE 5–18 **ESTIMATE** **Geosynchronous satellite, simplified.** A *geosynchronous* satellite of the Earth (as mentioned in Example 5–15) is one that stays above the same point on the equator of the Earth. Estimate the height above the Earth's surface needed for a geosynchronous weather satellite. (This is to be a "lunchtime" calculation done on a napkin, without calculator, as compared to our earlier calculation in Example 5–15.)

SOLUTION To use Kepler's third law we must compare the satellite to some other object that orbits Earth. The simplest choice is the Moon because we know its period and distance. The Moon's period is about $T_M \approx 27\,\text{d}$ and its distance from the earth about $r_{ME} \approx 380,000\,\text{km}$. The period of the weather satellite needs to be $T_{Sat} = 1\,\text{d}$ so that it stays above the same place on the Earth. Hence,

$$r_{Sat} = r_{ME}\left(\frac{T_{Sat}}{T_M}\right)^{2/3} = r_{ME}\left(\frac{1\,\text{d}}{27\,\text{d}}\right)^{2/3} = r_{ME}\left(\frac{1}{3}\right)^2 = \frac{r_{ME}}{9}.$$

(How nice the Moon's approximate period turns out to be a perfect cube.) A geosynchronous satellite must be $\frac{1}{9}$ the distance to the Moon, which is 42,000 km from the center of the Earth or 36,000 km above the Earth's surface. This is about 6 Earth radii high.

Accurate measurements on the orbits of the planets indicated that they did not precisely follow Kepler's laws. For example, slight deviations from perfectly elliptical orbits were observed. Newton was aware that this was to be expected from the law of universal gravitation ("every body in the universe attracts every other body...") because each planet exerts a gravitational force on the other planets. Since the mass of the Sun is much greater than that of any planet, the force on one planet due to any other planet will be small in comparison to the force on it due to the Sun. (The derivation of perfectly elliptical orbits ignores the forces due to other planets.) But because of this small force, each planetary orbit should depart from a perfect ellipse, especially when a second planet is fairly close to it. Such deviations, or **perturbations**, as they are called, from perfect ellipses are indeed observed. In fact, Newton's recognition of perturbations in the orbit of Saturn was a hint that helped him formulate the law of universal gravitation, that all bodies attract gravitationally. Observation of other perturbations later led to the discovery of Neptune and Pluto. Deviations in the orbit of Uranus, for example, could not be accounted for by perturbations due to the other known planets. Careful calculation in the nineteenth century indicated that these deviations could be accounted for if there were another planet farther out in the solar system. The position of this planet was predicted from the deviations in the orbit of Uranus, and telescopes focused on that region of the sky quickly found it; the new planet was called Neptune. Similar but much smaller perturbations of Neptune's orbit led to the discovery of Pluto in 1930.

More recently, in 1996, planets revolving about distant stars (Fig. 5–29) were inferred from the regular "wobble" of each star due to the gravitational attraction of the revolving planet.

The development by Newton of the law of universal gravitation and the three laws of motion was a major intellectual achievement. For with these laws, Newton was able to describe the motion of objects on Earth and in the heavens. The motions of heavenly bodies and bodies on Earth were seen to follow the same laws (something not previously recognized generally, although Galileo and Descartes had argued in its favor). For this reason (and also because Newton integrated the results of earlier workers into his system), we sometimes speak of Newton's "synthesis."

Newton's work was so encompassing that it constituted a theory of the universe, and influenced philosophy and other fields. The laws formulated by Newton are referred to as **causal laws**. By **causality** we mean the idea that one occurrence can cause another. We have repeatedly observed, for example, that when a rock strikes a window, the window almost immediately breaks. We infer that the rock *caused* the window to break. This idea of "cause and effect" took on more forceful meaning with Newton's laws. For the motion—or rather the acceleration—of any object was seen to be *caused* by the net force acting on it. As a result, the universe came to be pictured by many scientists and philosophers as a big machine whose parts

FIGURE 5–29 Our solar system (a), compared to recently discovered planets orbiting (b) the star 51 Pegasi, and (c) the star 47 Ursae Majoris.

Sun   Mercury Venus Earth   Mars   Jupiter

(a)

51 Pegasi   Planet (0.6 Jupiter's mass)

(b)

47 Ursae Majoris   Planet (3.5 Jupiter's mass)

(c)

move in a predetermined way—according to natural laws. However, this *deterministic* view of the universe had to be modified by scientists in the twentieth century, as we shall see in Chapters 27 and 28.

5–10 Types of Forces in Nature

We have already discussed that Newton's law of universal gravitation, Eq. 5–4, describes how a particular type of force—gravity—depends on the distance between, and masses of, the objects involved. Newton's second law, $\Sigma \mathbf{F} = m\mathbf{a}$, on the other hand, tells how a body will accelerate due to *any* type of force. But what are the types of forces that occur in nature besides gravity?

In the twentieth century, physicists came to recognize four different fundamental forces in nature: (1) the gravitational force; (2) the electromagnetic force (we shall see later that electric and magnetic forces are intimately related); (3) the strong nuclear force; and (4) the weak nuclear force. In this chapter, we discussed the gravitational force in detail. The nature of the electromagnetic force will be discussed in detail in Chapters 16 to 22. The strong and weak nuclear forces, which are discussed in Chapters 30 to 32, operate at the level of the atomic nucleus, and although they manifest themselves in such phenomena as radioactivity and nuclear energy, they are much less obvious in our daily lives.

Physicists have been working on theories that would unify these four forces—that is, to consider some or all of these forces as different manifestations of the same basic force. So far, the electromagnetic and weak nuclear forces have been theoretically united to form *electroweak* theory, in which the electromagnetic and weak forces are seen as two different manifestations of a single *electroweak force*. Attempts to further unify the forces, such as in *grand unified theories* (GUT), are hot research topics today.

Electroweak and GUT

But where do everyday forces fit into this scheme? Ordinary forces, other than gravity, such as pushes, pulls, and other contact forces like the normal force and friction, are today considered to be due to the electromagnetic force acting at the atomic level. For example, the force your fingers exert on a pencil is the result of electrical repulsion between the outer electrons of the atoms of your finger and those of the pencil.

Everyday forces are gravity and electromagnetic

➡ PROBLEM SOLVING   **Uniform Circular Motion and Gravity**

1. Draw a free-body diagram, showing all the forces acting on the object under consideration. If more than one object is involved, draw a separate free-body diagram for each. Be sure you can identify the source of each force (tension in a cord, Earth's gravity, friction, normal force, and so on), so you don't put in something that isn't there.

2. Determine which of these forces, or which of their components, act to provide the centripetal (radial) acceleration—that is, all the forces or components that act radially, toward or away from the center of the circular path. The sum of these forces (or components) provides the centripetal acceleration, $a_R = v^2/r$.

3. Choose a coordinate system, and positive and negative directions, and apply Newton's second law to the radial direction:

$$\Sigma F_R = ma_R = m\frac{v^2}{r}.$$

4. For the gravitational force, use Newton's law of universal gravitation (if the object is close to the Earth's surface, you can simply use $m\mathbf{g}$); be sure to use the correct value for r. Remember that for large objects, such as the Earth or Moon, r is measured from the *center* of the sphere, not from the surface.

SUMMARY

An object moving in a circle of radius r with constant speed v is said to be in **uniform circular motion**. It has a **centripetal acceleration** a_R that is directed radially toward the center of the circle (also called **radial acceleration**), whose magnitude is

$$a_R = \frac{v^2}{r}.$$

The direction of the velocity vector and that of the acceleration $\mathbf{a}_R$ are continually changing in direction, but are perpendicular to each other at each moment.

A force is needed to keep a particle revolving in a circle, and the direction of this force is toward the center of the circle. This force may be due to gravity, to tension in a cord, to a component of the normal force, or other type of force.

Newton's **law of universal gravitation** states that every particle in the universe attracts every other particle with a force proportional to the product of their masses and inversely proportional to the square of the distance between them:

$$F = G\frac{m_1 m_2}{r^2}.$$

The direction of this force is along the line joining the two particles. It is this gravitational force that keeps the Moon revolving around the Earth, and the planets revolving around the Sun.

Satellites revolving around the Earth are acted on by gravity, but "stay up" because of their high tangential speed.

QUESTIONS

1. It is sometimes said that water is removed from clothes in a spin dryer by centrifugal force throwing the water outward. Is this correct?

2. Will the acceleration of a car be the same when it travels around a sharp curve at 60 km/h as when it travels around a gentle curve at the same speed? Explain.

3. Suppose a car moves at constant speed along a mountain road. At what places does it exert the greatest and least forces on the road: (a) at the top of a hill, (b) at a dip between two hills, (c) on a level stretch near the bottom of a hill?

4. Describe all the forces acting on a child riding a horse on a merry-go-round. Which of these forces provides the centripetal acceleration of the child?

5. A bucket of water can be whirled in a vertical circle without the water spilling out, even at the top of the circle when the bucket is upside down. Explain.

6. Does an apple exert a gravitational force on the Earth? If so, how large a force? Consider an apple (a) attached to a tree, and (b) falling.

7. If the Earth's mass were double what it is, in what ways would the Moon's orbit be different?

8. Describe how careful measurements of the variation in g in the vicinity of an ore deposit might be used to estimate the amount of ore present.

9. When will your apparent weight be the greatest, as measured by a scale in a moving elevator: when the elevator (a) accelerates downward, (b) accelerates upward, (c) is in free fall, (d) moves upward at constant speed? In which case would your weight be the least? When would it be the same as when you are on the ground?

10. The Earth is not completely spherical but bulges outward at the equator. Why?

11. An antenna loosens and becomes detached from a satellite in a circular orbit around the Earth. Describe the antenna's motion subsequently. If it will land on Earth, describe where; if not, describe how it could be made to land on Earth.

12. Astronauts who spend long periods in outer space could be adversely affected by weightlessness. One way to simulate gravity is to shape the spaceship like a cylindrical shell that rotates, with the astronauts walking on the inside surface (Fig. 5–30). Explain how this simulates gravity. Consider (a) how objects fall, (b) the force we feel on our feet, and (c) any other aspects of gravity you can think of.

FIGURE 5–30 Question 12 and Problem 41.

13. Which pulls harder gravitationally, the Earth on the Moon, or the Moon on the Earth? Which accelerates more?

14. How many "accelerators" do you have in your car? There are at least three controls in the car which can be used to cause the car to accelerate. What are they? What accelerations do they produce?

15. A child on a sled comes flying over the crest of a small hill, as shown in Fig. 5–31. His sled does not leave the ground (he does not achieve "air"), but he feels the normal force between his chest and the sled decrease as he goes over the hill. Explain this decrease using Newton's second law.

16. People sometimes ask, "What keeps a satellite up in its orbit around the Earth?" How would you respond?

17. Explain how a runner experiences "free fall" or "apparent weightlessness" between steps.

FIGURE 5–31   Question 15.

18. The Earth moves faster in its orbit around the Sun in winter than in summer. Is it closer to the Sun in summer or in winter? Does this affect the seasons? Explain. [*Note*: This is not much of a factor in explaining the seasons—the main factor is the tilt of the Earth's axis relative to the plane of its orbit.]

PROBLEMS

SECTIONS 5–1 TO 5–3

1. (I) A jet plane traveling 1800 km/h (500 m/s) pulls out of a dive by moving in an arc of radius 6.00 km. What is the plane's acceleration in g's?

2. (I) A child on a merry-go-round is moving with a speed of 1.35 m/s when 1.20 m from the center of the merry-go-round. Calculate (*a*) the centripetal acceleration of the child, and (*b*) the net horizontal force exerted on the child (mass = 25.0 kg).

3. (I) Calculate the centripetal acceleration of the Earth in its orbit around the Sun and the net force exerted on the Earth. What exerts this force on the Earth? Assume that the Earth's orbit is a circle of radius 1.50×10^{11} m.

4. (I) A horizontal force of 280 N is exerted on a 2.0-kg discus as it is rotated uniformly in a horizontal circle (at arms length) of radius 1.00 m. Calculate the speed of the discus.

5. (II) A flat puck (mass M) is rotated in a circle on a frictionless air hockey tabletop, and is held in this orbit by a light cord which is connected to a dangling mass (mass m) through the central hole as shown in Fig. 5–32. Show that the speed of the puck is given by

$$v = \sqrt{\frac{mgR}{M}}.$$

6. (II) A 0.40-kg ball, attached to the end of a horizontal cord, is rotated in a circle of radius 1.3 m on a frictionless horizontal surface. If the cord will break when the tension in it exceeds 60 N, what is the maximum speed the ball can have? How would your answer be affected if there were friction?

7. (II) What is the maximum speed with which a 1050-kg car can round a turn of radius 70 m on a flat road if the coefficient of friction between tires and road is 0.80? Is this result independent of the mass of the car?

FIGURE 5–32   Problem 5.

FIGURE 5–33   Problem 8.

FIGURE 5–34   Problem 12.

8. (II) A ball on the end of a string is cleverly revolved at a uniform rate in a vertical circle of radius 85.0 cm, as shown in Fig. 5–33. If its speed is 4.15 m/s and its mass is 0.300 kg, calculate the tension in the string when the ball is (a) at the top of its path, and (b) at the bottom of its path.

9. (II) How large must the coefficient of friction be between the tires and the road if a car is to round a level curve of radius 85 m at a speed of 95 km/h?

10. (II) A device for training astronauts and jet fighter pilots is designed to rotate the trainee in a horizontal circle of radius 10.0 m. If the force felt by the trainee is 7.75 times her own weight, how fast is she rotating? Express your answer in both m/s and rev/s.

11. (II) A coin is placed 11.0 cm from the axis of a rotating turntable of variable speed. When the speed of the turntable is slowly increased, the coin remains fixed on the turntable until a rate of 36 rpm is reached, at which point the coin slides off. What is the coefficient of static friction between the coin and the turntable?

12. (II) At what minimum speed must a roller coaster be traveling when upside down at the top of a circle (Fig. 5–34) if the passengers are not to fall out? Assume a radius of curvature of 8.6 m.

13. (II) A 1000-kg sports car moving at 20 m/s crosses the rounded top of a hill (radius = 100 m). Determine (a) the normal force on the car, (b) the normal force on the 70-kg driver, and (c) the car speed at which the normal force equals zero.

14. (II) How many revolutions per minute would a 15-m-diameter Ferris wheel need to make for the passengers to feel "weightless" at the topmost point of the trip?

15. (II) Use dimensional analysis (see Appendix B) to check the form for the centripetal acceleration, $a_R = v^2/r$.

16. (II) Two masses, m_1 and m_2, connected to each other and to a central post by cords, as shown in Fig. 5–35, rotate about the post at a frequency f (revolutions per second) on a frictionless horizontal surface at distances r_1 and r_2, respectively, from the post. Derive an algebraic expression for the tension in each segment of the cord.

17. (II) A 1200-kg car rounds a curve of radius 70 m banked at an angle of 12°. If the car is traveling at 90 km/h, will a friction force be required? If so, how much and in what direction?

FIGURE 5–35   Problem 16.

FIGURE 5–36 Problem 18.

18. (II) In a "Rotor-ride" at a carnival, people pay money to be rotated in a vertical cylindrically walled "room." (See Fig. 5–36.) If the room radius is 5.0 m, and the rotation frequency is 0.50 revolutions per second when the floor drops out, what is the minimum coefficient of static friction so that the people will not slip down? People describe this ride by saying they were being "pressed against the wall." Is this true? Is there really an outward force pressing them against the wall? If so, what is its source? If not, what is the proper description of their situation (besides "scary")? [*Hint*: First draw the free-body diagram for a person.]

19. (III) Redo Example 5–3, precisely this time, by not ignoring the weight of the ball. In particular, find the magnitude of $\mathbf{F}_T$, and the angle it makes with the horizontal. [*Hint*: Set the horizontal component of $\mathbf{F}_T$ equal to ma_R; also, since there is no vertical motion, what can you say about the vertical component of $\mathbf{F}_T$?]

20. (III) If a curve with a radius of 80 m is perfectly banked for a car traveling 70 km/h, what must be the coefficient of static friction for a car not to skid when traveling at 90 km/h?

21. (III) A pilot performs an evasive maneuver by diving vertically at 310 m/s. If he can withstand an acceleration of 9.0 *g*'s without blacking out, at what altitude must he begin to pull out of the dive to avoid crashing into the sea?

*SECTION 5–4

* 22. (I) Determine the tangential and centripetal components of the net force exerted on the car (by the ground) in Example 5–9 when its speed is 30 m/s. The car's mass is 1000 kg.

* 23. (II) A car at the Indianapolis-500 accelerates uniformly from the pit area, going from rest to 320 km/h in a semicircular arc with a radius of 200 m. Determine the tangential and radial acceleration of the car when it is halfway through the turn, assuming constant tangential acceleration. If the curve were flat, what would the coefficient of static friction have to be between the tires and the roadbed to provide this acceleration with no slipping or skidding?

* 24. (III) A particle revolves in a horizontal circle of radius 2.70 m. At a particular instant, its acceleration is 1.05 m/s², in a direction that makes an angle of 32.0° to its direction of motion. Determine its speed (*a*) at this moment, and (*b*) 2.00 s later, assuming constant tangential acceleration.

SECTIONS 5–6 AND 5–7

25. (I) Calculate the force of gravity on a spacecraft 12,800 km (2 earth radii) above the Earth's surface if its mass is 1400 kg.

26. (I) Calculate the acceleration due to gravity on the Moon. The Moon's radius is about 1.74×10^6 m and its mass is 7.35×10^{22} kg.

27. (I) A hypothetical planet has a radius 2.5 times that of Earth, but has the same mass. What is the acceleration due to gravity near its surface?

28. (I) A hypothetical planet has a mass 2.5 times that of Earth, but the same radius. What is *g* near its surface?

29. (I) At the surface of a certain planet, the gravitational acceleration *g* has a magnitude of 12.0 m/s². A 2.10-kg brass ball is transported to this planet. What is (*a*) the mass of the brass ball on the Earth and on the planet, and (*b*) the weight of the brass ball on the Earth and on the planet?

30. (II) You are explaining to friends why astronauts feel weightless orbiting in the space shuttle, and they respond that they thought gravity was just a lot weaker up there. Convince them and yourself that it isn't so by calculating how much weaker gravity is 300 km above the Earth's surface.

31. (II) An exotic finish to massive stars is that of a neutron star, which might have as much as five times the mass of our Sun packed into a sphere about 10 km in radius! Estimate the surface gravity on this monster.

32. (II) What is the distance from the Earth's center to a point outside the Earth where the gravitational acceleration due to the Earth is $\frac{1}{10}$ of its value at the Earth's surface?

33. (II) A typical white dwarf star, which once was an average star like our Sun but is now in the last stage of its evolution, is the size of our Moon but has the mass of our Sun. What is the surface gravity on this star?

34. (II) Calculate the effective value of *g*, the acceleration of gravity, at (*a*) 3200 m, and (*b*) 3200 km, above the Earth's surface.

35. (II) Four 7.5-kg spheres are located at the corners of a square of side 0.60 m. Calculate the magnitude and direction of the gravitational force on one sphere due to the other three.

36. (II) Every few hundred years most of the planets line up on the same side of the Sun. Calculate the total force on the Earth due to Venus, Jupiter, and Saturn, assuming all four planets are in a line, Fig. 5–37. The masses are respectively $M_V = 0.815\,M_E$, $M_J = 318\,M_E$, $M_S = 95.1\,M_E$, and their mean distances from the Sun are 108, 150, 778, and 1430 million km.

FIGURE 5–37 Problem 36 (not to scale).

37. (II) Given that the acceleration of gravity at the surface of Mars is 0.38 of what it is on Earth, and that Mars' radius is 3400 km, determine the mass of Mars.

38. (III) Determine the mass of the Sun using the known value for the period of the Earth and its distance from the Sun. [*Note*: Compare your answer to that obtained using Kepler's laws, Example 5–17.]

SECTION 5–8

39. (I) Calculate the velocity of a satellite moving in a stable circular orbit about the Earth at a height of 3600 km.

40. (II) A 17.0-kg monkey hangs from a cord suspended from the ceiling of an elevator. The cord can withstand a tension of 220 N and breaks as the elevator accelerates. What was the elevator's minimum acceleration (magnitude and direction)?

41. (II) At what rate must the cylindrical spaceship of Fig. 5–30 rotate (see Question 12), if occupants are to experience simulated gravity of $\frac{1}{2}g$? Assume the spaceship's diameter is 32 m, and give your answer as the time needed for one revolution.

42. (II) Determine the time it takes for a satellite to orbit the Earth in a circular "near-Earth" orbit. The definition of "near-Earth" orbit is one which is at a height above the surface of the Earth which is small compared to the radius of the Earth, so that you may take the acceleration due to gravity as essentially the same as that on the surface. Does your result depend on the mass of the satellite?

43. (II) During an *Apollo* lunar landing mission, the command module continued to orbit the Moon at an altitude of about 100 km. How long did it take to go around the Moon once?

44. (II) What will a spring scale read for the weight of a 58-kg woman in an elevator that moves (*a*) with constant upward speed of 6.0 m/s, (*b*) with constant downward speed of 6.0 m/s, (*c*) with upward acceleration of 0.33 *g*, (*d*) with downward acceleration 0.33 *g*, and (*e*) in free fall?

45. (II) The rings of Saturn are composed of chunks of ice that orbit the planet. The inner radius of the rings is 73,000 km, while the outer radius is 170,000 km. Find the period of an orbiting chunk of ice at the inner radius and the period of a chunk at the outer radius. Compare your numbers with Saturn's mean rotation period of 10 hours and 39 minutes. The mass of Saturn is 5.69×10^{26} kg.

46. (II) A Ferris wheel, 24.0 m in diameter, rotates once every 12.5 s (see Fig. 5–10). What is the fractional change in a person's apparent weight (*a*) at the top, and (*b*) at the bottom, as compared to her weight at rest?

47. (II) What is the apparent weight of a 70-kg astronaut 4200 km from the center of the Earth's Moon in a space vehicle (*a*) moving at constant velocity, and (*b*) accelerating toward the Moon at 2.9 m/s²? State "direction" in each case.

48. (II) Describe a general procedure to determine the mass of a planet from observations on the orbit of one of its satellites.

49. (II) Suppose that a binary star system consists of two stars of equal mass. They are observed to be separated by 360 million km and to take 5.0 Earth years to orbit about a point midway between them. What is the mass of each?

50. (III) (*a*) Show that if a satellite orbits very near the surface of a planet with period *T*, the density (mass/volume) of the planet is $\rho = 3\pi/GT^2$. (*b*) Estimate the density of the Earth, given that a satellite near the surface orbits with a period of about 90 minutes.

*SECTION 5–9

* 51. (I) Use Kepler's laws and the period of the Moon (27.4 d) to determine the period of an artificial satellite orbiting very near the Earth's surface.

* 52. (I) The asteroid Icarus, though only a few hundred meters across, orbits the Sun like the other planets. Its period is about 410 d. What is its mean distance from the Sun?

* 53. (I) Neptune is an average distance of 4.5×10^9 km from the Sun. Estimate the length of the Neptunian year given that the Earth is 1.50×10^8 km from the Sun on the average.

* 54. (II) Determine the mass of the Earth from the known period and distance of the Moon.

FIGURE 5–38
Problem 55.

* 55. (II) Halley's comet orbits the Sun roughly once every 76 years. It comes very close to the surface of the Sun on its closest approach (Fig. 5–38). Roughly how far out from the Sun is it at its farthest? Is it still "in" the Solar System? What planet's orbit is nearest when it is out there? [*Hint*: The "mean distance" in Kepler's third law is $\frac{1}{2}$ the sum of the nearest and farthest distance from the Sun.]

* 56. (II) The Sun rotates about the center of the Milky Way Galaxy (Fig. 5–39) at a distance of about 30,000 light years from the center (1 ly = 9.5×10^{15} m). If it takes about 200 million years to make one rotation, estimate the mass of our galaxy. Assume that the mass distribution of our galaxy is concentrated mostly in a central uniform sphere. If all the stars had about the mass of our Sun (2×10^{30} kg), how many stars would there be in our galaxy?

FIGURE 5–39 Edge-on view of our galaxy. Problem 56.

TABLE 5–2 Principal Moons of Jupiter (Problems 57 and 58)

Moon	Mass (kg)	Period (Earth days)	Mean distance from Jupiter (km)
Io	8.9×10^{22}	1.77	422×10^3
Europa	4.9	3.55	671
Ganymede	15	7.16	1070
Callisto	11	16.7	1883

*57. (II) Table 5–2 gives the mean distance, period, and mass for the four largest moons of Jupiter (those discovered by Galileo in 1609). (a) Determine the mass of Jupiter using the data for Io. (b) Determine the mass of Jupiter using data for each of the other three moons. Are the results consistent?

*58. (II) Determine the mean distance from Jupiter for each of Jupiter's moons, using the distance of Io and the periods given in Table 5–2. Compare to the values in the Table.

*59. (II) The asteroid belt between Mars and Jupiter consists of many fragments (which some space scientists think came from a planet that once orbited the Sun but was destroyed). (a) If the center of mass of the asteroid belt is about 3 times farther from the Sun than the Earth is, how long would it have taken this hypothetical planet to orbit the Sun? (b) Can we use this data to deduce the mass of this planet?

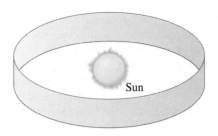

FIGURE 5–40   Problem 61.

*60. (III) (a) Use Kepler's second law to show that the ratio of the speeds of a planet at its nearest and farthest points from the Sun is equal to the inverse ratio of the near and far distances: $v_N/v_F = d_F/d_N$. (b) Given that the Earth's distance from the Sun varies from 1.47 to 1.52×10^{11} m, determine the minimum and maximum velocities of the Earth in its orbit around the Sun.

*61. (III) A science fiction tale describes an artificial "planet" in the form of a band completely encircling a sun, the inhabitants living on the inside surface (Fig. 5–40) (where it is always noon). Imagine the sun is exactly like our own, that the distance to the band is the same as the Earth–Sun distance (to make the climate temperate), and that the ring rotates quickly enough to produce an apparent gravity of one g as on Earth. What will be the period of revolution, this planet's year, in Earth days?

GENERAL PROBLEMS

62. How far above the Earth's surface will the acceleration of gravity be half what it is on the surface?

63. Tarzan plans to cross a gorge by swinging in an arc from a hanging vine (Fig. 5–41). If his arms are capable of exerting a force of 1400 N on the rope, what is the maximum speed he can tolerate at the lowest point of his swing? His mass is 80 kg and the vine is 4.8 m long.

64. Is it possible to whirl a bucket of water fast enough in a vertical circle so the water won't fall out? If so, what is the minimum speed?

65. On an ice rink two skaters of equal mass grab hands and spin in a mutual circle once every three seconds. If we assume their arms are each 0.80 m long, how hard are they pulling on one another, assuming their individual masses are 60.0 kg?

66. Because the Earth rotates once per day, the effective acceleration of gravity at the equator is slightly less than it would be if the Earth didn't rotate. Estimate the magnitude of this effect. What fraction of g is this?

FIGURE 5–41   Problem 63.

FIGURE 5-42   Problem 68.

FIGURE 5-44   Problem 72.

67. At what distance from the Earth will a spacecraft traveling directly from the Earth to the Moon experience zero net force because the Earth and Moon pull with equal and opposite forces?

68. A projected space station consists of a circular tube that is set rotating about its center (like a tubular bicycle tire) (Fig. 5–42). The circle formed by the tube has a diameter of about 1.1 km. (a) On which wall inside the tube will people be able to walk? (b) What must be the rotation speed (revolutions per day) if an effect equal to gravity at the surface of the Earth (1.0 g) is to be felt?

69. You know your mass is 60 kg, but when you stand on a bathroom scale in an elevator, it says your mass is 80 kg. What is the acceleration of the elevator, and in which direction?

70. A jet pilot takes his aircraft in a vertical loop (Fig. 5–43). (a) If the jet is moving at a speed of 1500 km/h at the lowest point of the loop, determine the minimum radius of the circle so that the centripetal acceleration at the lowest point does not exceed 6.0 g's. (b) Calculate also the 80-kg pilot's effective weight (the force with which the seat pushes up on him) at the bottom of the circle, and (c) at the top of the circle (assume the same speed).

71. Derive a formula for the mass of a planet in terms of its radius, r, the acceleration due to gravity at its surface, g_P, and the gravitational constant, G.

FIGURE 5-43   Problem 70.

72. A plumb bob is deflected from the vertical by an angle θ due to a massive mountain nearby (Fig. 5–44). (a) Find an approximate formula for θ in terms of the mass of the mountain, M_M, the distance to its center, D_M, and the radius and mass of the Earth. (b) Make a rough estimate of the mass of Mt. Everest, assuming it has the shape, say, of an equilateral pyramid (or cone) 4000 m high above its base, and then (c) estimate the angle θ of the pendulum bob if it is 5 km from the center of Mt. Everest.

73. A curve of radius 60 m is banked for a design speed of 100 km/h. If the coefficient of static friction is 0.30 (wet pavement), at what range of speeds can a car safely make the curve?

74. How long would a day be if the Earth were rotating so fast that objects at the equator were apparently weightless?

75. Two equal-mass stars maintain a constant distance apart of 8.0×10^{10} m and rotate about a point midway between them at a rate of one revolution every 12.6 yr. (a) Why don't the two stars crash into one another due to the gravitational force between them? (b) What must be the mass of each star?

76. A train traveling at a constant speed rounds a curve of radius 275 m. A chandelier suspended from the ceiling swings out to an angle of 17.5° throughout the turn. What is the speed of the train?

77. The planet Jupiter is about 320 times as massive as the Earth. Thus, it has been claimed that a person would be crushed by the force of gravity on Jupiter since people can't survive more than a few g's. Calculate the number of g's a person would experience if she could stand on the equator of Jupiter. Use the following astronomical data for Jupiter: mass = 1.9×10^{27} kg, equatorial radius = 7.1×10^4 km, rotation period 9 hr 55 min. Take the centripetal acceleration into account.

78. Astronomers using the Hubble Space Telescope have recently deduced the presence of an extremely massive core in the distant galaxy M87, so dense that it could well be a black hole (from which no light escapes). They did this by measuring the speed of gas clouds orbiting the core to be 780 km/s at a distance of 60 light-years (5.7×10^{17} m) from the core. Deduce the mass of the core, and compare it to the mass of our Sun.

A roller coaster at the highest point of its journey has its maximum potential energy (PE). As it rolls downhill, it loses PE and gains in kinetic energy (KE). Total energy is conserved. So if there is no friction, the loss in PE equals the gain in KE. If there is friction, the loss in PE equals the gain in KE plus the thermal energy produced by the work done by friction.

WORK AND ENERGY 6

Until now we have been studying the motion of an object in terms of Newton's three laws of motion. In that analysis, *force* played a central role as the quantity determining the motion. In this chapter and the next, we discuss an alternative analysis of the motion of an object in terms of the quantities *energy* and *momentum*. The importance of these quantities is that they are *conserved*. That is, in quite general circumstances they remain constant. That conserved quantities exist not only gives us a deeper insight into the nature of the world, but also gives us another way to approach practical problems. We still consider only translational motion, without rotation, in this chapter.

The conservation laws of energy and momentum are especially valuable in dealing with systems of many objects in which a detailed consideration of the forces involved would be difficult.

This chapter is devoted to the very important concept of *energy* and the closely related concept of *work*, which are scalar quantities and thus have no direction associated with them. Since these two quantities are scalars, they are often easier to deal with than are vector quantities such as force and acceleration. Energy derives its importance from two sources. First, it is a conserved quantity. Second, energy is a concept that is useful not only in the study of motion, but in all areas of physics and other sciences as well. But before discussing energy itself, we first examine the concept of work.

6–1 Work Done by a Constant Force

The word *work* has a variety of meanings in everyday language. But in physics, work is given a very specific meaning to describe what is accomplished by the action of a force when it acts on an object as the object moves through a distance. Specifically, the **work** done on an object by a constant force (constant in both magnitude and direction) is defined to be *the product of the magnitude of the displacement times the component of the force parallel to the displacement*. In equation form, we can write

$$W = F_{\parallel}d$$

where $F_{\parallel}$ is the component of the constant force **F** parallel to the displacement **d**. We can also write

Work defined

$$W = Fd \cos \theta, \tag{6–1}$$

where F is the magnitude of the constant force, d is the magnitude of the displacement of the object, and θ is the angle between the directions of the force and the displacement. The $\cos \theta$ factor appears in Eq. 6–1 because $F \cos \theta \, (= F_{\parallel})$ is the component of **F** parallel to **d** (Fig. 6–1). Work is a scalar quantity—it has only magnitude.

Let's first consider the case in which the motion and the force are in the same direction, so $\theta = 0$ and $\cos \theta = 1$, and then $W = Fd$. For example, if you push a loaded grocery cart a distance of 50 m by exerting a horizontal force of 30 N on the cart, you do 30 N $\times$ 50 m = 1500 N·m of work on the cart.

Units for work: the joule

As this example shows, in SI units, work is measured in newton-meters. A special name is given to this unit, the **joule** (J): 1 J = 1 N·m. In the cgs system, the unit of work is called the *erg* and is defined as 1 erg = 1 dyne·cm. In British units, work is measured in foot-pounds. It is easy to show that 1 J = 10^7 erg = 0.7376 ft·lb.

Force without work

A force can be exerted on an object and yet do no work. For example, if you hold a heavy bag of groceries in your hands at rest, you do no work on it. A force is exerted, but the displacement is zero, so the work $W = 0$. You also do no work on the bag of groceries if you carry it as you walk

FIGURE 6–1 A person pulling a crate along the floor. The work done by the force **F** is $W = Fd \cos \theta$, where **d** is the displacement.

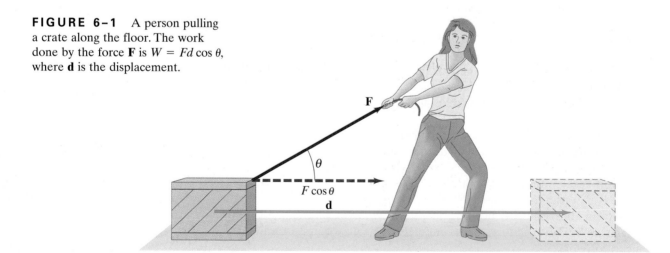

horizontally across the floor at constant velocity, as shown in Fig. 6–2. No horizontal force is required to move the package at a constant velocity. However, you do exert an upward force **F** on the package equal to its weight. But this upward force is perpendicular to the horizontal motion of the package and thus has nothing to do with that motion. Hence, the upward force is doing no work. This conclusion comes from our definition of work, Eq. 6–1: $W = 0$, because $\theta = 90°$ and $\cos 90° = 0$. Thus, when a particular force is perpendicular to the motion, no work is done by that force. (When you start or stop walking, there is a horizontal acceleration and you do briefly exert a horizontal force, and thus do work.)

When dealing with work, as with force, it is necessary to specify whether you are talking about work done *by* a specific object or done *on* a specific object. It is also important to specify whether the work done is due to one particular force (and which one), or work done by the total *net force* on the object.

FIGURE 6–2 Work done on the bag of groceries in this case is zero since **F** is perpendicular to the displacement **d**.

EXAMPLE 6–1 **Work done on a crate.** A 50-kg crate is pulled 40 m along a horizontal floor by a constant force exerted by a person, $F_P = 100$ N, which acts at a 37° angle as shown in Fig. 6–3. The floor is rough and exerts a friction force $F_{fr} = 50$ N. Determine the work done by each force acting on the crate, and the net work done on the crate.

SOLUTION We choose our coordinate system so that **x** can be the vector that represents the 40-m displacement (that is, along the x axis). There are four forces acting on the crate, as shown in Fig. 6–3: the force exerted by the person $\mathbf{F}_P$; the friction force $\mathbf{F}_{fr}$; the crate's weight $m\mathbf{g}$; and the normal force $\mathbf{F}_N$ exerted upward by the floor. The work done by the gravitational and normal forces is zero, since they are perpendicular to the displacement **x** ($\theta = 90°$ in Eq. 6–1):

$$W_G = mgx \cos 90° = 0$$

$$W_N = F_N x \cos 90° = 0.$$

The work done by $\mathbf{F}_P$ is

$$W_P = F_P x \cos \theta = (100 \text{ N})(40 \text{ m}) \cos 37° = 3200 \text{ J}.$$

The work done by the friction force is

$$W_{tr} = F_{fr} x \cos 180°$$

$$= (50 \text{ N})(40 \text{ m})(-1) = -2000 \text{ J}.$$

The angle between the displacement **x** and $\mathbf{F}_{fr}$ is 180° because they point in opposite directions. Since the force of friction is opposing the motion, it does *negative* work on the crate.

FIGURE 6–3 Example 6–1: 50-kg crate being pulled along a floor.

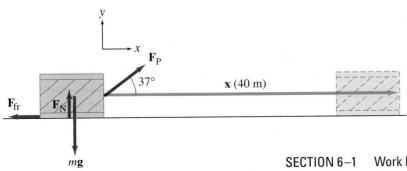

Finally, the net work can be calculated in two equivalent ways. (1) The net work done on an object is the algebraic sum of the work done by each force, since work is a scalar:

$$W_{net} = W_G + W_N + W_P + W_{fr}$$

$$= 0 + 0 + 3200 \text{ J} - 2000 \text{ J} = 1200 \text{ J}.$$

(2) The net work can also be calculated by first determining the net force on the object and then taking its component along the displacement: $(F_{net})_x = F_P \cos \theta - F_{fr}$. Then the net work is

$$W_{net} = (F_{net})_x x = (F_P \cos \theta - F_{fr})x$$

$$= (100 \text{ N} \cos 37° - 50 \text{ N})(40 \text{ m}) = 1200 \text{ J}.$$

In the vertical (y) direction, there is no displacement and no work done.

*W*_{net} *is the work done by all the forces acting on the body*

Negative work   In Example 6–1 we saw that friction did negative work. In general, the work done by a force is negative whenever the force (or the component of the force, $F_\parallel$) acts in the direction opposite to the direction of motion.

FIGURE 6–4 Example 6–2.

EXAMPLE 6–2 **Work on a backpack.** (*a*) Determine the work a hiker must do on a 15.0-kg backpack to carry it up a hill of height $h = 10.0$ m, as shown in Fig. 6–4a. Determine also (*b*) the work done by gravity on the backpack, and (*c*) the net work done on the backpack. For simplicity, assume the motion is smooth and at constant velocity (i.e., there is negligible acceleration).

SOLUTION (*a*) The forces on the backpack are shown in Fig. 6–4b: the force of gravity, $m\mathbf{g}$, acting downward; and $\mathbf{F}_H$, the force the hiker must exert upward to support the pack. Since we assume there is negligible acceleration, horizontal forces are negligible. In the vertical (y) direction, we choose up as positive. Newton's second law applied to the backpack gives

$$\Sigma F_y = ma_y$$

$$F_H - mg = 0.$$

Hence,

$$F_H = mg = (15.0 \text{ kg})(9.80 \text{ m/s}^2) = 147 \text{ N}.$$

To calculate the work done by the hiker on the backpack, Eq. 6–1 can be written

$$W_H = F_H(d \cos \theta),$$

and we note from Fig. 6–4a that $d \cos \theta = h$. So the work done by the hiker can be written:

$$W_H = F_H(d \cos \theta) = F_H h = mgh$$

$$= (147 \text{ N})(10.0 \text{ m}) = 1470 \text{ J}.$$

Note that the work done depends only on the change in elevation and not on the angle of the hill, θ. The same work would be done to lift the pack vertically the same height h.

(b) The work done by gravity is (from Eq. 6–1 and Fig. 6–4c):

$$W_G = (F_G)(d) \cos (180° - \theta).$$

Since $\cos (180° - \theta) = -\cos \theta$, we have

$$W_G = (F_G)(d)(-\cos \theta) = mg(-d \cos \theta)$$

$$= -mgh$$

$$= -(15.0 \text{ kg})(9.80 \text{ m/s}^2)(10.0 \text{ m}) = -1470 \text{ J}.$$

Note that the work done by gravity doesn't depend on the angle of the incline but only on the vertical height h of the hill. This is because gravity does work only in the vertical direction. We will make use of this important result later.

Work done by gravity depends on the height of the hill and not on the angle of incline

(c) The *net* work done on the backpack is $W_{net} = 0$, since the net force on the backpack is zero (it is assumed not to accelerate significantly). We can also determine the net work done by writing

$$W_{net} = W_G + W_H = -1470 \text{ J} + 1470 \text{ J} = 0$$

which is, as it should be, the same result.

Note in this example that even though the *net* work on the backpack is zero, the hiker nonetheless does do work on the backpack equal to 1470 J.

Hiker does work on pack, but the net work = 0

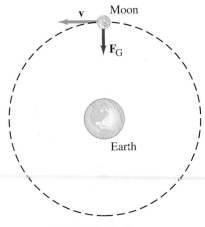

CONCEPTUAL EXAMPLE 6–3 **Does Earth do work on the Moon?** The Moon revolves around the Earth in a circular orbit, kept there by the gravitational force exerted by the Earth. Does gravity do (a) positive work, (b) negative work, or (c) no work at all on the Moon?

RESPONSE The gravitational force on the Moon (Fig. 6–5) acts toward the Earth (as a centripetal force), inward along the radius of the Moon's orbit. The Moon's displacement at any moment is along the circle, in the direction of its velocity, perpendicular to the radius and perpendicular to the force of gravity. Hence the angle θ between the force and the instantaneous displacement of the Moon is 90°, and the work done by gravity is therefore zero ($\cos 90° = 0$).

FIGURE 6–5
Conceptual Example 6–3.

➡ **PROBLEM SOLVING** Work

1. Choose an *xy* coordinate system. If the body is in motion, it may be convenient to choose the direction of motion as one of the coordinate directions. [Thus, for an object on an incline, you might choose one coordinate axis to be parallel to the incline.]

2. Draw a free-body diagram showing all the forces acting on the body.

3. Determine any unknown forces using Newton's laws.

4. Find the work done *by* a specific force *on* the body by using $W = Fd \cos \theta$. Note that the work done is negative when a force tends to oppose the displacement.

5. To find the *net* work done on the body, either (a) find the work done by each force and add the results algebraically; or (b) find the net force on the object, F_{net}, and then use it to find the net work done:

$$W_{net} = F_{net} \, d \cos \theta.$$

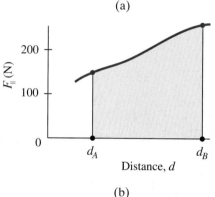

FIGURE 6–6 The work done by a force F can be calculated by taking: (a) the sum of the areas of the rectangles; (b) the area under the curve of $F_\parallel$ vs. d.

Work Done by a Varying Force

If the force acting on an object is constant, the work done by that force can be calculated using Eq. 6–1. But in many cases, the force varies in magnitude or direction during a process. For example, as a rocket moves away from Earth, work is done to overcome the force of gravity, which varies as the inverse square of the distance from the Earth's center. Other examples are the force exerted by a spring, which increases with the amount of stretch, or the work done by a varying force in pulling a box or cart up an uneven hill.

The work done by a varying force can be determined graphically. The procedure is like that for determining displacement when the velocity is known as a function of time (Section 2–8). To determine the work done by a variable force, we plot $F_\parallel(= F\cos\theta$, the component of **F** parallel to the direction of motion at any point) as a function of distance d, as in Fig. 6–6a. We divide the distance into small segments Δd. For each segment, we indicate the average of $F_\parallel$ by a horizontal dashed line. Then the work done for each segment is $\Delta W = F_\parallel \Delta d$, which is the area of a rectangle (Δd) wide and ($F_\parallel$) high. The total work done to move the object a total distance $d = d_B - d_A$ is the sum of the areas of the rectangles (five in the case shown in Fig. 6–6a). Usually, the average value of ($F_\parallel$) for each segment must be estimated, and a reasonable approximation of the work done can then be made. If we subdivide the distance into many more segments, Δd can be made smaller and our estimate of the work done more accurate. In the limit as Δd approaches zero, the total area of the many narrow rectangles approaches the area under the curve, Fig. 6–6b. That is, *the work done by a variable force in moving an object between two points is equal to the area under the $F_\parallel$ vs. d curve between those two points.*

6-3 Kinetic Energy, and the Work-Energy Principle

Energy is one of the most important concepts in science. Yet we cannot give a simple general definition of energy in only a few words. Nonetheless, each specific type of energy can be defined fairly simply. In this chapter, we define translational kinetic energy and potential energy. In later chapters, we will examine other types of energy, such as that related to heat (Chapters 14 and 15). The crucial aspect of all the types of energy is that the sum of all types, the *total energy*, remains the same after any process occurs as it was before: that is, the quantity "energy" can be defined so that it is a conserved quantity. More on this shortly.

For the purposes of this chapter, we can define energy in the traditional way as "the ability to do work." This simple definition is not very precise, nor is it really valid for all types of energy.[†] However, for mechanical energy which we discuss in this chapter, it serves to underscore the fundamental connection between work and energy. We now define and discuss one of the basic types of energy, kinetic energy.

A moving object can do work on another object it strikes. A flying cannonball does work on a brick wall it knocks down; a moving hammer does work on a nail it strikes. In either case, a moving object exerts a force on a

[†]Energy associated with heat is often not available to do work, as we will discuss in detail in Chapter 15.

second object and moves it through a distance. An object in motion has the ability to do work and thus can be said to have energy. The energy of motion is called **kinetic energy**, from the Greek word *kinetikos*, meaning "motion."

In order to obtain a quantitative definition for kinetic energy, let us consider an object of mass m that is moving in a straight line with an initial speed v_1. To accelerate it uniformly to a speed v_2, a constant net force F_{net} is exerted on it parallel to its motion over a distance d, Fig. 6–7. Then the net work done on the object is $W_{net} = F_{net} d$. We apply Newton's second law, $F_{net} = ma$, and use Eq. 2–10c, which we now write as $v_2^2 = v_1^2 + 2ad$, with v_1 as the initial speed and v_2 the final speed. We solve for a in Eq. 2–10c,

$$a = \frac{v_2^2 - v_1^2}{2d},$$

then substitute this into $F_{net} = ma$, and determine the work done:

$$W_{net} = F_{net} d = mad = m\left(\frac{v_2^2 - v_1^2}{2d}\right)d$$

or

$$W_{net} = \tfrac{1}{2} mv_2^2 - \tfrac{1}{2} mv_1^2. \tag{6–2}$$

We *define* the quantity $\tfrac{1}{2}mv^2$ to be the **translational kinetic energy** (KE) of the object:

$$KE = \tfrac{1}{2} mv^2. \tag{6–3}$$ *Kinetic energy defined*

(We call this "translational" kinetic energy to distinguish it from rotational kinetic energy, which we will discuss later, in Chapter 8.) Equation 6–2, derived here for one-dimensional motion, is valid in general for translational motion in three dimensions and even if the force varies. We can rewrite Eq. 6–2 as:

$$W_{net} = KE_2 - KE_1$$

or

$$W_{net} = \Delta KE. \tag{6–4}$$ WORK-ENERGY PRINCIPLE

Equation 6–4 (or Eq. 6–2) is an important result. It can be stated in words:

The net work done on an object is equal to the change in its kinetic energy.   WORK-ENERGY PRINCIPLE

This is known as the **work-energy principle**. Notice, however, that we made use of Newton's second law, $F_{net} = ma$, where F_{net} is the *net* force—the sum of all forces acting on the object. Thus, the work-energy principle is valid only if W is the *net work* done on the object—that is, the work done by all forces acting on the object.

The work-energy principle tells us that if (positive) net work W is done on a body, its kinetic energy increases by an amount W. The principle

FIGURE 6–8 A moving hammer strikes a nail and comes to rest. The hammer exerts a force F on the nail; the nail exerts a force $-F$ on the hammer (Newton's third law). The work done on the nail is positive ($W_n = Fd > 0$). The work done on the hammer is negative ($W_h = -Fd$).

Work-energy principle summarized

Energy units: the joule

also holds true for the reverse situation: if negative net work W is done on the body, the body's kinetic energy decreases by an amount W. That is, a net force exerted on a body opposite to the body's direction of motion reduces its speed and its kinetic energy. An example is a moving hammer (Fig. 6–8) striking a nail. The net force on the hammer ($-F$ in the figure, where F is assumed constant for simplicity) acts toward the left, whereas the displacement d is toward the right. So the net work done on the hammer, $W_h = (F)(d)(\cos 180°) = -Fd$, is negative and the hammer's kinetic energy decreases (usually to zero). Also note in this example that the hammer, as it slows down, does positive work on the nail: if the nail exerts a force $-F$ on the hammer to slow it down, the hammer exerts a force $+F$ on the nail (Newton's third law) through the distance d. Hence the net work done on the nail is $W_n = (+F)(+d) = Fd = -W_h$, and W_n is positive. Thus the decrease in kinetic energy of the hammer is also equal to the work the hammer can do on another object—which is consistent with energy being the ability to do work.

Note that whereas the translational kinetic energy ($=\frac{1}{2}mv^2$) is directly proportional to the mass of the object, it is proportional to the *square* of the speed. Thus, if the mass is doubled, the kinetic energy is doubled. But if the speed is doubled, the object has four times as much kinetic energy and is therefore capable of doing four times as much work.

To summarize, the connection between work and kinetic energy (Eq. 6–4) operates both ways. If the net work W done on an object is positive, then the object's kinetic energy increases. If the net work W done on an object is negative, its kinetic energy decreases. If the net work done on the object is zero, its kinetic energy remains constant (which also means its speed is constant).

Because of the direct connection between work and kinetic energy (Eq. 6–4), energy is measured in the same units as work: joules in SI units, ergs in the cgs, and foot-pounds in the British system. Like work, kinetic energy is a scalar quantity. The kinetic energy of a group of objects is the (algebraic) sum of the kinetic energies of the individual objects.

EXAMPLE 6–4 **KE and work done on a baseball.** A 145-g baseball is thrown with a speed of 25 m/s. (*a*) What is its kinetic energy? (*b*) How much work was done on the ball to make it reach this speed, if it started from rest?

SOLUTION (*a*) The kinetic energy is

$$\text{KE} = \tfrac{1}{2}mv^2 = \tfrac{1}{2}(0.145 \text{ kg})(25 \text{ m/s})^2 = 45 \text{ J}.$$

(*b*) Since the initial kinetic energy was zero, the net work done is just equal to the final kinetic energy, 45 J.

FIGURE 6–9 Example 6–5.

$v_1 = 20$ m/s $\qquad$ $v_2 = 30$ m/s

EXAMPLE 6–5 **Work on a car, to increase its KE.** How much work is required to accelerate a 1000-kg car from 20 m/s to 30 m/s (Fig. 6–9)?

SOLUTION The net work needed is equal to the increase in kinetic energy:

$$W = \text{KE}_2 - \text{KE}_1$$
$$= \tfrac{1}{2}mv_2^2 - \tfrac{1}{2}mv_1^2$$
$$= \tfrac{1}{2}(1000 \text{ kg})(30 \text{ m/s})^2 - \tfrac{1}{2}(1000 \text{ kg})(20 \text{ m/s})^2$$
$$= 2.5 \times 10^5 \text{ J}.$$

FIGURE 6–10
Conceptual Example 6–6.

CONCEPTUAL EXAMPLE 6–6 | **Work to stop a car.** An automobile traveling 60 km/h can brake to a stop within a distance of 20 m (Fig. 6–10a). If the car is going twice as fast, 120 km/h, what is its stopping distance (Fig. 6–10b)? The maximum braking force is approximately independent of speed.

RESPONSE Since the stopping force F is approximately constant, the work needed to stop the car, Fd, is proportional to the distance traveled. We apply the work-energy principle, noting that **F** and **d** are in opposite directions and that the final speed of the car is zero:

$$W_{net} = Fd \cos 180° = -Fd$$
$$= \Delta KE = 0 - \tfrac{1}{2}mv^2.$$

Thus, since the force and mass are constant, we can see that the stopping distance, d, increases with the square of the speed:

$$d \propto v^2.$$

Stopping distance ∝ initial speed squared

If the car's initial speed is doubled, the stopping distance is $(2)^2 = 4$ times as great, or 80 m.

6–4 Potential Energy

We have just discussed how an object is said to have energy by virtue of its motion, which we call kinetic energy. But it is also possible to have **potential energy**, which is the energy associated with forces that depend on the position or configuration of a body (or bodies) and the surroundings. Various types of potential energy (PE) can be defined, and each type is associated with a particular force.

Potential energy

A wound-up clock spring is an example of potential energy. The clock spring acquired its potential energy because work was done *on* it by the person winding the clock. As the spring unwinds, it exerts a force and does work to move the clock hands around.

Perhaps the most common example of potential energy is *gravitational potential energy*. A heavy brick held high in the air has potential energy because of its position relative to the Earth. It has the ability to do work, for if it is released, it will fall to the ground due to the gravitational force, and can do work on, say, a stake, driving it into the ground. Let us determine the gravitational potential energy of an object near the surface of the Earth. In order to

PE of gravity

FIGURE 6-11 A person exerts an upward force $F_{ext} = mg$ to lift a brick from y_1 to y_2.

lift an object of mass m vertically, an upward force at least equal to its weight, mg, must be exerted on it, say, by a person's hand. In order to lift it without acceleration a height h, from position y_1 to y_2 in Fig. 6-11 (upward direction chosen positive), a person must do work equal to the product of the needed external force, $F_{ext} = mg$ upward, and the vertical distance h. That is,

$$W_{ext} = F_{ext}\, d \cos 0° = mgh$$
$$= mg\,(y_2 - y_1). \qquad \text{(6-5a)}$$

Gravity is also acting on the object as it moves from y_1 to y_2, and does work on it equal to

$$W_G = F_G d \cos \theta = mgh \cos 180°,$$

where $\theta = 180°$ because $\mathbf{F}_G$ and $\mathbf{d}$ point in opposite directions. So

$$W_G = -mgh$$
$$= -mg(y_2 - y_1). \qquad \text{(6-5b)}$$

If we now allow the object to start from rest and fall freely under the action of gravity, it acquires a velocity given by $v^2 = 2gh$ (Eq. 2-10c) after falling a height h. It then has kinetic energy $\frac{1}{2}mv^2 = \frac{1}{2}m(2gh) = mgh$, and if it strikes a stake it can do work on the stake equal to mgh (work-energy principle). Thus, to raise an object of mass m to a height h *requires* an amount of work equal to mgh (Eq. 6-5a). And once at height h, the object has the *ability* to do an amount of work equal to mgh.

We therefore define the **gravitational potential energy** of a body as the product of its weight mg and its height y above some reference level (such as the ground):

Gravitational PE

$$\text{PE}_{grav} = mgy. \qquad \text{(6-6)}$$

The higher an object is above the ground, the more gravitational potential energy it has. We combine Eq. 6-5a with Eq. 6-6:

$$W_{ext} = mg\,(y_2 - y_1)$$
$$W_{ext} = \text{PE}_2 - \text{PE}_1 = \Delta\text{PE}. \qquad \text{(6-7a)}$$

That is, the work done by an external force to move the mass m from point 1 to point 2 (without acceleration) is equal to the change in potential energy of the object between points 1 and 2.

We can also write ΔPE in terms of the work done by gravity itself, starting from Eq. 6-5b, and obtain

$$W_G = -mg\,(y_2 - y_1)$$
$$W_G = -\Delta\text{PE}. \qquad \text{(6-7b)}$$

That is, the work done by gravity as the mass m moves from point 1 to point 2 is equal to the negative of the difference in potential energy between points 1 and 2.

Notice that the gravitational potential energy depends on the *vertical height* of the object *above some reference level* (Eq. 6-6), and in some situations, you may wonder from what point to measure the height y. The gravitational potential energy of a book held high above a table, for example, depends on whether we measure y from the top of the table, from the floor, or from some other reference point. What is physically important in any situation is the *change* in potential energy, ΔPE, because that is what is related to the work done, Eqs. 6–7, and it is ΔPE that can be measured. We can thus choose to measure y from any reference point that is convenient, but we must choose it at the start and be consistent throughout a calculation. The *change* in potential energy between any two points does not depend on this choice.

Change in PE *is what is physically meaningful*

An important result we discussed earlier (see Example 6–2 and Fig. 6–4) is that, since the gravity force does work only in the vertical direction, the work done by gravity depends only on the vertical height h, and not on the path taken, whether it be purely vertical motion or, say, motion along an incline. Thus, from Eqs. 6–7, we see that changes in gravitational potential energy depend only on the change in vertical height and not on the path taken.

EXAMPLE 6–7 **Potential energy changes for a roller coaster.** A 1000-kg roller-coaster car moves from point A, Fig. 6–12, to point B and then to point C. (*a*) What is its gravitational potential energy at B and C relative to point A? That is, take $y = 0$ at point A. (*b*) What is the change in potential energy when it goes from B to C? (*c*) Repeat parts (*a*) and (*b*), but take the reference point ($y = 0$) to be at point C.

SOLUTION (*a*) We take upward as the positive direction, and measure the heights from point A, which means initially that the potential energy is zero. At point B, where $y_B = 10\,\text{m}$,

$$\text{PE}_B = mgy_B = (1000\ \text{kg})(9.8\ \text{m/s}^2)(10\ \text{m}) = 9.8 \times 10^4\ \text{J}.$$

At point C, $y_C = -15\,\text{m}$, since C is below A. Therefore,

$$\text{PE}_C = mgy_C = (1000\ \text{kg})(9.8\ \text{m/s}^2)(-15\ \text{m}) = -1.5 \times 10^5\ \text{J}.$$

(*b*) In going from B to C, the potential energy change ($\text{PE}_{\text{final}} - \text{PE}_{\text{initial}}$) is

$$\text{PE}_C - \text{PE}_B = (-1.5 \times 10^5\ \text{J}) - (9.8 \times 10^4\ \text{J})$$

$$= -2.5 \times 10^5\ \text{J}.$$

The gravitational potential energy decreases by $2.5 \times 10^5\,\text{J}$.
(*c*) In this instance, $y_A = +15\,\text{m}$ at point A, so the potential energy initially (at A) is equal to

$$\text{PE}_A = (1000\ \text{kg})(9.8\ \text{m/s}^2)(15\ \text{m}) = 1.5 \times 10^5\ \text{J}.$$

At B, $y_B = 25\,\text{m}$, so the potential energy is

$$\text{PE}_B = 2.5 \times 10^5\ \text{J}.$$

At C, $y_C = 0$, so the potential energy is zero. The change in potential energy going from B to C is

$$\text{PE}_C - \text{PE}_B = 0 - 2.5 \times 10^5\ \text{J} = -2.5 \times 10^5\ \text{J},$$

which is the same as in part (*b*).

FIGURE 6–12 Example 6–7.

FIGURE 6–13 A spring (a) can store energy (elastic PE) when compressed as in (b), and can do work when released, as in (c).

(a) (b) (c)

PE defined in general

There are other kinds of potential energy besides gravitational. Each form of potential energy is associated with a particular force, and can be defined analogously to gravitational potential energy. In general, the *change in potential energy associated with a particular force, is equal to the negative of the work done by that force if the object is moved from one point to a second point* (as in Eq. 6–7b for gravity). Alternatively, we can define the *change in potential energy as the work required of an external force to move the object without acceleration between the two points*, as in Eq. 6–7a.

PE of elastic spring

We now consider another type of potential energy, that associated with elastic materials. This includes a great variety of practical applications. To take a simple example, consider the simple coil spring shown in Fig. 6–13. The spring has potential energy when compressed (or stretched), for when it is released, it can do work on a ball as shown. For a person to hold a spring either stretched or compressed an amount x from its normal (unstretched) length requires a force F_P that is directly proportional to x. That is,

$$F_P = kx,$$

where k is a constant, called the *spring constant*, and is a measure of the stiffness of the particular spring. The spring itself exerts a force in the opposite direction (Fig. 6–14),

$$F_S = -kx. \tag{6–8}$$

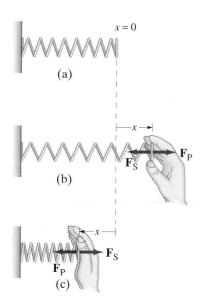

FIGURE 6–14 (a) Spring in normal (unstretched) position. (b) Spring is stretched by a person exerting a force $\mathbf{F}_P$ to the right (positive direction). The spring pulls back with a force $\mathbf{F}_S$ where $F_S = -kx$. (c) Person compresses the spring ($x < 0$), and the spring pushes back with a force $F_S = -kx$ where $F_S > 0$ because $x < 0$.

This force is sometimes called a "restoring force" because the spring exerts its force in the direction opposite the displacement (hence the minus sign), acting to return it to its normal length. Equation 6–8 is known as the **spring equation** and also as **Hooke's law** (see Chapter 9), and is accurate for springs as long as x is not too great.

In order to calculate the potential energy of a stretched spring, let us calculate the work required to stretch it (Fig. 6–14b). We might expect to use Eq. 6–1 for the work done on it, $W = Fx$, where x is the amount it is stretched from its normal length. But this would be incorrect since the force F_P ($= kx$) is not constant but varies over this distance, becoming greater the more the spring is stretched, as shown graphically in Fig. 6–15. So let us use the average force, $\bar{F}$. Since F_P varies linearly—from zero at the unstretched position to kx when stretched to x—the average force is $\bar{F} = \frac{1}{2}[0 + kx] = \frac{1}{2}kx$, where x here is the final amount stretched (shown as x_f in Fig. 6–15 for clarity). The work done is then

$$W = \bar{F}_P x = (\tfrac{1}{2}kx)(x) = \tfrac{1}{2}kx^2.$$

Hence the *elastic potential energy* is proportional to the square of the amount stretched[†]:

Elastic PE

$$\text{elastic PE} = \tfrac{1}{2}kx^2. \tag{6–9}$$

If a spring is *compressed* a distance x from its normal length, the force is still $F_P = kx$, and again the potential energy is given by this equation.

[†]We can also obtain Eq. 6–9 using Section 6–2. The work done, and hence ΔPE, equals the area under the F vs. x graph of Fig. 6–15. This area is a triangle (colored in Fig. 6–15) of altitude kx and base x, and hence of area (for a triangle) equal to $\frac{1}{2}(kx)(x) = \frac{1}{2}kx^2$.

Thus x can be either the amount compressed or amount stretched from the normal position.

In each of the above examples of potential energy—from a brick held at a height y, to a stretched or compressed spring—an object has the capacity or *potential* to do work even though it is not yet actually doing it. From these examples, we can also see that energy can be *stored*, for later use, in the form of potential energy (Fig. 6–13, for example, for a spring). It is also worth noting that although there is a single universal formula for the translational kinetic energy of an object, $\frac{1}{2}mv^2$, there is no single formula for potential energy. Instead, the mathematical form of the potential energy depends on the force involved.

Potential energy belongs to a system, and not to a single object alone. Potential energy is associated with a force, and a force on one object is always exerted by some other object. Thus potential energy is a property of the system as a whole. For a particle raised to a height y above the Earth's surface, the potential energy change is mgy. The system here is the particle plus the Earth, and properties of both are involved: particle (m) and Earth (g).

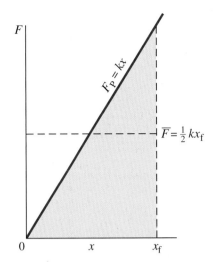

FIGURE 6–15 As a spring is stretched (or compressed), the force increases linearly as x increases: graph of $F = kx$ vs. x from $x = 0$ to $x = x_f$.

6–5 Conservative and Nonconservative Forces

The work done against gravity in moving an object from one point to another does not depend on the path taken. For example, it takes the same work ($= mgy$) to lift an object of mass m vertically a certain height as to carry it up an incline of the same vertical height, as in Fig. 6–4 (see Example 6–2). Forces such as gravity, for which the work done does not depend on the path taken but only on the initial and final positions, are called **conservative forces**. The elastic force of a spring (or other elastic material) in which $F = -kx$, is also a conservative force. Friction, on the other hand, is a **nonconservative** force since the work it does, for example, when a crate is moved across a floor from one point to another depends on whether the path taken is straight, or is curved or zigzag. As shown in Fig. 6–16, if a crate is pushed from point 1 to point 2 via the larger semicircular path rather than in the straight path, more work is done against friction, since the distance is greater and, unlike the gravitational force, the friction force is always directed opposite to the direction of motion. (The $\cos\theta$ term in Eq. 6–1 is always $\cos 180° = -1$ at all points on the path for the friction force.) Other forces that are nonconservative include the force exerted by a person and tension in a rope (see Table 6–1).

TABLE 6–1 Conservative and Nonconservative Forces

Conservative Forces	Nonconservative Forces
Gravitational	Friction
Elastic	Air resistance
Electric	Tension in cord
	Motor or rocket propulsion
	Push or pull by person

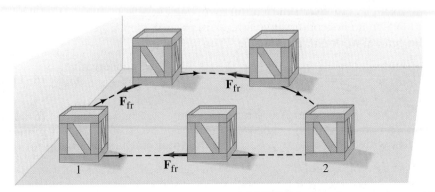

FIGURE 6–16 A crate is pulled across the floor from position 1 to position 2 via two paths, one straight and one curved. The friction force is always in the direction exactly opposed to the direction of motion. Hence, for a constant magnitude friction force, $W_{fr} = -F_{fr}d$, so if d is greater (as for the curved path), then W is greater.

Because potential energy is energy associated with the position or configuration of bodies, potential energy can only make sense if it can be stated uniquely for a given point. This cannot be done with nonconservative forces since the work done between two points depends not only on the two points but also on what path is taken (as in Fig. 6–16). Hence, *potential energy can be defined only for a conservative force.* Thus, although potential energy is always associated with a force, not all forces have a potential energy—for example, there is no potential energy for friction.

PE can be defined only for a conservative force

We can now extend the **work-energy principle** (discussed earlier in Section 6–3) to include potential energy. Suppose several forces act on an object which can undergo translational motion. And suppose only some of these forces are conservative, and we can write a potential-energy function for these conservative forces. We write the total (net) work W_{net} as a sum of the work done by conservative forces, W_C, and the work done by nonconservative forces, W_{NC}:

$$W_{net} = W_C + W_{NC}.$$

Then, from the work-energy principle, Eq. 6–4, we have

$$W_{net} = \Delta KE$$

$$W_C + W_{NC} = \Delta KE$$

where $\Delta KE = KE_2 - KE_1$, and so

$$W_{NC} = \Delta KE - W_C.$$

Work done by a conservative force can be written in terms of potential energy, as we saw in Eq. 6–7b for gravitational potential energy:

$$W_C = -\Delta PE.$$

We substitute this into the last equation above:

WORK-ENERGY PRINCIPLE (general form)

$$W_{NC} = \Delta KE + \Delta PE. \qquad \textbf{(6–10)}$$

Thus, *the work W_{NC} done by the nonconservative forces acting on an object is equal to the total change in kinetic and potential energy.*

It must be emphasized that *all* the forces acting on a body must be included in Eq. 6–10, either in the potential energy term on the right (if it is a conservative force), or in the work term, W_{NC}, on the left (but not in both!).

6–6 Mechanical Energy and Its Conservation

If only conservative forces are acting on a system, we arrive at a particularly simple and beautiful relation involving energy.

When no nonconservative forces are present, then $W_{NC} = 0$ in Eq. 6–10, the general work-energy principle. Then we have

$$\Delta KE + \Delta PE = 0 \qquad \begin{bmatrix} \text{conservative} \\ \text{forces only} \end{bmatrix} \qquad \textbf{(6–11a)}$$

or

$$(KE_2 - KE_1) + (PE_2 - PE_1) = 0. \qquad \begin{bmatrix} \text{conservative} \\ \text{forces only} \end{bmatrix} \qquad \textbf{(6–11b)}$$

We now define a quantity E, called the **total mechanical energy** of our system, as the sum of the kinetic and potential energies at any moment

$$E = \text{KE} + \text{PE}.$$

Total mechanical energy defined

Now we can rewrite Eq. 6–11b as

$$\text{KE}_2 + \text{PE}_2 = \text{KE}_1 + \text{PE}_1 \qquad \begin{bmatrix} \text{conservative} \\ \text{forces only} \end{bmatrix} \quad \textbf{(6–12a)}$$

or

$$E_2 = E_1 = \text{constant.} \qquad \begin{bmatrix} \text{conservative} \\ \text{forces only} \end{bmatrix} \quad \textbf{(6–12b)}$$

CONSERVATION OF MECHANICAL ENERGY

Equations 6–12 express a useful and profound principle regarding the total mechanical energy—namely, that it is a **conserved quantity**. The total mechanical energy E remains constant as long as no nonconservative forces act: (KE + PE) at some initial point 1 is equal to the (KE + PE) at any later point 2. To say it another way, consider Eq. 6–11a which tells us $\Delta \text{PE} = -\Delta \text{KE}$; that is, if the kinetic energy KE increases, then the potential energy PE must decrease by an equivalent amount to compensate. Thus, the total, KE + PE, remains constant. This is called the **principle of conservation of mechanical energy** for conservative forces:

> **If only conservative forces are acting, the total mechanical energy of a system neither increases nor decreases in any process. It stays constant—it is conserved.**

CONSERVATION OF MECHANICAL ENERGY

We now see the reason for the term "conservative force"—because for such forces, mechanical energy is conserved.

In the next Section we shall see the great usefulness of the conservation of mechanical energy principle in a variety of situations, and how it is often easier to use than the kinematic equations or Newton's laws. After that we will discuss how other forms of energy can be included in a grander conservation of energy law that includes energy associated with nonconservative forces.

6–7 Problem Solving Using Conservation of Mechanical Energy

A simple example of the conservation of mechanical energy is a rock allowed to fall from a height h under gravity (neglecting air resistance), as shown in Fig. 6–17. At the instant it is dropped, the rock, starting at rest, initially has only potential energy. As it falls, its potential energy decreases (because y decreases), but its kinetic energy increases to compensate, so that the sum of the two remains constant. At any point along the path, the total mechanical energy is given by

$$E = \text{KE} + \text{PE} = \tfrac{1}{2}mv^2 + mgy$$

where y is the rock's height above the ground at a given instant and v is its speed at that point. If we let the subscript 1 represent the rock at one

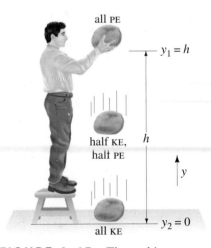

FIGURE 6–17 The rock's potential energy changes to kinetic energy as it falls.

point along its path (for example, the initial point), and 2 represent it at some other point, then we can write

total mechanical energy at point 1 = total mechanical energy at point 2

or (see also Eq. 6–12a)

Conservation of energy when only gravity acts

$$\tfrac{1}{2}mv_1^2 + mgy_1 = \tfrac{1}{2}mv_2^2 + mgy_2. \qquad \text{[grav. PE only]} \quad (6\text{–}13)$$

Just before the rock hits the ground, all of its initial potential energy will have been transformed into kinetic energy. We can see this from Eq. 6–13: initially (point 1), we set $y_1 = h$ and $v_1 = 0$ (the rock starts from rest). Just before hitting the ground (point 2), we have $y_2 = 0$, so from Eq. 6–13 we have

$$0 + mgh = \tfrac{1}{2}mv_2^2 + 0,$$

or $\text{KE}_2 = \tfrac{1}{2}mv_2^2 = mgh = \text{PE}_1$; the original potential energy has become kinetic energy.

EXAMPLE 6–8 **Falling rock.** If the original height of the stone in Fig. 6–17 is $y_1 = h = 3.0$ m, calculate the stone's speed when it has fallen to 1.0 m above the ground.

SOLUTION Since $v_1 = 0$ (the moment of release), $y_2 = 1.0$ m, and $g = 9.8$ m/s^2, Eq. 6–13 gives

$$\tfrac{1}{2}mv_1^2 + mgy_1 = \tfrac{1}{2}mv_2^2 + mgy_2$$
$$0 + (m)(9.8 \text{ m/s}^2)(3.0 \text{ m}) = \tfrac{1}{2}mv_2^2 + (m)(9.8 \text{ m/s}^2)(1.0 \text{ m}).$$

The m's cancel out, and solving for v_2^2 (which we see doesn't depend on m), we find

$$v_2^2 = 2[(9.8 \text{ m/s}^2)(3.0 \text{ m}) - (9.8 \text{ m/s}^2)(1.0 \text{ m})] = 39.2 \text{ m}^2/\text{s}^2,$$

and

$$v_2 = \sqrt{39.2} \text{ m/s} = 6.3 \text{ m/s}.$$

FIGURE 6–18 Energy buckets (for Example 6–8). Kinetic energy is red and potential energy is blue. The total (KE + PE) is the same for the three points shown. Note that the speed at $y = 0$, just before the rock hits, is $\sqrt{2(9.8 \text{ m/s}^2)(3.0 \text{ m})} = 7.7$ m/s.

$v = 0$ $y = 3.0$ m   PE

$v = 6.3$ m/s $y = 1.0$ m   KE PE

$v = 7.7$ m/s $y = 0$ y KE

A simple way to visualize energy conservation is with an "energy bucket" as shown in Fig. 6–18. At each point in the fall of the stone, the amount of kinetic energy and potential energy are shown as if they were two differently colored materials in the bucket. The total amount of material in the bucket (= total mechanical energy), remains constant.

Equation 6–13 can be applied to any object moving without friction under the action of gravity. For example, Fig. 6–19 shows a roller-coaster car starting from rest at the top of a hill, and coasting without friction to the bottom and up the hill on the other side.[†] Initially, the car has only potential energy. As it coasts down the hill, it loses potential energy and gains in kinetic energy, but the sum of the two remains constant. At the bottom of the hill it has its maximum kinetic energy, and as it climbs up the other side the kinetic energy changes back to potential energy. When the car comes to rest again, all of its energy will be potential energy. Given that the potential energy is proportional to the vertical height, energy conservation tells us that (in the

[†]The forces on the car are: gravity, the normal force exerted by the road, and friction (here assumed zero). The normal force acts perpendicular to the road, and so is always perpendicular to the motion and does no work. Thus $W_{\text{NC}} = 0$ in Eq. 6–10 (so mechanical energy is conserved) and we can use Eq. 6–13 with the potential energy being only gravitational potential energy. We will see how to deal with friction, for which $W_{\text{NC}} \neq 0$, in Section 6–9.

absence of friction) the car comes to rest at a height equal to its original height. If the two hills are the same height, the car will just barely reach the top of the second hill when it stops. If the second hill is lower than the first, not all of the car's kinetic energy will be transformed to potential energy and the car can continue over the top and down the other side. If the second hill is higher, the car will only reach a height on it equal to its original height on the first hill. This is true (in the absence of friction) no matter how steep the hill is, since potential energy depends only on the vertical height.

FIGURE 6–19 A roller-coaster car moving without friction illustrates the conservation of mechanical energy.

EXAMPLE 6–9 **Roller-coaster speed using energy conservation.** Assuming the height of the hill in Fig. 6–19 is 40 m, and the roller-coaster car starts from rest at the top, calculate (a) the speed of the roller-coaster car at the bottom of the hill, and (b) at what height it will have half this speed. Take $y = 0$ at the bottom of the hill.

SOLUTION (a) We use Eq. 6–13 with $v_1 = 0, y_1 = 40$ m, and $y_2 = 0$. Then

$$\tfrac{1}{2}mv_1^2 + mgy_1 = \tfrac{1}{2}mv_2^2 + mgy_2$$

$$0 + (m)(9.8 \text{ m/s}^2)(40 \text{ m}) = \tfrac{1}{2}mv_2^2 + 0.$$

The m's cancel out and we find $v_2 = \sqrt{2(9.8 \text{ m/s}^2)(40 \text{ m})} = 28$ m/s.
(b) We use the same equation, but now $v_2 = 14$ m/s (half of 28 m/s) and y_2 is unknown:

$$\tfrac{1}{2}mv_1^2 + mgy_1 = \tfrac{1}{2}mv_2^2 + mgy_2$$

$$0 + (m)(9.8 \text{ m/s}^2)(40 \text{ m}) = \tfrac{1}{2}(m)(14 \text{ m/s})^2 + (m)(9.8 \text{ m/s}^2)(y_2).$$

We cancel the m's and solve for y_2 and find $y_2 = 30$ m. That is, the car has a speed of 14 m/s when it is 30 *vertical* meters above the lowest point, both when descending the left-hand hill and when ascending the right-hand hill.

The mathematics of this Example is almost the same as that in Example 6–8. But there is an important difference between them. Example 6–8 could have been solved using force and acceleration. But here, where the motion is not vertical, using $F = ma$ would have been very difficult, whereas energy conservation readily gives us the answer.

CONCEPTUAL EXAMPLE 6–10 **Speeds on two water slides.** Two water slides at a pool are shaped differently, but have the same length and start at the same height h (Fig. 6–20). Two riders, Paul and Kathleen, start from rest at the same time on different slides. (a) Which rider, Paul or Kathleen, is traveling faster at the bottom? (b) Which rider makes it to the bottom first? Ignore friction.

RESPONSE (a) Each rider's initial potential energy mgh gets transformed to kinetic energy, so the speed v at the bottom is obtained from $\tfrac{1}{2}mv^2 = mgh$. The mass cancels in this equation and so the speed will be the same, regardless of the mass of the rider. Since they descend the same vertical height, they will finish with the same speed.
(b) Note that Kathleen is consistently at a lower elevation than Paul for the entire trip. This means she has converted her potential energy to kinetic energy earlier. Consequently, she is traveling faster than Paul for the whole trip, except toward the end where Paul finally gets up to the same speed. Since she was going faster for the whole trip, and the distance is the same, Kathleen gets to the bottom first.

FIGURE 6–20
Conceptual Example 6–10.

Paul PE KE

Kathleen PE KE

h

KE

FIGURE 6–21 Transformation of energy during a pole vault.

FIGURE 6–22 By bending their bodies, pole vaulters can keep their center of mass so low that it may even pass below the bar. By changing their kinetic energy (of running) into gravitational potential energy ($=mgy$) in this way, vaulters can cross over a higher bar than if the change in potential energy were accomplished without carefully bending the body.

There are many interesting examples of the conservation of energy in sports, one of which is the pole vault illustrated in Fig. 6–21. We often have to make approximations, but the sequence of events in broad outline for this case is as follows. The kinetic energy of the running athlete is transformed into elastic potential energy of the bending pole and, as the athlete leaves the ground, into gravitational potential energy. When the vaulter reaches the top and the pole has straightened out again, the energy has all been transformed into gravitational potential energy (if we ignore the vaulter's low horizontal speed over the bar). The pole does not supply any energy, but it acts as a device to *store* energy and thus aid in the transformation of kinetic energy into gravitational potential energy, which is the net result. The energy required to pass over the bar depends on how high the center of mass[†] (CM) of the vaulter must be raised. By bending their bodies, pole vaulters keep their CM so low that it can actually pass slightly beneath the bar (Fig. 6–22), thus enabling them to cross over a higher bar than would otherwise be possible.

EXAMPLE 6–11 ESTIMATE Pole vault. Estimate the kinetic energy and the speed required for a 70-kg pole vaulter to just pass over a bar 5.0 m high. Assume the vaulter's center of mass is initially 0.90 m off the ground and reaches its maximum height at the level of the bar itself.

SOLUTION We equate the total energy just before the vaulter places the end of the pole onto the ground (and the pole begins to bend and store potential energy) with the vaulter's total energy when passing over the bar (we ignore the small amount of kinetic energy at this point). We choose the initial position of the vaulter's center of mass to be $y_1 = 0$. The vaulter's body must then be raised to a height $y_2 = 5.0 \text{ m} - 0.9 \text{ m} = 4.1 \text{ m}$. Thus, using Eq. 6–13,

$$\tfrac{1}{2}mv_1^2 + 0 = 0 + mgy_2$$

and

$$\text{KE}_1 = \tfrac{1}{2}mv_1^2 = mgy_2$$
$$= (70 \text{ kg})(9.8 \text{ m/s}^2)(4.1 \text{ m}) = 2.8 \times 10^3 \text{ J}.$$

The speed is (solving for v_1 in $\text{KE}_1 = \tfrac{1}{2}mv_1^2$):

$$v_1 = \sqrt{\frac{2\text{KE}_1}{m}} = \sqrt{\frac{2(2800 \text{ J})}{70 \text{ kg}}} = 8.9 \text{ m/s}.$$

This is an approximation because we have ignored such things as the vaulter's speed while crossing over the bar, mechanical energy transformed when the pole is planted in the ground, and work done by the vaulter on the pole.

[†]The center of mass (CM) of a body is that point at which the entire mass of the body can be considered as concentrated, for purposes of describing its translational motion. (This is discussed in Chapter 7.) In Eq. 6–13, y is the position of the CM.

As another example of the conservation of mechanical energy, let us consider a mass m connected to a spring whose own mass can be neglected and whose stiffness constant is k. The mass m has speed v at any moment and the potential energy of the system is $\frac{1}{2}kx^2$, where x is the displacement of the spring from its unstretched length. If neither friction nor any other force is acting, the conservation-of-energy principle tells us that

$$\tfrac{1}{2}mv_1^2 + \tfrac{1}{2}kx_1^2 = \tfrac{1}{2}mv_2^2 + \tfrac{1}{2}kx_2^2 \qquad \text{[elastic PE only]} \quad \textbf{(6–14)}$$

Conservation of energy when PE is elastic

where the subscripts 1 and 2 refer to the velocity and displacement at two different points.

EXAMPLE 6–12   Toy dart gun. A dart of mass 0.100 kg is pressed against the spring of a toy dart gun as shown in Fig. 6–23. The spring (with spring constant $k = 250$ N/m) is compressed 6.0 cm and released. If the dart detaches from the spring when the latter reaches its normal length ($x = 0$), what speed does the dart acquire?

SOLUTION In the horizontal direction, the only force on the dart (neglecting friction) is the force exerted by the spring. Vertically, gravity is counterbalanced by the normal force exerted on the dart by the gun barrel. (After the dart leaves the barrel it will follow a projectile's path under gravity.) We use Eq. 6–14 with point 1 being at the maximum compression of the spring, so $v_1 = 0$ (dart not yet released) and $x_1 = -0.060$ m. Point 2 we choose to be the instant the dart flies off the end of the spring (Fig. 6–23b), so $x_2 = 0$ and we want to find v_2. Thus Eq. 6–14 can be written

$$0 + \tfrac{1}{2}kx_1^2 = \tfrac{1}{2}mv_2^2 + 0.$$

Then

$$v_2^2 = kx_1^2/m = (250 \text{ N/m})(0.060 \text{ m})^2/(0.100 \text{ kg}) = 9.0 \text{ m}^2/\text{s}^2$$

so $v_2 = \sqrt{v_2^2} = 3.0$ m/s.

(a)

(b)

|←6.0 cm→|

FIGURE 6–23 Example 6–12. (a) A dart is pushed against a spring, compressing it 6.0 cm. The dart is then released, and in (b) it leaves the spring at high velocity (v_2).

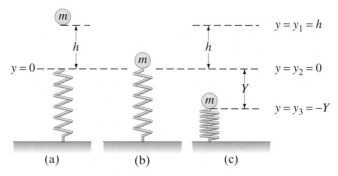

FIGURE 6–24
Example 6–13.

(a)   (b)   (c)

EXAMPLE 6–13 **Two kinds of PE.** A ball of mass $m = 2.60\,\text{kg}$, starting from rest, falls a vertical distance $h = 55.0\,\text{cm}$ before striking a vertical coiled spring, which it compresses (see Fig. 6–24) an amount $Y = 15.0\,\text{cm}$. Determine the spring constant of the spring. Assume the spring has negligible mass. Measure all distances from the point where the ball first touches the uncompressed spring ($y = 0$ at this point).

SOLUTION Since the motion is vertical, we use y instead of x (y positive upward). We divide this solution into two parts. (See also alternate solution below.)
Part 1: Let us first consider the energy changes of the ball as it falls from a height $y_1 = h = 0.55\,\text{m}$, Fig. 6–24a, to $y_2 = 0$, just as it touches the spring, Fig. 6–24b. Our system is the ball acted on by gravity (up to here the spring doesn't do anything) so

$$\tfrac{1}{2}mv_1^2 + mgy_1 = \tfrac{1}{2}mv_2^2 + mgy_2$$

$$0 + mgh = \tfrac{1}{2}mv_2^2 + 0$$

and $v_2 = \sqrt{2\,gh} = \sqrt{2(9.80\,\text{m/s}^2)(0.550\,\text{m})} = 3.28\,\text{m/s}$.
Part 2: Let's see what happens as the ball compresses the spring, Fig. 6–24b to c. Now there are two conservative forces on the ball—gravity and the spring force. So our energy equation becomes

$$E(\text{ball touches spring}) = E(\text{spring compressed})$$

Conservation of energy: gravity and elastic PE

$$\tfrac{1}{2}mv_2^2 + mgy_2 + \tfrac{1}{2}ky_2^2 = \tfrac{1}{2}mv_3^2 + mgy_3 + \tfrac{1}{2}ky_3^2.$$

We take point 2 to be the instant when the ball just touches the spring, so $y_2 = 0$ and $v_2 = 3.28\,\text{m/s}$. Point 3 we take to be when the ball comes to rest and the spring is fully compressed, so $v_3 = 0$ and $y_3 = -Y = -0.150\,\text{m}$ (given). Putting these into the above energy equation we get

$$\tfrac{1}{2}mv_2^2 + 0 + 0 = 0 - mgY + \tfrac{1}{2}kY^2.$$

We know m, v_2, and Y, so we can solve for k:

$$k = \frac{2}{Y^2}\left[\tfrac{1}{2}mv_2^2 + mgY\right]$$

$$= \frac{m}{Y^2}\left[v_2^2 + 2gY\right]$$

$$= \frac{(2.60\,\text{kg})}{(0.150\,\text{m})^2}\left[(3.28\,\text{m/s})^2 + 2(9.80\,\text{m/s}^2)(0.150\,\text{m})\right] = 1580\,\text{N/m}.$$

➡ **P R O B L E M   S O L V I N G**

Alternate Solution

Alternate solution: Instead of dividing the solution into two parts, we can do it all at once. After all, we get to choose what two points are used on the left and right of the energy equation. Let us write the energy equation for points 1 and 3 (Fig. 6–24). Point 1 is the initial point just before the ball starts to fall (Fig. 6–24a), so $v_1 = 0$, $y_1 = h = 0.550\,\text{m}$; and point 3 is when the spring is fully compressed (Fig. 6–24c), so $v_3 = 0$, $y_3 = -Y = -0.150\,\text{m}$. The forces on the ball in this process are gravity and (at least part of the time) the spring. So conservation of energy tells us

$$\tfrac{1}{2}mv_1^2 + mgy_1 + \tfrac{1}{2}k(0)^2 = \tfrac{1}{2}mv_3^2 + mgy_3 + \tfrac{1}{2}ky_3^2$$
$$0 + mgh + 0 = 0 - mgY + \tfrac{1}{2}kY^2$$

where we have set $y = 0$ for the spring at point 1 because it is not acting and is not compressed or stretched at this point. We solve for k:

$$k = \frac{2\,mg\,(h + Y)}{Y^2} = \frac{2(2.60\,\text{kg})(9.80\,\text{m/s}^2)(0.550\,\text{m} + 0.150\,\text{m})}{(0.150\,\text{m})^2} = 1580\,\text{N/m}$$

just as in our first method of solution.

EXAMPLE 6–14 **Bungee jump.** Dave jumps off a bridge with a bungee cord (a heavy stretchable cord) tied around his ankle (Fig. 6–25). He falls for 15 meters before the bungee cord begins to stretch. Dave's mass is 75 kg and we assume the cord obeys Hooke's law, $F = -kx$, with $k = 50\,\text{N/m}$. If we neglect air resistance, estimate how far below the bridge Dave will fall before coming to a stop. Ignore the mass of the cord (not realistic, however).

SOLUTION We first recognize that this problem is too hard to solve with the kinematic equations of Chapter 2. Those equations assumed a constant acceleration. But here the force the cord exerts on Dave gets stronger the farther he drops. But we can treat it readily using conservation of energy. Dave starts out with gravitational potential energy. The gravitational potential energy is tranformed into both kinetic energy and elastic potential energy. Assuming no frictional forces act on our system, the original total energy must be the same as the total energy at the end. If we define our coordinate system such that $y = 0$ at the lowest point in Dave's dive, and let the stretch of the cord at this point be represented by Δy, then the total fall is (see Fig. 6–25)

$$h = 15\,\text{m} + \Delta y.$$

Conservation of energy then gives:

$$\text{KE}_1 + \text{PE}_1 = \text{KE}_2 + \text{PE}_2$$
$$0 + mg(15\,\text{m} + \Delta y) = 0 + \tfrac{1}{2}k(\Delta y)^2.$$

To solve for Δy, we rewrite this in standard quadratic form ($ax^2 + bx + c = 0$):

$$(\Delta y)^2 - \frac{2\,mg}{k}\,\Delta y - \frac{2\,mg}{k}\,(15\,\text{m}) = 0.$$

Using the quadratic formula, $\Delta y = (-b \pm \sqrt{b^2 - 4ac})/2a$, with $a = 1$, $b = (-2\,mg/k) = -29\,\text{m}$, and $c = -2\,mg(15\,\text{m})/k = -440\,\text{m}^2$, we get two solutions:

$$\Delta y = 40\,\text{m} \quad \text{and} \quad \Delta y = -11\,\text{m}.$$

The negative solution is nonphysical, so the distance that Dave drops in his fall is:

$$h = 15\,\text{m} + 40\,\text{m} = 55\,\text{m}.$$

➡ **PHYSICS APPLIED**

Bungee jumping

FIGURE 6–25 Example 6–14. (a) Bungee jumper about to jump. (b) Bungee cord at its unstretched length. (c) Maximum stretch of cord.

6-8 Other Forms of Energy; Energy Transformations and the Law of Conservation of Energy

Besides the kinetic energy and potential energy of ordinary objects, other forms of energy can be defined as well. These include electric energy, nuclear energy, thermal energy, and the chemical energy stored in food and fuels. With the advent of the atomic theory, these other forms of energy have come to be considered as kinetic or potential energy at the atomic or molecular level. For example, according to the atomic theory, thermal energy is interpreted as the kinetic energy of rapidly moving molecules—when an object is heated, the molecules that make up the object move faster. On the other hand, the energy stored in food and fuel such as gasoline can be regarded as potential energy stored by virtue of the relative positions of the atoms within a molecule due to electric forces between the atoms (referred to as chemical bonds). For the energy in chemical bonds to be used to do work, it must be released, usually through chemical reactions. This is analogous to a compressed spring which, when released, can do work. Enzymes in our bodies allow the release of energy stored in food molecules. The violent spark of a spark plug in an automobile allows the mixture of gas and air to react chemically, releasing the stored energy which can then do work against the piston to propel the car forward. Electric, magnetic, and nuclear energy also can be considered examples of kinetic and potential (or stored) energy. We will deal with these other forms of energy in detail in later chapters.

Energy can be transformed from one form to another, and we have already encountered several examples of this. A stone held high in the air has potential energy; as it falls, it loses potential energy, since its height above the ground decreases. At the same time, it gains in kinetic energy, since its velocity is increasing. Potential energy is being transformed into kinetic energy.

Often the transformation of energy involves a transfer of energy from one body to another. The potential energy stored in the spring of Fig. 6–13b is transformed into kinetic energy of the ball, Fig. 6–13c. The kinetic energy of a running pole vaulter is transformed into elastic potential energy of the bending pole which in turn is transformed into the increasing potential energy of the rising athlete, Fig. 6–21. Water at the top of a dam has potential energy, which is transformed into kinetic energy as the water falls. At the base of the dam, the kinetic energy of the water can be transferred to turbine blades and further transformed into electric energy, as we shall see in a later chapter. The potential energy stored in a bent bow can be transformed into kinetic energy of the arrow (Fig. 6–26).

In each of these examples, the transfer of energy is accompanied by the performance of work. The spring of Fig. 6–13 does work on the ball. Water does work on turbine blades. A bow does work on an arrow. This observation gives us a further insight into the relation between work and energy: *work is done when energy is transferred from one object to another.*[†] A person throwing a ball or pushing a grocery cart provides another example.

FIGURE 6–26 Potential energy of a bent bow about to be transformed into kinetic energy of an arrow.

Work is done when energy is transferred

[†]If the objects are at different temperatures, heat can flow between them instead, or in addition. See Chapters 14 and 15.

The work done is a manifestation of energy being transferred from the person (ultimately derived from the chemical energy of food) to the ball or cart.

One of the great results of physics is that whenever energy is transferred or transformed, it is found that no energy is gained or lost in the process.

This is the **law of conservation of energy**, one of the most important principles in physics; it can be stated as:

The total energy is neither increased nor decreased in any process. Energy can be transformed from one form to another, and transferred from one body to another, but the total amount remains constant.

LAW OF CONSERVATION OF ENERGY

We have already discussed the conservation of energy for mechanical systems involving conservative forces, and we saw how it could be derived from Newton's laws and thus is equivalent to them. But in its full generality, the validity of the law of conservation of energy, encompassing all forms of energy including those associated with nonconservative forces like friction, rests on experimental observation. Even though Newton's laws are found to fail in the submicroscopic world of the atom, the law of conservation of energy has been found to hold in every experimental situation so far tested.

The law of conservation of energy is so wide-ranging that it plays an important role in other areas of physics (as we will see throughout this book), as well as in other sciences. We discuss another example of its usefulness in the next Section.

6-9 Energy Conservation with Dissipative Forces: Solving Problems

In our applications of energy conservation in Section 6–7, we neglected friction, a nonconservative force, but in many situations it cannot be ignored. In a real situation, the roller-coaster car in Fig. 6–19, for example, will not in fact reach the same height on the second hill as it had on the first hill because of friction. In this, and in other natural processes, the mechanical energy (sum of the kinetic and potential energies) does not remain constant but decreases. Because frictional forces reduce the total mechanical energy, they are called **dissipative forces**. Historically, the presence of dissipative forces hindered the formulation of a comprehensive conservation of energy law until well into the nineteenth century. It was not until then that heat, which is always produced when there is friction (try rubbing your hands together), was interpreted as a form of energy. Quantitative studies by nineteenth-century scientists (discussed in Chapters 14 and 15) demonstrated that if heat is considered as energy (properly called **thermal energy**), then the total energy is conserved in any process. For example, if the roller-coaster car in Fig. 6–19 is subject to frictional forces, then the initial total energy of the car will be equal to the kinetic plus potential energy of the car at any subsequent point along its path plus the amount of thermal energy produced in the process. The thermal energy produced by a constant friction force F_{fr} is equal to

Dissipative forces

the work done by this force. We now apply the general form of the work-energy principle, Eq. 6–10:

$$W_{NC} = \Delta KE + \Delta PE.$$

We can write $W_{NC} = -F_{fr}d$, where d is the distance over which the force acts. (**F** and **d** are in opposite directions, hence the minus sign.) Thus, with $KE = \frac{1}{2}mv^2$ and $PE = mgy$, we have

$$-F_{fr}d = \frac{1}{2}mv_2^2 - \frac{1}{2}mv_1^2 + mgy_2 - mgy_1$$

or

Conservation of energy with gravity and friction

$$\frac{1}{2}mv_1^2 + mgy_1 = \frac{1}{2}mv_2^2 + mgy_2 + F_{fr}d \quad \begin{bmatrix} \text{gravity and} \\ \text{friction acting} \end{bmatrix} \quad \textbf{(6–15)}$$

where d is the distance along the path traveled by the object in going from point 1 to point 2. Equation 6–15 can be seen to be Eq. 6–13 modified to include friction. It can be interpreted in a simple way: the initial mechanical energy of the car (point 1) equals the (reduced) final mechanical energy of the car plus the energy transformed by friction into thermal energy.

FIGURE 6–27 Example 6–15. Because of friction, a roller coaster car does not reach the original height on the second hill.

EXAMPLE 6–15 **Friction on the roller coaster.** The roller-coaster car in Example 6–9 is found to reach a vertical height of only 25 m on the second hill before coming to a stop (Fig. 6–27). It traveled a total distance of 400 m. Estimate the average friction force (assume constant) on the car, whose mass is 1000 kg.

SOLUTION We use conservation of energy, here in the form of Eq. 6–15, taking point 1 to be the instant when the car started coasting and point 2 to be the instant it stopped. Then $v_1 = 0$, $y_1 = 40\,\text{m}$, $v_2 = 0$, $y_2 = 25\,\text{m}$, and $d = 400\,\text{m}$. Thus

$$0 + (1000\,\text{kg})(9.8\,\text{m/s}^2)(40\,\text{m}) = 0 + (1000\,\text{kg})(9.8\,\text{m/s}^2)(25\,\text{m}) + F_{fr}(400\,\text{m}).$$

We solve this for F_{fr} to find $F_{fr} = 370\,\text{N}$.

Another example of the transformation of kinetic energy into thermal energy occurs when an object, such as the rock of Fig. 6–17, strikes the ground. Mechanical energy is not conserved at this encounter, but the total energy is. Kinetic energy is transformed into thermal energy (and probably some sound energy) and both the rock and the ground will be slightly warmer as a result of their collision. A more apparent example of this transformation of kinetic energy into thermal energy can be observed by vigorously striking a nail several times with a hammer, and then gently touching the nail with your finger to see how hot it is.

When other forms of energy are involved, such as chemical or electrical energy, the total amount of energy is always found to be conserved. Hence the law of conservation of energy is believed to be universally valid.

1. Draw a picture.
2. Determine the system for which energy will be conserved: the object or objects and the forces acting.
3. Ask yourself what quantity you are looking for, and decide what are the initial (point 1) and final (point 2) locations.
4. If the body under investigation changes its height during the problem, then choose a $y = 0$ level for gravitational potential energy. This may be chosen for convenience; the lowest point in the problem is often a good choice.
5. If springs are involved, choose the unstretched spring position to be x (or y) $= 0$.

6. If no friction or other nonconservative forces act, then apply conservation of mechanical energy:
$$KE_1 + PE_1 = KE_2 + PE_2,$$
7. Solve for the unknown quantity.
8. If friction or other nonconservative forces are present, then an additional term (W_{NC}) will be needed:
$$W_{NC} = \Delta KE + \Delta PE.$$
To be sure which sign to give W_{NC}, or on which side of the equation to put it, use your intuition: is the total mechanical energy increased or decreased in the process?

Problem solving is not a process that can be done by following a set of rules. The box above is thus not a prescription, but is a *summary* of steps to help you get started in solving problems involving energy.

6-10 Power

Average **power** is defined as the *rate at which work is done* (= work done divided by the time to do it), or as the *rate at which energy is transformed.* That is:

Power defined

$$\overline{P} = \text{average power} = \frac{\text{work}}{\text{time}} = \frac{\text{energy transformed}}{\text{time}}. \quad \textbf{(6-16)}$$

Average power

The power of a horse refers to how much work it can do per unit time. The power rating of an engine refers to how much chemical or electrical energy can be transformed into mechanical energy per unit time. In SI units, power is measured in joules per second, and this unit is given a special name, the **watt** (W): $1\ W = 1\ J/s$. We are most familiar with the watt for measuring the rate at which an electric lightbulb or heater changes electric energy into light or heat energy, but it is used for other types of energy transformations as well. In the British system, the unit of power is the foot-pound per second (ft·lb/s). For practical purposes, a larger unit is often used, the **horsepower**. One horsepower[†] (hp) is defined as 550 ft·lb/s, which equals 746 W.

Power units: the watt

It is very important to see the distinction between energy and power. To help make this distinction consider the following example. A person is limited in the work he or she can do, not only by the total energy required, but also by how fast this energy is transformed: that is, by power. For example, a person may be able to walk a long distance or climb many flights of stairs before having to stop because so much energy has been expended. On the other hand, a person who runs very quickly upstairs may fall exhausted after only a flight or two. He or she is limited in this case by power, the rate at which his or her body can transform chemical energy into mechanical energy.

Power and energy distinguished

[†]The unit was first chosen by James Watt (1736–1819), who needed a way to specify the power of his newly developed steam engines. He found by experiment that a good horse can work all day at an average rate of about 360 ft·lb/s. So as not to be accused of exaggeration in the sale of his steam engines, he multiplied this by $1\frac{1}{2}$ when he defined the hp.

EXAMPLE 6–16 **Stair-climbing power.** A 70-kg jogger runs up a long flight of stairs in 4.0 s. The vertical height of the stairs is 4.5 m. (*a*) Estimate the jogger's power output in watts and horsepower. (*b*) How much energy did this require?

SOLUTION (*a*) The work done is against gravity, and equals $W = mgy$. Then the average power output was

$$\overline{P} = \frac{W}{t} = \frac{mgy}{t} = \frac{(70 \text{ kg})(9.8 \text{ m/s}^2)(4.5 \text{ m})}{4.0 \text{ s}} = 770 \text{ W}.$$

Since there are 746 W in 1 hp, the jogger is doing work at a rate of just over 1 hp. It is worth noting that a human cannot do work at this rate for very long.

(*b*) The energy required is $E = \overline{P}t$ (Eq. 6–16). Since $\overline{P} = 770 \text{ W} = 770 \text{ J/s}$, then $E = (770 \text{ J/s})(4.0 \text{ s}) = 3100 \text{ J}$. [Note that the person had to transform more energy than this. The total energy transformed by a person or an engine always includes some thermal energy (recall how hot you get running up stairs).]

CONCEPTUAL EXAMPLE 6–17 **Energy of a powerful laser.** The Nova laser at Lawrence Livermore National Lab has ten beams, each of which has a power output greater than that of all the power plants in the United States. Where does this power come from?

RESPONSE Power companies do not really sell power. They sell energy. When you get your bill at the end of each month, you are charged for kilowatt-hours. A kilowatt is a unit of energy per time (power), so multiplying this by time will yield energy. Power stations are rated by power output, or how quickly they can deliver the energy. The Nova laser buys energy during the course of an afternoon and stores it. It then dumps the energy into the laser beams in a very short time interval, on the order of 1 ns (10^{-9} s). The *energy* delivered to the target is not so great, on the order of 10^5 J or what you get eating a doughnut. But it is delivered in so short a time that the power is extremely high, on the order of 10^{14} W.

Automobiles do work to overcome the force of friction (and air resistance), to climb hills, and to accelerate. A car is limited by the rate it can do work, which is why automobile engines are rated in horsepower. A car needs power most when it is climbing hills and when accelerating. In the next Example, we will calculate how much power is needed in these situations for a car of reasonable size. Even when a car travels on a level road at constant speed, it needs some power just to do work to overcome the retarding forces of internal friction and air resistance. These forces depend on the conditions and speed of the car, but are typically in the range 400 N to 1000 N.

It is often convenient to write the power in terms of the net force F applied to an object and its speed v. This is readily done since $\overline{P} = W/t$ and $W = Fd$ where d is the distance traveled. Then

$$\overline{P} = \frac{W}{t} = \frac{Fd}{t} = F\overline{v} \tag{6–17}$$

where $\overline{v} = d/t$ is the average speed of the object.

EXAMPLE 6–18 **Power needs of a car.** Calculate the power required of a 1400-kg car under the following circumstances: (*a*) the car climbs a 10° hill (a fairly steep hill) at a steady 80 km/h; and (*b*) the car accelerates along a level road from 90 to 110 km/h in 6.0 s to pass another car. Assume the retarding force on the car is $F_R = 700$ N throughout. See Fig. 6–28. (Be careful not to confuse $\mathbf{F_R}$, which is due to air resistance and friction that retards the motion, with the force $\mathbf{F}$ needed to accelerate the car, which is the frictional force exerted by the road on the tires—the reaction to the motor-driven tires pushing against the road.)

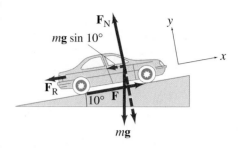

FIGURE 6–28 Example 6–18: calculation of power needed for a car (*a*) to climb a hill, (*b*) to pass another car.

SOLUTION (*a*) To move at a steady speed up the hill, the car must exert a force equal to the sum of the retarding force, 700 N, and the component of gravity parallel to the hill, $mg \sin 10° = (1400\,\text{kg})(9.80\,\text{m/s}^2)(0.174) = 2400$ N. Since $\bar{v} = 80$ km/h $= 22$ m/s and is parallel to $\mathbf{F}$, then (Eq. 6–17):

$$\overline{P} = F\bar{v}$$

$$= (2400\,\text{N} + 700\,\text{N})(22\,\text{m/s}) = 6.80 \times 10^4\,\text{W}$$

$$= 91\,\text{hp}.$$

(*b*) The car accelerates from 25.0 m/s to 30.6 m/s (90 to 110 km/h). Thus the car must exert a force that overcomes the 700 N retarding force plus that required to give it the acceleration $\bar{a}_x = (30.6\,\text{m/s} - 25.0\,\text{m/s})/6.0\,\text{s} = 0.93\,\text{m/s}^2$. We apply Newton's second law with x being the direction of motion:

$$ma_x = \Sigma F_x = F - F_R.$$

Then the force required, F, is

$$F = ma_x + F_R$$

$$= (1400\,\text{kg})(0.93\,\text{m/s}^2) + 700\,\text{N}$$

$$= 1300\,\text{N} + 700\,\text{N} = 2000\,\text{N}.$$

Since $\overline{P} = F\bar{v}$, the required power increases with speed and the motor must be able to provide a maximum power output of

$$\overline{P} = (2000\,\text{N})(30.6\,\text{m/s})$$

$$= 6.12 \times 10^4\,\text{W}$$

$$= 82\,\text{hp}.$$

Even taking into account the fact that only 60 to 80 percent of the engine's power output reaches the wheels, it is clear from these calculations that an engine of 100 to 150 hp is quite adequate from a practical point of view.

SUMMARY

Work is done on an object by a force when the object moves through a distance, d. If the direction of a constant force F makes an angle θ with the direction of motion, the work done by this force is

$$W = Fd \cos \theta.$$

Energy can be defined as the ability to do work. In SI units, work and energy are measured in **joules** ($1 \text{ J} = 1 \text{ N·m}$).

Kinetic energy (KE) is energy of motion. A body of mass m and speed v has translational kinetic energy

$$\text{KE} = \tfrac{1}{2} mv^2.$$

Potential energy (PE) is energy associated with forces that depend on the position or configuration of bodies. Gravitational potential energy is

$$\text{PE}_{\text{grav}} = mgy,$$

where y is the height of the object of mass m above an arbitrary reference point. Elastic potential energy is equal to $\tfrac{1}{2}kx^2$ for a stretched or compressed spring, where x is the displacement from the unstretched position. Other potential energies include chemical, electrical, and nuclear energy. The change in potential energy when an object changes position is equal to the external work needed to take the object from one position to the other.

The **work-energy principle** states that the *net* work done on a body (by the *net* force) equals the change in kinetic energy of that body:

$$W_{\text{net}} = \Delta \text{KE} = \tfrac{1}{2} mv_2^2 - \tfrac{1}{2} mv_1^2.$$

The **law of conservation of energy** states that energy can be transformed from one type to another, but the total energy remains constant. It is valid even when friction is present, since the heat generated can be considered a form of energy. In the absence of friction and other nonconservative forces, the total mechanical energy is conserved:

$$\text{KE} + \text{PE} = \text{constant}.$$

When nonconservative forces act, then

$$W_{\text{NC}} = \Delta \text{KE} + \Delta \text{PE},$$

where W_{NC} is the work done by nonconservative forces.

Power is defined as the rate at which work is done, or the rate at which energy is transformed. The SI unit of power is the **watt** ($1 \text{ W} = 1 \text{ J/s}$).

QUESTIONS

1. In what ways is the word "work" as used in everyday language the same as defined in physics? In what ways is it different?

2. Can a centripetal force ever do work on an object? Explain.

3. Can the normal force on an object ever do work? Explain.

4. A woman swimming upstream is not moving with respect to the shore. Is she doing any work? If she stops swimming and merely floats, is work done on her?

5. Is the work done by kinetic friction forces always negative? [*Hint:* Consider what happens to the dishes when pulling a tablecloth from under your mom's best china.]

6. Why is it tiring to push hard against a solid wall even though no work is done?

7. You have two springs that are identical except that spring 1 is stiffer than spring 2 ($k_1 > k_2$). On which spring is more work done (*a*) if they are stretched using the same force, (*b*) if they are stretched the same distance?

8. Can kinetic energy ever be negative? Explain.

9. A hand exerts a constant horizontal force on a block that is free to slide on a frictionless surface, as shown below (Fig. 6–29). The block starts from rest at point A, and by the time it has traveled a distance d to point B it is traveling with speed v_B. When the block has traveled another distance d to point C, will its speed be greater than, less than, or equal to $2v_B$? Explain your reasoning.

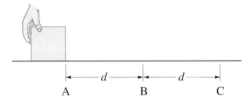

FIGURE 6–29 Question 9.

10. By approximately how much does your gravitational potential energy change when you jump as high as you can?

FIGURE 6–30 Question 11.

11. Describe precisely what is "wrong" physically in the famous Escher drawing shown in Fig. 6–30.

12. In Fig. 6–31, water balloons are tossed from the roof of a building, all with the same speed but with different launch angles. Which one has the highest speed on impact?

FIGURE 6–31 Question 12.

13. A pendulum is launched from a point that is a height h above its lowest point in 2 different ways (Fig. 6–32). During both launches, the bob is given an initial speed of 3.0 m/s. On the first launch, the initial velocity of the bob is directed upward along the trajectory, and on the second launch it is directed downward along the trajectory. Which launch will cause the pendulum to swing the largest angle from the equilibrium position?

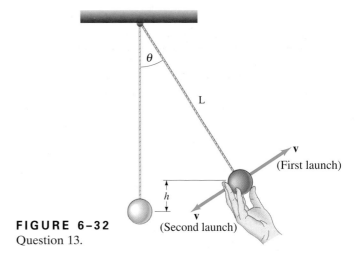

FIGURE 6–32 Question 13.

14. A coil spring of mass m rests upright on a table. If you compress the spring by pressing down with your hand and then release it, can the spring actually leave the table? Explain, using the law of conservation of energy.

15. In a well-known lecture demonstration, a bowling ball is hung from the ceiling by a steel wire (Fig. 6–33). The lecturer pulls the ball back and stands against the side wall of the lecture hall with the ball against his nose. To avoid injury the lecturer is supposed to release the ball, but not push it. Why?

FIGURE 6–33 Question 15.

16. What happens to the gravitational potential energy when water at the top of a waterfall falls to the pool below?

17. Describe the energy transformations when a child hops around on a pogo stick.

18. Describe the energy transformations that take place when a skier starts skiing down a hill, but after a time is brought to rest by striking a snowdrift.

19. A hill has a height h. A child on a sled (total mass m) slides down starting from rest at the top. Does the velocity at the bottom depend on the angle of the hill if (a) it is icy and there is no friction, and (b) there is friction (deep snow)?

20. Seasoned hikers prefer to step over a fallen log in their path rather than stepping on top and jumping down on the other side. Explain.

21. (*a*) Where does the kinetic energy come from when a car accelerates uniformly starting from rest? (*b*) How is the increase in kinetic energy related to the friction force the road exerts on the tires?

22. Two identical arrows, one with twice the speed of the other, are fired into a bale of hay. Assuming the hay exerts a constant frictional force on the arrows, the faster arrow will penetrate how much farther than the slower arrow? Explain.

23. Analyze the motion of a simple swinging pendulum in terms of energy, (*a*) ignoring friction, and (*b*) taking it into account. Explain why a grandfather clock has to be wound up.

24. When a "superball" is dropped, can it rebound to a height greater than its original height?

25. Suppose you lift a suitcase from the floor to a table. Does the work you do on the suitcase depend on (*a*) whether you lift it straight up or along a more complicated path, (*b*) the time it takes, (*c*) the height of the table, and (*d*) the weight of the suitcase?

26. Repeat the previous question for the *power* needed rather than the work.

27. Why is it easier to climb a mountain via a zigzag trail rather than to climb straight up?

28. Recall from Chapter 4, Example 4–12, that you can use a pulley and ropes to decrease the force needed to raise a heavy load (see Fig. 6–34). But for every meter the load is raised, how much rope must be pulled up? Account for this, using energy concepts.

FIGURE 6–34   Question 28.

PROBLEMS

SECTION 6–1

1. (I) A 75.0-kg firefighter climbs a flight of stairs 10.0 m high. How much work is required?

2. (I) A 900-N crate rests on the floor. How much work is required to move it at constant speed (*a*) 6.0 m along the floor against a friction force of 180 N, and (*b*) 6.0 m vertically?

3. (I) How much work did the movers do (horizontally) pushing a 150-kg crate 12.3 m across a rough floor without acceleration, if the effective coefficient of friction was 0.70?

4. (I) A car does 7.0×10^4 J of work in traveling 2.8 km at constant speed. What was the average retarding force (from all sources) acting on the car?

5. (I) How high will a 0.325-kg rock go if thrown straight up by someone who does 115 J of work on it? Neglect air resistance.

6. (I) A hammerhead with a mass of 2.0 kg is allowed to fall onto a nail from a height of 0.40 m. What is the maximum amount of work it could do on the nail? Why do people not just "let it fall" but add their own force to the hammer as it falls?

7. (II) What is the minimum work needed to push a 1000-kg car 300 m up a 17.5° incline? (*a*) Ignore friction. (*b*) Assume the effective coefficient of friction is 0.25.

8. (II) A grocery cart with mass of 18 kg is pushed at constant speed along an aisle by a force $F = 12$ N. The applied force acts at a 20° angle to the horizontal. Find the work done by each of the forces on the cart if the aisle is 15 m long.

9. (II) Eight books, each 4.6 cm thick with mass 1.8 kg, lie flat on a table. How much work is required to stack them one on top of another?

10. (II) A 280-kg piano slides 4.3 m down a 30° incline and is kept from accelerating by a man who is pushing back on it *parallel to the incline* (Fig. 6–35). The effective coefficient of kinetic friction is 0.40. Calculate: (*a*) the force exerted by the man, (*b*) the work done by the man on the piano, (*c*) the work done by the friction force, (*d*) the work done by the force of gravity, and (*e*) the net work done on the piano.

FIGURE 6–35
Problem 10.

11. (II) (a) Find the force required to give a helicopter of mass M an acceleration of $0.10\,g$ upward. (b) Find the work done by this force as the helicopter moves a distance h upward.

*SECTION 6-2

* 12. (II) In Fig. 6–6a, assume the distance axis is linear and that $d_A = 10.0\,\text{m}$ and $d_B = 35.0\,\text{m}$. Estimate the work done by this force in moving a 2.80-kg object from d_A to d_B.

* 13. (II) The x component of the force on an object varies as shown in Fig. 6–36. Determine the work done by this force to move the object (a) from $x = 0.0$ to $x = 10.0\,\text{m}$, and (b) from $x = 0.0$ to $x = 15.0\,\text{m}$.

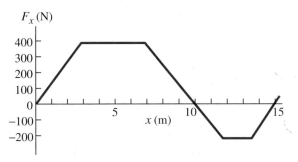

FIGURE 6-36   Problem 13.

* 14. (II) A spring has $k = 88\,\text{N/m}$. Use a graph to determine the work needed to stretch it from $x = 3.8\,\text{cm}$ to $x = 5.8\,\text{cm}$, where x is the displacement from its unstretched length.

* 15. (III) The x component of force exerted on a particle increases linearly from zero at $x = 0$, to $24.0\,\text{N}$ at $x = 3.0\,\text{m}$. It remains constant at $24.0\,\text{N}$ from $x = 3.0\,\text{m}$ to $x = 8.0\,\text{m}$, and then decreases linearly to zero at $x = 11.0\,\text{m}$. Determine the work done to move the particle from $x = 0$ to $x = 11.0\,\text{m}$ graphically by determining the area under the F_x vs. x graph.

* 16. (III) A 1300-kg space vehicle falls from a vertical height of 2500 km above the Earth's surface. Use Eq. 5–4 to estimate how much work is done by the force of gravity in bringing the vehicle to the Earth's surface. (First construct an F vs. r graph, where r is the distance from the Earth's center; then determine the work graphically.)

SECTION 6-3

17. (I) At room temperature, an oxygen molecule, with mass of $5.31 \times 10^{-26}\,\text{kg}$, typically has a KE of about $6.21 \times 10^{-21}\,\text{J}$. How fast is it moving?

18. (I) (a) If the KE of an arrow is doubled, by what factor has its speed increased? (b) If its speed is doubled, by what factor does its KE increase?

19. (I) How much work is required to stop an electron ($m = 9.11 \times 10^{-31}\,\text{kg}$) which is moving with a speed of $1.90 \times 10^{6}\,\text{m/s}$?

20. (I) How much work must be done to stop a 1000-kg car traveling at 110 km/h?

21. (II) An automobile is traveling along a highway at 90 km/h. If it travels instead at 100 km/h, what is the percent increase in the automobile's kinetic energy?

22. (II) An 80-g arrow is fired from a bow whose string exerts an average force of 95 N on the arrow over a distance of 80 cm. What is the speed of the arrow as it leaves the bow?

23. (II) A baseball ($m = 140\,\text{g}$) traveling 35 m/s moves a fielder's glove backward 25 cm when the ball is caught. What was the average force exerted by the ball on the glove?

24. (II) If the speed of a car is increased by 50%, by what factor will its minimum braking distance be increased, assuming all else is the same? Ignore the driver's reaction time.

25. (II) At an accident scene on a level road, investigators measure a car's skid mark to be 88 m long. It was a rainy day and the coefficient of friction was estimated to be 0.42. Use these data to determine the speed of the car when the driver slammed on (and locked) the brakes. (Why does the car's mass not matter?)

26. (II) A softball having a mass of 0.25 kg is pitched at 95 km/h. By the time it reaches the plate, it may have slowed by 10 percent. Neglecting gravity, estimate the average force of air resistance during a pitch, if the distance between the plate and the pitcher is about 15 m.

27. (III) One car has twice the mass of a second car, but only half as much kinetic energy. When both cars increase their speed by 5.0 m/s, they then have the same kinetic energy. What were the original speeds of the two cars?

28. (III) A 220-kg load is lifted 21.0 m vertically with an acceleration $a = 0.150\,g$ by a single cable. Determine (a) the tension in the cable, (b) the net work done on the load, (c) the work done by the cable on the load, (d) the work done by gravity on the load, and (e) the final speed of the load assuming it started from rest.

SECTIONS 6-4 AND 6-5

29. (I) A spring has a spring constant, k, of 440 N/m. How much must this spring be stretched to store 25 J of potential energy?

30. (I) A 6.0-kg monkey swings from one branch to another 1.2 m higher. What is the change in potential energy?

31. (I) By how much does the gravitational potential energy of a 64-kg pole vaulter change if his center of mass rises about 4.0 m during the jump?

32. (II) In starting an exercise, a 1.60-m tall person lifts a 2.10-kg book on the ground so it is 2.20 m above the ground. What is the potential energy of the book relative to (a) the ground, and (b) the top of the person's head? (c) How is the work done by the person related to the answers in parts (a) and (b)?

33. (II) A 55-kg hiker starts at an elevation of 1600 m and climbs to the top of a 3100-m peak. (a) What is the hiker's change in potential energy? (b) What is the minimum work required of the hiker? (c) Can the actual work done be more than this? Explain why.

34. (II) (a) A spring of spring constant k is initially compressed a distance x_0 from its unstretched length. What is the change in potential energy if it is then compressed to an amount x from its unstretched length? (b) Suppose the spring is then *stretched* a distance x_0 from the unstretched length. What is the change in potential energy as compared to when it is compressed by an amount x_0?

SECTIONS 6–6 AND 6–7

35. (I) Jane, looking for Tarzan, is running at top speed (5.6 m/s) and grabs a vine hanging vertically from a tall tree in the jungle. How high can she swing upward? Does the length of the vine (or rope) affect your answer?

36. (I) A novice skier, starting from rest, slides down a frictionless 35.0° incline whose vertical height is 125 m. How fast is she going when she reaches the bottom?

37. (I) A sled is initially given a shove up a frictionless 25.0° incline. It reaches a maximum vertical height 1.35 m higher than where it started. What was its initial speed?

38. (II) In the high jump, the kinetic energy of an athlete is transformed into gravitational potential energy without the aid of a pole. With what minimum speed must the athlete leave the ground in order to lift his center of mass 2.10 m and cross the bar with a speed of 0.70 m/s?

39. (II) A 75-kg trampoline artist jumps vertically upward from the top of a platform with a speed of 5.0 m/s. (a) How fast is he going as he lands on the trampoline, 3.0 m below (Fig. 6–37)? (b) If the trampoline behaves like a spring of spring constant 5.2×10^4 N/m, how far does he depress it?

FIGURE 6–37 Problem 39.

40. (II) A roller coaster, shown in Fig. 6–38, is pulled up to point A where it and its screaming occupants are released from rest. Assuming no friction, calculate the speed at points B, C, D.

FIGURE 6–38 Problems 40 and 53.

41. (II) A projectile is fired at an upward angle of 45.0° from the top of a 265-m cliff with a speed of 185 m/s. What will be its speed when it strikes the ground below? (Use conservation of energy.)

42. (II) A 60-kg bungee jumper jumps from a bridge. She is tied to a 12-m-long bungee cord and falls a total of 31 m. (a) Calculate the spring constant k of the bungee cord. (b) Calculate the maximum acceleration experienced by the jumper.

43. (II) A vertical spring (ignore its mass), whose spring constant is 900 N/m, is attached to a table and is compressed 0.150 m. (a) What speed can it give to a 0.300-kg ball when released? (b) How high above its original position (spring compressed) will the ball fly?

44. (II) A small mass m slides without friction along the looped apparatus shown in Fig. 6–39. If the object is to remain on the track, even at the top of the circle (whose radius is r), from what minimum height h must it be released?

FIGURE 6–39 Problems 44 and 79.

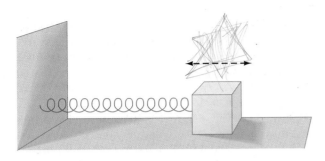

FIGURE 6–40   Problems 45, 55, 56, and 78.

45. (II) A mass m is attached to the end of a spring (constant k), Fig. 6–40. The mass is given an initial displacement x_0, after which it oscillates back and forth. Write a formula for the total mechanical energy (ignore friction and mass of the spring) in terms of position x and speed v.

46. (III) An engineer is designing a spring to be placed at the bottom of an elevator shaft. If the elevator cable should happen to break at a height h above the top of the spring, calculate the value that the spring constant k should have so that passengers undergo an acceleration of no more than $5.0\,g$ when brought to rest. Let M be the total mass of the elevator and passengers.

47. (III) What should be the spring constant k of a spring designed to bring a 1200-kg car to rest from a speed of 100 km/h so that the occupants undergo a maximum acceleration of $5.0\,g$?

48. (III) A cyclist intends to cycle up a 7.50° hill whose vertical height is 120 m. Assuming the mass of bicycle plus person is 75.0 kg, (a) calculate how much work must be done against gravity. (b) If each complete revolution of the pedals moves the bike 5.10 m along its path, calculate the average force that must be exerted on the pedals tangent to their circular path. Neglect work done by friction and other losses. The pedals turn in a circle of diameter 36.0 cm.

SECTIONS 6–8 AND 6–9

49. (I) Two railroad cars, each of mass 6500 kg and traveling 95 km/h, collide head-on and come to rest. How much thermal energy is produced in this collision?

50. (II) A 17-kg child descends a slide 3.5 m high and reaches the bottom with a speed of 2.3 m/s. How much thermal energy due to friction was generated in this process?

51. (II) A ski starts from rest and slides down a 20° incline 100 m long. (a) If the coefficient of friction is 0.090, what is the ski's speed at the base of the incline? (b) If the snow is level at the foot of the incline and has the same coefficient of friction, how far will the ski travel along the level? Use energy methods.

52. (II) A 90-kg crate, starting from rest, is pulled across a floor with a constant horizontal force of 350 N. For the first 15 m the floor is frictionless, and for the next 15 m the coefficient of friction is 0.30. What is the final speed of the crate?

53. (II) Suppose the roller coaster in Fig. 6–38 passes point A with a speed of 1.70 m/s. If the average force of friction is equal to one fifth of its weight, with what speed will it reach point B? The distance traveled is 45.0 m.

54. (II) A skier traveling 12.0 m/s reaches the foot of a steady upward 18° incline and glides 12.2 m up along this slope before coming to rest. What was the average coefficient of friction?

55. (III) A 0.520-kg wood block is firmly attached to a very light horizontal spring ($k = 180\,\text{N/m}$) as shown in Fig. 6–40. It is noted that the block-spring system, when compressed 5.0 cm and released, stretches out 2.3 cm beyond the equilibrium position before stopping and turning back. What is the coefficient of kinetic friction between the block and the table?

56. (III) A 180-g wood block is firmly attached to a very light horizontal spring, Fig. 6–40. The block can slide along a table where the coefficient of friction is 0.30. A force of 20 N compresses the spring 18 cm. If the spring is released from this position, how far beyond its equilibrium position will it stretch on its first swing?

57. (III) In early test flights for the space shuttle using a "glider" (mass of 1000 kg including pilot), it was noted that after a horizontal launch at 500 km/h at a height of 3500 m, the glider eventually landed at a speed of 200 km/h. (a) What would its landing speed have been in the absence of air resistance? (b) What was the average force of air resistance exerted on it if it came in at a constant glide of 10° to the Earth?

SECTION 6–10

58. (I) How long will it take a 1750-W motor to lift a 285-kg piano to a sixth-story window 16.0 m above?

59. (I) If a car generates 18 hp when traveling at a steady 90 km/h, what must be the average force exerted on the car due to friction and air resistance?

60. (I) (a) Show that a British horsepower (550 ft·lb/s) is equal to 746 W. (b) What is the horsepower rating of a 100 W lightbulb?

61. (II) Electric energy units are often expressed in the form of "kilowatt-hours." (a) Show that one kilowatt-hour (kWh) is equal to 3.6×10^6 J. (b) If the typical family of four in the United States uses electric energy at an average rate of 500 W, how many kWh would their electric bill be for one month, and (c) how many joules would this be? (d) At a cost of $0.12 per kWh, what would their monthly bill be in dollars? Does the monthly bill depend on the *rate* at which they use the electric energy?

62. (II) A driver notices that her 1000-kg car slows down from 90 km/h to 70 km/h in about 6.0 s on the level when it is in neutral. Approximately what power (watts and hp) is needed to keep the car traveling at a constant 80 km/h?

63. (II) How much work can a 3.0-hp motor do in 1.0 h?

64. (II) A shot-putter accelerates a 7.3-kg shot from rest to 14 m/s. If this motion takes 2.0 s, what average power was developed?

65. (II) A pump is to lift 8.00 kg of water per minute through a height of 3.50 m. What output rating (watts) should the pump motor have?

66. (II) During a workout, the football players at State U ran up the stadium stairs in 61 s. The stairs are 140 m long and inclined at an angle of 30°. If a typical player has a mass of 105 kg, estimate the average power output on the way up. Ignore friction and air resistance.

67. (II) How fast must a cyclist climb a 6.0° hill to maintain a power output of 0.25 hp? Neglect work done by friction and assume the mass of cyclist plus bicycle is 70 kg.

68. (II) A 1000-kg car has a maximum power output of 120 hp. How steep a hill can it climb at a constant speed of 70 km/h if the frictional forces add up to 600 N?

69. (II) Squaw Valley ski area in California claims that its lifts can move 47,000 people per hour. If the average lift carries people about 200 m (vertically) higher, estimate the maximum total power needed.

70. (III) A bicyclist coasts down a 7.0° hill at a steady speed of 5.0 m/s. Assuming a total mass of 75 kg (bicycle plus rider), what must be the cyclist's power output to climb the same hill at the same speed?

GENERAL PROBLEMS

71. A paratrooper fell 370 m after jumping from an aircraft without his parachute opening. He landed in a snowbank, creating a crater 1.1 m deep, but survived with only minor injuries. Assuming the paratrooper's mass was 80 kg and his terminal velocity was 30 m/s, estimate: (a) the work done by the snow in bringing him to rest; (b) the average force exerted on him by the snow to stop him; and (c) the work done on him by air resistance as he fell.

72. Designers of today's cars have built "5 mi/h (8 km/h) bumpers" that are designed to elastically compress and rebound without any physical damage at speeds below 8 km/h. If the material of the bumpers permanently deforms after a compression of 1.5 cm, but remains like an elastic spring up to that point, what must the effective spring constant of the bumper material be, assuming the car has a mass of 1400 kg and is tested by ramming into a solid wall?

73. In a certain library the first shelf is 10.0 cm off the ground, and the remaining 4 shelves are each spaced 30.0 cm above the previous one. If the average book has a mass of 1.5 kg with a height of 20 cm, and an average shelf holds 25 books, how much work is required to fill this bookshelf from scratch, assuming the books are all laying flat on the floor to start?

74. In a film of Jesse Owens's famous long jump in the 1936 Olympics, it is observed that his center of mass rose 1.1 m from launch point to the top of the arc. What minimum speed did he need at launch if he was also noted to be traveling at 6.5 m/s at the top of the arc?

75. A 0.20-kg pinecone falls from a branch 18 m above the ground. (a) With what speed would it hit the ground if air resistance could be ignored? (b) If it actually hits the ground with a speed of 10.0 m/s, what was the average force of air resistance exerted on it?

FIGURE 6-41   Problem 76.

76. A ball is attached to a horizontal cord of length L whose other end is fixed, Fig. 6–41. (a) If the ball is released, what will be its speed at the lowest point of its path? (b) A peg is located a distance h directly below the point of attachment of the cord. If $h = 0.80 L$, what will be the speed of the ball when it reaches the top of its circular path about the peg?

77. A 65-kg hiker climbs to the top of a 3900-m-high mountain. The climb is made in 5.0 h starting at an elevation of 2200 m. Calculate (a) the work done against gravity, (b) the average power output in watts and in horsepower, and (c) assuming the body is 15% efficient, what rate of energy input was required.

78. A mass m is attached to the end of a spring (constant k) as shown in Fig. 6–40. The mass is given an initial displacement x_0 from equilibrium, and an initial speed v_0. Ignoring friction and the mass of the spring, use energy methods to find (a) its maximum speed, and (b) its maximum stretch from equilibrium, in terms of the given quantities.

79. The small mass m sliding without friction along the looped apparatus shown in Fig. 6–39 is to remain on the track at all times, even at the very top of the loop of radius r. (a) Calculate, in terms of the given quantities, the minimum release height h (as in Problem 44). Next, if the actual release height is $2h$, calculate (b) the normal force exerted by the track at the bottom of the loop, (c) the normal force exerted by the track at the top of the loop, and (d) the normal force exerted by the track after the block exits the loop onto the flat section.

80. An elevator cable breaks when a 900-kg elevator is 30 m above a huge spring ($k = 4.0 \times 10^5$ N/m) at the bottom of the shaft. Calculate (a) the work done by gravity on the elevator before it hits the spring, (b) the speed of the elevator just before striking the spring, and (c) the amount the spring compresses (note that work is done by both the spring and gravity in this part).

81. Water flows over a dam at the rate of 550 kg/s and falls vertically 80 m before striking the turbine blades. Calculate: (a) the speed of the water just before striking the turbine blades (neglect air resistance), and (b) the rate at which mechanical energy is transferred to the turbine blades, assuming 60% efficiency.

82. A bicyclist of mass 75 kg (including the bicycle) can coast down a 4.0° hill at a steady speed of 10 km/h. Pumping hard, the cyclist can descend the hill at a speed of 30 km/h. Using the same power, at what speed can the cyclist climb the same hill? Assume the force of friction is proportional to the square of the speed v; that is, $F_{fr} = bv^2$, where b is a constant.

83. Show that on a roller coaster with a circular vertical loop (Fig. 6–42), the difference in your apparent weight at the top of the loop and the bottom of the loop is $6\,g$'s—that is, six times your weight. Ignore friction. Show also that as long as your speed is above the minimum needed, this answer doesn't depend on the size of the loop or how fast you go through it.

FIGURE 6–42 Problem 83.

84. If you stand on a bathroom scale, the spring inside the scale compresses 0.50 mm, and it tells you your weight is 700 N. Now if you jump on the scale from a height of 1.0 m, what does the scale read at its peak?

85. A 75-kg student runs at 5.0 m/s, grabs a rope, and swings out over a lake (Fig. 6–43). He releases the rope when his velocity is zero. (a) What is the angle θ when he releases the rope? (b) What is the tension in the rope just before he releases it? (c) What is the maximum tension in the rope?

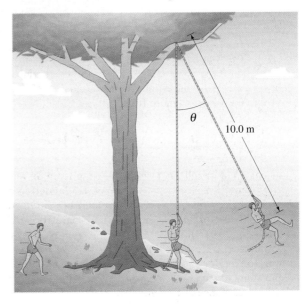

FIGURE 6–43 Problem 85.

86. In the rope climb, a 70-kg athlete climbs a vertical distance of 5.0 m in 9.0 s. What minimum power output was used to accomplish this feat?

The faster a bumper car goes, and the more mass it has, the more momentum it has. When two bumper cars collide, the collision may be nearly elastic: that is, momentum and kinetic energy are conserved.

$\mathbf{p} = m\mathbf{v}$

7 LINEAR MOMENTUM

The law of conservation of energy, which we discussed in the previous chapter, is one of several great conservation laws in physics. Among the other quantities found to be conserved are linear momentum, angular momentum, and electric charge. We will eventually discuss all of these because the conservation laws are among the most important in all of science. In this chapter, we discuss linear momentum and its conservation. We then make use of the laws of conservation of linear momentum and of energy to analyze collisions. Indeed, the law of conservation of momentum is particularly useful when dealing with two or more bodies that interact with each other, as in collisions.

Our focus up to now has been mainly on the motion of a single object, often thought of as a "particle" in the sense that we have ignored any rotation or internal motion. In this chapter we will deal with systems of two or more particles (and with extended bodies which can be considered as collections of particles). An important concept for this study is that of center of mass, which we discuss later in the chapter.

7–1 Momentum and Its Relation to Force

The **linear momentum** (or "momentum" for short) of a body is defined as the product of its mass and its velocity. Momentum (plural is momenta) is usually represented by the symbol **p**. If we let m represent the mass of a body and **v** represent its velocity, then its momentum **p** is

$$\mathbf{p} = m\mathbf{v}. \qquad\qquad (7\text{–}1) \qquad \textit{Linear momentum}$$

Since velocity is a vector, momentum is a vector. The direction of the momentum is the direction of the velocity, and the magnitude of the momentum is $p = mv$. Since **v** depends on the reference frame, this frame must be specified. The unit of momentum is simply that of mass $\times$ velocity, which in SI units is kg·m/s. There is no special name for this unit.

Everyday usage of the term *momentum* is in accord with the definition above. For according to Eq. 7–1, a fast-moving car has more momentum than a slow-moving car of the same mass, and a heavy truck has more momentum than a small car moving with the same speed. The more momentum an object has, the harder it is to stop it, and the greater effect it will have if it is brought to rest by impact or collision. A football player is more likely to be stunned if tackled by a heavy opponent running at top speed than by a lighter or slower-moving tackler. A heavy, fast-moving truck can do more damage than a slow-moving motorcycle.

A force is required to change the momentum of an object, whether it is to increase the momentum, to decrease it (such as to bring a moving object to rest), or to change its direction. Newton originally stated his second law in terms of momentum (although he called the product *mv* the "quantity of motion"). Newton's statement of the **second law of motion**, translated into modern language, is as follows:

The rate of change of momentum of a body is equal to the net force applied to it.

We can write this as an equation,

$$\Sigma\mathbf{F} = \frac{\Delta\mathbf{p}}{\Delta t}, \qquad\qquad (7\text{–}2) \qquad \boxed{\textit{NEWTON'S SECOND LAW}}$$

where $\Sigma\mathbf{F}$ is the net force applied to the object (the vector sum of all forces acting on it) and $\Delta\mathbf{p}$ is the resulting momentum change that occurs during the time interval[†] Δt. We can readily derive the familiar form of the second law, $\Sigma\mathbf{F} = m\mathbf{a}$, from Eq. 7–2 for the case of constant mass. If $\mathbf{v}_0$ is the initial velocity of an object and **v** is its velocity after a time Δt has elapsed, then

$$\Sigma\mathbf{F} = \frac{\Delta\mathbf{p}}{\Delta t} = \frac{m\mathbf{v} - m\mathbf{v}_0}{\Delta t} = \frac{m(\mathbf{v} - \mathbf{v}_0)}{\Delta t}$$

$$= m\frac{\Delta\mathbf{v}}{\Delta t}$$

$$= m\mathbf{a} \qquad\qquad \text{[constant mass]}$$

because, by definition, $\mathbf{a} = \Delta\mathbf{v}/\Delta t$. Newton's statement, Eq. 7–2, is actually more general than the more familiar one because it includes the situation in which the mass may change. This is important in certain circumstances, such as for rockets which lose mass as they burn fuel.

[†]Normally we think of Δt as being a small time interval. If it is not small, then Eq. 7–2 is valid if $\Sigma\mathbf{F}$ is constant over that time interval, or if $\Sigma\mathbf{F}$ is the average net force over that time interval.

Here is a simple Example of momentum change.

Washing a car: momentum change and force. Water leaves a hose at a rate of 1.5 kg/s with a speed of 20 m/s and is aimed at the side of a car, which stops it, Fig. 7–1. (That is, we ignore any splashing back.) What is the force exerted by the water on the car?

SOLUTION We take the x direction positive to the right. In each second, water with a momentum of $p_x = mv_x = (1.5\ \text{kg})(20\ \text{m/s}) = 30\ \text{kg}\cdot\text{m/s}$ is brought to rest at the moment it hits the car. The magnitude of the force (assumed constant) that the car must exert to change the momentum of the water by this amount is

$$F = \frac{\Delta p}{\Delta t} = \frac{p_{\text{final}} - p_{\text{initial}}}{\Delta t} = \frac{0 - 30\ \text{kg}\cdot\text{m/s}}{1.0\ \text{s}} = -30\ \text{N}.$$

The minus sign indicates that the force on the water is opposite to the water's original velocity. The car exerts a force of 30 N to the left to stop the water, so by Newton's third law, the water exerts a force of 30 N on the car.

FIGURE 7–1 Example 7–1.

FIGURE 7–2 Conceptual Example 7–2. Momentum of water before and after splashing back, and $\Delta \mathbf{p}$.

CONCEPTUAL EXAMPLE 7–2 | **The water splashes back.** What if the water splashes back from the car in Example 7–1? Would the force on the car be greater or less?

RESPONSE If the water splashes back toward the hose, the change in momentum will be greater in magnitude, and so the force on the car will be greater in magnitude. Note that $\mathbf{p}_{\text{final}}$ will now point in the negative x direction, as shown in Fig. 7–2 (instead of being zero as in Example 7–1). So the result for F (see displayed equation in Example 7–1), will be minus something more than $-30\ \text{N}$ (i.e., -35 to $-40\ \text{N}$, depending on the water's rebound speed). To put it simply, the car exerts not only a force to stop the water, but also an additional force to give it momentum in the opposite direction.

7–2 Conservation of Momentum

The concept of momentum is particularly important because, under certain circumstances, momentum is a conserved quantity. In the mid-seventeenth century, shortly before Newton's time, it had been observed that the vector sum of the momenta of two colliding objects remains constant. Consider, for example, the head-on collision of two billiard balls, as shown in Fig. 7–3. We assume the net external force on this system of two balls is zero—that is, the only significant forces are those that each ball exerts on the other during the collision. Although the momentum of each of the two balls changes as a result of the collision, the *sum* of their momenta is found to be the same before as after the collision. If $m_1\mathbf{v}_1$ is the momentum of ball number 1 and $m_2\mathbf{v}_2$ the momentum of ball 2, both measured before the collision, then the total momentum of the two balls before the collision is $m_1\mathbf{v}_1 + m_2\mathbf{v}_2$. After the collision, the balls each have a different velocity and momentum, which we will designate by a "prime" on the velocity: $m_1\mathbf{v}_1'$

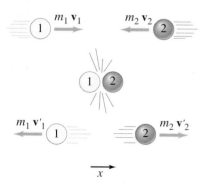

FIGURE 7–3 Momentum is conserved in a collision of two balls.

and $m_2\mathbf{v}_2'$. The total momentum after the collision is $m_1\mathbf{v}_1' + m_2\mathbf{v}_2'$. No matter what the velocities and masses involved are, it is found that the total momentum before the collision is the same as afterwards, whether the collision is head-on or not, as long as no net external force acts:

momentum before = momentum after

$$m_1\mathbf{v}_1 + m_2\mathbf{v}_2 = m_1\mathbf{v}_1' + m_2\mathbf{v}_2'. \qquad \textbf{(7–3)}$$

That is, the total vector momentum of the system of two balls is conserved: it stays constant.

Although the law of conservation of momentum was discovered experimentally, it is closely connected to Newton's laws of motion and they can be shown to be equivalent. We will do a simple derivation for the one-dimensional case illustrated in Fig. 7–3. We assume the force F that one ball exerts on the other during the collision is constant over the time of collision Δt. We use Newton's second law as expressed in Eq. 7–2, and rewrite it by multiplying both sides by Δt:

$$\Delta\mathbf{p} = \mathbf{F}\Delta t. \qquad \textbf{(7–4)}$$

We apply this to ball 2, noting that the force $\mathbf{F}_{21}$ on ball 2 due to ball 1 during the collision is to the right ($+x$ direction—see Fig. 7–4):

$$\Delta\mathbf{p}_2 = m_2\mathbf{v}_2' - m_2\mathbf{v}_2 = \mathbf{F}_{21}\Delta t.$$

By Newton's third law, the force $\mathbf{F}_{12}$ on ball 1 due to ball 2 is $\mathbf{F}_{12} = -\mathbf{F}_{21}$ and acts to the left, so

$$\Delta\mathbf{p}_1 = m_1\mathbf{v}_1' - m_1\mathbf{v}_1 = \mathbf{F}_{12}\Delta t = -\mathbf{F}_{21}\Delta t.$$

FIGURE 7–4 Forces on the balls during the collision of Fig. 7–3.

We can combine these last two equations (their right sides differ only by a minus sign):

$$m_1\mathbf{v}_1' - m_1\mathbf{v}_1 = -(m_2\mathbf{v}_2' - m_2\mathbf{v}_2)$$

or

$$m_1\mathbf{v}_1 + m_2\mathbf{v}_2 = m_1\mathbf{v}_1' + m_2\mathbf{v}_2'$$

which is Eq. 7–3, the conservation of momentum.

The above derivation can be extended to include any number of interacting bodies. To show this in a simple way, we let $\mathbf{p}$ in Eq. 7–2 represent the total momentum of a system—that is, the vector sum of the momenta of all objects in the system. (For our two-body system above, $\mathbf{p} = m_1\mathbf{v}_1 + m_2\mathbf{v}_2$.) If the net force $\Sigma\mathbf{F}$ on the system is zero [as it was above for our two-body system, $\mathbf{F} + (-\mathbf{F}) = 0$], then from Eq. 7–2, $\Delta\mathbf{p} = \mathbf{F}\Delta t = 0$, so the total momentum doesn't change. Thus the general statement of the **law of conservation of momentum** is

The total momentum of an isolated system of bodies remains constant.

By a **system**, we simply mean a set of objects that interact with each other. An **isolated system** is one in which the only forces present are those between the objects of the system. The sum of all these forces will be zero because of Newton's third law. If there are *external forces*—by which we mean forces exerted by objects outside the system—and they don't add up to zero (vectorially), then the total momentum won't be conserved. However, if the "system" can be redefined so as to include the other objects exerting these forces, then the conservation of momentum principle can apply. For example, if we take as our system a falling rock, it does not conserve momentum since an external force, the force of gravity exerted by the Earth, is acting on it and its momentum changes. However, if we include the Earth in the system, the total momentum of rock plus Earth is conserved. (This of course means that the Earth comes up to meet the ball. Since the Earth's mass is so great, its upward velocity is very tiny.)

(a) Before collision

(b) After collision

FIGURE 7–5 Example 7–3.

EXAMPLE 7–3 **Railroad cars collide: momentum conserved.** A 10,000-kg railroad car traveling at a speed of 24.0 m/s strikes an identical car at rest. If the cars lock together as a result of the collision, what is their common speed afterward? See Fig. 7–5.

SOLUTION The initial total momentum is

$$m_1 v_1 + m_2 v_2 = (10{,}000 \text{ kg})(24.0 \text{ m/s}) + (10{,}000 \text{ kg})(0 \text{ m/s})$$
$$= 2.40 \times 10^5 \text{ kg·m/s}.$$

and is to the right in the $+x$ direction. After the collision, the total momentum will be the same, and it will be shared by both cars. Since the two cars become attached, they will have the same speed, call it v'. Then:

$$(m_1 + m_2) v' = 2.40 \times 10^5 \text{ kg·m/s}$$

$$v' = \frac{2.40 \times 10^5 \text{ kg·m/s}}{2.00 \times 10^4 \text{ kg}} = 12.0 \text{ m/s},$$

to the right. Their mutual speed after collision is half the initial speed of car 1.

$p = 0$

(a)

$\mathbf{p}_{gas}$ $\mathbf{p}_{rocket}$

(b)

FIGURE 7–6 (a) A rocket, containing fuel, at rest in some reference frame. (b) In the same reference frame, the rocket fires and gases are expelled at high speed out the rear. The total vector momentum, $\mathbf{p}_{gas} + \mathbf{p}_{rocket}$, remains zero.

The law of conservation of momentum is particularly useful when we are dealing with fairly simple systems such as collisions and certain types of explosions. For example, *rocket propulsion*, which we saw in Chapter 4 can be understood on the basis of action and reaction, can also be explained on the basis of the conservation of momentum. Before a rocket is fired, the total momentum of rocket plus fuel is zero. As the fuel burns, the total momentum remains unchanged: the backward momentum of the expelled gases is just balanced by the forward momentum gained by the rocket itself (see Fig. 7–6). Thus, a rocket can accelerate in empty space. There is no need for the expelled gases to push against the Earth or the air (as is sometimes erroneously thought), as we already discussed in Chapter 4. Similar examples are the recoil of a gun and the throwing of a package from a boat.

(a) Before shooting

(b) After shooting

FIGURE 7-7 Example 7-4.

EXAMPLE 7-4 **Rifle recoil.** Calculate the recoil velocity of a 5.0-kg rifle that shoots a 0.050-kg bullet at a speed of 120 m/s, Fig. 7-7.

SOLUTION The total momentum of the system is conserved. We let the subscripts B represent the bullet and R the rifle; the final velocities are indicated by primes. Then conservation of momentum in the x direction gives

$$m_B v_B + m_R v_R = m_B v'_B + m_R v'_R$$

$$0 + 0 = (0.050 \text{ kg})(120 \text{ m/s}) + (5.0 \text{ kg})(v'_R)$$

$$v'_R = -\frac{(0.050 \text{ kg})(120 \text{ m/s})}{(5.0 \text{ kg})} = -1.2 \text{m/s}.$$

Since the rifle has a much larger mass, its (recoil) velocity is much less than that of the bullet. The minus sign indicates that the velocity (and momentum) of the rifle is in the negative x direction, opposite to that of the bullet. Notice that it is the *vector sum* of the momenta that is conserved.

7-3 Collisions and Impulse

Conservation of momentum is a very useful tool for dealing with collision processes, as we already saw in the examples of the previous section. Collisions are a common occurrence in everyday life: a tennis racket or a baseball bat striking a ball, two billiard balls colliding, one railroad car striking another, a hammer hitting a nail. At the subatomic level, scientists learn about the structure of nuclei and their constituents, and about the nature of the forces involved, by careful study of collisions between nuclei and/or elementary particles.

In a collision of two ordinary objects, both objects are deformed, often considerably, because of the large forces involved (Fig. 7 8). When the collision occurs, the force usually jumps from zero at the moment of contact to a very large value within a very short time, and then abruptly returns to zero again. A graph of the magnitude of the force one object exerts on the other during a collision, as a function of time, is something like that shown by the red curve in Fig. 7-9. The time interval Δt is usually very distinct and usually very small.

From Newton's second law, Eq. 7-2, the *net* force on an object is equal to the rate of change of its momentum:

$$\mathbf{F} = \frac{\Delta \mathbf{p}}{\Delta t}.$$

(We have written **F** instead of Σ**F** for the net force, which we assume is entirely due to the brief but large force that acts during the collision.) This equation applies, of course, to *each* of the objects in a collision. If we multiply both sides of this equation by the time interval Δt, we obtain

$$\text{Impulse} = \mathbf{F} \Delta t = \Delta \mathbf{p}. \qquad (7-5)$$

The quantity on the left, the product of the force **F** times the time Δt over which the force acts, is called the **impulse**. We see that the total change in momentum is equal to the impulse. The concept of impulse is of help

FIGURE 7-8 Tennis racket striking a ball. Note the deformation of both ball and racket due to the large force each exerts on the other.

FIGURE 7-9 Force as a function of time during a typical collision.

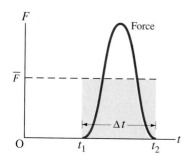

FIGURE 7–10 The average force $\bar{F}$ acting over an interval of time Δt gives the same impulse ($\bar{F}\Delta t$) as the actual force.

FIGURE 7–11 Period during which impulse acts (Example 7–5).

$v = 7.7$ m/s

$v = 0$

mainly when dealing with forces that act over a short time, as when a bat hits a baseball. The force is generally not constant and often its variation in time is like that graphed in Fig. 7–10. It is often sufficient to approximate such a varying force by an average force $\bar{F}$ acting over a time Δt, as indicated by the dashed line in Fig. 7–10. $\bar{F}$ is chosen so that the area shown shaded in Fig. 7–10 (equal to $\bar{F} \times \Delta t$) is equal to the area under the actual curve of F vs. t (which area represents the impulse). Note from Eq. 7–5 that the same impulse, and same change in momentum, can be given to an object by a smaller force F if the time Δt over which it acts is greater, as long as the product, $F \times \Delta t$, remains the same.

EXAMPLE 7–5 **Bend your knees when landing.** (*a*) Calculate the impulse experienced when a 70-kg person lands on firm ground after jumping from a height of 3.0 m. Then estimate the average force exerted on the person's feet by the ground, if the landing is (*b*) stiff-legged, and (*c*) with bent legs. In the former case, assume the body moves 1.0 cm during impact, and in the second case, when the legs are bent, about 50 cm.

SOLUTION (*a*) Although we don't know F and thus can't calculate the impulse $F\Delta t$ directly, we can use the fact that the impulse equals the change in momentum of the object. We need to determine the velocity of the person just before striking the ground, which we can do using conservation of energy (Eq. 6–11a):

$$\Delta \text{KE} = -\Delta \text{PE}$$
$$\tfrac{1}{2}mv^2 - 0 = -mg(y - y_0),$$

where we assume he started from rest ($v_0 = 0$), and $y_0 = 3.0$ m and $y = 0$. Thus, after falling 3.0 m, the person's velocity just before hitting the ground will be

$$v = \sqrt{2g(y_0 - y)} = \sqrt{2(9.8 \text{ m/s}^2)(3.0 \text{ m})} = 7.7 \text{ m/s}.$$

As the person strikes the ground, the momentum is quickly brought to zero, Fig. 7–11. The impulse on the person is

$$\bar{F}\Delta t = \Delta p = p - p_0$$
$$= 0 - (70 \text{ kg})(7.7 \text{ m/s}) = -540 \text{ N·s}.$$

The negative sign tells us that the force is opposed to the original momentum—that is, the force acts upward.

(*b*) In coming to rest, the body decelerates from 7.7 m/s to zero in a distance $d = 1.0 \text{ cm} = 1.0 \times 10^{-2}$ m. The average speed during this brief period is

$$\bar{v} = \frac{(7.7 \text{ m/s} + 0 \text{ m/s})}{2} = 3.8 \text{ m/s},$$

so the collision lasts a time

$$\Delta t = \frac{d}{\bar{v}} = \frac{(1.0 \times 10^{-2} \text{ m})}{(3.8 \text{ m/s})} = 2.6 \times 10^{-3} \text{ s}.$$

Since the magnitude of the impulse is $\bar{F}\Delta t = 540$ N·s, and $\Delta t = 2.6 \times 10^{-3}$ s, the average net force $\bar{F}$ has magnitude

$$\bar{F} = \frac{540 \text{ N·s}}{2.6 \times 10^{-3} \text{ s}} = 2.1 \times 10^5 \text{ N}.$$

The force $\bar{F}$ is the *net* force upward on the person (we calculated it from

Newton's second law). $\bar{F}$ equals the vector sum of the average force upward on the legs exerted by the ground, F_{grd}, which we take as positive, plus the downward force of gravity, $-mg$ (see Fig. 7–12):

$$\bar{F} = F_{grd} - mg.$$

Since $mg = (70\,\text{kg})(9.8\,\text{m/s}^2) = 690\,\text{N}$, then

$$F_{grd} = \bar{F} + mg = 2.1 \times 10^5\,\text{N} + 0.690 \times 10^3\,\text{N} \approx 2.1 \times 10^5\,\text{N}.$$

(c) This is just like part (b), except $d = 0.50\,\text{m}$, so $\Delta t = (0.50\,\text{m})/(3.8\,\text{m/s}) = 0.13\,\text{s}$, and

$$\bar{F} = \frac{540\,\text{N·s}}{0.13\,\text{s}} = 4.2 \times 10^3\,\text{N}.$$

The upward force exerted on the person's feet by the ground is, as in part (b):

$$F_{grd} = \bar{F} + mg = 4.2 \times 10^3\,\text{N} + 0.69 \times 10^3\,\text{N} = 4.9 \times 10^3\,\text{N}.$$

Clearly, the force on the feet and legs is much less when the knees are bent. In fact, the ultimate strength of the leg bone (see Chapter 9, Table 9–2) is not great enough to support the force calculated in part (b), so the leg would likely break in such a stiff landing, whereas it probably wouldn't in part (c).

FIGURE 7–12 When the person lands on the ground, the average net force during impact is $\bar{F} = F_{grd} - mg$, where F_{grd} is the force the ground exerts upward on the person.

7–4 Conservation of Energy and Momentum in Collisions

During most collisions, we usually don't know how the collision force varies over time, and so analysis using Newton's second law becomes difficult or impossible. But we can still determine a lot about the motion after a collision, given the initial motion, by making use of the conservation laws for momentum and energy. We saw in Section 7–2 that in the collision of two objects such as billiard balls, the total momentum is conserved. If the two objects are very hard and no heat is produced in the collision, then kinetic energy is conserved as well. By this we mean that the sum of the kinetic energies of the two objects is the same after the collision as before. Of course, for the brief moment during which the two objects are in contact, some (or all) of the energy is stored momentarily in the form of elastic potential energy. But if we compare the total kinetic energy before the collision with the total after the collision, they are found to be the same. Such a collision, in which the total kinetic energy is conserved, is called an **elastic collision**. If we use the subscripts 1 and 2 to represent the two objects, we can write the equation for conservation of total kinetic energy as

$$\tfrac{1}{2}m_1v_1^2 + \tfrac{1}{2}m_2v_2^2 = \tfrac{1}{2}m_1v_1'^2 + \tfrac{1}{2}m_2v_2'^2. \qquad \text{[elastic collision]} \quad (7\text{–}6)$$

Elastic collision

KE *conserved in elastic collisions*

Here, primed quantities ($'$) mean after the collision and unprimed mean before the collision, just as in Eq. 7–3 for conservation of momentum.

Although at the atomic level the collisions of atoms and molecules are often elastic, in the "macroscopic" world of ordinary objects, an elastic collision is an ideal that is never quite reached, since at least a little thermal energy (and perhaps sound and other forms of energy) is always produced during a collision. The collision of two hard elastic balls, such as billiard

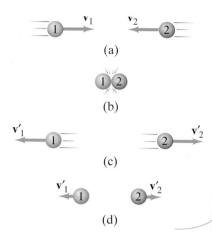

FIGURE 7–13 Two equal mass objects (a) approach each other with equal speeds, (b) collide, and then (c) bounce off with equal speeds in the opposite directions if the collision is elastic, or (d) bounce back much less or not at all, if the collision is inelastic.

Momentum conservation

KE *conservation*

FIGURE 7–14 Two particles, of masses m_1 and m_2, (a) before the collision, and (b) after the collision.

(a)

(b)

balls, however, is very close to perfectly elastic, and we often treat it as such. Even when the kinetic energy is not conserved, the *total* energy is, of course, always conserved.

Collisions in which kinetic energy is not conserved are said to be **inelastic collisions**. The kinetic energy that is lost is changed into other forms of energy, often thermal energy, so that the total energy (as always) is conserved. In this case, we can write that

$$\text{KE}_1 + \text{KE}_2 = \text{KE}_1' + \text{KE}_2' + \text{thermal and other forms of energy.}$$

See Fig. 7–13.

7–5 Elastic Collisions in One Dimension— Solving Problems Using Energy and Momentum Conservation

We now apply the conservation laws for momentum and kinetic energy to an elastic collision between two small objects (particles) that collide head-on, so all the motion is along a line. Let us assume that both particles are initially moving with velocities v_1 and v_2 along the x axis, Fig. 7–14a. After the collision, their velocities are v_1' and v_2', Fig. 7–14b. For any $v > 0$, the particle is moving to the right (increasing x), whereas for $v < 0$, the particle is moving to the left (toward decreasing values of x).

From conservation of momentum, we have

$$m_1 v_1 + m_2 v_2 = m_1 v_1' + m_2 v_2'.$$

Because the collision is assumed to be elastic, kinetic energy is also conserved:

$$\tfrac{1}{2} m_1 v_1^2 + \tfrac{1}{2} m_2 v_2^2 = \tfrac{1}{2} m_1 v_1'^2 + \tfrac{1}{2} m_2 v_2'^2.$$

We have two equations, so we can solve for two unknowns. If we know the masses and initial velocities, then we can solve these two equations for the velocities after the collision, v_1' and v_2'. We will do this in a moment in some Examples, but first we derive a useful result. To do so we rewrite the momentum equation as

$$m_1(v_1 - v_1') = m_2(v_2' - v_2), \qquad \textbf{(i)}$$

and we rewrite the KE equation as

$$m_1(v_1^2 - v_1'^2) = m_2(v_2'^2 - v_2^2)$$

or [noting that $(a - b)(a + b) = a^2 - b^2$] we write this as

$$m_1(v_1 - v_1')(v_1 + v_1') = m_2(v_2' - v_2)(v_2' + v_2). \qquad \textbf{(ii)}$$

We divide Eq. (ii) by Eq. (i), and (assuming $v_1 \neq v_1'$ and $v_2 \neq v_2'$) and obtain

$$v_1 + v_1' = v_2' + v_2.$$

We can rewrite this equation as

$$v_1 - v_2 = v_2' - v_1'$$

$$= -(v_1' - v_2'). \qquad \text{[head-on elastic collision]} \quad \textbf{(7–7)}$$

This is an interesting result: it tells us that for any elastic head-on collision, the relative speed of the two particles after the collision has the same magnitude as before (but opposite direction), no matter what the masses are.

EXAMPLE 7–6 **Pool or billiards.** A billiard ball of mass m moving with speed v collides head-on with a second ball of equal mass at rest ($v_2 = 0$). What are the speeds of the two balls after the collision, assuming it is elastic?

SOLUTION Since $v_1 = v$ and $v_2 = 0$, and $m_1 = m_2 = m$, then conservation of momentum gives

$$mv = mv_1' + mv_2'$$

or

$$v = v_1' + v_2'$$

since the m's cancel out. We have two unknowns (v_1' and v_2'), so we need a second equation, which could be the conservation of kinetic energy, or the simpler Eq. 7–7 we derived from it, which gives

$$v_1 - v_2 = v_2' - v_1'$$

or

$$v = v_2' - v_1'.$$

We subtract this equation from our momentum equation ($v = v_1' + v_2'$) and obtain

$$0 = 2v_1',$$

so $v_1' = 0$. This is one of our desired unknowns, and we can now solve for the other:

$$v_2' = v + v_1' = v + 0 = v.$$

To summarize, before the collision we have

$$v_1 = v, \qquad v_2 = 0$$

and after the collision

$$v_1' = 0, \qquad v_2' = v.$$

That is, ball 1 is brought to rest by the collision, whereas ball 2 acquires the original velocity of ball 1. This result is often observed by billiard and pool players, and is valid only if the two balls have equal masses (and no spin is given to the balls). See Fig. 7–15.

EXAMPLE 7–7 **A nuclear collision.** A proton of mass 1.01 u (unified atomic mass units) traveling with a speed of 3.60×10^4 m/s has an elastic head-on collision with a helium (He) nucleus ($m_{He} = 4.00$ u) initially at rest. What are the velocities of the proton and helium nucleus after the collision? (As mentioned in Chapter 1, 1 u = 1.66×10^{-27} kg, but we won't need this fact.)

SOLUTION Call the initial direction of motion the $+x$ direction. We have $v_2 = v_{He} = 0$ and $v_1 = v_p = 3.60 \times 10^4$ m/s. We want to find the velocities v'_p and v'_{He} after the collision. From conservation of momentum we have

$$m_p v_p + 0 = m_p v'_p + m_{He} v'_{He}.$$

Because the collision is elastic, kinetic energy is conserved and we can use Eq. 7–7, which becomes

$$v_p - 0 = v'_{He} - v'_p.$$

Thus

$$v'_p = v'_{He} - v_p,$$

and substituting this into the momentum equation we get

$$m_p v_p = m_p v'_{He} - m_p v_p + m_{He} v'_{He}.$$

Solving for v'_{He}, we obtain

$$v'_{He} = \frac{2 m_p v_p}{m_p + m_{He}} = \frac{2(1.01 \text{ u})(3.60 \times 10^4 \text{ m/s})}{5.01 \text{ u}} = 1.45 \times 10^4 \text{ m/s}.$$

The other unknown is v'_p, which we can now obtain from

$$v'_p = v'_{He} - v_p$$

$$= 1.45 \times 10^4 \text{ m/s} - 3.60 \times 10^4 \text{ m/s} = -2.15 \times 10^4 \text{ m/s}.$$

The minus sign tells us that the proton reverses direction upon collision, and we see that its speed is less than its initial speed (see Fig. 7–16). This makes sense from ordinary experience: the lighter proton would be expected to "bounce back" somewhat from the more massive helium nucleus, but not with its full original velocity as it would from a rigid wall (which would correspond to extremely large, or infinite, mass).

FIGURE 7–16 Example 7–7: (a) before collision, (b) after collision.

7–6 Inelastic Collisions

Collisions in which kinetic energy is not conserved are called inelastic collisions. Some of the initial kinetic energy in such collisions is transformed into other types of energy, such as thermal or potential energy, so the total final kinetic energy is less than the total initial kinetic energy. The inverse can also happen when potential energy (such as chemical or nuclear) is released, in which case the total final kinetic energy can be greater than the initial kinetic energy. Explosives are examples of this type. Typical macroscopic collisions are inelastic, at least to some extent, and often to a large extent. If two objects stick together as a result of a collision, the collision is said to be **completely inelastic**. Two colliding balls of putty that stick together or two railroad cars that couple together when they collide are examples of completely inelastic collisions. The kinetic energy in some cases is all transformed to other forms of energy in an inelastic collision, but in other cases only part of it is. In Example 7–3, for instance, we saw that when a traveling railroad car collided with a stationary one, the coupled cars traveled off with some kinetic energy.

Completely inelastic collision

In a completely inelastic collision, the maximum amount of kinetic energy is transformed to other forms consistent with conservation of momentum. Even though kinetic energy is not conserved in inelastic collisions, the total energy is conserved, and the total vector momentum is also always conserved.

EXAMPLE 7–8 **Railroad cars again.** For the completely inelastic collision of two railroad cars that we considered in Example 7–3, calculate how much of the initial kinetic energy is transformed to thermal or other forms of energy.

SOLUTION Initially, the total kinetic energy is
$$\tfrac{1}{2} m_1 v_1^2 = \tfrac{1}{2}(10{,}000 \text{ kg})(24.0 \text{ m/s})^2 = 2.88 \times 10^6 \text{ J}.$$
After the collision, the total kinetic energy is
$$\tfrac{1}{2}(20{,}000 \text{ kg})(12.0 \text{ m/s})^2 = 1.44 \times 10^6 \text{ J}.$$
Hence the energy transformed to other forms is
$$2.88 \times 10^6 \text{ J} - 1.44 \times 10^6 \text{ J} = 1.44 \times 10^6 \text{ J},$$
which is just half the original KE.

EXAMPLE 7–9 **Ballistic pendulum.** The *ballistic pendulum* is a device *Ballistic pendulum* used to measure the speed of a projectile, such as a bullet. The projectile, of mass m, is fired into a large block (of wood or other material) of mass M, which is suspended like a pendulum. (Usually, M is somewhat greater than m.) As a result of the collision, the pendulum-projectile system swings up to a maximum height h, Fig. 7–17. Determine the relationship between the initial speed of the projectile, v, and the height h.

SOLUTION We analyze this process by dividing it into two parts: (1) the collision itself, and (2) the subsequent motion of the pendulum from the vertical hanging position to height h. In part (1), Fig. 7–17a, we assume the collision time is very short, and so the projectile comes to rest in the block before the block has moved significantly from its position directly below its support. Thus there is no net external force and momentum is conserved:

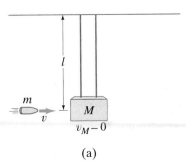

$$mv = (m + M)v', \qquad \textbf{(i)}$$

where v' is the speed of the block and embedded projectile just after the collision, before they have moved significantly. Once the pendulum begins to move (part 2, Fig. 7–17b), there will be a net external force (gravity, tending to pull it back to the vertical position). So, for part (2), we cannot use conservation of momentum. But we can use conservation of mechanical energy since the kinetic energy immediately after the collision is changed entirely to gravitational potential energy when the pendulum reaches its maximum height, h. Therefore (letting $y = 0$ for the pendulum in the vertical position):

$$\text{KE}_1 + \text{PE}_1 = \text{KE}_2 + \text{PE}_2$$

or

$$\tfrac{1}{2}(m + M)v'^2 + 0 = 0 + (m + M)gh, \qquad \textbf{(ii)}$$

so $v' = \sqrt{2gh}$. We combine equations (i) and (ii) to obtain

FIGURE 7–17 Ballistic pendulum (Example 7–9).

$$v = \frac{m + M}{m}v' = \frac{m + M}{m}\sqrt{2gh},$$

which is the final result. To obtain this result, we had to be opportunistic, in that we used whichever conservation laws we could: in (1) we could use only conservation of momentum, since the collision is inelastic and conservation of mechanical energy is not valid[†]; and in (2), conservation of mechanical energy is valid, but not conservation of momentum. In part (1), if there were significant motion of the pendulum during the deceleration of the projectile in the block, then there *would* be an external force during the collision—so conservation of momentum would not be valid, and this would have to be taken into account.

* 7–7 Collisions in Two or Three Dimensions

Conservation of momentum and energy can also be applied to collisions in two or three dimensions, and the vector nature of momentum is especially important. One common type of non-head-on collision is that in which a moving particle (called the "projectile") strikes a second particle initially at rest (the "target" particle). This is the common situation in games such as billiards, and for experiments in atomic and nuclear physics (the projectiles, from radioactive decay or a high-energy accelerator, strike a stationary target nucleus).

Figure 7–18 shows particle 1 (the projectile, m_1) heading along the x axis toward particle 2 (the target, m_2), which is initially at rest. If these are billiard balls, m_1 strikes m_2 and they go off at the angles θ_1' and θ_2', respectively, which are measured relative to m_1's initial direction (the x axis). The particles may begin to deflect even before they touch if electric, magnetic, or nuclear forces act between them.[‡]

Let us apply the law of conservation of momentum to a collision like that of Fig. 7–18. We choose the xy plane to be the plane in which the initial and final momenta lie. Because momentum is a vector, and is conserved, its components in the x and y directions remain constant. In the x direction,

$$p_{1x} + p_{2x} = p_{1x}' + p_{2x}'$$

p_x conserved

or

$$m_1 v_1 = m_1 v_1' \cos \theta_1' + m_2 v_2' \cos \theta_2'. \qquad \textbf{(7–8a)}$$

Because there is no motion in the y direction initially, the y component of

[†]Total energy of course is conserved.

[‡]You might think, for example, of two magnets oriented so that they repel each other: when one moves toward the other, the second moves away before the first one touches it.

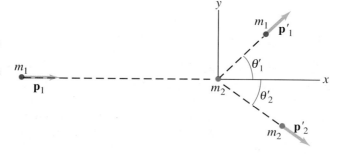

FIGURE 7–18 Particle 1, the projectile, collides with particle 2, the target. They move off, after the collision, with momenta $\mathbf{p}_1'$ and $\mathbf{p}_2'$ at angles θ_1' and θ_2'.

the total momentum is zero:

$$p_{1y} + p_{2y} = p'_{1y} + p'_{2y}$$

or

$$0 = m_1 v'_1 \sin \theta'_1 + m_2 v'_2 \sin \theta'_2. \qquad \textbf{(7-8b)} \qquad p_y \; conserved$$

EXAMPLE 7-10 **Billiard ball collision in 2-D.** A billiard ball moving with speed $v_1 = 3.0$ m/s in the $+x$ direction (Fig. 7-18) strikes an equal-mass ball initially at rest. The two balls are observed to move off at 45°, ball 1 above the x axis and ball 2 below. That is $\theta'_1 = 45°$ and $\theta'_2 = -45°$ in Fig. 7-18. What are the speeds of the two balls?

SOLUTION From symmetry we might guess that the two balls have the same speed. But let us not assume that now. Even though we aren't told whether the collision is elastic or inelastic, we can still always use conservation of momentum. So we can apply Eqs. 7-8a and b, and we want to solve for v'_1 and v'_2. We are given $m_1 = m_2 (= m)$ so

$$mv_1 = mv'_1 \cos(45°) + mv'_2 \cos(-45°)$$

and

$$0 = mv'_1 \sin(45°) + mv'_2 \sin(-45°).$$

The m's cancel out in both equations. The second equation yields [recall $\sin(-\theta) = -\sin \theta$]:

$$v'_2 = -v'_1 \frac{\sin(45°)}{\sin(-45°)} = -v'_1 \left(\frac{\sin 45°}{-\sin 45°} \right) = v'_1,$$

so they do have equal speeds as we guessed at first. The x component equation gives [recall $\cos(-\theta) = \cos \theta$]:

$$v_1 = v'_1 \cos(45°) + v'_2 \cos(45°) = 2v'_1 \cos(45°)$$

so

$$v'_1 = v'_2 = \frac{v_1}{2 \cos(45°)} = \frac{3.0 \text{ m/s}}{2(0.707)} = 2.1 \text{ m/s}.$$

When we have two independent equations, we can solve for, at most, two unknowns.

If we know that a collision is elastic, we can then apply conservation of kinetic energy and obtain a third equation:

$$\text{KE}_1 + \text{KE}_2 = \text{KE}'_1 + \text{KE}'_2$$

or, for the collision shown in Fig. 7-18,

$$\tfrac{1}{2}m_1 v_1^2 = \tfrac{1}{2}m_1 v'^2_1 + \tfrac{1}{2}m_2 v'^2_2. \qquad \text{[elastic collision]} \qquad \textbf{(7-8c)} \qquad \text{KE} \; conserved$$

If the collision is elastic, we have three independent equations and we can solve for three unknowns. If we are given m_1, m_2, v_1 (and v_2, if it is not zero), we cannot, for example, predict the final variables, $v'_1, v'_2, \theta'_1,$ and θ'_2, because there are four of them. However, if we measure one of these variables, say θ'_1, then the other three variables (v'_1, v'_2, and θ'_2) are uniquely determined, and we can determine them using Eqs. 7-8a, b, c.

1. Be sure no significant external force acts on your chosen system. That is, the forces that act between the interacting bodies must be the only significant ones if momentum conservation is to be used. [Note: If this is valid for a portion of the problem, you can use momentum conservation for that portion only.]

2. Draw a diagram of the initial situation, just before the interaction (collision, explosion) takes place, and represent the momentum of each object with an arrow and label. Do the same for the final situation, just after the interaction.

3. Choose a coordinate system and "+" and "−" directions. (For a head-on collision, you will need only an x axis.) It is often convenient to choose the +x axis in the direction of one object's initial velocity.

4. Write momentum conservation equation(s):

   total initial momentum = total final momentum.

   You have one equation for each component (x, y, z); only one equation for a head-on collision. [Don't forget that it is the *total* momentum, not the individual momenta, that is conserved.]

5. If the collision is elastic, you can also write down a conservation of kinetic energy equation:

   total initial KE = total final KE.

   [Alternately, you could use Eq. 7–7: $v_1 - v_2 = v_2' - v_1'$, if the collision is one dimensional (head-on).]

6. Solve algebraically for the unknown(s).

7–8 Center of Mass (CM)

Until now, we have been mainly concerned with the motion of single particles. When we have dealt with an extended body (that is, a body that has size), we have assumed that it could be approximated as a point particle or that it underwent only translational motion. Real "extended" bodies, however, can undergo rotational and other types of motion as well. For example, the diver in Fig. 7–19a undergoes only translational motion (all parts of the body follow the same path), whereas the diver in Fig. 7–19b undergoes

FIGURE 7–19 The motion of the diver is pure translation in (a), but is translation plus rotation in (b).

(a)

(b)

FIGURE 7–20 Translation plus rotation: a wrench moving over a horizontal surface. The CM, marked with a **+**, moves in a straight line.

both translational and rotational motion. We will refer to motion that is not pure translation as *general motion*.

Observations of the motion of bodies indicate that even if a body rotates, or there are several bodies that move relative to one another, there is one point that moves in the same path that a particle would if subjected to the same net force. This point is called the **center of mass** (abbreviated CM). The general motion of an extended body (or system of bodies) can be considered as *the sum of the translational motion of the CM, plus rotational, vibrational, or other types of motion about the CM.*

Center of mass and general motion

As an example, consider the motion of the center of mass of the diver in Fig. 7–19: the CM follows a parabolic path even when the diver rotates, as shown in Fig. 7–19b. This is the same parabolic path that a projected particle follows when acted on only by the force of gravity (that is, projectile motion). Other points in the rotating diver's body follow more complicated paths.

Figure 7–20 shows a wrench translating and rotating along a horizontal surface—note that its CM, marked by a red **+**, moves in a straight line, as shown by the dashed white line.

The CM is defined in the following way. We can consider any extended body as being made up of many tiny particles. But first we consider a system made up of only two particles, of mass m_1 and m_2. We choose a coordinate system so that both particles lie on the x axis at positions x_1 and x_2, Fig. 7–21. The center of mass of this system is defined to be at the position, x_{CM}, given by

FIGURE 7–21 The center of mass of a two-particle system lies on the line joining the two masses.

$$x_{CM} = \frac{m_1 x_1 + m_2 x_2}{m_1 + m_2} = \frac{m_1 x_1 + m_2 x_2}{M}, \qquad \textbf{(7–9a)}$$

Center of mass (x coordinate)

where $M = m_1 + m_2$ is the total mass of the system. The center of mass lies on the line joining m_1 and m_2. If the two masses are equal ($m_1 = m_2 = m$), x_{CM} is midway between them, since in this case

$$x_{CM} = \frac{m(x_1 + x_2)}{2m} = \frac{(x_1 + x_2)}{2}. \qquad \text{[equal masses]}$$

If one mass is greater than the other, say, $m_1 > m_2$, then the CM is closer to the larger mass. If there are more than two particles along a line, there will be additional terms in Eq. 7–9a, as the following Example shows.

FIGURE 7–22
Example 7–11.

$x = 0$ 1.0 m   5.0 m   6.0 m

EXAMPLE 7–11 **CM of three guys on a raft.** Three people of roughly equivalent mass m on a lightweight (air-filled) banana boat sit along the x axis at positions $x_1 = 1.0\,\text{m}$, $x_2 = 5.0\,\text{m}$, and $x_3 = 6.0\,\text{m}$ (Fig. 7–22). Find the position of the CM.

SOLUTION We use Eq. 7–9a with a third term:

$$x_{\text{CM}} = \frac{mx_1 + mx_2 + mx_3}{m + m + m} = \frac{m(x_1 + x_2 + x_3)}{3m}$$

$$= \frac{(1.0\,\text{m} + 5.0\,\text{m} + 6.0\,\text{m})}{3} = \frac{12.0\,\text{m}}{3} = 4.0\,\text{m}.$$

If the particles are spread out in two or three dimensions, then we need to specify not only the x coordinate of the CM (x_{CM}), but also the y and z coordinates, which will be given by formulas just like Eq. 7–9a. For example, for two particles of mass m_1 and m_2, whose y coordinates are y_1 and y_2, respectively, the y coordinate of their CM, y_{CM}, will be:

y coordinate of center of mass

$$y_{\text{CM}} = \frac{m_1 y_1 + m_2 y_2}{m_1 + m_2} = \frac{m_1 y_1 + m_2 y_2}{M}. \qquad \textbf{(7–9b)}$$

For more particles, there would be more terms in this formula.

Center of gravity

A concept similar to *center of mass* is **center of gravity** (CG). The CG of a body is that point at which the force of gravity can be considered to act. Of course, the force of gravity actually acts on all the different parts or particles of a body, but for purposes of determining the translational motion of a body as a whole, we can assume that the entire weight of the body (which is the sum of the weights of all its parts) acts at the CG. Strictly speaking, there is a conceptual difference between the center of gravity and the center of mass, but for practical purposes, they are generally at the same point.[†]

It is often easier to determine the CM or CG of an extended body experimentally rather than analytically. If a body is suspended from any point, it will swing (Fig. 7–23) unless it is placed so its CG lies on a vertical line directly below the point from which it is suspended. If the object is two-dimensional, or has a plane of symmetry, it need only be hung from two

---

[†]There would be a difference between the two points only if a body were large enough so that the acceleration due to gravity was different at different parts of the body.

FIGURE 7–23 The force of gravity, considered to act at the CG, causes the body to rotate about the pivot point unless the CG is on a vertical line directly below the pivot, in which case the body remains at rest.

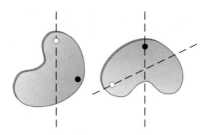

FIGURE 7–24 Finding the CG.

different pivot points and the respective vertical (plumb) lines drawn. Then the center of gravity will be at the intersection of the two lines, as in Fig. 7–24. If the object doesn't have a plane of symmetry, the CG with respect to the third dimension is found by suspending the object from at least three points whose plumb lines do not lie in the same plane. For symmetrically shaped bodies such as uniform cylinders (wheels), spheres, and rectangular solids, the CG is located at the geometric center of the body.

* 7–9 CM for the Human Body

If we have a group of extended bodies, each of whose CM is known, we can find the CM of the group using Eqs. 7–9a and b. As an example, we consider the human body. Table 7–1 indicates the CM and hinge points (joints) for the different components of a "representative" person. Of course, there are wide variations among people, so these data represent only a very rough average. Note that the numbers represent a percentage of the total height, which is regarded as 100 units; similarly the total mass is 100 units. Thus, for example, if a person is 1.70 m tall, his or her shoulder joint would be (1.70 m)(81.2/100) = 1.38 m above the floor.

TABLE 7–1
Center of Mass of Parts of Typical Human Body
(full height and mass = 100 units)

Distance of Hinge Points (%)	Hinge Points (•) (Joints)	Center of Mass (×) (% Height Above Floor)		Percent Mass
91.2	Base of skull on spine	Head	93.5	6.9
81.2	Shoulder joint	Trunk and neck	71.1	16.1
		Upper arms	71.7	6.6
	elbow 62.2	Lower arms	55.3	4.2
	wrist 46.2	Hands	43.1	1.7
52.1	Hip	Upper legs (thighs)	42.5	21.5
28.5	Knee			
		Lower legs	18.2	9.6
4.0	Ankle	Feet	1.8	3.4
			58.0	100.0

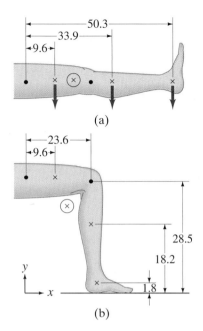

(a)

(b)

FIGURE 7–25 Example 7–12: finding the CM of a leg in two different positions (⊗ represents the calculated CM).

CM can be outside a body

EXAMPLE 7–12 **A leg's CM.** Determine the position of the CM of a whole leg (*a*) when stretched out, and (*b*) when bent at 90°, as shown in Fig. 7–25. Assume the person is 1.70 m tall.

SOLUTION (*a*) Table 7–1 uses percentage units, meaning the person has a mass of 100 units and a height of 100 units. At the end we can multiply by (1.70 m/100). We measure the distance from the hip joint using Table 7–1 and obtain the numbers shown in Fig. 7–25a. Using Eq. 7–9a, we obtain

$$x_{CM} = \frac{(21.5)(9.6) + (9.6)(33.9) + (3.4)(50.3)}{21.5 + 9.6 + 3.4} = 20.4 \text{ units.}$$

Thus, the center of mass of the leg and foot is 20.4 units from the hip joint, or 52.1 − 20.4 = 31.7 units from the base of the foot. Since the person is 1.70 m tall, this is (1.70 m)(31.7/100) = 0.54 m.

(*b*) In this part, we have a two-dimensional problem. We use an *xy* coordinate system, as shown in Fig. 7–25b. First, we calculate how far to the right of the hip joint the CM lies:

$$x_{CM} = \frac{(21.5)(9.6) + (9.6)(23.6) + (3.4)(23.6)}{21.5 + 9.6 + 3.4} = 14.9 \text{ units.}$$

For our 1.70 m tall person, this is (1.70 m)(14.9/100) = 0.25 m. Next, we calculate the distance, y_{CM}, of the CM above the floor:

$$y_{CM} = \frac{(3.4)(1.8) + (9.6)(18.2) + (21.5)(28.5)}{21.5 + 9.6 + 3.4} = 23.1 \text{ units,}$$

or (1.70 m)(23.1/100) = 0.39 m. Thus, the CM is located 39 cm above the floor and 25 cm to the right of the hip joint.

Note in this last Example that the CM can actually lie *outside* the body. Another example is a doughnut, whose CM is at the center of the hole.

Knowing the CM of the body when it is in various positions is of great use in studying body mechanics. One simple example from athletics is shown in Fig. 7–26. If high jumpers can get into the position shown, their CM can actually pass below the bar which their bodies go over, which means that for a particular takeoff speed, they can clear a higher bar. This is indeed what they try to do.

FIGURE 7–26 The CM of a high jumper may actually pass beneath the bar.

* 7-10 Center of Mass and Translational Motion

As mentioned in Section 7–8, a major reason for the importance of the concept of center of mass is that the motion of the CM for a system of particles (or an extended body) is directly related to the net force acting on the system as a whole. We now show this, taking the simple case of one-dimensional motion (x direction) and only three particles, but the extension to more bodies and to three dimensions follows the same lines.

Suppose the three particles lie on the x axis and have masses m_1, m_2, m_3, and positions x_1, x_2, x_3. From Eq. 7–9a for the CM, we can write

$$Mx_{CM} = m_1x_1 + m_2x_2 + m_3x_3,$$

where $M = m_1 + m_2 + m_3$ is the total mass of the system. If these particles are in motion, say along the x axis with velocities v_1, v_2, and v_3, respectively, then in a short time Δt they each will have traveled a distance:

$$\Delta x_1 = x_1' - x_1 = v_1 \Delta t$$
$$\Delta x_2 = x_2' - x_2 = v_2 \Delta t$$
$$\Delta x_3 = x_3' - x_3 = v_3 \Delta t,$$

where x_1', x_2', and x_3' represent their new positions after a time Δt. The position of the new CM is given by

$$Mx_{CM}' = m_1x_1' + m_2x_2' + m_3x_3'.$$

If we subtract the two CM equations, we get

$$M \Delta x_{CM} = m_1 \Delta x_1 + m_2 \Delta x_2 + m_3 \Delta x_3.$$

During the time Δt, the CM will have moved a distance

$$\Delta x_{CM} = x_{CM}' - x_{CM} = v_{CM} \Delta t,$$

where v_{CM} is the velocity of the CM. Into the equation just before the last one, we substitute the relations for all the Δx's:

$$Mv_{CM} \Delta t = m_1v_1 \Delta t + m_2v_2 \Delta t + m_3v_3 \Delta t.$$

We divide out Δt and get

$$Mv_{CM} = m_1v_1 + m_2v_2 + m_3v_3. \tag{7–10}$$

Since $m_1v_1 + m_2v_2 + m_3v_3$ is the sum of the momenta of the particles of the system, it represents the *total momentum* of the system. Thus we see from Eq. 7–10 that *the total (linear) momentum of a system of particles is equal to the product of the total mass M and the velocity of the center of mass of the system.* Or, *the linear momentum of an extended body is the product of the body's mass and the velocity of its CM.*

Total momentum

If there are forces acting on the particles, then the particles may be accelerating. In a short time Δt, each particle's velocity will change by an amount

$$\Delta v_1 = a_1 \Delta t, \qquad \Delta v_2 = a_2 \Delta t, \qquad \Delta v_3 = a_3 \Delta t.$$

If we now use the same reasoning as we did to derive Eq. 7–10, we obtain

$$Ma_{CM} = m_1a_1 + m_2a_2 + m_3a_3.$$

According to Newton's second law, $m_1a_1 = F_1$, $m_2a_2 = F_2$, and $m_3a_3 = F_3$, where F_1, F_2, and F_3 are the net forces on the three particles, respectively. Thus we get for the system as a whole:

*Newton's second law
for a system of particles
or an extended body*

$$Ma_{CM} = F_1 + F_2 + F_3 = F_{net}. \qquad (7\text{–}11)$$

That is, *the sum of all the forces acting on the system is equal to the total mass of the system times the acceleration of its center of mass.* This is **Newton's second law** for a system of particles, and it also applies to an extended body (which can be thought of as a collection of particles). Thus we conclude that the *center of mass of a system of particles (or of an extended body) with total mass M moves like a single particle of mass M acted upon by the same net external force.* That is, the system moves as if all its mass were concentrated at the CM and all the external forces acted at that point. We can thus treat the translational motion of any body or system of bodies as the motion of a particle (see Figs. 7–19 and 7–20). This theorem clearly simplifies our analysis of the motion of complex systems and extended bodies. Although the motion of various parts of the system may be complicated, we may often be satisfied with knowing the motion of the CM. This theorem also allows us to solve certain types of problems very easily, as illustrated by the following Example.

CONCEPTUAL EXAMPLE 7–13 | **A two-stage rocket.** A rocket is shot into the air as shown in Fig. 7–27. At the moment it reaches its highest point, a horizontal distance d from its starting point, a prearranged explosion separates it into two parts of equal mass. Part I is stopped in midair and falls vertically to Earth. Where does part II land? Assume **g** = constant.

SOLUTION After the rocket is fired, the path of the CM of the system continues to follow the parabolic trajectory of a projectile acted on by only a constant gravitational force. The CM will thus arrive at a point $2d$ from the starting point. Since the masses of I and II are equal, the CM must be midway between them. Therefore, II lands a distance $3d$ from the starting point. (If part I had been given a kick up or down, instead of merely falling, the solution would have been somewhat more complicated.)

FIGURE 7–27
Example 7–13.

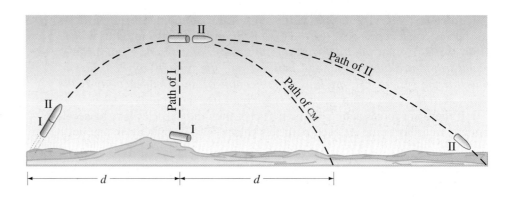

SUMMARY

The **momentum**, **p**, of a body is defined as the product of its mass times its velocity,

$$\mathbf{p} = m\mathbf{v}.$$

In terms of momentum, **Newton's second law** can be written as

$$\Sigma\mathbf{F} = \frac{\Delta\mathbf{p}}{\Delta t}.$$

That is, the rate of change of momentum equals the net applied force.

The **law of conservation of momentum** states that the total momentum of an isolated system of objects remains constant. An isolated system is one on which the net external force is zero.

The law of conservation of momentum is very useful in dealing with **collisions**. In a collision, two (or more) bodies interact with each other for a very short time, and the force between them during this time is very large.

The **impulse** of a force on a body is defined as **F** Δt, where **F** is the average force acting during the (usually short) time Δt. The impulse is equal to the change in momentum of the body:

$$\text{Impulse} = \mathbf{F}\,\Delta t = \Delta\mathbf{p}.$$

Total momentum is conserved in *any* collision. If $m_1\mathbf{v}_1$ and $m_2\mathbf{v}_2$ are the momenta of two objects before the collision and $m_1'\mathbf{v}_1'$ and $m_2'\mathbf{v}_2'$ are their momenta after, then

$$m_1\mathbf{v}_1 + m_2\mathbf{v}_2 = m_1\mathbf{v}_1' + m_2\mathbf{v}_2'.$$

Total energy is also conserved, but this may not be helpful in problem solving unless the only type of energy transformation involves kinetic energy. In that case kinetic energy is conserved and the collision is called an **elastic collision**, and we can write

$$\tfrac{1}{2}mv_1^2 + \tfrac{1}{2}mv_2^2 = \tfrac{1}{2}m_1v_1'^2 + \tfrac{1}{2}m_2v_2'^2.$$

If kinetic energy is not conserved, the collision is called **inelastic**. A **completely inelastic** collision is one in which the colliding bodies stick together after the collision.

The **center of mass** (CM) of an object (or group of objects) is that point at which the net force can be considered to act for purposes of determining the translational motion of the body as a whole. The complete motion of a body can be described as the translational motion of its center of mass plus rotation (or other internal motion) about its center of mass.

QUESTIONS

1. We claim that momentum is conserved. Yet most moving objects eventually slow down and stop. Explain.
2. When a person jumps from a tree to the ground, what happens to the momentum of the person upon striking the ground?
3. Why, when you release an inflated but untied balloon, does it fly across the room?
4. It is said that in ancient times a rich man with a bag of gold coins froze to death stranded on the surface of a frozen lake. Because the ice was frictionless, he could not push himself to shore. What could he have done to save himself had he not been so miserly?
5. How can a rocket change direction when it is far out in space and is essentially in a vacuum?
6. According to Eq. 7–5, the longer the impact time of an impulse, the smaller the force can be for the same momentum change, and hence the smaller the deformation of the object on which the force acts. Explain on this basis the value of "air bags," which are intended to inflate during an automobile collision and reduce the possibility of fracture or death.

7. It used to be common wisdom to build cars to be as rigid as possible to withstand collisions. Today, though, cars are designed to have "crumple zones" that collapse upon impact. What advantage does this have?
8. Why is it easier to hit a home run from a pitched ball than from one tossed in the air by the batter?
9. The speed of a tennis ball on the return of a serve can be just as fast as the serve, even though the racket isn't swung very fast. How can this be?
10. Is it possible for a body to receive a larger impulse from a small force than from a large force?
11. A light body and a heavy body have the same kinetic energy. Which has the greater momentum?
12. Is it possible for an object to have momentum without having kinetic energy? Can it have kinetic energy but no momentum? Explain.
13. At a hydroelectric power plant, water is directed at high speed against turbine blades on an axle that turns an electric generator. Do you think the turbine blades should be designed so that the water is brought to a dead stop, or so that the water rebounds?

14. A superball is dropped from a height h onto a hard steel plate (fixed to the Earth), from which it rebounds at very nearly its original speed. (*a*) Is the momentum of the ball conserved during any part of this process? (*b*) If we consider the ball and Earth as our system, during what parts of the process is momentum conserved? (*c*) Answer part (*b*) for a piece of putty that falls and sticks to the steel plate.

15. Why do you tend to lean backward when carrying a heavy load in your arms?

16. Why is the CM of a 1-m length of pipe at its midpoint, whereas this is not true for your arm or leg?

17. Show on a diagram how your CM shifts when you change from a lying position to a sitting position.

18. Describe an analytic way of determining the CM of any triangular-shaped, thin uniform plate.

* **19.** A rocket following a parabolic path through the air suddenly explodes into many pieces. What can you say about the motion of this system of pieces?

* **20.** If only an external force can change the momentum of the center of mass of an object, how can the internal force of the engine accelerate a car?

PROBLEMS

SECTIONS 7–1 AND 7–2

1. (I) What is the magnitude of the momentum of a 22-g sparrow flying with a speed of 8.1 m/s?

2. (II) A child in a boat throws a 5.40-kg package out horizontally with a speed of 10.0 m/s, Fig. 7–28. Calculate the velocity of the boat immediately after, assuming it was initially at rest. The mass of the child is 26.0 kg and that of the boat is 55.0 kg.

$v = 10$ m/s

FIGURE 7–28   Problem 2.

3. (II) Calculate the force exerted on a rocket, given that the propelling gases are expelled at a rate of 1300 kg/s with a speed of 40,000 m/s (at the moment of takeoff).

4. (II) A halfback on an apparent breakaway for a touchdown is tackled from behind. If the halfback has a mass of 95 kg and was moving at 4.1 m/s when he was tackled by an 85-kg cornerback running at 5.5 m/s in the same direction, what was their mutual speed immediately after the touchdown-saving tackle?

5. (II) A 12,500-kg railroad car travels alone on a level frictionless track with a constant speed of 18.0 m/s. A 5750-kg additional load is dropped onto the car. What then will be the car's speed?

6. (II) A 9500-kg boxcar traveling at 16 m/s strikes a second car. The two stick together and move off with a speed of 6.0 m/s. What is the mass of the second car?

7. (II) A gun is fired vertically into a 1.40-kg block of wood at rest directly above it. If the bullet has a mass of 21.0 g and a speed of 210 m/s, how high will the block rise into the air after the bullet becomes embedded in it?

8. (II) A 15-g bullet strikes and becomes embedded in a 1.10-kg block of wood placed on a horizontal surface just in front of the gun. If the coefficient of kinetic friction between the block and the surface is 0.25, and the impact drives the block a distance of 9.5 m before it comes to rest, what was the muzzle speed of the bullet?

9. (II) An atomic nucleus at rest decays radioactively into an alpha particle and a smaller nucleus. What will be the speed of this recoiling nucleus if the speed of the alpha particle is 3.8×10^5 m/s? Assume the recoiling nucleus has a mass 57 times greater than that of the alpha particle.

10. (II) An atomic nucleus initially moving at 420 m/s emits an alpha particle in the direction of its velocity, and the new nucleus slows to 350 m/s. If the alpha particle has a mass of 4.0 u and the original nucleus has a mass of 222 u, what speed does the alpha particle have when it is emitted?

11. (II) A 13-g bullet traveling 230 m/s penetrates a 2.0-kg block of wood and emerges going 170 m/s. If the block is stationary on a frictionless surface when hit, how fast does it move after the bullet emerges?

12. (II) A 975-kg two-stage rocket is traveling at a speed of 5.80×10^3 m/s with respect to Earth when a predesigned explosion separates the rocket into two sections of equal mass that then move with a relative speed (relative to each other) of 2.20×10^3 m/s along the original line of motion. (*a*) What is the speed and direction of each section (relative to Earth) after the explosion? (*b*) How much energy was supplied by the explosion? [*Hint:* What is the change in KE as a result of the explosion?]

13. (III) A rocket of total mass 3180 kg is traveling in outer space with a velocity of 115 m/s toward the Sun. It wishes to alter its course by 35.0°, and can do this by firing its rockets briefly in a direction perpendicular to its original motion. If the rocket gases are expelled at a speed of 1750 m/s, how much mass must be expelled?

14. (I) A tennis ball may leave the racket of a top player on the serve with a speed of 65.0 m/s. If the ball's mass is 0.0600 kg and it is in contact with the racket for 0.0300 s, what is the average force on the ball? Would this force be large enough to lift a 60-kg person?

15. (I) A 0.145-kg baseball pitched at 39.0 m/s is hit on a horizontal line drive straight back toward the pitcher at 52.0 m/s. If the contact time between bat and ball is 1.00×10^{-3} s, calculate the average force between the ball and bat during contact.

16. (II) A golf ball of mass 0.045 kg is hit off the tee at a speed of 45 m/s. The golf club was in contact with the ball for 5.0×10^{-3} s. Find (a) the impulse imparted to the golf ball, and (b) the average force exerted on the ball by the golf club.

17. (II) A tennis ball of mass $m = 0.060$ kg and speed $v = 25$ m/s strikes a wall at a 45° angle and rebounds with the same speed at 45° (Fig. 7–29). What is the impulse given the wall?

FIGURE 7-29
Problem 17.

18. (II) A 115-kg fullback is running at 4.0 m/s to the east and is stopped in 0.75 s by a head-on tackle by a tackler running due west. Calculate (a) the original momentum of the fullback, (b) the impulse exerted on the fullback, (c) the impulse exerted on the tackler, and (d) the average force exerted on the tackler.

19. (II) Suppose the force acting on a tennis ball (mass 0.060 kg) points in the $+x$ direction and is given by the graph of Fig. 7–30 as a function of time. Use graphical methods to estimate (a) the total impulse given the ball, and (b) the velocity of the ball after being struck, assuming the ball is being served so it is nearly at rest initially.

20. (III) From what maximum height can a 75-kg person jump without breaking the lower leg bone on either leg? Ignore air resistance and assume the CM of the person moves a distance of 0.60 m from the standing to the seated position (that is, in breaking the fall). Assume the breaking strength (force per unit area) of bone is 170×10^6 N/m^2, and its smallest cross-sectional area is 2.5×10^{-4} m^2.

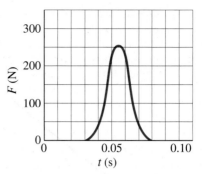

FIGURE 7-30   Problem 19.

SECTIONS 7-4 AND 7-5

21. (II) A ball of mass 0.440 kg moving east ($+x$ direction) with a speed of 3.70 m/s collides head-on with a 0.220-kg ball at rest. If the collision is perfectly elastic, what will be the speed and direction of each ball after the collision?

22. (II) A 0.450-kg ice puck, moving east with a speed of 3.00 m/s, has a head-on collision with a 0.900-kg puck initially at rest. Assuming a perfectly elastic collision, what will be the speed and direction of each object after the collision?

23. (II) Two billiard balls of equal mass undergo a perfectly elastic head-on collision. If the speed of one ball was initially 2.00 m/s, and of the other 3.00 m/s in the opposite direction, what will be their speeds after the collision?

24. (II) A 0.060-kg tennis ball, moving with a speed of 2.50 m/s, has a head-on collision with a 0.090-kg ball initially moving away from it at a speed of 1.00 m/s. Assuming a perfectly elastic collision, what is the speed and direction of each ball after the collision?

25. (II) A softball of mass 0.220 kg that is moving with a speed of 5.5 m/s collides head-on and elastically with another ball initially at rest. Afterward it is found that the incoming ball has bounced backward with a speed of 3.7 m/s. Calculate (a) the velocity of the target ball after the collision, and (b) the mass of the target ball.

26. (II) A pair of bumper cars in an amusement park ride collide elastically as one approaches the other directly from the rear (Fig. 7–31). One has a mass of 450 kg and the other 550 kg, owing to differences in passenger mass. If the lighter one approaches at 4.50 m/s and the other is moving at 3.70 m/s, calculate (a) their velocities after the collision, and (b) the change in momentum of each.

FIGURE 7-31   Problem 26: (a) before collision, (b) after collision.

203

27. (III) A 0.280-kg croquet ball makes an elastic head-on collision with a second ball initially at rest. The second ball moves off with half the original speed of the first ball. (a) What is the mass of the second ball? (b) What fraction of the original kinetic energy (ΔKE/KE) gets transferred to the second ball?

28. (III) In a physics lab, a small cube slides down a frictionless incline as shown in Fig. 7–32, and elastically strikes a cube at the bottom that is only one-half its mass. If the incline is 30 cm high and the table is 90 cm off the floor, where does each cube land?

FIGURE 7–32   Problem 28.

29. (III) Take the general case of a body of mass m_1 and velocity v_1 elastically striking a stationary ($v_2 = 0$) body of mass m_2 head-on. (a) Show that the final velocities v_1' and v_2' are given by

$$v_1' = \left(\frac{m_1 - m_2}{m_1 + m_2}\right) v_1,$$

$$v_2' = \left(\frac{2m_1}{m_1 + m_2}\right) v_1.$$

(b) What happens in the extreme case when m_1 is much smaller than m_2? Cite a common example of this. (c) What happens in the extreme case when m_1 is much larger than m_2? Cite a common example of this. (d) What happens in the case when $m_1 = m_2$? Cite a common example of this.

SECTION 7–6

30. (II) An 18-g rifle bullet traveling 230 m/s buries itself in a 3.6-kg pendulum hanging on a 2.8-m-long string, which makes the pendulum swing upward in an arc. Determine the horizontal component of the pendulum's displacement.

31. (II) (a) Derive a formula for the fraction of kinetic energy lost, ΔKE/KE, for the ballistic pendulum collision of Example 7–9. (b) Evaluate for $m = 14.0$ g and $M = 380$ g.

32. (II) An explosion breaks an object into two pieces, one of which has 1.5 times the mass of the other. If 7500 J were released in the explosion, how much kinetic energy did each piece acquire?

33. (II) A 1.0×10^3-kg Toyota collides into the rear end of a 2.2×10^3-kg Cadillac stopped at a red light. The bumpers lock, the brakes are locked, and the two cars skid forward 2.8 m before stopping. The police officer, knowing that the coefficient of kinetic friction between tires and road is 0.40, calculates the speed of the Toyota at impact. What was that speed?

34. (II) A measure of inelasticity in a head-on collision of two bodies is the *coefficient of restitution*, e, defined as

$$e = \frac{v_1' - v_2'}{v_2 - v_1},$$

where $v_1' - v_2'$ is the relative velocity of the two bodies after and $v_2 - v_1$ is their relative velocity before the collision. (a) Show that for a perfectly elastic collision, $e = 1$, and for a completely inelastic collision, $e = 0$. (b) A simple method for measuring the coefficient of restitution for a body colliding with a very hard surface like steel is to drop the body onto a heavy steel plate, as shown in Fig. 7–33. Determine a formula for e in terms of the original height h and the maximum height reached after collision h'.

FIGURE 7–33   Problem 34. Measurement of coefficient of restitution.

35. (III) A wooden block is cut into two pieces, one with three times the mass of the other. A depression is made in both faces of the cut, so that a firecracker can be placed in it with the block reassembled. The reassembled block is set on a rough-surfaced table and the fuse is lit. When the firecracker explodes, the two blocks separate. What is the ratio of distances each block travels?

36. (III) A 5.0-kg body moving in the $+x$ direction at 5.5 m/s collides head-on with a 3.0-kg body moving in the $-x$ direction at 4.0 m/s. Find the final velocity of each mass if: (a) the bodies stick together; (b) the collision is elastic; (c) the 5.0-kg body is at rest after the collision; (d) the 3.0-kg body is at rest after the collision; (e) the 5.0-kg body has a velocity of 4.0 m/s in the $-x$ direction after the collision. Are the results in (c), (d), and (e) "reasonable"? Explain.

*SECTION 7–7

* 37. (II) A radioactive nucleus at rest decays into a second nucleus, an electron, and a neutrino. The electron and neutrino are emitted at right angles and have momenta of 9.30×10^{-23} kg·m/s, and 5.40×10^{-23} kg·m/s, respectively. What is the magnitude and direction of the momentum of the second (recoiling) nucleus?

* 38. (II) An eagle ($m_1 = 4.3$ kg) moving with speed $v_1 = 7.8$ m/s is on a collision course with a second eagle ($m_2 = 5.6$ kg) moving at $v_2 = 10.2$ m/s in a direction at right angles to the first. After they collide, they hold onto one another. In what direction, and with what speed, are they moving after the collision?

* 39. (II) A billiard ball of mass $m_A = 0.400$ kg moving with speed $v_A = 1.80$ m/s strikes a second ball, initially at rest, of mass $m_B = 0.500$ kg. As a result of the collision, the first ball is deflected off at an angle of 30.0° with a speed $v'_A = 1.10$ m/s. (a) Taking the x axis to be the original direction of motion of ball A, write down the equations expressing the conservation of momentum for the components in the x and y directions separately. (b) Solve these equations for the speed, v'_B, and angle, θ', of ball B. Do not assume the collision is elastic.

* 40. (III) An atomic nucleus of mass m traveling with speed v collides elastically with a target particle of mass $2m$ (initially at rest) and is scattered at 90°. (a) At what angle does the target particle move after the collision? (b) What are the final speeds of the two particles? (c) What fraction of the initial KE (ΔKE/KE) is transferred to the target particle?

* 41. (III) After a completely inelastic collision between two objects of equal mass, each having initial speed v, the two move off together with speed $v/3$. What was the angle between their initial directions?

* 42. (III) In order to convert a tough split in bowling, it is necessary to strike the pin a glancing blow as shown in Fig. 7–34. Assume that the bowling ball, initially traveling at 12.0 m/s, has five times the mass of a pin and that the pin goes off at 80° from the original direction of the ball. Calculate (a) the final speed of the pin, (b) the final speed of the ball, and (c) the angle through which the ball was deflected. Assume the collision is elastic.

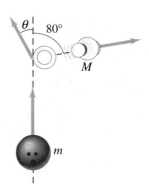

FIGURE 7–34   Problem 42.

* 43. (III) A neutron collides elastically with a helium nucleus (at rest initially) whose mass is four times that of the neutron. The helium nucleus is observed to rebound at an angle $\theta'_2 = 45°$. Determine the angle of the neutron, θ'_1, and the speeds of the two particles, v'_n and v'_{He}, after the collision. The neutron's initial speed is 6.2×10^5 m/s.

* 44. (III) Two billiard balls of equal mass move at right angles and meet at the origin of an xy coordinate system. One is moving upward along the y axis at 2.0 m/s, and the other is moving to the right along the x axis with speed 3.7 m/s. After the collision (assumed elastic), the second ball is moving along the positive y axis (Fig. 7–35). What is the final direction of the first ball, and what are their two speeds?

* 45. (III) Prove that in the elastic collision of two objects of identical mass, with one being a target initially at rest, the angle between their final velocity vectors is always 90°.

FIGURE 7–35   Problem 44. (Ball 1 after the collision is not shown.)

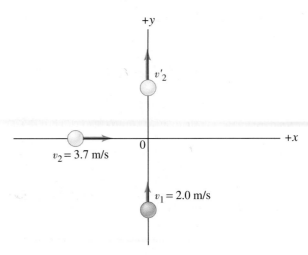

46. (I) The distance between a carbon atom ($m = 12$ u) and an oxygen atom ($m = 16$ u) in the CO molecule is 1.13×10^{-10} m. How far from the carbon atom is the center of mass of the molecule?

47. (I) An empty 1050-kg car has its CM 2.50 m behind the front of the car. How far from the front of the car will the CM be when two people sit in the front seat 2.80 m from the front of the car, and three people sit in the back seat 3.90 m from the front? Assume that each person has a mass of 70.0 kg.

48. (II) Three cubes, of side l_0, $2l_0$, and $3l_0$, are placed next to one another (in contact) with their centers along a straight line and the $l = 2l_0$ cube in the center (Fig. 7–36). What is the position, along this line, of the CM of this system? Assume the cubes are made of the same uniform material.

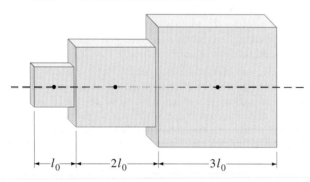

FIGURE 7–36   Problem 48.

49. (II) A square uniform raft, 18 m by 18 m, of mass 6200 kg, is used as a ferryboat. If three cars, each of mass 1200 kg, occupy the NE, SE, and SW corners, determine the CM of the loaded ferryboat.

50. (II) A (light) pallet has a load of cases of tomato paste (see Fig. 7–37), each of which is a cube of length l and has identical mass. Find the center of gravity in the horizontal plane, so that the crane operator can pick up the load without tipping it.

FIGURE 7–37
Problem 50.

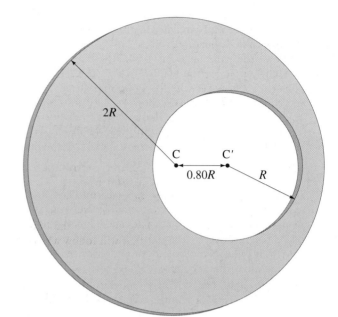

FIGURE 7–38   Problem 51.

51. (III) A uniform circular plate of radius $2R$ has a circular hole of radius R cut out of it. The center of the smaller circle is a distance $0.80R$ from the center of the larger circle, Fig. 7–38. What is the position of the center of mass of the plate? [Hint: Try subtraction.]

*SECTION 7–9

* 52. (I) Assume that your proportions are the same as those in Table 7–1, and calculate the mass of one of your legs.

* 53. (I) Determine the CM of an outstretched arm using Table 7–1.

* 54. (II) Use Table 7–1 to calculate the position of the CM of an arm bent at a right angle. Assume that the person is 155 cm tall.

* 55. (II) Calculate how far below the torso's median line the CM will be when a jumper is in a position such that his arms and legs are hanging vertically, and his trunk and head are horizontal. Will this be outside the body? Use Table 7–1.

*SECTION 7–10

* 56. (II) The masses of the Earth and Moon are 5.98×10^{24} kg and 7.35×10^{22} kg, respectively, and their centers are separated by 3.84×10^8 m. (a) Where is the CM of this system located? (b) What can you say about the motion of the Earth-Moon system about the Sun, and of the Earth and Moon separately about the Sun?

* 57. (II) A 55-kg woman and a 90-kg man stand 10.0 m apart on frictionless ice. (a) How far from the woman is their CM? (b) If they hold on to the two ends of a rope, and the man pulls on the rope so that he moves 2.5 m, how far from the woman will he be now? (c) How far will the man have moved when he collides with the woman?

FIGURE 7–39
Problem 58.

24.0 cm

8.00 cm

*58. (II) A mallet consists of a uniform cylindrical head of mass 2.00 kg and a diameter 0.0800 m mounted on a uniform cylindrical handle of mass 0.500 kg and length 0.240 m, as shown in Fig. 7–39. If this mallet is tossed, spinning, into the air, how far above the bottom of the handle is the point that will follow a parabolic trajectory?

*59. (II) (a) Suppose that in Example 7–13 (Fig. 7–27), $m_{II} = 3m_I$. Where then would m_{II} land? (b) What if $m_I = 3m_{II}$?

*60. (III) A helium balloon and its gondola, of mass M, are in the air and stationary with respect to the ground. A passenger, of mass m, then climbs out and slides down a rope with speed v, measured with respect to the balloon. With what speed and direction (relative to Earth) does the balloon then move? What happens if the passenger stops?

GENERAL PROBLEMS

61. During a Chicago storm, winds can whip horizontally at speeds of 100 km/h. If the air strikes a person at the rate of 40 kg/s per square meter and is brought to rest, calculate the force of the wind on a person. Assume the person's area to be 1.50 m high and 0.50 m wide. Compare to the typical maximum force of friction ($\mu \approx 1.0$) between the person and the ground, if the person has a mass of 70 kg.

62. A 5800-kg open railroad car coasts along with a constant speed of 8.60 m/s on a level track. Snow begins to fall vertically and fills the car at a rate of 3.50 kg/min. Ignoring friction with the tracks, what is the speed of the car after 90.0 min?

63. A 0.145 kg pitched baseball moving horizontally at 35.0 m/s strikes a bat and is popped straight up to a height of 55.6 m before turning around. If the contact time is 0.50 ms, calculate the average force on the ball during the contact.

64. A rocket of mass m traveling with speed v_0 along the x axis suddenly shoots out fuel, equal to one third of its mass, parallel to the y axis with speed $2v_0$. Give the components of the final velocity of the rocket.

65. A novice pool player is faced with the corner pocket shot shown in Fig. 7–40. The relative size of some of the dimensions (the units aren't important, only their ratios) are also shown. Should the player be worried about this being a "scratch shot," one where the cue ball will also fall into a pocket? Give details.

66. A 140-kg astronaut (including space suit) acquires a speed of 2.50 m/s by pushing off with his legs from an 1800-kg space capsule. (a) What is the change in speed of the space capsule? (b) If the push lasts 0.500 s, what is the average force exerted by each on the other? As the reference frame, use the position of the capsule before the push.

4.0

√3.0

1.0

Cue ball

FIGURE 7–40   Problem 65.

67. A golf ball rolls off the top of a flight of concrete stairs of total vertical height 4.00 m. The ball hits four times on the way down, each time striking the horizontal part of a different step. If all collisions are perfectly elastic, what is the bounce height on the fifth bounce when the ball reaches the bottom of the stairs?

68. A ball of mass m makes a head-on elastic collision with a second ball (at rest) and rebounds with a speed equal to one-fourth its original speed. What is the mass of the second ball?

69. You have been hired as an expert witness in a court case involving an automobile accident. The accident involved a car of mass 2000 kg (car A) which approached a stationary car of mass 1000 kg (car B). The driver of car A applied his brakes 15 m before he crashed into car B. After the collision, car A slid 15 m while car B slid 30 m. The coefficient of kinetic friction between the locked wheels and the road was measured to be 0.60. Prove to the court that the driver of car A was exceeding the 55-mph speed limit before applying the brakes.

* 70. Two people, one of mass 75 kg and the other of mass 60 kg, sit in a rowboat of mass 80 kg. With the boat initially at rest, the two people, who have been sitting at opposite ends of the boat 2.0 m apart from each other, now exchange seats. How far and in what direction will the boat move?

71. A meteor whose mass was about 10^8 kg struck the Earth ($m = 6.0 \times 10^{24}$ kg) with a speed of about 15 km/s and came to rest in the Earth. (a) What was the Earth's recoil speed? (b) What fraction of the meteor's kinetic energy was transformed to KE of the Earth? (c) By how much did the Earth's KE change as a result of this collision?

72. An explosion breaks an object, originally at rest, into two fragments. One fragment acquires twice the kinetic energy of the other. What is the ratio of their masses?

73. The force on a bullet is given by the formula $F = 580 - 1.8 \times 10^5 t$ over the time interval $t = 0$ to $t = 3.0 \times 10^{-3}$ s. In this formula, t is in seconds and F is in newtons. (a) Plot a graph of F vs. t for $t = 0$ to $t = 3.0$ ms. (b) Estimate, using graphical methods, the impulse given the bullet. (c) If the bullet achieves a speed of 220 m/s as a result of this impulse, given to it in the barrel of a gun, what must its mass be?

3.6 m

30°

M

FIGURE 7–41 Problems 74 and 75.

75. In Problem 74 (Fig. 7–41), what is the upper limit on the mass m if it is to rebound from M, slide up the incline, stop, slide down the incline, and collide with M again?

76. A 0.25-kg skeet (clay target) is fired at an angle of 30° to the horizon with a speed of 30 m/s (Fig. 7–42). When it reaches the maximum height, it is hit from below by a 15-g pellet traveling vertically upward at a speed of 200 m/s. The pellet is embedded in the skeet. (a) How much higher did the skeet go up? (b) How much extra distance, Δx, does the skeet travel because of the collision?

FIGURE 7–42 Problem 76.

Skeet

h'

$v = 200$ m/s

Pellet

h

y

x

Skeet $v_0 = 30$ m/s

30°

Δx

74. A block of mass $m = 2.20$ kg slides down a 30.0° incline which is 3.60 m high. At the bottom, it strikes a block of mass $M = 7.00$ kg which is at rest on a horizontal surface, Fig. 7–41. (Assume a smooth transition at the bottom of the incline.) If the collision is elastic, and friction can be ignored, determine (a) the speeds of the two blocks after the collision, and (b) how far back up the incline the smaller mass will go.

77. **The gravitational slingshot effect.** Figure 7–43 shows the planet Saturn moving in the negative x direction at its orbital speed (with respect to the sun) of 9.6 km/s. The mass of Saturn is 5.69×10^{26} kg. A spacecraft with mass 825 kg approaches Saturn, moving initially in the $+x$ direction at 10.4 km/s. The gravitational attraction of Saturn (a conservative force) causes the spacecraft to swing around it (orbit shown as dashed line) and head off in the opposite direction. Estimate the final speed of the spacecraft after it is far enough away to be nearly free of Saturn's gravitational pull.

FIGURE 7–43
Problem 77.

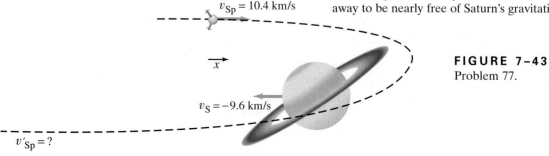

$v_{Sp} = 10.4$ km/s

x

$v_S = -9.6$ km/s

$v'_{Sp} = ?$

ROTATIONAL MOTION

We have, until now, been concerned mainly with translational motion. In this chapter, we will deal with rotational motion. We will mainly be concerned with rigid bodies. By a **rigid body** we mean a body with a definite shape that doesn't change, so that the particles composing it stay in fixed positions relative to one another. Of course, any real body is capable of vibrating or deforming when a force is exerted on it. But these effects are often very small, so the concept of an ideal rigid body is very useful as a good approximation.

The motion of a rigid body (as mentioned in Chapter 7) can be analyzed as the translational motion of its center of mass, plus rotational motion about its center of mass. We have already discussed translational motion in detail, so now we focus our attention on purely rotational motion. By *purely rotational motion*, we mean that all points in the body move in circles, such as the point P in the rotating wheel of Fig. 8–1, and that the centers of these circles all lie on a line called the **axis of rotation**, which in Fig. 8–1 is perpendicular to the page and passes through point O.

8–1 Angular Quantities

To describe rotational motion, we make use of angular quantities, such as angular velocity and angular acceleration. These are defined in analogy to the corresponding quantities in linear motion.

Every point in a body rotating about a fixed axis moves in a circle (shown dashed in Fig. 8–1 for point P) whose center is on the axis and

FIGURE 8–1 Looking down on a wheel that is rotating counterclockwise about an axis through the wheel's center at O (perpendicular to the page). The dashed circular line is the path of point P.

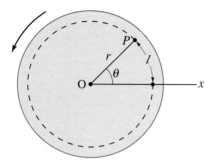

FIGURE 8–1 (Repeated.)

whose radius is r, the distance of that point from the axis of rotation. A straight line drawn from the axis to any point sweeps out the same angle θ in the same time.

To indicate the angular position of the body, or how far it has rotated, we specify the angle θ of some particular line in the body with respect to some reference line, such as the x axis (see Fig. 8–1). A point in the body (such as P in Fig. 8–1) moves through an angle θ when it travels the distance l measured along the circumference of its circular path. Angles are commonly measured in degrees, but the mathematics of circular motion is much simpler if we use the *radian* for angular measure. One **radian** (rad) is defined as the angle subtended by an arc whose length is equal to the radius. For example, in Fig. 8–1, point P is a distance r from the axis of rotation, and it has moved a distance l along the arc of a circle. The arc length l is said to "subtend" the angle θ. If $l = r$, then θ is exactly equal to 1 rad. In general, any angle θ is given by

θ in radians

$$\theta = \frac{l}{r}, \tag{8–1}$$

where r is the radius of the circle, and l is the arc length subtended by the angle θ which is specified in radians. Radians can be related to degrees in the following way. In a complete circle there are 360°, which of course must correspond to an arc length equal to the circumference of the circle, $l = 2\pi r$. Thus $\theta = l/r = 2\pi r/r = 2\pi$ rad in a complete circle, so

$$360° = 2\pi \text{ rad.}$$

1 rad ≈ 57.3° One radian is therefore $360°/2\pi \approx 360°/6.28 \approx 57.3°$.

FIGURE 8–2 (a) Example 8–1. (b) For small angles, arc length and the chord length (straight line) are nearly equal.

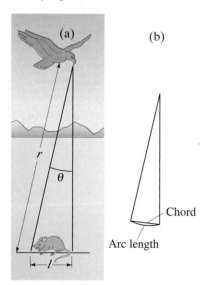

EXAMPLE 8–1 **Birds of prey—in radians.** A particular bird's eye can just distinguish objects that subtend an angle no smaller than about 3×10^{-4} rad. (a) How many degrees is this? (b) How small an object can the bird just distinguish when flying at a height of 100 m (Fig. 8–2a)?

SOLUTION (a) One radian is $360°/2\pi$, so 3×10^{-4} rad is

$$(3 \times 10^{-4} \text{ rad})\left(\frac{360°}{2\pi \text{ rad}}\right) = 0.017°.$$

(b) From Eq. 8–1, $l = r\theta$. For small angles, the arc length and the chord length are approximately[†] the same (Fig. 8–2b). Since $r = 100$ m and $\theta = 3 \times 10^{-4}$ rad, we find

$$l = (100 \text{ m})(3 \times 10^{-4} \text{ rad}) = 3 \times 10^{-2} \text{ m} = 3 \text{ cm.}$$

Had the angle been given in degrees, we would first have had to change it to radians to make this calculation.

Note in this Example that we used the fact that the radian is dimensionless (has no units) since it is the ratio of two lengths.

[†]Even for an angle as large as 15°, the error in making this estimate is only 1 percent, but for larger angles the error increases rapidly.

When an object, such as the bicycle wheel in Fig. 8–3, rotates from some initial position, specified by θ_0, to some final position, θ, its angular displacement is $\Delta\theta = \theta - \theta_0$. The *angular velocity* is defined in analogy with ordinary linear velocity. Instead of linear displacement, we use the angular displacement. Thus the **average angular velocity** (denoted by ω, the Greek lowercase letter omega) is defined as

$$\bar{\omega} = \frac{\Delta\theta}{\Delta t}, \qquad \textbf{(8–2a)}$$

where $\Delta\theta$ is the angle through which the body has rotated in the time Δt. We define the **instantaneous angular velocity** as the very small angle, $\Delta\theta$, through which the body turns in the very short time interval Δt:

$$\omega = \lim_{\Delta t \to 0} \frac{\Delta\theta}{\Delta t}. \qquad \textbf{(8–2b)}$$

(a)

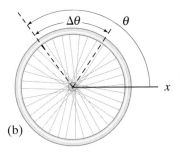

(b)

Angular velocity

Angular velocity is generally specified in radians per second (rad/s). Note that *all points in a rigid body rotate with the same angular velocity*, since every position in the body moves through the same angle in the same time interval.

Angular acceleration, in analogy to ordinary linear acceleration, is defined as the change in angular velocity divided by the time required to make this change. The **average angular acceleration** (denoted by α, the Greek lowercase letter alpha) is defined as

$$\bar{\alpha} = \frac{\omega - \omega_0}{\Delta t} = \frac{\Delta\omega}{\Delta t}, \qquad \textbf{(8–3a)}$$

where ω_0 is the angular velocity initially, and ω is the angular velocity after a time Δt. **Instantaneous angular acceleration** is defined in the usual way:

$$\alpha = \lim_{\Delta t \to 0} \frac{\Delta\omega}{\Delta t}. \qquad \textbf{(8–3b)}$$

FIGURE 8–3 A wheel rotates from (a) initial position θ_0, to (b) final position θ. The angular displacement is $\Delta\theta = \theta - \theta_0$.

Angular acceleration

Since ω is the same for all points of a rotating body, Eq. 8–3 tells us that α also will be the same for all points. Thus, ω and α are properties of the rotating body as a whole. With ω measured in radians per second and t in seconds, α will be expressed as radians per second squared (rad/s²).

Each particle or point of a rotating rigid body has, at any moment, a linear velocity v and a linear acceleration a. We can relate these linear quantities, v and a, of each particle, to the angular quantities, ω and α, of the rotating body as a whole. Consider a particle located a distance r from the axis of rotation, as in Fig. 8–4. If the body rotates with angular velocity ω, any particle will have a linear velocity whose direction is tangent to its circular path. The magnitude of its linear velocity, v, is $v = \Delta l/\Delta t$. From Eq. 8–1, a change in rotation angle $\Delta\theta$ is related to the linear distance traveled by $\Delta l = r\,\Delta\theta$. Hence

$$v = \frac{\Delta l}{\Delta t} = r\frac{\Delta\theta}{\Delta t}$$

or

$$v = r\omega. \qquad \textbf{(8–4)}$$

Thus, although ω is the same for every point in the rotating body at any

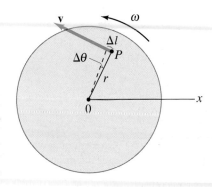

FIGURE 8–4 A particle P on a rotating wheel has a linear velocity **v** at any moment.

Linear and angular velocity related

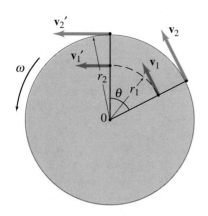

FIGURE 8–5 A wheel rotating uniformly counterclockwise. Two points on the wheel, at distances r_1 and r_2 from the center, have different linear velocities because they travel different distances in the same time interval. Since $r_2 > r_1$, then $v_2 > v_1$ ($v = r\omega$). But the two points have the same angular velocity ω because they travel through the same angle θ in the same time interval.

Tangential acceleration

Centripetal (or radial) acceleration

FIGURE 8–6 On a rotating wheel whose rotation speed is increasing, a point P has both tangential and radial (centripetal) components of acceleration. (See also Chapter 5.)

instant, the linear velocity v is greater for points farther from the axis (Fig. 8–5). Note that Eq. 8–4 is valid both instantaneously and on the average.

CONCEPTUAL EXAMPLE 8–2 **Is the lion faster than the horse?** A rotating carousel has one child sitting on a horse near the outer edge and another child seated on a lion halfway out from the center. (*a*) Which child has the greater linear speed? (*b*) Which child has the greater angular speed?

RESPONSE (*a*) The *linear* speed is the distance traveled divided by the time interval. In one rotation the child on the outer edge travels a longer distance than the child near the center, but the time interval is the same for both. Thus the child at the outer edge has the greater linear speed.
(*b*) The *angular* speed is the angle of rotation divided by the time interval. In one rotation both children rotate through the same angle ($360° = 2\pi$ radians). Thus the two children have the same angular speed.

We can use Eq. 8–4 to show that the angular acceleration α is related to the tangential linear acceleration a_{tan} of a particle in the rotating body by

$$a_{tan} = \frac{\Delta v}{\Delta t} = r\frac{\Delta \omega}{\Delta t}$$

or

$$a_{tan} = r\alpha. \tag{8–5}$$

In this equation, r is the radius of the circle in which the particle is moving, and the subscript tan in a_{tan} stands for "tangential" since the acceleration considered here is along the circle (that is, tangent to it).

The total linear acceleration of a particle is the vector sum of two components:

$$\mathbf{a} = \mathbf{a}_{tan} + \mathbf{a}_R,$$

where the radial[†] component, $\mathbf{a}_R$, is the radial or "centripetal" acceleration and points toward the center of the particle's circular path (Fig. 8–6). We saw in Chapter 5 (Eq. 5–1) that $a_R = v^2/r$, and we rewrite this in terms of ω using Eq. 8–4:

$$a_R = \frac{v^2}{r} = \frac{(\omega r)^2}{r} = \omega^2 r. \tag{8–6}$$

Thus the centripetal acceleration is greater the farther you are from the axis of rotation: the children farthest out on a merry-go-round feel the greatest acceleration. Equations 8–4, 8–5, and 8–6 relate the angular quantities describing the rotation of a body to the linear quantities for each particle of the body.

We can relate the angular velocity ω to the frequency of rotation, f, where by **frequency** we mean the number of complete revolutions (rev)

[†]"Radial" means along the radius—that is, toward or away from the center or axis.

per second. One revolution (of a wheel, say) corresponds to an angle of 2π radians, and thus $1\ \mathrm{rev/s} = 2\pi\ \mathrm{rad/s}$. Hence, in general, the frequency f is related to the angular velocity ω by

$$f = \frac{\omega}{2\pi}$$ *Frequency*

or

$$\omega = 2\pi f. \tag{8–7}$$

The unit for frequency, revolutions per second (rev/s), is given the special name, the hertz (Hz). That is

$$1\ \mathrm{Hz} = 1\ \mathrm{rev/s}.$$

Note that "revolution" is not really a unit, so we can also write $1\ \mathrm{Hz} = 1\ \mathrm{s}^{-1}$.

The time required for one complete revolution is called the **period**, T, and it is related to the frequency by

$$T = \frac{1}{f}. \tag{8–8}$$ *Period*

For example, if a particle rotates at a frequency of three revolutions per second, then each revolution takes $\frac{1}{3}$ s.

EXAMPLE 8–3 **Speed and acceleration on a merry-go-round.** (*a*) What is the linear speed of a child seated 1.2 m from the center of a steadily rotating merry-go-round (Fig. 8–7) that makes one complete revolution in 4.0 s? (*b*) What is her acceleration?

SOLUTION (*a*) First, we find the angular velocity in radians per second: the period is given as 4.0 s, so

$$f = \frac{1}{T} = \frac{1\ \mathrm{rev}}{4.0\ \mathrm{s}} = 0.25\ \mathrm{rev/s} = 0.25\ \mathrm{Hz}.$$

Then

$$\omega = 2\pi f = \left(2\pi\, \frac{\mathrm{rad}}{\mathrm{rev}}\right)\!\left(0.25\, \frac{\mathrm{rev}}{\mathrm{s}}\right) = 1.6\ \mathrm{rad/s}.$$

The radius r is 1.2 m, so the speed v is

$$v = r\omega = (1.2\ \mathrm{m})(1.6\ \mathrm{rad/s}) = 1.9\ \mathrm{m/s}.$$

(*b*) Since $\omega = 1.6\ \mathrm{rad/s} = \text{constant}$, then $\alpha = 0$ and the tangential component of the linear acceleration (Eq. 8–5) is

$$a_{\mathrm{tan}} = r\alpha = 0.$$

From Eq. 8–6, the radial component is

$$a_{\mathrm{R}} = \omega^2 r = (1.6\ \mathrm{rad/s})^2 (1.2\ \mathrm{m}) = 3.1\ \mathrm{m/s}^2.$$

Or we can solve instead: $a_{\mathrm{R}} = v^2/r = (1.9\ \mathrm{m/s})^2/(1.2\ \mathrm{m}) = 3.0\ \mathrm{m/s}^2$. (The difference is due to rounding off.) What force causes this acceleration? Is it a force of friction exerted by the merry-go-round?

FIGURE 8–7 Merry-go-round.

PHYSICS APPLIED

EXAMPLE 8–4 **Hard drive.** The platter of the hard disk of a computer rotates at 5400 rpm (revolutions per minute). (*a*) What is the angular velocity of the disk? (*b*) If the reading head of the drive is located 3.0 cm from the rotation axis, what is the speed of the disk below it? (*c*) What is the linear acceleration of this point? (*d*) If a single bit requires 5 μm of length along the motion direction, how many bits per second can the writing head write when it is 3.0 cm from the axis?

SOLUTION (*a*) The angular velocity

$$\omega = 2\pi f = (2\pi \, \text{rad/rev})\frac{(5400 \, \text{rev/min})}{(60 \, \text{s/min})} = 570 \, \text{rad/s}.$$

(*b*) The speed of a point 3.0 cm out from the axis is

$$v = r\omega = (3.0 \times 10^{-2} \, \text{m})(570 \, \text{rad/s}) = 17 \, \text{m/s}.$$

(*c*) The linear acceleration has two components, tangential and radial. Since ω = constant, then $\alpha = 0$, so $a_{\text{tan}} = r\alpha = 0$. The radial acceleration is

$$a_R = \omega^2 r = (570 \, \text{rad/s})^2(0.030 \, \text{m}) = 9700 \, \text{m/s}^2$$

toward the axis.
(*d*) Each bit requires 5.0×10^{-6} m, so at a speed of 17 m/s, the number of bits passing the head per second is

$$\frac{17 \, \text{m/s}}{5.0 \times 10^{-6} \, \text{m}} = 3.4 \times 10^6 \, \text{bits per second}.$$

EXAMPLE 8–5 **Centrifuge acceleration.** A centrifuge rotor is accelerated from rest to 20,000 rpm in 5.0 min. What is its average angular acceleration?

SOLUTION To calculate $\bar{\alpha} = \Delta\omega/\Delta t$ we need the initial and final angular velocities. The initial angular velocity is $\omega = 0$. The final angular velocity is

$$\omega = 2\pi f = (2\pi \, \text{rad/rev})\frac{(20{,}000 \, \text{rev/min})}{(60 \, \text{s/min})} = 2100 \, \text{rad/s}.$$

Then, since $\bar{\alpha} = \Delta\omega/\Delta t$ and $\Delta t = 5.0 \, \text{min} = 300 \, \text{s}$, we have

$$\bar{\alpha} = \frac{\omega - \omega_0}{\Delta t} = \frac{2100 \, \text{rad/s} - 0}{300 \, \text{s}} = 7.0 \, \text{rad/s}^2.$$

That is, every second the rotor's angular velocity increases by 7.0 rad/s, or by $(7.0/2\pi) = 1.1$ revolutions per second.

8–2 Kinematic Equations for Uniformly Accelerated Rotational Motion

In Chapter 2, we derived the important equations (2–10) that relate acceleration, velocity, and distance for the situation of uniform linear acceleration. Those equations were derived from the definitions of linear velocity and acceleration, assuming constant acceleration. The definitions of angular velocity and angular acceleration are the same as for their linear counterparts, except that θ has replaced the linear displacement x, ω has replaced v, and α has replaced a. Therefore, the angular equations for **constant angular acceleration** will be analogous to Eqs. 2–10 with x replaced by θ, v by ω, and a by α, and they can be derived in exactly the same way. We summarize them here, opposite their linear equivalents (we've chosen $x_0 = 0$, and $\theta_0 = 0$):

Angular	Linear			
$\omega = \omega_0 + \alpha t$	$v = v_0 + at$	[constant α, a]	**(8–9a)**	*Uniformly*
$\theta = \omega_0 t + \frac{1}{2}\alpha t^2$	$x = v_0 t + \frac{1}{2}at^2$	[constant α, a]	**(8–9b)**	*accelerated*
$\omega^2 = \omega_0^2 + 2\alpha\theta$	$v^2 = v_0^2 + 2ax$	[constant α, a]	**(8–9c)**	*rotational*
$\overline{\omega} = \dfrac{\omega + \omega_0}{2}$	$\overline{v} = \dfrac{v + v_0}{2}$	[constant α, a]	**(8–9d)**	*motion*

Note that ω_0 represents the angular velocity at $t = 0$, whereas θ and ω represent the angular position and velocity, respectively, at time t. Since the angular acceleration is constant, $\alpha = \overline{\alpha}$. These equations are of course also valid for constant angular velocity, for which case $\alpha = 0$ and we have $\omega = \omega_0$, $\theta = \omega_0 t$, and $\overline{\omega} = \omega$.

EXAMPLE 8–6 **Centrifuge revisited.** Through how many revolutions has the centrifuge rotor of Example 8–5 turned during its acceleration period? Assume constant angular acceleration.

SOLUTION We know that $\omega_0 = 0$, $\omega = 2100 \,\text{rad/s}$, $\alpha = \overline{\alpha} = 7.0 \,\text{rad/s}^2$, and $t = 300 \,\text{s}$. We could use either Eq. 8–9b or 8–9c to find θ. The former gives

$$\theta = 0 + \tfrac{1}{2}(7.0 \,\text{rad/s}^2)(300 \,\text{s})^2 = 3.15 \times 10^5 \,\text{rad},$$

where we have kept an extra digit because this is an intermediate result. To find the total number of revolutions, we divide by 2π and obtain

$$\frac{3.15 \times 10^5 \,\text{rad}}{2\pi \,\text{rad/rev}} = 5.0 \times 10^4 \,\text{rev}.$$

8–3 Rolling Motion

The rolling motion of a ball or wheel is familiar in everyday life: a ball rolling across the floor, or the wheels and tires of a car or bicycle rolling along the pavement. Rolling *without slipping* is readily analyzed and depends on static friction between the rolling object and the ground. The friction is static because the rolling object's point of contact with the ground is at rest at each moment. (Kinetic friction comes in if, for example, you brake too hard so the tires skid, or you accelerate so fast that you "burn rubber"— but these are more complicated situations.)

Rolling without slipping involves both rotation and translation. But there is a simple relation between the linear speed v of the axle and the

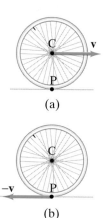

(a)

(b)

FIGURE 8–8 (a) A wheel rolling to the right. Its center C moves with velocity **v**. (b) The same wheel as seen from a reference frame in which the axle of the wheel C is at rest—that is, we are moving to the right with velocity **v** relative to part (a). Point P, which was at rest in (a), here in (b) moves to the left with velocity −**v** as shown. (See also Section 3–8 on relative velocity.)

angular velocity ω of the rotating wheel or sphere, namely $v = r\omega$ where r is the radius, as we now show. Figure 8–8a shows a wheel rolling to the right without slipping. At the moment shown, point P on the wheel is in contact with the ground and is momentarily at rest. The velocity of the axle at the wheel's center C is **v**. In Figure 8–8b we have put ourselves in the reference frame of the wheel—that is, we are moving to the right with velocity **v** relative to the ground. In this reference frame the axle C is at rest, whereas the ground and point P are moving to the left with velocity −**v** as shown. Here we are seeing pure rotation. So we can use Eq. 8–4 to obtain $v = r\omega$ where r is the radius of the wheel, and v is still equal to the translational speed of the wheel's center.

EXAMPLE 8–7 **Bicycle.** A bicycle slows down uniformly from $v_0 = 8.40$ m/s to rest over a distance of 115 m, Fig. 8–9a. Each wheel and tire has an overall diameter of 68.0 cm. Determine (a) the angular velocity of the wheels at the initial instant, (b) the total number of revolutions each wheel rotates in coming to rest, (c) the angular acceleration of the wheel, and (d) the time it took to come to a stop.

SOLUTION (a) Let us put ourselves in the reference frame of the bike—that is, as if we were riding the bike. Then the ground is going past us, initially, at a speed of 8.40 m/s, Fig. 8–9b. Since the tire is in contact with the ground at any moment, then a point on the rim of the tire (such as that touching the ground) moves at an initial speed of $v = 8.40$ m/s in

$v_0 = 8.40$ m/s

115 m

(a) Bike as seen from the ground ($t = 0$).

FIGURE 8–9 Example 8–7.

$v_{\text{ground}} = 8.40$ m/s

(b) From rider's reference frame, the ground is moving to the rear at an initial speed of 8.40 m/s ($t = 0$).

this reference frame. Hence the initial angular velocity of the wheel is

$$\omega_0 = \frac{v_0}{r} = \frac{8.40 \text{ m/s}}{0.340 \text{ m}} = 24.7 \text{ rad/s}.$$

(b) In coming to a stop, 115 m of ground passes beneath the tire. Because the tire is in firm contact with the ground, any point on the edge of the rotating tire travels 115 m total. Each revolution corresponds to a distance of $2\pi r$, so the number of revolutions the wheel makes in coming to a stop is

$$\frac{115 \text{ m}}{2\pi r} = \frac{115 \text{ m}}{(2\pi)(0.340 \text{ m})} = 53.8 \text{ rev}.$$

(c) The angular acceleration of the wheel can be obtained from Eq. 8–9c:

$$\alpha = \frac{\omega^2 - \omega_0^2}{2\theta} = \frac{0 - (24.7 \text{ rad/s})^2}{2(2\pi)(53.8 \text{ rev})} = -0.902 \text{ rad/s}^2,$$

where we have set $\theta = 2\pi \text{ rad/rev} \times 53.8 \text{ rev} (= 338 \text{ rad})$ because each revolution corresponds to 2π radians. [Alternatively, we could have used Eq. 8–1 to get the total θ: $\theta = l/r = 115 \text{ m}/0.340 \text{ m} = 338 \text{ rad}$.]

(d) Eq. 8–9a or b allows us to solve for the time. The first is easier:

$$t = \frac{\omega - \omega_0}{\alpha} = \frac{0 - 24.7 \text{ rad/s}}{-0.902 \text{ rad/s}^2} = 27.4 \text{ s}.$$

8–4 Torque

We have so far discussed rotational kinematics—the description of rotational motion in terms of angle, angular velocity, and angular acceleration. Now we discuss the dynamics, or causes, of rotational motion. Just as we found analogies between linear and rotational motion for the description of motion, so rotational equivalents for dynamics exist as well.

To make an object start rotating about an axis clearly requires a force. But the direction of this force, and where it is applied, are also important. Take, for example, an ordinary situation such as the door in Fig. 8–10 (looking down from above). If you apply a force $\mathbf{F}_1$ perpendicular to the door as shown, you will find that the greater the magnitude, F_1, the more quickly the door opens. (We assume that only this one force acts—we ignore friction in the hinges, and so on.) But now if you apply the same magnitude force at a point closer to the hinge, say $\mathbf{F}_2$ in Fig. 8–10, you will find that the door will not open so quickly. The effect of the force is less. Indeed, it is found that the angular acceleration of the door is proportional not only to the magnitude of the force, but it is also directly proportional to *the perpendicular distance from the axis of rotation to the line along which the force acts*. This distance is called the **lever arm**, or **moment arm**, of the force, and is labeled r_1 and r_2 for the two forces in Fig. 8–10. Thus, if r_1 in Fig. 8–10 is three times as large as r_2, then the angular acceleration of the door will be three times as great, assuming that the magnitudes of the forces are the same. To say it another way, if $r_1 = 3r_2$, then F_2 must be three times as large as F_1 to give the same angular acceleration. (Figure 8–11 shows two examples of tools whose long lever arms help produce large torque.)

FIGURE 8–10 Applying the same force with different lever arms, r_1 and r_2. If $r_1 = 3r_2$, then to create the same effect (angular acceleration), F_2 needs to be three times F_1, or $F_1 = \frac{1}{3}F_2$.

FIGURE 8–11 (a) A plumber can exert greater torque using a wrench with a long lever arm. (b) A tire iron too can have a long lever arm.

(a)   (b)

The angular acceleration, then, is proportional to the product of the *force times the lever arm*. This product is called the *moment of the force* about the axis, or, more commonly, it is called the **torque**, and is abbreviated τ (Greek lowercase letter tau). Thus, the angular acceleration α of an object is directly proportional to the net applied torque, τ:

$$\alpha \propto \tau.$$

This is the rotational analog of Newton's second law for linear motion, $a \propto F$.

We defined the lever arm as the *perpendicular* distance of the axis of rotation to the line of action of the force—that is, the distance which is perpendicular both to the axis of rotation and to an imaginary line drawn along the direction of the force. We do this to take into account the effect of forces acting at an angle. It is clear that a force applied at an angle, such as $\mathbf{F}_3$ in Fig. 8–12, will be less effective than the same magnitude force applied straight on, such as $\mathbf{F}_1$ (Fig. 8–12a). And if you push on the end of the door so that the force is directed at the hinge (the axis of rotation), as indicated by $\mathbf{F}_4$, the door will not rotate at all.

The lever arm for a force such as $\mathbf{F}_3$ is found by drawing a line along the direction of $\mathbf{F}_3$ (this is the "line of action" of $\mathbf{F}_3$). Then we draw another line, perpendicular to this "line of action", that goes to the axis and is perpendicular to it. The length of this second line is the lever arm for $\mathbf{F}_3$ and is labeled r_3 in Fig. 8–12b. The lever arm is perpendicular both to the line of action of the force and, at its other end, perpendicular to the rotation axis.

The magnitude of the torque associated with $\mathbf{F}_3$ is then $r_3 F_3$. This short lever arm and the corresponding smaller torque associated with $\mathbf{F}_3$ is consistent with the observation that $\mathbf{F}_3$ is less effective in accelerating the door than is $\mathbf{F}_1$. When the lever arm is defined in this way, experiment shows that the relation $\alpha \propto \tau$ is valid in general. Notice in Fig. 8–12 that the line of action of the force $\mathbf{F}_4$ passes through the hinge and hence its lever arm is zero. Consequently, zero torque is associated with $\mathbf{F}_4$ and it gives rise to no angular acceleration, in accord with everyday experience.

In general, then, we can write the torque about a given axis as

$$\tau = r_\perp F, \tag{8–10a}$$

where $r_\perp$ is the lever arm, and the perpendicular symbol ($\perp$) reminds us that we must use the distance from the axis of rotation that is perpendicular to the line of action of the force (Fig. 8–13a).

An alternate but equivalent way of determining the torque associated with a force is to resolve the force into components parallel and perpendicular to the line joining the point of application of the force to the axis, as shown in Fig. 8–13b. The component $F_\parallel$ exerts no torque since it is directed

(a)

(b)

FIGURE 8–12 (a) Forces acting at different angles at the doorknob. (b) The lever arm is defined as the perpendicular distance from the axis of rotation (the hinge) to the line of action of the force.

FIGURE 8–13
Torque $= r_\perp F = r F_\perp$

(a)

(b)

at the rotation axis (its moment arm is zero). Hence the torque will be equal to $F_\perp$ times the distance r from the axis to the point of application of the force:

$$\tau = rF_\perp. \qquad\qquad (8\text{--}10\text{b})$$

That this gives the same result as Eq. 8–10a can be seen from the fact that $F_\perp = F\sin\theta$ and $r_\perp = r\sin\theta$. [Note that θ is the angle between the directions of $\mathbf{F}$ and r (radial line from the axis to where $\mathbf{F}$ acts)]. So

$$\tau = rF\sin\theta \qquad\qquad (8\text{--}10\text{c})$$ *Magnitude of a torque*

in either case. We can use any of Eqs. 8–10 to calculate the torque, whichever is easiest.

Since torque is a distance times a force, it is measured in units of m·N in SI units,[†] cm·dyne in the cgs system, and ft·lb in the English system.

EXAMPLE 8–8 **Biceps torque.** The biceps muscle exerts a vertical force on the lower arm as shown in Figs. 8–14a and b. For each case, calculate the torque about the axis of rotation through the elbow joint, assuming the muscle is attached 5.0 cm from the elbow as shown.

SOLUTION (a) $F = 700$ N and $r_\perp = 0.050$ m, so

$$\tau = r_\perp F = (0.050\ \text{m})(700\ \text{N}) = 35\ \text{m·N}.$$

(b) Because the arm is at an angle, the lever arm is shorter (Fig. 8–14c): $r_\perp = (0.050\ \text{m})(\sin 60°)$. F is still 700 N, so

$$\tau = (0.050\ \text{m})(0.866)(700\ \text{N}) = 30\ \text{m·N}.$$

The arm can exert less torque at this angle. Weight machines at gyms are often designed to take this variation with angle into account.

CONCEPTUAL EXAMPLE 8–9 **The chimp is no wimp.** Adult chimpanzees have muscle mass of only about one third of an adult human male, but have been shown to be over twice as strong in some movements. Can you account for this?

RESPONSE The performance difference can be traced to anatomy. The attachment point of the biceps muscle to the forearm, for example, is much farther away from the elbow in chimps than in humans. The increase in the lever arm means that the same muscle force exerted by chimpanzee biceps produces a greater torque. A chimp has more leverage.

700 N

5.0 cm   (a)

700 N

30°

60°

(b)

$r_\perp$ **F**

Axis

30°

r 60°

(c)

FIGURE 8–14 Example 8–8.

When more than one torque acts on a body, the acceleration α is found to be proportional to the *net* torque. If all the torques acting on a body tend to rotate it in the same direction, the net torque is the sum of the torques. But if, say, one torque acts to rotate a body in one direction,

---

[†]Note that the units for torque are the same as those for energy. We write the unit for torque here as m·N (in SI) to help distinguish it from energy (N·m) because the two quantities are very different. An obvious difference is that energy is a scalar, whereas torque has a direction and is a vector. The special name *joule* (1 J = 1 N·m) is used only for energy (and for work), *never* for torque.

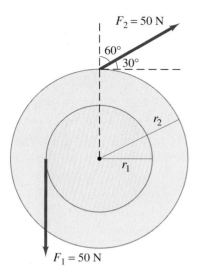

$F_2 = 50$ N

$60°$

$30°$

r_2

r_1

$F_1 = 50$ N

FIGURE 8–15 Example 8–10. The torque due to F_1 tends to accelerate the wheel counterclockwise, whereas the torque due to F_2 tends to accelerate the wheel clockwise.

FIGURE 8–16 Only the component of F that acts in the plane perpendicular to the rotation axis, $F_\perp$, acts to turn the wheel about the axis. The component parallel to the axis, $F_\parallel$, would tend to move the axis itself, which we assume is fixed.

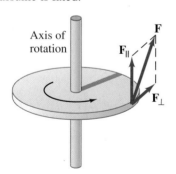

Axis of rotation

$F_\parallel$

F

$F_\perp$

and a second torque acts to rotate the body in the opposite direction (as in Fig. 8–15), the net torque is the difference of the two torques. We can assign a positive sign to torques that act to rotate the body in one direction (say counterclockwise) and a negative sign to torques that act to rotate the body in the opposite direction (clockwise).

EXAMPLE 8–10 **Torque on a compound wheel.** Two thin cylindrical wheels, of radii $r_1 = 30$ cm and $r_2 = 50$ cm, are attached to each other on an axle that passes through the center of each, as shown in Fig. 8–15. Calculate the net torque on this compound wheel due to the two forces shown, each of magnitude 50 N.

SOLUTION The force F_1 acts to rotate the system counterclockwise, whereas F_2 acts to rotate it clockwise. So the two forces act in opposition to each other. We must choose one direction of rotation to be positive—say, counterclockwise. Then F_1 exerts a positive torque, $\tau_1 = r_1 F_1$, since the lever arm is r_1. F_2, on the other hand, produces a negative (clockwise) torque and does not act perpendicular to r_2, so we must use its perpendicular component to calculate the torque it produces: $\tau_2 = -r_2 F_{2\perp} = -r_2 F_2 \sin\theta$, where $\theta = 60°$. (Note that θ must be the angle between F_2 and a radial line from the axis.) Hence the net torque is

$$\tau = r_1 F_1 - r_2 F_2 \sin 60°$$

$$= (0.30 \text{ m})(50 \text{ N}) - (0.50 \text{ m})(50 \text{ N})(0.866) = -6.7 \text{ m·N}.$$

This net torque acts to accelerate the rotation of the wheel in the clockwise direction. Note that the two forces have the same magnitude, yet produce a net torque because their lever arms are different.

[Since we are interested only in rotation about a fixed axis, we consider only forces that act in a plane perpendicular to the axis of rotation. If there is a force (or component of a force) acting parallel to the axis of rotation, it will tend to turn the axis of rotation—the component $F_\parallel$ in Fig. 8–16 is an example. Since we are assuming the axis remains fixed in direction, either there can be no such forces or else the axis must be an axle (or the like) which is mounted in bearings or hinges that hold the axis fixed. Thus, only a force, or component of a force ($F_\perp$ in Fig. 8–16), in a plane perpendicular to the axis will give rise to rotation about the axis, and it is only these that we consider.]

8–5 Rotational Dynamics; Torque and Rotational Inertia

We have discussed that the angular acceleration α of a rotating body is proportional to the net torque τ applied to it:

$$\alpha \propto \Sigma\tau$$

where we write $\Sigma\tau$ to remind us[†] that it is the net torque (sum of all torques acting on the body) that is proportional to α. This corresponds to Newton's second law for translational motion, $a \propto \Sigma F$, but here torque has taken the place of force, and, correspondingly, the angular accelera-

[†]Recall from Chapter 4 that Σ (Greek letter sigma) means "sum of."

tion α takes the place of the linear acceleration a. In the linear case, the acceleration is not only proportional to the net force, but it is also inversely proportional to the inertia of the body, which we call its mass, m. Thus we could write $a = \Sigma F/m$. But what plays the role of mass for the rotational case? That is what we now set out to determine. At the same time, we will see that the relation $\alpha \propto \Sigma \tau$ follows directly from Newton's second law, $\Sigma F = ma$.

We first consider a very simple case: a particle of mass m rotating in a circle of radius r at the end of a string or rod whose mass we can ignore (Fig. 8–17), and we assume a single force F acts on it as shown. The torque that gives rise to the angular acceleration is $\tau = rF$. If we make use of Newton's second law for linear quantities, $\Sigma F = ma$, and Eq. 8–5 relating the angular acceleration to the tangential linear acceleration, $a_{\text{tan}} = r\alpha$, we have

$$F = ma$$
$$= mr\alpha.$$

When we multiply both sides by r, we find that the torque $\tau = rF$ is given by

$$\tau = mr^2\alpha. \qquad \text{[single particle]} \quad \textbf{(8–11)}$$

FIGURE 8–17 A mass m rotating in a circle of radius r about a fixed point.

Here at last we have a direct relation between the angular acceleration and the applied torque τ. The quantity mr^2 represents the *rotational inertia* of the particle and is called its *moment of inertia*.

Now let us consider a rotating rigid body, such as a wheel rotating about an axis through its center, such as an axle. We can think of the wheel as consisting of many particles located at various distances from the axis of rotation. We can apply Eq. 8–11 to each particle of the body, and then sum over all the particles. The sum of the various torques is just the total torque, $\Sigma \tau$, so we obtain:

$$\Sigma \tau = (\Sigma mr^2)\alpha \qquad \textbf{(8–12)}$$

where we factored out the α since it is the same for all the particles of the body. The sum, Σmr^2, represents the sum of the masses of each particle in the body multiplied by the square of the distance of that particle from the axis of rotation. If we give each particle a number (1, 2, 3, ...), then $\Sigma mr^2 = m_1 r_1^2 + m_2 r_2^2 + m_3 r_3^2 + \cdots$. This quantity is called the **moment of inertia** (or *rotational inertia*) of the body, I:

$$I = \Sigma mr^2 = m_1 r_1^2 + m_2 r_2^2 + \cdots. \qquad \textbf{(8–13)}$$ *Moment of inertia*

Combining Eqs. 8–12 and 8–13, we can write

$$\Sigma \tau = I\alpha. \qquad \textbf{(8–14)}$$

NEWTON'S SECOND LAW FOR ROTATION

This is the rotational equivalent of Newton's second law. It is valid for the rotation of a rigid body about a fixed axis.[†]

We see that the moment of inertia, I, which is a measure of the rotational inertia of a body, plays the same role for rotational motion that mass does for translational motion. As can be seen from Eq. 8–13, the rotational inertia of an object depends not only on its mass, but also on how that mass is distributed with respect to the axis. For example, a large-diameter

[†]It can be shown that Eq. 8–14 is valid also when the body is translating with acceleration, as long as I and α are calculated about the center of mass of the body, and the rotation axis through the CM doesn't change direction.

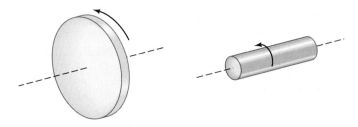

FIGURE 8–18 A large-diameter cylinder has greater rotational inertia than one of smaller diameter but equal mass.

Careful:
Mass can not be considered concentrated at CM for rotational motion

cylinder will have greater rotational inertia than one of equal mass but smaller diameter (and therefore greater length), Fig. 8–18. The former will be harder to start rotating, and harder to stop. When the mass is concentrated farther from the axis of rotation, the rotational inertia is greater. For rotational motion, the mass of a body *cannot* be considered as concentrated at its center of mass.

8–6 Solving Problems in Rotational Dynamics

Whenever you use Eq. 8–14, remember to use a consistent set of units, which in SI is: α in rad/s^2; τ in m·N; and the moment of inertia, I, in kg·m^2.

EXAMPLE 8–11 **Two weights on a bar: different axis, different *I*.** Two "weights" of mass 5.0 kg and 7.0 kg are mounted 4.0 m apart on a light rod (whose mass can be ignored), as shown in Fig. 8–19. Calculate the moment of inertia of the system (*a*) when rotated about an axis halfway between the weights, Fig. 8–19a, and (*b*) when the system rotates about an axis 0.50 m to the left of the 5.0-kg mass (Fig. 8–19b).

SOLUTION (*a*) Both weights are the same distance, 2.0 m, from the axis of rotation. Thus

$$I = \Sigma mr^2 = (5.0\ \text{kg})(2.0\ \text{m})^2 + (7.0\ \text{kg})(2.0\ \text{m})^2$$
$$= 20\ \text{kg·m}^2 + 28\ \text{kg·m}^2 = 48\ \text{kg·m}^2.$$

(*b*) The 5.0-kg mass is now 0.50 m from the axis and the 7.0-kg mass is 4.50 m from the axis. Then

$$I = \Sigma mr^2 = (5.0\ \text{kg})(0.50\ \text{m})^2 + (7.0\ \text{kg})(4.5\ \text{m})^2$$
$$= 1.3\ \text{kg·m}^2 + 142\ \text{kg·m}^2 = 143\ \text{kg·m}^2.$$

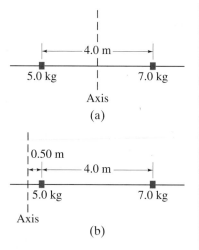

FIGURE 8–19 Example 8–11: calculating the moment of inertia.

The above Example illustrates two important points. First, the moment of inertia of a given system is different for different axes of rotation. Second, we see in part (*b*) that mass close to the axis of rotation contributes little to the total moment of inertia; in this example, the 5.0-kg object contributed less than 1 percent to the total.

I depends on axis of rotation and on distribution of mass

For most ordinary bodies, the mass is distributed continuously, and the calculation of the moment of inertia, $\Sigma\ mr^2$, can be difficult. Expressions can, however, be worked out (using calculus) for the moments of inertia of regularly shaped bodies in terms of their dimensions. Figure 8–20 gives these expressions for a number of solids rotated about the axes specified. The only one for which the result is obvious is that for the thin hoop or ring rotated about an axis passing through its center perpendicular to the plane of the hoop (Fig. 8–20a). For this object, all the mass is concentrated at the same distance from the axis, R. Thus $\Sigma mr^2 = (\Sigma m)R^2 = MR^2$, where M is the total mass of the hoop.

Object	Location of axis		Moment of inertia
(a)	Thin hoop of radius R	Through center	MR^2
(b)	Thin hoop of radius R and width W	Through central diameter	$\frac{1}{2}MR^2 + \frac{1}{12}MW^2$
(c)	Solid cylinder of radius R	Through center	$\frac{1}{2}MR^2$
(d)	Hollow cylinder of inner radius R_1 and outer radius R_2	Through center	$\frac{1}{2}M(R_1^2 + R_2^2)$
(e)	Uniform sphere of radius R	Through center	$\frac{2}{5}MR^2$
(f)	Long uniform rod of length L	Through center	$\frac{1}{12}ML^2$
(g)	Long uniform rod of length L	Through end	$\frac{1}{3}ML^2$
(h)	Rectangular thin plate, of length L and width W	Through center	$\frac{1}{12}M(L^2 + W^2)$

FIGURE 8–20 Moments of inertia for various objects of uniform composition.

➡ **PROBLEM SOLVING** Rotational Motion

1. As always, draw a clear and complete diagram.
2. Draw a free-body diagram for the body under consideration (or for each body if more than one), showing only (and all) the forces acting on that body and exactly where they act, so you can determine the torque due to each. Gravity acts at the CG of the body (Section 7–8).
3. Identify the axis of rotation and calculate the torques about it. Choose positive and negative directions of rotation (counterclockwise and clockwise), and assign the correct sign to each torque.
4. Apply Newton's second law for rotation, $\Sigma\tau = I\alpha$. If the moment of inertia is not given, and it is not the unknown sought, you need to determine it first. Use consistent units, which in SI are: α in rad/s²; τ in m·N; and I in kg·m².
5. Solve the resulting equation(s) for the unknown(s).
6. As always, do a rough estimate to determine if your answer is reasonable: does it make sense?

33.0 cm
(=R)

F_T

FIGURE 8–21 Example 8–12.

EXAMPLE 8–12 A heavy pulley. A 15.0-N force (represented by $\mathbf{F}_T$) is applied to a cord wrapped around a pulley of mass $M = 4.00\,\text{kg}$ and radius $R = 33.0\,\text{cm}$, Fig. 8–21. The pulley is observed to accelerate uniformly from rest to reach an angular speed of 30.0 rad/s in 3.00 s. If there is a frictional torque (at the axle), $\tau_{\text{fr}} = 1.10\,\text{m·N}$, determine the moment of inertia of the pulley. The pulley is assumed to rotate about its center. Its free-body diagram is shown in Fig. 8–21, although the friction force is not shown since we are given only its torque.

SOLUTION We can calculate the moment of inertia from Eq. 8–14, $\Sigma\tau = I\alpha$, since from the measurements given we can determine $\Sigma\tau$ and α. The net torque is the applied torque due to $\mathbf{F}_T$ minus the frictional torque; we take positive to be counterclockwise:

$$\Sigma\tau = (0.330\,\text{m})(15.0\,\text{N}) - 1.10\,\text{m·N} = 3.85\,\text{m·N}.$$

The angular acceleration is

$$\alpha = \frac{\Delta\omega}{\Delta t} = \frac{30.0\,\text{rad/s} - 0}{3.00\,\text{s}} = 10.0\,\text{rad/s}^2.$$

Hence

$$I = \frac{\Sigma\tau}{\alpha} = \frac{3.85\,\text{m·N}}{10.0\,\text{rads/s}^2} = 0.385\,\text{kg·m}^2.$$

FIGURE 8–22 Example 8–13, with free-body diagram for the falling bucket of mass m shown in (b).

R

F_T

(a)

F_T

$m\mathbf{g}$

(b)

EXAMPLE 8–13 Pulley and bucket: raising water from a well. Consider again the pulley in Fig. 8–21. But this time, suppose that, instead of a constant 15.0-N force being exerted on the cord, we now have a bucket of weight 15.0 N (mass $m = 1.53\,\text{kg}$) hanging from the cord, which we assume not to stretch or slip on the pulley. See Fig. 8–22a. (*a*) Calculate the angular acceleration α of the pulley and the linear acceleration a of the bucket. (*b*) Determine the angular velocity ω of the pulley and the linear velocity v of the bucket at $t = 3.00\,\text{s}$ if the pulley (and bucket) start from rest at $t = 0$.

SOLUTION (*a*) Let F_T be the tension in the cord. Then a force F_T acts at the edge of the pulley, and we have for the rotation of the pulley:

$$I\alpha = \Sigma\tau = F_T R - \tau_{\text{fr}}. \qquad \text{[pulley]}$$

Next we look at the (linear) motion of the bucket of mass m. Figure 8–22b shows a free-body diagram for the bucket. Two forces act on the bucket: the force of gravity mg acts downward, and the tension of the cord F_T pulls upward. So by $\Sigma F = ma$, for the bucket we have (taking downward as positive):

$$mg - F_T = ma. \qquad \text{[bucket]}$$

Note that the tension F_T, which is the force exerted on the edge of the pulley, is *not* equal to the weight of the bucket ($=mg = 15.0\,\text{N}$). There must be a net force (so $F_T < mg$) if the bucket is accelerating. Indeed, by the last equation above, $F_T = mg - ma$. To obtain α, we use Eq. 8–5,

$$a = R\alpha,$$

which is valid since the tangential acceleration of a point on the edge of the pulley is the same as the acceleration of the bucket if the cord doesn't stretch or slip. Substituting $F_T = mg - ma$ into the first equation above, we obtain

$$I\alpha = \Sigma\tau = F_T R - \tau_{fr} = (mg - mR\alpha)R - \tau_{fr} = mgR - mR^2\alpha - \tau_{fr}.$$

Now α appears on both sides of this last relation, so we solve for α:

$$\alpha(I + mR^2) = mgR - \tau_{fr},$$

and then

$$\alpha = \frac{mgR - \tau_{fr}}{I + mR^2}.$$

Then, since $I = 0.385$ kg·m² (Example 8–12),

$$\alpha = \frac{(15.0\ \text{N})(0.330\ \text{m}) - 1.10\ \text{m·N}}{0.385\ \text{kg·m}^2 + (1.53\ \text{kg})(0.330\ \text{m})^2} = 6.98\ \text{rad/s}^2.$$

The angular acceleration is somewhat less in this case than the 10.0 rad/s² of Example 8–12. Why? Because F_T ($ = mg - ma$) is somewhat less than the weight of the bucket, mg. The linear acceleration of the bucket is

$$a = R\alpha = (0.330\ \text{m})(6.98\ \text{rad/s}^2) = 2.30\ \text{m/s}^2.$$

(*b*) Since the angular acceleration is constant,

$$\omega = \omega_0 + \alpha t = 0 + (6.98\ \text{rad/s}^2)(3.00\ \text{s}) = 20.9\ \text{rad/s}$$

after 3.00 s. The linear velocity of the bucket is the same as that of a point on the wheel's edge:

$$v = R\omega = (0.330\ \text{m})(20.9\ \text{rad/s}) = 6.90\ \text{m/s}.$$

The same result can also be obtained by using the linear equation $v = v_0 + at = 0 + (2.30\ \text{m/s}^2)(3.00\ \text{s}) = 6.90\ \text{m/s}.$

8–7 Rotational Kinetic Energy

The quantity $\frac{1}{2}mv^2$ is the kinetic energy of a body undergoing translational motion. A body rotating about an axis is said to have **rotational kinetic energy**. By analogy with translational KE, we would expect this to be given by the expression $\frac{1}{2}I\omega^2$ where I is the moment of inertia of the body and ω is its angular velocity. We can indeed show that this is true. Consider any rigid rotating object as made up of many tiny particles, each of mass m. If we let r represent the distance of any one particle from the axis of rotation, then its linear velocity is $v = r\omega$. The total kinetic energy of the whole body will be the sum of the kinetic energies of all its particles:

$$\text{KE} = \Sigma(\tfrac{1}{2}mv^2) = \Sigma(\tfrac{1}{2}mr^2\omega^2)$$
$$= \tfrac{1}{2}\Sigma(mr^2)\omega^2,$$

where we have factored out the $\frac{1}{2}$ and the ω^2 since they are the same for every particle of a rigid body. Since $\Sigma mr^2 = I$, the moment of inertia, we see that the kinetic energy of a rigid rotating object, as expected, is

$$\text{rotational KE} = \tfrac{1}{2}I\omega^2. \qquad\qquad \textbf{(8–15)} \quad \textit{Rotational KE}$$

The units are joules, as with all other forms of energy.

An object that rotates while its center of mass (CM) undergoes translational motion will have both translational and rotational KE. Equation 8–15 gives the rotational KE if the rotation axis is fixed. If the object is moving (such as a wheel rolling down a hill), this equation is still valid as long as the rotation axis is fixed in direction. Then the total kinetic energy is

Total KE

$$\text{KE} = \tfrac{1}{2} M v_{\text{CM}}^2 + \tfrac{1}{2} I_{\text{CM}} \omega^2,$$

where v_{CM} is the linear velocity of the CM, I_{CM} is the moment of inertia about an axis through the CM, ω is the angular velocity about this axis, and M is the total mass of the body.

EXAMPLE 8–14 Sphere rolling down an incline. What will be the speed of a solid sphere of mass M and radius R when it reaches the bottom of an incline if it starts from rest at a vertical height H and rolls without slipping? See Fig. 8–23. Ignore losses due to dissipative forces, and compare your result to that for an object *sliding* down a frictionless incline.

SOLUTION We use the law of conservation of energy, and we must now include rotational kinetic energy. The total energy at any point a vertical distance y above the base of the incline is

$$\tfrac{1}{2} M v^2 + \tfrac{1}{2} I_{\text{CM}} \omega^2 + M g y,$$

where v is the speed of the CM. We equate the total energy at the top ($y = H$ and $v = \omega = 0$) to the total energy at the bottom ($y = 0$):

$$0 + 0 + M g H = \tfrac{1}{2} M v^2 + \tfrac{1}{2} I_{\text{CM}} \omega^2 + 0.$$

From Fig. 8–20, the moment of inertia of a solid sphere about an axis through its CM is $I_{\text{CM}} = \tfrac{2}{5} M R^2$. Since the sphere rolls without slipping, the speed, v, of the center of mass with respect to the point of contact (which is momentarily at rest at any instant) is equal to the speed of a point on the edge relative to the center, as we saw in Section 8–3 (Fig. 8–8). We therefore have $\omega = v/R$. Hence

$$\tfrac{1}{2} M v^2 + \tfrac{1}{2} (\tfrac{2}{5} M R^2) \left(\frac{v^2}{R^2} \right) = M g H.$$

Canceling the M's and R's, we obtain

$$(\tfrac{1}{2} + \tfrac{1}{5}) v^2 = g H$$

or

$$v = \sqrt{\tfrac{10}{7} g H}.$$

Note first that v is independent of both the mass M and the radius R of the sphere. Also, we can compare this result for the speed of a rolling sphere to that for an object sliding down a plane without rotating and without friction (see Section 6–7, $\tfrac{1}{2} m v^2 = m g H$), in which case $v = \sqrt{2 g H}$, which is greater. An object sliding without friction transforms its initial potential energy entirely into translational KE (none into rotational KE), so its speed is greater.

➡ **PROBLEM SOLVING**

Rotational energy adds to other forms of energy to get the total energy which is conserved

FIGURE 8–23 A sphere rolling down a hill has both translational and rotational kinetic energy. Example 8–14.

CONCEPTUAL EXAMPLE 8–15 Who's fastest? Several objects roll without slipping down an incline of vertical height H, all starting from rest at the same moment. The objects are a thin hoop (or a plain wedding band), a marble, a solid cylindrical battery (D-cell), an empty soup can, and an unopened soup can. In addition a greased box slides down without friction. In what order do they reach the bottom of the incline?

RESPONSE The sliding box wins every time! As we saw in Example 8–14, the speed of a rolling sphere at the bottom of the incline is less than that for a sliding box (without friction) because the potential energy loss (MgH) is transformed completely into translational KE for the box, whereas for rolling objects, the initial PE is shared between translational and rotational kinetic energy. For each of the rolling objects we can state that the loss in potential energy equals the increase in kinetic energy:

$$MgH = \tfrac{1}{2}Mv^2 + \tfrac{1}{2}I_{CM}\omega^2.$$

First we note that for all our rolling objects, the moment of inertia I_{CM} is a numerical factor times the mass M and the radius R^2 (Fig. 8–20). The mass M is in each term, so the translational speed v doesn't depend on M, nor does it depend on the radius R since $\omega = v/R$ so R^2 cancels out for all the rolling objects, just as in Example 8–14. Thus the speed v at the bottom depends only on that numerical factor in I_{CM} which expresses how the mass is distributed. Consequently, the hoop, with all its mass concentrated at radius R ($I_{CM} = MR^2$), will have the lowest speed and will arrive at the bottom behind the D-cell ($I_{CM} = \tfrac{1}{2}MR^2$) which in turn will be behind the marble ($I_{CM} = \tfrac{2}{5}MR^2$). The empty can, which is mainly a hoop plus a small disk, has its mass concentrated almost at R; so it will be a bit faster than the pure hoop but slower than the D-cell. See Fig. 8–24. The unopened soup can is much more complicated. It cannot be considered a solid cylinder because the soup can move about inside, and that will dissipate some energy; so we expect it to be slower than the D-cell, but that's about all we can safely say. For all the other objects, note that the speed at the bottom does not depend on the object's mass M or radius R, but only on its shape (and the height of the hill H).

If there had been no friction between the sphere (and other rolling objects) and the plane in these Examples, the sphere would have slid rather than rolled. Friction must be present to make a round object roll. We did not need to take friction into account in the energy equation because it is *static* friction and does no work. If we assume the sphere is perfectly rigid and thus is in contact with the surface at a point, then the force of friction acts parallel to the plane. But the point of contact of the sphere at each instant does not slide—it moves perpendicular to the plane (first down and then up) as the sphere rolls (Fig. 8–25). Thus, no work is done by the friction force because the force and the motion are perpendicular. The reason the rolling objects in Examples 8–14 and 8–15 move down the slope more slowly than if they were sliding is *not* because friction is doing work. Rather it is because some of the gravitional PE is converted to rotational KE, leaving less for the translational KE.

The work done on a body rotating about a fixed axis, such as the wheel or pulley in Fig. 8–21 and 8–22, can be written using angular quantities. As shown in Fig. 8–26, a force F exerting a torque $\tau = rF$ on the wheel does work $W = F\,\Delta l$ in rotating the wheel a small distance Δl. The wheel has rotated through a small angle $\Delta\theta = \Delta l/r$ (see Eq. 8–1). Hence

$$W = F\,\Delta l = Fr\,\Delta\theta.$$

Since $\tau = rF$, then

$$W = \tau\,\Delta\theta \qquad\qquad \textbf{(8–16)} \qquad \textit{Work done by torque}$$

is the work done by the torque τ in rotating the wheel through an angle $\Delta\theta$.

FIGURE 8–24   Example 8–15.

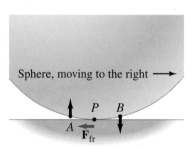

FIGURE 8–25   A sphere rolling to the right on a plane surface. The point in contact with the ground at any moment, point P, is momentarily at rest. Point A on the left of P is moving nearly vertically upward at the instant shown, and point B on the right is moving nearly vertically downward. (An instant later, point B will touch the plane and be at rest momentarily.)

FIGURE 8–26   Torque $\tau = rF$ does work in rotating a wheel equal to $W = F\Delta l = Fr\Delta\theta = \tau\Delta\theta$.

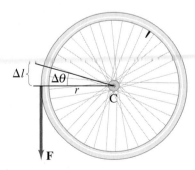

8–8 Angular Momentum and Its Conservation

Throughout this chapter we have seen that if we use the appropriate angular variables, the kinematic and dynamic equations for rotational motion are analogous to those for ordinary linear motion. In the last section (Section 8–7) we saw, for example, that rotational kinetic energy can be written as $\frac{1}{2}I\omega^2$, which is analogous to the translational $\text{KE} = \frac{1}{2}mv^2$. In like manner, the linear momentum, $p = mv$, has a rotational analog. It is called **angular momentum**, L, and for a body rotating about a fixed axis it is defined as

Angular momentum

$$L = I\omega, \tag{8–17}$$

where I is the moment of inertia, and ω is the angular velocity. The SI units for L are $\text{kg·m}^2/\text{s}$.

We saw in Chapter 7 (Section 7–1) that Newton's second law can be written not only as $\Sigma F = ma$, but also more generally in terms of momentum (Eq. 7–2), $\Sigma F = \Delta p/\Delta t$. In a similar way, the rotational equivalent of Newton's second law, which we saw in Eq. 8–14 can be written as $\Sigma \tau = I\alpha$, can also be written in terms of angular momentum:

$$\Sigma \tau = \frac{\Delta L}{\Delta t}, \tag{8–18}$$

where $\Sigma \tau$ is the net torque acting to rotate the body, and ΔL is the change in angular momentum in the time Δt. Equation 8–14, $\Sigma \tau = I\alpha$, is a special case of Eq. 8–18 when the moment of inertia is constant. This can be seen as follows. If a body has angular velocity ω_0 at time $t = 0$, and angular velocity ω at a time Δt later, then its angular acceleration (Eq. 8–3) is

$$\alpha = \frac{\Delta \omega}{\Delta t} = \frac{\omega - \omega_0}{\Delta t}.$$

Then from Eq. 8–18, we have

$$\Sigma \tau = \frac{\Delta L}{\Delta t} = \frac{I\omega - I\omega_0}{\Delta t} = \frac{I(\omega - \omega_0)}{\Delta t} = I\frac{\Delta \omega}{\Delta t} = I\alpha,$$

which is Eq. 8–14.

Angular momentum is an important concept in physics because, under certain conditions, it is a conserved quantity. We can see from Eq. 8–18 that if the net torque $\Sigma \tau$ on a body is zero, then $\Delta L/\Delta t$ equals zero. That is, L does not change. This, then, is the **law of conservation of angular momentum** for a rotating body:

The total angular momentum of a rotating body remains constant if the net torque acting on it is zero.

The law of conservation of angular momentum is one of the great conservation laws of physics.

When there is zero net torque acting on a body, and the body is rotating about a fixed axis or about an axis through its CM such that its direction doesn't change, we can write

$$I\omega = I_0\omega_0 = \text{constant.}$$

I_0 and ω_0 are the moment of inertia and angular velocity, respectively, about that axis at some initial time ($t = 0$), and I and ω are their values at some other time. The parts of the body may alter their positions relative to one another, so that I changes. But then ω changes as well and the product $I\omega$ remains constant.

Many interesting phenomena can be understood on the basis of conservation of angular momentum. Consider a skater doing a spin on the tips of her skates, Fig. 8–27. She rotates at a relatively low speed when her arms are outstretched, but when she brings her arms in close to her body, she suddenly spins much faster. By remembering the definition of moment of inertia as $I = \Sigma mr^2$, it is clear that when she pulls her arms in closer to the axis of rotation, r is reduced for the arms so her moment of inertia is reduced. Since the angular momentum $I\omega$ remains constant (we ignore the small torque due to friction), if I decreases, then the angular velocity ω must increase. If the skater reduces her moment of inertia by a factor of 2, she will then rotate with twice the angular velocity.

A similar example is the diver shown in Fig. 8–28. The push as she leaves the board gives her an initial angular momentum about her CM. When she curls herself into the tuck position, she rotates quickly one or more times. She then stretches out again, increasing her moment of inertia, which reduces the angular velocity to a small value, and then she enters the water. The change in moment of inertia from the straight position to the tuck position can be a factor of as much as $3\frac{1}{2}$.

Note that for angular momentum to be conserved, the net torque must be zero, but the net force does not necessarily have to be zero. The net force on the diver in Fig. 8–28, for example, is not zero (gravity is acting), but the net torque on her is zero.

FIGURE 8–27 A skater doing a spin on ice, illustrating conservation of angular momentum: in (a), I is large and ω is small; in (b), I is smaller so ω is larger.

FIGURE 8–28 A diver rotates faster when arms and legs are tucked in than when they are outstretched. Angular momentum is conserved.

I large, ω small

I small, ω large

(a)   (b)

FIGURE 8–29 Example 8–16.

EXAMPLE 8–16 **Object rotating on a string of changing length.** A mass m attached to the end of a string revolves in a circle on a frictionless tabletop. The other end of the string passes through a hole in the table (Fig. 8–29). Initially, the mass revolves with a speed $v_1 = 2.4$ m/s in a circle of radius $r_1 = 0.80$ m. The string is then pulled slowly through the hole so that the radius is reduced to $r_2 = 0.48$ m. What is the speed, v_2, of the mass now?

SOLUTION The force exerted by the string on the mass m does not alter its angular momentum about the axis of rotation, because the force is exerted toward the axis so the lever arm is zero and $\tau = 0$. Hence, from conservation of angular momentum:

$$I_1\omega_1 = I_2\omega_2.$$

Our small mass is essentially a particle whose moment of inertia is $I = mr^2$ (Section 8–5, Eq. 8–11), so we have

$$mr_1^2\omega_1 = mr_2^2\omega_2,$$

or

$$\omega_2 = \omega_1\left(\frac{r_1^2}{r_2^2}\right).$$

Then, since $v = r\omega$, we can write:

$$v_2 = r_2\omega_2 = r_2\omega_1\left(\frac{r_1^2}{r_2^2}\right) = r_2\frac{v_1}{r_1}\left(\frac{r_1^2}{r_2^2}\right) = v_1\frac{r_1}{r_2}$$

$$= (2.4 \text{ m/s})\left(\frac{0.80 \text{ m}}{0.48 \text{ m}}\right) = 4.0 \text{ m/s}.$$

FIGURE 8–30 (a) Rotating wheel. (b) Right-hand rule for obtaining the direction of $\boldsymbol{\omega}$.

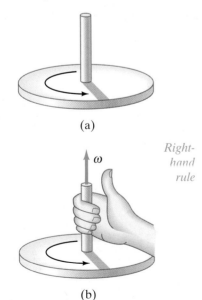

(a)

Right-hand rule

(b)

* 8–9 Vector Nature of Angular Quantities

Up to now we have considered only the magnitudes of angular variables such as ω, α, and L. But they can be treated as vectors, and now we consider in what directions these vectors point. In fact, we have to *define* the directions for rotational quantities, and we take first the angular velocity, $\boldsymbol{\omega}$.

Consider the rotating wheel shown in Fig. 8–30a. The linear velocities of different particles of the wheel point in all different directions. The only unique direction in space associated with the rotation is along the axis of rotation, perpendicular to the actual motion. We therefore choose the axis of rotation to be the direction of the angular velocity vector, $\boldsymbol{\omega}$. Actually, there is still an ambiguity since $\boldsymbol{\omega}$ could point in either direction along the axis of rotation (up or down in Fig. 8–30a). The convention we use, called **the right-hand rule**, is the following: When the fingers of the right hand are curled around the rotation axis and point in the direction of the rotation, then the thumb points in the direction of $\boldsymbol{\omega}$. This is shown in Fig. 8–30b. Note that $\boldsymbol{\omega}$ points in the direction a right-handed screw would move when turned in the direction of rotation. Thus, if the rotation of the wheel in Fig. 8–30a is counterclockwise, the direction of $\boldsymbol{\omega}$ is upward as shown in Fig. 8–30b. If the wheel rotates clockwise, then $\boldsymbol{\omega}$ points in the opposite direction, downward. Note that no part of the rotating body moves in the direction of $\boldsymbol{\omega}$.

If the axis of rotation is fixed, then $\boldsymbol{\omega}$ can change only in magnitude. Thus $\boldsymbol{\alpha} = \Delta\boldsymbol{\omega}/\Delta t$ must also point along the axis of rotation. If the rotation is counterclockwise as in Fig. 8–30a, and if the magnitude ω is increasing, then $\boldsymbol{\alpha}$ points upward; but if ω is decreasing (the wheel is slowing down), $\boldsymbol{\alpha}$ points downward. If the rotation is clockwise, $\boldsymbol{\alpha}$ will point downward if ω is increasing, and point upward if ω is decreasing.

Angular momentum, like linear momentum, is a vector quantity. For a symmetrical body (such as a wheel, cylinder, hoop, sphere) rotating about a symmetry axis, we can write the vector angular momentum as

$$\mathbf{L} = I\boldsymbol{\omega}. \qquad (8\text{–}19)$$

The angular velocity vector $\boldsymbol{\omega}$ (and therefore also $\mathbf{L}$) points along the axis of rotation in the direction given by the right-hand rule (Fig. 8–30b).

The vector nature of angular momentum can be used to explain a number of interesting (and sometimes surprising) phenomena. For example, consider a person standing at rest on a circular platform capable of rotating without friction about an axis through its center (that is, a simplified merry-go-round). If the person now starts to walk along the edge of the platform, Fig. 8–31a, the platform starts rotating in the opposite direction. Why? First of all because the person's foot exerts a force on it; but also because (and this is the most useful analysis here) this is an example of the conservation of angular momentum. If the person starts walking counterclockwise, the person's angular momentum will be pointed upward along the axis of rotation (remember how we defined the direction of $\boldsymbol{\omega}$ using the right-hand rule). The magnitude of the person's angular momentum will be $L = I\omega = (mr^2)(v/r)$, where v is the person's speed (relative to Earth, not the platform), r is his distance from the rotation axis, m is his mass, and his moment of inertia is mr^2 if we consider him a particle (mass concentrated at one point). The platform rotates in the opposite direction, so its angular momentum points downward. If the initial total angular momentum was zero (person and platform at rest), it will remain zero after the person starts walking—that is, the upward angular momentum of the person just balances the oppositely directed downward angular momentum of the platform (Fig. 8–31b), so the total vector angular momentum remains zero. Even though the person exerts a force (and torque) on the platform, and vice versa, these are internal torques (internal to the system consisting of platform plus person). There are no external torques (assuming friction-free bearings of the platform), so the angular momentum remains constant.

CONCEPTUAL EXAMPLE 8–17 **Spinning bicycle wheel.** Your physics teacher is holding a spinning bicycle wheel while standing on a stationary frictionless turntable (Fig. 8–32). What will happen if the teacher suddenly flips the bicycle wheel over so that it is spinning in the opposite direction?

RESPONSE The total angular momentum initially is $\mathbf{L}$ vertically upward, and that's what the system's angular momentum must be afterward since $\mathbf{L}$ is conserved. Thus, if the wheel's angular momentum afterward is $-\mathbf{L}$ downward, then the angular momentum of teacher plus turntable will have to be $+2\mathbf{L}$ upward. We can safely predict that the teacher will begin spinning around in the same direction the wheel was spinning originally.

(b)

FIGURE 8–31 (a) A person standing on a circular platform, both initially at rest, begins walking along the edge at speed v. The platform, assumed to be mounted on friction-free bearings, begins rotating in the opposite direction, so that the total angular momentum remains zero, as shown in (b).

FIGURE 8–32 Example 8–17.

SUMMARY

When a rigid body rotates about a fixed axis, each point of the body moves in a circular path. Lines drawn perpendicularly from the rotation axis to various points in the body all sweep out the same angle θ in any given time interval.

Angles are conveniently measured in **radians**, where one radian is the angle subtended by an arc whose length is equal to the radius, or

$$2\pi \, \text{rad} = 360°$$

$$1 \, \text{rad} \approx 57.3°.$$

Angular velocity, ω, is defined as the rate of change of angular position:

$$\omega = \frac{\Delta\theta}{\Delta t}.$$

All parts of a rigid body rotating about a fixed axis have the same angular velocity at any instant.

Angular acceleration, α, is defined as the rate of change of angular velocity:

$$\alpha = \frac{\Delta\omega}{\Delta t}.$$

The linear velocity v and acceleration a of a point fixed at a distance r from the axis of rotation are related to ω and α by

$$v = r\omega, \qquad a_{\text{tan}} = r\alpha, \qquad a_{\text{R}} = \omega^2 r,$$

where a_{tan} and a_{R} are the tangential and radial (centripetal) components of the linear acceleration, respectively.

The frequency f is related to ω by $\omega = 2\pi f$, and to the period T by $T = 1/f$.

The equations describing uniformly accelerated rotational motion ($\alpha = $ constant) have the same form as for uniformly accelerated linear motion:

$$\omega = \omega_0 + \alpha t; \qquad \theta = \omega_0 t + \tfrac{1}{2}\alpha t^2;$$

$$\omega^2 = \omega_0^2 + 2\alpha\theta; \qquad \bar{\omega} = \frac{\omega + \omega_0}{2}.$$

The dynamics of rotation is analogous to the dynamics of linear motion. Force is replaced by **torque**, τ, which is defined as the product of force times lever arm (perpendicular distance from the line of action of the force to the axis of rotation). Mass is replaced by **moment of inertia**, I, which depends not only on the mass of the body, but also on how the mass is distributed about the axis of rotation. Linear acceleration is replaced by angular acceleration. The rotational equivalent of Newton's second law is then

$$\Sigma\tau = I\alpha.$$

The **rotational kinetic energy** of a body rotating about a fixed axis with angular velocity ω is

$$\text{KE} = \tfrac{1}{2}I\omega^2.$$

For a body both translating and rotating, the total kinetic energy is the sum of the translational KE of the body's CM plus the rotational KE of the body about its CM:

$$\text{KE} = \tfrac{1}{2}Mv_{\text{CM}}^2 + \tfrac{1}{2}I_{\text{CM}}\omega^2$$

as long as the rotation axis is fixed in direction.

The **angular momentum**, L, of a body about a fixed rotation axis is given by

$$L = I\omega.$$

Newton's second law, in terms of angular momentum, becomes

$$\Sigma\tau = \frac{\Delta L}{\Delta t}.$$

If the net torque on the body is zero, $\Delta L/\Delta t = 0$, so $L = $ constant. This is the **law of conservation of angular momentum** for a rotating body.

1. You are standing a known distance from the Statue of Liberty. Describe how you could determine its height using only a meter stick, and without moving from your place.

2. A bicycle odometer (which measures distance traveled) is attached near the wheel hub and is designed for 27-inch wheels. What happens if you use it on a bicycle with 24-inch wheels?

3. Suppose a record turntable rotates at constant angular velocity. Does a point on the rim have radial and/or tangential acceleration? If the turntable's angular velocity increases uniformly, does the point have radial and/or tangential acceleration? For which cases would the magnitude of either of the components of linear acceleration change?

4. If the angular quantities θ, ω, and α were specified in terms of degrees rather than radians, how would Eqs. 8–9 for uniformly accelerated rotational motion have to be altered?

5. Can a small force exert a greater torque than a larger force? Explain.

6. If a force **F** acts on a body such that its lever arm is zero, does it have any effect on the body's motion?

7. Why is it more difficult to do a sit-up with your hands behind your head than when they are outstretched in front of you? A diagram may help you to answer this.

8. Expert bicyclists use very lightweight "sew-up" (tubular) tires. They claim that reducing the mass of the tires is far more significant than an equal reduction in mass elsewhere on the bicycle. Explain why.

9. A 21-speed bicycle has seven sprockets at the rear wheel and three at the pedal cranks. In which gear is it harder to pedal, a small rear sprocket or a large rear sprocket? Why? In which gear is it harder to pedal, a small front sprocket or a large front sprocket? Why?

10. Mammals that depend on being able to run fast have slender lower legs with flesh and muscle concentrated high, close to the body (Fig. 8–33). On the basis of rotational dynamics, explain why this distribution of mass is advantageous.

FIGURE 8–33 A gazelle. Question 10.

FIGURE 8–34 Question 11.

11. Why do tightrope walkers carry a long, narrow beam (Fig. 8–34)?

12. If the net force on a system is zero, is the net torque also zero? If the net torque on a system is zero, is the net force zero?

13. A stick stands vertically on its end on a frictionless surface. Describe the motion of its CM, and of each end, when it is tipped slightly to one side and falls.

14. Two inclines have the same height but make different angles with the horizontal. The same steel ball is rolled down each incline. On which incline will the speed of the ball at the bottom be greatest? Explain.

15. Two solid spheres simultaneously start rolling (from rest) down an incline. One sphere has twice the radius and twice the mass of the other. Which reaches the bottom of the incline first? Which has the greater speed there? Which has the greater total kinetic energy at the bottom?

16. A sphere and a cylinder have the same radius and the same mass. They start from rest at the top of an incline. Which reaches the bottom first? Which has the greater speed at the bottom? Which has the greater total kinetic energy at the bottom? Which has the greater rotational KE?

17. A cyclist rides over the top of a hill. Is the bicycle's motion rotational, translational, or a combination of both?

18. We claim that momentum and angular momentum are conserved. Yet most moving or rotating bodies eventually slow down and stop. Explain.

19. If there were a great migration of people toward the equator, how would this affect the length of the day?

20. Can the diver of Fig. 8–28 do a somersault without having any initial rotation when she leaves the board?

*21. In what direction is the Earth's angular velocity for its daily rotation on its axis?

*22. The angular velocity of a wheel rotating on a horizontal axle points west. In what direction is the linear velocity of a point on the top of the wheel? If the angular acceleration points east, describe the tangential linear acceleration of this point. Is the angular speed increasing or decreasing?

*23. When a motorcyclist leaves the ground on a jump, if the throttle is left on (so that the rear wheel spins), why does the front of the cycle rise up?

*24. Suppose you are standing on the edge of a large rotating turntable. What happens if you walk toward the center?

*25. A quarterback leaps into the air to throw a forward pass. As he throws the ball, the upper part of his body rotates. If you look quickly you will notice that his hips and legs rotate in the opposite direction (Fig. 8–35). Explain.

*26. Look at the face of a clock with a second hand. In what direction is the angular momentum of the second hand?

FIGURE 8–35
Quarterback in the air, throwing a pass. Question 25.

*27. On the basis of the law of conservation of angular momentum, discuss why a helicopter must have more than one rotor (or propeller). Discuss one or more ways the second propeller can operate in order to keep the body stable.

PROBLEMS

SECTION 8–1

1. (I) What are the following angles expressed in radians: (a) 30°, (b) 57°, (c) 90°, (d) 360°, and (e) 420°? Give as numerical values and as fractions of π.

2. (I) The Sun subtends an angle of about 0.5° to us on the Earth, 150 million km away. What is the radius of the Sun?

3. (I) Eclipses happen on Earth because of an amazing coincidence. Calculate, using the information inside the front cover, the angular diameter (in radians) of the Sun and the angular diameter of the Moon, as seen on Earth.

4. (I) The Eiffel Tower is 300 m tall. When you are standing at a certain place in Paris, it subtends an angle of 6°. How far are you, then, from the Eiffel Tower?

5. (I) A laser beam is directed at the Moon, 380,000 km from Earth. The beam diverges at an angle θ (Fig. 8–36) of 1.8×10^{-5} rad. What diameter spot will it make on the Moon?

6. (I) A 0.35-m diameter grinding wheel rotates at 1800 rpm. Calculate its angular velocity in rad/s.

7. (I) What is the linear speed and acceleration of a point on the edge of the grinding wheel in Problem 6?

8. (I) A 33-rpm phonograph record reaches its rated speed 1.8 s after it is turned on. What was the angular acceleration?

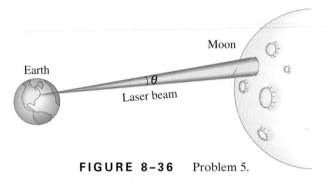

FIGURE 8–36   Problem 5.

9. (I) Calculate the angular velocity of (a) the second hand, (b) the minute hand, and (c) the hour hand, of a clock. State in rad/s. (d) What is the angular acceleration in each case?

10. (I) The blades in a blender rotate at a rate of 7500 rpm. When the motor is turned off during operation, the blades slow to rest in 3.0 s. What is the angular acceleration as the blades slow down?

11. (II) A child rolls a ball on a level floor 4.5 m to another child. If the ball makes 15.0 revolutions, what is its diameter?

12. (II) A bicycle with 68-cm-diameter tires travels 7.0 km. How many revolutions do the wheels make?

13. (II) Estimate the angle subtended by the Moon using a ruler and your finger or other object to just blot out the Moon. Describe your measurement and the result obtained and then use it to estimate the diameter of the Moon. The Moon is about 380,000 km from the Earth.

14. (II) Calculate the angular velocity of the Earth (a) in its orbit around the Sun, and (b) about its axis.

15. (II) What is the linear speed of a point (a) on the equator, (b) on the Arctic Circle (latitude 66.5° N), and (c) at a latitude of 45.0° N, due to the Earth's rotation?

16. (II) How fast (in rpm) must a centrifuge rotate if a particle 7.0 cm from the axis of rotation is to experience an acceleration of 100,000 g's?

17. (II) A 70-cm-diameter wheel accelerates uniformly from 160 rpm to 280 rpm in 4.0 s. Determine (a) its angular acceleration, and (b) the radial and tangential components of the linear acceleration of a point on the edge of the wheel 2.0 s after it has started accelerating.

18. (II) A record player turntable of radius R_1 is turned by a circular rubber roller of radius R_2 in contact with it at their outer edges. What is the ratio of their angular velocities, ω_1/ω_2?

19. (II) In traveling to the Moon astronauts aboard the Apollo spacecraft put themselves into a slow rotation in order to distribute the Sun's energy evenly. At the start of their trip, they accelerated from no rotation to one revolution every minute during a 10-min time interval. The spacecraft can be thought of as a cylinder with a diameter of 8.5 m. Determine (a) the angular acceleration, and (b) the radial and tangential components of the linear acceleration of a point on the skin of the ship 5.0 min after it started this acceleration.

SECTIONS 8–2 AND 8–3

20. (I) A phonograph turntable reaches its speed of 33 rpm after making 1.7 revolutions. What was its angular acceleration?

21. (I) A centrifuge accelerates from rest to 15,000 rpm in 220 s. Through how many revolutions did it turn in this time?

22. (I) An automobile engine slows down from 4000 rpm to 1200 rpm in 3.5 s. Calculate (a) its angular acceleration, assumed uniform, and (b) the total number of revolutions the engine makes in this time.

23. (II) Pilots can be tested for the stresses of flying high-speed jets in a whirling "human centrifuge" which takes 1.0 min to turn through 20 complete revolutions before reaching its final speed. (a) What was its angular acceleration (assume constant), and (b) what was its final speed in rpm?

24. (II) A 40-cm-diameter wheel accelerates uniformly from 240 rpm to 360 rpm in 6.5 s. How far will a point on the edge of the wheel have traveled in this time?

25. (II) Starting from the definitions of ω and α, derive Eqs. 8–9 assuming constant angular acceleration.

26. (II) A small rubber wheel is used to drive a large pottery wheel, and they are mounted so that their circular edges touch. If the small wheel has a radius of 2.0 cm and accelerates at the rate of 7.2 rad/s², and it is in contact with the pottery wheel (radius 25.0 cm) without slipping, calculate (a) the angular acceleration of the pottery wheel, and (b) the time it takes the pottery wheel to reach its required speed of 65 rpm.

27. (II) The tires of a car make 65 revolutions as the car reduces its speed uniformly from 100 km/h to 50 km/h. The tires have a diameter of 0.80 m. (a) What was the angular acceleration? (b) If the car continues to decelerate at this rate, how much more time is required for it to stop?

28. (III) A wheel, starting from rest, undergoes uniform angular acceleration α about its fixed axle. (a) Write the components of the linear acceleration, a_{tan} and a_R, for a point P which is a distance r from the axle in terms of α, r, and time t. (b) Let ϕ be the angle between the linear acceleration vector, $\mathbf{a}$, and the line drawn between P and the axis. Express ϕ in terms of the total number of revolutions of the wheel, N.

SECTION 8–4

29. (I) What is the maximum torque exerted by a 55-kg person riding a bike if the rider puts all her weight on each pedal when climbing a hill? The pedals rotate in a circle of radius 17 cm.

30. (I) A person exerts a force of 45 N on the end of a door 84 cm wide. What is the magnitude of the torque if the force is exerted (a) perpendicular to the door, and (b) at a 60° angle to the face of the door?

31. (II) Calculate the net torque about the axle of the wheel shown in Fig. 8–37. Assume that a friction torque of 0.40 m·N opposes the motion.

FIGURE 8–37
Problem 31.

32. (II) If the coefficient of static friction between tires and pavement is 0.75, calculate the minimum torque that must be applied to the 66-cm-diameter tire of a 1080-kg automobile in order to "lay rubber" (make the wheels spin, slipping as the car accelerates). Assume each wheel supports an equal share of the weight.

FIGURE 8–38   Problem 33.

33. (II) The bolts on the cylinder head of an engine require tightening to a torque of 80 m·N. If a wrench is 30 cm long, what force perpendicular to the wrench must the mechanic exert at its end? If the six-sided bolt head is 15 mm in diameter, estimate the force applied near each of the six points by a socket wrench (Fig. 8–38).

SECTIONS 8–5 AND 8–6

34. (I) Calculate the moment of inertia of a 12.2-kg sphere of radius 0.623 m when the axis of rotation is through its center.

35. (I) Calculate the moment of inertia of a 66.7-cm-diameter bicycle wheel. The rim and tire have a combined mass of 1.25 kg. The mass of the hub can be ignored (why?).

36. (II) Calculate the moment of inertia of the array of point objects shown in Fig. 8–39 about (a) the vertical axis, and (b) the horizontal axis. Assume the objects are wired together by very light rigid pieces of wire. About which axis would it be harder to accelerate this array? In Fig. 8–39, $m = 1.8$ kg and $M = 3.1$ kg. The array is rectangular and it is split through the middle by the horizontal axis.

FIGURE 8–39   Problem 36.

37. (II) An oxygen molecule consists of two oxygen atoms whose total mass is 5.3×10^{-26} kg and whose moment of inertia about an axis perpendicular to the line joining the two atoms, midway between them, is 1.9×10^{-46} kg·m^2. Estimate, from these data, the effective distance between the atoms.

38. (II) A small 1.05-kg ball on the end of a light rod is rotated in a horizontal circle of radius 0.900 m. Calculate (a) the moment of inertia of the system about the axis of rotation, and (b) the torque needed to keep the ball rotating at constant angular velocity if air resistance exerts a force of 0.0800 N on the ball.

FIGURE 8–40
Problem 39.

39. (II) In order to get a flat uniform cylindrical satellite spinning at the correct rate, engineers fire four tangential rockets as shown in Fig. 8–40. If the satellite has a mass of 2600 kg and a radius of 3.0 m, what is the required steady force of each rocket if the satellite is to reach 30 rpm in 5.0 min?

40. (II) A grinding wheel is a uniform cylinder with a radius of 8.50 cm and a mass of 0.580 kg. Calculate (a) its moment of inertia about its center, and (b) the applied torque needed to accelerate it from rest to 1500 rpm in 5.00 s if it is known to slow down from 1500 rpm to rest in 55.0 s.

41. (II) A softball player swings a bat, accelerating it from rest to 3.0 rev/s in a time of 0.20 s. Approximate the bat as a 2.2-kg uniform rod of length 0.95 m, and compute the torque the player applies to one end of it.

42. (II) A day-care worker pushes tangentially on a small hand-driven merry-go-round and is able to accelerate it from rest to a spinning rate of 20 rpm in 10.0 s. Assume the merry-go-round is a disk of radius 2.5 m and has a mass of 800 kg, and two children (each with a mass of 25 kg) sit opposite each other on the edge. Calculate the torque required to produce the acceleration, neglecting frictional torque. What force is required?

43. (II) A centrifuge rotor rotating at 10,000 rpm is shut off and is eventually brought to rest by a frictional torque of 1.20 m·N. If the mass of the rotor is 4.80 kg and it can be approximated as a solid cylinder of radius 0.0710 m, through how many revolutions will the rotor turn before coming to rest, and how long will it take?

44. (II) The forearm in Fig. 8–41 accelerates a 3.6-kg ball at 7.0 m/s^2 by means of the triceps muscle, as shown. Calculate (a) the torque needed, and (b) the force that must be exerted by the triceps muscle. Ignore the mass of the arm.

FIGURE 8–41
Problems 44 and 45.

45. (II) Assume that a 1.50-kg ball is thrown solely by the action of the forearm, which rotates about the elbow joint under the action of the triceps muscle, Fig. 8–41. The ball is accelerated from rest to 10.0 m/s in 0.350 s, at which point it is released. Calculate (a) the angular acceleration of the arm, and (b) the force required of the triceps muscle. Assume that the forearm has a mass of 3.70 kg and rotates like a uniform rod about an axis at its end.

46. (II) A helicopter rotor blade can be considered a long thin rod, as shown in Fig. 8–42. If each of the three rotor helicopter blades is 3.75 m long and has a mass of 160 kg, calculate the moment of inertia of the three rotor blades about the axis of rotation. How much torque must the motor apply to bring the blades up to a speed of 5.0 rev/s in 8.0 s?

FIGURE 8–42
Problem 46.

47. (II) The radius of the roll of paper shown in Fig. 8–43 is 7.6 cm and its moment of inertia is $I = 2.9 \times 10^{-3}$ kg·m². A force of 3.2 N is exerted on the end of the roll for 1.3 s, but the paper does not tear so it begins to unroll. A constant friction torque of 0.11 m·N is exerted on the roll which gradually brings it to a stop. Assuming that the paper's thickness is negligible, calculate (a) the length of paper that unrolls during the time that the force is applied (1.3 s) and (b) the length of paper that unrolls from the time the force ends to the time when the roll has stopped moving.

FIGURE 8–43
Problem 47.

48. (III) An *Atwood's machine* consists of two masses, m_1 and m_2, which are connected by a massless inelastic cord that passes over a pulley, Fig. 8–44. If the pulley has radius R and moment of inertia I about its axle, determine the acceleration of the masses m_1 and m_2, and compare to the situation in which the moment of inertia of the pulley is ignored. [*Hint:* The tensions F_{T1} and F_{T2} are not necessarily equal.]

49. (III) A hammer thrower accelerates the hammer (mass = 7.30 kg) from rest within four full turns (revolutions) ,and releases it at a speed of 28.0 m/s. Assuming a uniform rate of increase in angular velocity

FIGURE 8–44
Atwood's machine.
Problems 48 and 55.

and a radius of 1.20 m, calculate (a) the angular acceleration, (b) the (linear) tangential acceleration, (c) the centripetal acceleration just before release, (d) the net force being exerted on the hammer by the athlete just before release, and (e) the angle of this force with respect to the radius of the circular motion.

SECTION 8–7

50. (I) A bowling ball of mass 7.3 kg and radius 9.0 cm rolls without slipping down a lane at 4.3 m/s. Calculate its total kinetic energy.

51. (I) A centrifuge rotor has a moment of inertia of 3.15×10^{-2} kg·m². How much energy is required to bring it from rest to 8000 rpm?

52. (II) Estimate the kinetic energy of the Earth with respect to the Sun as the sum of two terms, (a) that due to its daily rotation about its axis, and (b) that due to its yearly revolution about the Sun. [Assume the Earth is a uniform sphere, mass = 6.0×10^{24} kg, radius = 6.4×10^6 m, and is 1.5×10^8 km from the Sun.]

53. (II) A merry-go-round has a mass of 1640 kg and a radius of 8.20 m. How much net work is required to accelerate it from rest to a rotation rate of one revolution in 8.00 s? (Assume it is a solid cylinder.)

54. (II) (a) Calculate the translational and rotational speeds of a sphere (radius 20.0 cm and mass 1.20 kg), that rolls without slipping down a 30.0° incline that is 10.0 m long, when it reaches the bottom. Assume it started from rest. (b) What is its ratio of translational to rotational KE at the bottom? Try to avoid putting in numbers until the end so you can answer: (c) do your answers in (a) and (b) depend on the radius of the sphere or its mass?

55. (III) Two masses, $m_1 = 18.0$ kg and $m_2 = 26.5$ kg, are connected by a rope that hangs over a pulley (as in Fig. 8–44). The pulley is a uniform cylinder of radius 0.260 m and mass 7.50 kg. Initially, m_1 is on the ground and m_2 rests 3.00 m above the ground. If the system is now released, use conservation of energy to determine the speed of m_2 just before it strikes the ground. Assume the pulley is frictionless.

56. (III) A 3.30-m-long pole is balanced vertically on its tip. It is given a tiny push. What will be the speed of the upper end of the pole just before it hits the ground? Assume the lower end does not slip. **237**

57. (I) What is the angular momentum of a 0.210-kg ball rotating on the end of a string in a circle of radius 1.10 m at an angular speed of 10.4 rad/s?

58. (I) A person stands, hands at the side, on a platform that is rotating at a rate of 1.30 rev/s. If the person now raises his arms to a horizontal position, Fig. 8-45, the speed of rotation decreases to 0.80 rev/s. (a) Why does this occur? (b) By what factor has the moment of inertia of the person changed?

FIGURE 8-45
Problem 58.

59. (I) A diver (such as the one shown in Fig. 8-28) can reduce her moment of inertia by a factor of about 3.5 when changing from the straight position to the tuck position. If she makes two rotations in 1.5 s when in the tuck position, what is her angular speed (rev/s) when in the straight position?

60. (I) A figure skater during her finale can increase her rotation rate from an initial rate of 1.0 rev every 2.0 s to a final rate of 3.0 rev/s. If her initial moment of inertia was 4.6 kg·m², what is her final moment of inertia? How does she physically accomplish this change?

61. (II) Hurricanes can involve winds in excess of 120 km/h. Make a crude estimate of (a) the energy, and (b) the angular momentum, of such a hurricane, approximating it as a rigidly rotating uniform cylinder of air (density 1.3 kg/m³) of radius 100 km and height 4.0 km.

62. (II) (a) What is the angular momentum of a figure skater spinning (with arms in close to her body) at 3.5 rev/s, assuming her to be a uniform cylinder with a height of 1.5 m, a radius of 15 cm, and a mass of 55 kg. (b) How much torque is required to slow her to a stop in 5.0 s, assuming she does *not* move her arms?

63. (II) Determine the angular momentum of the Earth (a) about its rotation axis (assume the Earth is a uniform sphere), and (b) in its orbit around the Sun (treat the Earth as a particle orbiting the Sun). The Earth has mass = 6.0×10^{24} kg, radius = 6.4×10^6 m, and is 1.5×10^8 km from the Sun.

64. (II) A nonrotating cylindrical disk of moment of inertia I is dropped onto an identical disk rotating at angular speed ω. Assuming no external torques, what is the final common angular speed of the two disks?

65. (II) A uniform disk, such as a record turntable, turns at 7.0 rev/s around a frictionless spindle. A nonrotating rod, of the same mass as the disk and length equal to the disk's diameter, is dropped onto the freely spinning disk. They then both turn around the spindle with their centers superposed, Fig. 8-46. What is the angular velocity in rev/s of the combination?

FIGURE 8-46
Problem 65.

66. (II) A person of mass 75 kg stands at the center of a rotating merry-go-round platform of radius 3.0 m and moment of inertia 1000 kg·m². The platform rotates without friction with angular velocity 2.0 rad/s. The person walks radially to the edge of the platform. (a) Calculate the angular velocity when the person reaches the edge. (b) Compare the rotational kinetic energies of the system of platform plus person before and after the person's walk.

67. (II) An asteroid of mass 1.0×10^5 kg, traveling at a speed of 30 km/s relative to the Earth, hits the Earth at the equator. It hits the Earth tangentially and in the direction of Earth's rotation. Use angular momentum to estimate the fractional change in the angular speed of the Earth as a result of the collision.

68. (II) A 4.2-m-diameter merry-go-round is rotating freely with an angular velocity of 0.80 rad/s. Its total moment of inertia is 1760 kg·m². Four people standing on the ground, each of mass 65 kg, suddenly step onto the edge of the merry-go-round. What is the angular velocity of the merry-go-round now? What if the people were on it initially and then jumped off in a radial direction (relative to the merry-go-round)?

69. (II) Suppose our Sun eventually collapses into a white dwarf, in the process losing about half its mass and winding up with a radius 1.0 percent of its existing radius. Assuming the lost mass carries away no angular momentum, what would the Sun's new rotation rate be? (Take the Sun's current period to be about 30 days.) What would be its final KE in terms of its initial KE of today?

*SECTION 8-9

*70. (II) Suppose a 55-kg person stands at the edge of a 6.5-m-diameter merry-go-round turntable that is mounted on frictionless bearings and has a moment of inertia of 1700 kg·m². The turntable is at rest initially, but when the person begins running at a speed of 3.8 m/s (with respect to the turntable) around its edge, the turntable begins to rotate in the opposite direction. Calculate the angular velocity of the turntable.

*71. (II) A person stands on a platform, initially at rest, that can rotate freely without friction. The moment of inertia of the person plus the platform is I_P. The person holds a spinning bicycle wheel with axis horizontal. The wheel has moment of inertia I_W and angular velocity ω_W. What will be the angular velocity ω_P of the platform if the person moves the axis of the wheel so that it points (a) vertically upward, (b) at a 60° angle to the vertical, (c) vertically downward? (d) What will ω_P be if the person reaches up and stops the wheel in part (a)?

GENERAL PROBLEMS

72. A large spool of rope stands on the ground with the end of the rope lying on the top edge of the spool. A person grabs the end of the rope and walks a distance L, holding onto it, Fig. 8–47. The spool rolls behind the person without slipping. What length of rope unwinds from the spool? How far does the spool's CM move?

FIGURE 8–47   Problem 72.

73. The Moon orbits the Earth so that the same side always faces the Earth. Determine the ratio of its spin angular momentum (about its own axis) to its orbital angular momentum. (In the latter case, treat the Moon as a particle orbiting the Earth.)

74. A cyclist accelerates from rest at a rate of 1.00 m/s^2. How fast will a point on the rim of the tire (diameter = 68 cm) at the top be moving after 3.0 s? [Hint: At any moment, the lowest point on the tire is in contact with the ground and is at rest—see Fig. 8–48.]

$v = ?$

$a = 1.00 \text{ m/s}^2$

This point on tire at rest momentarily

FIGURE 8–48   Problem 74.

75. (a) A yo-yo is made of two solid cylindrical disks, each of mass 0.050 kg and diameter 0.075 m, joined by a (concentric) thin solid cylindrical hub of mass 0.0050 kg and diameter 0.010 m. Use conservation of energy to calculate the linear speed of the yo-yo when it reaches the end of its 1.0-m-long string, if it is released from rest. (b) What fraction of its kinetic energy is rotational?

ω_R

Rear sprocket

R_R

v

v

R_F

ω_F

Front sprocket

FIGURE 8–49
Problem 76.

76. (a) How is the angular speed of the rear wheel (ω_R) of a bicycle related to that of the pedals and front sprocket (ω_F)? That is, derive a formula for ω_R/ω_F. Let N_F and N_R be the number of teeth on the front and rear sprockets, respectively. The teeth are spaced equally on all sprockets so that the chain meshes properly. See Fig. 8–49. Then evaluate the ratio ω_R/ω_F when (b) the front and rear sprockets have 52 and 13 teeth, respectively, and (c) when they have 42 and 28.

77. Suppose a star the size of our Sun, but with mass 8.0 times as great, were rotating at a speed of 1.0 revolution every 10 days. If it were to undergo gravitational collapse to a neutron star of radius 10 km, losing $\frac{3}{4}$ of its mass in the process, what would its rotation speed be? Assume the star is a uniform sphere at all times, and that the thrown-off mass carries off no angular momentum.

78. One possibility for a low-pollution automobile is for it to use energy stored in a heavy rotating flywheel. Suppose such a car has a total mass of 1400 kg, uses a 1.50 m diameter uniform cylindrical flywheel of mass 240 kg, and should be able to travel 300 km without needing a flywheel "spinup." (a) Make reasonable assumptions (average frictional retarding force = 500 N, twenty acceleration periods from rest to 90 km/h, equal uphill and downhill—assuming during downhill, energy can be put back into the flywheel), and show that the total energy needed to be stored in the flywheel is about $1.6 \times 10^8 \text{ J}$. (b) What is the angular velocity of the flywheel when it has a full "energy charge"? (c) About how long would it take a 150-hp motor to give the flywheel a full energy charge before a trip?

FIGURE 8-50 Problem 80.

79. A hollow cylinder (hoop) is rolling on a horizontal surface at speed $v = 4.3$ m/s when it reaches a 15° incline. (a) How far along the incline will it go? (b) How long will it be on the incline before it arrives back at the bottom?

80. A uniform rod of mass M and length L can pivot freely (i.e., we ignore friction) about a hinge attached to a wall, as in Fig. 8–50. The rod is held horizontally and then released. At the moment of release, determine (a) the angular acceleration of the rod and (b) the linear acceleration of the tip of the rod. Assume that the force of gravity acts at the center of mass of the rod, as shown. [*Hint*: See Fig. 8–20g.]

81. A wheel of mass M has radius R. It is standing vertically on the floor, and we want to exert a horizontal force F at its axle so that it will climb a step against which it rests (Fig. 8–51). The step has height h, where $h < R$. What minimum force F is needed?

FIGURE 8-51 Problem 81.

82. A bicyclist traveling with speed $v = 4.2$ m/s on a flat road is making a turn with a radius $r = 6.4$ m. The forces acting on the cyclist and cycle are the normal force ($\mathbf{F}_N$) and friction force ($\mathbf{F}_{fr}$) exerted by the road on the tires, and $m\mathbf{g}$, the total weight of the cyclist and cycle (see Fig. 8–52). (a) Explain carefully why the angle θ the bicycle makes with the vertical (Fig. 8–52) must be given by $\tan\theta = F_{fr}/F_N$ if the cyclist is to maintain balance. (b) Calculate θ for the values given. [*Hint*: Consider the "circular" translational motion of the bicycle and rider.] (c) If the coefficient of static friction between tires and road is $\mu_s = 0.70$, what is the minimum turning radius?

FIGURE 8-52 Problem 82.

83. A marble of mass m and radius r rolls along the looped rough track of Fig. 8–53. What is the minimum value of h if the marble is to reach the highest point of the loop without leaving the track? Assume $r \ll R$, and ignore frictional losses.

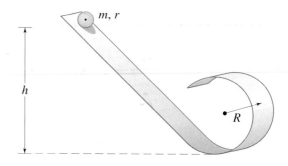

FIGURE 8-53 Problems 83 and 84.

84. Repeat Problem 83, but do not assume $r \ll R$.

85. A thin uniform stick of mass M and length L is positioned vertically, with its tip on a frictionless table. It is released and allowed to slip and fall (Fig. 8–54). Determine the speed of its center of mass just before it hits the table.

FIGURE 8-54 Problem 85.

The Pont du Gard, in southern France, was built by the Romans nearly 2000 years ago, and was built to last.

9

BODIES IN EQUILIBRIUM; ELASTICITY AND FRACTURE

I n this chapter, we will study a special case of motion—when the net force and the net torque on an object, or system of objects, are both zero. In this case the object or system is either at rest, or its center of mass is moving at constant velocity. We will be concerned mainly with the first case, when the object or objects are all at rest. Now you may think that the study of objects at rest is not very interesting since the objects will have neither velocity nor acceleration; and the net force and the net torque will be zero. But this does not imply that no forces at all act on the objects. In fact it is virtually impossible to find a body on which no forces act at all. Sometimes the forces may be so great that the object is seriously deformed, or it may even fracture (break)—and avoiding such problems gives this field of **statics** great importance.

This chapter deals with forces within objects at rest

FIGURE 9–1 The book is in equilibrium; the net force on it is zero.

9–1 Statics—The Study of Forces in Equilibrium

Objects within our experience have at least one force acting on them (gravity), and if they are at rest then there must be other forces acting on them as well so that the net force is zero. An object at rest on a table, for example, has two forces acting on it, the downward force of gravity and the normal force the table exerts upward on it (Fig. 9–1). Since the net

Normal force

Gravity

FIGURE 9–2 Elevated walkway collapse in a Kansas City hotel in 1981. How a simple physics calculation could have prevented the tragic loss of 100 lives is considered in Example 9–14.

force is zero, the upward force exerted by the table must be equal in magnitude to the force of gravity acting downward. (Do not confuse these two forces with the equal and opposite forces of Newton's third law which act on different bodies; here both forces act on the same body.) Such a body is said to be in **equilibrium** (Latin for "equal forces" or "balance") under the action of these two forces.

The subject of statics is concerned with the calculation of the forces acting on and within structures that are in equilibrium. Determination of these forces, which occupies us in the first part of this chapter, then allows a determination of whether the structures can sustain the forces without significant deformation or fracture, subjects we discuss later in this chapter. These techniques can be applied in a wide range of fields. Architects and engineers must be able to calculate the forces on the structural components of buildings, bridges, machines, vehicles, and other structures, since any material will break or buckle if too much force is applied (Fig. 9–2). In the human body, a knowledge of the forces in muscles and joints is of great value for medicine and physical therapy, and is also valuable for the study of athletic activity.

Let us take a simple Example of the addition of forces applied to orthodonture.

EXAMPLE 9–1 **Straightening teeth.** The wire band shown in Fig. 9–3a has a tension F_T of 2.0 N along it. It therefore exerts forces of 2.0 N on the tooth (to which it is attached) in the two directions shown. Calculate the resultant force on the tooth due to the wire, F_W.

SOLUTION Since the two forces are equal, their sum will be directed along the line that bisects the angle between them, which we have labeled the y axis. The x components of the two forces add up to zero. The y component of each force is $(2.0\,\text{N})(\cos 70°) = 0.68\,\text{N}$; adding the two together we get a total force F_W of 1.36 N as shown in Fig. 9–3b. Note that if the wire is firmly attached to the tooth, the tension to the right, say, can be made larger than that to the left, and the resultant force would correspondingly be directed more toward the right.

FIGURE 9–3 Forces on a tooth. Example 9–1.

(a)

(b)

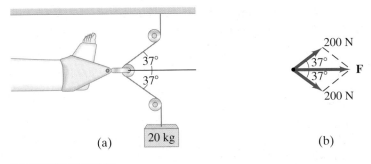

(a)

20 kg

200 N

$37°$
$37°$
F
$37°$

200 N

(b)

FIGURE 9–4 Traction apparatus exerts force on a leg.

EXAMPLE 9–2 **Traction.** Calculate the force exerted on the leg by the traction apparatus shown in Fig. 9–4. Assume the pulleys are frictionless.

SOLUTION There is a tension of 20 kg × 9.8 m/s² = 200 N all along the cord. Thus there are two 200-N forces acting at 37° angles on the central pulley and on the leg, Fig. 9–4b. So the resultant force on the leg is $F = 2(200 \text{ N}) \cos 37° = 320$ N acting to the right. (The leg is in equilibrium, so there must be another 320-N force acting on the leg to keep it at rest. What exerts this force?)

9–2 The Conditions for Equilibrium

For a body to be at rest, the sum of the forces acting on it must add up to zero. Since force is a vector, the components of the net force must each be zero. Hence, a condition for equilibrium is that

$$\Sigma F_x = 0, \qquad \Sigma F_y = 0, \qquad \Sigma F_z = 0. \qquad \textbf{(9–1)}$$

First condition for equilibrium: the sum of all forces is zero

We will mainly be dealing with forces that act in a plane, so we usually need only the x and y components. We must remember that if a particular force component points along the negative x or y axis, it must have a negative sign. Equation 9–1 is called the **first condition for equilibrium**.

EXAMPLE 9–3 **Pull-ups on a scale.** A 90-kg weakling cannot do even one pull-up. By standing on a scale (Fig. 9–5), he can determine how close he gets. His best effort results in a scale reading of 23 kg. What force is he exerting?

SOLUTION There are three forces acting on our nonathlete, as shown in Fig. 9–5: gravity, $mg = (90 \text{ kg})(9.8 \text{ m/s}^2)$ downward, and two upward forces which are (1) the force the bar pulls upward on him, F_B (equal and opposite to the force he exerts on the bar), and (2) the force the scale exerts on his feet, F_S. At best, $F_S = (23 \text{ kg})(g)$. The person doesn't move, so the sum of these forces is zero:

$$\Sigma F_y = 0$$

$$F_B + F_S - mg = 0.$$

We solve for F_B:

$$F_B = mg - F_S$$

$$= (90 \text{ kg} - 23 \text{ kg})(g) = (67 \text{ kg})(9.8 \text{ m/s}^2) = 660 \text{ N}.$$

That is, he could lift himself if his mass were only 67 kg.

F_B

y

F_B F_S

mg

F_S

mg

(a)   (b)

FIGURE 9–5 Example 9–3: (a) A person trying to do a pull-up while standing on a scale. (b) Simple free-body diagram.

FIGURE 9-6 Example 9-4.

FIGURE 9-7 Although the net force on it is zero, the ruler will move (rotate). A pair of equal forces acting in opposite directions but at different points on a body (as shown here) is referred to as a *couple*.

Second condition for equilibrium: the sum of all torques is zero

EXAMPLE 9–4 **Chandelier cord tension.** Calculate the tensions $\mathbf{F}_1$ and $\mathbf{F}_2$ in the two cords that are connected to the cord supporting the 200-kg chandelier in Fig. 9–6.

SOLUTION The three forces, $\mathbf{F}_1$, $\mathbf{F}_2$, and the weight of the 200-kg chandelier, act at the point where the three cords join. We choose this junction point (it could be a knot) as the object for which we write $\Sigma F_x = 0$, $\Sigma F_y = 0$. (We don't bother considering the chandelier itself since only two forces act on it, gravity downward and the equal but opposite force exerted upward by the cord, both of which equal $mg = (200 \text{ kg})(9.8 \text{ m/s}^2) = 1960 \text{ N}$.) There are two unknowns (F_1 and F_2)[†] and we can solve for them using Eqs. 9–1. We first resolve $\mathbf{F}_1$ into its horizontal (x) and vertical (y) components. Although we don't know the value of F_1, we can write $F_{1x} = -F_1 \cos 60°$ and $F_{1y} = F_1 \sin 60°$. $\mathbf{F}_2$ has only an x component. In the vertical direction, we have only the weight of the chandelier $= (200 \text{ kg})(g)$ acting downward and the vertical component of $\mathbf{F}_1$ upward. Since $\Sigma F_y = 0$, we have

$$\Sigma F_y = F_1 \sin 60° - (200 \text{ kg})(g) = 0$$

so

$$F_1 = \frac{(200 \text{ kg})g}{\sin 60°} = \frac{(200 \text{ kg})g}{0.866} = (231 \text{ kg})g = 2260 \text{ N}.$$

In the horizontal direction,

$$\Sigma F_x = F_2 - F_1 \cos 60° = 0.$$

Thus

$$F_2 = F_1 \cos 60° = (231 \text{ kg})(g)(0.500) = (115 \text{ kg})g = 1130 \text{ N}.$$

The magnitudes of $\mathbf{F}_1$ and $\mathbf{F}_2$ determine the strength of cord or wire that must be used. In this case, the wire must be able to hold at least 231 kg. Note in this example that we didn't insert the value of g, the acceleration due to gravity, until the end. In this way we found the magnitude of the force in terms of the number of kilograms (which may be a more familiar quantity than newtons) times g.

Although Eqs. 9–1 must be true if an object is to be in equilibrium, they are not a sufficient condition. Figure 9–7 shows an object on which the net force is zero. Although the two forces labeled $\mathbf{F}$ add up to give zero net force on the object, they do give rise to a net torque that will rotate the object. Referring to Eq. 8–14, $\Sigma \tau = I\alpha$, we see that if a body is to remain at rest, the net torque applied to it (calculated about *any* axis) must be zero. Thus we have the **second condition for equilibrium**: that the sum of the torques acting on a body must be zero:

$$\Sigma \tau = 0. \qquad \textbf{(9 2)}$$

This will assure that the angular acceleration, α, about any axis will be zero. If the body is not rotating initially ($\omega = 0$), it will not start rotating. Equations 9–1 and 9–2 are the only requirements for a body to be in equilibrium.

We will consider cases in which the forces all act in a plane (we call it the xy plane). In this case the torque is calculated about an axis that is per-

---

[†]The directions of $\mathbf{F}_1$ and $\mathbf{F}_2$ are known, since tension in a rope can only be along the rope—any other direction would cause the rope to bend, as already pointed out in Chapter 4. Thus, our unknowns are the magnitudes F_1 and F_2.

pendicular to the *xy* plane. *The choice of this axis is arbitrary.* If the body is at rest, then $\Sigma\tau = 0$ about any axis whatever. Therefore we can choose any axis that makes our calculation easier.

CONCEPTUAL EXAMPLE 9–5 **A lever.** The bar in Fig. 9–8 is being used as a lever to pry up a large rock. The small rock acts as a fulcrum. The force F_P required at the long end of the bar can be quite a bit smaller than the rock's weight *mg*, since it is the *torques* that balance in the rotation about the fulcrum. If, however, the leverage isn't quite good enough, and the rock isn't budged, what are two ways to increase the leverage?

RESPONSE One way is to increase the lever arm of the force F_P by slipping a pipe over the end of the bar and thereby pushing with a longer lever arm. A second, perhaps handier, way is to move the fulcrum closer to the large rock. This may change the long lever arm *R* only a little, but it changes the short lever arm *r* by a good fraction and therefore changes the ratio of *R/r* dramatically. In order to pry the rock, the torque due to F_P must at least balance the torque due to *mg*, so $mgr = F_P R$ and

$$\frac{r}{R} = \frac{F_P}{mg}.$$

With *r* smaller, the weight *mg* can be balanced with less force F_P.

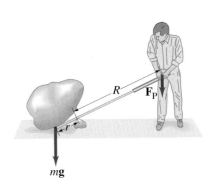

FIGURE 9–8 Example 9–5

9–3 Solving Statics Problems

This subject of statics is important because it allows us to calculate certain forces on (or within) a structure when some of the forces on it are already known. There is no single technique for attacking such statics problems, but the following procedure may be helpful:

➡ **PROBLEM SOLVING** **Statics**

1. Choose one body at a time for consideration, and make a careful *free-body diagram* for it by showing all the forces acting on that body and the points at which these forces act.

2. Choose a convenient coordinate system, and resolve the forces into their components.

3. Using letters to represent unknowns, write down equations for

$$\Sigma F_x = 0, \qquad \Sigma F_y = 0, \qquad \text{and} \qquad \Sigma\tau = 0.$$

4. For the $\Sigma\tau = 0$ equation, choose any axis perpendicular to the *xy* plane that you like. (For example, you can reduce the number of unknowns in the resulting equation by choosing the axis so that one of the unknown forces passes through

the axis; then this force will have zero lever arm and produce zero torque and so won't appear in the equation.) Pay careful attention to determining the lever arm for each force correctly. Give each torque a + or − sign. If torques that would tend to rotate the object counterclockwise are given a + sign, then torques that would tend to rotate it clockwise are negative.

5. Solve these equations for the unknowns. Three equations allow a maximum of three unknowns to be solved for; they can be forces, distances, even angles. (If an unknown force comes out negative in your solution, it means the direction you originally chose for that force is actually the opposite.)

Probably the hardest step is (1): *all* the forces *on* the body must be included, but the forces exerted *by* this body on other objects must *not* be included.

It is useful to refer back to Example 9–4 to see how this procedure was followed. Notice that we considered the forces on the chandelier only briefly (gravity downward and the rope pulling upward with an equal magnitude force). We wanted to know the tension in the two cords, so we chose the junction point (or knot) of the cords as our object. Since the object is essentially a point, there would be no torques, so the torque equation was not used. Of course, you do not have to use all three equations if they are not needed. In many situations we must make use of the torque equation, as we shall soon see. Just as the components of a force are positive or negative depending on their direction, so too are torques. If a torque that tends to rotate the object counterclockwise is considered positive, then a torque that tends to rotate it clockwise must be considered negative.

Alternate form of $\Sigma\tau = 0, \Sigma\mathbf{F} = 0$

Another way of stating the torque equilibrium condition is that the sum of all clockwise torques is equal to the sum of all counterclockwise torques. And for the forces, the sum of the upward forces is equal to the sum of the downward forces, and the sum of horizontal forces to the left is equal to the sum of horizontal forces to the right.

One of the forces that acts on bodies is the force of gravity. Our analysis in this chapter is greatly simplified if we use the concept of center of gravity (CG) or center of mass (CM), which for practical purposes are the same point. As we discussed in Section 7–8, we can consider the force of gravity on the body as a whole as acting at its CG. For uniform symmetrically shaped bodies, the CG is at the geometric center. For more complicated bodies, the CG can be determined as discussed in Section 7–8.

EXAMPLE 9–6 **Balancing a seesaw.** A 2.0-kg board serves as a seesaw for two children, as shown in Fig. 9–9a. One child has a mass of 30 kg and sits 2.5 m from the pivot point (i.e., his CG is 2.5 m from the pivot). At what distance, x, from the pivot must a 25-kg child place herself to balance the seesaw? Assume the board is uniform and centered over the pivot.

SOLUTION The free-body diagram for the board is as shown in Fig. 9–9b. The forces acting on the board are the forces exerted downward on it by each child, $\mathbf{F}_1$ and $\mathbf{F}_2$, the upward force exerted by the

FIGURE 9–9 (a) Two children on a seesaw, Example 9–6. (b) Free-body diagram of the board.

pivot, $\mathbf{F}_N$, and the force of gravity (its weight), which acts at the center of the uniform board. Let us calculate the torque about the pivot point, P: the lever arms for F_N and for the weight of the board are then zero, and they will not appear in the torque equation. Thus the torque equation will involve only the forces $\mathbf{F}_1$ and $\mathbf{F}_2$, which are equal to the weights of the children. The torque exerted by each child will be mg times the appropriate lever arm, which here is the distance of each child from the pivot point. Hence the torque equation is

$$\Sigma\tau = (30 \text{ kg})(g)(2.5 \text{ m}) - (25 \text{ kg})(g)(x) = 0,$$

where a torque tending to rotate the board counterclockwise has been chosen to be a positive torque, and a torque tending to rotate it clockwise to be negative. The acceleration of gravity, g, appears in both terms and cancels out. We solve for x and find:

$$x = \frac{(30 \text{ kg})(2.5 \text{ m})}{25 \text{ kg}} = 3.0 \text{ m}.$$

To balance the seesaw, the second child must sit so that her CG is 3.0 m from the pivot. This makes sense: since she is lighter, she must sit farther from the pivot.

EXAMPLE 9–7 **Forces on a beam and supports.** A uniform 1500-kg beam, 20.0 m long, supports a 15,000-kg printing press 5.0 m from the right support column (Fig. 9–10). Calculate the force on each of the vertical support columns.

SOLUTION We analyze the forces on the beam, since the force the beam exerts on the columns is equal and opposite to the forces exerted by the columns on the beam. We call the latter $\mathbf{F}_1$ and $\mathbf{F}_2$ in Fig. 9–10. The weight of the beam itself acts at its center of gravity, 10.0 m from either end. Since it doesn't matter which point we choose as the axis for writing the torque equation, we can choose one that is convenient. If we calculate the torques about the point of application of $\mathbf{F}_1$, then $\mathbf{F}_1$ will not enter the equation (its lever arm will be zero) and we will have an equation in only one unknown, F_2. We choose the counterclockwise direction as positive, and $\Sigma\tau = 0$ gives

$$\Sigma\tau = (10.0 \text{ m})(1500 \text{ kg})g - (15.0 \text{ m})(15,000 \text{ kg})g + (20.0 \text{ m})F_2 = 0.$$

Solving for F_2 we find $F_2 = (12,000 \text{ kg})g = 118,000 \text{ N}$. To find F_1, we use $\Sigma F_y = 0$:

$$\Sigma F_y = F_1 - (1500 \text{ kg})g - (15,000 \text{ kg})g + F_2 = 0.$$

Putting in $F_2 = (12,000 \text{ kg})g$ we find that $F_1 = (4500 \text{ kg})g = 44,100 \text{ N}$.

To confirm that the results don't depend on where the axis is chosen, let us choose a different axis, say, at the other end of the beam, where F_2 acts. In this case, the torque equation is

$$\Sigma\tau = -(20.0 \text{ m})F_1 + (10.0 \text{ m})(1500 \text{ kg})g + (5.0 \text{ m})(15,000 \text{ kg})g = 0.$$

We solve this for F_1 and find $F_1 = (4500 \text{ kg})g$, just as before. The ΣF_y equation is the same as before and gives $F_2 = (12,000 \text{ kg})g$. So our results are the same, no matter which axis we choose to calculate the torques about.

FIGURE 9–10 A 1500-kg beam supports a 15,000-kg machine. Example 9–7.

FIGURE 9–11   A cantilever.

➡ **PROBLEM SOLVING**

If a force comes out negative

Figure 9–11 shows a beam that extends beyond its support like a diving board. Such a beam is called a *cantilever*. The forces acting on the beam in this figure are those due to the supports, F_1 and F_2, and the force of gravity, which acts at the CG, 5.0 m to the right of the right-hand support. If you follow the procedure of the last Example to calculate F_1 and F_2, assuming they point upward as shown in the figure, you will find that F_1 comes out negative. If the beam has a mass of 1200 kg, then $F_2 = 15,000$ N and $F_1 = -3000$ N (see Problem 15). Whenever an unknown force comes out negative, it merely means that the force actually points in the opposite direction from what you assumed. Thus in Fig. 9–11, F_1 actually points downward. With a little reflection it should become obvious that the left-hand support must indeed pull downward on the beam (by means of bolts, screws, fasteners and/or glue) if the beam is to be in equilibrium; otherwise the sum of the torques about the CG (or about the point where F_2 acts) could not be zero.

Our next Example involves a beam attached to a wall by a hinge and supported by a guy wire (Fig. 9–12). It is important to remember that a flexible cord or wire can support a force only along its length. (If there were a component of force perpendicular to the cord or wire, it would bend because it is flexible.) For a rigid device, such as the hinge in Fig. 9–12, the force can be in any direction and we can know the direction only after solving the problem.

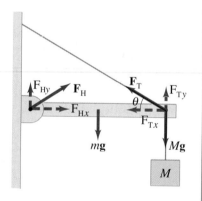

FIGURE 9–12   Example 9–8.

EXAMPLE 9–8 **Beam and wire.** A uniform beam, 2.20 m long with mass $m = 25.0$ kg, is mounted by a hinge on a wall as shown in Fig. 9–12. The beam is held in a horizontal position by a wire that makes an angle $\theta = 30.0°$ as shown. The beam supports a mass $M = 280$ kg suspended from its end. Determine the components of the force F_H that the hinge exerts on the beam, and the components of the tension F_T in the supporting wire.

SOLUTION Figure 9–12 is the free-body diagram for the beam, showing all the forces acting on the beam; it also shows the components of F_H and F_T. We have three unknowns, F_{Hx}, F_{Hy}, and F_T (we are given θ), so we will need all three equations, $\Sigma F_x = 0$, $\Sigma F_y = 0$, $\Sigma \tau = 0$. The sum of the forces in the vertical (y) direction is

$$\Sigma F_y = F_{Hy} + F_{Ty} - mg - Mg = 0. \tag{i}$$

In the horizontal (x) direction, the sum of the forces is

$$\Sigma F_x = F_{Hx} - F_{Tx} = 0. \tag{ii}$$

For the torque equation, we choose the axis at the point where F_T and $M\mathbf{g}$ act (so our equation then contains only one unknown, F_{Hy}, and we can solve it more quickly); we choose torques that tend to rotate the beam counterclockwise as positive. The weight mg of the (uniform) beam acts at its center, so we have:

$$\Sigma \tau = -(F_{Hy})(2.20 \text{ m}) + (mg)(1.10 \text{ m}) = 0$$

or

$$F_{Hy} = \tfrac{1}{2}mg = \tfrac{1}{2}(25.0 \text{ kg})(9.80 \text{ m/s}^2) = 123 \text{ N}.$$

Next, since the tension F_T in the wire acts along the wire ($\theta = 30°$),

$$F_{Ty} = F_{Tx} \tan \theta = 0.577 F_{Tx}. \tag{iii}$$

From Eqs. i, ii, and iii we get

$$F_{Ty} = (m + M)g - F_{Hy} = (305 \text{ kg})(9.80 \text{ m/s}^2) - 123 \text{ N} = 2870 \text{ N}$$

$$F_{Tx} = F_{Ty}/0.577 = 4970 \text{ N}$$

$$F_{Hx} = F_{Tx} = 4970 \text{ N}.$$

The components of $\mathbf{F_H}$ are $F_{Hy} = 123 \text{ N}$ and $F_{Hx} = 4970 \text{ N}$. The tension in the wire is $F_T = \sqrt{F_{Tx}^2 + F_{Ty}^2} = 5740 \text{ N}$.

EXAMPLE 9–9 **Ladder.** A 5.0-m-long ladder leans against a wall at a point 4.0 m above the ground as shown in Fig. 9–13. The ladder is uniform and has mass 12.0 kg. Assuming the wall is frictionless (but the ground is not), determine the forces exerted on the ladder by the ground and the wall.

SOLUTION Figure 9–13 shows the free-body diagram for the ladder, showing all the forces acting on the ladder. The wall, since it is frictionless, can exert a force only perpendicular to the wall, and we label that force $\mathbf{F_W}$. The ground, however, can exert both horizontal, F_{Gx}, and vertical, F_{Gy}, force components, the former being frictional and the latter the normal force. Finally, gravity exerts a force $mg = (12.0 \text{ kg})(9.80 \text{ m/s}^2) = 118 \text{ N}$ on the ladder at its midpoint, since the ladder is uniform. The y component of the force equation is

$$\Sigma F_y = F_{Gy} - mg = 0,$$

so immediately we have $F_{Gy} = mg = 118 \text{ N}$. The x component of the force equation is

$$\Sigma F_x = F_{Gx} - F_W = 0.$$

To determine F_{Gx} and F_W, which are both unknowns, we need another equation, namely a torque equation, which we calculate about the point where the ladder touches the ground. This point is a distance $x_0 = \sqrt{(5.0 \text{ m})^2 - (4.0 \text{ m})^2} = 3.0 \text{ m}$ from the wall. The lever arm for mg is half this, or 1.5 m, and the lever arm for F_W is 4.0 m. Since $\mathbf{F_G}$ acts at the axis, its lever arm is zero and so doesn't enter the equation (we planned it like that), and we get

$$\Sigma \tau = (4.0 \text{ m})F_W - (1.5 \text{ m})mg = 0.$$

Thus

$$F_W = \frac{(1.5 \text{ m})(12.0 \text{ kg})(9.8 \text{ m/s}^2)}{4.0 \text{ m}} = 44 \text{ N}.$$

Then, from the x component of the force equation,

$$F_{Gx} = F_W = 44 \text{ N}.$$

Since the components of $\mathbf{F_G}$ are $F_{Gx} = 44 \text{ N}$ and $F_{Gy} = 118 \text{ N}$, then

$$F_G = \sqrt{(44 \text{ N})^2 + (118 \text{ N})^2} = 126 \text{ N} \approx 130 \text{ N}$$

(rounded off to two significant figures) and it acts at an angle

$$\theta = \tan^{-1}(118 \text{ N}/44 \text{ N}) = 70°$$

to the ground. Note the force $\mathbf{F_G}$ does *not* have to act along the ladder's direction because the ladder is rigid and not flexible like a cord or cable.

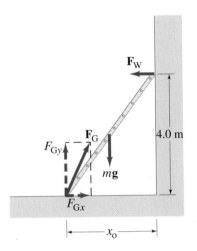

FIGURE 9–13 A ladder leaning against a wall. Example 9–9.

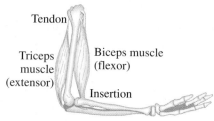

Tendon

Triceps muscle (extensor)

Biceps muscle (flexor)

Insertion

FIGURE 9–14 Diagram showing the biceps (flexor) and triceps (extensor) muscles in the human arm.

➡ **PHYSICS APPLIED**

Forces in muscles and joints

F_M

F_J

$(2.0 \text{ kg})(g)$

$(5.0 \text{ kg})(g)$

5.0 cm

←15 cm→

←————— 35 cm —————→

(a)

30°

(b)

FIGURE 9–15 Example 9–10.

➡ **PHYSICS APPLIED**

Muscle insertion and lever arm

* 9–4 Applications to Muscles and Joints

The techniques we have been discussing for calculating forces on bodies in equilibrium can readily be applied to the human (or animal) body. This is of great use in studying the forces on muscles, bones, and joints for organisms in motion or at rest. Generally a muscle is attached, via tendons, to two different bones (see Fig. 9–14). The points of attachment are called *insertions*. The two bones are flexibly connected at a *joint*, such as those at the elbow, knee, and hip. A muscle exerts a pull when its fibers contract under stimulation by a nerve, but it cannot exert a push. Muscles that tend to bring two limbs closer together, such as the biceps muscle in the upper arm (Fig. 9–14) are called *flexor* muscles; those that act to extend a limb outward, such as the triceps, are called *extensor* muscles. The flexor muscle in the upper arm is used when lifting an object in your hand; the extensor muscle is used when throwing a ball.

EXAMPLE 9–10 **Force exerted by biceps muscle.** How much force must the biceps muscle exert when a 5.0-kg mass is held in the hand (a) with the arm horizontal as in Fig. 9–15a, and (b) when the arm is at a 30° angle as in Fig. 9–15b? Assume that the mass of forearm and hand together is 2.0 kg and their CG is as shown.

SOLUTION (a) The forces acting on the forearm are shown in Fig. 9–15a and include the upward force F_M exerted by the muscle and a force F_J exerted at the joint by the bone in the upper arm (both assumed to act vertically). We wish to find F_M, which is done most easily by using the torque equation and by choosing our axis through the joint so that F_J does not enter:

$$\Sigma\tau = (0.050 \text{ m})(F_M) - (0.15 \text{ m})(2.0 \text{ kg})(g) - (0.35 \text{ m})(5.0 \text{ kg})(g) = 0.$$

We solve this for F_M and find $F_M = (41 \text{ kg})(g) = 400 \text{ N}$.
(b) The lever arm, as calculated about the joint, is reduced by the factor cos 30° for all three forces. So our torque equation will look like the one just above, except that each term will have a "cos 30°." The latter will cancel out so the same result will be obtained, $F_M = 400 \text{ N}$.

Note in this Example that the force required of the muscle (400 N) is quite large compared to the weight of the object lifted (49 N). Indeed, the muscles and joints of the body are generally subjected to quite large forces. The point of insertion of a muscle varies from person to person. A slight increase in the point of insertion of the biceps muscle from 5.0 cm to 5.5 cm can be a considerable advantage for lifting and other exertions. Indeed, champion athletes are often found to have muscle insertions farther from the joint than the average person, and if this applies to one muscle, it usually applies to all.

As another example of the large forces acting within the human body, we consider the muscles used to support the trunk when a person bends forward (Fig. 9–16a). The lowest vertebra on the spinal column (fifth lumbar vertebra) acts as a fulcrum for this bending position. The "erector spinae" muscles in the back that support the trunk act at an effective angle of about 12° to the axis of the spine. Figure 9–16b is a simplified schematic drawing showing the forces on the upper body. We assume the trunk makes an angle of 30° with the horizontal. The force exerted by the back muscles is represented by $\mathbf{F}_M$, the force exerted on the base of the spine at the lowest vertebra is $\mathbf{F}_V$, and $\mathbf{w}_1$, $\mathbf{w}_2$, and $\mathbf{w}_3$ represent the weights of the head, freely hanging arms, and trunk, respectively. The values shown are approximations taken from Table 7–1. The distances (in cm) refer to a person 180 cm tall, but are approximately in the same ratio of 1:2:3 for an average person of any height, and the result in the following Example is then independent of the height of the person.

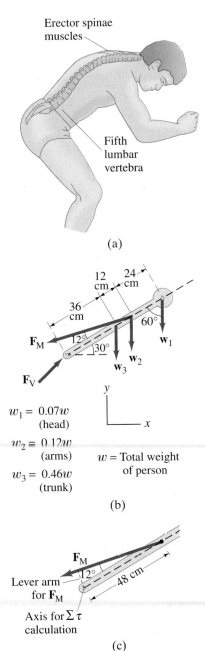

(a)

(b)

(c)

FIGURE 9–16 (a) A person bending over. (b) Forces on the back exerted by the back muscles ($\mathbf{F}_M$) and by the vertebrae ($\mathbf{F}_V$) when a person bends over.

EXAMPLE 9–11 **Forces on your back.** Calculate the magnitude and direction of the force $\mathbf{F}_V$ acting on the fifth lumbar vertebra for the example shown in Fig. 9–16b.

SOLUTION First we calculate F_M using the torque equation, taking the axis at the base of the spine. To figure the lever arms, we need to use trigonometric functions. For $\mathbf{F}_M$, the lever arm (perpendicular distance from axis to line of action of the force) will be the real distance to where the force acts (48 cm) times sin 12° as shown in Fig. 9–16c. The lever arms for $\mathbf{w}_1$, $\mathbf{w}_2$, and $\mathbf{w}_3$ can be seen from Fig. 9–16b to be their respective distances times sin 60°. Thus $\Sigma\tau = 0$ gives

$$(0.48\ \text{m})(\sin 12°)(F_M) - (0.72\ \text{m})(\sin 60°)(w_1)$$
$$- (0.48\ \text{m})(\sin 60°)(w_2) - (0.36\ \text{m})(\sin 60°)(w_3) = 0,$$

where we chose the positive sign for counterclockwise torque. Putting in the values for w_1, w_2, w_3 given in the figure, we find $F_M = 2.2w$, where w is the total weight of the body. To get the components of $\mathbf{F}_V$ we use the x and y components of the force equation (noting that $30° - 12° = 18°$):

$$\Sigma F_y = F_{Vy} - F_M \sin 18° - w_1 - w_2 - w_3 = 0$$

$$F_{Vy} = 1.3w$$

$$\Sigma F_x = F_{Vx} - F_M \cos 18° = 0$$

$$F_{Vx} = 2.1w.$$

Then

$$F_V = \sqrt{F_{Vx}^2 + F_{Vy}^2} = 2.5w.$$

The angle θ that F_V makes with the horizontal is given by $\tan\theta = F_{Vy}/F_{Vx} = 0.62$, so $\theta = 32°$.

$w_1 = 0.07w$ (head)

$w_2 = 0.12w$ (arms)

$w_3 = 0.46w$ (trunk)

w = Total weight of person

The force on the lowest vertebra is thus $2\frac{1}{2}$ times the body weight! This force is transmitted from the "sacral" bone at the base of the spine, through the fluid-filled and somewhat flexible *intervertebral disc*. The discs at the base of the spine are clearly being compressed under very large forces. [If the body was less bent over (say the 30° angle in Fig. 9–16b becomes 60° or 70°) then the stress on the lower back will be less (see Problem 40).]

If the person in Fig. 9–16 has a mass of 90 kg, and is holding 20 kg in his hands (this increases w_2 to $0.33w$), then F_V is increased to nearly five times the person's weight ($5w$)! (For this 200-lb person, the force on the disc would be 1000 lb!) With such strong forces acting, it is little wonder that so many people suffer from low back pain at one time or another in their lives.

* 9–5 Stability and Balance

A body in static equilibrium, if left undisturbed, will undergo no translational or rotational acceleration since the sum of all the forces and the sum of all the torques acting on it are zero. However, if the object is displaced slightly, three different outcomes are possible: (1) the object returns to its original position, in which case it is said to be in **stable equilibrium**; (2) the object moves even farther from its original position, in which case it is said to be in **unstable equilibrium**; or (3) the object remains in its new position, in which case it is said to be in **neutral equilibrium**.

Consider the following examples. A ball suspended freely from a string is in stable equilibrium, for if it is displaced to one side, it will quickly return to its original position (Fig. 9–17a). On the other hand, a pencil standing on its point is in unstable equilibrium. If its CG is directly over its tip (Fig. 9–17b), the net force and net torque on it will be zero. But if it is displaced ever so slightly—say by a slight vibration or tiny air current—there will be a torque on it, and it will continue to fall in the direction of the original displacement. Finally, an example of an object in neutral equilibrium is a sphere resting on a horizontal tabletop. If it is placed slightly to one side, it will remain in its new position.

In most situations, such as in the design of structures and in working with the human body, we are interested in maintaining stable equilibrium or *balance*, as we sometimes say. In general, an object whose CG is below its point of support, such as a ball on a string, will be in stable equilibrium. If the CG is above the base of support, we have a more complicated situation. Consider a standing refrigerator (Fig. 9–18a). If it is tipped slightly, it will return to its original position due to the torque on it as shown in Fig. 9–18b. But if it is tipped too far, Fig. 9–18c, it will fall over. The critical point is reached when the CG is no longer above the base of support. In general, *a body whose CG is above its base of support will be stable if a vertical line projected downward from the CG falls within the base of support*. This is because the normal force upward on the object (which balances out gravity) can be exerted only within the area of contact, so that if the force of gravity acts beyond this area, a net torque will act to topple the object. Stability, then, can be relative. A brick lying on its widest face is more stable than a brick standing on its end, for it will take more of an effort to tip it over. In the extreme case of the pencil in Fig. 9–17b, the base is practically a point and the slightest disturbance will topple it. In general, the larger the base and the lower the CG, the more stable the object.

Stable and unstable equilibrium

Net force

FIGURE 9–17 (a) Stable equilibrium, and (b) unstable equilibrium.

FIGURE 9–18 Equilibrium of a refrigerator resting on a surface.

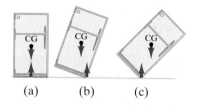

CG   CG   CG

(a)   (b)   (c)

In this sense, humans are much less stable than four-legged mammals, which not only have a larger base of support because of their four legs, but also have a lower center of gravity. The human species has had to develop special apparatus, such as certain very strong muscles, in order to deal with the problem of keeping a person upright and at the same time stable. Because of their upright position, humans suffer from numerous ailments such as low back pain due to the large forces involved, as we saw in Example 9–11. When walking and performing other kinds of movement, a person continually shifts the body so that its CG is over the feet, although in the normal adult this requires no conscious thought. Even as simple a movement as bending over requires moving the hips backward so that the CG remains over the feet, and this repositioning is done without thinking about it. To see this, position yourself with your heels and back to a wall and try to touch your toes. You won't be able to do it without falling. Persons carrying heavy loads automatically adjust their posture so that the CG of the total mass is over their feet, Fig. 9–19.

FIGURE 9–19 Humans adjust their posture to achieve stability when carrying loads.

9–6 Elasticity; Stress and Strain

In the first part of this chapter we studied how to calculate the forces on objects in equilibrium. In this Section we study the effects of these forces: any object changes shape under the action of applied forces. In Section 9–7 we will see that if the forces are great enough, the object will break or *fracture*.

If a force is exerted on an object, such as the vertically suspended metal rod shown in Fig. 9–20, the length of the object changes. If the amount of elongation, ΔL, is small compared to the length of the object, experiment shows that ΔL is proportional to the weight or force exerted on the object. This proportionality, as we saw in Section 6–4, can be written as an equation:

$$F = k \, \Delta L. \tag{9–3}$$

Here F represents the force (or weight) pulling on the object, ΔL is the change in length, and k is a proportionality constant. Equation 9–3, which is sometimes called **Hooke's law**[†], after Robert Hooke (1635–1703) who first noted it, is found to be valid for almost any solid material from iron to bone, but it is valid only up to a point. For if the force is too great, the object stretches excessively and eventually breaks. Figure 9–21 shows a typical graph of elongation versus applied force. Up to a point called the **proportional limit**, Eq. 9–3 is a good approximation for many common materials, and the curve is a straight line. Beyond this point, the graph deviates from a straight line, and no simple relationship exists between F and ΔL. Nonetheless, up to a point farther along the curve called the **elastic limit**, the object will return to its original length if the applied force is removed. The region from the origin to the elastic limit is called the *elastic region*. If the object is stretched beyond the elastic limit, it enters the *plastic region*: it does not return to the original length upon removal of the external force, but remains permanently deformed (like bending a paper clip). The maxi-

FIGURE 9–20 Hooke's law: $\Delta L \propto$ applied force.

Hooke's law (again)

FIGURE 9–21 Applied force vs. elongation for a typical metal under tension.

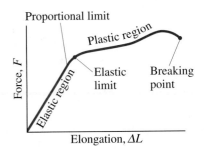

[†]The term "law" applied to this relation is not really appropriate, since first of all, it is only an approximation, and secondly, it refers only to a limited set of phenomena. Most physicists prefer to reserve the word "law" for those relations that are deeper and more encompassing and precise, such as Newton's laws of motion or the law of conservation of energy.

mum elongation is reached at the *breaking point*. The maximum force that can be applied without breaking is called the **ultimate strength** of the material (actually force per unit area as discussed in Section 9–7).

The amount of elongation of an object, such as the rod shown in Fig. 9–20, depends not only on the force applied to it, but also on the material from which it is made and on its dimensions. That is, the constant k in Eq. 9–3 can be written in terms of these factors. If we compare rods made of the same material but of different lengths and cross-sectional areas, it is found that for the same applied force, the amount of stretch (again assumed small compared to the total length) is proportional to the original length and inversely proportional to the cross-sectional area. That is, the longer the object, the more it elongates for a given force; and the thicker it is, the less it elongates. These findings can be combined with Eq. 9–3 to yield

Young's modulus

$$\Delta L = \frac{1}{E} \frac{F}{A} L_0, \tag{9–4}$$

where L_0 is the original length of the object, A is the cross-sectional area, and ΔL is the change in length due to the applied force F. E is a constant of proportionality[†] known as the **elastic modulus**, or **Young's modulus**, and its value depends only on the material. The value of Young's modulus for various materials is given in Table 9–1 (note: shear and bulk modulus in

TABLE 9–1 Elastic Moduli

Material	Elastic Modulus, E (N/m²)	Shear Modulus, G (N/m²)	Bulk Modulus, B (N/m²)
Solids			
Iron, cast	100×10^9	40×10^9	90×10^9
Steel	200×10^9	80×10^9	140×10^9
Brass	100×10^9	35×10^9	80×10^9
Aluminum	70×10^9	25×10^9	70×10^9
Concrete	20×10^9		
Brick	14×10^9		
Marble	50×10^9		70×10^9
Granite	45×10^9		45×10^9
Wood (pine)			
(parallel to grain)	10×10^9		
(perpendicular to grain)	1×10^9		
Nylon	5×10^9		
Bone (limb)	15×10^9	80×10^9	
Liquids			
Water			2.0×10^9
Alcohol (ethyl)			1.0×10^9
Mercury			2.5×10^9
Gases[†]			
Air, H_2, He, CO_2			1.01×10^5

[†]At normal atmospheric pressure; no variation in temperature during process.

[†]The fact that E is in the denominator, so that $1/E$ is the actual proportionality constant, is merely a convention. When we rewrite Eq. 9–4 to get Eq. 9–5, E is found in the numerator.

this Table are discussed later in this section). Because E is a property only of the material and is independent of the object's size or shape, Eq. 9–4 is far more useful for practical calculation than Eq. 9–3.

From Eq. 9–4, we see that the change in length of an object is directly proportional to the product of the object's length L_0 and the force per unit area F/A applied to it. It is general practice to define the force per unit area as the **stress**:

$$\text{stress} = \frac{\text{force}}{\text{area}} = \frac{F}{A},$$ *Stress*

which has units of N/m^2. Also, the **strain** is defined to be the ratio of the change in length to the original length:

$$\text{strain} = \frac{\text{change in length}}{\text{original length}} = \frac{\Delta L}{L_0},$$ *Strain*

and is dimensionless (no units). Strain is thus the fractional change in length of the object, and is a measure of how much the rod has been deformed. Stress is applied to the material by external agents, whereas strain is the material's response to the stress. Equation 9–4 can be rewritten as

$$\frac{F}{A} = E \frac{\Delta L}{L_0}$$ **(9–5)**

or

$$E = \frac{F/A}{\Delta L/L_0} = \frac{\text{stress}}{\text{strain}}.$$ *Young's modulus (again)*

Thus we see that the strain is directly proportional to the stress, in the linear (elastic) region of Fig. 9–21.

EXAMPLE 9–12 **Tension in piano wire.** A 1.60-m-long steel piano wire has a diameter of 0.20 cm. How great is the tension in the wire if it stretches 0.30 cm when tightened?

SOLUTION We solve for F in Eq. 9–5 and note that the area $A = \pi r^2 = (3.14)(0.0010 \text{ m})^2 = 3.1 \times 10^{-6} \text{ m}^2$. Then

$$F = E \frac{\Delta L}{L_0} A$$

$$= (2.0 \times 10^{11} \text{ N/m}^2)\left(\frac{0.0030 \text{ m}}{1.60 \text{ m}}\right)(3.1 \times 10^{-6} \text{ m}^2) = 1200 \text{ N}.$$

where we obtained the value for E from Table 9–1. The strong tension in all the wires in a piano must be supported by a strong frame.

The rod shown in Fig. 9–20 is said to be under *tension* or **tensile** *Tension* **stress**. For not only is there a force pulling down on the rod at its lower end, but since the rod is in equilibrium we know that the support at the top is exerting an equal[†] upward force on the rod at its upper end,

---

[†]If we ignore the weight of the bar.

FIGURE 9-22 Stress exists *within* the material.

FIGURE 9-23 This Greek temple, in Agrigento, Sicily, built 2500 years ago, shows the post-and-beam construction.

Fig. 9–22a. In fact, this tensile stress exists throughout the material. Consider, for example, the lower half of a suspended rod as shown in Fig. 9–22b. This lower half is in equilibrium, so there must be an upward force on it to balance the downward force at its lower end. What exerts this upward force? It must be the upper part of the rod. Thus we see that external forces applied to an object give rise to internal forces, or stress, within the material itself. (Recall also the discussion of tension in a cord, page 92.)

Compression   Strain or deformation due to tensile stress is but one type of stress to which materials can be subjected. There are two other common types of stress: compressive and shear. **Compressive stress** is the exact opposite of tensile stress. Instead of being stretched, the material is compressed: the forces act inwardly on the body. Columns that support a weight, such as the columns of a Greek temple (Fig. 9–23), or those that support the beam in Fig. 9–10, are subjected to compressive stress. Equations 9–4 and 9–5 apply equally well to compression and tension, and the values for the elastic modulus E are usually the same.

Shear   Figure 9–24 compares tensile and compressive stresses as well as the third type, shear stress. An object under **shear stress** has equal and opposite forces applied *across* its opposite faces. An example is a book or brick firmly attached to a tabletop, on which a force is exerted parallel to the

FIGURE 9-24 The three types of stress for rigid bodies.

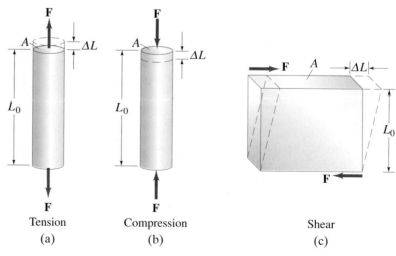

Tension
(a)

Compression
(b)

Shear
(c)

top surface. The table exerts an equal and opposite force along the bottom surface. Although the dimensions of the object do not change significantly, the shape of the object does change as shown in the figure. An equation similar to Eq. 9–4 can be applied to calculate shear strain:

$$\Delta L = \frac{1}{G}\frac{F}{A}L_0, \qquad\qquad (9\text{--}6)$$

but ΔL, L_0, and A must be reinterpreted as indicated in Fig. 9–24c. Note that A is the area of the surface *parallel* to the applied force (and not perpendicular as for tension and compression), and ΔL is *perpendicular* to L_0. The constant of proportionality, G, is called the **shear modulus** and is generally one half to one third the value of the elastic modulus, E (see Table 9–1). Figure 9–25 illustrates why $\Delta L \propto L_0$: the fatter book shifts more for the same shearing force.

Shear modulus

    The rectangular object undergoing shear in Fig. 9–24c would not actually be in equilibrium under the forces shown, for a net torque would exist. If the object is in fact in equilibrium, there must be two more forces acting on it which balance out this torque. One acts vertically upward on the right, and the other acts vertically downward on the left, as shown in Fig. 9–26. This is generally true of shear forces. If the object is a brick or book lying on a table, these two additional forces can be exerted by the table and by whatever exerts the other horizontal force (such as a hand pushing across the top of a book).

    If an object is subjected to inward forces from all sides, its volume will decrease. A common situation is a body submerged in a fluid; for in this case, the fluid exerts a pressure on the object in all directions, as we shall see in Chapter 10. Pressure is defined as force per area, and thus is the equivalent of stress. For this situation the change in volume, ΔV, is found to be proportional to the original volume, V_0, and to the increase in the pressure, ΔP. We thus obtain a relation of the same form as Eq. 9–4 but with a proportionality constant called the **bulk modulus**, B:

$$\frac{\Delta V}{V_0} = -\frac{1}{B}\Delta P \qquad\qquad (9\text{--}7)$$

or

$$B = -\frac{\Delta P}{\Delta V/V_0}.$$

The minus sign is included to indicate that the volume *decreases* with an

FIGURE 9–25 The fatter book (a) shifts more than the thinner book (b) with the same applied shear force.

FIGURE 9–26 Balance of forces and torques for shear stress.

(a)

(b)

increase in pressure. Values for the bulk modulus are given in Table 9–1. Since liquids and gases do not have a fixed shape, only the bulk modulus applies to them.

9–7 Fracture

FIGURE 9–27 Fracture as a result of the three types of stress.

If the stress on a solid object is too great, the object fractures or breaks (Fig. 9–27). Table 9–2 lists the ultimate tensile strength, compressive strength, and shear strength for a variety of materials. These values give the maximum force per unit area that an object can withstand under each of these three types of stress. They are, however, representative values only, and the actual value for a given specimen can differ considerably. It is therefore necessary to maintain a "safety factor" of from 3 to perhaps 10 or more—that is, the actual stresses on a structure should not exceed one tenth to one third of the values given in the table. You may encounter tables of the "allowable stresses" in which appropriate safety factors have already been included.

TABLE 9–2 Ultimate Strengths of Materials (force/area)

Material	Tensile Strength (N/m²)	Compressive Strength (N/m²)	Shear Strength (N/m²)
Iron, cast	170×10^6	550×10^6	170×10^6
Steel	500×10^6	500×10^6	250×10^6
Brass	250×10^6	250×10^6	200×10^6
Aluminum	200×10^6	200×10^6	200×10^6
Concrete	2×10^6	20×10^6	2×10^6
Brick		35×10^6	
Marble		80×10^6	
Granite		170×10^6	
Wood (pine)			
(parallel to grain)	40×10^6	35×10^6	5×10^6
(perpendicular to grain)		10×10^6	
Nylon	500×10^6		
Bone (limb)	130×10^6	170×10^6	

EXAMPLE 9–13 **Size and compression of support columns.** (*a*) What minimum cross-sectional area should the two columns have to support the beam of Example 9–7 (Fig. 9–10) assuming the columns are made of concrete and a safety factor of 6 is required? We saw in Example 9–7 that the column on the left supports 4.4×10^4 N and that on the right supports 1.2×10^5 N. (*b*) How much will the chosen supports compress under the given load?

SOLUTION (*a*) The right-hand column receives the larger force, 1.2×10^5 N. It is clearly under compression, and from Table 9–2, we see that the ultimate compressive strength of concrete is 2.0×10^7 N/m². Using a safety factor of 6, the maximum allowable stress is $\frac{1}{6}(2.0 \times 10^7$ N/m²) = 3.3×10^6 N/m², which equals F/A. Since $F = 1.2 \times 10^5$ N, we can solve for A, and we find:

$$A = \frac{1.2 \times 10^5 \text{ N}}{3.3 \times 10^6 \text{ N/m}^2} = 3.6 \times 10^{-2} \text{ m}^2, \text{ or } 360 \text{ cm}^2.$$

A support 18 cm $\times$ 20 cm will be adequate.
(*b*) We solve for

$$\frac{\Delta L}{L_0} = \frac{1}{E}\frac{F}{A} = \left(\frac{1}{2.0 \times 10^{10} \text{ N/m}^2}\right)(3.3 \times 10^6 \text{ N/m}^2) = 1.7 \times 10^{-4}.$$

Thus, if the support has a length $L_0 = 5.0$ m, $\Delta L = 0.85 \times 10^{-3}$ m, or about 1 mm. This calculation was for the right-hand support. If the left-hand support is made of the same cross-sectional area, it will compress less and this should be taken into account.

As can be seen in Table 9–2, concrete (like stone and brick) is reasonably strong under compression but extremely weak under tension. Thus concrete can be used as vertical columns placed under compression but is of little value as a beam since it cannot withstand the tensile forces that arise (see Fig. 9–28). *Reinforced concrete*, in which iron rods are embedded in the concrete, is much stronger (Fig. 9–29). But the concrete on the lower edge of a loaded beam still tends to crack because of its weakness under tension. This problem is solved with **prestressed concrete**, which also contains iron rods or a wire mesh, but during the pouring of the concrete, the rods or wire are held under tension. After the concrete dries, the tension on the iron is released, putting the concrete under compression. The amount of compressive stress is carefully predetermined so that when the design loads are applied to the beam, they reduce the compression on the lower edge but never put the concrete into tension.

➡ **PHYSICS APPLIED**

Reinforced concrete
and
prestressed concrete

FIGURE 9–28 A beam sags, at least a little (but is exaggerated here), even under its own weight. The beam thus changes shape so that the upper portion is compressed, and the lower portion is under tension (elongated). Shearing stress also occurs within the beam.

FIGURE 9–29 Steel rods waiting for concrete to be poured around them to form a new highway.

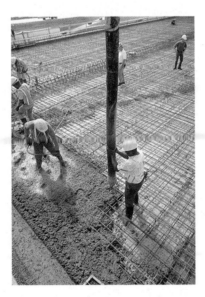

Let's consider one more instructive Example.

CONCEPTUAL EXAMPLE 9–14 **A tragic substitution.** Two walkways, one above the other, are suspended from vertical rods attached to the ceiling of a high hotel lobby, Fig. 9–30a (only one rod is shown). The original design called for single rods 14 m long, but when such long rods proved to be unwieldy to install, it was decided to replace each long rod with two shorter ones as shown in Fig. 9–30b. Determine the net force exerted by the rods on the supporting pin A (assumed the same size) for each design. Assume each vertical rod supports a mass m of each bridge.

RESPONSE The single long vertical rod in Fig. 9–30a exerts an upward force equal to mg on pin A to support the mass m of the upper bridge. Why? Because the pin is in equilibrium, and the other force that balances this is the downward force mg exerted on it by the upper bridge. (There is thus a shear stress on the pin.) See Fig. 9–30c. The situation when two shorter rods support the bridges (Fig. 9–30b) is shown in Fig. 9–30d in which only the connections at the upper bridge are shown. The lower rod exerts a force of mg downward on the lower of the two pins because it supports the lower bridge. The upper rod exerts a force of $2mg$ on the upper pin (pin A) because the upper rod supports both bridges. Thus we see that when the builders substituted two shorter rods for each single long one, the stress in the supporting pin A was *doubled*. What perhaps seemed like a simple substitution did, in fact, lead to a tragic collapse in 1981 with a loss of life of over 100 people (see Fig. 9–2). Having a feel for physics, and being able to make simple calculations based on physics, can have a great effect, literally, on people's lives.

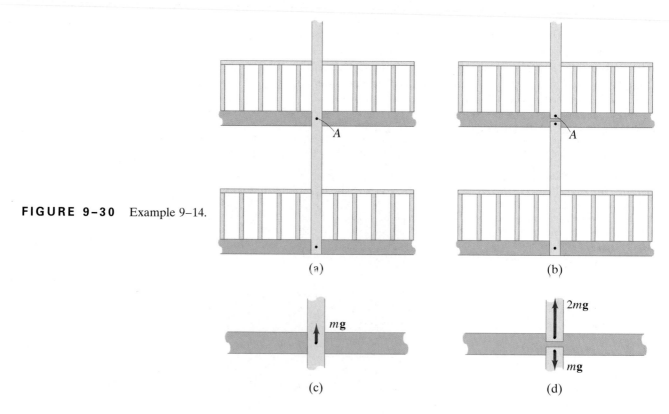

FIGURE 9–30 Example 9–14.

(a)   (b)

(c)   (d)

*9–8 Spanning a Space: Arches and Domes

There are a great many areas where the arts and humanities overlap the sciences, and this is especially clear in architecture, where the forces in the materials that make up a structure need to be understood to avoid excessive deformation and collapse. Many of the features we admire in the architecture of the past were introduced not simply for their decorative effect, but often for technical reasons. One example is the development of methods to span a space, from the simple beam to arches and domes.

It might be said that the first important architectural invention was the post-and-beam (or post-and-lintel) construction, in which two upright posts support a horizontal beam. Before steel was introduced in the nineteenth century, the length of a beam was quite limited because the strongest building materials were then stone and brick. Hence the width of a span was limited by the size of available stones. Equally important, stone and brick, though strong under compression—are very weak under tension and shear; all three types of stress occur in a beam, as shown in Fig. 9–28. The minimal space that could be spanned using stone is shown by the closely spaced columns of the great Greek temples (Fig. 9–23).

The introduction of the semicircular **arch** by the Romans (Fig. 9–31), aside from its aesthetic appeal, was a tremendous technological innovation. It had been preceded by the so-called "triangle arch" and the "corbeled arch," but these were relatively small improvements over the post-and-beam (see Fig. 9–32). The advantage of the "true" or semicircular arch is that, if well designed, its wedge-shaped stones experience stress which is mainly compressive even when supporting a large load such as the wall and roof of a cathedral. Because the stones are forced to squeeze against one another, they are mainly under compression (see Fig. 9–33). Note, however, that the arch transfers horizontal as well as vertical forces to the supports. A round arch consisting of many well-shaped stones could span a very wide space. However, considerable buttressing on the sides was needed to support the horizontal components of the forces.

➡ **PHYSICS APPLIED**

Architecture: Beams, arches and domes

FIGURE 9–31 Round arches in the Roman Forum. The one in the background is the Arch of Titus.

FIGURE 9–33 Stones in a round (or "true") arch are mainly under compression.

FIGURE 9–32 Various means to span a space.

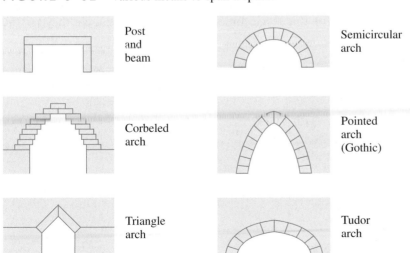

Post and beam

Semicircular arch

Corbeled arch

Pointed arch (Gothic)

Triangle arch

Tudor arch

FIGURE 9–34 Flying buttresses (on the cathedral of Notre Dame, in Paris).

The pointed arch came into use about A.D. 1100 and became the hallmark of the great Gothic cathedrals. It too was an important technical innovation, and was first used to support heavy loads such as the tower of a cathedral, and as the central arch. Apparently the builders realized that, because of the steepness of the pointed arch, the forces due to the weight above could be brought down more nearly vertically, so less horizontal buttressing would be needed. The pointed arch reduced the load on the walls, so there could be more openness and light. The smaller buttressing needed was provided on the outside by graceful flying buttresses (Fig. 9–34).

The technical innovation of the pointed arch was achieved not through calculation but through experience and intuition, for it was not until much later that detailed calculations, such as those presented earlier in this chapter, came into use. To make an accurate analysis of a stone arch is quite difficult in practice. But if we make some simplifying assumptions, we can show why the horizontal component of the force at the base is less for a pointed arch than for a round one. Figure 9–35 shows a round arch and a pointed arch, each with an 8.0-m span. The height of the round arch is thus 4.0 m, whereas that of the pointed arch is larger and has been chosen to be 8.0 m. Each arch supports a weight of 12.0×10^4 N ($= 12{,}000$ kg $\times g$) which, for simplicity, we have divided into two parts (each 6.0×10^4 N) acting on the two halves of each arch as shown. To be in equilibrium, each of the supports must exert an upward force of 6.0×10^4 N. Each support also exerts a horizontal force, F_H, at the base of the arch, and it is this we want to calculate. We focus only on the right half of each arch. We set equal to zero the total torque calculated about the apex of the arch due to the forces exerted on that half arch, as if there were a hinge at the apex. For the round arch, the torque equation is

$$(4.0 \text{ m})(6.0 \times 10^4 \text{ N}) - (2.0 \text{ m})(6.0 \times 10^4 \text{ N}) - (4.0 \text{ m})(F_H) = 0.$$

Thus $F_H = 3.0 \times 10^4$ N. For the pointed arch, the torque equation is

$$(4.0 \text{ m})(6.0 \times 10^4 \text{ N}) - (2.0 \text{ m})(6.0 \times 10^4 \text{ N}) - (8.0 \text{ m})(F_H) = 0.$$

Solving, we find that $F_H = 1.5 \times 10^4$ N—only half as much! From this calculation we can see that the horizontal buttressing force required for a pointed arch is less because the arch is higher, and there is therefore a longer lever arm for this force. Indeed, the steeper the arch, the less the

FIGURE 9–35 Forces in a round arch (a), compared with those in a pointed arch (b).

(a)

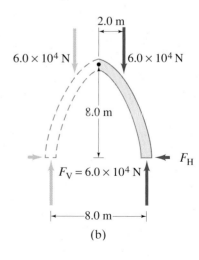

(b)

horizontal component of the force needs to be, and hence the more nearly vertical is the force exerted at the base of the arch.

The further development of the arch was one of decline. For the subsequent flattened arches, such as the Tudor arch (Fig. 9–32), were structurally weaker than the simple pointed arch. However, with the coming of advanced methods of calculation in the nineteenth and twentieth centuries, it became possible to calculate the best shape of arch for a given load condition. For example, if the load is uniform across the span, it can be shown that the stresses within the arch will be purely compressive if the arch has a parabolic shape.

Whereas an arch spans a two-dimensional space, a **dome**—which is basically an arch rotated about a vertical axis—spans a three-dimensional space. The Romans built the first large domes. Their shape was hemispherical and some still stand, such as that of the Pantheon in Rome (Fig. 9–36). By the time of the Renaissance, the technique for constructing large domes seems to have been lost. Indeed, the dome of the Pantheon was a source of wonder to Renaissance architects. The problem came to the fore in fifteenth-century Florence with the designing of a new cathedral that was to have a dome 43 m in diameter to rival that of the Pantheon. In 1418, after the cathedral was finished except for the dome, a competition for the design of the dome was held and was won by Filippo Brunelleschi (1377–1446). One problem that had to be dealt with was that the dome was to rest on a "drum" that had been completed with no external abutments; and there was no place to put any. Hence the dome needed to exert a minimum of horizontal force. Brunelleschi solved this by designing a pointed dome (Fig. 9–37), since a pointed dome, like a pointed arch, exerts a smaller side thrust against its base.

The other major problem was how to support the dome during construction. A dome, like an arch, is not stable until all the stones are in place. It had been the custom to support a dome during construction with a wooden framework. But no trees big enough or strong enough could be found to span the 43-m space required for the cathedral in Florence. Instead of using a wooden framework, Brunelleschi built the dome in horizontal layers. Each

FIGURE 9–36 Interior of the Pantheon in Rome, built in the first century. This view, showing the great dome and its central opening for light, was painted about 1740 by Panini. Photographs do not capture its grandeur as well as this painting does.

FIGURE 9–37 The skyline of Florence, showing Brunelleschi's famous dome on the cathedral.

FIGURE 9–38 The dome of the Small Sports Palace in Rome, built for the 1960 Olympics.

layer was bonded to the previous one, which held it in place until the last stone of the circle was placed, and then that circle was stable just as a completed arch is stable. Each closed ring was then strong enough to support the next layer. It was an amazing feat.

To end this Section we will consider the forces necessary to support a modern dome, that of the Small Sports Palace in Rome (Fig. 9–38). A dome, like an arch, is statically most stable when under compression. The 36 buttresses which support the 1.2×10^6-kg dome are positioned at a 38° angle and connect smoothly with the dome.

EXAMPLE 9–15 **A modern dome.** Calculate the components of the force, F_V and F_H, that each buttress exerts on the dome so that the force acts compressively—that is, at a 38° angle (Fig. 9–39).

SOLUTION The vertical load on *each* buttress is $\frac{1}{36}$ of the total weight. Thus

$$F_V = \frac{(1.2 \times 10^6 \text{ kg})(9.8 \text{ m/s}^2)}{36} = 3.4 \times 10^5 \text{ N}.$$

The force must act at a 38° angle at the base of the dome in order to be purely compressive. Thus

$$\tan 38° = \frac{F_V}{F_H} = \frac{340{,}000 \text{ N}}{F_H}$$

$$F_H = \frac{340{,}000 \text{ N}}{\tan 38°} = 430{,}000 \text{ N}.$$

In order that each of the buttresses be able to exert this 430,000-N horizontal force, a prestressed-concrete tension ring surrounds the base of the buttresses beneath the ground (see Problem 63 and Fig. 9–71).

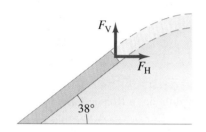

FIGURE 9–39 Example 9–15.

SUMMARY

A body at rest [or one in uniform motion at constant velocity] is said to be in **equilibrium**. The subject concerned with the determination of the forces within a structure at rest is called **statics**.

The two necessary conditions for a body to be in equilibrium are (1) the vector sum of all the forces on it must be zero, and (2) the sum of all the torques (calculated about any arbitrary axis) must also be zero:

$$\Sigma F_x = 0 \qquad \Sigma F_y = 0 \qquad \Sigma \tau = 0.$$

It is important when doing statics problems to apply the equilibrium conditions to only one body at a time.

A body in static equilibrium is said to be in (a) **stable**, (b) **unstable**, or (c) **neutral equilibrium**, depending on whether a slight displacement leads to (a) a return to the original position, (b) further movement away from the original position, or

(c) rest in the new position. An object in stable equilibrium is also said to be in **balance**.

Hooke's law applies to many elastic solids, and states that the change in length of an object is proportional to the applied force:

$$F = k \, \Delta L.$$

If the force is too great, the object will exceed its **elastic limit**, which means it will no longer return to its original shape when the distorting force is removed. If the force is even greater, the **ultimate strength** of the material can be exceeded and the object fractures.

The force per unit area acting on a body is called the **stress**, and the resulting fractional change in length is called the **strain**.

The stress on a body is present within the body and can be of three types: **compression**, **tension**, and **shear**.

The ratio of stress to strain is called the **elastic modulus** of the material. **Young's modulus** applies for compression and tension, and the **shear modulus** for shear; **bulk modulus** applies to an object whose volume changes as a result of pressure on all sides. All three moduli are constants for a given material when distorted within the elastic region.

QUESTIONS

1. Describe several situations where a body is not in equilibrium, even though the net force on it is zero.

2. A bungee jumper momentarily comes to rest at the bottom of the dive before he springs back upward. At that moment, is the bungee jumper in equilibrium? Explain.

3. You can find the center of gravity of a meter stick by resting it horizontally on your index fingers, and then slowly drawing your fingers together. First the meter stick will slip on one finger, and then on the other, but eventually the fingers meet at the CG. Why does this work?

4. Your doctor's scale has arms on which weights slide to counter your weight, Fig. 9–40. These weights are obviously much lighter than you are. How does this work?

FIGURE 9–40   Question 4.

5. A bear sling, Fig. 9–41, is used in some national parks for placing backpackers' food out of the reach of bears. Explain why the force needed to pull the backpack up increases as the backpack gets higher and higher. Is it possible to pull the rope hard enough so that it doesn't sag at all?

6. A ladder, leaning against a wall, makes a 60° angle with the ground. When is it more likely to slip: when a person stands near the top or near the bottom? Explain.

7. Explain why touching the toes while seated on the floor with outstretched legs produces less stress on the lower spinal column than when touching the toes from a standing position. Use a diagram.

FIGURE 9–41   Question 5, Problems 24 and 29.

8. An earthen retaining wall is shown in Fig. 9–42a. The earth, particularly when wet, can exert a significant force F on the wall. (a) What force produces the torque to keep the wall upright? (b) Explain why the retaining wall in Fig. 9–42b would be much less likely to overturn.

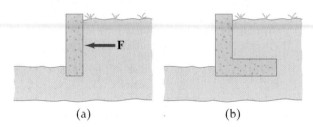

(a)   (b)

FIGURE 9–42   Question 8.

9. Figure 9–43 shows a cone. Draw how to lay it on a flat table so that it is in (a) stable equilibrium, (b) unstable equilibrium, (c) neutral equilibrium.

FIGURE 9–43   Question 9.

FIGURE 9-44   Question 10.

10. A uniform meter stick supported at the 25-cm mark is in equilibrium when a 1-kg rock is suspended at the 0-cm end (as shown in Fig. 9–44). Is the mass of the meter stick greater than, equal to, or less than the mass of the rock? Explain your reasoning.

11. Which of the configurations of brick, (a) or (b) of Fig. 9–45, is the more likely to be stable? Why?

(a)   (b)

FIGURE 9-45   The dots indicate the CG of each brick. The fractions $\frac{1}{4}$ and $\frac{1}{2}$ indicate what portion of each brick is hanging beyond its support. Question 11.

FIGURE 9-46   Question 12.

12. Name the type of equilibrium for each position of the ball in Fig. 9–46.

13. Why do you tend to lean backward when carrying a heavy load in your arms?

14. Place yourself facing the edge of an open door. Position your feet astride the door with your nose and abdomen touching the door's edge. Try to rise on your tiptoes. Why can't this be done?

15. Why is it not possible to sit upright in a chair and rise to one's feet without first leaning forward?

16. Why is it more difficult to do sit-ups when your knees are bent than when your legs are stretched out?

17. Examine how a pair of scissors or shears cuts through a piece of cardboard. Is the name "shears" justified?

18. Materials such as ordinary concrete and stone are very weak under tension or shear. Would it be wise to use such a material for either of the supports of the cantilever shown in Fig. 9–11? If so, which one(s)?

PROBLEMS

SECTIONS 9–1 TO 9–3

1. (I) Three forces are applied to a tree sapling, as shown in Fig. 9–47, to stabilize it. If $\mathbf{F}_1 = 282$ N and $\mathbf{F}_2 = 355$ N, find $\mathbf{F}_3$ in magnitude and direction.

FIGURE 9-47   Problem 1.

2. (I) What should be the tension in the wire if the net force exerted on the tooth in Fig. 9–3 is to be 0.75 N? Assume that the angle between the two forces is 155° rather than the 140° in the figure.

3. (I) Calculate the torque about the front support of a diving board, on the right in Fig. 9–48, exerted by a 60-kg person 3.0 m from that support.

FIGURE 9-48   Problems 3, 4, 19, and 20.

4. (I) How far out on the diving board (Fig. 9–48) would a 60-kg diver have to be to exert a torque of 1000 m·N on the board, relative to the left support post?

5. (I) Two cords support a chandelier in the manner shown in Fig. 9–6 except that the upper wire makes an angle of 45° with the ceiling. If the cords can sustain a force of 1300 N without breaking, what is the maximum chandelier weight that can be supported?

FIGURE 9–49 Problems 6 and 21.

6. (I) Calculate the mass m needed in order to suspend the leg shown in Fig. 9–49. Assume the leg (with cast) has a mass of 15.0 kg, and its CG is 35.0 cm from the hip joint; the sling is 80.5 cm from the hip joint.

7. (II) A 160-kg horizontal beam is supported at each end. A 300-kg piano rests a quarter of the way from one end. What is the vertical force on each of the supports?

8. (II) A uniform steel beam has a mass of 1000 kg. On it is resting half of an identical beam, as shown in Fig. 9–50. What is the vertical support force at each end?

FIGURE 9–50 Problem 8.

9. (II) Suppose that the net force as calculated in Example 9–1 is 10° to the left of where it should point if the tooth is to move correctly. If the tension to the left is 2.0 N, what should the tension to the right be to make the net force act in the correct direction?

10. (II) A 70-kg adult sits at one end of a 10-m board, on the other end of which sits his 30-kg child. Where should the pivot be placed so the board (ignore its mass) is balanced?

11. (II) Repeat Problem 10 taking into account the board's 15-kg mass.

12. (II) Find the tension in the two cords shown in Fig. 9–51. Neglect the mass of the cords, and assume that the angle θ is 30° and the mass m is 200 kg.

FIGURE 9–51
Problem 12.

FIGURE 9–52
Problems 13 and 48.

13. (II) Find the tension in the two wires supporting the traffic light shown in Fig. 9–52.

14. (II) Determine the force F_N that the pivot exerts on the seesaw board in Fig. 9–9.

15. (II) Calculate F_1 and F_2 for the uniform cantilever shown in Fig. 9–11 whose mass is 1200 kg.

16. (II) A 0.60 kg sheet hangs from a massless clothesline as shown in Fig. 9–53. The line on either side of the sheet makes an angle of 3.5° with the horizontal. Calculate the tension in the clothesline on either side of the sheet. Why is the tension so much greater than the weight of the sheet?

FIGURE 9–53 Problem 16.

17. (II) A door, 2.30 m high and 1.30 m wide, has a mass of 13.0 kg. A hinge 0.40 m from the top and another hinge 0.40 m from the bottom each support half the door's weight (Fig. 9–54). Assume that the center of gravity is at the geometrical center of the door, and determine the horizontal and vertical force components exerted by each hinge on the door.

FIGURE 9–54
Problem 17.

FIGURE 9–55 Problem 18.

18. (II) Three boys are trying to balance on a seesaw, which consists of a fulcrum rock, acting as a pivot at the center, and a very light board 3.6 m long (Fig. 9–55). Two boys are already on either end. One has a mass of 50 kg, and the other a mass of 35 kg. Where should the third boy, whose mass is 25 kg, place himself so as to balance the seesaw?

19. (II) Calculate the forces F_1 and F_2 that the supports exert on the diving board of Fig. 9–48 when a 60-kg person stands at its tip. Ignore the weight of the board.

20. (II) Repeat the last problem, taking into account the board's mass of 35 kg. Assume the board's CG is at its center.

21. (II) Calculate the mass m required in Fig. 9–49 to support the leg (without cast), using the result of Example 7–12 and the values given in Table 7–1, assuming a 60.0-kg person 160 cm tall. The leg pivots about the hip joint and the support acts at the ankle joint.

22. (II) Calculate F_1 and F_2 for the beam shown in Fig. 9–56. Assume it is uniform and has a mass of 250 kg.

FIGURE 9–56 Problem 22.

23. (II) Calculate the tension F_T in the wire that supports the 30-kg beam shown in Fig. 9–57, and the force $\mathbf{F}_W$ exerted by the wall on the beam (give magnitude and direction).

24. (II) The two trees in Fig. 9–41 are 7.6 m apart. Calculate the magnitude of the force $\mathbf{F}$ a backpacker must exert to hold a 16-kg backpack so that the rope sags at its midpoint by (a) 1.5 m, (b) 0.15 m.

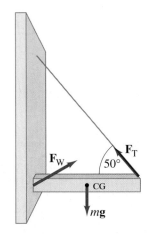

FIGURE 9–57 Problem 23.

25. (II) A 170-cm-tall person lies on a light (massless) board which is supported by two scales, one under the feet and one beneath the top of the head (Fig. 9–58). The two scales read, respectively, 31.6 and 35.1 kg. Where is the center of gravity of this person?

FIGURE 9–58 Problem 25.

26. (II) A shop sign weighing 215 N is supported by a uniform 135-N beam as shown in Fig. 9–59. Find the tension in the guy wire and the horizontal and vertical forces exerted by the hinge on the beam.

FIGURE 9–59 Problem 26.

FIGURE 9–60   Problem 27.

FIGURE 9–62   Problem 30.

27. (II) A traffic light hangs from a structure as shown in Fig. 9–60. The uniform aluminum pole AB is 7.5 m long and has a mass of 8.0 kg. The mass of the traffic light is 12.0 kg. Determine the tension in the horizontal massless cable CD, and the vertical and horizontal components of the force exerted by the pivot A on the aluminum pole.

28. (II) A uniform ladder of mass m and length L leans at an angle θ against a frictionless wall, Fig. 9–61. If the coefficient of static friction between the ladder and the ground is μ, what is the minimum angle at which the ladder will not slip?

30. (III) A meter stick with a mass of 230 g is supported horizontally by two vertical strings, one at the 0-cm mark and the other at the 90-cm mark (Fig. 9–62). (a) What is the tension in the string at 0 cm? (b) What is the tension in the string at 90 cm?

31. (III) Consider again the ladder of Example 9–9 but with a painter climbing up. If the mass of the ladder is 12.0 kg, the mass of the painter is 60.0 kg, and the ladder begins to slip at its base when she is 70 percent of the way up the length of the ladder, what is the coefficient of static friction between the ladder and the floor? Again assume the wall is frictionless. A free-body diagram is shown in Fig. 9–63.

FIGURE 9–61   Problems 28, 76, and 77.

29. (III) A 23.0 kg backpack is suspended midway between two trees by a light cord as in Fig. 9–41. A bear grabs the backpack and pulls vertically downward with a constant force, so that each section of cord makes an angle of 30° below the horizontal. Initially, without the bear pulling, the angle was 15°; the tension in the cord with the bear pulling is double what it was when he was not. Calculate the force the bear is exerting on the backpack.

FIGURE 9–63   Problem 31.

FIGURE 9-64   Problem 32.

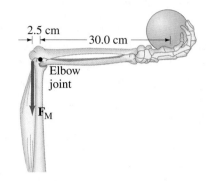

FIGURE 9-66   Problem 35.

32. (III) A person wants to push a lamp (mass 7.2 kg) across the floor. (*a*) Assuming the person pushes at a height of 60 cm above the ground and the coefficient of friction is 0.20, determine whether the lamp will slide or tip over (Fig. 9–64). (*b*) Calculate the maximum height above the floor at which the person can push the lamp so that it slides rather than tips.

33. (III) Two guy wires run from the top of a pole 2.6 m tall that supports a volleyball net. The two wires are anchored to the ground 2.0 m apart and each is 2.0 m from the pole (Fig. 9–65). The tension in each wire is 95 N. What is the tension in the net, assumed horizontal and attached at the top of the pole?

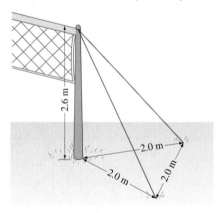

FIGURE 9-65   Problem 33.

*SECTION 9-4

* 34. (I) If the point of insertion of the biceps muscle into the lower arm shown in Fig. 9–15a is 6.0 cm, how much mass could the person hold with a muscle exertion of 400 N?

* 35. (I) Approximately what force, F_M, must the extensor muscle in the upper arm exert on the lower arm to hold a 7.3-kg shot put (Fig. 9–66)? Assume the lower arm has a mass of 2.8 kg and its CG is 12 cm from the pivot point.

* 36. (II) Calculate the force required of the "deltoid" muscle, F_M, to hold up the outstretched arm shown in Fig. 9–67. The total mass of the arm is 3.3 kg.

FIGURE 9-67   Problems 36, 37, and 38.

* 37. (II) Suppose the hand in the last problem holds a 15-kg mass. What force, F_M, is required of the deltoid muscle, assuming the mass is 52 cm from the shoulder joint?

* 38. (II) Calculate the magnitude of the force F_J exerted by the shoulder on the upper arm at the joint for Problems 36 and 37.

* 39. (II) The Achilles tendon is attached to the rear of the foot as shown in Fig. 9–68. When a person elevates himself just barely off the floor on the "ball of one foot," estimate the tension in the Achilles tendon (pulling upward), and the (downward) force exerted by the lower leg bone on the foot. Assume the person has a mass of 70 kg, and that D is twice as long as d.

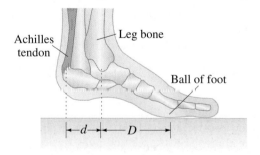

FIGURE 9-68   Problem 39.

40. (II) Redo Example 9–11 assuming now that the person is less bent over so that the 30° in Fig. 9–16b is instead 60°. What will the magnitude of F_V on the vertebra be?

*41. (III) Calculate the magnitude of the force at the base of the spine, $\mathbf{F}_V$ in Example 9–11 (Fig. 9–16b), if a mass of 20 kg is held in the person's hands, with arms hanging vertically. Assume the person's mass is 70 kg.

*SECTION 9–5

*42. (II) The Leaning Tower of Pisa is 55 m tall and about 7.0 m in diameter. The top is 4.5 m off center. Is the tower in stable equilibrium? If so, how much farther can it lean before it becomes unstable? Assume the tower is of uniform composition.

*43. (III) Four bricks are to be stacked at the edge of a table, each brick overhanging the one below it, so that the top brick extends as far as possible beyond the edge of the table. (a) To achieve this, show that successive bricks must extend no more than (starting at the top) 1/2, 1/4, 1/6, and 1/8 of their length beyond the one below (Fig. 9–69). (b) Is the top brick completely beyond the base? (c) Determine a general formula for the maximum total distance spanned by n bricks if they are to remain stable. (d) A builder wants to construct a corbeled arch (Fig. 9–32) based on the principle of stability discussed in (a) and (c) above. What minimum number of bricks, each 0.30 m long, is needed if the arch is to span 1.0 m?

FIGURE 9–69
Problem 43.

SECTION 9 6

44. (I) A nylon tennis string on a racket is under a tension of 250 N. If its diameter is 1.00 mm, by how much is it lengthened from its untensioned length of 30.0 cm?

45. (I) A marble column of cross-sectional area 2.0 m² supports a mass of 25,000 kg. (a) What is the stress within the column? (b) What is the strain?

46. (I) By how much is the column in the previous problem shortened if it is 12 m high?

47. (I) A vertical steel girder with a cross-sectional area of 0.15 m² has a sign (mass 2000 kg) hanging from its end. (a) What is the stress within the girder? (b) What is the strain on the girder? (c) If the girder is 9.50 m long, how much is it lengthened? (Ignore the mass of the girder itself.)

48. (II) If the two wires in Fig. 9–52 (Problem 13) are made of steel wire 1.0 mm in diameter, what is the percentage stretch of each because of the load?

49. (II) One liter of alcohol (1000 cm³) in a flexible container is carried to the bottom of the sea, where the pressure is 2.6×10^6 N/m². What will be its volume there?

50. (II) A 15-cm-long animal tendon was found to stretch 3.7 mm by a force of 13.4 N. The tendon was approximately round with an average diameter of 8.5 mm. Calculate the elastic modulus of this tendon.

51. (II) How much pressure is needed to compress the volume of an iron block by 0.10 percent? Express answer in N/m², and compare it to atmospheric pressure (1.0×10^5 N/m²).

52. (II) At depths of 2000 m in the sea, the pressure is about 200 times atmospheric pressure (1.0×10^5 N/m²). By what percentage does an iron bathysphere's volume change at this depth?

53. (III) A scallop forces open its shell with an elastic material called abductin, whose elastic modulus is about 2.0×10^6 N/m². If this piece of abductin is 3.0 mm thick and has a cross-sectional area of 0.50 cm², how much potential energy does it store when compressed 1.0 mm?

54. (III) A pole projects horizontally from the front wall of a shop. A 5.1-kg sign hangs from the pole at a point 2.2 m from the wall (Fig. 9–70). (a) What is the torque due to this sign calculated about the point where the pole meets the wall? (b) If the pole is not to fall off, there must be another torque exerted to balance it. What exerts this torque? Use a diagram to show how this torque must act. (c) Discuss whether compression, tension, and/or shear play a role in part (b).

FIGURE 9–70   Problem 54.

SECTION 9–7

55. (I) The femur bone in the leg has a minimum effective cross section of about 3.0 cm² ($= 3.0 \times 10^{-4}$ m²). How much compressive force can it withstand before breaking?

56. (II) What is the maximum tension possible in a 1.00-mm-diameter nylon tennis racket string? If you want tighter strings, what do you do to prevent breakage: go to thinner or thicker strings? Why? What causes strings to break when they are hit by the ball?

57. (II) If a compressive force of 3.6×10^4 N is exerted on the end of a 20-cm-long bone of cross-sectional area 3.6 cm^2, (a) will the bone break, and (b) if not, by how much does it shorten?

58. (II) (a) What is the minimum cross-sectional area required of a vertical steel cable from which is suspended a 320-kg chandelier? Assume a safety factor of 7.0. (b) If the cable is 7.5 m long, how much does it elongate?

59. (II) Assume the supports of the cantilever shown in Fig. 9–11 (mass = 2600 kg) are made of wood. Calculate the minimum cross-sectional area required of each, assuming a safety factor of 8.5.

60. (II) An iron bolt is used to connect two iron plates together. The bolt must withstand shear forces up to about 3200 N. Calculate the minimum diameter for the bolt, based on a safety factor of 6.0.

61. (III) A steel cable is to support an elevator whose total (loaded) mass is not to exceed 3100 kg. If the maximum acceleration of the elevator is 1.2 m/s^2, calculate the diameter of cable required. Assume a safety factor of 7.0.

*SECTION 9–8

* 62. (II) How high must a pointed arch be if it is to span a space 8.0 m wide and exert one third the horizontal force at its base that a round arch would?

* 63. (II) The subterranean tension ring that exerts the balancing horizontal force on the abutments for the dome in Fig. 9–38 is 36-sided, so each segment makes a 10° angle with the adjacent one (Fig. 9–71). Calculate the tension F that must exist in each segment so that the required force of 4.3×10^5 N can be exerted at each corner (Example 9–15).

FIGURE 9–71
Problem 63.   430,000 N

GENERAL PROBLEMS

64. The mobile in Fig. 9–72 is in equilibrium. The object B has mass of 0.735 kg. Determine the masses of the objects A, C, and D. (Neglect the weights of the crossbars.)

FIGURE 9–72   Problem 64.

65. A 50-story building is being planned. It is to be 200 m high with a base 40 m by 70 m. Its total mass will be about 1.8×10^7 kg and its weight therefore about 1.8×10^8 N. Suppose a 200-km/h wind exerts a force of 950 N/m^2 over the 70-m-wide face (Fig. 9–73). Calculate the torque about the potential

pivot point, the rear edge of the building (where $\mathbf{F_E}$ acts in Fig. 9–73), and determine whether the building will topple. Assume the total force of the wind acts at the midpoint of the building's face, and that the building is not anchored in bedrock. [*Hint:* $\mathbf{F_E}$ in Fig. 9–73 represents the force that the Earth exerts on the building in the case where the building is just beginning to tip.]

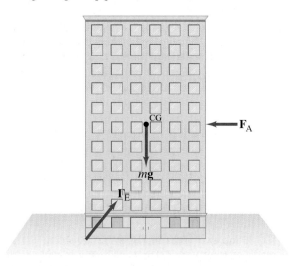

FIGURE 9–73   Force on a building subjected to wind ($\mathbf{F_A}$) and gravity ($m\mathbf{g}$); $\mathbf{F_E}$ is the force on the building due to the Earth in the situation when the building is just about to tip. Problem 65.

66. A tightly stretched "high wire" is 46 m long. It sags 3.4 m when a 60-kg tightrope walker stands at its center. What is the tension in the wire? Is it possible to increase the tension in the wire so that there is no sag?

67. What minimum horizontal force F is needed to pull a wheel of radius R and mass M over a step of height h as shown in Fig. 9–74 ($R > h$)? (a) Assume the force is applied at the top edge as shown. (b) Assume the force is applied instead at the wheel's center.

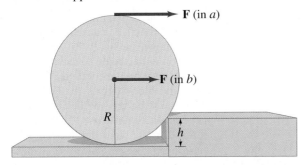

FIGURE 9–74 Problem 67.

68. The center of gravity of a loaded truck depends on how it is packed. If a truck is 4.0 m high and 2.4 m wide, and its CG is 2.2 m above the ground, how steep a slope can the truck be parked on without tipping over sideways (Fig. 9–75)?

FIGURE 9–75 Problem 68.

69. In Example 7–5 in Chapter 7, we calculated the impulse and average force on the leg of a person who jumps 3.0 m down to the ground. If the legs are not bent upon landing, so that the body moves a distance d of only 1.0 cm during collision, determine (a) the stress in the tibia bone (area $= 3.0 \times 10^{-4}\,\text{m}^2$), and (b) whether or not the bone will break. (c) Repeat for a bent-knees landing ($d = 50.0$ cm).

70. The roof of a 9.0 m $\times$ 10.0 m room in a school has a total mass of 12,600 kg. The roof is to be supported by "2 $\times$ 4s" (actually about 4.0 cm $\times$ 9.0 cm) along the 10.0-m sides. How many supports are required on each side and how far apart must they be? Consider only compression and assume a safety factor of 12.

71. In Fig. 9–76, consider the right-hand (northernmost) section of the Golden Gate Bridge, which has a length $d_1 = 343$ m. Assume the CG of this span is halfway between the tower and anchor. Determine F_{T1} and F_{T2} (which act on the northernmost cable) in terms of mg, the weight of the northernmost span, and calculate the tower height h needed for equilibrium. Assume the roadway is supported only by the suspension cables and neglect the mass of the cables. [*Hint*: F_{T3} does not act on this section.]

FIGURE 9–76 Problems 71 and 75.

72. A 20.0-m-long uniform beam weighing 600 N is supported on walls A and B, as shown in Fig. 9–77. (a) Find the maximum weight a person can be to walk to the extreme end D without tipping the beam. Find the forces that the walls A and B exert on the beam when the person is standing: (b) at D; (c) at a point 2.0 m to the right of B; (d) 2.0 m to the right of A.

FIGURE 9–77 Problem 72.

73. A 36-kg round table is supported by three legs placed equal distances apart on the edge. What minimum mass, placed on the table's edge, will cause the table to overturn?

FIGURE 9–78 Problem 74.

FIGURE 9–79
Problem 80.

74. A uniform flexible steel cable of weight mg is suspended between two equal elevation points as shown in Fig. 9–78, where $\theta = 60°$. Determine the tension in the cable (a) at its lowest point, and (b) at the points of attachment. (c) What is the direction of the tension force in each case?

75. Assume that a single-span suspension bridge such as the Golden Gate Bridge has the configuration indicated in Fig. 9–76. Assume that the roadway is uniform over the length of the bridge and that each segment of the suspension cable provides the sole support for the roadway directly below it. The ends of the cable are anchored to the ground only, not to the roadway. What must the ratio of d_2 to d_1 be so that the suspension cable exerts no net horizontal force on the towers? Neglect the mass of the cables and the fact that the roadway isn't precisely horizontal.

76. A uniform 7.0-m long ladder of mass 15.0 kg leans against a smooth wall (so the force exerted by the wall, F_W, is perpendicular to the wall). The ladder makes an angle of 20° with the vertical wall (see Fig. 9–61); and the ground is rough. (a) Calculate the components of the force exerted by the ground on the ladder at its base, and (b) determine what the coefficient of friction at the base of the ladder must be if the ladder is not to slip when a 70-kg person stands three fourths of the way up the ladder.

77. If the coefficient of friction between the ladder and the ground in the situation described in the preceding problem is 0.30, how far up the ladder can the person climb before the ladder starts to slip?

78. There is a maximum height of a uniform vertical column made of any material that can support itself without buckling, and it is independent of the cross-sectional area (why?). Calculate this height for (a) steel (density $7.8 \times 10^3 \, \text{kg/m}^3$), and (b) granite (density $2.7 \times 10^3 \, \text{kg/m}^3$).

79. From what minimum height must a 1.2-kg rectangular brick 15.0 cm × 6.0 cm × 4.0 cm be dropped above a rigid steel floor in order to break the brick? Assume the brick strikes the floor directly on its largest face, and that the compression of the brick is much greater than that of the steel (that is, ignore compression of the steel). State other simplifying assumptions that may be necessary.

80. A cube of side L rests on a rough floor. It is subjected to a steady horizontal pull, F, exerted a distance h above the floor as shown in Fig. 9–79. As F is increased, the block will either begin to slide, or begin to tip over. (a) What must be the coefficient of static friction μ_s so that the block begins to slide rather than tip? (b) What must be the coefficient of static friction so that the block begins to tip? [*Hint*: Where will the normal force on the block act if it tips?]

81. A man doing push-ups pauses in the position shown in Fig. 9–80. His mass $m = 70$ kg. Determine the normal force exerted by the floor (a) on each hand; (b) on each foot.

FIGURE 9–80 Problem 81.

82. A 60-kg painter is on a scaffold supported from above by ropes (Fig. 9–81). The scaffold has a mass of 25 kg and is uniformly constructed. There is a 4.0-kg pail of paint off to one side, as shown in the figure. Can the painter walk safely to both ends of the scaffold? If not, which end(s) is dangerous and how close to the end can he approach safely?

FIGURE 9–81 Problem 82.

Under water, fish and scuba divers experience a buoyant force that almost exactly balances their weight.

FLUIDS 10

The three common states, or **phases**, of matter are solid, liquid, and gas. We can distinguish these three phases as follows. A **solid** maintains a fixed shape and a fixed size; even if a large force is applied to a solid, it does not readily change its shape or volume. A **liquid** does not maintain a fixed shape—it takes on the shape of its container—but like a solid it is not readily compressible, and its volume can be changed significantly only by a very large force. A **gas** has neither a fixed shape nor a fixed volume—it will expand to fill its container. For example, when air is pumped into an automobile tire, the air does not all run to the bottom of the tire as a liquid would; it spreads out to fill the whole volume of the tire. Since liquids and gases do not maintain a fixed shape, they both have the ability to flow; they are thus often referred to collectively as **fluids**.

Phases of matter

The division of matter into three states is not always simple. How, for example, should butter be classified? Furthermore, a fourth state of matter can be distinguished, the **plasma** state, which occurs only at very high temperatures and consists of ionized atoms (electrons separated from the nuclei). Some scientists believe that so-called colloids (suspensions of tiny particles in a liquid) should also be considered a separate state of matter. However, for our present purposes we will mainly be interested in the three ordinary states of matter. In Chapter 9 we discussed some of the properties of solid materials. In this chapter we will discuss the properties of fluids.

10–1 Density and Specific Gravity

It is sometimes said that iron is "heavier" than wood. This cannot really be true since a large log clearly weighs more than an iron nail. What we should say is that iron is more *dense* than wood.

TABLE 10-1
Densities of Substances†

Substance	Density, ρ (kg/m³)
Solids	
Aluminum	2.70×10^3
Iron and steel	7.8×10^3
Copper	8.9×10^3
Lead	11.3×10^3
Gold	19.3×10^3
Concrete	2.3×10^3
Granite	2.7×10^3
Wood (typical)	$0.3–0.9 \times 10^3$
Glass, common	$2.4–2.8 \times 10^3$
Ice	0.917×10^3
Bone	$1.7–2.0 \times 10^3$
Liquids	
Water (4° C)	1.00×10^3
Blood, plasma	1.03×10^3
Blood, whole	1.05×10^3
Sea water	1.025×10^3
Mercury	13.6×10^3
Alcohol, ethyl	0.79×10^3
Gasoline	0.68×10^3
Gases	
Air	1.29
Helium	0.179
Carbon dioxide	1.98
Water (steam) (100° C)	0.598

†Densities are given at 0°C and 1 atm pressure unless otherwise specified.

The **density**, ρ, of an object (ρ is the lowercase Greek letter "rho") is defined as its mass per unit volume:

$$\rho = \frac{m}{V}, \qquad (10\text{--}1)$$

where m is the mass of the object and V its volume. Density is a characteristic property of any pure substance. Objects made of a given pure substance, such as pure gold, can have any size or mass, but the density will be the same for each. (Sometimes we will find Eq. 10–1 useful for writing the mass of an object as $m = \rho V$, and the weight of an object, mg, as $\rho V g$.)

The SI unit for density is kg/m³. Sometimes densities are given in g/cm³. Note that since $1 \text{ kg/m}^3 = 1000 \text{ g}/(100 \text{ cm})^3 = 10^{-3} \text{g/cm}^3$, then a density given in g/cm³ must be multiplied by 1000 to give the result in kg/m³. Thus the density of aluminum is $\rho = 2.70 \text{ g/cm}^3$, which is equal to 2700 kg/m³. The densities of a variety of substances are given in Table 10–1. The table specifies temperature and pressure because they affect the density of substances (although the effect is slight for liquids and solids).

EXAMPLE 10–1 **Mass, given volume and density.** What is the mass of a solid iron wrecking ball of radius 18 cm?

SOLUTION The volume of any sphere is $V = \frac{4}{3}\pi r^3$ so we have

$$V = \tfrac{4}{3}\pi r^3 = \tfrac{4}{3}(3.14)(0.18 \text{ m})^3 = 0.024 \text{ m}^3.$$

From Table 10–1, the density of iron is $\rho = 7800 \text{ kg/m}^3$, so we have from Eq. 10–1,

$$m = \rho V = (7800 \text{ kg/m}^3)(0.024 \text{ m}^3) = 190 \text{ kg}.$$

The **specific gravity** of a substance is defined as the ratio of the density of that substance to the density of water at 4.0°C. Specific gravity (abbreviated SG) is a number, without dimensions or units. Since the density of water is $1.00 \text{ g/cm}^3 = 1.00 \times 10^3 \text{ kg/m}^3$, the specific gravity of any substance will be equal numerically to its density specified in g/cm³, or 10^{-3} times its density specified in kg/m³. For example (see Table 10–1), the specific gravity of lead is 11.3, and that of alcohol is 0.79.

10-2 Pressure in Fluids

Pressure is defined as force per unit area, where the force F is understood to be acting perpendicular to the surface area A:

$$\text{pressure} = P = \frac{F}{A}. \qquad (10\text{--}2)$$

The pascal

The SI unit of pressure is N/m². This unit has the official name **pascal** (Pa), in honor of Blaise Pascal (see Section 10–4); that is, $1 \text{ Pa} = 1 \text{ N/m}^2$. However, for simplicity, we will often use N/m². Other units sometimes used are dynes/cm², lb/in.² (sometimes abbreviated "psi"). We will meet several other units shortly, and will discuss conversions between them in Section 10–5 (see also the Table inside the front cover).

As an example of calculating pressure, a 60-kg person whose two feet cover an area of 500 cm² will exert a pressure of

$$F/A = mg/A = (60 \text{ kg})(9.8 \text{ m/s}^2)/(0.050 \text{ m}^2) = 12 \times 10^3 \text{ N/m}^2$$

on the ground. If the person stands on one foot, the force is the same but the area will be half, so the pressure will be twice as much: $24 \times 10^3 \text{ N/m}^2$.

The concept of pressure is particularly useful in dealing with fluids. It is an experimental fact that *a fluid exerts a pressure in all directions*. This is well known to swimmers and divers who feel the water pressure on all parts of their bodies. At any point in a fluid at rest, the pressure is the same in all directions. This is illustrated in Fig. 10–1. Consider a tiny cube of the fluid which is so small that we can ignore the force of gravity on it. The pressure on one side of it must equal the pressure on the opposite side. If this weren't true, there would be a net force on the cube and it would start moving. If the fluid is not flowing, then the pressures must be equal.

Fluids exert pressure in all directions

Another important property of a fluid at rest is that the force due to fluid pressure always acts *perpendicular* to any surface it is in contact with. If there were a component of the force parallel to the surface as shown in Fig. 10–2, then according to Newton's third law, the surface would exert a force back on the fluid that also would have a component parallel to the surface. Such a component would cause the fluid to flow, in contradiction to our assumption that the fluid is at rest. Thus the force due to the pressure is perpendicular to the surface.

FIGURE 10–1 Pressure is the same in every direction in a fluid at a given depth; if it weren't, the fluid would be in motion.

Let us now calculate quantitatively how the pressure in a liquid of uniform density varies with depth. Consider a point which is at a depth h below the surface of the liquid (that is, the surface is a height h above this point), as shown in Fig. 10–3. The pressure due to the liquid at this depth h is due to the weight of the column of liquid above it. Thus the force acting on the area is $F = mg = \rho Ahg$, where Ah is the volume of the column, ρ is the density of the liquid (assumed to be constant), and g is the acceleration of gravity. The pressure, P, is then

$$P = \frac{F}{A} = \frac{\rho Ahg}{A}$$

$$P = \rho gh. \qquad \text{[liquid]} \quad \textbf{(10–3a)}$$

Pressure variation with depth

Thus the pressure is directly proportional to the density of the liquid, and to the depth within the liquid. In general, the *pressure at equal depths within a uniform liquid is the same*. (Equation 10–3a gives the pressure due to the liquid itself. If an external pressure is exerted at the surface of the liquid, this must be taken into account, as we will discuss in Section 10–4.)

FIGURE 10–2 If there were a component of force parallel to the solid surface, the liquid would move in response to it; for a liquid at rest, $F_{\parallel} = 0$.

FIGURE 10–3 Calculating the pressure at a depth h in a liquid.

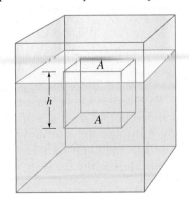

Equation 10–3a is extremely useful. It is valid for fluids whose density is constant and does not change with depth—that is, if the fluid is *incompressible*. This is usually a good approximation for liquids (although at great depths in the ocean the density of water is increased substantially by compression due to the great weight of water above). Gases, on the other hand, are very compressible, and density can vary significantly with depth. If the density varies only slightly, Eq. 10–3a can be used to determine the difference in pressure ΔP at different heights with ρ being the average density:

Change in pressure with change in depth in a fluid

$$\Delta P = \rho g\, \Delta h. \tag{10–3b}$$

PHYSICS APPLIED

Water supply

FIGURE 10–4 Example 10–2.

EXAMPLE 10–2 **Pressure at a faucet.** The surface of the water in a storage tank is 30 m above a water faucet in the kitchen of a house, Fig. 10–4. Calculate the water pressure at the faucet.

SOLUTION Atmospheric pressure acts both at the surface of the water in the storage tank, and on the water leaving the faucet. The pressure difference between the inside and outside of the faucet is

$$\Delta P = \rho g h = (1.0 \times 10^3 \text{ kg/m}^3)(9.8 \text{ m/s}^2)(30 \text{ m})$$
$$= 2.9 \times 10^5 \text{ N/m}^2.$$

The height h is sometimes called the **pressure head**. In this example, the head of the water is 30 m. Note that the very different diameters of the tank and faucet don't affect the result—only pressure does.

10–3 | Atmospheric Pressure and Gauge Pressure

The pressure of the Earth's atmosphere, as in any fluid, changes with depth. But the Earth's atmosphere is somewhat complicated because not only does the density of air vary greatly with altitude, but there is no distinct top surface to the atmosphere from which h (in Eq. 10–3a) can be measured. We can, however, calculate the approximate difference in pressure between two altitudes using Eq. 10–3b.

The pressure of the air at a given place varies slightly according to the weather. At sea level, the pressure of the atmosphere on the average is $1.013 \times 10^5 \text{ N/m}^2$ (or 14.7 lb/in.²). This value is used to define another unit of pressure in common use, the **atmosphere** (abbreviated atm):

One atmosphere (unit)

$$1 \text{ atm} = 1.013 \times 10^5 \text{ N/m}^2 = 101.3 \text{ kPa.}$$

The bar (unit)

Another unit of pressure sometimes used (in meteorology and on weather maps) is the **bar**, which is defined as $1 \text{ bar} = 1.00 \times 10^5 \text{ N/m}^2 = 100 \text{ kPa}$. Thus standard atmospheric pressure is slightly more than 1 bar.

The pressure due to the weight of the atmosphere is exerted on all objects immersed in this great sea of air, including our bodies. How does a human body withstand the enormous pressure on its surface? The answer is that living cells maintain an internal pressure that closely equals the external pressure, just as the pressure inside a balloon only slightly exceeds the out-

278   CHAPTER 10   Fluids

side pressure of the atmosphere. An automobile tire, because of its rigidity, can maintain internal pressures much greater than the external pressure.

It is important to note that tire gauges, and most other pressure gauges, register the pressure over and above atmospheric pressure. This is called **gauge pressure**. Thus, to get the absolute pressure, P, one must add the atmospheric pressure, P_A, to the gauge pressure, P_G:

$$P = P_A + P_G.$$

Gauge pressure

Absolute pressure = atmospheric pressure + gauge pressure

For example, if a tire gauge registers 220 kPa, the absolute pressure within the tire is 220 kPa + 101 kPa = 321 kPa. This is equivalent to about 3.2 atm (2.2 atm gauge pressure).

| **CONCEPTUAL EXAMPLE 10–3** | **Finger holds water in a straw.** |

You insert a straw of length L into a tall glass of your favorite beverage. You place your finger over the top of the straw so that no air can get in or out, and then lift the straw from the liquid. You find that the straw retains the liquid such that the distance from the bottom of your finger to the top of the liquid is h. (See Fig. 10–5.) Does the air in the space between your finger and the top of the liquid have a pressure P that is (*a*) greater than, (*b*) equal to, or (*c*) less than, the atmospheric pressure P_A outside the straw?

RESPONSE Consider the forces on the column of liquid. Atmospheric pressure on the outside of the straw pushes upward on the liquid at the bottom of the straw, gravity pulls the liquid downward, and the air pressure inside the top of the straw pushes downward on the liquid. Since the liquid is in equilibrium, the upward force due to atmospheric pressure must balance the two downward forces. The only way this is possible is for the air pressure inside the straw to be rather less than the atmosphere pressure outside the straw.

FIGURE 10–5 Example 10–3.

10–4 Pascal's Principle

The Earth's atmosphere exerts a pressure on all objects with which it is in contact, including other fluids. External pressure acting on a fluid is transmitted throughout that fluid. For instance, according to Eq. 10–3a, the pressure due to the water pressure at a depth of 100 m below the surface of a lake is $P = \rho g h = (1000 \text{ kg/m}^3)(9.8 \text{ m/s}^2)(100 \text{ m}) = 9.8 \times 10^5 \text{ N/m}^2$, or 9.7 atm. However, the total pressure at this point is due to the pressure of water plus the pressure of the air above it. Hence the total pressure (if the lake is near sea level) is 9.7 atm + 1.0 atm = 10.7 atm. This is just one example of a general principle attributed to the French philosopher and scientist Blaise Pascal (1623–1662). **Pascal's principle** states that *pressure applied to a confined fluid increases the pressure throughout by the same amount.*

A number of practical devices make use of Pascal's principle. Two examples, hydraulic brakes in an automobile and the hydraulic lift, are illus-

Pascal's principle

➡ **PHYSICS APPLIED**

Brakes and hydraulic lift

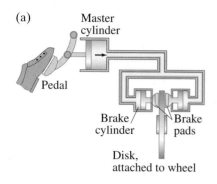

(a)

Master cylinder

Pedal

Brake cylinder

Brake pads

Disk, attached to wheel

(b)

F_{out}

F_{in}

P_{in} A_{in}

A_{out} P_{out}

FIGURE 10–6 Applications of Pascal's principle: (a) hydraulic brakes in a car; (b) hydraulic lift.

trated in Fig. 10–6. In the case of a hydraulic lift, a small force can be used to exert a large force by making the area of one piston (the output) larger than the area of the other (the input). To see how this works, we assume the input and output pistons are at the same height (at least approximately). Then the external input force F_{in}, by Pascal's principle, increases the pressure equally throughout so that at the same level (see Fig. 10–6b):

$$P_{out} = P_{in}$$

where the input quantities are represented by the subscript "in" and the output by "out." Thus

$$\frac{F_{out}}{A_{out}} = \frac{F_{in}}{A_{in}},$$

or, finally,

$$\frac{F_{out}}{F_{in}} = \frac{A_{out}}{A_{in}}.$$

The quantity F_{out}/F_{in} is called the "mechanical advantage" of the hydraulic lift, and is equal to the ratio of the areas. For example, if the area of the output piston is 20 times that of the input cylinder, the force is multiplied by a factor of 20: thus a force of 200 lb could lift a 4000-lb car.

10–5 Measurement of Pressure; Gauges and the Barometer

Many devices have been invented to measure pressure, some of which are shown in Fig. 10–7. The simplest is the open-tube *manometer* (Fig. 10–7a) which is a U-shaped tube partially filled with a liquid, usually mercury or water. The pressure P being measured is related to the difference in height h of the two levels of the liquid by the relation (see Eq. 10–3a)

Pressure beneath the surface of an open liquid

$$P = P_0 + \rho g h, \tag{10–3c}$$

where P_0 is atmospheric pressure (acting on the top of the fluid in the left-hand tube), and ρ is the density of the liquid. Note that the quantity $\rho g h$ is the "gauge pressure"—the amount by which P exceeds atmospheric

FIGURE 10–7 Pressure gauges: (a) open-tube manometer, (b) aneroid gauge, and (c) common tire pressure gauge.

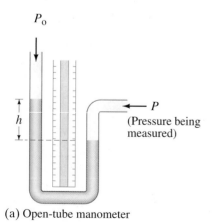

P_0

h

P
(Pressure being measured)

(a) Open-tube manometer

Flexible chamber

(b) Aneroid gauge (used mainly for air pressure and then called an aneroid barometer)

Scale reading

Atmospheric pressure

Spring

Pressure of air in tire

(c) Tire gauge

pressure. If the liquid in the left-hand column were lower than that in the right-hand column, this would indicate that P was less than atmospheric pressure (and h would be negative).

Instead of calculating the product ρgh, it is common to simply specify the height h. In fact, pressures are sometimes specified as so many "millimeters of mercury" (mm-Hg), and sometimes as so many "mm of water" (mm-H_2O). The unit mm-Hg is equivalent to a pressure of 133 N/m², since 1.00 mm = 1.00×10^{-3} m and the density of mercury is 13.6×10^3 kg/m³:

$$\rho gh = (13.6 \times 10^3 \text{ kg/m}^3)(9.80 \text{ m/s}^2)(1.00 \times 10^{-3} \text{ m})$$

$$= 1.33 \times 10^2 \text{ N/m}^2.$$

The unit mm-Hg is also called the **torr** in honor of Evangelista Torricelli (1608–1647), who invented the barometer (see below). Conversion factors among the various units of pressure (an incredible nuisance!) are given in Table 10–2. It is important that only N/m² = Pa, the proper SI unit, be used in calculations involving other quantities specified in SI units.

The torr (unit)

➡ PROBLEM SOLVING

Use SI unit in calculations:
1 Pa = 1 N/m²

TABLE 10–2
Conversion Factors Between Different Units of Pressure

In Terms of 1 Pa = 1 N/m²	Related to 1 atm
1 atm = 1.013×10^5 N/m²	1 atm = 1.013×10^5 N/m²
= 1.013×10^5 Pa = 101.3 kPa	
1 bar = 1.000×10^5 N/m²	1 atm = 1.013 bar
1 dyne/cm² = 0.1 N/m²	1 atm = 1.013×10^6 dyne/cm²
1 lb/in² = 6.90×10^3 N/m²	1 atm = 14.7 lb/in²
1 lb/ft² = 47.9 N/m²	1 atm = 2.12×10^3 lb/ft²
1 cm-Hg = 1.33×10^3 N/m²	1 atm = 76 cm-Hg
1 mm-Hg = 133 N/m²	1 atm = 760 mm-Hg
1 torr = 133 N/m²	1 atm = 760 torr
1 mm H_2O (4°C) = 9.81 N/m²	1 atm = 1.03×10^4 mm-H_2O (4°C)

Another type of pressure gauge is the aneroid gauge (Fig. 10–7b) in which the pointer is linked to the flexible ends of an evacuated thin metal chamber. In an electronic gauge, the pressure may be applied to a thin metal diaphragm whose resulting distortion is detected electrically. How a common tire gauge is constructed is shown in Fig. 10–7c.

Atmospheric pressure is often measured by a modified kind of mercury manometer with one end closed, called a mercury **barometer** (Fig. 10–8). The glass tube is completely filled with mercury and then inverted into the bowl of mercury. If the tube is long enough, the level of the mercury will drop, leaving a vacuum at the top of the tube, since atmospheric pressure can support a column of mercury only about 76 cm high (exactly 76.0 cm at standard atmospheric pressure). That is, a column of mercury 76 cm high exerts the same pressure as the atmosphere; we can see this using the formula $P = \rho gh$, with $\rho = 13.6 \times 10^3$ kg/m³ for mercury and $h = 76.0$ cm:

$$P = (13.6 \times 10^3 \text{ kg/m}^3)(9.80 \text{ m/s}^2)(0.760 \text{ m})$$

$$= 1.013 \times 10^5 \text{ N/m}^2 = 1.00 \text{ atm.}$$

Household barometers are usually of the aneroid type (Fig. 10–7b), either mechanical or electronic.

FIGURE 10–8 Diagram of a mercury barometer, when the air pressure is 76 cm-Hg.

A calculation similar to that above will show that atmospheric pressure can maintain a column of water 10.3 m high in a tube whose top is under vacuum (Fig. 10–9). A few centuries ago, it was a source of wonder and frustration that no matter how good a vacuum pump was, it could not lift water more than about 10 m. The only way to pump water out of deep mine shafts, for example, was to use multiple stages for depths greater than 10 m. Galileo studied this problem, and his student Torricelli was the first to explain it. The point is that a pump does not really suck water up a tube—it merely reduces the pressure at the top of the tube. Atmospheric air pressure *pushes* the water up the tube if the top end is at low pressure (under a vacuum), just as it is air pressure that pushes (or maintains) the mercury 76 cm high in a barometer.

FIGURE 10–9 A water barometer: water was added at the top and the spigot at the top then closed; the water level dropped, leaving a vacuum between its upper surface and the spigot. Why? Because air pressure could not support a column of water more than 10 m high.

CONCEPTUAL EXAMPLE 10–4 **Suction.** You sit in a meeting where a novice NASA engineer proposes suction cup shoes for Space Shuttle astronauts working on the exterior of the spacecraft. Having just studied this chapter, you gently remind him of the fallacy of this plan. What is it?

RESPONSE Suction cups work by pushing out the air underneath the cup. What holds the cup in place is the air pressure outside the cup. (This can be a substantial force on Earth. For example, a 10 cm diameter cup has an area of $7.8 \times 10^{-3} \, \text{m}^2$. The force of the atmosphere on the outside of the cup is then 800 N, about 180 lbs!) With no air pressure in outer space, nothing can hold the suction cup on the spacecraft.

We sometimes mistakenly think of suction as something we actively do. For example, we intuitively think that we pull the soda up through a straw. Instead, all we do is lower the pressure at the top of the straw, and the atmosphere *pushes* the soda up the straw.

10–6 Buoyancy and Archimedes' Principle

Objects submerged in a fluid appear to weigh less than they do when outside the fluid. For example, a large rock that you would have difficulty lifting off the ground can often be easily lifted from the bottom of a stream. When the rock breaks through the surface of the water, it suddenly seems to be much heavier. Many objects, such as wood, float on the surface of water. These are two examples of *buoyancy*. In each example, the force of gravity is acting downward. But in addition, an upward *buoyant force* is exerted by the liquid. The upward buoyant force on fish and underwater divers (chapter opening photo) almost exactly balances the force of gravity downward.

The buoyant force occurs because the pressure in a fluid increases with depth. Thus the upward pressure on the bottom surface of a submerged object is greater than the downward pressure on its top surface. To see the effect of this, consider a cylinder of height h whose top and bottom ends have an area A and which is completely submerged in a fluid of density ρ_F, as shown in Fig. 10–10. The fluid exerts a pressure $P_1 = \rho_F g h_1$ at the top surface of the cylinder. The force due to this pressure on top of the cylinder is $F_1 = P_1A = \rho_F g h_1 A$, and it is directed downward. Similarly, the fluid exerts an upward force on the bottom of the cylinder equal to $F_2 = P_2A = \rho_F g h_2 A$. The net force due to the fluid pressure, which is the **buoyant force**, $\mathbf{F}_B$, acts upward and has the magnitude

$$F_B = F_2 - F_1$$
$$= \rho_F g A(h_2 - h_1)$$
$$= \rho_F g A h$$
$$= \rho_F g V,$$

where $V = Ah$ is the volume of the cylinder. Since ρ_F is the density of the fluid, the product $\rho_F g V = m_F g$ is the weight of fluid which takes up a volume equal to the volume of the cylinder. Thus the buoyant force on the cylinder is equal to the weight of fluid displaced[†] by the cylinder. This result is valid no matter what the shape of the object. Its discovery is credited to Archimedes (287?–212 B.C.), and it is called **Archimedes' principle**: *the buoyant force on a body immersed in a fluid is equal to the weight of the fluid displaced by that object.*

We can derive Archimedes' principle in general by the following simple but elegant argument. The irregularly shaped object D shown in Fig. 10–11a is acted on by the force of gravity (its weight, $\mathbf{w}$, downward) and the buoyant force, F_B, upward. We wish to determine F_B. To do so, we next consider a body, this time made of our same fluid (D' in Fig. 10–11b) with the same shape and size as the original object, and located at the same depth. You might think of this body of fluid as being separated from the rest of the fluid by an imaginary transparent membrane. The buoyant force F_B on this body of fluid will be exactly the same as that on the original object since the surrounding fluid, which exerts F_B, is in exactly the same configuration. Now the body of fluid D' is in equilibrium (the fluid as a whole is at rest); therefore, $F_B = w'$, where w' is the weight of the body of fluid. Hence the buoyant force F_B is equal to the weight of the body of fluid whose volume equals the volume of the original submerged object, which is Archimedes' principle.

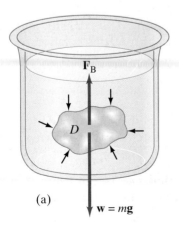

FIGURE 10–10 Determination of buoyant force.

Archimedes' principle

[†]By "fluid displaced," we mean a volume of fluid equal to the volume of the object, or that part of the object submerged if it floats or is only partly submerged (the fluid that used to be where the object is). If the object is placed in a glass or tub initially filled to the brim with water, the water that flows over the top represents the water displaced by the object.

(a)

(b)

FIGURE 10–11
Archimedes' principle.

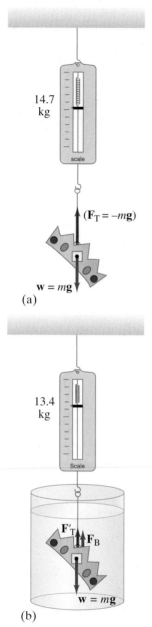

14.7 kg

$(\mathbf{F}_T = -m\mathbf{g})$

$\mathbf{w} = m\mathbf{g}$

(a)

13.4 kg

$\mathbf{F'}_T$ $\mathbf{F}_B$

$\mathbf{w} = m\mathbf{g}$

(b)

FIGURE 10–12 (a) A scale reads the mass of an object in air—in this case the crown of Example 10–6. All objects are at rest, so the tension F_T in the connecting cord or wire equals the weight w of the object: $F_T = mg$. Note that we show the free-body diagram of the crown, and that F_T is what causes the scale reading (it's equal to the net downward force on the scale). (b) The object submerged has an additional force on it, the buoyant force F_B. The net force is zero, so $F'_T + F_B = w$. The scale now reads $m' = 13.4$ kg, where m' is related to the effective weight by $w' = m'g$, where $F'_T = w' = w - F_B$.

EXAMPLE 10–5 **Recovering a submerged statue.** A 70-kg ancient statue lies at the bottom of the sea. Its volume is $3.0 \times 10^4\ \text{cm}^3$. How much force is needed to lift it?

SOLUTION The buoyant force on the statue due to the water is equal to the weight of $3.0 \times 10^4\ \text{cm}^3 = 3.0 \times 10^{-2}\ \text{m}^3$ of water (for seawater $\rho = 1.025 \times 10^3\ \text{kg/m}^3$):

$$F_B = m_{\text{H}_2\text{O}}g = \rho_{\text{H}_2\text{O}}gV$$
$$= (1.025 \times 10^3\ \text{kg/m}^3)(9.8\ \text{m/s}^2)(3.0 \times 10^{-2}\ \text{m}^3)$$
$$= 3.0 \times 10^2\ \text{N}.$$

The weight of the statue is $mg = (70\ \text{kg})(9.8\ \text{m/s}^2) = 6.9 \times 10^2\ \text{N}$. Hence the force needed to lift it is $690\ \text{N} - 300\ \text{N} = 390\ \text{N}$. It is as if the statue had a mass of only $(390\ \text{N})/(9.8\ \text{m/s}^2) = 40\ \text{kg}$.

Archimedes is said to have discovered his principle in his bath while thinking how he might determine whether the king's new crown was pure gold or a fake. Gold has a specific gravity of 19.3, somewhat higher than that of most metals, but a determination of specific gravity or density is not readily done directly because, even if the mass is easily known, the volume of an irregularly shaped object is not easily calculated. However, if the object is weighed in air ($=w$) and also "weighed" while it is under water ($=w'$), the density can be determined using Archimedes' principle, as the following Example shows. The quantity w' is called the *apparent weight* in water, and is what a scale reads when the object is submerged in water (see Fig. 10–12); w' is equal to the true weight ($w = mg$) minus the buoyant force.

EXAMPLE 10–6 **Archimedes: Is the crown gold?** When a crown of mass 14.7 kg is submerged in water, an accurate scale reads only 13.4 kg. Is the crown made of gold?

SOLUTION See analysis in Fig. 10–12. The apparent weight of the submerged object, w' ($= F'_T$ in Fig. 10–12b), equals its actual weight w minus the buoyant force F_B as shown:

$$w' = F'_T$$
$$= w - F_B = \rho_O gV - \rho_F gV,$$

where V is the volume of the object, ρ_O its density, and ρ_F the density of the fluid (water in this case). From this relation, we can see that $w - w' = \rho_F gV$. Then we can write

$$\frac{w}{w - w'} = \frac{\rho_O gV}{\rho_F gV} = \frac{\rho_O}{\rho_F}.$$

[Thus $w/(w - w')$ is equal to the specific gravity of the object if the fluid in which it is submerged is water.] For the crown we have

$$\frac{\rho_O}{\rho_{\text{H}_2\text{O}}} = \frac{w}{w - w'} = \frac{(14.7\ \text{kg})g}{(14.7\ \text{kg} - 13.4\ \text{kg})g} = \frac{14.7\ \text{kg}}{1.3\ \text{kg}} = 11.3.$$

This corresponds to a density of $11{,}300\ \text{kg/m}^3$. The crown seems to be made of lead (see Table 10–1)!

$\mathbf{F_B} = (2000 \text{ kg})\mathbf{g}$

$m_O = 1200 \text{ kg}$
$V = 2.0 \text{ m}^3$

a

$\mathbf{w} = (1200 \text{ kg})\mathbf{g}$

(a)

$\mathbf{F_B}$

$\mathbf{w}$

(b)

FIGURE 10–13 (a) The fully submerged log accelerates upward because $F_B > w$. It comes to equilibrium (b) when $\Sigma F = 0$, so $F_B = w = m_O g = (1200 \text{ kg})g$. Thus 1200 kg, or 1.2 m³, of water is displaced.

Archimedes' principle applies equally well to objects that float, such as wood. In general, *an object floats on a fluid if its density is less than that of the fluid.* At equilibrium—that is, when floating—the buoyant force on an object has magnitude equal to the weight of the object. For example, a log whose specific gravity is 0.60 and whose volume is 2.0 m³ has a mass $m_O = \rho_O V = (0.60 \times 10^3 \text{ kg/m}^3)(2.0 \text{ m}^3) = 1200 \text{ kg}$. If the log is fully submerged, it will displace a mass of water $m_F = \rho_F V = (1000 \text{ kg/m}^3)(2.0 \text{ m}^3) = 2000 \text{ kg}$. Hence the buoyant force on the log will be greater than its weight, and it will float upward to the surface (Fig. 10–13). It will come to equilibrium when it displaces 1200 kg of water, which means that 1.2 m³ of its volume will be submerged. This 1.2 m³ corresponds to 60 percent of the volume of the log $(1.2/2.0 = 0.60)$, so 60 percent of the log is submerged. In general when an object floats, we have $F_B = w$, which we can write as (see Fig. 10–14)

$$\rho_F V_{\text{displ}} \, g = \rho_O V_O \, g$$

where V_O is the full volume of the object and V_{displ} is the volume of fluid it displaces (= volume submerged). Thus

$$\frac{V_{\text{displ}}}{V_O} = \frac{\rho_O}{\rho_F}.$$

That is, the fraction of the object submerged is given by the ratio of the object's density to that of the fluid.

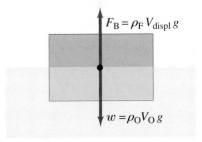

$F_B = \rho_F V_{\text{displ}} \, g$

$w = \rho_O V_O \, g$

FIGURE 10–14 An object floating in equilibrium: $F_B = w$.

FIGURE 10–15 A hydrometer. Example 10–7.

1.00

25.0 cm

x

EXAMPLE 10–7 **Hydrometer calibration.** A **hydrometer** is a simple instrument used to indicate specific gravity of a liquid by measuring how deeply it sinks in the liquid. A particular hydrometer (Fig. 10–15) consists of a glass tube, weighted at the bottom, which is 25.0 cm long, 2.00 cm² in cross-sectional area, and has a mass of 45.0 g. How far from the end should the 1.000 mark be placed?

SOLUTION The hydrometer has a density

$$\rho = \frac{m}{V} = \frac{45.0 \text{ g}}{(2.00 \text{ cm}^2)(25.0 \text{ cm})} = 0.900 \text{ g/cm}^3.$$

Thus, when placed in water, it will come to equilibrium when 0.900 of its volume is submerged. Since it is of uniform cross section, $(0.900)(25.0 \text{ cm}) = 22.5 \text{ cm}$ of its length will be submerged. Since the specific gravity of water is defined to be 1.000, the mark should be placed 22.5 cm from the end.

Archimedes' principle is also useful in geology. According to the theory of plate tectonics and continental drift, the continents can be considered to be floating on a fluid "sea" of slightly deformable rock (mantle rock). Some interesting calculations can be done using very simple models, which we consider here and in the problems at the end of the chapter.

FIGURE 10–16 Model of a continent "floating" in the Earth's crust. The buoyant force $\mathbf{F}_B$ has magnitude equal to the weight $\mathbf{w}$ for a body that floats (is in equilibrium). Example 10–8.

EXAMPLE 10–8 **ESTIMATE** **Floating continents.** A simple model (Fig. 10–16) considers a continent as a block (density = $2800\,\text{kg/m}^3$) floating in the mantle rock around it (density = $3300\,\text{kg/m}^3$). Assuming the continent is 35 km thick (the average thickness of the Earth's crust), estimate the height of the continent above the surrounding rock.

SOLUTION As discussed above, the fraction of continent submerged is equal to the ratio of the densities:

$$\frac{(2800\,\text{kg/m}^3)}{(3300\,\text{kg/m}^3)} = 0.85.$$

Hence 0.15 of the continent's height is above the surrounding rock, which is $(0.15)(35\,\text{km}) = 5.3\,\text{km}$. This result, 5.3 km, represents a rough estimate of the average depth of the ocean bottom relative to the height of the continents. (It is rough since it ignores such things as the weight of the oceans.)

Air is a fluid and it too exerts a buoyant force. Ordinary objects weigh less in air than they do if weighed in a vacuum. Because the density of air is so small, the effect for ordinary solids is slight. There are objects, however, that float in air—helium-filled balloons, for example, because the density of helium is less than the density of air.

FIGURE 10–17 Example 10–9.

EXAMPLE 10–9 **Helium balloon.** What volume V of helium is needed if a balloon is to lift a load of 800 kg (including the weight of the empty balloon)?

SOLUTION The buoyant force on the helium, F_B, which is equal to the weight of displaced air, must be at least equal to the weight of the helium plus the load (Fig. 10–17):

$$F_B = (m_{He} + 800\,\text{kg})g,$$

where g is the acceleration due to gravity. This equation can be written in terms of density:

$$\rho_{air}Vg = (\rho_{He}V + 800\,\text{kg})g.$$

Solving now for V, we find

$$V = \frac{800\,\text{kg}}{\rho_{air} - \rho_{He}} = \frac{800\,\text{kg}}{(1.29\,\text{kg/m}^3 - 0.18\,\text{kg/m}^3)}$$

$$= 720\,\text{m}^3.$$

This is the volume needed near the Earth's surface, where $\rho_{air} = 1.29\,\text{kg/m}^3$. To reach a high altitude, a greater volume would be needed since the density of air decreases with altitude.

10–7 Fluids in Motion; Flow Rate and the Equation of Continuity

We now turn from the study of fluids at rest to the more complex subject of fluids in motion, which is called **fluid dynamics**, or (especially if the fluid is water) **hydrodynamics**. Many aspects of fluid motion are still being studied today (for example, turbulence as a manifestation of chaos is a "hot" topic today). Nonetheless, with certain simplifying assumptions, a good understanding of this subject can be obtained.

(a)

(b)

FIGURE 10–18   (a) Streamline or laminar flow; (b) turbulent flow.

To begin with, we can distinguish two main types of fluid flow. If the flow is smooth, such that neighboring layers of the fluid slide by each other smoothly, the flow is said to be **streamline** or **laminar flow**.[†] In this kind of flow, each particle of the fluid follows a smooth path, and these paths do not cross over one another (Fig. 10–18a). Above a certain speed, which depends on a number of factors, as we shall see later, the flow becomes turbulent. **Turbulent flow** is characterized by erratic, small, whirlpool-like circles called eddy currents or eddies (Fig. 10–18b). Eddies absorb a great deal of energy, and although a certain amount of internal friction, called **viscosity**, is present even during streamline flow, it is much greater when the flow is turbulent. A few tiny drops of ink or food coloring dropped into a moving liquid can quickly reveal whether the flow is streamline or turbulent.

[†]The word laminar means "in layers."

FIGURE 10–19 Fluid flow through a pipe of varying diameter.

Let us consider the steady laminar flow of a fluid through an enclosed tube or pipe as shown in Fig. 10–19. First we determine how the speed of the fluid changes when the size of the tube changes. The mass **flow rate** is defined as the mass Δm of fluid that passes a given point per unit time Δt: mass flow rate = $\Delta m / \Delta t$. In Fig. 10–19, the volume of fluid passing point 1 (that is, through area A_1) in a time Δt is just $A_1 \Delta l_1$, where Δl_1 is the distance the fluid moves in time Δt. Since the velocity† of fluid passing point 1 is $v_1 = \Delta l_1 / \Delta t$, the mass flow rate $\Delta m_1 / \Delta t$ through area A_1 is

$$\frac{\Delta m_1}{\Delta t} = \frac{\rho_1 \Delta V_1}{\Delta t} = \frac{\rho_1 A_1 \Delta l_1}{\Delta t} = \rho_1 A_1 v_1,$$

where $\Delta V_1 = A_1 \Delta l_1$ is the volume of mass Δm_1, and ρ_1 is the fluid density. Similarly, at point 2 (through area A_2), the flow rate is $\rho_2 A_2 v_2$. Since no fluid flows in or out the sides, the flow rates through A_1 and A_2 must be equal. Thus, since:

$$\frac{\Delta m_1}{\Delta t} = \frac{\Delta m_2}{\Delta t},$$

then

$$\rho_1 A_1 v_1 = \rho_2 A_2 v_2.$$

Equation of continuity

This is called the **equation of continuity**. If the fluid is incompressible (ρ doesn't change with pressure), which is an excellent approximation for liquids under most circumstances (and sometimes for gases as well), then $\rho_1 = \rho_2$, and the equation of continuity becomes

$$A_1 v_1 = A_2 v_2. \qquad [\rho = \text{constant}] \quad \textbf{(10–4)}$$

Notice that the product Av represents the *volume rate of flow* (volume of fluid passing a given point per second), since $\Delta V / \Delta t = A\,\Delta l/\Delta t = Av$, which in SI units is m^3/s. Equation 10–4 tells us that where the cross-sectional area is large the velocity is small, and where the area is small the velocity is large. That this is reasonable can be seen by looking at a river. A river flows slowly through a meadow where it is broad, but speeds up to torrential speed when passing through a narrow gorge.

Equation 10–4 can be applied to the flow of blood in the body. Blood flows from the heart into the aorta, from which it passes into the major arteries. These branch into the small arteries (arterioles), which in turn branch into myriads of tiny capillaries. The blood returns to the heart via the veins (Fig. 10–20).

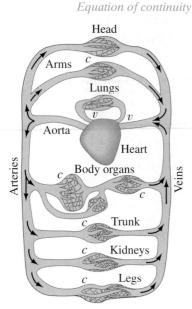

v = valves
c = capillaries

FIGURE 10–20 Human circulatory system.

➡ **PHYSICS APPLIED**

Blood flow

EXAMPLE 10–10 **ESTIMATE** **Blood flow.** The radius of the aorta is about 1.0 cm and the blood passing through it has a speed of about 30 cm/s. A typical capillary has a radius of about 4×10^{-4} cm, and blood flows through it at a speed of about 5×10^{-4} m/s. Estimate how many capillaries there are in the body.

---

†If there were no viscosity, the velocity would be the same across a cross section of the tube. Real fluids have viscosity, and this internal friction causes different layers of the fluid to flow at different speeds. In this case v_1 and v_2 represent the average speeds at each cross section.

SOLUTION Let A_1 be the area of the aorta, and A_2 be the area of *all* the capillaries through which blood flows. Then $A_2 = N\pi r_{cap}^2$ where N is the number of capillaries and $r_{cap} \approx 4 \times 10^{-4}$ cm is the estimated average radius of one capillary. From Eq. 10–4 we have

$$v_2 A_2 = v_1 A_1$$
$$v_2 N \pi r_{cap}^2 = v_1 \pi r_{aorta}^2$$

so

$$N = \frac{v_1}{v_2} \frac{r_{aorta}^2}{r_{cap}^2} = \left(\frac{0.30 \text{ m/s}}{5 \times 10^{-4} \text{ m/s}} \right) \left(\frac{1.0 \times 10^{-2} \text{ m}}{4 \times 10^{-6} \text{ m}} \right)^2 \approx 4 \times 10^9$$

or about 4 billion capillaries.

Another Example that makes use of the equation of continuity and the argument leading up to it is the following.

EXAMPLE 10–11 **Heating duct to a room.** How large must a heating duct be if air moving 3.0 m/s along it can replenish the air every 15 minutes in a room of 300-m^3 volume? (See Fig. 10–21.) Assume the air's density remains constant.

SOLUTION We can apply Eq. 10–4 if we consider the room (call it point 2) as a large section of the duct. Reasoning in the same way we did to obtain Eq. 10–4 (changing Δt to t), we see that $A_2 v_2 = A_2 l_2 / t = V_2 / t$ where V_2 is the volume of the room. Then $A_1 v_1 = A_2 v_2 = V_2 / t$ and

$$A_1 = \frac{V_2}{v_1 t} = \frac{300 \text{ m}^3}{(3.0 \text{ m/s})(900 \text{ s})} = 0.11 \text{ m}^2.$$

If the duct is square, then each side has length $l = \sqrt{A} = 0.33$ m or 33 cm. A rectangular duct 20 cm $\times$ 55 cm will also do.

➡ **PHYSICS APPLIED**
Heating duct

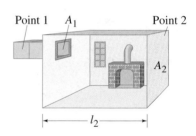

Point 1 A_1 Point 2

A_2

l_2

FIGURE 10–21
Example 10–11.

10–8 Bernoulli's Equation

Have you ever wondered how air can circulate in a prairie dog's burrow, why smoke goes up a chimney, or why a car's convertible top bulges upward at high speeds? These are examples of a principle worked out by Daniel Bernoulli (1700–1782) in the early eighteenth century. In essence, **Bernoulli's principle** states that *where the velocity of a fluid is high, the pressure is low, and where the velocity is low, the pressure is high.* For example, if the pressures at points 1 and 2 in Fig. 10–19 are measured, it will be found that the pressure is lower at point 2, where the velocity is greater, than it is at point 1, where the velocity is smaller. At first glance, this might seem strange; you might expect that the greater speed at point 2 would imply a higher pressure. But this cannot be the case. For if the pressure at point 2 were higher than at 1, this higher pressure would slow the fluid down, whereas in fact it has speeded up in going from point 1 to point 2. Thus the pressure at point 2 must be less than at point 1, to be consistent with the fact that the fluid accelerates.

Bernoulli's principle

(a)

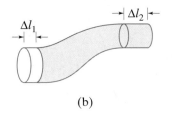

(b)

FIGURE 10–22 Fluid flow: for derivation of Bernoulli's equation.

Bernoulli developed an equation that expresses this principle quantitatively. To derive Bernoulli's equation, we assume the flow is steady and laminar, the fluid is incompressible, and the viscosity is small enough to be ignored. To be general, we assume the fluid is flowing in a tube of nonuniform cross section that varies in height above some reference level, Fig. 10–22. We will consider the amount of fluid shown in color and calculate the work done to move it from the position shown in (a) to that shown in (b). In this process, fluid at point 1 flows a distance Δl_1 and forces the fluid at point 2 to move a distance Δl_2. The fluid to the left of point 1 exerts a pressure P_1 on our section of fluid and does an amount of work

$$W_1 = F_1 \Delta l_1 = P_1 A_1 \Delta l_1.$$

At point 2, the work done on our section of fluid is

$$W_2 = -P_2 A_2 \Delta l_2;$$

the negative sign is present because the force exerted on the fluid is opposite to the motion (thus the fluid shown in color does work on the fluid to the right of point 2). Work is also done on the fluid by the force of gravity. Since the net effect of the process shown in Fig. 10–22 is to move a mass m of volume $A_1 \Delta l_1$ ($= A_2 \Delta l_2$, since the fluid is incompressible) from point 1 to point 2, the work done by gravity is

$$W_3 = -mg(y_2 - y_1),$$

where y_1 and y_2 are heights of the center of the tube above some (arbitrary) reference level. Notice that in the case shown in Fig. 10–22, this term is negative since the motion is uphill against the force of gravity. The net work W done on the fluid is thus:

$$W = W_1 + W_2 + W_3$$

$$W = P_1 A_1 \Delta l_1 - P_2 A_2 \Delta l_2 - mgy_2 + mgy_1.$$

According to the work energy principle (Section 6–3), the net work done on a system is equal to its change in kinetic energy. Thus

$$\tfrac{1}{2} mv_2^2 - \tfrac{1}{2} mv_1^2 = P_1 A_1 \Delta l_1 - P_2 A_2 \Delta l_2 - mgy_2 + mgy_1.$$

The mass m has volume $A_1 \Delta l_1 = A_2 \Delta l_2$. Thus we can substitute $m = \rho A_1 \Delta l_1 = \rho A_2 \Delta l_2$, and also divide through by $A_1 \Delta l_1 = A_2 \Delta l_2$, to obtain:

$$\tfrac{1}{2} \rho v_2^2 - \tfrac{1}{2} \rho v_1^2 = P_1 - P_2 - \rho gy_2 + \rho gy_1$$

which we rearrange to get

Bernoulli's equation

$$P_1 + \tfrac{1}{2} \rho v_1^2 + \rho gy_1 = P_2 + \tfrac{1}{2} \rho v_2^2 + \rho gy_2. \tag{10–5}$$

This is **Bernoulli's equation**. Since points 1 and 2 can be any two points along a tube of flow, Bernoulli's equation can be written:

$$P + \tfrac{1}{2} \rho v^2 + \rho gy = \text{constant}$$

at every point in the fluid.

Bernoulli's equation is an expression of the law of energy conservation, since we derived it from the work-energy principle.

EXAMPLE 10–12 **Flow and pressure in hot-water heating systems.** Water circulates throughout a house in a hot-water heating system. If the water is pumped at a speed of 0.50 m/s through a 4.0-cm-diameter pipe in the basement under a pressure of 3.0 atm, what will be the flow speed and pressure in a 2.6-cm-diameter pipe on the second floor 5.0 m above?

SOLUTION We first calculate the flow speed on the second floor, calling it v_2, using the equation of continuity, Eq. 10–4. Noting that the areas are proportional to the radii squared ($A = \pi r^2$), we call the basement point 1 and obtain

$$v_2 = \frac{v_1 A_1}{A_2} = \frac{v_1 \pi r_1^2}{\pi r_2^2} = (0.50 \text{ m/s})\frac{(0.020 \text{ m})^2}{(0.013 \text{ m})^2} = 1.2 \text{ m/s}.$$

To find the pressure, we use Bernoulli's equation:

$$\begin{aligned}
P_2 &= P_1 + \rho g(y_1 - y_2) + \tfrac{1}{2}\rho(v_1^2 - v_2^2) \\
&= (3.0 \times 10^5 \text{ N/m}^2) + (1.0 \times 10^3 \text{ kg/m}^3)(9.8 \text{ m/s}^2)(-5.0 \text{ m}) \\
&\quad + \tfrac{1}{2}(1.0 \times 10^3 \text{ kg/m}^3)[(0.50 \text{ m/s})^2 - (1.2 \text{ m/s})^2] \\
&= 3.0 \times 10^5 \text{ N/m}^2 - 4.9 \times 10^4 \text{ N/m}^2 - 6.0 \times 10^2 \text{ N/m}^2 \\
&= 2.5 \times 10^5 \text{ N/m}^2,
\end{aligned}$$

or 2.5 atm. Notice that the velocity term contributes very little in this case.

10–9 Applications of Bernoulli's Principle: From Torricelli to Sailboats, Airfoils, and TIA

Bernoulli's equation can be applied to many situations. One example is to calculate the velocity, v_1, of a liquid flowing out of a spigot at the bottom of a reservoir, Fig. 10–23. We choose point 2 in Eq. 10–5 to be the top surface of the liquid. Assuming the diameter of the reservoir is large compared to that of the spigot, v_2 will be almost zero. Points 1 (the spigot) and 2 (top surface) are open to the atmosphere so the pressure at both points is equal to atmospheric pressure: $P_1 = P_2$. Then Bernoulli's equation becomes

$$\tfrac{1}{2}\rho v_1^2 + \rho g y_1 = \rho g y_2$$

or

$$v_1 = \sqrt{2g(y_2 - y_1)}. \qquad \textbf{(10–6)}$$

This result is called **Torricelli's theorem.** Although it is seen to be a special case of Bernoulli's equation, it was discovered a century before Bernoulli by Evangelista Torricelli, a student of Galileo, hence its name. Equation 10–6 tells us that the liquid leaves the spigot with the same speed that a freely falling object would attain falling the same height. This should not be too surprising since the derivation of Bernoulli's equation relies on the conservation of energy.

Another special case of Bernoulli's equation arises when the fluid is flowing but there is no appreciable change in height; that is, $y_1 = y_2$. Then Eq. 10–5 becomes

$$P_1 + \tfrac{1}{2}\rho v_1^2 = P_2 + \tfrac{1}{2}\rho v_2^2. \qquad \textbf{(10–7)}$$

This tells us quantitatively that where the speed is high the pressure is low, and vice versa. It explains many common phenomena, some of which are

FIGURE 10–23 Torricelli's theorem: $v_1 = \sqrt{2g(y_2 - y_1)}$.

Torricelli's theorem

FIGURE 10-24 Examples of Bernoulli's principle: (a) atomizer, (b) Ping-Pong ball in jet of air, (c) airplane wing, (d) sailboat, (e) carburetor barrel.

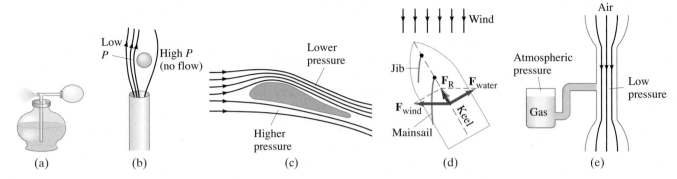

(a)   (b)   (c)   (d)   (e)

➡ **PHYSICS APPLIED**

Perfume atomizer

➡ **PHYSICS APPLIED**

Airplanes and dynamic lift

➡ **PHYSICS APPLIED**

Sailing against the wind

FIGURE 10-25 A sailboat moving against the wind makes use of Bernoulli's principle.

illustrated in Fig. 10-24. The pressure in the air blown at high speed across the top of the vertical tube of a perfume atomizer (Fig. 10-24a) is less than the normal air pressure acting on the surface of the liquid in the bowl. Thus perfume is pushed up the tube because of the reduced pressure at the top. A Ping-Pong ball can be made to float above a blowing jet of air (some vacuum cleaners can blow air), Fig 10-24b; if the ball begins to leave the jet of air, the higher pressure in the still air outside the jet (Bernoulli's principle) pushes the ball back in.

Airplane wings and other airfoils moving rapidly relative to the air deflect the air so that although streamline flow is largely maintained, the streamlines are crowded together above the wing, Fig. 10-24c. Just as the flow lines are crowded together in a pipe constriction where the velocity is high (see Fig. 10-19), so the crowded streamlines above the wing indicate that the air speed is greater above the wing than below it. Hence the air pressure above the wing is less than that below and there is thus a net upward force, called **dynamic lift**. Bernoulli's principle is only one aspect of the lift on a wing. Wings are usually tilted slightly upward so that air striking the bottom surface is deflected downward; the change in momentum of the rebounding air molecules results in an additional upward force on the wing. Turbulence also plays an important role.

A sailboat can move against the wind, Figs. 10-24d and Fig. 10-25, and the Bernoulli effect aids in this considerably if the sails are arranged so that the air velocity increases in the narrow constriction between the two sails. The normal atmospheric pressure behind the mainsail is larger than the reduced pressure in front of it (due to the fast moving air in the narrow slot between the sails), and this pushes the boat forward. When going against the wind, the mainsail is set at an angle approximately midway between the wind direction and the boat's axis (keel line) as shown in Fig. 10-24d. The net force on the sail (wind and Bernoulli) acts nearly perpendicular to the sail ($\mathbf{F}_{wind}$). This would tend to make the boat move sideways if it weren't for the keel that extends vertically downward beneath the water—for the water exerts a force ($\mathbf{F}_{water}$) on the keel nearly perpendicular to the keel. The resultant of these two forces ($\mathbf{F}_R$) is almost directly forward as shown.

A **venturi tube** is essentially a pipe with a narrow constriction (the throat). One example of a venturi tube is the barrel of a carburetor in a car, (Fig. 10–24e). The flowing air speeds up as it passes this constriction (Eq. 10–4) and so the pressure is lower. Because of the reduced pressure, gasoline under atmospheric pressure in the carburetor reservoir is forced into the air stream and mixes with the air before entering the cylinders.

The venturi tube is also the basis of the *venturi meter*, which is used to measure the flow speed of fluids, Fig. 10–26. Venturi meters can be used to measure the flow velocities of gases and liquids and have even been designed to measure blood velocity in arteries.

Why does smoke go up a chimney? It's partly because hot air rises (it's less dense and therefore buoyant). But Bernoulli's principle also plays a role. Because wind blows across the top of a chimney, the pressure is less there than inside the house. Hence, air and smoke are pushed up the chimney. Even on an apparently still night there is usually enough ambient air flow at the top of a chimney to assist upward flow of smoke.

If gophers, prairie dogs, rabbits, and other animals that live underground are to avoid suffocation, the air must circulate in their burrows. The burrows are always made to have at least two entrances (Fig. 10–27). The speed of air flow across different holes will usually be slightly different. This results in a slight pressure difference, which forces a flow of air through the burrow a la Bernoulli. The flow of air is enhanced if one hole is higher than the other (and this is often contrived by animals) since wind speed tends to increase with height.

In medicine, one of many applications of Bernoulli's principle is to explain a TIA, a transient ischemic attack (meaning a temporary lack of blood supply to the brain), caused by the so-called "subclavian steal syndrome." A person suffering a TIA may experience symptoms such as dizziness, double vision, headache, and weakness of the limbs. A TIA can occur as follows. Blood normally flows up to the brain at the back of the head via the two vertebral arteries—one going up each side of the neck—which meet to form the basilar artery just below the brain, as shown in Fig. 10–28. The vertebral arteries issue from the subclavian arteries, as shown, before the latter pass to the arms. When an arm is exercised vigorously, blood flow increases to meet the needs of the arm's muscles. If the subclavian artery on one side of the body is partially blocked, however, say by arteriosclerosis, the blood velocity will have to be higher on that side to supply the needed blood. (Recall the equation of continuity: smaller area means larger velocity for the same flow rate, Eq. 10–4.) The increased blood velocity past the opening to the vertebral artery results in lower pressure (Bernoulli's principle). Thus, blood rising in the vertebral artery on the "good" side at normal pressure can be *diverted down* into the other vertebral artery because of the low pressure on that side (like the Venturi effect), instead of passing upward into the basilar artery and the brain. Hence the blood supply to the brain is reduced due to "subclavian steal syndrome": the fast-moving blood in the subclavian artery "steals" the blood away from the brain. The resulting dizziness or weakness usually causes the person to stop the exertions, followed by a return to normal.

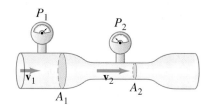

P_1 P_2 v_1 v_2 A_1 A_2

FIGURE 10–26 Venturi meter.

⮕ **PHYSICS APPLIED**

Smoke up a chimney

FIGURE 10–27 Bernoulli's principle is responsible for air flow in underground burrows.

Wind

⮕ **PHYSICS APPLIED**

Medicine—TIA

FIGURE 10–28 Rear of the head and shoulders showing arteries leading to the brain and to the arms. High blood velocity past the constriction in the left subclavian artery causes low pressure in the left vertebral artery, in which a reverse (downward) blood flow can then result: so-called "subclavian steal syndrome," resulting in a TIA (see text).

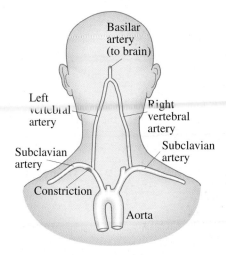

Basilar artery (to brain)

Left vertebral artery

Right vertebral artery

Subclavian artery

Subclavian artery

Constriction

Aorta

Bernoulli's equation ignores the effects of friction (viscosity) and the compressibility of the fluid. The energy that is transformed to internal (or potential) energy due to compression and to thermal energy by friction can be taken into account by adding terms to the right side of Eq. 10–5. These terms are difficult to calculate theoretically and are normally determined empirically. We will not pursue them here, but merely note that they do not significantly alter the explanations for the phenomena described above.

* 10–10 Viscosity

As already mentioned, real fluids have a certain amount of internal friction which is called **viscosity**. Viscosity exists in both liquids and gases, and is essentially a frictional force between adjacent layers of fluid as the layers move past one another. In liquids, viscosity is mainly due to the cohesive forces between the molecules. In gases, it arises from collisions between the molecules.

Different fluids possess different amounts of viscosity: syrup is more viscous than water; grease is more viscous than engine oil; liquids in general are much more viscous than gases. The viscosity of different fluids can be expressed quantitatively by a *coefficient of viscosity*, η (the Greek lowercase letter eta), which is defined in the following way. A thin layer of fluid is placed between two flat plates. One plate is stationary and the other is made to move at constant speed, Fig. 10–29. The fluid directly in contact with each plate is held to the surface by the adhesive force between the molecules of the liquid and those of the plate. Thus the upper surface of the fluid moves with the same speed v as the upper plate, whereas the fluid in contact with the stationary plate remains stationary. The stationary layer of fluid retards the flow of the layer just above it, which in turn retards the flow of the next layer, and so on. Thus the velocity varies continuously from 0 to v, as shown. The change in velocity divided by the distance over which this change is made—equal to v/l—is called the *velocity gradient*. To move the upper plate requires a force, which you can verify by moving a flat plate across a puddle of syrup on a table. For a given fluid, it is found that the force required, F, is proportional to the area of fluid in contact with each plate, A, and to the speed, v, and is inversely proportional to the separation, l, of the plates: $F \propto vA/l$. For different fluids, the more viscous the fluid, the greater is the required force. Hence the proportionality constant for this equation is defined as the *coefficient of viscosity*, η:

$$F = \eta A \frac{v}{l}. \qquad (10\text{–}8)$$

Solving for η, we find $\eta = Fl/vA$. The SI unit for η is $\text{N·s/m}^2 = \text{Pa·s}$ (pascal·second). In the cgs system, the unit is dyne·s/cm^2 and this unit is called a *poise* (P). Viscosities are often stated in *centipoise* (cP), which is one-hundredth of a poise. Table 10–3 lists the coefficient of viscosity for various fluids. The temperature is also specified, since it has a strong effect; the viscosity of liquids such as motor oil, for example, decreases rapidly as temperature increases.[†]

FIGURE 10–29 Determination of viscosity.

[†]The Society of Automotive Engineers assigns numbers to represent the viscosity of oils: 30-weight (SAE 30) is more viscous than 10-weight. Multigrade oils, such as 20–50, are designed to maintain viscosity as temperature increases; 20–50 means the oil is 20-wt when cool but is like a 50-wt pure oil when it is hot (engine-running temperature).

TABLE 10–3   Coefficient of Viscosity for Various Fluids

Fluid	Temperature (°C)	Coefficient of Viscosity, η (Pa·s)[†]
Water	0	1.8×10^{-3}
	20	1.0×10^{-3}
	100	0.3×10^{-3}
Whole blood	37	$\approx 4 \times 10^{-3}$
Blood plasma	37	$\approx 1.5 \times 10^{-3}$
Ethyl alcohol	20	1.2×10^{-3}
Engine oil (SAE 10)	30	200×10^{-3}
Glycerine	20	1500×10^{-3}
Air	20	0.018×10^{-3}
Hydrogen	0	0.009×10^{-3}
Water vapor	100	0.013×10^{-3}

[†]1 Pa·s = 10 P = 1000 cP

*10–11   Flow in Tubes: Poiseuille's Equation, Blood Flow

If a fluid had no viscosity, it could flow through a level tube or pipe without a force being applied. Because of viscosity, a pressure difference between the ends of a tube is necessary for the steady flow of any real fluid, be it water or oil in a pipe, or blood in the circulatory system of a human, even when the tube is level.

The rate of flow of a fluid in a round tube depends on the viscosity of the fluid, the pressure difference, and the dimensions of the tube. The French scientist J. L. Poiseuille (1799–1869), who was interested in the physics of blood circulation (and after whom the "poise" is named), determined how the variables affect the flow rate of an incompressible fluid undergoing laminar flow in a cylindrical tube. His result, known as *Poiseuille's equation*, is:

$$Q = \frac{\pi r^4 (P_1 - P_2)}{8 \eta L},$$ **(10–9)**

Poiseuille's equation for flow rate in a tube

where r is the inside radius of the tube, L is its length, $P_1 - P_2$ is the pressure difference between the ends, η is the coefficient of viscosity, and Q is the volume rate of flow (volume of fluid flowing past a given point per unit time which in SI has units of m^3/s). Equation 10–9 applies to laminar flow. There is no such simple mathematical relation if the flow is turbulent.

EXAMPLE 10–13   **Pressure in car engine oil tube.**   Engine oil (assume SAE 10, Table 10–3) passes through a fine 1.80-mm-diameter tube in a prototype engine. The tube is 5.5 cm long. What pressure difference is needed to maintain a flow rate of 5.6 mL/min?

SOLUTION The flow rate in SI units is $Q = 5.6 \times 10^{-6}\,\text{m}^3/60\,\text{s} = 9.3 \times 10^{-8}\,\text{m}^3/\text{s}$. We solve for $P_1 - P_2$ in Eq. 10–9 and put all terms in SI units:

$$P_1 - P_2 = \frac{8\eta L Q}{\pi r^4}$$

$$= \frac{8(2.0 \times 10^{-1}\,\text{N}\cdot\text{s/m}^2)(5.5 \times 10^{-2}\,\text{m})(9.3 \times 10^{-8}\,\text{m}^3/\text{s})}{3.14(0.90 \times 10^{-3}\,\text{m})^4}$$

$$= 4.0 \times 10^3\,\text{N/m}^2$$

or about 0.040 atm.

Poiseuille's equation tells us that the flow rate Q is directly proportional to the "pressure gradient," $(P_1 - P_2)/L$, and it is inversely proportional to the viscosity of the fluid. This is just what we might expect. It may be surprising, however, that Q also depends on the *fourth* power of the tube's radius. This means that for the same pressure gradient, if the tube radius is halved, the flow rate is decreased by a factor of 16! Thus the rate of flow, or alternately the pressure required to maintain a given flow rate, is greatly affected by only a small change in tube radius.

An interesting example of this r^4 dependence is *blood flow* in the human body. Poiseuille's equation is valid only for the streamline flow of an incompressible fluid with constant viscosity η. So it cannot be precisely accurate for blood whose flow is not without turbulence and that contains corpuscles (whose diameter is almost equal to that of a capillary). Hence η depends to a certain extent on the blood flow speed v. Nonetheless, Poiseuille's equation does give a reasonable first approximation. The body controls the flow of blood by means of tiny bands of muscle surrounding the arteries. Contraction of these muscles reduces the diameter of an artery and, because of the r^4 in Eq. 10–9, the flow rate is greatly reduced for only a small change in radius. Very small actions by these muscles can thus control precisely the flow of blood to different parts of the body. Another aspect is that the radius of arteries is reduced as a result of arteriosclerosis (hardening of the arteries) and by cholesterol buildup; when this happens, the pressure gradient must be increased to maintain the same flow rate. If the radius is reduced by half, the heart would have to increase the pressure by a factor of about $2^4 = 16$ in order to maintain the same blood-flow rate. The heart must work much harder under these conditions, but usually cannot maintain the original flow rate. Thus, high blood pressure is an indication both that the heart is working harder and that the blood-flow rate is reduced.

* 10–12 Surface Tension and Capillarity

Up to now in this chapter, we have mainly been studying what happens to fluids as a whole. But the *surface* of a liquid at rest also behaves in an interesting way. A number of common observations suggest that the surface of a liquid acts like a stretched membrane under tension. For example, a drop of water on the end of a dripping faucet, or hanging from a thin branch in the early morning dew (Fig. 10–30), forms into a nearly spherical shape as if it were a tiny balloon filled with water. A steel needle can be made to float on the surface of water even though it is denser than the water. The surface of a liquid acts as if it is under tension, and this tension, acting parallel to the surface, arises from the attractive forces between the molecules. This effect is called *surface tension*. More specifically, a quantity called the *surface tension*, γ (the Greek letter gamma), is defined as the force F per unit length L that acts across any line in a surface, tending to pull the surface closed:

$$\gamma = \frac{F}{L}.$$

(10–10)

To understand this, consider the U-shaped apparatus shown in Fig. 10–31 which encloses a thin film of liquid. Because of surface tension, a force F is required to pull the movable wire and thus increase the surface area of the liquid. The liquid contained by the wire apparatus is a thin film having both a top and a bottom surface. Hence the length of the surface being increased is $2l$, and the surface tension is $\gamma = F/2l$. A delicate apparatus of this type can be used to measure the surface tension of various liquids. The surface tension of water is 0.072 N/m at 20°C. Table 10–4 gives the values for other liquids. Note that temperature has a considerable effect on the surface tension.

We can see how surface tension arises by examining the process from the molecular point of view. The molecules of a liquid exert attractive forces on each other: these attractive forces are shown acting, in Fig. 10–32, on a molecule deep within the liquid and on a second molecule at the surface. The molecule inside the liquid is in equilibrium due to the forces of other molecules acting in all directions. The molecule at the surface is also normally in equilibrium (the liquid is at rest). This is true even though the forces on a molecule at the surface can be exerted only by molecules below it (or alongside it). Hence there is a net attractive force downward, which tends to compress the surface layer slightly—but only to the point where this downward force is balanced by an upward (repulsive) force due to close contact or collision with the molecules below.[†] This compression

[†]The air molecules above the surface exert a force, but this is a small effect since the air molecules are so far apart. The surface tension therefore does depend on the material above the surface, but the effect is generally small if the latter is a dilute gas. But because of this small effect, surface tensions are specified for the boundary between two materials. If the second material is not mentioned, it is assumed to be air at atmospheric pressure.

FIGURE 10–30 Spherical water droplets, dew on a grass blade.

FIGURE 10–31 U-shaped wire apparatus holding a film of liquid to measure surface tension ($\gamma = F/2l$).

(a) Top view

(b) Edge view (magnified)

TABLE 10–4
Surface Tension of Some Substances

Substance	Surface Tension (N/m)
Mercury (20°C)	0.44
Blood, whole (37°C)	0.058
Blood, plasma (37°C)	0.073
Alcohol, ethyl (20°C)	0.023
Water (0°C)	0.076
(20°C)	0.072
(100°C)	0.059
Benzene (20°C)	0.029
Soap solution (20°C)	≈0.025
Oxygen (−193°C)	0.016

FIGURE 10–32
Molecular theory of surface tension, showing attractive forces (only) on a molecule at the surface, and on one deep inside the liquid.

FIGURE 10–33 Forces on a molecule being brought to the surface because of an increase in the surface area. The molecules at the surface must exert a strong force on the molecule being pulled up, and they in turn feel a strong force back (Newton's third law). This is the surface-tension force.

of the surface means that, in essence, the liquid minimizes its surface area. This is why water tends to form spherical droplets (Fig. 10–30), for a sphere represents the minimum surface area for a given volume.

In order to increase the surface area of a liquid, a force is required and work must be done to bring molecules from the interior to the surface (Fig. 10–33). This work increases the potential energy of the molecules and is sometimes called *surface energy*. The greater the surface area, the greater the surface energy. The amount of work needed to increase the surface area can be calculated from Fig. 10–31 and Eq. 10–10:

$$W = F \, \Delta x$$
$$= \gamma L \, \Delta x$$
$$= \gamma \, \Delta A,$$

where Δx is the change in distance and $\Delta A \, (= L \, \Delta x)$ is the total increase in area (at both surfaces in Fig. 10–31). So we can write

$$\gamma = \frac{W}{\Delta A}.$$

Thus, the surface tension γ is not only equal to the force per unit length; it is also equal to the work done per unit increase in surface area. Hence, γ can be specified in N/m or J/m^2, which are the same thing.

Because of surface tension, insects can walk on water; and objects more dense than water, such as a steel needle, can actually float on the surface. Figure 10–34a shows how the surface tension can support the weight w of an object. Actually, the object sinks slightly into the fluid, so w is the "effective weight" of the object—its true weight less the buoyant force. If the object is spherical in shape, the surface tension acts at all points around a horizontal circle of approximately radius r (Fig. 10–34a). Only the vertical component, $\gamma \cos \theta$, acts to balance w. We set the length L equal to the circumference of the circle, $L \approx 2\pi r$, so the net upward force due to surface tension is $F \approx (\gamma \cos \theta)L \approx 2\pi r \gamma \cos \theta$.

FIGURE 10–34 Surface tension acting on (a) a sphere, and (b) an insect leg. Example 10–14.

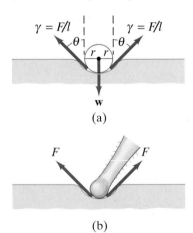

EXAMPLE 10–14 ESTIMATE A water strider. The base of an insect's leg is approximately spherical in shape, with a radius of about 2.0×10^{-5} m. The 0.0030 g mass of the insect is supported equally by the six legs. Estimate the angle θ (see Fig. 10–34) for an insect on the surface of water. Assume the water temperature is 20°C.

SOLUTION Since the insect is in equilibrium, the upward surface tension force is equal to the effective pull of gravity downward on each leg:

$$2\pi r \gamma \cos \theta \approx w,$$

where w is one sixth the weight of the insect (since it has six legs). Then

$$(6.28)(2.0 \times 10^{-5}\,\text{m})(0.072\,\text{N/m}) \cos \theta \approx \tfrac{1}{6}(3.0 \times 10^{-6}\,\text{kg})(9.8\,\text{m/s}^2)$$

$$\cos \theta \approx \frac{0.49}{0.90} = 0.54.$$

So $\theta \approx 57°$. Notice that if $\cos \theta$ were greater than 1, this would indicate that the surface tension would not be great enough to support the weight.

Calculations such as that in Example 10–14 are often only approximations, since the radius r of the surface depression is not precisely equal to the radius of the object. However, reasonable estimates can be made as to whether an object will or will not remain on the surface.

Soaps and detergents have the effect of lowering the surface tension of water. This is desirable for washing and cleaning since the high surface tension of pure water prevents it from penetrating easily between the fibers of material and into tiny crevices. Substances that reduce the surface tension of a liquid are called *surfactants*.

Surface tension plays a role in another interesting phenomenon, capillarity. It is a common observation that water in a glass container rises up slightly where it touches the glass, Fig. 10–35a. The water is said to "wet" the glass. Mercury, on the other hand, is depressed when it touches the glass, Fig. 10–35b; the mercury does not wet the glass. Whether or not a liquid wets a solid surface is determined by the relative strength of the cohesive forces between the molecules of the liquid compared to the adhesive forces between the molecules of the liquid and those of the container. (*Cohesion* refers to the force between molecules of the same type and *adhesion* to the force between molecules of different types.) Water wets glass because the water molecules are more strongly attracted to the glass molecules than they are to other water molecules. The opposite is true for mercury: the cohesive forces are stronger than the adhesive forces.

In tubes having very small diameters, liquids are observed to rise or fall relative to the level of the surrounding liquid. This phenomenon is called **capillarity**, and such thin tubes are called **capillaries**. Whether the liquid rises or falls (Fig. 10–36) depends on the relative strengths of the adhesive and cohesive forces. Thus water rises in a glass tube whereas mercury falls. The actual amount of rise (or fall) depends on the surface tension—which is what keeps the liquid surface from breaking apart.

FIGURE 10–35 Water (a) "wets" the surface of glass, whereas (b) mercury does not "wet" the glass.

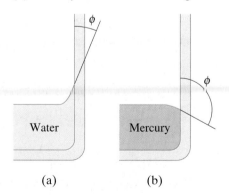

(a)   (b)

FIGURE 10–36 Capillarity.

(a)
Glass tube
in water

(b)
Glass tube
in mercury

FIGURE 10–37 An example of one kind of pump. As the diagram indicates, the intake valve opens and air (or fluid that is being pumped) fills the empty space when the piston moves to the left. When the piston moves to the right (not shown), the outlet valve opens and fluid is forced out.

* **10–13** **Pumps; the Heart and Blood Pressure**

We conclude this chapter with a brief discussion of pumps of various types, including the heart. Pumps can be classified into categories according to their function. A *vacuum pump* is designed to reduce the pressure (usually of air) in a given vessel. A *force pump*, on the other hand, is a pump that is intended to increase the pressure—for example, to lift a liquid (such as water from a well) or to push a fluid through a pipe. Figure 10–37 illustrates the principle behind a simple reciprocating pump. It could be a vacuum pump, in which case the intake is connected to the vessel to be evacuated. A similar mechanism is used in some force pumps, and in this case the fluid is forced under increased pressure through the outlet. Other kinds of pumps are illustrated in Fig. 10–38. The centrifugal pump, or any force pump, can be used as a *circulating pump*—that is, to circulate a fluid around a closed path, such as the cooling water or lubricating oil in an automobile.

FIGURE 10–38 (a) Centrifugal pump: the rotating blades force fluid through the outlet pipe; this kind of pump is used in vacuum cleaners and as a water pump in automobiles. (b) Rotary oil pump, used to obtain vacuums as low as 10^{-4} mm-Hg: gas (usually air) from the vessel to be evacuated diffuses into the space G via the intake pipe I; the rotating off-center cylinder C traps the gas in G and pushes it out the exhaust valve E, in the meantime allowing more gas to diffuse into G for the next cycle. The sliding valve V is kept in contact with C by a spring S, and this prevents the exhaust gas from returning to G. (c) Diffusion pump, used to obtain vacuums as low as 10^{-8} mm-Hg: air molecules from the vessel to be evacuated diffuse into the jet, where the rapidly moving jet of oil sweeps the molecules away. A "forepump" is needed, which is a mechanical pump, such as the rotary type (b), and acts as a first stage in reducing the pressure.

(a)

(b)

(c)

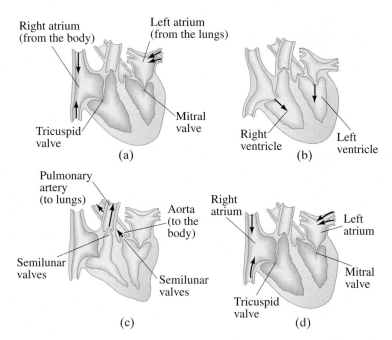

Right atrium
(from the body)

Left atrium
(from the lungs)

Tricuspid
valve

Mitral
valve

(a)

Right
ventricle

Left
ventricle

(b)

Pulmonary
artery
(to lungs)

Aorta
(to the
body)

Semilunar
valves

Semilunar
valves

(c)

Right
atrium

Left
atrium

Mitral
valve

Tricuspid
valve

(d)

FIGURE 10–39 (a) In the diastole phase, the heart relaxes between beats. Blood moves into the heart; both atria are filled rapidly. (b) When the atria contract, the systole, or pumping, phase begins. The contraction pushes the blood through the mitral and tricuspid valves into the ventricles. (c) The contraction of the ventricles forces the blood through the semilunar valves into the pulmonary artery (which leads to the lungs), and to the aorta (the body's largest artery) which leads to the arteries serving all the body (Figure 10–20). (d) When the heart relaxes, the semilunar valves close; blood fills the atria, beginning the cycle again.

The heart of a human (and of other animals as well) is essentially a circulating pump. The action of a human heart is shown in Fig. 10–39. There are actually two separate paths for blood flow. The longer path takes blood to the parts of the body, via the arteries, bringing oxygen to body tissues and picking up carbon dioxide which it carries back to the heart via veins. This blood is then pumped into the lungs (the second path), where the carbon dioxide is released and oxygen is taken up. The oxygen-laden blood is returned to the heart, where it is again pumped to the tissues of the body.

➡ **PHYSICS APPLIED**

Heart as a pump

Blood pressure is measured using either a mercury-filled manometer or one of the other types of gauge mentioned earlier (Section 10–5), and it is usually calibrated in mm-Hg. The gauge is attached to a closed air-filled jacket that is wrapped around the upper arm at the level of the heart, Fig. 10–40. Two values of blood pressure are measured: the maximum pressure when the heart is pumping, called *systolic pressure*; and the pressure when the heart is in the resting part of the cycle, called *diastolic pressure*. Initially, the air pressure in the jacket is increased high above the systolic pressure by means of a hand pump, and this compresses the main (brachial) artery in the arm and cuts off the flow of blood. The air pressure is then reduced slowly to the point where blood again begins to flow into the arm; it is detected by listening with a stethoscope to the characteristic tapping sound[†] of the blood returning to the forearm. At this point, systolic pressure is just equal to the air pressure in the jacket which can be read off the gauge. The air pressure is subsequently reduced further and the tapping sound disappears when blood at low pressure can enter the artery. At this point, the gauge indicates the diastolic pressure. Normal systolic pressure is around 120 mm-Hg, whereas normal diastolic pressure is around 80 mm-Hg.

➡ **PHYSICS APPLIED**

Blood pressure

FIGURE 10–40 Device for measuring blood pressure.

Jacket

Hand
pump

Gauge

[†]When the blood flows through the constriction caused by the tight jacket, its velocity is high and the flow is turbulent. It is the turbulence that causes the tapping sound.

The three common phases of matter are **solid**, **liquid**, and **gas**. Liquids and gases are collectively called **fluids**, meaning they have the ability to flow. The **density** of a material is defined as its mass per unit volume. **Specific gravity** is the ratio of the density of the material to the density of water (at 4°C).

Pressure is defined as force per unit area. The pressure at a depth h in a liquid is given by

$$P = \rho g h,$$

where ρ is the density of the liquid and g is the acceleration due to gravity.

Pascal's principle says that an external pressure applied to a confined fluid is transmitted throughout the fluid.

Pressure is measured using a manometer or other type of gauge. A **barometer** is used to measure atmospheric pressure. Standard atmospheric pressure (average at sea level) is $1.013 \times 10^5 \, \text{N/m}^2$. **Gauge pressure** is the total pressure less atmospheric pressure.

Archimedes' principle states that an object submerged wholly or partially in a fluid is buoyed up by a force equal to the weight of fluid it displaces.

Fluid flow rate is the mass or volume of fluid that passes a given point per unit time. The **equation of continuity** states that for an incompressible fluid flowing in an enclosed tube, the product of the velocity of flow and the cross-sectional area of the tube remains constant:

$$Av = \text{constant}.$$

Bernoulli's principle tells us that where the velocity of a fluid is high, the pressure in it is low, and where the velocity is low, the pressure is high. **Bernoulli's equation** is

$$P_1 + \tfrac{1}{2}\rho v_1^2 + \rho g y_1 = P_2 + \tfrac{1}{2}\rho v_2^2 + \rho g y_2.$$

for two points in a tube of flow.

Fluid flow can be characterized either as **streamline** (sometimes called **laminar**), in which the layers of fluid move smoothly and regularly along paths called streamlines, or as **turbulent**, in which case the flow is not smooth and regular but is characterized by irregularly shaped whirlpools.

Viscosity refers to friction within a fluid that prevents the fluid from flowing freely and is essentially a frictional force between adjacent layers of fluid as they move past one another.

Q U E S T I O N S

1. If one material has a higher density than another, does this mean the molecules of the first must be heavier than those of the second? Explain.

2. Airplane travelers often note that their cosmetics bottles and other containers have leaked after a trip. What might cause this?

3. The three containers in Fig. 10–41 are filled with water to the same height and have the same surface area at the base; hence the water pressure, and the total force on the base of each, is the same. Yet the total weight of water is different for each. Explain this "hydrostatic paradox."

FIGURE 10–41   Question 3.

4. Consider what happens when you push both a pin and the blunt end of a pen against your skin with the same force. Decide what determines whether your skin suffers a cut—the net force applied to it or the pressure.

5. It is often said that water seeks its own level. Explain.

6. A small amount of water is boiled in a one-gallon gasoline can. The can is removed from the heat and the lid put on. Shortly thereafter the can collapses. Explain.

7. Explain how the tube in Fig. 10–42, known as a *siphon*, can transfer liquid from one container to a lower one even though the liquid must flow uphill for part of its journey. (Note that the tube must be filled with liquid to start with.)

FIGURE 10–42
A siphon. Question 7.

8. An ice cube floats in a glass of water filled to the brim. As the ice melts, will the glass overflow?

9. Will an ice cube float in a glass of alcohol? Why or why not?

10. A barge filled high with sand approaches a low bridge over the river and cannot quite pass under it. Should sand be added to, or removed from, the barge?

11. Will a helium-filled balloon rise indefinitely in the air? Explain.

12. Will an empty balloon have precisely the same apparent weight on a scale as one that is filled with air? Explain.

13. Does the buoyant force on a diving bell deep beneath the ocean have precisely the same value as when the bell is just beneath the surface? Explain.

14. A small wooden boat floats in a swimming pool, and the level of the water at the edge of the pool is marked. Consider the following situations and determine whether the level of the water will rise, fall, or stay the same. (a) The boat is removed from the water. (b) The boat holds an iron anchor which is removed from the boat and placed on the shore. (c) The iron anchor is removed from the boat and dropped in the pool.

15. Explain why helium weather balloons, which are used to measure atmospheric conditions at high altitude, are normally released while filled to only 10%–20% of their maximum volume.

16. Why do you float more easily in salt water than in fresh?

17. When measuring blood pressure, why must the jacket be held at the level of the heart?

18. Roofs of houses are sometimes "blown" off (or are they pushed off?) during a tornado or hurricane. Explain, using Bernoulli's principle.

19. If you dangle two pieces of paper vertically, a few inches apart (Fig. 10–43), and blow between them, how do you think the papers will move? Try it and see. Explain.

FIGURE 10–43 Question 19.

20. Why does the canvas top of a convertible bulge out when the car is traveling at high speed?

21. Children are told to avoid standing too close to a rapidly moving train because they might get sucked under it. Is this possible? Explain.

22. Why does a sailboat need a keel? In a small sailboat, the keel (a vertical "board" that extends below the boat into the water) is removed when the boat is anchored. Why?

23. A tall Styrofoam cup is filled with water. Two holes are punched in the cup near the bottom, and water begins rushing out. If the cup is dropped so it falls freely, will the water continue to flow from the holes? Explain.

24. With a little effort, you can blow across a dime on a table and make it land in a cup as shown in Fig. 10–44, without touching either cup or dime. Explain (and try it).

FIGURE 10–44 Question 24.

25. Why do airplanes normally take off into the wind?

26. Hummingbirds expend 20 times as much energy to hover in front of a flower as they do in normal flight. Explain.

27. Why does the stream of water from a faucet become narrower as it falls (Fig. 10–45)?

FIGURE 10–45 Water coming from a faucet. Question 27 and Problem 78.

Explain why the speed of wind increases (on average) with increased height above the Earth's surface. [*Hint:* See Fig. 10–29.] Explain how this affects a mole that builds its burrow with one of the two entrances higher than the other.

* 29. Identical steel ball bearings are dropped into tubs of water at 10°C and 40°C. In which tub will the ball bearing reach the bottom more quickly?

PROBLEMS

SECTION 10–1

1. (I) The approximate volume of the granite monolith known as El Capitan in Yosemite National Park is about 10^8 m^3. What is its approximate mass?

2. (I) What is the approximate mass of air in a living room $5.8 \text{ m} \times 3.8 \text{ m} \times 2.8 \text{ m}$?

3. (I) If you tried to nonchalantly smuggle gold bricks by filling your backpack, whose dimensions are $60 \text{ cm} \times 25 \text{ cm} \times 15 \text{ cm}$, what would its mass be?

4. (I) Estimate your volume. [*Hint:* Because you can swim on or just under the surface of the water in a swimming pool, you have a pretty good idea of your density.]

5. (II) A bottle has a mass of 35.00 g when empty and 98.44 g when filled with water. When filled with another fluid, the mass is 88.78 g. What is the specific gravity of this other fluid?

6. (II) If 5.0 L of antifreeze solution (specific gravity = 0.80) is added to 4.0 L of water to make a 9.0-L mixture, what is the specific gravity of the mixture?

SECTIONS 10–2 TO 10–5

7. (I) What is the difference in blood pressure (mm-Hg) between the top of the head and bottom of the feet of a 1.60-m-tall person standing vertically?

8. (I) Estimate the pressure exerted on a floor by (a) a 50-kg model standing momentarily on a single spiked heel (area = 0.05 cm^2), and compare it (b) to the pressure exerted by a 1500-kg elephant standing on one foot (area = 800 cm^2).

9. (I)(a) Calculate the total force of the atmosphere acting on the top of a table that measures $1.6 \text{ m} \times 1.9 \text{ m}$. (b) What is the total force acting upward on the underside of the table?

10. (II) In a movie, Tarzan is shown evading his captors by hiding underwater for many minutes while breathing through a long thin reed. Assuming the maximum pressure difference lungs can manage and still breathe is −80 mm-Hg, calculate the deepest he could have been.

11. (II) The gauge pressure in each of the four tires of an automobile is 240 kPa. If each tire has a "footprint" of 200 cm^2, estimate the mass of the car.

12. (II) The maximum gauge pressure in a hydraulic lift is 18 atm. What is the largest size vehicle (kg) it can lift if the diameter of the output line is 22 cm?

13. (II) How high would the level be in an alcohol barometer at normal atmospheric pressure?

14. (II) What is the total force and the absolute pressure on the bottom of a swimming pool 22.0 m by 12.0 m whose uniform depth is 2.0 m? What will be the pressure against the *side* of the pool near the bottom?

15. (II) How high would the atmosphere extend if it were of uniform density, equal to that at sea level throughout?

16. (II) Water and then oil are poured into a U-shaped tube, open at both ends, and do not mix. They come to equilibrium as shown in Fig. 10–46. What is the density of the oil? [*Hint:* Pressures at points a and b are equal. Why?]

FIGURE 10–46 Problem 16.

17. (II) Determine the water gauge pressure at a house at the bottom of a hill fed by a full tank of water 5.0 m deep and connected to the house by a pipe that is 100 m long at an angle of 60° from the horizontal (Fig. 10–47). Neglect turbulence, and frictional and viscous effects. How high would the water shoot if it came vertically out of a broken pipe in front of the house?

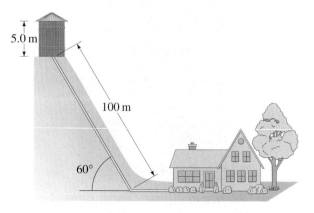

FIGURE 10–47 Problem 17.

18. (II) Determine the minimum gauge pressure needed in the water pipe leading into a building if water is to come out of a faucet on the twelfth floor, 41 m above.

19. (II) In working out his principle, Pascal showed dramatically how force can be multiplied with fluid pressure. He placed a long thin tube of 0.30-cm radius vertically into a 20-cm-radius wine barrel, Fig. 10–48. He found that when the barrel was filled with water and the tube filled to a height of 12 m, the barrel burst. Calculate (a) the mass of fluid in the tube, and (b) the net force on the lid exerted by the water in the barrel just before rupture.

r = 0.30 cm
12 m
R = 20 cm

FIGURE 10–48   Problem 19. (Not to scale)

20. (III) Estimate the density of the water 6.0 km deep in the sea. (See Section 9–6 and Table 9–1.) By what fraction does it differ from the density at the surface?

SECTION 10–6

21. (I) The hydrometer of Example 10–7 sinks to a depth of 22.9 cm when placed in a fermenting vat. What is the density of the brewing liquid?

22. (I) A geologist finds that a moon rock whose mass is 8.20 kg has an apparent mass of 6.18 kg when submerged in water. What is the density of the rock?

23. (I) What fraction of a piece of aluminum will be submerged when it floats in mercury?

24. (II) A spherically shaped balloon has a radius of 9.5 m, and is filled with helium. How large a cargo can it lift, assuming that the skin and structure of the balloon have a mass of 1000 kg? Neglect the buoyant force on the cargo volume itself.

25. (II) A 78-kg person has an apparent mass of 54 kg (because of buoyancy) when standing in water that comes up to the hips. Estimate the mass of each leg. Assume the body has SG = 1.00.

26. (II) What is the likely identity of a metal (see Table 10–1) if a sample has a mass of 63.5 g when measured in air and an apparent mass of 56.4 g when submerged in water?

27. (II) Archimedes' principle can be used not only to determine the specific gravity of a solid using a known liquid (Example 10–6). The reverse can be done as well. (a) As an example, a 3.40-kg aluminum ball has an apparent mass of 2.10 kg when submerged in a particular liquid: calculate the density of the liquid. (b) Derive a simple formula for determining the density of a liquid using this procedure.

28. (II) (a) Show that the buoyant force on a partially submerged object acts at the center of gravity of the fluid before it is displaced. This point is called the *center of buoyancy*. (b) For a ship to be stable, should its center of buoyancy be above, below, or at the same point as, its center of gravity? Explain.

29. (II) A 0.48-kg piece of wood floats in water but is found to sink in alcohol (specific gravity = 0.79) in which it has an apparent mass of 0.035 kg. What is the SG of the wood?

30. (II) The specific gravity of ice is 0.917, whereas that for seawater is 1.025. What fraction of an iceberg is above the surface of the water?

31. (III) A polar bear partially supports herself by pulling part of her body out of the water onto a rectangular slab of ice. The ice sinks down so that only half of what was once exposed now is exposed, and the bear has 70 percent of her volume (and weight) out of the water. Estimate the bear's mass, assuming that the total volume of the ice is 10 m³, and the bear's specific gravity is 1.0.

32. (III) A 2.52-kg piece of wood (SG = 0.50) floats on water. What minimum mass of lead, hung from it by a string, will cause it to sink?

33. (III) If an object floats in water, its density can be determined by tying a sinker on it so that both the object and the sinker are submerged. Show that the specific gravity is given by $w/(w_1 - w_2)$, where w is the weight of the object alone in air, w_1 is the apparent weight when a sinker is tied to it and the sinker only is submerged, and w_2 is the apparent weight when both the object and the sinker are submerged.

SECTIONS 10–7 TO 10–9

34. (I) Using the data of Example 10–10, calculate the average speed of blood flow in the major arteries of the body which have a total cross-sectional area of about 2.0 cm².

35. (I) A 17-cm-radius air duct is used to replenish the air of a room 9.2 m × 5.0 m × 4.5 m every 10 min. How fast does the air flow in the duct?

36. (II) A $\frac{5}{8}$-inch (inside) diameter garden hose is used to fill a round swimming pool 7.2 m in diameter. How long will it take to fill the pool to a depth of 1.5 m if water issues from the hose at a speed of 0.28 m/s?

37. (II) What gauge pressure in the water mains is necessary if a firehose is to spray water to a height of 12.0 m?

38. (II) What is the volume rate of flow of water from a 1.60-cm-diameter faucet if the pressure head is 12.0 m?

39. (II) If wind blows at 30 m/s over your house, what is the net force on the flat roof if its area is 240 m²?

40. (II) What is the lift (in newtons) due to Bernoulli's principle on a wing of area 80 m² if the air passes over the top and bottom surfaces at speeds of 340 m/s and 290 m/s, respectively?

41. (II) Estimate the air pressure at the center of a category 5 hurricane with a wind speed at the center of 300 km/h.

42. (II) Show that the power needed to drive a fluid through a pipe with uniform cross-section is equal to the volume rate of flow, Q, times the pressure difference, $P_1 - P_2$.

43. (II) Water at a pressure of 3.8 atm at street level flows into an office building at a speed of 0.60 m/s through a pipe 5.0 cm in diameter. The pipes taper down to 2.6 cm in diameter by the top floor, 20 m above (Fig. 10–49). Calculate the flow velocity and the pressure in such a pipe on the top floor. Ignore viscosity. Pressures are gauge pressures.

FIGURE 10–49   Problem 43.

44. (III) (a) Show that the flow velocity measured by a venturi meter is given by the relation $v_1 = A_2\sqrt{2(P_1 - P_2)/\rho(A_1^2 - A_2^2)}$. See Fig. 10–26. (b) A venturi tube is measuring the flow of water; it has a main diameter of 3.0 cm tapering down to a throat diameter of 1.0 cm; if the pressure difference is measured to be 18 mm-Hg, what is the velocity of the water?

*SECTION 10–10

* **45.** (II) A viscometer consists of two concentric cylinders, 10.20 cm and 10.60 cm in diameter. A particular liquid fills the space between them to a depth of 12.0 cm. The outer cylinder is fixed, and a torque of 0.024 N·m keeps the inner cylinder turning at a steady rotational speed of 62 rev/min. What is the viscosity of the liquid?

*SECTION 10–11

* **46.** (I) A gardener feels it is taking him too long to water a garden with a $\frac{3}{8}$-in-diameter hose. By what factor will his time be cut if he uses a $\frac{5}{8}$-in-diameter hose? Assume nothing else is changed.

* **47.** (II) What must be the pressure difference between the two ends of a 1.9-km section of pipe, 29 cm in diameter, if it is to transport oil ($\rho = 950 \text{ kg/m}^3$, $\eta = 0.20$ Pa·s) at a rate of 450 cm³/s?

* **48.** (II) Blood from an animal is placed in a bottle 1.70 m above a 3.8-cm-long needle of inside diameter 0.40 mm from which it flows at a rate of 4.1 cm³/min. What is the viscosity of this blood?

* **49.** (II) What diameter must a 21-m-long air duct have if the pressure of a ventilation and heating system is to replenish the air in a room 9.0 m × 14 m × 4.0 m every 10 min? Assume the pump can exert a gauge pressure of 0.71×10^{-3} atm.

* **50.** (II) Assuming a constant pressure gradient, by what factor must a blood vessel decrease in radius if the blood flow is reduced by 75 percent?

* **51.** (II) Calculate the pressure drop per cm along the aorta using the data of Example 10–10 and Table 10–3.

* **52.** (II) Poiseuille's equation does not hold if the flow velocity is high enough that turbulence sets in. The onset of turbulence occurs when the so-called *Reynolds number*, *Re*, exceeds approximately 2000. *Re* is defined as

$$Re = \frac{2\bar{v}r\rho}{\eta},$$

where $\bar{v}$ is the average speed of the fluid, ρ is its density, η is its viscosity, and r is the radius of the tube in which the fluid is flowing. (a) Determine if blood flow through the aorta is laminar or turbulent, when the average speed of blood in the aorta ($r = 1.0$ cm) during the resting part of the heart's cycle is about 30 cm/s. (b) During exercise, the blood-flow speed approximately doubles. Calculate the Reynolds number in this case and determine if the flow is laminar or turbulent.

* **53.** (III) According to the plate tectonic model of the Earth, the plates supporting the continents move very slowly on the hot deformable rock below. Make a rough estimate of the Reynolds number (see Problem 52) to show that this flow is laminar using the following data: speed $v = 50$ mm/yr, density and viscosity of deformable rock below are $\rho = 3200 \text{ kg/m}^3$ and $\eta = 4 \times 10^{19}$ Pa·s (notice the magnitude!), with thickness ≈ 100 km.

FIGURE 10–50 Problems 54 and 61.

* 54. (III) A patient is to be given a blood transfusion. The blood is to flow through a tube from a raised bottle to a needle inserted in the vein (Fig. 10–50). The inside diameter of the 4.0-cm-long needle is 0.40 mm and the required flow rate is 4.0 cm³ of blood per minute. How high should the bottle be placed above the needle? Obtain ρ and η from the Tables. Assume the blood pressure is 18 torr above atmospheric pressure.

* 55. (I) If the force F needed to move the wire in Fig. 10–31 is 5.1×10^{-3} N, calculate the surface tension γ of the enclosed fluid. Assume $l = 0.070$ m.

* 56. (I) Calculate the force needed to move the wire in Fig. 10–31 if it is immersed in a soapy solution and the wire is 18.2 cm long.

* 57. (II) If the base of an insect's leg has a radius of about 3.0×10^{-5} m and its mass is 0.016 g, would you expect the six-legged insect to remain on top of the water?

* 58. (II) The surface tension of a liquid can be determined by measuring the force F needed to just lift a circular platinum ring of radius r from the surface of the liquid. (a) Find a formula for γ in terms of F and r. (b) At 30°C, if $F = 840 \times 10^{-3}$ N and $r = 2.8$ cm, calculate γ for the tested liquid.

* 59. (II) A small pool of water on a table is broken into 100 droplets. By what factor does the surface energy change, assuming the original pool is flat and of depth h and the droplets are hemispheres of radius h?

* 60. (III) Show that inside a soap bubble, there must be a pressure ΔP in excess of that outside equal to $\Delta P = 4\gamma/r$, where r is the radius of the bubble and γ is the surface tension. [*Hint*: Think of the bubble as two hemispheres in contact with each other; and remember that there are two surfaces to the bubble. Note that this result applies to any kind of membrane, where $T = 2\gamma$ is the tension per unit length in that membrane.]

GENERAL PROBLEMS

61. Intravenous infusions are often made under gravity, as shown in Fig. 10–50. Assuming the fluid has a density of 1.00 g/cm³, at what height h should the bottle be placed so the liquid pressure is (a) 65 mm-Hg, (b) 550 mm-H$_2$O? (c) If the blood pressure is 18 mm-Hg above atmospheric pressure, how high should the bottle be placed so that the fluid just barely enters the vein?

62. A 2.4-N force is applied to the plunger of a hypodermic needle. If the diameter of the plunger is 1.3 cm and that of the needle 0.20 mm, (a) with what force does the fluid leave the needle? (b) What force on the plunger would be needed to push fluid into a vein where the gauge pressure is 18 mm-Hg? Answer for the instant just before the fluid starts to move.

63. A bicycle pump is used to inflate a tire. The initial tire pressure is 210 kPa (30 psi). At the end of the pumping process, the final pressure is 310 kPa (45 psi). If the diameter of the plunger in the cylinder of the pump is 3.0 cm, what is the range of the force that needs to be applied to the pump handle from beginning to end?

64. What is the approximate difference in air pressure between the top and the bottom of the World Trade Center buildings in New York City? They are 410 m tall and are located at sea level. Express as a fraction of atmospheric pressure at sea level.

65. Estimate the pressure on the mountains underneath the Antarctic ice pack which is typically 4 km thick.

66. Giraffes are a wonder of cardiovascular engineering. Calculate the difference in pressure (in atmospheres) that the blood vessels in a giraffe's head have to accommodate as it lowers its head from a full upright position to ground level for a drink. The height of an average giraffe is about 6 meters.

67. When you drive up into the mountains, or descend rapidly from the mountains, your ears "pop," which means that the pressure behind the eardrum is being equalized to that outside. If this did not happen, what would be the approximate force on an eardrum of area 0.50 cm² if a change in altitude of 1000 m takes place?

68. A small animal is found to remain suspended in a mixture of 18.0 percent (by weight) alcohol and 82.0 percent water. What is the density of the animal?

69. The contraction of the left ventricle (chamber) of the heart pumps blood to the body. Assuming that the inner surface of the left ventricle has an area of 82 cm² and the maximum pressure in the blood is 120 mm-Hg, estimate the force exerted by the ventricle at maximum pressure.

70. Estimate the total mass of the Earth's atmosphere using the known value of atmospheric pressure at sea level.

71. Suppose a person can reduce the pressure in the lungs to −80 mm-Hg gauge pressure. How high can water then be sucked up a straw?

72. How high should the pressure head be if water is to come from a faucet at a speed of 7.2 m/s? Ignore viscosity.

73. A ship, carrying freshwater to a desert island in the Caribbean, has a horizontal cross-sectional area of 2650 m² at the waterline. When unloaded, the ship rises 8.50 m higher in the sea. How much water was delivered?

74. A raft is made of 10 logs lashed together. Each is 33 cm in diameter and has a length of 6.1 m. How many people can the raft hold before they start getting their feet wet, assuming the average person has a mass of 70 kg? Do *not* neglect the weight of the logs. Assume the specific gravity of wood is 0.60.

75. During each heartbeat, approximately 70 cm³ of blood is pushed from the heart at an average pressure of 105 mm-Hg. Calculate the power output of the heart, in watts, assuming 70 beats per minute.

76. A bucket of water is accelerated upward at 3.5 g. What is the buoyant force on a 3.0-kg granite rock (SG = 2.7) submerged in the water? Will the rock float? Why or why not?

77. The drinking fountain outside your classroom shoots water about 12 cm up in the air from a nozzle of diameter 0.60 cm. The pump at the base of of the unit (1.1 m below the nozzle) pushes water into a 1.2-cm-diameter supply pipe that goes up to the nozzle. What gauge pressure does the pump have to provide? Ignore the viscosity; your answer will therefore be an underestimate.

78. The stream of water from a faucet decreases in diameter as it falls (Fig. 10–45). Derive an equation for the diameter of the stream as a function of the distance y below the faucet, given that the water has speed v_0 when it leaves the faucet, whose diameter is D.

79. A race car has 12 cylinders with a total displacement of 5.0 liters. When turbocharged, the air is forced in at about 1.5 atmospheres, meaning that about 7.5 liters of air (plus gas vapor) is brought into the 12 cylinders per revolution of the engine. At cruising speeds, the engines may run at 3000 rpm, and the fuel injection mechanisms (four for each cylinder) each have effective openings of diameter 6.0 mm. (a) What is the speed of the air flow through each injector, and (b) what is the pressure in the air as it travels through the injector into each cylinder?

FIGURE 10–51   Problem 80.

80. You need to siphon water from a clogged sink. The sink has an area of 0.375 m² and is filled to a height of 4.0 cm. Your siphon tube rises 50 cm above the bottom of the sink and then descends 100 cm to a pail. The siphon tube has a diameter of 2.0 cm. (a) Assuming that the water enters the siphon tube with almost zero velocity, calculate its velocity when it enters the pail. (b) Estimate how long it will take to empty the sink.

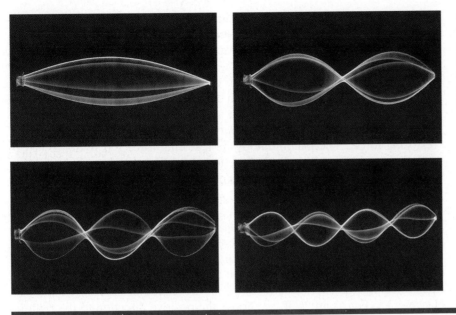

A stretched cord is made to vibrate, and in the four cases shown here, the frequency has been adjusted so that very nice standing waves are produced. Can you guess the relation between the frequencies for each case?

VIBRATIONS AND WAVES

Many objects vibrate or oscillate—an object on the end of a spring, a tuning fork, the balance wheel of an old watch, a pendulum, a plastic ruler held firmly over the edge of a table and gently struck, the strings of a guitar or piano. Spiders detect prey by the vibrations of their webs, cars oscillate up and down when they hit a bump, buildings and bridges vibrate when heavy trucks pass or the wind is fierce. Indeed, because most solids are elastic (see Chapter 9), most material objects vibrate (at least briefly) when given an impulse. Electrical oscillations occur in radio and television sets. At the atomic level, atoms vibrate within a molecule, and the atoms of a solid vibrate about their relatively fixed positions.

Vibrations and wave motion are intimately related subjects. Waves—whether ocean waves, waves on a string, earthquake waves, or sound waves in air—have as their source a vibration. In the case of sound, not only is the source a vibrating object, but so is the detector—the eardrum or the membrane of a microphone. Indeed, the medium through which a wave travels itself vibrates (such as air for sound waves). In the second half of this chapter, after we discuss vibrations, we will discuss simple waves such as those on water and on a string. In Chapter 12 we will study sound waves, and in later chapters we will encounter other forms of wave motion, including electromagnetic waves and light.

11–1 Simple Harmonic Motion

When a **vibration** or an **oscillation** repeats itself, back and forth, over the same path, the motion is **periodic**. The simplest form of periodic motion is represented by an object oscillating on the end of a coil spring. Because many other types of vibrational motion closely resemble this system, we will

(a)

$x = 0$

(b)

F

x (> 0)

(c)

F

x (< 0)

FIGURE 11–1 Mass vibrating at the end of a spring.

look at it in detail. We assume that the mass of the spring can be ignored, and that the spring is mounted horizontally, as shown in Fig. 11–1a, so that the object of mass m slides without friction on the horizontal surface. Any spring has a natural length at which it exerts no force on the mass m, and the position of the mass at this point is called the **equilibrium position**. If the mass is moved either to the left, which compresses the spring, or to the right, which stretches it, the spring exerts a force on the mass that acts in the direction of returning the mass to the equilibrium position; hence it is called a "restoring force." The magnitude of the restoring force F is found to be directly proportional to the displacement x the spring has been stretched or compressed from the equilibrium position (Fig. 11–1b and c):

$$F = -kx. \tag{11–1}$$

Note that the equilibrium position is at $x = 0$. Equation 11–1, which is often referred to as Hooke's law (see Sections 6–4 and 9–6), is accurate as long as the spring is not compressed to the point where the coils are about to touch, or stretched beyond the elastic region (see Fig. 9–21). The minus sign indicates that the restoring force is always in the direction opposite to the displacement x. For example, if we choose the positive direction to the right in Fig. 11–1, x is positive when the spring is stretched, but the direction of the restoring force is to the left (negative direction). If the spring is compressed, x is negative (to the left) but the force F acts toward the right (Fig. 11–1c).

The proportionality constant k in Eq. 11–1 is called the "spring constant." In order to stretch the spring a distance x, one has to exert an (external) force on the spring at least equal to $F = +kx$. The greater the value of k, the greater the force needed to stretch a spring a given distance. That is, the stiffer the spring, the greater the spring constant k.

Note that the force F in Eq. 11–1 is *not* a constant, but varies with position. Therefore the acceleration of the mass m is not constant, so we *cannot* use the equations for constant acceleration developed in Chapter 2.

Let us examine what happens when the spring is initially stretched a distance $x = A$, as shown in Fig. 11–2a, and then released. The spring exerts a force on the mass that pulls it toward the equilibrium position. But because the mass has been accelerated by the force, it passes the equilibrium position with considerable speed. Indeed, as the mass reaches the equilibrium position, the force on it decreases to zero, but its speed at this point is a maximum, Fig. 11–2b. As it moves farther to the left, the force on it acts to slow it down, and it stops momentarily at $x = -A$, Fig. 11–2c. It then begins moving back in the opposite direction, Fig. 11–2d, until it reaches the original starting point, $x = A$, Fig. 11–2e. It then repeats the motion, moving back and forth symmetrically between $x = A$ and $x = -A$.

To discuss vibrational motion, we need to define a few terms. The distance x of the mass from the equilibrium point at any moment is called the **displacement**. The maximum displacement—the greatest distance from the equilibrium point—is called the **amplitude**, A. One **cycle** refers to the complete to-and-fro motion from some initial point back to that same point, say from $x = A$ to $x = -A$ back to $x = A$. The **period**, T, is defined as the time required for one complete cycle. Finally, the **frequency**, f, is the number of complete cycles per second. Frequency is generally specified in hertz (Hz), where 1 Hz = 1 cycle per second (s^{-1}). It is easy to see, from their definitions, that frequency and period are inversely related:

$$f = \frac{1}{T} \qquad \text{and} \qquad T = \frac{1}{f}; \qquad\qquad \textbf{(11–2)}$$

for example, if the frequency is 5 cycles per second, then each cycle takes $\frac{1}{5}$ s.

  The oscillation of a spring hung vertically is essentially the same as that of a horizontal spring. Because of the force of gravity, the length of the vertical spring at equilibrium will be longer than when it is horizontal, as shown in Fig. 11–3. The spring is in equilibrium when $\Sigma F = 0 = kx_0 - mg$, so the spring stretches an extra amount $x_0 = mg/k$ to be in equilibrium. If x is measured from this new equilibrium position, Eq. 11–1 can be used directly with the same value of k.

EXAMPLE 11–1   **Car springs.**   When a family of four people with a total mass of 200 kg step into their 1200-kg car, the car's springs compress 3.0 cm. (*a*) What is the spring constant of the car's springs (Fig. 11–4), assuming they act as a single spring? (*b*) How far will the car lower if loaded with 300 kg?

SOLUTION   (*a*) The added force of $(200\,\text{kg})(9.8\,\text{m/s}^2) = 1960\,\text{N}$ causes the springs to compress 3.0×10^{-2} m. Therefore, by Eq. 11–1, the spring constant is

$$k = \frac{F}{x} = \frac{1960\,\text{N}}{3.0 \times 10^{-2}\,\text{m}} = 6.5 \times 10^4\,\text{N/m}.$$

(*b*) If the car is loaded with 300 kg,

$$x = \frac{F}{k} = \frac{(300\,\text{kg})(9.8\,\text{m/s}^2)}{(6.5 \times 10^4\,\text{N/m})} = 4.5 \times 10^{-2}\,\text{m},$$

or 4.5 cm. We could have obtained this answer without solving for k: since x is proportional to F, if 200 kg compresses the spring 3.0 cm, then 1.5 times the force will compress the spring 1.5 times as much, or 4.5 cm.

(a)

(b)

(c)

(d)

(e)

FIGURE 11–2   Force on, and velocity of, mass at different positions of its oscillation.

FIGURE 11–3 (a) Free spring, hung vertically. (b) Mass m attached to spring in new equilibrium position, which occurs when $\Sigma F = 0 = mg - kx_0$.

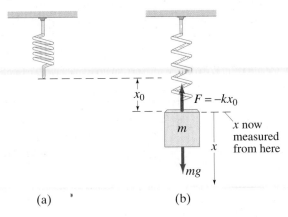

(a)   (b)

FIGURE 11–4   Photo of a car's spring. (Also visible is the shock absorber, in red—see Section 11–5.)

Any vibrating system for which the restoring force is directly proportional to the negative of the displacement (as in Eq. 11–1, $F = -kx$) is said to exhibit **simple harmonic motion** (SHM).[†] Such a system is often called a **simple harmonic oscillator** (SHO). We saw in Section 9–6 that most solid materials stretch or compress according to Eq. 11–1 as long as the displacement is not too great. Because of this, many natural vibrations are simple harmonic, or sufficiently close to it that they can be treated using this SHM model.

SHM
SHO

11–2 Energy in the Simple Harmonic Oscillator

When dealing with forces that are not constant, such as here with simple harmonic motion, it is often convenient and useful to use the energy approach, as we saw in Chapter 6.

To stretch or compress a spring, work has to be done. Hence potential energy is stored in a stretched or compressed spring. Indeed, we have already seen in Section 6–4 that elastic potential energy is given by

$$\text{PE} = \tfrac{1}{2}kx^2.$$

Thus, since the total mechanical energy E of a mass–spring system is the sum of the kinetic and potential energies, we have

$$E = \tfrac{1}{2}mv^2 + \tfrac{1}{2}kx^2, \tag{11–3}$$

where v is the velocity of the mass m when it is a distance x from the equilibrium position. As long as there is no friction, the total mechanical energy E remains constant. As the mass oscillates back and forth, the energy continuously changes from potential energy to kinetic energy, and back again (Fig. 11–5). At the extreme points, $x = A$ and $x = -A$, all the energy is stored in the spring as potential energy (and is the same whether the spring is compressed or stretched to the full amplitude). At these extreme points, the mass stops momentarily as it changes direction, so $v = 0$ and:

$$E = \tfrac{1}{2}m(0)^2 + \tfrac{1}{2}kA^2 = \tfrac{1}{2}kA^2. \tag{11–4a}$$

Thus, the **total mechanical energy of a simple harmonic oscillator is proportional to the square of the amplitude.** At the equilibrium point, $x = 0$, all the energy is kinetic:

$$E = \tfrac{1}{2}mv_0^2 + \tfrac{1}{2}k(0)^2 = \tfrac{1}{2}mv_0^2, \tag{11–4b}$$

where v_0 represents the *maximum* velocity during the motion (which occurs at $x = 0$). At intermediate points, the energy is part kinetic and part potential. By combining Eq. 11–4a with Eq. 11–3, we can find a useful equation for the velocity as a function of the position x:

$$\tfrac{1}{2}mv^2 + \tfrac{1}{2}kx^2 = \tfrac{1}{2}kA^2.$$

Solving for v^2, we have

$$v^2 = \frac{k}{m}(A^2 - x^2) = \frac{k}{m}A^2\left(1 - \frac{x^2}{A^2}\right).$$

From Eqs. 11–4a and 11–4b, we have $\tfrac{1}{2}mv_0^2 = \tfrac{1}{2}kA^2$, so $v_0^2 = (k/m)A^2$.

[†]The word "harmonic" refers to the motion being sinusoidal, which we discuss in Section 11–4. It is "simple" when there is pure sinusoidal motion of a single frequency (rather than a mixture of frequencies as, for example, discussed in Section 11–12 and in Chapter 12).

Total energy of SHO

FIGURE 11–5 Energy changes from kinetic energy to potential energy and back again as the spring oscillates.

(a)

$E = \tfrac{1}{2}kA^2$

$x = -A \quad x = 0 \quad x = A$

(b)

$E = \tfrac{1}{2}mv_0^2$

$x = -A \quad x = 0 \quad x = A$

(c) $E = \tfrac{1}{2}kA^2$

$x = -A \quad x = 0 \quad x = A$

(d)

$E = \tfrac{1}{2}mv^2 + \tfrac{1}{2}kx^2$

$x = -A \quad x = 0 \quad x = A$
x

Inserting this in the equation above and taking the square root, we have

$$v = \pm v_0 \sqrt{1 - x^2/A^2}. \qquad (11\text{-}5)$$

This gives the velocity of the object at any position x. The object moves back and forth, so its velocity can be either in the $+$ or $-$ direction, but its magnitude depends only on the magnitude of x.

CONCEPTUAL EXAMPLE 11–2 **Doubling the amplitude.** Suppose the spring in Fig. 11–5 is stretched twice as far (to $x = 2A$). What happens to (*a*) the energy of the system, (*b*) the maximum velocity, (*c*) the maximum acceleration?

RESPONSE (*a*) From Eq. 11–4a, the energy is related to the square of the amplitude, so stretching it twice as far quadruples the energy. [You may protest: "I did work stretching the spring from $x = 0$ to $x = A$. Don't I do the same work stretching it from A to $2A$?" No. The force you have to exert for the second leg ($x = A$ to $2A$) is more than for the first leg (0 to A), so the work done is more too.]
(*b*) From Eq. 11–4b, we can see that since the energy is quadrupled, the maximum velocity must be double what it was before.
(*c*) Since the force is twice as great when we stretch it twice as far, the acceleration here is also twice as great.

EXAMPLE 11–3 **Spring calculations.** A spring (Fig. 11–6a) stretches 0.150 m when a 0.300-kg mass is hung from it (Fig. 11–6b). The spring is then stretched an additional 0.100 m from this equilibrium point, and released (Fig. 11–6c). Determine (*a*) the spring constant k, (*b*) the amplitude of the oscillation A, (*c*) the maximum velocity v_0, (*d*) the magnitude of the velocity, v, when the mass is 0.050 m from equilibrium, and (*e*) the magnitude of the maximum acceleration of the mass.

SOLUTION (*a*) Since the spring stretches 0.150 m when 0.300 kg is hung from it, we find k from Eq. 11–1 to be

$$k = \frac{F}{x} = \frac{mg}{x} = \frac{(0.300 \text{ kg})(9.80 \text{ m/s}^2)}{0.150 \text{ m}} = 19.6 \text{ N/m}.$$

(*b*) Since the spring is stretched 0.100 m from equilibrium (Fig. 11–6c) and is given no initial speed, $A = 0.100$ m.

0.150 m

0.100 m

m

m

(a)   (b)   (c)

FIGURE 11–6
Examples 11–3 and 11–4.

(c) The maximum velocity v_0 is attained as the mass passes through the equilibrium point where all the energy is kinetic. By conservation of energy (see Eq. 11–3):

$$\tfrac{1}{2} m v_0^2 + 0 = 0 + \tfrac{1}{2} k A^2,$$

where $A = 0.100$ m. Solving for v_0, we have

$$v_0 = A \sqrt{\frac{k}{m}} = (0.100 \text{ m}) \sqrt{\frac{19.6 \text{ N/m}}{0.300 \text{ kg}}} = 0.808 \text{ m/s}.$$

(d) We use Eq. 11–5 and find that

$$v = v_0 \sqrt{1 - x^2/A^2}$$

$$= (0.808 \text{ m/s}) \sqrt{1 - \frac{(0.050 \text{ m})^2}{(0.100 \text{ m})^2}} = 0.700 \text{ m/s}.$$

(e) By Newton's second law, $F = ma$. So the maximum acceleration occurs where the force is greatest—that is, when $x = A = 0.100$ m. Thus

$$a = \frac{kA}{m} = \frac{(19.6 \text{ N/m})(0.100 \text{ m})}{0.300 \text{ kg}} = 6.53 \text{ m/s}^2.$$

EXAMPLE 11–4 **More spring calculations—energy.** For the simple harmonic oscillator of Example 11–3, determine (a) the total energy, and (b) the kinetic and potential energies at half amplitude ($x = \pm A/2$).

SOLUTION (a) Since $k = 19.6$ N/m and $A = 0.100$ m, the total energy E from Eq. 11–4a is

$$E = \tfrac{1}{2} k A^2 = \tfrac{1}{2} (19.6 \text{ N/m})(0.100 \text{ m})^2 = 9.80 \times 10^{-2} \text{ J}.$$

(b) At $x = A/2 = 0.050$ m, we have

$$\text{PE} = \tfrac{1}{2} k x^2 = 2.5 \times 10^{-2} \text{ J},$$

and

$$\text{KE} = E - \text{PE} = 7.3 \times 10^{-2} \text{ J}.$$

11–3 The Period and Sinusoidal Nature of SHM

The period of a simple harmonic oscillator is found to depend on the stiffness of the spring and also on the mass m that is oscillating. But—strange as it may seem—the *period does not depend on the amplitude*. You can find this out for yourself by using a watch and timing 10 or 20 cycles of an oscillating spring for a small amplitude and then for a large amplitude.

We can derive a formula for the period of simple harmonic motion (SHM), and this can be done by comparing SHM to an object rotating in a circle. From this same "reference circle" we can obtain a second useful result—a formula for the position of an oscillating mass as a function of time. There is, of course, nothing rotating in a circle when a spring oscillates linearly, but it is the mathematical similarity here that we find useful.

Consider now a mass m revolving counterclockwise in a circle of radius A, with constant speed v_0, on top of a table as shown in Fig. 11–7. As viewed from above, the motion is a circle in the xy plane. But a person who looks at the motion from the edge of the table sees an oscillatory motion back and forth, and this one-dimensional motion corresponds precisely to simple harmonic motion, as we shall now see. What the person sees, and what we are interested in, is the projection of the circular motion onto the x axis (Fig. 11–7b). To see that this x-motion is analogous to SHM, let us calculate the magnitude of the x component of the velocity v_0, which is labeled v in Fig. 11–7. The two triangles involving θ in Fig. 11–7 are similar, so

$$\frac{v}{v_0} = \frac{\sqrt{A^2 - x^2}}{A}$$

or

$$v = v_0 \sqrt{1 - \frac{x^2}{A^2}}.$$

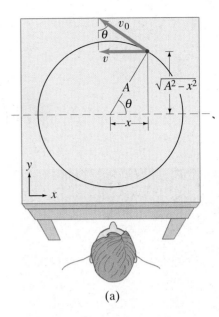

(a)

This is exactly the equation for the speed of a mass oscillating with SHM, as we saw in Eq. 11–5. Thus the projection on the x axis of an object revolving in a circle has the same motion as a mass at the end of a spring.

We can now determine the period of SHM because it is equal to that of the revolving object making one complete revolution. First we note that the velocity v_0 is equal to the circumference of the circle (distance) divided by the period T:

$$v_0 = \frac{2\pi A}{T} = 2\pi A f. \tag{11–6}$$

We solve for the period T:

$$T = \frac{2\pi A}{v_0}.$$

From Eqs. 11–4a and b, we have $\frac{1}{2}kA^2 = \frac{1}{2}mv_0^2$, so $A/v_0 = \sqrt{m/k}$. Thus

$$T = 2\pi \sqrt{\frac{m}{k}}. \tag{11–7a}$$

(b)

FIGURE 11–7 Analysis of simple harmonic motion as a side view (b) of circular motion (a).

Period T of SHO

This is the formula we were looking for. The period depends on the mass m and the spring constant k, but not on the amplitude. We see from Eq. 11–7a that the greater the mass, the longer the period; and the stiffer the spring, the shorter the period. This makes sense since a greater mass means more inertia and therefore slower response (smaller acceleration). And larger k means greater force and therefore quicker response (larger acceleration). Notice that Eq. 11–7a is not a direct proportion: the period varies as the *square root* of m/k. For example, the mass must be quadrupled to double the period. Equation 11–7a is fully in accord with experiment and is valid not only for a spring, but for all kinds of simple harmonic motion that obey Eq. 11–1.

We can also write, since $f = 1/T$ (Eq. 11–2), that

$$f = \frac{1}{T} = \frac{1}{2\pi} \sqrt{\frac{k}{m}}. \tag{11–7b}$$

EXAMPLE 11–5 **Car springs again.** What are the period and frequency of the car in Example 11–1 after hitting a bump? Assume the shock absorbers are poor, so the car really oscillates up and down.

SOLUTION From Eq. 11–7a,

$$T = 2\pi \sqrt{\frac{m}{k}} = 6.28 \sqrt{\frac{1400 \text{ kg}}{6.5 \times 10^4 \text{ N/m}}} = 0.92 \text{ s},$$

or slightly less than a second. The frequency $f = 1/T = 1.09$ Hz (Eq. 11–2).

EXAMPLE 11–6 **Spider web.** A small insect of mass 0.30 g is caught in a spiderweb of negligible mass (Fig. 11–8). The web vibrates predominantly with a frequency of 15 Hz. (a) Estimate the value of the spring constant k for the web. (b) At what frequency would you expect the web to vibrate if an insect of mass 0.10 g were trapped?

SOLUTION (a) We use Eq. 11–7b,

$$f = \frac{1}{2\pi} \sqrt{\frac{k}{m}},$$

and solve for k:

$$k = (2\pi f)^2 m$$

$$= (6.28 \times 15 \text{ s}^{-1})^2 (3.0 \times 10^{-4} \text{ kg}) = 2.7 \text{ N/m}.$$

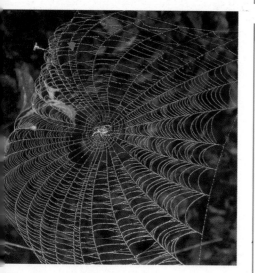

FIGURE 11–8 A spider waits for its prey (Example 11–6) at the center of its web.

(b) We could substitute $m = 1.0 \times 10^{-4}$ kg in the above equation for f. Instead, we notice that the frequency decreases with the square root of the mass. Since the new mass is one-third the first mass, the frequency increases by a factor of $\sqrt{3}$. That is, $f = (15 \text{ Hz})(\sqrt{3}) = 26$ Hz.

We now use the reference circle to find the position of a mass undergoing simple harmonic motion as a function of time. From Fig. 11–7, we see that $\cos\theta = x/A$, so the projection of the ball's position on the x axis is

$$x = A \cos\theta.$$

Since the mass is rotating with angular velocity ω, we can write $\theta = \omega t$ (see Section 8–1), where θ is in radians. Thus

Position as a function of time (SHM)

$$x = A \cos\omega t. \tag{11–8a}$$

Furthermore, since the angular velocity ω (specified in radians per second) can be written as $\omega = 2\pi f$, where f is the frequency (Eq. 8–7), we also can write

$$x = A \cos 2\pi ft, \tag{11–8b}$$

or in terms of the period T:

$$x = A \cos\frac{2\pi t}{T}. \tag{11–8c}$$

Notice in Eq. 11–8c that when $t = T$ (that is, after a time equal to one period), we have the cosine of 2π, which is the same as the cosine of zero. This makes sense since the motion repeats itself after a time T.

As we have seen, the projection on the x axis of the rotating object corresponds precisely to the motion of a simple harmonic oscillator. Thus,

Eqs. 11–8 give the position of an object oscillating under simple harmonic motion. Since the cosine function varies between 1 and −1, then x varies between A and −A, as it must. If a pen is attached to a vibrating mass as shown in Fig. 11–9 and a sheet of paper is moved at a steady rate beneath it, a curve will be drawn that accurately follows Eqs. 11–8. Figure 11–10a shows a graph of x vs. t, Eq. 11–8.

We can also find the velocity v as a function of time from Fig. 11–7a. For the point shown (red dot), we see that the magnitude of v is $v_0 \sin \theta$, but **v** points to the left, so $v = -v_0 \sin \theta$. Again setting $\theta = \omega t = 2\pi f t = 2\pi t/T$, we have

$$v = -v_0 \sin 2\pi f t = -v_0 \sin \frac{2\pi t}{T}. \qquad \textbf{(11–9)}$$

At $t = 0$ the velocity is negative (points to the left) and remains so until $t = \frac{1}{2}T$ (corresponding to $\theta = 180° = \pi$ radians). Then from $t = \frac{1}{2}T$ to $t = T$ the velocity is positive. The velocity as a function of time (Eq. 11–9) is plotted in Fig. 11–10b. Recall that $v_0 = 2\pi A f$, Eq. 11–6, which we can rewrite using Eq. 11–7b as

$$v_0 = 2\pi A f = A\sqrt{\frac{k}{m}}.$$

The acceleration as a function of time is readily obtained from Newton's second law, $F = ma$:

$$a = \frac{F}{m} = \frac{-kx}{m} = -\left(\frac{kA}{m}\right)\cos 2\pi f t = -a_0 \cos 2\pi f t$$
$$= -a_0 \cos \frac{2\pi}{T} t \qquad \textbf{(11–10)}$$

where the maximum acceleration is $a_0 = kA/m$. Equation 11–10 is plotted in Fig. 11–10c. Because the acceleration of a SHO is *not* constant, the equations for uniformly accelerated motion do *not* apply to SHM.

FIGURE 11–9 Sinusoidal nature of SHM as a function of time; in this case, $x = A \cos (2\pi t/T)$.

(a)

(b)

(c)

FIGURE 11–10 Graphs showing (a) displacement x as a function of time t: $x = A \cos (2\pi t/T)$; (b) velocity as a function of time: $v = -v_0 \sin (2\pi t/T)$; (c) acceleration as a function of time, $a = -(kA/m)\cos (2\pi t/T)$.

Other equations for simple harmonic motion are also possible, depending on the initial conditions (or on when you choose t to be zero). For example, if at $t = 0$ the oscillations are begun by giving the mass a push when it is at the equilibrium position, the equation would be

$$x = A \sin \frac{2\pi t}{T}.$$

This curve (Fig. 11–11) has precisely the same shape as the cosine curve shown in Fig. 11–10a, except it is shifted to the right by a quarter cycle so that it starts out at $x = 0$ instead of at $x = A$.

SHM is sinusoidal

Both of the curves, sine and cosine, are referred to as being **sinusoidal** (having the shape of a sine function). Thus simple harmonic motion[†] is said to be sinusoidal because the position varies as a sinusoidal function of time.

FIGURE 11–11 Sinusoidal nature of SHM as a function of time; in this case, $x = A \sin (2\pi t/T)$ because at $t = 0$ the mass is at the equilibrium position $x = 0$, but it also has (or is given) an initial speed at $t = 0$ that carries it to $x = A$ at $t = \frac{1}{4}T$.

EXAMPLE 11–7 **Loudspeaker.** The cone of a loudspeaker vibrates in SHM at a frequency of 262 Hz ("middle C"). The amplitude at the center of the cone is $A = 1.5 \times 10^{-4}$ m, and at $t = 0$, $x = A$. (a) What is the equation describing the motion of the center of the cone? (b) What is its maximum velocity and maximum acceleration? (c) What is the position of the cone at $t = 1.00$ ms ($= 1.00 \times 10^{-3}$ s)?

SOLUTION (a) The amplitude $A = 1.5 \times 10^{-4}$ m and $2\pi f = (6.28 \text{ rad})(262 \text{ s}^{-1}) = 1650$ rad/s. The motion begins ($t = 0$) with the cone at its maximum displacement ($x = A$ at $t = 0$), so we use the cosine function:

$$x = A \cos 2\pi ft = (1.5 \times 10^{-4} \text{ m}) \cos (1650\, t).$$

(b) From Eq. 11–6,

$$v_0 = 2\pi Af = 2\pi(1.5 \times 10^{-4} \text{ m})(262 \text{ s}^{-1}) = 0.25 \text{ m/s}.$$

From Eqs. 11–10 and 11–7b

$$a_0 = \left(\frac{k}{m}\right)A = (2\pi f)^2 A$$

$$= 4\pi^2(262 \text{ s}^{-1})^2(1.5 \times 10^{-4} \text{ m}) = 410 \text{ m/s}^2,$$

which is more than 40 g's.
(c) At $t = 1.00 \times 10^{-3}$ s

$$x = (1.5 \times 10^{-4} \text{ m}) \cos [(1650 \text{ rad/s})(1.00 \times 10^{-3} \text{ s})]$$

$$= (1.5 \times 10^{-4} \text{ m}) \cos (1.65 \text{ rad}) = -1.2 \times 10^{-5} \text{ m}.$$

[†]Simple harmonic motion can be *defined* as motion that is sinusoidal. This definition is fully consistent with our earlier definition in Section 11–1.

11–4 The Simple Pendulum

A **simple pendulum** consists of a small object (the pendulum bob) suspended from the end of a lightweight cord, Fig. 11–12. We assume that the cord doesn't stretch and that its mass can be ignored relative to that of the bob. The motion of a simple pendulum moving back and forth (Fig. 11–13) with negligible friction resembles simple harmonic motion: the pendulum oscillates along the arc of a circle with equal amplitude on either side of its equilibrium point (where it hangs vertically) and as it passes through the equilibrium point it has its maximum speed. But is it really undergoing SHM? That is, is the restoring force proportional to its displacement? Let us find out.

The displacement of the pendulum along the arc is given by $x = L\theta$, where θ is the angle the cord makes with the vertical and L is the length of the cord (Fig. 11–12). Thus, if the restoring force is proportional to x or to θ, the motion will be simple harmonic. The restoring force is the component of the weight, mg, tangent to the arc:

$$F = -mg \sin \theta,$$

where the minus sign, as in Eq. 11–1, means the force is in the direction opposite to the angular displacement θ. Since F is proportional to the sine of θ and not to θ itself, the motion is *not* SHM. However, if θ is small, then $\sin \theta$ is very nearly equal to θ when the latter is specified in radians. This can be seen by looking at the trigonometry table inside the back cover, or by noting in Fig. 11–12 that the arc length x $(= L\theta)$ is nearly the same length as the chord $(= L \sin \theta)$ indicated by the straight dashed line, *if θ is small*. For angles less than 15°, the difference between θ (in radians) and $\sin \theta$ is less than 1 percent. Thus, to a very good approximation for small angles,

$$F = -mg \sin \theta \approx -mg\theta.$$

Using $x = L\theta$, we have

$$F \approx -\frac{mg}{L}x.$$

Thus, for small displacements, the motion is essentially simple harmonic, since this equation fits Hooke's law, $F = -kx$, where the effective force constant is $k = mg/L$. The period of a simple pendulum can be found using Eq. 11–7, where for k we substitute mg/L:

$$T = 2\pi \sqrt{\frac{m}{mg/L}}$$

$$T = 2\pi \sqrt{\frac{L}{g}}; \qquad \text{[}\theta \text{ small]} \quad \textbf{(11–11a)}$$

and the frequency[†] is

$$f = \frac{1}{T} = \frac{1}{2\pi}\sqrt{\frac{g}{L}}. \qquad \text{[}\theta \text{ small]} \quad \textbf{(11–11b)}$$

A surprising result is that the period does not depend on the mass of the pendulum bob! You may have noticed this if you pushed a small child and a large one on the same swing.

[†]Equations 11–11 apply to a simple pendulum—a concentrated mass at the end of a string of negligible mass—but not to the oscillation of, say, a baseball bat suspended from one end.

FIGURE 11–12 Simple pendulum.

FIGURE 11–13 Strobe-light photo of an oscillating pendulum.

FIGURE 11-14 The swinging motion of this lamp, hanging by a very long cord from the ceiling of the cathedral at Pisa, is said to have been observed by Galileo and to have inspired him to the conclusion that the period of a pendulum does not depend on amplitude.

EXAMPLE 11–8 **Grandfather clock.** (*a*) Estimate the length of the pendulum in a grandfather clock that ticks once per second. (*b*) What would be the period of a clock with a 1.0-m-long pendulum?

SOLUTION If we assume the clock "ticks" once per cycle, then the period $T = 1.0\,\text{s}$. We solve Eq. 11–11a for L:

$$L = \frac{T^2}{4\pi^2}\,g = \frac{(1.0\,\text{s})^2}{4\pi^2}(9.8\,\text{m/s}^2) = 0.25\,\text{m}$$

(*b*) From Eq. 11–11a, $T \propto \sqrt{L}$. So if L is four times as great ($4 \times 0.25\,\text{m} = 1.0\,\text{m}$), then T will be $\sqrt{4} = 2$ times larger, or $2.0\,\text{s}$.

We saw in Section 11–3 that the period of an object undergoing SHM, including a pendulum, does not depend on the amplitude. Galileo is said to have first noted this fact while watching a swinging lamp in the cathedral at Pisa (Fig. 11–14). This discovery led to the pendulum clock,[†] the first really precise timepiece, which became the standard for centuries.

Because a pendulum does not undergo *precisely* SHM, the period does depend slightly on the amplitude, the more so for large amplitudes. The accuracy of a pendulum clock would be affected, after many swings, by the decrease in amplitude due to friction; but the mainspring in a pendulum clock (or the falling weight in a grandfather clock) supplies energy to compensate for the friction and to maintain the amplitude constant, so that the timing remains accurate.

11–5 Damped Harmonic Motion

The amplitude of any real oscillating spring or swinging pendulum slowly decreases in time until the oscillations stop altogether. Figure 11–15 shows a typical graph of the displacement as a function of time. This is called **damped harmonic motion**. The damping[‡] is generally due to the resistance of air and to internal friction within the oscillating system. The energy that is thus dissipated to thermal energy is reflected in a decreased amplitude of oscillation.

Since natural oscillating systems are damped in general, why do we even talk about (undamped) simple harmonic motion? The answer is that SHM is much easier to deal with mathematically. And if the damping is not large, the oscillations can be thought of as simple harmonic motion on which the damping is superposed—that is, the decrease in amplitude rep-

[†] It is not known whether Galileo actually built a pendulum clock. Christiaan Huygens (1629–1695) did build one more than 10 years after Galileo's death.
[‡] To "damp" means to diminish, restrain, or extinguish, as to "dampen one's spirits."

FIGURE 11-15 Damped harmonic motion.

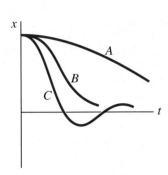

FIGURE 11–16 Graphs that represent (A) overdamped, (B) critically damped, and (C) underdamped oscillatory motion.

Piston

Viscous fluid

FIGURE 11–17 Automobile spring and shock absorber to provide damping so that car won't bounce up and down endlessly.

resented by the dashed curves in Fig. 11–15. Although frictional damping does alter the frequency of vibration, the effect is usually small unless the damping is large; thus Eqs. 11–7 can still be used in most cases.

Sometimes the damping is so large that the motion no longer resembles simple harmonic motion. Three common cases of heavily damped systems are shown in Fig. 11–16. Curve A represents the **overdamped** situation, when the damping is so large that it takes a long time to reach equilibrium. Curve C represents an **underdamped** situation in which the system makes several swings before coming to rest. Curve B represents **critical damping**; in this case equilibrium is reached the quickest. These terms all derive from the use of practical damped systems such as door-closing mechanisms and shock absorbers in a car (Fig. 11–17). Such devices are usually designed to give critical damping; but as they wear out, underdamping occurs: a door slams or a car bounces up and down several times each time it hits a bump.

11–6 Forced Vibrations; Resonance

When a vibrating system is set into motion, it vibrates at its natural frequency (Eqs. 11–7b and 11–11b). However, a system may have an external force applied to it that has its own particular frequency and then we have a **forced vibration**. For example, we might pull the mass on the spring of Fig. 11–1 back and forth at a frequency f. The mass then vibrates at the frequency f of the external force, even if this frequency is different from the **natural frequency** of the spring, which we will now denote by f_0 where (see Eq. 11–7b)

$$f_0 = \frac{1}{2\pi} \sqrt{\frac{k}{m}}.$$

For a forced vibration, the amplitude of vibration is found to depend on the difference between f and f_0, and is a maximum when the

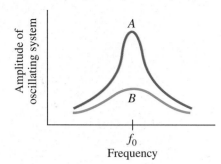

FIGURE 11–18 Resonance for lightly damped (*A*) and heavily damped (*B*) systems.

➡ **PHYSICS APPLIED**

Swinging

➡ **PHYSICS APPLIED**

Resonant collapse

frequency of the external force equals the natural frequency of the system—that is, when $f = f_0$. The amplitude is plotted in Fig. 11–18 as a function of the external frequency f. Curve *A* represents light damping and curve *B* heavy damping. The amplitude can become large when the driving frequency f is near the natural frequency, $f \approx f_0$, as long as the damping is not too large. When the damping is small, the increase in amplitude near $f = f_0$ is very large (and often dramatic). This effect is known as **resonance**. The natural vibrating frequency f_0 of a system is called its **resonant frequency**.

A simple illustration of resonance is pushing a child on a swing. A swing, like any pendulum, has a natural frequency of oscillation. If you push on the swing at a random frequency, the swing bounces around and reaches no great amplitude. But if you push with a frequency equal to the natural frequency of the swing, the amplitude increases greatly. This clearly illustrates that at resonance, relatively little effort is required to obtain a large amplitude.

The great tenor Enrico Caruso was said to be able to shatter a crystal goblet by singing a note of just the right frequency at full voice. This is an example of resonance, for the sound waves emitted by the voice act as a forced vibration on the glass. At resonance, the resulting vibration of the goblet may be large enough in amplitude that the glass exceeds its elastic limit and breaks.

Since material objects are, in general, elastic, resonance is an important phenomenon in a variety of situations. It is particularly important in building, although the effects are not always foreseen. For example, it has been reported that a railway bridge collapsed because a nick in one of the wheels of a crossing train set up a resonant vibration in the bridge. Indeed, marching soldiers break step when crossing a bridge to avoid the possibility of a similar catastrophe. And the famous collapse of the Tacoma Narrows Bridge (Fig. 11–19a) in 1940 was due in part to resonance of the bridge (in tune with wind gusts), as was the Oakland freeway collapse in the 1989 California earthquake (Fig. 11–19b).

We will meet important examples of resonance later in this chapter and in succeeding chapters. We will also see that vibrating objects often have not one, but many resonant frequencies.

FIGURE 11–19 (a) Large-amplitude oscillations of the Tacoma Narrows Bridge, due to gusty winds, led to its collapse (November 7, 1940). (b) Collapse of freeway in California, due to the 1989 earthquake, in which resonance played a part.

(a)

(b)

11–7 Wave Motion

When you throw a stone into a lake or pool of water, circular waves form and move outward, Fig. 11–20. Waves will also travel along a cord (or a "slinky") that is stretched out straight on a table if you vibrate one end back and forth as shown in Fig. 11–21. Water waves and waves on a cord are two common examples of wave motion. We will meet other kinds of wave motion later, but for now we will concentrate on these **mechanical waves**.

If you have ever watched ocean waves moving toward shore (and before they break), you may have wondered if the waves were carrying water from out at sea into the beach. This is, in fact, not the case.[†] Water waves move with a recognizable velocity. But each particle of the water itself merely oscillates about an equilibrium point. This is clearly demonstrated by observing leaves on a pond as waves move by. The leaves (or a cork) are not carried forward by the waves, but simply oscillate about an equilibrium point because this is the motion of the water itself.

| **CONCEPTUAL EXAMPLE 11-9** | **Wave vs. particle velocity.** Is the velocity of a wave moving along a cord the same as the velocity of a particle of the cord? See Fig. 11–21.

SOLUTION No. The two velocities are different, both in magnitude and direction. The wave on the rope of Fig. 11–21 moves to the right, but each piece of the rope only vibrates to and fro. (The rope clearly does not travel in the direction that the wave on it does.)

Waves can move over large distances, but the medium (the water or the rope) itself has only a limited movement. Thus, although a wave is not matter, the wave pattern can travel in matter. A wave consists of oscillations that move without carrying matter with them.

Waves are moving oscillations, not carrying matter along

[†]Do not be confused by the "breaking" of ocean waves, which occurs when a wave interacts with the ground in shallow water and hence is no longer a simple wave.

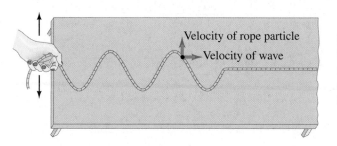

Velocity of rope particle
Velocity of wave

FIGURE 11-21 Wave traveling on a cord. The wave travels to the right along the cord. Particles of the cord oscillate back and forth on the tabletop.

Waves carry energy from one place to another. Energy is given to a water wave, for example, by a rock thrown into the water, or by wind far out at sea. The energy is transported by waves to the shore. The oscillating hand in Fig. 11–21 transfers energy to the rope, which is then transported down the rope and can be transferred to an object at the other end. All forms of traveling waves transport energy.

Wave "pulse"

Let us look a little more closely at how a wave is formed and how it comes to "travel." We first look at a single wave bump or **pulse**. A single pulse can be formed on a rope by a quick up-and-down motion of the hand, Fig. 11–22. The hand pulls up on one end of the rope, and because the end piece is attached to adjacent pieces, these also feel an upward force and they, too, begin to move upward. As each succeeding piece of rope moves upward, the wave crest moves outward along the rope. Meanwhile, the end piece of rope has been returned to its original position by the hand, and as each succeeding piece of rope reaches its peak position, it, too, is pulled back down again by the adjacent section of rope. Thus the source of a traveling wave pulse is a disturbance, and cohesive forces between adjacent pieces of rope cause the pulse to travel outward. Waves in other media are created and propagate outward in a similar fashion.

Periodic wave

A **continuous** or **periodic wave**, such as that shown in Fig. 11–21, has as its source a disturbance that is continuous and oscillating; that is, the source is a *vibration* or *oscillation*. In Fig. 11–21, a hand oscillates one end of the rope. Water waves may be produced by any vibrating object placed at the surface, such as your hand; or the water itself is made to vibrate when wind blows across it or a rock is thrown into it. A vibrating tuning fork or drum membrane gives rise to sound waves in air. And we will see later that oscillating electric charges give rise to light waves. Indeed, almost any vibrating object sends out waves.

The source of any wave, then, is a vibration. And it is a *vibration* that propagates outward and thus constitutes the wave. If the source vibrates sinusoidally in SHM, then the wave itself—if the medium is perfectly elastic—will have a sinusoidal shape both in space and in time. (1) In space: if you take a picture of the wave spread throughout space at a given instant of time, the wave will have the shape of a sine or cosine function. (2) In time: if you look at the motion of the medium at one place over a long period of time—for example, if you look between two closely spaced posts

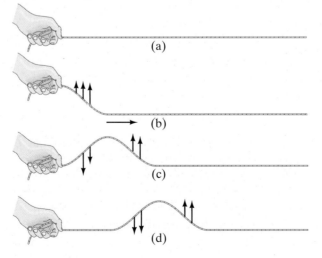

FIGURE 11–22 Motion of a wave pulse to the right. Arrows indicate velocity of cord particles.

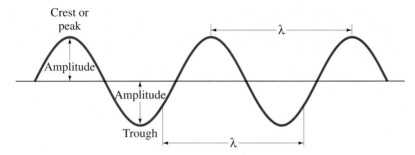

FIGURE 11–23
Characteristics of a single-frequency continuous wave.

of a pier or out of a ship's porthole as water waves pass by—the up-and-down motion of that small segment of water will be simple harmonic motion—the water moves up and down sinusoidally in time.

Some of the important quantities used to describe a periodic sinusoidal wave are shown in Fig. 11–23. The high points on a wave are called crests, the low points troughs. The **amplitude** is the maximum height of a crest, or depth of a trough, relative to the normal (or equilibrium) level. The total swing from a crest to a trough is twice the amplitude. The distance between two successive crests is called the **wavelength**, λ (the Greek letter lambda). The wavelength is also equal to the distance between *any* two successive identical points on the wave. The **frequency**, f, is the number of crests—or complete cycles—that pass a given point per unit time. The period, T, of course, is just $1/f$, and is the time elapsed between two successive crests passing by the same point in space.

Amplitude

Wavelength, λ

Frequency, f

The **wave velocity**, v, is the velocity at which wave crests (or any other part of the waveform) move. The wave velocity must be distinguished from the velocity of a particle of the medium itself. For example, for a wave traveling along a cord as in Fig. 11–21, the wave velocity is to the right, along the cord, whereas the velocity of a particle of the cord is perpendicular to it.

Wave velocity (don't confuse with the velocity of a particle)

A wave crest travels a distance of one wavelength, λ, in one period, T. Thus the wave velocity is equal to λ/T: $v = \lambda/T$. Then, since $1/T = f$:

$$v = \lambda f. \qquad (11\text{–}12)$$

$v = \lambda f$ (sinusoidal waves)

For example, suppose a wave has a wavelength of 5 m and a frequency of 3 Hz. Since three crests pass a given point per second, and the crests are 5 m apart, the first crest (or any other part of the wave) must travel a distance of 15 m during the 1 s. So its speed is 15 m/s.

The velocity of a wave depends on the properties of the medium in which it travels. The velocity of a wave on a stretched string, for example, depends on the tension in the string, F_T, and on the string's mass per unit length, m/L. For waves of small amplitude, the relationship is

$$v = \sqrt{\frac{F_T}{m/L}}. \qquad \text{[wave on a string]} \qquad (11\text{–}13)$$

This formula makes sense qualitatively on the basis of Newtonian mechanics. That is, we expect the tension to be in the numerator and the mass per unit length in the denominator. Why? Because when the tension is greater, we expect the velocity to be greater since each segment of string is in tighter contact with its neighbor; and the greater the mass per unit length, the more inertia the string has and the more slowly the wave would be expected to propagate.

EXAMPLE 11–10 **Wave on a wire.** A wave whose wavelength is 0.30 m is traveling down a 300-m-long wire whose total mass is 15 kg. If the wire is under a tension of 1000 N, what is the velocity and frequency of this wave?

SOLUTION From Eq. 11–13, the velocity is

$$v = \sqrt{\frac{1000 \text{ N}}{(15 \text{ kg})/(300 \text{ m})}} = 140 \text{ m/s}.$$

The frequency then is (Eq. 11–12)

$$f = \frac{v}{\lambda} = \frac{140 \text{ m/s}}{0.30 \text{ m}} = 470 \text{ Hz}.$$

Note that a higher tension would increase both v and f, whereas a thicker, denser wire would reduce v and f.

11–8 Types of Waves: Transverse and Longitudinal

We saw earlier that although waves may travel over long distances, the particles of the medium vibrate only over a limited region of space. When a wave travels down a rope, say from left to right, the particles of the rope vibrate up and down in a direction transverse (or perpendicular) to the motion of the wave itself. Such a wave is called a **transverse wave**. There exists another type of wave known as a **longitudinal wave**. In a longitudinal wave, the vibration of the particles of the medium is along the *same* direction as the motion of the wave. Longitudinal waves are readily formed on a stretched spring or "slinky" by alternately compressing and expanding one end. This is shown in Fig. 11–24b, and can be compared to the transverse wave in Fig. 11–24a. A series of compressions and expansions propagate along the spring. The *compressions* are those areas where the coils are momentarily close together. *Expansions* (sometimes called *rarefactions*) are regions where the coils are momentarily far apart. Compressions and expansions correspond to the crests and troughs of a transverse wave.

Transverse and longitudinal waves

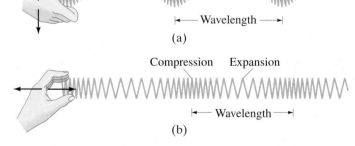

FIGURE 11–24 (a) Transverse wave; (b) longitudinal wave.

An important example of a longitudinal wave is a sound wave in air. A vibrating drumhead, for example, alternately compresses and rarefies the air, producing a longitudinal wave that travels outward in the air, as shown in Fig. 11–25.

As in the case of transverse waves, each section of the medium in which a longitudinal wave passes oscillates over a very small distance, whereas the wave itself can travel large distances. Wavelength, frequency, and wave velocity all have meaning for a longitudinal wave. The wavelength is the distance between successive compressions (or between successive expansions), and frequency is the number of compressions that pass a given point per second. The wave velocity is the velocity with which each compression appears to move; and it is equal to the product of wavelength and frequency (Eq. 11–12).

A longitudinal wave can be represented graphically by plotting the density of air molecules (or coils of a slinky) versus position at a given instant, as shown in Fig. 11–26. We will often use such a graphical representation because it is much easier to illustrate what is happening. Note that the graph looks much like a transverse wave.

The velocity of a longitudinal wave has a form similar to that for a transverse wave on a string (Eq. 11–13); that is

$$v = \sqrt{\frac{\text{elastic force factor}}{\text{inertia factor}}}.$$

In particular, for a longitudinal wave traveling down a long solid rod,

$$v = \sqrt{\frac{E}{\rho}}, \qquad\qquad\qquad \textbf{(11–14a)}$$

where E is the elastic modulus (Section 9–6) of the material and ρ is its density. For a longitudinal wave traveling in a liquid or gas,

$$v = \sqrt{\frac{B}{\rho}}, \qquad\qquad\qquad \textbf{(11–14b)}$$

where B is the bulk modulus (Section 9–6) and ρ the density.

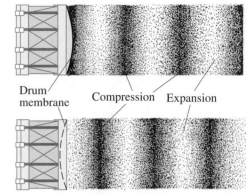

FIGURE 11–25 Production of a sound wave, which is longitudinal, shown at two moments in time, about a half period ($\frac{1}{2}T$) apart.

Drum membrane   Compression   Expansion

(a)

(b)

FIGURE 11–26 (a) A longitudinal wave with (b) its graphical representation, at a particular instant in time.

Density of air

High

Normal

Low

x

Sound velocity in a steel rail. You can often hear a distant train approaching by putting your ear to the track. How long does it take for the wave to travel down the steel track if the train is 1.0 km away?

SOLUTION Referring to Tables 9–1 and 10–1 for the elastic modulus and density of steel, respectively, we have

$$v = \sqrt{\frac{2.0 \times 10^{11} \text{ N/m}^2}{7.8 \times 10^3 \text{ kg/m}^3}} = 5.1 \times 10^3 \text{ m/s}.$$

Then the time $t = \text{distance/velocity} = (1.0 \times 10^3 \text{ m})/(5.1 \times 10^3 \text{ m/s}) = 0.20$ s.

⇒ **PHYSICS APPLIED**

Earthquake waves

Both transverse and longitudinal waves are produced when an *earthquake* occurs. The transverse waves that travel through the body of the Earth are called *S* waves (*S* for shear) and the longitudinal waves are called *P* waves (*P* for pressure). Both longitudinal and transverse waves can travel through a solid since the atoms or molecules can vibrate about their relatively fixed positions in any direction. But in a fluid, only longitudinal waves can propagate, because any transverse motion would experience no restoring force since a fluid can flow. This fact was used by geophysicists to infer that the Earth's outer core is molten: longitudinal waves are detected diametrically across the Earth, but not transverse waves. The accepted explanation is that the core of the Earth must be liquid.

Besides these two types of waves that can pass through the body of the Earth (or other substance), there can also be *surface waves* that travel along the boundary between two materials. A wave on water is actually a surface wave that moves on the boundary between water and air. The motion of each particle of water at the surface is circular or elliptical (Fig. 11–27), so it is a combination of transverse and longitudinal motions. Below the surface, there is also transverse plus longitudinal wave motion, as shown. At the bottom, the motion is only longitudinal. (When a wave approaches shore, the water drags at the bottom and is slowed down, while the crests move ahead at higher speed (Fig. 11–28) and "spill" over the top.)

Surface waves are also set up on the Earth when an earthquake occurs. The waves that travel along the surface are mainly responsible for the damage caused by earthquakes.

FIGURE 11–27 A water wave is an example of a *surface wave*, which is a combination of transverse and longitudinal wave motions.

FIGURE 11–28 How a wave breaks. The green arrows represent the local velocity of water molecules.

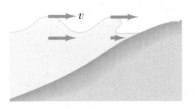

11–9 Energy Transported by Waves

Waves transport energy from one place to another. As waves travel through a medium, the energy is transferred as vibrational energy from particle to particle of the medium. For a sinusoidal wave of frequency f, the particles move in SHM as a wave passes, so each particle has an energy $E = \frac{1}{2}kA^2$, where A is the amplitude of its motion, either transversely or longitudinally. (See Eq. 11–4a.)

Thus, we have the important result that the **energy transported by a wave is proportional to the square of the amplitude**. The **intensity** I of a wave is defined as the power (energy per unit time) transported across unit area perpendicular to the direction of energy flow:

Wave energy $\propto$ (Amplitude)²

$$I = \frac{\text{energy/time}}{\text{area}} = \frac{\text{power}}{\text{area}}.$$

Intensity

Since the energy is proportional to the wave amplitude squared as we just saw, so too is the intensity:

$$I \propto A^2. \tag{11–15}$$

If a wave flows out from the source in all directions, it is a three-dimensional wave. Examples are sound traveling in the open air, earthquake waves, and light waves. If the medium is isotropic (same in all directions), the wave is said to be a *spherical wave* (Fig. 11–29). As the wave moves outward, the energy it carries is spread over a larger and larger area since the surface area of a sphere of radius r is $4\pi r^2$. Thus the intensity of a wave is

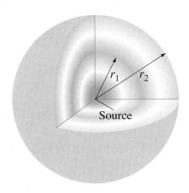

$$I = \frac{\text{power}}{\text{area}} = \frac{P}{4\pi r^2}. \tag{11–16a}$$

If the power output P of the source is constant, then the intensity decreases as the inverse square of the distance from the source:

$$I \propto \frac{1}{r^2}. \tag{11–16b}$$

FIGURE 11–29 Wave traveling outward from source has spherical shape. Two different crests (or compressions) are shown, of radius r_1 and r_2.

If we consider two points at distances r_1 and r_2 from the source, as in Fig. 11–29, then $I_1 = P/4\pi r_1^2$ and $I_2 = P/4\pi r_2^2$, so

$$\frac{I_2}{I_1} = \frac{r_1^2}{r_2^2}. \tag{11–16c}$$

$I \propto \dfrac{1}{r^2}$

Thus, for example, when the distance doubles ($r_2/r_1 = 2$), then the intensity is reduced to $\frac{1}{4}$ its earlier value: $I_2/I_1 = (\frac{1}{2})^2 = \frac{1}{4}$.

The amplitude of a wave also decreases with distance. Since the intensity is proportional to the square of the amplitude (Eq. 11–15), then the amplitude A must decrease as $1/r$ so that $I \propto A^2$ will be proportional to $1/r^2$ (as in Eq. 11–16b), so:

$$A \propto \frac{1}{r}.$$

If we consider again two different distances from the source, r_1 and r_2, then:

$$\frac{A_2}{A_1} = \frac{r_1}{r_2}.$$

When the wave is twice as far from the source, the amplitude is half as large, and so on (ignoring damping due to friction).

Earthquake intensity. If the intensity of an earthquake P wave 100 km from the source is 1.0×10^6 W/m², what is the intensity 400 km from the source?

SOLUTION The intensity decreases as the square of the distance from the source. Therefore, at 400 km, the intensity will be $\left(\frac{1}{4}\right)^2 = \frac{1}{16}$ of its value at 100 km, or 6.2×10^4 W/m². Alternatively, Eq. 11–16b could be used: $I_2 = I_1 r_1^2 / r_2^2 = (1.0 \times 10^6 \text{ W/m}^2)(100 \text{ km})^2 / (400 \text{ km})^2 = 6.2 \times 10^4$ W/m².

The situation is different for a one-dimensional wave, such as a transverse wave on a string or a longitudinal wave pulse traveling down a thin uniform metal rod. The area remains constant, so the amplitude A also remains constant (ignoring friction). Thus the amplitude and the intensity do not decrease with distance.

In practice, frictional damping is generally present, and some of the energy is transformed into thermal energy. Thus the amplitude and intensity of a one-dimensional wave decrease with distance from the source, and for a three-dimensional wave the decrease will be greater than that discussed above, although the effect may often be small.

* 11–10 Intensity Related to Amplitude and Frequency

We can obtain an explicit relation between the energy carried by a wave, or the wave's intensity I, and the amplitude and frequency of the wave. For a sinusoidal wave of frequency f, the particles move in SHM as a wave passes, so each particle has an energy $E = \frac{1}{2}kA^2$, where A is the amplitude of its motion, either transversely or longitudinally. Using Eq. 11–7, we can write k in terms of the frequency, $k = 4\pi^2 m/T^2 = 4\pi^2 mf^2$, where m is the mass of a particle (or small volume) of the medium. Then

$$E = 2\pi^2 mf^2 A^2.$$

The mass $m = \rho V$, where ρ is the density of the medium and V the volume of a small slice of the medium. The volume $V = Sl$, where S is the cross-sectional surface area through which the wave travels (Fig. 11–30), and we can write l as the distance the wave travels in a time t as $l = vt$, where v is the speed of the wave. Thus $m = \rho V = \rho Sl = \rho Svt$ and

$$E = 2\pi^2 \rho Svt f^2 A^2. \tag{11–17a}$$

From this equation, we see again the important result that the energy transported by a wave is proportional to the square of the amplitude. The power transported $P = E/t$ is

$$P = \frac{E}{t} = 2\pi^2 \rho Sv f^2 A^2. \tag{11–17b}$$

Finally, the **intensity** I of a wave is the power transported across unit area

FIGURE 11–30 Calculating the energy carried by a wave moving with velocity v.

$l = vt$

perpendicular to the direction of energy flow:

$$I = \frac{P}{S} = 2\pi^2 v \rho f^2 A^2. \tag{11-18}$$

This relation shows explicitly that the intensity of a wave is proportional to the square of the wave amplitude A at any point and to the square of the frequency.

11–11 Reflection and Interference of Waves

Reflection.   When a wave strikes an obstacle, or comes to the end of the medium it is traveling in, at least a part of the wave is reflected. You have probably seen water waves reflect off a rock or the side of a swimming pool. And you may have heard a shout reflected from a distant cliff—which we call an "echo."

A wave pulse traveling down a rope is reflected as shown in Fig. 11–31. You can observe this for yourself (try it with a rope lying on a table) and see that the reflected pulse returns inverted as in Fig. 11–31a if the end of the rope is fixed; and returns right side up if the end is free as in Fig. 11–31b. When the end is fixed to a support, as in Fig. 11–31a, the pulse reaching that fixed end exerts a force (upward) on the support. The support exerts an equal but opposite force (Newton's third law) downward on the cord. This downward force on the cord is what "generates" the inverted reflected pulse. In Fig. 11–31b, the free end is constrained by neither a support nor by additional cord. It therefore tends to overshoot—its displacement being momentarily greater than that of the traveling pulse. The overshooting end exerts an upward pull on the cord, and this is what generates the reflected pulse, which is not inverted.

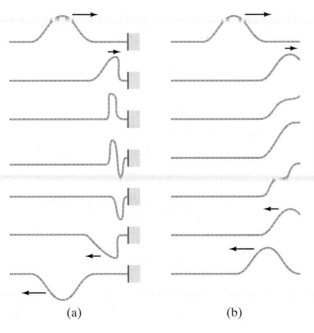

(a)   (b)

FIGURE 11–31 Reflection of a wave pulse on a cord when the end of the cord is (a) fixed and (b) free.

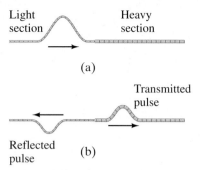

(a)

Transmitted
pulse

Reflected
pulse (b)

FIGURE 11–32 When a wave pulse traveling to the right (a) reaches a discontinuity, then (b) part is reflected and part is transmitted.

Wave fronts

Plane waves

Law of reflection

When the wave pulse on the rope in Fig. 11–31a reaches the wall, not all of the energy is reflected. Some of it is absorbed by the wall. Part of the absorbed energy is transformed into thermal energy, and part continues to propagate through the material of the wall. This is more clearly illustrated by considering a pulse that travels down a rope which consists of a light section and a heavy section, as shown in Fig. 11–32. When the wave reaches the boundary between the two sections, part of the pulse is reflected and part is transmitted, as shown. The heavier the second section, the less is transmitted; and when the second section is a wall or rigid support, very little is transmitted.

For a two- or three-dimensional wave, such as a water wave, we are concerned with **wave fronts**, by which we mean the whole width of a wave crest (what we usually refer to simply as a "wave" when we are at the seashore). A line drawn in the direction of motion, perpendicular to the wave front, is called a **ray**, as shown in Fig. 11–33. Note in Fig. 11–33b that wave fronts far from the source have lost almost all their curvature and are nearly straight, as ocean waves often are; they are then called **plane waves**. For reflection of a two- or three-dimensional plane wave, as shown in Fig. 11–34, the angle that the incoming or *incident wave* makes with the reflecting surface is equal to the angle made by the reflected wave. This is the law of reflection: **the angle of reflection equals the angle of incidence**. The "angle of incidence" is defined as the angle the incident ray makes with the perpendicular to the reflecting surface (or the wave front makes with a tangent to the surface), and the "angle of reflection" is the corresponding angle for the reflected wave.

FIGURE 11–33 Rays, signifying the direction of motion, are always perpendicular to the wave fronts (wave crests). (a) Circular or spherical waves near the source; (b) far from the source the wave fronts are nearly straight or flat, and are called plane waves.

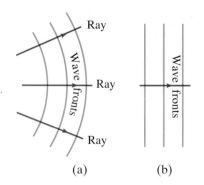

(a)   (b)

FIGURE 11–34 Law of reflection.

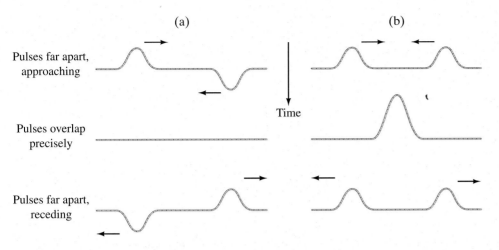

(a)   (b)

Pulses far apart, approaching

Time

Pulses overlap precisely

Pulses far apart, receding

FIGURE 11–35
Two wave pulses pass each other. Where they overlap, interference occurs: (a) destructive, and (b) constructive.

Interference.   **Interference** refers to what happens when two waves pass through the same region of space at the same time. Consider, for example, the two wave pulses on a string traveling toward each other as shown in Fig. 11–35. In part (a) the two pulses have the same amplitude, but one is a crest and the other a trough; in part (b) they are both crests. In both cases, the waves meet and pass right on by each other. However, in the region where they overlap, the resultant displacement is the *algebraic sum of their separate displacements* (a crest is considered positive and a trough negative). This is called the **principle of superposition**. In Fig. 11–35a, the two waves oppose one another as they pass by and the result is called **destructive interference**. In Fig. 11–35b, the resultant displacement is greater than that of either pulse and the result is **constructive interference**.

Superposition principle

Constructive and destructive interference

When two rocks are thrown into a pond simultaneously, the two sets of circular waves interfere with one another as shown in Fig. 11–36. In some areas of overlap, crests of one wave repeatedly meet crests of the other (and troughs meet troughs); this is constructive interference and the water continuously oscillates up and down with greater amplitude than either wave separately. In other areas, destructive interference occurs where the water actually does not move up and down at all over time—this is where crests of

FIGURE 11–36   Interference of water waves.

Constructive interference

Destructive interference

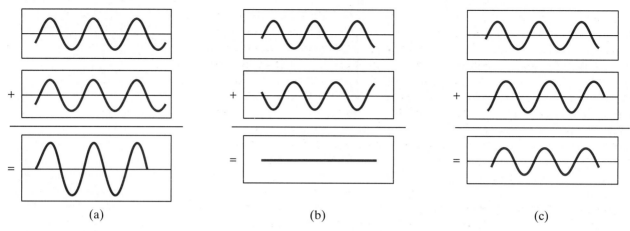

FIGURE 11–37 Graphs showing two waves, and their sum, as a function of time at three locations. In (a) the two waves interfere constructively, in (b) destructively, and in (c) partially destructively.

(a)   (b)   (c)

FIGURE 11–38 Standing waves corresponding to three resonant frequencies.

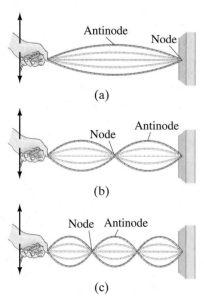

Antinode
Node

(a)

Node   Antinode

(b)

Node   Antinode

(c)

one wave meet troughs of the other, and vice versa. Figure 11–37a shows the displacement of the two waves as a function of time, as well as their sum, for the case of constructive interference. For any two such waves, we use the term **phase** to describe the relative positions of their crests. When the crests and troughs are aligned as in Fig. 11–37a for constructive interference, the two waves are **in phase**. At points where destructive interference occurs—see Fig. 11–37b—crests of one wave repeatedly meet troughs of the other wave, and the two waves are said to be completely **out of phase** or, more precisely, out of phase by one-half wavelength (that is, the crests of one wave occur a half wavelength behind the crests of the other wave). Of course, the relative phase of the two water waves in Fig. 11–36 will at most points be intermediate between these two extremes, which results in *partially* destructive interference, as illustrated in Fig. 11–37c.

11–12 Standing Waves; Resonance

If you shake one end of a cord (or slinky) and the other end is kept fixed, a continuous wave will travel down to the fixed end and be reflected back, inverted (Fig. 11–31). As you continue to vibrate the cord, there will be waves traveling in both directions, and the wave traveling down the cord will interfere with the reflected wave coming back. Usually there will be quite a jumble. But if you vibrate the cord at just the right frequency, the two traveling waves will interfere in such a way that a large-amplitude **standing wave** will be produced, Fig. 11–38. It is called a "standing wave" because it doesn't appear to be traveling. The cord simply oscillates up and down with a fixed pattern. The points of destructive interference, where the cord remains still, are called **nodes;** points of constructive interference, where the cord oscillates with maximum amplitude, are called **antinodes**. The nodes and antinodes remain in fixed positions for a given frequency. Standing waves can occur at more than one frequency. The lowest frequency of vibration that produces a standing wave gives rise to the pattern shown in Fig. 11–38a. The standing waves shown in parts (b) and (c) are produced at precisely two and three times the lowest frequency, respectively, assuming the tension in the cord is the same. The cord can also vibrate with four loops at four times the lowest frequency, and so on.

The frequencies at which standing waves are produced are the **natural frequencies** or **resonant frequencies** of the cord, and the different standing wave patterns shown in Fig. 11–38 are different "resonant modes of vibration." For although a standing wave is the result of the interference of two waves traveling in opposite directions, it is also an example of a vibrating object at resonance (Section 11–6). Standing waves then represent the same phenomenon as the resonance of a vibrating spring or pendulum, which we discussed earlier. The only difference is that a spring or pendulum has only one resonant frequency, whereas the cord has a large number of resonant frequencies, each of which is a whole-number multiple of the lowest resonant frequency.

Now let us consider a cord stretched between two supports that is plucked like a guitar or violin string, Fig. 11–39a. Waves of a great variety of frequencies will travel in both directions along the string, will be reflected at the ends, and will travel back in the opposite direction. Most of these waves interfere in a random way with each other and quickly die away. However, those waves that correspond to the resonant frequencies of the string will persist. The ends of the string, since they are fixed, will be nodes. There may be other nodes as well. Some of the possible resonant modes of vibration (standing waves) are shown in Fig. 11–39b. Generally, the motion will be a combination of these different resonant modes; but only those frequencies that correspond to a resonant frequency will be present.

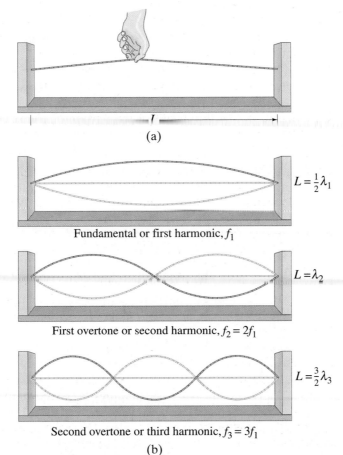

(a)

$L = \frac{1}{2}\lambda_1$

Fundamental or first harmonic, f_1

$L = \lambda_2$

First overtone or second harmonic, $f_2 = 2f_1$

$L = \frac{3}{2}\lambda_3$

Second overtone or third harmonic, $f_3 = 3f_1$

(b)

FIGURE 11–39
(a) A string is plucked.
(b) Only standing waves corresponding to resonant frequencies persist for long.

To determine the resonant frequencies, we first note that the wavelengths of the standing waves bear a simple relationship to the length L of the string. The lowest frequency, called the **fundamental frequency**, corresponds to one antinode (or loop). And as can be seen in Fig. 11–39b, the whole length corresponds to one-half wavelength. Thus $L = \frac{1}{2}\lambda_1$, where λ_1 stands for the wavelength of the fundamental. The other natural frequencies are called **overtones**; when they are integral multiples of the fundamental (as they are for a simple string), they are also called **harmonics**, with the fundamental being referred to as the **first harmonic**.[†] The next mode after the fundamental has two loops and is called the **second harmonic** (or first overtone); the length of the string L corresponds to one complete wavelength: $L = \lambda_2$. For the third and fourth harmonics, $L = \frac{3}{2}\lambda_3$ and $L = 2\lambda_4$, respectively, and so on. In general, we can write

Fundamental frequency

Overtones and harmonics

$L = \frac{1}{2}\lambda_1$

Fundamental or first harmonic, f_1

$L = \lambda_2$

First overtone or second harmonic, $f_2 = 2f_1$

$L = \frac{3}{2}\lambda_3$

Second overtone or third harmonic, $f_3 = 3f_1$

(b)

FIGURE 11–39b   Fundamental and first two overtones. (Repeated.)

$$L = \frac{n\lambda_n}{2}, \qquad \text{where } n = 1, 2, 3, \cdots .$$

The integer n labels the number of the harmonic: $n = 1$ for the fundamental, $n = 2$ for the second harmonic, and so on. We solve for λ_n and find

$$\lambda_n = \frac{2L}{n}, \qquad n = 1, 2, 3, \cdots . \qquad \textbf{(11–19)}$$

In order to find the frequency f of each vibration we use Eq. 11–12, $f = v/\lambda$, and we see that

$$f_n = \frac{v}{\lambda_n} = \frac{nv}{2L} = nf_1,$$

where $f_1 = v/\lambda_1 = v/2L$ is the fundamental frequency. That is, each of the resonant frequencies is an integral multiple ($2\times$, $3\times$, etc.) of the fundamental frequency.

Because a standing wave is equivalent to two traveling waves moving in opposite directions, the concept of wave velocity still makes sense and is given by Eq. 11–13 in terms of the tension F_T in the string and its mass per unit length (m/L): that is, $v = \sqrt{F_T/(m/L)}$.

EXAMPLE 11–13 **Piano string.**   A piano string is 1.10 m long and has a mass of 9.00 g. (*a*) How much tension must the string be under if it is to vibrate at a fundamental frequency of 131 Hz? (*b*) What are the frequencies of the first four harmonics?

SOLUTION   (*a*) The wavelength of the fundamental is $\lambda = 2L = 2.20$ m (Eq. 11–19). The velocity is then $v = \lambda f = (2.20 \text{ m})(131 \text{ s}^{-1}) = 288$ m/s. Then, from Eq. 11–13, we have

$$F_T = \frac{m}{L} v^2$$

$$= \left(\frac{9.00 \times 10^{-3} \text{ kg}}{1.10 \text{ m}} \right) (288 \text{ m/s})^2 = 679 \text{ N}.$$

(*b*) The frequencies of the second, third, and fourth harmonics are two, three, and four times the fundamental frequency: 262, 393, and 524 Hz.

[†]The term "harmonic" comes from music, because such integral multiples of frequencies "harmonize."

A standing wave does appear to be standing in place (and a traveling wave appears to move). The term "standing" wave is also meaningful from the point of view of energy. Since the string is at rest at the nodes, no energy flows past these points. Hence the energy is not transmitted down the string, but "stands" in place in the string.

Standing waves are produced not only on strings, but on any object that is set into vibration. Even when a rock or a piece of wood is struck with a hammer, standing waves are set up that correspond to the natural resonant frequencies of the object. In general, the resonant frequencies depend on the dimensions of the object, just as for a string they depend on its length. For example, a small object does not have as low resonant frequencies as does a large object made of the same material. All musical instruments depend on standing waves to produce their musical sounds, from stringed instruments to wind instruments (in which a column of air vibrates as a standing wave) to drums and other percussion instruments, as we discuss in the next chapter.

* 11–13 Refraction and Diffraction

In Section 11–11, we discussed two important aspects of wave behavior: reflection and interference. Now we discuss two others, refraction and diffraction. All four of these aspects of wave behavior are important in understanding light, as we will see in Chapters 23, 24, and 25.

Refraction.   When any wave strikes a boundary, some of the energy is reflected and some is transmitted or absorbed. When a two- or three-dimensional wave traveling in one medium crosses a boundary into a medium where its velocity is different, the transmitted wave may move in a different direction than the incident wave, as shown in Fig. 11–40. This phenomenon is known as **refraction**. One example is a water wave; the velocity decreases in shallow water and the waves refract, Fig. 11–41. [When the wave velocity changes gradually, as in Fig. 11–41, without a sharp boundary, the waves change direction (refract) gradually.] In Fig. 11–40, the velocity of the wave in medium 2 is less than in medium 1. In this case, the direction of the wave bends so it travels more nearly perpendicular to the boundary. That is, the *angle of refraction*, θ_r, is less than the *angle of incidence*, θ_i. To see why this is so, and to help us get a quantitative relation between θ_r and θ_i, let us think of each wave front as a row of soldiers. The soldiers are marching from firm ground (medium 1) into mud (medium 2) and hence are slowed

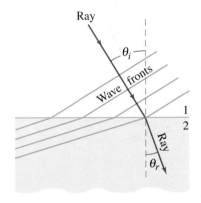

FIGURE 11–40   Refraction of waves passing a boundary.

Angle of refraction

FIGURE 11–41   Water waves refracting as they approach the shore, where their velocity is less. There is no distinct boundary, as in Fig. 11–40, because the wave velocity changes gradually.

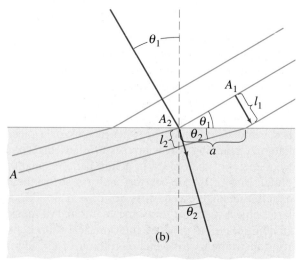

FIGURE 11–42 Soldier analogy (a), to derive law of refraction for waves (b).

down. The soldiers that reach the mud first are slowed down first and the row bends as shown in Fig. 11–42a. Let us consider the wave front (or row of soldiers) labeled A in Fig. 11–42b. In the same time t that A_1 moves a distance $l_1 = v_1 t$, we see that A_2 moves a distance $l_2 = v_2 t$. The two triangles shown have the side labeled a in common. Thus

$$\sin \theta_1 = \frac{l_1}{a} = \frac{v_1 t}{a}$$

and

$$\sin \theta_2 = \frac{l_2}{a} = \frac{v_2 t}{a}.$$

Dividing these two equations, we find that

Law of refraction

$$\frac{\sin \theta_2}{\sin \theta_1} = \frac{v_2}{v_1}. \tag{11–20}$$

Since θ_1 is the angle of incidence (θ_i), and θ_2 is the angle of refraction (θ_r), Eq. 11–20 gives the quantitative relation between the two. Of course, if the wave were going in the opposite direction, the argument would not be changed. Only θ_1 and θ_2 would change roles: θ_1 would be the angle of refraction and θ_2 the angle of incidence. Clearly then, if the wave travels into a medium where it can move faster, it will bend in the opposite way, $\theta_r > \theta_i$. We see from Eq. 11–20 that if the velocity increases, the angle increases, and vice versa.

Earthquake waves refract within the Earth as they travel through rock layers of different densities (and therefore the velocity is different) just as water waves do. Light waves refract as well, and when we discuss light we shall find Eq. 11–20 very useful.

EXAMPLE 11–14 **Refraction of earthquake wave.** An earthquake P wave passes across a boundary in rock where its velocity increases from 6.5 km/s to 8.0 km/s. If it strikes this boundary at 30°, what is the angle of refraction?

SOLUTION Since sin 30° = 0.50, Eq. 11–20 yields

$$\sin \theta_2 = \frac{(8.0 \text{ m/s})}{(6.5 \text{ m/s})} (0.50) = 0.62$$

so $\theta_2 = 38°$.

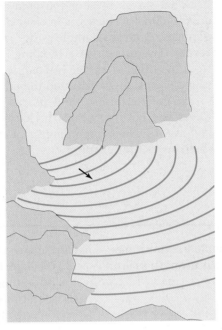

FIGURE 11–43 Wave diffraction. The waves come from the upper left. Note how the waves, as they pass between the two great rocks, bend around behind each of the rock cliffs. (Each rock can be considered an obstacle in the sense of Fig. 11–44. The two rocks can also be considered as a "slit" through which the waves pass and spread out—compare to Fig. 24–2c for light.)

Diffraction. Waves spread as they travel, and when they encounter an obstacle they bend around it somewhat and pass into the region behind as shown in Fig. 11–43 for water waves. This phenomenon is called **diffraction**.

The amount of diffraction depends on the wavelength of the wave and on the size of the obstacle, as shown in Fig. 11–44. If the wavelength is much larger than the object, as with the grass blades of Fig. 11–44a, the wave bends around them almost as if they were not there. For larger objects—parts (b) and (c)—there is more of a "shadow" region behind the obstacle where we might not expect the waves to penetrate (but they do, at least a little). But notice in part (d), where the obstacle is the same as in part (c) but the wavelength is longer, that there is more diffraction into the shadow region. As a rule of thumb, *only if the wavelength is smaller than the size of the object will there be a significant shadow region.* It is worth noting that this rule applies to *reflection* from an obstacle as well. Very little of a wave is reflected unless the wavelength is smaller than the size of the obstacle.

A rough guide to the amount of diffraction is

$$\theta \text{ (radians)} \approx \frac{\lambda}{L}$$

where θ is roughly the angular spread of waves behind an opening of width L or around an obstacle of width L.

FIGURE 11–44 Water waves passing objects of various sizes. Note that the larger the wavelength compared to the size of the object, the more diffraction there is into the "shadow region."

(a) Water waves passing blades of grass

(b) Stick in water

(c) Short-wavelength waves passing log

(d) Long-wavelength waves passing log

That waves can bend around obstacles, and thus can carry energy to areas behind obstacles, is in clear distinction to energy carried by material particles. A clear example is the following: if you are standing around a corner on one side of a building, you can't be hit by a baseball thrown from the other side, but you can hear a shout or other sound because the sound waves diffract around the edges.

Both interference and diffraction occur only for energy carried by waves and not for energy carried by material particles. This distinction is important for understanding the nature of light, and of matter itself, as we shall see in later chapters.

■ SUMMARY

A vibrating object undergoes **simple harmonic motion** (SHM) if the restoring force is proportional to the displacement,

$$F = -kx.$$

The maximum displacement is called the **amplitude**.

The **period**, T, is the time required for one complete cycle (back and forth), and the **frequency**, f, is the number of cycles per second; they are related by

$$f = \frac{1}{T}.$$

The period of vibration for a mass m on the end of a spring is given by

$$T = 2\pi \sqrt{m/k}.$$

SHM is **sinusoidal**, which means that the displacement as a function of time follows a sine or cosine curve.

During SHM, the total energy $E = \frac{1}{2}mv^2 + \frac{1}{2}kx^2$ is continually changing from potential to kinetic and back again.

A **simple pendulum** of length L approximates SHM if its amplitude is small and friction can be ignored. Its period is then given by (for small amplitudes)

$$T = 2\pi \sqrt{L/g},$$

where g is the acceleration of gravity.

When friction is present (for all real springs and pendulums), the motion is said to be **damped**. The maximum displacement decreases in time, and the energy is eventually all transformed to heat.

If an oscillating force is applied to a system capable of vibrating, the system's amplitude of vibration can be very large if the frequency of the applied force matches the **natural** (or **resonant**) **frequency** of the oscillator. This is called **resonance**.

Vibrating objects act as sources of **waves** that travel outward from the source. Waves on water and on a string are examples. The wave may be a **pulse** (a single crest) or it may be continuous (many crests and troughs).

The **wavelength** of a continuous sinusoidal wave is the distance between two successive crests.

The **frequency** is the number of full wavelengths (or crests) that pass a given point per unit time.

The **wave velocity** (how fast a crest moves) is equal to the product of wavelength and frequency,

$$v = \lambda f.$$

The **amplitude** of a wave is the maximum height of a crest, or depth of a trough, relative to the normal (or equilibrium) level.

In a **transverse wave**, the oscillations are perpendicular to the direction in which the wave travels. An example is a wave on a string.

In a **longitudinal wave**, the oscillations are along (parallel to) the line of travel; sound is an example.

Waves reflect off objects in their path. When the **wave front** (of a two- or three-dimensional wave) strikes an object, the angle of reflection is equal to the angle of incidence. When a wave strikes a boundary between two materials in which it can travel, part of the wave is reflected and part is transmitted.

When two waves pass through the same region of space at the same time, they **interfere**. The resultant displacement at any point and time is the sum of their separate displacements; this can result in **constructive interference**, **destructive interference**, or something in between, depending on the amplitudes and relative phases of the waves.

Waves traveling on a string (or other medium) of fixed length interfere with waves that have reflected off the end and are traveling back in the opposite direction. At certain frequencies, **standing waves** can be produced in which the waves seem to be standing still rather than traveling. The string (or other medium) is vibrating as a whole. This is a resonance phenomenon and the frequencies at which standing waves occur are called **resonant frequencies**. The points of destructive interference (no vibration) are called **nodes**. Points of constructive interference (maximum amplitude of vibration) are called **antinodes**.

QUESTIONS

1. Give some examples of everyday vibrating objects. Which exhibit SHM, at least approximately?

2. Is the acceleration of a simple harmonic oscillator ever zero? If so, where?

3. Is the motion of a piston in an automobile engine simple harmonic? Explain.

4. Real springs have mass. How will the true period and frequency differ from those given by the equations for a mass oscillating on the end of an idealized massless spring?

5. How could you double the maximum speed of a simple harmonic oscillator (SHO)?

6. Does a car bounce on its springs faster when it is empty or when it is fully loaded?

7. If you double the amplitude of a SHO, how does this change the frequency, maximum velocity, maximum acceleration, and total mechanical energy?

8. If a pendulum clock is accurate at sea level, will it gain or lose time when taken to high altitude?

9. A tire swing hangs from a branch nearly to the ground. How could you estimate the height of the branch using only a stopwatch?

10. Why can you make water slosh back and forth in a pan only if you shake the pan at a certain frequency?

11. A tuning fork of natural frequency 264 Hz sits on a table at the front of a room. At the back of the room, two tuning forks, one of natural frequency 260 Hz and one of 420 Hz, are initially silent; but when the tuning fork at the front of the room is set into vibration, the 260-Hz fork spontaneously begins to vibrate, but the 420-Hz fork does not. Explain.

12. Give several everyday examples of resonance.

13. Is a rattle in a car ever a resonance phenomenon? Explain.

14. Over the years, buildings have been built out of lighter and lighter materials. How has this affected the natural vibration frequencies of buildings and the problems of resonance due to passing trucks, airplanes, or natural sources of vibration?

15. Is the frequency of a simple periodic wave equal to the frequency of its source? Why or why not?

16. Explain the difference between the speed of a transverse wave traveling down a cord and the speed of a tiny piece of the cord.

17. Why do the strings used for the lowest-frequency notes on a piano normally have wire wrapped around them?

18. What kind of waves do you think will travel down a horizontal metal rod if you strike its end (a) vertically from above and (b) horizontally parallel to its length?

19. Since the density of air decreases with an increase in temperature, but the bulk modulus B is nearly independent of temperature, how would you expect the speed of sound waves in air to vary with temperature?

20. Give two reasons why circular water waves decrease in amplitude as they travel away from the source.

21. Two linear waves have the same amplitude and otherwise are identical, except one has half the wavelength of the other. Which transmits more energy? How much more energy does it transfer?

22. When a sinusoidal wave crosses the boundary between two sections of cord as in Fig. 11–32, the frequency does not change (although the wavelength and velocity do change). Explain why.

23. If we knew that energy was being transmitted from one place to another, how might we determine whether the energy was being carried by particles (material bodies) or by waves?

24. AM radio signals can usually be heard behind a hill, but FM often cannot. That is, AM signals bend more than FM. Explain. (Radio signals, as we shall see, are carried by electromagnetic waves whose wavelength for AM is typically 200 to 600 m and for FM about 3 m.)

25. If a string is vibrating in three segments, are there any places one can touch it with a knife blade without disturbing the motion?

26. When a standing wave exists on a string, the vibrations of incident and reflected waves cancel at the nodes. Does this mean that energy was destroyed? Explain.

PROBLEMS

SECTIONS 11–1 TO 11–3

1. (I) When a 65-kg person climbs into a 1000-kg car, the car's springs compress vertically by 2.8 cm. What will be the frequency of vibration when the car hits a bump? Ignore damping.

2. (I) An elastic cord is 65 cm long when a weight of 55 N hangs from it and is 85 cm long when a weight of 80 N hangs from it. What is the "spring" constant of this elastic cord?

3. (I) If a particle undergoes SHM with amplitude 0.25 m, what is the total distance it travels in one period?

4. (I) A fisherman's scale stretches 3.9 cm when a 2.7-kg fish hangs from it. (a) What is the spring constant and (b) what will be the amplitude and frequency of vibration if the fish is pulled down 2.5 cm more and released so that it vibrates up and down?

5. (II) Construct a table indicating the position of the mass in Fig. 11–2 at the times $t = 0, \frac{1}{4}T, \frac{1}{2}T, \frac{3}{4}T, T$, and $\frac{5}{4}T$, where T is the period of oscillation. On a graph of x versus t, plot these six points. Now connect these points with a smooth curve. Does your curve, based on these simple considerations, resemble that of a cosine or sine wave (Fig. 11–9 or 11–11)?

6. (II) A small fly of mass 0.15 g is caught in a spider's web. The web vibrates predominately with a frequency of 4.0 Hz. (a) What is the value of the effective spring constant k for the web? (b) At what frequency would you expect the web to vibrate if an insect of mass 0.50 g were trapped?

7. (II) A balsa wood block of mass 50 g floats on a lake, bobbing up and down at a frequency of 2.5 Hz. (a) What is the value of the effective spring constant of the water? (b) A partially filled water bottle of mass 0.25 kg and almost the same size and shape of the balsa block is tossed into the water. At what frequency would you expect the bottle to bob up and down? Assume SHM.

8. (II) An elastic cord vibrates with a frequency of 3.0 Hz when a mass of 0.60 kg is hung from it. What will its frequency be if only 0.38 kg hangs from it?

9. (II) A 0.50-kg mass at the end of a spring vibrates 3.0 times per second with an amplitude of 0.15 m. Determine (a) the velocity when it passes the equilibrium point, (b) the velocity when it is 0.10 m from equilibrium, (c) the total energy of the system, and (d) the equation describing the motion of the mass, assuming that at $t = 0, x$ was a maximum.

10. (II) A mass m at the end of a spring vibrates with a frequency of 0.88 Hz. When an additional 600-g mass is added to m, the frequency is 0.60 Hz. What is the value of m?

FIGURE 11–45
Problem 11.

11. (II) A block of mass m is supported by two identical parallel vertical springs, each with spring constant k (Fig. 11–45). What will be the frequency of vibration?

12. (II) A mass of 1.62 kg stretches a vertical spring 0.315 m. If the spring is stretched an additional 0.130 m and released, how long does it take to reach the (new) equilibrium position again?

13. (II) It takes a force of 80.0 N to compress the spring of a toy popgun 0.200 m to "load" a 0.150-kg ball. With what speed will the ball leave the gun?

14. (II) At $t = 0$, a 750-g mass at rest on the end of a horizontal spring ($k = 124$ N/m) is struck by a hammer, which gives it an initial speed of 2.76 m/s. Determine (a) the period and frequency of the motion, (b) the amplitude, (c) the maximum acceleration, (d) the position as a function of time, and (e) the total energy.

15. (II) A mass sitting on a horizontal, frictionless surface is attached to one end of a spring; the other end is fixed to a wall. 3.0 J of work is required to compress the spring by 0.12 m. If the mass is released from rest with the spring compressed, it experiences a maximum acceleration of 15 m/s². Find the value of (a) the spring constant and (b) the mass.

16. (II) A mass attached to the end of a spring is stretched a distance x_0 from equilibrium and released. At what distance from equilibrium will it have (a) velocity equal to half its maximum velocity and (b) acceleration equal to half its maximum acceleration?

17. (II) A 0.50-kg mass vibrates according to the equation $x = 0.45 \cos 8.40t$, where x is in meters, and t in seconds. Determine (a) the amplitude, (b) the frequency, (c) the total energy, and (d) the kinetic energy and potential energy when $x = 0.30$ m.

18. (II) A 400-g mass vibrates according to the equation $x = 0.35 \sin (5.50t)$ where x is in meters and t is in seconds. Determine (a) the amplitude, (b) the frequency, (c) the period, (d) the total energy, and (e) the KE and PE when x is 10 cm. (f) Draw a careful graph of x vs. t showing correct amplitude and period.

19. (II) A spring of force constant 210 N/m vibrates with an amplitude of 28.0 cm when 0.250 kg hangs from it. (a) What is the equation describing this motion as a function of time? Assume that the mass passes through the equilibrium point with positive velocity at $t = 0$. (b) At what times will the spring have its maximum and minimum extensions?

20. (II) A 2.00-kg pumpkin oscillates from a vertically hanging light spring once every 0.55 s. (a) Write down the equation giving the pumpkin's position y (+ upward) as a function of time t, assuming it started by being compressed 10 cm from the equilibrium position (where $y = 0$), and released. (b) How long will it take to get to the equilibrium position for the first time? (c) What will be its maximum speed? (d) What will be its maximum acceleration, and where will it first be attained?

21. (II) A 25.0-g bullet strikes a 0.600-kg block attached to a fixed horizontal spring whose spring constant is 6.70×10^3 N/m and sets it into vibration with an amplitude of 21.5 cm. What was the speed of the bullet before impact if the two objects move together after impact?

22. (II) If one vibration has 10 times the energy of a second, but their frequencies and masses are the same, how do their amplitudes compare?

23. (II) (a) At what displacement from equilibrium is the energy of a SHO half KE and half PE? (b) At what displacement from equilibrium is the speed half the maximum value?

24. (II) If a mass m hangs from a vertical spring, as shown in Fig. 11–3, show that $F = -kx$ holds for extension or compression of the spring, where x is the displacement from the (vertical position) equilibrium point.

25. (III) Show that the conservation of energy (Eq. 11–3) holds also for a vertical spring (see Fig. 11–3) where x is measured from the vertical equilibrium position ($x_0 = mg/k$) for a mass m.

26. (III) A bungee jumper with mass 65.0 kg jumps from a high bridge. After reaching his lowest point, he oscillates up and down, hitting a low point eight more times in 34.7 s. He finally comes to rest 25.0 m below the level of the bridge. Calculate the spring constant and the unstretched length of the bungee cord.

27. (III) A mass m is connected to two springs, with spring constants k_1 and k_2, in two different ways, as shown in Fig. 11–46a and b. Show that the period for the configuration shown in part (a) is given by

$$T = 2\pi \sqrt{m\left(\frac{1}{k_1} + \frac{1}{k_2}\right)}$$

and for that in part (b) is given by

$$T = 2\pi \sqrt{\frac{m}{k_1 + k_2}}.$$

Ignore friction.

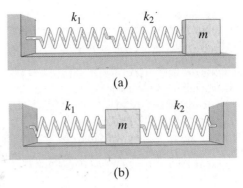

(a)

(b)

FIGURE 11–46   Problem 27.

SECTION 11–4

28. (I) A pendulum makes 36 vibrations in 50 s. What is its (a) period, and (b) frequency?

29. (I) You want to build a grandfather clock with a pendulum (a weight on the end of a light cable) that has one second between its "tick" (swinging to) and its "tock" (swinging fro). (a) How long do you make the cable? (b) Suppose the cable stretches a bit from the weight on the end. Will the clock run fast or slow?

30. (II) What is the period of a simple pendulum 50 cm long (a) on the Earth, and (b) when it is in a freely falling elevator?

31. (II) The length of a simple pendulum is 0.66 m, the pendulum bob has a mass of 310 grams, and it is released at an angle of 12° to the vertical. (a) With what frequency does it vibrate? Assume SHM. (b) What is the pendulum bob's speed when it passes through the lowest point of the swing? (c) What is the total energy stored in this oscillation, assuming no losses?

32. (II) Derive a formula for the maximum speed v_0 of a simple pendulum bob in terms of g, the length L, and the angle of swing θ_0.

33. (III) A clock pendulum oscillates at a frequency of 2.0 Hz. At $t = 0$, it is released from rest starting at an angle of 15° to the vertical. Ignoring friction, what will be the position (angle) of the pendulum at (a) $t = 0.25$ s, (b) $t = 1.60$ s, and (c) $t = 500$ s? [Hint: Do not confuse the angle of swing θ of the pendulum with the angle that appears as the argument of the cosine.]

SECTIONS 11–7 AND 11–8

34. (I) A fisherman notices that wave crests pass the bow of his anchored boat every 3.0 s. He measures the distance between two crests to be 8.5 m. How fast are the waves traveling?

35. (I) A sound wave in air has a frequency of 262 Hz and travels with a speed of 330 m/s. How far apart are the wave crests (compressions)?

36. (I) AM radio signals have frequencies between 550 kHz and 1600 kHz (kilohertz) and travel with a speed of 3.00×10^8 m/s. What are the wavelengths of these signals? On FM, the frequencies range from 88.0 MHz to 108 MHz (megahertz) and travel at the same speed; what are their wavelengths?

37. (I) Calculate the speed of longitudinal waves in (a) water, (b) granite, and (c) steel.

38. (I) Two solid rods have the same elastic modulus, but one is twice as dense as the other. In which rod will the speed of longitudinal waves be greater, and by what factor?

39. (II) A cord of mass 0.55 kg is stretched between two supports 30 m apart. If the tension in the cord is 150 N, how long will it take a pulse to travel from one support to the other?

40. (II) Determine the wavelength of a 6,000-Hz sound wave traveling along an iron rod. [*Hint*: See Table 9–1.]

41. (II) A sailor strikes the side of his ship just below the surface of the sea. He hears the echo of the wave reflected from the ocean floor directly below 3.0 s later. How deep is the ocean at this point?

42. (II) S and P waves from an earthquake travel at different speeds, and this difference helps in the determination of the earthquake "epicenter" (where the disturbance took place). (a) Assuming typical speeds of 8.5 km/s and 5.5 km/s for P and S waves, respectively, how far away did the earthquake occur if a particular seismic station detects the arrival of these two types of waves 2.0 min apart? (b) Is one seismic station sufficient to determine the position of the epicenter? Explain.

43. (II) An earthquake-produced surface wave can be approximated by a sinusoidal transverse wave. Assuming a frequency of 0.50 Hz (typical of earthquakes, which actually include a mixture of frequencies), what amplitude is needed so that objects begin to leave contact with the ground?

SECTION 11–9

44. (II) Compare (a) the intensities and (b) the amplitudes of an earthquake wave as it passes two points 10 km and 20 km from the source.

45. (II) The intensity of a particular earthquake wave is measured to be 2.0×10^6 J/m²·s at a distance of 50 km from the source. (a) What was the intensity when it passed a point only 1.0 km from the source? (b) What was the rate energy passed through an area of 10.0 m² at 1.0 km?

46. (III) Show that the amplitude A of circular water waves decreases as the square root of the distance r from the source: $A \propto 1/\sqrt{r}$. Ignore damping.

*SECTION 11–10

* 47. (I) Two earthquake waves of the same frequency travel through the same portion of the Earth, but one is carrying twice the energy. What is the ratio of the amplitudes of the two waves?

* 48. (I) Two waves traveling along a stretched string have the same frequency, but one transports three times the power of the other. What is the ratio of the amplitudes of the two waves?

* 49. (II) A bug on the surface of a pond is observed to move up and down a total vertical distance of 6.0 cm, lowest to highest point, as a wave passes. If the ripples decrease to 4.5 cm, by what factor does the bug's maximum KE change?

SECTION 11–11

50. (I) The two pulses shown in Fig. 11–47 are moving toward each other. (a) Sketch the shape of the string at the moment they directly overlap. (b) Sketch the shape of the string a few moments later. (c) In Fig. 11–35a, at the moment the pulses pass each other, the string is straight. What has happened to the energy at this moment?

FIGURE 11–47   Problem 50.

SECTION 11–12

51. (I) If a violin string vibrates at 440 Hz as its fundamental frequency, what are the frequencies of the first four harmonics?

52. (I) A violin string vibrates at 294 Hz when unfingered. At what frequency will it vibrate if it is fingered one third of the way down from the end?

53. (I) A particular string resonates in four loops at a frequency of 280 Hz. Name at least three other frequencies at which it will resonate.

54. (I) In an earthquake, it is noted that a footbridge oscillated up and down in a one loop (fundamental standing wave) pattern once every 2.5 s. What other possible resonant periods of motion are there for this bridge? What frequencies do they correspond to?

55. (II) The velocity of waves on a string is 92 m/s. If the frequency of standing waves is 475 Hz, how far apart are two adjacent nodes?

56. (II) If two successive overtones of a vibrating string are 280 Hz and 350 Hz, what is the frequency of the fundamental?

57. (II) A guitar string is 90 cm long and has a mass of 3.6 g. From the bridge to the support post ($= L$) is 60 cm, and the string is under a tension of 520 N. What are the frequencies of the fundamental and first two overtones?

58. (II) A particular guitar string is supposed to vibrate at 200 Hz, but it is measured to actually vibrate at 205 Hz. By what percent should the tension in the string be changed to get the frequency to the correct value?

59. (II) Show that the frequency of standing waves on a string of length L and linear density μ, which is stretched to a tension F_T, is given by

$$f = \frac{n}{2L}\sqrt{\frac{F_T}{\mu}},$$

where n is an integer.

60. (II) One end of a horizontal string of linear density 4.3×10^{-4} kg/m is attached to a small-amplitude mechanical 60-Hz vibrator. The string passes over a pulley, a distance $L = 1.50$ m away, and weights are hung from this end, Fig. 11–48. What mass must be hung from this end of the string to produce (a) one loop, (b) two loops, and (c) five loops of a standing wave? Assume the string at the vibrator is a node, which is nearly true. Why can the amplitude of the standing wave be much greater than the vibrator amplitude?

Vibrator

FIGURE 11–48
Problems 60 and 61.

61. (II) In Problem 60, the length of the string may be adjusted by moving the pulley. If the hanging mass is fixed at 0.080 kg, how many different standing wave patterns may be achieved by varying L between 10 cm and 1.5 m?

62. (III) (a) Show that if the tension in a stretched string is changed by a small amount ΔF_T, the frequency of the fundamental is changed by an amount $\Delta f = \frac{1}{2}(\Delta F_T / F_T) f$. (b) By what percent must the tension in a piano string be increased or decreased to raise the frequency from 438 Hz to 442 Hz? (c) Does the formula in part (a) apply to the overtones as well?

*SECTION 11–13

* 63. (I) An earthquake P wave traveling 8.0 km/s strikes a boundary within the earth between two kinds of material. If it approaches the boundary at an incident angle of 50° and the angle of refraction is 31°, what is the speed in the second medium?

* 64. (I) Water waves approach an underwater "shelf" where the velocity changes from 2.8 m/s to 2.1 m/s. If the incident wave crests make a 34° angle with the shelf, what will be the angle of refraction?

* 65. (II) A longitudinal earthquake wave strikes a boundary between two types of rock at a 35° angle. As it crosses the boundary, the specific gravity of the rock changes from 3.7 to 2.8. Assuming that the elastic modulus is the same for both types of rock, determine the angle of refraction.

* 66. (II) It is found for any type of wave, say an earthquake wave, that if it reaches a boundary beyond which its speed is increased, there is a maximum incident angle if there is to be a transmitted refracted wave. This maximum incident angle θ_{iM} corresponds to an angle of refraction equal to 90°. If $\theta_i > \theta_{iM}$, all the wave is reflected at the boundary and none is refracted (because this would correspond to $\sin \theta_r > 1$, where θ_r is the angle of refraction, which is impossible); this is referred to as *total internal reflection*. (a) Find a formula for θ_{iM} using Eq. 11–20. (b) At what angles of incidence will there be only reflection and no transmission for an earthquake P wave traveling 7.2 km/s when it reaches a different kind of rock where its speed is 8.4 km/s?

GENERAL PROBLEMS

67. When you walk with a cup of coffee (diameter 8 cm) at just the right pace of about 1 step per second, the coffee builds up its "sloshing" until eventually, after a few steps, it starts to spill over the top. What is the speed of the waves in the coffee?

68. A 70-kg person jumps from a window to a fire net 20 m below, which stretches the net 1.1 m. Assume that the net behaves like a simple spring, and calculate how much it would stretch if the same person were lying in it. How much would it stretch if the person jumped from 35 m?

69. A crane has hoisted a 1200-kg car at the junkyard. The steel crane cable is 20 m long and has a diameter of 6.4 mm. A breeze starts the car bouncing at the end of the cable. What is the period of the bouncing? [*Hint*: Refer to Table 9–1.]

70. An energy-absorbing car bumper has a spring constant of 500 kN/m. Find the maximum compression of the bumper if the car, with mass 1500 kg, collides with a wall at a speed of 2 m/s (approximately 5 mi/h).

FIGURE 11–49
Problem 71.

71. A block of Jell-O rests on a cafeteria plate as shown in Fig. 11–49 (which also gives the dimensions of the block). You push it sideways as shown, and then you let go. The Jell-O springs back and begins to vibrate. In analogy to a mass vibrating on a spring, estimate the frequency of this vibration, given that the shear modulus of Jell-O is 520 N/m^2 and its density is 1300 kg/m^3.

72. A simple pendulum oscillates with frequency f. What is its frequency if it accelerates at $\frac{1}{2}g$ (a) upward, and (b) downward?

73. A 250-kg wooden raft floats on a lake. When a 75-kg man stands on the raft, it sinks 4.0 cm deeper into the water. When he steps off, the raft vibrates for a while. (a) What is the frequency of vibration? (b) What is the total energy of vibration (ignoring damping)?

74. The ripples in a certain groove 12.8 cm from the center of a 33-rpm phonograph record have a wavelength of 1.70 mm. What will be the frequency of the sound emitted?

75. Two strings on a musical instrument are tuned to play at 392 Hz (G) and 440 Hz (A). (a) What are the first two overtones for each string? (b) If the two strings have the same length and are under the same tension, what must be the ratio of their masses (M_G/M_A)? (c) If the strings, instead, have the same mass per unit length and are under the same tension, what is the ratio of their lengths (L_G/L_A)? (d) If their masses and lengths are the same, what must be the ratio of the tensions in the two strings?

76. A 900-kg car strikes a huge spring at a speed of 20 m/s (Fig. 11–50), compressing it 5.0 m. (a) What is the spring constant of the spring? (b) How long is the car in contact with the spring before it bounces off in the opposite direction?

900 kg

FIGURE 11–50   Problem 76.

77. Consider a sine wave traveling down the stretched two-part cord of Fig. 11–32. Determine a formula (a) for the ratio of the speeds of the wave in the two sections, v_2/v_1, and (b) for the ratio of the wavelengths in the two sections. (The frequency is the same in both sections. Why?) (c) Is the wavelength greater in the heavier cord or the lighter?

78. A diving board oscillates with simple harmonic motion of frequency 3.5 cycles per second. What is the maximum amplitude with which the end of the board can vibrate in order that a pebble placed there (Fig. 11–51) not lose contact with the board during the oscillation?

FIGURE 11–51   Problem 78.

79. A rectangular block of wood floats in a calm lake. Show that, if friction is ignored, when the block is pushed gently down into the water, it will then vibrate with SHM. Also, determine an equation for the force constant.

80. A mass m is gently placed on the end of a freely hanging spring. The mass then falls 30 cm before it stops and begins to rise. What is the frequency of the oscillation?

81. The pendulum of a clock consists of a small heavy bob at the end of a brass rod. The clock keeps accurate time at 20°C, at which point the pendulum has a period of 0.5520 s. When the temperature is 35°C, will the clock be fast or slow? By how much will it be in error after one day at 35°C? [*Hint:* Consult Table 13–1 and Eq. 13–1.]

82. The water in a U-shaped tube is displaced an amount Δx from equilibrium. (The level in one side is $2 \Delta x$ above the level in the other side.) If friction is neglected, will the water oscillate harmonically? Determine a formula for the equivalent spring constant k. Does k depend on the density of the liquid, the cross section of the tube, or the length of the water column?

If music be the food of physics, play on.

SOUND 12

Sound is associated with our sense of hearing and, therefore, with the physiology of our ears and the psychology of our brain which interprets the sensations that reach our ears. The term *sound* also refers to the physical sensation that stimulates our ears: namely, longitudinal waves.

We can distinguish three aspects of any sound. First, there must be a *source* for a sound; and as with any wave, the source of a sound wave is a vibrating object. Second, the energy is transferred from the source in the form of longitudinal sound *waves*. And third, the sound is *detected* by an ear or an instrument. We will discuss sources and detectors of sound later in this chapter, as well as some important applications, but now we look at some aspects of sound waves themselves.

12–1 Characteristics of Sound

We already saw in Chapter 11, Fig. 11–25, how a vibrating drumhead produces a sound wave in air. Indeed, we usually think of sound waves traveling in the air, for normally it is the vibrations of the air that force our eardrums to vibrate. But sound waves can also travel in other materials.

TABLE 12–1
Speed of Sound in Various Materials, at 20°C and 1 atm

Material	Speed (m/s)
Air	343
Air (0°C)	331
Helium	1005
Hydrogen	1300
Water	1440
Seawater	1560
Iron and steel	≈ 5000
Glass	≈ 4500
Aluminum	≈ 5100
Hardwood	≈ 4000

Two stones struck together under water can be heard by a swimmer beneath the surface, for the vibrations are carried to the ear by the water. When you put your ear flat against the ground, you can hear an approaching train or truck. In this case the ground does not actually touch your eardrum, but the longitudinal wave transmitted by the ground is called a sound wave just the same, for its vibrations cause the outer ear and the air within it to vibrate. Clearly, sound cannot travel in the absence of matter. For example, a bell ringing inside an evacuated jar cannot be heard, nor does sound travel through the empty reaches of outer space.

The **speed of sound** is different in different materials. In air at 0°C and 1 atm, sound travels at a speed of 331 m/s. We saw in Eq. 11–14b ($v = \sqrt{B/\rho}$) that the speed depends on the elastic modulus, B, and the density, ρ, of the material. Thus for helium, whose density is much less than that of air but whose elastic modulus is not greatly different, the speed is about three times as great as in air. In liquids and solids, which are much less compressible and therefore have much greater elastic moduli, the speed is larger still. The speed of sound in various materials is given in Table 12–1. The values depend somewhat on temperature, but this is significant mainly for gases. For example, in air, the speed increases approximately 0.60 m/s for each Celsius degree increase in temperature:

Speed of sound in air

$$v \approx (331 + 0.60\,T)\ \text{m/s},$$

where T is the temperature in °C. Unless stated otherwise, we will assume in this chapter that $T = 20°C$, so that $v = [\,331 + (0.60)(20)\,]$ m/s $= 343$ m/s.

➥ PHYSICS APPLIED

How far away is the lightning?

| **CONCEPTUAL EXAMPLE 12–1** | **Distance from a lightning strike.** A rule of thumb that tells how close lightning has hit is: "one mile for every five seconds before the thunder is heard." Justify, noting that the speed of light is so high (3×10^8 m/s) that the time for light to travel is negligible compared to the time for sound. |

RESPONSE The speed of sound in air is about 340 m/s, so to travel 1 km = 1000 m takes about 3 seconds. One mile is about 1.6 kilometers, so the time for the thunder to travel a mile is about $(1.6)(3) \approx 5$ seconds.

Two aspects of any sound are immediately evident to a human listener. These are "loudness" and "pitch," and each refers to a sensation in the consciousness of the listener. But to each of these subjective sensations there corresponds a physically measurable quantity. **Loudness** is related to the energy in the sound wave, and we shall discuss it in the next section.

Loudness

Pitch

The **pitch** of a sound refers to whether it is high, like the sound of a piccolo or violin, or low, like the sound of a bass drum or string bass. The physical quantity that determines pitch is the frequency, as was first noted by Galileo. The lower the frequency, the lower the pitch, and the higher the frequency, the higher the pitch. The human ear responds to frequencies in the range from about 20 Hz to about 20,000 Hz. (Recall that 1 Hz is 1 cycle per second.) This is called the **audible range**. These limits vary somewhat from one individual to another. One general trend is that as people age, they are less able to hear the high frequencies, so that the high-frequency limit may be 10,000 Hz or less.

Audible frequency range

Sound waves whose frequencies are outside the audible range may reach the ear, but we are not generally aware of them. Frequencies above

20,000 Hz are called **ultrasonic** (do not confuse with *supersonic*, which is used for an object moving with a speed faster than the speed of sound). Many animals can hear ultrasonic frequencies; dogs, for example, can hear sounds as high as 50,000 Hz, and bats can detect frequencies as high as 100,000 Hz. Ultrasonic waves have a number of applications in medicine and other fields, which we will discuss later in this chapter.

➡ PHYSICS APPLIED

Autofocusing camera

EXAMPLE 12–2 **Autofocusing with sound waves.** Autofocusing cameras emit a pulse of very high frequency (ultrasonic) sound that travels to the object being photographed, and include a sensor that detects the returning reflected sound, as shown in Fig. 12–1. To get an idea of the time sensitivity of the detector, calculate the travel time of the pulse for an object (*a*) 1.0 m away, (*b*) 20 m away.

SOLUTION We assume the temperature is about 20°C, so the speed of sound, as calculated above, is 343 m/s. (*a*) The pulse travels 1.0 m to the object and 1.0 m back, for a total of 2.0 m. Since speed = distance/time, we have

$$t = \frac{\text{distance}}{\text{speed}} = \frac{2.0 \text{ m}}{343 \text{ m/s}} = 0.0059 \text{ s} = 5.9 \text{ ms}.$$

(*b*) The total distance now is $2 \times 20 \text{ m} = 40 \text{ m}$, so

$$t = \frac{40 \text{ m}}{343 \text{ m/s}} = 0.12 \text{ s} = 120 \text{ ms}.$$

FIGURE 12–1 Example 12–2. Autofocusing camera emits an ultrasonic pulse. Solid lines represent the moving wave front of the outgoing wave pulse moving to the right; dashed lines represent the wave front of the pulse reflected off the person's face, returning to the camera. The time information allows the camera mechanism to adjust the lens to focus on the face.

Sound waves whose frequencies are below the audible range (that is, less than 20 Hz) are called **infrasonic**. Sources of infrasonic waves include earthquakes, thunder, volcanoes, and waves produced by vibrating heavy machinery. This last source can be particularly troublesome to workers, for infrasonic waves—even though inaudible—can cause damage to the human body. These low-frequency waves act in a resonant fashion, causing considerable motion and irritation of internal organs of the body.

We often describe sound waves in terms of the vibration of the molecules of the medium—that is, in terms of the motion or displacement of the molecules. But sound waves can also be analyzed from the point of view of pressure. Indeed, longitudinal waves are often called **pressure waves**. The pressure variation is usually easier to measure than the displacement. As can be seen in Figure 12–2, in a wave "compression" (where molecules are closest together), the pressure is higher than normal,

FIGURE 12–2 The membrane of a drum, as it vibrates, alternately compresses the air and then as it recedes (moves to the left) leaves a rarefaction or expansion of air. See also Fig. 11–25.

Expansion (pressure lower)   Compression (pressure higher)

Drum membrane   ◄►► Motion of a molecule

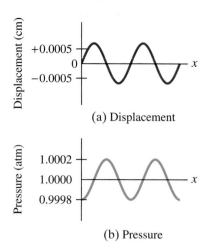

(a) Displacement

(b) Pressure

FIGURE 12–3 Representation of a sound wave in space at a given instant in terms of (a) displacement, and (b) pressure.

whereas in an expansion (or rarefaction) the pressure is less than normal. Figure 12–3 shows a graphical representation of a sound wave in air in terms of (a) displacement and (b) pressure. Note that the displacement wave is a quarter wavelength out of phase with the pressure wave: where the pressure is a maximum or minimum, the displacement from equilibrium is zero; and where the pressure variation is zero, the displacement is a maximum or minimum.

12–2 Intensity of Sound: Decibels

Like pitch, *loudness* is a sensation in the consciousness of a human being. It too is related to a physically measurable quantity, the **intensity** of the wave. Intensity is defined as the energy transported by a wave per unit time across unit area and, as we saw in the previous chapter (Section 11–9), is proportional to the square of the wave amplitude. Since energy per unit time is power, intensity has units of power per unit area, or watts/meter^2 (W/m^2).

The human ear can detect sounds with an intensity as low as 10^{-12} W/m^2 and as high as 1 W/m^2 (and even higher, although above this it is painful). This is an incredibly wide range of intensity, spanning a factor of a trillion (10^{12}) from lowest to highest. Presumably because of this wide range, what we perceive as loudness is not directly proportional to the intensity. To produce a sound that sounds about twice as loud requires a sound wave that has about 10 times the intensity. This is roughly valid at any sound level for frequencies near the middle of the audible range. For example, a sound wave of intensity 10^{-2} W/m^2 sounds to an average human being like it is about twice as loud as one whose intensity is 10^{-3} W/m^2, and four times as loud as 10^{-4} W/m^2.

Because of this relationship between the subjective sensation of loudness and the physically measurable quantity "intensity," it is usual to specify sound-intensity levels using a logarithmic scale. The unit on this scale is a **bel**, after Alexander Graham Bell (1847–1922), the inventor of the telephone, or much more commonly, the **decibel** (dB), which is $\frac{1}{10}$ bel (10 dB = 1 bel). The **intensity level**, β, of any sound is defined in terms of its intensity, I, as follows:

Intensity level (decibels)

$$\beta \text{ (in dB)} = 10 \log \frac{I}{I_0}, \tag{12–1}$$

where I_0 is the intensity of some reference level, and the logarithm is to the base 10. I_0 is usually taken as the minimum intensity audible to an average person, the "threshold of hearing," which is $I_0 = 1.0 \times 10^{-12}$ W/m^2. The intensity level of a sound whose intensity is $I = 1.0 \times 10^{-10}$ W/m^2, for example, will be

$$\beta = 10 \log \left(\frac{1.0 \times 10^{-10}}{1.0 \times 10^{-12}} \right) = 10 \log 100 = 20 \text{ dB},$$

since log 100 is equal to 2.0 (logarithms are discussed in Appendix A). Notice that the intensity level at the threshold of hearing is 0 dB; that is, $\beta = 10 \log (10^{-12}/10^{-12}) = 10 \log 1 = 0$ since log 1 = 0. Notice, too, that an increase in intensity by a factor of 10 corresponds to a level increase of 10 dB. An increase in intensity by a factor of 100 corresponds to a level increase of 20 dB. Thus a 50-dB sound is 100 times more intense than a 30-dB sound.

The intensities and intensity levels for a number of common sounds are listed in Table 12–2.

TABLE 12–2 Intensity of Various Sounds

Source of the Sound	Intensity Level (dB)	Intensity (W/m^2)
Jet plane at 30 m	140	100
Threshold of pain	120	1
Loud indoor rock concert	120	1
Siren at 30 m	100	1×10^{-2}
Auto interior, moving at 90 km/h	75	3×10^{-5}
Busy street traffic	70	1×10^{-5}
Ordinary conversation, at 50 cm	65	3×10^{-6}
Quiet radio	40	1×10^{-8}
Whisper	20	1×10^{-10}
Rustle of leaves	10	1×10^{-11}
Threshold of hearing	0	1×10^{-12}

EXAMPLE 12–3 **Loudspeaker response.** A high-quality loudspeaker is advertised to reproduce, at full volume, frequencies from 30 Hz to 18,000 Hz with uniform intensity ± 3 dB. That is, over this frequency range, the intensity level does not vary by more than 3 dB from the average. By what factor does the intensity change for the maximum intensity-level change of 3 dB?

SOLUTION Let us call the average intensity I_1 and the average level β_1. Then the maximum intensity, I_2, corresponds to a level $\beta_2 = \beta_1 + 3$ dB. Thus

$$\beta_2 - \beta_1 = 10 \log \frac{I_2}{I_0} - 10 \log \frac{I_1}{I_0}$$

$$3 \text{ dB} = 10 \left(\log \frac{I_2}{I_0} - \log \frac{I_1}{I_0} \right)$$

$$= 10 \log \frac{I_2}{I_1}$$

because $(\log a - \log b) = \log a/b$ (see Appendix A). Then

$$\log \frac{I_2}{I_1} = 0.30, \qquad \text{or} \qquad \frac{I_2}{I_1} = 10^{0.30}.$$

Using a calculator we calculate 10^x with $x = 0.30$, or from a log table, we find what number has a logarithm equal to 0.30. The result is 2.0, so

$$\frac{I_2}{I_1} = 2.0,$$

or I_2 is twice as intense as I_1.

It is worth noting that a sound-level difference of 3 dB (which corresponds to a doubled intensity as we just saw) corresponds to only a very small change in the subjective sensation of apparent loudness. Indeed, the average human can distinguish a difference in level of only about 1 or 2 dB.

Normally, the loudness or intensity of a sound decreases as you get farther from the source of the sound. In interior rooms, this effect is reduced because of reflections from the walls. However, if a source is in the open so that sound can radiate out freely in all directions, the intensity decreases as the inverse square of the distance,

$$I \propto \frac{1}{r^2},$$

as we saw in Section 11–9. Of course, if there is significant reflection from structures or the ground, the situation will be more complicated.

EXAMPLE 12–4 **Airplane roar.** The intensity level of the sound from a jet plane at a distance of 30 m is 140 dB. What is the intensity level at 300 m? (Ignore reflections from the ground.)

SOLUTION The intensity I at 30 m is found from Eq. 12–1:

$$140 \text{ dB} = 10 \log \left(\frac{I}{10^{-12} \text{ W/m}^2} \right).$$

Reversing the log equation to solve for I we have:

$$10^{14} = \frac{I}{10^{-12} \text{ W/m}^2},$$

so $I = (10^{14})(10^{-12} \text{ W/m}^2) = 10^2 \text{ W/m}^2$. At 300 m, 10 times as far away, the intensity will be $(\frac{1}{10})^2 = 1/100$ as much, or 1 W/m². Hence, the intensity level is

$$\beta = 10 \log \left(\frac{1 \text{ W/m}^2}{10^{-12} \text{ W/m}^2} \right) = 120 \text{ dB}.$$

Even at 300 m, the sound is at the threshold of pain. This is why workers at airports wear ear covers to protect their ears from damage (Fig. 12–4).

FIGURE 12–4 Airport worker with sound-intensity reducing ear covers.

* ### 12–3 Amplitude Related to Intensity

In this brief section we illustrate how marvelously sensitive the human ear is: it can detect displacements of air molecules that are actually less than the diameter of atoms (about 10^{-10} m).

The intensity I of a wave is proportional to the square of the wave amplitude, A, as discussed in Sections 11–9 and 11–10. Starting from Eq. 11–18, we can relate the amplitude quantitatively to the intensity I or level β, as the following Example shows.

➡ P H Y S I C S   A P P L I E D

Incredible sensitivity of the ear

EXAMPLE 12–5 **How tiny the displacement is.** Calculate the displacement of air molecules for a sound having a frequency of 1000 Hz at the threshold of hearing.

SOLUTION At the threshold of hearing, $I = 1 \times 10^{-12}$ W/m². We use Eq. 11–18 and solve for the amplitude A:

$$A = \frac{1}{\pi f} \sqrt{\frac{I}{2\rho v}}$$

$$= \frac{1}{(3.14)(1.0 \times 10^3 \text{ s}^{-1})} \sqrt{\frac{1.0 \times 10^{-12} \text{ W/m}^2}{(2)(1.29 \text{ kg/m}^3)(343 \text{ m/s})}},$$

where we have taken the density of air to be 1.29 kg/m³ and the speed of sound in air (assumed 20°C) as 343 m/s. When we do the arithmetic, we find that $A = 1.1 \times 10^{-11}$ m.

* 12–4 The Ear and Its Response; Loudness

The human ear, as we have seen, is a remarkably sensitive detector of sound. Mechanical detectors of sound, namely microphones, can barely match the ear in detecting low-intensity sounds.

PHYSICS APPLIED
Human ear

The function of the ear is to efficiently transform the vibrational energy of waves into electrical signals which are carried to the brain by way of nerves. A microphone performs a similar task. Sound waves striking the diaphragm of a microphone set it into vibration, and these vibrations are transformed into an electrical signal with the same frequencies, which can then be amplified and sent to a loudspeaker or tape recorder. We shall discuss the operation of microphones when we study electricity and magnetism in later chapters. Here we shall discuss the structure and response of the ear.

Figure 12–5 is a diagram of the human ear. The ear is conveniently divided into three main divisions: the outer ear, middle ear, and inner ear. In the outer ear, sound waves from the outside travel down the ear canal to the eardrum (the tympanum), which vibrates in response to the impinging waves. The middle ear consists of three small bones known as the hammer, anvil, and stirrup, which transfer the vibrations of the eardrum to the inner ear at the oval window. This delicate system of levers, coupled with the relatively large area of the eardrum compared to the area of the oval window, results in the pressure being amplified by a factor of about 40.

Structure of the ear

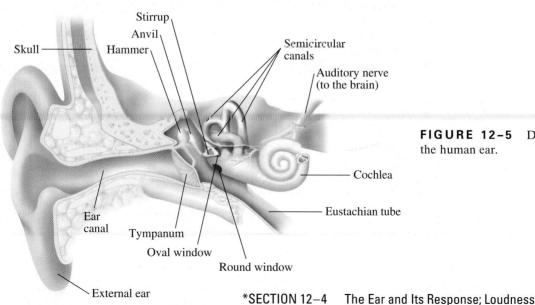

FIGURE 12–5 Diagram of the human ear.

SECTION 12–4 The Ear and Its Response; Loudness **353**

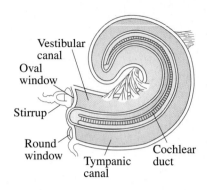

Vestibular canal
Oval window
Stirrup
Round window
Tympanic canal
Cochlear duct

FIGURE 12–6 Diagram of the cochlea.

The inner ear consists of the semicircular canals, which are important for controlling balance, and the liquid-filled cochlea, where the vibrational energy of sound waves is transformed into electrical energy and sent to the brain. Figure 12–6 is a diagrammatic representation of the cochlea. The sound vibration travels from the oval window down the vestibular canal and back up the tympanic canal. Because of the viscosity of the liquid, considerable damping occurs, but any remaining energy is dissipated at the round window at the end of the tympanic canal. Between these two canals is a third, known as the cochlear duct. On the membrane separating the cochlear duct from the tympanic canal (the basilar membrane) is the "organ of Corti," which contains some 30,000 nerve endings. As a pressure wave passes along the tympanic canal, it causes ripples in the basilar membrane and the attached organ of Corti; this is where the energy is transformed into electrical impulses and sent to the brain by way of auditory nerves. The basilar membrane is under tension, but is less taut and becomes thicker as it goes from the middle ear to the apex of the cochlea. From our earlier considerations, we might expect that the thicker, less taut end would be more sensitive to low frequencies and the tighter, thinner end more sensitive to the higher frequencies. Careful experiments indicate that this is indeed true, and this fact is important for our sensing of pitch.

Sensitivity of the ear

The ear is not equally sensitive to all frequencies. To hear the same loudness for sounds of different frequencies requires different intensities. Studies averaged over large numbers of people have produced the curves shown in Fig. 12–7. On this graph, each curve represents sounds that seemed to be equally loud. The number labeling each curve represents the *loudness level* (the units are called *phons*), which is numerically equal to the intensity level in dB at 1000 Hz. For example, the curve labeled 40 represents sounds that are

Loudness (in "phons")

FIGURE 12–7 Sensitivity of the human ear as a function of frequency (see text). Note that the frequency scale is "logarithmic" in order to cover a wide range of frequencies.

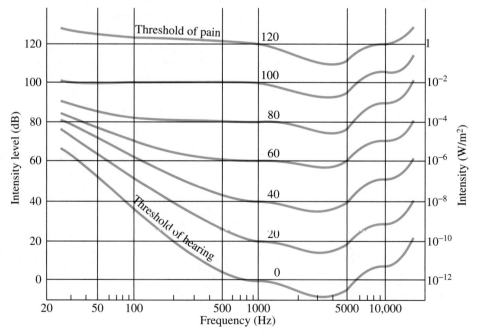

heard to have the same loudness as a 1000-Hz sound with an intensity level of 40 dB. From this 40-phon curve we see that a 100-Hz tone must have an intensity of about 62 dB to sound as loud (to an average person) as a 1000-Hz tone of only 40 dB. The lowest curve in Fig. 12–7 (labeled 0) represents the intensity level, as a function of frequency, for the softest sound that is just audible by a very good ear. Note that the ear is most sensitive to sounds of frequency between 2000 and 4000 Hz. Note too that whereas a 1000-Hz sound is audible at a level of 0 dB, a 100-Hz sound must be at least 40 dB to be heard.

The top curve in Fig. 12–7, labeled 120, represents the "threshold of feeling or pain." Sounds above this level can actually be felt and cause pain. As can be seen, it does not vary much with frequency.

Figure 12–7 shows that at lower intensity levels, our ears are less sensitive to the high and low frequencies relative to middle frequencies. The "loudness" control on stereo systems is intended to compensate for this. As the volume is turned down, the loudness control boosts the high and low frequencies relative to the middle frequencies so that the sound will have a more "normal-sounding" frequency balance. Many listeners, however, find the sound more pleasing or natural without the loudness control.

12–5 Sources of Sound: Vibrating Strings and Air Columns

The source of any sound is a vibrating object. Almost any object can vibrate and hence be a source of sound. We now discuss some simple sources of sound, particularly musical instruments. In musical instruments, the source is set into vibration by striking, plucking, bowing, or blowing. Standing waves are produced and the source vibrates at its natural resonant frequencies. The vibrating source is in contact with the air (or other medium) and pushes on it to produce sound waves that travel outward. The frequencies of the waves are the same as the source, but the speed and wavelengths can be different. A drum has a stretched membrane that vibrates. Xylophones and marimbas have metal or wood bars that can be set into vibration. Bells, cymbals, and gongs also make use of a vibrating metal. The most widely used instruments make use of vibrating strings, such as the violin, guitar, and piano, or make use of vibrating columns of air, such as the flute, trumpet, and pipe organ. We have already seen that the pitch of a pure sound is determined by the frequency. Typical frequencies for musical notes on the so-called "equally tempered chromatic scale" are given in Table 12–3 for the octave beginning with middle C. Note that one octave corresponds to a doubling of frequency. For example, middle C has frequency of 262 Hz whereas C′ (C above middle C) has twice the frequency, 524 Hz.

TABLE 12–3
Equally Tempered Chromatic Scale[†]

Note	Frequency (Hz)
C	262
C# or D♭	277
D	294
D# or E♭	311
E	330
F	349
F# or G♭	370
G	392
G# or A♭	415
A	440
A# or B♭	466
B	494
C′	524

[†]Only one octave is included.

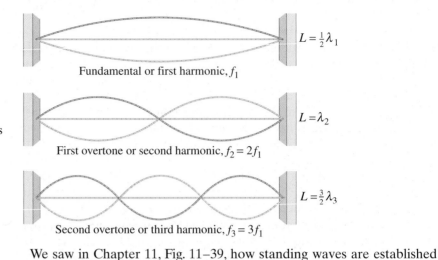

$L = \frac{1}{2}\lambda_1$

Fundamental or first harmonic, f_1

$L = \lambda_2$

First overtone or second harmonic, $f_2 = 2f_1$

$L = \frac{3}{2}\lambda_3$

Second overtone or third harmonic, $f_3 = 3f_1$

FIGURE 12–8 Standing waves on a string—only the lowest three frequencies are shown.

➡ **PHYSICS APPLIED**

Stringed instruments

FIGURE 12–9 The wavelength of a fingered string (b) is shorter than that of an unfingered string (a). Hence, the frequency of the fingered string is higher. Only one string is shown on this guitar, and only the simplest standing wave, the fundamental, is shown.

(a)

(b)

We saw in Chapter 11, Fig. 11–39, how standing waves are established on a string, and we show this again here in Fig. 12–8. This is the basis for all stringed instruments. The pitch is normally determined by the lowest resonant frequency, the **fundamental**, which corresponds to nodes occurring only at the ends. The wavelength of the fundamental on the string is equal to twice the length of the string. Therefore, the fundamental frequency is $f = v/\lambda = v/2L$, where v is the velocity of the wave on the string. When a finger is placed on the string of, say, a guitar or violin, the effective length of the string is shortened. So its fundamental frequency, and pitch, is higher since the wavelength of the fundamental is shorter (Fig. 12–9). The strings on a guitar or violin are all the same length. They sound at a different pitch because the strings have different mass per unit length, m/L, which affects the velocity as seen in Eq. 11–13, $v = \sqrt{F_T/(m/L)}$. (The tension may also be different; adjusting the tension is the means for tuning the instrument.) Thus the velocity on a heavier string is less and the frequency will be less for the same wavelength. In pianos and harps, the strings are each of different length. For the lower notes the strings are not only longer, but heavier as well, and the reason why is illustrated in the following Example.

EXAMPLE 12–6 **Piano strings.** The highest key on a piano corresponds to a frequency about 150 times that of the lowest key. If the string for the highest note is 5.0 cm long, how long would the string for the lowest note have to be if it had the same mass per unit length and was under the same tension?

SOLUTION The velocity would be the same on each string, so the frequency is inversely proportional to the length L of the string ($f = v/\lambda = v/2L$). Thus

$$\frac{L_L}{L_H} = \frac{f_H}{f_L},$$

where the subscripts L and H refer to the lowest and highest notes, respectively. Thus $L_L = L_H(f_H/f_L) = (5.0 \text{ cm})(150) = 750 \text{ cm}$, or 7.5 m. This would be ridiculously long for a piano. The longer lower strings are made heavier partly to avoid this, so that even on grand pianos the strings are no longer than about 3 m.

EXAMPLE 12–7 **Frequencies and wavelengths in the violin.** A 0.32-m-long violin string is tuned to play A above middle C at 440 Hz. (a) What is the wavelength of the fundamental string vibration, and (b) what are the frequency and wavelength of the sound wave produced? (c) Why is there a difference?

SOLUTION (*a*) From Fig. 12–8 we see that the wavelength of the fundamental is

$$\lambda = 2L = 0.64 \text{ m} = 64 \text{ cm}.$$

This is the wavelength of the standing wave on the string.
(*b*) The sound wave that travels outward in the air (to reach our ears) has the same frequency, 440 Hz (why?). Its wavelength is

$$\lambda = \frac{v}{f} = \frac{343 \text{ m/s}}{440 \text{ Hz}} = 0.78 \text{ m},$$

where v is the speed of sound in air (assumed at 20°C).
(*c*) The wavelength of the sound wave is different from that of the standing wave on the string because the speed of sound in air (343 m/s at 20°C) is different from the speed of the wave on the string ($= f\lambda = 440 \text{ Hz} \times 0.64 \text{ m} = 280 \text{ m/s}$), which of course depends on the tension in the string and its mass per unit length.

Stringed instruments would not be very loud if they relied on their vibrating strings to produce the sound waves since the strings are simply too thin to compress and expand much air. Stringed instruments therefore make use of a kind of mechanical amplifier known as a *sounding board* (piano) or *sounding box* (guitar, violin), which acts to amplify the sound by putting a greater surface area in contact with the air (Fig. 12–10). When the strings are set into vibration, the sounding board or box is set into vibration as well. Since it has much greater area in contact with the air, it can produce a much stronger sound wave. On an electric guitar, the sounding box is not so important since the vibrations of the strings are amplified electronically.

Instruments such as woodwinds, the brasses, and the pipe organ produce sound from the vibrations of standing waves in a column of air within a tube or pipe (Fig. 12–11). Standing waves can occur in the air of any cavity, but the frequencies present are complicated for any but very simple shapes such as a long, narrow tube. In some instruments, a vibrating reed or the vibrating lip of the player helps to set up vibrations of the air column. In others, a stream of air is directed against one edge of the opening or mouthpiece, leading to turbulence which sets up the vibrations. Because of the disturbance, whatever its source, the air within the tube vibrates with a variety of frequencies; but only frequencies that correspond to standing waves will persist.

For a string fixed at both ends, Fig. 12–8, the standing waves have nodes (no movement) at the two ends, and one or more antinodes (large amplitude of vibration) in between; a node separates successive antinodes. The lowest-frequency standing wave, the *fundamental*, corresponds to a single antinode. The higher-frequency standing waves are called **overtones** or **harmonics**. Specifically, the first harmonic is the fundamental, the second harmonic has twice the frequency of the fundamental,[†] and so on. (See also Section 11–12.)

[†]When the resonant frequencies above the fundamental (that is, the overtones) are integral multiples of the fundamental, they are called harmonics. But if the overtones are not integral multiples of the fundamental, as is the case for a vibrating drumhead, for example, they are not harmonics.

(a)

(b)

FIGURE 12–10 (a) Sounding box (guitar); (b) sounding board (to which strings are attached, inside of piano).

➡ **PHYSICS APPLIED**

Wind instruments

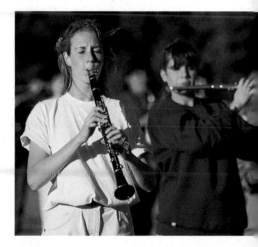

FIGURE 12–11 Wind instruments: clarinet (left) and flute.

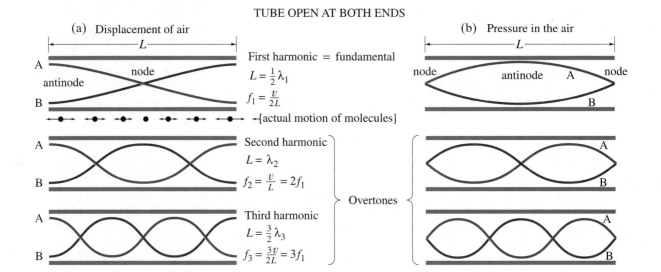

(a) Displacement of air

(b) Pressure in the air

A

antinode   node

B

←[actual motion of molecules]

First harmonic = fundamental

$L = \frac{1}{2}\lambda_1$

$f_1 = \frac{v}{2L}$

node   antinode A node

B

A

B

Second harmonic

$L = \lambda_2$

$f_2 = \frac{v}{L} = 2f_1$

A

B

A

B

Third harmonic

$L = \frac{3}{2}\lambda_3$

$f_3 = \frac{3v}{2L} = 3f_1$

Overtones

A

B

FIGURE 12–12 Modes of vibration (standing waves) for a tube open at both ends ("open tube"). The simplest modes of vibration are shown in (a), on the left, in terms of the motion of the air (displacement), and in (b), on the right, in terms of air pressure. These are graphs, shown placed within the tube, and are labeled A and B where B represents the wave form $\frac{1}{2}$ period after the moment when it has the shape labeled A. The actual motion of molecules for one case is shown just below the tube at upper left.

Open tubes produce all harmonics

The situation is similar for a column of air, but we must remember that it is now air itself that is vibrating. We can describe the waves either in terms of the flow of the air—that is, in terms of the *displacement* of air— or in terms of the *pressure* in the air (see Figs. 12–2 and 12–3). In terms of displacement, the air at the closed end of a tube is a displacement node since the air is not free to move there, whereas at the open end of a tube there will be an antinode since the air can move freely. The air within the tube vibrates in the form of longitudinal standing waves. The possible modes of vibration for a tube open at both ends (called an **open tube**), are shown *graphically* in Fig. 12–12. They are shown for a tube that is open at one end but closed at the other (called a **closed tube**) in Fig. 12–13. [A tube closed at *both* ends, having no connection to the outside air, would be useless as an instrument.] The graphs in part (a) of each Figure (left sides) represent the displacement amplitude of the vibrating air in the tube. Note that these are graphs, and that the air molecules themselves oscillate *horizontally*, parallel to the tube length, as shown by the small arrows in the top diagram of Fig. 12–12a (on the left). The exact position of the antinode near the open end of a tube depends on the diameter of the tube, but if the diameter is small compared to the length, which is the usual case, the antinode occurs very close to the end as shown. We assume this is the case in what follows. (The position of the antinode may also depend slightly on the wavelength and other factors.)

Let us look in detail at the open tube, in Fig. 12–12a, which might be a flute. An open tube has displacement antinodes at both ends. Notice that there must be at least one node within an open tube if there is to be a standing wave at all. A single node corresponds to the fundamental frequency of the tube. Since the distance between two successive nodes, or between two successive antinodes, is $\frac{1}{2}\lambda$, there is one-half a wavelength within the length of the tube in this case: $L = \frac{1}{2}\lambda$, or $\lambda = 2L$. So the fundamental frequency is $f_1 = v/\lambda = v/2L$, where v is the velocity of sound in air. The standing wave with two nodes is the first overtone or second harmonic and has half the wavelength ($L = \lambda$) and twice the frequency. Indeed, the frequency of each overtone is an integral multiple of the fundamental frequency. This is just what is found for a string.

(a) Displacement of air

A
B

First harmonic = fundamental
$$L = \frac{1}{4}\lambda_1$$
$$f_1 = \frac{v}{4L}$$

A
B

Third harmonic
$$L = \frac{3}{4}\lambda_3$$
$$f_3 = \frac{3v}{4L} = 3f_1$$

} Overtones {

A
B

Fifth harmonic
$$L = \frac{5}{4}\lambda_5$$
$$f_5 = \frac{5v}{4L} = 5f_1$$

(b) Pressure in the air

A
B

A
B

A
B

For a closed tube, shown in Fig. 12–13a, which could be a clarinet, there is always a displacement node at the closed end (because the air is not free to move) and an antinode at the open end (where the air can move freely). Since the distance between a node and the nearest antinode is $\frac{1}{4}\lambda$, we see that the fundamental in a closed tube corresponds to only one–fourth of a wavelength within the length of the tube: $L = \lambda/4$, and $\lambda = 4L$. The fundamental frequency is thus $f_1 = v/4L$, or half what it is for an open pipe of the same length. There is another difference, for as we can see from Fig. 12–13a, only the odd harmonics are present in a closed pipe: the overtones have frequencies equal to 3, 5, 7, $\cdots$ times the fundamental frequency. There is no way for waves with 2, 4, 6, $\cdots$ times the fundamental frequency to have a node at one end and an antinode at the other, and thus they cannot exist as standing waves in a closed tube.

If this description in terms of displacement seems hard to understand, or you want to understand it from another point of view, then consider a description in terms of the *pressure* in the air, shown in part (b) of Figs. 12–12 and 12–13 (right sides). Where the air in a wave is compressed, the pressure is higher, whereas in a wave expansion (or rarefaction), the pressure is less than normal. The open end of a tube is open to the atmosphere. Hence the pressure at an open end must be a *node:* the pressure doesn't alternate, but remains at the outside atmospheric pressure. If a tube has a closed end, the pressure at that closed end can readily alternate to be above or below atmospheric pressure. Hence there is a pressure *antinode* at a closed end of a tube. Of course there can be pressure nodes and antinodes within the tube, and some of the possible vibrational modes in terms of pressure for an open tube are shown in Fig. 12–12b, and for a closed tube are shown in Fig. 12–13b.

Pipe organs (Fig. 12–14) make use of both open and closed pipes. Notes of different pitch are sounded using different pipes with different lengths from a few centimeters to 5 m or more. Other musical instruments act either like a closed tube or an open tube. A flute, for example, is an open tube, for it is open not only where you blow into it, but also at the opposite end as well. The different notes on a flute and many other instruments are obtained by shortening the length of the tube—that is, by uncovering holes along its length. In a trumpet, on the other hand, pushing down on the valves opens additional lengths of tube. In all these instruments, the longer the length of the vibrating air column, the lower the frequency.

FIGURE 12–13 Modes of vibration (standing waves) for a tube closed at one end ("closed tube"). See caption for Fig. 12–12.

Closed tubes produce only odd harmonics

FIGURE 12–14 Pipe organ.

EXAMPLE 12–8 **Open and closed organ pipes.** What will be the fundamental frequency and first three overtones for a 26-cm-long organ pipe at 20°C if it is (a) open and (b) closed?

SOLUTION At 20°C, the speed of sound in air is 343 m/s (Section 12–1). (a) For the open pipe, the fundamental frequency is

$$f_1 = \frac{v}{2L} = \frac{343 \text{ m/s}}{2(0.26 \text{ m})} = 660 \text{ Hz}.$$

The overtones, which include all harmonics, are 1320 Hz, 1980 Hz, 2640 Hz, and so on.
(b) Referring to Fig. 12–13, we see that for a closed pipe

$$f_1 = \frac{v}{4L} = \frac{343 \text{ m/s}}{4(0.26 \text{ m})} = 330 \text{ Hz}.$$

But only the odd harmonics will be present, so the first three overtones will be 990 Hz, 1650 Hz, and 2310 Hz. (The closed pipe plays 330 Hz, which, from Table 12–3, is E above middle C, whereas the open pipe of the same length plays 660 Hz, an octave higher.)

EXAMPLE 12–9 **Flute.** A flute is designed to play middle C (262 Hz) as the fundamental frequency when all the holes are covered. Approximately how long should the distance be from the mouthpiece to the far end of the flute? (Note: This is only approximate since the antinode does not occur precisely at the mouthpiece.) Assume the temperature is 20°C.

SOLUTION The speed of sound in air at 20°C is 343 m/s. Then, from Fig. 12–12 (a flute is an open tube, as mentioned earlier), the fundamental frequency f_1 is related to the length of the vibrating air column by $f = v/2L$. Solving for L, we find

$$L = \frac{v}{2f} = \frac{343 \text{ m/s}}{2(262 \text{ s}^{-1})} = 0.655 \text{ m}.$$

EXAMPLE 12–10 **A cold flute.** If the temperature is only 10°C, what will be the frequency of the note played when all the openings are covered in the flute of Example 12–9?

SOLUTION The length L is still 65.5 cm. But now the velocity of sound is less since it changes by 0.60 m/s per each C°. For a drop of 10 C°, the velocity decreases by 6 m/s to 337 m/s. The frequency will be

$$f = \frac{v}{2L} = \frac{337 \text{ m/s}}{2(0.655 \text{ m})} = 257 \text{ Hz}.$$

This example illustrates why players of wind instruments take time to "warm up" their instruments so they will be in tune. The effect of temperature on stringed instruments is much smaller.

*12–6 Quality of Sound, and Noise

Whenever we hear a sound, particularly a musical sound, we are aware of its loudness, its pitch, and also of a third aspect called "quality." For example, when a piano and then a flute play a note of the same loudness and pitch (say middle C), there is a clear difference in the overall sound. We would never mistake a piano for a flute. This is what is meant by the **quality** of a sound. For musical instruments, the terms *timbre* or *tone color* are also used.

Just as loudness and pitch can be related to physically measurable quantities, so too can quality. The quality of a sound depends on the presence of overtones—their number and their relative amplitudes. Generally, when a note is played on a musical instrument, the fundamental as well as overtones are present simultaneously. Figure 12–15 illustrates how the superposition of three wave forms, in this case the fundamental and first two overtones (with particular amplitudes), would combine to give a composite *waveform*. Of course, more than two overtones are usually present.

The relative amplitudes of the various overtones are different for different musical instruments, and this is what gives each instrument its characteristic quality or timbre. A graph showing the relative amplitudes of the harmonics produced by an instrument is called a "sound spectrum." Several typical examples for different instruments are shown in Fig. 12–16. Normally, the fundamental has the greatest amplitude and its frequency is what is heard as the pitch.

The manner in which an instrument is played strongly influences the sound quality. Plucking a violin string, for example, makes a very different sound than pulling a bow across it. The sound spectrum at the very start (or end) of a note (as when a hammer strikes a piano string) can be very different from the subsequent sustained tone. This too affects the subjective tone quality of an instrument.

An ordinary sound, like that made by striking two stones together, is a noise that has a certain quality, but a clear pitch is not discernible. A noise such as this is a mixture of many frequencies which bear little relation to one another. If a sound spectrum were made of this noise, it would not show discrete lines like those of Fig. 12–16. Instead it would show a continuous, or nearly continuous, spectrum of frequencies. Such a sound we call "noise" in comparison with the more harmonious sounds which contain frequencies that are simple multiples of the fundamental.

Noise affects us in various ways, particularly psychologically. Sometimes it is a mere annoyance, but loud noise can cause loss of hearing, and this is particularly a problem in factories and other industrial workplaces, where the sound level may be high for long periods. Rock musicians, too, can suffer hearing loss, for intensity levels as high as 120 dB are commonly produced. Hearing loss due to excessive noise levels was recognized by the ancient Romans. Regardless of source, hearing loss due to noise is particularly serious in the frequency range from about 2000 to 5000 Hz, an important region for speech and music.

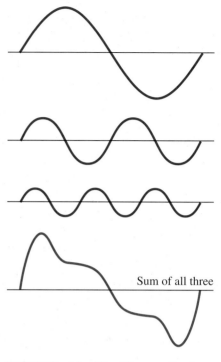

FIGURE 12–15 The amplitudes of the fundamental and first two overtones are added at each point to get the "sum," or composite waveform.

FIGURE 12–16 Sound spectra. The shapes of the spectra change as the instruments play different notes.

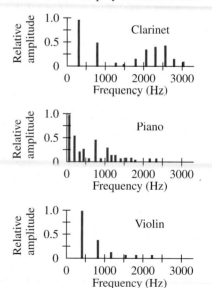

The problem of noise control is a difficult one. Isolation of a source of sound with barriers is helpful but expensive and not always convenient. Attacking the source of the sound is often beneficial. Reducing the area of vibration of machinery is important, for the greater the area, the more air that can be "pushed" and the louder the sound. Constructing surfaces out of stiffer materials, thus reducing the amplitude, or coating surfaces with energy-absorbing material, can be effective. The placement of machinery is important, for the floor, a wall, or other object may resonate with the machine's vibration, significantly increasing the amplitude. Careful maintenance is important, for lack of lubrication, loose bolts, and worn parts can cause vibrations. The noise produced by jet planes is also a serious problem and can affect a whole community.

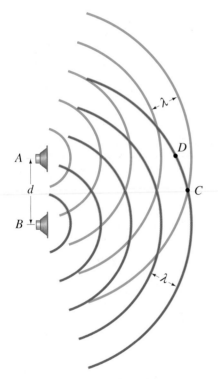

FIGURE 12–17 Sound waves from two loudspeakers interfere.

12–7 Interference of Sound Waves; Beats

We saw in Section 11–11 that when two waves simultaneously pass through the same region of space, they interfere with one another. Since this can occur for any kind of wave, we should expect that interference will occur with sound waves, and indeed it does.

As a simple example, consider two large loudspeakers, A and B, a distance d apart on the stage of an auditorium as shown in Fig. 12–17. Let us assume the two speakers are emitting sound waves of the same single frequency and that they are in phase: that is, when one speaker is forming a compression, so is the other. (We ignore reflections from walls, floor, etc.) The curved lines in the diagram represent the crests of sound waves from each speaker. Of course, we must remember that for a sound wave, a crest is a compression in the air whereas a trough—which falls between two crests—is a rarefaction. A person or detector at a point such as C, which is the same distance from each speaker, will experience a loud sound because the interference will be constructive. On the other hand, at a point such as D in the diagram, little if any sound will be heard because destructive interference occurs—compressions of one wave meet rarefactions of the other and vice versa (see Fig. 11–36 and related discussion on water waves in Section 11–11).

An analysis of this situation is perhaps clearer if we graphically represent the wave forms as in Fig. 12–18. In Fig. 12–18a it can be seen that at point C, constructive interference occurs since both waves simultaneously have crests or simultaneously have troughs. In Fig. 12–18b we see the situation for point D. The wave from speaker B must travel a greater distance than the wave from A. Thus the wave from B lags behind that from A. In this diagram, point E is chosen so that the distance ED is equal to AD. Thus we see that if the distance BE is equal to precisely one-half the wavelength of the sound, the two waves will be exactly out of phase when they reach D, and destructive interference occurs. This then is the criterion for determining at what points destructive interference occurs: destructive interference occurs at any point whose distance from one speaker is greater than its distance from the other speaker by exactly one-half wavelength. Notice that if this extra distance (BE in Fig. 12–18b) is equal to a whole wavelength (or 2, 3, $\cdots$ wavelengths), then the two waves will be in phase and *constructive interference* occurs. If the distance BE equals $\frac{1}{2}$, $1\frac{1}{2}$, $2\frac{1}{2}$, $\cdots$ wavelengths, *destructive interference* occurs.

It is important to realize that a person sitting at point D hears nothing at all (or nearly so), yet sound is coming from both speakers. Indeed, if one of the speakers is turned off, the sound from the other speaker would be clearly heard.

If a loudspeaker emits a whole range of frequencies, not all wavelengths will destructively interfere at any one point such as D. Rather, only specific wavelengths will destructively interfere completely, according to the criteria above.

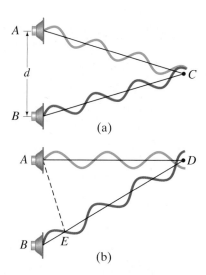

FIGURE 12–18 Sound waves of a single frequency from loudspeakers A and B (see Fig. 12–17) constructively interfere at C and destructively interfere at D.

EXAMPLE 12–11 **Loudspeakers' interference.** The two loudspeakers in Fig. 12–17 are 1.00 m apart. A person stands 4.00 m from one speaker. How far must this person be from the second speaker in order to detect destructive interference when the speakers emit an 1150-Hz sound? Assume the temperature is 20°C.

SOLUTION The wavelength of this sound is

$$\lambda = \frac{v}{f} = \frac{343 \text{ m/s}}{1150 \text{ Hz}} = 0.30 \text{ m.}$$

For destructive interference to occur, the person must be one-half wavelength farther from one loudspeaker than from the other, or 0.15 m. Thus the person must be 4.13 m (or 3.85 m) from the second speaker. Notice in this example that if the speakers were less than 0.15 m apart, there would be no point that was 0.15 m farther from one speaker than the other, and there would be no point where destructive interference would occur.

We have been discussing interference of sound waves that takes place in space. An interesting and important example of interference that occurs in time is the phenomenon known as **beats**. This is the phenomenon that occurs if two sources of sound—say, two tuning forks—are close in frequency but not exactly the same. Sound waves from the two sources interfere with each other and the sound level at a given position alternately rises and falls; the regularly spaced intensity changes are called beats.

Beats

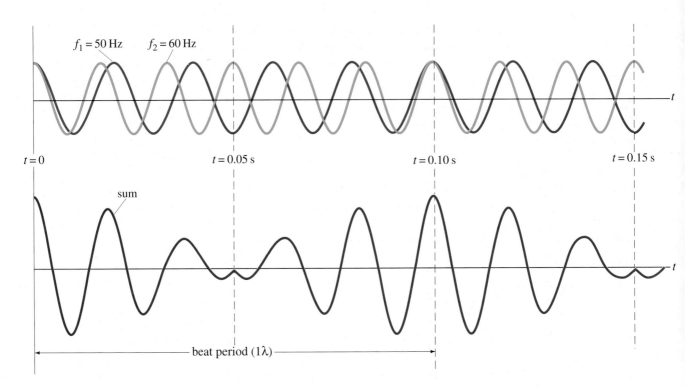

FIGURE 12–19 Beats occur as a result of the superposition of two sound waves of slightly different frequency.

Beat frequency = difference in the two wave frequencies

➡ **PHYSICS APPLIED**

Tuning a piano

To see how beats arise, consider two equal-amplitude sound waves of frequency $f_1 = 50$ Hz and $f_2 = 60$ Hz, respectively. In 1.00 s, the first source makes 50 vibrations whereas the second makes 60. We now examine the waves at one point in space equidistant from the two sources. The waveforms for each wave as a function of time are shown on the top graph of Fig. 12–19; the magenta line represents the 50 Hz wave, the blue line represents the 60 Hz wave. The lower graph in Fig. 12–19 shows the sum of the two waves. At time $t = 0$ the two waves are shown to be in phase and interfere constructively. Because the two waves vibrate at different rates, at time $t = 0.05$ s they are completely out of phase and destructive interference occurs as shown in the figure. At $t = 0.10$ s, they are again in phase and the resultant amplitude again is large. Thus the resultant amplitude is large every 0.10 s and in between it drops drastically. This rising and falling of the intensity is what is heard as beats.[†] In this case the beats are 0.10 s apart. That is, the **beat frequency** is ten per second or 10 Hz. This result, that the beat frequency equals the difference in frequency of the two waves, is valid in general.

The phenomenon of beats can occur with any kind of wave and is a very sensitive method for comparing frequencies. For example, to tune a piano, a piano tuner listens for beats produced between his standard tuning fork and that of a particular string on the piano, and knows it is in tune when the beats disappear. The members of an orchestra can tune up by listening for beats between their instruments and that of a standard tone (usually A above middle C at 440 Hz) produced by a piano or an oboe.

[†]Beats will be heard even if the amplitudes are not equal, as long as the difference is not great.

EXAMPLE 12–12 **Beats.** A tuning fork produces a steady 400-Hz tone. When this tuning fork is struck and held near a vibrating guitar string, twenty beats are counted in five seconds. What are the possible frequencies produced by the guitar string?

SOLUTION The beat frequency is

$$f_{\text{beat}} = \frac{20 \text{ vibrations}}{5 \text{ s}} = 4 \text{ Hz}.$$

This is the difference of the frequencies of the two waves, and because one wave is known to be 400 Hz, the other must be either 404 Hz or 396 Hz.

12–8 Doppler Effect

You may have noticed that the pitch of the siren on a speeding firetruck drops abruptly as it passes you. Or you may have noticed the change in pitch of a blaring horn on a fast-moving car as it passes by. The pitch of the sound from the engine of a race car changes as it passes an observer. When a source of sound is moving toward an observer, the pitch is higher than when the source is at rest; and when the source is traveling away from the observer, the pitch is lower. This phenomenon is known as the **Doppler effect**[†] and occurs for all types of waves. Let us now see why it occurs, and calculate the change in frequency for sound waves.

To be concrete, consider the siren of a firetruck at rest, which is emitting sound of a particular frequency in all directions as shown in Fig. 12–20a. The wave velocity depends only on the medium in which it is traveling, and is independent of the velocity of the source or observer. If our source, the firetruck, is moving, the siren emits sound at the same frequency as it does at rest. But the sound waves it emits forward are closer together than normal, as shown in Fig. 12–20b. This is because the fire engine, as it moves, is "catching up" with the previously emitted wavefronts. Thus an observer on the sidewalk will detect more wave crests passing per second, so the frequency is higher. The wavefronts emitted behind the truck, on the other hand, are farther apart than normal because the truck is speeding away from them. Hence, fewer wave crests per second pass by an observer behind the truck and the pitch is lower.

[†]After J. C. Doppler (1803–1853), who enunciated it in 1842.

(a) At rest

(b) Firetruck moving

FIGURE 12–20
(a) Both observers on the sidewalk hear the same frequency from the firetruck at rest. (b) Doppler effect: observer toward whom the firetruck moves hears a higher-frequency sound, and observer behind the firetruck hears a lower frequency.

Frequency change—
moving source

To calculate the change in frequency, we make use of Fig. 12–21, and we assume the air (or other medium) is at rest in our reference frame. In Fig. 12–21a, the source of the sound, shown as a dot, is at rest; two successive wave crests are shown, the second of which is just in the process of being emitted. The distance between these crests is λ, the wavelength. If the frequency of the source is f, then the time between emissions of wave crests is

$$T = \frac{1}{f}.$$

In Fig. 12–21b, the source is moving with a velocity v_s. In a time T (as just defined), the first wave crest has moved a distance $d = vT$, where v is the velocity of the sound wave in air (which is, of course, the same whether the source is moving or not). In this same time, the source has moved a distance $d_s = v_s T$. Then the distance between successive wave crests, which is the new wavelength $λ'$, is (since $d = λ$)

$$λ' = d - d_s$$

$$= λ - v_s T$$

$$= λ - v_s \frac{λ}{v}$$

$$= λ\left(1 - \frac{v_s}{v}\right).$$

The change in wavelength, $Δλ$, is

$$Δλ = λ' - λ = -v_s \frac{λ}{v}.$$

So the shift in wavelength is directly proportional to the speed v_s of the source. The new frequency, on the other hand, is given by

$$f' = \frac{v}{λ'} = \frac{v}{λ\left(1 - \dfrac{v_s}{v}\right)}$$

or, since $v/λ = f$,

$$f' = \frac{f}{\left(1 - \dfrac{v_s}{v}\right)}. \qquad \begin{bmatrix} \text{source moving toward} \\ \text{stationary observer} \end{bmatrix} \quad \textbf{(12–2a)}$$

Because the denominator is less than 1, $f' > f$. For example, if a source emits a sound of frequency 400 Hz when at rest, then when the source moves toward a fixed observer with a speed of 30 m/s, the observer hears a frequency (at 20°C) of

$$f' = \frac{400 \text{ Hz}}{1 - \dfrac{30 \text{ m/s}}{343 \text{ m/s}}} = 438 \text{ Hz}.$$

For a source that is moving *away* from the observer at a speed v_s, the

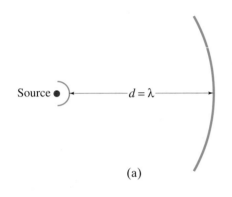

Source ● $d = λ$

(a)

Crest emitted when source was at point 1.

Crest emitted when source was at point 2.

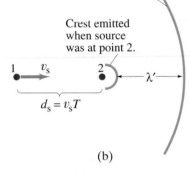

1 v_s 2 $λ'$

$d_s = v_s T$

(b)

FIGURE 12–21
Determination of frequency change in the Doppler effect (see text). Red dot is the source.

new wavelength will be

$$\lambda' = d + d_s,$$

and the change in wavelength will be

$$\Delta\lambda = \lambda' - \lambda = +v_s\frac{\lambda}{v}.$$

The frequency of the wave will be

$$f' = \frac{f}{\left(1 + \dfrac{v_s}{v}\right)}. \qquad \begin{bmatrix}\text{source moving away from}\\ \text{stationary observer}\end{bmatrix} \qquad \textbf{(12–2b)}$$

In this case, if a source vibrating at 400 Hz is moving away from a fixed observer at 30 m/s, the latter will hear a frequency of about 368 Hz.

The Doppler effect also occurs when the source is at rest and the observer is in motion. If the observer is traveling toward the source, the pitch is higher; and if the observer is traveling away from the source, the pitch is lower. Quantitatively, the change in frequency is slightly different than for the case of a moving source. With a fixed source and a moving observer, the distance between wave crests, the wavelength λ, is not changed. But the velocity of the crests with respect to the observer is changed. If the observer is moving toward the source, Fig. 12–22, the speed of the waves relative to the observer is $v' = v + v_o$, where v is the velocity of sound in air (we assume the air is still) and v_o is the velocity of the observer. Hence, the new frequency is

Frequency change— moving observer

$$f' = \frac{v'}{\lambda} = \frac{v + v_o}{\lambda}$$

or, since $\lambda = v/f$,

$$f' = \left(1 + \frac{v_o}{v}\right)f. \qquad \begin{bmatrix}\text{observer moving toward}\\ \text{stationary source}\end{bmatrix} \qquad \textbf{(12–3a)}$$

If the observer is moving away from the source, the relative velocity is $v' = v - v_o$ and

$$f' = \left(1 - \frac{v_o}{v}\right)f. \qquad \begin{bmatrix}\text{observer moving away}\\ \text{from stationary source}\end{bmatrix} \qquad \textbf{(12–3b)}$$

Source
$v_s = 0$
λ
v
Observer
v_o

FIGURE 12–22 Observer moving with speed v_o toward stationary source "sees" wave crests pass at speed $v' = v + v_o$ where v is the speed of the sound waves in air.

EXAMPLE 12–13 **A moving siren.** The siren of a police car at rest emits at a predominant frequency of 1600 Hz. What frequency will you hear if you are at rest and the police car moves at 25.0 m/s (*a*) toward you, (*b*) away from you?

SOLUTION (*a*) We use Eq. 12–2a:

$$f' = \frac{f}{\left(1 - \dfrac{v_s}{v}\right)} = \frac{1600\ \text{Hz}}{\left(1 - \dfrac{25.0\ \text{m/s}}{343\ \text{m/s}}\right)} = 1726\ \text{Hz}.$$

(*b*) We use Eq. 12–2b:

$$f' = \frac{f}{\left(1 + \dfrac{v_s}{v}\right)} = \frac{1600\ \text{Hz}}{\left(1 + \dfrac{25.0\ \text{m/s}}{343\ \text{m/s}}\right)} = 1491\ \text{Hz}.$$

When a sound wave is reflected from a moving obstacle, the frequency of the reflected wave will, because of the Doppler effect, be different from that of the incident wave. This is illustrated in the following Example.

EXAMPLE 12–14 **Two Doppler shifts.** A 5000-Hz sound wave is directed toward an object moving 3.50 m/s toward the (stationary) source (Fig. 12–23). What is the frequency of the reflected wave?

SOLUTION There are actually two Doppler shifts in this situation. First, the object acts like a moving observer (Fig. 12–23a) and "detects" a sound wave of frequency (Eq. 12–3a)

$$f' = \left(1 + \frac{v_0}{v}\right)f = \left(1 + \frac{3.50\ \text{m/s}}{343\ \text{m/s}}\right)(5000\ \text{Hz}) = 5051\ \text{Hz}.$$

Second, the object acts like a moving source (Eq. 12–2a) in reemitting (reflecting) the sound, so the reflected frequency is

$$f'' = \frac{f'}{\left(1 - \dfrac{v_s}{v}\right)} = \frac{5051\ \text{Hz}}{\left(1 - \dfrac{3.50\ \text{m/s}}{343\ \text{m/s}}\right)} = 5103\ \text{Hz}.$$

Thus the frequency shifts by 103 Hz.

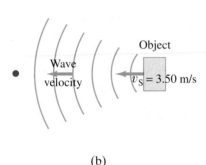

FIGURE 12–23 Example 12–14.

The incident wave and the reflected wave, when mixed together (say, electronically), interfere with one another and beats are produced. The beat frequency is equal to the difference in the two frequencies, and in Example 12–14 would be 103 Hz. This Doppler technique is used in a variety of medical applications, usually with ultrasonic waves in the megahertz frequency range. For example, ultrasonic waves reflected from red blood cells can be used to determine the velocity of blood flow. Similarly, the technique can be used to detect the movement of the chest of a young fetus and to monitor its heartbeat.

For convenience, we can write Eqs. 12–2 and 12–3 as a single equation that covers all cases of both source and observer in motion:

$$f' = f\left(\frac{v \pm v_0}{v \mp v_s}\right). \tag{12–4}$$

⮕ **PHYSICS APPLIED**

Medicine

Source and observer moving

⮕ **PROBLEM SOLVING**

Getting the signs right

The upper signs apply if source and/or observer move toward each other; the lower signs apply if they are moving apart.

The Doppler effect occurs for other types of waves as well. Light and other types of electromagnetic waves exhibit the Doppler effect: although the formulas for the frequency shift are not identical to Eqs. 12–2 and 12–3, the effect is similar. An important application is to astronomy where, for example, the velocities of distant galaxies can be determined from the Doppler shift. Light from such galaxies is shifted toward lower frequencies, indicating that the galaxies are moving away from us. (This is called the **red shift** since red has the lowest frequency of visible light.) The greater the frequency shift, the greater the velocity of recession. It is found that the farther the galaxies are from us, the faster they move away. This observation is the basis for the idea that the universe is expanding, as discussed in Chapter 33.

* 12–9 Shock Waves and the Sonic Boom

An object such as an airplane traveling faster than the speed of sound is said to have a **supersonic speed**. Such a speed is often given as a **Mach**[†] **number**, which is defined as the ratio of the object's speed to that of sound in the medium at that location. For example, a plane traveling 900 m/s high in the atmosphere, where the speed of sound is only 300 m/s, has a speed of Mach 3.

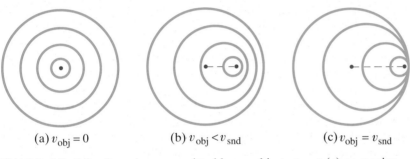

(a) $v_{obj} = 0$ (b) $v_{obj} < v_{snd}$ (c) $v_{obj} = v_{snd}$

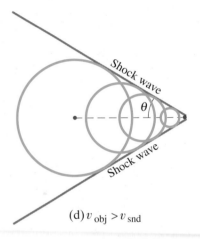

(d) $v_{obj} > v_{snd}$

FIGURE 12–24 Sound waves emitted by an object at rest (a) or moving (b, c, and d). If the object's velocity is less than the velocity of sound, the Doppler effect occurs (b); if its velocity is greater than the velocity of sound, a shock wave is produced (d).

When a source of sound moves at subsonic speeds, the pitch of the sound is altered, as we have seen (the Doppler effect); see also Fig. 12–24a and b. But if a source of sound moves faster than the speed of sound, a more dramatic effect known as a **shock wave** occurs. In this case the source is actually "outrunning" the waves it produces. As shown in Fig. 12–24c, when the source is traveling at the speed of sound, the wave fronts it emits in the forward direction "pile up" directly in front of it. When the object moves at a supersonic speed, the wave fronts pile up on one another along the sides, as shown in Fig. 12–24d. The different wave crests overlap one another and form a single very large crest which is the shock wave. Behind this very large crest there is usually a very large trough. A shock wave is essentially the result of constructive interference of a large number of wave fronts. A shock wave in air is analogous to the bow wave of a boat traveling faster than the speed of the water waves it produces, Fig. 12–25.

FIGURE 12–25 Bow waves produced by a boat.

[†]After the Austrian physicist Ernst Mach (1838–1916).

(a)   (b)

FIGURE 12–26 (a) The (double) sonic boom has already been heard by person A on the left. It is just being heard by person B in the center. And it will shortly be heard by person C on the right. (b) Special photo of supersonic aircraft showing shock waves produced in the air. (Several closely spaced shock waves are produced by different parts of the aircraft.)

➡ **PHYSICS APPLIED**

Sonic boom

When an airplane travels at supersonic speeds, the noise it makes and its disturbance of the air form into a shock wave containing a tremendous amount of sound energy. When the shock wave passes a listener, it is heard as a loud "sonic boom." A sonic boom lasts only a fraction of a second, but the energy it contains is often sufficient to break windows and cause other damage. It can be psychologically unnerving as well. Actually, a sonic boom is made up of two or more booms since major shock waves can form at the front and the rear of the aircraft, as well as at the wings, etc. (Fig. 12–26). Bow waves of a boat are also multiple, as can be seen in Fig. 12–25.

*Breaking the
sound barrier*

When an aircraft approaches the speed of sound, it encounters a barrier of sound waves in front of it (see Fig. 12–24c). In order to exceed the speed of sound, extra thrust is needed to pass through this "sound barrier." This is called "breaking the sound barrier." Once a supersonic speed is attained, this barrier no longer impedes the motion. It is sometimes erroneously thought that a sonic boom is produced only at the moment an aircraft is breaking through the sound barrier. Actually, a shock wave follows the aircraft at all times it is traveling at supersonic speeds. A series of observers on the ground will each hear a loud "boom" as the shock wave passes, Fig. 12–26. The shock wave consists of a cone whose apex is at the aircraft. The angle of this cone, θ, (see Fig. 12–24d) is given by

$$\sin \theta = \frac{v_{\text{snd}}}{v_{\text{obj}}}, \qquad \textbf{(12–5)}$$

where v_{obj} is the velocity of the object (the aircraft) and v_{snd} is the velocity of sound in the medium (the proof is left as a problem).

*12–10 Applications; Ultrasound and Medical Imaging

➡ **PHYSICS APPLIED**

Sonar

The reflection of sound is used in many applications to determine distance. The **sonar**[†] or pulse-echo technique is used to locate underwater objects. (It is also used in medical applications, as we shall see shortly.) A transmitter sends out a sound pulse through the water, and a detector receives its reflection, or echo, a short time later. This time interval is carefully measured, and from it the distance to the reflecting object can be determined since the

[†]Sonar stands for "*so*und *na*vigation *r*anging."

speed of sound in water is known. The depth of the sea and the location of reefs, sunken ships, submarines, or schools of fish can be determined in this way. The interior structure of the Earth is studied in a similar way by detecting reflections of waves traveling through the Earth whose source was a deliberate explosion (called "soundings"). An analysis of waves reflected from various structures and boundaries within the Earth reveals characteristic patterns that are also useful in the exploration for oil and minerals.

Sonar generally makes use of **ultrasonic** frequencies: that is, waves whose frequencies are above 20 kHz, beyond the range of human detection. For sonar, the frequencies are typically in the range 20 kHz to 100 kHz. One reason for using ultrasound waves, other than the fact that they are inaudible, is that for shorter wavelengths there is less diffraction, so the beam spreads less and smaller objects can be detected. For as we saw in Chapter 11, particularly in Fig. 11–44, an obstacle intercepts and reflects a portion of a wave significantly only if the wavelength is less than the size of the object. Indeed, the smallest-sized objects that can be detected are on the order of the wavelength of the wave used. With the higher frequencies of ultrasound, the wavelength is smaller, so smaller objects can be detected.

In medicine, ultrasonic waves are used both in diagnosis and in treatment. Treatment involves destruction of unwanted tissue or objects in the body (such as tumors or kidney stones) using ultrasonic waves of very high intensity (as high as 10^7 W/m^2) that are focused on the undesired material. Ultrasound is also used in physical therapy, to provide local heating of injured muscles.

The diagnostic use of ultrasound in medicine is a more complicated and very interesting application of physical principles. A **pulse-echo technique** is used, much like sonar. A high-frequency sound pulse is directed into the body, and its reflections from boundaries or interfaces between organs and other structures and lesions in the body are then detected. By using this technique, tumors and other abnormal growths, or pockets of fluid, can be distinguished; the action of heart valves and the development of a fetus can be examined; and information about various organs of the body, such as the brain, heart, liver, and kidneys, can be obtained. Although ultrasound does not replace X-rays, for certain kinds of diagnosis it is more helpful. Some kinds of tissue or fluid are not detected in X-ray photographs, but ultrasound waves are reflected from their boundaries. It is also possible to produce "real-time" ultrasound images, as if one were watching a movie of a section of the interior of the body. Furthermore, at the low levels used for diagnosis ($< 3 \times 10^4$ W/m^2), no adverse effects have been reported. So ultrasound is considered a noninvasive method for probing the body.

The frequencies used in ultrasonic diagnosis are in the range 1 to 10 MHz (1 megahertz $= 10^6$ Hz). The speed of sound waves in the tissues of the human body averages about 1540 m/s (close to that for water); so the wavelength of a 1-MHz wave is about

$$\lambda = \frac{v}{f} = \frac{(1540 \text{ m/s})}{(10^6 \text{ s}^{-1})} \approx 1.5 \times 10^{-3} \text{ m} = 1.5 \text{ mm,}$$

and this sets the limit to the smallest-sized objects that can be detected. Higher frequencies mean shorter wavelengths, and thus in principle visualization of finer detail. However, the higher the frequency, the more the wave is absorbed by the body, and reflections from deeper inside the body are lost.

FIGURE 12–27
(a) Ultrasound pulse passes through abdomen, reflecting from surfaces in its path.
(b) Reflected pulses plotted as a function of time (A-scan) when received by transducer; time is proportional to distance of travel. The vertical dashed lines point out which reflected pulse goes with which surface.
(c) B-mode display for the same echoes: brightness of each dot is related to signal strength.

The pulse-echo technique for medical imaging works as follows. A brief pulse of ultrasound is emitted by a transducer, which is a device that transforms an electrical pulse into a sound-wave pulse. (This is also what a loudspeaker does, as discussed in Chapter 20). Part of the pulse is reflected at various interface surfaces in the body, and most (usually) continue on (see Fig. 11–32 for the case of a pulse on a rope). The detection of the reflected pulses is done by the same transducer, which transforms the sound pulses into electrical pulses (as a microphone does—Chapter 21), and these pulses can then be displayed on the screen of a display terminal or monitor. As an example, consider a sound pulse passing through the abdomen, as shown in Fig. 12–27a. At various interfaces in the body, part of the pulse is reflected. The time elapsed from when the pulse is emitted to when its reflection (echo) is received is proportional to the distance to the reflecting surface. For example, if the distance from transducer to the vertebra is 25 cm, the pulse travels a round-trip distance of 2×25 cm $= 0.50$ m, and the time taken is

$$t = \frac{d}{v} = \frac{0.50 \text{ m}}{1540 \text{ m/s}} = 3.2 \times 10^{-4} \text{ s} = 320 \ \mu\text{s}.$$

A pulse reflected from an object only 10 cm from the transducer would be received after only 130 μs. Figure 12–27b plots the reflected pulses as a function of the time they are received by the transducer for the situation shown in part (a).

The *strength* of a reflected pulse depends mainly on the difference in density of the two materials on either side of the interface. It also depends on the sound velocity in each material, but this normally has little effect since in most tissue the velocity is within a few percent of the average 1540 m/s. [Exceptions are bone (4000 m/s) and air (340 m/s). At interfaces involving bone or the lungs, most of the sound pulse is reflected, so ultrasound cannot be used as a probe beyond such interfaces.[†]]

A trace such as that shown in Fig. 12–27b can be displayed directly on a monitor as shown in Fig. 12–28. Such a display is called an *A-scan* or A-mode scan. Far more common today is a *B-scan*, which can be used to form two-dimensional images of a cross section of the body. In a B-mode scan, each echo is represented as a point, whose position is given by the time delay and whose brightness depends on the strength of the echo. Figure 12–27c shows these points for the echoes produced in part (a). A two-dimensional image can be formed from a series of B-scans. The transducer is moved, and at each position it sends out a pulse and receives echoes as shown in Fig. 12–29. Each B-scan trace can be plotted, spaced appropriately one below the other, to form an image on a display terminal as shown in Fig. 12–29b. Only 10 lines are shown in Fig. 12–29, so the image

FIGURE 12–28 An A-scan of an eye.

[†]Generally the transducer is placed in contact with a person's body, or the body and transducer are immersed in water, so as to avoid the high reflectivity at the first air-body interface, which would greatly reduce the amount of sound waves passing into the body.

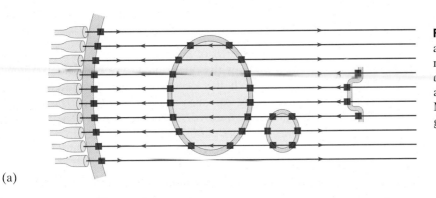

(a)

(b)

FIGURE 12–29 (a) Ten traces are made across the abdomen by moving the transducer, or by using an array of transducers. (b) The echoes are plotted to produce the image. More closely spaced traces would give a more detailed image.

(a)

(b)

FIGURE 12–30
(a) Ultrasound image of a human fetus (with head at the left) within the uterus. (b) False-color high-resolution ultrasound image of fetus (different colors represent different intensities of reflected pulses).

is crude. More lines give a more precise image.[†] Photographs of ultrasound images are shown in Fig. 12–30.

A faster scan can be obtained by using an array of transducers, or by rotating a single transducer about a point so it "looks" in a sequence of different directions through the body.

Ultrasound imaging has been an important advance in medicine. It, along with other types of medical imaging, which we will discuss in later chapters (Chapters 25 and 31), has in many cases replaced exploratory surgery and other risky, painful, or costly procedures. There is no evidence of harmful effects with ultrasound imaging, as there is with X-ray imaging. It is thus considered noninvasive. Nonetheless it is not a technique that replaces all others. Beam spreading limits the sharpness of images. And the way sound reflects from matter is not the same as for light or X-rays, so different kinds of information can be obtained from different imaging techniques. A new mode of ultrasound imaging is now being developed based on *transmitted* waves (instead of reflected) making use of computer-assisted tomography techniques (normally done with X-rays—see Chapter 25). Another technique, which we already discussed in Section 12–8, employs the Doppler shift of ultrasound echoes to measure velocities within the human body, such as blood flow and fetal heartbeat.

[†]*Radar* used for aircraft involves a similar pulse-echo technique except that it uses electromagnetic (EM) waves which, like light, travel with a speed of 3×10^8 m/s. This works for the large distances in the sky, but the high speed precludes the use of EM wave pulse-echo techniques in the body since the picosecond times involved [$\approx (0.1 \text{ m})/(3 \times 10^8 \text{ m/s}) \approx 3 \times 10^{-10} \text{ s} = 300 \text{ ps}$] between reflections cannot at present be separated electronically. Other techniques using EM waves will be discussed later.

SUMMARY

Sound travels as a longitudinal wave in air and other materials. In air, the speed of sound increases with temperature; at 20°C, it is about 343 m/s.

The **pitch** of a sound is determined by the frequency; the higher the frequency, the higher the pitch.

The **audible range** of frequencies for humans is roughly 20 to 20,000 Hz (1 Hz = 1 cycle per second).

The **loudness** or **intensity** of a sound is related to the amplitude of the wave. Because the human ear can detect sound intensities from 10^{-12} W/m² to over 1 W/m², intensity levels are specified on a logarithmic scale. The **intensity level** β, specified in decibels, is defined in terms of intensity I as

$$\beta = 10 \log (I/I_0),$$

where the reference intensity I_0 is usually taken to be 10^{-12} W/m².

Musical instruments are simple sources of sound in which *standing waves* are produced.

The strings of a stringed instrument may vibrate as a whole with nodes only at the ends; the frequency at which this occurs is called the **fundamental**. The string can also vibrate at higher frequencies, called **overtones** or **harmonics**, in which there are one or more additional nodes. The frequency of each harmonic is a whole-number multiple of the fundamental.

In wind instruments, standing waves are set up in the column of air within the tube.

The vibrating air in an **open tube** (open at both ends) has displacement antinodes at both ends. The fundamental frequency corresponds to a wavelength equal to twice the tube length. The harmonics have frequencies that are 2, 3, 4, ⋯ times the fundamental frequency, just as for strings.

For a **closed tube** (closed at one end), the fundamental corresponds to a wavelength four times the length of the tube. Only the odd harmonics are present, equal to 1, 3, 5, 7, ⋯ times the fundamental frequency.

Sound waves from different sources can interfere with each other. If two sounds are at slightly different frequencies, **beats** can be heard at a frequency equal to the difference in frequency of the two sources.

The **Doppler effect** refers to the change in pitch of a sound due to the motion either of the source or of the listener. If they are approaching each other, the pitch is higher; if they are moving apart, the pitch is lower.

QUESTIONS

1. What is the evidence that sound travels as a wave?

2. What is the evidence that sound is a form of energy?

3. Children sometimes play with a homemade "telephone" by attaching a string to the bottoms of two paper cups. When the string is stretched and a child speaks into one of the cups, the sound can be heard at the other cup. Explain clearly how the sound wave travels from one cup to the other. (See Fig. 12–31.)

FIGURE 12–31   Question 3 and Problem 4.

4. When a sound wave passes from air into water, do you expect the frequency or wavelength to change?

5. What evidence can you give that the speed of sound in air does not depend significantly on frequency?

6. The voice of a person who has inhaled helium sounds very high-pitched. Why?

*7. Explain why the larger area of the eardrum, as compared to that of the oval window, leads to an amplification of the pressure.

8. What is the reason that catgut strings on some musical instruments are wrapped with fine wire?

9. Explain how a tube might be used as a filter to reduce the amplitude of sounds in various frequency ranges. (An example is a car muffler.)

10. How will the air temperature in a room affect the pitch of organ pipes?

11. Noise control is an important goal today. One mode of attack is to reduce the area of vibration of noisy machinery, for example by keeping it as small as possible or isolating it (acoustically) from the floor and walls. A second method is to make the surface out of a thicker material. Explain how each of these can reduce the noise level.

12. Why are the frets on a guitar spaced closer together as you move up the fingerboard toward the bridge?

13. Standing waves can be said to be due to "interference in space," whereas beats can be said to be due to "interference in time." Explain.

14. In Fig. 12–17, if the frequency of the speakers were lowered, would the points D and C (where destructive and constructive interference occur) move farther apart or closer together?

15. Traditional methods of protecting the hearing of people who work in areas with very high noise levels have consisted mainly of efforts to block or reduce noise levels. With a relatively new technology, headphones are worn that do not block the ambient noise. Instead, a device is used which detects the noise, inverts it electronically, then feeds it to the headphones *in addition to* the ambient noise. How could adding *more* noise actually reduce the sound levels reaching the ears?

16. Suppose a source of sound moves at right angles to the line of sight of a listener at rest in still air. Will there be a Doppler effect? Explain.

17. Figure 12–32 shows various positions of a child in motion on a swing. A monitor is blowing a whistle in front of the child on the ground. At which position will the child hear the highest frequency for the sound of the whistle? Explain your reasoning.

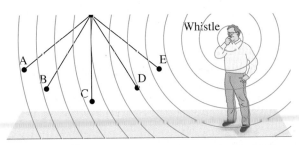

FIGURE 12–32   Question 17.

18. If a wind is blowing, will this alter the frequency of the sound heard by a person at rest with respect to the source? Is the wavelength or velocity changed?

*19. A sonic boom sounds much like an explosion. Explain the similarity between the two.

[Unless stated otherwise, assume $T = 20°C$ and $v_{sound} = 343$ m/s.]

SECTION 12–1

1. (I) A hiker determines the length of a lake by listening for the echo of her shout reflected by a cliff at the far end of the lake. She hears the echo 1.5 s after shouting. Estimate the length of the lake.

2. (I) A sailor strikes the side of his ship just below the waterline. He hears the echo of the sound reflected from the ocean floor directly below 2.0 s later. How deep is the ocean at this point?

3. (I) (a) Calculate the wavelengths in air at 20°C for sounds in the maximum range of human hearing, 20 Hz to 20,000 Hz. (b) What is the wavelength of a 10-MHz ultrasonic wave?

4. (II) Two children use a homemade "telephone" consisting of two paper cups attached by an aluminum wire which is 8.4 m long, as in Fig. 12–31. Determine the time for the "sound" to travel from one cup to the other. How does this compare with the time for sound to travel the same distance through air?

5. (II) A person sees a heavy stone strike the concrete pavement. A moment later two sounds are heard from the impact: one travels in the air and the other in the concrete, and they are 1.4 s apart. How far away did the impact occur?

6. (II) A fishing boat is drifting just above a school of tuna on a foggy day. Without warning, an engine backfire occurs on another boat 1.0 km away (Fig. 12–33). How much time elapses between the instant when the backfire is heard (a) by the fish and (b) by the fishermen?

FIGURE 12–33   Problem 6.

7. (II) The sound from a very high burst of fireworks takes 4.5 s to arrive at your eardrums. The burst occurred 1500 m above you and traveled vertically through two stratified layers of air, the top one at 0°C and the bottom one at 20°C. How thick is each layer of air?

8. (II) If the camera of Example 12–2 focuses precisely at 20°C, what percent error will there be in distances when used at 0°C?

SECTION 12–2

9. (I) What is the intensity of sound at the pain level of 120 dB? Compare it to that of a whisper at 20 dB.

10. (I) What is the intensity level of a sound whose intensity is 2.0×10^{-6} W/m²?

11. (II) Human beings can typically detect a difference in sound intensity level of 2.0 dB. What is the ratio of the amplitudes of two sounds whose levels differ by this amount?

12. (II) A person standing a certain distance from an airplane with four equally noisy jet engines is feeling a sound level intensity bordering on pain, 120 dB. What sound level intensity would this person experience if the captain shut down all but one engine?

13. (II) A stereo tape recorder is said to have a signal-to-noise ratio of 58 dB. What is the ratio of intensities of the signal and the background noise?

14. (II) (a) Estimate the power output of sound from a person speaking in normal conversation. Use Table 12–2. Assume the sound spreads roughly uniformly over a hemisphere in front of the mouth. (b) How many people would produce a total sound output of 100 W of ordinary conversation?

15. (II) A 50-dB sound wave strikes an eardrum whose area is 5.0×10^{-5} m². (a) How much energy is absorbed by the eardrum per second? (b) At this rate, how long would it take your eardrum to receive a total energy of 1.0 J?

16. (II) Expensive stereo amplifier A is rated at 250 W per channel, while the more modest amplifier B is rated at 40 W per channel. (a) Estimate the intensity level in decibels you would expect at a point 3.0 m from a loudspeaker connected in turn to each amp. (b) Will the expensive amp sound twice as loud as the cheaper one?

17. (II) At a recent rock concert, a dB meter registered 130 dB when placed 2.5 m in front of a loudspeaker on the stage. (a) What was the power output of the speaker, assuming uniform spherical spreading of the sound and neglecting absorption in the air? (b) How far away would the intensity level be a somewhat reasonable 90 dB?

18. (II) If the amplitude of a sound wave is tripled, (a) by what factor will the intensity increase? (b) By how many dB will the intensity level increase?

19. (III) A jet plane emits 5.0×10^5 J of sound energy per second. (a) What is the intensity level 30 m away? Air absorbs sound at a rate of about 7.0 dB/km; calculate what the intensity level will be (b) 1.0 km and (c) 5.0 km away from this jet plane, taking into account air absorption.

*SECTION 12–3

*20. (II) Two sound waves have equal displacement amplitudes, but one has twice the frequency of the other. What is the ratio of their intensities?

*21. (II) What would be the intensity level (in dB) of a sound wave in air that corresponds to a displacement amplitude of vibrating air molecules of 1.3 mm at 260 Hz?

*22. (II) Calculate the maximum displacement of air molecules when a 131-Hz sound wave passes whose intensity is at the threshold of pain (120 dB).

*SECTION 12–4

*23. (I) A 6000-Hz tone must have what intensity level to seem as loud as a 100-Hz tone that has a 50-dB intensity level? (See Fig. 12–7.)

*24. (I) What are the lowest and highest frequencies that an ear can detect when the intensity level is 30 dB? (See Fig. 12–7.)

*25. (II) Your auditory system can accommodate a huge range of sound levels. What is the ratio of highest to lowest intensity at (a) 100 Hz, (b) 5000 Hz? (See Fig. 12–7.)

SECTION 12–5

26. (I) The A string on a violin has a fundamental frequency of 440 Hz. The length of the vibrating portion is 32 cm and has a mass of 0.35 g. Under what tension must the string be placed?

27. (I) An unfingered guitar string is 0.70 m long and is tuned to play E above middle C (330 Hz). How far from the end of this string must the finger be placed to play A above middle C (440 Hz)?

28. (I) Determine the length of an open organ pipe that emits middle C (262 Hz) when the temperature is 21°C.

29. (I) (a) What resonant frequency would you expect from blowing across the top of an empty soda bottle that is 15 cm deep? (b) How would that change if it was one-third full of soda?

30. (I) If you were to build a pipe organ with open-tube pipes spanning the range of human hearing (20 Hz to 20 kHz), what would be the range of the lengths of pipes required?

31. (I) An organ pipe is 112 cm long. What are the fundamental and first three audible overtones if the pipe is (a) closed at one end, and (b) open at both ends?

32. (II) How far from the end of the flute in Example 12–9 should the hole be that must be uncovered to play D above middle C at 294 Hz?

33. (II) A highway overpass was observed to resonate as one full loop when a small earthquake shook the ground vertically at 4.0 Hz. The highway department put a support at the center of the overpass, anchoring it to the ground as shown in Fig. 12–34. What resonant frequency would you now expect for the overpass? It is noted that earthquakes rarely do significant shaking above 5 or 6 Hz. Did the modifications do any good?

Before modification

Added support

After modification

FIGURE 12–34   Problem 33.

34. (II) An organ is in tune at 20°C. By what percent will the frequency be off at 5.0°C?

35. (II) (a) At $T = 15$°C, how long must a closed organ pipe be if it is to have a fundamental frequency of 294 Hz? (b) If this pipe were filled with helium, what would its fundamental frequency be?

36. (II) A particular organ pipe can resonate at 264 Hz, 440 Hz, and 616 Hz, but not at any other frequencies between 264 and 616 Hz. (a) Is this an open or closed pipe? (b) What is the fundamental frequency of this pipe?

37. (II) A uniform narrow tube 1.80 m long is open at both ends. It resonates at two successive harmonics of frequency 275 Hz and 330 Hz. What is the speed of sound in the gas in the tube?

38. (II) A pipe in air at 20°C is to be designed to produce two successive harmonics at 240 Hz and 280 Hz. How long must the pipe be, and is it open or closed?

39. (II) How many overtones are present within the audible range for a 2.44-m-long organ pipe at 20°C (a) if it is open, and (b) if it is closed?

40. (II) The human ear is most sensitive at about 3500 Hz, approximately independent of the loudness level (Fig. 12–7). What does this imply about the length of the ear canal from the outer opening to the eardrum?

*SECTION 12–6

*41. (II) Approximately what are the intensities of the first two overtones of a violin compared to the fundamental? How many decibels softer than the fundamental are the first and second overtones? (See Fig. 12–16.)

SECTION 12–7

42. (I) A piano tuner hears one beat every 2.0 s when trying to adjust two strings, one of which is sounding 440 Hz, so that they sound the same tone. How far off in frequency is the other string?

43. (I) What will be the "beat frequency" if middle C (262 Hz) and C♯ (277 Hz) are played together? Will this be audible? What if each is played two octaves lower (each frequency reduced by a factor of 4)?

44. (I) A certain dog whistle operates at 23.5 kHz, while another (brand X) operates at an unknown frequency. If neither whistle can be heard by humans when played separately, but a shrill whine of frequency 5000 Hz occurs when they are played simultaneously, estimate the operating frequency of brand X.

45. (II) A guitar string produces 4 beats/s when sounded with a 350-Hz tuning fork and 9 beats/s when sounded with a 355-Hz tuning fork. What is the vibrational frequency of the string? Explain your reasoning.

46. (II) Two piano strings are supposed to be vibrating at 132 Hz, but a piano tuner hears three beats every 2.0 s when they are played together. (a) If one is vibrating at 132 Hz, what must be the frequency of the other (is there only one answer)? (b) By how much (in percent) must the tension be increased or decreased to bring them in tune?

47. (II) Two violin strings are tuned to the same frequency, 294 Hz. The tension in one string is then decreased by 1.5 percent. What will be the beat frequency heard when the two strings are played together?

48. (II) Two loudspeakers are 2.5 m apart. A person stands 3.0 m from one speaker and 3.5 m from the other. (a) What is the lowest frequency at which destructive interference will occur at this point? (b) Calculate two other frequencies that also result in destructive interference at this point (give the next two highest).

49. (III) A source emits sound of wavelengths 2.64 m and 2.76 m in air. (a) How many beats per second will be heard (assume $T = 20°C$)? (b) How far apart in space are the regions of maximum intensity?

50. (III) Show that the two speakers in Fig. 12–17 must be separated by at least a distance d equal to one-half the wavelength λ of sound if there is to be any place where complete destructive interference occurs. The speakers are in phase.

SECTION 12–8

51. (I) The predominant frequency of a certain police car's siren is 1800 Hz when at rest. What frequency do you detect if you move with a speed of 30.0 m/s (a) toward the car, and (b) away from the car.

52. (I) A bat at rest sends out ultrasonic sound waves at 50,000 Hz and receives them returned from an object moving radially away from it at 25.0 m/s. What is the received sound frequency?

53. (II) In one of the original Doppler experiments, one tuba was played on a moving platform car at a frequency of 75 Hz, and a second identical one was played on the same tone while at rest in the railway station. What beat frequency was heard if the train approached the station at a speed of 10.0 m/s?

54. (II) Two automobiles are equipped with the same single-frequency horn. When one is at rest and the other is moving toward an observer at 15 m/s, a beat frequency of 5.5 Hz is heard. What is the frequency the horns emit? Assume $T = 20°C$.

55. (II) Two trains emit whistles of the same frequency, 277 Hz. If one train is at rest and the other is traveling at 40 km/h away from an observer at rest, what beat frequency will the observer detect?

56. (II) Compare the shift in frequency if a 2000-Hz source is moving toward you at 15 m/s versus if you are moving toward it at 15 m/s. Are the two frequencies exactly the same? Are they close? Repeat the calculation for 150 m/s and then again for 300 m/s. What can you conclude about the asymmetry of the Doppler formulas? Show that at low speeds (relative to the speed of sound), the two formulas—source approaching and detector approaching—yield the same result.

57. (II) Ultrasound waves are used to measure blood-flow speeds. Suppose a device emits sound at 500 kHz, and the speed of sound in human tissue is taken to be 1540 m/s. What is the expected beat frequency if blood is flowing normally in large leg arteries at 2.0 cm/s directly away from the sound source?

58. (III) The Doppler effect using ultrasonic waves of frequency 2.25×10^6 Hz is used to monitor the heartbeat of a fetus. A (maximum) beat frequency of 600 Hz is observed. Assuming that the speed of sound in tissue is 1.54×10^3 m/s, calculate the maximum velocity of the surface of the beating heart.

59. (III) In Problem 58, the beat frequency is found to appear and then disappear 180 times per minute, which reflects the fact that the heart is beating and its surface changes speed. What is the heartbeat rate?

*60. (I) You are flying above the Gulf of Mexico in a po-
lice helicopter when you see a fast-moving boat
below. The wake from the boat forms an angle of 20°
on either side of the straight-line motion. Knowing
that the speed of the water waves is about 2.0 m/s,
what is the speed of the boat?

*61. (I) (a) How fast is an object moving on land if it is
moving at Mach 0.33? (b) A high-flying Concorde
passenger jet displays its Mach number on a screen
while cruising at 3000 km/h to be 3.2. What is the
speed of sound at that altitude?

*62. (II) Show that the angle θ a sonic boom makes with
the path of a supersonic object is given by Eq. 12–5.

*63. (II) An airplane travels at Mach 2.3 where the speed
of sound is 310 m/s. (a) What is the angle the shock
wave makes with the direction of the airplane's motion? (b) If the plane is flying at a height of 7100 m,
how long after it is directly overhead will a person
on the ground hear the shock wave?

*64. (II) A space probe enters the thin atmosphere of an-
other planet where the speed of sound is only about
35 m/s. (a) What is the probe's Mach number if its
initial speed is 15,000 km/h? (b) What is the apex
angle of the shock wave it produces?

*65. (II) A meteorite traveling 8000 m/s strikes the ocean.
Determine the shock wave angle it produces (a) in
the air just before entering the ocean, and (b) in the
water just after entering. Assume $T = 20°C$.

*66. (II) You look directly overhead and see a plane ex-
actly 1.5 km above the ground flying faster than the
speed of sound. By the time you hear the sonic
boom, the plane has traveled a horizontal distance of
2.0 km. See Fig. 12–35. (a) Find the angle of the shock
cone, θ. (b) Find the speed of the plane (the Mach
number). Assume the speed of sound is 330 m/s.

FIGURE 12–35   Problem 66.

GENERAL PROBLEMS

67. Approximately how many octaves are there in the
human audible range?

68. A stone is dropped from the top of a cliff. The splash
it makes when striking the water below is heard 3.5 s
later. How high is the cliff?

69. A single mosquito 5.0 m from a person makes a
sound close to the threshold of human hearing
(0 dB). What will be the intensity level of 1000 such
mosquitoes?

70. At the Indianapolis 500, you can measure the speed
of cars just by listening to the difference in pitch of
the engine noise between approaching and receding
cars. Suppose the sound of a certain car drops by a
full octave as it goes by on the straightaway. How
fast is it going?

71. A tight guitar string has a frequency of 600 Hz as its
third harmonic. What will be its fundamental frequency if it is fingered at a length of only 60 percent
of its original length?

72. Each string on a violin is tuned to a frequency $1\frac{1}{2}$
times that of its neighbor. If all the strings are to be
placed under the same tension, what must be the
mass per unit length of each string relative to that of
the lowest string?

73. The A string of a violin is 32 cm long between fixed
points with a fundamental frequency of 440 Hz and a
linear density of 5.5×10^{-4} kg/m. (a) What are the
wave speed and tension in the string? (b) What is the
length of the tube of a simple wind instrument (say,
an organ pipe) closed at one end whose fundamental
is also 440 Hz if the speed of sound is 343 m/s in air?
(c) What is the frequency of the first overtone of
each instrument?

74. A stereo amplifier is rated at 100 W output at 1000 Hz.
The power output drops by 10 dB at 15 kHz. What is
the power output in watts at 15 kHz?

75. A tuning fork is set into vibration above a vertical
open tube filled with water (Fig. 12–36). The water
level is allowed to drop slowly. As it does so, the air
in the tube above the water level is heard to resonate
with the tuning fork when the distance from the tube
opening to the water level is 0.125 m and again at
0.395 m. What is the frequency of the tuning fork?

0.125m

0.395m

FIGURE 12–36
Problem 75.

76. In audio and communications systems, the *gain*, β, in decibels is defined as

$$\beta = 10 \log \frac{P_{out}}{P_{in}},$$

where P_{in} is the power input to the system and P_{out} is the power output. A particular stereo amplifier puts out 100 W of power for an input of 1 mW. What is its gain in dB?

77. Two loudspeakers are at opposite ends of a railroad car as it moves past a stationary observer at 10.0 m/s, as shown in Fig. 12–37. If they have identical sound frequencies of 200 Hz, what is the beat frequency heard by the observer when (*a*) he observes from the position in A, in front of the car, (*b*) he is between the speakers as in B, and (*c*) he observes the speakers after they have passed him, as in C?

FIGURE 12–37 Problem 77.

78. The frequency of a steam train whistle as it approaches you is 522 Hz. After it passes you, its frequency is measured as 486 Hz. How fast was the train moving (assume constant velocity)?

79. A 75-cm-long guitar string of mass 1.80 g is placed near a tube open at one end, and also 75 cm long. How much tension should be in the string if it is to produce resonance (in its fundamental mode) with the third harmonic in the tube?

80. If the velocity of blood flow in the aorta is normally about 0.32 m/s, what beat frequency would you expect if 5.50-MHz ultrasound waves were directed along the flow and reflected from the red blood cells? Assume that the waves travel with a speed of 1.54×10^3 m/s.

81. A source of sound waves (wavelength λ) is a distance l from a detector. Sound reaches the detector directly, and also by reflecting off an obstacle, as shown in Fig. 12–38. The obstacle is equidistant from source and detector. When the obstacle is a distance d to the right of the line of sight between source and detector, as shown, the two waves arrive in phase. How much farther to the right must the obstacle be moved if the two waves are to be out of phase by $\frac{1}{2}$ wavelength, so destructive interference occurs? (Assume $\lambda \ll l, d$.)

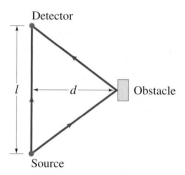

FIGURE 12–38 Problem 81.

82. A factory whistle emits sound of frequency 645 Hz. On a day when the wind velocity is 9.0 m/s from the north, what frequency will observers hear who are located, at rest, (*a*) due north, (*b*) due south, (*c*) due east, and (*d*) due west, of the whistle? What frequency is heard by a cyclist heading (*e*) north, or (*f*) west, toward the whistle at 13.0 m/s? ($T = 20°C$.)

*83. A person hears a pure tone coming from two sources that seems to be in the 500–1000 Hz range. The sound is loudest at points equidistant from the two sources. In order to determine exactly what the frequency is, the person moves about and finds that the sound level is minimal at a point 0.22 m farther from one source than the other. What is the frequency of the sound?

84. *Echolocation* is a form of sensory perception used by most bats, toothed whales and porpoises, as well as a few birds. The animal emits a pulse of sound which is reflected from objects; the reflected pulse is then detected by the animal to learn about its environment or interact with other animals. Echolocation waves emitted by whales have frequencies of about 200,000 Hz. (*a*) What is the wavelength of the whale's echolocation wave? (*b*) If an obstacle is 100 m from the whale, how long after the whale emits a wave will the reflected wave return to him?

85. A bat emits a series of high frequency sound pulses as it approaches a moth. The pulses are approximately 70.0 ms apart and each is about 3.0 ms long. How far away can the moth be detected by the bat, so that the echo from one chirp returns before the next chirp is emitted?

86. A bat flies toward a moth at speed 8.0 m/s while the moth is flying toward the bat at speed 5.0 m/s. The bat emits a sound wave of 51.35 kHz. What is the frequency of the wave detected by the bat after it reflects off the moth?

Heating the air within a hot-air balloon, and thus raising the air's temperature, forces some air to escape (Charles's law). The air's density (mass per unit volume) within the balloon is thus reduced and the balloon can "float" upward.

TEMPERATURE AND 13
KINETIC THEORY

This chapter is the first of three (Chapters 13, 14, and 15) that are devoted to the subject of temperature, heat, and thermodynamics. Much of this chapter will be devoted to an investigation of the theory that matter is made up of atoms and that these atoms are in continuous random motion. This theory is called the *kinetic theory*. ("Kinetic," you may recall, is Greek for "moving.")

We also discuss the concept of temperature as well as the experimentally measured properties of gases, for these serve as a foundation for testing the kinetic theory.

13–1 Atomic Theory of Matter

The idea that matter is made up of atoms dates back to the ancient Greeks. According to the Greek philosopher Democritus, if a pure substance—say, a piece of iron—were cut into smaller and smaller bits, eventually a smallest piece of that substance would be obtained which could not be divided further. This smallest piece was called an **atom**, which in Greek means "indivisible."[†] The only real alternative to the atomic theory of matter was the idea that matter is continuous and can be subdivided indefinitely.

Today the atomic theory is generally accepted by scientists. The experimental evidence in its favor, however, came mainly in the eighteenth, nineteenth, and twentieth centuries, and much of it was obtained from the analysis of chemical reactions. A crucial piece of evidence was the *law of definite*

[†]Today, of course, we don't consider the atom as indivisible, but rather as consisting of a nucleus (containing protons and neutrons) and electrons.

proportions, which is a summation of experimental results collected during the half century prior to 1800. It states that when two or more elements[†] combine to form a compound, they always do so in the same proportions by weight. For example, table salt is always formed from 23 parts sodium and 35 parts chlorine; and water is formed from one part hydrogen and eight parts oxygen, by weight. A continuous theory of matter could hardly account for the law of definite proportions, but, as John Dalton (1766–1844) pointed out, the atomic theory could: the weight proportions of each element required to form a compound correspond to the relative weights of the combining atoms. One atom of sodium (Na), for example, could combine with one atom of chlorine (Cl) to form one molecule of salt (NaCl), and one atom of sodium would have a mass 23/35 times as large as one of chlorine. By measuring the relative amounts of each element needed to form a large variety of compounds, experimenters established the relative weights of atoms. Hydrogen, the lightest atom, was arbitrarily assigned the relative weight of 1. On this scale, carbon was about 12, oxygen 16, sodium 23, and so on.[‡]

Today, we speak of the relative masses of atoms and molecules—what we call the **atomic mass** or **molecular mass**, respectively.[§] These are based on assigning the abundant carbon atom, ^{12}C, the value of exactly 12.0000 **unified atomic mass units** (u). In terms of kilograms,

$$1 \text{ u} = 1.66 \times 10^{-27} \text{ kg}.$$

The atomic mass of hydrogen is then 1.0078 u, and the values for other atoms are as listed in the periodic table inside the back cover of this book, and also in Appendix F. The molecular mass of a compound is the sum of atomic masses of the atoms making up the molecule.

Another important piece of evidence for the atomic theory is the so-called **Brownian movement**, named after the biologist Robert Brown, who is credited with its discovery in 1827. While he was observing tiny pollen grains suspended in water under his microscope, Brown noticed that the tiny grains moved about in tortuous paths (Fig. 13–1), even though the water appeared to be perfectly still. The atomic theory easily explains Brownian movement if the further reasonable assumption is made that the atoms of any substance are continually in motion. Then Brown's tiny pollen grains are jostled about by the vigorous barrage of rapidly moving molecules of water.

In 1905, Albert Einstein examined[††] Brownian movement from a theoretical point of view and was able to calculate from the experimental

FIGURE 13–1 Path of a tiny particle (pollen grain, for example) suspended in water. The straight lines connect observed positions of the particle at equal time intervals. [The plot has the same general shape independent of the time interval—whether it be every 60s or 0.1s, a phenomenon referred to as *fractal* behavior.]

[†]An *element* is a substance, such as gold, iron, or copper, that cannot be broken down into simpler substances by chemical means. *Compounds* are substances made up of elements, and can be broken down into them; examples are carbon dioxide and water. The smallest piece of an element is an atom; the smallest piece of a compound is a molecule. Molecules are made up of atoms; a molecule of water is made up of two atoms of hydrogen and one of oxygen; its chemical formula is H_2O.

[‡]It was not quite so simple, however. For example, from the various compounds oxygen formed, its relative weight was judged to be 16. But this was inconsistent with the weight ratio in water of oxygen to hydrogen (only 8 to 1). This difficulty was explained by assuming two H atoms combine with one O atom to form a water molecule. To make a self-consistent scheme, many other molecules also had to be judged as containing more than one atom of a given type.

[§]The terms *atomic weight* and *molecular weight* are popularly used for these quantities, but properly speaking we are comparing masses.

[††]It is possible that Einstein was unaware of Brown's work, and so independently predicted, from theoretical ideas, the existence of Brownian movement.

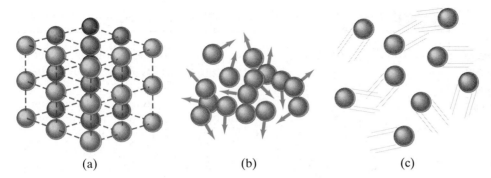

FIGURE 13–2
Atomic arrangements
in (a) a crystalline solid,
(b) a liquid, and (c) a gas.

(a)   (b)   (c)

data the approximate size and mass of atoms and molecules. His calculations showed that the diameter of a typical atom is about 10^{-10} m.

At the start of Chapter 10, we distinguished the three common states of matter—solid, liquid, gas—based on **macroscopic**, or "large-scale," properties. Now let us see how these three phases of matter differ from the atomic, or **microscopic**, point of view. Clearly, atoms and molecules must exert attractive forces on each other. For how else could a brick or a piece of aluminum stay together in one piece? The attractive forces between molecules are of an electrical nature (more on this in later chapters). If the molecules come too close together, the force between them becomes repulsive (electric repulsion between their outer electrons). Thus molecules maintain a minimum distance from each other. In a solid material, the attractive forces are strong enough that the atoms or molecules are held in more or less fixed positions, often in an array known as a crystal lattice, as shown in Fig. 13–2a. The atoms or molecules in a solid are in motion—they vibrate about their nearly fixed positions. In a liquid, the atoms or molecules are moving more rapidly, or the forces between them are weaker, so that they are sufficiently free to roll over one another, as in Fig. 13–2b. In a gas, the forces are so weak, or the speeds so high, that the molecules do not even stay close together. They move rapidly every which way, Fig. 13–2c, filling any container and occasionally colliding with one another. On the average, the speeds are sufficiently high in a gas that when two molecules collide, the force of attraction is not strong enough to keep them close together and they fly off in new directions.

Macroscopic vs. microscopic properties

EXAMPLE 13–1 ESTIMATE Distance between atoms. The density of copper is 8.9×10^3 kg/m³ and each copper atom has a mass of 63 u. Estimate the average distance between neighboring atoms.

SOLUTION The mass of 1 copper atom is $63 \times 1.66 \times 10^{-27}$ kg $= 1.04 \times 10^{-25}$ kg. This means that in a cube of copper 1 m on a side (volume $= 1$ m³), there are

$$\frac{8.9 \times 10^3 \text{ kg/m}^3}{1.04 \times 10^{-25} \text{ kg/atom}} = 8.5 \times 10^{28} \text{ atoms/m}^3.$$

On one side of the 1-m-long cube there are $(8.5 \times 10^{28})^{1/3}$ atoms $= 4.4 \times 10^9$ atoms. Hence the distance between neighboring atoms is

$$\frac{1 \text{ m}}{4.4 \times 10^9 \text{ atoms}} = 2.3 \times 10^{-10} \text{ m between atoms.}$$

FIGURE 13–3 Expansion joint on a bridge.

FIGURE 13–4 (a) Model of Galileo's original idea for a thermometer. (b) Actual thermometers built by the Accademia del Cimento (1657–1667) in Florence are among the earliest known. These sensitive and exquisite instruments contained alcohol, sometimes colored, like many thermometers today. (c) Clinical thermometers in the shape of a frog, also built by the Accademia del Cimento, could be tied to a patient's wrist. The small spheres suspended in the liquid each have a slightly different density. The number of spheres that would sink was a measure of the patient's fever.

In everyday life, **temperature** is a measure of how hot or cold an object is. A hot oven is said to have a high temperature, whereas the ice of a frozen lake is said to have a low temperature.

Many properties of matter change with temperature. For example, most materials expand when heated.[†] An iron beam is longer when hot than when cold. Concrete roads and sidewalks expand and contract slightly according to temperature, which is why compressible spacers or expansion joints (Fig. 13–3) are placed at regular intervals. The electrical resistance of matter changes with temperature (see Chapter 18). So too does the color radiated by objects, at least at high temperatures: you may have noticed that the heating element of an electric stove glows with a red color when hot. At higher temperatures, solids such as iron glow orange or even white. The white light from an ordinary incandescent lightbulb comes from an extremely hot tungsten wire. The surface temperatures of the Sun and other stars can be measured by the predominant color (more precisely, wavelengths) of light they emit.

Instruments designed to measure temperature are called **thermometers**. There are many kinds of thermometers, but their operation always depends on some property of matter that changes with temperature. Most common thermometers rely on the expansion of a material with an increase in temperature. The first idea for a thermometer (Fig. 13–4a), by Galileo, made use of the expansion of a gas. Common thermometers today consist of a hollow glass tube filled with mercury or with alcohol colored with a red dye, as were the earliest usable thermometers (Fig. 13–4b). Figure 13–4c shows an early clinical thermometer of a different type, also based on a change in density with temperature (see caption).

[†]Most materials expand when their temperature is raised, but not all. Water, for example, in the range 0°C to 4°C contracts with an increase in temperature (see Section 13–5).

(a)

(b)

(c)

FIGURE 13–5 (a) Mercury- or alcohol-in-glass thermometer; (b) bimetallic strip; (c) photo of bimetallic-strip thermometer.

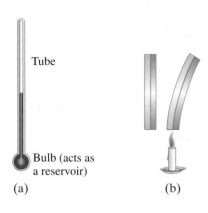

Tube

Bulb (acts as a reservoir)

(a)   (b)

(c)

In the common liquid-in-glass thermometer, the liquid expands more than the glass when the temperature is increased, so the liquid level rises in the tube (Fig. 13–5a). Although metals also expand with temperature, the change in length of a metal rod, say, is generally too small to measure accurately for ordinary changes in temperature. However, a useful thermometer can be made by bonding together two dissimilar metals whose rates of expansion are different (Fig. 13–5b). When the temperature is increased, the different amounts of expansion cause the bimetallic strip to bend. Often the bimetallic strip is in the form of a coil, one end of which is fixed while the other is attached to a pointer (Fig. 13–5c). This kind of thermometer is used as ordinary air thermometers, oven thermometers, automatic-off switch in electric coffeepots, and in room thermostats for determining when the heater or air conditioner should go on or off. Very precise thermometers make use of electrical properties (Chapter 18), such as resistance thermometers, thermocouples, and thermistors.

In order to measure temperature quantitatively, some sort of numerical scale must be defined. The most common scale today is the **Celsius** scale, sometimes called the **centigrade** scale. In the United States, the **Fahrenheit** scale is also common. The most important scale in scientific work is the absolute, or Kelvin, scale, and it will be discussed later in this chapter.

One way to define a temperature scale is to assign arbitrary values to two readily reproducible temperatures. For both the Celsius and Fahrenheit scales these two fixed points are chosen to be the freezing point and the boiling point[†] of water, both taken at atmospheric pressure. On the Celsius scale, the freezing point of water is chosen to be 0°C ("zero degrees Celsius") and the boiling point 100°C. On the Fahrenheit scale, the freezing point is defined as 32°F and the boiling point 212°F. A practical

Temperature scales

[†]The freezing point of a substance is defined as that temperature at which the solid and liquid phases coexist in equilibrium—that is, without any net liquid changing into the solid or vice versa. Experimentally, this is found to occur at only one definite temperature, for a given pressure. Similarly, the boiling point is defined as that temperature at which the liquid and gas coexist in equilibrium. Since these points vary with pressure, the pressure must be specified (usually it is 1 atm).

FIGURE 13–6 Celsius and Fahrenheit scales compared.

thermometer is calibrated by placing it in carefully prepared environments at each of the two temperatures and marking the position of the mercury or pointer. For a Celsius scale, the distance between the two marks is then divided into one hundred equal intervals separated by small marks representing each degree between 0°C and 100°C (hence the name "centigrade scale" meaning "hundred steps"). For a Fahrenheit scale, the two points are labeled 32°F and 212°F and the distance between them is divided into 180 equal intervals. For temperatures below the freezing point of water and above the boiling point of water the scales can be extended using the same equally spaced intervals. However, ordinary thermometers can be used only over a limited temperature range because of their own limitations—for example, the mercury in a mercury-in-glass thermometer solidifies at some point, below which the thermometer will be useless. It is also rendered useless above temperatures where the fluid vaporizes. For very low or very high temperatures, specialized thermometers are required, some of which we will mention later.

Every temperature on the Celsius scale corresponds to a particular temperature on the Fahrenheit scale, Fig. 13–6. It is easy to convert from one to the other if you remember that 0°C corresponds to 32°F and that a range of 100° on the Celsius scale corresponds to a range of 180° on the Fahrenheit scale. Thus, one Fahrenheit degree (1 F°) corresponds to $100/180 = \frac{5}{9}$ of a Celsius degree (1 C°). That is, $1\,F° = \frac{5}{9}\,C°$. (Notice that when we refer to a specific temperature, we say "degrees Celsius," as in 20°C; but when we refer to a change in temperature or a temperature interval, we say "Celsius degrees," as in "1 C°.") The conversion between the two temperature scales can be written

$$T(°C) = \frac{5}{9}[T(°F) - 32] \qquad \text{or} \qquad T(°F) = \frac{9}{5}T(°C) + 32.$$

Rather than memorizing these relations (it would be easy to confuse them), it is usually easier to simply remember that 0°C = 32°F and 5 C° = 9 F° (or 100°C = 212°F).

EXAMPLE 13–2 **Taking your temperature.** Normal body temperature is 98.6°F. What is this on the Celsius scale?

SOLUTION First we note that 98.6°F is 98.6 − 32.0 = 66.6 F° above the freezing point of water. Since each F° is equal to $\frac{5}{9}$ C°, this corresponds to $66.6 \times \frac{5}{9} = 37.0$ Celsius degrees above the freezing point. Since the freezing point is 0°C, the temperature is 37.0°C.

Different materials do not expand in quite the same way over a wide temperature range. Consequently, if we calibrate different kinds of thermometers exactly as described above, they will not usually agree precisely. Because of how we calibrated them, they will agree at 0°C and at 100°C. But because of different expansion properties, they may not agree precisely at intermediate temperatures (remember we arbitrarily divided the thermometer scale into 100 equal divisions between 0°C and 100°C). Thus a carefully calibrated mercury-in-glass thermometer might register 52.0°C, whereas a carefully calibrated thermometer of another type might read 52.6°C.

Because of this discrepancy, some standard kind of thermometer must be chosen so that these intermediate temperatures can be precisely defined. The chosen standard for this purpose is the so-called **constant-volume gas**

thermometer. As shown in the simplified diagram of Fig. 13–7, this thermometer consists of a bulb filled with a dilute gas connected by a thin tube to a mercury manometer. The volume of the gas is kept constant by raising or lowering the right-hand tube of the manometer so that the mercury in the left tube coincides with the reference mark. An increase in temperature causes a proportional increase in pressure in the bulb. Thus the tube must be lifted higher to keep the gas volume constant. The height of the mercury in the right-hand column is then a measure of the temperature. This thermometer can be calibrated and gives the same results for all gases in the limit of reducing the gas pressure in the bulb toward zero. The resulting scale is defined as the standard temperature scale.

FIGURE 13–7 Constant-volume gas thermometer.

*13–3 Thermal Equilibrium and the Zeroth Law of Thermodynamics

We are all familiar with the fact that if two objects at different temperatures are placed in thermal contact (so that heat energy can transfer from one to the other), the two objects will eventually reach the same temperature. They will then be said to be in **thermal equilibrium**. For example, an ice cube placed in a large glass of hot water melts to water, all of which eventually comes to the same temperature. If you put your hand in the water of an icy lake, you can *feel* the temperature of your hand drop, as energy flows out of your hand into the cold water. (It's best to pull your hand out before thermal equilibrium is reached!) Two objects are defined to be in *thermal equilibrium* if, when placed in thermal contact, no energy flows from one to the other, and their temperatures don't change.

Suppose you wanted to determine if two systems, A and B, were in thermal equilibrium, but without putting them in contact. You could do so by making use of a third system, C (which could be considered a thermometer). Suppose that C and A are in thermal equilibrium and that C and B are in thermal equilibrium. Does this imply that A and B are necessarily in thermal equilibrium with each other? Actually, it isn't completely obvious. Nonetheless, a great deal of experiment indicates that *if two systems are in thermal equilibrium with a third system, then they are in thermal equilibrium with each other*. This postulate is called the **zeroth law of thermodynamics**. (It has this rather odd name since it was not until after the great first and second laws of thermodynamics (Chapter 15) were worked out that scientists realized that this apparently obvious postulate needed to be stated first.)

Temperature is a property of a system that determines whether the system will be in thermal equilibrium with other systems. When two systems are in thermal equilibrium, their temperatures are, by definition, equal. This is consistent with our everyday notion of temperature, since when a hot body and a cold one are put into contact, they eventually come to the same temperature. Thus the importance of the zeroth law is that it allows a useful definition of temperature. Suppose that the zeroth law were not valid. That is, suppose A and C are in thermal equilibrium and so are B and C, but A and B are not in thermal equilibrium; this would imply that $T_A = T_C$ and $T_B = T_C$, but that $T_A \neq T_B$, which clearly would not be consistent, so temperature would be a useless variable. But no experiment has contradicted the zeroth law, so we assume it is valid, and therefore temperature is a useful quantity.

13–4 Thermal Expansion

Most substances expand when heated and contract when cooled. However, the amount of expansion or contraction varies, depending on the material.

Experiments indicate that the change in length ΔL of almost all solids is, to a very good approximation, directly proportional to the change in temperature ΔT. As might be expected, the change in length is also proportional to the original length of the object, L_0, Fig. 13–8. That is, for the same temperature change, a 4-m-long iron rod will increase in length twice as much as a 2-m-long iron rod. We can write this proportionality as an equation:

Linear expansion

$$\Delta L = \alpha L_0 \, \Delta T, \tag{13–1a}$$

where α, the proportionality constant, is called the *coefficient of linear expansion* for the particular material and has units of $(\text{C}°)^{-1}$. This equation can also be written as

$$L = L_0(1 + \alpha \Delta T), \tag{13–1b}$$

where L_0 is the length initially, at temperature T_0, and L is the length after heating or cooling to a temperature T. If the temperature change $\Delta T = T - T_0$ is negative, then $\Delta L = L - L_0$ is also negative; thus the length shortens.

The values of α for various materials at 20°C are listed in Table 13–1. It should be noted that α does vary slightly with temperature (which is why thermometers made of different materials do not agree precisely). However, if the temperature range is not too great, the variation can usually be ignored.

FIGURE 13–8 A thin rod of length L_0 at temperature T_0 is heated to a new uniform temperature T and acquires length L, where $L = L_0 + \Delta L$.

TABLE 13–1 Coefficients of Expansion, at 20°C

Material	Coefficient of Linear Expansion, α $(\text{C}°)^{-1}$	Coefficient of Volume Expansion, β $(\text{C}°)^{-1}$
Solids		
Aluminum	25×10^{-6}	75×10^{-6}
Brass	19×10^{-6}	56×10^{-6}
Iron or steel	12×10^{-6}	35×10^{-6}
Lead	29×10^{-6}	87×10^{-6}
Glass (Pyrex)	3×10^{-6}	9×10^{-6}
Glass (ordinary)	9×10^{-6}	27×10^{-6}
Quartz	0.4×10^{-6}	1×10^{-6}
Concrete and brick	$\approx 12 \times 10^{-6}$	$\approx 36 \times 10^{-6}$
Marble	$1.4\text{–}3.5 \times 10^{-6}$	$4\text{–}10 \times 10^{-6}$
Liquids		
Gasoline		950×10^{-6}
Mercury		180×10^{-6}
Ethyl alcohol		1100×10^{-6}
Glycerin		500×10^{-6}
Water		210×10^{-6}
Gases		
Air (and most other gases at atmospheric pressure)		3400×10^{-6}

EXAMPLE 13–3 **Bridge expansion.** The steel bed of a suspension bridge is 200 m long at 20°C. If the extremes of temperature to which it might be exposed are −30°C to +40°C, how much will it contract and expand?

➡ **PHYSICS APPLIED**
Expansion in structures

SOLUTION From Table 13–1, we find that $\alpha = 12 \times 10^{-6}(\text{C}°)^{-1}$. The increase in length when it is at 40°C will be

$$\Delta L = (12 \times 10^{-6}/\text{C}°)(200 \text{ m})(40°\text{C} - 20°\text{C}),$$
$$= 4.8 \times 10^{-2} \text{ m},$$

or 4.8 cm. When the temperature decreases to −30°C, $\Delta T = -50 \text{ C}°$. So the change in length is

$$\Delta L = (12 \times 10^{-6}/\text{C}°)(200 \text{ m})(-50 \text{ C}°) = -12.0 \times 10^{-2} \text{ m},$$

or a decrease in length of 12 cm.

CONCEPTUAL EXAMPLE 13–4 **Do holes expand or contract?** A circular hole is cut from a cookie sheet, as in Fig. 13–9. When the cookie sheet is heated in the oven, does the hole get larger or smaller?

RESPONSE Instead of cutting a hole, imagine just drawing a circle on the cookie sheet with a pencil. When the cookie sheet expands, it's easy to see that the circle will expand along with the rest of the sheet. Cutting the cookie sheet where the circle is means the hole grows as well. Note that the material does *not* expand inward to fill the hole. In a solid object, *all* sections expand with increased temperature.

FIGURE 13–9 Example 13–4.

EXAMPLE 13–5 **Ring on a rod.** An iron ring is to fit snugly on a cylindrical iron rod. At 20°C, the diameter of the rod is 6.445 cm and the inside diameter of the ring is 6.420 cm. To slip over the rod, the ring must be slightly larger than the rod diameter by about 0.008 cm. To what temperature must the ring be brought if its hole is to be large enough so it will slip over the rod?

SOLUTION The hole in the ring must be increased from a diameter of 6.420 cm to 6.445 cm + 0.008 cm = 6.453 cm. The ring must be heated since the hole diameter will increase linearly with temperature (as in Example 13–4). We solve for ΔT in Eq. 13–1a and find

$$\Delta T = \frac{\Delta L}{\alpha L_0} = \frac{6.453 \text{ cm} - 6.420 \text{ cm}}{(12 \times 10^{-6} \text{ C}°^{-1})(6.420 \text{ cm})} = 430 \text{ C}°.$$

So it must be raised at least to $T = (20°\text{C} + 430 \text{ C}°) = 450°\text{C}.$

CONCEPTUAL EXAMPLE 13–6 **Opening a tight jar lid.** When the lid of a glass jar is tight, holding it under hot water for a short time will often make it easier to open. Why?

➡ **PHYSICS APPLIED**
Opening a tight lid

RESPONSE The lid may be struck by the hot water more directly than the glass and so expand sooner. But even if not, metals generally expand more than glass for the same temperature change (α is greater—see Table 13–1).

The change in *volume* of a material which undergoes a temperature change is given by a relation similar to Eq. 13–1, namely,

Volume expansion

$$\Delta V = \beta V_0 \,\Delta T, \tag{13–2}$$

where ΔT is the change in temperature, V_0 is the original volume, ΔV is the change in volume, and β is the *coefficient of volume expansion*. The units of β are $(\text{C}°)^{-1}$. The value of β for a number of materials is given in Table 13–1. Notice that for solids, β is normally equal to approximately 3α

$\beta \approx 3\alpha$

(why this should be so is the subject of Problem 21). This is not true, however, for solids that are not isotropic (isotropic means having the same properties in all directions). Note also that linear expansion has no meaning for liquids and gases since they do not have fixed shapes.

Equations 13–1 and 13–2 are accurate only if ΔL (or ΔV) is small compared to L_0 (or V_0). This is of particular concern for liquids and even more so for gases because of the large values of β. Furthermore, β itself varies substantially with temperature for gases. Therefore, a more convenient way of dealing with gases is needed, and this will be discussed starting in Section 13–7.

EXAMPLE 13–7 **Gas tank in the sun.** The 70-L steel gas tank of a car is filled to the top with gasoline at 20°C. The car is then left to sit in the sun, and the tank reaches a temperature of 40°C (104°F). How much gasoline do you expect to overflow from the tank?

SOLUTION The gasoline expands (Eq. 13–2) by

$$\Delta V = \beta V_0 \,\Delta T = (950 \times 10^{-6}\ \text{C}°^{-1})(70\ \text{L})(20\ \text{C}°) = 1.3\ \text{L}.$$

The tank also expands. We can think of it as a steel shell that undergoes volume expansion ($\beta \approx 3\alpha = 36 \times 10^{-6}\ \text{C}°^{-1}$). If the tank were solid, the exterior surface layer (the shell) would expand just the same. Thus the tank increases in volume by

$$\Delta V = (36 \times 10^{-6}\ \text{C}°^{-1})(70\ \text{L})(20\ \text{C}°) = 0.050\ \text{L},$$

so the tank expansion has little effect. If this full tank were left in the sun, over a liter of gas could spill out into the road.

Want to save a few pennies? Fill your gas tank early in the morning when it is cool and the gas is denser—but don't fill the tank quite all the way.

➡ **PHYSICS APPLIED**

Gas tank overflow

13–5 Anomalous Behavior of Water Below 4°C

Water is unusual: it expands *when cooled from 4°C to 0°C*

Most substances expand more or less uniformly with an increase in temperature (as long as no phase change occurs). Water, however, does not follow the usual pattern. If water at 0°C is heated, it actually *decreases* in volume until it reaches 4°C. Above 4°C water behaves normally and expands in volume as the temperature is increased, Fig. 13–10. Water thus has its greatest density at 4°C. This anomalous behavior of water is of great importance for the survival of aquatic life during cold winters. When the water in a lake or river is above 4°C and begins to cool by contact with cold air, the water at the surface sinks because of its greater density and it is replaced by warmer water from below. This mixing continues until the temperature reaches 4°C. As the surface water cools further, it remains on the surface because it is less dense than the 4°C water below. Water then

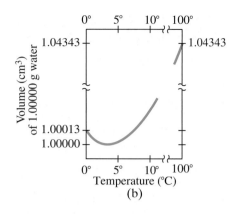

FIGURE 13–10 Behavior of water as a function of temperature near 4°C. (a) Density vs. temperature. (b) Volume, of 1.00000 gram of water, as a function of temperature. [Note the break in each axis.]

freezes first at the surface, and the ice remains on the surface since ice (specific gravity = 0.917) is less dense than water. The water at the bottom remains at 4°C until almost the whole body of water is frozen. If water were like most substances, becoming more dense as it cools, the water at the bottom of a lake would be frozen first. Lakes would freeze solid more easily since circulation would bring the warmer water to the surface to be efficiently cooled. The complete freezing of a lake would cause severe damage to its plant and animal life. Because of the unusual behavior of water below 4°C, it is rare for any large body of water to freeze completely, and this is helped by the layer of ice on the surface, which acts as an insulator to reduce the flow of heat out of the water into the cold air above. Without this peculiar but wonderful property of water, life on this planet as we know it might not have been possible.

Not only does water expand as it cools from 4°C to 0°C, it expands even more as it freezes to ice. This is why ice cubes float in water and pipes break when water inside them freezes.

➡ **PHYSICS APPLIED**

Life under ice

13–6 Thermal Stresses

In many situations, such as in buildings and roads, the ends of a beam or slab of material are rigidly fixed, which greatly limits expansion or contraction. If the temperature should change, large compressive or tensile stresses, called *thermal stresses*, will occur. The magnitude of such stresses can be calculated using the concept of elastic modulus developed in Chapter 9. To calculate the internal stress, we can think of this process as occurring in two steps. The rod expands (or contracts) by an amount ΔL given by Eq. 13–1, and then a force is applied to compress (or expand) the material back to its original length. The force F required is given by Eq. 9–4:

$$\Delta L = \frac{1}{E}\frac{F}{A}L_0,$$

where E is Young's modulus for the material. To calculate the internal stress, F/A, we then set ΔL in Eq. 13–1a equal to ΔL in the equation above and find

$$\alpha L_0\,\Delta T = \frac{1}{E}\frac{F}{A}L_0.$$

Hence, $F = \alpha EA\,\Delta T$.

EXAMPLE 13–8 **Stress in concrete on a hot day.** A highway is to be made of blocks of concrete 10 m long placed end to end with no space in between them to allow for expansion. If the blocks were placed at a temperature of 10°C, what force of compression would occur if the temperature reached 40°C? The contact area between each block is 0.20 m². Will fracture occur?

➡ **PHYSICS APPLIED**

Highway buckling

SOLUTION We solve for F in the equation above and use the value of E found in Table 9–1:

$$F = \alpha \, \Delta T \, E \, A$$

$$= (12 \times 10^{-6}/\text{C}°)(30 \text{ C}°)(20 \times 10^9 \text{ N/m}^2)(0.20 \text{ m}^2)$$

$$= 1.4 \times 10^6 \text{ N}.$$

The stress, F/A, is $(1.4 \times 10^6 \text{ N})/(0.20 \text{ m}^2) = 7.0 \times 10^6 \text{ N/m}^2$. This is not far from the ultimate strength of concrete (Table 9–2) under compression and exceeds it for tension and shear. Hence, assuming the concrete is not perfectly aligned, part of the force will act in shear, and fracture is likely.

13–7 The Gas Laws and Absolute Temperature

Equation 13–2 is not very useful for describing the expansion of a gas, partly because the expansion can be so great, and partly because gases generally expand to fill whatever container they are in. Indeed, Eq. 13–2 is meaningful only if the pressure is kept constant. The volume of a gas depends very much on the pressure as well as on the temperature. It is therefore valuable to determine a relation between the volume, the pressure, the temperature, and the mass of a gas. Such a relation is called an **equation of state**.[†] (By the word *state*, we mean the physical condition of the system.)

If the state of a system is changed, we will always wait until the pressure and temperature have reached the same values throughout. We thus consider only **equilibrium states** of a system—when the variables that describe it (such as temperature and pressure) are the same throughout the system and are not changing in time. We also note that the results of this Section are accurate only for gases that are not too dense (the pressure is not too high, on the order of an atmosphere or so) and not close to the liquefaction (boiling) point.

For a given quantity of gas it is found experimentally that, to a good approximation, *the volume of a gas is inversely proportional to the pressure applied to it when the temperature is kept constant.* That is,

Boyle's law

$$V \propto \frac{1}{P}. \qquad\qquad [\text{constant } T]$$

where P is the absolute pressure (not "gauge pressure"—see Chapter 10). For example, if the pressure on a gas is doubled, the volume is reduced to half its original volume. This relation is known as **Boyle's law**, after Robert Boyle (1627–1691), who first stated it on the basis of his own experiments.

[†]An equation of state can also be sought for solids and liquids since for them, too, the volume depends on the mass, temperature, and external pressure, although the temperature and pressure have much less effect than on gases. However, the situation with solids and liquids is much more complicated.

A graph of P vs. V for a fixed temperature is shown in Fig. 13–11. Boyle's law can also be written

$$PV = \text{constant.} \qquad \text{[constant } T\text{]}$$

That is, at constant temperature, if either the pressure or volume of the gas is allowed to vary, the other variable also changes so that the product PV remains constant.

Temperature also affects the volume of a gas, but a quantitative relationship between V and T was not found until more than a century after Boyle's work. The Frenchman Jacques Charles (1746–1823) found that when the pressure is not too high and is kept constant, the volume of a gas increases with temperature at a nearly constant rate, as in Fig. 13–12a. However, all gases liquefy at low temperatures (for example, oxygen liquefies at $-183°C$) and so the graph cannot be extended below the liquefaction point. Nonetheless, the graph is essentially a straight line and if projected to lower temperatures, as shown by the dashed line, it crosses the axis at about $-273°C$.

Such a graph can be drawn for any gas, and the straight line always projects back to $-273°C$ at zero volume. This seems to imply that if a gas could be cooled to $-273°C$ it would have zero volume, and at lower temperatures a negative volume, which makes no sense, of course. It could be argued that $-273°C$ is the lowest temperature possible, and many other more recent experiments indicate that it is so. This temperature is called the **absolute zero** of temperature. Its value has been determined to be $-273.15°C$.

Absolute zero forms the basis of a temperature scale known as the **absolute** or **Kelvin scale**, and it is used extensively in scientific work. On this scale the temperature is specified as degrees Kelvin or, preferably, simply as kelvins (K) without the degree sign. The intervals are the same as for the Celsius scale, but the zero on this scale (0 K) is chosen as absolute zero itself. Thus the freezing point of water (0°C) is 273.15 K and the boiling point of water is 373.15 K. Indeed, any temperature on the Celsius scale can be changed to kelvins by adding 273.15 to it:

$$T(\text{K}) = T(°\text{C}) + 273.15.$$

Now let us look at Fig. 13–12b, where we see that the graph of the volume of a gas versus absolute temperature is a straight line that passes through the origin. Thus, to a good approximation, *the volume of a given amount of gas is directly proportional to the absolute temperature when the pressure is kept constant.* This is known as **Charles's law**, and is written

$$V \propto T. \qquad \text{[constant } P\text{]}$$

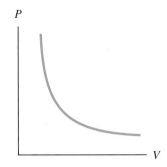

FIGURE 13–11 Pressure vs. volume of a gas at a constant temperature, showing the inverse relationship as given by Boyle's law: as the volume increases, the pressure decreases.

Absolute zero

Kelvin scale

Charles's law

FIGURE 13–12 Volume of a gas as a function of (a) Celsius temperature, and (b) Kelvin temperature, when the pressure is kept constant.

(a)

(b)

A third gas law, known as **Gay-Lussac's law**, after Joseph Gay-Lussac (1778–1850), states that *at constant volume, the pressure of a gas is directly proportional to the absolute temperature*:

Gay-Lussac's law

$$P \propto T. \qquad \text{[constant } V\text{]}$$

A familiar example is that a closed jar, or aerosol can, thrown into a fire will explode due to the increase in gas pressure inside.

The laws of Boyle, Charles, and Gay-Lussac are not really laws in the sense that we use this term today (precise, deep, wide-ranging validity). They are really only approximations that are accurate for real gases only as long as the pressure and density of the gas are not too high, and the gas is not too close to condensation. The term *law* applied to these three relationships has become traditional, however, so we have stuck with that usage.

CONCEPTUAL EXAMPLE 13–9 | **Does warm air rise?** Warm air is less dense than cool air. Therefore, because of buoyant forces, warm air should rise in the atmosphere. Why, then, is the air invariably *colder* at the top of a mountain, compared to the air at the bottom of a mountain?

RESPONSE It is true that warm air rises in the atmosphere. However, as it rises, it moves toward a region of lower pressure. This effect is particularly pronounced where a moving air mass is forced upward by a mountain. According to Gay-Lussac's law the temperature of the air decreases in direct proportion to the pressure, so it cools. Also, air moving in the opposite direction, downslope from higher to lower altitudes, creates the *chinook* winds of the Rockies and the Santa Anas of southern California. The compressional heating can raise the temperature in Denver to 60°F in January, and to 100°F in Los Angeles in November.

13–8 The Ideal Gas Law

The gas laws of Boyle, Charles, and Gay-Lussac were obtained by means of a technique that is very useful in science: namely, to hold one or more variables constant in order to see clearly the effects of changing only one of the variables. These laws can now be combined into a single more general relation between the pressure, volume, and temperature of a fixed quantity of gas:

$$PV \propto T.$$

This relation indicates how any of the quantities P, V, or T will vary when the other two quantities change. This relation reduces to Boyle's, Charles's, or Gay-Lussac's law when either the temperature, the pressure, or the volume, respectively, is held constant.

Finally, we must incorporate the effect of the amount of gas present. Anyone who has blown up a balloon knows that the more air forced into the balloon, the bigger it gets (Figure 13–13). Indeed, careful experiments show that at constant temperature and pressure, the volume V of an en-

FIGURE 13–13 Blowing up a balloon means putting more air (more air molecules) into the balloon, which increases its volume.

closed gas increases in direct proportion to the mass m of gas present. Hence we write

$$PV \propto mT.$$

This proportion can be made into an equation by inserting a constant of proportionality. Experiment shows that this constant has a different value for different gases. However, the constant of proportionality turns out to be the same for all gases if, instead of the mass m, we use the number of moles. One **mole** (abbreviated mol) is defined as the amount of substance that contains as many atoms or molecules as there are in 12.00 grams of carbon 12 (whose atomic mass is exactly 12 u). A simpler but equivalent definition is: 1 mol is that number of grams of a substance numerically equal to the molecular mass (Section 13–1) of the substance. For example, the molecular mass of hydrogen gas (H_2) is 2.0 u (since each molecule contains two atoms of hydrogen and each atom has an atomic mass of 1.0 u). Thus 1 mol of H_2 has a mass of 2.0 g. Similarly, 1 mol of neon gas has a mass of 20 g, and 1 mol of CO_2 has a mass of $[12 + (2 \times 16)] = 44$ g. The mole is the official unit in the SI system. Sometimes the kilomole (kmol) is used, which is that number of kilograms numerically equal to the molecular mass of the substance. In general, the number of moles, n, in a given sample of a pure substance is equal to its mass in grams divided by its molecular mass specified as grams per mole:

Mole (unit)

$$n \text{ (mol)} = \frac{\text{mass (grams)}}{\text{molecular mass (g/mol)}}.$$

For example, the number of moles in 132 g of CO_2 is

$$n = \frac{132 \text{ g}}{44 \text{ g/mol}} = 3.0 \text{ mol.}$$

We can now write the proportion discussed above as an equation:

$$PV = nRT, \tag{13–3}$$

IDEAL GAS LAW

where n represents the number of moles and R is the constant of proportionality. R is called the **universal gas constant** because its value is found experimentally to be the same for all gases. The value of R, in several sets of units (only the first is the proper SI unit), is:

$$R = 8.315 \text{ J/(mol·K)} \qquad \text{[SI units]}$$
$$= 0.0821 \text{ (L·atm)/(mol·K)}$$
$$= 1.99 \text{ calories/(mol·K).}^{\dagger}$$

Equation 13–3 is called the **ideal gas law**, or the **equation of state for an ideal gas**. We use the term "ideal" because real gases do not follow Eq. 13–3 precisely, particularly at high pressure (and density) or when the gas is near the liquefaction point (= boiling point). However, at pressures less than an atmosphere or so, and when T is not close to the liquefaction point of the gas, Eq. 13–3 is quite accurate and useful for real gases.

---

†Calories are defined in the next chapter, Section 14–1, and sometimes it is useful to use R as given in terms of calories.

13–9 Problem Solving with the Ideal Gas Law

STP = 273 K, 1 atm

The ideal gas law is an extremely useful tool, and we now consider some Examples. We will often refer to "standard conditions" or "standard temperature and pressure" (STP), which means $T = 273$ K (0°C) and $P = 1.00$ atm $= 1.013 \times 10^5$ N/m^2 = 101.3 kPa.

EXAMPLE 13–10 **Volume of one mol at STP.** Determine the volume of 1.00 mol of any gas at STP, assuming it behaves like an ideal gas.

SOLUTION We solve for V in Eq. 13–3:

$$V = \frac{nRT}{P} = \frac{(1.00 \text{ mol})(8.315 \text{ J/mol·K})(273 \text{ K})}{(1.013 \times 10^5 \text{ N/m}^2)}$$

$$= 22.4 \times 10^{-3} \text{ m}^3.$$

1 mol of gas at STP has volume = 22.4 L

Since 1 liter is 1000 cm^3 = 1×10^{-3} m^3, 1 mol of any gas has a volume of 22.4 L at STP. [How big is 22.4 L? About the size of a cube one foot on a side; more precisely, $\sqrt[3]{22.4 \times 10^{-3} \text{ m}^3} = 0.28$ m $= 28$ cm on a side.]

The value for the volume of 1 mol of an ideal gas at STP (22.4 L) is worth remembering, for it sometimes makes calculation simpler, as the following Example shows.

EXAMPLE 13–11 **Easy way to get gas mass.** A flexible container of oxygen (O_2, molecular mass = 32.0 u) at STP has a volume of 10.0 m^3. What is the mass of gas enclosed?

SOLUTION Since 1 mol occupies a volume of 22.4×10^{-3} m^3, the 10.0 m^3 of oxygen corresponds to

$$n = \frac{(10.0 \text{ m}^3)}{(22.4 \times 10^{-3} \text{ m}^3/\text{mol})} = 446 \text{ mol}.$$

Since 1 mol has a mass of 0.0320 kg, the mass of oxygen is

$$m = (446 \text{ mol}) \times (0.0320 \text{ kg/mol}) = 14.3 \text{ kg}.$$

EXAMPLE 13–12 **Helium balloon.** A helium party balloon, assumed to be a perfect sphere, has a radius of 18.0 cm. At room temperature (20°C), its internal pressure is 1.05 atm. Find the number of moles of helium in the balloon and the mass of helium needed to inflate the balloon to these values.

SOLUTION We use the ideal gas law to find n. We first find the volume of the balloon:

$$V = \tfrac{4}{3} \pi r^3 = 0.0244 \text{ m}^3.$$

The pressure is given as 1.05 atm $= 1.064 \times 10^5$ N/m^2. The temperature must always be expressed in kelvins, so we change 20°C to $(20 + 273)$K $= 293$ K. Finally we choose the value of R to be $R = 8.315$ J/(mol·K) because we are using SI units. Thus

$$n = \frac{PV}{RT} = \frac{(1.064 \times 10^5 \text{ N/m}^2)(0.0244 \text{ m}^3)}{(8.315 \text{ J/mol·K})(293 \text{ K})} = 1.066 \text{ mol}$$

The mass of helium (molecular mass = 4.00 g/mol as given in Appendix F or the periodic table) can be obtained from:

$$\text{mass} = n \times \text{molecular mass} = (1.066 \text{ mol})(4.00 \text{ g/mol}) = 4.26 \text{ g}.$$

Frequently, volume is specified in liters and pressure in atmospheres. Rather than convert these to SI units, we can instead use the value of R given above as 0.0821 L·atm/mol·K.

In many situations it is not necessary to use the value of R at all. For example, many problems involve a change in the pressure, temperature, and volume of a fixed amount of gas. In this case, $PV/T = nR$ = constant, since n and R remain constant. If we now let P_1, V_1, and T_1 represent the appropriate variables initially, and P_2, V_2, T_2 represent the variables after the change is made, then we can write

$$\frac{P_1 V_1}{T_1} = \frac{P_2 V_2}{T_2}.$$

> **► PROBLEM SOLVING**
> *Using the ideal gas law as a ratio*

If we know any five of the quantities in this equation, we can solve for the sixth. Or, if one of the three variables is constant ($V_1 = V_2$, or $P_1 = P_2$, or $T_1 = T_2$) then we can use this equation to solve for one unknown when given the other three quantities.

EXAMPLE 13–13 **Check tires cold.** An automobile tire is filled to a gauge pressure of 200 kPa at 10°C. After driving 100 km, the temperature within the tire rises to 40°C. What is the pressure within the tire now?

> **► PHYSICS APPLIED**
> *Pressure in a hot tire*

SOLUTION Since the volume remains essentially constant, $V_1 = V_2$, and therefore

$$\frac{P_1}{T_1} = \frac{P_2}{T_2}.$$

This is, incidentally, a statement of Gay-Lussac's law. Since the pressure given is the gauge pressure, we must add atmospheric pressure ($= 101$ kPa) to get the absolute pressure $P_1 = (200\text{ kPa} + 101\text{ kPa}) = 301$ kPa. And we convert temperatures to kelvin by adding 273:

$$P_2 = \frac{P_1}{T_1} T_2 = \frac{(3.01 \times 10^5\text{ Pa})(313\text{ K})}{(283\text{ K})} = 333\text{ kPa}.$$

Subtracting atmospheric pressure, we find the resulting gauge pressure to be 232 kPa, which is a 15 percent increase. This Example shows why car manuals suggest checking the pressure when the tires are cold.

Always remember, when using the ideal gas law, that temperatures must be given in kelvins (K) and that the pressure P must always be *absolute* pressure, not gauge pressure.

> **► PROBLEM SOLVING**
> *Always give T in K, and P as absolute, not gauge, pressure*

13–10 Ideal Gas Law in Terms of Molecules: Avogadro's Number

The fact that the gas constant, R, has the same value for all gases is a remarkable reflection of simplicity in nature. It was first recognized, although in a slightly different form, by the Italian scientist Amedeo Avogadro (1776–1856). Avogadro stated that *equal volumes of gas at the same pressure and temperature contain equal numbers of molecules*. This is sometimes called **Avogadro's hypothesis**. That this is consistent with R

> *Avogadro's hypothesis*

being the same for all gases can be seen as follows. First of all, from Eq. 13–3 we see that for the same number of moles, n, and the same pressure and temperature, the volume will be the same for all gases as long as R is the same. Second, the number of molecules in 1 mole is the same for all gases (this follows[†] directly from the definition of a mole). Thus Avogadro's hypothesis is equivalent to R being the same for all gases.

The number of molecules in a mole is known as **Avogadro's number**, N_A. Although Avogadro conceived the notion, he was not able to actually determine the value of N_A. Indeed, precise measurements were not done until the twentieth century. A number of methods have been devised to measure N_A and the accepted value is

Avogadro's number

$$N_A = 6.02 \times 10^{23}. \qquad \text{[molecules/mol]}$$

Since the total number of molecules N in a gas is equal to the number per mole times the number of moles $(N = nN_A)$, then the ideal gas law (Eq. 13–3) can be written in terms of the number of molecules present:

$$PV = nRT = \frac{N}{N_A} RT,$$

or

IDEAL GAS LAW
(in terms of number of molecules)

$$PV = NkT. \qquad \textbf{(13–4)}$$

The constant $k = R/N_A$ is called **Boltzmann's constant** and has the value

$$k = \frac{R}{N_A} = \frac{8.315 \text{ J/mol·K}}{6.02 \times 10^{23}/\text{mol}} = 1.38 \times 10^{-23} \text{ J/K}.$$

EXAMPLE 13–14 **Hydrogen atom mass.** Use Avogadro's number to determine the mass of a hydrogen atom.

SOLUTION One mole of hydrogen (atomic mass = 1.008 u) has a mass of 1.008×10^{-3} kg and contains 6.02×10^{23} atoms. Thus one atom has a mass

$$m = \frac{1.008 \times 10^{-3} \text{ kg}}{6.02 \times 10^{23}} = 1.67 \times 10^{-27} \text{ kg}.$$

Historically, the reverse process was one method used to obtain N_A: that is, from the measured mass of the hydrogen atom.

[†]For example, the molecular mass of H_2 gas is 2.0 atomic mass units (u), whereas that of O_2 gas is 32.0 u. Thus 1 mol of H_2 has a mass of 0.0020 kg and 1 mol of O_2 gas, 0.032 kg. The number of molecules in a mole is equal to the total mass M of a mole divided by the mass m of one molecule; since this ratio (M/m) is the same for all gases by definition of the mole, a mole of any gas must contain the same number of molecules.

EXAMPLE 13-15 ESTIMATE | **How many molecules in one breath?**
Estimate how many molecules you breathe in with a 1.0-L breath of air.

SOLUTION One mole corresponds to 22.4 L (Example 13–10), so 1.0 L of air is $1/22.4 = 0.045$ mol. Then 1.0 L of air contains

$$(0.045 \text{ mol})(6.02 \times 10^{23} \text{ molecules/mole}) = 2.7 \times 10^{22} \text{ molecules}.$$

13-11 Kinetic Theory and the Molecular Interpretation of Temperature

The concept that matter is made up of atoms which are in continual random motion is called the **kinetic theory**. We now investigate the properties of a gas from the point of view of kinetic theory, and in particular we will calculate the pressure in a gas in terms of its molecular properties. We will also arrive at an important relation between the average kinetic energy of molecules in a gas and the absolute temperature.

We make the following assumptions about the molecules in a gas. These assumptions reflect a simple view of a gas, but nonetheless correspond well to the essential features of real gases that are at low pressures and far from the liquefaction point. Under these circumstances gases follow the ideal gas law closely, and indeed we shall refer to such a gas as an **ideal gas**. The assumptions, which represent the basic postulates of the kinetic theory, are:

Postulates of kinetic theory

1. There are a large number of molecules, N, each of mass m, moving in random directions with a variety of speeds. This assumption is in accord with our observation that a gas fills its container and, in the case of air on the earth, is kept from escaping only by the force of gravity.

2. The molecules are, on the average, far apart from one another. That is, their average separation is much greater than the diameter of each molecule.

3. The molecules are assumed to obey the laws of classical mechanics, and are assumed to interact with one another only when they collide. Although molecules exert weak attractive forces on each other between collisions, the potential energy associated with these forces is small compared to the kinetic energy, and we ignore it for now.

4. Collisions with another molecule or the wall of the vessel are assumed to be perfectly elastic, like the collisions of perfectly elastic billiard balls (Chapter 7).

We can see immediately how this kinetic view of a gas can explain Boyle's law (Section 13–7). The pressure exerted on a wall of a container of gas is due to the constant bombardment of molecules. If the volume is reduced by (say) half, the molecules are closer together and twice as many will be striking a given area of the wall per second. Hence we expect the pressure to be twice as great, which is Boyle's law.

Boyle's law explained

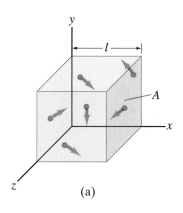

Calculating the pressure in an ideal gas due to its molecules

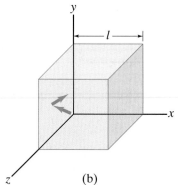

(a)

(b)

FIGURE 13–14 (a) Molecules of a gas moving about in a cubical container. (b) Arrows indicate the momentum of one molecule as it rebounds from the end wall.

Now let us calculate quantitatively the pressure in a gas based on kinetic theory. For purposes of argument, we imagine that the molecules are contained in a rectangular vessel whose ends have area A and whose length is l, as shown in Fig. 13–14a. The pressure exerted by the gas on the walls of its container is, according to our model, due to the collisions of the molecules with the walls. Let us focus our attention on the wall, of area A, at the left end of the container and examine what happens when one molecule strikes this wall, as shown in Fig. 13–14b. This molecule exerts a force on the wall, and the wall exerts an equal and opposite force back on the molecule. The magnitude of this force, according to Newton's second law, is equal to the molecule's rate of change of momentum (Section 7–1). Assuming the collision is elastic, only the x component of the molecule's momentum changes, and it changes from $-mv_x$ (it is moving in the negative x direction) to $+mv_x$. Thus the change in momentum, $\Delta(mv)$, which is the final momentum minus the initial momentum, is

$$\Delta(mv) = mv_x - (-mv_x) = 2mv_x$$

for one collision. This molecule will make many collisions with the wall, each separated by a time Δt, which is the time it takes the molecule to travel across the box and back again, a distance equal to $2l$. Thus $2l = v_x \, \Delta t$ or $\Delta t = 2l/v_x$. The time Δt between collisions is very small, so the number of collisions per second is very large. Thus the average force—averaged over many collisions—will be equal to the force exerted during one collision divided by the time between collisions (Newton's second law):

$$F = \frac{\Delta(mv)}{\Delta t} = \frac{2mv_x}{2l/v_x} = \frac{mv_x^2}{l}. \qquad \text{[due to one molecule]}$$

Of course, the actual force due to one molecule is intermittent, but because a huge number of molecules are striking the wall per second, the force is, on average, pretty steady. To calculate the force due to *all* the molecules in the box, we have to add the contributions of each. Thus the net force on the wall is

$$F = \frac{m}{l}(v_{x1}^2 + v_{x2}^2 + \cdots + v_{xN}^2),$$

where v_{x1} means v_x for molecule number 1 (we arbitrarily assign each molecule a number) and the sum extends over the total of N molecules. Now the average value of the square of the x component of velocity is

$$\overline{v_x^2} = \frac{v_{x1}^2 + v_{x2}^2 + \cdots + v_{xN}^2}{N}. \qquad \textbf{(13–5)}$$

Thus we can write the force as

$$F = \frac{m}{l} N \overline{v_x^2}.$$

We know that the square of any vector is equal to the sum of the squares of its components (theorem of Pythagoras). Thus $v^2 = v_x^2 + v_y^2 + v_z^2$ for any velocity v. Taking averages, we obtain

$$\overline{v^2} = \overline{v_x^2} + \overline{v_y^2} + \overline{v_z^2}.$$

Since the velocities of the molecules in our gas are assumed to be random, there is no preference to one direction or another. Hence

$$\overline{v_x^2} = \overline{v_y^2} = \overline{v_z^2}.$$

Combining this relation with the one just above, we get

$$\overline{v^2} = 3\overline{v_x^2}.$$

We substitute this into the equation for the force F:

$$F = \frac{m}{l} N \frac{\overline{v^2}}{3}.$$

The pressure on the wall is then

$$P = \frac{F}{A} = \frac{1}{3} \frac{Nm\overline{v^2}}{Al}$$

or

$$P = \frac{1}{3} \frac{Nm\overline{v^2}}{V}, \tag{13-6}$$

Pressure in a gas

where $V = lA$ is the volume of the container. This is the result we were seeking, the pressure in a gas expressed in terms of molecular properties.

Equation 13–6 can be rewritten in a clearer form by multiplying both sides by V and slightly rearranging the right side:

$$PV = \tfrac{2}{3} N(\tfrac{1}{2} m\overline{v^2}). \tag{13-7}$$

The quantity $\tfrac{1}{2} m\overline{v^2}$ is the average kinetic energy ($\overline{\mathrm{KE}}$) of the molecules in the gas. If we compare Eq. 13–7 with Eq. 13–4, the ideal gas law, we see that the two agree if

$$\tfrac{2}{3}(\tfrac{1}{2} m\overline{v^2}) = kT,$$

or

$$\boxed{\overline{\mathrm{KE}} = \tfrac{1}{2} m\overline{v^2} = \tfrac{3}{2} kT.} \tag{13-8}$$

TEMPERATURE RELATED TO $\overline{KE}$ OF MOLECULES

This equation tells us that

the average translational kinetic energy of molecules in a gas is directly proportional to the absolute temperature.

The higher the temperature, according to kinetic theory, the faster the molecules are moving on the average. This relation is one of the triumphs of the kinetic theory.

EXAMPLE 13–16 **Molecular KE.** What is the average translational kinetic energy of molecules in a gas at 37°C?

SOLUTION We use Eq. 13–8 and change 37°C to 310 K:

$$\overline{\mathrm{KE}} = \tfrac{3}{2} kT$$
$$= \tfrac{3}{2}(1.38 \times 10^{-23} \text{ J/K})(310 \text{ K}) = 6.42 \times 10^{-21} \text{J}.$$

Equation 13–8 holds not only for gases, but also applies reasonably accurately to liquids and solids. Thus the result of Example 13–16 would apply to molecules within living cells at body temperature (37°C).

SECTION 13–11   Kinetic Theory and the Molecular Interpretation of Temperature   **401**

We can also use Eq. 13–8 to calculate how fast molecules are moving on the average. Notice that the average in Eqs. 13–5 through 13–8 is over the *square* of the velocity. The square root of $\overline{v^2}$ is called the **root-mean-square** velocity, v_{rms} (since we are taking the square *root* of the *mean* of the *square* of the velocity):

Root-mean-square (rms) velocity

rms speed of molecules

$$v_{rms} = \sqrt{\overline{v^2}} = \sqrt{\frac{3kT}{m}}. \tag{13–9}$$

EXAMPLE 13–17 **Speeds of air molecules.** What is the rms speed of air molecules (O_2 and N_2) at room temperature (20°C)?

SOLUTION We must apply Eq. 13–9 to oxygen and nitrogen separately since they have different masses. Using the result of Example 13–14, the masses of one molecule of O_2 (molecular mass = 32 u) and N_2 (molecular mass = 28 u) are

$$m(O_2) = (32)(1.67 \times 10^{-27}\ kg) = 5.3 \times 10^{-26}\ kg$$

$$m(N_2) = (28)(1.67 \times 10^{-27}\ kg) = 4.7 \times 10^{-26}\ kg.$$

Thus, for oxygen

$$v_{rms} = \sqrt{\frac{3kT}{m}} = \sqrt{\frac{(3)(1.38 \times 10^{-23}\ J/K)(293\ K)}{(5.3 \times 10^{-26}\ kg)}}$$

$$= 480\ m/s,$$

and for nitrogen the result is $v_{rms} = 510\ m/s$. (These speeds are more than 1500 km/h or 1000 mi/h.)

*13–12 Distribution of Molecular Speeds

The molecules in a gas are assumed to be in random motion, which means that many molecules have speeds less than the rms speed and others have greater speeds. In 1859, James Clerk Maxwell (1831–1879) derived, on the basis of kinetic theory, that the speeds of molecules in a gas are distributed according to the graph shown in Fig. 13–15. This is known as the **Maxwell distribution of speeds.**[†] The speeds vary from zero up to many times the rms speed, but as can be seen from the graph, most molecules have speeds that are not far from the average. Less than one percent of the molecules exceed four times v_{rms}.

Maxwell distribution of speeds of molecules in a gas

Experiments to determine the distribution in real gases, starting in the 1920s, confirmed with considerable accuracy the Maxwell distribution and the direct proportion between average kinetic energy and absolute temperature, Eq. 13–8.

Figure 13–16 shows the distribution for two different temperatures, just as v_{rms} increases with temperature, so the whole distribution curve shifts to the right at higher temperatures. This figure illustrates how kinetic theory can be used to explain why many chemical reactions, including those in bio-

FIGURE 13–15 Distribution of speeds of molecules in an ideal gas. Note that v_{rms} is not at the peak of the curve (that speed is called the "most probable speed," v_P). This is because the curve is skewed to the right: it is not symmetrical.

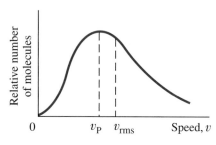

[†]Mathematically, the distribution is given by $\Delta N = Cv^2 \exp\left(-\frac{1}{2}mv^2/kT\right)\Delta v$, where ΔN is the number of molecules with speed between v and $v + \Delta v$, C is a constant, and exp means the expression in parentheses is an exponent on the natural number $e = 2.718\ldots$.

logical cells, take place more rapidly as the temperature increases. Most chemical reactions take place in a liquid solution, and the molecules in a liquid have a distribution of speeds close to the Maxwell distribution. Two molecules may chemically react only if their kinetic energy is great enough so that when they collide, they penetrate into each other somewhat. The minimum energy required is called the *activation energy*, E_A, and it has a specific value for each chemical reaction. The molecular speed corresponding to a kinetic energy of E_A for a particular reaction is indicated in Fig. 13–16. The relative number of molecules with energy greater than this value is given by the area under the curve beyond E_A. In Fig. 13–16, the respective areas for two different temperatures are indicated by the two different shadings in the figure. It is clear that the number of molecules that have kinetic energies in excess of E_A increases greatly for only a small increase in temperature. The rate at which a chemical reaction occurs is proportional to the number of molecules with energy greater than E_A, and thus we see why reaction rates increase rapidly with increased temperature.

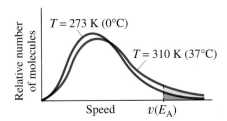

FIGURE 13–16 Distribution of molecular speeds for two different temperatures.

* 13–13 Real Gases and Changes of Phase

The ideal gas law is an accurate description of the behavior of a real gas as long as the pressure is not too high and as long as the temperature is far from the liquefaction point. But what happens when these two criteria are not satisfied?

Let us look at a graph of pressure plotted against volume for a given amount of gas. On such a "*PV* diagram," Fig. 13–17, each point represents an equilibrium state of the given substance. The various curves (labeled *A*, *B*, *C*, and *D*) show how the pressure varies as the volume is changed at constant temperature for several different values of the temperature. The dashed curve *A'* represents the behavior of a gas as predicted by the ideal gas law; that is, $PV = $ constant. The solid curve *A* represents the behavior of a real gas at the same temperature. Notice that at high pressure, the volume of a real gas is less than that predicted by the ideal gas law. The curves *B* and *C* in Fig. 13–17 represent the gas at successively lower temperatures, and we see that the behavior deviates even more from the curves predicted by the ideal gas law (for example, *B'*), and the deviation is greater the closer the gas is to liquefying.

To explain this, we note that at higher pressure we expect the molecules to be closer together. And, particularly at lower temperatures, the potential energy associated with the attractive forces between the molecules (which we ignored before) is no longer negligible compared to the now reduced kinetic energy of the molecules. These attractive forces tend to pull the molecules closer together so that at a given pressure, the volume is less than expected from the ideal gas law. At still lower temperatures, these forces cause liquefaction, and the molecules become very close together.

Curve *D* represents the situation when liquefaction occurs. At low pressure on curve *D* (on the right in Fig. 13–17), the substance is a gas and occupies a large volume. As the pressure is increased, the volume decreases until point *b* is reached. Beyond *b*, the volume decreases with no change in pressure; the substance is gradually changing from the gas to the liquid

PV diagram

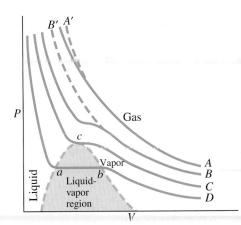

FIGURE 13–17 *PV* diagram for a real substance. Curves *A, B, C,* and *D* represent the same gas at different temperatures.

phase. At point *a*, all of the substance has changed to liquid. Further increase in pressure reduces the volume only slightly—liquids are nearly incompressible—so the curve is very steep as shown. The area within the tongue-shaped dashed line represents the region where the gas and liquid phases exist together in equilibrium.

Curve *C* in Fig. 13–17 represents the behavior of the substance at its **critical temperature**; and the point *c* (the one point where this curve is horizontal) is called the **critical point**. At temperatures less than the critical temperature (and this is the definition of the term), a gas will change to the liquid phase if sufficient pressure is applied. Above the critical temperature, no amount of pressure can cause a gas to change phase and become a liquid. What happens instead is that the gas becomes denser and denser as the pressure is increased and gradually it acquires properties resembling a liquid, but no liquid surface forms. The critical temperatures for various gases are given in Table 13–2. Scientists tried for many years to liquefy oxygen without success. Only after the discovery of the behavior of substances associated with the critical point was it realized that oxygen can be liquefied only if first cooled below its critical temperature of −118°C.

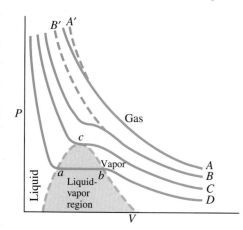

FIGURE 13–17 (Repeated) *PV* diagram for a real substance. Curves *A, B, C*, and *D* represent the same gas at different temperatures.

TABLE 13–2 Critical Temperatures and Pressures

Substance	Critical Temperature		Critical Pressure (atm)
	°C	K	
Water	374	647	218
Carbon dioxide	31	304	72.8
Oxygen	−118	155	50
Nitrogen	−147	126	33.5
Hydrogen	−239.9	33.3	12.8
Helium	−267.9	5.3	2.3

Often a distinction is made between the terms "gas" and "vapor": a substance below its critical temperature in the gaseous state is called a **vapor**; when above the critical temperature, it is called a **gas**. This is indicated in Fig. 13–17.

The behavior of a substance can be diagrammed not only on a *PV* diagram but also on a *PT* diagram. A *PT* diagram, often called a **phase diagram**, is particularly convenient for comparing the different phases of a substance. Figure 13–18 is the phase diagram for water. The curve labeled *l-v* represents those points where the liquid and vapor phases are in equilibrium—it is thus a graph of the boiling point versus pressure. Note that the curve correctly shows that at a pressure of 1 atm the boiling point is 100°C and that the boiling point is lowered for a decreased pressure. The curve *s-l* represents points where solid and liquid exist in equilibrium and thus is a graph of the freezing point versus pressure. At 1 atm, the freezing point of water is, of course, 0°C, as shown. Notice also in Fig. 13–18 that at a pressure of 1 atm, the substance is in the liquid phase if the temperature is between 0°C and 100°C, but is in the solid or vapor phase if the temperature is below 0°C or above 100°C. The curve labeled *s-v* is the *sublimation point* versus pressure curve. **Sublimation** refers to the process whereby at low pressures a solid changes directly into the vapor phase without passing through the liquid phase. For water, this occurs at pressures less than

FIGURE 13–18 Phase diagram for water (note that the scales are not linear).

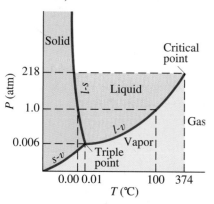

TABLE 13-3 Triple-Point Data

Substance	Temperature (K)	Pressure N/m²	Pressure atm
Water	273.16 (0.01°C)	6.10×10^2	6.03×10^{-3}
Carbon dioxide	216.6	5.16×10^5	5.11
Ammonia	195.40	6.06×10^3	6.00×10^{-2}
Nitrogen	63.2	1.25×10^4	1.24×10^{-1}
Oxygen	54.4	1.52×10^2	1.50×10^{-3}
Hydrogen	13.8	7.03×10^3	6.95×10^{-2}

0.0060 atm. Carbon dioxide, which in the solid phase is called dry ice, sublimates even at atmospheric pressure.

Triple point

The intersection of the three curves (in Fig. 13–18) is the **triple point**, and it is only at this point that the three phases can exist together in equilibrium. Because the triple point corresponds to a unique value of temperature and pressure, it is precisely reproducible and is often used as a point of reference. For example, the standard of temperature is usually specified as exactly 273.16 K at the triple point of water (see Table 13–3), rather than 273.15 K at the freezing point of water at 1 atm.

Notice that the s-l curve for water slopes upward to the left. This is true only of substances that *expand* upon freezing; for at a higher pressure, a lower temperature is needed to cause the liquid to freeze. More commonly, substances contract upon freezing and the s-l curve slopes upward to the right, as shown for CO_2 in Fig. 13–19.

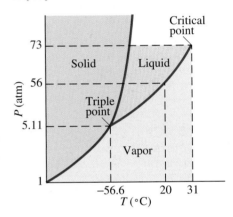

FIGURE 13-19 Phase diagram for carbon dioxide.

* 13-14 Vapor Pressure and Humidity

If a glass of water is left out overnight, the water level will have dropped by morning. We say the water has evaporated, meaning that some of the water has changed to the vapor or gas phase.

Evaporation

This process of **evaporation** can be explained on the basis of kinetic theory. The molecules in a liquid move past one another with a variety of speeds that follow, approximately, the Maxwell distribution. There are strong attractive forces between these molecules, which is what keeps them close together in the liquid phase. A molecule in the upper regions of the liquid may, because of its speed, leave the liquid momentarily. But just as a rock thrown into the air returns to the Earth, so the attractive forces of the other molecules can pull the vagabond molecule back to the liquid surface—that is, if its velocity is not too large. A molecule with a high enough velocity, however, will escape from the liquid entirely, like a rocket escaping the Earth, and become part of the gas phase. Only those molecules that have kinetic energy above a particular value can escape to the gas phase. We have already seen that kinetic theory predicts that the relative number of molecules with kinetic energy above a particular value (such as E_A in Fig. 13–16) increases with temperature. This is in accord with the well-known observation that the evaporation rate is greater at higher temperatures.

Because it is the fastest molecules that escape from the surface, the average speed of those remaining is less. When the average speed is less,

Evaporation cools

FIGURE 13–20 Vapor appears above a liquid in a closed container.

**TABLE 13–4
Saturated Vapor Pressure
of Water**

Temp-erature (°C)	Saturated Vapor Pressure	
	torr (= mmHg)	Pa (= N/m²)
− 50	0.030	4.0
− 10	1.95	2.60×10^2
0	4.58	6.11×10^2
5	6.54	8.72×10^2
10	9.21	1.23×10^3
15	12.8	1.71×10^3
20	17.5	2.33×10^3
25	23.8	3.17×10^3
30	31.8	4.24×10^3
40	55.3	7.37×10^3
50	92.5	1.23×10^4
60	149	1.99×10^4
70	234	3.12×10^4
80	355	4.73×10^4
90	526	7.01×10^4
100	760	1.01×10^5
120	1489	1.99×10^5
150	3570	4.76×10^5

the absolute temperature is less. Thus kinetic theory predicts that *evaporation is a cooling process*. You have no doubt noticed this effect when you stepped out of a warm shower and felt cold as the water on your body began to evaporate; and after working up a sweat on a hot day, even a slight breeze makes you feel cool through evaporation.

Air normally contains water vapor (water in the gas phase) and it comes mainly from evaporation. To look at this process in a little more detail, consider a closed container that is partially filled with water (it could just as well be any other liquid) and from which the air has been removed (Fig. 13–20). The fastest moving molecules quickly evaporate into the space above. As they move about, some of these molecules strike the liquid surface and again become part of the liquid phase; this is called **condensation**. The number of molecules in the vapor increases for a time, until a point is reached where the number returning to the liquid equals the number leaving in the same time interval. Equilibrium then exists, and the space is said to be *saturated*. The pressure of the vapor when it is saturated is called the **saturated vapor pressure** (or sometimes simply the vapor pressure).

The saturated vapor pressure does not depend on the volume of the container. If the volume above the liquid were reduced suddenly, the density of molecules in the vapor phase would be increased temporarily. More molecules would then be striking the liquid surface per second. There would be a net flow of molecules back to the liquid phase until equilibrium was again reached, and this would occur at the same value of the saturated vapor pressure.

The saturated vapor pressure of any substance depends on the temperature. At higher temperatures, more molecules have sufficient kinetic energy to break from the liquid surface into the vapor phase. Hence equilibrium will be reached at a higher pressure. The saturated vapor pressure of water at various temperatures is given in Table 13–4. Notice that even solids—for example, ice—have a measurable saturated vapor pressure.

In everyday situations, evaporation from a liquid takes place into the air above it rather than into a vacuum. This does not materially alter the discussion above relating to Fig. 13–20. Equilibrium will still be reached when there are sufficient molecules in the gas phase that the number reentering the liquid equals the number leaving. The concentration of molecules in the gas phase is not affected by the presence of air, although collisions with air molecules may lengthen the time needed to reach equilibrium. Thus equilibrium occurs at the same value of the saturated vapor pressure as if air weren't there.

Of course, if the container is large or is not closed, all the liquid may evaporate before saturation is reached. And if the container is not sealed—as, for example, a room in your house—it is not likely that the air will become saturated with water vapor; unless of course, it is raining outside.

The saturated vapor pressure of a liquid increases with temperature. When the temperature is raised to the point where the saturated vapor pressure at that temperature equals the external pressure, **boiling** occurs (Fig. 13–21). As the boiling point is approached, tiny bubbles tend to form in the liquid, which indicate a change from the liquid to the gas phase. However, if the vapor pressure inside the bubbles is less than the external pressure, the bubbles immediately are crushed. As the temperature is increased, the saturated vapor pressure inside a bubble eventually becomes equal to or exceeds the external air pressure. The bubble will then not collapse but will increase in size and rise to the surface. Boiling has then begun. *A liquid boils when its saturated vapor pressure equals the external pressure.* This occurs for water under 1 atm (760 torr) of pressure at 100°C, as can be seen from Table 13–4.

The boiling point of a liquid clearly depends on the external pressure. At high elevations, the boiling point of water is somewhat less than at sea level since the air pressure is less. For example, on the summit of Mt. Everest (8850 m) the air pressure is about one-third of what it is at sea level, and from Table 13–4 we can see that water will boil at about 70°C. Cooking food by boiling takes longer at high elevations, since the temperature is less. Pressure cookers, however, reduce cooking time, since they build up a pressure as high as 2 atm, allowing higher temperatures to be attained.

When we refer to the weather as being dry or humid, we are referring to the water vapor content of the air. In a gas such as air, which is a mixture of several types of gases, the total pressure is the sum of the *partial pressures* of each gas present.[†] By **partial pressure**, we mean the pressure each gas would exert if it alone occupied the whole volume. The partial pressure of water in the air can be as low as zero and can vary up to a maximum equal to the saturated vapor pressure of water at the given temperature. Thus, at 20°C, the partial pressure of water cannot exceed 17.5 torr (see Table 13–4). The **relative humidity** is defined as the ratio of the partial pressure to the saturated vapor pressure at a given temperature. It is usually expressed as a percentage:

FIGURE 13–21 Boiling: bubbles of water vapor float upward from the bottom of the pot (where the temperature is highest).

Relative humidity

$$\text{Relative humidity} = \frac{\text{partial pressure of H}_2\text{O}}{\text{saturated vapor pressure of H}_2\text{O}} \times 100\%.$$

Thus, when the humidity is close to 100 percent, the air holds nearly all the water vapor it can.

EXAMPLE 13–18 **Relative humidity.** On a particular hot day, the temperature is 30°C and the partial pressure of water vapor in the air is 21.0 torr. What is the relative humidity?

SOLUTION From Table 13–4, the saturated vapor pressure of water at 30°C is 31.8 torr. Hence the relative humidity is

$$\frac{21.0 \text{ torr}}{31.8 \text{ torr}} \times 100\% = 66\%.$$

Humans are sensitive to humidity. A relative humidity of 40–50 percent is generally optimum for both health and comfort. High humidity, particularly on a hot day, reduces the evaporation of moisture from the skin, which is one of the body's vital mechanisms for regulating body temperature. Very low humidity, on the other hand, can have a drying effect on the skin and mucous membranes.

Proper humidity must be maintained to prevent damage to paintings, tape recordings, and a wide variety of sensitive objects. Thus the design of heating and air-conditioning systems for buildings must take into account not only heating and cooling but the control of relative humidity as well.

➡ **PHYSICS APPLIED**

Humidity and comfort

[†]For example, 78 percent (by volume) of air molecules are nitrogen and 21 percent oxygen, with much smaller amounts of water vapor, argon, and other gases. At an air pressure of 1 atm, oxygen exerts a partial pressure of 0.21 atm and nitrogen 0.78 atm.

Air is saturated with water vapor when the partial pressure of water in the air is equal to the saturated vapor pressure at that temperature. If the partial pressure of water exceeds the saturated vapor pressure, the air is said to be **supersaturated**. This situation can occur when a temperature decrease occurs. For example, suppose the temperature is 30°C and the partial pressure of water is 21 torr, which represents a humidity of 66 percent as we saw above. Suppose now that the temperature falls to, say, 20°C, as might happen at nightfall. From Table 13–4 we see that the saturated vapor pressure of water at 20°C is 17.5 torr. Hence the relative humidity would be greater than 100 percent, and the supersaturated air cannot hold this much water. The excess water condenses and appears as dew. This process is also responsible for the formation of fog, clouds, and rain.

When air containing a given amount of water is cooled, a temperature is reached where the partial pressure of water equals the saturated vapor pressure. This is called the **dew point**. Measurement of the dew point is the most accurate means of determining the relative humidity. One method uses a polished metal surface in contact with air, which is gradually cooled down. The temperature at which moisture begins to appear on the surface is the dew point, and the partial pressure of water can then be obtained from saturated vapor pressure tables. If, for example, on a given day the temperature is 20°C, and the dew point is 5°C, then the partial pressure of water (Table 13–4) in the original air was 6.54 torr, whereas its saturated vapor pressure was 17.5 torr; hence the relative humidity was 6.54/17.5 = 37 percent.

A more convenient but less accurate method for measuring relative humidity is the so-called wet-bulb–dry-bulb technique, which makes use of two thermometers. One thermometer bulb is fitted with a snug cloth jacket that is soaking wet. The apparatus is usually swung in the air: the lower the humidity, the more evaporation takes place from the wet bulb, causing its temperature reading to be less. A comparison of the temperature readings on the wet-bulb thermometer and the dry (ordinary) thermometer can then be checked against special tables that have been compiled to obtain the relative humidity.

* 13–15 Diffusion

If you carefully place a few drops of food coloring in a container of water as in Fig. 13–22, you will find that the color spreads throughout the water. The process may take several hours (assuming you don't shake the glass), but eventually the color will become uniform. This mixing, known as **diffusion**, takes place because of the random movement of the molecules. Diffusion occurs in

FIGURE 13–22 A few drops of food coloring spreads slowly throughout the water, eventually becoming uniform.

(a)

(b)

(c)

gases too. Common examples include perfume or smoke (or the odor of something cooking on the stove) diffusing in air, although convection (moving air currents) often plays a greater role in spreading odors than does diffusion. Diffusion depends on concentration, by which we mean the number of molecules or moles per unit volume. In general, *the diffusing substance moves from a region where its concentration is high to one where its concentration is low.*

Diffusion can be readily understood on the basis of kinetic theory and the random motion of molecules. Consider a tube of cross-sectional area A containing molecules in a higher concentration on the left than on the right, Fig. 13–23. We assume the molecules are in random motion. Yet there will be a net flow of molecules to the right. To see why this is true, let us consider the small section of tube of length Δx as shown. Molecules from both regions 1 and 2 cross into this central section as a result of their random motion. The more molecules there are in a region, the more will strike a given area or cross a boundary. Since there is a greater concentration of molecules in region 1 than in region 2, more molecules cross into the central section from region 1 than from region 2. There is, then, a net flow of molecules from left to right, from high concentration toward low concentration. The flow stops only when the concentrations become equal.

You might expect that the greater the difference in concentration, the greater the flow rate. This is indeed the case. In 1855, the physiologist Adolf Fick (1829–1901) determined experimentally that the rate of diffusion (J) is directly proportional to the change in concentration per unit distance $(C_1 - C_2)/\Delta x$ (which is called the **concentration gradient**), and to the cross-sectional area A (see Fig. 13–23):

$$J = DA\frac{C_1 - C_2}{\Delta x}. \tag{13–10}$$

D is a constant of proportionality called the **diffusion constant**. Equation 13–10 is known as the **diffusion equation**, or **Fick's law**. If the concentrations are given in mol/m^3, then J is the number of moles passing a given point per second; if the concentrations are given in kg/m^3, then J is the mass movement per second (kg/s). The length Δx, of course, is given in meters.

Equation 13–10 applies not only to the simple situation of a gas diffusing as shown in Fig. 13–23, which is called *self-diffusion*, but also to a gas diffusing in a second gas (perfume vapor in air) or to a substance dissolved in a liquid, which are the more common situations. The rate of diffusion will be slower, particularly in liquids, because there will be collisions with the other molecules. Thus, the diffusion constant D will depend on the properties of the substances involved, and also on the temperature and the external pressure. The values of D for a variety of substances are given in Table 13–5.

Diffusion occurs from high to low concentration

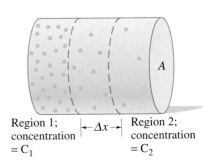

Region 1; concentration $= C_1$ |←Δx→| Region 2; concentration $= C_2$

FIGURE 13–23 Diffusion occurs from a region of high concentration to one of lower concentration (only one type of molecule is shown).

Diffusion equation

TABLE 13–5 Diffusion Constants, D (20°C, 1 atm)

Diffusing Molecules	Medium	D (m²/s)
H_2	Air	6.3×10^{-5}
O_2	Air	1.8×10^{-5}
O_2	Water	100×10^{-11}
Blood hemoglobin	Water	6.9×10^{-11}
Glycine (an amino acid)	Water	95×10^{-11}
DNA (molecular mass 6×10^6 u)	Water	0.13×10^{-11}

EXAMPLE 13–19 **ESTIMATE** **Diffusion of ammonia in air.** To get an idea of the time required for diffusion, estimate how long it might take for ammonia (NH_3) to be detected 10 cm from a bottle after it is opened, assuming only diffusion.

SOLUTION This will be an order-of-magnitude calculation. The rate of diffusion J can be set equal to the number of molecules N diffusing across area A in a time t: $J = N/t$. We solve for t:

$$t = \frac{N}{J}$$

and then use Eq. 13–10:

$$t = \frac{N}{AD} \frac{\Delta x}{\Delta C}.$$

The average concentration (midway between bottle and nose) can be approximated by $\overline{C} \approx N/V$, where V is the volume over which the molecules move and is roughly of the order of $V \approx A \, \Delta x$ where Δx is 10 cm. We substitute $N = \overline{C} A \, \Delta x$ into the above equation:

$$t \approx \frac{(\overline{C} A \, \Delta x) \, \Delta x}{AD \, \Delta C} = \frac{\overline{C}}{\Delta C} \frac{(\Delta x)^2}{D}.$$

The concentration of ammonia is high near the bottle and low near the detecting nose, so $\overline{C} \approx \Delta C/2$ or $(\overline{C}/\Delta C) \approx \frac{1}{2}$. Since NH_3 molecules have a size somewhere between H_2 and O_2, from Table 13–5 we can estimate $D \approx 4 \times 10^{-5} \, m^2/s$. Then

$$t \approx \frac{1}{2} \frac{(0.10 \, m)^2}{(4 \times 10^{-5} \, m^2/s)} \approx 100 \, s,$$

or about a minute or two. This seems rather long from experience, suggesting that air currents (convection) are more important than diffusion for transmitting odors.

Diffusion is extremely important for living organisms. Within cells, for example, molecules produced in certain chemical reactions must diffuse through water to other areas where they take part in other reactions.

Gas diffusion is important too. Plants require carbon dioxide for photosynthesis. The CO_2 diffuses into leaves from the outside through tiny openings called stomata. As CO_2 is utilized by the cells, its concentration drops below that in the air outside, and more diffuses inward as described by Fick's law. Water vapor and oxygen produced by the cells diffuse outward into the air.

Animals also exchange oxygen and CO_2 with the environment. Oxygen is required for energy-producing reactions and must diffuse into cells. CO_2 is produced as an end product of many metabolic reactions and must diffuse out of cells. Because of the slowness of diffusion over longer distances, only the smallest animals could survive without having developed complex respiratory and circulatory systems. In humans, oxygen is taken into the lungs where it diffuses across lung tissue and into the blood, which distributes it to cells throughout the body. The blood also carries carbon dioxide produced by the cells back to the lungs, where it diffuses outward.

The atomic theory of matter postulates that all matter is made up of tiny entities called **atoms**, which are typically 10^{-10} m in diameter.

Atomic and **molecular masses** are specified on a scale where ordinary carbon (^{12}C) is arbitrarily given the value 12.0000 u (atomic mass units).

The distinction between solids, liquids, and gases can be attributed to the strength of the attractive forces between the atoms or molecules and to their average speed.

Temperature is a measure of how hot or cold a body is. **Thermometers** are used to measure temperature on the **Celsius** (°C), **Fahrenheit** (°F), and **Kelvin** (K) scales. Two standard points on each scale are the freezing point of water (0°C, 32°F, 273.15 K) and the boiling point of water (100°C, 212°F, 373.15 K). A one kelvin change in temperature equals a change of one Celsius degree or $\frac{9}{5}$ Fahrenheit degrees.

The change in length, ΔL, of a solid, when its temperature changes by an amount ΔT, is directly proportional to the temperature change and to its original length L_0. That is,

$$\Delta L = \alpha L_0 \, \Delta T,$$

where α is the *coefficient of linear expansion*.

The change in volume of most solids, liquids, and gases is proportional to the temperature change and to the original volume V_0: $\Delta V = \beta V_0 \Delta T$. The *coefficient of volume expansion, β,* is approximately equal to 3α for solids.

Water is unusual because, unlike most materials whose volume increases with temperature, its volume actually decreases as the temperature increases in the range from 0°C to 4°C.

The **ideal gas law**, or **equation of state for an ideal gas**, relates the pressure P, volume V, and temperature T (in kelvins) of n moles of gas by the equation

$$PV = nRT,$$

where $R = 8.315$ J/mol·K for all gases. Real gases obey the ideal gas law quite accurately if they are not at too high a pressure or near their liquefaction point.

One **mole** of a substance is defined as the number of grams which is numerically equal to the atomic or molecular mass.

Avogadro's number, $N_A = 6.02 \times 10^{23}$, is the number of atoms or molecules in 1 mol of any pure substance.

The ideal gas law can be written in terms of the number of molecules N in the gas as

$$PV = NkT,$$

where $k = R/N_A = 1.38 \times 10^{-23}$ J/K is Boltzmann's constant.

According to the **kinetic theory** of gases, which is based on the idea that a gas is made up of molecules that are moving rapidly and at random, the average kinetic energy of molecules is proportional to the Kelvin temperature T:

$$\overline{KE} = \frac{3}{2} kT,$$

where k is Boltzmann's constant. At any moment, there exists a wide distribution of molecular speeds within a gas.

Q U E S T I O N S

1. Which has more atoms: 1 kg of iron or 1 kg of aluminum (see Appendix F)?

2. Name several properties of materials that could be exploited to make a thermometer.

3. Which is larger, 1C° or 1F°?

4. In a foreign country, a thermometer tells you that you have a fever of 40.0°C. What is this in Fahrenheit?

5. Figure 13–24 shows a diagram of a typical *thermostat* used to control a furnace (or other heating or cooling system). The bimetallic strip consists of two strips of different metals bonded together. Explain why this strip bends when the temperature changes, and how this controls the furnace.

Wires to heater

Rotate this cam to adjust for desired temperature   Contact

FIGURE 13–24   A typical thermostat.

6. A flat bimetallic strip consists of aluminum riveted to a strip of iron. When heated, which metal will be on the outside of the curve?

7. In the relation $\Delta L = \alpha L_0 \, \Delta T$, should L_0 be the initial length, the final length, or does it matter?

8. Long steam pipes often have a section in the shape of a ∪. Why?

9. The units for the coefficients of expansion α are $(C°)^{-1}$, and there is no mention of a length unit such as meters. Would the expansion coefficient change if we used feet or millimeters instead of meters?

10. A flat, uniform cylinder of lead floats in mercury at 0°C. Will the lead float higher or lower when the mercury temperature is raised?

11. Explain why it is advisable to add water to an overheated automobile engine only slowly, and only with the engine running.

12. A glass container may break if one part of it is heated or cooled more rapidly than adjacent parts. Explain.

13. When a cold mercury-in-glass thermometer is first placed in a hot tub of water, the mercury initially descends a bit and then rises. Explain.

14. The principal virtue of Pyrex glass is that its coefficient of linear expansion is much smaller than that for ordinary glass (Table 13–1). Explain why this gives rise to the increased heat resistance of Pyrex.

15. Will a grandfather clock, accurate at 20°C, run fast or slow on a hot day (30°C)? The clock uses a pendulum supported on a long thin brass rod.

16. Freezing a can of soda will cause its bottom and top to bulge so badly the can will not stand up. What's happened?

17. Why doesn't the size of different molecules enter into the gas laws?

18. When a gas is rapidly compressed (say, by pushing down a piston) its temperature increases. When a gas expands against a piston, it cools. Explain these changes in temperature using the kinetic theory, in particular noting what happens to the momentum of molecules when they strike the moving piston.

19. Explain in words how Charles's law follows from kinetic theory and the relation between average kinetic energy and the absolute temperature.

20. Explain in words how Gay-Lussac's law follows from kinetic theory.

21. As you go higher in the Earth's atmosphere, the ratio of N_2 molecules to O_2 molecules increases. Why?

22. Escape velocity for the Earth refers to the minimum speed an object must have to leave the Earth and never return. The escape velocity for the Moon is about one tenth what it is for the Earth since the Moon is smaller. Explain, then, why the Moon has practically no atmosphere.

23. Discuss why the Maxwell distribution of speeds (Fig. 13–15) is not a symmetrical curve.

24. A thermal gradient precipitator is a device for removing particulate matter from polluted air. It consists, basically, of two solid surfaces, one hot and one cold, close to each other. Particles in air that passes between the two surfaces tend to collect on the cold surface and thus are removed. Explain. (This effect can be noticed on the wall behind a heating radiator, particularly if the wall is an outside wall and therefore quite cool.)

* 25. Alcohol evaporates more quickly than water at room temperature. What can you infer about the molecular properties of one relative to the other?

* 26. Explain why a hot humid day is far more uncomfortable than a hot dry day at the same temperature.

* 27. Is it possible to boil water at room temperature (20°C) without heating it? Explain.

PROBLEMS

SECTION 13–1

1. (I) How does the number of atoms in a 30.0-gram gold ring compare to the number in a silver ring of the same mass?

2. (I) How many atoms are there in a 3.4-gram penny?

SECTION 13–2

3. (I) (a) "Room temperature" is often taken to be 68°F; what is this on the Celsius scale? (b) The temperature of the filament in a lightbulb is about 1800°C; what is this on the Fahrenheit scale?

4. (I) (a) 15° below zero on the Celsius scale is what Fahrenheit temperature? (b) 15° below zero on the Fahrenheit scale is what Celsius temperature?

5. (I) The highest and lowest recorded temperatures were 136°F (in the Libyan desert) and −129°F (in Antarctica). What are these temperatures on the Celsius scale?

6. (I) In an alcohol-in-glass thermometer, the alcohol column has length 10.70 cm at 0.0°C and length 22.85 cm at 100.0°C. What is the temperature if the column has length (a) 16.70 cm, and (b) 20.50 cm?

7. (I) The original Celsius temperature scale [due to Anders Celsius (1701–1744)] defined the freezing point of water as 100° and the boiling point as 0°. What temperature on this scale corresponds to 35°C on our present Celsius scale?

8. (II) At what temperature will the Fahrenheit and Centigrade scales yield the same numerical value?

SECTION 13–4

9. (I) A concrete highway is built of slabs 14 m long (20°C). How wide should the expansion cracks be (at 20°C) between the slabs to prevent buckling if the range of temperature is −30°C to +50°C?

10. (I) Super Invar, an alloy of iron and nickel, is a strong material with a very low coefficient of thermal expansion [$0.2 \times 10^{-6}(\text{C}°)^{-1}$]. A 2.0-m-long tabletop of this alloy is used for sensitive laser measurements where extremely high tolerances are required. How much will this alloy table expand along its length if the temperature increases 5.0°C?

11. (I) By what factor would the coefficients of linear expansion in Table 13–1 change if British units (ft, lb, °F) were used?

12. (II) To make a secure fit, rivets that are larger than the rivet hole are often used and the rivet is cooled (usually in dry ice) before it is placed in the hole. A steel rivet 1.871 cm in diameter is to be placed in a hole 1.869 cm in diameter. To what temperature must the rivet be cooled if it is to fit in the hole (at 20°C)?

13. (II) A uniform rectangular plate of length l and width w has coefficient of linear expansion α. Show that, if we neglect very small quantities, the change in area of the plate due to a temperature change ΔT is $\Delta A = 2\alpha l w\, \Delta T$. Fig. 13–25.

FIGURE 13–25 Rectangular plate is heated. Problem 13.

14. (II) Estimate the percent change in density of iron when it is still a solid, but deep in the Earth where the temperature is 2000°C and it is under 5000 atm of pressure. Take into account both thermal expansion and changes due to increased outside pressure. Assume both the bulk modulus and the volume coefficient of expansion do not vary with temperature and are the same as at normal room temperature. The bulk modulus for iron is about $90 \times 10^9\,\text{N/m}^2$.

15. (II) An ordinary glass is filled to the brim with 350.0 mL of water at 100°C. If the temperature decreased to 20°C, how much water could be added to the glass?

16. (II) It is observed that 65.50 mL of water at 20°C completely fills a container to the brim. When the container and the water are heated to 60°C, 0.35 g of water is lost. (a) What is the coefficient of volume expansion of the container? (b) What is the most likely material of the container? Density of water at 60°C is 0.98324 g/mL.

17. (II) A quartz sphere is 14.5 cm in diameter. What will be its change in volume if it is heated from 30°C to 200°C?

18. (II) A brass plug is to be placed in a ring made of iron. At room temperature, the diameter of the plug is 8.753 cm and that of the inside of the ring is 8.743 cm. They must both be brought to what common temperature in order to fit?

19. (II) If a fluid is contained in a long narrow vessel so it can expand in essentially one direction only, show that the effective coefficient of linear expansion α is approximately equal to the coefficient of volume expansion β.

20. (II) (a) Show that the change in the density ρ of a substance, when the temperature changes by ΔT, is given by $\Delta\rho = -\beta\rho\,\Delta T$. (b) What is the fractional change in density of a lead sphere whose temperature decreases from 25°C to −40°C?

21. (III) Show that for an isotropic solid, $\beta = 3\alpha$, if the amount of expansion is small. β and α are the coefficients of volume and linear expansion, respectively. [Hint: consider a cubical solid and neglect very small quantities. See also Problem 13.]

22. (III) The pendulum in a grandfather clock is made of brass and keeps perfect time at 17°C. How much time is gained or lost in a year if the clock is kept at 25°C? (Assume the frequency dependence on length for a simple pendulum applies.)

23. (III) A 23.4-kg solid aluminum cylindrical wheel of radius 0.52 m is rotating about its axle in frictionless bearings with angular velocity $\omega = 32.8$ rad/s. If its temperature is now raised from 20.0°C to 75.0°C, what is the fractional change in ω?

SECTION 13–6

24. (I) At what temperature will the ultimate compressive strength of concrete be exceeded for the blocks discussed in Example 13–8?

25. (I) An aluminum bar has the precisely desired length when at 15°C. How much stress is required to keep it at this length if the temperature increases to 35°C?

26. (II) (a) A horizontal steel I-beam of cross-sectional area 0.031 m^2 is rigidly connected to two vertical steel girders. If the beam was installed when the temperature was 30°C, what stress is developed in the beam when the temperature drops to -30°C? (b) Is the ultimate strength of the steel exceeded? (c) What stress is developed if the beam is concrete and has a cross-sectional area of 0.13 m^2? Will it fracture?

27. (III) A wine barrel of diameter 134.122 cm at 20°C is to be enclosed by an iron band. The circular band has an inside diameter of 134.110 cm at 20°C. It is 8.9 cm wide and 0.65 cm thick. (a) To what temperature must the band be heated so that it will fit over the barrel? (b) What will be the tension in the band when it cools to 20°C?

SECTION 13–7

28. (I) What are the following temperatures on the Kelvin scale: (a) 86°C, (b) 78°F, (c) -100°C, (d) 5500°C?

29. (I) Absolute zero is what temperature on the Fahrenheit scale?

30. (II) Typical temperatures in the interior of the Earth and Sun are about 4000°C and 15×10^6°C, respectively. (a) What are these temperatures in kelvins? (b) What percent error is made in each case if a person forgets to change °C to K?

SECTIONS 13–8 AND 13–9

31. (I) If 3.00 m^3 of a gas initially at STP is placed under a pressure of 4.00 atm, the temperature of the gas rises to 38.0°C. What is the volume?

32. (II) Calculate the density of oxygen at STP using the ideal gas law.

33. (II) A storage tank contains 21.6 kg of nitrogen (N$_2$) at an absolute pressure of 3.65 atm. What will the pressure be if the nitrogen is replaced by an equal mass of CO$_2$?

34. (II) A storage tank at STP contains 18.5 kg of nitrogen (N$_2$). (a) What is the volume of the tank? (b) What is the pressure if an additional 15.0 kg of nitrogen is added?

35. (II) If 25.50 mol of helium gas is at 10°C and a gauge pressure of 0.350 atm, calculate (a) the volume of the helium gas under these conditions, and (b) the temperature if the gas is compressed to precisely half the volume at a gauge pressure of 1.00 atm.

36. (II) What is the pressure inside a 50.0-L container holding 105.0 kg of argon gas at 20°C?

37. (II) A tire is filled with air at 15°C to a gauge pressure of 220 kPa. If the tire reaches a temperature of 38°C, what fraction of the original air must be removed if the original pressure of 220 kPa is to be maintained?

38. (II) If 55.0 L of oxygen at 18.0°C and an absolute pressure of 2.45 atm are compressed to 48.8 L and at the same time the temperature is raised to 50.0°C, what will the new pressure be?

39. (III) Compare the value for the density of water vapor at 100°C and 1 atm (Table 10–1) with the value predicted from the ideal gas law. Why would you expect a difference?

40. (III) A child's helium-filled balloon escapes at sea level and 20.0°C. When it reaches an altitude of 3000 m, where the temperature is 5.0°C and the pressure only 0.70 atm, how will its volume compare to that at sea level?

41. (III) An air bubble at the bottom of a lake 43.5 m deep has a volume of 1.00 cm^3. If the temperature at the bottom is 5.5°C and at the top 21.0°C, what is the volume of the bubble just before it reaches the surface?

SECTION 13–10

42. (I) Calculate the number of molecules/m^3 in an ideal gas at STP.

43. (I) How many moles of water are there in 1.000 L? How many molecules?

44. (II) Estimate the number of (a) moles, and (b) molecules of water in all the Earth's oceans. Assume water covers 75 percent of the Earth to an average depth of 3 km.

45. (II) A cubic box of volume 3.9×10^{-2} m^3 is filled with air at atmospheric pressure at 20°C. The box is closed and heated to 180°C. What is the net force on each side of the box?

46. (III) Estimate how many molecules of air are in each 2.0-L breath you take that were also in the last breath Einstein took. [Hint: assume the atmosphere is about 10 km high and of constant density.]

SECTION 13–11

47. (I) Calculate the rms speed of helium atoms near the surface of the Sun at a temperature of about 6000 K.

48. (I) (a) What is the average translational kinetic energy of a nitrogen molecule at STP? (b) What is the total translational kinetic energy of 1.00 mol of N$_2$ molecules at 20°C?

49. (I) Twelve molecules have the following speeds, given in arbitrary units: 6, 2, 4, 6, 0, 4, 1, 8, 5, 3, 7, and 8. Calculate their rms speed.

50. (I) A gas is at 0°C. To what temperature must it be raised to double the rms speed of its molecules?

51. (II) The rms speed of molecules in a gas at 20.0°C is to be increased by 2.0 percent. To what temperature must it be raised?

52. (II) If the pressure in a gas is doubled while its volume is held constant, by what factor does v_{rms} change?

53. (II) Show that for a mixture of two gases at the same temperature, the ratio of the rms speeds of their molecules is equal to the inverse ratio of the square roots of their molecular masses.

54. (II) (a) Calculate the approximate rms speed of an amino acid whose molecular mass is 89 u in a living cell at 37°C. (b) What would be the rms speed of a protein of molecular mass 50,000 u at 37°C?

55. (II) Show that the pressure P in a gas can be written $P = \frac{1}{3}\rho v^2$, where ρ is the density of the gas and v is the rms speed of the molecules.

56. (III) The two isotopes of uranium, ^{235}U and ^{238}U (the superscripts refer to their atomic mass), can be separated by a gas-diffusion process by combining them with fluorine to make the gaseous compound UF_6. Calculate the ratio of the rms speeds of these molecules for the two isotopes, at constant T.

*SECTION 13–13

*57. (I) (a) At atmospheric pressure, in what phases can CO_2 exist? (b) For what range of pressures and temperatures can CO_2 be a liquid? Refer to Fig. 13–19.

*58. (I) CO_2 exists in what phase when the pressure is 30 atm and the temperature is 30°C (Fig. 13–19)?

*59. (I) Water is in which phase when the pressure is 0.01 atm and the temperature is (a) 90°C, (b) −20°C?

*SECTION 13–14

*60. (I) What is the dew point (approximately) if the humidity is 50 percent on a day when the temperature is 25°C?

*61. (I) What is the air pressure at a place where water boils at 90°C?

*62. (I) If the air pressure at a particular place in the mountains is 0.85 atm, estimate the temperature at which water boils.

*63. (I) What is the temperature on a day when the partial pressure of water is 530 Pa and the relative humidity is 40 percent?

*64. (II) What is the approximate pressure inside a pressure cooker if the water is boiling at a temperature of 120°C? Assume no air escaped during the heating process, which started at 20°C.

*65. (II) If the humidity in a room of volume 680 m³ at 25°C is 80 percent, what mass of water can still evaporate from an open pan?

*66. (II) In humid climates, people constantly *dehumidify* their cellars in order to prevent rot and mildew. If the cellar in a house (kept at 20°C) has 95 m² of floor space and a ceiling height of 2.8 m, what is the mass of water that must be removed from it in order to drop the humidity from 95 percent to a more reasonable 30 percent?

*67. (II) An *autoclave* is a device used to sterilize laboratory instruments. It is essentially a high-pressure steam boiler and operates on the same principle as a pressure cooker. However, because hot steam under pressure is more effective in killing microorganisms than moist air at the same temperature and pressure, the air is removed and replaced by steam. Typically, the gauge pressure inside the autoclave is 1.0 atm. What is the temperature of the steam? Assume the steam is in equilibrium with boiling water.

*68. (III) Air that is at its dew point of 5°C is drawn into a building where it is heated to 25°C. What will be the relative humidity at this temperature? Assume constant pressure of 1.0 atm. Take into account the expansion of the air.

*69. (III) On a day when the temperature is 30°C, the wet-bulb thermometer of a humidity tester falls to 10°C. What is the relative humidity?

*SECTION 13–15

*70. (I) Approximately how long would it take for the ammonia of Example 13–19 to be detected 2.0 m from the bottle after it is opened? What does this suggest about the relative importance of diffusion and convection for carrying odors?

*71. (II) What is the time needed for a glycine molecule (see Table 13–5) to diffuse a distance of 15 μm in water at 20°C if its concentration varies over that distance from 1.0 mol/m³ to 0.40 mol/m³? Compare this "speed" to its average thermal speed. The molecular mass of glycine is about 75 u.

*72. (II) Oxygen diffuses from the surface of insects to the interior through tiny tubes called tracheae. An average trachea is about 2 mm long and has cross-sectional area of 2×10^{-9} m². Assuming the concentration of oxygen inside is half what it is outside in the atmosphere, (a) show that the concentration of oxygen in the air (assume 21 percent is oxygen) at 20°C is about 8.7 mol/m³, then (b) calculate the diffusion rate J, and (c) estimate the average time for a molecule to diffuse in. Assume the diffusion constant is 1×10^{-5} m²/s.

*73. (III) (a) Derive *Graham's law*, which states that "the rate of diffusion of gas molecules is inversely proportional to the square root of the molecular mass." (b) Which would diffuse faster, N_2 gas or O_2 gas, and by how much (percentage)?

74. A precise steel tape measure has been calibrated at 20°C. At 34°C, (a) will it read high or low, and (b) what will be the percentage error?

75. A Pyrex measuring cup was calibrated at normal room temperature. How much error will be made in a recipe calling for 300 mL of cool water, if the water and the cup are hot, at 80°C, instead of at room temperature? Neglect the glass expansion.

76. The pressure in a helium gas cylinder is initially 28 atmospheres. After many balloons have been blown up, the pressure has decreased to 5 atm. What fraction of the original gas remains in the cylinder?

77. Write the ideal gas law in terms of density of the gas.

78. Estimate the number of air molecules in a room of length 5 m, width 3 m, and height 2.5 m. Assume the temperature is 20°C. How many moles does that correspond to?

79. The lowest pressure attainable using the best available vacuum techniques is about $10^{-12} \, \text{N/m}^2$. At such a pressure, how many molecules are there per cm^3 at 0°C?

80. In outer space the density of matter is about one atom per cm^3, mainly hydrogen atoms, and the temperature is about 2.7 K. Calculate the rms speed of these hydrogen atoms, and the pressure (in atmospheres).

81. If a scuba diver fills his lungs to full capacity of 5.5 L when 10 m below the surface, to what volume would his lungs expand if he quickly rose to the surface? Is this advisable?

82. A scuba tank has a volume of 3500 cm^3. For very deep dives, the tank is filled with 50 percent (by volume) pure oxygen and 50 percent pure helium. (a) How many molecules are there of each type in the tank if it is filled at 20°C to a gauge pressure of 10 atm? (b) What is the ratio of the average kinetic energies of the two types of molecule? (c) What is the ratio of the rms speeds of the two types of molecule?

83. A space vehicle returning from the Moon enters the atmosphere at a speed of about 40,000 km/h. Molecules (assume nitrogen) striking the nose of the vehicle with this speed correspond to what temperature? (Because of this high temperature, the nose of a space vehicle must be made of special materials; indeed, part of it does vaporize, and this is seen as a bright blaze upon reentry.)

84. The temperature of an ideal gas is increased from 120°C to 360°C while the volume and the number of moles stay constant. By what factor does the pressure change? By what factor does v_{rms} change?

85. A house has a volume of 770 m^3. (a) What is the total mass of air inside the house at 20°C? (b) If the temperature drops to -10°C, what mass of air enters or leaves the house?

86. (a) Use the ideal gas law to show that, for an ideal gas at constant pressure, the coefficient of volume expansion is equal to $\beta = 1/T$ where T is the kelvin temperature. Compare to Table 13–1 for gases at $T = 293$ K. (b) Show that the bulk modulus (Section 9–6) for an ideal gas held at constant temperature is $B = P$ where P is the pressure.

87. From the known value of atmospheric pressure at the surface of the Earth, estimate the total number of air molecules in the Earth's atmosphere.

88. What is the rms speed of nitrogen molecules contained in a 7.6 m^3 volume at 4.2 atm if the total amount of nitrogen is 1800 mol?

89. (a) The tube of a mercury thermometer has an inside diameter of 0.140 mm. The bulb has a volume of 0.255 cm^3. How far will the thread of mercury move when the temperature changes from 11.5°C to 33.0°C? Take into account expansion of the Pyrex glass. (b) Determine a formula for the length of the mercury column in terms of relevant variables.

90. What is the average distance between oxygen molecules at STP?

91. Calculate approximately the total translational kinetic energy of all the molecules in an *E. coli* bacterium of mass 2.0×10^{-15} kg at 37°C. Assume 70 percent of the cell, by weight, is water, and the other molecules have an average molecular weight on the order of 10^5.

92. An iron cube floats in a bowl of liquid mercury at 0°C. (a) If the temperature is raised to 25°C, will the cube float higher or lower in the mercury? (b) By what percent will the fraction of volume submerged change?

93. If a steel band were to fit snugly around the Earth's equator at 20°C, but then was heated to 30°C, how high above the Earth would the band be (assume equal everywhere)?

On a cold night, heat radiation from a campfire can warm you and also dry your socks. Warm clothes act as insulators to reduce heat loss from the body to the outside by conduction and convection.

HEAT 14

When a pot of cold water is placed on a hot burner of a stove, the temperature of the water increases. We say that heat flows from the hot burner to the cold water. When two objects at different temperatures are put in contact, heat spontaneously flows from the hotter one to the colder one. The spontaneous flow of heat is always in the direction tending to equalize the temperature. If the two objects are kept in contact long enough for their temperatures to become equal, the two bodies are said to be in thermal equilibrium, and there is no further heat flow between them. For example, when a fever thermometer is first put in a patient's mouth, heat is flowing from the patient's mouth to the thermometer; when the temperature reading stops increasing, the thermometer is then in equilibrium with the person's mouth, and they are at the same temperature.

14–1 Heat as Energy Transfer

We use the term "heat" in everyday life as if we knew what we meant. But the term is often used inconsistently, so it is important for us to define heat clearly, and to clarify the phenomena and concepts related to heat.

We commonly speak of the "flow" of heat—heat flows from a stove burner to a pot of coffee, from the Sun to the Earth, from a person's mouth into a fever thermometer. Heat flows spontaneously from an object at higher temperature to one at lower temperature. Indeed, an eighteenth-century model of heat pictured heat flow as movement of a fluid substance called *caloric*. However, the caloric fluid was never able to be detected. In the nineteenth century, it was found that the various phenomena associated with heat could be described consistently without the need to use the fluid model. The new model viewed heat as being akin to work and energy, as we will discuss in a moment. First we note that a common unit for heat, still in use today, is named after caloric. It is called the **calorie** (cal) and is defined as *the amount of heat necessary to raise the temperature of 1 gram of water by 1 Celsius degree.* [To be precise, the particular temperature range from 14.5°C to 15.5°C is specified because the heat required is very slightly different at different temperatures. The difference is less than 1 percent over the range 0 to 100°C and we will ignore it for most purposes.] More often used than the calorie is the **kilocalorie** (kcal), which is 1000 calories. Thus 1 *kcal is the heat needed to raise 1 kg of water by 1 C°.* Sometimes a kilocalorie is called a **Calorie** (with a capital C), and it is by this unit that the energy value of food is specified (we might also call it the "dietary calorie"). In the British system of units, heat is measured in British thermal units (Btu). One Btu is defined as the heat needed to raise the temperature of 1 lb of water by 1 F°. It can be shown (Problem 5) that 1 Btu = 0.252 kcal = 1055 J.

Calorie

Kilocalorie (= dietary Calorie)

BTU

The idea that heat is related to energy was pursued by a number of scientists in the 1800s, particularly by an English brewer, James Prescott Joule (1818–1889). Joule performed a number of experiments that were crucial for establishing our present-day view that heat, like work, represents a transfer of energy. One of Joule's experiments is shown (simplified) in Fig. 14–1. The falling weight causes the paddle wheel to turn. The friction between the water and the paddle wheel causes the temperature of the water to rise slightly (barely measurable, in fact, by Joule). Of course, the same temperature rise could also be obtained by heating the water on a hot stove. In this and a great many other experiments (some involving electrical energy), Joule determined that a given amount of work done was always equivalent to a particular amount of heat input. Quantitatively, 4.186 joules (J) of work was found to be equivalent to 1 calorie (cal) of heat. This is known as the **mechanical equivalent of heat**:

Weight

FIGURE 14–1 Joule's experiment on the mechanical equivalent of heat.

$$4.186 \text{ J} = 1 \text{ cal}$$

$$4.186 \times 10^3 \text{ J} = 1 \text{ kcal}.$$

Mechanical equivalent of heat

As a result of these and other experiments, scientists came to interpret heat not as a substance, and not even as a form of energy. Rather, heat refers to a *transfer of energy*: when heat flows from a hot object to a cooler one, it is energy that is being transferred from the hot to the cold object. Thus, **heat** is *energy that is transferred from one body to another because of a difference in temperature.* In SI units, the unit for heat, as for any form of energy, is the joule. Nonetheless, calories and kcal are still sometimes used. Today the calorie is *defined* in terms of the joule (via the mechanical equivalent of heat, above), rather than in terms of the properties of water, as given previously. The latter is still handy to remember: 1 cal raises 1 g of water by 1 C°, or 1 kcal raises 1 kg of water by 1 C°.

Heat is energy transferred because of a ΔT

EXAMPLE 14–1 **ESTIMATE** **Working off the extra calories.** A young couple throw caution to the wind one afternoon, eating too much ice cream and cake. They realize that they each overate by about 500 Calories, and to compensate they want to do an equivalent amount of work climbing stairs. How much total height must each person walk up stairs? Each has a mass of 60 kg.

▶ **PHYSICS APPLIED**
Working off Calories

SOLUTION 500 Calories is 500 kcal, which in joules is

$$(500 \text{ kcal})(4.186 \times 10^3 \text{ J/kcal}) = 2.1 \times 10^6 \text{ J.}$$

The work done to climb a vertical height h is $W = mgh$. We want to solve for h given that $W = 2.1 \times 10^6 \text{ J}$:

$$h = \frac{W}{mg} = \frac{2.1 \times 10^6 \text{ J}}{(60 \text{ kg})(9.80 \text{ m/s}^2)} = 3600 \text{ m.}$$

They need to climb a very high mountain (over 11,000 ft) or many flights of stairs. (The human body does not transform energy with 100 percent efficiency—more like 20 percent—just as no engine does. As we'll see in the next chapter, some energy is always "wasted," so the couple would actually lose weight by doing this much work.)

EXAMPLE 14–2 **Kinetic energy transformed to heat.** When a 3.0-g bullet, traveling with a speed of 400 m/s, passes through a tree its speed is reduced to 200 m/s. How much heat Q is produced and shared by the bullet and the tree?

SOLUTION Energy conservation tells us that

$$\text{KE}_i = \text{KE}_f + Q$$

where Q is the heat produced, and the subscripts i and f stand for initial and final, respectively. Hence

$$Q = \tfrac{1}{2}m(v_i^2 - v_f^2) = \tfrac{1}{2}(3.0 \times 10^{-3} \text{ kg})[(400 \text{ m/s})^2 - (200 \text{ m/s})^2] = 180 \text{ J}$$

$$= \frac{180 \text{ J}}{4.186 \text{ J/cal}} = 43 \text{ cal.}$$

This is distributed between the bullet and the tree according to the relative abilities of the materials to transmit and store heat energy.

14–2 Distinction Between Temperature, Heat, and Internal Energy

We introduce the concept of internal energy now since it will help clarify ideas about heat. The sum total of all the energy of all the molecules in an object is called its **thermal energy** or **internal energy**. (We will use the two terms interchangeably.) Occasionally, the term "heat content" of a body is used for this purpose, but it is not a good term because it can be confused with heat itself. Heat, as we have seen, is not the energy a body contains, but rather refers to the amount of energy transferred from one body to another at a different temperature.

Using the kinetic theory, we can make a clear distinction between temperature, heat, and internal energy. Temperature (in kelvins) is a measure of the *average* kinetic energy of individual molecules. Thermal energy and internal energy refer to the *total* energy of all the molecules in the object.

Heat vs. internal energy vs. temperature

(Thus two equal-mass hot ingots of iron may have the same temperature, but two of them have twice as much thermal energy as one does.) Heat, finally, refers to a *transfer* of energy (such as thermal energy) from one object to another because of a difference in temperature.

Direction of heat flow depends on temperature

Notice that the direction of heat flow between two objects depends on their temperatures, not on how much internal energy each has. Thus, if 50 g of water at 30°C is placed in contact (or mixed) with 200 g of water at 25°C, heat flows *from* the water at 30°C *to* the water at 25°C even though the internal energy of the 25°C water is much greater because there is so much more of it.

14–3 Internal Energy of an Ideal Gas

Let us calculate the internal energy of n moles of an ideal monatomic (one atom per molecule) gas. The internal energy, U, is the sum of the translational kinetic energies of all the atoms. This sum is just equal to the average kinetic energy per molecule times the total number of molecules, N:

$$U = N(\tfrac{1}{2} m\overline{v^2}).$$

From Eq. 13–8, we then have

$$U = \tfrac{3}{2} NkT$$

or

$$U = \tfrac{3}{2} nRT, \qquad \text{[monatomic ideal gas]} \quad \textbf{(14–1)}$$

where n is the number of moles. Thus, the internal energy of an ideal gas depends only on temperature and the number of moles of gas.

If the gas molecules contain more than one atom, then the rotational and vibrational energy of the molecules (Fig. 14–2) must also be taken into account. The internal energy will be greater at a given temperature than for a monatomic gas, but it will still be a function only of temperature for an ideal gas.

The internal energy of real gases also depends mainly on temperature, but where they deviate from ideal gas behavior, the internal energy also depends somewhat on pressure and volume.

The internal energy of liquids and solids is quite complicated, for it includes electrical potential energy associated with the forces (or "chemical" bonds) between atoms and molecules.

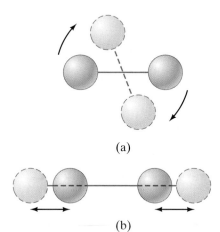

(a)

(b)

FIGURE 14–2 Molecules can have (a) rotational and (b) vibrational energy, as well as translational energy.

14–4 Specific Heat

If heat is put into an object, its temperature rises. But how much does the temperature rise? That depends. As early as the eighteenth century, experimenters had recognized that the amount of heat Q required to change the temperature of a given material is proportional to the mass m of the material and to the temperature change ΔT. This remarkable simplicity in nature can be expressed in the equation

Relation between heat flow and temperature change

$$Q = mc\,\Delta T, \qquad \textbf{(14–2)}$$

Specific heat

where c is a quantity characteristic of the material called its **specific heat**. Because $c = Q/m\,\Delta T$, specific heat is specified in units of J/kg·C° (the proper SI unit) or kcal/kg·C°. For water at 15°C and a constant pressure of 1 atm, $c = 1.00$ kcal/kg·C° or 4.19×10^3 J/kg·C°, since, by definition of the cal and the joule, it takes 1 kcal of heat to raise the temperature of

TABLE 14–1   Specific Heats
(at 1 atm constant pressure and 20°C unless otherwise stated)

Substance	Specific Heat, c kcal/kg·C°	J/kg·C°	Substance	Specific Heat, c kcal/kg·C°	J/kg·C°
Aluminum	0.22	900	Alcohol (ethyl)	0.58	2400
Copper	0.093	390	Mercury	0.033	140
Glass	0.20	840	Water		
Iron or steel	0.11	450	Ice (−5°C)	0.50	2100
Lead	0.031	130	Liquid (15°C)	1.00	4186
Marble	0.21	860	Steam (110°C)	0.48	2010
Silver	0.056	230	Human body (average)	0.83	3470
Wood	0.4	1700	Protein	0.4	1700

1 kg of water by 1 C°. Table 14–1 gives the values of specific heat for other substances at 20°C. The values of c depend to some extent on temperature (as well as slightly on pressure), but for temperature changes that are not too great, c can often be considered constant.

EXAMPLE 14–3   How heat depends on specific heat. (*a*) How much heat is required to raise the temperature of an empty 20-kg vat made of iron from 10°C to 90°C? (*b*) What if the vat is filled with 20 kg of water?

SOLUTION (*a*) From Table 14–1, the specific heat of iron is 450 J/kg·C°. The change in temperature is (90°C − 10°C) = 80 C°. Thus,

$$Q = mc\,\Delta T = (20\text{ kg})(450\text{ J/kg·C}°)(80\text{ C}°) = 7.2 \times 10^5\text{ J} = 720\text{ kJ}.$$

(Note: 1 kJ = 10^3 J = 0.239 kcal since 1 kcal = 4.19×10^3 J.)
(*b*) The water alone would require

$$Q = mc\,\Delta T = (20\text{ kg})(4186\text{ J/kg·°C})(80\text{ C}°) = 6.7 \times 10^6\text{ J} = 6700\text{ kJ}$$

or almost 10 times what an equal mass of iron requires. The total, for the vat plus the water, is 720 kJ + 6700 kJ = 7400 kJ.

If the iron vat in part (*a*) had been *cooled* from 90°C to 10°C, 720 kJ of heat would have flowed *out* of the iron. In other words, Eq. 14–2 is valid for heat flow either in or out, with a corresponding increase or decrease in temperature. We saw in part (*b*) of Example 14–3 that water requires almost 10 times as much heat as an equal mass of iron to make the same temperature change. Water has one of the highest specific heats of all substances, which makes it an ideal substance for hot-water space-heating systems and other uses that require a minimal drop in temperature for a given amount of heat transfer. It is the water content, too, that causes the apples rather than the crust in hot apple pie to burn our tongues, through heat transfer.

CONCEPTUAL EXAMPLE 14–4   A very hot skillet. You accidentally let an empty skillet get very hot on the stove (200°C or even more). What happens when you dunk it into a couple of inches of cool water in the bottom of the sink? Will the final temperature be midway between the initial temperatures of water and skillet? Will the water start boiling?

RESPONSE Experience may tell you that the water warms up—perhaps by a few degrees or perhaps by 10 or 20 degrees. The water doesn't come close to boiling. The water's temperature increase is a lot less than the skillet's temperature decrease. Why? Because the mass of water is roughly equal to that of the skillet, and metals like iron and aluminum have specific heats 5 to 10 times smaller than that for water (Table 14–1). As heat leaves the skillet and enters the water, the water's temperature change will be 5 to 10 times less than that of the skillet. If, instead, you let a few drops of water fall onto the hot skillet, that very small mass of water will sizzle and boil away (the skillet's mass may be 100× that of the water).

[Specific heats for gases are more complicated than for solids and liquids, which change in volume only slightly with a change in temperature (Section 13–4). Gases change strongly in volume with a change in temperature at constant pressure, as we saw in Chapter 13 with the gas laws; or, if kept at constant volume, the pressure in a gas changes strongly with temperature. The specific heat of a gas depends very much on how the process of changing its temperature is carried out. Most commonly, we deal with the specific heats of gases kept (a) at constant pressure (c_p) or (b) at constant volume (c_v). Some values are given in Table 14–2, where we see that c_p is always greater than c_v. For liquids and solids, this distinction is usually negligible.]

[Optional paragraph]

TABLE 14–2
Specific Heats of Gases (kcal/kg·C°)

Gas	c_p (constant pressure)	c_v (constant volume)
Steam (100°C)	0.482	0.350
Oxygen	0.218	0.155
Helium	1.15	0.75
Carbon dioxide	0.199	0.153
Nitrogen	0.248	0.177

14–5 Calorimetry—Solving Problems

When different parts of an isolated system are at different temperatures, heat will flow from the part at higher temperature to the part at lower temperature. If the system is completely isolated, no energy can flow into or out of it. So, again, the *conservation of energy* plays an important role for us: the heat lost by one part of the system is equal to the heat gained by the other part:

Energy conservation

$$\text{heat lost} = \text{heat gained.}$$

Let us take an Example.

EXAMPLE 14–5 **The cup cools the tea.** If 200 cm^3 of tea at 95°C is poured into a 150-g glass cup initially at 25°C, what will be the final temperature T of the mixture when equilibrium is reached, assuming no heat flows to the surroundings?

SOLUTION Since tea is mainly water, its specific heat is 4186 J/kg·C° (Table 14–1) and its mass m is its density times its volume ($V = 200 \text{ cm}^3 = 200 \times 10^{-6} \text{ m}^3$): $m = \rho V = (1.0 \times 10^3 \text{ kg/m}^3)(200 \times 10^{-6} \text{ m}^3) = 0.20 \text{ kg}$. Applying conservation of energy, we can set

$$\text{heat lost by tea} = \text{heat gained by cup}$$

$$m_{tea}c_{tea}(95°C - T) = m_{cup}c_{cup}(T - 25°C)$$

where T is the as yet unknown final temperature. Putting in numbers and using Table 14–1, we solve for T, and find

$$(0.20 \text{ kg})(4186 \text{ J/kg·C°})(95°C - T) = (0.15 \text{ kg})(840 \text{ J/kg·C°})(T - 25°C)$$

$$79{,}400 \text{ J} - (836 \text{ J/C°})T = (126 \text{ J/C°})T - 6300 \text{ J}$$

$$T = 89°C.$$

The tea drops in temperature by 6C° by coming into equilibrium with the cup.

Notice in this Example that the ΔT (of Eq. 14–2, $Q = mc\,\Delta T$) is a positive quantity on both sides of our conservation of energy equation. On the left we have "heat lost," and ΔT is the initial minus final temperature ($95°C - T$), whereas on the right, the "heat gained," ΔT, is the final minus the initial temperature.

The exchange of energy, as exemplified in Example 14–5, is the basis for a technique known as **calorimetry**, which is the quantitative measurement of heat exchange. To make such measurements, a **calorimeter** is used; a simple water calorimeter is shown in Fig. 14–3. It is very important that the calorimeter be well insulated so that only a minimal amount of heat is exchanged with the outside. One important use of the calorimeter is in the determination of specific heats of substances. In the technique known as the "method of mixtures," a sample of the substance is heated to a high temperature, which is accurately measured, and then quickly placed in the cool water of the calorimeter. The heat lost by the sample will be gained by the water and the calorimeter. By measuring the final temperature of the mixture, the specific heat can be calculated, as illustrated in the following Example.

⇒ PHYSICS APPLIED
Calorimetry

FIGURE 14–3 Simple water calorimeter.

EXAMPLE 14–6 **Unknown specific heat determined by calorimetry.** We wish to determine the specific heat of a new alloy. A 0.150-kg sample of the alloy is heated to 540°C. It is then quickly placed in 400 g of water at 10.0°C, which is contained in a 200-g aluminum calorimeter cup. (We do not need to know the mass of the insulating jacket since we assume the air space between it and the cup insulates it well, so that its temperature does not change significantly.) The final temperature of the mixture is 30.5°C. Calculate the specific heat of the alloy.

SOLUTION Again we apply the conservation of energy and write that the heat lost equals the heat gained:

$$\left(\begin{array}{c}\text{heat lost}\\\text{by sample}\end{array}\right) = \left(\begin{array}{c}\text{heat gained}\\\text{by water}\end{array}\right) + \left(\begin{array}{c}\text{heat gained by}\\\text{calorimeter cup}\end{array}\right)$$

$$m_s c_s \Delta T_s = m_w c_w\,\Delta T_w + m_{cal} c_{cal}\,\Delta T_{cal}$$

where the subscripts s, w, and cal refer to the sample, water, and calorimeter, respectively. When we put in values and use Table 14–1, this equation becomes

$$(0.150\,\text{kg})(c_s)(540°C - 30.5°C) = (0.40\,\text{kg})(4186\,\text{J/kg·C}°)(30.5°C - 10.0°C)$$
$$+ (0.20\,\text{kg})(900\,\text{J/kg·C}°)(30.5°C - 10.0°C)$$
$$76.4 c_s = (34{,}300 + 3700)\,\text{J/kg·C}°$$
$$c_s = 500\,\text{J/kg·C}°.$$

In making this calculation, we have ignored any heat transferred to the thermometer and the stirrer (which is needed to quicken the heat transfer process and thus reduce heat loss to the outside). It can be taken into account by adding additional terms to the right side of the above equation and will result in a slight correction to the value of c_s (see Problem 20). It should be noted that the quantity $m_{cal} c_{cal}$ is often called the **water equivalent** of the calorimeter—that is, $m_{cal} c_{cal}$ is numerically equal to the mass of water (in kilograms) that would absorb the same amount of heat.

A **bomb calorimeter** is used to measure the heat released when a substance burns. Important applications are the burning of foods to determine their Calorie content, and burning of seeds and other substances to determine their "energy content," or heat of combustion. A carefully weighed sample of the substance, together with an excess amount of oxygen at high pressure, is placed in a sealed container (the "bomb"). The bomb is placed in the water of the calorimeter and a fine wire passing into the bomb is then heated briefly, which causes the mixture to ignite.

EXAMPLE 14–7 Measuring Calorie content of a brownie. Determine the Calorie content of 100 g of Mullaney's fudge brownie from the following measurements. A 10-g sample of the brownie is allowed to dry before putting it in the bomb. The aluminum bomb has a mass of 0.615 kg and is placed in 2.00 kg of water contained in an aluminum calorimeter cup of mass 0.524 kg. The initial temperature of the mixture is 15.0°C and its temperature after ignition is 36.0°C.

SOLUTION In this case, the heat Q released in the burning is absorbed by the entire mixture of bomb, calorimeter, and water:

$$Q = (m_w c_w + m_{cal} c_{cal} + m_{bomb} c_{bomb})\,\Delta T$$
$$= [(2.00 \text{ kg})(1.0 \text{ kcal/kg·C°}) + (0.524 \text{ kg})(0.22 \text{ kcal/kg·C°})$$
$$+ (0.615 \text{ kg})(0.22 \text{ kcal/kg·C°})]\,[36.0°C - 15.0°C]$$
$$= 47 \text{ kcal.}$$

Since 47 kcal is released in the burning of 10 g of brownie, a 100-g portion would contain 470 "Calories."

14–6 Latent Heat

When a material changes phase from solid to liquid, or from liquid to gas (see also Section 13–13), a certain amount of energy is involved in this **change of phase**. For example, let us trace what happens when a 1.0-kg block of ice at $-40°C$ has heat added to it at a steady rate until all the ice has changed to water, then the (liquid) water is heated to 100°C and changed to steam above 100°C, all at 1 atm pressure. As shown in the graph of Fig. 14–4, as heat is added to the ice, its temperature rises at a rate of about 2 C°/kcal of heat added (since for ice, $c \approx 0.50$ kcal/kg·C°). However, when 0°C is reached, the temperature stops increasing even though heat is still being added. Now as heat is added, the ice gradually changes to water in the liquid state without any change in temperature. After about 40 kcal have been added at 0°C, half the ice remains and half has changed to water. After about 80 kcal, or 330 kJ, has been added, all the ice has changed to water, still at 0°C. Further addition of heat causes the water's temperature to again increase, now at a rate of 1 C°/kcal. When 100°C is reached, the temperature again remains constant as the heat added changes the liquid water to vapor (steam). About 540 kcal (2260 kJ) is required to change the 1.0 kg of water completely to steam, after which the curve rises again, indicating that the temperature of the steam now rises as heat is added.

The heat required to change 1.0 kg of a substance from the solid to the liquid state is called the **heat of fusion**; it is denoted by L_F. The heat of fusion of water is 79.7 kcal/kg or, in proper SI units, 333 kJ/kg ($=3.33 \times 10^5$ J/kg).

Heat of fusion

FIGURE 14–4 Temperature as a function of the heat added to bring 1.0 kg of ice at −40°C to steam above 100°C. Note the scale break between 200 and 740 kcal (the page isn't wide enough to fit it).

The heat required to change a substance from the liquid to the vapor phase is called the **heat of vaporization**, L_V, and for water it is 539 kcal/kg or 2260 kJ/kg. Other substances follow graphs similar to Fig. 14–4, although the melting-point and boiling-point temperatures are different, as are the specific heats and heats of fusion and vaporization. Values for the heats of fusion and vaporization, which are also called the **latent heats**, are given in Table 14–3 for a number of substances.

Heat of vaporization

The heats of vaporization and fusion also refer to the amount of heat *released* by a substance when it changes from a gas to a liquid, or from a liquid to a solid. Thus, steam releases 2260 kJ/kg when it changes to water, and water releases 333 kJ/kg when it becomes ice.

Of course, the heat involved in a change of phase depends not only on the latent heat, but also on the total mass of the substance. That is,

$$Q = mL,$$

where L is the latent heat of the particular process and substance, m is the mass of the substance, and Q is the heat required or given off during

TABLE 14–3 Latent Heats (at 1 atm)

Substance	Melting Point	Heat of Fusion		Boiling Point	Heat of Vaporization	
	(°C)	kcal/kg[†]	J/kg	(°C)	kcal/kg[†]	J/kg
Oxygen	−218.8	3.3	0.14×10^5	−183	51	2.1×10^5
Nitrogen	−210.0	6.1	0.26×10^5	−195.8	48	2.00×10^5
Ethyl alcohol	−114	25	1.04×10^5	78	204	8.5×10^5
Ammonia	−77.8	8.0	0.33×10^5	−33.4	33	1.37×10^5
Water	0	79.7	3.33×10^5	100	539	22.6×10^5
Lead	327	5.9	0.25×10^5	1750	208	8.7×10^5
Silver	961	21	0.88×10^5	2193	558	23×10^5
Iron	1808	69.1	2.89×10^5	3023	1520	63.4×10^5
Tungsten	3410	44	1.84×10^5	5900	1150	48×10^5

[†]Numerical values in kcal/kg are the same in cal/g.

the phase change. For example, when 5.00 kg of water freezes at 0°C, $(5.00 \text{ kg})(3.33 \times 10^5 \text{ J/kg}) = 1.67 \times 10^6 \text{ J}$ of energy is released.

Calorimetry sometimes involves a change of state, as the following Examples show. Indeed, latent heats are often measured using calorimetry.

EXAMPLE 14–8 **Making ice.** How much energy does a refrigerator have to remove from 1.5 kg of water at 20°C to make ice at $-12°C$?

SOLUTION Heat must flow out to reduce the water from 20°C to 0°C, to change it to ice, and then to lower the ice from 0°C to $-12°C$,

$$Q = mc_w(20°C - 0°C) + mL_F + mc_{ice}[0° - (-12°C)]$$

$$= (1.5 \text{ kg})(4186 \text{ J/kg·C°})(20 \text{ C°}) + (1.5 \text{ kg})(3.33 \times 10^5 \text{ J/kg})$$

$$+ (1.5 \text{ kg})(2100 \text{ J/kg·C°})(12 \text{ C°})$$

$$= 6.6 \times 10^5 \text{ J} = 660 \text{ kJ}.$$

EXAMPLE 14–9 **Will all the ice melt?** At a party, a 0.50-kg chunk of ice at $-10°C$ is placed in 3.0 kg of "iced" tea at 20°C. At what temperature and in what phase will the final mixture be? The tea can be considered as water.

➡ **PROBLEM SOLVING**

First determine (or estimate) the final state

SOLUTION In this situation, before we can write down an equation, we must first check to see if the final state will be all ice, a mixture of ice and water at 0°C, or all water. To bring the 3.0 kg of water at 20°C down to 0°C would require an energy release of

$$m_w c_w(20°C - 0°C) = (3.0 \text{ kg})(4186 \text{ J/kg·C°})(20 \text{ C°}) = 250 \text{ kJ}.$$

On the other hand, to raise the ice from $-10°C$ to 0°C would require

$$m_{ice} c_{ice}[0°C - (-10°C)] = (0.50 \text{ kg})(2100 \text{ J/kg·C°})(10 \text{ C°}) = 10.5 \text{ kJ},$$

and to change the ice to water at 0°C would require

$$m_{ice} L_F = (0.50 \text{ kg})(333 \text{ kJ/kg}) = 167 \text{ kJ},$$

for a total of $10.5 \text{ kJ} + 167 \text{ kJ} = 177 \text{ kJ}$. This is not enough energy to bring the 3.0 kg of water at 20°C down to 0°C, so we know that the mixture must end up all water, somewhere between 0°C and 20°C. Now we can determine the final temperature T by applying the conservation of energy and writing

Then determine the final temperature

$$\begin{pmatrix} \text{heat to raise} \\ \text{0.50 kg of ice} \\ \text{from } -10°C \\ \text{to } 0°C \end{pmatrix} + \begin{pmatrix} \text{heat to change} \\ \text{0.50 kg} \\ \text{of ice} \\ \text{to water} \end{pmatrix} + \begin{pmatrix} \text{heat to raise} \\ \text{0.50 kg of water} \\ \text{from } 0°C \\ \text{to } T \end{pmatrix} = \begin{pmatrix} \text{heat lost by} \\ \text{3.0 kg of} \\ \text{water cooling} \\ \text{from } 20°C \text{ to } T \end{pmatrix}$$

so

$$10.5 \text{ kJ} + 167 \text{ kJ} + (0.50 \text{ kg})(4186 \text{ J/kg·C°})(T)$$

$$= (3.0 \text{ kg})(4186 \text{ J/kg·C°})(20°C - T)$$

or

$$14,600 \, T = 73,800$$

$$T = 5.1°C.$$

EXAMPLE 14–10 **Determining a latent heat.** The specific heat of mercury is 0.033 kcal/kg·C°. When 1.0 kg of solid mercury at its melting point of −39°C is placed in a 0.50-kg aluminum calorimeter filled with 1.2 kg of water at 20.0°C, the final temperature of the mixture is found to be 16.5°C. What is the heat of fusion of mercury in kcal/kg?

SOLUTION The heat gained by the mercury (Hg) equals the heat lost by the water and calorimeter:

$$m_{Hg}L_{Hg} + m_{Hg}c_{Hg}[16.5°C - (-39°C)]$$
$$= m_w c_w(20°C - 16.5°C) + m_{Al}c_{Al}(20.0°C - 16.5°C)$$

or

$$(1.0 \text{ kg})(L_{Hg}) + (1.0 \text{ kg})(0.033 \text{ kcal/kg·C°})(55.5 \text{ C°})$$
$$= (1.2 \text{ kg})(1.0 \text{ kcal/kg·C°})(3.5 \text{ C°})$$
$$+ (0.50 \text{ kg})(0.22 \text{ kcal/kg·C°})(3.5 \text{ C°}).$$

Thus

$$L_{Hg} = (4.2 + 0.4 - 1.8) \text{ kcal/kg} = 2.8 \text{ kcal/kg}.$$

→ **P R O B L E M   S O L V I N G**   **Calorimetry**

1. Be sure you have enough information to apply energy conservation. Ask yourself: is the system isolated (or very nearly so, enough to get a good estimate)? Do we know or can we calculate all significant sources of heat energy flow?

2. Apply conservation of energy:

   heat gained = heat lost.

   For each substance in the system, a heat (energy) term will appear on either the left or right of this equation.

3. If no phase changes occur, each term in the energy conservation equation (above) will have the form

   $$Q(\text{gain}) = mc(T_f - T_i)$$
   or
   $$Q(\text{lost}) = mc(T_i - T_f)$$

where T_i and T_f are the initial and final temperatures of the substance, and m and c are its mass and specific heat.

4. If phase changes do or might occur, there may be terms in the energy conservation equation of the form $Q = mL$, where L is the latent heat. But *before* applying energy conservation, determine (or estimate) in which phase the final state will be, as we did in Example 14–9 by calculating the different contributing values for heat Q.

5. Be sure each term appears on the correct side of the energy equation (heat gained or heat lost) and that each ΔT is positive.

6. Note that when the system reaches thermal equilibrium, the final temperature of each substance will have the *same* value. There is only one T_f.

7. Solve your energy equation for the unknown.

The latent heat to change a liquid to a gas is needed not only at the boiling point. Water can change from the liquid to the gas phase even at room temperature. This process is called **evaporation** (see also Section 13–14). The value of the heat of vaporization increases slightly with a decrease in temperature: at 20°C, for example, it is 2450 kJ/kg (585 kcal/kg) compared to 2260 kJ/kg (539 kcal/kg) at 100°C. When water evaporates, it cools, since the energy required (the latent heat of vaporization) comes from the water itself; so its internal energy, and therefore its temperature, must drop.[†]

Evaporation of water from the skin is one of the most important methods the body uses to control its temperature. When the temperature of the blood rises slightly above normal, the hypothalamus gland detects this temperature increase and sends a signal to the sweat glands to increase their production. The energy required to vaporize this water comes from the body, and hence the body cools.

We can make use of kinetic theory to see why energy is needed to melt or vaporize a substance. At the melting point, the latent heat of fusion does not increase the kinetic energy (and the temperature) of the molecules in the solid, but instead is used to overcome the potential energy associated with the forces between the molecules. That is, work must be done against these attractive forces to break the molecules loose from their relatively fixed positions in the solid so they can freely roll over one another in the liquid phase. Similarly, energy is required for molecules held close together in the liquid phase to escape into the gaseous phase. This process is a more violent reorganization of the molecules than is melting (the average distance between the molecules is greatly increased), and hence the heat of vaporization is generally much greater than the heat of fusion for a given substance.

14–7 Heat Transfer: Conduction

Heat is transferred from one place or body to another in three different ways: by *conduction*, *convection*, and *radiation*. We now discuss each of these in turn; but in practical situations, any two or all three may be operating at the same time. We start with conduction.

When a metal poker is put in a hot fire, or a silver spoon is placed in a hot bowl of soup, the exposed end of the poker or spoon soon becomes hot as well, even though it is not directly in contact with the source of heat. We say that heat has been conducted from the hot end to the cold end.

Heat **conduction** in many materials can be visualized as the result of molecular collisions. As one end of the object is heated, the molecules there move faster and faster. As they collide with their slower-moving neighbors, they transfer some of their energy to these molecules whose speeds thus increase. These in turn transfer some of their energy by collision with molecules still farther along the object. Thus the energy of thermal motion is transferred by molecular collision along the object. In metals, according to modern theory, it is collisions of free electrons within the metal with each other and with metal atoms that are visualized as being mainly responsible for conduction.

Heat conduction takes place only if there is a difference in temperature. Indeed, it is found experimentally that the rate of heat flow through a substance is proportional to the difference in temperature between its ends. The rate of heat flow also depends on the size and shape of the ob-

---

[†]According to kinetic theory, evaporation is a cooling process because it is the fastest-moving molecules that escape from the surface (Section 13–14). Hence the average speed of the remaining molecules is less, so by Eq. 13–8 the temperature is less.

ject, and to investigate this quantitatively, let us consider the heat flow through a uniform object, as illustrated in Fig. 14–5. It is found experimentally that the heat flow ΔQ per time interval Δt is given by the relation

$$\frac{\Delta Q}{\Delta t} = kA \frac{T_1 - T_2}{l} \qquad \textbf{(14–3)}$$

Rate of heat flow by conduction

where A is the cross-sectional area of the object, l is the distance between the two ends, which are at temperatures T_1 and T_2, and k is a proportionality constant called the **thermal conductivity**, which is characteristic of the material. From Eq. 14–3, we see[†] that the rate of heat flow (units of J/s) is directly proportional to the cross-sectional area and to the temperature gradient $(T_1 - T_2)/l$.

The thermal conductivities, k, for a variety of substances are given in Table 14–4. Substances for which k is large conduct heat rapidly and are said to be good **conductors**. Most metals fall in this category, although there is a wide range even among them as you may observe by holding the ends of a silver spoon and a stainless-steel spoon immersed in the same hot cup of soup. Substances for which k is small, such as wool, fiberglass, polyurethane, and down, are poor conductors of heat and are therefore good **insulators**. The relative magnitudes of k can explain simple phenomena

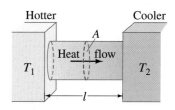

FIGURE 14–5 Heat conduction between areas at temperatures T_1 and T_2. If T_1 is greater than T_2, the heat flows to the right; the rate is given by Eq. 14–3.

[†]This is quite similar to the relations describing diffusion (Chapter 13) and the flow of fluids through a pipe (Chapter 10). In those cases, the flow of matter was found to be proportional to the concentration gradient $(C_1 - C_2)/l$, or to the pressure gradient $(P_1 - P_2)/l$. This close similarity is one reason we speak of the flow of heat. Yet we must keep in mind that no substance is flowing in this case—it is energy that is being transferred.

TABLE 14–4 Thermal Conductivities

Substance	Thermal Conductivity, k	
	kcal/s·m·C°	J/s·m·C°
Silver	10×10^{-2}	420
Copper	9.2×10^{-2}	380
Aluminum	5.0×10^{-2}	200
Steel	1.1×10^{-2}	40
Ice	5×10^{-4}	2
Glass (typical)	2.0×10^{-4}	0.84
Brick and concrete	2.0×10^{-4}	0.84
Water	1.4×10^{-4}	0.56
Human tissue (excluding blood)	0.5×10^{-4}	0.2
Wood	0.2–0.4×10^{-4}	0.08–0.16
Fiberglass insulation	0.12×10^{-4}	0.048
Cork and glass wool	0.1×10^{-4}	0.042
Wool	0.1×10^{-4}	0.040
Goose down	0.06×10^{-4}	0.025
Polyurethane foam	0.06×10^{-4}	0.024
Air	0.055×10^{-4}	0.023

such as why a tile floor is much colder on the feet than a rug-covered floor at the same temperature. Tile is a better conductor of heat than the rug; heat transferred from your foot to the rug is not conducted away rapidly, so the rug's surface quickly heats up to the temperature of your foot. But the tile conducts the heat away rapidly and thus can take more heat from your foot quickly, so your foot's surface temperature drops.

➡ PHYSICS APPLIED

Heat loss through windows

FIGURE 14–6 Example 14–11.

EXAMPLE 14–11 **Heat loss through windows.** A major source of heat loss from a house is through the windows. Calculate the rate of heat flow through a glass window 2.0 m × 1.5 m in area and 3.2 mm thick, if the temperatures at the inner and outer surfaces are 15.0°C and 14.0°C, respectively (Fig. 14–6).

SOLUTION Since $A = (2.0 \text{ m})(1.5 \text{ m}) = 3.0 \text{ m}^2$, $l = 3.2 \times 10^{-3} \text{ m}$, and using Table 14–3 to get k, we have from Eq. 14–3

$$\frac{\Delta Q}{\Delta t} = kA\frac{T_1 - T_2}{l} = \frac{(0.84 \text{ J/s·m·C°})(3.0 \text{ m}^2)(15.0°C - 14.0°C)}{(3.2 \times 10^{-3} \text{ m})}$$

$$= 790 \text{ J/s}.$$

This is equivalent to $(790 \text{ J/s})/(4.19 \times 10^3 \text{ J/kcal}) = 0.19 \text{ kcal/s}$, or $(0.19 \text{ kcal/s})·(3600 \text{ s/h}) = 680 \text{ kcal/h}$.

You might notice in this Example that 15°C is not very warm for the living room of a house. The room itself may indeed be much warmer, and the outside might be colder than 14°C. But the temperatures of 15°C and 14°C were specified as those at the window surfaces, and there is usually a considerable drop in temperature of the air in the vicinity of the window both on the inside and the outside. That is, the layer of air on either side of the window acts as an insulator, and normally the major part of the temperature drop between the inside and outside of the house takes place across the air layer. If there is a heavy wind, the air outside a window will constantly be replaced with cold air; the temperature gradient across the glass will be greater and there will be a much greater rate of heat loss. Increasing the width of the air layer, such as using two panes of glass separated by an air gap, will reduce the heat loss more than simply increasing the glass thickness, since the thermal conductivity of air is much less than for glass.

The insulating properties of clothing come from the insulating properties of air. Without clothes, our bodies would heat the air in contact with the skin and would soon become reasonably comfortable because air is a very good insulator. But since air moves—there are breezes and drafts, and people themselves move about—the warm air would be replaced by cold air, thus increasing the temperature difference and the heat loss from the body. Clothes keep us warm by holding air so it cannot move readily. It is not the cloth that insulates us, but the air that the cloth traps. Down is a very good insulator because even a small amount of it fluffs up and traps a great amount of air. On this basis, can you see one reason why drapes in front of a window reduce heat loss from a house?

Wind can cause much greater heat loss

➡ PHYSICS APPLIED

Thermal windows

➡ PHYSICS APPLIED

How clothes keep us warm

Clothing traps air

For practical purposes, the thermal properties of building materials, particularly when considered as insulation, are usually specified by R-values (or "thermal resistance"), defined for a given thickness l of material as:

$$R = \frac{l}{k}.$$

The R-value of a given piece of material thus combines the thickness l and the thermal conductivity k in one number. In the United States, R-values are given in British units (although often not stated at all!), as $ft^2 \cdot h \cdot F° /$Btu. Table 14–5 gives R-values for some common building materials: note that R-values increase directly with material thickness. For example, 2 inches of fiberglass has $R = 6 \ ft^2 \cdot h \cdot F°/$Btu, half that for 4 inches.

TABLE 14–5 *R*-values

Material	Thickness	R-value (ft²·h·F°/Btu)
Glass	$\frac{1}{8}$ inch	1
Brick	$3\frac{1}{2}$ inch	0.6–1
Plywood	$\frac{1}{2}$ inch	0.6
Fiberglass insulation	4 inches	12

| CONCEPTUAL EXAMPLE 14–12 | **Boil the pasta?** When cooking pasta, why do you turn the heat down once the pasta is in and the water boiling? Would it cook faster if you kept the heat up high?

RESPONSE Once the pasta is in the water and the water is boiling, the pasta is cooking at 100°C. Leaving the heat on high will boil the water faster, but will not raise the temperature, and so will *not* speed the cooking process. 100°C is all you can get. Cooking time depends on the rate of heat transfer, which varies with the temperature (Eq. 14–3). Boiling the water fast wastes fuel and runs up your bill; it also reduces the volume of water more quickly as more boils away. (Moderately fast boiling has one advantage: to help keep the pasta from sticking together—but that can be remedied by putting a teaspoon of cooking oil in the water.) Of course to bring the water from room temperature to boiling requires large heat input and is most efficiently done with high heat. The higher the heat the more quickly the water boils. But, if the heat is not high enough, the water never boils no matter how long you wait. Why is that?

→ **PHYSICS APPLIED**

Cooking speed

14–8 Heat Transfer: Convection

Although liquids and gases are generally not very good conductors of heat, they can transfer heat quite rapidly by convection. **Convection** is the process whereby heat is transferred by the mass movement of molecules from one place to another. Whereas conduction involves molecules (and/or electrons) moving only over small distances and colliding, convection involves the movement of molecules over large distances.

A forced-air furnace, in which air is heated and then blown by a fan into a room, is an example of *forced convection. Natural convection* occurs as well, and one familiar example is that hot air rises. For instance, the air above a radiator (or other type of heater) expands as it is heated, and hence its density decreases; because its density is less, it rises, just as a log submerged in water floats upward because its density is less than that of water. Warm or cold ocean currents, such as the balmy Gulf Stream, represent natural convection on a large scale. Wind is another example of convection, and weather in general is a result of convective air currents.

When a pot of water is heated (Fig. 14–7), convection currents are set up as the heated water at the bottom of the pot rises because of its reduced density and is replaced by cooler water from above. This principle is used in many heating systems, such as the hot-water radiator system shown in

FIGURE 14–7 Convection currents in a pot of water being heated on a stove.

Cooler water

Hotter water

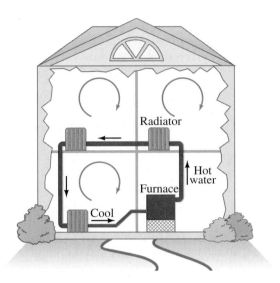

FIGURE 14–8 Convection plays a role in heating a house. The circular arrows show the convective air currents in the rooms.

➡ **PHYSICS APPLIED**

House heating

FIGURE 14–9 Convection on a hiking trail: (a) upward movement of air in the morning because it is heated; (b) downward movement in the evening because the air is cooled.

Fig. 14–8. Water is heated in the furnace and as its temperature increases, it expands and rises, as shown. This causes the water to circulate in the system. Hot water then enters the radiators, heat is transferred by conduction to the air, and the cooled water returns to the furnace. Thus, the water circulates because of convection; pumps are sometimes used to improve circulation. The air throughout the room also becomes heated as a result of convection. The air heated by the radiators rises and is replaced by cooler air, resulting in convective air currents, as shown.

Other types of furnaces also depend on convection. Hot-air furnaces with registers (openings) near the floor often do not have fans but depend on natural convection, which can be appreciable. In other systems, a fan is used. In either case, it is important that cold air can return to the furnace so that convective currents circulate throughout the room if the room is to be uniformly heated.

Another example of convection and its effects is given in the following excerpt from the early environmentalist François Matthes's "The Winds of Yosemite Valley"[†]:

> It happens to be so ordained in nature that the sun shall heat the ground more rapidly than the air. And so it comes that every slope or hillside basking in the morning sun soon becomes itself a source of heat. It gradually warms the air immediately over it, and the latter, becoming lighter, begins to rise. But not vertically upward, for above it is still the cool air pressing down. Up along the warm slope it ascends, much as shown by the arrows in the accompanying diagram [Fig. 14–9a]. Few visitors to the valley but will remember toiling up some never ending zigzags on a hot and breathless day, with the sun on their backs and their own dust floating upward with them in an exasperating, choking cloud. Perhaps they thought it was simply their misfortune that the dust should happen to rise on that particular day. It always does on a sun-warmed slope.
>
> But again, memories may arise of another occasion when, on coming down a certain trail the dust ever descended with the travelers, wafting down upon them from zigzag to zigzag as if with malicious pleasure. That, however, undoubtedly happened on the shady side of the valley. For there the conditions are exactly reversed. When the sun leaves a slope the latter begins at

[†]Reprinted from the *Sierra Club Bulletin*, June 1911, pp. 91–92.

once to lose its heat by radiation, and in a short time is colder than the air. The layer next to the ground then gradually chills by contact, and, becoming heavier as it condenses, begins to creep down along the slope [Fig. 14–9b]. There is, thus, normally a warm updraft on a sunlit slope and a cold downdraft on a shaded slope—and that rule one may depend on almost any day in a windless region like the Yosemite. Indeed, one might readily take advantage of it and plan his trips so as to have a dust-free journey.

➡ **PHYSICS APPLIED**

Body heat:
convection by blood

The human body produces a great deal of thermal energy. Of the food energy transformed within the body, at best 20 percent is used to do work, so over 80 percent appears as thermal energy. During light activity, for example, if this thermal energy were not dissipated, the body temperature would rise about $3\,C°$ per hour. Clearly, the heat generated by the body must be transferred to the outside. Is the heat transferred by conduction? The temperature of the skin in a comfortable environment is 33 to 35°C, whereas the interior of the body is at 37°C. A simple calculation (see Problem 56) shows that, because of this small temperature difference, plus the low thermal conductivity of tissue, direct conduction is responsible for very little of the heat that must be dissipated. Instead, the heat is carried to the surface by the blood. In addition to all its other important responsibilities, blood acts as a convective fluid to transfer heat to just beneath the surface of the skin. It is then conducted (over a very short distance) to the surface. Once at the surface, the heat is transferred to the environment by convection, evaporation, and radiation (see the next Section).

14–9 Heat Transfer: Radiation

Convection and conduction require the presence of matter as a medium to carry the heat from the hotter to the colder region. But a third type of heat transfer occurs without any medium at all. All life on earth depends on the transfer of energy from the Sun, and this energy is transferred to the Earth over empty (or nearly empty) space. This form of energy transfer is heat—since the Sun's temperature is much higher (6000 K) than Earth's—and is referred to as **radiation**. The warmth we receive from a fire is mainly radiant energy (most of the air heated by a fire rises by convection up the chimney and does not reach us).

As we shall see in later chapters, radiation consists essentially of electromagnetic waves. Suffice it to say for now that radiation from the Sun consists of visible light plus many other wavelengths that the eye is not sensitive to, including infrared (IR) radiation which is mainly responsible for heating the Earth.

The rate at which an object radiates energy has been found to be proportional to the fourth power of the Kelvin temperature, T. That is, a body at 2000 K as compared to one at 1000 K radiates energy at a rate $2^4 = 16$ times as much. The rate of radiation is also proportional to the area A of the emitting object, so the rate at which energy leaves the object, $\Delta Q/\Delta t$, is

$$\frac{\Delta Q}{\Delta t} = e\sigma A T^4. \tag{14–4}$$

Radiation $\propto T^4$

This is called the **Stefan-Boltzmann equation**, and σ is a universal constant called the **Stefan-Boltzmann constant** which has the value

$$\sigma = 5.67 \times 10^{-8}\ \mathrm{W/m^2 \cdot K^4}.$$

The factor e, called the **emissivity**, is a number between 0 and 1 that is characteristic of the material. Very black surfaces, such as charcoal, have emissivity close to 1, whereas shiny surfaces have e close to zero and thus emit correspondingly less radiation. The value of e depends somewhat on the temperature of the body.

Not only do shiny surfaces emit less radiation, but they absorb little of the radiation that falls upon them (most is reflected). Black and very dark objects, on the other hand, absorb nearly all the radiation that falls on them—which is why light-colored clothing is usually preferable to dark clothing on a hot day. Thus, **a good absorber is also a good emitter**.

Any object not only emits energy by radiation, but also absorbs energy radiated by other bodies. If an object of emissivity e and area A is at a temperature T_1, it radiates energy at a rate $e\sigma A T_1^4$. If the object is surrounded by an environment at temperature T_2 and high emissivity (≈ 1), the rate the surroundings radiate energy is proportional to T_2^4, and the rate that energy is absorbed by the object is proportional to T_2^4. The *net* rate of radiant heat flow from the object is given by the equation

Good absorber is good emitter

Net flow rate of heat radiation

$$\frac{\Delta Q}{\Delta t} = e\sigma A(T_1^4 - T_2^4), \tag{14-5}$$

where A is the surface area of the object, T_1 its temperature and e its emissivity (at temperature T_1), and T_2 is the temperature of the surroundings. Notice in this equation that the rate of heat absorption by an object was taken to be $e\sigma A T_2^4$; that is, the proportionality constant is the same for both emission and absorption. This must be true to correspond with the experimental fact that equilibrium between object and surroundings is reached when they come to the same temperature. That is, $\Delta Q/\Delta t$ must equal zero when $T_1 = T_2$, so the coefficients of emission and absorption terms must be the same. This confirms the idea that a good emitter is a good absorber.

Because both the object and its surroundings radiate energy, there is a net transfer of energy from one to the other unless everything is at the same temperature. From Eq. 14–5 it is clear that if $T_1 > T_2$, the net flow of heat is from the body to the surroundings, so the body cools. But if $T_1 < T_2$, the net heat flow is from the surroundings into the body, and its temperature rises. If different parts of the surroundings are at different temperatures, Eq. 14–5 becomes more complicated.

➡ **PHYSICS APPLIED**

Body's radiative heat loss

➡ **PROBLEM SOLVING**

Must use Kelvin temperature

EXAMPLE 14–13 **ESTIMATE** **Cooling by radiation.** An athlete is sitting unclothed in a locker room whose dark walls are at a temperature of 15°C. Estimate the rate of heat loss by radiation assuming a skin temperature of 34°C and $e = 0.70$. Take the surface area of the body not in contact with the chair to be 1.5 m².

SOLUTION From Eq. 14–5, we have

$$\frac{\Delta Q}{\Delta t} = e\sigma A(T_1^4 - T_2^4)$$

$$= (0.70)(5.67 \times 10^{-8} \text{ W/m}^2 \cdot \text{K}^4)(1.5 \text{ m}^2)[(307 \text{ K})^4 - (288 \text{ K})^4]$$

$$= 120 \text{ W}.$$

A resting person naturally produces heat internally at a rate of about 100 W (Chapter 15), less than the heat loss by radiation as calculated in this Example. Hence, the person's temperature would drop, causing considerable discomfort. The body responds to excessive heat loss by increasing its metabolic rate (Section 15–3) and shivering is one method by which the body increases its metabolism. Naturally, clothes help a lot. Example 14–13 illustrates that a person may be uncomfortable even if the temperature of the air is, say, 25°C, which is quite a warm room. If the walls or floor are cold, radiation to them occurs no matter how warm the air is. Indeed, it is estimated that radiation accounts for about 50 percent of the heat loss from a sedentary person in a normal room. Rooms are most comfortable when the walls and floor are warm and the air not so warm. Floors and walls can be heated by means of hot-water conduits or electric heating elements. Although such first-rate heating systems are not common today, it is interesting to note that 2000 years ago the Romans, even in houses in the remote province of Great Britain, made use of hot-water and steam conduits in the floor to heat their houses.

EXAMPLE 14–14 | **ESTIMATE** | **Two teapots.** A ceramic teapot ($e = 0.70$) and a shiny one ($e = 0.10$) each hold 0.75 L of tea at 95°C. (a) Estimate the rate of heat loss from each, and (b) estimate the temperature drop after 30 min for each. Consider only radiation and assume the surroundings are at 20°C.

SOLUTION (a) A teapot that holds 0.75 L can be approximated by a cube 10 cm on a side, with five sides exposed, so its surface area would be about $5 \times 10^{-2} \, m^2$. The rate of heat loss would be about

$$\frac{\Delta Q}{\Delta t} = e\sigma A(T_1^4 - T_2^4)$$

$$= e(5.67 \times 10^{-8} \, W/m^2 \cdot K^4)(5 \times 10^{-2} \, m^2)[(368 \, K)^4 - (293 \, K)^4]$$

$$= e(30) \, W,$$

or about 20 W for the ceramic pot ($e = 0.70$) and 3 W for the shiny one ($e = 0.10$).

(b) To estimate the temperature drop, we use the concept of specific heat and ignore the contribution of the pots compared to that of the 0.75 L of water. Then, using Eq. 14–2,

$$\frac{\Delta T}{\Delta t} = \frac{\Delta Q/\Delta t}{mc} = \frac{e(30) \, J/s}{(0.75 \, kg)(4.19 \times 10^3 \, J/kg \cdot C°)} = e(0.01) \, C°/s,$$

which for 30 min (1800 s) represents about 12 C° for the ceramic pot and about 2 C° for the shiny one. The shiny one clearly has an edge, at least as far as radiation is concerned. However, convection and conduction could play a greater role than radiation.

Heating of an object by radiation from the Sun cannot be calculated using Eq. 14–5 since this equation assumes a uniform temperature, T_2, of the environment surrounding the object, whereas the Sun is essentially a point source. Hence the Sun must be treated as a separate source of energy. Heating by the Sun is calculated using the fact that about 1350 J of energy strikes the atmosphere of the Earth from the Sun per second per

Radiation from the Sun

FIGURE 14–10 Radiant energy striking a body at an angle θ.

square meter of area at right angles to the Sun's rays. This number, 1350 W/m², is called the **solar constant**. The atmosphere may absorb as much as 70 percent of this energy before it reaches the ground, depending on the cloud cover. On a clear day, about 1000 W/m² reaches the Earth's surface. An object of emissivity e with area A facing the Sun absorbs heat at a rate, in watts, of about

$$\frac{\Delta Q}{\Delta t} = (1000 \text{ W/m}^2)eA \cos \theta,$$

where θ is the angle between the Sun's rays and a line perpendicular to the area A (Fig. 14–10). That is, $A \cos \theta$ is the "effective" area, at right angles to the Sun's rays. The explanations for the seasons and the polar ice caps (Fig. 14–11), and why the Sun heats the Earth more at midday than at sunrise or sunset, are also related to this $\cos \theta$ factor.

FIGURE 14–11 June sun makes an angle of about 23° with the equator. Thus (a) θ in the southern United States is near 0° (direct summer sun), whereas (b) in the southern hemisphere, θ is 50° or 60°, and less heat can be absorbed—hence it is winter. (c) At the poles, there is never strong direct sun; $\cos \theta$ varies from about $\frac{1}{2}$ in summer to 0 in winter; so with little heating, ice can form.

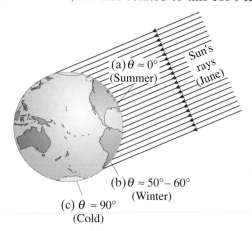

EXAMPLE 14–15 **Getting a tan—energy absorption** What is the rate of energy absorption from the Sun by a person lying flat on the beach on a clear day if the Sun makes a 30° angle with the vertical? Assume that $e = 0.70$, the area of the body exposed to the Sun is 0.80 m², and that 1000 W/m² reaches the Earth's surface.

SOLUTION Since $\cos 30° = 0.866$, we have

$$\frac{\Delta Q}{\Delta t} = (1000 \text{ W/m}^2)eA \cos \theta$$

$$= (1000 \text{ W/m}^2)(0.70)(0.80 \text{ m}^2)(0.866)$$

$$= 490 \text{ W}.$$

Notice that if a person wears light-colored clothing, e is much smaller, so the energy absorbed is less.

An interesting application of thermal radiation to diagnostic medicine is **thermography**. A special instrument, the thermograph, scans the body, measuring the intensity of radiation from many points and forming a picture that resembles an X-ray (Fig. 14–12). Areas where metabolic activity is high, such as in tumors, can often be detected on a thermogram as a result of their higher temperature and consequent increased radiation.

(a)　　　　　　　　　(b)

FIGURE 14–12
Thermograms of a healthy person's arms and hands (a) before smoking, (b) after smoking a cigarette, showing temperature decrease due to impaired blood circulation associated with smoking. The thermograms have been color-coded according to temperature; the scale on the right goes from blue (cold) to white (hot).

SUMMARY

Thermal energy, or **internal energy**, U, refers to the total energy of all the molecules in a body. For an ideal monatomic gas

$$U = \tfrac{3}{2} NkT = \tfrac{3}{2} nRT.$$

Heat refers to the transfer of energy from one body to another because of a difference of temperature. Heat is thus measured in energy units, such as joules.

Heat and thermal energy are also sometimes specified in calories or kilocalories, where

$$1 \text{ cal} = 4.186 \text{ J}$$

is the amount of heat needed to raise the temperature of 1 g of water by 1 C°.

The **specific heat**, c, of a substance is defined as the energy (or heat) required to change the temperature of unit mass of substance by 1 degree; as an equation,

$$Q = mc\,\Delta T,$$

where Q is the heat absorbed or given off, ΔT is the temperature increase or decrease, and m is the mass of the substance.

When heat flows within an isolated system, conservation of energy tells us that the heat gained by one part of the system is equal to the heat lost by the other part of the system; this is the basis of **calorimetry**, which is the quantitative measurement of heat exchange.

Exchange of energy occurs, without a change in temperature, whenever a substance changes phase. The **heat of fusion** is the heat required to melt 1 kg of a solid into the liquid phase; it is also equal to the heat given off when the substance changes from liquid to solid. The **heat of vaporization** is the energy required to change 1 kg of a substance from the liquid to the vapor phase; it is also the energy given off when the substance changes from vapor to liquid.

Heat is transferred from one place (or body) to another in three different ways. In **conduction**, energy is transferred from higher-KE molecules or electrons to lower-KE neighbors when they collide.

Convection is the transfer of energy by the mass movement of molecules over considerable distances.

Radiation, which does not require the presence of matter, is energy transfer by electromagnetic waves, such as from the Sun. All bodies radiate energy in an amount that is proportional to the fourth power of their Kelvin temperature (T^4) and to their surface area. The energy radiated (or absorbed) also depends on the nature of the surface (dark surfaces absorb and radiate more than bright shiny ones), which is characterized by the emissivity, e.

1. What happens to the work done when a jar of orange juice is vigorously shaken?

2. When a hot object warms a cooler object, does temperature flow between them? Are the temperature changes of the two objects equal?

3. (a) If two objects of different temperature are placed in contact, will heat naturally flow from the object with higher internal energy to the object with lower internal energy? (b) Is it possible for heat to flow even if the internal energies of the two objects are the same? Explain.

4. In warm zones where tropical plants grow but the temperature may drop below freezing a few times in the winter, the destruction of sensitive plants due to freezing can be reduced by watering them in the evening. Explain.

5. The specific heat of water is quite large. Explain why this fact makes water particularly good for heating systems (that is, hot-water radiators).

6. Why does water in a canteen stay cooler if the cloth jacket surrounding the canteen is kept moist?

7. Explain why burns caused by steam on the skin are often more severe than burns caused by water at 100°C.

8. Explain, using the concepts of latent heat and internal energy, why water cools (its temperature drops) when it evaporates.

9. Will potatoes cook faster if the water is boiling faster?

10. Does an ordinary electric fan cool the air? Why or why not? If not, why use it?

11. The temperature very high in the Earth's atmosphere can be 700°C. Yet an animal there would freeze to death rather than roast. Explain.

12. Explorers on failed Arctic expeditions have survived by covering themselves with snow. Why on earth would they do that?

13. Why is wet sand at the beach cooler to walk on than dry sand?

14. Homeless people know that sleeping under a blanket of newspapers is a good way to keep warm. Why are thin newspapers so effective?

15. Why is it important, when hot-air furnaces are used to heat a house, that there be a vent for air to return to the furnace? What happens if this vent is blocked by a bookcase?

16. Ceiling fans are sometimes reversible, so that they drive the air down in one season and pull it up in another season. Which way should you set the fan for summer? For winter?

17. Down sleeping bags and parkas are often specified as so many inches or centimeters of *loft*, the actual thickness of the garment when it is fluffed up. Explain.

18. Microprocessor chips nowadays have a "heat sink" glued on top that looks like a series of fins. Why is it shaped like that?

19. Sea breezes are often encountered on sunny days at the shore of a large body of water. Explain in light of the fact that the temperature of the land rises more rapidly than that of the nearby water.

20. The floor of a house on a foundation under which the air can flow is often cooler than a floor that rests directly on the ground (such as a concrete slab foundation). Explain.

21. The Earth cools off at night much more quickly when the weather is clear than when cloudy. Why?

22. Why is the liner of a thermos bottle silvered, and why does it have a vacuum between its two walls?

23. Explain why air-temperature readings arc always taken with the thermometer in the shade.

24. A premature baby in an incubator can be dangerously cooled even when the air temperature in the incubator is warm. Explain.

25. Suppose you are designing one of the following (choose one): house, concert hall, medical office building. List as many sources of heat as you can think of. Estimate the heat produced by each.

26. Heat loss occurs through windows by the following processes: (1) ventilation around edges; (2) through the frame, particularly if it is metal; (3) through the glass panes; and (4) radiation. (a) For the first three, what is (are) the mechanism(s): conduction, convection, or radiation? (b) Heavy curtains reduce which of these heat losses? Explain in detail.

27. A piece of wood lying in the Sun absorbs more heat than a piece of shiny metal. Yet the wood feels less hot than the metal when you pick it up. Explain.

28. People often ask: Which will cool the coffee faster, (1) adding cold milk immediately after pouring the coffee into a mug or (2) waiting a few minutes until you are ready to drink the coffee?

PROBLEMS

SECTION 14–1

1. (I) How much heat (joules) is required to raise the temperature of 20.0 kg of water from 15°C to 95°C?

2. (I) How much work must a person do to offset eating a 750-Cal piece of cake?

3. (I) To what temperature will 7700 J of work raise 3.0 kg of water initially at 10.0°C?

4. (I) An average active person consumes about 2500 Cal a day. (*a*) What is this in joules? (*b*) What is this in kilowatt-hours? (*c*) Your power company charges about a dime per kilowatt-hour. How much would your energy cost per day if you bought from the power company? Could you feed yourself on this much money per day?

5. (II) A British thermal unit (Btu) is a unit of heat in the British system of units. One Btu is defined as the heat needed to raise 1 pound of water by 1 F°. Show that 1 Btu = 0.252 kcal = 1055 J.

6. (II) A small immersion heater is rated at 350 W. Estimate how long it will take to heat a cup of soup (assume this is 250 mL of water) from 20°C to 50°C.

7. (II) A water heater can generate 7200 kcal/h. How much water can it heat from 15°C to 50°C per hour?

8. (II) How many kilocalories of heat are generated when the brakes are used to bring a 1000-kg car to rest from a speed of 100 km/h?

SECTIONS 14–4 AND 14–5

9. (I) What is the specific heat of a metal substance if 135 kJ of heat is needed to raise 5.1 kg of the metal from 20°C to 30°C?

10. (I) An automobile cooling system holds 18 L of water. How much heat does it absorb if its temperature rises from 20°C to 90°C?

11. (I) Samples of copper, aluminum, and water experience the same temperature rise when they absorb the same amount of heat. What is the ratio of their masses?

12. (I) What is the specific heat capacity of water in Btu/lb·F°?

13. (I) What is the water equivalent of 4.00 kg of lead?

14. (II) A 30-g glass thermometer reads 21.6°C before it is placed in 135 mL of water. When the water and thermometer come to equilibrium, the thermometer reads 39.2°C. What was the original temperature of the water?

15. (II) The 1.20-kg head of a hammer has a speed of 8.0 m/s just before it strikes a nail (Fig. 14–13) and is brought to rest. Estimate the temperature rise of a 14-g iron nail generated by ten such hammer blows done in quick succession. Assume the nail absorbs all the energy.

16. (II) What will be the equilibrium temperature when a 270-g block of copper at 300°C is placed in a 150-g aluminum calorimeter cup containing 820 g of water at 12.0°C?

FIGURE 14–13   Problem 15.

17. (II) An old way of heating cider in the winter was to put a hot iron poker from the fireplace into a ceramic mug containing the cider. Suppose we have half a liter of cider (assume water) initially at 15°C and an effective mass of iron of 0.55 kg initially at 700°C. If we assume that the entire heat transfer takes place between the poker and the cider—that is, the effect of the mug and the air in the room surrounding the cider can be neglected—what will be the final temperature of the cider?

18. (II) A hot iron horseshoe (mass = 0.40 kg) which has just been forged, is dropped into 1.60 L of water in a 0.30-kg iron pot initially at 20°C. If the final equilibrium temperature is 25°C, estimate the initial temperature of the hot horseshoe.

19. (II) When a 290-g piece of iron at 180°C is placed in a 100-g aluminum calorimeter cup containing 250 g of glycerin at 10°C, the final temperature is observed to be 38°C. What is the specific heat of glycerin?

20. (II) When 195 g of a substance is heated to 330°C and then plunged into a 100-g aluminum calorimeter cup containing 150 g of water at 12.5°C, the final temperature, as registered by a 17-g glass thermometer, is 35.0°C. What is the specific heat of the substance?

21. (II) How long does it take a 750-W coffeepot to bring to a boil 0.60 L of water initially at 8.0°C? Assume that the part of the pot which is heated with the water is made of 360 g of aluminum, and that no water boils away.

SECTION 14–6

22. (I) How much heat is needed to melt 16.50 kg of silver that is initially at 20°C?

23. (I) During exercise, a person may give off 180 kcal of heat in 30 min by evaporation of water from the skin. How much water has been lost?

24. (II) A 30-g ice cube at its melting point is dropped into an insulated container of liquid nitrogen. How much nitrogen evaporates if it is at its boiling point of 77 K and has a latent heat of vaporization of 200 kJ/kg? Assume for simplicity that the specific heat of ice is a constant and is equal to its value near its melting point.

25. (II) A cube of ice is taken from the freezer at $-8.5°C$ and placed in a 100-g aluminum calorimeter filled with 300 g of water at room temperature of 20°C. The final situation is observed to be all water at 17°C. What was the mass of the ice cube?

26. (II) An iron boiler of mass 230 kg contains 830 kg of water at 20°C. A heater supplies energy at the rate of 52,000 kJ/h. How long does it take for the water (a) to reach the boiling point, and (b) to all have changed to steam?

27. (II) On a hot day's race, a bicyclist consumes 8.0 L of water over the span of four hours. Making the approximation that all of the cyclist's energy goes into evaporating this water as sweat, how much energy in kcal did the rider use during the ride? (Since the efficiency of the rider is only about 20 percent, most of the energy consumed does go to heat, so our approximation is not far off.)

28. (II) What mass of steam at 100°C must be added to 1.00 kg of ice at 0°C to yield liquid water at 20°C?

29. (II) The specific heat of mercury is 138 J/kg·C°. Determine the latent heat of fusion of mercury using the following calorimeter data: 1.00 kg of solid Hg at its melting point of $-39.0°C$ is placed in a 0.620-kg aluminum calorimeter with 0.400 kg of water at 12.80°C; the resulting equilibrium temperature is 5.06°C.

30. (II) A 54.0-kg ice-skater moving at 6.4 m/s glides to a stop. Assuming the ice is at 0°C and that 50 percent of the heat generated by friction is absorbed by the ice, how much ice melts?

31. (II) At a crime scene, the forensic investigator notes that the 8.2-g lead bullet that was stopped in a doorframe apparently melted completely on impact. Assuming the bullet was fired at room temperature (20°C), what does the investigator calculate the *minimum* muzzle velocity of the gun was?

SECTIONS 14–7 TO 14–9

32. (I) Calculate the rate of heat flow by conduction in Example 14–11 assuming there are strong gusty winds and that the external temperature is $-5°C$.

33. (I) (a) How much power is radiated by a tungsten sphere (emissivity $e = 0.35$) of radius 22 cm at a temperature of 25°C? (b) If the sphere is enclosed in a room whose walls are kept at $-5°C$, what is the *net* flow rate of energy out of the sphere?

34. (I) Over what distance must there be heat flow by conduction from the blood capillaries beneath the skin to the surface if the temperature difference is 0.50 C°? Assume 200 W must be transferred through the whole body's surface area of 1.5 m².

35. (I) The cool red giant star Betelgeuse has a radius $r = 3.1 \times 10^{11}$ m. (If it were centered on our Sun, it would extend out beyond Mars.) Its surface temperature is 2800 K (about half that of our Sun). Assuming it is a perfect emitter ($e = 1.0$), what is its power output? Compare to the power output of our Sun.

36. (I) At approximately what rate does the top of a person's head absorb the sun's energy on a clear day if: (a) it is covered with hair ($e = 0.75$), has a surface area of 225 square centimeters (assumed flat), and the person stands straight with the Sun making an angle of 40° with the vertical; or if (b) it is bald with $e = 0.20$ under the same conditions. See Fig. 14–14.

FIGURE 14–14 Problem 36.

37. (II) Two rooms, each a cube 4.0 m on a side, share a 12–cm thick brick wall. Because of a number of 100–W light-bulbs in one room, the air is at 30°C, while in the other room it is at 10°C. How many of the 100–W light-bulbs are needed to maintain the temperature difference across the wall?

38. (II) How long does it take the Sun to melt a block of ice at 0°C with an area of 1.0 m² and thickness 1.0 cm? Assume that the Sun's rays make an angle of 30° with the normal to the area, and that the emissivity of ice is 0.050.

39. A copper rod and an aluminum rod of the same length and cross-sectional area are attached end to end (Fig. 14–15). The copper end is placed in a furnace which is maintained at a constant temperature of 250°C. The aluminum end is placed in an ice bath held at constant temperature of 0.0°C. Calculate the temperature at the point where the two rods are joined.

FIGURE 14–15 Problem 39.

40. (II) The Earth receives about 430 W/m^2 from the Sun, averaged over the whole surface, and radiates an equal amount back into space (that is, the Earth is in equilibrium). Assuming the Earth is a perfect emitter ($e = 1.0$), estimate its average surface temperature.

41. (II) A 100-W lightbulb generates 95 W of heat, which is dissipated through a glass bulb that has a radius of 3.0 cm and is 1.0 mm thick. What is the difference in temperature between the inner and outer surfaces of the glass?

42. (II) Write an equation for the total rate of heat flow through the wall of a house if the wall consists of material with thermal conductivity k_1, total area A_1, and thickness l_1, and of windows with thermal conductivity k_2, area A_2, and thickness l_2. The temperature difference is ΔT.

43. (III) Suppose the insulating qualities of the wall of a house come mainly from a 4.0-inch layer of brick and an R-19 layer of insulation, as shown in Fig. 14–16. What is the total rate of heat loss through such a wall, if its total area is 240 ft^2 and the temperature difference across it is 10 F°?

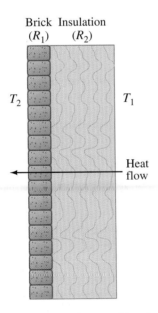

Brick Insulation
(R_1)   (R_2)

T_2 T_1

Heat flow

FIGURE 14–16 Two layers of insulation of a wall of a building. Problem 43.

Glass

T_1 Air T_2

l_1 l_2 l_3

FIGURE 14–17   Problem 44.

44. (III) A double-glazed window is one with two panes of glass separated by an air space, Fig. 14–17. (a) Show that the rate of heat flow by conduction is given by

$$\frac{\Delta Q}{\Delta t} = \frac{A(T_2 - T_1)}{l_1/k_1 + l_2/k_2 + l_3/k_3},$$

where k_1, k_2, and k_3 are the thermal conductivities for glass, air, and glass, respectively. (b) Generalize this expression for any number of materials placed next to one another.

45. (III) Approximately how long should it take 11.0 kg of ice at 0°C to melt when it is placed in a carefully sealed Styrofoam icebox of dimensions 25 cm × 35 cm × 50 cm whose walls are 1.5 cm thick? Assume that the conductivity of Styrofoam is double that of air and that the outside temperature is 30°C.

46. (III) A house thermostat is normally set to 22°C, but at night it is turned down to 12°C for 7.0 h. Estimate how much more heat would be needed (state as a percentage of daily usage) if the thermostat were not turned down at night. Assume that the outside temperature averages 0°C for the 7.0 h at night and 8°C for the remainder of the day, and that the heat loss from the house is proportional to the difference in temperature inside and out. To obtain an estimate from the data, you will have to make other simplifying assumptions; state what these are.

GENERAL PROBLEMS

47. If coal gives off 7000 kcal/kg when it is burned, how much coal will be needed to heat a house that requires 4.8×10^7 kcal for the whole winter? Assume that an additional 30 percent of the heat is lost up the chimney.

48. A 15-g lead bullet is tested by firing it into a fixed block of wood with a mass of 1.05 kg. If the block and imbedded bullet together absorb all the heat energy generated and, after thermal equilibrium has been reached, the system has a temperature rise measured as 0.020 C°, estimate the entering speed of the bullet.

FIGURE 14–18 Problem 49.

49. (a) Find the total power radiated into space by the Sun, assuming it to be a perfect emitter at $T = 5500$ K. The Sun's radius is 7.0×10^8 m. (b) From this, determine the power per unit area arriving at the Earth, 1.5×10^{11} m away (Fig. 14–18).

50. During light activity, a 70-kg person may generate 200 kcal/h. Assuming that 20 percent of this goes into useful work and the other 80 percent is converted to heat, calculate the temperature rise of the body after 1.00 h if none of this heat were transferred to the environment.

51. A 340-kg marble boulder rolls off the top of a cliff and falls a vertical height of 140 m before striking the ground. Estimate the temperature rise of the rock if 50 percent of the heat generated remains in the rock.

52. The *heat capacity*, C, of an object is defined as the amount of heat needed to raise its temperature by 1 C°. Thus, to raise the temperature by ΔT requires heat Q given by

$$Q = C \Delta T.$$

(a) Write the heat capacity C in terms of the specific heat, c, of the material. (b) What is the heat capacity of 1 kg of water? (c) Of 50 kg of water?

53. A very long 2.00-cm-diameter lead rod absorbs 320 kJ of heat. By how much does its length change? What would happen if the rod were only 2.0 cm long?

54. A mountain climber wears down clothing 3.5 cm thick with total surface area 1.7 m². The temperature at the surface of the clothing is -20°C and at the skin is 34°C. Determine the rate of heat flow by conduction through the clothing (a) assuming it is dry and that the thermal conductivity, k, is that of down, and (b) assuming the clothing is wet, so that k is that of water and the jacket has matted down to 0.50 cm thickness.

55. A marathon runner has an average metabolism rate for the race of about 1000 kcal/h. If the runner has a mass of 65.0 kg, how much water would the runner lose to evaporation from the skin for a race that lasts 2.5 h?

56. Estimate the rate at which heat can be conducted from the interior of the body to the surface. Assume that the thickness of tissue is 4.0 cm, that the skin is at 34°C and the interior at 37°C, and that the surface area is 1.5 m². Compare this to the measured value of about 230 W that must be dissipated by a person working lightly. This clearly shows the necessity of convective cooling by the blood.

57. *Newton's law of cooling* states that for small temperature differences, if a body at a temperature T_1 is in surroundings at a temperature T_2, the body cools at a rate given by

$$\frac{\Delta Q}{\Delta t} = K(T_1 - T_2)$$

where K is a constant. It includes the effects of conduction, convection, and radiation. That this linear relationship should hold is obvious if only conduction is considered. Show that it is also approximately true for radiation by showing that Eq. 14–5 reduces to

$$\Delta Q/\Delta t = 4\sigma e A T_2^3(T_1 - T_2)$$
$$= \text{constant} \times (T_1 - T_2)$$

if $(T_1 - T_2)$ is small.

58. A house has well-insulated walls 17.5 cm thick (assume conductivity of air) and area 410 m², a roof of wood 6.5 cm thick and area 280 m², and uncovered windows 0.65 cm thick and total area 33 m². (a) Assuming that the heat loss is only by conduction, calculate the rate at which heat must be supplied to this house to maintain its temperature at 23°C if the outside temperature is -10°C. (b) If the house is initially at 10°C, estimate how much heat must be supplied to raise the temperature to 23°C within 30 min. Assume that only the air needs to be heated and that its volume is 750 m³. (c) If natural gas costs $0.080 per kilogram and its heat of combustion is 5.4×10^7 J/kg, how much is the monthly cost to maintain the house as in part (a) for 24 h each day assuming 90 percent of the heat produced is used to heat the house? Take the specific heat of air to be 0.24 kcal/kg·C°.

59. A 15-g lead bullet traveling at 220 m/s passes through a thin iron wall and emerges at a speed of 160 m/s. If the bullet absorbs 50 percent of the heat generated, (a) what will be the temperature rise of the bullet? (b) If the ambient temperature is 20°C, will any of the bullet melt, and if so, how much?

60. A leaf of area 40 cm² and mass 4.5×10^{-4} kg directly faces the Sun on a clear day. The leaf has an emissivity of 0.85 and a specific heat of 0.80 kcal/kg·K. (a) Estimate the rate of rise of the leaf's temperature. (b) Calculate the temperature the leaf would reach if it lost all its heat by radiation (the surroundings are at 20°C). (c) In what other ways can the heat be dissipated by the leaf?

61. Use the result of part (a) in the previous problem and take into account radiation from the leaf to calculate how much water must be transpired (evaporated) by the leaf per hour to maintain a temperature of 35°C.

62. An iron meteorite melts when it enters the Earth's atmosphere. If its initial temperature, outside of the atmosphere, was -125°C, calculate the minimum velocity the meteorite must have had before it entered Earth's atmosphere.

This old train locomotive is an example of a steam engine, one type of heat engine whose development helped lead to the second law of thermodynamics.

THE LAWS OF 15 THERMODYNAMICS

Thermodynamics is the name we give to the study of processes in which energy is transferred as heat and as work.

In Chapter 6 we saw that work is done when energy is transferred from one body to another by mechanical means. In Chapter 14 we saw that heat is a transfer of energy from one body to a second body at a lower temperature. Thus, heat is much like work. To distinguish them, *heat* is defined as a *transfer of energy due to a difference in temperature*, whereas work is a transfer of energy that is not due to a temperature difference.

Heat distinguished from work

In discussing thermodynamics, we shall often refer to particular systems. A **system** is any object or set of objects that we wish to consider. Everything else in the universe we will refer to as its "environment." There are several kinds of systems. A **closed system** is one for which no mass enters or leaves (but energy may be exchanged with the environment). In an **open system**, mass may enter or leave (as well as energy). Many (idealized) systems we study in physics are closed systems. But many systems, including plants and animals, are open systems since they exchange materials (food, oxygen, waste products) with the environment. A closed system is said to be **isolated** if no energy in any form passes across its boundaries; otherwise it is not isolated.

Open and closed systems

15–1 The First Law of Thermodynamics

In Section 14–2, we defined the internal energy of a system as the sum total of all the energy of the molecules of the system. We would expect that the internal energy of a system would be increased if work were done on the system, or if heat were added to it. Similarly the internal energy would be decreased if heat flowed out of the system or if work were done by the system on something else.

Thus, from conservation of energy, it is reasonable to propose an important law: the change in internal energy of a closed system, ΔU, will be equal to the heat added to the system minus the work done by the system; in equation form:

$$\Delta U = Q - W, \tag{15–1}$$

FIRST LAW OF THERMODYNAMICS

where Q is the net heat *added* to the system and W is the net work done *by* the system. We must be careful and consistent in following the sign conventions for Q and W. Because W in Eq. 15–1 is the work done *by* the system, then if work is done *on* the system, W will be negative and U will increase. [Of course, we could have defined W as the work done *on* the system, in which case there would be a plus sign in Eq. 15–1; but it is conventional to define W and Q as we have done.] Similarly, Q is positive for heat added to the system, so if heat leaves the system, Q is negative. Equation 15–1 is known as the **first law of thermodynamics**. It is one of the great laws of physics, and its validity rests on experiments (such as Joule's) in which no exceptions have been seen. Since Q and W represent energy transferred into or out of the system, the internal energy changes accordingly. Thus, the first law of thermodynamics is a statement of the *law of conservation of energy*. It is worth noting that the conservation of energy law was not formulated until the nineteenth century, for it depended on the interpretation of heat as a transfer of energy.

Heat added is +
Heat lost is −
Work on system is −
Work by system is +

First law of thermodynamics is conservation of energy

Equation 15–1 applies to a closed system. It also applies to an open system if we take into account the change in internal energy due to the increase or decrease in the amount of matter. For an isolated system, no work is done and no heat enters or leaves the system, so $W = Q = 0$, and hence $\Delta U = 0$.

A given system, in a particular state, can be said to have a certain amount of internal energy, U. This cannot be said of heat or work. A system in a given state does not "have" a certain amount of heat or work. Rather, when work is done on a system (such as compressing a gas), or when heat is added or removed from a system, the state of the system *changes*. Thus, work and heat are involved in *thermodynamic processes* that can change the system from one state to another; they are not characteristic of the state itself, as are pressure P, volume V, temperature T, mass m or number of moles n, and internal energy U.

Internal energy is a property of a system; work and heat are not

EXAMPLE 15–1 Using the first law. (*a*) An amount of heat equal to 2500 J is added to a system, and 1800 J of work is done on the system. What is the change in internal energy of the system? (*b*) What would be the internal energy change if 2500 J of heat is added to the system and 1800 J of work is done *by* the system (i.e., as output)?

SOLUTION (*a*) We use the first law of thermodynamics, Eq. 15–1. The heat added to the system is $Q = 2500$ J. The work done by the system W is -1800 J. Why the minus sign? Because 1800 J done *on* the system (as given) equals $-1800\,\text{J}$ done *by* the system, and it is the latter we need to put in Eq. 15–1 by the conventions given above. Hence

$$\Delta U = 2500\,\text{J} - (-1800\,\text{J}) = 2500\,\text{J} + 1800\,\text{J} = 4300\text{J}.$$

You may have intuitively thought that the $2500\,\text{J}$ and the $1800\,\text{J}$ would need to be added together, since both refer to energy added to the system. You would have been right. We did this exercise in detail to emphasize the importance of keeping careful track of signs.

(*b*) Heat is still being added to the system, so $Q = +2500\,\text{J}$, but now the work is being done *by* the system. Hence

$$\Delta U = 2500\,\text{J} - 1800\,\text{J} = 700\,\text{J}.$$

The change in internal energy is much less because the work was an output.

15–2 First Law of Thermodynamics Applied to Some Simple Systems

Let us analyze some simple processes in the light of the first law of thermodynamics.

First we consider an idealized process that is carried out at constant temperature. Such a process is called an **isothermal** process (from the Greek meaning "same temperature"). If the system is an ideal gas, then $PV = nRT$ (Eq. 13–3), so for constant temperature $PV = $ constant. Thus the process follows a curve like AB on the PV diagram shown in Fig. 15–1, which is a curve for $PV = $ constant (see Fig. 13–11). Each point on the curve, such as point A, represents the state of the system at a given moment—that is, its pressure P and volume V. At a lower temperature, another isothermal process would be represented by a curve like $A'B'$ in Fig. 15–1 (the product $PV = nRT = $ constant is less when T is less). The curves shown in Fig. 15–1 are referred to as *isotherms*. Let us assume that the gas is enclosed in a container fitted with a movable piston, Fig. 15–2, and that the gas is in contact with a **heat reservoir** (a body whose mass is so large that, ideally, its temperature does not change significantly when heat is exchanged with our system). We also assume that the process of compression (volume decreases) or expansion (volume increases) is done very slowly to make certain that all of the gas stays in equilibrium at the same constant temperature. If the gas is initially in a state represented by point A in Fig. 15–1, and an amount of heat Q is added to the system, the system will move to another point, B, on the diagram. If the temperature is to remain constant, the gas must expand and do an amount of

Isothermal process ($\Delta T = 0$)

Heat reservoir

FIGURE 15–1 *PV* diagram for an ideal gas undergoing isothermal processes at two different temperatures.

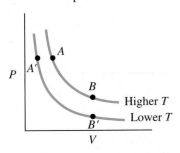

FIGURE 15–2 An ideal gas in a cylinder fitted with a movable piston.

Movable cylinder

Ideal gas

445

work W on the environment (it exerts a force on the piston and moves it through a distance). The temperature is kept constant so, from Eq. 14–1, the internal energy does not change: $\Delta U = \frac{3}{2}nR\Delta T = 0$. Hence, by the first law of thermodynamics (Eq. 15–1), $\Delta U = Q - W = 0$, so $W = Q$: the work done by the gas in an isothermal process equals the heat added to the gas.

An **adiabatic** process is one in which no heat is allowed to flow into or out of the system: $Q = 0$. This situation can occur if the system is extremely well insulated, or the process happens so quickly that heat—which flows slowly—has no time to flow in or out. The very rapid expansion of gases in an internal combustion engine is one example of a process that is very nearly adiabatic. A slow adiabatic expansion of an ideal gas follows a curve like that labeled AC in Fig. 15–3. Since $Q = 0$, we have from Eq. 15–1 that $\Delta U = -W$. That is, the internal energy decreases if the gas expands; hence the temperature decreases as well (because $\Delta U = \frac{3}{2}nR\Delta T$). This is evident in Fig. 15–3 where the product PV ($= nRT$) is less at point C than at point B (curve AB is for an isothermal process, for which $\Delta U = 0$ and $\Delta T = 0$). In an adiabatic compression (going from C to A, for example), work is done *on* the gas, and hence the internal energy increases and the temperature rises. In a diesel engine, the air-fuel mixture is rapidly compressed adiabatically by a factor of 15 or more; the temperature rise is so great that the mixture ignites spontaneously.

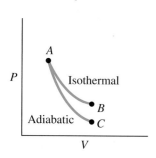

FIGURE 15–3 *PV* diagram for adiabatic (*AC*) and isothermal (*AB*) processes on an ideal gas.

Isothermal and adiabatic processes are just two possible processes that can occur. Two other simple thermodynamic processes are illustrated on the *PV* diagram of Fig. 15–4: (a) an **isobaric** process is one in which the pressure is kept constant, so the process is represented by a straight line on the *PV* diagram (Fig. 15–4a); (b) an **isochoric** or isovolumetric process is one in which the volume does not change (Fig. 15–4b). In these, and in all other processes, the first law of thermodynamics holds.

It is often valuable to calculate the work done in a process. If the pressure is kept constant during a process (isobaric), the work done is easily calculated. For example, if the gas in Fig. 15–5 expands slowly against the piston, the work done by the gas to raise the piston is the force F times the distance d. But the force is just the pressure P of the gas times the area A of the piston, $F = PA$. Thus,

$$W = Fd = PAd$$

or

$$W = P\,\Delta V, \qquad\qquad \text{[constant pressure]} \quad \textbf{(15–2)}$$

where $\Delta V = Ad$ is the change in volume of the gas. This equation also

FIGURE 15–4 (a) Isobaric ("same pressure") process. (b) Isochoric (same volume) process.

(a) Isobaric

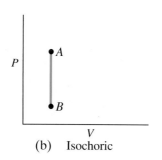

(b) Isochoric

FIGURE 15–5 Work is done on the piston when the gas expands, moving the piston a distance d.

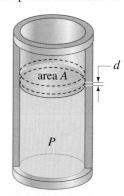

holds if the gas is *compressed* at constant pressure, in which case (since V decreases) ΔV is negative; W is then negative, which indicates that work is done *on* the gas. Equation 15–2 is also valid for liquids and solids, as long as the pressure is constant during the process. In an isochoric process (Fig. 15–4b) the volume does not change, so no work is done, $W = 0$.

Figure 15–6 shows the isotherm AB we saw in Fig. 15–1 as well as another possible process represented by the path ADB. In going from A to D, the gas does no work since the volume does not change. But in going from D to B, the gas does work equal to $P_B(V_B - V_A)$, and this is the total work done in the process ADB.

If the pressure varies during a process, such as for the isothermal process AB in Fig. 15–1, Eq. 15–2 cannot be used directly to determine the work. A rough estimate can be obtained, however, by using an "average" value for P in Eq. 15–2. More accurately, the work done is equal to the area under the PV curve. This is obvious when the pressure is constant, for as can be seen from Fig. 15–7a, the shaded area is just $P_B(V_B - V_A)$ and this is the work done. Similarly, the work done during an isothermal process is equal to the shaded area shown in Fig. 15–7b. The calculation of work done in this case can be carried out using calculus, or by estimating the area on graph paper.

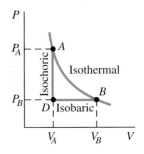

FIGURE 15–6 *PV* diagram for different processes (see the text).

Work =
area under PV curve

FIGURE 15–7 Work done by a gas is equal to the area under the *PV* curve.

(a)

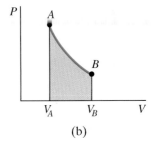

(b)

| **CONCEPTUAL EXAMPLE 15–2** | **Work in isothermal and adiabatic processes.** |

In Fig. 15–3 we saw the *PV* diagrams for a gas expanding in two ways, isothermally and adiabatically. The initial volume V_A was the same in each case, and the final volumes were the same ($V_B = V_C$). In which process was more work done by the gas?

RESPONSE More work was done in the isothermal process. We can see this in two simple ways by looking at Fig. 15–3. First, the "average" pressure was higher during the isothermal process AB, so $W = P_{ave} \Delta V$ was greater (ΔV is the same for both processes). Secondly, we can look at the area under each curve: the area under curve AB, which represents the work done, was greater (since curve AB is higher) than that under AC.

| **CONCEPTUAL EXAMPLE 15–3** | **Simple adiabatic process.** |

Here is an example of an adiabatic process that you can do with just a rubber band. Hold the rubber band loosely with two hands and gauge the temperature with your lips. Stretch the rubber band suddenly and again touch it lightly to your lips. You should notice an increase in temperature. Explain clearly why the temperature increases.

RESPONSE Stretching the rubber band *suddenly* makes the process adiabatic because there is no time for heat to enter or leave the system, so $Q = 0$. You do work on the system, representing an energy input, so W in Eq. 15–1 is negative. Hence ΔU must be positive. An increase in internal energy corresponds to an increase in temperature (Eq. 14–1), so the rubber band heats up.

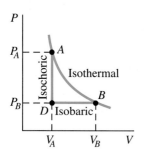

FIGURE 15–6 (Repeated for Example 15–4.)

EXAMPLE 15–4 **First law in isobaric and isochoric processes.** An ideal gas is slowly compressed at a constant pressure of 2.0 atm from 10.0 L to 2.0 L. This process is represented in Fig. 15–6 as the path B to D. (In this process, some heat flows out and the temperature drops.) Heat is then added to the gas, holding the volume constant, and the pressure and temperature are allowed to rise until the temperature reaches its original value. This process is D to A in Fig. 15–6. Calculate (*a*) the total work done by the gas in the process BDA, and (*b*) the total heat flow into the gas.

SOLUTION (*a*) Work is done only in the first part, the compression (BD):

$$W = P\,\Delta V = (2.0 \times 10^5\,\text{N/m}^2)(2.0 \times 10^{-3}\,\text{m}^3 - 10.0 \times 10^{-3}\,\text{m}^3)$$
$$= -1.6 \times 10^3\,\text{J}.$$

From D to A no work is done ($\Delta V = 0$); so the total work done by the gas is $-1.6 \times 10^3\,\text{J}$, where the minus means that $+1.6 \times 10^3\,\text{J}$ of work is done *on* the gas.

(*b*) Since the temperature at the beginning and at the end of the process is the same, there is no change in internal energy: $\Delta U = 0$. From the first law of thermodynamics we have

$$0 = \Delta U = Q - W,$$

so

$$Q = W$$
$$= -1.60 \times 10^3\,\text{J}.$$

Since Q is negative, we know that 1600 J of heat flow out of the gas. Note that this is the total heat loss for the whole process, BDA.

EXAMPLE 15–5 **Work done in an engine.** In an engine, 0.25 moles of gas in the cylinder expands rapidly and adiabatically against the piston. In the process, the temperature drops from 1150 K to 400 K. How much work does the gas do? Assume the gas is ideal.

SOLUTION The pressure is not constant so we can't use Eq. 15-2. Instead, we can use the first law of thermodynamics if we can determine ΔU, since we know $Q = 0$ because the process is adiabatic. We determine ΔU from Eq. 14–1 for the internal energy of an ideal monatomic gas:

$$\Delta U = U_\text{f} - U_\text{i} = \tfrac{3}{2}nR(T_\text{f} - T_\text{i})$$
$$= \tfrac{3}{2}(0.25\,\text{mol})(8.315\,\text{J/mol·K})(400\,\text{K} - 1150\,\text{K})$$
$$= -2300\,\text{J}.$$

Then, from Eq. 15–1,

$$W = Q - \Delta U = 0 - (-2300\,\text{J}) = 2300\,\text{J}.$$

EXAMPLE 15–6 **ΔU for boiling water to steam.** Determine the change in internal energy of 1.0 liter of water at 100°C when it is fully boiled, which results in 1671 liters of steam at 100°C. Assume the process is done at atmospheric pressure.

SOLUTION 1.0 L of water has a mass of 1.0 kg, and its latent heat of vaporization (Table 14–3) is $L_V = 22.6 \times 10^5$ J/kg. So the heat input required for this process is

$$Q = mL = (1.0 \text{ kg})(22.6 \times 10^5 \text{ J/kg}) = 22.6 \times 10^5 \text{ J}.$$

The work done by the water is (Eq. 15–2)

$$W = P \, \Delta V = (1.0 \times 10^5 \text{ N/m}^2)(1671 \times 10^{-3} \text{ m}^3 - 1 \times 10^{-3} \text{ m}^3)$$

$$= 1.7 \times 10^5 \text{ J},$$

where we used 1 atm $= 1.0 \times 10^5$ N/m² and 1 L $= 10^3$ cm³ $= 10^{-3}$ m³. Then

$$\Delta U = Q - W = 22.6 \times 10^5 \text{ J} - 1.7 \times 10^5 \text{ J}$$

$$= 21 \times 10^5 \text{ J}.$$

Note that most of the heat added goes to increasing the internal energy of the water (increasing molecular energy to overcome the attraction that held the molecules close together in the liquid state), and only a small part (< 10 percent) goes into doing work.

*15–3 Human Metabolism and the First Law

Human beings and other animals do work. Work is done when a person walks or runs, or lifts a heavy object. Work requires energy. Energy is also needed for growth—to make new cells, and to replace old cells that have died. A great many energy-transforming processes occur within an organism, and they are referred to as *metabolism*.

We can apply the first law of thermodynamics,

$$\Delta U = Q - W,$$

to an organism, say the human body. Work W is done by the body in its various activities, and if this is not to result in a decrease in the body's internal energy (and temperature), energy must somehow be added to compensate. The body's internal energy is not maintained by a flow of heat Q into the body, however. Normally, the body is at a higher temperature than its surroundings, so heat usually flows *out* of the body. Even on a very hot day when heat is absorbed, the body has no way of utilizing this heat to support its vital processes. What then is the source of energy? It is the internal energy (chemical potential energy) stored in foods. Now in a closed system, the internal energy changes only as a result of heat flow or work done. In an open system, such as an animal, internal energy itself can flow into or out of the system. When we eat food, we are bringing internal energy into our bodies directly, which thus increases the total internal energy U in our bodies. This energy eventually goes into work and heat flow from the body according to the first law.

The metabolic rate is the rate at which internal energy is transformed within the body. It is usually specified in kcal/h or in watts. Typical metabolic rates for a variety of human activities are given in Table 15–1 for an "average" 65-kg adult.

PHYSICS APPLIED

Energy in the human body

TABLE 15–1 Metabolic Rates (65-kg human)

Activity	Metabolic Rate (approximate)	
	kcal/h	watts
Sleeping	60	70
Sitting upright	100	115
Light activity (eating, dressing, household chores)	200	230
Moderate work (tennis, walking)	400	460
Running (15 km/h)	1000	1150
Bicycling (race)	1100	1270

EXAMPLE 15–7 **Energy transformation in the body.** How much energy is transformed in 24 h by a 65-kg person who spends 8.0 h sleeping, 1.0 h at moderate physical labor, 4.0 h in light activity, and 11.0 h working at a desk or relaxing?

SOLUTION Table 15–1 gives the metabolic rate in watts (J/s). Since there are 3600 s in an hour, the total energy transformed is

$$\left[\begin{array}{l} (8.0\,\text{h})(70\,\text{J/s}) + (1.0\,\text{h})(460\,\text{J/s}) \\ + (4.0\,\text{h})(230\,\text{J/s}) + (11.0\,\text{h})(115\,\text{J/s}) \end{array} \right](3600\,\text{s/h}) = 1.15 \times 10^7\,\text{J}.$$

Since 4.19×10^3 J = 1 kcal, this is equivalent to 2800 kcal; so a food intake of 2800 Cal would compensate for this energy output. A person who wanted to lose weight would have to eat less than 2800 Cal a day, or increase his or her level of activity.

15–4 The Second Law of Thermodynamics—Introduction

The first law of thermodynamics states that energy is conserved. There are, however, many processes we can imagine that conserve energy but are not observed to occur in nature. For example, when a hot object is placed in contact with a cold object, heat flows from the hotter one to the colder one, never spontaneously the reverse. If heat were to leave the colder object and pass to the hotter one, energy could still be conserved. Yet it doesn't happen spontaneously.[†] As a second example, consider what happens when you drop a rock and it hits the ground. The initial potential energy of the rock changes to kinetic energy as the rock falls, and when the rock hits the ground this energy in turn is transformed into internal energy of the rock and the ground in the vicinity of the impact; the molecules move faster and the temperature rises slightly. But have you seen the reverse happen—a rock at rest on the ground suddenly rise up in the air because the thermal energy of molecules is transformed into kinetic energy of the rock as a whole? Energy could be conserved in this process, yet we never see it happen.

There are many other examples of processes that occur in nature but whose reverse does not. Here are two. (1) If you put a layer of salt in a jar and cover it with a layer of similar-sized grains of pepper, when you shake it you get a thorough mixture. But no matter how long you shake it, the mixture is unlikely to separate into two layers again. (2) Coffee cups and glasses break spontaneously if you drop them. But they don't go back together spontaneously (Fig. 15–8).

The first law of thermodynamics, conservation of energy, would not be violated if any of these processes occurred in reverse. To explain this lack of reversibility, scientists in the latter half of the nineteenth century came to formulate a new principle known as the **second law of thermodynamics**. This law is a statement about which processes occur in nature and which

[†]By spontaneously, we mean by itself without input of work of some sort. (A refrigerator does move heat from a cold environment to a warmer one, but only by doing work.)

(a)

(b)

(c)

FIGURE 15-8 Have you ever observed this process (left to right)?

do not. It can be stated in a variety of ways, all of which are equivalent. One statement, due to R. J. E. Clausius (1822–1888), is that

> **heat flows naturally from a hot object to a cold object; heat will not flow spontaneously from a cold object to a hot object.**

SECOND LAW OF THERMODYNAMICS (Clausius statement)

Since this statement applies to one particular process, it is not obvious how it applies to other processes. A more general statement is needed that will include other possible processes in a more obvious way.

The development of a general statement of the second law was based partly on the study of heat engines. A **heat engine** is any device that changes thermal energy into mechanical work, such as steam engines and automobile engines. We now examine heat engines, both from a practical point of view and to show their importance in developing the second law of thermodynamics.

15–5 Heat Engines

It is easy to produce thermal energy by doing work—for example, by simply rubbing your hands together briskly, or indeed by any frictional process. But to get work from thermal energy is more difficult, and the invention of a practical device to do this came only about 1700 with the development of the steam engine.

The basic idea behind any heat engine is that mechanical energy can be obtained from thermal energy only when heat is allowed to flow from a high temperature to a low temperature. In the process, some of the heat can then be transformed to mechanical work, as diagrammed schematically in Fig. 15–9. That is, a heat input Q_H at a high temperature T_H is partly transformed into work W and partly exhausted as heat Q_L at a lower temperature T_L. By conservation of energy, $Q_H = W + Q_L$. The high and low temperatures, T_H and T_L, are called the **operating temperatures** of the engine. We will be interested only in engines that run in a repeating *cycle* (that is, the system returns repeatedly to its starting point) and thus can run continuously. [Note carefully that we are now using a new sign convention: we take Q_H, Q_L, and W as always positive. Their directions are found from the applicable diagram, such as Fig. 15–9.]

The operation of two practical engines, the steam engine and the internal combustion engine (used in most automobiles), is illustrated in

FIGURE 15–9 Schematic diagram of energy transfers for a heat engine.

High temperature

Steam

Water

Boiler

Intake valve
(open during
expansion)

Exhaust valve
(closed during
expansion)

Pump

Water

Condenser

Low temperature

High-pressure steam,
from boiler

Low-pressure steam,
exhausted to condenser

FIGURE 15–10
Steam engines.

(a) Reciprocating type

(b) Turbine (boiler and
condenser not shown)

➡ **PHYSICS APPLIED**

Engines

Figs. 15–10 and 15–11. Steam engines are of two main types, each making use of steam heated by combustion of coal, oil, or gas (or nuclear energy). In the so-called reciprocating type, Fig. 15–10a, the heated steam passes through the intake valve and expands against a piston, forcing it to move. As the piston returns to its original position, it forces the gases out the exhaust valve. In a steam turbine, Fig. 15–10b, everything is essentially the same, except that the reciprocating piston is replaced by a rotating turbine that resembles a paddlewheel with many sets of blades. Most of our electricity today is generated using steam turbines. In a steam engine, the high temperature is obtained by burning coal, oil, or other fuel to heat the steam. In an internal combustion engine, the high temperature is achieved by burning the gasoline–air mixture in the cylinder itself (ignited by the spark plug), Fig. 15–11.

FIGURE 15–11 (below)
Four-cycle internal combustion engine: (a) the gasoline–air mixture flows into the cylinder as the piston moves down; (b) the piston moves upward and compresses the gas; (c) firing of the spark plug ignites the gasoline–air mixture, raising it to a high temperature; (d) the gases, now at high temperature and pressure, expand against the piston in this, the power stroke; (e) the burned gases are pushed out to the exhaust pipe; the intake valve then opens, and the whole cycle repeats.

Intake valve (open)   Exhaust valve (closed)

Gas-air mixture from carburetor

Connecting rod

Crankshaft

Both valves closed

Cylinder

Piston

Both valves closed

Spark plug (firing)

Both valves closed

Rings

Exhaust valve (open)

To exhaust pipe

(a) Intake

(b) Compression

(c) Ignition

(d) Expansion
(power stroke)

(e) Exhaust

To see why a *temperature difference* is required to run an engine, let us examine the steam engine. In the reciprocating engine, for example (Fig. 15–10a), suppose there were no condenser or pump, and that the steam was at the same temperature throughout the system. This would mean that the pressure of the gas being exhausted would be the same as that on intake. Thus, although work would be done by the gas *on* the piston when it expanded, an equal amount of work would have to be done *by* the piston to force the steam out the exhaust; hence, no net work would be done. In a real engine, the exhausted gas is cooled to a lower temperature and condensed so that the exhaust pressure is less than the intake pressure. Thus, although the piston must do work on the gas to expel it on the exhaust stroke, it is less than the work done by the gas on the piston during the intake. So a net amount of work can be obtained—but only if there is a difference of temperature. Similarly, in the gas turbine if the gas were not cooled, the pressure on each side of the blades would be the same; by cooling the gas on the exhaust side, the pressure on the front side of the blade is greater and hence the turbine turns.

The efficiency, e, of any heat engine can be defined as the ratio of the work it does, W, to the heat input at the high temperature, Q_H (Fig. 15–9):

$$e = \frac{W}{Q_H}.$$

This is a sensible definition since W is the output (what you get from the engine), whereas Q_H is what you put in and pay for in burned fuel. Since energy is conserved, the heat input Q_H must equal the work done plus the heat that flows out at the low temperature (Q_L):

$$Q_H = W + Q_L.$$

Thus $W = Q_H - Q_L$, and the efficiency of an engine is

$$e = \frac{W}{Q_H} = \frac{Q_H - Q_L}{Q_H} = 1 - \frac{Q_L}{Q_H}. \qquad (15\text{–}3)$$

Note that to give the efficiency as a percent, Eq. 15–3 must be multiplied by 100.

EXAMPLE 15–8 **Car efficiency.** An automobile engine has an efficiency of 20 percent and produces an average of 23,000 J of mechanical work per second during operation. How much heat is discharged from this engine per second?

SOLUTION The output heat is Q_L. We are given $e = 0.20$, so from Eq. 15–3 we find that

$$\frac{Q_L}{Q_H} = 1 - e = 0.80.$$

We also know that $e = W/Q_H$, by definition, so in one second

$$Q_H = \frac{W}{e} = \frac{23{,}000 \text{ J}}{0.20} = 1.15 \times 10^5 \text{ J}.$$

Thus

$$Q_L = 0.80 Q_H = (0.80)(1.15 \times 10^5 \text{ J}) = 9.2 \times 10^4 \text{ J}.$$

The engine discharges 9.2×10^4 J/s = 92,000 watts.

FIGURE 15–12 The Carnot cycle. Heat engines work in a cycle, and the cycle for the Carnot engine begins at point a on this PV diagram. (1) The gas is first expanded isothermally, with the addition of heat Q_H, along the path ab at temperature T_H. (2) Next the gas expands adiabatically from b to c—no heat is exchanged, but the temperature drops to T_L. (3) The gas is then compressed at constant temperature T_L, path c to d, and heat Q_L flows out. (4) Finally, the gas is compressed adiabatically, path da, back to its original state. No Carnot engine actually exists, but as a theoretical engine it played an important role in the development of thermodynamics.

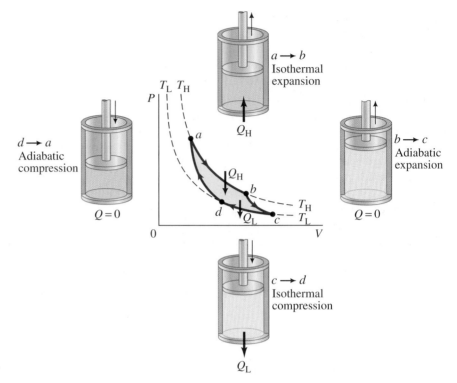

To see how to increase efficiency, the French scientist Sadi Carnot (1796–1832) examined the characteristics of an ideal engine (now called a **Carnot engine**). Each of the processes of heat addition and exhaust, of gas expansion or compression, were considered to be done **reversibly**. That is, each of the processes (say, during expansion of the gases against a piston) was done so slowly that the process could be considered a series of equilibrium states, and the whole process could be done in reverse with no change in the magnitude of work done or heat exchanged. A real process, on the other hand, would occur more quickly; there would be turbulence present in the gas, friction would be present, and so on. Because of these factors, a real process cannot be done precisely in reverse—the turbulence would be different and the heat lost to friction would not reverse itself. Thus, real processes are called **irreversible**. An important result is that for an idealized reversible engine, the heats Q_H and Q_L are proportional to the operating temperatures T_H and T_L (in kelvins) so that the efficiency can be written as:

Carnot (ideal) engine

$$e_{\text{ideal}} = \frac{T_H - T_L}{T_H} = 1 - \frac{T_L}{T_H}. \qquad \begin{bmatrix} \text{Carnot (ideal)} \\ \text{efficiency} \end{bmatrix} \quad \textbf{(15–4)}$$

This is the theoretical limit to the efficiency. Real engines cannot have an efficiency even this high because of losses due to friction and the like. Real engines that are well designed reach 60 to 80 percent of the Carnot efficiency. [Figure 15–12 shows the operating cycle of a Carnot engine.]

EXAMPLE 15–9 **Steam engine efficiency.** A steam engine operates between 500°C and 270°C. What is the maximum possible efficiency of this engine?

SOLUTION We must first change the temperature to kelvins. Thus, $T_H = 773$ K and $T_L = 543$ K. Then from Eq. 15–4,

$$e_{ideal} = 1 - \frac{543}{773} = 0.30.$$

To get the efficiency in percent, we multiply by 100. Thus, the maximum (or Carnot) efficiency is 30 percent. Realistically, an engine might attain 0.70 of this value or 21 percent. Note in this Example that the exhaust temperature is still rather high, 270°C. Steam engines are often arranged in series so that the exhaust of one engine is used as intake by a second or third engine.

EXAMPLE 15–10 **A phony claim?** An engine manufacturer makes the following claims: The heat input per second of the engine is 9.0 kJ at 375 K. The heat output per second is 4.0 kJ at 225 K. Do you believe these claims?

SOLUTION The efficiency of the engine is (Eq. 15–3)

$$e = \frac{Q_H - Q_L}{Q_H} = \frac{9.0\,\text{kJ} - 4.0\,\text{kJ}}{9.0\,\text{kJ}} = 0.56.$$

However, the maximum possible efficiency is given by the Carnot efficiency, Eq. 15–4:

$$e_{ideal} = \frac{T_H - T_L}{T_H} = \frac{375\,\text{K} - 225\,\text{K}}{375\,\text{K}} = 0.40.$$

So the manufacturer's claims violate the second law of thermodynamics and cannot be believed.

It is quite clear from Eq. 15–4 that at normal temperatures, a 100 percent efficient engine is not possible. Only if the exhaust temperature, T_L, were at absolute zero could 100 percent efficiency be obtained. But reaching absolute zero is a practical (as well as theoretical) impossibility.[†] Thus it is possible to state that

no device is possible whose sole effect is to transform a given amount of heat completely into work.

That is, there can be no perfect (100 percent efficient) heat engine such as the one diagrammed in Fig. 15–13. This is another way of expressing the *second law of thermodynamics*, and is known as the **Kelvin-Planck statement of the second law of thermodynamics**. If the second law were not true, so that a perfect engine could be built, some rather remarkable things could happen. For example, if the engine of a ship did not need a low-temperature reservoir to exhaust heat into, the ship could sail across the ocean using the vast resources of the internal energy of the ocean water. Indeed, we would have no fuel problems at all!

[†]It seems, from careful experimentation, that absolute zero is unattainable. This result is known as the **third law of thermodynamics**.

FIGURE 15–13 Diagram of hypothetical perfect heat engine in which all the heat input is used to do work.

15–6 Refrigerators, Air Conditioners, and Heat Pumps

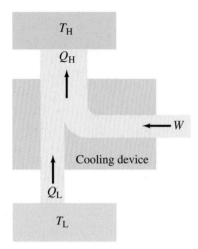

FIGURE 15–14 Schematic diagram of energy transfers for a refrigerator or air conditioner.

The operating principle of refrigerators, air conditioners, and heat pumps is just the reverse of a heat engine. Each operates to transfer heat out of a cool environment into a warm environment. As diagrammed in Fig. 15–14, by doing work W, heat is taken from a low-temperature region, T_L (inside a refrigerator, say), and a greater amount of heat is exhausted at a high temperature, T_H (the room). You can often feel this heat blowing out beneath a refrigerator. The work W is usually done by a compressor motor which compresses a fluid, as illustrated in Fig. 15–15.

A perfect refrigerator—one in which no work is required to take heat from the low-temperature region to the high-temperature region—is not possible. This is **the Clausius statement of the second law of thermodynamics**, which we already mentioned in Section 15–4: heat does not flow spontaneously from a cold object to a hot object. To accomplish such a task, work must be done. Thus, *there can be no perfect refrigerator.*

The coefficient of performance (CP) of a refrigerator is defined as the heat Q_L removed from the low-temperature area (inside a refrigerator) divided by the work W done to remove the heat (Fig. 15–14):

$$\text{CP} = \frac{Q_L}{W}. \qquad (15\text{–}5a)$$

This makes sense since the more heat, Q_1, that can be removed from the

FIGURE 15–15 Typical refrigerator system. The compressor motor forces a gas at high pressure through a heat exchanger (condenser) on the outside walls of the refrigerator where Q_H is given off and the gas cools to become liquid. The liquid passes from a high-pressure region, via a valve, to low-pressure tubes on the inside walls of the refrigerator; the liquid evaporates at this lower pressure and thus absorbs heat (Q_L) from the inside of the refrigerator. The fluid returns to the compressor where the cycle begins again.

inside of the refrigerator for a given amount of work, the better (more efficient) the refrigerator. Energy is conserved, so from the first law of thermodynamics we can write (see Fig. 15–14) $Q_L + W = Q_H$, or $W = Q_H - Q_L$. Then Eq. 15–5a becomes

$$\text{CP} = \frac{Q_L}{W} = \frac{Q_L}{Q_H - Q_L}. \qquad \textbf{(15–5b)}$$

For an ideal refrigerator (not a perfect one, which is impossible), the best one could do would be

$$\text{CP}_{\text{ideal}} = \frac{T_L}{T_H - T_L}, \qquad \textbf{(15–5c)}$$

as for an ideal (Carnot) engine (Eq. 15–4).

An air conditioner works very much like a refrigerator, although the actual construction details are different because an air conditioner takes heat Q_L from inside a room or building at a low temperature, and deposits heat Q_H outside to the environment at a higher temperature. Equations 15–5 also describe the coefficient of performance for an air conditioner.

➡ **PHYSICS APPLIED**

Air conditioner

Heat naturally flows from high temperature to low temperature. Refrigerators and air conditioners do work to accomplish the opposite: to make heat flow from cold to hot. We might say they "pump" heat from cold areas to hotter areas, against the natural tendency of heat to flow from hot to cold, just as water can be pumped uphill, against the natural tendency to flow downhill. The term **heat pump** is usually reserved for a device that can heat a house in winter by taking heat Q_L from the outside at low temperature and delivering heat Q_H to the warmer inside of the house, by doing work W; see Fig. 15–16. The operating principle is like that for a refrigerator or air conditioner; but the objective of a heat pump is to heat (deliver Q_H), rather than to cool (remove Q_L). Thus, the coefficient of performance of a heat pump is defined differently than for an air conditioner because it is the heat Q_H delivered to the inside of the house that is important now:

➡ **PHYSICS APPLIED**

Heat pump

$$\text{CP} = \frac{Q_H}{W}. \qquad \textbf{(15–6)}$$

Most heat pumps can, however, be "turned around" and used as air conditioners in the summer.

Cold outside   Inside of house

FIGURE 15–16 A heat pump "pumps" heat from the cold outside to the warm inside of a house.

EXAMPLE 15–11 **Heat pump.** A heat pump has a coefficient of performance of 3.0 and is rated to do work at 1500 W. (*a*) How much heat can it add to a room per second? (*b*) If the heat pump were turned around to act as an air conditioner in the summer, what would you expect its coefficient of performance to be, assuming all else stays the same?

SOLUTION (*a*) We use Eq. 15–6 for the heat pump and, since our device does 1500 J of work per second, it can pour heat into the room at a rate of

$$Q_H = CP \times W = 3.0 \times 1500 \text{ J} = 4500 \text{ J}$$

per second, or at a rate of 4500 W.
(*b*) If our device is turned around in summer, it can take heat Q_L from inside the house and do 1500 J of work per second, and then dumps $Q_H = 4500$ J per second to the hot outside. Energy is conserved, so $Q_L + W = Q_H$ (see Fig. 15–16, but reverse inside and outside of house), and

$$Q_L = Q_H - W = 4500 \text{ J} - 1500 \text{ J} = 3000 \text{ J}.$$

The coefficient of performance as an air conditioner would thus be (Eq. 15–5a)

$$CP = \frac{Q_L}{W} = \frac{3000 \text{ J}}{1500 \text{ J}} = 2.0.$$

A good heat pump can be a money saver and an energy saver. Compare, for example, our heat pump in this Example to, say, a 1500-watt electric heater. We plug the latter into the wall, it draws 1500 watts of electricity and delivers 1500 watts of heat to the room. Our heat pump when plugged into the wall also draws 1500 W of electricity (which is what we pay for), but it delivers 4500 W of heat!

15–7 Entropy and the Second Law of Thermodynamics

We have seen several aspects of the second law of thermodynamics; and the different statements of it that we have discussed can be shown to be completely equivalent. But what we would really like is a general statement of the second law of thermodynamics. It was not until the latter half of the nineteenth century that the *second law of thermodynamics* was finally stated in a general way, namely in terms of a quantity called **entropy**, introduced by Clausius in the 1860s. Entropy, unlike heat, is a function of the state of a system. In the next section, we will see that entropy can be interpreted as a measure of the order or disorder of a system.

When we deal with entropy—as with potential energy—it is the *change* in entropy during a process that is important, not the absolute amount. According to Clausius, the change in entropy S of a system, when an amount of heat Q is *added* to it by a reversible process at constant temperature, is given by

Entropy change
$$\Delta S = \frac{Q}{T},$$ (15–7)

where T is the Kelvin temperature.

EXAMPLE 15-12 **Entropy change in melting.** An ice cube of mass 60 g is taken from a storage compartment at 0°C and placed in a paper cup. After a few minutes, exactly half of the mass of the ice cube has melted, becoming water at 0°C. Find the change in entropy of the ice/water.

SOLUTION The heat required to melt 30 g of ice (half of the ice cube) is found from the latent heat of fusion (see Section 14–6):

$$Q = mL = (30 \text{ g})(79.7 \text{ cal/g}) = 2400 \text{ cal}$$

or 2.4 kcal. Since the temperature remains constant in our process, the change in entropy can be found from Eq. 15–7,

$$\Delta S = \frac{Q}{T} = \frac{2400 \text{ cal}}{273 \text{ K}} = +8.8 \text{ cal/K}.$$

Note that the change in entropy of the surroundings (cup, air) has not been computed.

The temperature in Example 15–12 was constant, and the calculation easy. If the temperature varies during a process, a summation of the heat flow over the changing temperature can often be calculated using calculus or a computer. However, if the temperature change is not too great, a reasonable approximation can be made using the average value of the temperature, as indicated in the next Example.

EXAMPLE 15-13 **ESTIMATE** **Entropy change when mixing water.** A sample of 50.0 kg of water at 20.0°C is mixed with 50.0 kg of water at 24.0°C. Estimate the change in entropy.

SOLUTION The final temperature of the mixture will be 22.0°C, since we started with equal amounts of water. A quantity of heat,

$$Q = mc \, \Delta T = (50.0 \text{ kg})(1.00 \text{ kcal/kg·K})(2.0 \text{ C°}) = 100 \text{ kcal},$$

flows out of the hot water as it cools down from 24°C to 22°C, and this heat flows into the cold water as it warms from 20°C to 22°C. The total change in entropy, ΔS, will be the sum of the changes in entropy of the hot water, ΔS_H, and that of the cold water, ΔS_C:

$$\Delta S = \Delta S_H + \Delta S_C.$$

We now make use of Eq. 15–7; for the hot water we use an average temperature of 23.0°C (296 K), and for the cold water an average temperature of 21.0°C (294 K). Thus

$$\Delta S_H = -\frac{100 \text{ kcal}}{296 \text{ K}} = -0.338 \text{ kcal/K}$$

and

$$\Delta S_C = \frac{100 \text{ kcal}}{294 \text{ K}} = 0.340 \text{ kcal/K}.$$

Note that the entropy of the hot water (S_H) decreases since heat flows out of the hot water. But the entropy of the cold water (S_C) increases by a greater amount. The total change in entropy is

$$\Delta S = \Delta S_H + \Delta S_C = -0.338 \text{ kcal/K} + 0.340 \text{ kcal/K}$$

$$= +0.002 \text{ kcal/K}.$$

In this last Example (15–13), we see that although the entropy of one part of the system decreased, the entropy of the other part increased by a greater amount so that the net change in entropy of the whole system is positive. This result, which we have calculated for a specific case in Example 15–13, has been found to hold in all other cases tested. That is, the total entropy is found to increase in all natural processes. The second law of thermodynamics can be stated in terms of entropy as: *The entropy of an isolated system never decreases. It can only stay the same or increase.* Entropy can remain the same only for an idealized (reversible) process. For any real process, the change in entropy ΔS is greater than zero:

Entropy of isolated system never decreases

$$\Delta S > 0. \qquad \qquad (15\text{--}8)$$

If the system is not isolated, then the change in entropy of the system, ΔS_s, plus the change in entropy of the environment, ΔS_{env}, must be greater than or equal to zero:

$$\Delta S = \Delta S_s + \Delta S_{env} \geq 0. \qquad \qquad (15\text{--}9)$$

Only idealized processes have $\Delta S = 0$. Real processes have $\Delta S > 0$. This, then, is the *general statement of the second law of thermodynamics*:

SECOND LAW OF THERMODYNAMICS (general statement)

The total entropy of any system plus that of its environment increases as a result of any natural process.

Although the entropy of one part of the universe may decrease in any process (see Example 15–13), the entropy of some other part of the universe always increases by a greater amount, so the total entropy always increases.

Now that we finally have a quantitative general statement of the second law of thermodynamics, we can see that it is an unusual law. It differs considerably from other laws of physics, which are typically equalities (such as $F = ma$) or conservation laws (such as for energy and momentum). The second law of thermodynamics introduces a new quantity, the entropy, S, but does not tell us it is conserved. Quite the opposite. Entropy is not conserved in natural processes; it always increases in time.

15–8 Order to Disorder

The concept of entropy, as we have discussed it so far, may seem rather abstract. To get a feel for the concept of entropy, we can relate it to the concepts of *order* and *disorder*. In fact, the entropy of a system can be considered a *measure of the disorder of the system*. Then the second law of thermodynamics can be stated simply as:

SECOND LAW OF THERMODYNAMICS (general statement)

Natural processes tend to move toward a state of greater disorder.

Exactly what we mean by disorder may not always be clear, so we now consider a few examples. Some of these will show us how this very general statement of the second law actually applies beyond what we usually consider as thermodynamics.

Let us first look at the simple processes mentioned in Section 15–4. A jar containing separate layers of salt and pepper is more orderly than when the salt and pepper are all mixed up. Shaking a jar containing separate layers results in a mixture, and no amount of shaking brings the orderly layers back again. The natural process is from a state of relative order (layers) to one of relative disorder (a mixture), not the reverse. That is, disorder increases. Similarly, a solid coffee cup is a more "orderly" object than the pieces of a broken cup. Cups break when they fall, but they do not spontaneously mend themselves. Again, the normal course of events is an increase of disorder.

When a hot object is put in contact with a cold object, heat flows from the high temperature to the low until the two objects reach the same intermediate temperature. At the beginning of the process we can distinguish two classes of molecules: those with a high average kinetic energy and those with a low average kinetic energy. After the process, all the molecules are in one class with the same average kinetic energy, and we no longer have the more orderly arrangement of molecules in two classes. Order has gone to disorder. Furthermore, note that the separate hot and cold objects could serve as the hot- and cold-temperature regions of a heat engine, and thus could be used to obtain useful work. But once the two objects are put in contact and reach the same temperature, no work can be obtained. Disorder has increased, since a system that has the ability to perform work must surely be considered to have a higher order than a system no longer able to do work.

These examples illustrate the general concept that an increase in entropy corresponds to an increase in disorder. (We discuss this connection quantitatively in Section 15–11.) In general, we associate disorder with randomness: salt and pepper in layers is more orderly than a random mixture; a neat stack of numbered pages is more orderly than pages strewn randomly about on the floor. We can also say that a more orderly arrangement is one that requires more **information** to specify or classify it. When we have one hot and one cold body, we have two classes of molecules and two pieces of information; when the two bodies come to the same temperature, there is only one class and one piece of information. When salt and pepper are mixed, there is only one (uniform) class; when they are in layers, there are two classes. In this sense, information is connected to order, or low entropy. This is the foundation upon which the modern field of **information theory** is built.

The remaining example of those we discussed earlier is that of a stone falling to the ground, its kinetic energy being transformed to thermal energy. (We noted that the reverse never happens; a stone never absorbs thermal energy and rises into the air of its own accord.) This is another example of order changing to disorder. For although thermal energy is associated with the disorderly random motion of molecules, the molecules in the falling stone all have the same velocity downward in addition to their own random velocities. Thus, the more orderly kinetic energy of the stone is changed to disordered thermal energy when it strikes the ground. Disorder increases in this process, as it does in all processes that occur in nature.

15–9   Unavailability of Energy; Heat Death

In the process of heat conduction from a hot body to a cold one, we have seen that entropy increases and that order goes to disorder. The separate hot and cold objects could serve as the high- and low-temperature regions for a heat engine and thus could be used to obtain useful work. But after the two objects are put in contact with each other and reach the same uniform temperature, no work can be obtained from them. With regard to being able to do useful work, order has gone to disorder in this process.

The same can be said about a falling rock that comes to rest upon striking the ground. Just before hitting the ground, all the kinetic energy of the rock could have been used to do useful work. But once the rock's mechanical kinetic energy becomes thermal energy, this is no longer possible.

Both these examples illustrate another important aspect of the second law of thermodynamics—*in any natural process, some energy becomes unavailable to do useful work*. In any process, no energy is ever lost (it is always conserved). Rather, it becomes less useful—it can do less useful work. As time goes on, **energy is degraded**, in a sense; it goes from more orderly forms (such as mechanical) eventually to the least orderly form, internal or thermal energy. Entropy is a factor here because the amount of energy that becomes unavailable to do work is proportional to the change in entropy during any process.

Degradation of energy

A natural outcome of this is the prediction that as time goes on, the universe will approach a state of maximum disorder. Matter will become a uniform mixture, heat will have flowed from high-temperature regions to low-temperature regions until the whole universe is at one temperature. No work can then be done. All the energy of the universe will have become degraded to thermal energy. All change will cease. This, the so-called **heat death** of the universe, has been much discussed by philosophers. This final state seems an inevitable consequence of the second law of thermodynamics, although it lies very far in the future. Yet it is based on the assumption that the universe is finite, which cosmologists are not really sure of. Furthermore, there is some question as to whether the second law of thermodynamics, as we know it, actually applies in the vast reaches of the universe. The answers are not yet in.

"Heat death"

* 15–10 Evolution and Growth; "Time's Arrow"

➡ PHYSICS APPLIED

Biological evolution

An interesting example of the increase in entropy relates to biological evolution and to growth of organisms. Clearly, a human being is a highly ordered organism. The process of evolution from the early macromolecules and simple forms of life to *Homo sapiens* is a process of increasing order. So, too, the development of an individual from a single cell to a grown person is a process of increasing order. Do these processes violate the second law of thermodynamics? No, they do not. In the processes of evolution and growth, and even during the mature life of an individual, waste products are eliminated. These small molecules that remain as a result of metabolism are simple molecules without much order. Thus they represent relatively higher disorder or entropy. Indeed, the total entropy of the molecules cast aside by organisms during the processes of evolution and growth is greater than the decrease in entropy associated with the order of the growing individual or evolving species.

Another aspect of the second law of thermodynamics is that it tells us in which *direction* processes go. If you were to see a film being run backward, you would undoubtedly be able to tell that it *was* run backward. For you would see odd occurrences, such as a broken coffee cup rising from the floor and reassembling on a table, or a torn balloon suddenly becoming whole again and filled with air. We know these things don't happen in real life; they are processes in which order increases—or entropy decreases. They violate the second law of thermodynamics. When watching a movie (or imagining that time could go backward), we are tipped off to a reversal of time by observing whether entropy is increasing or decreasing. Hence, entropy has been called **time's arrow**, for it can tell us in which direction time is going.

*15-11 Statistical Interpretation of Entropy and the Second Law

The ideas of entropy and disorder are made clearer with the use of a statistical or probabilistic analysis of the molecular state of a system. This statistical approach, which was first applied toward the end of the nineteenth century by Ludwig Boltzmann (1844–1906), makes a clear distinction between the "macrostate" and the "microstate" of a system. The **microstate** of a system would be specified when the position and velocity of every particle (or molecule) is given. The **macrostate** of a system is specified by giving the macroscopic properties of the system—the temperature, pressure, number of moles, and so on. In reality, we can know only the macrostate of a system. There are generally far too many molecules in a system to be able to know the velocity and position of every one at a given moment. Nonetheless, it is important to recognize that a great many different microstates can correspond to the *same* macrostate.

Let us take a simple example. Suppose you repeatedly shake four coins in your hand and drop them on the table. Specifying the number of heads and the number of tails that appear on a given throw is the macrostate of this system. Specifying each coin as being a head or a tail is the microstate of the system. In the following table we see the number of microstates that correspond to each macrostate:

Macrostate	Possible Microstates (H = heads, T = tails)	Number of Microstates
4 heads	H H H H	1
3 heads, 1 tail	H H H T, H H T H, H T H H, T H H H	4
2 heads, 2 tails	H H T T, H T H T, T H H T, H T T H, T H T H, T T H H	6
1 head, 3 tails	T T T H, T T H T, T H T T, H T T T	4
4 tails	T T T T	1

A basic assumption behind the statistical approach is that *each microstate is equally probable*. Thus the number of microstates that give the same macrostate corresponds to the relative probability of that macrostate occurring. The macrostate of two heads and two tails is the most probable one in our case of tossing four coins; out of the total of 16 possible microstates, six correspond to two heads and two tails, so the probability of throwing two heads and two tails is 6 out of 16, or 38 percent. The probability of throwing one head and three tails is 4 out of 16, or 25 percent. The probability of four heads is only 1 in 16, or 6 percent. Of course if you threw the coins 16 times, you might not find that two heads and two tails appear exactly 6 times, or four tails exactly once. These are only probabilities or averages. But if you made 1600 throws, very nearly 38 percent of them would be two heads and two tails. The greater the number of tries, the closer are the percentages to the calculated probabilities.

If we consider tossing more coins, say 100 all at the same time, the relative probability of throwing all heads (or all tails) is greatly reduced. There is only one microstate corresponding to all heads. For 99 heads and 1 tail, there are 100 microstates since each of the coins could be the one tail. The relative

TABLE 15–2
Probabilities of Various Macrostates for 100 Coin Tosses

Macrostate		Number of Microstates, W
Heads	Tails	
100	0	1
99	1	1.0×10^2
90	10	1.7×10^{13}
80	20	5.4×10^{20}
60	40	1.4×10^{28}
55	45	6.1×10^{28}
50	50	1.0×10^{29}
45	55	6.1×10^{28}
40	60	1.4×10^{28}
20	80	5.4×10^{20}
10	90	1.7×10^{13}
1	99	1.0×10^2
0	100	1

FIGURE 15–17 (a) Most probable distribution of molecular speeds in a gas (Maxwellian, or random); (b) orderly, but highly unlikely, distribution of speeds in which all molecules have nearly the same speed.

(a)

(b)

probabilities for other macrostates are given in Table 15–2. There are a total of about 10^{30} microstates possible.[†] Thus the relative probability of finding all heads is 1 in 10^{30}, an incredibly unlikely event! The probability of obtaining 50 heads and 50 tails (see Table 15–2) is $1.0 \times 10^{29}/10^{30} = 0.10$, or 10 percent. The probability of obtaining between 45 and 55 heads is 90 percent.

Thus we see that as the number of coins increases, the probability of obtaining the most orderly arrangement (all heads or all tails) becomes extremely unlikely. The least orderly arrangement (half heads, half tails) is the most probable and the probability of being within a certain percentage (say, 5 percent) of the most probable arrangement greatly increases as the number of coins increases. These same ideas can be applied to the molecules of a system. For example, the most probable state of a gas (say, the air in a room) is one in which the molecules take up the whole space and move about randomly; this corresponds to the Maxwellian distribution, Fig. 15–17a (and see Chapter 13). On the other hand, the very orderly arrangement of all the molecules located in one corner of the room and all moving with the same velocity (Fig. 15–17b) is extremely unlikely.

From these examples, it is clear that probability is directly related to disorder and hence to entropy. That is, the most probable state is the one with greatest entropy, or greatest disorder and randomness.

In terms of probability, the second law of thermodynamics—which tells us that entropy increases in any process—reduces to the statement that those processes occur which are most probable. The second law thus becomes a trivial statement. However, there is an additional element now. The second law in terms of probability does not *forbid* a decrease in entropy. Rather, it says the probability is extremely low. It is not impossible that salt and pepper should separate spontaneously into layers, or that a broken teacup should mend itself. It is even possible that a lake should freeze over on a hot summer day (that is, for heat to flow out of the cold lake into the warmer surroundings). But the probability for such events occurring is extremely small. In our coin examples, we saw that increasing the number of coins from 4 to 100 reduced drastically the probability of large deviations from the average, or most probable, arrangement. In ordinary systems, we are dealing not with 100 molecules, but with incredibly large numbers of molecules: in 1 mole alone there are 6×10^{23} molecules. Hence the probability of deviation far from the average is incredibly tiny. For example, it has been calculated that the probability that a stone resting on the ground could transform 1 cal of thermal energy into mechanical energy and rise up into the air, is much less likely than the probability that a group of monkeys typing randomly would by chance produce the complete works of Shakespeare.

* 15–12 Energy Resources: Thermal Pollution

When we speak in everyday life of energy usage, we are speaking of the *transformation* of energy from one form to another form that is more useful to us. For space heating of homes and buildings, the direct burning of

[†]Each coin has two possibilities, either heads or tails. Then the possible number of microstates is $2 \times 2 \times 2 \times \cdots = 2^{100}$, which, using a calculator, gives 1.27×10^{30}.

fuels such as gas, oil, or coal releases energy stored as potential energy of the molecules. For many transformations, however, a *heat engine* is required, such as to run cars, aircraft, and other vehicles and, very importantly, to generate electricity.

Before discussing various means for producing electricity, we first look at two types of pollution associated with heat engines: air pollution and thermal pollution. Air pollution can result from burning any fossil fuels (coal, oil, gas) such as in cars, in industrial furnaces for smelting, and in electric generating plants. The internal combustion engines of automobiles are especially polluting because the burning takes place so quickly that complete combustion does not take place and more noxious gases are thus produced. To help reduce air pollution, special devices are used (such as catalytic converters). Even when combustion is complete, the CO_2 thrown into the atmosphere absorbs some of the natural infrared radiation emitted by the warm Earth that would otherwise escape. The buildup of atmospheric CO_2, and the consequent heating of the atmosphere is referred to as the **greenhouse effect** since a similar effect is partly responsible for warming an enclosed greenhouse; and it is projected to raise the average temperature of the atmosphere by several degrees C within the next century, causing shifts in rainfall patterns, melting of polar ice caps that would raise sea levels (thus flooding low-lying areas), and turning forests into deserts. An alternative scenario, equally troubling, projects a layer of CO_2 blocking some of the Sun's rays and causing a new Ice Age. In either case, experts agree on the urgency for conservation: limit the burning of fossil fuels.

Another type of environmental pollution is **thermal pollution**. Every heat engine, from automobiles to power plants, exhausts heat to the environment (Q_L in Fig. 15–9). Most electricity-producing power plants today use a heat engine to transform thermal energy into electricity, and the exhaust heat is generally absorbed by a coolant such as water. If the engine is run efficiently (at best, 30 to 40 percent today), the temperature T_L (see Eq. 15–4) must be kept as low as possible. Hence a great deal of water must flow as coolant through a power plant. The water is usually obtained from a nearby river or lake, or from the ocean. As a result of the transfer of heat to the water, its temperature rises, which can cause significant damage to aquatic life, in large part because the warmed water holds less dissolved oxygen. Another way that heat is exhausted is by discharging it into the atmosphere by means of large cooling towers, Fig. 15–18. Unfortunately, this method can also have environmental effects, for the heated air can alter the weather of a region. Although careful controls may eventually reduce air pollution to an acceptable level, thermal pollution cannot be avoided. What we can do, in light of the second law of thermodynamics, is use less energy and try to build more efficient engines.

Most of the electricity produced in the U.S. at the present time makes use of a heat engine coupled with an electric generator. Electric generators are devices that transform mechanical energy, usually rotational kinetic energy of turbines containing many blades (Fig. 15–19), into electric energy. Details of how electric generators transform mechanical energy to electrical energy will be discussed in Chapter 21. The various means to turn the turbine will be discussed now, along with some of the advantages and disadvantages of each.

FIGURE 15–18 Large cooling towers at an electric generating plant.

⇒ **PHYSICS APPLIED**
Thermal pollution

FIGURE 15–19 Mechanical or heat energy is transformed to electric energy with a turbine and generator.

⇒ **PHYSICS APPLIED**
Electric energy generation (energy transformation)

Fossil-fuel steam plants. At a steam plant, coal, oil, or natural gas is burned to boil water and produce high-pressure steam that turns the turbine. The basic principles of this sort of steam engine were covered in Section 15–5. The advantages of such plants are that we know how to build them and they are not too expensive to run (yet). The disadvantages are: the products of combustion create air pollution; like all heat engines, their efficiency is limited (30 to 40 percent is typical); the waste heat produces thermal pollution; the extraction of the raw materials can be devastating to the land, especially in coal strip mining and oil-shale recovery; accidents are possible, such as oil spills at sea; and there may not be a great deal of fossil fuel left (estimates range from a supply lasting a few centuries to only a few decades). At present, 90 percent of all energy (not only electrical) in the United States is produced with fossil fuels.

Nuclear energy. Two nuclear processes release energy: fission and fusion. In fission, the nuclei of uranium or plutonium atoms are made to split ("fission") with the release of energy. In fusion, energy is released when small nuclei, such as those of hydrogen, combine ("fuse"). These processes are described in Chapter 31. The fission process is used in all present nuclear power plants, since fusion has not yet been controlled. Nuclear energy is used to heat steam just as fossil fuels are. A nuclear power plant is thus essentially a steam engine using uranium as its fuel. It suffers from the low efficiency characteristic of all heat engines (Section 15–5) and the accompanying thermal pollution. Nuclear power plants in normal operation produce practically no air pollution. But accidents can release high levels of damaging radioactivity such as happened at Chernobyl in 1986 and, earlier though less drastically, at Three Mile Island in 1979. Nuclear plants present other problems as well: radioactive substances produced are difficult to dispose of; nuclear material may possibly be diverted by terrorists and used to make bombs; and the nuclear fuel supply is limited. However, the energy produced per kilogram of fuel is very large, and extraction is less damaging to the land than for fossil fuels. The fusion process, on the other hand, seems to have fewer disadvantages and has the advantage of a vast supply of fuel—the hydrogen in the water molecules (H_2O) in the oceans. Unfortunately, fusion cannot yet be controlled sufficiently, but holds promise for the future. Both nuclear processes release no CO_2 and thus do not contribute to the greenhouse effect.

Geothermal energy. Both fossil-fuel and nuclear plants heat water to steam for the steam turbine. Natural steam can be obtained from the Earth itself (Fig. 15–20). In many places, water beneath the ground is in contact with the hot interior of the Earth and is raised to high temperature and pressure. It comes to the surface as hot springs, geysers, or steam vents. We can use not only natural vents but can drill down to trapped steam beds, or pass cold water down into contact with hot dry rock to heat it to steam. The largest geothermal plant at present operates at the Geysers in northern California. A similar plant has been successfully operating for almost a century in Italy, and a number of others are functioning in various parts of the world. In Iceland, geothermal energy is a major player. Geothermal energy appears to be clean in that it produces little air pollution, although there is some (nonsteam) gas emission. There is also the thermal pollution of the spent hot water, and the mineral content of the water (often high) may not only be environmentally polluting but can be corrosive to the parts of the

Nuclear power uses a steam engine

FIGURE 15–20 A geothermal power plant.

FIGURE 15–21 The Glen Canyon Dam (in Arizona) created an artificial lake that inundated an area of great scenic beauty.

apparatus itself. Nonetheless, geothermal is a reasonably inexpensive means of production and holds considerable promise, although naturally produced steam can, presumably, run out, leaving a site dry.

Hydroelectric power plants. Hydroelectric plants use falling water to directly turn the turbines of the generators so no heat engine is needed. They are usually located at the base of a dam. Hydroelectric power plants produce about 5 percent of all U.S. energy used. They produce practically no air or water pollution. Furthermore, they are nearly 100 percent efficient, since very little waste heat is produced. But they may not always be reliable because of drought, and there are not many good locations remaining in which to build dams. The reservoirs behind dams inundate land that may be fertile or of great scenic beauty (Fig. 15–21), and therefore involve a serious detriment to the environment.

Tidal energy. The Earth's tides are another form of water power. As shown in Fig. 15–22, a basin behind a dam is filled at high tide and the water is released at low tide to drive turbines. At the next high tide, the reservoir is filled again, and the inrushing water also turns turbines. A tidal plant is in operation in France, and a possible future site is in Canada's Bay of Fundy, where the tidal rise reaches 11 m (as compared to an average of 0.3 m elsewhere). This remarkable tide is attributed to a resonance effect (Section 11–6): the resonant frequency for water "sloshing" back and forth within the Bay of Fundy has a period of about 13 hours, as compared to 12.4 h for the tides (which is the driving force). One project for generating electricity proposes a dam to shorten the bay so as to reduce the resonant period slightly and increase the amplitude even more. Other good sites for tidal power (where there is a large difference between high and low tides) are not plentiful and would require large dams across natural or artificial bays. The rather abrupt changes of water level could have an effect on wildlife, but otherwise tidal power would seem to have a minimal environmental effect. Unfortunately, reasonable estimates of available sites indicate that tidal power could at best produce only a small fraction of the world's energy needs.

High tide

Low tide: Water is beginning to flow out of basin to ocean, driving turbines

Water level equalized (at low tide).

High tide: Water is allowed to flow back into basin, driving turbines.

Water level equalized (at high tide).

FIGURE 15–22 Tidal power plant. Turbines are located inside the dam.

(a)   (b)

FIGURE 15–23 Two types of windmill: (a) propeller type and (b) "eggbeater" type.

Wind power. Windmills were once widely used. Their comeback, on a much grander scale as a means of turning a generator to produce electricity, has already begun. There are already some 17,000 windmills in California alone (Fig. 15–23) capable of producing 1700 MW, enough power for a city of 300,000 people; they are supplying over 1 percent of the state's energy needs. Windmills of various sizes are being produced: from small 3-kW models (say, for a remote house), to large models producing several megawatts, with vanes 50 m wide. Windmills are generally "clean," although a large array producing a significant portion of U.S. energy needs might affect the weather and be considered an eyesore.

Solar energy. Many kinds of solar energy are already in use: fossil fuels are the remains of plant life that grew by photosynthesis of light from the Sun; hydroelectric power depends on the Sun to evaporate

FIGURE 15–24 (a) Solar heating system for a house. Water-carrying tubes are located on the roof in contact with a large black surface that absorbs the Sun's radiant energy and heats the water. The surface is covered with a piece of glass to prevent loss by convection, and the other sides of the tubes are well insulated to reduce conductive heat losses. The heated water is circulated to a large well-insulated reservoir (kept in the basement perhaps), where it is stored and recirculated to heaters in the house. The reservoir can also serve as the source for the hot-water supply. Either natural convection or forced convection (pumps) can move the water in the two parts of the system. Some form of backup system is needed in many climates to serve when there are prolonged periods of heavy clouds. (b) Solar panels on top of a house.

(a)   Insulation   (b)

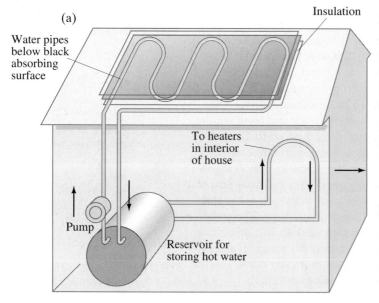

Water pipes below black absorbing surface

To heaters in interior of house

Pump

Reservoir for storing hot water

FIGURE 15–25 A passive
solar energy system: south-facing
windows allow passage of sunlight in winter when the Sun is
low in the sky. The floor is made
of high-heat-capacity material
(brick, tile) to store the energy
received from the sunlight and
release it when the temperature
drops. The overhanging sun shield
shades the window in summer
when the Sun is high, helping to
keep the interior cooler.

water that later comes down as rain to fill the reservoirs; and wind power
depends on convection currents produced by the Sun heating the atmosphere. *Active solar heating* systems, such as that described in Fig. 15–24,
can be used for space heating (keeping interiors warm) and for heating
hot water. So-called *passive solar heating* refers to architectural devices to
wisely use the Sun's energy, such as placing windows along the southern
exposure of a building to catch the Sun's rays in the winter, but with a
sunshade above to keep the Sun's rays out in the summer (Fig. 15–25).
The Sun's rays can also be used to produce electrical energy. For example, an array of large mirrors can focus the Sun's rays onto a boiler atop a
tall tower, as in a small pilot plant in California (Fig. 15–26). The focused
light can heat water to steam to drive a turbine. Such a system could be
used for a home generating system, although a backup system would
probably be needed for cloudy days. On a larger scale, large areas of land
would be needed to collect sufficient sunlight, roughly 1 square mile for
100 MW output. Although the ubiquitous thermal pollution would exist
and the climate might be affected, there would be essentially no air or
water pollution and no radioactivity, and the technology would not be too
difficult. Another direct user of sunlight is the **solar cell**, or more correctly **photovoltaic cell**, which converts sunlight directly into electricity without a heat engine being involved. The efficiency of solar cells, through
continuing research, can now be over 30 percent. Solar cells have been
very expensive, but recently the cost has dropped significantly and may
soon become competitive, making them very desirable since thermal (and
other) pollution would be very low (no heat engine is involved). Chemical pollution produced in their manufacture in large numbers could, however, be serious. Solar cells might be placed on roofs for home use. On a
large scale, they would require a large land area since the Sun's energy is
not very concentrated.

It should be clear that all forms of energy production have undesir-
able side effects. Some are worse than others, and not all problems can be
anticipated. New forms of energy production will be needed in the future
as old reserves of fuel are used up. Many different methods—including
those we have just discussed—may have to be used in the future to satisfy
the needs of our energy-consuming society. It is clear that conserving our
limited fuel supplies, by avoiding wasteful use of energy, should be of great
concern to society.

➡ **PHYSICS APPLIED**

Solar heating

➡ **PHYSICS APPLIED**

Solar-generated electricity

FIGURE 15–26 Array of mirrors
focus sunlight on a boiler to produce
steam at a solar energy installation
near Barstow, California.

1. Define the system you are dealing with; be careful to distinguish the system under study from its surroundings.

2. Be careful of signs associated with work and heat. In the first law work done *by* the system is positive; work done *on* the system is negative. Heat added to the system is positive, but heat removed from it is negative. With heat engines, we usually consider heat and work as positive and write energy conservation equations with + and − signs taking into account directions.

3. Watch the units used for work and heat; work is most often expressed in joules, and heat in calories or kilocalories, but choose only one unit for a given problem.

4. Temperatures must generally be expressed in kelvins; temperature **differences** may be expressed in C° or K.

5. Efficiency (or coefficient of performance) is a ratio of two energies or powers: useful output divided by required input. Efficiency (but *not* coefficient of performance) is always less than 1 in value, and hence is often stated as a percentage.

6. The entropy of a system increases when heat is added to the system, and decreases when heat is removed. If heat is transferred from system A to system B, the change in entropy of A is negative and the change in entropy of B is positive.

SUMMARY

The **first law of thermodynamics** states that the change in internal energy ΔU of a system is equal to the heat *added* to the system, Q, minus the work, W, done *by* the system:

$$\Delta U = Q - W.$$

This is a statement of the conservation of energy, and is found to hold for all types of processes.

An **isothermal** process is a process carried out at constant temperature.

In an **adiabatic** process, no heat is exchanged ($Q = 0$).

The work W done by a gas at constant pressure, P, is given by

$$W = P \, \Delta V,$$

where ΔV is the change in volume of the gas.

A **heat engine** is a device for changing thermal energy, by means of heat flow, into useful work.

The **efficiency** of a heat engine is defined as the ratio of the work W done by the engine to the heat input Q_H. Because of conservation of energy, the work output equals $Q_H - Q_L$, where Q_L is the heat exhausted to the environment; hence the efficiency

$$e = \frac{W}{Q_H} = 1 - \frac{Q_L}{Q_H}.$$

The upper limit on the efficiency can be written in terms of the higher and lower operating temperatures (in kelvins) of the engine, T_H and T_L, as

$$e_{\text{ideal}} = 1 - \frac{T_L}{T_H}.$$

The operation of **refrigerators** and air conditioners is the reverse of that of a heat engine: work is done to extract heat from a cool region and exhaust it to a region at a higher temperature.

The **second law of thermodynamics** can be stated in several equivalent ways:

1. heat flows spontaneously from a hot object to a cold one, but not the reverse;

2. there can be no 100 percent efficient heat engine—that is, one that can change a given amount of heat completely into work;

3. natural processes tend to move toward a state of greater disorder or greater **entropy**.

Statement (3) is the most general statement of the second law of thermodynamics, and can be restated as: the total entropy, S, of any system plus that of its environment increases as a result of any natural process:

$$\Delta S > 0.$$

Entropy is a quantitative measure of the disorder of a system. As time goes on, energy is degraded to less useful forms—that is, it is less available to do useful work.

QUESTIONS

1. What happens to the internal energy of water vapor in the air that condenses on the outside of a cold glass of water? Is work done or heat exchanged? Explain.

2. Use the conservation of energy to explain why the temperature of a gas increases when it is compressed—say, by pushing down on a cylinder—whereas the temperature decreases when the gas expands.

3. In an isothermal process, 3700 J of work is done by an ideal gas. Is this enough information to tell how much heat has been added to the system? If so, how much?

4. For the processes illustrated in Fig. 15–6, is more work or less work done in the isothermal process AB or in the process ADB?

5. Is it possible for the temperature of a system to remain constant even though heat flows into or out of it? If so, give examples.

6. Explain why the temperature of a gas increases when it is adiabatically compressed.

7. Can mechanical energy ever be transformed completely into heat or internal energy? Can the reverse happen? In each case, if your answer is no, explain why not; if yes, give examples.

8. Is it possible to cool down a room on a hot summer day by leaving the refrigerator door open?

9. Would a definition of heat engine efficiency as $e = W/Q_L$ be a useful one? Explain.

10. What plays the role of high-temperature and low-temperature reservoirs in (a) an internal combustion engine, and (b) a steam engine?

11. Which will give the greater improvement in the efficiency of a Carnot engine, a $10\,C°$ increase in the high-temperature reservoir, or a $10\,C°$ decrease in the low-temperature reservoir?

12. The oceans contain a tremendous amount of thermal energy. Why, in general, is it not possible to put this energy to useful work?

13. A gas is allowed to expand (a) adiabatically and (b) isothermally. In each process, does the entropy increase, decrease, or stay the same?

14. Give three examples, other than those mentioned in this chapter, of naturally occurring processes in which order goes to disorder. Discuss the observability of the reverse process.

15. Which do you think has the greater entropy, 1 kg of solid iron or 1 kg of liquid iron? Why?

16. What happens if you remove the lid of a bottle containing chlorine gas? Does the reverse process ever happen? Why or why not? Can you think of other examples of irreversibility?

17. Think up several processes (other than those already mentioned) that would obey the first law of thermodynamics, but, if they actually occurred, would violate the second law.

18. Suppose you collect a lot of papers strewn all over the floor and put them in a neat stack; does this violate the second law of thermodynamics? Explain.

19. The first law of thermodynamics is sometimes whimsically stated as, "You can't get something for nothing," and the second law as, "You can't even break even." Explain how these statements could be equivalent to the formal statements.

20. Give three examples of naturally occurring processes that illustrate the degradation of usable energy into internal energy.

* 21. Entropy is often called "time's arrow" because it tells us in which direction natural processes occur. If a movie film were run backward, name some processes that you might see that would tell you that time was "running backward."

* 22. Living organisms convert relatively simple food molecules into a complex structure as they grow. Is this a violation of the second law of thermodynamics?

PROBLEMS

SECTIONS 15–1 AND 15–2

1. (I) In Example 15–4, if the heat lost from the gas in the process BD is 3.42×10^3 J, what is the change in internal energy of the gas?

2. (I) An ideal gas expands isothermally, performing 4.40×10^3 J of work in the process. Calculate (a) the change in internal energy of the gas, and (b) the heat absorbed during this expansion.

3. (II) Sketch a PV diagram of the following process: 2.0 L of ideal gas at atmospheric pressure are cooled at constant pressure to a volume of 1.0 L, and then expanded isothermally back to 2.0 L, whereupon the pressure is increased at constant volume until the original pressure is reached.

4. (II) 1.0 L of air initially at 6.5 atm of (absolute) pressure is allowed to expand isothermally until the pressure is 1.0 atm. It is then compressed at constant pressure to its initial volume, and lastly is brought back to its original pressure by heating at constant volume. Draw the process on a PV diagram, including numbers and labels for the axes.

5. (II) One liter of air is cooled at constant pressure until its volume is halved, and then it is allowed to expand isothermally back to its original volume. Draw the process on a PV diagram.

6. (II) An ideal gas has its pressure cut in half slowly, while being kept in a container with rigid walls. In the process, 265 kJ of heat left the gas. (*a*) How much work was done during this process? (*b*) What was the change in internal energy of the gas during this process?

7. (II) In an engine, an almost ideal gas is compressed adiabatically to half its volume. In doing so, 1350 J of work is done on the gas. (*a*) How much heat flows into or out of the gas? (*b*) What is the change in internal energy of the gas? (*c*) Does its temperature rise or fall?

8. (II) An ideal gas expands at a constant pressure of 5.0 atm from 400 mL to 660 mL. Heat then flows out of the gas, at constant volume, and the pressure and temperature are allowed to drop until the temperature reaches its original value. Calculate (*a*) the total work done by the gas in the process, and (*b*) the total heat flow into the gas.

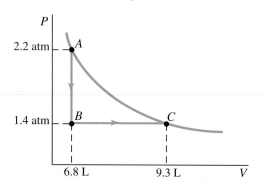

FIGURE 15–27 Problem 9.

9. (II) Consider the following two-step process. Heat is allowed to flow out of an ideal gas at constant volume so that its pressure drops from 2.2 atm to 1.4 atm. Then the gas expands at constant pressure, from a volume of 6.8 L to 9.3 L, where the temperature reaches its original value. See Fig. 15–27. Calculate (*a*) the total work done by the gas in the process, (*b*) the change in internal energy of the gas in the process, and (*c*) the total heat flow into or out of the gas.

10. (II) The PV diagram in Fig. 15–28 shows two possible states of a system containing 1.25 moles of a monatomic ideal gas. ($P_1 = P_2 = 450$ N/m^2, $V_1 = 2.00$ m^3, $V_2 = 8.00$ m^3). (*a*) Draw the process which depicts an isobaric expansion from state 1 to state 2 and label this process (A). (*b*) Find the work done by the gas and the change in internal energy of the gas in process (A).

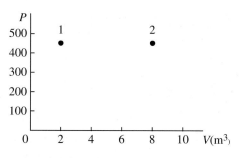

FIGURE 15–28 Problem 10.

(*c*) Draw the process which depicts an isothermal expansion from state 1 to the volume V_2 followed by an isochoric increase in temperature to state 2 and label this two-step process (B). (*d*) Find the change in internal energy of the gas for the two-step process (B).

11. (II) When a gas is taken from *a* to *c* along the curved path in Fig. 15–29, the work done by the gas is $W = -35$ J and the heat added to the gas is $Q = -63$ J. Along path *abc*, the work done is $W = -48$ J. (*a*) What is Q for path *abc*? (*b*) If $P_c = \frac{1}{2}P_b$, what is W for path *cda*? (*c*) What is Q for path *cda*? (*d*) What is $U_a - U_c$? (*e*) If $U_d - U_c = 5$ J, what is Q for path *da*?

12. (III) In the process of taking a gas from state *a* to state *c* along the curved path shown in Fig. 15 29, 80 J of heat leave the system and 55 J of work are done *on* the system. (*a*) Determine the change in internal energy, $U_a - U_c$. (*b*) When the gas is taken along the path *cda*, the work done by the gas is $W = 38$ J. How much heat Q is added to the gas in the process *cda*? (*c*) If $P_a = 2.5P_d$, how much work is done by the gas in the process *abc*? (*d*) What is Q for path *abc*? (*e*) If $U_a - U_b = 10$ J, what is Q for the process *bc*? Here is a summary of what is given:

$$Q_{a \to c} = -80 \text{ J} \qquad U_a - U_b = 10 \text{ J}$$
$$W_{a \to c} = -55 \text{ J} \qquad P_a = 2.5P_d$$
$$W_{cda} = 38 \text{ J}$$

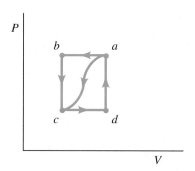

FIGURE 15–29 Problems 11, 12, and 13.

13. (III) Suppose a gas is taken clockwise around the rectangular cycle shown in Fig. 15–29, starting at b, then to a, to d, to c, and returning to b. Using the values given in Problem 12, calculate (a) the net work done during the cycle, (b) the net heat flow during the cycle, and (c) the total internal energy change during the cycle. (d) What percentage of the *intake* heat was turned into useable work: i.e., how efficient is this "rectangular" cycle (give as a percentage)?

*SECTION 15–3

* 14. (I) How much energy would the person of Example 15–7 transform if instead of working 11.0 h she took a noontime break and ran for 1.0 h?

* 15. (I) Calculate the average metabolic rate of a person who sleeps 8.0 h, sits at a desk 8.0 h, engages in light activity 4.0 h, watches television 2.0 h, plays tennis 1.5 h, and runs 0.5 h daily.

* 16. (II) A person decides to lose weight by sleeping one hour less per day, using the time for light activity. How much weight (or mass) can this person expect to lose in 1 year, assuming no change in food intake? Assume that 1 kg of fat stores about 9000 kcal of energy.

SECTION 15–5

17. (I) A heat engine produces 8200 J of heat while performing 3200 J of useful work. What is the efficiency of this engine?

18. (I) A heat engine does 7200 J of work in each cycle while absorbing 12.0 kcal of heat from a high-temperature reservoir. What is the efficiency of this engine?

19. (I) What is the maximum efficiency of a heat engine whose operating temperatures are 580°C and 320°C?

20. (I) The exhaust temperature of a heat engine is 230°C. What must be the high temperature if the Carnot efficiency is to be 28 percent?

21. (II) A nuclear power plant operates at 75 percent of its maximum theoretical (Carnot) efficiency between temperatures of 600° and 350°C. If the plant produces electric energy at the rate of 1.3 GW, how much exhaust heat is discharged per hour?

22. (II) A Carnot engine performs work at the rate of 440 kW while using 680 kcal of heat per second. If the temperature of the heat source is 570°C, at what temperature is the waste heat exhausted?

23. (II) A 100-hp car engine operates at about 15 percent efficiency. Assume the engine's water temperature of 85°C is its cold-temperature (exhaust) reservoir and 500°C is its thermal "intake" temperature (the temperature of the exploding gas/air mixture). (a) Calculate its efficiency relative to its maximum possible (Carnot) efficiency. (b) Estimate how much power (in watts) goes into moving the car, and how much heat, in joules and in kcal, is exhausted to the air in 1 h.

24. (II) A heat engine utilizes a heat source at 550°C and has an ideal (Carnot) efficiency of 30 percent. To increase the ideal efficiency to 40 percent, what must be the temperature of the heat source?

25. (II) A heat engine exhausts its heat at 350°C and has a Carnot efficiency of 39 percent. What exhaust temperature would enable it to achieve a Carnot efficiency of 50 percent?

26. (III) At a steam power plant, steam engines work in pairs, the output of heat from one being the approximate heat input of the second. The operating temperatures of the first are 670°C and 440°C, and of the second 430°C and 290°C. If the heat of combustion of coal is 2.8×10^7 J/kg, at what rate must coal be burned if the plant is to put out 1000 MW of power? Assume the efficiency of the engines is 60 percent of the ideal (Carnot) efficiency.

27. (III) Water is used to cool the power plant in Problem 26. If the water temperature is allowed to increase by no more than 6.0 C°, estimate how much water must pass through the plant per hour.

*SECTION 15–6

* 28. (I) The low temperature of a freezer cooling coil is −15°C and the discharge temperature is 30°C. What is the maximum theoretical coefficient of performance?

* 29. (I) If an ideal refrigerator keeps its contents at −15°C when the house temperature is 22°C, what is its coefficient of performance?

* 30. (II) A restaurant refrigerator has a coefficient of performance of 5.0. If the temperature in the kitchen outside the refrigerator is 29°C, what is the lowest temperature that could be obtained inside the refrigerator if it were ideal?

* 31. (II) A heat pump is used to keep a house warm at 22°C. How much work is required of the pump to deliver 2800 J of heat into the house if the outdoor temperature is (a) 0°C, (b) −15°C? Assume ideal (Carnot) behavior.

* 32. (II) An ideal engine has an efficiency of 35 percent. If it were run backward as a heat pump, what would be its coefficient of performance?

* 33. (II) The engine of Example 15–9 is run in reverse. How long would it take to freeze a tray of a dozen 40-g compartments of liquid water at room temperature into a dozen ice cubes at the freezing point, assuming that it takes 450 W of input electric power to run it? Assume ideal (Carnot) behavior.

SECTION 15–7

34. (I) What is the change in entropy of 100 g of steam at 100°C when it is condensed to water at 100°C?

35. (I) One kilogram of water is heated from 0°C to 100°C. Estimate the change in entropy of the water.

36. (I) What is the change in entropy of 1.00 m³ of water at 0°C when it is frozen to ice at 0°C?

37. (II) If the water in Problem 36 were frozen by being in contact with a great deal of ice at −10°C, what would be the total change in entropy of the process?

38. (II) An aluminum rod conducts 7.50 cal/s from a heat source maintained at 240°C to a large body of water at 27°C. Calculate the rate entropy increases per unit time in this process.

39. (II) 1.0 kg of water at 30°C is mixed with 1.0 kg of water at 60°C in a well-insulated container. Estimate the net change in entropy of the system.

40. (II) A 3.8-kg piece of aluminum at 30°C is placed in 1.0 kg of water in a Styrofoam container at room temperature (20°C). Calculate the approximate net change in entropy of the system.

41. (III) A real heat engine working between heat reservoirs at 970 K and 650 K produces 550 J of work per cycle for a heat input of 2200 J. (a) Compare the efficiency of this real engine to that of an ideal (Carnot) engine. (b) Calculate the total entropy change of the universe per cycle of the real engine. (c) Calculate the total entropy change of the universe per cycle of a Carnot engine operating between the same two temperatures.

*SECTION 15–11

* 42. (II) Calculate the probabilities, when you throw two dice, of obtaining (a) a 5, and (b) an 11.

* 43. (II) Rank the following five-card hands in order of increasing probability: (a) four aces and a king; (b) six of hearts, eight of diamonds, queen of clubs, three of hearts, jack of spades; (c) two jacks, two queens, and an ace; and (d) any hand having no two equal-value cards. Explain your ranking using microstates and macrostates.

* 44. (II) Suppose that you repeatedly shake six coins in your hand and drop them on the table. Construct a table showing the number of microstates that correspond to each macrostate. What is the probability of obtaining (a) three heads and three tails, and (b) six heads?

* 45. (III) Estimate the probability that a bridge player will be dealt (a) all four aces (among 13 cards), and (b) all 13 cards of one suit.

*SECTION 15–12

* 46. (I) Solar cells can produce about 40 W of electricity per square meter of surface area if directly facing the Sun. How large an area is needed to supply the needs of a house that requires 22 kWh/day? Would this fit on the roof of an average house? (Assume that the Sun shines 12 h/day.)

* 47. (II) Water is stored in an artificial lake created by a dam. The water depth is 45 m at the dam, and a steady flow rate of 35 m³/s is maintained through hydroelectric turbines installed near the base of the dam. How much electrical power can be produced?

* 48. (II) Energy may be stored for use during peak demand by pumping water to a high reservoir when demand is low and then releasing it to drive turbines when needed. Suppose water is pumped to a lake 135 m above the turbines at a rate of 1.00×10^5 kg/s for 10.0 h at night. (a) How much energy (kWh) is needed to do this each night? (b) If all this energy is released during a 14-h day, at 75 percent efficiency, what is the average power output?

* 49. (II) The basin of the tidal power plant at the mouth of the Rance River in France covers an area of 23 km². The average difference in water height between high and low tide is 8.5 m. Estimate how much work the falling water can do on the turbines per day, assuming there are two high and two low tides per day. Assume the basin is flat, and neglect any KE of water before and after its fall. See Fig. 15–22.

GENERAL PROBLEMS

50. To get an idea of how much thermal energy is contained in the world's oceans, estimate the heat liberated when a cube, 1 km on a side, of ocean water is cooled by 1 K. (Approximate the ocean water by pure water for this estimate.)

51. One and one-half moles of an ideal monatomic gas expand adiabatically, performing 7500 J of work in the process. What is the change in temperature of the gas during this expansion?

52. An inventor claims to have designed and built an engine that produces 1.50 MW (megawatts) of usable work while taking in 3.00 MW of thermal energy at 425 K, and rejecting 1.50 MW of thermal energy at 215 K. Is there anything fishy about his claim?

53. When 5.30×10^4 J of heat are added to a gas enclosed in a cylinder fitted with a light frictionless piston maintained at atmospheric pressure, the volume is observed to increase from 1.9 m³ to 4.1 m³. Calculate (a) the work done by the gas, and (b) the change in internal energy of the gas. (c) Graph this process on a PV diagram.

54. A 4-cylinder gasoline engine has an efficiency of 0.25 and delivers 200 J of work per cycle per cylinder. When the engine fires at 25 cycles per second, (a) what is the work done per second? (b) What is the total heat input per second from the fuel? (c) If the energy content of gasoline is 130 MJ per gallon, how long does one gallon last?

55. It's not necessary that a heat engine's hot environment be hotter than ambient temperature. Liquid nitrogen (90 K) is about as cheap as bottled water. What would be the efficiency of an engine that made use of heat transferred from air at room temperature (293 K) to the liquid nitrogen "fuel"?

56. It has been suggested that a heat engine could be developed that made use of the temperature difference between water at the surface of the ocean and that several hundred meters deep. In the tropics, the temperatures may be 27°C and 4°C, respectively. What is the maximum efficiency such an engine could have? Why might such an engine be feasible in spite of the low efficiency? Can you imagine any adverse environmental effects that might occur?

57. Two 1100-kg cars are traveling 100 km/h in opposite directions when they collide and are brought to rest. Estimate the change in entropy of the universe as a result of this collision. Assume $T = 20°C$.

58. A 120-g insulated aluminum cup at 15°C is filled with 140 g of water at 50°C. After a few minutes, equilibrium is reached. (a) Determine the final temperature, and (b) estimate the total change in entropy.

*59. (a) What is the coefficient of performance of an ideal heat pump that extracts heat from 6°C air outside and deposits heat inside your house at 24°C? (b) If this heat pump operates on 1000 W of electrical energy, what is the maximum heat it can deliver into your house each hour?

60. The burning of gasoline in a car releases about 3.0×10^4 kcal/gal. If a car averages 41 km/gal when driving 90 km/h, which requires 25 hp, what is the efficiency of the engine under those conditions?

61. A 15-kg box having an initial speed of 2.4 m/s slides along a rough table and comes to rest. Estimate the total change in entropy of the universe. Assume all objects are at room temperature (293 K).

62. A vertical steel I-beam at the base of a building is 7.5 m tall, has a mass of 300 kg, and supports a load of 4.3×10^5 N. If the beam's temperature decreases by 6.0 C°, calculate the change in its internal energy using the facts that for steel, the specific heat is 0.11 kcal/kg·C° and the coefficient of linear expansion is 11×10^{-6} C°$^{-1}$.

63. A 33 percent efficient power plant puts out 800 MW (megawatts) of work (electrical energy). Cooling towers are used to take away the exhaust heat. If the air temperature is allowed to rise 7.0 C°, what volume of air (km³) is heated per day? Will the local climate be heated significantly? If the heated air were to form a layer 200 m thick, how large an area would it cover for 24 h of operation? (The heat capacity of air is about 7.0 cal/mol·C° at constant pressure.)

64. Suppose a power plant delivers energy at 900 MW using steam turbines. The steam goes into the turbines superheated at 600 K and deposits its unused heat in river water at 285 K. Assume that the turbine operates as an ideal Carnot engine. (a) If the river flow rate is 37 m³/s, calculate the average temperature increase of the river water downstream from the power plant. (b) What is the entropy increase per kilogram of the downstream river water in J/kg·K?

This comb has been rubbed by a cloth or paper towel to give it a static electric charge. Because the comb is electrically charged, it induces a separation of charge in all those scraps of paper, and thus attracts them.

16 ELECTRIC CHARGE AND ELECTRIC FIELD

The word "electricity" may evoke an image of complex modern technology: computers, lights, motors, electric power. But the electric force would seem to play an even deeper role in our lives: according to atomic theory, the forces that act between atoms and molecules to hold them together to form liquids and solids are electrical forces, and electric forces are also involved in the metabolic processes that occur within our bodies. Many of the forces we have dealt with so far, such as elastic forces, the normal force, and other contact forces (pushes and pulls) are now considered to result from electric forces acting at the atomic level. This does not include gravity, however, which is a separate force.[†]

The earliest studies on electricity date back to the ancients, but it has been only in the past two centuries that electricity was studied in detail. We will discuss the development of ideas about electricity, including practical devices, as well as the relation to magnetism, in the next seven chapters.

[†]As we discussed in Section 5–10, physicists in this century came to recognize four different fundamental forces in nature: (1) gravitational force, (2) electromagnetic force (we will see later that electric and magnetic forces are intimately related), (3) strong nuclear force, and (4) weak nuclear force. The last two forces operate at the level of the nucleus of an atom. The electromagnetic and weak nuclear forces are now thought to have a common origin known as the electroweak force. We will discuss these forces in later chapters.

(a)   (b)

16–1 Static Electricity; Electric Charge and Its Conservation

The word *electricity* comes from the Greek word *elektron*, which means "amber." Amber is petrified tree resin, and the ancients knew that if you rub an amber rod with a piece of cloth, the amber attracts small pieces of leaves or dust. A piece of hard rubber, a glass rod, or a plastic ruler rubbed with a cloth will also display this "amber effect," or **static electricity** as we call it today. You can readily pick up small pieces of paper with a plastic comb or ruler that you've just vigorously rubbed with even a paper towel. See the photo on the previous page and Fig. 16–1. You have probably experienced static electricity when combing your hair or when taking a synthetic blouse or shirt from a clothes dryer. And you may have felt a shock when you touched a metal doorknob after sliding across a car seat or walking across a nylon carpet. In each case, an object becomes "charged" due to a rubbing process and is said to possess a net **electric charge**.

Is all electric charge the same, or is it possible that there is more than one type? In fact, there are two types of electric charge, as the following simple experiments show. A plastic ruler is suspended by a thread and rubbed vigorously with a cloth to charge it. When a second ruler, which has also been charged in the same way, is brought close to the first, it is found that the one ruler *repels* the other. This is shown in Fig. 16–2a. Similarly, if a rubbed glass rod is brought close to a second charged glass rod, again a repulsive force is seen to act, Fig. 16–2b. However, if the charged glass rod is brought close to the charged plastic ruler, it is found that they *attract* each other, Fig. 16–2c. The charge on the glass must therefore be different from that on the plastic. Indeed, it is found experimentally that all charged objects fall into one of two categories. Either they are attracted to the plastic and repelled by the glass, just as glass is; or they are repelled by the plastic and attracted to the glass, just as the plastic ruler is. Thus there seem to be two, and only two, types of electric charge. Each type of charge repels the same type but attracts the opposite type. That is: **unlike charges attract; like charges repel**.

The two types of electric charge were referred to as *positive* and *negative* by the American statesman, philosopher, and scientist Benjamin Franklin (1706–1790). The choice of which name went with which type of charge was of course arbitrary. Franklin's choice sets the charge on the rubbed glass rod to be positive charge, so the charge on a rubbed plastic ruler (or amber) is called negative charge. We still follow this convention today.

Franklin argued that whenever a certain amount of charge is produced on one body in a process, an equal amount of the opposite type of charge is produced on another body. The positive and negative are to be treated *algebraically*, so that during any process, the net change in the

(a) Two charged plastic rulers repel

(b) Two charged glass rods repel

(c) Charged glass rod attracts charged plastic ruler

Likes repel;
unlikes attract

amount of charge produced is zero. For example, when a plastic ruler is rubbed with a paper towel, the plastic acquires a negative charge and the towel an equal amount of positive charge. The charges are separated, but the sum of the two is zero. This is an example of a law that is now well established: the **law of conservation of electric charge**, which states that

the net amount of electric charge produced in any process is zero.

If one object or one region of space acquires a positive charge, then an equal amount of negative charge will be found in neighboring areas or objects. No violations have ever been found, and this conservation law is as firmly established as those for energy and momentum.

16–2 Electric Charge in the Atom

Only within the past century has it become clear that electricity starts inside the atom itself. In later chapters we will discuss atomic structure and the ideas that led to our present view of the atom in more detail. But it will help our understanding of electricity if we discuss it briefly now.

Today's view, somewhat simplified, shows the atom as having a heavy, positively charged nucleus surrounded by one or more negatively charged electrons (Fig. 16–3). The nucleus contains protons, which are positively charged, and neutrons, which have no net electric charge. The magnitude of the charge on protons and electrons is exactly the same, but their signs are opposite. Hence, neutral atoms contain equal numbers of protons and electrons. Sometimes, however, an atom may lose one or more of its electrons, or may gain extra electrons. In this case the atom will have a net positive or negative charge, and is called an **ion**.

In solid materials the nuclei tend to remain close to fixed positions, whereas some of the electrons move quite freely. The charging of a solid object by rubbing is explained mainly by the transfer of electrons from one material to the other. When a plastic ruler becomes negatively charged by rubbing with a paper towel, the transfer of electrons from the towel to the plastic leaves the towel with a positive charge equal in magnitude to the negative charge acquired by the plastic. (In liquids and gases, nuclei or ions can move as well as electrons.)

Normally when objects are charged by rubbing, they hold their charge only for a limited time and eventually return to the neutral state. Where does the charge go? In some cases it is neutralized by charged ions in the air (formed, for example, by collisions with charged particles known as cosmic rays that reach the Earth from space). Often more importantly, the charge can "leak off" onto water molecules in the air. This is because water molecules are **polar**—that is, even though they are neutral, their charge is not distributed uniformly, Fig. 16–4. Thus the extra electrons on, say, a charged plastic ruler can "leak off" into the air because they are attracted to the positive end of water molecules. A positively charged object, on the other hand, can be neutralized by transfer of loosely held electrons from water molecules in the air. On dry days, static electricity is much more noticeable since the air contains fewer water molecules to allow leakage. On humid or rainy days, it is difficult to make any object hold its charge for long.

FIGURE 16–3 Simple model of the atom.

FIGURE 16–4 Diagram of a water molecule. Because it has opposite charges on different ends, it is called a "polar" molecule.

Charged   Neutral

Metal

Wood

(a)   (b)   (c)

16–3 Insulators and Conductors

Suppose we have two metal spheres, one highly charged and the other electrically neutral (Fig. 16–5a). If we now place an iron nail so that it touches both the spheres (Fig. 16–5b), it is found that the previously uncharged sphere quickly becomes charged. If, instead, we connect the two spheres together by a wooden rod or a piece of rubber (Fig. 16–5c), the uncharged ball does not become noticeably charged. Materials like the iron nail are said to be **conductors** of electricity, whereas wood and rubber are **nonconductors** or **insulators**.

Metals are generally good conductors whereas most other materials are insulators (although even insulators conduct electricity very slightly). It is interesting that nearly all natural materials fall into one or the other of these two quite distinct categories. There are a few materials, however (notably silicon, germanium, and carbon), that fall into an intermediate (but distinct) category known as **semiconductors**.

Metals are good conductors

From the atomic point of view, the electrons in an insulating material are bound very tightly to the nuclei. In a good conductor, on the other hand, some of the electrons are bound very loosely and can move about freely within the material (although they cannot *leave* the metal easily) and are often referred to as *free electrons* or *conduction electrons*. When a positively charged object is brought close to or touches a conductor, the free electrons are attracted by this positive charge and move quickly toward it. On the other hand, the free electrons move swiftly away from a negative charge that is brought close. In a semiconductor, there are very few free electrons, and in an insulator, almost none.

16–4 Induced Charge; the Electroscope

Suppose a positively charged metal object is brought close to an uncharged metal object. If the two touch, the free electrons in the neutral one are attracted to the positively charged object and some will pass over to it, Fig. 16–6. Since the second object is now missing some of its negative electrons, it will have a net positive charge. This process is called "charging by conduction," or "by contact," and the two objects end up with the same sign of charge.

Now suppose a positively charged object is brought close to a neutral metal rod, but does not touch it. Although the electrons of the metal rod do not leave the rod, they still move within the metal toward the charged object, which leaves a positive charge at the opposite end, Fig. 16–7. A charge is said to have been *induced* at the two ends of the metal rod. Of course no net charge has been created in the rod; charges have merely been *separated*. The net charge on the metal rod is still zero. However, if the metal were broken into two pieces, we could have two charged objects, one charged positively and one charged negatively.

(a) Neutral metal rod

(b) Metal rod acquires charge by contact

FIGURE 16–6 (a) Neutral metal rod acquires a charge (b) when placed in contact with a charged metal object.

FIGURE 16–7 Charging by induction.

(a)   Neutral metal rod

(b)   Metal rod still neutral, but with a separation of charge

(a)

(b)

(c)

FIGURE 16–8 Inducing a charge on an object connected to ground.

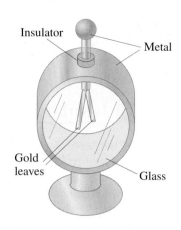

Insulator

Metal

Gold leaves

Glass

FIGURE 16–9 Electroscope.

(a)   (b)

FIGURE 16–10 Electroscope charged (a) by induction, (b) by conduction.

FIGURE 16–11 A previously charged electroscope can be used to determine the sign of a given charge.

(a)   (b)   (c)

Another way to induce a net charge on a metal object is to connect it with a conducting wire to the ground (or a conducting pipe leading into the ground) as shown in Fig. 16–8a ($\equiv$ means "ground"). The object is then said to be "grounded" or "earthed." Now the Earth, since it is so large and can conduct, can easily accept or give up electrons; hence it acts like a reservoir for charge. If a charged object—let's say negative this time—is brought up close to the metal, free electrons in the metal are repelled and many of them move down the wire into the Earth, Fig. 16–8b. This leaves the metal positively charged. If the wire is now cut, the metal will have a positive induced charge on it (Fig. 16–8c). If the wire were cut after the negative object is moved away, the electrons would all have moved back into the metal and it would be neutral.

An **electroscope** is a device that can be used for detecting charge. As shown in Fig. 16–9, inside of a case are two movable leaves, often made of gold. (Sometimes only one leaf is movable.) The leaves are connected by a conductor to a metal ball on the outside of the case, but are insulated from the case itself. If a positively charged object is brought close to the knob, a separation of charge is induced, as electrons are attracted up into the ball, leaving the leaves positively charged, Fig. 16–10a. The two leaves repel each other as shown. If, instead, the knob is charged by conduction, the whole apparatus acquires a net charge as shown in Fig. 16–10b. In either case, the greater the amount of charge, the greater the separation of the leaves.

Note, however, that you cannot tell the sign of the charge in this way, since a negative charge will cause the leaves to separate just as much as an equal-magnitude positive charge—in either case the two leaves repel each other. An electroscope can, however, be used to determine the sign of the charge if it is first charged by conduction, say negatively, as in Fig. 16–11a. Now if a negative object is brought close, as in Fig. 16–11b, more electrons are induced to move down into the leaves and they separate further. On the other hand, if a positive charge is brought close, the electrons are induced to flow upward, leaving the leaves less negative and their separation is reduced, Fig. 16–11c.

The electroscope was much used in the early studies of electricity. The same principle, aided by some electronics, is used in much more sensitive modern **electrometers**.

16–5 Coulomb's Law

We have seen that an electric charge exerts a force on other electric charges. What factors affect the magnitude of this force? To answer this, the French physicist Charles Coulomb (1736–1806) investigated electric forces in the 1780s using a torsion balance (Fig. 16–12) much like that used by Cavendish for his studies of the gravitational force (Section 5–6).

Although precise instruments for the measurement of electric charge were not available in Coulomb's time, he was able to prepare small spheres with different magnitudes of charge in which the *ratio* of the charges was known. He reasoned that if a charged conducting sphere is placed in contact with an identical uncharged sphere, the charge on the first would be shared equally by the two of them because of symmetry. He thus had a way to produce charges equal to $\frac{1}{2}$, $\frac{1}{4}$, and so on, of the original charge. Although he had some difficulty with induced charges, Coulomb was able to argue that the force one tiny charged object exerted on a second tiny charged object is directly proportional to the charge on each of them. That is, if the charge on either one of the objects was doubled, the force was doubled; and if the charge on both of the objects was doubled, the force increased to four times the original value. This was the case when the distance between the two charges remained the same. If the distance between them was allowed to increase, he found that the force decreased with the *square of the distance* between them. That is, if the distance was doubled, the force fell to one-fourth of its original value. Thus, Coulomb concluded, the force one tiny charged object exerts on a second one is proportional to the product of the magnitude of the charge on one, Q_1, times the magnitude of the charge on the other, Q_2, and inversely proportional to the square of the distance r between them (Fig. 16–13). As an equation, we can write **Coulomb's law** as

$$F = k\frac{Q_1 Q_2}{r^2}, \qquad (16\text{–}1)$$

where k is a proportionality constant. The validity of Coulomb's law today rests on precision measurements that are much more sophisticated than Coulomb's original experiment.

Since we are dealing here with a new quantity (electric charge), we could choose its unit so that the proportionality constant k in Eq. 16–1 would be one. Indeed, such a system of units was once common.[†] However, the most widely used unit now is the **coulomb** (C), which is the SI unit. The precise definition of the coulomb today is in terms of electric current and magnetic field, and will be discussed later (Section 20–7). In SI units, k has the value

$$k = 8.988 \times 10^9 \ \text{N} \cdot \text{m}^2/\text{C}^2 \approx 9.0 \times 10^9 \ \text{N} \cdot \text{m}^2/\text{C}^2.$$

Thus, 1 C is that amount of charge which, if placed on each of two point objects 1.0 m apart, will result in each object exerting a force of $(9.0 \times 10^9 \ \text{N} \cdot \text{m}^2/\text{C}^2)(1.0 \ \text{C})(1.0 \ \text{C})/(1.0 \ \text{m})^2 = 9.0 \times 10^9 \ \text{N}$ on the other. This would be an enormous force, equal to the weight of almost a million tons. We don't normally encounter charges as large as a coulomb.

Charges produced by rubbing ordinary objects (such as a comb or plastic ruler) are typically around a microcoulomb ($1 \ \mu\text{C} = 10^{-6} \ \text{C}$) or

FIGURE 16–12 Schematic diagram of Coulomb's apparatus. It is similar to Cavendish's, which was used for the gravitational force. When a charged sphere is placed close to the one on the suspended bar, the bar rotates slightly. The suspending fiber resists the twisting motion and the angle of twist is proportional to the force applied. By the use of this apparatus, Coulomb investigated how the electric force varies as a function of the magnitude of the charges and of the distance between them.

COULOMB'S LAW

FIGURE 16–13 Coulomb's law, Eq. 16–1, gives the force between two point charges, Q_1 and Q_2, a distance r apart.

Q_1 Q_2
●———r———●

Unit for charge:
the coulomb

[†]This is a cgs system of units, and the unit of electric charge is called the *electrostatic unit* (esu) or the statcoulomb. One esu is defined as that charge, on each of two point objects 1 cm apart, that gives rise to a force of 1 dyne.

less. The magnitude of the charge on one electron, on the other hand, has been determined to be about 1.602×10^{-19} C, and its sign is negative. This is the smallest known charge,[†] and because of its fundamental nature, it is given the symbol e and is often referred to as the *elementary charge*:

$$e = 1.602 \times 10^{-19} \text{ C}.$$

Note that e is defined as a positive number, so the charge on the electron is $-e$. (The charge on a proton, on the other hand, is $+e$.) Since an object cannot gain or lose a fraction of an electron, the net charge on any object must be an integral multiple of this charge. Electric charge is thus said to be **quantized** (existing only in discrete amounts: $1e, 2e, 3e$, etc.). Because e is so small, however, we normally don't notice this discreteness in macroscopic charges (1 μC requires about 10^{13} electrons), which thus seem continuous.

Equation 16–1 gives the *magnitude* of the electric force that either object exerts on the other, when the magnitudes of the charges Q_1 and Q_2 are given. The *direction* of the electric force *is always along the line joining the two objects*. If the two charges have the same sign, the force on either object is directed away from the other. If the two charges have opposite signs, the force on one is directed toward the other, Fig. 16–14. Notice that the force one charge exerts on the second is equal but opposite to that exerted by the second on the first, in accord with Newton's third law.

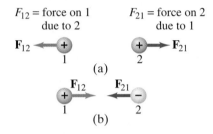

F_{12} = force on 1 due to 2 F_{21} = force on 2 due to 1

(a)

(b)

FIGURE 16–14 Direction of the force depends on whether the charges have (a) the same sign, or (b) opposite sign.

[Note the similarity of Coulomb's law to the law of universal gravitation, Eq. 5–4. Both are inverse square laws ($F \propto 1/r^2$). Both also have a proportionality to a product of a property of each body—mass for gravity, electric charge for electricity. A major difference between the two laws is that gravity is always an attractive force, whereas the electric force can be either attractive or repulsive.]

The constant k in Eq. 16–1 is often written in terms of another constant, ϵ_0, called the **permittivity of free space**. It is related to k by $k = 1/4\pi\epsilon_0$. Coulomb's law can then be written

$$F = \frac{1}{4\pi\epsilon_0} \frac{Q_1 Q_2}{r^2}, \tag{16–2}$$

where

$$\epsilon_0 = \frac{1}{4\pi k} = 8.85 \times 10^{-12} \text{ C}^2/\text{N·m}^2.$$

Equation 16–2 looks more complicated than Eq. 16–1, but other fundamental equations we haven't seen yet are simpler in terms of ϵ_0 rather than k. It doesn't matter which form we use, of course, since Eqs. 16–1 and 16–2 are equivalent.

It should be recognized that Eqs. 16–1 and 16–2 apply to objects whose size is much smaller than the distance between them. Ideally, it is precise for **point charges** (spatial size negligible compared to other distances). For finite-sized objects, it is not always clear what value to use for r, particularly since the charge may not be distributed uniformly on the objects. If the two objects are spheres and the charge is known to be distributed uniformly on each, then r is the distance between their centers.

Coulomb's law describes the force between two charges when they are at rest. Additional forces come into play when charges are in motion, and these will be discussed in later chapters. In this chapter we discuss only charges at rest, the study of which is called **electrostatics**.

[†] According to the standard model of elementary particle physics, subnuclear particles called quarks have a smaller charge than that on the electron, equal to $\frac{1}{3}e$ or $\frac{2}{3}e$. Quarks have not been detected directly, and theory indicates that free quarks may not be detectable.

When calculating with Coulomb's law, we usually ignore the signs of the charges and determine direction based on whether the force is attractive or repulsive.

EXAMPLE 16–1 **Electric force on electron by proton.** Determine the magnitude of the electric force on the electron of a hydrogen atom exerted by the single proton ($Q_2 = +e$) that is its nucleus. Assume the electron "orbits" the proton at its average distance of $r = 0.53 \times 10^{-10}$ m, Fig. 16–15.

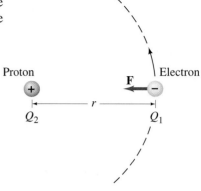

SOLUTION We use Eq. 16–1 with $r = 0.53 \times 10^{-10}$ m, and $Q_1 = Q_2 = 1.6 \times 10^{-19}$ C (ignoring the signs of the charges):

$$F = \frac{(9.0 \times 10^9 \text{ N·m}^2/\text{C}^2)(1.6 \times 10^{-19} \text{ C})(1.6 \times 10^{-19} \text{ C})}{(0.53 \times 10^{-10} \text{ m})^2}$$

$$= 8.2 \times 10^{-8} \text{ N}.$$

FIGURE 16–15 Example 16–1.

The direction of the force on the electron is toward the proton, since the charges have opposite signs and the force is attractive.

CONCEPTUAL EXAMPLE 16–2 **Which charge exerts the greater force?**
Two positive point charges, $Q_1 = 50\,\mu\text{C}$ and $Q_2 = 1\,\mu\text{C}$, are separated by a distance l, Fig. 16–16. Which is larger in magnitude, the force that Q_1 exerts on Q_2, or the force that Q_2 exerts on Q_1?

RESPONSE From Coulomb's law, the force on Q_1 exerted by Q_2 is:

$$F_{12} = k\frac{Q_1 Q_2}{l^2}.$$

FIGURE 16–16 Example 16–2.

The force on Q_2 exerted by Q_1 is the same except that Q_1 and Q_2 are reversed. The equation is symmetric with respect to the two charges, so $F_{21} = F_{12}$. Newton's third law also tells us that these two forces must have equal magnitude.

It is very important to keep in mind that Eq. 16–1 (or 16–2) gives the force on a charge due to only *one* other charge. If several (or many) charges are present, the *net force on any one of them will be the vector sum of the forces due to each of the others*.

Electric forces add as vectors

16–6 Solving Problems Involving Coulomb's Law and Vectors

The electric force between charged particles at rest (sometimes referred to as the **electrostatic force** or as the **Coulomb force**) is, like all forces, a vector: it has both magnitude and direction. When several forces act on an object (call them $\mathbf{F}_1$, $\mathbf{F}_2$, etc.), the net force $\mathbf{F}_{net}$ on the object is the vector sum of all the forces acting on it:

$$\mathbf{F}_{net} = \mathbf{F}_1 + \mathbf{F}_2 + \cdots.$$

(This is sometimes referred to as the principle of superposition for forces). We studied how to add vectors in Chapter 3, and in Chapter 4 we applied the rules for adding vectors to forces. It might be a good idea now to review Sections 3–2, 3–3, 3–4, as well as Section 4–9 on general problem-solving techniques. Here is a brief review of vectors.

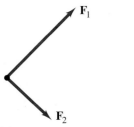

(a) Two forces acting on an object.

(b) The total, or net, force is $\mathbf{F} = \mathbf{F}_1 + \mathbf{F}_2$ by the tail-to-tip method of adding vectors.

(c) $\mathbf{F} = \mathbf{F}_1 + \mathbf{F}_2$ by the parallelogram method.

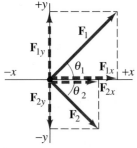

(d) $\mathbf{F}_1$ and $\mathbf{F}_2$ resolved into their x and y components.

FIGURE 16–17 Review of vector addition.

Given two vector forces, $\mathbf{F}_1$ and $\mathbf{F}_2$, acting on a body (Fig. 16–17a), they can be added using the tail-to-tip method (Fig. 16–17b) or by the parallelogram method (Fig. 16–17c), as discussed in Section 3–2. These two methods are useful for *understanding* a given problem (for getting a picture in your mind of what is going on), but for *calculating* the direction and magnitude of the resultant sum, it is more precise to use the method of adding components. Figure 16–17d shows the components of our $\mathbf{F}_1$ and $\mathbf{F}_2$ resolved into components along chosen x and y axes (for more details, see Section 3–4). From the definitions of the trigonometric functions (Figs. 3–11 and 3–12), we have

$$F_{1x} = F_1 \cos \theta_1 \qquad F_{2x} = F_2 \cos \theta_2$$
$$F_{1y} = F_1 \sin \theta_1 \qquad F_{2y} = -F_2 \sin \theta_2.$$

We add up the x and y components separately to obtain the components of the resultant force $\mathbf{F}$, which are

$$F_x = F_{1x} + F_{2x} = F_1 \cos \theta_1 + F_2 \cos \theta_2,$$
$$F_y = F_{1y} + F_{2y} = F_1 \sin \theta_1 - F_2 \sin \theta_2.$$

The magnitude of $\mathbf{F}$ is

$$F = \sqrt{F_x^2 + F_y^2}.$$

The direction of $\mathbf{F}$ is specified by the angle θ that $\mathbf{F}$ makes with the x axis, which is given by

$$\tan \theta = \frac{F_y}{F_x}.$$

This review has been necessarily brief; a rereading of the appropriate parts of Chapters 3 and 4 will give more details.

When dealing with several charges, it is often helpful to use subscripts on each of the forces involved. The first subscript refers to the particle *on* which the force acts; the second refers to the particle that exerts the force. For example, if we have three charges, $\mathbf{F}_{31}$ means the force exerted *on* particle 3 *by* particle 1.

As in all problem solving, it is very important to draw a diagram, in particular a free-body diagram for each body (Chapter 4), showing all the forces acting on that body. In applying Coulomb's law, we usually deal with charge magnitudes only (leaving out minus signs) to get the magnitude of each force. Then determine the direction of the force physically (like charges repel, unlike attract), and show the force on the diagram. Finally, add the forces on one object together as vectors.

EXAMPLE 16–3 **Three charges in a line.** Three charged particles are arranged in a line, as shown in Fig. 16–18a. Calculate the net electrostatic force on particle 3 (the $-4.0\ \mu C$ on the right) due to the other two charges.

SOLUTION The net force on particle 3 will be the vector sum of the force $\mathbf{F}_{31}$ exerted by particle 1 and the force $\mathbf{F}_{32}$ exerted by particle 2: $\mathbf{F} = \mathbf{F}_{31} + \mathbf{F}_{32}$. The magnitudes of these two forces are

$$F_{31} = \frac{(9.0 \times 10^9\ \mathrm{N \cdot m^2/C^2})(4.0 \times 10^{-6}\ \mathrm{C})(8.0 \times 10^{-6}\ \mathrm{C})}{(0.50\ \mathrm{m})^2} = 1.2\ \mathrm{N}.$$

$$F_{32} = \frac{(9.0 \times 10^9\ \mathrm{N \cdot m^2/C^2})(4.0 \times 10^{-6}\ \mathrm{C})(3.0 \times 10^{-6}\ \mathrm{C})}{(0.20\ \mathrm{m})^2} = 2.7\ \mathrm{N}.$$

Since we were calculating the magnitudes of the forces, we omitted the signs of the charges; but we must be aware of them to get the direction of each force. Let the line joining the particles be the x axis, and we take it positive to the right. Then, because $\mathbf{F}_{31}$ is repulsive and $\mathbf{F}_{32}$ is attractive, the directions of the forces are as shown in Fig. 16–18b: $\mathbf{F}_{31}$ points in the positive x direction and $\mathbf{F}_{32}$ points in the negative x direction. The net force on particle 3 is then

$$F = -F_{32} + F_{31} = -2.7\,\text{N} + 1.2\,\text{N} = -1.5\,\text{N}.$$

The magnitude of the net force is 1.5 N, and it points to the left.

Notice in this Example that the charge in the middle (Q_2) in no way blocks the effect of the other charge (Q_1); Q_2 does exert its own force, of course.

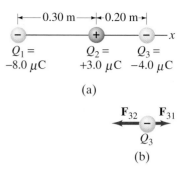

FIGURE 16–18 Diagram for Example 16–3.

EXAMPLE 16–4 **Electric force using vector components.** Calculate the net electrostatic force on charge Q_3 shown in Fig. 16–19a due to the charges Q_1 and Q_2.

SOLUTION The forces $\mathbf{F}_{31}$ and $\mathbf{F}_{32}$ have the directions shown in the diagram since Q_1 exerts an attractive force and Q_2 a repulsive force. The magnitudes of $\mathbf{F}_{31}$ and $\mathbf{F}_{32}$ are (ignoring signs since we know the directions)

$$F_{31} = \frac{(9.0 \times 10^9\,\text{N}\cdot\text{m}^2/\text{C}^2)(6.5 \times 10^{-5}\,\text{C})(8.6 \times 10^{-5}\,\text{C})}{(0.60\,\text{m})^2} = 140\,\text{N},$$

$$F_{32} = \frac{(9.0 \times 10^9\,\text{N}\cdot\text{m}^2/\text{C}^2)(6.5 \times 10^{-5}\,\text{C})(5.0 \times 10^{-5}\,\text{C})}{(0.30\,\text{m})^2} = 330\,\text{N}.$$

We resolve $\mathbf{F}_{31}$ into its components along the x and y axes, as shown:

$$F_{31x} = F_{31} \cos 30° = 120\,\text{N},$$

$$F_{31y} = -F_{31} \sin 30° = -70\,\text{N}.$$

The force $\mathbf{F}_{32}$ has only a y component. So the net force $\mathbf{F}$ on Q_3 has components

$$F_x = F_{31x} = 120\,\text{N}$$

$$F_y = F_{32} + F_{31y} = 330\,\text{N} - 70\,\text{N} = 260\,\text{N}.$$

Thus the magnitude of the net force is

$$F = \sqrt{F_x^2 + F_y^2} = \sqrt{(120\,\text{N})^2 + (260\,\text{N})^2} = 290\,\text{N};$$

and it acts at an angle θ (see Fig. 16–19b) given by $\tan \theta = F_y/F_x = 260\,\text{N}/120\,\text{N} = 2.2$, so $\theta = 65°$.

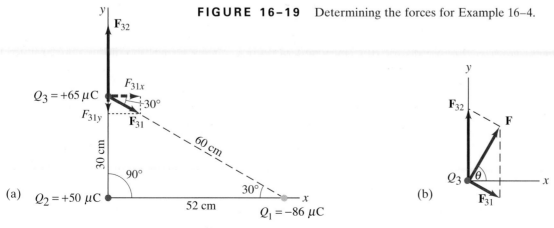

FIGURE 16–19 Determining the forces for Example 16–4.

The Electric Field

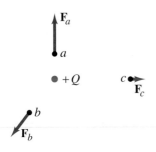

FIGURE 16–20 An electric field surrounds every charge. *P* is an arbitrary point.

FIGURE 16–21 Force exerted by charge $+Q$ on a small test charge, *q*, placed at points *a*, *b*, and *c*.

Many common forces might be referred to as "contact forces," such as your hands pushing or pulling a cart, or a tennis racket hitting a tennis ball.

In contrast, both the gravitational force and the electrical force act over a distance: there is a force even when the two objects are not touching. The idea of a force *acting at a distance* was a difficult one for early thinkers. Newton himself felt uneasy with this idea when he published his law of universal gravitation. A helpful way to look at the situation uses the idea of the **field**, developed by the British scientist Michael Faraday (1791–1867). In the electrical case, according to Faraday, an *electric field* extends outward from every charge and permeates all of space (Fig. 16–20). When a second charge is placed near the first charge, it feels a force because of the electric field that is there (say, at point *P* in Fig. 16–20). The electric field at the location of the second charge is considered to interact directly with this charge to produce the force. It must be emphasized, however, that a field, as we think of it here, is *not* a kind of matter.

We can investigate the electric field surrounding a charge or group of charges by measuring the force on a small positive **test charge**. By a test charge we mean a charge so small that the force it exerts does not significantly alter the distribution of the charges that create the field being measured. The force on a tiny positive test charge *q* placed at various locations in the vicinity of a single positive charge *Q* would be as shown in Fig. 16–21. The force at *b* is less than at *a* because the distance is greater (Coulomb's law); and the force at *c* is smaller still. In each case, the force is directed radially outward from *Q*. The electric field is defined in terms of the force on such a positive test charge. In particular, the **electric field**, **E**, at any point in space is defined as the force **F** exerted on a tiny positive test charge at that point divided by the magnitude of the test charge *q*:

Definition of electric field

$$\mathbf{E} = \frac{\mathbf{F}}{q}. \tag{16–3}$$

Ideally, **E** is defined as the limit of $\mathbf{F}/q$ as *q* is taken smaller and smaller, approaching zero. From this definition (Eq. 16–3), we see that the electric field at any point in space is a vector whose direction is the direction of the force on a positive test charge at that point, and whose magnitude is the *force per unit charge*. Thus **E** is measured in units of newtons per coulomb (N/C).

The reason for defining **E** as $\mathbf{F}/q$ (with $q \rightarrow 0$) is so that **E** does not depend on the magnitude of the test charge *q*. This means that **E** describes only the effect of the charges creating the electric field at that point.

The electric field at any point in space can be measured, based on the definition, Eq. 16–3. For simple situations involving one or several point charges, we can calculate what **E** will be. For example, the electric field at a distance *r* from a single point charge *Q* would have magnitude

$$E = \frac{kqQ/r^2}{q}$$

Electric field due to one point charge

$$= k\frac{Q}{r^2}; \qquad \text{[single point charge]} \tag{16–4a}$$

or, in terms of ϵ_0 (Eq. 16–2)

$$E = \frac{1}{4\pi\epsilon_0}\frac{Q}{r^2}. \qquad \text{[single point charge]} \quad \textbf{(16–4b)}$$

This relation for the electric field due to a single point charge is also (in addition to Eq. 16–1) referred to as Coulomb's law. Notice that E is independent of q—that is, it depends only on the charge Q which produces the field, and not on the value of the test charge q.

EXAMPLE 16–5 **Electrostatic copier.** An electrostatic copier works by selectively arranging positive charges (in a pattern to be copied) on the surface of a nonconducting drum, then gently sprinkling negatively charged dry toner (ink) particles onto the drum. The toner particles temporarily stick to the pattern on the drum and are later transferred to paper and "melted" to produce the copy. Suppose each toner particle has a mass of 9.0×10^{-16} kg and carries an average of 20 extra electrons to provide an electric charge. Assuming that the electric force on a toner particle must exceed twice its weight in order to ensure sufficient attraction, compute the required electric field strength near the surface of the drum. See Fig. 16–22.

SOLUTION The minimum value of electric field satisfies the relation

$$qE = 2\,mg$$

where $q = 20e$. Hence

$$E = \frac{2\,mg}{q} = \frac{2(9.0 \times 10^{-16}\,\text{kg})(9.8\,\text{m/s}^2)}{20(1.6 \times 10^{-19}\,\text{C})}$$

$$= 5.5 \times 10^3\,\text{N/C}.$$

Surface of drum

E

Toner particles held to drum surface by electric field **E**

FIGURE 16–22 Example 16–5.

EXAMPLE 16–6 **Electric field of a single point charge.** Calculate the magnitude and direction of the electric field at a point P which is 30 cm to the right of a point charge $Q = -3.0 \times 10^{-6}$ C.

SOLUTION The magnitude of the electric field due to a single point charge is given by Eq. 16–4:

$$E = k\frac{Q}{r^2} = \frac{(9.0 \times 10^9\,\text{N·m}^2/\text{C}^2)(3.0 \times 10^{-6}\,\text{C})}{(0.30\,\text{m})^2} = 3.0 \times 10^5\,\text{N/C}.$$

The direction of the electric field is *toward* the charge Q as shown in Fig. 16–23a since we defined the direction as that of the force on a positive test charge. If Q had been positive, the electric field would have pointed away, as in Fig. 16–23b.

This Example illustrates a general result: The electric field due to a positive charge points away from the charge, whereas **E** due to a negative charge points toward that charge.

FIGURE 16–23 Example 16–6. Electric field at point P (a) due to a negative charge Q, and (b) due to a positive charge Q.

|← 30 cm →|

$Q = -3.0 \times 10^{-6}$ C $E = 3.0 \times 10^5$ N/C

(a)

$Q = +3.0 \times 10^{-6}$ C P $E = 3.0 \times 10^5$ N/C

(b)

If the field is due to more than one charge, the individual fields (call them $\mathbf{E}_1$, $\mathbf{E}_2$, etc.) due to each charge are added vectorially to get the total field at any point:

Superposition principle for electric fields

$$\mathbf{E} = \mathbf{E}_1 + \mathbf{E}_2 + \cdots .$$

The validity of this **superposition principle** for electric fields is fully confirmed by experiment.[†]

EXAMPLE 16–7 **E in between two point charges.** Two point charges are separated by a distance of 10.0 cm. One has a charge of $-25\ \mu C$ and the other $+50\ \mu C$. (*a*) What is the direction and magnitude of the electric field at a point P in between them, that is 2.0 cm from the negative charge (Fig. 16–24a)? (*b*) If an electron is placed at rest at P, what will its acceleration (direction and magnitude) be initially?

SOLUTION (*a*) The field will be a combination of two fields both pointing to the left: the field due to the negative charge Q_1 points toward Q_1, and the field due to the positive charge Q_2 points away from Q_2, again to the left, Fig. 16–24b. Thus, we can add the magnitudes of the two fields together algebraically, ignoring the signs of the charges:

$$E = k\frac{Q_1}{r_1^2} + k\frac{Q_2}{r_2^2} = k\left(\frac{Q_1}{r_1^2} + \frac{Q_2}{r_2^2}\right) = k\frac{Q_1}{r_1^2}\left[1 + \frac{(Q_2/Q_1)}{(r_2^2/r_1^2)}\right]$$

where in the last step we factored out (Q_1/r_1^2). We substitute in $r_1 = 2.0\ \text{cm} = 2.0 \times 10^{-2}\ \text{m}$ and $r_2 = 8.0 \times 10^{-2}\ \text{m}$:

$$E = (9.0 \times 10^9\ \text{N·m}^2/\text{C}^2)\frac{(25 \times 10^{-6}\ \text{C})}{(2.0 \times 10^{-2}\ \text{m})^2}\left[1 + \frac{(50/25)}{(8.0/2.0)^2}\right]$$

$$= 5.6 \times 10^8\ [1 + \tfrac{1}{8}]\ \text{N/C} = 6.3 \times 10^8\ \text{N/C}.$$

Notice how factoring out Q_1/r_1^2 on the first line allowed us to see the relative strengths of the two contributing fields—namely that Q_2's field is only $\tfrac{1}{8}$ of Q_1's (or $\tfrac{1}{9}$ of the total).
(*b*) The electron will feel a force to the *right* since it is negatively charged and the acceleration will therefore be to the right, with a magnitude of

$$a = \frac{F}{m} = \frac{qE}{m} = \frac{(1.60 \times 10^{-19}\ \text{C})(6.3 \times 10^8\ \text{N/C})}{9.1 \times 10^{-31}\ \text{kg}} = 1.1 \times 10^{20}\ \text{m/s}^2.$$

FIGURE 16–24 Example 16–7. In (b), we don't know the relative lengths of $\mathbf{E}_1$ and $\mathbf{E}_2$ until we do the calculation.

[†]A more general form of Coulomb's law, which allows calculation of the electric field in some useful situations, is Gauss's law, discussed in Appendix D.

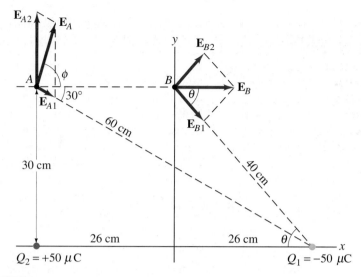

FIGURE 16-25 Calculation of the electric field at points A and B for Example 16–8.

EXAMPLE 16–8 **E above two point charges.** Calculate the total electric field (a) at point A and (b) at point B in Fig. 16–25 due to both charges, Q_1 and Q_2.

SOLUTION (a) The calculation is much like that of Example 16–4, but now we are dealing with electric fields. The electric field at A is the vector sum of the fields $\mathbf{E}_{A1}$ due to Q_1, and $\mathbf{E}_{A2}$ due to Q_2; using Eq. 16–4, $E = kQ/r^2$, they have magnitudes:

$$E_{A1} = \frac{(9.0 \times 10^9 \ \text{N·m}^2/\text{C}^2)(50 \times 10^{-6} \ \text{C})}{(0.60 \ \text{m})^2} = 1.25 \times 10^6 \ \text{N/C},$$

$$E_{A2} = \frac{(9.0 \times 10^9 \ \text{N·m}^2/\text{C}^2)(50 \times 10^{-6} \ \text{C})}{(0.30 \ \text{m})^2} = 5.0 \times 10^6 \ \text{N/C}.$$

➡ **PROBLEM SOLVING**

Ignore signs of charges and determine direction physically, showing directions on diagram

The directions are as shown, so the total electric field at A, $\mathbf{E}_A$, has components

$$E_{Ax} = E_{A1} \cos 30° = 1.1 \times 10^6 \ \text{N/C},$$

$$E_{Ay} = E_{A2} - E_{A1} \sin 30° = 4.4 \times 10^6 \ \text{N/C}.$$

Thus the magnitude of $\mathbf{E}_A$ is

$$E_A = \sqrt{(1.1)^2 + (4.4)^2} \times 10^6 \ \text{N/C} = 4.5 \times 10^6 \ \text{N/C},$$

and its direction is ϕ given by $\tan \phi = E_{Ay}/E_{Ax} = 4.4/1.1 = 4.0$, so $\phi = 76°$.

(b) Because B is equidistant (40 cm by the Pythagorean theorem) from the two equal charges, the magnitudes of E_{B1} and E_{B2} are the same; that is,

➡ **PROBLEM SOLVING**

Use symmetry to save work, when possible

$$E_{B1} = E_{B2} = \frac{kQ}{r^2} = \frac{(9.0 \times 10^9 \ \text{N·m}^2/\text{C}^2)(50 \times 10^{-6} \ \text{C})}{(0.40 \ \text{m})^2}$$

$$= 2.8 \times 10^6 \ \text{N/C}.$$

Also, because of the symmetry, the y components are equal and opposite. Hence the total field E_B is horizontal and equals $E_{B1} \cos \theta + E_{B2} \cos \theta = 2E_{B1} \cos \theta$; from the diagram, $\cos \theta = 26 \ \text{cm}/40 \ \text{cm} = 0.65$. Then

$$E_B = 2E_{B1} \cos \theta = 2(2.8 \times 10^6 \ \text{N/C})(0.65) = 3.6 \times 10^6 \ \text{N/C},$$

and the direction of $\mathbf{E}_B$ is along the $+x$ direction.

Solving electrostatics problems follows, to a large extent, the general problem-solving procedure discussed in Section 4–9. In particular,

1. Draw a careful diagram—namely, a free-body diagram for each object, showing all the forces acting on that object, or the electric field at a point due to all sources.

2. Apply Coulomb's law to get the magnitude of each force on a charged object, or the electric field at a point. Deal only with magnitudes of charges (leaving out minus signs), and obtain the magnitude of each force or electric field. Then determine the direction of each force or electric field physically (like charges repel each other, unlike charges attract). Show and label each vector force or field on your diagram. Then add vectorially all the forces on an object, or the contributing fields at a point, to get the resultant.

3. Use symmetry (say, in the geometry) whenever possible.

FIGURE 16–26 (a) Electric field at a given point in space. (b) Force on a positive charge. (c) Force on a negative charge.

If we are given the electric field **E** at a given point in space, then we can calculate the force **F** on a charge q placed at that point by writing (see Eq. 16–3):

$$\mathbf{F} = q\mathbf{E}.$$

If q is positive, **F** and **E** will point in the same direction. If q is negative, **F** and **E** point in opposite directions. See Fig. 16–26.

16–8   Field Lines

Since the electric field is a vector, it is sometimes referred to as a *vector field*. We could indicate the electric field with arrows at various points in a given situation, such as at *a*, *b*, and *c* in Fig. 16–27. The directions of $\mathbf{E}_a$, $\mathbf{E}_b$, and $\mathbf{E}_c$ are the same as that of the forces shown earlier in Fig. 16–21, but the lengths (magnitudes) are different since we divide by q. However, the relative lengths of $\mathbf{E}_a$, $\mathbf{E}_b$, and $\mathbf{E}_c$ are the same as for the forces since we divide by the same q each time. To indicate the electric field in such a way at *many* points, however, would result in many arrows, which might appear complicated or confusing. To avoid this, we use another technique, that of field lines.

In order to visualize the electric field, we draw a series of lines to indicate the direction of the electric field at various points in space. These **electric field lines** (sometimes called *lines of force*) are drawn so that they indicate the direction of the force due to the given field on a positive test charge. The lines of force due to a single positive charge are shown in Fig. 16–28a and for a single negative charge in Fig. 16–28b. In part (a) the

FIGURE 16–27 Electric field vector shown at three points, due to a single point charge Q. (Compare to Fig. 16–21.)

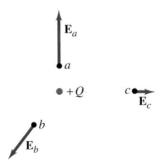

FIGURE 16–28 Electric field lines (a) near a single positive point charge, (b) near a single negative point charge.

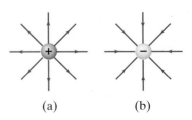

(a)   (b)

lines point radially outward from the charge, and in part (b) they point radially inward toward the charge because that is the direction the force would be on a positive test charge in each case (as in Fig. 16–23). Only a few representative lines are shown. One could just as well draw lines in between those shown since the electric field exists there as well. However, we can always draw the lines so that the *number of lines starting on a positive charge, or ending on a negative charge, is proportional to the magnitude of the charge.* Notice that near the charge, where the force is greatest, the lines are closer together. This is a general property of electric field lines: *the closer the lines are together, the stronger the electric field in that region.* In fact the lines can always be drawn so that the number of lines crossing unit area perpendicular to **E** is proportional to the magnitude of the electric field.

Figure 16–29a shows the electric field lines surrounding two charges of opposite sign. The electric field lines are curved in this case and they are directed from the positive charge to the negative charge. The direction of the field at any point is directed tangentially as shown by the arrow at point P. To satisfy yourself that this is the correct pattern for the electric field lines, you can make a few calculations such as those done in Example 16–8 for just this case (see Fig. 16–25). Figures 16–29b and c show the electric field lines surrounding two equal positive charges (b), and (c) for unequal charges, $+2Q$ and $-Q$; note that twice as many lines leave $+2Q$ as there are lines entering $-Q$ (number of lines is proportional to magnitude of Q). Finally, in (d), we see the field between two oppositely charged parallel plates. Notice that the electric field lines between the two plates start out perpendicular to the surface of the metal plates (we'll see why this is always true in the next Section) and go directly from one plate to the other, as we expect because a positive test charge placed between the plates would feel a strong repulsion from the positive plate and a strong attraction to the negative plate. The field lines between the plates are parallel and equally spaced, except near the edges. Thus, in the central region, the electric field has the same magnitude at all points and we can write

$$E = \text{constant.} \quad \text{[between two closely spaced parallel plates]} \quad \textbf{(16–5)}$$

Although the field fringes near the edges (the lines curve), we can often ignore this, particularly if the separation of the plates is small compared to their size. This should be compared to the field of a single point charge, where the field decreases as the square of the distance, Eq. 16–4.

We summarize the properties of field lines as follows:

1. The field lines indicate the direction of the electric field; the field points in the direction tangent to the field line at any point.

2. The lines are drawn so that the magnitude of the electric field, E, is proportional to the number of lines crossing unit area perpendicular to the lines. The closer the lines, the stronger the field.

3. Electric field lines start on positive charges and end on negative charges; and the number starting or ending is proportional to the magnitude of the charge.

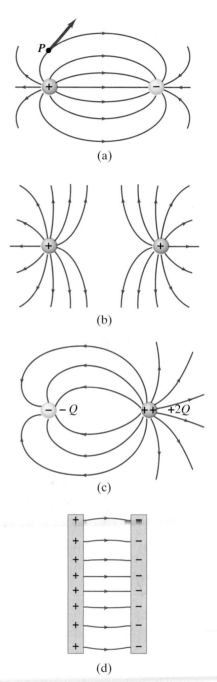

(a)

(b)

(c)

(d)

FIGURE 16–29 Electric field lines for four arrangements of charges.

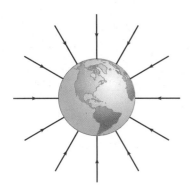

FIGURE 16–30 The Earth's gravitational field.

The field concept can also be applied to the gravitational force. Thus we can say that a **gravitational field** exists for every object that has mass. One object attracts another by means of the gravitational field. The Earth, for example, can be said to possess a gravitational field (Fig. 16–30) which is responsible for the gravitational force on objects. The *gravitational field intensity* is defined as the *force per unit mass*. The magnitude of the Earth's gravitational field intensity at any point is then (GM_E/r^2), where M_E is the mass of the Earth, r is the distance of the point from the Earth's center, and G is the gravitational constant (Chapter 5). At the Earth's surface, r is simply the radius of the Earth and the gravitational field intensity is simply equal to g, the acceleration due to gravity (since $F/m = mg/m = g$). Beyond the Earth, the gravitational field intensity can be calculated at any point as a sum of terms due to Earth, Sun, Moon, and other bodies that contribute significantly.

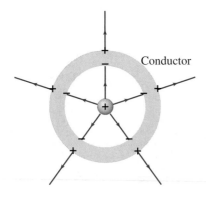

FIGURE 16–31 A charge placed inside a spherical shell. Charges are induced on the conductor surfaces. The electric field exists even beyond the shell but not within the conductor itself.

FIGURE 16–32 If the electric field **E** at the surface of a conductor had a component parallel to the surface, $E_\parallel$, the latter would accelerate electrons into motion. In the static case (no charges are in motion), $E_\parallel$ must be zero, and so the electric field must be perpendicular to the conductor's surface: $E = E_\perp$.

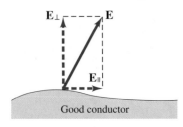

16–9 Electric Fields and Conductors

We now discuss some properties of good conductors. First, *the electric field inside a good conductor is zero in the static situation*—that is, when the charges are at rest. If there were an electric field within a conductor, there would be a force on its free electrons since $F = qE$. The electrons would move until they reached positions where the electric field, and therefore the electric force on them, did become zero.

This reasoning has some interesting consequences. For one, *any net charge on a good conductor distributes itself on the surface*. For a negatively charged conductor, you can imagine that the negative charges repel one another and race to the surface to get as far from one another as possible. Another consequence is the following. Suppose that a positive charge Q is surrounded by an isolated uncharged metal conductor whose shape is a spherical shell, Fig. 16–31. Because there can be no field within the metal, the lines leaving the positive charge must end on negative charges on the inner surface of the metal. Thus an equal amount of negative charge, $-Q$, is induced on the inner surface of the spherical shell. Then, since the shell is neutral, a positive charge, $+Q$, of the same magnitude must exist on the outer surface of the shell. Thus, although no field exists in the metal itself, an electric field exists outside of it, as shown in Fig. 16–31, as if the metal were not even there.

A related property of static electric fields and conductors is that *the electric field is always perpendicular to the surface outside of a conductor*. If there were a component of **E** parallel to the surface (Fig. 16–32), electrons at the surface would move along the surface in response to this force until they reached positions where no force was exerted on them—that is, until the electric field was perpendicular to the surface.

These properties pertain only to conductors. Inside a nonconductor, which does not have free electrons, an electric field can exist (Section 17–8). And the electric field outside a nonconductor does not necessarily make an angle of 90° to the surface.

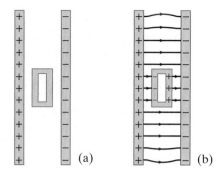

FIGURE 16–33 Example 16–9.

➡ **PHYSICS APPLIED**

Shielding
Safety in a storm

CONCEPTUAL EXAMPLE 16–9 **Shielding, and safety in a storm.** A hollow metal box is placed between two parallel charged plates as shown in Fig. 16–33a. What's the field like inside the box?

RESPONSE If our metal box were solid, and not hollow, the electrons in the box, even if it were neutral overall, would redistribute themselves along the surface so that the field lines would not penetrate the conducting metal of the box. For a hollow box, the external field is not changed since the electrons in the metal can move just as freely as before to the surface. Hence we conclude that the field inside the hollow metal box is zero. So the field lines are something like those shown in Fig. 16–33b. A conducting box used in this way is an effective device for shielding delicate instruments and electronic circuits from unwanted external electric fields. We also can see that a relatively safe place to be during a lightning storm is inside a car, surrounded by metal.

*16–10 Electric Forces in Molecular Biology: DNA Structure and Replication

The study of the structure and functioning of a living cell at the molecular level is known as molecular biology. It is an important area for application of physics. Since the interior of a cell is mainly water, we can imagine it as a vast sea of molecules continually in motion (as in kinetic theory, Chapter 13), colliding with one another with various amounts of kinetic energy. These molecules interact with one another in various ways—chemical reactions (making and breaking of bonds between atoms) and more brief interactions or unions that occur because of *electrostatic attraction* between molecules.

The many processes that occur within the cell are now considered to be the result of *random ("thermal") molecular motion plus the ordering effect of the electrostatic force*. We now use these ideas to analyze some basic cellular processes involving macromolecules (large molecules). The picture we present here has not been seen "in action." Rather, it is a model of what happens based on presently accepted physical theories and a great variety of experimental results.

The genetic information that is passed on from generation to generation in all living objects is contained in the chromosomes, which are made up of genes. Each gene contains the information needed to produce a particular type of protein molecule. The genetic information contained in a gene is built into the principal molecule of a chromosome, the DNA (deoxyribonucleic acid). A DNA molecule consists of a long chain of many small molecules known as nucleotide bases. There are only four types of bases: adenine (A), cytosine (C), guanine (G), and thymine (T).

The DNA in a chromosome generally consists of two long DNA chains wrapped about one another in the shape of a "double helix." The

➡ **PHYSICS APPLIED**

Inside a cell:
Kinetic theory plus
electrostatic force

➡ **PHYSICS APPLIED**

DNA structure

FIGURE 16–34 (a) Section of a DNA double helix. (b) "Close-up" view of the helix, showing how A and T attract each other and how G and C attract each other through electrostatic forces. The + and − signs indicated on certain atoms represent net charges, usually a fraction of e, due to uneven sharing of electrons. The red dots are used to indicate the electrostatic attraction (often called a "weak bond" or "hydrogen bond"). Note that there are two weak bonds between A and T, and three between C and G. The distance unit is the angstrom (1 Å = 10^{-10} m).

(a)

(b)

gcnctic information is contained in the specific order of the 4 bases (A, C, G, T) along the chain. As shown in Fig. 16–34, the two strands are attracted by electrostatic forces—that is, by the attraction of positive charges to negative charges. We see in part (a) that an A (adenine) on one strand is always opposite a T on the other strand; similarly, a G is always opposite a C. This important ordering effect occurs because the shapes of the four molecules A, T, C, and G are such that a T fits closely only into an A, and a G into a C; and only in the case of this close proximity of the charged portions is the electrostatic force great enough to hold them together even for a short time (Fig. 16–34b), forming what are often referred to as "weak bonds." The electrostatic force between A and T, and between C and G, exists because these molecules have charged parts. These charges are due to some electrons in each of these molecules spending more time orbiting one atom than another. For example, the electron normally on the H atom of adenine spends some of its time orbiting the adjacent N atom (more on this in Chapter 29), so the N has a net negative charge and the H a positive charge (upper part of Fig. 16–34b). This H^+ atom of adenine is then attracted to the O^- atom of thymine. These net + and – charges usually have magnitudes of a fraction of e (charge on the electron) such as $0.2e$ or $0.4e$. See Problem 47.

How does the arrangement shown in Fig. 16–34 come about? It occurs when the chromosome replicates (duplicates) itself just before cell division. Indeed, the arrangement of A opposite T and G opposite C is crucial for ensuring that the genetic information is passed on accurately to the next generation. The process of replication is shown in a simplified form in Fig. 16–35. The two strands of the DNA chain separate (with the help of enzymes, which also operate via the electrostatic force), leaving the charged parts of the bases exposed. Once replication starts, let us see how the correct order of bases occurs by focusing our attention on the G molecule

⇒ **PHYSICS APPLIED**

DNA replication

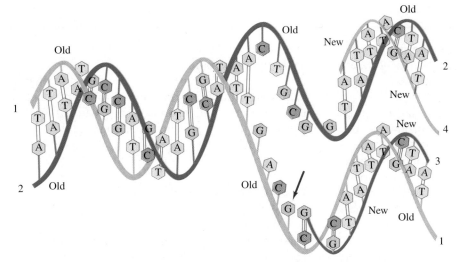

FIGURE 16–35
Replication of DNA.

indicated by the arrow on the lowest strand in the figure. There are many un-attached nucleotide bases of all four kinds bouncing around in the cellular fluid. The only one of the four bases that will experience attraction to our G, if it bounces close to it, will be a C. The charges on the other three bases are not arranged so that they can get close to those on the G, and thus there will be no significant attractive force exerted on them—remember that the force decreases rapidly with distance. Because the G attracts an A, T, or G almost not at all, an A, T, or G will be knocked away by collisions with other mole-cules before enzymes can attach it to the growing (number 3) chain. But the electrostatic force will often hold a C opposite our G long enough so that an enzyme can attach the C to the growing end of the new chain.

Thus we see that electrostatic forces are responsible for selecting the bases in the proper order during replication, so the genetic information is passed on accurately to the next generation. Note in Fig. 16–35 that the new number 4 strand has the same order of bases as the old number 1 strand; and the new number 3 strand is the same as the old number 2. So the two new helixes, 1–3 and 2–4, are identical to the original 1–2 helix. The error rate—say a T being incorporated in a new chain opposite a G—is on the order of 1 in 10^4 [and is kept even lower (1 in 10^8 to 10^9) with the aid of special enzymatic "proofreading and repair" mechanisms]. Such an error constitutes a spontaneous mutation and a possible change in some characteristic of the organism. It is important for the survival of the organism that the error rate be low, but it cannot be zero if evolution (which can only occur through mutation) is to take place.

This process of DNA replication is often presented as if it occurred in clockwork fashion—as if each molecule knew its role and went to its as-signed place, like bees in a hive. But this is not the case. The forces of at-traction between the electric charges of the molecules are rather weak and become significant only when the molecules can come close together and several "weak bonds" can be made. Indeed, if the shapes are not just right, there is almost no electrostatic attraction, which is why there are few mis-takes. The point is that there are many molecules in the cell, all jostling about, but only that one type which has the proper shape will be attracted sufficiently so as to remain long enough to become attached to the grow-ing chain. Thus, out of the random motion of the molecules, the electro-static force acts to bring order out of chaos.

SUMMARY

There are two kinds of **electric charge**, positive and negative. These designations are to be taken algebraically—that is, any charge is plus or minus so many coulombs (C), in SI units.

Electric charge is **conserved**: if a certain amount of one type of charge is produced in a process, an equal amount of the opposite type is also produced; thus the *net* charge produced is zero.

According to the atomic theory, electricity originates in the atom, which consists of a positively charged nucleus surrounded by negatively charged electrons. Each electron has a charge $-e = -1.6 \times 10^{-19}\,\mathrm{C}$.

Conductors are those materials in which many electrons are relatively free to move, whereas electric **insulators** are those in which very few electrons are free to move.

An object is negatively charged when it has an excess of electrons, and positively charged when it has less than its normal amount of electrons. The charge on any object is thus a whole number times $+e$ or $-e$; that is, charge is **quantized**.

An object can become charged by rubbing (in which electrons are transferred from one material to another), by conduction (which is transfer of charge from one charged object to another by touching), or by induction (the separation of charge within an object because of the close approach of another charged object but without touching).

Electric charges exert a force on each other. If two charges are of opposite types, one positive and one negative, they each exert an attractive force on the other. If the two charges are the same type, each repels the other.

The magnitude of the force one point charge exerts on another is proportional to the product of their charges, and inversely proportional to the square of the distance between them:

$$F = k\frac{Q_1 Q_2}{r^2};$$

this is **Coulomb's law**. In SI units, k is often written as $1/4\pi\epsilon_0$.

We think of an **electric field** as existing in space around any charge or group of charges. The force on another charged object is then said to be due to the electric field present at its location.

The *electric field*, **E**, at any point in space due to one or more charges, is defined as the force per unit charge that would act on a test charge q placed at that point:

$$\mathbf{E} = \frac{\mathbf{F}}{q}.$$

Electric fields are represented by **electric field lines** that start on positive charges and end on negative charges. Their direction indicates the direction the force would be on a tiny positive test charge placed at a point. The lines can be drawn so that the number per unit area is proportional to the magnitude of E.

The static electric field (that is, no charges moving) inside a good conductor is zero, and the electric field lines just outside a charged conductor are perpendicular to its surface.

QUESTIONS

1. If you charge a pocket comb by rubbing it with a silk scarf, how can you determine if the comb is positively or negatively charged?
2. Why does a shirt or blouse taken from a clothes dryer sometimes cling to your body?
3. Explain why fog or rain droplets tend to form around ions or electrons in the air.
4. A positively charged rod is brought close to a neutral piece of paper, which it attracts. Draw a diagram showing the separation of charge and explain why attraction occurs.
5. Why does a plastic ruler that has been rubbed with a cloth have the ability to pick up small pieces of paper? Why is this difficult to do on a humid day?
6. Contrast the *net charge* on a conductor to the "free charges" in the conductor.
7. Figures 16–7 and 16–8 show how a charged rod placed near an uncharged metal object can attract (or repel) electrons. There are a great many electrons in the metal, yet only some of them move as shown. Why not all of them?
8. When an electroscope is charged, the two leaves repel each other and remain at an angle. What balances the electric force of repulsion so that the leaves don't separate further?
9. The form of Coulomb's law is very similar to that for Newton's law of universal gravitation. What are the differences between these two laws? Compare also gravitational mass and electric charge.
10. We are not normally aware of the gravitational or electrical force between two ordinary objects. What is the reason in each case? Give an example where we are aware of each one and why.

11. Is the electric force a conservative force? Why or why not? (See Chapter 6.)

12. What experimental observations mentioned in the text rule out the possibility that the numerator in Coulomb's law contains the sum $(Q_1 + Q_2)$ rather than the product Q_1Q_2?

13. When a charged ruler attracts small pieces of paper, sometimes a piece jumps quickly away after touching the ruler. Explain.

14. Explain why we use *small* test charges when measuring electric fields.

15. When determining an electric field, must we use a *positive* test charge, or would a negative one do as well? Explain.

16. Draw the electric field lines surrounding two negative electric charges a distance l apart.

17. Assume that the two opposite charges in Fig. 16–29a are 12.0 cm apart. Consider the magnitude of the electric field 2.5 cm from the positive charge. On which side of this charge—top, bottom, left, or right—is the electric field the strongest? The weakest?

18. Consider the electric field at the three points indicated by the letters A, B, and C in Fig. 16–36. First draw an arrow at each point indicating the direction of the net force that a positive test charge would experience if placed at that point, then list the letters in order of *decreasing* field strength (strongest first).

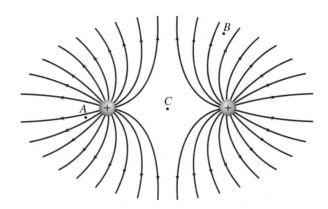

FIGURE 16-36 Question 18.

19. Why can electric field lines never cross?

20. Consider a small positive test charge located on an electric field line at some point, such as point P in Fig. 16–29a. Is the direction of the velocity and/or acceleration of the test charge along this line? Discuss.

21. We wish to determine the electric field at a point near a positively charged metal sphere (a good conductor). We do so by bringing a small test charge, q_0, to this point and measure the force F_0 on it. Will F_0/q_0 be greater than, less than, or equal to, the electric field $\mathbf{E}$ as it was at that point before the test charge was present?

PROBLEMS

SECTIONS 16-5 AND 16-6

1. (I) How many electrons make up a charge of $-30.0 \mu C$?

2. (I) Two charged smoke particles exert a force of 4.2×10^{-2} N on each other. What will be the force if they are moved so they are only one eighth as far apart?

3. (I) Two charged balls are 20.0 cm apart. They are moved, and the force on each of them is found to have been tripled. How far apart are they now?

4. (I) Two charged Ping-Pong balls separated by a distance of 1.50 m exert an electric force of 0.0200 N on each other. What will be the force if the objects are brought closer, to a separation of only 30.0 cm?

5. (I) What is the magnitude of the electric force of attraction between an iron nucleus ($q = +26e$) and its innermost electron if the distance between them is 1.5×10^{-12} m?

6. (I) What is the repulsive electrical force between two protons in a nucleus that are 5.0×10^{-15} m apart from each other?

7. (I) What is the magnitude of the force a $+15$-μC charge exerts on a $+3.0$-mC charge 40 cm away? ($1 \mu C = 10^{-6}$ C, 1 mC $= 10^{-3}$ C.)

8. (II) A person scuffing her feet on a wool rug on a dry day accumulates a net charge of $-60 \mu C$. How many excess electrons does this person get, and by how much does her mass increase?

9. (II) Imagine that space invaders could deposit extra electrons in equal amounts on the Earth and on your car, which has a mass of 1050 kg. Note that the rubber tires would provide some insulation. How much charge Q would need to be placed on your car (same amount on the Earth) in order to levitate it (overcome gravity)? [*Hint*: Assume that the Earth's charge is spread uniformly so it acts as if it were located at the Earth's center, and then the separation distance is the radius of the Earth.]

10. (II) What is the total charge of all the electrons in 1.0 kg of H_2O?

11. (II) Particles of charge $+70$, $+48$, and $-80 \mu C$ are placed in a line (Fig. 16–37). The center one is 0.35 m from each of the others. Calculate the net force on each charge due to the other two.

FIGURE 16-37 Problem 11.

FIGURE 16–38 Problem 12.

12. (II) Three positive particles of charges $11.0 \, \mu C$ are located at the corners of an equilateral triangle of side 15.0 cm (Fig. 16–38). Calculate the magnitude and direction of the net force on each particle.

13. (II) A charge of 6.00 mC is placed at each corner of a square 1.00 m on a side. Determine the magnitude and direction of the force on each charge.

14. (II) Repeat Problem 13 for the case when two of the positive charges, on opposite corners, are replaced by negative charges of the same magnitude (Fig. 16–39).

FIGURE 16–39 Problem 14.

15. (II) Compare the electric force holding the electron in orbit ($r = 0.53 \times 10^{-10}$ m) around the proton nucleus of the hydrogen atom, with the gravitational force between the same electron and proton. What is the ratio of these two forces?

16. (II) Suppose that electrical attraction, rather than gravity, were responsible for holding the Moon in orbit around the Earth. If equal and opposite charges Q were placed on the Earth and the Moon, what should be the value of Q to maintain the present orbit? Use these data: mass of Earth = 5.97×10^{24} kg, mass of Moon = 7.35×10^{22} kg, radius of orbit = 3.84×10^8 m. Treat the Earth and Moon as point particles.

17. (II) Two positive point charges are a fixed distance apart. The sum of their charges is Q_T. What charge must each have in order to (a) maximize the electric force between them, and (b) minimize it?

18. (II) In one model of the hydrogen atom, the electron revolves in a circular orbit around the proton with a speed of 1.1×10^6 m/s. Determine the radius of the electron's orbit.

19. (III) A $+5.7 \, \mu C$ and a $-3.5 \, \mu C$ charge are placed 25 cm apart. Where can a third charge be placed so that it experiences no net force?

20. (III) Two small nonconducting spheres have a total charge of $80.0 \, \mu C$. When placed 1.06 m apart, the force each exerts on the other is 12.0 N and is repulsive. What is the charge on each? What if the force were attractive?

SECTIONS 16–7 AND 16–8

21. (I) What is the magnitude of the acceleration experienced by an electron in an electric field of 600 N/C? How does the direction of the acceleration depend on the direction of the field at that point? How does the direction of the acceleration depend on the electron's velocity at that point?

22. (I) What is the magnitude and direction of the electric force on an electron in a uniform electric field of strength 3500 N/C that points due east?

23. (I) A proton is released in a uniform electric field, and it experiences an electric force of 3.2×10^{-14} N toward the south. What are the magnitude and direction of the electric field?

24. (I) A force of 8.4 N is exerted on a -8.8-μC charge in a downward direction. What is the magnitude and direction of the electric field at this point?

25. (I) What is the magnitude and direction of the electric field 30.0 cm directly above a 33.0×10^{-6}-C charge?

26. (II) What is the magnitude and direction of the electric field at a point midway between a -8.0-μC and a $+6.0$-μC charge 4.0 cm apart?

27. (II) An electron is released from rest in a uniform electric field and accelerates to the north at a rate of 125 m/s². What is the magnitude and direction of the electric field?

28. (II) The electric field midway between two equal but opposite point charges is 1750 N/C, and the distance between the charges is 16.0 cm. What is the magnitude of the charge on each?

29. (II) Determine the direction and magnitude of the electric field at the point P shown in Fig. 16–40. The two charges are separated by a distance of $2a$ and the point P is a distance x out on the perpendicular bisector of the line joining them. Express your answers in terms of Q, x, a, and k.

FIGURE 16–40 Problem 29.

FIGURE 16-41 Problem 30.

30. (II) Use Coulomb's law to determine the magnitude and direction of the electric field at points A and B in Fig. 16–41 due to the two positive charges ($Q = 9.0 \, \mu C$) shown. Is your result consistent with Fig. 16–29b?

31. (II) Calculate the electric field at the center of a square 60 cm on a side if one corner is occupied by a $+45.0$-μC charge and the other three are occupied by -31.0-μC charges.

32. (II) Calculate the electric field at one corner of a square 1.00 m on a side if the other three corners are occupied by 2.80×10^{-6}-C charges.

33. (II) (a) Determine the electric field $\mathbf{E}$ at the origin O in Fig. 16–42 due to the two charges at A and B. (b) Repeat, but let the charge at B be reversed in sign.

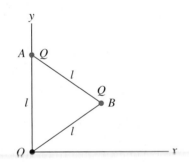

FIGURE 16-42
Problem 33.

34. (II) Draw, approximately, the electric field lines about two point charges, $+Q$ and $-3Q$, which are a distance l apart.

35. (II) What is the electric field strength at a point in space where a proton ($m = 1.67 \times 10^{-27}$ kg) experiences an acceleration of 1 million "g's"?

36. (II) A spacecraft makes a trip from the Earth to the Moon, 380,000 km away. At what point in the trip will the gravitational field be zero? The mass of the Moon is about $\frac{1}{81}$ that of the Earth.

FIGURE 16-43
Problem 37.

37. (III) You are given two unknown point charges, Q_1 and Q_2. At a point on the line joining them, one third of the way from Q_1 to Q_2, the electric field is zero (Fig. 16–43). What can you say about these two charges?

38. (III) An electron (mass $m = 9.11 \times 10^{-31}$ kg) is accelerated in the uniform field $\mathbf{E}$ ($E = 1.85 \times 10^4$ N/C) between two parallel charged plates. The separation of the plates is 1.20 cm. The electron is accelerated from rest near the negative plate and passes through a tiny hole in the positive plate, Fig. 16–44. (a) With what speed does it leave the hole? (b) Show that the gravitational force can be ignored.

FIGURE 16-44
Problem 38.

39. (III) An electron moving at 1 percent the speed of light to the right enters a uniform electric field region where the field is known to be parallel to its direction of motion. If the electron is to be brought to rest in the space of 5.0 cm, (a) what direction is required for the electric field, and (b) what is the strength of the field?

*SECTION 16–10

* 40. (III) The two strands of the helix-shaped DNA molecule are held together by electrostatic forces as shown in Fig. 16–34. Assume that the net average charge indicated on H and N atoms is $0.2e$, and on the indicated C and O atoms is $0.4e$, that atoms on each molecule are separated by 1.0×10^{-10} m, and that all relevant angles are 120°. Estimate the net force between: (a) a thymine and an adenine; and (b) a cytosine and a guanine. (c) Estimate the total force for a DNA molecule containing 10^5 pairs of such molecules.

GENERAL PROBLEMS

41. How close must two electrons be if the electric force between them is equal to the weight of either at the Earth's surface?

42. A 3.0-g copper penny has a positive charge of 42 μC. What fraction of its electrons has it lost?

43. A proton ($m = 1.67 \times 10^{-27}$ kg) is suspended at rest in a uniform electric field $\mathbf{E}$. Take into account gravity and determine $\mathbf{E}$.

44. Measurements indicate that there is an electric field surrounding the Earth. Its magnitude is about 150 N/C at the Earth's surface and points inward toward the Earth's center. What is the magnitude of the electric charge on the Earth? Is it positive or negative? [*Hint*: The electric field outside a uniformly charged sphere is the same as if all the charge were concentrated at its center.]

45. A water droplet of radius 0.018 mm remains stationary in the air. If the electric field of the Earth is 150 N/C, how many excess electron charges must the water droplet have?

46. Calculate the magnitude of the electric field at the center of a square with sides 25 cm long if the corners, taken in rotation, have charges of $1.0 \mu C$, $2.0 \mu C$, $3.0 \mu C$, and $4.0 \mu C$ (all positive).

47. Estimate the net force between the CO group and the HN group shown in Fig. 16–45. The C and O have charges $\pm 0.40e$ and the H and N have charges $\pm 0.20e$ where $e = 1.6 \times 10^{-19}$ C. [*Hint*: Do not include the "internal" forces between C and O, or between H and N.]

FIGURE 16–45   Problem 47.

48. Two charges, $-Q_0$ and $-3Q_0$, are a distance l apart. These two charges are free to move along the line passing through them both, but do not because there is a third charge nearby. What must be the magnitude of the third charge and its placement in order for the first two to be in equilibrium?

49. A point charge ($m = 1.0$ g) at the end of an insulating string of length 50 cm (Fig. 16–46) is observed to be in equilibrium in a known uniform horizontal electric field, $E = 9200$ N/C, when the pendulum has swung so it is 1.0 cm high. If the field points to the right in Fig. 16–46, determine the magnitude and sign of the point charge.

FIGURE 16–46
Problem 49.

50. A positive point charge $Q_1 = 2.5 \times 10^{-5}$ C is fixed at the origin of coordinates, and a negative charge $Q_2 = -5.0 \times 10^{-6}$ C is fixed to the x axis at $x = +2.0$ m. Find the location of the place(s) along the x axis where the electric field due to these two charges is zero.

51. An electron with speed $v_0 = 1.5 \times 10^6$ m/s is traveling parallel to an electric field ($\mathbf{v}_0 \parallel \mathbf{E}$) of magnitude $E = 7.7 \times 10^3$ N/C. (*a*) How far will it travel before it stops? (*b*) How much time will elapse before it returns to its starting point?

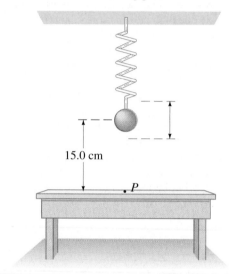

FIGURE 16–47   Problem 52.

52. A small lead ball is encased in insulating plastic and suspended vertically from an ideal spring ($k = 126$ N/m) above a lab table, Fig. 16–47. The total mass of the coated ball is 0.800 kg, and its center lies 15.0 cm above the tabletop when in equilibrium. The ball is pulled down 5.00 cm below equilibrium, an electric charge $Q = -3.00 \times 10^{-6}$ C is deposited on the ball, and the system is released. Using what you know about harmonic oscillation, write an expression for the electric field strength as a function of time that would be measured at the point on the tabletop (P) directly below the ball.

53. A large electroscope is made with "leaves" that are 70-cm-long wires with 24-g balls at the ends. When charged, nearly all the charge resides on the balls. If the wires each make a 30° angle with the vertical (Fig. 16–48), what total charge Q must have been applied to the electroscope?

FIGURE 16–48
Problem 53.

54. (II) Dry air will break down and generate a spark if the electric field exceeds about 3×10^6 N/C. How much charge could be packed onto a green pea (diameter 0.75 cm) before the pea spontaneously discharges? [*Hint*: Eq. 16–4 works outside a sphere if r is measured from its center.]

55. Two point charges, $Q_1 = -6.7\ \mu C$ and $Q_2 = 1.3\ \mu C$ are located between two oppositely charged parallel plates, as shown in Fig. 16–49. The two point charges are separated by a distance of $x = 0.34$ m. Assume that the electric field produced by the charged plates is uniform and equal to $E = 73{,}000$ N/C. Calculate the net electrostatic force on Q_1 and give its direction.

FIGURE 16–49   Problem 55.

56. A point charge of mass 0.210 kg, and net charge $+0.340\ \mu C$, hangs at rest at the end of an insulating string above a single sheet of charge. The horizontal sheet of charge creates a uniform vertical electric field in the vicinity of the point charge. The tension in the string is measured to be 5.67 N. Calculate the magnitude and direction of the electric field due to the sheet of charge (Fig. 16–50).

Uniform sheet of charge

FIGURE 16–50   Problem 56.

Lightning: The potential difference (voltage) between clouds and the Earth can become so high that electrons are pulled off atoms of the air by the large electric field. The air becomes a conductor as the ionized atoms and freed electrons flow rapidly, colliding with more atoms, and causing more ionization. The massive flow of charge reduces the potential difference and the "discharge" quickly ceases. The light represents energy released when the ions and electrons recombine to form atoms (Chapter 27).

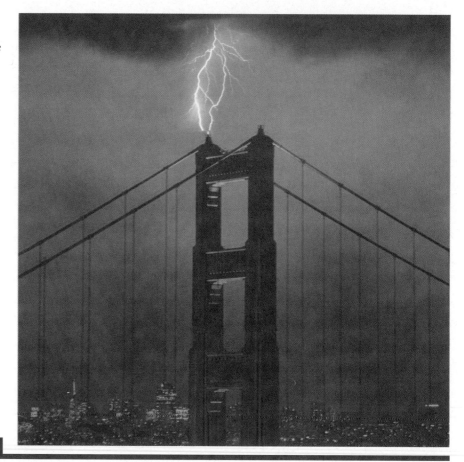

17 ELECTRIC POTENTIAL AND ELECTRIC ENERGY; CAPACITANCE

We saw in Chapter 6 that the concept of energy was extremely valuable in dealing with mechanical problems. For one thing, energy is a conserved quantity and is thus an important aspect of nature. Furthermore, we saw that many problems could be solved using the energy concept even though a detailed knowledge of the forces involved was not possible, or when a calculation involving Newton's laws would have been too difficult.

The energy point of view can be used in electricity, and it is especially useful. It not only extends the law of conservation of energy, but it gives us another way to view electrical phenomena; and it is a tool in solving problems more easily, in many cases, than by using forces and electric fields.

17–1 Electric Potential and Potential Difference

To apply conservation of energy, we need to define electric potential energy as for other types of potential energy (Chapter 6). That is, we define the change in electric potential energy, $PE_a - PE_b$, when a charge q moves from some point b to a second point a, as the negative of the work done by the electric force to move the charge from b to a. For example, consider the electric field between two equally but oppositely charged parallel plates whose separation is small compared to their width and height, so the field will be uniform over most of the region, Fig. 17–1. Now consider a small positive point charge q placed at point b very near the positive plate as shown. If the charge is released, the electric force will do work on the charge and accelerate it toward the negative plate. In the process, the charged particle will have its kinetic energy increased. The potential energy will be decreased by an equal amount, equal to the negative of the work done by the electric force. In accord with the conservation of energy, electric potential energy is transformed into kinetic energy, and the total energy is conserved. Note that the positive charge q has its greatest potential energy at point b, near the positive plate,[†] so $(\text{PE}_b - \text{PE}_a) > 0$. The reverse is true for a negative charge: its potential energy is greatest near the negative plate.

We defined the electric field (Chapter 16) as the force per unit charge. Similarly, it is useful to define the **electric potential** (or simply the **potential** when "electric" is understood) as the *potential energy per unit charge.* Electric potential is given the symbol V. If a point charge q has electric potential energy PE_a at some point a, the electric potential V_a at this point is

$$V_a = \frac{\text{PE}_a}{q}.$$

FIGURE 17–1 Work is done by the electric field in moving the positive charge from position b to position a.

As we discussed in Chapter 6 (Section 6–4), only differences in potential energy are physically measurable. Hence only the **difference in potential**, or the **potential difference**, between two points a and b (such as between a and b in Fig. 17–1) is measurable. Since the difference in potential energy, $\text{PE}_a - \text{PE}_b$, is equal to the negative of the work, W_{ba}, done by the electric force to move the charge from point b to point a, we have that the potential difference V_{ab} is

Potential difference

$$V_{ab} = V_a - V_b = -\frac{W_{ba}}{q}$$

The unit of electric potential, and of potential difference, is joules/coulomb and is given a special name, the **volt**, in honor of Alessandro Volta (1745–1827; he is best known for having invented the electric battery, as discussed in Chapter 18). The volt is abbreviated V, so $1\,\text{V} = 1\,\text{J/C}$. Note from our definition that the positive plate in Fig. 17–1 is at a higher potential than the negative plate. Thus a positively charged object moves naturally from a high potential to a low potential. A negative charge does the reverse. Potential difference, since it is measured in volts, is often referred to as **voltage**.

The volt ($1\,\text{V} = 1\,\text{J/C}$)

If we wish to speak of the potential, V_a, at some point a, we must be aware that V_a depends on where the potential is chosen to be zero. The zero point for electric potential in a given situation, just as for potential energy, can be chosen arbitrarily since only differences in potential energy can be measured. Often the ground, or a conductor connected directly to the ground, is taken as zero potential, and other potentials are given with respect to ground. (Thus, a point where the voltage is 50 V is one where the difference of potential between it and ground is 50 V.) In other cases, as we shall see, we may choose the potential to be zero at infinity (see Section 17–5).

Voltage = potential difference

V = 0 chosen arbitrarily

[†]At this point it has its greatest ability to do work (on some other object or system).

Since the electric potential is defined as the potential energy per unit charge, then the change in potential energy of a charge q when moved between two points a and b is

$$\Delta PE = PE_b - PE_a = qV_{ba}. \qquad (17\text{--}1)$$

That is, if an object with charge q moves through a potential difference V_{ba}, its potential energy changes by an amount qV_{ba}. For example, if the potential difference between the two plates in Fig. 17–1 is 6 V, then a 1-C charge moved (say by an external force) from a to b will gain $(1\,C)(6\,V) = 6\,J$ of electric potential energy. (And it will lose 6 J of electric PE if it moves from b to a.) Similarly, a 2-C charge will gain 12 J, and so on. Thus, electric potential difference is a measure of how much energy an electric charge can acquire in a given situation. And, since energy is the ability to do work, the electric potential difference is also a measure of how much work a given charge can do. The exact amount depends both on the potential difference and on the charge.

To better understand electric potential, let's make a comparison to the gravitational case when a rock falls from the top of a cliff. The greater the height, h, of a cliff, the more potential energy ($= mgh$) the rock has at the top of the cliff, relative to the bottom, and the more kinetic energy it will have when it reaches the bottom. The actual amount of kinetic energy it will acquire, and the amount of work it can do, depends both on the height of the cliff and the mass m of the rock. A large rock and a small rock can be at the same height h (Fig. 17–2a) and thus have the same "gravitational potential," but the larger rock has the greater potential energy. Similarly, in the electrical case (Fig. 17–2b): the potential energy change, or the work that can be done, depends both on the potential difference (corresponding to the height of the cliff) and on the charge (corresponding to mass), Eq. 17–1.

Practical sources of electrical energy such as batteries and electric generators are meant to maintain a particular potential difference. The actual amount of energy used or transformed depends on how much charge flows. For example, consider an automobile headlight connected to a 12.0-V battery. The amount of energy transformed (into light, and of course thermal energy) is proportional to how much charge flows, which in turn depends on how long the light is on. If over a given period 5.0 C of charge flows through the light, the total energy transformed is $(5.0\,C)(12.0\,V) = 60\,J$. If the headlight is left on twice as long, 10.0 C of charge will flow and the energy transformed is $(10.0\,C)(12.0\,V) = 120\,J$.

Table 17–1 presents some typical voltages.

TABLE 17–1
Some Typical Voltages

Source	Voltage (approx.)
Thundercloud to ground	10^8 V
High-voltage power line	10^6 V
Power supply for TV tube	10^4 V
Automobile ignition	10^4 V
Household outlet	10^2 V
Automobile battery	12 V
Flashlight battery	1.5 V
Resting potential across nerve membrane	10^{-1} V
Potential changes on skin (EKG and EEG)	10^{-4} V

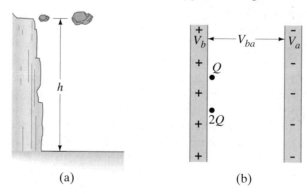

FIGURE 17–2 (a) Two rocks are at the same height. The larger rock has more PE. (b) Two charges have the same electric potential. The $2Q$ charge has more PE.

(a)

(b)

EXAMPLE 17–1 **Electron in TV tube.** Suppose an electron in the picture tube of a television set is accelerated from rest through a potential difference $V_{ba} = +5000\text{ V}$ (Fig. 17–3). (a) What is the change in potential energy of the electron? (b) What is the speed of the electron ($m = 9.1 \times 10^{-31}$ kg) as a result of this acceleration? (c) Repeat for a proton ($m = 1.67 \times 10^{-27}$ kg) that accelerates through a potential difference of $V_{ba} = -5000\text{ V}$.

SOLUTION (a) The charge on an electron is $e = -1.6 \times 10^{-19}$ C. Therefore its change in potential energy (Eq. 17–1) is equal to

$$\Delta\text{PE} = qV_{ba} = (-1.6 \times 10^{-19}\text{ C})(+5000\text{ V})$$

$$= -8.0 \times 10^{-16}\text{ J}.$$

FIGURE 17–3 Electron accelerated in TV picture tube. Example 17–1.

The minus sign in the result indicates that the PE decreases. (The potential difference, V_{ba}, has a positive sign since the final potential is higher than the initial potential; that is, negative electrons are attracted from a negative electrode to a positive one.)

(b) The potential energy lost by the electron becomes kinetic energy. From conservation of energy (see Eq. 6–11), $\Delta\text{KE} + \Delta\text{PE} = 0$, so

$$\Delta\text{KE} = -\Delta\text{PE}$$

$$\tfrac{1}{2}mv^2 - 0 = -qV_{ba},$$

where the initial KE = 0 since we assume the electron started from rest. We solve for v and put in the mass of the electron $m = 9.1 \times 10^{-31}$ kg:

$$v = \sqrt{-\frac{2qV_{ba}}{m}}$$

$$= \sqrt{-\frac{2(-1.6 \times 10^{-19}\text{ C})(5000\text{ V})}{9.1 \times 10^{-31}\text{ kg}}}$$

$$= 4.2 \times 10^7\text{ m/s}.$$

(Note: For such a high speed, which is $\frac{1}{7}$ the speed of light, we should use the theory of relativity, Chapter 26.)

(c) The proton has the same magnitude of charge as the electron, though of opposite sign. Hence for the same magnitude of V_{ba} we expect the same change in PE, but a lesser speed since the proton's mass is greater. Thus:

$$\Delta\text{PE} = qV_{ba} = (+1.6 \times 10^{-19}\text{ C})(-5000\text{ V}) = -8.0 \times 10^{-16}\text{ J},$$

and

$$v = \sqrt{-\frac{2qV_{ba}}{m}}$$

$$= \sqrt{-\frac{2(1.6 \times 10^{-19}\text{ C})(-5000\text{ V})}{(1.67 \times 10^{-27}\text{ kg})}}$$

$$= 9.8 \times 10^5\text{ m/s}.$$

Note that the energy doesn't depend on the mass, only on the charge and voltage. The speed *does* depend on m.

17–2 Relation Between Electric Potential and Electric Field

The effects of any charge distribution can be described either in terms of electric field or in terms of electric potential. Electric potential is often easier to use since it is a scalar, whereas electric field is a vector. There is an intimate connection between the potential and the field. Let us examine this relation for the case of a uniform electric field, such as that between the parallel plates of Fig. 17–1 whose difference of potential is V_{ba}. We won't worry about signs. The work done by the electric field to move a positive charge q from b to a is

$$W = qV_{ba}.$$

We can also write the work done as the force times distance and recall that the force on q is $F = qE$, where E is the uniform electric field between the plates. Thus

$$W = Fd = qEd$$

where d is the distance (parallel to the field lines) between points a and b. We now set these two expressions for W equal and find $qV_{ba} = qEd$, or

$$V_{ba} = Ed. \qquad \text{[E uniform]} \qquad \textbf{(17–2a)}$$

If we solve for E, we find that

$$E = V_{ba}/d. \qquad \text{[E uniform]} \qquad \textbf{(17–2b)}$$

From this equation we can see that the units for electric field can be written as volts per meter (V/m) as well as newtons per coulomb (N/C). These are equivalent in general, since $1\,\text{N/C} = 1\,\text{N·m/C·m} = 1\,\text{J/C·m} = 1\,\text{V/m}$.

V related to uniform E

FIGURE 17–4 Example 17–2.

[Optional paragraph: More general relation between E and V]

EXAMPLE 17–2 **Electric field obtained from voltage.** Two parallel plates are charged to a voltage of 50 V. If the separation between the plates is 0.050 m, calculate the electric field between them (Fig. 17–4).

SOLUTION We have from Eq. 17–2,

$$E = \frac{V}{d} = \frac{50\,\text{V}}{0.050\,\text{m}} = 1000\,\text{V/m}.$$

[In a region where E is not uniform, the connection between E and V takes on a different form than Eq. 17–2. In general, it is possible to show that the electric field in a given direction at any point in space is equal to the *rate at which the electric potential changes over distance in that direction*. Actually, if we take into account direction, this gives the negative of the electric field. For example, the x component of the electric field is given by $E_x = -\Delta V/\Delta x$, where ΔV is the change in potential over the very short distance Δx. Note that this relation resembles Eq. 17–2b except that the distance Δx must be very small—so small that E does not change appreciably over this distance. Similar relations apply for the y and z components of **E**. Another way of stating the relation is this: if we plot V on a graph versus x, the slope of the graph at any point equals the magnitude of the x component of the electric field at that point. And we must insert a minus sign if we want the direction to come out right.]

17-3 Equipotential Lines

The electric potential can be represented diagrammatically by drawing **equipotential lines** or, in three dimensions, **equipotential surfaces**. An equipotential surface is one on which all points are at the same potential. That is, the potential difference between any two points on the surface is zero, and no work is required to move a charge from one point to the other. An *equipotential surface must be perpendicular to the electric field* at any point. If this were not so—that is, if there were a component of **E** parallel to the surface—it would require work to move the charge along the surface against this component of **E**; and this would contradict the idea that it is an *equi*potential surface.

*Equipotentials ⊥ **E***

The fact that the electric field lines and equipotential surfaces are mutually perpendicular helps us locate the equipotentials when the electric field lines are known. In a normal two-dimensional drawing, we show equipotential *lines*, which are the intersections of equipotential surfaces with the plane of the drawing. In Fig. 17–5, a few of the equipotential lines are drawn (dashed green lines) for the electric field (red lines) between two parallel plates at a potential difference of 20 V. The negative plate is arbitrarily chosen to be zero volts and the potential of each equipotential line is indicated. Note that **E** points toward lower values of *V*. The equipotential lines for the case of two equal but oppositely charged particles are shown in Fig. 17–6 as green dashed lines. Equipotential lines and surfaces, unlike field lines, are always continuous and never end, and so continue beyond the borders

FIGURE 17–5 Equipotential lines (the green dashed lines) between two charged parallel plates; note that they are perpendicular to the electric field (solid red lines).

20 V 15 V 10 V 5 V 0 V

FIGURE 17–6 Equipotential lines (green, dashed) are always perpendicular to the electric field lines (solid red) shown here for two equal but oppositely charged particles.

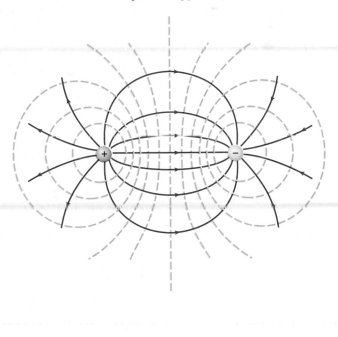

FIGURE 17–7 A topographic map (here, a portion of the Sierra Nevada in California) shows continuous contour lines, each of which is at a fixed height above sea level. Here they are at 80 ft (25 m) intervals. If you walk along one contour line, you neither climb nor descend. If you cross lines, and especially (maximally), if you climb perpendicular to the lines, you will be changing your gravitational potential (rapidly, if the lines are close together).

of Figs. 17–5 and 17–6. A useful analogy is a topographic map: the contour lines are essentially gravitational equipotential lines (Fig. 17–7).

We saw in Section 16–9 that there can be no electric field within a conductor in the static case, for otherwise the free electrons would feel a force and would move. Indeed *a conductor must be entirely at the same potential in the static case*, and the surface of a conductor is then an equipotential surface. (If it weren't, the free electrons at the surface would move, since whenever there is a potential difference between two points, work can be done on charged particles to move them.) This is fully consistent with our result, discussed earlier, that the electric field at the surface of a conductor must be perpendicular to the surface.

Conductors are equipotential surfaces

17–4 The Electron Volt, a Unit of Energy

The joule is a very large unit for dealing with energies of electrons, atoms, or molecules, whether in atomic and nuclear physics or in chemistry or molecular biology (see Example 17–1). For this purpose, the **electron volt** (eV) is used. One electron volt is defined as the energy acquired by a particle carrying a charge equal to that on the electron ($q = e$) as a result of moving through a potential difference of 1 V. Since the charge on an electron has magnitude 1.6×10^{-19} C, and since the change in potential energy equals qV, 1 eV is equal to $(1.6 \times 10^{-19} \text{C}) \cdot (1.0 \text{ V}) = 1.6 \times 10^{-19}$ J:

Electron volt

$$1 \text{ eV} = 1.6 \times 10^{-19} \text{ J}.$$

An electron that accelerates through a potential difference of 1000 V will lose 1000 eV of potential energy and will thus gain 1000 eV or 1 keV (kilo-electron volt) of kinetic energy. On the other hand, if a particle has a charge equal to twice the charge on the electron ($= 2e = 3.2 \times 10^{-19}$ C), when it moves through a potential difference of 1000 V its energy will change by 2000 eV.

Although the electron volt is handy for *stating* the energies of molecules and elementary particles, it is *not* a proper SI unit. For calculations, electron volts should be converted to joules using the conversion factor just given. In Example 17–1, for example, the electron acquired a kinetic energy of 8.0×10^{-16} J. We normally would quote this energy as 5000 eV ($= 8.0 \times 10^{-16}$ J$/1.6 \times 10^{-19}$ J/eV). But in determining the speed in SI units, we have to use the KE in joules (J).

17–5 Electric Potential Due to Point Charges

The electric potential at a distance r from a single point charge Q can be derived from the expression for its electric field (Eq. 16–4) using calculus. The potential in this case is usually taken to be zero at infinity (∞); this is also where the electric field ($E = kQ/r^2$) is zero. The result is

$$V = k\frac{Q}{r}$$

[single point charge]   (17–3)

$$= \frac{1}{4\pi\epsilon_0}\frac{Q}{r}.$$

Electric potential of point charge ($V = 0$ at $r = \infty$)

We can think of V here as representing the absolute potential, where $V = 0$ at $r = \infty$, or we can think of V as the potential difference between r and infinity. Notice that the potential V decreases with the first power of the distance, whereas the electric field (Eq. 16–4) decreases as the *square* of the distance. The potential near a positive charge is large, and it decreases toward zero at very large distances. For a negative charge, the potential is negative and increases toward zero at large distances (Fig. 17–8).

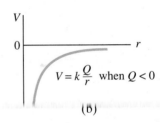

FIGURE 17–8 Potential V as a function of distance r from a single point charge Q when the charge is (a) positive, (b) negative.

EXAMPLE 17–3 **Work to force two + charges close together.** What minimum work is required by an external force to bring a charge $q = 3.00\ \mu C$ from a great distance away (take $r = \infty$) to a point 0.500 m from a charge $Q = 20.0\ \mu C$?

SOLUTION The work required is equal to the change in potential energy:

$$W = qV_{ba} = q\left(\frac{kQ}{r_b} - \frac{kQ}{r_a}\right),$$

where $r_b = 0.500$ m and $r_a = \infty$. The second term in parentheses is zero ($1/\infty - 0$) so

$$W = (3.00 \times 10^{-6}\ C)\frac{(9.00 \times 10^9\ \text{N·m}^2/\text{C}^2)(2.00 \times 10^{-5}\ C)}{(0.500\ \text{m})} = 1.08\ \text{J}.$$

[Note that we could not calculate the work done by multiplying force times distance because the force is not constant.]

To determine the electric field surrounding a collection of two or more point charges requires adding up the electric fields due to each charge. Since the electric field is a vector, this can often be a chore. To find the electric potential due to a collection of point charges is far easier, since the electric potential is a scalar, and hence you only need to add numbers together without concern for direction. This is a major advantage in using electric potential. We do have to include the signs of charges, however.

Potentials add as scalars (Fields add as vectors)

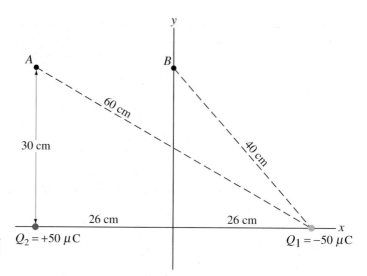

FIGURE 17–9 Example 17–4.
(See also Example 16–8, Fig. 16–25
in the previous chapter.)

(See also Example 16–8, Fig. 16–25
in the previous chapter.)

EXAMPLE 17–4 **Potential above two charges.** Calculate the electric potential at points A and B in Fig. 17–9 due to the two charges shown. (This is the same situation as Example 16–8, Fig. 16–25, where we calculated the electric field at these points.)

SOLUTION The potential at point A is the sum of the potentials due to the $+$ and $-$ charges, and we use Eq. 17–3 for each:

$$V_A = V_{A2} + V_{A1}$$

$$= \frac{(9.0 \times 10^9\ \text{N}\cdot\text{m}^2/\text{C}^2)(5.0 \times 10^{-5}\ \text{C})}{0.30\ \text{m}}$$

$$+ \frac{(9.0 \times 10^9\ \text{N}\cdot\text{m}^2/\text{C}^2)(-5.0 \times 10^{-5}\ \text{C})}{0.60\ \text{m}}$$

$$= 1.50 \times 10^6\ \text{V} - 0.75 \times 10^6\ \text{V}$$

$$= 7.5 \times 10^5\ \text{V}.$$

At point B:

$$V_B = V_{B2} + V_{B1}$$

$$= \frac{(9.0 \times 10^9\ \text{N}\cdot\text{m}^2/\text{C}^2)(5.0 \times 10^{-5}\ \text{C})}{0.40\ \text{m}}$$

$$+ \frac{(9.0 \times 10^9\ \text{N}\cdot\text{m}^2/\text{C}^2)(-5.0 \times 10^{-5}\ \text{C})}{0.40\ \text{m}}$$

$$= 0\ \text{V}.$$

It should be clear that the potential will be zero everywhere on the plane equidistant between the two charges. Thus this plane is an equipotential surface with $V = 0$.

A simple summation like these can easily be performed for any number of point charges.

CONCEPTUAL EXAMPLE 17–5 **Potential energies.** Consider the three pairs of charges in Fig. 17–10. (*a*) Which set has a positive potential energy? (*b*) Which set has the most negative potential energy? (*c*) Which set requires the most work to separate the charges to infinity? Assume the charges all have the same magnitude.

RESPONSE We can combine Eqs. 17–1 and 17–3, calling the two charges Q_1 and Q_2:

$$\text{PE} = k\frac{Q_1 Q_2}{r}.$$

(*a*) Set (iii) has a positive potential energy because the charges have the same sign. (*b*) Set (i) has the most negative potential energy because the charges are of opposite sign and their separation is less than that for set (ii). That is, r is smaller for (i). (*c*) Set (i) will require the most work for separation to infinity. The more negative the potential energy, the more work required to separate the charges and bring the PE up to zero ($r = \infty$).

FIGURE 17–10 Example 17–5.

* **17–6 Electric Dipoles**

Two equal point charges Q, of opposite sign, separated by a distance l, are called an **electric dipole**. The two charges we saw in Fig. 17–9 constitute an electric dipole. The electric field lines and equipotential surfaces for a dipole were shown in Fig. 17–6. Because electric dipoles occur often in physics, as well as in other fields such as molecular biology, it is useful to examine them more closely.

Let us calculate the electric potential at an arbitrary point P due to a dipole, as shown in Fig. 17–11. Since V is the sum of the potentials due to each of the two charges, we have

$$V = \frac{kQ}{r} + \frac{k(-Q)}{r + \Delta r} = kQ\left(\frac{1}{r} - \frac{1}{r + \Delta r}\right) = kQ\frac{\Delta r}{r(r + \Delta r)},$$

where r is the distance from P to the positive charge and $r + \Delta r$ is the distance to the negative charge. This equation becomes simpler if we consider points P whose distance from the dipole is much larger than the separation of the two charges—that is, for $r \gg l$. From the diagram we can see that in this case, $\Delta r \approx l\cos\theta$; and since $r \gg \Delta r = l\cos\theta$, we can neglect Δr in the denominator as compared to r. Therefore, we obtain

$$V = \frac{kQl\cos\theta}{r^2}. \qquad \text{[dipole; } r \gg l] \quad \textbf{(17–4a)}$$

When θ is between $0°$ and $90°$, V is positive. If θ is between $90°$ and $180°$, V is negative (since $\cos\theta$ is then negative). This makes sense since in the first case P is closer to the positive charge and in the second case it is closer to the negative charge. At $\theta = 90°$, the potential is zero ($\cos 90° = 0$), in agreement with the result of Example 17–4 (point B). From Eq. 17–4a, we see that the potential decreases as the *square* of the distance from the dipole, whereas for a single point charge the potential decreases with the first power of the distance (Eq. 17–3). It is not surprising that the potential should fall off faster for a dipole; for when you are far from a dipole, the two equal but opposite charges appear so close together as to tend to neutralize each other.

FIGURE 17–11 Electric dipole. Calculation of potential V at point P.

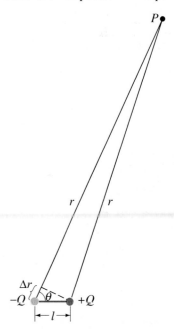

Potential far from a dipole

The product Ql which occurs in Eq. 17–4a is referred to as the **dipole moment**, p, of the dipole. Equation 17–4a can be written in terms of the dipole moment as

$$V = \frac{kp \cos \theta}{r^2}.$$ [dipole; $r \gg l$]   **(17–4b)**

A dipole moment has units of coulomb-meters (C·m), although for molecules a smaller unit called a *debye* is sometimes used: 1 debye = 3.33×10^{-30} C·m.

In many molecules, even though they are electrically neutral, the electrons spend more time in the vicinity of one atom than another, which results in a separation of charge. Such molecules have a dipole moment and are called **polar molecules**. We have already seen that water (Fig. 16–4) is a polar molecule, and we have encountered others in our discussion of molecular biology (Section 16–10). Table 17–2 gives the dipole moments for several molecules. The $+$ and $-$ signs indicate on which atoms these charges lie. The last two entries are a part of many organic molecules and play an important role in molecular biology.

TABLE 17–2
Dipole Moments of Selected Molecules

Molecule	Dipole Moment (C·m)
$H_2^{(+)}O^{(-)}$	6.1×10^{-30}
$H^{(+)}Cl^{(-)}$	3.4×10^{-30}
$N^{(-)}H_3^{(+)}$	5.0×10^{-30}
$>N^{(-)}{-}H^{(+)\ddagger}$	$\approx 3.0 \times 10^{-30}$
$>C^{(+)}{=}O^{(-)\ddagger}$	$\approx 8.0 \times 10^{-30}$

‡These groups often appear on larger molecules; hence the value for the dipole moment will vary somewhat, depending on the rest of the molecule.

EXAMPLE 17–6 **The C═O group dipole.** The distance between the carbon $(+)$ and oxygen $(-)$ atoms in the group C═O is about 1.2×10^{-10} m. Calculate (*a*) the net charge Q on the C (carbon) and O (oxygen) atoms, and (*b*) the potential 9.0×10^{-10} m from the dipole along its axis, with the oxygen being the nearer atom (that is, to the left in Fig. 17–11, so $\theta = 180°$). (*c*) What would the potential be at this point if only the oxygen (O) were charged?

SOLUTION (*a*) The dipole moment $p = Ql$. Therefore $Q = p/l$, and from Table 17–2:

$$Q = \frac{p}{l} = \frac{8.0 \times 10^{-30} \text{ C·m}}{1.2 \times 10^{-10} \text{ m}} = 6.7 \times 10^{-20} \text{ C}.$$

(*b*) Since $\theta = 180°$, we have, using Eq. 17–4:

$$V = \frac{kp \cos \theta}{r^2}$$

$$= \frac{(9.0 \times 10^9 \text{ N·m}^2/\text{C}^2)(8.0 \times 10^{-30} \text{ C·m})(-1.00)}{(9.0 \times 10^{-10} \text{ m})^2}$$

$$= -0.089 \text{ V}.$$

(*c*) If we assume that the oxygen has charge $Q = -6.7 \times 10^{-20}$ C [as in part (*a*) above] and that the carbon is not charged, we use the formula for a single charge, Eq. 17–3:

$$V = \frac{kQ}{r} = \frac{(9.0 \times 10^9 \text{ N·m}^2/\text{C}^2)(-6.7 \times 10^{-20} \text{ C})}{9.0 \times 10^{-10} \text{ m}} = -0.67 \text{ V}.$$

Of course, we expect the potential of a single charge to have greater magnitude than that of a dipole of equal charge at the same distance.

(a)

Insulator

(b)

(c)

FIGURE 17–12 Capacitors:
Diagrams of (a) parallel plate,
(b) cylindrically shaped (rolled
up parallel plate). (c) Photo of
some real capacitors.

17–7 Capacitance

A **capacitor**, sometimes called a *condenser*, is a device that can store electric charge, and consists of two conducting objects (usually plates or sheets) placed near each other but not touching. Capacitors are widely used in electronic circuits: they store charge for later use, as in a camera flash, and as energy backup in computers if the power fails; capacitors block surges of charge and energy to protect circuits; very tiny capacitors serve as memory for the "ones" and "zeroes" of the binary code in the random access memory (RAM) of computers; and capacitors serve many other applications, some of which we will discuss. A typical capacitor consists of a pair of parallel plates of area A separated by a small distance d (Fig. 17–12a). Often the two plates are rolled into the form of a cylinder with paper or other insulator separating the plates (Fig. 17–12b; Fig. 17–12c is a photo of some actual capacitors used for various applications). In a diagram, a capacitor is represented by the symbol

$$\dashv\vdash.$$ [capacitor symbol]

If a voltage is applied to a capacitor, say by connecting the capacitor to a battery as in Fig. 17–13, it quickly becomes charged. One plate acquires a negative charge, and the other an equal amount of positive charge. For a given capacitor, the amount of charge Q acquired by each plate is proportional to the potential difference V:

$$Q = CV. \qquad (17\text{–}5)$$

The constant of proportionality, C, in this relation is called the **capacitance** of the capacitor. The unit of capacitance is coulombs per volt, and this unit is called a **farad** (F). Most capacitors have capacitance in the range 1 pF (picofarad $= 10^{-12}$ F) to 1 μF (microfarad $= 10^{-6}$ F). The relation, Eq. 17–5, was first suggested by Volta in the late eighteenth century.

The capacitance C is a constant for a given capacitor: it does not depend on Q or V. Its value depends only on the structure and dimensions of the capacitor itself. For a parallel-plate capacitor whose plates have area A and are separated by a distance d of air (Fig. 17–12a), the capacitance is given by

$$C = \epsilon_0 \frac{A}{d}. \qquad \text{[parallel-plate capacitor]} \quad (17\text{–}6)$$

This relation makes sense intuitively: a larger area A means that for a

→ **PHYSICS APPLIED**

Uses of capacitors

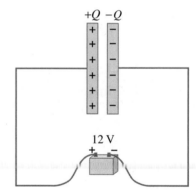

$+Q$ $-Q$

12 V

FIGURE 17–13 Parallel-plate capacitor connected to a battery.

Capacitance

Unit is farad (1 F = 1 C/V)

given number of charges, there will be less repulsion between them (they're farther apart), so we expect that more charge can be held on each plate. And a greater separation d means the charge on each plate exerts less attractive force on the other plate, so less charge is drawn from the battery, and the capacitance is less.[†] The constant ϵ_0 is the *permittivity of free space* which, as we saw in Chapter 16, has the value $8.85 \times 10^{-12} \, C^2/N \cdot m^2$.

EXAMPLE 17–7 **Capacitor calculations.** (*a*) Calculate the capacitance of a capacitor whose plates are $20 \, cm \times 3.0 \, cm$ and are separated by a 1.0-mm air gap. (*b*) What is the charge on each plate if the capacitor is connected to a 12-V battery? (*c*) What is the electric field between the plates?

SOLUTION (*a*) The area $A = (20 \times 10^{-2} \, m)(3.0 \times 10^{-2} \, m) = 6.0 \times 10^{-3} \, m^2$. The capacitance C is then

$$C = \epsilon_0 \frac{A}{d} = (8.85 \times 10^{-12} \, C^2/N \cdot m^2) \frac{6.0 \times 10^{-3} \, m^2}{1.0 \times 10^{-3} \, m} = 53 \, pF.$$

(*b*) The charge on each plate is

$$Q = CV = (53 \times 10^{-12} \, F)(12 \, V) = 6.4 \times 10^{-10} \, C.$$

(*c*) From Eq. 17–2 for a uniform electric field

$$E = \frac{V}{d} = \frac{12 \, V}{1.0 \times 10^{-3} \, m} = 1.2 \times 10^4 \, V/m.$$

17–8 Dielectrics

In most capacitors there is an insulating sheet (such as paper or plastic) called a **dielectric** between the plates. This serves several purposes. First, because higher voltages can be applied without charge passing across the gap, dielectrics break down (charge suddenly starts to flow through them when the voltage is high enough) less readily than air. Furthermore, a dielectric allows the plates to be placed closer together without touching, thus allowing an increased capacitance because d is less in Eq. 17–6. Finally, it is found experimentally that if the dielectric fills the space between the two conductors, the capacitance is increased by a factor K which is known as the **dielectric constant** (Table 17–3). Thus, for a parallel-plate capacitor,

$$C = K\epsilon_0 \frac{A}{d}. \qquad \text{[parallel-plate capacitor]} \quad \textbf{(17–7)}$$

TABLE 17–3
Dielectric Constants (20°C)

Material	Dielectric Constant, K
Vacuum	1.0000
Air (1 atm)	1.0006
Paraffin	2.2
Rubber, hard	2.8
Vinyl (plastic)	2.8–4.5
Paper	3–7
Quartz	4.3
Glass	4–7
Porcelain	6–8
Mica	7
Ethyl alcohol	24
Water	80

[†]Equation 17–6 is readily derived using the result from Appendix D on Gauss's law, namely that the electric field between two parallel plates is given by Eq. D–4,

$$E = \frac{Q/A}{\epsilon_0}.$$

We combine this with Eq. 17–2, $V = Ed$, to obtain

$$V = \left(\frac{Q}{A\epsilon_0}\right)d.$$

Thus, from Eq. 17–5, the definition of capacitance,

$$C = \frac{Q}{V} = \frac{Q}{(Q/A\epsilon_0)d} = \epsilon_0 \frac{A}{d}$$

which is Eq. 17–6.

This can also be written

$$C = \epsilon \frac{A}{d},$$

where

$$\epsilon = K\epsilon_0$$

is the *permittivity of the material*.

Let us now examine, from the molecular point of view, why the capacitance of a capacitor should increase when a dielectric is inserted between the plates. Consider a capacitor whose plates are separated by an air gap. This capacitor has a charge $+Q$ on one plate and $-Q$ on the other (Fig. 17–14a). The capacitor is isolated (not connected to a battery) so charge cannot flow to or from the plates. The potential difference between the plates, V_0, is given by Eq. 17–5: $Q = C_0 V_0$; the subscripts ($_0$) refer to the situation when only air is between the plates. Now we insert a dielectric between the plates (Fig. 17–14b). The molecules of the dielectric may be *polar*. That is, although the molecules are neutral, the electrons may not be evenly distributed, so that one part of the molecule is positive and another part negative. Because of the electric field between the plates, the molecules will tend to become oriented as shown. Even if the molecules are not polar, the electric field between the plates will induce some separation of charge in the molecules. Although the electrons do not leave the molecules, they will move slightly within the molecules toward the positive plate. So the situation is still as illustrated in Fig. 17–14b. The net effect in either case is as if there were a net negative charge on the outer edge of the dielectric facing the positive plate, and a net positive charge on the opposite side, as shown in Fig. 17–14c.

Now imagine a positive test charge within the dielectric. The force that it feels is reduced by a factor K, the dielectric constant. This is reflected by the fact that some of the electric field lines actually do not pass through the dielectric, but end (and restart) on the charges induced on the surface of the dielectric (Fig. 17–14c). Because the force on our test charge is reduced by a factor K, the work needed to move it from one plate to the other is reduced by a factor K. (We assume that the dielectric fills all the space between the plates.) The voltage, which is the work done per unit charge, must therefore also have decreased by the factor K. That is, the voltage between the plates is now

$$V = \frac{V_0}{K}.$$

Now the charge Q on the plates has not changed, because they are isolated. So we have

$$Q = CV,$$

where C is the capacitance when the dielectric is present. When we combine this with the relation, $V = V_0/K$, we obtain

$$C = \frac{Q}{V} = \frac{Q}{V_0/K} = \frac{QK}{V_0} = KC_0,$$

since $C_0 = Q/V_0$. Thus we see, from an atomic point of view, why the capacitance is increased by the factor K.

Molecular description of dielectrics

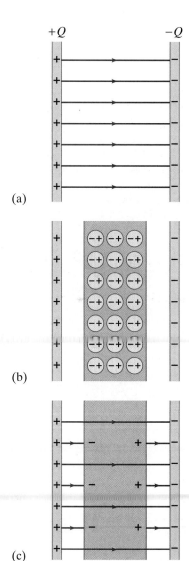

(a)

(b)

(c)

FIGURE 17–14 Molecular view of the effects of a dielectric.

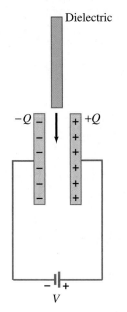

FIGURE 17–15 Conceptual Example 17–8.

FIGURE 17–16 Key on a computer keyboard. Pressing the key reduces the capacitor spacing thus increasing the capacitance which can be detected electronically.

CONCEPTUAL EXAMPLE 17–8 | Inserting a dielectric at constant *V*. A capacitor consisting of two plates separated by a distance d is connected to a battery of voltage V and acquires a charge Q. While it is still connected to the battery, a slab of dielectric material is inserted between the plates of the capacitor (Fig. 17–15). Will Q increase, decrease, or stay the same?

RESPONSE Since the capacitor remains connected to the battery, the voltage stays constant and equal to the V. From the discussion above, we know that the capacitance C must increase when the dielectric material is inserted. From the relation $Q = CV$, if V stays constant, but C increases, Q must increase as well. As the dielectric is inserted, more charge will be pulled from the battery and deposited onto the plates of the capacitor as its capacitance increases.

One type of computer keyboard operates by capacitance. As shown in Fig. 17–16, each key is connected to the upper plate of a capacitor. The upper plate moves down when the key is pressed, reducing the spacing between the capacitor plates, and increasing the capacitance (Eq. 17–7: smaller d, larger C). The *change* in capacitance becomes an electric signal that is detected by an electronic circuit.

17–9 Storage of Electric Energy

A charged capacitor stores electric energy. The energy stored in a capacitor will be equal to the work done to charge it. The net effect of charging a capacitor is to remove charge from one plate and add it to the other plate. This is what a battery does when it is connected to a capacitor. A capacitor does not become charged instantly. It takes time (Section 19–7). When some charge is on each plate, it requires work to add more charge of the same sign. The more charge already on a plate, the more work is required to add more. The work needed to add a small amount of charge Δq, when a potential difference V is across the plates, is $\Delta W = V \Delta q$. Initially, when the capacitor is uncharged, no work is required to move the first bit of charge over. By the end of the charging process, however, the work needed to add a charge Δq will be equal to $V_f \Delta q$ where V_f is the final voltage ($V_f = Q/C$). If the voltage across the capacitor were constant, the work needed to move charge Q would be $W = QV$. But the voltage across the capacitor is proportional to how much charge it already has accumulated (Eq. 17–5), and so the voltage increases during the charging process from zero to its final value, V_f, at the end. Then the total work done, W, will be equivalent to moving all the charge Q at once across a voltage equal to the average voltage during the whole process. (This is just like calculating the work done to compress a spring, Section 6–4.) The average voltage is $(V_f - 0)/2 = V_f/2$, so

$$W = Q \frac{V_f}{2}.$$

Thus we can say that the energy, U, stored in a capacitor is

$$U = \text{energy} = \tfrac{1}{2}QV,$$

where V is the potential difference between the plates (we have dropped the subscript), and Q is the charge on each plate. Since $Q = CV$, we can also write

$$U = \tfrac{1}{2}QV = \tfrac{1}{2}CV^2 = \tfrac{1}{2}\frac{Q^2}{C}. \qquad \textbf{(17–8)}$$

Energy stored in capacitor

EXAMPLE 17–9 **Energy stored in a capacitor.** A camera flash unit stores energy in a $150\,\mu\text{F}$ capacitor at $200\,\text{V}$. How much electric energy can be stored?

SOLUTION From Eq. 17–8, we have

$$U = \text{energy} = \tfrac{1}{2}CV^2 = \tfrac{1}{2}(150 \times 10^{-6}\,\text{F})(200\,\text{V})^2 = 3.0\,\text{J}.$$

Notice how the units work out: $\text{FV}^2 = \left(\dfrac{\text{C}}{\text{V}}\right)(\text{V}^2) = \text{CV} = \text{C}\left(\dfrac{\text{J}}{\text{C}}\right) = \text{J}.$

If this energy could be released in $\frac{1}{1000}$ of a second ($10^{-3}\,\text{s}$) the power output would be equivalent to $3000\,\text{W}$.

⇒ **PHYSICS APPLIED**

Camera flash

Energy is not a substance and does not have a definite location. Nonetheless, it is often useful to think of it as being stored in the electric field between the plates. As an example, let us calculate the energy stored in a parallel-plate capacitor in terms of the electric field.

We saw in Eq. 17–2 that the electric field E between two large but close parallel plates is uniform and is related to the potential difference by $V = Ed$, where d is the separation. Also, Eq. 17–6 tells us that $C = \epsilon_0 A/d$. Thus

$$U = \tfrac{1}{2}CV^2 = \tfrac{1}{2}\left(\frac{\epsilon_0 A}{d}\right)(E^2 d^2)$$

$$= \tfrac{1}{2}\epsilon_0 E^2 Ad.$$

The quantity Ad is simply the volume between the plates in which the electric field E exists. If we divide both sides by the volume, we obtain an expression for the energy per unit volume or **energy density**:

$$u = \text{energy density} = \frac{\text{energy}}{\text{volume}} = \tfrac{1}{2}\epsilon_0 E^2. \qquad \textbf{(17–9)}$$

Energy stored per unit volume in electric field

The energy stored per unit volume is proportional to the square of the electric field in that region. If a dielectric is present, ϵ_0 is replaced by ϵ. We derived Eq. 17–9 for the special case of a capacitor. But it can be shown to be valid for any region of space where there is an electric field.

Cathode   Anode

– +
Battery

FIGURE 17–17 If the cathode inside the evacuated glass tube is heated to glowing, negatively charged "cathode rays" (electrons) are "boiled off" and flow across to the anode (+) to which they are attracted.

➡ **PHYSICS APPLIED**

CRT

* 17–10 Cathode Ray Tube: TV and Computer Monitors, Oscilloscope

An important device that makes use of voltage, and that allows us to "visualize" voltages in the sense of displaying graphically how a voltage changes in time, is the **cathode ray tube (CRT)**. A CRT used in this way is an *oscilloscope*—but an even more common use of a CRT is as the picture tube of television sets and computer monitors.

The operation of a CRT depends first of all on the phenomenon of **thermionic emission**, discovered by Thomas Edison (1847–1931) in the course of experiments on developing the electric light bulb. To understand how thermionic emission occurs, consider two small plates (electrodes) inside an evacuated "bulb" or "tube" as shown in Fig. 17–17, to which is applied a potential difference (by a battery, say). The negative electrode is called the **cathode**[†], the positive one the **anode**. If the negative cathode is heated (usually by an electric current, as in a lightbulb) so that it becomes hot and glowing, it is found that negative charge leaves the cathode and flows to the positive anode. These negative charges are now called electrons, but originally they were called **cathode rays** since they seemed to come from the cathode (see Section 27–1 on the discovery of the electron).

We can understand how electrons might be "boiled off" a hot metal plate if we treat electrons like molecules in a gas. This makes sense if electrons are relatively free to move about inside a metal, which is consistent with metals being good conductors. However, electrons don't readily escape from the metal. If an electron were to escape outside the metal surface, a net positive charge would remain behind, and this would attract the electron back. To escape, an electron needs a certain minimum kinetic energy, just as molecules in a liquid must have a minimum KE to "evaporate" into the gaseous state. We saw in Chapter 13 that the average kinetic energy ($\overline{KE}$) of molecules in a gas is proportional to the absolute temperature T. We can apply this idea, but only very roughly, to free electrons in a metal as if they made up an "electron gas." Of course, some electrons have more KE than average and others less. At room temperature, very few electrons would have sufficient energy to escape. At high temperature, $\overline{KE}$ is larger and many electrons escape—just as molecules evaporate from liquids, which occurs more readily at high temperatures. Thus, significant thermionic emission occurs only at elevated temperatures.

The **cathode-ray tube** (CRT) derives its name from the fact that inside an evacuated glass tube, a beam of cathode rays (electrons) is directed to various parts of a screen to produce a "picture." A simple CRT is diagrammed in Fig. 17–18. Electrons emitted by the heated cathode are accelerated by a high voltage (5,000–50,000 V) applied to the anode. The electrons pass out of this "electron gun" through a small hole in the anode. The inside of the tube face is coated with a fluorescent material that glows

---

[†]These terms were coined by Michael Faraday and come from the Greek words meaning, respectively, "descent" and "a way up."

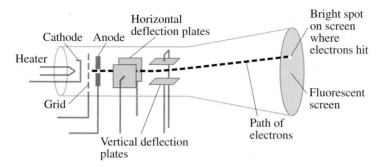

FIGURE 17–18   A cathode-ray tube. Magnetic deflection coils are often used in place of the electric deflection plates. The relative positions of the elements have been exaggerated for clarity.

when struck by electrons. A tiny bright spot is thus visible where the electron beam strikes the screen. Two horizontal and two vertical plates deflect the beam of electrons when a voltage is applied to them. The electrons are deflected toward whichever plate is positive. By varying the voltage on the deflection plates, the bright spot can be placed at any point on the screen. Today it is more usual for CRTs to make use of magnetic deflection coils (Chapter 20) instead of electric plates.

In the picture tube or monitor for a computer or television set, the electron beam is made to sweep over the screen in the manner shown in Fig. 17–19. The beam is swept horizontally by the horizontal deflection plates or coils. When the horizontal deflecting field is maximum in one direction, the beam is at one edge of the screen. As the field decreases to zero, the beam moves to the center; and as the field increases to a maximum in the opposite direction, the beam approaches the opposite edge. When the beam reaches this edge, the voltage or current abruptly changes to return the beam to the opposite side of the screen. Simultaneously, the beam is deflected downward slightly by the vertical deflection plates (or coils), and then another horizontal sweep is made. For television in the United States, 525 lines constitutes a complete sweep over the entire screen. (High-definition TV will provide more than double this number of lines, giving greater picture sharpness. Some European systems already provide significantly more lines than the present U.S. standard.) The complete picture of 525 lines is swept out in $\frac{1}{30}$ s. Actually, a single vertical sweep takes $\frac{1}{60}$ s and involves every other line. The lines in between are then swept out over the next $\frac{1}{60}$ s (called interlacing). We see a picture because the image is retained by the fluorescent screen and by our eyes for about $\frac{1}{20}$ s. The picture we see consists of the varied brightness of the spots on the screen. The brightness at any point is controlled by the grid (a "porous" electrode, such as a wire grid, that allows passage of electrons) which can limit the flow of electrons by means of the voltage applied to it: the more negative this voltage, the more electrons are repelled and the fewer pass through. The voltage on the grid is determined by the video signal (a voltage) sent out by the TV station and received by the TV set. Accompanying this signal are signals that synchronize the grid voltage to the horizontal and vertical sweeps.

➡ PHYSICS APPLIED

TV and computer monitors

FIGURE 17–19   Electron beam sweeps across a television screen in a succession of horizontal lines.

FIGURE 17–20 An electrocardio-gram (ECG) trace displayed on a CRT.

An **oscilloscope** is a device for amplifying, measuring, and visually observing an electrical signal (a "signal" is usually a time-varying voltage), especially rapidly changing signals. The signal is displayed on the screen of a CRT. In normal operation, the electron beam is swept horizontally at a uniform rate in time by the horizontal deflection plates. The signal to be displayed is applied, after amplification, to the vertical deflection plates. The visible "trace" on the screen, which could be an ECG (Fig. 17–20), a voltage in an electronic device being repaired, or a signal from an experiment on nerve conduction, is thus a plot of the signal voltage (vertically) versus time (horizontally).

* 17–11 The Electrocardiogram (ECG or EKG)

Each time the heart beats, changes in electrical potential occur on its surface that can be detected using metal contacts, called "electrodes," which are attached to the skin. The changes in potential are small, on the order of millivolts (mV), and must be amplified. They are displayed either with a chart recorder on paper, or on a cathode-ray-tube oscilloscope screen (Fig. 17–20). The record of the potential changes for a given person's heart is called an electrocardiogram (EKG or ECG). An example is shown in Fig. 17–21. The instrument itself is called an electrocardiograph. We are not so interested now in the electronics, but in the source of these potential changes and their relation to heart activity.

Muscle cells and nerve cells are similar in that both have an electric dipole layer across the cell wall. That is, in the normal situation there is a net positive charge on the exterior surface and a net negative charge on the interior surface, as shown in Fig. 17–22a. The amount of charge depends on the size of the cell, but is approximately $10^{-3}\,C/m^2$ of surface. For a cell whose surface area is $10^{-5}\,m^2$, the total charge on either surface is thus $\approx 10^{-8}\,C$. Just before the contraction of heart muscles, changes occur in the cell wall, so that positive ions on the exterior of the cell are able to pass through the wall and neutralize those on the inside, or even make the inside surface slightly positive compared to the exterior, as shown in Fig. 17–22b. This depolarization, as it is called, starts at one end of the cell and progresses toward the opposite end, as indicated by the arrow in part (b), until the whole muscle is depolarized. The muscle then slowly repolarizes to its original state (Fig. 17–22a). The whole process requires less than a second. Figure 17–22c shows rough graphs of the potential V as a function of time at the two points P and P' (on either side of this cell) as the depolarization moves across the cell.

In the heart, the path of depolarization is complicated. Furthermore, after depolarization, the muscles repolarize to the resting state (Fig. 17–22a). Thus the potential difference as a function of time is quite complicated (Fig. 17–21).

FIGURE 17–21 Typical ECG. Two heart beats are shown.

It is standard procedure to divide a typical electrocardiogram into regions corresponding to the various deflections (or "waves" as they are called), as shown in Fig. 17–21. Each of the deflections corresponds to the activity of a particular part of the heart beat (Fig. 10–39). The P wave corresponds to contraction of the atria. The QRS group corresponds to contraction of the ventricles; this group has three main phases because the depolarization follows a complicated path from left to right, and toward the front, then downward to the left and toward the rear. The T wave corresponds to recovery (repolarization) of the heart in preparation for the next cycle.

Electrocardiograms make use of three basic electrodes, one placed on either side of the heart on the hands, and one on the left foot. Sometimes six additional electrodes are placed at other locations. The measurement of so many potential differences provides additional information (some of it redundant), since the heart is a three-dimensional object and depolarization takes place in all three dimensions. A complete electrocardiogram may include as many as 12 graphs.

The ECG is a powerful tool in identifying heart defects. For example, the right side of the heart enlarges if the right ventricle must push against an abnormally large load (as when blood vessels become hardened or clogged). This problem is readily observed on an ECG, since the S wave becomes very large (negatively). *Infarcts*, which are dead regions of the heart muscle that result from heart attacks, are also detected on an ECG because they reflect the depolarization wave.

The interpretation of an ECG depends to a great extent on experience obtained with many patients rather than on theoretical understanding. A good deal of scientific research remains to be done.

(a)

(b)

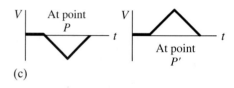
(c)

FIGURE 17–22 Heart muscle cell showing: (a) charge dipole layer in resting state; (b) depolarization of cell progressing as muscle begins to contract; and (c) potential V at points P and P' as a function of time.

▌ S U M M A R Y

The **electric potential** at any point in space is defined as the electric potential energy per unit charge.

The **electric potential difference** between any two points is defined as the work done to move a 1-C electric charge between the two points. Potential difference is measured in volts (1 V = 1 J/C) and is sometimes referred to as **voltage**.

The change in PE of a charge q when it moves through a potential difference V_{ba} is

$$\Delta\text{PE} = qV_{ba}.$$

The potential difference V between two points where a uniform electric field E exists is given by

$$V = Ed,$$

where d is the distance between the two points.

An **equipotential line** or **surface** is all at the same potential, and is perpendicular to the electric field at all points.

The electric potential due to a single point charge Q, relative to zero potential at infinity, is given by

$$V = \frac{kQ}{r}.$$

A **capacitor** is a device used to store charge and consists of two nontouching conductors. The two conductors generally hold equal and opposite charges, Q, and the ratio of this charge to the potential difference V between the conductors is called the **capacitance**, C; so

$$Q = CV.$$

The capacitance of a parallel-plate capacitor is proportional to the area of each plate and inversely proportional to their separation.

The space between the two conductors of a capacitor contains a nonconducting material such as air, paper, or plastic; these materials are referred to as **dielectrics**, and the capacitance is proportional to a property of dielectrics called the *dielectric constant*, K (nearly equal to 1 for air).

A charged capacitor stores an amount of electric energy given by

$$\tfrac{1}{2}QV = \tfrac{1}{2}CV^2 = \tfrac{1}{2}\frac{Q^2}{C}.$$

This energy can be thought of as stored in the electric field between the plates.

The energy stored in any electric field E has a density (energy per unit volume) of $\tfrac{1}{2}\epsilon_0 E^2$.

QUESTIONS

1. If two points are at the same potential, does this mean that no work is done in moving a test charge from one point to the other? Does this imply that no force must be exerted?

2. Can two equipotential lines cross? Explain.

3. Draw a few equipotential lines in Fig. 16–29b.

4. Is there a point along the line joining two equal positive charges where the electric field is zero? Where the electric potential is zero? Explain.

5. An electron is accelerated by a potential difference of, say, 100 V. How much greater would its final speed be if it were accelerated with four times as much voltage?

6. If a negative charge is initially at rest in an electric field, will it move toward a region of higher potential or lower potential? What about a positive charge? How does the potential energy of the charge change in each case?

7. State clearly the difference between: (a) electric potential and electric field, (b) electric potential and electric potential energy.

8. If the potential at a point is zero, must the electric field also be zero? Give an example.

9. What can you say about the electric field in a region of space that has the same potential throughout?

10. How does the Earth's gravitational field change with distance? What about its gravitational potential?

11. Can a particle ever move from a region of low electric potential to one of high potential and yet have its electric potential energy decrease? Explain.

12. When dealing with practical devices, we often take the ground (the earth) to be 0 V. If, instead, we said the ground was -10 V, how would this affect (a) V, and (b) E, at other points?

13. When a battery is connected to a capacitor, why do the two plates acquire charges of the same magnitude? Will this be true if the two conductors are different sizes or shapes?

14. We have seen that the capacitance C depends on the size, shape, and position of the two conductors, as well as on the dielectric constant K. What then did we mean when we said that C is a constant in Eq. 17–5?

15. How does the energy stored in a capacitor change when a dielectric is inserted if (a) the capacitor is isolated so Q doesn't change, (b) the capacitor remains connected to a battery so V doesn't change?

PROBLEMS

SECTIONS 17–1 TO 17–4

1. (I) How much work is needed to move a -8.6-μC charge from ground to a point whose potential is $+75$ V?

2. (I) How much work is needed to move a proton from a point with a potential of $+100$ V to a point where it is -50 V? Express your answer both in joules and electron volts.

3. (I) How much kinetic energy will an electron gain (in joules and eV) if it falls through a potential difference of 21,000 V in a TV picture tube?

4. (I) An electron acquires 3.45×10^{-16} J of kinetic energy when it is accelerated by an electric field in a computer monitor from plate A to plate B. What is the potential difference between the plates, and which plate is at the higher potential?

5. (I) How strong is the electric field between two parallel plates 5.2 mm apart if the potential difference between them is 220 V?

6. (I) An electric field of 640 V/m is desired between two parallel plates 11.0 mm apart. How large a voltage should be applied?

7. (I) What potential difference is needed to give a helium nucleus ($Q = 2e$) 65.0 keV of KE?

8. (II) Two parallel plates, connected to a 100-V power supply, are separated by an air gap. How small can the gap be if the air is not to exceed its breakdown value of $E = 3 \times 10^6$ V/m?

9. (II) The work done by an external force to move a -7.50-μC charge from point a to point b is 25.0×10^{-4} J. If the charge was started from rest and had 4.82×10^{-4} J of kinetic energy when it reached point b, what must be the potential difference between a and b?

10. (II) What is the speed of (a) a 750-eV, and (b) a 3.5 keV, electron?

11. (II) What is the speed of a proton whose kinetic energy is 28.0 MeV?

12. (II) An alpha particle (which is a helium nucleus, $Q = +2e$, $m = 6.64 \times 10^{-27}$ kg) is emitted in a radioactive decay with KE = 5.53 MeV. What is its speed?

SECTION 17–5

13. (I) What is the electric potential 15.0 cm from a 4.00-μC point charge?

14. (I) A charge Q creates an electric potential of $+125$ V at a distance of 15 cm. What is Q?

15. (II) A $+30$-μC charge is placed 32 cm from an identical $+30$-μC charge. How much work would be required to move a $+0.50$-μC test charge from a point midway between them to a point 10 cm closer to either of the charges?

16. (II) (*a*) What is the electric potential a distance of 2.5×10^{-15} m away from a proton? (*b*) What is the electric potential energy of a system that consists of two protons 2.5×10^{-15} m apart—as might occur inside a typical nucleus?

17. How much voltage must be used to accelerate a proton (radius 1.2×10^{-15} m) so that it has sufficient energy to just penetrate a silicon nucleus? A silicon nucleus has a charge of $+14e$ and its radius is about 3.6×10^{-15} m. Assume the potential is that for point charges.

18. (II) How much work must be done to bring three electrons from a great distance apart to within 1.0×10^{-10} m from one another?

19. (II) Consider point *a* which is 70 cm north of a -3.8-μC point charge, and point *b* which is 80 cm west of the charge (Fig. 17–23). Determine (*a*) $V_{ba} = V_b - V_a$, and (*b*) $\mathbf{E}_b - \mathbf{E}_a$ (magnitude and direction).

FIGURE 17–23   Problem 19.

20. (II) An electron starts from rest 72.5 cm from a fixed point charge with $Q = -0.125\ \mu$C. How fast will the electron be moving when it is very far away?

21. (II) Two identical $+7.5$-μC point charges are initially spaced 5.5 cm from each other. If they are released at the same instant from rest, how fast will they be moving when they are very far away from each other? Assume they have identical masses of 1.0 mg.

22. (III) In the Bohr model of the hydrogen atom, an electron orbits a proton (the nucleus) in a circular orbit of radius 0.53×10^{-10} m. (*a*) What is the electric potential at the electron's orbit due to the proton? (*b*) What is the kinetic energy of the electron? (*c*) What is the total energy of the electron in its orbit? (*d*) What is the *ionization energy*—that is, the energy required to remove the electron from the atom and take it to $r = $ infinity, at rest?

23. (III) Two equal but opposite charges are separated by a distance *d*, as shown in Fig. 17–24. Determine a formula for $V_{BA} = V_B - V_A$ for points *B* and *A* on the line between the charges situated as shown.

FIGURE 17–24   Problem 23.

*SECTION 17–6

*24. (II) An electron and a proton are 0.53×10^{-10} m apart. (*a*) What is their dipole moment if they are at rest? (*b*) What is the average dipole moment if the electron revolves about the proton in a circular orbit?

*25. (II) Calculate the electric potential due to a dipole whose dipole moment is 4.8×10^{-30} C·m at a point 1.1×10^{-9} m away if this point is: (*a*) along the axis of the dipole nearer the positive charge; (*b*) 45° above the axis but nearer the positive charge; (*c*) 45° above the axis but nearer the negative charge.

*26. (II) (*a*) In Example 17–6, part (*b*), calculate the electric potential without using the dipole approximation, Eq. 17–4; that is, don't assume $r \gg l$. (*b*) What is the percent error in this case when the dipole approximation is used?

*27. (III) The dipole moment, considered as a vector, points from the negative to the positive charge. The water molecule, Fig. 17–25, has a dipole moment **p** which can be considered as the vector sum of the two dipole moments, $\mathbf{p}_1$ and $\mathbf{p}_2$, as shown. The distance between each H and the O is about 0.96×10^{-10} m. The lines joining the center of the O atom with each H atom make an angle of 104°, as shown, and the net dipole moment has been measured to be $p = 6.1 \times 10^{-30}$ C·m. Determine the charge *q* on each H atom.

FIGURE 17–25   Problem 27.

*28. (III) Show that if two dipoles with dipole moments p_1 and p_2 are in line with one another (Fig. 17–26), the potential energy of one in the presence of the other (their "interaction energy") is given by

$$\text{PE} = -\frac{2kp_1p_2}{r^3},$$

where *r* is the distance between the two dipoles. [*Hint:* Assume that *r* is much greater than the length of either dipole.]

FIGURE 17–26   Problem 28.

*29. (III) Show that if an electric dipole is placed in a uniform electric field, then a torque is exerted on it equal to $pE \sin \phi$, where ϕ is the angle between the dipole moment vector and the direction of the electric field as shown in Fig. 17–27. What is the net force on the dipole? How are your answers affected if the field is nonuniform? Note that the dipole moment vector $\mathbf{p}$ is defined so that its magnitude is Ql and its direction is pointing from the negative end to the positive end as shown.

FIGURE 17–27   Problem 29.

SECTIONS 17–7 AND 17–8

30. (I) The two plates of a capacitor hold $+2500 \, \mu C$ and $-2500 \, \mu C$ of charge, respectively, when the potential difference is 950 V. What is the capacitance?

31. (I) The potential difference between two parallel wires in air is 120 V. They carry equal and opposite charge of magnitude 95 pC. What is the capacitance of the two wires?

32. (I) A 7500-pF capacitor holds $16.5 \times 10^{-8} \, C$ of charge. What is the voltage across the capacitor?

33. (I) How much charge flows from a 12.0-V battery when it is connected to a 9.00-μF capacitor?

34. (I) A 0.20-F capacitor is desired. What area must the plates have if they are to be separated by a 2.2-mm air gap?

35. (I) What is the capacitance of a pair of circular plates with a radius of 5.0 cm separated by 3.2 mm of mica?

36. (II) The charge on a capacitor increases by 15 μC when the voltage across it increases from 97 V to 121 V. What is the capacitance of the capacitor?

37. (II) An electric field of $8.50 \times 10^5 \, V/m$ is desired between two parallel plates each of area 35.0 cm^2 and separated by 2.45 mm of air. What charge must be on each plate?

38. (II) If a capacitor has 4.2 μC of charge on it and an electric field of 2.0 kV/mm is desired between the two plates which are separated by 4.0 mm of air, what must each plate's area be?

39. (II) How strong is the electric field between the plates of a 0.80-μF air-gap capacitor if they are 2.0 mm apart and each has a charge of 72 μC?

40. (II) The electric field between the plates of a paper-separated ($K = 3.75$) capacitor is $9.21 \times 10^4 \, V/m$. The plates are 1.95 mm apart and the charge on each plate is 0.775 μC. Determine the capacitance of this capacitor and the area of each plate.

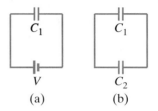

FIGURE 17–28   Problems 41 and 49.

41. (III) A 7.7-μF capacitor is charged by a 125-V battery and then is disconnected from the battery. When this capacitor (C_1) is then connected (Fig. 17–28) to a second (initially uncharged) capacitor, C_2, the voltage on the first drops to 15 V. What is the value of C_2? [*Hint*: Charge is conserved.]

42. (III) A 2.50-μF capacitor is charged to 1000 V and a 6.80-μF capacitor is charged to 650 V. These capacitors are then disconnected from their batteries, and the positive plates are now connected to each other and the negative plates are connected to each other. What will be the potential difference across each and the charge on each? [*Hint*: Charge is conserved.]

SECTION 17–9

43. (I) 550 V is applied to a 7200-pF capacitor. How much energy is stored?

44. (I) A cardiac defibrilator is used to shock a heart that is beating erratically. A capacitor in this device is charged to 6000 V and stores 200 J of energy. What is its capacitance?

45. (II) A homemade capacitor is assembled by placing two 9-in pie pans 10 cm apart and connecting them to the opposite terminals of a 9-V battery. Estimate (*a*) the capacitance, (*b*) the charge on each plate, (*c*) the electric field halfway between the plates, (*d*) the work done by the battery to charge the plates. (*e*) Which of the above values change if a dielectric is inserted?

46. (II) A parallel-plate capacitor has fixed charges $+Q$ and $-Q$. The separation of the plates is then doubled. By what factor does the energy stored in the electric field change?

47. (II) How does the energy stored in a capacitor change if (*a*) the potential difference is doubled, (*b*) the charge on each plate is doubled, and (*c*) the separation of the plates is doubled, as the capacitor remains connected to a battery?

48. (II) A parallel-plate capacitor is isolated with a charge $\pm Q$ on each plate. If the separation of the plates is halved and a dielectric (constant K) is inserted in place of air, by what factor does the energy storage change? To what do you attribute the change in stored potential energy? How does the new value of the electric field between the plates compare with the original value?

49. (III) A 2.70-μF capacitor is charged by a 45.0-V battery. It is disconnected from the battery and then connected to an uncharged 4.00-μF capacitor (Fig. 17–28). Determine the total stored energy (a) before the two capacitors are connected, and (b) after they are connected. (c) What is the change in energy?

SECTION 17–10

***50.**(I) Use the ideal gas as a model to estimate the rms speed of a free electron in a metal at 300 K, and at 2500 K (the typical temperature of the cathode in a tube).

***51.**(III) In a given CRT, electrons are accelerated horizontally by 15 kV. They then pass through a uniform electric field E for a distance of 2.8 cm which deflects them upward so they reach the top of the screen 22 cm away, 11 cm above the center. Estimate the value of E.

***52.**(III) Electrons are accelerated by 14 kV in a CRT. The screen is 30 cm wide and is 34 cm from the 2.6-cm-long deflection plates. Over what range must the horizontally deflecting electric field vary to sweep the beam fully across the screen?

GENERAL PROBLEMS

53. There is an electric field near the Earth's surface whose intensity is about 150 V/m. How much energy is stored per cubic meter in this field?

54. A lightning flash transfers 4.0 C of charge and 4.2 MJ of energy to the Earth. (a) Between what potential difference did it travel? (b) How much water could this boil, starting from room temperature?

55. Calculate the average translational kinetic energy in eV for (a) an oxygen molecule at room temperature, (b) a nitrogen molecule at room temperature, (c) an iron atom in the Sun's corona where the temperature is about 2 million K, and (d) a carbon dioxide molecule in the lower atmosphere of Mars where the temperature is $-50°C$.

56. In a television picture tube, electrons are accelerated by thousands of volts through a vacuum. If a television set were laid on its back, would electrons be able to move upward against the force of gravity? What potential difference, acting over a distance of 3.0 cm, would be needed to balance the downward force of gravity so that an electron would remain stationary? Assume that the electric field is uniform.

57. It takes 8.5 J of energy to move a 3.0-mC charge from one plate of a 9.0-μF capacitor to the other. How much charge is on each plate?

58. An electron starting from rest acquires 5.2 keV of KE in moving from point A to point B. (a) How much KE would a proton acquire, starting from rest at B and moving to point A? (b) Determine the ratio of their speeds at the end of their respective trajectories.

59. A 2600-pF air-gap capacitor is connected to a 9.0-V battery. If a piece of mica is placed between the plates, how much charge will then flow from the battery?

60. A huge 4.0-F capacitor has enough stored energy to heat 2.5 kg of water from 20°C to 95°C. What is the potential difference across the plates?

61. An uncharged capacitor is connected to a 24.0-V battery until it is fully charged, after which it is disconnected from the battery. A slab of paraffin is then inserted between the plates. What will now be the voltage between the plates?

62. Dry air will break down if the electric field exceeds 3.0×10^6 V/m. What amount of charge can be placed on a capacitor if the area of each plate is 56 cm²?

63. A 3.4-μC and a -2.0-μC charge are placed 1.5 cm apart. At what points along the line joining them is (a) the electric field zero, and (b) the potential zero?

64. Three charges are at the corners of an equilateral triangle (side l) as shown in Fig. 17–29. Determine the potential at the midpoint of each of the sides.

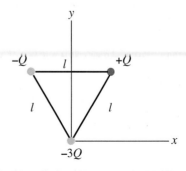

FIGURE 17–29 Problem 64.

65. A capacitor C_1 carries a charge Q_0. It is then connected directly to a second, uncharged, capacitor C_2, as shown in Fig. 17–30. What charge will each carry now? What will be the potential difference across each?

FIGURE 17–30   Problem 65.

66. An electron is accelerated horizontally from rest in a television picture tube by a potential difference of 25,000 V. It then passes between two horizontal plates 6.5 cm long and 1.3 cm apart that have a potential difference of 250 V (Fig. 17–31). At what angle θ will the electron be traveling after it passes between the plates?

FIGURE 17–31   Problem 66.

67. In a photocell, ultraviolet (UV) light provides enough energy to some electrons in barium metal to eject them from a surface at high speed. See Fig. 17–32. To measure the maximum energy of the electrons, another plate above the barium surface is kept at a negative enough potential that the emitted electrons are slowed down and stopped, and return to the barium surface. If the plate voltage is -3.02 V (compared to the barium) when the fastest electrons are stopped, what was the speed of these electrons when they were emitted?

FIGURE 17–32   Problem 67.

68. To get an idea how big a farad is, suppose you wanted to make a 1-F air-filled parallel-plate capacitor for a circuit you were building. (*a*) To make it a reasonable size, suppose you limited the plate area to 1.0 cm^2. What would the gap have to be between the plates? Is this practically achievable? (*b*) Suppose you instead chose the gap between the plates to be 1.0 mm. What would be the area of the plates? Is this a practical solution for your circuit?

69. Near the surface of the Earth there is an electric field of about 150 V/m which points downward. Two identical balls with mass $m = 0.540$ kg are dropped from a height of 2.00 m, but one of the balls is positively charged with $q_1 = 550\ \mu\text{C}$, and the second is negatively charged with $q_2 = -550\ \mu\text{C}$. Use conservation of energy to determine the difference in the speed of the two balls when they hit the ground. (Neglect air resistance.)

70. The power supply for a pulsed nitrogen laser has a $0.050\ \mu\text{F}$ capacitor with a maximum voltage rating of 30 kilovolts. (*a*) Estimate how much energy could be stored in this capacitor. (*b*) If 10 percent of this stored electrical energy is converted to light energy in a pulse that is 10 microseconds long, what is the power of the laser pulse?

71. In lightning storms, the potential difference between the Earth and the bottom of the thunderclouds can be as high as 35,000,000 V. The bottoms of the thunderclouds are typically 1500 m above the Earth, and can have an area of 110 km^2. For the purposes of this problem, model the Earth-cloud system as a huge capacitor and calculate (*a*) the capacitance of the Earth-cloud system, (*b*) the charge stored in the "capacitor," and (*c*) the energy stored in the "capacitor."

The glow of the thin wire filament of a light bulb is caused by the electric current passing through it. Electric energy is transformed to thermal energy (via collisions between moving electrons and atoms of the wire), which causes the wire's temperature to become so high that it glows.

ELECTRIC CURRENTS 18

In the previous two chapters we have been studying static electricity: electric charges at rest. In this chapter we begin our study of charges in motion as an electric current. Until the year 1800, the technical development of electricity consisted mainly of producing a static charge by friction—for example by machines that could produce rather large potentials such as the one shown in Fig. 18–1. Large sparks could be produced by such machines, but they had little practical value.

In nature itself there were grander displays of electricity such as lightning and "St. Elmo's fire," which is a glow that appeared around the yardarms of ships during storms. That these phenomena were electrical in origin was not recognized until the eighteenth century. For example, it was only in 1752 that Franklin, in his famous kite experiment, showed that lightning was an electric discharge—a giant electric spark.

Finally, in 1800, an event of great practical importance occurred: Alessandro Volta (1745–1827; Fig. 18–2) invented the electric battery, and with it produced the first steady flow of electric charge—that is, a steady electric current. This discovery opened a new era, which transformed our civilization, for today's electrical technology is based on electric current.

FIGURE 18–1 Early electrostatic generator.

FIGURE 18–2 Alessandro Volta. In this portrait, Volta exhibits his battery to Napoleon in 1801.

18–1 The Electric Battery

The events that led to the discovery of the battery are interesting. For not only was this an important discovery, but it also gave rise to a famous scientific debate between Volta and Luigi Galvani (1737–1798), eventually involving many others in the scientific world.

In the 1780s, Galvani, a professor at the University of Bologna (thought to be the world's oldest university still in existence), carried out a long series of experiments on the contraction of a frog's leg muscle through electricity produced by a static-electricity machine. Galvani found that contraction of the muscle could be produced by other means as well: when a brass hook was pressed into the frog's spinal cord and then hung from an iron railing that also touched the frog, the leg muscles again would contract. He found that this phenomenon occurred for other pairs of metals as well.

What was the source of this unusual phenomenon? Galvani believed that the source of the electric charge was in the frog muscle or nerve itself, and that the wire merely transmitted the charge to the proper points; and when he published his work in 1791, he termed it "animal electricity." Many wondered, including Galvani himself, if he had discovered the long-sought "life-force."

Volta, at the University of Pavia 200 km away, was skeptical of Galvani's results. Although he soon confirmed and extended those experiments, Volta still doubted Galvani's idea of animal electricity. Instead he came to believe that the source of the electricity was not in the animal itself, but rather in the *contact between the two metals*. Volta made public his views and soon had many followers, although others still sided with Galvani.

Volta soon realized that a moist conductor, such as a frog muscle or moisture at the contact point of the two dissimilar metals, was necessary in the circuit if it was to be effective. He also saw that the contracting frog muscle was a sensitive instrument for detecting electric "tension" or "electromotive force" (his words for what we now call potential), in fact more sensitive than the best available electroscopes that he and others had developed. Most important, he recognized that a decisive answer to Galvani could be given only if the sensitive frog leg was replaced by an inorganic detector. That is, to cement his view that it was the contact of two dissimilar metals that caused the frog muscle to contract, Volta would have to connect the two dissimilar metals directly to an electroscope and observe a separation of the leaves representing a potential difference. This proved difficult since his most sensitive[†] electroscopes were much less sensitive than the frog muscle. But the eventual success of this experiment vindicated Volta's theory.

Volta's research found that certain combinations of metals produced a greater effect than others, and, using his measurements, he listed them in order of effectiveness. (This "electrochemical series" is still used by chemists today.) And he found that carbon could be used in place of one of the metals.

Volta then conceived what is his greatest contribution to science. Between a disc of zinc and one of silver, he placed a piece of cloth or paper soaked in salt solution or dilute acid and piled a "battery" of such couplings, one on top of another, as shown in Fig. 18–3a. This "pile" or "battery" produced a much increased potential difference. Indeed, when strips

[†]Volta's most sensitive electroscope measured about 40 V per degree (angle of leaf separation). Nonetheless, he was finally able to estimate the potential differences produced by dissimilar metals in contact: for a silver–zinc contact he got about 0.7 V, remarkably close to today's value of 0.78 V.

FIGURE 18–3 Two types of voltaic battery: (a) a pile; (b) a "crown of cups." Z stands for zinc and A for silver (*argentum* in Latin). Taken from Volta's original publication.

(a)

(b)

of metal connected to the two ends of the pile were brought close, a spark was produced. Volta had designed and built the first electric battery. A second design, known as the "crown of cups," is shown in Fig. 18–3b. Volta made public this great discovery in 1800.

The potential produced by Volta's battery was still weak compared to that produced by the best friction machines of the time, although it could produce considerable charge. (The electrostatic machines were high-potential, low-charge devices.) But it had a great advantage: it was "self-renewing"—it could produce a flow of electric charge continuously for a relatively long period of time. It was not long before even more powerful batteries were constructed.

After Volta's discovery of the electric battery, scientists eventually recognized that a battery produces electricity by transforming chemical energy into electrical energy. Today a great variety of electric cells and batteries are available, from flashlight batteries (sometimes called "dry cells") to the storage battery of a car. The simplest batteries contain two plates or rods made of dissimilar metals (one can be carbon) called **electrodes**. The electrodes are immersed in a solution, such as a dilute acid, called the **electrolyte**. (In a dry cell, the electrolyte is absorbed in a powdery paste.) Such a device is properly called an **electric cell**, and several cells connected together is a **battery** (although today even a single cell is called a battery). The chemical reactions involved in most electric cells are quite complicated. Here we describe how one very simple cell works, emphasizing the physical aspects.

Cells and batteries

The simple cell shown in Fig. 18–4 uses dilute sulfuric acid as the electrolyte. One of the electrodes is made of carbon, the other of zinc. That part of each electrode remaining outside the solution is called the **terminal**, and connections to wires and circuits are made here. The acid attacks the zinc electrode and tends to dissolve it. But each zinc atom leaves two electrons behind, so it enters the solution as a positive ion. The zinc electrode thus acquires a negative charge. As more zinc ions enter solution, the electrolyte can momentarily become positively charged. Because of this, and through other chemical reactions, electrons are pulled off the carbon electrode. Thus the carbon electrode becomes positively charged. Because there is an opposite charge on the two electrodes, there is a potential difference between the two terminals. In a cell whose terminals are not connected, only a small amount of the zinc is dissolved, for as the zinc electrode becomes increasingly negative, any new positive zinc ions produced are attracted back to the electrode. Thus, a particular potential difference or voltage is maintained between the two terminals. If charge is allowed to flow between the terminals, say, through a wire (or a lightbulb),

How a simple battery works

FIGURE 18–4 Simple electric cell.

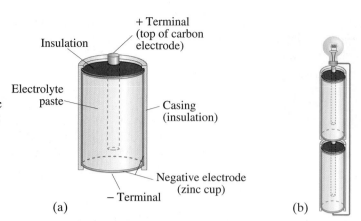

FIGURE 18–5 (a) Diagram of an ordinary dry cell (like a D-cell or AA). The cylindrical zinc cup is covered on the sides; its flat bottom is the negative terminal. (b) Two dry cells (AA type) connected in series. Note that the positive terminal of one cell pushes against the negative terminal of the other.

then more zinc can be dissolved. After a time, one or the other electrode is used up and the cell becomes "dead."

The voltage that exists between the terminals of a battery depends on what the electrodes are made of and their relative ability to be dissolved or give up electrons. When two or more cells are connected so that the positive terminal of one is connected to the negative terminal of the next, they are said to be connected in *series* and their voltages add up. Thus, the voltage between the ends of two 1.5-V flashlight batteries connected in series is 3.0 V, whereas the six 2-V cells of an automobile storage battery give 12 V. Figure 18–5 shows (a) a diagram of a common dry cell (or "flashlight battery") used in portable radios, Walkmans, flashlights, etc., and (b) shows two of them in series.

18–2 Electric Current

Electric circuit

When a continuous conducting path is connected between the terminals of a battery, we have an electric **circuit**, Fig. 18–6a. On any diagram of a circuit, as in Fig. 18–6b, we will represent a battery by the symbol

Battery

$$\dashv\vdash \qquad . \qquad \text{[battery symbol]}$$

The longer line on this symbol represents the positive terminal, and the shorter line, the negative terminal. The device powered by the battery could be a lightbulb (which is just a fine wire inside an evacuated glass bulb), a heater, a radio, or whatever. When such a circuit is formed, charge can flow through the wires of the circuit, from one terminal of the battery to the other. A flow of charge such as this is called an **electric current**.

FIGURE 18–6 (a) A simple electric circuit. (b) Schematic drawing of the same circuit.

More precisely, the electric current in a wire is defined as the net amount of charge that passes through it per unit time at any point. Thus, the average current I is defined as

$$I = \frac{\Delta Q}{\Delta t},$$

(18–1) *Electric current.*

where ΔQ is the amount of charge that passes through the conductor at any location during the time interval Δt. Electric current is measured in coulombs per second; this is given a special name, the **ampere** (abbreviated amp or A), after the French physicist André Ampère (1775–1836). Thus, 1 A = 1 C/s. Smaller units of current are often used such as the milliampere (1 mA = 10^{-3} A) and microampere (1 μA = 10^{-6} A).

The ampere (1 A = 1 C/s)

In any single circuit, such as in Fig. 18–6, the current at any instant is the same at one point (say point A) as at any other point (such as B). This follows from the conservation of electric charge (charge doesn't disappear).

EXAMPLE 18–1 **Current is flow of charge.** A steady current of 2.5 A flows in a wire for 4.0 min. (*a*) How much charge passed through any point in the circuit? (*b*) How many electrons would this be?

SOLUTION (*a*) Since the current was 2.5 A, or 2.5 C/s, then in 4.0 minutes (= 240 seconds) the total charge that flowed was, from Eq. 18–1,

$$\Delta Q = I \, \Delta t$$

$$= (2.5 \text{ C/s})(240 \text{ s}) = 600 \text{ C}.$$

(*b*) The charge on one electron is 1.60×10^{-19} C, so 600 C would consist of

$$\frac{600 \text{ C}}{1.6 \times 10^{-19} \text{ C/electron}} = 3.8 \times 10^{21} \text{ electrons}.$$

CONCEPTUAL EXAMPLE 18–2 **How to connect a battery.** What's wrong with each of the schemes shown in Fig. 18–7 for lighting a flashlight with a flashlight battery and a single wire?

RESPONSE (a) There is no loop for current to flow around. Charges might briefly start to flow from the battery toward the lightbulb, but there they run into a "dead end," and the flow would immediately come to a stop. (b) Now there is a loop passing to and from the lightbulb, but there is no potential difference, and so there will be nothing to push the current. (c) Nothing is wrong here. This way will light the bulb.

FIGURE 18–7 Example 18–2.

(a)　　　　　(b)　　　　　(c)

FIGURE 18-8 Conventional current from + to − is equivalent to a negative (electron) current flowing from − to +.

Conventional current

We saw in Chapter 16 (Section 16–3) that conductors contain many free electrons. Thus, when a conducting wire is connected to the terminals of a battery as in Fig. 18–6, it is actually the negatively charged electrons that flow in the wire. When the wire is first connected, the potential difference between the terminals of the battery sets up an electric field inside the wire[†] and parallel to it. Thus free electrons at one end of the wire are attracted into the positive terminal, and at the same time, electrons leave the negative terminal of the battery and enter the wire at the other end. There is a continuous flow of electrons through the wire that begins as soon as the wire is connected to *both* terminals. However, when the conventions of positive and negative charge were invented two centuries ago, it was assumed that positive charge flowed in a wire. Actually, for nearly all purposes, positive charge flowing in one direction is exactly equivalent to negative charge flowing in the opposite direction,[‡] as shown in Fig. 18–8. Today, we still use the historical convention of positive current flow when discussing the direction of a current. So when we speak of the current flowing in a circuit, we mean the direction positive charge would flow. This is sometimes referred to as **conventional current**. When we want to speak of the direction of electron flow, we will specifically state it is the electron current. In liquids and gases, both positive and negative charges (ions) can move.

18–3 Ohm's Law: Resistance and Resistors

To produce an electric current in a circuit, a difference in potential is required. One way of producing a potential difference is by a battery. It was Georg Simon Ohm (1787–1854) who established experimentally that the current in a metal wire is proportional to the potential difference V applied to its ends:

Current ∝ voltage

$$I \propto V.$$

If, for example, we connect a wire to a 6-V battery, the current flow will be twice what it would be if the wire were connected to a 3-V battery.

Water analogy

It is helpful to compare an electric current to the flow of water in a river or a pipe acted on by gravity. If the pipe (or river) is nearly level, the flow rate is small. But if one end is somewhat higher than the other, the flow rate—or current—is greater. The greater the difference in height, the greater the current. We saw in Chapter 17 that electric potential is analogous, in the gravitational case, to the height of a cliff; and this applies in the present case to the height through which the fluid flows. Just as an increase in height causes a greater flow of water, so a greater electric potential difference, or voltage, causes a greater electric current flow.

Exactly how much current flows in a wire depends not only on the voltage, but also on the resistance the wire offers to the flow of electrons. The walls of a pipe, or the banks of a river and rocks in the middle, offer resistance to the flow of current. Similarly, electrons are slowed down because of interactions with the atoms of the wire. The higher this resistance, the less the current for a given voltage V. We then define *resistance* so that

---

[†]This does not contradict what was said in Section 16–9 that in the *static* case, there can be no electric field within a conductor since otherwise the charges would move. Indeed, when there is an electric field in a conductor, charges do move, and we get an electric current.

[‡]An exception is discussed in Section 20–11.

the current is inversely proportional to the resistance. When we combine this with the above proportion, we have

$$I = \frac{V}{R},$$ (18–2)

where R is the **resistance** of a wire or other device, V is the potential difference across the device, and I is the current that flows through it. This relation (Eq. 18–2) is often written

$$V = IR,$$

and is known as **Ohm's law**. Many physicists would say that it is not a law, but rather is the *definition of resistance*. If we want to call something Ohm's law, it would be the statement that the current through a *metal conductor* is proportional to the applied voltage, $I \propto V$. That is, R is a constant, independent of V, for metal conductors. But this relation does *not* apply generally for other substances and devices such as diodes, vacuum tubes, transistors, and so on. Thus "Ohm's law" is not a fundamental law, but rather a description of a certain class of materials (metal conductors). The habit of calling it Ohm's law is so ingrained that we won't quibble with continuing this usage, as long as we keep in mind its limitations. Materials or devices that do not follow Ohm's law are said to be *nonohmic*. See Fig. 18–9. The definition of resistance,

$$R = V/I$$

(Eq. 18–2), can be applied also to nonohmic cases, but in these cases, R would not be constant and would depend on the applied voltage.

The unit for resistance is called the **ohm** and is abbreviated Ω (Greek capital omega). Because $R = V/I$, we see that $1.0\ \Omega$ is equivalent to $1.0\ V/A$.

OHM'S LAW

(a)

(b)

FIGURE 18–9 Graphs of current vs. voltage for (a) a metal conductor which obeys Ohm's law, and (b) for a nonohmic device, in this case a semiconductor diode.

The ohm $(1\ \Omega = 1\ V/A)$

FIGURE 18–10 Flashlight (Example 18–3). Note how circuit is completed along the side strip.

EXAMPLE 18–3 Flashlight bulb resistance. A small flashlight bulb (Fig. 18–10) draws 300 mA from its 1.5-V battery. (*a*) What is the resistance of the bulb? (*b*) If the voltage dropped to 1.2 V, how would the current change?

SOLUTION (*a*) We use Ohm's law, Eq. 18–2, and find

$$R = \frac{V}{I} = \frac{1.5\ V}{0.30\ A} = 5.0\ \Omega.$$

(*b*) If the resistance stayed constant, the current would be approximately

$$I = \frac{V}{R} = \frac{1.2\ V}{5.0\ \Omega} = 0.24\ A,$$

or a drop of 60 mA. Actually, resistance depends on temperature (Section 18–4), so this is only an approximation.

All electric devices, from heaters to lightbulbs to stereo amplifiers, offer resistance to the flow of current. The filaments of lightbulbs and electric heaters are special types of wires whose resistance results in their becoming very hot. Generally, the connecting wires have very low resistance in comparison to the resistance of the wire filaments or coils. In many circuits, particularly in electronic devices, **resistors** are used to control the

Resistors

FIGURE 18–11 Photo of resistors (mostly).

First digit
Second digit
Multiplier
Tolerance

FIGURE 18–12 The resistance value of a given resistor is written on the exterior, or may be given as a color code, as shown above and in the table: the first two colors represent the first two digits in the value of the resistance, the third color represents the power of ten that it must be multiplied by, and the fourth is the manufactured tolerance. For example, a resistor whose four colors are red, green, orange, and silver has a resistance of 25,000 Ω (25 kΩ), give or take 10 percent.

Resistor Color Code			
Color	Number	Multi-plier	Toler-ance (%)
Black	0	1	
Brown	1	10^1	
Red	2	10^2	
Orange	3	10^3	
Yellow	4	10^4	
Green	5	10^5	
Blue	6	10^6	
Violet	7	10^7	
Gray	8	10^8	
White	9	10^9	
Gold		10^{-1}	5%
Silver		10^{-2}	10%
No color			20%

amount of current. Resistors have resistances from less than an ohm to millions of ohms (see Figs. 18–11 and 18–12). The two main types are "wire-wound" resistors, which consist of a coil of fine wire, and "composition" resistors, which are usually made of the semiconductor carbon.

When we draw a diagram of a circuit, we indicate a resistance with the symbol

Resistor —/\/\/— . [resistor symbol]

Wires whose resistance is negligible, however, are shown simply as straight lines.

18–4 Resistivity

We might expect that the resistance of a thick wire would be less than that of a thin one because a thicker wire has more area for the electrons to pass through. And you might expect the resistance to be greater if the length is greater since there would be more obstacles to electron flow. Indeed, it is found experimentally that the resistance R of a metal wire is directly proportional to its length L and inversely proportional to the cross-sectional area A. That is,

$$R = \rho \frac{L}{A}, \tag{18–3}$$

Resistivity where ρ, the constant of proportionality, is called the **resistivity** and depends on the material used. Typical values of ρ, whose units are $\Omega \cdot m$ (see Eq. 18–3), are given for various materials in the middle column of Table 18–1. The values depend somewhat on purity, heat treatment, temperature, and other factors. Notice that silver has the lowest resistivity and is thus the best conductor (although it is expensive). Copper is not far behind, so it is clear why most wires are made of copper. Aluminum, although it has a higher resistivity, is much less dense than copper; it is thus preferable to copper in some situations, such as transmission lines, because its resistance for the same weight is less than that for copper.

TABLE 18–1 Resistivity and Temperature Coefficients (at 20°C)

Material	Resistivity, ρ ($\Omega\cdot$m)	Temperature Coefficient, α (C°)$^{-1}$
Conductors		
Silver	1.59×10^{-8}	0.0061
Copper	1.68×10^{-8}	0.0068
Gold	2.44×10^{-8}	0.0034
Aluminum	2.65×10^{-8}	0.00429
Tungsten	5.6×10^{-8}	0.0045
Iron	9.71×10^{-8}	0.00651
Platinum	10.6×10^{-8}	0.003927
Mercury	98×10^{-8}	0.0009
Nichrome (alloy of Ni, Fe, Cr)	100×10^{-8}	0.0004
Semiconductors†		
Carbon (graphite)	$(3–60) \times 10^{-5}$	-0.0005
Germanium	$(1–500) \times 10^{-3}$	-0.05
Silicon	$0.1 - 60$	-0.07
Insulators		
Glass	$10^{9} - 10^{12}$	
Hard rubber	$10^{13} - 10^{15}$	

†Values depend strongly on presence of even slight amounts of impurities.

EXAMPLE 18–4 Speaker wires. Suppose you want to connect your stereo to remote speakers (Fig. 18–13). (*a*) If each wire must be 20 m long, what diameter copper wire should you use to keep the resistance less than $0.10\,\Omega$ per wire? (*b*) If the current to each speaker is 4.0 A, what is the voltage drop across each wire?

SOLUTION (*a*) We solve Eq. 18–3 for the area A and use Table 18–1.

$$A = \rho \frac{L}{R} = \frac{(1.68 \times 10^{-8}\ \Omega\cdot\text{m})(20\ \text{m})}{(0.10\ \Omega)} = 3.4 \times 10^{-6}\ \text{m}^2.$$

The cross-sectional area A of a circular wire is related to its diameter d by $A = \pi d^2/4$. The diameter must then be at least

$$d = \sqrt{\frac{4A}{\pi}} = 2.1 \times 10^{-3}\ \text{m} = 2.1\ \text{mm}.$$

(*b*) From Ohm's law,

$$V = IR = (4.0\ \text{A})(0.10\ \Omega) = 0.40\ \text{V}.$$

FIGURE 18–13 Example 18–4

CONCEPTUAL EXAMPLE 18–5 **Stretching changes resistance.** A wire of resistance R is stretched uniformly until it is twice its original length. What happens to its resistance?

RESPONSE If the length L doubles then the cross-sectional area A halves, so that the volume ($V = AL$) of the wire remains the same. From Eq. 18–3 we see that the resistance would increase by a factor of four ($2/\frac{1}{2} = 4$).

The resistivity of a material depends somewhat on temperature. In general, the resistance of metals increases with temperature. This is not surprising, for at higher temperatures, the atoms are moving more rapidly and are arranged in a less orderly fashion; so they might be expected to interfere more with the flow of electrons. If the temperature change is not too great, the resistivity of metals usually increases nearly linearly with temperature. That is,

Effect of temperature

$$\rho_T = \rho_0[1 + \alpha(T - T_0)] \qquad \text{(18–4)}$$

where ρ_0 is the resistivity at some reference temperature T_0 (such as 0°C or 20°C), ρ_T is the resistivity at a temperature T and α is the *temperature coefficient of resistivity*. Values for α are given in Table 18–1. Note that the temperature coefficient for semiconductors can be negative. Why? It seems that at higher temperatures, some of the electrons that are not normally free in a semiconductor become free and can contribute to the current. Thus, the resistance of a semiconductor can decrease with an increase in temperature, although this is not always the case.

➡ **PHYSICS APPLIED**

Resistance thermometer

EXAMPLE 18–6 **Resistance thermometer.** The variation in electrical resistance with temperature can be used to make precise temperature measurements. Platinum is usually used since it is relatively free from corrosive effects and has a high melting point. To be specific, suppose at 20°C the resistance of a platinum resistance thermometer is 164.2 Ω. When placed in a particular solution, the resistance is 187.4 Ω. What is the temperature of this solution?

SOLUTION Since the resistance R is directly proportional to the resistivity ρ, we can combine Eq. 18–3 with Eq. 18–4:

$$R = R_0[1 + \alpha(T - T_0)].$$

Here $R_0 = \rho_0 L/A$ is the resistance of the wire at $T_0 = 20°C$. We solve this equation for T and find

$$T = T_0 + \frac{R - R_0}{\alpha R_0} = 20°C + \frac{187.4\ \Omega - 164.2\ \Omega}{(3.927 \times 10^{-3}(C°)^{-1})(164.2\ \Omega)} = 56.0°C.$$

More convenient for some applications is a *thermistor*, which consists of a metal oxide or semiconductor whose resistance also varies in a repeatable way with temperature. They can be made quite small and respond very quickly to temperature changes. They have another advantage in that they can be used at very high or low temperatures where gas or liquid thermometers would be useless.

The value of α in Eq. 18–4 depends on temperature, so it is important to check the temperature range of validity of any value (say, in a handbook of physical data). If the temperature range is wide, Eq. 18–4 is not adequate and terms proportional to the square and cube of the temperature are needed, but they are generally very small except when $T - T_0$ is large.

Superconductivity

At very low temperatures, near absolute zero, the resistivity of certain metals and their compounds or alloys becomes zero as measured by the highest-precision techniques. Materials in such a state are said to be **superconducting**. This phenomenon was first observed by H. K. Onnes (1853–1926) in 1911 when he cooled mercury below 4.2 K ($-269°C$). He found that at this temperature, the resistance of mercury suddenly dropped to zero. In general, superconductors become superconducting only below a certain *transition temperature* T_C, which is usually within a few degrees of absolute zero. Current in a ring-shaped superconducting material has been observed to flow for years in the absence of a potential difference, with no measurable decrease. Measurements show that the resistivity ρ of superconductors is less than $4 \times 10^{-25} \, \Omega \cdot m$, which is over 10^{16} times smaller than that for copper, and is considered to be zero in practice.

Much research has been done on superconductivity in recent years to try to understand why it occurs,[†] and to find materials that superconduct at more reasonable temperatures to reduce the cost and inconvenience of refrigeration at the required very low temperature. Before 1986 the highest temperature at which a material was found to superconduct was 23 K, and this required liquid helium to keep the material cold. In 1987, a compound of yttrium, barium, copper, and oxygen was developed that can be superconducting at 90 K. Since this is above the temperature of liquid nitrogen, 77 K, boiling liquid nitrogen is sufficiently cold to keep the material superconducting. This was an important breakthrough since liquid nitrogen is much more easily and cheaply obtained than is the liquid helium needed for previous superconductors. Since then, superconductivity at temperatures in the vicinity of 160K have been reported, though in fragile compounds.

High-temperature superconductors

Considerable research is being done to develop high-T_C superconductors as wires that can carry currents strong enough to be practical. Most applications today use a bismuth-strontium-calcium-copper oxide, known (for short) as BSCCO. Applications of superconductivity that once seemed like science fiction may become real. A major use today of superconductors is for carrying the current in electromagnets (we shall see in Chapter 20 that electric currents produce magnetic fields). In large magnets, a great amount of energy is needed just to maintain the current, and this energy is wasted as heat.

The use of higher-temperature superconductors would enable motors and generators to be much smaller (perhaps $\frac{1}{10}$ the size today) if superconductors can be developed that can sustain large currents. Transmission of power over long distances using superconductors would similarly require much smaller and less costly transmission lines.

Superconductors could make electric cars more practical, make computers far faster than they are today, and have great potential in devices to store energy for use at peak demand. Superconductors are already being studied for use in high-speed ground transportation: the magnetic fields produced by superconducting magnets would be used to "levitate" vehicles over tracks so there is essentially no friction (Fig. 18–14). The levitation arises from the repulsive force between the magnet (on the train) and the eddy currents (Section 21–6) produced in the track below.

FIGURE 18–14 An experimental train in Japan, supported by the magnetic field produced by current in coils beneath the tracks (in the red containers).

[†]The first successful superconductivity theory was published by Bardeen, Cooper, and Schrieffer (the BCS theory) in 1957. It is based on quantum theory and cannot be explained on the basis of classical physics.

18–6 | Electric Power

Filament

Insulator

FIGURE 18–15
Incandescent lightbulb.

Electric energy is useful to us because it can be easily transformed into other forms of energy. Motors, whose operation we will examine in Chapter 20, transform electric energy into mechanical work.

In other devices such as electric heaters, stoves, toasters, and hair dryers, electric energy is transformed into thermal energy in a wire resistance known as a "heating element." And in an ordinary lightbulb, the tiny wire filament (Fig. 18–15) becomes so hot it glows; only a few percent of the energy is transformed into visible light, and the rest, over 90 percent, into thermal energy. Lightbulb filaments and heating elements in household appliances have resistances typically of a few ohms to a few hundred ohms.

Electric energy is transformed into thermal energy or light in such devices because the current is usually rather large, and there are many collisions between the moving electrons and the atoms of the wire. In each collision, part of the electron's kinetic energy is transferred to the atom with which it collides. As a result, the kinetic energy of the atoms increases and hence the temperature of the wire element increases (see Section 13–11). The increased thermal energy (internal energy) can be transferred as heat by conduction and convection to the air in a heater or to food in a pan, by radiation to toast in a toaster, or radiated as light.

To find the power transformed by an electric device recall that the energy transformed when a charge Q moves through a potential difference V is QV (Eq. 17–1). Then the power P, which is the rate energy is transformed, is

$$P = \text{power} = \frac{\text{energy transformed}}{\text{time}} = \frac{QV}{t}.$$

The charge that flows per second, Q/t, is simply the electric current I. Thus we have

Electric power (general)

$$P = IV. \tag{18–5}$$

This general relation gives us the power transformed by any device, where I is the current passing through it and V is the potential difference across it. It also gives the power delivered by a source such as a battery. The SI unit of electric power is the same as for any kind of power, the **watt** ($1\,\text{W} = 1\,\text{J/s}$).

The rate of energy transformation in a resistance R can be written, using Ohm's law ($V = IR$), in two other ways:

Electric power
(in resistance R)

$$P = IV \tag{18–6a}$$

$$= I(IR) = I^2R \tag{18–6b}$$

$$= \left(\frac{V}{R}\right)V = \frac{V^2}{R}. \tag{18–6c}$$

Equations 18–6b and c apply only to resistors, whereas Eq. 18–6a, $P = IV$, applies to any device.

EXAMPLE 18–7 **Headlights.** Calculate the resistance of a 40-W automobile headlight designed for 12 V (Fig. 18–16).

SOLUTION Since we are given $P = 40\,\text{W}$ and $V = 12\,\text{V}$, we can use Eq. 18–6c and solve for R:

$$R = \frac{V^2}{P} = \frac{(12\,\text{V})^2}{(40\,\text{W})} = 3.6\,\Omega.$$

This is the resistance when the bulb is burning brightly at 40 W. When the bulb is cold, the resistance is much lower, as we saw in Eq. 18–4. (Since the current is high when the resistance is low, lightbulbs burn out most often when first turned on.)

40-W Headlight

FIGURE 18–16 Example 18–7.

It is energy, not power, that you pay for on your electric bill. Since power is the *rate* energy is transformed, the total energy used by any device is simply its power consumption multiplied by the time it is on. If the power is in watts and the time is in seconds, the energy will be in joules since $1\,\text{W} = 1\,\text{J/s}$. Electric companies usually specify the energy with a much larger unit, the **kilowatt-hour** (kWh). One kWh $= (1000\,\text{W})(3600\,\text{s}) = 3.60 \times 10^6\,\text{J}$.

You pay for energy

Kilowatt-hour (unit of energy)

EXAMPLE 18–8 **Electric heater.** An electric heater draws 15.0 A on a 120-V line. How much power does it use and how much does it cost per month (30 days) if it operates 3.0 h per day and the electric company charges 10.5 cents per kWh? (For simplicity, assume the current flows steadily in one direction.)

SOLUTION The power is

$$P = IV = (15.0\,\text{A})(120\,\text{V}) = 1800\,\text{W}$$

or 1.80 kW. To operate it for $(3.0\,\text{h/d})(30\,\text{d}) = 90\,\text{h}$ would cost $(1.80\,\text{kW})(90\,\text{h})(\$0.105) = \$17$.

EXAMPLE 18–9 **ESTIMATE** **Lightning bolt.** Lightning is a spectacular example of electric current in a natural phenomenon. There is much variability to lightning bolts, but a typical event can transfer $10^9\,\text{J}$ of energy across a potential difference of perhaps $5 \times 10^7\,\text{V}$ during a time interval of about 0.2 s. Use this information to estimate the total amount of charge transferred, the current, and the average power over the 0.2 s.

SOLUTION From Eq. 17–1, energy $= QV$, so

$$Q \approx \frac{10^9\,\text{J}}{5 \times 10^7\,\text{V}} = 20\,\text{C}.$$

The current over the 0.2 s is about

$$I = \frac{Q}{t} \approx \frac{20\,\text{C}}{0.2\,\text{sec}} = 100\,\text{A}.$$

[Since most lightning bolts consist of several stages, it is possible that individual parts could carry currents much higher than this.] The average power delivered is

$$\overline{P} = \frac{\text{energy}}{\text{time}} = \frac{10^9\,\text{J}}{0.2\,\text{sec}} = 5 \times 10^9\,\text{W} = 5\,\text{GW}.$$

We can also use Eq. 18–6a:

$$P = IV = (100\,\text{A})(5 \times 10^7\,\text{V}) = 5\,\text{GW}.$$

539

The electric wires that carry electricity to lights and other electric appliances have some resistance, although usually it is quite small. Nonetheless, if the current is large enough, the wires will heat up and produce thermal energy at a rate equal to I^2R, where R is the wire's resistance. One possible hazard is that the current-carrying wires in the wall of a building may become so hot as to start a fire. Thicker wires, of course, have less resistance (see Eq. 18–3) and thus can carry more current without becoming too hot. When a wire carries more current than is safe, it is said to be "overloaded." A building should, of course, be designed with wiring heavy enough for any expected load. To prevent overloading, fuses or circuit breakers are installed in circuits. They are basically switches (Fig. 18–17) that open the circuit when the current exceeds some particular value. A 20-A fuse or circuit breaker, for example, opens when the current passing through it exceeds 20 A. If a circuit repeatedly burns out a fuse or opens a circuit breaker, there are two possibilities: there may be too many devices drawing current in that circuit; or there is a fault somewhere, such as a "short." A short, or "short circuit," means that two wires have crossed (perhaps because the insulation has worn down) so the path of the current is shortened. The resistance of the circuit is then very small, so the current will be very large. Short circuits, of course, should be remedied immediately.

FIGURE 18–17 (a) Fuses. When the current exceeds a certain value, the metallic ribbon melts and the circuit opens. Then the fuse must be replaced. (b) A circuit breaker. Electric current passes through a bimetallic strip. When the current is great enough (i.e., too great to be safe), the bimetallic strip is heated and bends so far to the left that the notch in the spring-loaded metal strip drops down over the end of the bimetallic strip; (c) the circuit then opens at the contact points (one is attached to the metal strip) and the outside switch is also flipped. As soon as the bimetallic strip cools down, it can be reset using the outside switch.

(a) Fuses   (b) Circuit breaker   (c) Circuit breaker

Household circuits are designed with the various devices connected so that each receives the standard voltage (usually 120 V in the United States) from the electric company (Fig. 18–18). (Circuits with the devices arranged as in Fig. 18–18 are called *parallel circuits*, as we will discuss more fully in the next chapter.) When a fuse blows or circuit breaker opens, the total current being drawn on that circuit should be checked.

EXAMPLE 18–10 **Will a fuse blow?** Determine the total current drawn by all the devices in the circuit of Fig. 18–18.

SOLUTION The circuit in Fig. 18–18 draws the following currents: the lightbulb draws $I = P/V = 100\,\text{W}/120\,\text{V} = 0.8\,\text{A}$; the heater draws $1800\,\text{W}/120\,\text{V} = 15.0\,\text{A}$; the stereo draws a maximum of $350\,\text{W}/120\,\text{V} = 2.9\,\text{A}$; and the hair dryer draws $1200\,\text{W}/120\,\text{V} = 10.0\,\text{A}$. The total current drawn, if all devices are used at the same time, is

$$0.8\,\text{A} + 15.0\,\text{A} + 2.9\,\text{A} + 10.0\,\text{A} = 28.7\,\text{A}.$$

Switch

Lightbulb
100 W

Electric
heater
1800 W

Stereo receiver
350 W

Hair dryer
1200 W

Fuse

120 V
(from electric company)

FIGURE 18–18 Connection of household appliances.

If the circuit in Fig. 18–18 has a 20-A fuse, it would be expected to blow, and we hope it will, to prevent overloaded wires from getting hot enough to start a fire. Something will have to be turned off to get this circuit below 20 A. (Houses and apartments usually have several circuits, each with its own fuse or circuit breaker; try moving one of the devices to another circuit.) If the circuit is designed for a 30-A fuse, it shouldn't blow, so if it does, a short may be the problem. (The most likely place is in the cord of one of the devices.) Proper fuse size is selected according to the wire used to supply the current; a properly rated fuse should *never* be replaced by a higher-rated one. A fuse blowing or a circuit breaker opening is acting like a switch, making an "open circuit." By an open circuit, we mean that there is no longer a complete conducting path, so no current can flow; it is as if $R = \infty$ (infinity).

In electric circuits, heat dissipation by resistors must be considered. The physical size of a resistor is a rough indicator of the maximum permissible power it can dissipate ($= I^2R$) without appreciable rise in temperature. Common values are $\frac{1}{4}$ W, $\frac{1}{2}$ W, and 1 W; the higher the wattage, the larger the physical size.

18–8 Alternating Current

When a battery is connected to a circuit, the current flows steadily in one direction. This is called a **direct current**, or **DC**. Electric generators at electric power plants, however, produce **alternating current**, or **AC**.[†] An alter

DC and AC

[†]Although it is redundant, we sometimes say "ac current" and "ac voltage," which really mean "alternating current" and "alternating voltage."

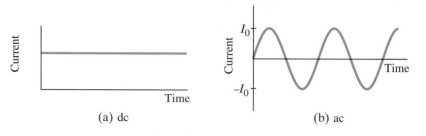

FIGURE 18-19 (a) Direct current. (b) Alternating current.

(a) dc

(b) ac

nating current reverses direction many times per second and is usually sinusoidal, as shown in Fig. 18–19. The electrons in a wire first move in one direction and then in the other. The current supplied to homes and businesses by electric companies is ac throughout virtually the entire world.

The voltage produced by an ac electric generator is sinusoidal, as we shall see in Chapter 21. The current it produces is thus sinusoidal (Fig. 18–19b). We can write the voltage as a function of time as

$$V = V_0 \sin 2\pi ft.$$

The potential V oscillates between $+V_0$ and $-V_0$. V_0 is referred to as the **peak voltage**. The frequency f is the number of complete oscillations made per second. In most areas of the United States and Canada, f is 60 Hz (the unit "hertz," as we saw in Chapter 11, means cycles per second). In some countries, 50 Hz is used.

From Ohm's law, if a voltage V exists across a resistance R, then the current I is

$$I = \frac{V}{R} = \frac{V_0}{R} \sin 2\pi ft = I_0 \sin 2\pi ft. \qquad \textbf{(18–7)}$$

The quantity $I_0 = V_0/R$ is the **peak current**. The current is considered positive when the electrons flow in one direction and negative when they flow in the opposite direction. It is clear from Fig. 18–19b that an alternating current is as often positive as it is negative. Thus, the average current is zero. This does not mean, however, that no power is needed or that no heat is produced in a resistor. Electrons do move back and forth, and do produce heat. Indeed, the power delivered to a resistance R at any instant is

$$P = I^2R = I_0^2 R \sin^2 2\pi ft.$$

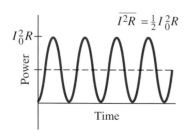

FIGURE 18-20 Power delivered to a resistor in an ac circuit.

$$\overline{I^2R} = \tfrac{1}{2}I_0^2 R$$

Because the current is squared, we see that the power is always positive, Fig. 18–20. The quantity $\sin^2 2\pi ft$ varies between 0 and 1; and it is not too difficult to show that its average value is $\tfrac{1}{2}$ as can be seen graphically in the figure. Thus, the *average power* developed, $\overline{P}$, is

$$\overline{P} = \tfrac{1}{2}I_0^2 R.$$

Since power can also be written $P = V^2/R = (V_0^2/R) \sin^2 2\pi ft$, we also have that the average power is

$$\overline{P} = \tfrac{1}{2}\frac{V_0^2}{R}.$$

The average or mean value of the *square* of the current or voltage is thus what is important for calculating average power: $\overline{I^2} = \tfrac{1}{2}I_0^2$ and $\overline{V^2} = \tfrac{1}{2}V_0^2$. The square root of each of these is the **rms** (root-mean-square) value of the current or voltage:

$$I_{rms} = \sqrt{\overline{I^2}} = \frac{I_0}{\sqrt{2}} = 0.707 I_0, \qquad \textbf{(18–8a)}$$ *rms current*

$$V_{rms} = \sqrt{\overline{V^2}} = \frac{V_0}{\sqrt{2}} = 0.707 V_0. \qquad \textbf{(18–8b)}$$ *rms voltage*

The rms values of V and I are sometimes called the "effective values." They are useful because they can be substituted directly into the power formulas, Eqs. 18–6, to get the average power:

$$\overline{P} = \tfrac{1}{2} I_0^2 R = I_{rms}^2 R \qquad \textbf{(18–9a)}$$

$$\overline{P} = \tfrac{1}{2} \frac{V_0^2}{R} = \frac{V_{rms}^2}{R}. \qquad \textbf{(18–9b)}$$

Thus, a direct current whose values of I and V equal the rms values of I and V for an alternating current will produce the same power. Hence it is usually the rms value of current that is specified or measured. For example, in the United States and Canada, standard line voltage[†] is 120 V ac. The 120 V is V_{rms}; the peak voltage V_0 is

$$V_0 = \sqrt{2} V_{rms} = 170 \text{ V}.$$

In most of Europe the rms voltage is 240 V, so the peak voltage is 340 V.

EXAMPLE 18–11 **Hair dryer.** (*a*) Calculate the resistance and the peak current in a 1000-W hair dryer (Fig. 18–21) connected to a 120-V line. (*b*) What happens if it is connected to a 240-V line in Britain?

SOLUTION (*a*) We can apply Eq. 18–6a using rms values. Then the rms current is

$$I_{rms} = \frac{\overline{P}}{V_{rms}} = \frac{1000 \text{ W}}{120 \text{ V}} = 8.33 \text{ A}.$$

Thus $I_0 = \sqrt{2} I_{rms} = 11.8$ A. From Ohm's law the resistance is

$$R = \frac{V_{rms}}{I_{rms}} = \frac{120 \text{ V}}{8.33 \text{ A}} = 14.4 \text{ } \Omega.$$

The resistance could equally well be calculated using peak values: $R = V_0/I_0 = 170 \text{ V}/11.8 \text{ A} = 14.4 \Omega.$
(*b*) When connected to a 240-V line, more current would flow and the resistance would change with the increased temperature (Section 18–4). But let us make an estimate based on the same 14.4 Ω resistance. The average power delivered would be

$$\overline{P} = \frac{V_{rms}^2}{R} = \frac{(240 \text{ V})^2}{(14.4 \text{ } \Omega)} = 4000 \text{ W}.$$

This is four times the dryer's power rating and would undoubtedly melt the heating element or the wire coils of the motor. Be sure your hair dryer (or electric shaver) has a 120/240 V switch before traveling too far from home, or carry a transformer (Section 21–7) for travelers.

FIGURE 18–21 A hair dryer. Most of the current goes through the heating coils, a pure resistance; a small part goes to the motor to turn the fan. Example 18–11.

[†]The line voltage can vary, depending on the total load; the frequency of 60 Hz, however, remains extremely steady.

EXAMPLE 18–12 **Stereo power.** Each channel of a stereo receiver is capable of an average power output of 100 W into an 8-Ω loudspeaker (see Fig. 18–13). What is the rms voltage and rms current fed to the speaker (*a*) at the maximum power of 100 W, and (*b*) at 1.0 W?

SOLUTION We assume that the loudspeaker can be treated as a simple resistance (not quite true—see Chapter 21) with $R = 8.0\ \Omega$. (*a*) From Eq. 18–9, with $\bar{P} = 100$ W

$$V_{rms} = \sqrt{\bar{P}R} = \sqrt{(100\text{ W})(8.0\ \Omega)} = 28\text{ V},$$

and

$$I_{rms} = \sqrt{\frac{\bar{P}}{R}} = \sqrt{\frac{100\text{ W}}{8.0\ \Omega}} = 3.5\text{ A},$$

or

$$I_{rms} = \frac{V_{rms}}{R} = \frac{28\text{ V}}{8.0\ \Omega} = 3.5\text{ A}.$$

(*b*) At $\bar{P} = 1.0$ W,

$$V_{rms} = \sqrt{(1.0\text{ W})(8.0\ \Omega)} = 2.8\text{ V}$$

$$I_{rms} = \frac{2.8\text{ V}}{8.0\ \Omega} = 0.35\text{ A}.$$

This section has given a brief introduction to the simpler aspects of alternating currents. We will discuss ac circuits in more detail in Chapter 21. In Chapter 19 we will deal with the details of dc circuits only.

* 18–9 Microscopic View of Electric Current

FIGURE 18–22 Electric field **E** in a wire gives electrons in random motion a drift velocity v_d.

Drift speed

FIGURE 18–23 Electrons in the volume Al will all pass through the cross-section indicated in a time Δt where $l = v_d\,\Delta t$.

It can be useful to analyze a simple model of electric current at the microscopic level of atoms and electrons. In a conducting wire, for example, we can imagine the free electrons as moving about randomly at high speeds, bouncing off the atoms of the wire (somewhat like the molecules of a gas—Sections 13–10 and ff). When an electric field exists in the wire, Fig. 18–22, the electrons feel a force and initially begin to accelerate. But they soon reach a more or less steady average speed (due to collisions with atoms in the wire), known as their **drift speed**, v_d. The drift speed is normally very much smaller than the electrons' average random speed.

We can relate v_d to the macroscopic current I in the wire. In a time Δt, the electrons will travel a distance $l = v_d\,\Delta t$ on average. Suppose the wire has cross-sectional area A. Then in time Δt, electrons in a volume $V = Al = Av_d\,\Delta t$ will pass through the cross section A of wire, as shown in Fig. 18–23. If there are n free electrons (each of charge e) per unit volume, ($n = N/V$) then the total charge ΔQ that passes through the area A in a time Δt is

$$\Delta Q = (\text{no. of charges, } N) \times (\text{charge per particle})$$
$$= (nV)(e) = (nAv_d\,\Delta t)(e).$$

The current I in the wire is thus

Current (microscopic variables)

$$I = \frac{\Delta Q}{\Delta t} = neAv_d. \tag{18–10}$$

EXAMPLE 18–13 **Electron speeds in a wire.** A copper wire, 3.2 mm in diameter, carries a 5.0-A current. Determine (*a*) the drift speed of the free electrons, and (*b*) the rms speed of electrons assuming they behave like an ideal gas at 20°C. Assume that one electron per Cu atom is free to move[†] (the others remain bound to the atom).

SOLUTION (*a*) The cross-sectional area of the wire is

$$A = \pi r^2 = (3.14)(1.60 \times 10^{-3} \text{ m})^2 = 8.0 \times 10^{-6} \text{ m}^2.$$

Since we assume there is one free electron per atom, the density of free electrons, *n*, is the same as the density of Cu atoms. The atomic mass of Cu is 63.5 u (see Periodic Table inside the back cover), so 63.5 g of Cu contains one mole or 6.02×10^{23} free electrons. The density of copper (Table 10–1) is $\rho = 8.9 \times 10^3 \text{ kg/m}^3$, where $\rho = m/V$. So the number of free electrons per unit volume is

$$n = \frac{N}{V} = \frac{N}{m/\rho} = \frac{N(1 \text{ mole})}{m(1 \text{ mole})}\rho$$

$$n = \left(\frac{6.02 \times 10^{23} \text{ electrons}}{63.5 \times 10^{-3} \text{ kg}}\right)(8.9 \times 10^3 \text{ kg/m}^3) = 8.4 \times 10^{28} \text{ m}^{-3}.$$

Then, by Eq. 18–10, the drift speed is

$$v_d = \frac{I}{neA} = \frac{5.0 \text{ A}}{(8.4 \times 10^{28} \text{ m}^{-3})(1.6 \times 10^{-19} \text{ C})(8.0 \times 10^{-6} \text{ m}^2)}$$

$$= 4.7 \times 10^{-5} \text{ m/s},$$

which is only about 0.05 mm/s.

(*b*) If we model the free electrons as an ideal gas (a rough approximation), we use Eq. 13–9 to estimate the random rms speed of an electron as it darts around:

$$v_{rms} = \sqrt{\frac{3kT}{m}} = \sqrt{\frac{3(1.38 \times 10^{-23} \text{ J/K})(293 \text{ K})}{9.11 \times 10^{-31} \text{ kg}}} = 1.2 \times 10^5 \text{ m/s}.$$

Thus we see that the drift speed (average speed in the direction of the current) is very much less than the rms thermal speed of the electrons (by a factor of about 10^9). [*Note*: The result in (*b*) is an underestimate. Quantum theory calculations, and experiments, give the rms speed in copper to be about $1.6 \times 10^6 \text{ m/s}$.]

The drift velocity of electrons in a wire is clearly very slow, only about 0.05 mm/s for the example above, which means it takes an electron 20×10^3 s, or $5\frac{1}{2}$ h, to travel only 1 m. This is not, of course, how *Electricity's "speed"* fast "electricity travels": when you flip a light switch, the light—even if many meters away—goes on nearly instantaneously, for electric fields travel essentially at the speed of light ($3 \times 10^8 \text{ m/s}$). We can think of electrons in a wire as being like a pipe full of water: when a little water enters one end of the pipe, almost immediately some water issues forth at the other end.

[†]The number of free electrons in many cases can be determined by measuring the drift speed using the Hall effect (Section 20–11)—that is, the reverse of what we are calculating here.

The Nervous System and Nerve Conduction

An interesting example of the flow of electric charge is our remarkable and complex nervous system, which provides us with the means for being aware of the world, for communication within the body, and for controlling the body's muscles. Although the detailed functioning of the nervous system is still not well understood, we do have a reasonably good understanding of how messages are transmitted within the nervous system: they are electrical signals passing along the basic element of the nervous system, the *neuron*.

Neurons are living cells of unusual shape (Fig. 18–24). Attached to the main cell body are several small appendages known as *dendrites* and a long tail called the *axon*. Signals are received by the dendrites and are propagated along the axon. When a signal reaches the nerve endings, it is transmitted to the next nerve or to a muscle at a connection called a *synapse*. (Some neurons have separate cells, called Schwann cells, wrapped around their axons; they form a layered sheath called a myelin sheath and help to insulate neurons from one another.) Neurons serve in three capacities. "Sensory neurons" carry messages from the eyes, ears, skin, and other organs to the central nervous system, which consists of the brain and spinal cord. "Motor neurons" carry signals from the central nervous system to particular muscles and can signal them to contract. These two types of neuron make up the "peripheral system" as distinguished from the central nervous system. The third type of neuron is the interneuron, which transmits signals between neurons. Interneurons are in the brain and spinal column, and often are connected in an incredibly complex array.

A neuron, before transmitting an electrical signal, is in the so-called "resting state." Like nearly all living cells, neurons have a net positive charge on the outer surface of the cell membrane and a negative charge on the inner surface. This was already mentioned in Section 17–11 with regard to heart muscles and the ECG. This difference in charge, or "dipole layer," means that a potential difference exists across the cell membrane. When a neuron is not transmitting a signal, this "resting potential," which is normally stated as

$$V_{\text{inside}} - V_{\text{outside}},$$

is typically $-60\,\text{mV}$ to $-90\,\text{mV}$, depending on the type of organism. The most common ions in a cell are K^+, Na^+, and Cl^-. There are large differences in the concentrations of these ions inside and outside a cell, as indicated by the typical values given in Table 18–2. Other ions are also present, so the fluids both inside and outside the axon are electrically neutral. Because of the differences in concentration, there is a tendency for ions to diffuse across the membrane (recall diffusion, Section 13–15). However, in the resting state the cell membrane prevents any net flow of Na^+ (through a mechanism of "active pumping" of Na^+ out of the cell). But it does allow the flow of Cl^- ions and less so of K^+ ions, and it is these

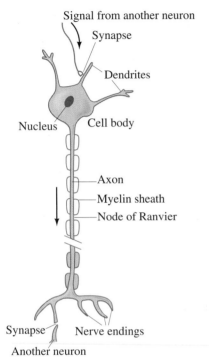

FIGURE 18–24 A neuron.

TABLE 18–2
Concentrations of Ions Inside and Outside a Typical Axon

	Concentration inside axon (mol/m³)	Concentration outside axon (mol/m³)
K^+	140	5
Na^+	15	140
Cl^-	9	125

two ions that produce the dipole charge layer on the membrane. Because there is a greater concentration of K^+ inside the cell than outside, more K^+ ions tend to diffuse outward across the membrane than diffuse inward. A K^+ ion that passes through the membrane becomes attached to the outer surface of the membrane, and leaves behind an equal negative charge that lies on the inner surface of the membrane (Fig. 18–25). The fluids themselves remain neutral. Indeed, what keeps the ions on the membrane is their attraction for each other across the membrane. Independent of this process, Cl^- ions tend to diffuse *into* the cell since their concentration outside is higher. Both K^+ and Cl^- diffusion tends to charge the interior surface of the membrane negatively and the outside positively. As charge accumulates on the membrane surface, it becomes increasingly difficult for more ions to diffuse: K^+ ions trying to move outward, for example, are repelled by the positive charge already there. Equilibrium is reached when the tendency to diffuse because of the concentration difference is just balanced by the electrical potential difference across the membrane. The greater the concentration difference, the greater the potential difference across the membrane, which, as mentioned above, is in the range $-60\,\text{mV}$ to $-90\,\text{mV}$.

The most important aspect of a neuron is not that it has a resting potential (most cells do), but rather that it can respond to a stimulus and conduct an electrical signal along its length. A nerve can be stimulated in various ways. The stimulus could be thermal (when you touch a hot stove) or chemical (as in taste buds); it could be pressure (as on the skin or at the eardrum), or light (as in the eye); or it could be the electric stimulus of a signal coming from the brain or another neuron. In the laboratory, the stimulus is usually electrical and is applied by a tiny probe at some point on the neuron. If the stimulus exceeds some threshold, a voltage pulse will travel down the axon. This voltage pulse can be detected at a point on the axon using a voltmeter or an oscilloscope connected as in Fig. 18–26. This voltage pulse has the shape shown in Fig. 18–27, and is called an *action potential*. As can be seen, the potential increases from a resting potential of about $-70\,\text{mV}$ and becomes a positive $30\,\text{mV}$ or $40\,\text{mV}$. The action potential lasts for about 1 ms and travels down an axon with a speed of $30\,\text{m/s}$ to $150\,\text{m/s}$.

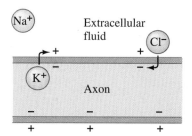

FIGURE 18–25 How a dipole layer of charge forms on a cell membrane.

FIGURE 18–26 Measuring the potential difference between the inside and outside of a nerve.

FIGURE 18–27 Action potential.

FIGURE 18-27
Action potential.

Resting potential | Action potential

FIGURE 18-28 Propagation of an action potential along an axon membrane.

Point of stimulation

Exterior | Membrane

(a)

(b)

(c)

Action potential moving to the right

(d)

But what causes the action potential? Apparently, the cell membrane has the ability to alter its permeability properties. At the point where the stimulation occurs, the membrane suddenly becomes much more permeable to Na^+ than to K^+ and Cl^- ions. Thus, Na^+ ions rush into the cell and the inner surface of the wall becomes positively charged, and the potential difference quickly swings positive ($\approx +30$ mV in Fig. 18–27). But then the membrane suddenly returns to its original characteristics: it becomes impermeable to Na^+ and in fact pumps out Na^+ ions. The diffusion of Cl^- and K^+ ions again predominates and the original resting potential is restored (-70 mV in Fig. 18–27).

What causes the action potential to travel along the axon? The action potential occurs at the point of stimulation, as shown in Fig. 18–28a. The membrane momentarily is positive on the inside and negative on the outside at this point. Nearby charges are attracted toward this region, as shown in Fig. 18–28b. The potential in these adjacent regions then drops, causing an action potential there. Thus, as the membrane returns to normal at the original point, nearby it experiences an action potential, so the action potential moves down the axon (Figs. 18–28c and d).

You may wonder if the number of ions that pass through the membrane would significantly alter the concentrations. The answer is no; and we can show why by treating the axon as a capacitor in the following Example.

EXAMPLE 18–14 **ESTIMATE** **Capacitance of an axon.** (a) Do an order-of-magnitude estimate for the capacitance of an axon 10 cm long of radius 10 μm. The thickness of the membrane is about 10^{-8} m and the dielectric constant is about 3. (b) By what factor does the concentration of Na^+ ions in the cell change as a result of one action potential?

SOLUTION (a) The membrane of an axon resembles a cylindrically shaped parallel-plate capacitor, with opposite charges on each side. The separation of the "plates" is the thickness of the membrane, $d \approx 10^{-8}$ m. The area A is the area of a cylinder of radius r and length l:

$$A = 2\pi rl \approx (6.28)(10^{-5} \text{ m})(0.1 \text{ m}) \approx 6 \times 10^{-6} \text{ m}^2.$$

From Eq. 17–7, we have

$$C = K\epsilon_0 \frac{A}{d} \approx (3)(8.85 \times 10^{-12} \text{ C}^2/\text{N·m}^2)\frac{6 \times 10^{-6} \text{ m}^2}{10^{-8} \text{ m}} \approx 10^{-8} \text{ F}.$$

(b) Since the voltage changes from $-70\,\mathrm{mV}$ to about $+30\,\mathrm{mV}$, the total change is about $100\,\mathrm{mV}$. The amount of charge that moves is then

$$Q = CV \approx (10^{-8}\,\mathrm{F})(0.1\,\mathrm{V}) = 10^{-9}\,\mathrm{C}.$$

Each ion carries a charge $e = 1.6 \times 10^{-19}\,\mathrm{C}$, so the number of ions that flow per action potential is $Q/e = (10^{-9}\,\mathrm{C})/(1.6 \times 10^{-19}\,\mathrm{C}) \approx 10^{10}$. The volume of our cylindrical axon is

$$V = \pi r^2 l \approx (3)(10^{-5}\,\mathrm{m})^2(0.1\,\mathrm{m}) = 3 \times 10^{-11}\,\mathrm{m}^3,$$

and the concentration of Na^+ ions inside the cell (Table 18–2) is $15\,\mathrm{mol/m^3} = 15 \times 6.02 \times 10^{23}\,\mathrm{ions/m^3} \approx 10^{25}\,\mathrm{ions/m^3}$. Thus, the cell contains $(10^{25}\,\mathrm{ions/m^3}) \times (3 \times 10^{-11}\,\mathrm{m^3}) \approx 3 \times 10^{14}\,Na^+$ ions. One action potential, then, will change the concentration of Na^+ ions by at most $10^{10}/(3 \times 10^{14}) = \frac{1}{3} \times 10^{-4}$ or 1 part in 30,000. This tiny change would not be measurable.

Thus, even 1000 action potentials will not alter the concentration significantly. The sodium pump does not, therefore, have to remove Na^+ ions quickly after an action potential, but can operate slowly over time to maintain a relatively constant concentration.

The propagation of a nerve pulse as described here applies to an unmyelinated axon. Myelinated axons, on the other hand, are insulated from the extracellular fluid by the myelin sheath except at the nodes of Ranvier (see Fig. 18–24). An action potential cannot be regenerated where there is a myelin sheath. Once such a neuron is stimulated, the pulse will still travel along the membrane, but there is resistance and the pulse becomes smaller as it moves down the axon. Nonetheless, the weakened signal can still stimulate a full-fledged action potential when it reaches a node of Ranvier. Thus, the signal is repeatedly amplified at these points. Compare this to an unmyelinated neuron, in which the signal is continually amplified by repeated action potentials all along its length. This naturally requires much more energy. Development of myelinated neurons was a significant evolutionary step, for it meant reliable transmission of nerve pulses with less energy expended. And the pulses travel more quickly, since ordinary conduction is faster than the repeated production of action potentials, whose speed depends on the flow of ions across the membrane.

SUMMARY

An electric battery serves as a source of potential difference by transforming chemical energy into electric energy. A simple battery consists of two electrodes made of different metals immersed in a solution or paste known as an electrolyte.

Electric current, I, refers to the rate of flow of electric charge and is measured in **amperes** (A): 1 A equals a flow of 1 C/s past a given point.

The direction of **conventional current** flow is that of positive charge. In a wire, it is actually negatively charged electrons that move, so they flow in a direction opposite to the direction of the conventional current. Positive conventional current always flows from a high potential to a low potential.

Ohm's law states that the current in a good conductor is proportional to the potential difference applied to its two ends. The proportionality constant is called the **resistance** R of the material, so

$$V = IR.$$

The unit of resistance is the **ohm** (Ω), where $1\,\Omega = 1\,\mathrm{V/A}$. See Table 18–3.

TABLE 18–3 Summary of Units

Current	$1\,\mathrm{A} = 1\,\mathrm{C/s}$
Potential difference	$1\,\mathrm{V} = 1\,\mathrm{J/C}$
Power	$1\,\mathrm{W} = 1\,\mathrm{J/s}$
Resistance	$1\,\Omega = 1\,\mathrm{V/A}$

The resistance R of a wire is inversely proportional to its cross-sectional area A, and directly proportional to its length L and to a property of the material called its resistivity: $R = \rho L/A$. The **resistivity**, ρ, increases with temperature for metals, but for semiconductors it may decrease.

At very low temperatures certain materials become **superconducting**, which means their electrical resistance becomes zero.

The rate at which energy is transformed in a resistance R from electric to other forms of energy (such as heat and light) is equal to the product of current and voltage. That is, the **power** transformed, measured in watts, is given by

$$P = IV$$

and for resistors can be written with the help of Ohm's law as

$$P = I^2R = \frac{V^2}{R}.$$

The total electric energy transformed in any device equals the product of power and the time during which the device is operated. In SI units, energy is given in joules ($1\,\text{J} = 1\,\text{W·s}$), but electric companies use a larger unit, the **kilowatt-hour** ($1\,\text{kWh} = 3.6 \times 10^6\,\text{J}$).

Electric current can be **direct current (dc)**, in which the current is steady in one direction; or it can be **alternating current (ac)**, in which the current reverses direction at a particular frequency, typically 60 Hz. Alternating currents are often sinusoidal in time, $I = I_0 \sin 2\pi ft$, and are produced by an alternating voltage.

The rms values of sinusoidally alternating currents and voltages are given by

$$I_{\text{rms}} = \frac{I_0}{\sqrt{2}} \quad \text{and} \quad V_{\text{rms}} = \frac{V_0}{\sqrt{2}},$$

respectively, where I_0 and V_0 are the peak values. The power relationship, $P = IV = I^2R = V^2/R$, is valid for the average power in alternating currents when the rms values of V and I are used.

QUESTIONS

1. Car batteries can be rated in ampere-hours (A·h). What quantity is being rated?

2. When an electric cell is connected to a circuit, electrons flow away from the negative terminal in the circuit. But within the cell, electrons flow *to* the negative terminal. Explain.

3. Develop an analogy between blood circulation and an electrical circuit. Discuss what plays the role of the heart for the electric case, and so on.

4. Design a circuit in which two different switches of the type shown in Fig. 18–29 can be used to operate the same lightbulb from opposite sides of a room.

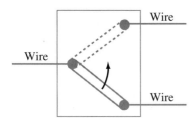

FIGURE 18–29 Question 4.

5. Can a copper wire and an aluminum wire of the same length have the same resistance? Explain.

6. If a rectangular solid made of carbon has sides a, $2a$, $3a$, how would you connect the wires from a battery so as to obtain (*a*) the least resistance, (*b*) the greatest resistance?

7. In a car, one terminal of the battery is said to be connected to "ground." Since it is not really connected to the ground, what is meant by this expression?

8. The equation $P = V^2/R$ indicates that the power dissipated in a resistor decreases if the resistance is increased, whereas the equation $P = I^2R$ implies the opposite. Is there a contradiction here? Explain.

9. What happens when a lightbulb burns out?

10. Explain why lightbulbs almost always burn out just as they are turned on and not after they have been on for some time.

11. Which draws more current, a 100-W lightbulb or a 75-W bulb? Which has the higher resistance?

12. Electric power is transferred over large distances at very high voltages. Explain how the high voltage reduces power losses in the transmission lines.

13. Why is it dangerous to replace a 15-A fuse that blows repeatedly with a 25-A fuse?

14. Electric lights operated on low-frequency ac (say, 10 Hz) flicker noticeably. Why?

15. Some lamps might have batteries connected in either of the two arrangements shown in Fig. 18–30. What would be the advantages of each scheme?

16. Driven by ac power, the same electrons pass back and forth through your reading lamp over and over again. Explain why the light stays lit instead of going out after the first pass of electrons.

17. The heating element in a toaster is made of Nichrome wire. Immediately after the toaster is turned on, is the current (I_{rms}) in the wire increasing, decreasing, or staying constant? Explain.

18. Is current used up in a resistor?

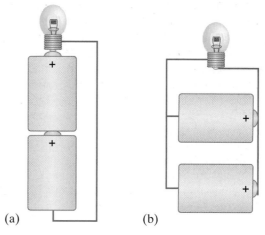

(a)   (b)

FIGURE 18–30 Question 15.

PROBLEMS

SECTIONS 18–2 AND 18–3

1. (I) A service station charges a battery using a current of 5.7 A for 7.0 h. How much charge passes through the battery?

2. (I) A current of 1.00 A flows in a wire. How many electrons are flowing past any point in the wire per second? The charge on one electron is 1.60×10^{-19} C.

3. (I) What is the current in amperes if 1000 Na$^+$ ions were to flow across a cell membrane in 6.5 μs? The charge on the sodium is the same as on an electron, but positive.

4. (I) What voltage will produce 0.25 A of current through a 3000-Ω resistor?

5. (I) What is the resistance of a toaster if 110 V produces a current of 3.1 A?

6. (II) An electrical device draws 5.50 A at 110 V. (a) If the voltage drops by 10 percent, what will be the current, assuming nothing else changes? (b) If the resistance of the device were reduced by 10 percent, what current would be drawn at 110 V?

7. (II) A 9.0-V battery is connected to a bulb whose resistance is 1.6 Ω. How many electrons leave the battery per minute?

8. (II) A hair dryer draws 9.0 A when plugged into a 120-V line. (a) What is its resistance? (b) How much charge passes through it in 15 min?

9. (II) If a 12-V battery pushes a current of 0.50 A through a resistor, what is its resistance, and how many joules of energy does the battery lose in a minute?

10. (II) A bird stands on an electric transmission line carrying 2500 A (Fig. 18–31). The line has 2.5×10^{-5} Ω resistance per meter and the bird's feet are 4.0 cm apart. What voltage does the bird feel?

FIGURE 18–31 Problem 10.

SECTION 18–4

11. (I) What is the resistance of a 3.0-m length of copper wire 1.5 mm in diameter?

12. (I) What is the diameter of a 1.00-m length of tungsten wire whose resistance is 0.22 Ω?

13. (II) Compare the resistance of 10.0 m of aluminum wire 2.0 mm in diameter with 15.0 m of copper filament wire 2.5 mm in diameter.

14. (II) Can a 2.5-mm-diameter copper wire have the same resistance as a tungsten wire of the same length? Give numerical details.

15. (II) A certain copper wire has a resistance of 10.0 Ω. At what point along the length must the wire be cut so that the resistance of one piece is 7.0 times the resistance of the other? Calculate the resistance of each piece.

16. (II) How much would you have to raise the temperature of a copper wire (originally at 20°C) to increase its resistance by 20 percent?

17. (II) A length of aluminum wire is connected to a precision 10.00-V power supply, and a current of 0.4212 A is precisely measured at 20.0°C. The wire is placed in a new environment of unknown temperature where the measured current is 0.3618 A. What is the unknown temperature?

18. (II) Estimate at what temperature copper will have the same resistivity as tungsten does at 20°C.

19. (II) A 100-W lightbulb has a resistance of about 12 Ω when cold (20°C) and 140 Ω when on (hot). Estimate the temperature of the filament when "on" assuming an average temperature coefficient of resistivity $\alpha = 0.0060\ (\text{C}°)^{-1}$.

20. (II) A rectangular solid made of carbon has sides lying along the x, y, and z axes, whose lengths are 1.0 cm, 2.0 cm, and 4.0 cm, respectively (Fig. 18–32). Determine the resistance for current that flows through the solid in (a) the x direction, (b) the y direction, and (c) the z direction. Assume the resistivity is $\rho = 3.0 \times 10^{-5}\ \Omega\cdot\text{m}$.

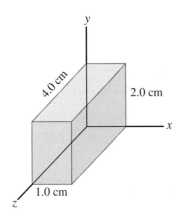

FIGURE 18–32 Problem 20.

21. (II) 10.0 m of wire consists of 5.0 m of copper followed by 5.0 m of aluminum of equal diameter (both 1.0 mm). A voltage difference of 80 V is placed across the composite wire. (a) What is the total resistance of the wire? (b) What is the current flow through the wire? (c) What are the voltages across the aluminum part and across the copper part?

22. (II) For some applications, it is important that the value of a resistance not change with temperature. For example, suppose you made a 4.70-kΩ resistor from a carbon resistor and a Nichrome wire-wound resistor connected together so the total resistance is the sum of their separate resistances. What value should each of these resistors have (at 0°C) so that the combination is temperature independent?

SECTIONS 18–6 AND 18–7

23. (I) The element of an electric oven is designed to produce 3.3 kW of heat when connected to a 240-V source. What must be the resistance of the element?

24. (I) What is the maximum power consumption of a 9.0-V portable cassette player that draws a maximum of 350 mA of current?

25. (I) What is the maximum voltage that can be applied to a 2.7-kΩ resistor rated at $\frac{1}{4}$ watt?

26. (I) A hair dryer has two settings: 600 W and 1200 W. (a) At which setting do you expect the resistance to be higher? After making a guess, determine the resistance at (b) the lower setting, (c) the higher setting.

27. (I) (a) What is the resistance and current through a 60-W lightbulb if it is connected to its proper source voltage of 120 V? (b) Repeat for a 440-W bulb.

28. (I) You buy a 60-W lightbulb in Europe, where electricity is delivered to homes at 240 V. If you use the lightbulb at home in the United States, how bright will it be relative to 60-W 120-V bulbs? Estimate how much power it will consume.

29. (II) How many kWh of energy does a 550-W toaster use in the morning if it is in operation for a total of 10 min? At a cost of 12 cents/kWh, how much would this add to your monthly electric energy bill if you made toast four mornings per week?

30. (II) At $0.110 per kWh, what does it cost to leave a 40-W porch light on day and night for a year?

31. (II) An ordinary flashlight uses 2 D-cell batteries (1.5 V each) and draws 350 mA when turned on. (a) Calculate the resistance of the bulb and the power dissipated. (b) By what factor would the power increase if 4 D-cells in series were used with the same bulb? (Neglect heating effects of the filament.) Why shouldn't you try this?

32. (II) What is the total amount of energy stored in a 12-V, 90-A·h car battery when it is fully charged?

33. (II) A transistor to be used in a circuit is rated for a maximum current of 25 mA if operated at 9.0 V. (a) What is the maximum power output of the transistor, and (b) what would be the limiting current if the voltage applied were actually only 7.0 V?

34. (II) How many 100-W lightbulbs, connected to 120 V as in Fig. 18–18, can be used without blowing a 2.5-A fuse?

35. (II) What is the efficiency of a 0.50-hp electric motor that draws 4.4 A from a 120-V line?

36. (II) A power station delivers 520 kW of power to a factory through wires of total resistance of 3.0 Ω. How much less power is wasted if the electricity is delivered at 50,000 V rather than 12,000 V?

37. (II) A 2200-W oven is hooked to a 240-V source. (a) What is the resistance of the oven? (b) How long will it take to boil 100 mL of water assuming 80 percent efficiency? (c) How much will this cost at 10 cents/kWh?

38. (III) The current in an electromagnet connected to a 240-V line is 14.5 A. At what rate must cooling water pass over the coils if the water temperature is to rise by no more than 7.50 C°?

39. (III) A small immersion heater can be used in a car to heat a cup of water for coffee. If the heater can heat 150 mL of water from 5°C to 95°C in 5.0 min, approximately how much current does it draw from the 12-V battery and what is its resistance? Assume the manufacturer's claim of 60 percent efficiency.

SECTION 18–8

40. (I) Calculate the peak current in a 2.2-kΩ resistor connected to a 120-V rms ac source.

41. (I) An ac voltage, whose peak value is 180 V, is across a 330-Ω resistor. What is the value of the rms and peak currents in the resistor?

42. (I) What is the resistance of your house when (a) everything electrical is turned off, and (b) there is a lone 75-W lightbulb burning on the porch?

43. (II) The peak value of an alternating current passing through a 1500-W device is 4.0 A. What is the rms voltage across it?

44. (II) Calculate the peak voltage and peak current through an 1800-W arc welder connected to a 450-V ac line.

45. (II) What is the maximum instantaneous power dissipated by, and maximum current passing through, a 3.0-hp pump connected to a 240-V ac power source?

46. (II) A heater coil connected to a 240-V ac line has a resistance of 34 Ω. What is the average power used? What are the maximum and minimum values of the instantaneous power?

*SECTION 18–9

* 47. (II) A 0.55-mm-diameter copper wire carries a tiny current of 2.5 μA. What is the electron drift speed in the wire?

* 48. (II) A 5.00-m length of 2.0-mm-diameter wire carries a 750-mA current when 22.0 mV is applied to its ends. If the drift velocity has been measured (by the Hall effect—Section 20–11) to be 1.7×10^{-5} m/s, determine (a) the resistance R of the wire, (b) the resistivity ρ, and (c) the number n of free electrons per unit volume.

* 49. (II) In a liquid solution, Na^+ ions in a concentration of 5.00 mol/m³ move with a drift speed of 5.00×10^{-4} m/s, and SO_4^{2-} ions move with a speed of 2.00×10^{-4} m/s in the same direction, with the result of producing zero net current. What is the concentration of SO_4^{2-} ions?

* 50. (III) At a point high in the Earth's atmosphere, He^{2+} ions in a concentration of 2.8×10^{12}/m³ are moving due north at a speed of 2.0×10^6 m/s. Also, an 8.0×10^{11}/m³ concentration of O_2^- ions is moving due south at a speed of 7.2×10^6 m/s. Determine the magnitude and direction of the net current passing through unit area (A/m²).

*SECTION 18–10

* 51. (I) What is the magnitude of the electric field across an axon membrane 1.0×10^{-8} m thick if the resting potential is −70 mV?

* 52. (II) A nerve is stimulated with an electric pulse. The action potential is detected at a point 3.40 cm down the axon 0.0052 s later. When the action potential is detected 7.20 cm from the point of stimulation, the time required is 0.0063 s. What is the speed of the electric pulse along the axon? (Why are two measurements needed instead of only one?)

* 53. (III) Estimate how much energy is required to transmit one action potential along the axon of Example 18–14. [Hint: One pulse is equivalent to charging and discharging the axon capacitance; see Section 17–9]. What minimum average power is required for 10^4 neurons transmitting 100 pulses per second?

* 54. (III) During the action potential, Na^+ ions move into the cell at a rate of about 3×10^{-7} mol/m²·s. How much power must be produced by the active transport system to produce this flow against a +30-mV potential difference? Assume that the axon is 10 cm long and 20 μm in diameter.

GENERAL PROBLEMS

55. How many coulombs are there in 1.00 ampere-hour?

56. What is the average current drawn by a 1.0-hp 120-V motor?

57. A person accidentally leaves a car with the lights on. If each of the two front lights uses 40 W and each of the two rear lights 6 W, for a total of 92 W, how long will a fresh 12-V battery last if it is rated at 90 A·h? Assume the full 12 V appears across each bulb.

58. The heating element of a 110-V, 1500-W heater is 5.4 m long. If it is made of iron, what must its diameter be?

59. The conductance G of an object is defined as the reciprocal of the resistance R; that is, $G = 1/R$. The unit of conductance is a mho ($= ohm^{-1}$), which is also called the siemens (S). What is the conductance (in siemens) of an object that draws 700 mA of current at 3.0 V?

60. (a) A particular household uses a 1.8-kW heater 3.0 h/day ("on" time), four 100-W lightbulbs 6.0 h/day, a 3.0-kW electric stove element for a total of 1.4 h, and miscellaneous power amounting to 2.0 kWh/day. If electricity costs $.105 per kWh, what will be their monthly bill (30 d)? (b) How much coal (which produces 7000 kcal/kg) must be burned by a 35-percent-efficient power plant to provide the yearly needs of this household?

61. A length of wire is cut in half and the two lengths are wrapped together side by side to make a thicker wire. How does the resistance of this new combination compare to the resistance of the original wire?

62. A 1200-W hair dryer is designed for 117 V. (a) What will be the percentage change in power output if the voltage drops to 105 V? Assume no change in resistance. (b) How would the actual change in resistivity with temperature affect your answer?

63. The wiring in a house must be thick enough so it doesn't become so hot as to start a fire. What diameter must a copper wire be if it is to carry a maximum current of 35 A and produce no more than 1.6 W of heat per meter of length?

64. Suppose a current is given by the equation $I = 1.80 \sin 210t$, where I is in amperes, and t in seconds. (a) What is the frequency? (b) What is the rms value of the current? (c) If this is the current through a 42.0-Ω resistor, what is the equation that describes the voltage as a function of time?

65. In a "brownout" situation, the voltage supplied by the electric company falls. Assuming the percent drop is small, show that the power output of a given appliance falls by approximately twice that percent, assuming the resistance does not change. How much of a voltage drop does it take for a 60-W lightbulb to begin acting like a 50-W bulb?

66. A microwave oven running at 60 percent efficiency delivers 900 W of energy per second to the interior. Find: (a) the power drawn from the source; (b) the current drawn. Assume a source voltage of 120 V.

67. A 1.00-Ω wire is drawn out to 3.00 times its original length. What is its resistance now?

68. 220 V is applied to two different conductors made of the same material. One conductor is twice as long and twice as thick as the second. What is the ratio of the power transformed in the first relative to the second?

69. An electric heater is used to heat a room of volume 62 m^3. Air is brought into the room at 5°C and is changed completely twice per hour. Heat loss through the walls amounts to approximately 850 kcal/h. If the air is to be maintained at 20°C, what minimum wattage must the heater have? (The specific heat of air is about 0.17 kcal/kg·C°.)

FIGURE 18-33 The EV-1 electric car. (Problem 70.)

70. The new EV-1 electric car (Fig. 18–33) makes use of storage batteries as its source of energy. Its mass is 1300 kg and it is powered by 26 batteries, each 12 V, 52 A·h. Assume that the car is driven on the level at an average speed of 40 km/h, and the average friction force is 240 N. Assume 100 percent efficiency and neglect energy used for acceleration. Note that no energy is consumed when the vehicle is stopped since the engine doesn't need to idle. (a) Determine the horsepower required. (b) After approximately how many kilometers must the batteries be recharged?

71. A 12.5-Ω resistor is made from a coil of copper wire whose total mass is 18.0 g. What is the diameter of the wire and how long is it?

72. A 100-W, 120-V lightbulb has a resistance of 12 Ω when cold (20°C) and 140 Ω when on (hot). Calculate its power consumption at (a) the instant it is turned on, and (b) after a few moments when it is hot.

73. A capacitor is often used in electronics to keep energy flowing even if there is a momentary loss of power from the electric company. What capacitance would be required for a 150-W television (plugged into a standard 120 V ac outlet) to protect it from a 0.10 s lapse in power?

*74. The Tevatron accelerator at Fermilab (Illinois) is designed to carry an 11 mA beam of protons traveling at very nearly the speed of light (3.0×10^8 m/s) around a ring 6300 m in circumference. How many protons are stored in the beam?

*75. How far does an average electron move along the wires of a 300-W toaster during an alternating current cycle? The power cord has copper wires of diameter 1.8 mm and is plugged into a standard ac outlet. [Hint: the maximum current in the cycle is related to the maximum drift velocity. The maximum velocity in an oscillation is related to the maximum displacement; see Section 11–3.]

Years ago when one bulb of a certain type of Christmas tree lights burned out, the whole string of lights went out. (Was this a series or a parallel circuit?) Today, if one bulb goes out, the rest stay lit. For other strings however (usually those with very tiny bulbs), if you *remove* a bulb (burned out or not), all the others go out. What kind of circuit could this be?

DC CIRCUITS 19

Electric circuits abound in today's world. They are basic parts of all electronic gear from radio and TV sets to computers and even automobiles. Scientific measurements, from physics to biology and medicine, make use of electric circuits. In Chapter 18, we discussed the basic principles of electric current. Now we will apply these principles to analyze dc circuits and to understand the operation of a number of useful instruments.[†]

When we draw a diagram for a circuit, we represent batteries, capacitors, and resistors by the symbols shown in Table 19–1. Wires whose resistance is negligible compared to other resistance in the circuit are drawn simply as straight lines. For the most part in this chapter, except in Section 19–7, we will be interested in circuits operating in their steady state—that is, we won't be looking at a circuit at the moment any change is made in it, such as when a battery or resistor is connected or disconnected, but rather a short time later when the currents have reached their steady values.

[†]Ac circuits that contain only a voltage source and resistors can be analyzed like the dc circuits in this chapter. However, ac circuits that contain capacitors and other circuit elements are more complicated, and we discuss them in Chapter 21.

**TABLE 19–1
Symbols for
Circuit Elements**

Symbol	Device
⊣⊢	Battery
⊣⊢	Capacitor
⌁⋀⋁⋀⌁	Resistor
——	Wire with negligible resistance

FIGURE 19–1 (a) Resistances connected in series: $R_{eq} = R_1 + R_2 + R_3$. (b) Resistances could be lightbulbs, or any other type of resistance. (c) Equivalent single resistance R_{eq} that draws the same current.

19–1 Resistors in Series and in Parallel

When two or more resistors are connected end to end as shown in Fig. 19–1, they are said to be connected in **series**. The resistors could be simple resistors as in Fig. 18–11, or they could be lightbulbs, heating elements, or other resistive devices. Any charge that passes through R_1 in Fig 19–1a will also pass through R_2 and then R_3. Hence the same current I passes through each resistor. (If it did not, this would imply that charge was accumulating at some point in the circuit, which does not happen in the steady state.) We let V represent the voltage across all three resistors. We assume all other resistance in the circuit can be ignored, and so V equals the voltage of the battery. We let V_1, V_2, and V_3 be the potential differences across each of the resistors, R_1, R_2, and R_3, respectively, as shown in Fig. 19–1a. By Ohm's law, $V_1 = IR_1$, $V_2 = IR_2$, and $V_3 = IR_3$. Because the resistors are connected end to end, energy conservation tells us that the total voltage V is equal to the sum of the voltages across each resistor:

Series circuit: voltages add; current the same in each R

$$V = V_1 + V_2 + V_3 = IR_1 + IR_2 + IR_3. \qquad \text{[series]} \quad \textbf{(19–1)}$$

[To see in more detail why this is true, note that an electric charge q passing through R loses potential energy by qV_1. In passing through R_2 and R_3, the PE decreases by qV_2 and qV_3, for a total $\Delta\text{PE} = qV_1 + qV_2 + qV_3$; this sum must equal the energy given to q by the battery, qV, so that energy is conserved. Hence $qV = q(V_1 + V_2 + V_3)$, and so $V = V_1 + V_2 + V_3$, which is Eq. 19–1.]

Now let us determine the equivalent single resistance R_{eq} that would draw the same current as our combination (Fig. 19–1c). Such a single resistance R_{eq} would be related to V by

$$V = IR_{eq}.$$

We equate this expression with Eq. 19–1, $V = I(R_1 + R_2 + R_3)$, and find

Resistances in series

$$R_{eq} = R_1 + R_2 + R_3. \qquad \text{[series]} \quad \textbf{(19–2)}$$

This is, in fact, what we expect. When we put several resistances in series, the total resistance is the sum of the separate resistances. This applies to any number of resistances in series. Note that when you add more resistance to the circuit, the current will decrease. For example, if a 12-V battery is connected to a 4-Ω resistor, the current will be 3 A. But if the 12-V battery is connected to three 4-Ω resistors in series, the total resistance is 12 Ω and the current will be only 1 A.

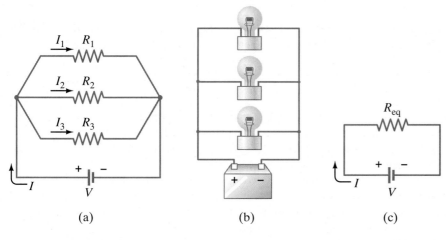

FIGURE 19–2
(a) Resistances connected
in parallel:
$1/R_{eq} = 1/R_1 + 1/R_2 + 1/R_3$,
which could be (b) lightbulbs;
(c) shows the equivalent circuit
with R_{eq} obtained from Eq. 19–3.

Another simple way to connect resistors is in **parallel**, so that the current from the source splits into separate branches, as shown in Fig. 19–2. The wiring in houses and buildings is arranged so all electric devices are in parallel, as we already saw in Fig. 18–18. With parallel wiring, if you disconnect one device (say R_1 in Fig. 19–2), the current to the others is not interrupted. But in a series circuit, if one device (say R_1 in Fig. 19–1) is disconnected, the current *is* stopped to all the others.

In a parallel circuit, Fig. 19–2a, the total current I that leaves the battery breaks into three branches. We let $I_1, I_2,$ and I_3 be the currents through each of the resistors, $R_1, R_2,$ and R_3, respectively. Because electric charge is conserved, the current flowing into a junction (where different wires or conductors meet) must equal the current flowing out of the junction. Thus, in Fig. 19–2a,

$$I = I_1 + I_2 + I_3. \qquad \text{[parallel]}$$

Parallel circuit:
currents add;
voltage the same across each R

When resistors are connected in parallel, each experiences the same voltage. (Indeed, any two points in a circuit connected by a wire of negligible resistance are at the same potential.) Hence the full voltage of the battery is applied to each resistor in Fig. 19–2a, so

$$I_1 = \frac{V}{R_1}, \qquad I_2 = \frac{V}{R_2}, \qquad \text{and} \qquad I_3 = \frac{V}{R_3}.$$

Let us now determine what single resistor R_{eq} (Fig. 19–2c) will draw the same current I as these three resistances in parallel. This equivalent resistance R_{eq} must satisfy

$$I = \frac{V}{R_{eq}}.$$

We now combine the equations above:

$$I = I_1 + I_2 + I_3,$$

$$\frac{V}{R_{eq}} = \frac{V}{R_1} + \frac{V}{R_2} + \frac{V}{R_3}.$$

When we divide out the V from each term, we have

$$\frac{1}{R_{eq}} = \frac{1}{R_1} + \frac{1}{R_2} + \frac{1}{R_3}. \qquad \text{[parallel]} \quad \textbf{(19–3)} \qquad \textit{Resistances in parallel}$$

For example, suppose you connect two 8-Ω loudspeakers to a single set of output terminals of your stereo amplifier or receiver. (Ignore the other

FIGURE 19–3 Water pipes in parallel—analogy to electric currents in parallel.

[Optional paragraph]

channel for a moment—our two speakers are both connected to the left channel, say.) The equivalent resistance will then be found from

$$\frac{1}{R} = \frac{1}{8\,\Omega} + \frac{1}{8\,\Omega} = \frac{2}{8\,\Omega} = \frac{1}{4\,\Omega}$$

and so $R = 4\,\Omega$. Thus the net resistance is *less* than that of each single resistance. This may at first seem surprising. But remember that when you put resistors in parallel, you are giving the current additional paths to follow. Hence the net resistance will be less.

[An analogy may help here. Consider two pipes taking in water near the top of a dam and releasing it below as shown in Fig. 19–3. The gravitational potential difference, proportional to the height h, is the same for both pipes, just as in the electrical case of parallel resistors. If both pipes are open, rather than only one, twice as much current will flow through. That is, with two equal pipes open, the net resistance to the flow of water will be reduced, by half. Note that if both pipes are closed, the dam offers infinite resistance to the flow of water. This corresponds in the electrical case to an open circuit—when no current flows—so the electrical resistance is infinite.]

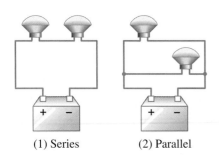

(1) Series   (2) Parallel

FIGURE 19–4 Example 19–1.

CONCEPTUAL EXAMPLE 19–1 **Series or Parallel?** (*a*) The lightbulbs in Fig 19–4 are identical and have identical resistance R. Which configuration produces more light? (*b*) Which way do you think the headlights of a car are wired?

RESPONSE (*a*) The parallel combination has lower resistance ($=R/2$) than the series combination ($=2R$). There will be more total current in configuration (2). The total power consumed, which is proportional to the light produced, is $P = IV$, so the greater current in (2) means more light.
(*b*) In parallel (2), because if one goes out, the other light can stay lit.

EXAMPLE 19–2 **Series and parallel resistors.** Two 100-Ω resistors are connected (*a*) in series, and (*b*) in parallel, to a 24.0-V battery. See Fig. 19–5. What is the current through each resistor and what is the equivalent resistance of each circuit?

FIGURE 19–5 Example 19–2.

(a)

(b)

SOLUTION (a) Every charge, every moving electron, passes first through R_1 and then R_2. So the current I is the same in both resistors; and the potential difference across the battery, V, equals the total change in potential across the two resistors:

$$V = V_1 + V_2 = IR_1 + IR_2.$$

Hence

$$I = \frac{V}{R_1 + R_2} = \frac{24.0 \text{ V}}{100 \text{ }\Omega + 100 \text{ }\Omega} = 0.120 \text{ A}.$$

The equivalent resistance, using Eq. 19–2, is $R_{eq} = R_1 + R_2 = 200 \text{ }\Omega$. We could also get R_{eq} by thinking from the point of view of the battery: the total resistance R_{eq} must equal the battery voltage divided by the current it puts out:

$$R_{eq} = \frac{V}{I} = \frac{24.0 \text{ V}}{0.120 \text{ A}} = 200 \text{ }\Omega.$$

[Note that the voltage across R_1 is $V_1 = IR_1 = (0.120 \text{ A})(100 \text{ }\Omega) = 12.0 \text{ V}$, and across R_2 is $V_2 = IR_2 = 12.0 \text{ V}$, each being half of the battery voltage. A simple circuit like Fig. 19–5a is thus often called a simple "voltage divider."]
(b) Any given charge (or electron) can flow through only one or the other of the two resistors. Just as a river may break into two streams when going around an island, here too the total current I from the battery (Fig. 19–5b) equals the sum of the separate currents through the two resistors:

$$I = I_1 + I_2.$$

The potential difference across each resistor is the battery voltage $V = 24.0 \text{ V}$. Hence

$$I = I_1 + I_2 = \frac{V}{R_1} + \frac{V}{R_2} = \frac{24.0 \text{ V}}{100 \text{ }\Omega} + \frac{24.0 \text{ V}}{100 \text{ }\Omega} = 0.24 \text{ A} + 0.24 \text{ A} = 0.48 \text{ A}.$$

The equivalent resistance is

$$R_{eq} = \frac{V}{I} = \frac{24.0 \text{ V}}{0.48 \text{ A}} = 50 \text{ }\Omega.$$

We could also have obtained this result from Eq. 19–3:

$$\frac{1}{R_{eq}} = \frac{1}{100 \text{ }\Omega} + \frac{1}{100 \text{ }\Omega} = \frac{2}{100 \text{ }\Omega} = \frac{1}{50 \text{ }\Omega},$$

and so $R_{eq} = 50 \text{ }\Omega$.

Note that whenever a group of resistors is replaced by the equivalent resistance, current and voltage and power in the rest of the circuit are unaffected.

EXAMPLE 19–3 **Series with parallel in a circuit.** How much current flows from the battery shown in Fig. 19–6a?

SOLUTION The current I that flows out of the battery all passes through the 400-Ω resistor, but then it splits into I_1 and I_2 passing through the 500-Ω and 700-Ω resistors. The latter two are in parallel. This seems to be a complicated circuit. We look for simplicity, something that we already know how to treat. So let's start by finding the equivalent resistance, R_P, of the parallel resistors, 500-Ω and 700-Ω:

$$\frac{1}{R_P} = \frac{1}{500\ \Omega} + \frac{1}{700\ \Omega} = 0.0020\ \Omega^{-1} + 0.0014\ \Omega^{-1} = 0.0034\ \Omega^{-1}.$$

➡ **PROBLEM SOLVING**

Remember to take the reciprocal

This is $1/R_P$, so we must take the reciprocal to find R_P. (It is a common mistake to forget to take this reciprocal. Notice that the units of reciprocal ohms, Ω^{-1}, help to remind us of this.) Thus

$$R_P = \frac{1}{0.0034\ \Omega^{-1}} = 290\ \Omega.$$

This 290 Ω is the equivalent resistance of the two parallel resistors, and is in series with the 400-Ω resistor as shown in the equivalent circuit of Fig. 19–6b. To find the total equivalent resistance R_{eq}, we add the 400-Ω and 290-Ω resistances together, since they are in series, and find

$$R_{eq} = 400\ \Omega + 290\ \Omega = 690\ \Omega.$$

The total current flowing from the battery is then

$$I = \frac{V}{R_{eq}} = \frac{12.0\ \text{V}}{690\ \Omega} = 0.017\ \text{A} = 17\ \text{mA}.$$

FIGURE 19–6 (a) Circuit for Examples 19–3 and 19–4. (b) Equivalent circuit, showing the equivalent resistance of 290 Ω for the two parallel resistors in (a).

EXAMPLE 19–4 **Current in one branch.** What is the current flowing through the 500-Ω resistor in Fig. 19–6a?

SOLUTION To solve this problem, we must find the voltage across the 500-Ω resistor, which is the voltage between points b and c in the diagram, and we call it V_{bc}. Once V_{bc} is known, we can apply Ohm's law to get the current. First we find the voltage across the 400-Ω resistor, V_{ab}, since we know that 17 mA passes through it; V_{ab} can be found using Ohm's law, $V = IR$:

$$V_{ab} = (0.017\ \text{A})(400\ \Omega) = 6.8\ \text{V}.$$

Since the total voltage across the network of resistors is $V_{ac} = 12.0\ \text{V}$, then V_{bc} must be $12.0\ \text{V} - 6.8\ \text{V} = 5.2\ \text{V}$. Then Ohm's law tells us that the current I_1 through the 500-Ω resistor is

$$I_1 = \frac{5.2\ \text{V}}{500\ \Omega} = 1.0 \times 10^{-2}\ \text{A} = 10\ \text{mA}.$$

This is the answer we wanted. However, we can also calculate the current I_2 through the 700-Ω resistor since the voltage across it is also 5.2 V:

$$I_2 = \frac{5.2\ \text{V}}{700\ \Omega} = 7\ \text{mA}.$$

Notice that when I_1 combines with I_2 to form the total current I (at point c in Fig. 19–6a), their sum is 10 mA + 7 mA = 17 mA. This is, of course, the total current as calculated in Example 19–3.

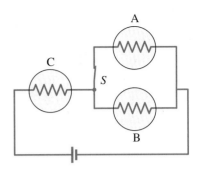

CONCEPTUAL EXAMPLE 19-5 **Bulb brightness in a circuit.** The circuit shown in Fig. 19–7 has three identical lightbulbs, each of resistance R. (a) When switch S is closed, how will the brightness of bulbs A and B compare with that of bulb C? (b) What happens when switch S is opened? Use a minimum of mathematics in your answers.

RESPONSE (a) With switch S closed, the current that passes through bulb C must then split into two equal parts when it reaches the junction leading to bulbs A and B. It splits into equal parts because the resistance of bulbs A and B are equal. Thus, bulbs A and B receive half the current, and will be less bright than bulb C.
(b) When the switch S is open, we have a simple one-loop series circuit, and we expect bulbs B and C to be equally bright. However, the equivalent resistance of this circuit $(=R + R)$ is greater than that of the circuit with the switch closed. So when we open the switch, we increase the resistance and reduce the current leaving the battery. Thus, bulb C will dim when we open the switch. But bulb B gets more current when the switch is open (you may have to use some mathematics here), and so it will brighten somewhat when the switch is opened.

FIGURE 19–7 Example 19–5, three identical lightbulbs.

EXAMPLE 19-6 **Resistor "ladder."** Estimate the equivalent resistance of the "ladder" of equal 100-Ω resistors shown in Fig. 19–8a. In other words, what would an ohmmeter read if connected between points A and B?

(a)

SOLUTION It may seem that none of these resistors is in series or parallel. But there is a place to start: the three resistors on the far right are definitely in series with one another. Each has resistance $R(= 100\,\Omega)$, so those last three have a net resistance of $3R(= 300\,\Omega)$.
   Next we can see that this combination is in parallel with the next resistor to the left, as shown in the dashed box in Fig. 19–8b. The equivalent resistance of the resistors in the dashed box (b), call it R_{eq1}, is given by

(b)

$$\frac{1}{R_{eq1}} = \frac{1}{R} + \frac{1}{3R} = \frac{4}{3R}, \qquad \text{so} \qquad R_{eq1} = \frac{3R}{4}$$

which numerically is $300\,\Omega/4 = 75\,\Omega$. Next, this equivalent resistance of $3R/4$ is in series with the next two (Fig. 19–8c).
The resistors in the dashed box (c) are in series, and are equivalent to $2R + 3R/4 = 11R/4$ (which equals $1100\,\Omega/4 = 275\,\Omega$). Now this $11R/4$ is in parallel with the next step on the ladder (Fig. 19–8d).
The resistance in the dashed box (d) is equivalent to

(c)

$$\frac{1}{R_{eq2}} = \frac{1}{R} + \frac{4}{11R} = \frac{15}{11R}, \qquad \text{so} \qquad R_{eq2} = \frac{11R}{15}.$$

This in turn is in series with two more, yielding the final equivalent resistance of

(d)

FIGURE 19–8 Example 19–6.

$$R_{eq} = \frac{11R}{15} + R + R = \frac{41}{15}R.$$

We are given that $R = 100\,\Omega$, so $R_{eq} = 273\,\Omega$. Notice again the value of using algebra all along: we can have an answer that enables us to calculate the overall resistance regardless of the value of the individual resistor.

19-2 EMF and Terminal Voltage

emf

A device such as a battery or an electric generator that transforms one type of energy (chemical, mechanical, light, and so on) into electric energy is called a **seat** or **source** of **electromotive force** or of **emf**. (The term "electromotive force" is a misnomer since it does not refer to a "force" that is measured in newtons. Hence, to avoid confusion, we prefer to use the abbreviation, emf.) The potential difference between the terminals of such a source, when no current flows to an external circuit, is called the emf of the source. The symbol $\mathscr{E}$ is usually used for emf (don't confuse it with E for electric field).

Why battery voltage isn't constant

You may have noticed in your own experience that when a current is drawn from a battery, the voltage across its terminals drops below its rated emf. For example, if you start a car with the headlights on, you may notice the headlights dim. This happens because the starter draws a large current, and the battery voltage drops as a result. The voltage drop occurs because the chemical reactions in a battery cannot supply charge fast enough to maintain the full emf. For one thing, charge must flow (within the electrolyte) between the electrodes of the battery, and there is always some hindrance to completely free flow. Thus, a battery itself has some resistance, which is called its **internal resistance**; it is usually designated r. A real battery is then modeled as if it were a perfect emf $\mathscr{E}$ in series with a resistor r, as shown in Fig. 19–9. Since this resistance r is inside the battery, we can never separate it from the battery. The two points a and b in the diagram represent the two terminals of the battery. What we measure is the **terminal voltage** V_{ab}. When no current is drawn from the battery, the terminal voltage equals the emf, which is determined by the chemical reactions in the battery: $V_{ab} = \mathscr{E}$. However, when a current I flows from the battery, there is an internal drop in voltage equal to Ir. Thus the terminal voltage (the actual voltage delivered) is[†]

Terminal voltage

$$V_{ab} = \mathscr{E} - Ir.$$

For example, if a 12-V battery has an internal resistance of 0.1 Ω, then when 10 A flows from the battery, the terminal voltage is 12 V $-$ (10 A)(0.1 Ω) $=$ 11 V. The internal resistance of a battery is usually small. For example, an ordinary flashlight battery when fresh may have an internal resistance of perhaps 0.05 Ω. (However, as it ages and the electrolyte dries out, the internal resistance increases to many ohms.) Car batteries have even lower internal resistance.

FIGURE 19–9 Diagram for an electric cell or battery.

[†]When a battery is being charged, a current is forced to pass through it (this happens in Fig. 19–14b), and we have to write

$$V_{ab} = \mathscr{E} + Ir.$$

See, for example, Problem 25 and Fig. 19–31.

EXAMPLE 19–7 **Analyzing a circuit.** A 9.0-V battery whose internal resistance r is 0.50 Ω is connected in the circuit shown in Fig. 19–10a. (a) How much current is drawn from the battery? (b) What is the terminal voltage of the battery? (c) What is the current in the 6.0-Ω resistor?

SOLUTION (a) First, we determine the equivalent resistance of the circuit. But where do we start? We note that the 4.0-Ω and 8.0-Ω resistors are in parallel, and so have an equivalent resistance R_{eq1} given by

$$\frac{1}{R_{eq1}} = \frac{1}{8.0\ \Omega} + \frac{1}{4.0\ \Omega} = \frac{3}{8.0\ \Omega};$$

(a)

so R_{eq1} = 2.7 Ω. This 2.7 Ω is in series with the 6.0-Ω resistor, as shown in Fig. 19–10b; so the net resistance of the lower arm of the circuit is

$$R_{eq2} = 6.0\ \Omega + 2.7\ \Omega = 8.7\ \Omega,$$

as shown in Fig. 19–10c. The equivalent resistance R_{eq3} of this 8.7-Ω and the 10.0-Ω resistances in parallel is given by

$$\frac{1}{R_{eq3}} = \frac{1}{10.0\ \Omega} + \frac{1}{8.7\ \Omega} = 0.21\ \Omega^{-1},$$

(b)

so R_{eq3} = $(1/0.21\ \Omega^{-1})$ = 4.8 Ω. This 4.8 Ω is in series with the 5.0-Ω resistor and the 0.50-Ω internal resistance of the battery (Fig. 19–10d), so the total equivalent resistance R_{eq} of the circuit is R_{eq} = 4.8 Ω + 5.0 Ω + 0.50 Ω = 10.3 Ω. Hence the current drawn is

$$I = \frac{\mathcal{E}}{R_{eq}} = \frac{9.0\ V}{10.3\ \Omega} = 0.87\ A.$$

(b) The terminal voltage of the battery is

$$V_{ab} = \mathcal{E} - Ir = 9.0\ V - (0.87\ A)(0.50\ \Omega) = 8.6\ V.$$

(c)

(c) Now we can work back and get the current in the 6.0-Ω resistor. It must be the same as the current through the 8.7 Ω shown in Fig. 19–10c (why?). The voltage across the 8.7 Ω shown will be the emf of the battery minus the voltage drops across r and the 5.0-Ω resistor. Thus the current will be (call it I')

$$I' = \frac{9.0\ V - (0.87\ A)(0.50\ \Omega + 5.0\ \Omega)}{8.7\ \Omega} = 0.48\ A,$$

and this is the current through the 6.0-Ω resistor.

(d)

FIGURE 19–10 Circuit for Example 19–7, where r is the internal resistance of the battery.

FIGURE 19–11 Currents can be calculated using Kirchhoff's rules.

19-3 Kirchhoff's Rules

In the last few examples we have been able to find the currents flowing in circuits by combining resistances in series and parallel, and using Ohm's law. This technique can be used for many circuits. However, we sometimes encounter a circuit that is too complicated for this analysis. For example, we cannot find the currents flowing in each part of the circuits shown in Fig. 19–11 simply by combining resistances as we did before.

To deal with such complicated circuits, we use Kirchhoff's rules, invented by G. R. Kirchhoff (1824–1887) in the mid-nineteenth century. There are two of them, and they are simply convenient applications of the laws of conservation of charge and energy. **Kirchhoff's first** or **junction rule** is based on the conservation of charge, and we already used it in deriving the rule for parallel resistors. It states that

Junction rule
(conservation of charge)

at any junction point, the sum of all currents entering the junction must equal the sum of all currents leaving the junction.

(That is, what goes in must come out.) For example, at the junction point a in Fig. 19–11, I_3 is entering whereas I_1 and I_2 are leaving. Thus Kirchhoff's junction rule states that $I_3 = I_1 + I_2$. We already saw an instance of this at the end of Example 19–4, where the currents passing through the 400-Ω and 700-Ω resistors were 10 mA and 7 mA, respectively, and they added at point c in Fig. 19–6a to give the outgoing current of 17 mA.

Kirchhoff's junction rule is based on the conservation of charge. Charges that enter a junction must also leave—none is lost or gained. **Kirchhoff's second** or **loop rule** is based on the conservation of energy. It states that

Loop rule
(conservation of energy)

the sum of the changes in potential around any closed path of a circuit must be zero.

To see why this should hold, consider the analogy of a roller coaster on its track. When it starts from the station, it has a particular potential energy. As it climbs the first hill, its potential energy increases and reaches a peak at the top. As it descends the other side, its potential energy decreases and reaches a local minimum at the bottom of the hill. As the roller coaster continues on its path, its potential energy goes through more changes. But when it arrives back at the starting point, it has exactly as much potential energy as it had when it started at this point. Another way of saying this is that there was as much uphill as there was downhill.

The same reasoning can be applied to an electric circuit. We will do the circuit of Fig. 19–11 shortly but first we consider the simple circuit in Fig. 19–12. We have chosen it to be the same as the equivalent circuit of Fig. 19–6b already discussed. The current flowing in this circuit is $I = (12.0\,\text{V})/(690\,\Omega) = 0.0174\,\text{A}$, as we calculated in Example 19–3, although now we calculate to three significant figures and assume all values are known to three significant figures. The positive side of the battery, point e in Fig. 19–12a, is at a high potential compared to point d at the negative side of the battery. That is, point e is like the top of a hill for a roller coaster. We can now follow the current around the circuit starting at any point we choose. Let us start at point e and follow a positive test charge completely around this circuit. As we go, we will note all changes in potential. When the test charge returns to point e, the potential there will be the same as when we started, so the total change in potential will be zero. It is helpful to plot the changes in voltage around the circuit, and we do this in Fig. 19–12b; point d is arbitrarily taken as zero. As our positive test charge goes from point e to point a, there is no change in potential since there is no source of potential nor any resistance. However, as the charge passes through the 400-Ω resistor to get to point b, there is a decrease in potential of $V = IR = (0.0174\,\text{A})(400\,\Omega) = 6.96\,\text{V}$. In effect, the positive test charge is flowing "downhill" since it is heading toward the negative terminal of the battery. This is indicated in the graph of Fig. 19–12b. The decrease in voltage between the two ends of a resistor ($= IR$) is called a **voltage drop**. Because this is a *decrease* in a voltage, we use a *negative* sign when applying Kirchhoff's loop rule; that is,

$$V_{ba} = V_b - V_a = -6.96\,\text{V}.$$

As the charge proceeds from b to c there is a further voltage drop of $(0.0174\,\text{A}) \times (290\,\Omega) = 5.04\,\text{V}$, and since this is a decrease, we write

$$V_{cb} = -5.04\,\text{V}.$$

There is no change in potential as our test charge moves from c to d. But when it moves from d, which is the negative or low potential side of the battery, to point e which is the positive terminal, the voltage *increases* by 12.0 V. That is,

$$V_{ed} = +12.0\,\text{V}.$$

The sum of all the changes in potential in going around the circuit of Fig. 19–12 is then

$$-6.96\,\text{V} - 5.04\,\text{V} + 12.0\,\text{V} = 0.$$

And this is exactly what Kirchhoff's loop rule said it would be.

We already studied the details of this circuit, but have studied it again to illustrate how the loop rule is applied. In the next Section, we will see how to use Kirchhoff's rules to determine the currents in more difficult circuits.

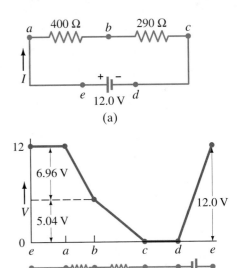

FIGURE 19–12 Changes in potential around the circuit in (a) are plotted in (b).

➡ **PROBLEM SOLVING**

Be consistent with signs

19–4 Solving Problems with Kirchhoff's Rules

When using Kirchhoff's rules, we will find it useful to designate the current in each separate branch of the given circuit by a different subscript, such as I_1, I_2, and I_3 in Fig. 19–13 (this is the same circuit as in Fig. 19–11). You do not have to know in advance in which direction these currents actually are moving. You make a guess and calculate the potentials around the circuit for that direction. If the current actually flows in the opposite direction, your answer will merely have a negative sign. This and other details of using Kirchhoff's rules will become clearer in the following Example.

➡ **PROBLEM SOLVING**

*Choose current
directions arbitrarily*

FIGURE 19–13 Currents can be calculated using Kirchhoff's rules. See Example 19–8.

EXAMPLE 19–8 **Using Kirchhoff's rules.** Calculate the currents I_1, I_2, and I_3 in each of the branches of the circuit in Fig. 19–13.

SOLUTION We choose the directions of the currents as shown in Fig. 19–13. Since (positive) current tends to move away from the positive terminal of a battery, we expect I_2 and I_3 to have the directions shown. It is hard to tell the direction of I_1 in advance, so we arbitrarily chose the direction indicated. We have three unknowns and therefore we need three equations. We first apply Kirchhoff's junction rule to the currents at point a, where I_3 enters and I_2 and I_1 leave:

$$I_3 = I_1 + I_2. \qquad (a)$$

This same equation holds at point d, so we get no new information there. We now apply Kirchhoff's loop rule to two different closed loops. First we apply it to the loop $ahdcba$. From a to h we have a voltage drop $V_{ha} = -(I_1)(30\,\Omega)$. From h to d there is no change, but from d to c the potential increases by 45 V: that is, $V_{cd} = +45$ V. From c to a the voltage drops through the two resistances by an amount $V_{ac} = -(I_3)(40\,\Omega + 1\,\Omega)$. Thus we have $V_{ha} + V_{cd} + V_{ac} = 0$, or

$$-30I_1 - 41I_3 + 45 = 0, \qquad (b)$$

where we have omitted the units. For our second loop, we take the complete circuit $ahdefga$. (We could have just as well taken $abcdefg$ instead.) Again we have $V_{ha} = -(I_1)(30\,\Omega)$, and $V_{dh} = 0$. But when we take our positive test charge from d to e, it actually is going uphill, against the flow of current—or at least against the *assumed* direction of the current, which is what counts in this calculation. Thus $V_{ed} = I_2(20\,\Omega)$ has a *positive* sign. Similarly, $V_{fe} = I_2(1\,\Omega)$. From f to g there is a decrease in potential of 80 V since we go from the high potential terminal of the battery to the low. Thus $V_{gf} = -80$ V. Finally, $V_{ag} = 0$, and the sum of the potentials around this loop is then

$$-30I_1 + 21I_2 - 80 = 0. \qquad (c)$$

The physics is now done. The rest is algebra. We have three equations—labeled (a), (b), and (c)—in three unknowns. From Eq. (c) we have

$$I_2 = \frac{80 + 30I_1}{21} = 3.8 + 1.4I_1. \qquad (d)$$

From Eq. (b) we have

$$I_3 = \frac{45 - 30I_1}{41} = 1.1 - 0.73I_1. \qquad (e)$$

We substitute these into Eq. (a) and solve for I_1:

$$I_1 = I_3 - I_2 = 1.1 - 0.73I_1 - 3.8 - 1.4I_1$$

$$I_1 = -2.7 - 2.1I_1,$$

so

$$3.1I_1 = -2.7$$

$$I_1 = -0.87 \text{ A}.$$

I_1 has magnitude 0.87 A. The negative sign indicates that its direction is actually opposite to that initially assumed and shown in Fig. 19–13. Note that the answer automatically comes out in amperes since all values were in volts and ohms. From Eq. (d) we have

$$I_2 = 3.8 + 1.4I_1 = 2.6 \text{ A},$$

and from Eq. (e)

$$I_3 = 1.1 - 0.73I_1 = 1.7 \text{ A}.$$

This completes the solution.

The unknowns in different situations are not necessarily currents. It might be that the currents are given and we have to solve for unknown resistance or voltage.

➡ **PROBLEM SOLVING** Kirchhoff's Rules

1. Label + and − for each battery. The long side of a battery symbol is +.
2. Label the current in each branch of the circuit with a symbol and an arrow (as in Fig. 19–13). The direction of the arrow can be chosen arbitrarily. If the current is actually in the opposite direction, it will come out with a minus sign in the solution.
3. Apply Kirchhoff's junction rule at each junction, and the loop rule for one or more loops. You will need as many independent equations as there are unknowns. You may write down more equations than this, but you will find that some of the equations will be redundant (that is, not be independent in the sense of providing new information). You may use $V = IR$ for each resistor, which sometimes will reduce the number of unknowns.
4. In applying the loop rule, follow each loop in one direction only (either clockwise or coun-

terclockwise—your choice). Pay careful attention to signs:

(a) For a resistor, the sign of the potential difference is negative if your chosen loop direction is the same as the chosen current direction through that resistor; the sign is positive if you are moving opposite to the chosen current direction.

(b) For a battery, the sign of the potential difference is positive if your loop direction moves from the negative terminal toward the positive; the sign will be negative if you are moving from the positive terminal toward the negative terminal.

5. Solve the equations algebraically for the unknowns. Be careful in manipulating equations not to err with signs. At the end, check your answers by plugging them into the original equations, or even by using any additional equations not used previously (either loop or junction rule equations).

FIGURE 19–14 Batteries in series, (a) and (b), and in parallel, (c).

When two or more sources of emf, such as batteries, are arranged in series, the total voltage is the algebraic sum of their respective voltages. For example, if two 1.5-V flashlight batteries are connected as shown in Fig. 19–14a, the voltage V_{ca} across the light bulb, represented by the resistance R, is 3.0 V. (To be absolutely correct, we should also take into account the internal resistance of the batteries, but we assume it to be small.) On the other hand, when a 20-V and a 12-V battery are connected oppositely, as shown in Fig. 19–14b, the net voltage V_{ca} is 8 V. That is, a positive test charge moved from a to b gains in potential by 20 V, but when it passes from b to c it drops by 12 V. So the net change is 20 V − 12 V = 8 V. You might think that connecting batteries in reverse like this would be wasteful. And for most purposes that would be true. But such a reverse arrangement is precisely how a battery charger works. In Fig. 19–14b, the 20-V source is charging up the 12-V battery. Because of its greater voltage, the 20-V source is forcing charge back into the 12-V battery: electrons are being forced into its negative terminal and removed from its positive terminal. An automobile alternator keeps the car battery charged in the same way. A voltmeter placed across the terminals of a (12-V) car battery with the engine running fairly fast can tell you whether or not the alternator is charging the battery. If it is, the voltmeter reads 13 or 14 V. If the battery is not being charged, the voltage will drop below 12 V because the battery is discharging. Car batteries can be recharged, but other batteries may not be rechargeable, since the chemical reactions in many cannot be reversed. In such cases, the arrangement of Fig. 19–14b would simply waste energy.

Sources of emf can also be arranged in parallel, Fig. 19–14c. A parallel arrangement is not used to increase voltage, but rather to provide more energy when large currents are needed (such as for starting some diesel engines). Each of the cells in parallel has to produce only a fraction of the total current, so the loss due to internal resistance is less than for a single cell; and the batteries will go dead less quickly.

19–6 Circuits Containing Capacitors in Series and in Parallel

Just as resistors can be placed in series or in parallel in a circuit, so can capacitors (Chapter 17). We first consider a **parallel** connection as shown in Fig. 19–15a. If a battery of voltage V is connected to points a and b, this voltage exists across each of the capacitors. Each acquires a charge given by $Q_1 = C_1V$, $Q_2 = C_2V$, and $Q_3 = C_3V$. The total charge Q that must leave the battery is then

$$Q = Q_1 + Q_2 + Q_3 = C_1V + C_2V + C_3V.$$

A single equivalent capacitor that will hold the same charge Q at the same voltage V will have a capacitance C_{eq} given by

$$Q = C_{eq}V.$$

Combining the previous two equations, we have

$$C_{eq}V = C_1V + C_2V + C_3V,$$

or

$$C_{eq} = C_1 + C_2 + C_3.$$ [parallel] **(19–4)**   *Capacitors in parallel*

The net effect of connecting capacitors in parallel is thus to increase the capacitance. This makes sense because we are essentially increasing the area of the plates where charge can accumulate (see Eq. 17–6).

If the capacitors are connected in **series**, as in Fig. 19–15b, a charge $+Q$ flows from the battery to one plate of C_1, and $-Q$ flows to one plate of C_3. The regions A and B between the capacitors were originally neutral, so the net charge there must still be zero. The $+Q$ on the left plate of C_1 attracts a charge of $-Q$ on the opposite plate. Because region A must have a zero net charge, there is thus $+Q$ on the left plate of C_2. The same considerations apply to the other capacitors, so we see that the charge on each capacitor is the same, namely Q. A single capacitor that could replace these three in series without affecting the circuit (that is, Q and V the same) would have a capacitance C_{eq} given by

$$Q = C_{eq}V.$$

Now the total voltage V across the three capacitors in series must equal the sum of the voltages across each capacitor:

$$V = V_1 + V_2 + V_3.$$

We also have $Q = C_1V_1$, $Q = C_2V_2$, and $Q = C_3V_3$, so we substitute for V_1, V_2, and V_3 into the last equation and get

$$\frac{Q}{C_{eq}} = \frac{Q}{C_1} + \frac{Q}{C_2} + \frac{Q}{C_3},$$

or

$$\frac{1}{C_{eq}} = \frac{1}{C_1} + \frac{1}{C_2} + \frac{1}{C_3}.$$ [series] **(19–5)**   *Capacitors in series*

Notice that the forms of the equations for capacitors in series or in parallel are just the reverse of their counterparts for resistance. That is, the formula for capacitors in series resembles the formula for resistors in parallel, and vice versa.

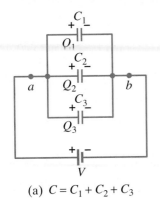

(a) $C = C_1 + C_2 + C_3$

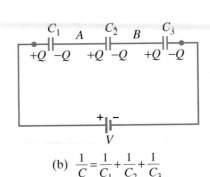

(b) $\dfrac{1}{C} = \dfrac{1}{C_1} + \dfrac{1}{C_2} + \dfrac{1}{C_3}$

FIGURE 19–15 Capacitors (a) in parallel, (b) in series.

FIGURE 19–16 Example 19–9.

Remember to take the reciprocal

EXAMPLE 19–9 **Equivalent capacitance.** Determine the capacitance of a single capacitor that will have the same effect as the combination shown in Fig. 19–16. Take $C_1 = C_2 = C_3 = C$.

SOLUTION C_2 and C_3 are connected in parallel, so they are equivalent to a single capacitor having capacitance

$$C_{23} = C_2 + C_3 = 2C.$$

C_{23} is in series with C_1, so the equivalent capacitance, C_{eq}, is given by

$$\frac{1}{C_{eq}} = \frac{1}{C_1} + \frac{1}{C_{23}} = \frac{1}{C} + \frac{1}{2C} = \frac{3}{2C}.$$

Hence $C_{eq} = \frac{2}{3}C$, which is the equivalent capacitance of the entire combination.

19–7 Circuits Containing a Resistor and a Capacitor

RC circuits

Charging the capacitor

Capacitors and resistors are often found together in a circuit. A simple example is shown in Fig. 19–17a. We now analyze this **RC circuit**. When the switch S is closed, current immediately begins to flow through the circuit. Electrons will flow out from the negative terminal of the battery, through the resistor R, and accumulate on the upper plate of the capacitor. And electrons will flow into the positive terminal of the battery, leaving a positive charge on the other plate of the capacitor. As charge accumulates on the capacitor, the potential difference across it increases, and the current is reduced until eventually the voltage across the capacitor equals the emf of the battery, $\mathscr{E}$. There is then no potential difference across the resistor, and no further current flow. The potential difference across the capacitor, which is proportional to the charge on the capacitor ($V = Q/C$, Eq. 17–5), thus increases in time, as shown in Fig. 19–17b. The actual shape of this curve is a type of exponential. It is given by the formula[†]

$$V = \mathscr{E}(1 - e^{-t/RC}),$$

where V is the voltage across the capacitor as a function of time t.

FIGURE 19–17 For the *RC* circuit shown in (a), the voltage across the capacitor increases with time, as shown in (b), after the switch S is closed.

(a)

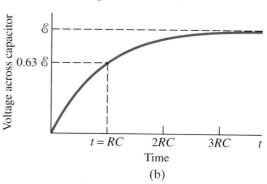

(b)

[†]The derivation is easy, using calculus. The exponential, e, has the value $e = 2.718\cdots$.

570   CHAPTER 19   DC Circuits

The product of the resistance R times the capacitance C, which appears in the exponent, is called the **time constant** τ of the circuit: $\tau = RC$. The time constant is a measure of how quickly the capacitor becomes charged. [The units of RC are $\Omega \cdot F = (V/A)(C/V) = C/(C/s) = s$.] Specifically, it can be shown that the product RC gives the time required for the capacitor to reach 63 percent of full voltage. This can be checked[†] using any calculator with an e^x key: $e^{-1} = 0.37$, so for $t = RC$, then $(1 - e^{-t/RC}) = (1 - e^{-1}) = (1 - 0.37) = 0.63$. In a circuit, for example, where $R = 200\,k\Omega$ and $C = 3.0\,\mu F$, the time constant is $(2.0 \times 10^5\,\Omega)(3.0 \times 10^{-6}\,F) = 0.60\,s$. If the resistance is much lower, the time constant is much smaller and the capacitor becomes charged almost instantly. This makes sense, since a lower resistance will retard the flow of charge less. All circuits contain some resistance (if only in the connecting wires), so a capacitor can never be charged instantaneously when connected to a battery.

Time constant = RC

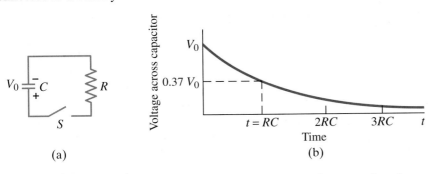

(a)

(b)

FIGURE 19–18 For the RC circuit shown in (a), the voltage V on the capacitor decreases with time, as shown in (b), after the switch S is closed. The charge on the capacitor follows the same curve since $Q \propto V$.

The circuit just discussed involved the *charging* of a capacitor by a battery through a resistance. Now let us look at another situation: when a capacitor is already charged (say to a voltage V_0), and it is allowed to *discharge* through a resistance R as shown in Fig. 19–18a. (In this case there is no battery.) When the switch S is closed, charge begins to flow through resistor R from one side of the capacitor toward the other side, until it is fully discharged. The voltage across the capacitor decreases, as shown in Fig. 19–18b. This "exponential decay" curve is given by

Capacitor discharges

$$V = V_0 e^{-t/RC},$$

where V_0 is the initial voltage across the capacitor. The voltage falls 63 percent of the way to zero (to $0.37\,V_0$) in a time $\tau = RC$.

The charging and discharging in an RC circuit can be used to produce voltage pulses at a regular frequency. The charge on the capacitor increases to a particular voltage, and then discharges. A simple way of initiating the discharge is by the use of a gas-filled tube that breaks down when the voltage across it reaches a certain value V_0. After the discharge is finished, the tube no longer conducts current and the recharging process repeats itself, starting at V_0'. Figure 19–19 shows a possible circuit, and the "sawtooth" voltage it produces.

FIGURE 19–19 (a) An RC circuit, coupled with a gas-filled tube as a switch, can produce a repeating "sawtooth" voltage, as shown in (b).

(a)

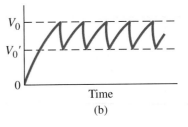

(b)

[†]More simply, since $e = 2.718\cdots$, then $e^{-1} = 1/e = 1/2.718 = 0.37$. Note that the exponential e is the inverse operation to the natural logarithm ln: $\ln(e) = 1$, and $\ln(e^x) = x$.

EXAMPLE 19–10 **An RC circuit.** If a charged capacitor, $C = 35\,\mu\text{F}$, is connected to a resistance, $R = 120\,\Omega$ (as in Fig. 19–18a), how much time will elapse until the voltage falls to 10 percent of its original (maximum) value?

SOLUTION The time constant for this circuit is given by

$$\tau = RC = (120\,\Omega)(35\,\mu\text{F}) = 4.2 \times 10^{-3}\,\text{s}.$$

After a time t the voltage across the capacitor will be

$$V = V_0\,(e^{-t/RC}).$$

We want to know the time t for which $V = (0.10)V_0$. Thus we set

$$(0.10)V_0 = V = V_0 e^{-t/RC}$$

so

$$e^{-t/RC} = 0.10.$$

The inverse operation to the exponential e is the natural log, ln. Thus

$$\ln(e^{-t/RC}) = -\frac{t}{RC} = \ln 0.10 = -2.3.$$

Thus the elapsed time is

$$t = -2.3(RC) = (2.3)(4.2 \times 10^{-3}\,\text{s}) = 9.7 \times 10^{-3}\,\text{s}$$

or 9.7 ms.

CONCEPTUAL EXAMPLE 19–11 **Bulb in RC circuit.** In the circuit of Fig. 19–20, the capacitor is originally uncharged. Describe the behavior of the lightbulb from the instant the switch is thrown until a long time later.

RESPONSE When the switch is first thrown, the charging current is high and the lightbulb burns brightly. As the capacitor charges, the voltage across it increases and the current is reduced causing the lightbulb to dim. As the capacitor approaches the same voltage as the battery, the current drops toward zero and the lightbulb goes out.

FIGURE 19–20 Example 19–11.

* 19–8 Heart Pacemakers

An interesting use of an RC circuit is the electronic pacemaker, which can make a stopped heart start beating again by applying an electric stimulus through electrodes attached to the chest. The stimulus can be repeated at the normal heartbeat rate if necessary.

The heart itself contains a *pacemaker*, which sends out tiny electric pulses at a rate of 60 to 80 per minute. These are the signals that induce the start of each heartbeat. In some forms of heart disease, the pacemaker cells fail to function properly, and the heart loses its beat. People suffering from this ailment now commonly make use of *electronic pacemakers*. These devices produce a regular voltage pulse that starts and controls the frequency of the heartbeat. The "fixed-rate" type produces signals continuously. The "demand" type operates only when the natural pacemaker fails. The electrodes are implanted in or near the heart and the circuit usually contains a capacitor and a resistor. The charge on the capacitor increases to a certain point and then discharges. Then it starts charging again. The pulsing rate depends on the values of R and C. Generally, the power source is a battery that must be replaced or recharged, depending on type. Some pacemakers obtain their energy from the heat produced by a radioactive element; the thermal energy is transformed to electricity by a thermocouple.

19–9 Electric Hazards; Leakage Currents

An electric shock can cause damage to the body or may even be fatal. The severity of a shock depends on the magnitude of the current, how long it acts, and through what part of the body it passes. A current passing through vital organs such as the heart or brain is especially serious for it can interfere with their operation. Electric current heats tissue and can cause burns. A current also stimulates nerves and muscles (whose operation, as we have seen in Sections 17–11 and 18–9, is electrical), and we feel a "shock."

Most people can "feel" a current of about 1 mA. Currents of a few mA cause pain but rarely much damage in a healthy person. However, currents above 10 mA cause severe contraction of the muscles, and a person may not be able to release the source of the current (say, a faulty appliance or wire). Death from paralysis of the respiratory system can occur. Artificial respiration, however, can often revive a victim. If a current above about 70 mA passes across the torso so that a portion passes through the heart for a second or more, the heart muscles will begin to contract irregularly and blood will not be properly pumped. This condition is called "ventricular fibrillation." If it lasts for long, death results. Strangely enough, if the current is much larger, on the order of 1 A, the damage may be less and death by heart failure may be less likely[†] under some conditions.

The seriousness of a shock depends on the effective resistance of the body. Living tissue has quite low resistance since the fluid of cells contains ions that can conduct quite well. However, the outer layers of skin, when dry, offer much resistance. The effective resistance between two points on opposite sides of the body when the skin is dry is in the range of 10^4 to $10^6 \, \Omega$. However, when the skin is wet, the resistance may be $10^3 \, \Omega$ or less. A person in good contact with the ground who touches a 120-V dc line with wet hands can suffer a current

$$I = \frac{120 \text{ V}}{1000 \, \Omega} = 120 \text{ mA.}$$

As we saw above, this could be lethal.

PHYSICS APPLIED

Electric shock

Beware of wet skin

[†]Apparently, larger currents bring the entire heart to a standstill. Upon release of the current, the heart returns to its normal rhythm. This may not happen when fibrillation occurs since it is often hard to stop once it starts. Fibrillation may also occur as a result of a heart attack or during heart surgery. A device known as a *defibrillator* can apply a brief high current to the heart; this causes complete heart stoppage and is often followed by resumption of normal beating.

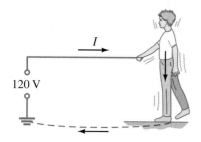

FIGURE 19–21 A person receives an electric shock when the circuit is completed.

Figure 19–21 shows how the circuit is completed when a person touches an electric wire. One side of a 120-V source is connected to ground by a wire connected to a buried conductor (say, a water pipe). Thus the current passes from the high-voltage wire, through the person, to the ground; it passes through the ground back to the other terminal of the source to complete the circuit. If the person in Fig. 19–21 stands on a good insulator—thick-soled shoes or a dry wood floor—there will be much more resistance in the circuit and consequently much less current will flow. However, if the person stands with bare feet on the ground, or is sitting in a bathtub, there is considerable danger because the resistance is much less. In a bathtub, not only are you wet, but the water is in contact with the drain pipe that leads to the ground. That is why it is strongly recommended not to touch anything electrical in such a situation.

(a)   (b)   (c)

FIGURE 19–22 (a) An electric oven operating normally with a two-prong plug. (b) Short to the case with ungrounded case: shock. (c) Short to the case with the case grounded with third prong.

➡ PHYSICS APPLIED

Grounding and shocks

A principal danger comes from touching a bare wire whose insulation has worn off, or from a bare wire inside an appliance when you're tinkering with it. (Always unplug an electrical device before investigating its insides!) Sometimes a wire inside a device breaks or loses its insulation and comes in contact with the case. If the case is metal, it will conduct electricity. A person could then suffer a severe shock merely by touching the case, as shown in Fig. 19–22b. To prevent an accident, metal cases are supposed to be connected directly to ground, so they cannot become "hot." Then if a "hot" wire touches the grounded case, a short circuit to ground immediately occurs internally, as shown in Fig. 19–22c; most of the current passes through the low resistance ground wire rather than through the person. Furthermore, the high current immediately opens the fuse or circuit breaker in the circuit. Grounding a metal case is best done by a separate ground wire connected to the third (round) prong of a 3-prong plug; it can also be done by connecting the case to the larger prong of a so-called "polarized" 2-prong plug. Of course not only the device, but also the outlets must be wired correctly to ground.

FIGURE 19–23 Human body modeled electrically as resistance and capacitor in parallel when a voltage is applied.

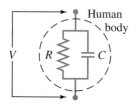

The human body acts as if it had capacitance in parallel with its resistance (Fig. 19–23). A dc current can pass through the resistance, but not the capacitance. An ac current, like the changing currents discussed in Section 18–8 (more on this in Chapter 21), can exist also in the capacitive branch. Because of the additional path allowing current flow, the ac current for a given V_{rms} will be greater than for the same dc voltage. Thus an ac voltage is more dangerous than an equal dc voltage.

Leakage current

Another danger is *leakage current*, by which we mean a current along an unintended path. Leakage currents are often capacitively coupled. For example, a wire in a lamp forms a capacitor with the metal case; charges moving in one conductor attract or repel charge in the other, so there is a current. Typical electrical codes limit leakage currents to 1 mA for any device. A 1-mA leakage current is usually harmless. It can be very dangerous, however, to a hospital patient with implanted electrodes connected to ground through the apparatus. This is because the current can pass directly through the heart as compared to the usual situation where the current enters at the hands and spreads out through the body. Although 70 mA may be needed to cause heart fibrillation when entering through the hands (very little of it actually passes through the heart), as little as 0.02 mA has been known to cause fibrillation when passing directly to the heart. Thus, a "wired" patient is in considerable danger from leakage current even from as simple an act as touching a lamp.

* 19–10 DC Ammeters and Voltmeters

An **ammeter** is used to measure current, and a **voltmeter** measures potential difference or voltage. The crucial part of an analog ammeter or voltmeter, in which the reading is by a pointer on a scale (Fig. 19–24), is a *galvanometer*. The galvanometer works on the principle of the force between a magnetic field and a current-carrying coil of wire, and will be discussed in Chapter 20. For now, we merely need to know that the deflection of the needle of a galvanometer is proportional to the current flowing through it. The *full-scale current sensitivity, I_m*, of a galvanometer is the current needed to make the needle deflect full scale. If the sensitivity I_m is 50 μA, a current of 50 μA will cause the needle to go to the end of the scale. A current of 25 μA will then make it go only half way. If there is no current, the needle should be on zero, and usually there is an adjustment screw to make it so.

Many meters we see in everyday life are galvanometers connected as ammeters or voltmeters, such as some of the meters on a car's instrument panel.

A galvanometer can be used directly to measure small dc currents. For example, a galvanometer whose sensitivity I_m is 50 μA can measure currents from about 1 μA (currents smaller than this would be hard to read on the scale) up to 50 μA. To measure larger currents, a resistor is placed in parallel with the galvanometer. Thus, an ammeter (represented by the symbol •-Ⓐ-•) consists of a galvanometer (•-Ⓖ-•) in parallel with a resistor called the **shunt resistor**. ("Shunt" is a synonym for "in parallel.") This is shown in Fig. 19–25. The shunt resistance is R, and the resistance of the galvanometer coil (which carries the current) is r. The value of R is chosen according to what full-scale deflection is desired and is normally very small, giving an ammeter a very small internal resistance.

➡ **PHYSICS APPLIED**

Meters

FIGURE 19–24 A multimeter used as a voltmeter.

Ammeter uses shunt resistor

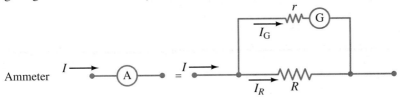

Ammeter

FIGURE 19–25 An ammeter is a galvanometer in parallel with a (shunt) resistor with low resistance, R.

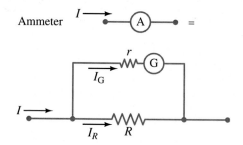

Ammeter

FIGURE 19–25 (Repeated.) An ammeter is a galvanometer in parallel with a (shunt) resistor with low resistance, R.

EXAMPLE 19–12 **Ammeter design.** Design an ammeter to read 1.0 A at full scale using a galvanometer with a full-scale sensitivity of 50 μA and a resistance $r = 30\,\Omega$. Check if the scale is linear.

SOLUTION When the total current I entering the ammeter is 1.0 A, we want the current I_G through the galvanometer to be precisely 50 μA (to give full-scale deflection). See Fig 19–25. Thus, when 1.0 A flows into the meter, we want 0.999950 A $(= I_R)$ to pass through the shunt resistor R. Since the potential difference across the shunt is the same as across the galvanometer,

$$I_R R = I_G r,$$

then

$$R = \frac{I_G r}{I_R} = \frac{(5.0 \times 10^{-5}\,\text{A})(30\,\Omega)}{(0.999950\,\text{A})} = 1.5 \times 10^{-3}\,\Omega,$$

or 0.0015 Ω. The shunt resistor must thus have a very low resistance so that most of the current passes through it.

If the current I into the meter is 0.50 A, this will produce a current to the galvanometer equal to $I_G = I_R R/r = (0.50\,\text{A})(1.5 \times 10^{-3}\,\Omega)/30\,\Omega = 25\,\mu$A, which gives a deflection half of full scale, as required.

Voltmeter uses series resistor

A voltmeter (•Ⓥ•) also consists of a galvanometer and a resistor. But the resistor R is connected in series, Fig. 19–26, and it is usually large, giving a voltmeter a high internal resistance.

FIGURE 19–26 A voltmeter is a galvanometer in series with a resistor with high resistance, R.

Voltmeter

EXAMPLE 19–13 **Voltmeter design.** Using the same galvanometer with internal resistance $r = 30\,\Omega$ and full-scale current sensitivity of 50 μA, design a voltmeter that reads from 0 to 15 V. Is the scale linear?

SOLUTION When a potential difference of 15 V exists across the terminals of our voltmeter, we want 50 μA to be passing through it so as to give a full-scale deflection. From Ohm's law we have (see Fig. 19–26)

$$15\,\text{V} = (50\,\mu\text{A})(r + R),$$

so

$$R = \frac{15\,\text{V}}{5.0 \times 10^{-5}\,\text{A}} - r = 300\,\text{k}\Omega - 30\,\Omega = 300\,\text{k}\Omega.$$

Notice that $r = 30\,\Omega$ is so small compared to the value of R that it doesn't influence the calculation significantly.

The scale will again be linear: if the voltage to be measured is 6.0 V, the current passing through the voltmeter will be $(6.0\,\text{V})/(3.0 \times 10^5\,\Omega) = 2.0 \times 10^{-5}\,\text{A}$, or 20 μA. This will produce two fifths of full-scale deflection, as required ($6.0\,\text{V}/15.0\,\text{V} = 2/5$).

The meters discussed above are for direct current. If an alternating current passed through these meters, the reading would fluctuate wildly, or possibly not move at all because the back and forth changes would be too

(a)   (b)   (c)

FIGURE 19–27 Measuring current and voltage.

rapid. A dc meter like those described above can be modified to measure ac with the addition of diodes (Chapter 29) which allow current to flow in one direction only. The resulting unidirectional current can be read by the meter. An ac meter can be calibrated to read rms or peak values.

Suppose you wish to determine the current I in the circuit shown in Fig. 19–27a and the voltage V across the resistor R_1. How exactly are ammeters and voltmeters connected to the circuit being measured?

How to connect meters

Because an ammeter is used to measure the current flowing in the circuit, it must be inserted directly into the circuit, in series with the other elements, as shown in Fig. 19–27b. The smaller its internal resistance, the less it will affect the circuit.

Ammeter is inserted into the circuit

A voltmeter, on the other hand, is connected in parallel with the circuit element across which the voltage is to be measured. It is used to measure the potential difference between two points and its two wire leads (connecting wires) are connected to the two points, as shown in Fig. 19–27c where the voltage across R_1 is being measured. The larger its internal resistance ($R + r$ in Fig. 19–26) the less it affects the circuit being measured.

Voltmeter is connected in parallel

Voltmeters and ammeters can have several series or shunt resistors to offer a choice of range. **Multimeters** can measure voltage, current, and resistance. Sometimes they are called VOMs (Volt-Ohm-Meter). Meters with digital readout are called digital voltmeters (DVM) or digital multimeters (DMM), Fig. 19–28.

FIGURE 19–28 A digital multimeter being used to measure resistance.

To measure resistance, the meter must contain a battery of known voltage connected in series to a resistor (R_{ser}) and to an ammeter (consisting of a galvanometer of resistance r and a shunt resistor R_{sh}). This makes an **ohmmeter** (see Fig. 19–29, and Problem 83). The resistor whose resistance is to be measured completes the circuit. The deflection in this case is inversely proportional to the resistance. Thus, if the resistance is small the deflection is large, and vice versa. The calibration of the scale depends on the value of the series resistor. Since an ohmmeter sends a current through the device whose resistance is to be measured, it should not be used on very delicate devices that could be damaged by the current.

FIGURE 19–29 An ohmmeter.

The **sensitivity** of a meter is generally specified on the face. It may be given as so many ohms per volt, which indicates how many ohms of resistance there are in the meter per volt of full-scale reading. For example, if the sensitivity is 30,000 Ω/V, this means that on the 10-V scale the meter has a resistance of 300,000 Ω. The full-scale current sensitivity, I_m, discussed earlier, is just the reciprocal of the sensitivity in Ω/V (note that the units Ω/V = A^{-1}). For example, a meter with sensitivity 30,000 Ω/V produces a full-scale deflection at 1.0 V when 30,000 Ω is in series with the galvanometer. Thus the current sensitivity is $(1.0 \text{ V})/(3.0 \times 10^4 \, \Omega) = 33 \, \mu\text{A}$ in this case.

Effects of Meter Resistance

⇨ **PHYSICS APPLIED**

Correcting for meter resistance

(a)

(c)

FIGURE 19–27 (a) and (c) repeated for Example 19–14.

It is important to know the sensitivity of a meter, for in many cases the resistance of the meter can seriously affect your results. Take the following Example.

EXAMPLE 19–14 **Voltage reading versus true voltage.** Suppose you are testing an electronic circuit which has two resistors, R_1 and R_2, each 15 kΩ, connected in series as shown in Fig. 19–27a. The battery maintains 8.0 V across them and has negligible internal resistance. A voltmeter whose sensitivity is 10,000 Ω/V is put on the 5.0-V scale. What voltage does the meter read when connected across R_1, and what error is caused by the finite resistance of the meter?

SOLUTION On the 5.0-V scale, the voltmeter has an internal resistance of $(5.0\,\text{V})(10{,}000\,\Omega/\text{V}) = 50{,}000\,\Omega$. When connected across R_1, as in Fig. 19–27c, we have this 50 kΩ in parallel with $R_1 = 15$ kΩ. The net resistance R_{eq} of these two is given by

$$\frac{1}{R_{eq}} = \frac{1}{50\text{ k}\Omega} + \frac{1}{15\text{ k}\Omega} = \frac{13}{150\text{ k}\Omega};$$

so $R_{eq} = 11.5$ kΩ. This $R_{eq} = 11.5$ kΩ is in series with $R_2 = 15$ kΩ, so the total resistance of the circuit is now 26.5 kΩ. Hence the current from the battery is

$$I = \frac{8.0\text{ V}}{26.5\text{ k}\Omega} = 0.30\text{ mA}.$$

Then the voltage drop across R_1, which is the same as that across the voltmeter, is $(3.0 \times 10^{-4}\,\text{A})(11.5 \times 10^3\,\Omega) = 3.5$ V. [The voltage drop across R_2 is $(3.0 \times 10^{-4}\,\text{A})(15 \times 10^3\,\Omega) = 4.5$ V, for a total of 8.0 V.] If we assume the meter is precise, it will read 3.5 V. In the normal circuit, without the meter, $R_1 = R_2$ so the voltage across R_1 is half that of the battery, or 4.0 V. Thus the voltmeter, because of its internal resistance, gives a low reading. In this case it is off by 0.5 V, or more than 10 percent.

Example 19–14 illustrates how seriously a meter can affect a circuit and give a misleading reading. If the resistance of a voltmeter is much higher than the resistance of the circuit, however, it will have little effect and its readings can be trusted, at least to the manufactured precision of the meter, which for ordinary analog meters is typically 3 to 4 percent of full-scale deflection. An ammeter also can interfere with a circuit, but the effect is minimal if its resistance is much less than that of the circuit as a whole. For both voltmeters and ammeters, the more sensitive the galvanometer the less effect it will have. A 50,000-Ω/V meter is far better than a 1,000-Ω/V meter.

Electronic voltmeters using transistors, including digital meters, have very high input resistance (usually specified in ohms) in the range 10^6 to $10^8\,\Omega$, and even higher. Hence they have very little effect on most circuits and their readings are reliable for nearly any circuit. The precision of digital meters is typically one part in 10^4 ($=0.01$ percent) or better. State-of-the-art instruments reach a precision of one part in 10^6.

SUMMARY

When resistances are connected in **series** (end to end), the equivalent resistance is the sum of the individual resistances:

$$R_{eq} = R_1 + R_2 + \cdots .$$

When resistors are connected in **parallel**, the reciprocal of the total resistance equals the sum of the reciprocals of the individual resistances:

$$\frac{1}{R_{eq}} = \frac{1}{R_1} + \frac{1}{R_2} + \cdots .$$

In a parallel connection, the net resistance is less than any of the individual resistances.

A device that transforms one type of energy into electrical energy is called a **seat** or **source** of **emf**. A battery behaves like a source of emf in series with an **internal resistance**. The emf is the potential difference determined by the chemical reactions in the battery and equals the terminal voltage when no current is drawn. When a current is drawn, the voltage at the battery's terminals is less than its emf by an amount equal to the *Ir* drop across the internal resistance.

Kirchhoff's rules are helpful in determining the currents and voltages in a complex circuit. Kirchhoff's **junction rule** is based on conservation of electric charge and states that the sum of all currents entering any junction equals the sum of all currents leaving that junction. The second, or **loop rule**, is based on conservation of energy and states that the algebraic sum of the voltage changes around any closed path of the circuit must be zero.

When capacitors are connected in **parallel**, the equivalent capacitance is the sum of the individual capacitances:

$$C_{eq} = C_1 + C_2 + \cdots .$$

When capacitors are connected in **series**, the reciprocal of the equivalent capacitance equals the sum of the reciprocals of the individual capacitances:

$$\frac{1}{C_{eq}} = \frac{1}{C_1} + \frac{1}{C_2} + \cdots .$$

If an **RC circuit** containing a resistor R in series with a capacitance C is connected to a dc source of emf, the voltage across the capacitor rises gradually in time characterized by the **time constant**

$$\tau = RC.$$

This is the time it takes for the voltage to reach 63 percent of its maximum value. A capacitor discharging through a resistor is characterized by the same time constant: in a time $\tau = RC$, the voltage across the capacitor drops to 37 percent of its initial value.

Electric shocks are caused by current passing through the body. To avoid shocks, the body must not become part of a circuit by allowing different parts of the body to touch objects at different potentials. Commonly, one part of the body may be touching ground and another part a high or low potential.

QUESTIONS

1. Explain why birds can sit on power lines safely, while leaning a metal ladder up against one to fetch a stuck kite is extremely dangerous.

2. Discuss the advantages and disadvantages of Christmas tree lights connected in parallel versus those connected in series.

3. If all you have is a 120-V line, would it be possible to light several 6-V lamps without burning them out? How?

4. Two lightbulbs of resistance R_1 and R_2 ($> R_1$) are connected in series. Which is brighter? What if they are connected in parallel?

5. Describe carefully the difference between emf and potential difference.

6. Household outlets are often double outlets. Are these connected in series or parallel? How do you know?

7. With two identical lightbulbs and two identical batteries, how would you arrange the bulbs and batteries in a circuit in order to get the maximum possible total power out. (Assume that the batteries have negligible internal resistance.)

8. Explain why Kirchhoff's first (junction) rule is equivalent to conservation of electric charge.

9. Explain why Kirchhoff's second (loop) rule is a result of the conservation of energy.

10. How does the overall resistance of your room's electric circuit change when instead of having a single 60-W lightbulb on, you turn on an additional 100-W bulb?

FIGURE 19–30
Question 11.

11. Given the circuit shown in Fig. 19–30 use the words "increases," "decreases," or "stays the same" to complete the following statements:
    (a) If R_7 increases, the potential difference between A and E (assume no resistance in Ⓐ and ℰ) ___.
    (b) If R_7 increases, the potential difference between A and E (assume Ⓐ and ℰ have resistance) ___.
    (c) If R_7 increases, the voltage drop across R_4 ___.
    (d) If R_2 decreases, the current through R_1 ___.
    (e) If R_2 decreases, the current through R_6 ___.
    (f) If R_2 decreases, the current through R_3 ___.
    (g) If R_5 increases, the voltage drop across R_2 ___.
    (h) If R_5 increases, the voltage drop across R_4 ___.
    (i) If R_2, R_5, and R_7 increase, ℰ _____.
12. Why are batteries connected in series? Why in parallel? Does it matter if the batteries are nearly identical or not in either case?
13. Describe a situation in which the terminal voltage of a battery is greater than its emf.
14. The 18-V source in Fig. 19–31 is "charging" the 12-V battery. Explain how it does this.

FIGURE 19–31
Question 14 and
Problem 25.

15. Explain in detail how you could measure the internal resistance of a battery.
16. Compare and discuss the formulas for resistors and for capacitors when connected in series and in parallel.
17. Suppose that three identical capacitors are connected to a battery. Will they store more energy if connected in series or in parallel?
18. Why is it more dangerous to turn on an electric appliance when you are standing outside in bare feet than when you are inside wearing shoes with thick soles?
19. Figure 19–32 is a diagram of a capacitor (or condenser) *microphone*. The changing air pressure in a sound wave causes one plate of the capacitor C to move back and forth. Explain how a current of the same frequency as the sound wave is produced.

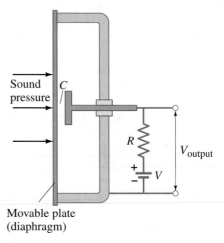

FIGURE 19–32 Diagram of a capacitor microphone. Question 19.

* 20. In an *RC* circuit, current flows from the battery until the capacitor is completely charged. Is the total energy supplied by the battery equal to the total energy stored by the capacitor? If not, where does the extra energy go?
* 21. What is the main difference between a voltmeter and an ammeter?
* 22. What would happen if you mistakenly used an ammeter where you needed to use a voltmeter?
* 23. Explain why an ideal ammeter would have zero resistance and an ideal voltmeter infinite resistance.

PROBLEMS

SECTION 19-1

In these Problems neglect the internal resistance of a battery unless the Problem refers to it.

1. (I) Four 140-Ω lightbulbs are connected in series. What is the total resistance of the circuit? What is their resistance if they are connected in parallel?

2. (I) Three 40-Ω lightbulbs and three 80-Ω lightbulbs are connected in series. (a) What is the total resistance of the circuit? (b) What is their resistance if all six are wired in parallel?

3. (I) Given only one 30-Ω and one 50-Ω resistor, list all possible values of resistance that can be obtained.

4. (I) Suppose that you have a 500-Ω, a 900-Ω, and a 1.40-kΩ resistor. What is (a) the maximum, and (b) the minimum resistance you can obtain by combining these?

5. (II) Suppose that you have a 6.0-V battery and you wish to apply a voltage of only 4.0 V. Given an unlimited supply of 1.0-Ω resistors, how could you connect them so as to make a "voltage divider" that produced a 4.0-V output for a 6.0-V input?

6. (II) Three 240-Ω resistors can be connected together in four different ways, making combinations of series and/or parallel circuits. What are these four ways and what is the net resistance in each case?

7. (II) What is the net resistance of the circuit connected to the battery in Fig. 19–33? Each resistance has $R = 2.8\,\text{k}\Omega$.

FIGURE 19-33
Problems 7 and 14.   12 V

8. (II) Eight lights are connected in series across a 110-V line. (a) What is the voltage across each bulb? (b) If the current is 0.40 A, what is the resistance of each bulb and the power dissipated in each?

9. (II) Eight lights are connected in parallel to a 110-V source by two leads of total resistance 1.5 Ω. If 240 mA flows through each bulb, what is the resistance of each, and what fraction of the total power is wasted in the leads?

10. (II) Eight 7.0-W Christmas tree lights are connected in series to each other and to a 110-V source. What is the resistance of each bulb?

11. (II) A close inspection of an electric circuit reveals that a 480-Ω resistor was inadvertently soldered in the place where a 320-Ω resistor is needed. How can this be fixed without removing anything from the existing circuit?

12. (II) Two resistors when connected in series to a 110-V line use one fourth the power that is used when they are connected in parallel. If one resistor is 2.2 kΩ, what is the resistance of the other?

13. (II) A 75-W, 110-V bulb is connected in parallel with a 40-W, 110-V bulb. What is the net resistance?

14. (II) Calculate the current through each resistor in Fig. 19–33 if each resistance $R = 2.20\,\text{k}\Omega$. What is the potential difference between points A and B?

15. (II) Consider the network of resistors shown in Fig. 19–34. Answer qualitatively: (a) What happens to the voltage across each resistor when the switch S is closed? (b) What happens to the current through each when the switch is closed? (c) What happens to the power output of the battery when the switch is closed? (d) Let $R_1 = R_2 = R_3 = R_4 = 100\,\Omega$ and $V = 45.0\,\text{V}$. Determine the current through each resistor before and after closing the switch. Are your qualitative predictions confirmed?

FIGURE 19-34   Problem 15.

16. (II) Three equal resistors (R) are connected to a power supply, as shown in Fig. 19–35. Qualitatively, what happens to (a) the voltage drop across each of these resistors, (b) the current flow through each, and (c) the terminal voltage of the battery, when the switch S is opened, after having been closed for a long time? (d) If the emf of the battery is 18.0 V, what is its terminal voltage when the switch is closed if the internal resistance is 0.50 Ω and $R = 5.50\,\Omega$? (e) What is the terminal voltage when the switch S is open?

FIGURE 19-35
Problem 16.

17. (III) A 2.8-kΩ and a 2.1-kΩ resistor are connected in parallel; this combination is connected in series with a 1.8-kΩ resistor. If each resistor is rated at $\frac{1}{2}$ W, what is the maximum voltage that can be applied across the whole network?

581

SECTION 19–2

18. (I) Calculate the terminal voltage for a battery with an internal resistance of $0.900\,\Omega$ and an emf of 8.50 V when the battery is connected in series with (a) an 81.0-Ω resistor, and (b) an 810-Ω resistor.

19. (I) Four 2.0-V cells are connected in series to a 12-Ω lightbulb. If the resulting current flow is 0.62 A, what is the internal resistance of each cell, assuming they are identical and neglecting the wires?

20. (II) A battery with an emf of 12.0 V shows a terminal voltage of 11.8 V when operating in a circuit with two lightbulbs rated at 3.0 W (at 12.0 V) which are connected in parallel with it. What is the battery's internal resistance?

21. (II) A 1.5-V dry cell can be tested by connecting it to a low-resistance ammeter. It should be able to supply at least 25 A. What is the internal resistance of the cell in this case?

22. (II) What is the internal resistance of a 12.0-V car battery whose terminal voltage drops to 8.8 V when the starter draws 60 A? What is the resistance of the starter?

23. (II) What is the current in the 8.0-Ω resistor in Fig. 19–10a?

SECTIONS 19–3 AND 19–4

24. (II) Calculate the current in the circuit of Fig. 19–36 and show that the sum of all the voltage changes around the circuit is zero.

FIGURE 19–36
Problem 24.

25. (II) Determine the terminal voltage of each battery in Fig. 19–31.

26. (II) What is the potential difference between points a and d in Fig. 19–13?

27. (II) What is the terminal voltage of each battery in Fig. 19–13?

28. (II) Determine the magnitudes and directions of the currents through R_1 and R_2 in Fig. 19–37.

29. (II) Repeat Problem 28, now assuming that each battery has an internal resistance $r = 1.2\,\Omega$.

FIGURE 19–37
Problems 28 and 29.

FIGURE 19–38
Problems 30 and 31.

30. (II) Determine the magnitudes and directions of the currents through each resistor shown in Fig. 19–38. The batteries have emfs of $\mathscr{E}_1 = 9.0$ V and $\mathscr{E}_2 = 12.0$ V and the resistors have values of $R_1 = 15\,\Omega$, $R_2 = 20\,\Omega$, and $R_3 = 30\,\Omega$.

31. (II) Repeat Problem 30 assuming each battery has internal resistance $r = 1.0\,\Omega$.

32. (III) Determine the current through each of the resistors in Fig. 19–39.

33. (III) If the 25-Ω resistor in Fig. 19–39 were shorted out (resistance = 0), what then would be the current through the 10-Ω resistor?

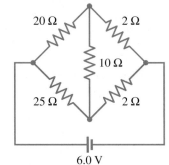

FIGURE 19–39
Problems 32 and 33.

34. (III) Determine the currents I_1, I_2, and I_3 in Fig. 19–40. Assume the internal resistance of each battery is $r = 1.0\,\Omega$. What is the terminal voltage of the 6.0-V battery?

35. (III) What would the current I_1 be in Fig. 19–40 if the 12-Ω resistor were shorted out ($r = 1.0\,\Omega$)?

FIGURE 19–40
Problems 34 and 35.

SECTION 19–5

36. (II) Suppose two batteries, with unequal emfs of 2.0V and 3.0 V, are connected as shown in Fig. 19–41. If each internal resistance is $r = 0.10\,\Omega$, and $R = 4.0\,\Omega$, what is the voltage across the resistor R?

FIGURE 19–41
Problem 36.

SECTION 19-6

37. (I) Six 3.7-μF capacitors are connected in parallel. What is the equivalent capacitance? What is their equivalent capacitance if connected in series?

38. (I) An electric circuit was accidentally constructed using a 5.0-μF capacitor instead of the required 16-μF value. What can a technician add to correct this circuit?

39. (I) The capacitance of a portion of a circuit is to be reduced from 4800 pF to 3300 pF. What capacitance can be added to the circuit to produce this effect without removing existing circuit elements? Must any existing connections be broken in the process?

40. (I) You have three capacitors, of capacitance 2000 pF, 7500 pF, and 0.0100 μF. What is the maximum and minimum capacitance that you can form from these? How do you make the connection in each case?

41. (II) A circuit contains a single 150-pF capacitor hooked across a battery. It is desired to store three times as much energy in a combination of two capacitors by adding a single capacitor to this one. How would you hook it up and what would its value be?

42. (II) (a) Determine the equivalent capacitance of the circuit shown in Fig. 19-42. (b) If $C_1 = C_2 = 2C_3 = 12.5\ \mu$F, how much charge is stored on each capacitor when $V = 45.0$ V?

FIGURE 19-42
Problems 42 and 43.

43. (II) In Fig. 19-42, let $V = 90$ V and $C_1 = C_2 = C_3 = 8.8\ \mu$F. How much energy is stored in the capacitor network?

44. (II) A 0.40-μF and a 0.50-μF capacitor are connected in series to a 9.0-V battery. Calculate (a) the potential difference across each capacitor, and (b) the charge on each. (c) Repeat parts (a) and (b) assuming the two capacitors are in parallel.

45. (II) Suppose three parallel-plate capacitors, whose plates have areas A_1, A_2, and A_3, and separations d_1, d_2, and d_3, are connected in parallel. Show, using Eq. 17-6, that Eq. 19-4 is valid.

46. (II) An air-filled parallel-plate capacitor of capacitance C_0 is filled with two identically sized dielectric slabs of dielectric constants K_1 and K_2 as shown in Fig. 19-43. What is the new capacitance? [*Hint:* Treat this as two capacitors in combination.]

FIGURE 19-43
Problem 46.

FIGURE 19-44
Problem 47.

47. (II) An air-filled parallel-plate capacitor of capacitance C_0 is filled with two identically sized dielectric slabs of dielectric constants K_1 and K_2 as shown in Fig. 19-44. What is the new capacitance? [*Hint:* Treat this as two capacitors in combination.]

48. (III) A 200-pF capacitor is connected in series with an unknown capacitor, and as a series combination they are connected to a battery with an emf of 25.0 V. If the 200-pF capacitor stores 125 pC of charge on its plates, what is the unknown capacitance?

49. (III) A 7.0-μF and a 3.0-μF capacitor are connected in series and this combination is connected in parallel with a 4.0-μF capacitor. (a) What is the net capacitance? (b) If 24 V is applied across the whole network, calculate the voltage across each capacitor.

SECTION 19-7

50. (I) Electrocardiographs are often connected as shown in Fig. 19-45. The leads are said to be capacitively coupled. A time constant of 3.0 s is typical and allows rapid changes in potential to be accurately recorded. If $C = 3.0\ \mu$F, what value must R have?

FIGURE 19-45   Problem 50.

51. (II) Suppose a 7.5-μF capacitor replaced the 10.0-Ω resistor in Fig. 19-10a. Calculate: (a) the charge on the capacitor in the steady state—that is, after the capacitor reaches its maximum charge; (b) the steady state current through the other resistors.

52. (II) In Fig. 19-17a, the total resistance is 15 kΩ, and the battery's emf is 24.0 V. If the time constant is measured to be 35 μs, calculate (a) the total capacitance of the circuit and (b) the time it takes for the voltage across the resistor to reach 16.0 V.

53. (II) The *RC* circuit of Fig. 19-18a has $R = 6.7$ kΩ and $C = 3.0\ \mu$F. The capacitor is at voltage V_0 at $t = 0$, when the switch is closed. How long does it take the capacitor to discharge to 1.0 percent of its initial voltage?

FIGURE 19–46
Problem 54.

54. (III) Two resistors and two uncharged capacitors are arranged as shown in Fig. 19–46. With a potential difference of 24 V across the combination, (a) what is the potential at point a with S open? (Let V = 0 at the negative terminal of the source.) (b) What is the potential at point b with the switch open? (c) When the switch is closed, what is the final potential of point b? (d) How much charge flows through the switch S after it is closed?

*SECTION 19–10

* 55. (I) What is the resistance of a voltmeter on the 250-V scale if the meter sensitivity is 30,000 Ω/V?

* 56. (II) An ammeter has a sensitivity of 10,000 Ω/V. What current passing through the galvanometer produces full-scale deflection?

* 57. (II) A galvanometer has an internal resistance of 30 Ω and deflects full scale for a 50-μA current. Describe how to use this galvanometer to make (a) an ammeter to read currents up to 30 A, and (b) a voltmeter to give a full-scale deflection of 1000 V.

* 58. (II) A galvanometer has a sensitivity of 35 kΩ/V and internal resistance 20.0 Ω. How could you make this into (a) an ammeter that reads 2.0 A full scale, or (b) a voltmeter reading 1.00 V full scale?

* 59. (II) A milliammeter reads 10 mA full scale. It consists of a 0.20-Ω resistor in parallel with a 30-Ω galvanometer. How can you change this ammeter to a voltmeter giving a full-scale reading of 10 V without taking the ammeter apart? What will be the sensitivity (Ω/V) of your voltmeter?

*SECTION 19–11

* 60. (II) A 45-V battery of negligible internal resistance is connected to a 37-kΩ and a 28-kΩ resistor in series. What reading will a voltmeter, of internal resistance 100 kΩ, give when used to measure the voltage across each resistor? What is the percent inaccuracy due to meter resistance for each case?

* 61. (II) An ammeter whose internal resistance is 60 Ω reads 5.25 mA when connected in a circuit containing a battery and two resistors in series whose values are 700 Ω and 400 Ω. What is the actual current when the ammeter is absent?

* 62. (II) A battery with $\mathcal{E}$ = 12.0 V and internal resistance r = 1.0 Ω is connected to two 9.0-kΩ resistors in series. An ammeter of internal resistance 0.50 Ω measures the current and at the same time a voltmeter with internal resistance 15 kΩ measures the voltage across one of the 9.0-kΩ resistors in the circuit. What do the ammeter and voltmeter read?

* 63. (III) The value of an unknown resistance, R, can be measured in at least two different ways using an ammeter and a voltmeter (Fig. 19–47). The measured value of R in each case is assumed to be R = V/I, where V is the voltmeter reading and I is the ammeter reading. Assume that the resistance of the ammeter is 1.00 Ω and that of the voltmeter is 10.0 kΩ in each case. Calculate the measured value for R for each setup, and determine which setup gives the most accurate value for R if the actual value of R is (a) R = 2.00 Ω, (b) R = 100 Ω, or (c) R = 5000 Ω.

Circuit 1   Circuit 2

FIGURE 19–47   Problem 63.

* 64. (III) Two 7.4-kΩ resistors are placed in series and connected to a battery. A voltmeter of sensitivity 1000 Ω/V is on the 3.0-V scale and reads 2.0 V when placed across either of the resistors. What is the emf of the battery? (Ignore its internal resistance.)

* 65. (III) What internal resistance should the voltmeter have to be in error by less than 3 percent for the situation of Example 19–14?

* 66. (III) The voltage across a 120-kΩ resistor in a circuit containing additional resistance (R_2) in series with a battery (V) is measured, by a 20,000-Ω/V meter on the 100-V scale, to be 25 V. On the 30-V scale, the reading is 23 V. What is the actual voltage in the absence of the voltmeter? What is R_2?

* 67. (III) A 12.0-V battery (assume the internal resistance = 0) is connected to two resistors in series. A voltmeter whose internal resistance is 15.0 kΩ measures 5.5 V and 4.0 V, respectively, when connected across each of the resistors. What is the resistance of each resistor?

GENERAL PROBLEMS

68. Suppose that you wish to apply a 0.25-V potential difference between two points on the body. The resistance is about 2000 Ω, and you only have a 9.0-V battery. How can you connect up one or more resistors so that you can produce the desired voltage?

69. A three-way lightbulb can produce 50 W, 100 W, or 150 W, at 120 V. Such a bulb contains two filaments that can be connected to the 120 V individually or in parallel. Describe how the connections to the two filaments are made to give each of the three wattages. What must be the resistance of each filament?

70. Suppose you want to run some apparatus that is 95 m from an electric outlet. Each of the wires connecting your apparatus to the 120-V source has a resistance per unit length of 0.0065 Ω/m. If your apparatus draws 3.0 A, what will be the voltage drop across the connecting wires and what voltage will be applied to your apparatus?

71. Electricity can be a hazard in hospitals, particularly to patients who are connected to electrodes, such as an ECG. For example, suppose that the motor of a motorized bed shorts out to the bed frame, and the bed frame's connection to a ground has broken (or was not there in the first place). If a nurse touches the bed and the patient at the same time, she becomes a conductor and a complete circuit can be made through the patient to ground through the ECG apparatus. This is shown schematically in Fig. 19–48. Calculate the current through the patient.

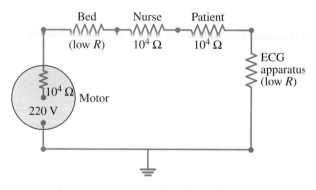

FIGURE 19–48 Problem 71.

72. How much energy must a 45-V battery expend to fully charge a 0.40-μF and a 0.60-μF capacitor when they are placed (a) in parallel, (b) in series? (c) How much charge flowed from the battery in each case?

73. A heart pacemaker is designed to operate at 72 beats/min using a 7.5-μF capacitor in a simple RC circuit. What value of resistance should be used if the pacemaker is to fire when the voltage reaches 63 percent of maximum?

74. The internal resistance of a 1.35-V mercury cell is 0.030 Ω, whereas that of a 1.5-V dry cell is 0.35 Ω. Explain why three mercury cells can more effectively power a 2-W hearing aid that requires 4.0 V than can three dry cells.

75. Suppose that a person's body resistance is 900 Ω. (a) What current passes through the body when the person accidentally is connected to 110 V? (b) If there is an alternative path to ground whose resistance is 40 Ω, what current passes through the person? (c) If the voltage source can produce at most 1.5 A, how much current passes through the person in case (b)?

76. The **Wheatstone bridge** is a circuit used to make precise measurements of resistance. The unknown resistance to be measured, R_x, is placed in the bridge circuit as shown in Fig. 19–49, and the variable resistor R_3 is adjusted until the galvanometer does not deflect when the switch is closed. For this setting, no current flows from B to D, so B and D are at the same potential. (a) Show that the unknown resistance R_x is given by

$$R_x = \frac{R_2}{R_1} R_3$$

when R_1, R_2, and R_3 are all accurately known. (b) A Wheatstone bridge is balanced when $R_1 = 630\,\Omega$, $R_2 = 972\,\Omega$, and $R_3 = 42.6\,\Omega$. What is the value of the unknown resistance in the fourth arm?

77. An unknown length of platinum wire 0.920 mm in diameter is placed as the unknown resistance in a Wheatstone bridge (see Problem 76). Arms 1 and 2 have resistance of 38.0 Ω and 46.0 Ω, respectively. Balance is achieved when R_3 is 3.48 Ω. How long is the platinum wire?

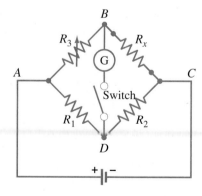

FIGURE 19–49 Wheatstone bridge. Problems 76 and 77.

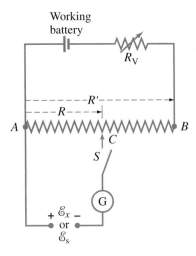

FIGURE 19–50 Potentiometer circuit. Problem 78.

FIGURE 19–51 Problems 79 and 86.

78. A **potentiometer** is a device to precisely measure potential differences or emf, using a "null" technique. In the simple potentiometer circuit shown in Fig. 19–50, R' represents the total resistance of the resistor from A to B (which could be a long uniform "slide" wire), whereas R represents the resistance of only the part from A to the movable contact at C. When the unknown emf to be measured, $\mathcal{E}_x$, is placed into the circuit as shown, the movable contact C is moved until the galvanometer G gives a null reading (i.e., zero) when the switch S is closed. The resistance between A and C for this situation we call R_x. Next, a standard emf, $\mathcal{E}_s$, which is known precisely, is inserted into the circuit in place of $\mathcal{E}_x$ and again the contact C is moved until zero current flows through the galvanometer when the switch S is closed. The resistance between A and C now is called R_s. (a) Show that the unknown emf is given by

$$\mathcal{E}_x = \left(\frac{R_x}{R_s}\right)\mathcal{E}_s$$

where R_x, R_s, and $\mathcal{E}_s$ are all precisely known. The working battery is assumed to be fresh and give a constant voltage. (b) A slide-wire potentiometer is balanced against a 1.0182-V standard cell when the slide wire is set at 25.4 cm out of a total length of 100.0 cm. For an unknown source, the setting is 45.8 cm. What is the emf of the unknown? (c) The galvanometer of a potentiometer has an internal resistance of $30\,\Omega$ and can detect a current as small as 0.015 mA. What is the minimum uncertainty possible in measuring an unknown voltage? (d) Explain the advantage of using this "null" method of measuring emf.

79. The variable capacitance of a radio tuner consists of four plates connected together placed alternately between four other plates, also connected together (Fig. 19–51). Each plate is separated from its neighbor by 2.0 mm of air. One set of plates can move so that the area of overlap varies from $2.0\,\text{cm}^2$ to $12.0\,\text{cm}^2$. (a) Are these seven capacitors connected in series or in parallel? (b) Determine the range of capacitance values.

80. A battery produces 40.8 V when 7.40 A are drawn from it and 47.3 V when 2.20 A are drawn. What is the emf and internal resistance of the battery?

81. How many $\frac{1}{2}$-W resistors, each of the same resistance, must be used to produce an equivalent 2.2-kΩ, 5-W resistor? What is the resistance of each, and how must they be connected?

82. The current through the 4.0-kΩ resistor in Fig. 19–52 is 3.50 mA. What is the terminal voltage V_{ba} of the "unknown" battery? (There are two answers. Why?) [*Hint:* Use conservation of energy or Kirchhoff's rules.]

FIGURE 19–52 Problem 82.

83. One type of *ohmmeter* consists of an ammeter connected to a series resistor and a battery, Fig. 19–29. The scale is different from ammeters and voltmeters. Zero resistance corresponds to full-scale deflection (because maximum current flows from the battery), whereas infinite resistance corresponds to no deflection (no current flow). Suppose a 3.0-V battery is used, and the galvanometer has internal resistance of $25\,\Omega$ and deflects full scale for a current of $35\,\mu A$. What values of shunt resistance, R_{sh}, and series resistance, R_{ser}, are needed to make an ohmmeter that registers a midscale deflection (that is, half of maximum) for a resistance of $30\,k\Omega$? [Note: An additional series resistor (variable) is also needed so the meter can be zeroed. The zero should be checked frequently by touching the leads together, since the battery voltage can vary. Because of battery voltage variation, such ohmmeters are not precision instruments, but they are useful to obtain approximate values.]

84. Some light dimmer switches use a variable resistor called a *potentiometer*, as shown in Fig. 19–53 (a variation of Fig. 19–50). The slide moves from position $x = 0$ to $x = 1$, and the resistance up to slide position x is proportional to x (the total resistance is $R_{pot} = 100\,\Omega$). What is the power expended in the lightbulb if (a) $x = 1$, (b) $x = \frac{1}{2}$, (c) $x = \frac{1}{4}$?

FIGURE 19–53   Problem 84.

Power supply   Power outage protector

FIGURE 19–54   Problem 85.

85. Electronic devices often use an *RC* circuit to protect against power outages as shown in Fig. 19–54. (a) If the device is supposed to keep the supply voltage at least 70 percent of nominal for as long as 0.20 s, how big a resistance is needed? The capacitor is $22\,\mu F$. (b) Between which two terminals should the device be connected, a and b, b and c, or a and c?

86. A high-voltage supply can be constructed from a variable capacitor with interleaving plates, one of which can be rotated, as in Fig. 19–51. A version of this type of capacitor with more plates has a capacitance which can be varied from $10\,pF$ to $1\,pF$ (a) Initially, this capacitor is connected to a 10,000-volt power supply when the capacitance is $10\,pF$ and charged. It is then disconnected from the power supply and the capacitance reduced to $1\,pF$ by rotating the plates. What is the voltage across the capacitor now? (b) What is a major disadvantage of this as a high-voltage power supply?

Magnets produce magnetic fields, but so do electric currents. An electric current flowing in this coil of wire produces a magnetic field which causes the tiny pieces of iron (iron "filings") to align in the field. Fig. 20–4 shows iron filings in the magnetic field produced by a magnet.

20 MAGNETISM

Today it is clear that magnetism and electricity are closely related. This relationship was not discovered, however, until the nineteenth century. The history of magnetism begins much earlier with the ancient civilizations in Asia Minor. It was in a region of Asia Minor known as Magnesia that rocks were found that would attract each other. These rocks were called "magnets" after their place of discovery.

20–1 Magnets and Magnetic Fields

Poles of a magnet

A magnet will attract paper clips, nails, and other objects made of iron. Any magnet, whether it is in the shape of a bar or a horseshoe, has two ends or faces, called **poles**, which is where the magnetic effect is strongest. If a magnet is suspended from a fine thread, it is found that one pole of the magnet will always point toward the north. It is not known for sure when this fact was discovered, but it is known that the Chinese were making use of it as an aid to navigation by the eleventh century and perhaps earlier. This is, of course, the principle of a compass. A compass needle is simply a magnet that is supported at its center of gravity so it can rotate freely. That pole of a freely suspended magnet which points toward the north is called the **north pole** of the magnet. The other pole points toward the south and is called the **south pole**.

It is a familiar fact that when two magnets are brought near one another, each exerts a force on the other. The force can be either attractive or repulsive and can be felt even when the magnets don't touch. If the north pole of one magnet is brought near the north pole of a second magnet, the force is repulsive. Similarly, if two south poles are brought close, the force is repulsive. But when a north pole is brought near a south pole, the force is attractive. These results are shown in Fig. 20–1, and are reminiscent of the force between electric charges; like poles repel, and unlike poles attract. *But do not confuse magnetic poles with electric charge.* They are not the same thing. One important difference is that a positive or negative electric charge can easily be isolated. But the isolation of a single magnetic pole seems impossible. If a bar magnet is cut in half, you do not obtain isolated north and south poles. Instead, two new magnets are produced, Fig. 20–2. If the cutting operation is repeated, more magnets are produced, each with a north and a south pole. Physicists have tried complicated means to isolate single magnetic poles (monopoles), but so far there is no firm experimental evidence for their existence.

Only iron and a few other materials such as cobalt, nickel, and gadolinium show strong magnetic effects. They are said to be **ferromagnetic** (from the Latin word *ferrum* for iron). All other materials show some slight magnetic effect, but it is extremely weak and can be detected only with delicate instruments. (We will look in more detail at ferromagnetism in Sections 20–13 and 20–15.)

We found it useful to speak of an electric field surrounding an electric charge. In the same way, we can imagine a **magnetic field** surrounding a magnet. The force one magnet exerts on another can then be described as the interaction between one magnet and the magnetic field of the other. Just as we drew electric field lines, we can also draw **magnetic field lines**. They can be drawn, as for electric field lines, so that (1) the direction of the magnetic field is tangent to a line at any point, and (2) the number of lines per unit area is proportional to the magnitude of the magnetic field.

The *direction* of the magnetic field at a given point can be defined as the direction that the north pole of a compass needle would point when placed at that point. Figure 20–3a shows how one magnetic field line around a bar magnet is found using compass needles. The magnetic field determined in this way for the field outside a bar magnet is shown in Fig. 20–3b. Notice that because of our definition, the lines always point from the north toward the south pole of a magnet (the north pole of a magnetic compass needle is attracted to the south pole of another magnet). Figure 20–4 shows how thin iron filings reveal the magnetic field lines by lining up like compass needles.

FIGURE 20–3 (a) Plotting a magnetic field line of a bar magnet. (b) Magnetic field lines outside of a bar magnet.

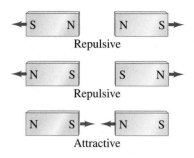

FIGURE 20–1 Like poles of a magnet repel; unlike poles attract.

Magnetic poles not found singly

FIGURE 20–2 If you break a magnet in half, you do not obtain isolated north and south poles; instead, two new magnets are produced, each with a north and a south pole.

Magnetic field lines

Magnetic field lines point from north to south magnetic poles

FIGURE 20–4 Thin iron filings indicate the magnetic field lines around a bar magnet.

(a)   (b)

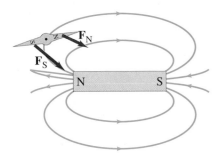

FIGURE 20–5 Forces on a compass needle that produce a torque to orient it parallel to the magnetic field lines. The net torque will be zero when the needle is parallel to the magnetic field line at that point. (Only the attractive forces are shown; try drawing in the repulsive forces and show that they produce a similar torque.)

➡ **PHYSICS APPLIED**

Use of a compass.

FIGURE 20–6 The Earth acts like a huge magnet but its magnetic poles are not at the geographic poles.

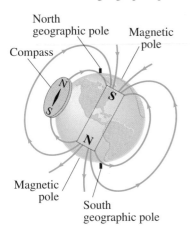

North geographic pole

Magnetic pole

Compass

S

N

Magnetic pole

South geographic pole

FIGURE 20–7 Using a map and compass in the wilderness. First you align the compass so the needle points away from true north (N) exactly the number of degrees of declination as stated on the (topographic) map: 15° in the case shown. Then align the map with true north, as shown, *not* with the compass needle.

We can define the magnetic field at any point as a vector, represented by the symbol **B**, whose direction is determined as discussed above, using a compass needle. The *magnitude* of **B** can be defined in terms of the torque exerted on a compass needle when it makes a certain angle with the magnetic field, as in Fig. 20–5. That is, the greater the torque, the greater the magnetic field strength. We can use this definition for now, but a more precise definition will be given in Section 20–3.

The Earth's magnetic field is shown in Fig. 20–6. The pattern of field lines is as if there were an (imaginary) bar magnet inside the Earth. Since the north pole of a compass needle points north, the magnetic pole which is in the geographic north is actually a south pole magnetically, as indicated in Fig. 20–6 by the S on the schematic bar magnet inside the Earth. (Remember that the north pole of one magnet is attracted to the south pole of a second.) Nonetheless, this pole is still often called the "north magnetic pole" simply because it is in the north. Similarly, the Earth's south magnetic pole, near the geographic south pole, is magnetically a north pole. The Earth's magnetic poles do not coincide with the geographic poles (which are on the Earth's axis of rotation). The north magnetic pole, for example, is in northern Canada, about 1300 km from the geographic north pole. This must be taken into account when using a compass (Fig. 20–7). The angular difference between magnetic north as indicated by a compass, and true (geographical) north, is called the **magnetic declination**. In the U.S. it varies from 0° to about 25°, depending on location.

Notice in Fig. 20–6 that the Earth's magnetic field is not tangent to the Earth's surface at all points. The angle that the Earth's magnetic field makes with the horizontal at any point is referred to as the **angle of dip**.

The simplest magnetic field is one that is uniform—it doesn't change from one point to another. A perfectly uniform field over a large area is not easy to produce. But the field between two flat parallel pole pieces of a magnet is nearly uniform if the area of the pole faces is large compared to their separation, as shown in Fig. 20–8. At the edges, the field "fringes" out somewhat and is no longer uniform. The parallel evenly spaced field lines in the drawing indicate that the field is uniform at points not too near the edge, much like the electric field between two parallel plates (Fig. 17–1).

FIGURE 20–8 Magnetic field between two large poles of a magnet is nearly uniform except at the edges.

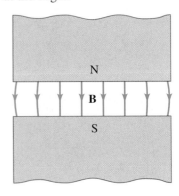

N

B

S

20-2 Electric Currents Produce Magnetism

During the eighteenth century, many natural philosophers sought to find a connection between electricity and magnetism. A stationary electric charge and a magnet were shown not to have any influence on each other. But in 1820, Hans Christian Oersted (1777–1851) found that when a compass needle is placed near an electric wire, the needle deflects as soon as the wire is connected to a battery and a current flows. As we have seen, a compass needle can be deflected by a magnetic field. What Oersted found was that **an electric current produces a magnetic field**. He had found a connection between electricity and magnetism.

A compass needle placed near a straight section of current-carrying wire aligns itself so it is tangent to a circle drawn around the wire, Fig. 20–9. Thus, the magnetic field lines produced by a current in a straight wire are in the form of circles with the wire at their center, Fig. 20–10a. The direction of these lines is indicated by the north pole of the compass in Fig. 20–9. There is a simple way to remember the direction of the magnetic field lines in this case. It is called a **right-hand rule**: you grasp the wire with your right hand so that your thumb points in the direction of the conventional (positive) current; then your fingers will encircle the wire in the direction of the magnetic field, Fig. 20–10b. The magnetic field lines due to a circular loop of current-carrying wire can be determined in a similar way using a compass. The result is shown in Fig. 20–11. Again the right-hand rule can be used, as shown in Fig. 20–12.

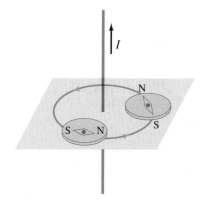

FIGURE 20-9 Deflection of a compass needle near a current-carrying wire, showing the presence and direction of the magnetic field.

FIGURE 20-10 (a) Magnetic field lines around an electric current in a straight wire. (b) Right-hand rule for remembering the direction of the magnetic field: when the thumb points in the direction of the conventional current, the fingers wrapped around the wire point in the direction of the magnetic field.

(a) (b)

FIGURE 20-11 Magnetic field due to a circular loop of wire.

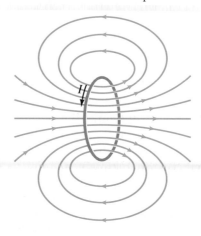

FIGURE 20-12 Right-hand rule for determining the direction of the magnetic field relative to the current.

Magnetic field

Force is up

B

Right-hand rule

(a)   (b)   (c)

FIGURE 20-13 (a) Force on a current-carrying wire placed in a magnetic field **B**; (b) same, but current reversed; (c) right-hand rule for setup in (b).

Magnet exerts a force on an electric current

Right-hand rule for force on current due to **B**

FIGURE 20-14 Current-carrying wire in a magnetic field. Force on the wire is directed into the page.

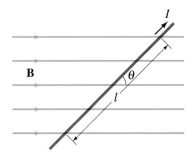

20-3 Force on an Electric Current in a Magnetic Field; Definition of B

In Section 20–2 we saw that an electric current exerts a force on a magnet, such as a compass needle. By Newton's third law, we might expect the reverse to be true as well: we should expect that *a magnet exerts a force on a current-carrying wire*. Experiments indeed confirm this effect, and it too was first observed by Oersted.

Let us look at the force exerted on a wire in detail. Suppose a straight wire is placed between the poles of a horseshoe magnet as shown in Fig. 20–13. When a current flows in the wire, a force is exerted on the wire. But this force is *not* toward one or the other poles of the magnet. Instead, the force is directed at right angles to the magnetic field direction. If the current is reversed in direction, the force is in the opposite direction. It is found that *the direction of the force is always perpendicular to the direction of the current and also perpendicular to the direction of the magnetic field*, **B**. This statement does not completely describe the direction, however: the force could be either up or down in Fig. 20–13b and still be perpendicular to both the current and to **B**. Experimentally, the direction of the force is given by another *right-hand rule*, as illustrated in Fig. 20–13c. First you orient your right hand so that the outstretched fingers point in the direction of the (conventional) current; from this position, bend your fingers, so that they point in the direction of the magnetic field lines (which point from the N toward the S pole outside a magnet); you may have to rotate your hand and arm about the wrist until they do point along **B** when bent, remembering that straightened fingers must point along the direction of the current first. When your hand is oriented in this way, then the extended thumb points in the direction of the force on the wire.

This describes the direction of the force. What about its magnitude? It is found experimentally that the magnitude of the force is directly proportional to the current I in the wire, to the length l of wire in the magnetic field (assumed uniform), and to the magnetic field B. The force also depends on the angle θ between the current direction and the magnetic field (Fig. 20–14). When the current is perpendicular to the field lines, the force is strongest. When the wire is parallel to the magnetic field lines, there is no force at all. At other angles, the force is proportional to $\sin \theta$ (Fig. 20–14).

Thus we have

$$F \propto IlB \sin \theta.$$

Up to now we have talked of magnetic field strengths in terms of the torque exerted by the field on a compass needle (Fig. 20–5). But we have not defined the magnetic field strength precisely. In fact, the magnetic field B is defined in terms of the above proportion so that the proportionality constant is precisely 1. Thus we have

$$F = IlB \sin \theta. \qquad \text{(20–1)}$$

FORCE ON ELECTRIC CURRENT IN A MAGNETIC FIELD

If the direction of the current is perpendicular to the field ($\theta = 90°$), then the force is

$$F_{max} = IlB. \qquad [\mathbf{I} \perp \mathbf{B}] \quad \text{(20–2)}$$

If the current is parallel to the field ($\theta = 0°$), the force is zero.[†]

In summary, the magnetic field vector $\mathbf{B}$ is defined as follows. The direction of $\mathbf{B}$ in a region of space is the direction that a straight section of current-carrying wire would have when placed in the field and the force on it is zero ($\theta = 0°$ in Eq. 20–1), and consistent with the right-hand rule when the wire is oriented in another direction. The magnitude of $\mathbf{B}$ is defined (from Eq. 20–2) as

Definition of magnetic field

$$B = \frac{F_{max}}{Il},$$

where F_{max} is the magnitude of the force on a straight length l of wire carrying a current I when the wire is perpendicular to $\mathbf{B}$.

The SI unit for magnetic field B is the **tesla** (T). From Eq. 20–1, it is clear that $1\,\text{T} = 1\,\text{N/A·m}$. An older name for the tesla is the "weber per meter squared" ($1\,\text{Wb/m}^2 = 1\,\text{T}$). Another unit commonly used to specify magnetic field is a cgs unit, the **gauss** (G): $1\,\text{G} = 10^{-4}\,\text{T}$. A field given in gauss should always be changed to teslas before using with other SI units. To get a "feel" for these units, we note that the magnetic field of the Earth at its surface is about $\frac{1}{2}$ G or 0.5×10^{-4} T. On the other hand, strong electromagnets can produce fields on the order of 2 T and superconducting magnets over 10 T.

The tesla and the gauss (units)

EXAMPLE 20–1 **Magnetic force on a current-carrying wire.** A wire carrying a 30-A current has a length $l = 12\,\text{cm}$ between the pole faces of a magnet at an angle $\theta = 60°$ (Fig. 20–14). The magnetic field is approximately uniform at 0.90 T. We ignore the field beyond the pole pieces. What is the force on the wire?

SOLUTION We use Eq. 20–1 and find that

$$F = IlB \sin \theta$$

$$= (30\,\text{A})(0.12\,\text{m})(0.90\,\text{T})(0.866) = 2.8\,\text{N}.$$

On a diagram, when we want to represent a magnetic field that is pointing out of the page (toward us) or into the page, we use $\odot$ or $\times$. The $\odot$ is meant to

---

[†] In our discussion, we have assumed that the magnetic field is uniform. If it is not, then B in Eqs. 20–1 and 20–2 is the average field over the length l of the wire. In practical cases, we can consider a wire as made up of many short segments Δl and the force on each segment is proportional to the length Δl of that segment and to the magnetic field B at that segment. The total force is the vector sum of the individual forces.

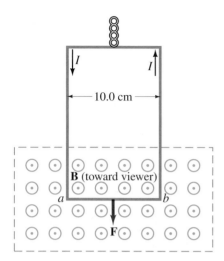

FIGURE 20–15 Measuring a magnetic field **B**. Example 20–2.

resemble the tip of an arrow pointing directly toward the reader, whereas the $\times$ or $\otimes$ resembles the tail of an arrow going away. (See Fig. 20–15.)

EXAMPLE 20–2 **Measuring a magnetic field.** A rectangular loop of wire hangs vertically as shown in Fig. 20–15. A magnetic field **B** is directed horizontally, perpendicular to the wire, and points out of the page at all points as represented by the symbol $\odot$. The magnetic field **B** is very nearly uniform along the horizontal portion of wire ab (length $l = 10.0$ cm) which is near the center of a large magnet producing the field. The top portion of the wire loop is free of the field. The loop hangs from a balance which measures a downward force (in addition to the gravitational force) of $F = 3.48 \times 10^{-2}$ N when the wire carries a current $I = 0.245$ A. What is the magnitude of the magnetic field B at the center of the magnet?

SOLUTION The magnetic forces on the two vertical sections of the wire loop point to the left and right, respectively. They are equal and in opposite directions and so add up to zero. Hence, the net magnetic force on the loop is that on the horizontal section ab whose length is $l = 0.100$ m (and $\theta = 90°$ so $\sin \theta = 1$); thus

$$B = \frac{F}{Il} = \frac{3.48 \times 10^{-2}\,\text{N}}{(0.245\,\text{A})(0.100\,\text{m})} = 1.42\,\text{T}.$$

This technique is a highly precise means of determining magnetic fields.

20–4 Force on an Electric Charge Moving in a Magnetic Field

We have seen that a current-carrying wire experiences a force when placed in a magnetic field. Since a current in a wire consists of moving electric charges, we might expect that freely moving charged particles (not in a wire) would also experience a force when passing through a magnetic field. Indeed, this is the case.

From what we already know, let us determine the force on a single moving electric charge. If N such particles of charge q pass by a given point in time t, they constitute a current $I = Nq/t$. We let t be the time for a charge q to travel a distance l in a magnetic field B; then $l = vt$, where v is the velocity of the particle. Thus, the force on these N particles is, by Eq. 20–1, $F = IlB \sin \theta = (Nq/t)(vt)B \sin \theta$. The force on *one* of the particles is found by dividing by N:

FORCE ON MOVING CHARGE IN MAGNETIC FIELD

$$F = qvB \sin \theta. \tag{20–3}$$

This equation gives the magnitude of the force on a particle of charge q moving with velocity v in a magnetic field of strength B, where θ is the angle between **v** and **B**. The force is greatest when the particle moves perpendicular to **B** ($\theta = 90°$):

$$F_{\text{max}} = qvB. \qquad [\text{v} \perp \text{B}] \quad (20–4)$$

The force is *zero* if the particle moves *parallel* to the field lines ($\theta = 0°$). The *direction* of the force is perpendicular to the magnetic field **B** and to the ve-

Right-hand rule locity **v** of the particle. It is again given by a *right-hand rule*: you orient your right hand so that your outstretched fingers point along the direction of motion of the particle (**v**), and when you bend your fingers they must point along the direction of **B**; then your thumb will point in the direction of the force. This is true only for *positively* charged particles, and will be "down" for the situation shown in Fig. 20–16. For negatively charged particles, the force is in exactly the opposite direction ("up" in Fig. 20–16).

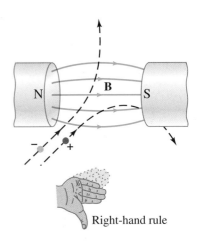

FIGURE 20–16 Force on charged particles due to a magnetic field is perpendicular to the magnetic field direction.

EXAMPLE 20–3 **Magnetic force on a proton.** A proton having a speed of 5.0×10^6 m/s in a magnetic field feels a force of 8.0×10^{-14} N toward the west when it moves vertically upward. When moving horizontally in a northerly direction, it feels zero force. What is the magnitude and direction of the magnetic field in this region? (The charge on a proton is $q = +e = 1.6 \times 10^{-19}$ C.)

SOLUTION Since the proton feels no force when moving north, the field must be in a north–south direction. The right-hand rule tells us that **B** must point toward the north in order to produce a force to the west when the proton moves upward. (Your thumb points west and the outstretched fingers of your right hand point upward only when your bent fingers point north.) The magnitude of **B**, from Eq. 20–3 with $\theta = 90°$, is

$$B = \frac{F}{qv} = \frac{8.0 \times 10^{-14} \text{ N}}{(1.6 \times 10^{-19} \text{ C})(5.0 \times 10^6 \text{ m/s})} = 0.10 \text{ T}.$$

The path of a charged particle moving in a plane perpendicular to a uniform magnetic field is a circle (or the arc of a circle if the particle later passes out of the magnetic field region). See Fig. 20–17, where the magnetic field is directed *into* the paper, as represented by ×'s. An electron at point P is moving to the right, and the force on it at this point is downward as shown (use the right-hand rule and reverse the direction for negative charge). The electron is thus deflected downward. A moment later, say when it reaches point Q, the force is still perpendicular to the velocity and is in the direction shown. Since the force is always perpendicular to **v**, the magnitude of **v** does not change but the particle changes direction and moves in a circular path, with a centripetal acceleration (we demonstrate this in Example 20–4). The force is directed toward the center of this circle at all points. Note that the electron moves clockwise in Fig. 20–17. A positive particle would feel a force in the opposite direction and would thus move counterclockwise.

FIGURE 20–17 Force exerted by a uniform magnetic field on a moving charged particle (in this case, an electron) produces a circular path.

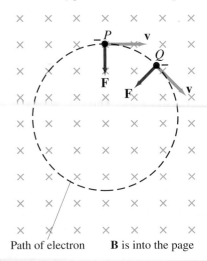

Path of electron   **B** is into the page

EXAMPLE 20–4 **Electron's path in a uniform magnetic field.** An electron travels at 2.0×10^7 m/s in a plane perpendicular to a 0.010-T magnetic field. Describe its path.

SOLUTION The electron moves at constant speed in a curved path whose radius of curvature is found using Newton's second law, $F = ma$. We have a centripetal acceleration $a = v^2/r$ (Eq. 5–1). The force is given by Eq. 20–4, $F = qvB$, so we have

$$F = ma$$

$$qvB = \frac{mv^2}{r}.$$

We solve for r and find

$$r = \frac{mv}{qB}.$$

Since **F** is perpendicular to **v**, the magnitude of **v** doesn't change. From this equation we see that if **B** = constant, then r = constant, and the curve must be a circle as we claimed above.

To get r, we put in the numbers:

$$r = \frac{(9.1 \times 10^{-31}\text{ kg})(2.0 \times 10^7\text{ m/s})}{(1.6 \times 10^{-19}\text{ C})(0.010\text{ T})} = 1.1 \times 10^{-2}\text{ m,}$$

or 1.1 cm.

FIGURE 20–18   Example 20–5.

CONCEPTUAL EXAMPLE 20–5 | **A spiral path.**   What is the path of a charged particle if its velocity is *not* perpendicular to the magnetic field?

RESPONSE   The velocity vector can be broken down into components parallel and perpendicular to the field. The velocity component parallel to the field lines experiences no force, and so this component remains constant. The velocity component perpendicular to the field results in circular motion about the field lines. Putting these two motions back together produces a helical (spiral) motion around the field lines as shown in Fig. 20–18.

➡ PHYSICS APPLIED

The aurora borealis

CONCEPTUAL EXAMPLE 20–6 | **Aurora borealis.**   Charged ions approach the Earth from the Sun (the "solar wind") and are drawn toward the poles, sometimes causing a phenomenon called the *aurora borealis* or "northern lights" in northern latitudes. Why toward the poles?

RESPONSE   A glance at Fig. 20–19 (see also Fig. 20–18) provides the answer. Imagine a stream of charged particles approaching the Earth as shown. The velocity component perpendicular to the field for each particle becomes a circular orbit around the field lines, whereas the velocity component parallel to the field carries the particle along the field lines toward the poles. The high concentration of charged particles ionizes the air, and the recombining of electrons with atoms emits light (Chapter 27), which is the aurora, especially during periods of high sun spot activity when the solar wind is greater.

FIGURE 20–19   (a) Diagram showing a charged particle approaching the Earth which is "captured" by the magnetic field of the Earth. Such particles follow the field lines toward the poles as shown. (b) Photo of aurora borealis.

(a)

(b)

20-5 Magnetic Field Due to a Straight Wire

We saw in Section 20-2, Fig. 20-10, that the magnetic field due to the electric current in a long straight wire is such that the field lines are circles with the wire at the center (Fig. 20-20). You might expect that the field strength at a given point would be greater if the current flowing in the wire were greater; and that the field would be less at points farther from the wire. This is indeed the case. Careful experiments show that the magnetic field B at a point near a long straight wire is directly proportional to the current I in the wire and inversely proportional to the distance r from the wire:

$$B \propto \frac{I}{r}.$$

This relation is valid as long as r, the perpendicular distance to the wire, is much less than the distance to the ends of the wire (i.e., the wire is long).

The proportionality constant is written[†] as $\mu_0/2\pi$; thus,

$$B = \frac{\mu_0}{2\pi}\frac{I}{r}. \qquad \text{[outside a long straight wire]} \quad (20\text{-}5)$$

The value of the constant μ_0, which is called the **permeability of free space**, is $\mu_0 = 4\pi \times 10^{-7}\,\text{T·m/A}$.

FIGURE 20-20 Same as Fig. 20-10a, magnetic field lines around a long straight wire carrying an electric current I.

Magnetic field due to current in straight wire

FIGURE 20-21 Example 20-7.

EXAMPLE 20-7 **Calculation of B near a wire.** A vertical electric wire in the wall of a building carries a dc current of 25 A upward. What is the magnetic field at a point 10 cm due north of this wire (Fig. 20-21)?

SOLUTION According to Eq. 20-5:

$$B = \frac{\mu_0 I}{2\pi r} = \frac{(4\pi \times 10^{-7}\,\text{T·m/A})(25\,\text{A})}{(2\pi)(0.10\,\text{m})} = 5.0 \times 10^{-5}\,\text{T},$$

or 0.50 G. By the right-hand rule (Fig. 20-10b), the field points to the west (into the page in Fig. 20-21) at this point. Since this field has about the same magnitude as Earth's, a compass would not point north but in a northwesterly direction.

➡ **PHYSICS APPLIED**

A compass, near a current, may not point North

20-6 Force Between Two Parallel Wires

We have seen that a wire carrying a current produces a magnetic field (magnitude given by Eq. 20-5 for a long straight wire), and furthermore that such a wire feels a force when placed in a magnetic field (Section 20-3, Eq. 20-1). Thus, we expect that two current-carrying wires would exert a force on each other.

[†]The constant is chosen in this complicated way so that Ampère's law (Section 20-8), which is considered more fundamental, will have a simple and elegant form.

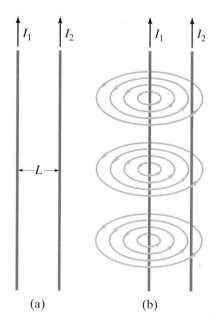

(a)   (b)

FIGURE 20–22 (a) Two parallel conductors carrying currents I_1 and I_2. (b) Magnetic field produced by I_1. (Field produced by I_2 is not shown.)

FIGURE 20–23 (a) Parallel currents in the same direction exert attractive force on each other. (b) Antiparallel currents (in opposite directions) exert repulsive force on each other.

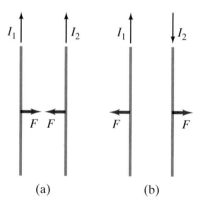

(a)   (b)

Consider two long parallel conductors separated by a distance L, as in Fig. 20–22a. They carry currents I_1 and I_2, respectively. Each current produces a magnetic field that is "felt" by the other so that each must exert a force on the other, as Ampère first pointed out. For example, the magnetic field B_1 produced by I_1 is given by Eq. 20–5. At the location of the second conductor, the magnitude of this field is

$$B_1 = \frac{\mu_0}{2\pi} \frac{I_1}{L}.$$

See Fig. 20–22b where the field due *only* to I_1 is shown. According to Eq. 20–2, the force F per unit length l on the conductor carrying current I_2 is

$$\frac{F}{l} = I_2 B_1.$$

Note that the force on I_2 is due only to the field produced by I_1. Of course I_2 also produces a field, but it does not exert a force on itself. We substitute in the above formula for B_1 and find

$$\frac{F}{l} = \frac{\mu_0}{2\pi} \frac{I_1 I_2}{L}. \tag{20–6}$$

If we use the right-hand rule of Fig. 20–10b, we see that the lines of B_1 are as shown in Fig. 20–22b. Then using the right-hand rule of Fig. 20–13c, we see that the force exerted on I_2 will be to the left in the figure. That is, I_1 exerts an attractive force on I_2 (Fig. 20–23a). This is true as long as the currents are in the same direction. If I_2 is in the opposite direction, the right-hand rule indicates that the force is in the opposite direction. That is, I_1 exerts a repulsive force on I_2 (Fig. 20–23b). Reasoning similar to that above shows that the magnetic field produced by I_2 exerts an equal but opposite force on I_1. We expect this to be true also, of course, from Newton's third law. Thus, as shown in Fig. 20–23, parallel currents in the same directions attract each other, whereas parallel currents in opposite directions repel.

EXAMPLE 20–8 **Force between two current carrying wires.** The two wires of a 2.0-m-long appliance cord are 3.0 mm apart and carry a current of 8.0 A dc. Calculate the force between these wires.

SOLUTION Equation 20–6 gives us

$$F = \frac{(2.0 \times 10^{-7}\,\text{T·m/A})(8.0\,\text{A})^2(2.0\,\text{m})}{(3.0 \times 10^{-3}\,\text{m})} = 8.5 \times 10^{-3}\,\text{N},$$

where we have written $\mu_0/2\pi = 2.0 \times 10^{-7}\,\text{T·m/A}$. Since the currents are in opposite directions, the force would tend to spread them apart.

EXAMPLE 20–9 **Suspending a current with a current.** A horizontal wire carries a current $I_1 = 80$ A dc. A second parallel wire 20 cm below it (Fig. 20–24) must carry how much current I_2 so that it doesn't fall due to gravity? The lower wire has a mass of 0.12 g per meter of length.

SOLUTION The force of gravity on the lower wire is downward and per each meter of length has magnitude

$$\frac{F}{l} = \frac{mg}{l} = \frac{(0.12 \times 10^{-3}\,\text{kg})(9.8\,\text{m/s}^2)}{1.0\,\text{m}} = 1.18 \times 10^{-3}\,\text{N/m}.$$

The magnetic force on wire 2 must be upward (hence I_2 must have the same direction as I_1) and with $L = 0.20$ m and $I_1 = 80$ A has magnitude

$$\frac{F}{l} = \frac{\mu_0}{2\pi}\frac{I_1 I_2}{L}.$$

We solve for I_2 and find

$$I_2 = \frac{2\pi L}{\mu_0 I_1}\left(\frac{F}{l}\right) = \frac{2\pi (0.20\,\text{m})}{(4\pi \times 10^{-7}\,\text{T·m/A})(80\,\text{A})}(1.18 \times 10^{-3}\,\text{N/m}) = 15\,\text{A}.$$

FIGURE 20–24 Example 20–9.

* 20–7 Definition of the Ampere and the Coulomb

You may have wondered how the constant μ_0 in Eq. 20–5 could be exactly $4\pi \times 10^{-7}$ T·m/A. Here is how it happened. With an older definition of the ampere, μ_0 was measured experimentally to be very close to this value. Today, however, μ_0 is *defined* to be exactly $4\pi \times 10^{-7}$ T·m/A. This, of course, could not be done if the ampere were defined independently. The ampere, the unit of current, is now defined in terms of the magnetic field B it produces using the defined value of μ_0.

In particular, we use the force between two parallel current-carrying wires, Eq. 20–6, to define the ampere precisely. If $I_1 = I_2 = 1$ A exactly, and the two wires are exactly 1 m apart, then

$$\frac{F}{l} = \frac{(4\pi \times 10^{-7}\,\text{T·m/A})}{(2\pi)}\frac{(1\,\text{A})(1\,\text{A})}{(1\,\text{m})} = 2 \times 10^{-7}\,\text{N/m}.$$

Thus, *one* **ampere** *is defined as that current flowing in each of two long parallel conductors 1 m apart, which results in a force of exactly 2×10^{-7} N/m of length of each conductor.*

Definitions of ampere and coulomb

This is the precise definition of the ampere. The **coulomb** is then defined as being *exactly* one ampere-second: $1\,\text{C} = 1\,\text{A·s}$. The value of k or ϵ_0 in Coulomb's law (Section 16–5) is obtained from experiment.

* 20–8 Ampère's Law

In Section 20–5, we saw that Eq. 20–5 gives the relation between the current in a long straight wire and the magnetic field it produces. This equation is valid only for a long straight wire. The following important question arises: Is there a general relation between a current in a wire of whatever shape and the magnetic field around it? The answer is yes: the French scientist André Marie Ampère (1775–1836) proposed such a relation shortly after Oersted's discovery. Consider any (arbitrary) closed path around a current, as shown in Fig. 20–25, and imagine this path as being made up of short segments each of length Δl. First, we take the product of the length of each segment times the component of **B** parallel to that segment. If we now sum all these terms, according to Ampère, the result will be equal to μ_0 times the net current I that passes through the surface enclosed by the path. This is known as **Ampère's law** and can be written mathematically as

$$\sum B_{\parallel} \Delta l = \mu_0 I. \qquad \text{(20–7)}$$

The symbol Σ means "the sum of" and $B_{\parallel}$ means the component of **B** parallel to that particular Δl. The lengths Δl are chosen so that $B_{\parallel}$ is essentially constant on each length. The sum must be made over a closed path; and I is the net current passing through the surface bounded by this closed path.

We can check Ampère's law by applying it to the simple case of a long straight wire carrying a current I, which we have already examined and which served as an inspiration for Ampère himself. Suppose that we want to find the magnitude of B at point A, a distance r from the wire in Fig. 20–26. We know that the magnetic field lines are circles with the wire at their center. We then choose a path to be used in Eq. 20–7: we choose a circle of radius r (the choice of path is ours—so we choose one that will be convenient). We choose this circular path because at any point on this path, **B** will be tangent to this circle. Thus, for any short segment of the circle (Fig. 20–26), **B** will be parallel to that segment, so $B_{\parallel} = B$. Suppose that we break the circular path down into 100 segments.[†] Then Ampère's law states that

$$(B \, \Delta l)_1 + (B \, \Delta l)_2 + (B \, \Delta l)_3 + \cdots + (B \, \Delta l)_{100} = \mu_0 I.$$

The dots represent all the terms we did not write down. Since all the segments are the same distance from the wire, we expect B to be the same at each segment. We can then factor out B from the sum:

$$B(\Delta l_1 + \Delta l_2 + \Delta l_3 + \cdots + \Delta l_{100}) = \mu_0 I.$$

The sum of the segment lengths is just the circumference of the circle, $2\pi r$. Thus we have

$$B(2\pi r) = \mu_0 I,$$

or

$$B = \frac{\mu_0 I}{2\pi r}.$$

This is just Eq. 20–5 for the field near a long straight wire, as discussed earlier.

Ampère's law thus works for this simple case. A great many experiments indicate that Ampère's law is valid in general. However, it can be used to calculate the magnetic field mainly for simple cases. Its importance is that it relates the magnetic field to the current in a direct and mathematically elegant way. Ampère's law is thus considered one of the basic laws of electricity and magnetism. It is valid for any situation where the currents and fields are not changing in time.

We now can see why the constant in Eq. 20–5 is written $\mu_0/2\pi$. This is done so that only μ_0 appears in Eq. 20–7 (rather than, say, $2\pi k$ if we had used k in Eq. 20–5). In this way, the more fundamental equation, Ampère's law, has the simpler form.

[†]Actually, Ampère's law is precisely accurate when there is an infinite number of infinitesimally short segments, but that leads into calculus.

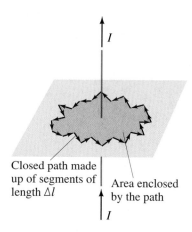

Closed path made up of segments of length Δl

Area enclosed by the path

FIGURE 20–25 Arbitrary path enclosing a current, for Ampère's law. The path is broken down into segments of equal length Δl.

FIGURE 20–26 Circular path of radius r.

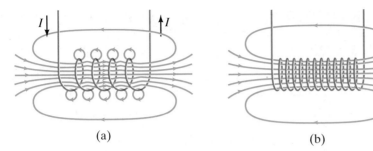

(a)   (b)

FIGURE 20–27 (a) Magnetic field due to several loops of a solenoid. (b) If the coils are closely spaced, the field is very nearly uniform.

We now use Ampère's law to calculate the magnetic field inside a long coil of wire with many loops, as shown in Fig. 20–27, which is known as a **solenoid**. Each coil produces a magnetic field as shown in Fig. 20–11, and the total field inside the solenoid will be the sum of the fields due to each current loop as shown in Fig. 20–27a. If the coils of the solenoid are very closely spaced, the field inside will be essentially parallel to the axis except at the ends, as shown in Fig. 20–27b. For applying Ampère's law, we choose the path *abcd* shown in Fig. 20–28, far from either end. We will consider this path as made up of four segments, the sides of the rectangle: *ab*, *bc*, *cd*, *da*. Then the left side of Eq. 20–7 becomes

$$(B_\parallel \Delta l)_{ab} + (B_\parallel \Delta l)_{bc} + (B_\parallel \Delta l)_{cd} + (B_\parallel \Delta l)_{da}.$$

The first term in this sum will be very small since the field outside the solenoid is so small as to be negligible compared to the field inside (the same number of lines inside the solenoid spread throughout space outside). Thus the first term will be zero. Furthermore, **B** is perpendicular to the segments *bc* and *da*, so these terms are zero, too. Therefore, the left side of Eq. 20–7 is simply $(B_\parallel \Delta l)_{cd} = Bl$, where B is the field inside the solenoid, and l is the length *cd*. Now we determine the current enclosed by our chosen rectangular loop, to use for the right side of Eq. 20–7. If a current I flows in the wire of the solenoid, the total current enclosed by our path *abcd* is NI, where N is the number of loops our path encircles (five in Fig. 20–28). Thus Ampère's law gives us

$$Bl = \mu_0 NI.$$

If we let $n = N/l$ be the *number of loops per unit length*, then

$$B = \mu_0 nI. \qquad\qquad \text{[solenoid]} \quad \textbf{(20–8)}$$

Magnetic field inside a solenoid

This is the magnitude of the magnetic field within a solenoid. Note that B depends only on the number of loops per unit length, n, and the current I. The field does not depend on the position within the solenoid, so B is uniform. This is strictly true only for an infinite solenoid, but it is a good approximation for real ones for points not close to the ends.

Current out of paper

Current into paper

FIGURE 20–28 Magnetic field inside a solenoid is straight except at the ends. Dashed lines indicate the path chosen for use in Ampère's law.

EXAMPLE 20–10 **Field inside a solenoid.** A thin 10-cm-long solenoid has a total of 400 turns of wire and carries a current of 2.0 A. Calculate the field inside near the center.

SOLUTION The number of turns per unit length is $n = 400/0.10\,\text{m} = 4.0 \times 10^3\,\text{m}^{-1}$. From Eq. 20–8:

$$B = \mu_0 nI = (12.57 \times 10^{-7}\,\text{T·m/A})(4.0 \times 10^3\,\text{m}^{-1})(2.0\,\text{A})$$

$$= 1.0 \times 10^{-2}\,\text{T}.$$

➥ **PHYSICS APPLIED**

Coaxial cable
(shielding)

CONCEPTUAL EXAMPLE 20–11 **Coaxial cable.** A *coaxial cable* is a single wire surrounded by a cylindrical metallic braid, as shown in Fig. 20–29. The two conductors are separated by an insulator. The central wire carries current to the other end of the cable, and the outer braid carries the return current and is usually considered ground. Describe the magnetic field (*a*) in the space between the conductors, and (*b*) outside the cable.

RESPONSE (*a*) In the space between the conductors, we can apply Ampère's law for a circular path around the center wire, just as we did for the case shown in Fig. 20–26 and the magnitude is as given by Eq. 20–5. The current in the outer conductor has no bearing on this result. (Ampère's law uses only the current enclosed *inside* the path; as long as the currents outside the path don't affect the symmetry of the field, they do not contribute to the field along the path at all).
(*b*) Outside the cable, we can draw a similar circular path, for we expect the field to have the same circular symmetry. Now, however, there are two currents enclosed by the path, and they add up to zero. The field outside the cable is zero.

The nice feature of coaxial cables is that they are self-shielding: no stray magnetic fields escape outside the cable. The outer cylindrical conductor also shields external electric fields from coming in (see also Example 16–9). This makes them ideal for carrying signals near sensitive equipment. Audiophiles use coaxial cables between stereo equipment components and even to the loudspeakers.

FIGURE 20–29 Coaxial cable. Example 20–11.

Insulating sleeve

Cylindrical braid

I

Solid wire

*20–9 Torque on a Current Loop; Magnetic Moment

When an electric current flows in a closed loop of wire placed in a magnetic field, as shown in Fig. 20–30, the magnetic force on the current can produce a torque. This is the basic principle behind a number of important practical devices, including meters and motors. (We discuss these applications in the next Section.) The interaction between a current and a magnetic field is important in other areas as well, including atomic physics.

When current flows through the loop in Fig. 20–30a, whose face we assume is parallel to **B** and is rectangular, the magnetic field exerts a force on both vertical sections of wire as shown, F_1 and F_2 (see also top view, Fig. 20–30b). Notice that, by the right-hand rule (Fig. 20–13c), the direction of the force on the upward current on the left is in the opposite direction from the equal magnitude force F_2 on the descending current on the right. These forces give rise to a net torque that tends to rotate the coil about its vertical axis.

Let us calculate the magnitude of this torque. From Eq. 20–2, the force $F = IaB$, where a is the length of the vertical arm of the coil. The lever arm for each force is $b/2$, where b is the width of the coil and the "axis" is at the midpoint. The total torque is the sum of the torques due to each of the forces, so

$$\tau = IaB\frac{b}{2} + IaB\frac{b}{2} = IabB = IAB,$$

where $A = ab$ is the area of the coil. If the coil consists of N loops of wire, the current is then NI, so the torque becomes

$$\tau = NIAB. \tag{20–9a}$$

If the coil makes an angle θ with the magnetic field, as shown in Fig. 20–30c, the forces are unchanged, but each lever arm is reduced from $\frac{1}{2}b$ to $\frac{1}{2}b\sin\theta$. Note that the angle θ is chosen to be the angle between **B** and the perpendicular to the face of the coil, Fig. 20–30c. So the torque becomes

$$\tau = NIAB \sin\theta. \tag{20–9b}$$

This formula, derived here for a rectangular coil, is valid for any shape of flat coil.

The quantity NIA is called the **magnetic dipole moment** of the coil and is considered a vector:

$$\mathbf{M} = NIA, \tag{20–10}$$

where the direction of **A** (and therefore of **M**) is *perpendicular* to the plane of the coil (the black arrow in Fig. 20–30c).

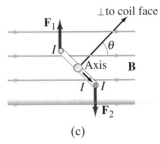

FIGURE 20–30 Calculating the torque on a current loop in a magnetic field **B**. (a) Loop face parallel to **B** field lines; (b) top view; (c) loop makes an angle to **B**, reducing the torque since the lever arm is reduced.

EXAMPLE 20-12 **Torque on a coil.** A circular coil of wire has a diameter of 20.0 cm and contains 10 loops. The current in each loop is 3.00 A, and the coil is placed in a 2.00-T magnetic field. Determine the maximum and minimum torque exerted on the coil by the field.

SOLUTION Equation 20–9 is valid for any shape of coil, including circular, where the area is

$$A = \pi r^2 = \pi(0.100 \text{ m})^2 = 3.14 \times 10^{-2} \text{ m}^2.$$

The maximum torque occurs when the coil's face is parallel to the magnetic field, so $\theta = 90°$ in Fig. 20–30c, and $\sin\theta = 1$ in Eq. 20–9b:

$$\tau = NIAB \sin\theta = (10)(3.00 \text{ A})(3.14 \times 10^{-2} \text{ m}^2)(2.00 \text{ T})(1) = 1.88 \text{ N·m}.$$

The minimum torque occurs if $\sin\theta = 0$, for which $\theta = 0°$, and then $\tau = 0$ from Eq. 20–9b.

---

* ## 20–10 Applications: Galvanometers, Motors, Loudspeakers

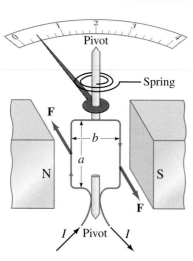

FIGURE 20–31 Galvanometer.

FIGURE 20–32 Galvanometer coil wrapped on an iron core.

The basic component of most meters, including ammeters, voltmeters, and ohmmeters, is a galvanometer. We have already seen how these meters are designed (Section 19–10), and now we can examine how the crucial element, a galvanometer, itself works. As shown in Fig. 20–31, a **galvanometer** consists of a coil of wire (with attached pointer) suspended in the magnetic field of a permanent magnet. When current flows through the loop of wire, which is usually rectangular, the magnetic field exerts a torque on the loop, as given by Eq. 20–9b, $\tau = NIAB \sin\theta$. This torque is opposed by a spring which exerts a torque τ_s approximately proportional to the angle ϕ through which it is turned (Hooke's law). That is,

$$\tau_s = k\phi,$$

where k is the stiffness constant of the spring. Thus the coil and the attached pointer will rotate only to the point where the spring torque balances the torque due to the magnetic field. From Eq. 20–9b we then have $k\phi = NIAB \sin\theta$ or

$$\phi = \frac{NIAB \sin\theta}{k}.$$

Thus the deflection of the pointer, ϕ, is directly proportional to the current I flowing in the coil. But it also depends on the angle θ the coil makes with **B**. For a useful meter we need ϕ to depend only on I, independent of θ. To solve this problem, curved pole pieces are used and the galvanometer coil is wrapped around a cylindrical iron core as shown in Fig. 20–32. The iron tends to concentrate the magnetic field lines so that **B** always points parallel to the face of the coil at the wire outside the core. The force is then always perpendicular to the face of the coil and the torque will not vary with angle. Thus ϕ will be proportional to I, as required.

A **chart recorder**, in which a pen graphs a signal such as an ECG (electrocardiogram–Section 17–11) on a moving roll of paper, is basically

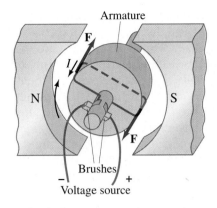

FIGURE 20-33 Diagram of a simple dc motor.

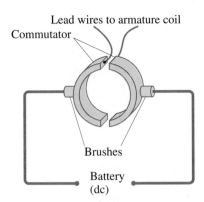

FIGURE 20-34 The commutator-brush arrangement in a dc motor assures alternation of the current in the armature to keep rotation continuous. The commutators are attached to the motor shaft and turn with it, whereas the brushes remain stationary.

a galvanometer. The pen is attached to an arm, which is connected to the galvanometer coil. The instrument could record either voltage or current, just as any galvanometer can be connected as a voltmeter or ammeter.

An **electric motor** changes electric energy into (rotational) mechanical energy. A motor works on the same principle as a galvanometer, except that there is no spring so the coil can rotate continuously in one direction. The coil is larger and is mounted on a large cylinder called the **rotor** or **armature**, Fig. 20-33. Actually, there are several coils, although only one is indicated in the figure. The armature is mounted on a shaft or axle. At the moment shown in Fig. 20-33, the magnetic field exerts forces on the current in the loop as shown. However, when the coil, which is rotating clockwise in Fig. 20-33, passes beyond the vertical position the forces would then act to return the coil back to vertical if the current remained the same. But if the current could somehow be reversed at that critical moment, the forces would reverse, and the coil would continue rotating in the same direction. Thus, alternation of the current is necessary if a motor is to turn continuously in one direction. This can be achieved in a **dc motor** with the use of **commutators** and **brushes**: as shown in Fig. 20-34, the brushes are stationary contacts that rub against the conducting commutators mounted on the motor shaft. At every half revolution, each commutator changes its connection to the other brush. Thus the current in the coil reverses every half revolution as required for continuous rotation. Most motors contain several coils, called "windings," each located in a different place on the armature, Fig. 20-35. Current flows through each coil only during a small part of a revolution, at the time when its orientation results in the maximum torque. In this way, a motor produces a much steadier torque than can be obtained from a single coil. An **ac motor**, with ac current as input, can work without commutators since the current itself alternates. Many motors use wire coils to produce the magnetic field (electromagnets) instead of a permanent magnet. Indeed the design of most practical motors is more complex than described here, but the general principles remain the same.

Electric motor

➡ **PHYSICS APPLIED**

DC motor

FIGURE 20-35 Motor with many windings.

➡ **PHYSICS APPLIED**

AC motor

Loudspeakers

Coil of wire
(attached to
speaker cone)

Rigid
metal
frame

Magnet

Lead-in
wires

Cone

FIGURE 20–36
Loudspeaker.

(a)

(b)

FIGURE 20–37 The Hall
effect. (a) Negative charges
moving to the right as the
current. (b) Positive charges
moving to the left as the current.

A **loudspeaker** also works on the principle that a magnet exerts a force on a current-carrying wire. The electrical output of a radio or TV set is connected to the wire leads of the speaker. The speaker leads are connected internally to a coil of wire, which is itself attached to the speaker cone, Fig. 20–36. The speaker cone is usually made of stiffened cardboard and is mounted so that it can move back and forth freely. A permanent magnet is mounted directly in line with the coil of wire. When the alternating current of an audio signal flows through the wire coil, the coil and the attached speaker cone experience a force due to the magnetic field of the magnet. As the current alternates at the frequency of the audio signal, the speaker cone moves back and forth at the same frequency, causing alternate compressions and rarefactions of the adjacent air, and sound waves are produced. A speaker thus changes electrical energy into sound energy, and the frequencies and intensities of the emitted sound waves can be an accurate reproduction of the electrical input.

* 20–11 The Hall Effect

When a current-carrying conductor is held firmly in a magnetic field, the field exerts a sideways force on the charges moving in the conductor. For example, if electrons move to the right in the rectangular conductor shown in Fig. 20–37a, the inward magnetic field will exert a downward force on the electrons $F_B = ev_d B$, where v_d is the drift velocity of the electrons (Section 18–9). So the electrons will tend to move nearer face S than face R. There will thus be a potential difference between faces R and S of the conductor. This potential difference builds up until the electric field $\mathbf{E}_H$ it produces exerts a force, $e\mathbf{E}_H$, on the moving charges that is equal and opposite to the magnetic force. This effect is called the **Hall effect** after E. H. Hall, who discovered it in 1879. The difference of potential produced is called the **Hall emf**.

The electric field due to the separation of charge is called the *Hall field*, $\mathbf{E}_H$, and points downward in Fig. 20–37a, as shown. In equilibrium, the force due to this electric field is balanced by the magnetic force $ev_d B$, so

$$eE_H = ev_d B.$$

Hence $E_H = v_d B$. The Hall emf is then (assuming the conductor is long and thin so E_H is uniform)

$$\mathcal{E}_H = E_H l = v_d B l, \tag{20–11}$$

where l is the width of the conductor.

A current of negative charges moving to the right is equivalent to positive charges moving to the left, at least for most purposes. But the Hall effect can distinguish these two. As can be seen in Fig. 20–37b, positive particles moving to the left are deflected downward, so that the bottom surface is positive relative to the top surface. This is the reverse of part (a). Indeed, the direction of the emf in the Hall effect first revealed that it is negative particles that move in metal conductors. In some semiconductors, however, the Hall effect reveals that the carriers of current are positive (more on this in Chapter 29).

The magnitude of the Hall emf is proportional to the strength of the magnetic field. The Hall effect can thus be used to measure magnetic field strengths. First the conductor, called a *Hall probe*, is calibrated with known magnetic fields. Then, for the same current, its emf output will be a measure of B. Hall probes can be made very small and are convenient and accurate to use.

The Hall effect can also be used to measure the drift velocity of charge carriers when the external magnetic field B is known. Such a measurement also allows us to determine the density of charge carriers in the material.

EXAMPLE 20–13 **Drift velocity using the Hall effect.** A long copper strip 1.8 cm wide and 1.0 mm thick is placed in a 1.2-T magnetic field as in Fig. 20–37a. When a steady current of 15 A passes through it, the Hall emf is measured to be 1.02 μV. Determine the drift velocity of the electrons and the density of free (conducting) electrons (number per unit volume) in the copper.

SOLUTION The drift velocity (Eq. 20–11) is

$$v_d = \frac{\mathscr{E}_H}{Bl} = \frac{1.02 \times 10^{-6} \text{ V}}{(1.2 \text{ T})(1.8 \times 10^{-2} \text{ m})} = 4.7 \times 10^{-5} \text{ m/s}.$$

The density of charge carriers n is obtained from Eq. 18–10, $I = nev_dA$, where A is the cross-sectional area through which the current I flows. Then

$$n = \frac{I}{ev_dA} = \frac{15 \text{ A}}{(1.6 \times 10^{-19} \text{ C})(4.7 \times 10^{-5} \text{ m/s})(1.8 \times 10^{-2} \text{ m})(1.0 \times 10^{-3} \text{ m})}$$
$$= 11 \times 10^{28} \text{ m}^{-3}.$$

This value for the density of free electrons in copper, $n = 11 \times 10^{28}$ per m³, is the experimentally measured value. It represents *more* than one free electron per atom, which as we saw in Example 18–13 is $8.4 \times 10^{28} \text{ m}^{-3}$.

* 20–12 Mass Spectrometer

Various methods were developed in the early part of this century to measure the masses of atoms. One of the most accurate was the **mass spectrometer**[†] of Fig. 20–38. Ions are produced by heating, or by an electric current, in the source S. Those that pass through slit S_1 enter a region where there are both electric and magnetic fields: the magnetic field points out of the page in Fig. 20–38, and the electric field points up (from the + plate toward the − plate). Ions will follow a straight-line path in this region, as shown, if the electric force qE (upward on a positive ion) is just balanced by the magnetic force qvB (downward on a positive ion): that is, if

$$qE = qvB$$

or

$$v = \frac{E}{B}.$$

In other words, those ions (and only those) whose speed is $v = E/B$ will pass through undeflected and emerge through slit S_2. (This arrangement is called a *velocity selector*.) In the second region, after S_2, there is only a

[†] The term *mass spectrograph* is also used.

➡ PHYSICS APPLIED

The mass spectrometer

FIGURE 20–38 Bainbridge mass spectrometer. The magnetic fields B and B' point out of the paper (indicated by the dots).

magnetic field B' so the ions follow a circular path. The radius of their path can be measured because the ions darken the photographic plate where they strike. Since $qvB' = mv^2/r$ and $v = E/B$, then we have

$$m = \frac{qB'r}{v} = \frac{qBB'r}{E}.$$

All the quantities on the right can be measured, and thus m can be determined. Note that for ions of the same charge, the mass of each is proportional to the radius of its path.

The masses of many atoms were measured in this way. When a pure substance was used, it was sometimes found that two or more closely spaced marks would appear on the film. For example, neon produced two marks whose radii corresponded to atoms of mass 20 and 22 atomic mass units (u). Impurities were ruled out and it was concluded that there must be two types of neon with different masses. These different forms were called **isotopes.** It was soon found that most elements are mixtures of isotopes. We shall see in Chapter 30 that the difference in mass is due to different numbers of neutrons.

Mass spectrometers can be used to separate not only different elements and isotopes, but different molecules as well. They are used in physics and chemistry, and in biological and biomedical laboratories.

EXAMPLE 20–14 **Mass spectrometry.** Carbon atoms of atomic mass 12.0 u are found to be mixed with another, unknown, element. In a mass spectrometer, the carbon traverses a path of radius 22.4 cm and the unknown's path has a 26.2 cm radius. What is the unknown element? Assume they have the same charge.

SOLUTION Since mass is proportional to the radius, we have

$$\frac{m_x}{m_C} = \frac{26.2\ \text{cm}}{22.4\ \text{cm}} = 1.17.$$

Thus $m_x = 1.17 \times 12.0\ \text{u} = 14.0\ \text{u}$. The other element is probably nitrogen (see the periodic table, inside the back cover). However, it could also be an isotope of carbon or oxygen. Further physical or chemical analysis would be needed.

* 20–13 Ferromagnetism; Domains

We saw in Section 20–1 that iron (and a few other materials) can be made into strong magnets. These materials are said to be **ferromagnetic**. We now look more deeply into the sources of ferromagnetism.

A bar magnet, with its two opposite poles at either end, resembles an electric dipole (equal-magnitude positive and negative charges separated by a distance). Indeed, a bar magnet is sometimes referred to as a "magnetic dipole." There are opposite "poles" separated by a distance. And the magnetic field lines of a bar magnet form a pattern much like that for the electric field of an electric dipole: compare Fig. 16–29a with Fig. 20–3b.

Domains in iron   Microscopic examination reveals that a magnet is actually made up of tiny regions known as **domains**, which are at most about 1 mm in length or width. Each domain behaves like a tiny magnet with a north and a south pole. In an unmagnetized piece of iron, these domains are arranged randomly, as shown in Fig. 20–39a. The magnetic effects of the domains cancel each other out, so this piece of iron is not a magnet. In a magnet, the domains are preferentially aligned in one direction as shown in Fig. 20–39b (downward in this case). A magnet can be made from an unmagnetized piece of iron by placing it in a strong magnetic field. (You can make a needle magnetic, for example, by stroking it with one pole of a strong magnet.) Careful observations show in this case that the magnetization of domains may actually rotate slightly so as to be more nearly parallel to the external field. Or, more commonly, the borders of domains move so that those domains whose magnetic orientation is parallel to the external field grow in size at the expense of other domains. This can be seen by comparing Figs. 20–39a and b. This explains how a magnet can pick up unmagnetized pieces of iron like paper clips or bobby pins. The magnet's field causes a slight alignment of the domains in the unmagnetized object so that the object becomes a temporary magnet with its north pole facing the south pole of the permanent magnet, and vice versa; thus, attraction results. In the same way, elongated iron filings will arrange themselves in a magnetic field just as a compass needle does, and will reveal the shape of the magnetic field, Fig. 20–40.

An iron magnet can remain magnetized for a long time, and thus it is referred to as a "permanent magnet." However, if you drop a magnet on the floor or strike it with a hammer, you may jar the domains into randomness. The magnet can thus lose some or all of its magnetism. Heating a magnet too can cause a loss of magnetism, for raising the temperature increases the random thermal motion of the atoms which tends to randomize the domains. Above a certain temperature known as the **Curie temperature** (1043 K for iron), a magnet cannot be made at all.[†]

There is a striking similarity between the fields produced by a bar magnet and by a loop of electric current or a solenoid (compare Fig. 20–3b with Figs. 20–11 and 20–27). This suggests that the magnetic field produced by a current may have something to do with ferromagnetism, an idea proposed by Ampère in the nineteenth century. According to modern atomic theory, the atoms that make up any material can be roughly visualized as containing electrons that orbit around a central nucleus. Since the electrons are charged, they constitute an electric current and therefore produce a magnetic field. But if there is no external field, the electron orbits in different atoms are arranged randomly, so the magnetic effects due to the many orbits in all the atoms in a material cancel out. However, electrons produce an additional magnetic field, almost as if they and their electric charge were spinning about their own axes. It is the magnetic field due to electron spin[‡] that is believed to produce ferromagnetism.

FIGURE 20–39 (a) An unmagnetized piece of iron is made up of domains that are randomly arranged. Each domain is like a tiny magnet; the arrows represent the magnetization direction, with the arrowhead being the N pole. (b) In a magnet, the domains are preferentially aligned in one direction, and may be altered in size by the magnetization process.

FIGURE 20–40 Iron filings line up along magnetic field lines.

[†]Iron, nickel, cobalt, gadolinium, and certain alloys are ferromagnetic at room temperature; several other elements and alloys have low Curie temperature and thus are ferromagnetic only at low temperatures.

[‡]The name "spin" comes from the early suggestion that the additional magnetic field arises from the electron "spinning" on its axis (as well as "orbiting" the nucleus) and this additional motion of the charge was supposed to produce the extra field. However this view of a spinning electron is oversimplified: see Chapter 28.

In most materials, the magnetic fields due to electron spin cancel out. But in iron and other ferromagnetic materials, a complicated cooperative mechanism seems to operate. The result is that the electrons contributing to the ferromagnetism in a domain "spin" in the same direction. Thus the tiny magnetic fields due to each of the electrons add up to give the magnetic field of a domain. And when the domains are aligned, as we have seen, a strong magnet results.

It is believed possible today that *all* magnetic fields are caused by electric currents. This would explain why no single magnetic pole has ever been found: there is no way to divide up a current and obtain a single magnetic pole. Of course if an isolated pole is found, we will have to alter the idea that all magnetic fields are produced by currents.

The lack of single magnetic poles means that magnetic field lines form closed loops, unlike electric field lines which begin on positive charges and end on negative charges.

Electromagnets and Solenoids

A long coil of wire consisting of many loops of wire, as discussed in Section 20–8, is called a solenoid. The magnetic field within a solenoid can be fairly large since it will be the sum of the fields due to the current in each loop (see Fig. 20–41). The solenoid acts like a magnet; one end can be considered the north pole and the other the south pole, depending on the direction of the current in the loops (use the right-hand rule). Since the magnetic field lines leave the north pole of a magnet, the north pole of the solenoid in Fig. 20–41 is on the right.

If a piece of iron is placed inside a solenoid, the magnetic field is increased greatly because the domains of the iron are aligned by the magnetic field produced by the current. The resulting magnetic field is the sum of that due to the current and that due to the iron, and can be hundreds or thousands of times that due to the current alone (see Section 20–15). This arrangement is called an **electromagnet**. The iron used in electromagnets acquires and loses its magnetism quite readily when the current is turned on or off, and so is referred to as "soft iron." (It is "soft" only in a magnetic sense.) Iron that holds its magnetism even when there is no externally applied field is called "hard iron." Hard iron is used in permanent magnets. Soft iron is usually used in electromagnets so that the field can be turned on and off readily. Whether iron is hard or soft depends on heat treatment and other factors.

Electromagnets find use in many practical applications, from use in motors and generators to producing large magnetic fields for research. Because the current flows continuously, a great deal of waste heat (I^2R power) is often produced. Cooling coils, which are tubes carrying water, must be used to absorb the heat in bigger installations. For some applications, superconducting magnets are coming into use. The current-carrying wires are made of superconducting material (Section 18–5) kept below the transition temperature. No electric power is needed to maintain large current, which means large savings of electricity. Of course, energy is needed to keep the superconducting coils at the necessary low temperature.

FIGURE 20–41 Magnetic field of a solenoid. The north pole of this solenoid, thought of as a magnet, is on the right, and the south pole is on the left.

➡ **PHYSICS APPLIED**

Electromagnets and solenoids

Another useful device consists of a solenoid into which a rod of iron is partially inserted. This combination is also referred to as a solenoid. One simple use is as a doorbell (Fig. 20–42). When the circuit is closed by pushing the button, the coil effectively becomes a magnet and exerts a force on the iron rod. The rod is pulled into the coil and strikes the bell. A larger solenoid is used in the starters of cars; when you engage the starter, you are closing a circuit that not only turns the starter motor, but activates a solenoid that first moves the starter into direct contact with the engine. Solenoids are used as switches in many other devices, such as tape recorders. They have the advantage of moving mechanical parts quickly and accurately.

FIGURE 20–42 Solenoid used as a doorbell.

*20–15 Magnetic Fields in Magnetic Materials; Hysteresis

The field of a long solenoid is directly proportional to the current. Indeed, Eq. 20–8 tells us that the field B_0 inside a solenoid is given by

$$B_0 = \mu_0 nI.$$

This is valid if there is only air inside the coil. If we put a piece of iron or other ferromagnetic material inside the solenoid, the field will be greatly increased, often by hundreds or thousands of times. This occurs because the domains in the iron become preferentially aligned by the external field. The resulting magnetic field is the sum of that due to the current and that due to the iron. It is sometimes convenient to write the total field in this case as a sum of two terms:

$$\mathbf{B} = \mathbf{B}_0 + \mathbf{B}_M. \tag{20–12}$$

Here, $\mathbf{B}_0$ refers to the field due only to the current in the wire (the "external field"). It is the field that would be present in the absence of a ferromagnetic material. Then $\mathbf{B}_M$ represents the additional field due to the ferromagnetic material itself, often $\mathbf{B}_M \gg \mathbf{B}_0$.

The total field inside a solenoid in such a case can also be written by replacing the constant μ_0 in Eq. 20–8 by another constant, μ, characteristic of the material inside the coil:

$$B = \mu nI; \tag{20–13}$$

μ is called the **magnetic permeability** of the material. For ferromagnetic materials, μ is much greater than μ_0. For all other materials, its value is very close to μ_0.[†] The value of μ, however, is not constant for ferromagnetic materials; it depends on the value of the external field B_0, as the following experiment shows.

[†]All materials are slightly magnetic. Nonferromagnetic materials fall into two principal classes: **paramagnetic**, in which μ is very slightly larger than μ_0; and **diamagnetic**, in which μ is very slightly less than μ_0. Paramagnetic materials apparently contain atoms that have a net magnetic dipole moment due to orbiting electrons, and these become slightly aligned with an external field just as the galvanometer coil in Fig. 20–31 experiences a torque that tends to align it. Atoms of diamagnetic materials have no net dipole moment. However, in the presence of an external field, electrons revolving in one direction are caused to increase in speed slightly, whereas those revolving in the opposite direction are reduced in speed. The result is a slight net magnetic effect that actually opposes the external field.

FIGURE 20–43 Iron-core torus.

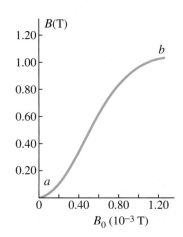

FIGURE 20–44 Total magnetic field B in an iron-core torus as a function of the external field B_0 (B_0 is caused by the current I in the coil).

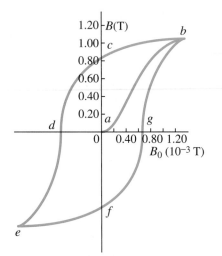

FIGURE 20–45 Hysteresis curve.

Hysteresis

Measurements on magnetic materials are generally done using a torus, which is essentially a long solenoid bent into the shape of a circle (Fig. 20–43), so that practically all the lines of **B** remain within the torus. Suppose the torus has an iron core that is initially unmagnetized and there is no current in the windings of the torus. Then the current I is slowly increased, and B_0 increases linearly with I. The total field B also increases, but follows the curved line shown in the graph of Fig. 20–44. (Note the different scales: $B \gg B_0$.) Initially, point a, the domains (Section 20–13) are randomly oriented. As B_0 increases, the domains become more and more aligned until at point b, nearly all are aligned. The iron is said to be approaching **saturation**. Point b is typically 70 percent of full saturation. (If B_0 is increased further, the curve continues to rise very slowly, and reaches 98 percent saturation only when B_0 reaches a value about a thousandfold above that at point b; the last few domains are very difficult to align.) Next, suppose the external field B_0 is reduced by decreasing the current in the coils. As the current is reduced to zero, point c in Fig. 20–45, the domains do not become completely random. Some permanent magnetism remains. If the current is then reversed in direction, enough domains can be turned around so $B = 0$ (point d). As the reverse current is increased further, the iron approaches saturation in the opposite direction (point e). Finally, if the current is again reduced to zero and then increased in the original direction, the total field follows the path *efgb*, again approaching saturation at point b.

Notice that the field did not pass through the origin (point a) in this cycle. The fact that the curves do not retrace themselves on the same path is called **hysteresis**. The curve *bcdefgb* is called a **hysteresis loop**. In such a cycle, much energy is transformed to thermal energy (friction) due to realigning of the domains. It can be shown that the energy dissipated in this way is proportional to the area of the hysteresis loop.

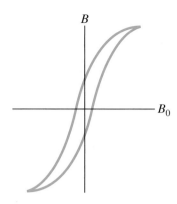

FIGURE 20–46 Hysteresis curve for soft iron.

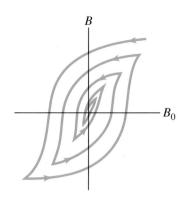

FIGURE 20–47 Successive hysteresis loops during demagnetization.

At points c and f, the iron core is magnetized even though there is no current in the coils. These points correspond to a permanent magnet. For a permanent magnet, it is desired that ac and af be as large as possible. Materials for which this is true are said to have high **retentivity**, and may be referred to as "hard." On the other hand, a hysteresis curve such as that in Fig. 20–46 occurs for so-called "soft iron" (soft from a magnetic point of view). This is preferred for *electromagnets* (Section 20–14) since the field can be more readily switched off, and the field can be reversed with less loss of energy.

A ferromagnetic material can be demagnetized—that is, made unmagnetized. This can be done by reversing the magnetizing current repeatedly while decreasing its magnitude. This results in the curve of Fig. 20–47. The heads of a tape recorder are demagnetized in this way. The alternating magnetic field acting at the heads due to a demagnetizer is strong when the demagnetizer is placed near the heads and decreases as it is moved slowly away. (Cassette tapes themselves can be erased and ruined by a magnetic field.)

Demagnetizing

> **PROBLEM SOLVING** **Magnetic Fields**

Magnetic fields are somewhat analogous to the electric fields of Chapter 16, but there are several important differences to recall:

1. The force experienced by a charged particle moving in a magnetic field is *perpendicular* to the direction of the magnetic field (and to the direction of the velocity of the particle), whereas the force exerted by an electric field is *parallel* to the direction of the field (and unaffected by the velocity of the particle).

2. The *right-hand rule*, in its many forms, is intended to help you determine the directions of magnetic field, and the forces they exert, and/or the directions of electric current or charged particle velocity. The right-hand rules are specifically designed to deal with the "perpendicular" nature of these quantities.

3. Note that the equations in this chapter are generally not printed as vector equations, but involve magnitudes only. The right-hand rule is to be used to find directions of vector quantities.

SUMMARY

A magnet has two **poles**, north and south. The north pole is that end which points toward the north when the magnet is freely suspended. Unlike poles of two magnets attract each other, whereas like poles repel.

We can imagine that a **magnetic field** surrounds every magnet. The SI unit for magnetic field is the **tesla** (T). The force one magnet exerts on another is said to be an interaction between one magnet and the magnetic field produced by the other.

Electric currents produce magnetic fields. For example, the lines of magnetic field due to a current in a straight wire form circles around the wire and the field exerts a force on magnets.

The magnitude of the magnetic field a distance r from a long straight wire carrying a current I is given by

$$B = \frac{\mu_0}{2\pi} \frac{I}{r}.$$

A magnetic field exerts a force on an electric current. For a straight wire of length l carrying a current I, the force has magnitude

$$F = IlB \sin \theta,$$

where θ is the angle between the magnetic field of strength B and the wire. The direction of the force is perpendicular to the wire and to the magnetic field, and is given by the right-hand rule.

Similarly, a magnetic field exerts a force on a charge q moving with velocity v of magnitude

$$F = qvB \sin \theta,$$

where θ is the angle between $\mathbf{v}$ and $\mathbf{B}$. The direction of $\mathbf{F}$ is perpendicular to $\mathbf{v}$ and to $\mathbf{B}$. The path of a charged particle moving perpendicular to a uniform magnetic field is a circle.

The force exerted on a current-carrying wire by a magnetic field is the basis for operation of many devices, such as meters, motors, and loudspeakers.

QUESTIONS

1. A compass needle is not always balanced parallel to the Earth's surface but one end may dip downward. Explain.

2. Draw the magnetic field lines around a straight section of wire carrying a current horizontally to the left.

3. In what direction are the magnetic field lines surrounding a straight wire carrying a current that is moving directly toward you?

4. The magnetic field due to current in wires in your home can affect a compass. Discuss the problem in terms of currents, including if they are ac or dc.

5. What kind of field or fields surround a moving electric charge?

6. Will a magnet attract any metallic object, or only those made of iron? (Try it and see.) Why is this so?

7. Two iron bars attract each other no matter which ends are placed close together. Are both magnets? Explain.

*8. Note that the pattern of magnetic field lines surrounding a bar magnet is similar to that of the electric field around an electric dipole. From this fact, predict how the magnetic field will change with distance (*a*) when near one pole of a very long bar magnet, and (*b*) when far from a magnet as a whole.

9. Suppose you have three iron rods, two of which are magnetized but the third is not. How would you determine which two are the magnets without using any additional objects?

10. How can you make a compass without using iron or other ferromagnetic material?

11. A horseshoe magnet is held vertically with the north pole on the left and south pole on the right. A wire passes between the poles, equidistant from them, and carries a current directly away from you. In what direction is the force on the wire?

12. Can you set a resting electron into motion with a magnetic field? With an electric field?

13. A charged particle is moving in a circle under the influence of a uniform magnetic field. If an electric field that points in the same direction as the magnetic field is turned on, describe the path the charged particle will take.

14. Each of the right-hand rules you learned in this chapter can be changed to *left-hand rules* if you are specifying the direction of movement of *negative* particles, such as electrons in a wire. Show, for each right-hand rule, that the same operations using the left hand give the same results if the direction of charge flow is for negative charges.

15. A charged particle moves in a straight line through a particular region of space. Could there be a nonzero magnetic field in this region? If so, give two possible situations.

16. If a moving charged particle is deflected sideways in some region of space, can we conclude for certain that $\mathbf{B} \neq 0$ in that region?

17. If a negatively charged particle enters a region of uniform magnetic field which is perpendicular to the particle's velocity, will the kinetic energy of the particle increase, decrease, or stay the same. Explain your answer. (Neglect gravity.)

18. In Fig. 20–48, charged particles move in the vicinity of a current carrying wire. For each charged particle the arrow indicates the direction of motion of the particle and the + or − indicates the sign of the charge. For each of the particles, indicate the direction of the magnetic force due to the magnetic field produced by the wire.

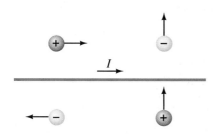

FIGURE 20–48 Question 18.

19. Explain why a strong magnet held near a television screen causes the picture to become distorted. Also, explain why the picture sometimes goes completely black where the field is the strongest.

20. In a particular region of space there is a uniform magnetic field **B**. Outside this region, $B = 0$. Can you inject an electron into the field perpendicularly so it will move in a closed circular path in the field?

21. How could you tell whether moving electrons in a certain region of space are being deflected by an electric field or by a magnetic field (or by both)?

22. A beam of electrons is directed perpendicularly toward a horizontal wire carrying a current from left to right. In what direction are the electrons deflected?

23. Two long wires carrying equal currents I are at right angles to each other, but don't quite touch. Describe the magnetic force one exerts on the other.

24. A horizontal current-carrying wire, free to move, is suspended directly above a second, parallel, current-carrying wire. (a) In what direction is the current in the lower wire? (b) Can the upper wire be held in stable equilibrium due to the magnetic force of the lower wire? Explain.

25. What factors determine the sensitivity of a galvanometer?

26. A rectangular piece of semiconductor is inserted in a magnetic field and a battery is connected to its ends as shown in Fig. 20–49. When a sensitive voltmeter is connected between points a and b, it is found that point a is at a higher potential than b. What is the sign of the charge carriers in this semiconductor material?

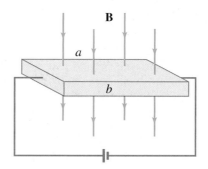

FIGURE 20–49 Question 26.

* 27. Two ions have the same mass, but one is singly ionized and the other is doubly ionized. How will their positions on the film of the mass spectrograph of Fig. 20–38 differ?

* 28. Why will either pole of a magnet attract an unmagnetized piece of iron?

* 29. An unmagnetized nail will not attract an unmagnetized paper clip. However, if one end of the nail is in contact with a magnet, the other end will attract a paper clip. Explain.

* 30. Another type of magnetic switch similar to a solenoid is a relay. A relay is an electromagnet (the iron rod inside the coil does not move) that, when activated, attracts a piece of soft iron on a pivot. Design a relay (a) to make a doorbell, and (b) to close an electrical switch. A relay is used in the latter case when you need to switch on a circuit carrying a very large current but you do not want that large current flowing through the main switch. For example, the starter switch of a car is connected to a relay so that the large currents needed for the starter do not pass to the dashboard switch.

PROBLEMS

SECTIONS 20–3 AND 20–4

1. (I) (a) What is the force per meter on a wire carrying a 9.80-A current when perpendicular to a 0.80-T magnetic field? (b) What if the angle between the wire and field is 45.0°?

2. (I) A 1.5-m length of wire carrying 6.5 A of current is oriented horizontally. At that point on the Earth's surface, the dip angle of the Earth's magnetic field makes an angle of 40° to the wire. Estimate the magnetic force on the wire due to the Earth's magnetic field of 5.5×10^{-5} T at this point.

3. (I) How much current is flowing in a wire 4.20 m long if the maximum force on it is 0.900 N when placed in a uniform 0.0800-T field?

4. (I) The force on a wire carrying 25.0 A is a maximum of 4.14 N when placed between the pole faces of a magnet. If the pole faces are 22.0 cm in diameter, what is the approximate strength of the magnetic field?

5. (I) Determine the magnitude and direction of the force on an electron traveling 3.58×10^6 m/s horizontally to the west in a vertically upward magnetic field of strength 1.30 T.

6. (I) Describe the path of an electron that is projected vertically upward with a speed of 1.80×10^6 m/s into a uniform magnetic field of 0.250 T that is directed away from the observer.

(a)   (b)   (c)   (d)

(e)   (f)

FIGURE 20–50   Problem 7.

7. (I) Find the direction of the force on a negative charge for each diagram shown in Fig. 20–50, where **v** is the velocity of the charge and **B** is the direction of the magnetic field. (⊗ means the vector points inward. ⊙ means it points outward, toward the viewer.)

8. (I) Determine the direction of **B** for each case in Fig. 20–51, where **F** represents the force on a positively charged particle moving with velocity **v**.

(a)   (b)   (c)

FIGURE 20–51   Problem 8.

9. (I) Alpha particles of charge $q = +2e$ and mass $m = 6.6 \times 10^{-27}$ kg are emitted from a radioactive source at a speed of 1.6×10^7 m/s. What magnetic field strength would be required to bend these into a circular path of radius $r = 0.25$ m?

10. (II) An electron experiences the greatest force as it travels 1.8×10^6 m/s in a magnetic field when it is moving southward. The force is upward and of magnitude 2.2×10^{-12} N. What is the magnitude and direction of the magnetic field?

11. (II) The magnetic force per meter on a wire is measured to be only 45 percent of its maximum possible value. Sketch the relationship of the wire and the field if the force were a maximum, and sketch the relationship as it actually is, calculating the angle between the wire and the magnetic field.

12. (II) The force on a wire is a maximum of 5.30 N when placed between the pole faces of a magnet. The current flows horizontally to the right and the magnetic field is vertical. The wire is observed to "jump" toward the observer when the current is turned on. (a) What type of magnetic pole is the top pole face? (b) If the pole faces have a diameter of 10.0 cm, estimate the current in the wire if the field is 0.15 T. (c) If the wire is tipped so that it now makes an angle of 10° with the horizontal, what force will it now feel?

13. (II) A proton moves in a circular path perpendicular to a 1.15-T magnetic field. The radius of its path is 8.40 mm. Calculate the energy of the proton in eV.

14. (II) For a particle of mass m and charge q moving in a circular path in a magnetic field B, show that its kinetic energy is proportional to r^2, the square of the radius of curvature of its path.

15. (II) A particle of charge q moves in a circular path of radius r in a uniform magnetic field B. Show that its momentum is $p = qBr$.

16. (II) For a particle of mass m and charge q moving in a circular orbit in a uniform magnetic field B, show that its angular momentum is given by $L = qBr^2$.

17. (II) A sort of "projectile launcher" is shown in Fig. 20–52. A large current moves in a closed loop composed of fixed rails, a power supply, and a very light, almost frictionless bar touching the rails. A magnetic field is perpendicular to the plane of the circuit. If the bar has a length of 20 cm, a mass of 1.5 g, and is placed in a field of 1.7 T, what constant current flow is needed in order for it to accelerate to 30 m/s in a distance of 1.0 m? In what direction must the field point?

FIGURE 20–52   Problem 17.

18. (III) A 3.80-g bullet moves with a speed of 180 m/s perpendicular to the Earth's magnetic field of 5.00×10^{-5} T. If the bullet possesses a net charge of 8.10×10^{-9} C, by what distance will it be deflected from its path due to the magnetic field after it has traveled 1.00 km?

SECTIONS 20–5 AND 20–6

19. (I) Jumper cables used to start a stalled vehicle often carry a 15-A current. How strong is the magnetic field 15 cm away? What percentage of the Earth's magnetic field is this?

20. (I) If a magnetic field no larger than that of the Earth (0.55×10^{-4} T) is to be allowed 30 cm from an electrical wire, what is the maximum current the wire can carry?

21. (I) What is the magnitude and direction of the force between two parallel wires 45 m long and 6.0 cm apart, each carrying 35 A in the same direction?

22. (I) A vertical straight wire carrying an upward 12-A current exerts an attractive force per unit length of 8.8×10^{-4} N/m on a second parallel wire 7.0 cm away. What current (magnitude and direction) flows in the second wire?

23. (II) What is the maximum current that a wire can carry if an experimenter is performing an experiment 1.0 m away that deals with the Earth's magnetic field, which she wishes to measure to ± 1 percent?

24. (II) What is the acceleration (in g's) of a 175-g model airplane charged to 18.0 C and traveling at 1.8 m/s as it passes within 8.6 cm of a wire, nearly parallel to its path, carrying a 30-A current?

25. (II) A horizontal compass is placed 20 cm due south from a straight vertical wire carrying a 30-A current downward. In what direction does the compass needle point at this location? Assume the horizontal component of the Earth's field at this point is 0.45×10^{-4} T and the magnetic declination is 0°.

26. (II) A long horizontal wire carries 12.0 A of current due north. What is the net magnetic field 20.0 cm due west of the wire if the Earth's field there points downward, 40° below the horizontal, and has magnitude 5.0×10^{-5} T?

27. (II) A stream of protons passes a given point in space at a rate of 10^9 protons/s. What magnetic field do they produce 2.0 m from the beam?

28. (II) Determine the magnetic field midway between two long straight wires 2.0 cm apart in terms of the current I in one when the other carries 15 A. Assume these currents are (a) in the same direction, and (b) in opposite directions.

29. (II) A long pair of wires serves to conduct 25.0 A of dc current to (and from) an instrument. If the wires are of negligible diameter but are 2.0 mm apart, what is the magnetic field 10.0 cm from their midpoint, in their plane (Fig. 20–53)? Compare to the magnetic field of the Earth.

FIGURE 20–53   Problem 29.

30. (II) A compass needle points 20° E of N outdoors. However, when it is placed 8.0 cm to the east of a vertical wire inside a building, it points 55° E of N. What is the magnitude and direction of the current in the wire? The Earth's field there is 0.50×10^{-4} T and is horizontal.

31. (II) Three long parallel wires are 38.0 cm from one another. (Looking along them, they are at three corners of an equilateral triangle.) The current in each wire is 8.00 A, but that in wire A is opposite to that in wires B and C (Fig. 20–54). Determine the magnetic force per unit length on each wire due to the other two.

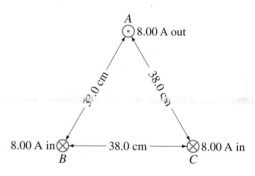

FIGURE 20–54   Problems 31 and 65.

32. (II) The magnetic field near the center of a single circular loop of radius r, carrying current I, is given by:

$$B = \frac{\mu_0 I}{2r}.$$

Assume the planetary model for the hydrogen atom, in which a single electron makes a circular orbit of radius 5.3×10^{-11} m about the nucleus. What magnitude of magnetic field would the orbiting electron produce at the nucleus?

FIGURE 20–55   Problem 33.

33. (II) Two long wires are oriented so that they are per-pendicular to each other, and at their closest, they are 20.0 cm apart (Fig. 20–55). What is the magnitude of the magnetic field at a point midway between them if the top one carries a current of 20.0 A and the bottom one carries 5.0 A?

34. (II) A long horizontal wire carries a current of 48 A. A second wire, made of 2.5-mm-diameter copper wire and parallel to the first but 15 cm below it, is held in suspension magnetically (Fig. 20–56). (a) What is the magnitude and direction of the current in the lower wire? (b) Is the lower wire in stable equilibrium? (c) Repeat parts (a) and (b) if the second wire is suspended 15 cm *above* the first due to the latter's field.

FIGURE 20–56   Problem 34.

* 35. (III) Two long parallel wires 6.00 cm apart carry 16.5-A currents in the same direction. Determine the magnetic field strength at a point 12.0 cm from one wire and 13.0 cm from the other. [*Hint*: Make a drawing in a plane containing the field lines, and recall the rules for vector addition.]

SECTION 20–8

* 36. (I) A 30.0-cm long solenoid 1.25 cm in diameter is to produce a field of 0.385 T at its center. How much current should the solenoid carry if it has 1000 turns of the wire?

* 37. (II) You have 1.0 kg of copper and want to make a practical solenoid that produces the greatest possible magnetic field. Should you make your copper wire long and thin, short and fat, or something else? Consider other variables, such as solenoid diameter, length, and so on.

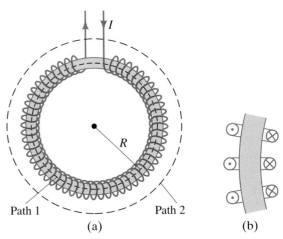

FIGURE 20–57   Problem 38. (a) A torus. (b) A section of the torus showing direction of the current for three loops: ⊙ means current toward viewer, and ⊗ means current away from viewer.

* 38. (II) A torus is a solenoid in the shape of a circle (Fig. 20–57). Use Ampère's law along the circular path, shown dashed in Fig. 20–57a, to determine that the magnetic field (a) inside the torus is $B = \mu_0 NI/2\pi R$, where N is the total number of turns, and (b) outside the torus is $B = 0$. (c) Is the field inside a torus uniform like a solenoid's? If not, how does it vary?

* 39. (II) Use Ampère's law to show that a uniform magnetic field, such as between the pole pieces of a magnet, Fig. 20–8, cannot drop abruptly to zero outside the magnet. [*Hint*: Take as your path a rectangle with one vertical side inside the field and one vertical side completely outside the field.]

* 40. (III) A current I, flowing in a long solid cylindrical wire of radius r_0, is uniform across the cross section (Fig. 20–58). (a) Use Ampère's law to show that the magnetic field inside the conductor at a distance r from the center of the conductor is

$$B = \frac{\mu_0 Ir}{2\pi r_0^2}.$$

Assume that the field lines are circles, just as they are outside the conductor. (b) Show that at the surface of the wire this agrees with the answer for the magnetic field outside of a long wire. (c) Where is the magnetic field a maximum, and what is its maximum value for a 1.0-mm-diameter wire carrying 15.0 A dc? (d) At what distance from the surface would the field be 10 percent of its maximum? [*Hint*: Make a plot of the magnetic field strength as a function of the distance perpendicularly out from the central axis of the wire.]

FIGURE 20–58
Problem 40.

*41. (I) A galvanometer needle deflects full scale for a 63.0-μA current. What current will give full-scale deflection if the magnetic field weakens to 0.860 of its original value?

*42. (I) If the restoring spring of a galvanometer weakens by 20 percent over the years, what current will give full-scale deflection if it originally required 36 μA?

*43. (I) If the current to a motor drops by 15 percent, by what factor does the output torque change?

*44. (I) A single square loop of wire 22.0 cm on a side is placed with its face parallel to the magnetic field between the pole pieces of a large magnet. When 6.30 A flows in the coil, the torque on it is 0.325 m·N. What is the magnetic field strength?

*45. (II) Show that the magnetic dipole moment M of an electron orbiting the proton nucleus of a hydrogen atom is related to the orbital momentum L of the electron by

$$M = \frac{e}{2m} L.$$

*46. (II) A circular coil 18.0 cm in diameter and containing eleven loops lies flat on the ground. The Earth's magnetic field at this location has magnitude 5.50×10^{-5} T and points into the Earth at an angle of 56.0° below a line pointing due north. If a 7.70-A counterclockwise current passes through the coil, (a) determine the torque on the coil, and (b) which edge of the coil rises up, north, east, south, or west?

*47. (II) A rectangular sample of a metal is 3.0 cm wide and 500 μm thick. When it carries a 30-A current and is placed in a 0.80-T magnetic field it produces a 6.5-μV Hall emf. Determine: (a) the Hall field in the conductor; (b) the drift speed of the conduction electrons; (c) the density of free electrons in the metal.

*48. (II) In a probe that uses the Hall effect to measure magnetic fields, a 12.0-A current passes through a 1.50-cm-wide 1.00-mm-thick strip of sodium metal. If the Hall emf is 2.42 μV, what is the magnitude of the magnetic field (take it perpendicular to the flat face of the strip)? Assume one free electron per atom of Na, and take its specific gravity to be 0.971.

*49. (II) The Hall effect can be used to measure blood flow rate because the blood contains ions that constitute an electric current. (a) Does the sign of the ions influence the emf? (b) Determine the flow velocity in an artery 3.3 mm in diameter if the measured emf is 0.10 mV and B is 0.070 T. (In actual practice, an alternating magnetic field is used.)

*50. (I) Protons move in a circle of radius 5.10 cm in a 0.566-T magnetic field. What value of electric field could make their paths straight? In what direction must it point?

*51. (I) In a mass spectrometer, germanium atoms have radii of curvature equal to 21.0, 21.6, 21.9, 22.2, and 22.8 cm. The largest radius corresponds to an atomic mass of 76 u. What are the atomic masses of the other isotopes?

*52. (II) Suppose the electric field between the electric plates in the mass spectrometer of Fig. 20–38 is 2.48×10^4 V/m and the magnetic fields $B = B' = 0.68$ T. The source contains carbon isotopes of mass numbers 12, 13, and 14 from a long-dead piece of a tree. (To estimate atomic masses, multiply by 1.67×10^{-27} kg.) How far apart are the lines formed by the singly charged ions of each type on the photographic film? What if the ions were doubly charged?

*53. (II) (a) What value of magnetic field would make a beam of electrons, traveling to the right at a speed of 4.8×10^6 m/s, go undeflected through a region where there is a uniform electric field of 10,000 V/m pointing vertically up? (b) What is the direction of the magnetic field if it is known to be perpendicular to the electric field? (c) What is the frequency of the circular orbit of the electrons if the electric field is turned off?

*54. (II) A mass spectrometer is being used to monitor air pollutants. It is difficult, however, to separate molecules with nearly equal mass such as CO (28.0106 u) and N_2 (28.0134 u). How large a radius of curvature must a spectrometer have if these two molecules are to be separated on the film by 0.50 mm?

*55. (II) One form of mass spectrometer accelerates ions by a voltage V before they enter a magnetic field B. The ions are assumed to start from rest. Show that the mass of an ion is $m = qB^2R^2/2V$, where R is the radius of the ions' path in the magnetic field and q is their charge.

*56. (II) An iron-core solenoid is 36 cm long, 1.5 cm in diameter, and has 600 turns of wire. A magnetic field of 1.8 T is produced when 40 A flows in the wire. What is the permeability μ at this high field strength?

57. Protons with momentum 4.8×10^{-16} kg·m/s are magnetically steered clockwise in a circular path 2.0 km in diameter at Fermi National Accelerator Laboratory in Illinois. What is the magnitude and direction of the field in the magnets surrounding the beam pipe?

58. A rectangular loop of wire is sitting next to a straight wire, as shown in Fig. 20–59. There is a current of 2.5 A in both wires. What is the magnitude and direction of the net force on the loop?

FIGURE 20–59   Problem 58.

59. A proton and an electron have the same kinetic energy upon entering a region of constant magnetic field. What is the ratio of the radii of their circular paths?

60. Near the equator, the Earth's magnetic field points almost horizontally to the north and has magnitude $B = 0.50 \times 10^{-4}$ T. What should be the magnitude and direction for the velocity of an electron if its weight is to be exactly balanced by the magnetic force?

61. Calculate the force on an airplane which has acquired a net charge of 155 C and moves with a speed of 120 m/s perpendicular to the Earth's magnetic field of 5.0×10^{-5} T.

62. The power cable for an electric trolley (Fig. 20–60) carries a horizontal current of 330 A toward the east. The Earth's magnetic field has a strength 5.0×10^{-5} T and makes an angle of dip of 22° at this location. Calculate the magnitude and direction of the magnetic force on a 10-m length of this cable.

63. A doubly charged helium atom, whose mass is 6.6×10^{-27} kg, is accelerated by a voltage of 2400 V. (a) What will be its radius of curvature in a uniform 0.240-T field? (b) What is its period of revolution?

64. A straight 1.00-mm-diameter copper wire can just "float" horizontally in air because of the force of the Earth's magnetic field **B** which is horizontal and of magnitude 5.00×10^{-5} T. What current does the wire carry?

FIGURE 20–60   Problem 62.

65. In Fig. 20–54 the top wire is 2.00-mm-diameter copper wire and is suspended in air due to the two magnetic forces from the bottom two wires. The current flow through the two bottom wires is 20.0 A in each. Calculate the required current flow in the suspended wire.

66. Two stiff parallel wires a distance l apart in a horizontal plane act as rails to support a light metal rod of mass m (perpendicular to each rail), Fig. 20–61. A magnetic field **B**, directed vertically upward (outward in diagram), acts throughout. At $t = 0$, wires connected to the rails are connected to a constant current source and a current I begins to flow through the system. Determine the speed of the rod as a function of time (a) assuming no friction between the rod and the rails, and (b) if the coefficient of friction is μ_k. (c) In which direction does the rod move, east or west, if the current through it heads north?

FIGURE 20–61   Looking down on a rod sliding on rails. Problem 66.

67. Estimate the approximate maximum deflection of the electron beam near the center of a TV screen due to the Earth's 5.0×10^{-5} T field. Assume the screen is 20 cm from the electron gun where the electrons are accelerated (a) by 2.0 kV, or (b) by 30 kV. Note that in color TV sets, the beam must be directed accurately to within less than 1 mm in order to strike the correct phosphor. Because the Earth's field is significant here, mu-metal shields are used to reduce the Earth's field in the CRT. (See Section 17–10.)

68. An electron enters a large solenoid at a 7.0° angle to the axis. If the field is a uniform 3.3×10^{-2} T, determine the radius and pitch (distance between loops) of the electron's helical path if its speed is 1.8×10^7 m/s.

69. The cyclotron (Fig. 20–62) is a device used to accelerate elementary particles such as protons to high speeds. Particles starting at point A with some initial velocity travel in circular orbits in the magnetic field B. The particles are accelerated to higher speeds each time they pass in the gap between the metal "dees," where there is an electric field E. (There is no electric field within the cavity of the metal dees.) The electric field changes direction each half-cycle, owing to an ac voltage $V = V_0 \sin 2\pi ft$, so that the particles are increased in speed at each passage through the gap. (a) Show that the frequency f of the voltage must be $f = Bq/2\pi m$, where q is the charge on the particles and m their mass. (b) Show that the kinetic energy of the particles increases by $2qV_0$ each revolution, assuming that the gap is small. (c) If the radius of the cyclotron is 2.0 m and the magnetic field strength is 0.50 T, what will be the maximum kinetic energy of accelerated protons in MeV? (d) How is a cyclotron like a swing?

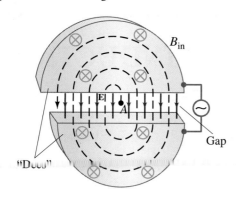

FIGURE 20-62   A cyclotron. Problem 69.

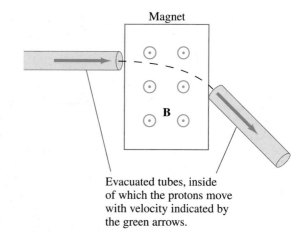

Evacuated tubes, inside of which the protons move with velocity indicated by the green arrows.

FIGURE 20-63   Problem 70.

70. Magnetic fields are very useful in particle accelerators for "beam steering"; that is, the magnetic fields can be used to change the beam's direction without altering its speed (Fig. 20–63). Show how this works with a beam of protons. What happens to protons that are not moving with the speed that the magnetic field is designed for? If the field extends over a region 5.0 cm wide and has a magnitude of 0.33 T, by approximately what angle will a beam of protons traveling at 1.0×10^7 m/s be bent?

71. A square loop of aluminum wire is 20.0 cm on a side. It is to carry 25.0 A and rotate in a 1.65-T magnetic field. (a) Determine the minimum diameter of the wire so that it will not fracture from tension or shear. Assume a safety factor of 10. (See Table 9–2.) (b) What is the resistance of a single loop of this wire?

72. The magnetic field B at the center of a circular coil of wire carrying a current I is

$$B = \frac{\mu_0 NI}{2r},$$

where N is the number of loops in the coil and r is its radius. Suppose that an electromagnet uses a coil 1.2 m in diameter made from square copper wire 1.6 mm on a side. The power supply produces 120 V at a maximum power output of 4.0 kW. (a) How many turns are needed to run the power supply at maximum power? (b) What is the magnetic field strength at the center of the coil? (c) If you use a greater number of turns and this same power supply (so the voltage remains at 120 V), will a greater magnetic field strength result? Explain.

Transmission lines carry electric power over great distances, at very high voltage for greater efficiency. To reduce high voltage to usable voltage, transformers are used, whose operation depends on electromagnetic induction. Induction is also the basis for electric generators, which produce the electric power in the first place.

21 ELECTROMAGNETIC INDUCTION AND FARADAY'S LAW; AC CIRCUITS

In Chapter 20, we discussed two ways in which electricity and magnetism are related: (1) an electric current produces a magnetic field; and (2) a magnetic field exerts a force on an electric current or moving electric charge. These discoveries were made in 1820–1821. Scientists then began to wonder: if electric currents produce a magnetic field, is it possible that a magnetic field can produce an electric current? Ten years later the American Joseph Henry (1797–1878) and the Englishman Michael Faraday (1791–1867) independently found that it was possible. Henry actually made the discovery first. But Faraday published his results earlier and investigated the subject in more detail. We now discuss this phenomenon and some of its world-changing applications.

21–1 Induced EMF

In his attempt to produce an electric current from a magnetic field, Faraday used an apparatus like that shown in Fig. 21–1. A coil of wire, X, was connected to a battery. The current that flowed through X produced a magnetic field that was intensified by the iron core. Faraday hoped that by using a

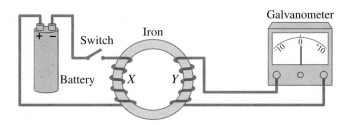

FIGURE 21–1 Faraday's experiment to induce an emf.

strong enough battery, a steady current in X would produce a great enough magnetic field to produce a current in a second coil Y. This second circuit, Y, contained a galvanometer to detect any current but contained no battery. He met no success with steady currents. But the long-sought effect was finally observed when Faraday saw the galvanometer in circuit Y deflect strongly at the moment he closed the switch in circuit X. And the galvanometer deflected strongly in the opposite direction when he opened the switch. A *steady* current in X had produced *no* current in Y. Only when the current in X was starting or stopping was a current produced in Y.

Constant **B** *induces no* emf

Faraday concluded that although a steady magnetic field produces no current, a *changing* magnetic field can produce an electric current! Such a current is called an **induced current**. When the magnetic field through coil Y changes, a current flows as if there were a source of emf in the circuit. We therefore say that an

induced emf is produced by a changing magnetic field.

Faraday did further experiments on **electromagnetic induction**, as this phenomenon is called. For example, Fig. 21–2 shows that if a magnet is moved quickly into a coil of wire, a current is induced in the wire. If the magnet is quickly removed, a current is induced in the opposite direction. Furthermore, if the magnet is held steady and the coil of wire is moved toward or away from the magnet, again an emf is induced and a current flows. Motion or change is required to induce an emf. It doesn't matter whether the magnet or the coil moves.

Changing **B** *induces an emf*

FIGURE 21–2 (a) A current is induced when a magnet is moved toward a coil. (b) The induced current is opposite when the magnet is moved away from the coil. Note that the galvanometer zero is at the center of the scale and the needle deflects left or right, depending on the direction of the current. In (c) no current is induced if the magnet does not move relative to the coil.

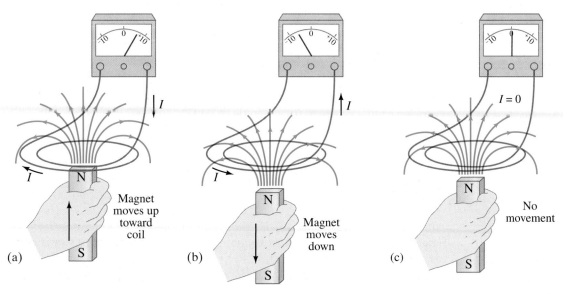

21–2 Faraday's Law of Induction; Lenz's Law

Faraday investigated quantitatively what factors influence the magnitude of the emf induced. He found first of all that it depends on time: the more rapidly the magnetic field changes, the greater the induced emf. But the emf is not simply proportional to the rate of change of the magnetic field, **B**. Rather it is proportional to the rate of change of the **magnetic flux**, Φ_B, passing through the loop of area A, which is defined as

Magnetic flux defined

$$\Phi_B = B_\perp A = BA \cos\theta. \tag{21–1}$$

Here $B_\perp$ is the component of the magnetic field **B** perpendicular to the face of the coil, and θ is the angle between **B** and a line drawn perpendicular to the face of the coil. These quantities are shown in Fig. 21–3 for a square coil of side l whose area $A = l^2$. When the face of the coil is parallel to **B**, $\theta = 90°$ and $\Phi_B = 0$. When **B** is perpendicular to the coil, $\theta = 0°$ and

$$\Phi_B = BA. \qquad\qquad [\textbf{B} \perp \text{coil face}]$$

As we saw earlier, the lines of **B** (like lines of **E**) can be drawn such that the number of lines per unit area is proportional to the field strength. Then the flux Φ_B can be thought of as being proportional to the *total number of lines passing through the coil*. This is illustrated in Fig. 21–4, where the coil is viewed from the side (on edge). For $\theta = 90°$, no lines pass through the coil and $\Phi_B = 0$, whereas Φ_B is a maximum when $\theta = 0°$. The unit of magnetic flux is the tesla-meter^2; this is called a **weber**: $1\ \text{Wb} = 1\ \text{T}\cdot\text{m}^2$.

With this definition of the flux, we can now write down the results of Faraday's investigations. If the flux through N loops of wire changes by an amount $\Delta\Phi_B$ during a time Δt, the average induced emf during this time is

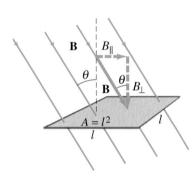

FIGURE 21–3 Determining the flux through a flat loop of wire. This loop is square, of side l and area $A = l^2$.

FARADAY'S LAW OF INDUCTION

$$\mathcal{E} = -N\frac{\Delta\Phi_B}{\Delta t}. \tag{21–2}$$

This fundamental result is known as **Faraday's law of induction**, and is one of the basic laws of electromagnetism.

The minus sign in Eq. 21–2 is placed there to remind us in which direction the induced emf acts. Experiments show that

Lenz's law

> **an induced emf always gives rise to a current whose magnetic field opposes the original change in flux.**

This is known as **Lenz's law**. Let us apply it to the case of relative motion between a magnet and a coil, Fig. 21–2. The changing flux induces an emf, which produces a current in the coil. And this induced current produces its own magnetic field. In Fig. 21–2a the distance between the coil and the magnet decreases. So the magnetic field, and therefore the flux, through the coil increases. The magnetic field of the magnet points upward. To oppose this upward increase, the field produced by the induced current points *downward*. Thus, Lenz's law tells us that the current moves as shown (use the right-hand rule). In Fig. 21–2b, the flux *decreases* (because the magnet is moved away), so the induced current produces an *upward* magnetic field that is "trying" to maintain the status quo. Thus the current is as shown.

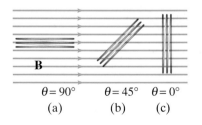

FIGURE 21–4 Magnetic flux Φ_B is proportional to the number of lines of **B** that pass through the loop.

$\theta = 90°$ $\theta = 45°$ $\theta = 0°$
(a)   (b)   (c)

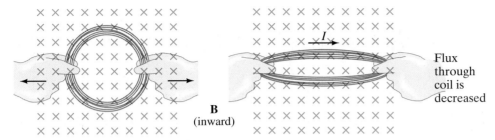

FIGURE 21–5 A current can be induced by changing the area of the coil. In both this case and that of Fig. 21–6, the flux through the coil is reduced. Here the brief induced current acts in the direction shown so as to try to maintain the original flux ($\Phi = BA$) by producing its own magnetic field into the page. That is, as the area A decreases, the current acts to increase B in the original (inward) direction.

Let us consider what would happen if Lenz's law were not true, but were just the reverse. The induced current in this imaginary situation would produce a flux in the same direction as the original change. This greater change in flux would produce an even larger current followed by a still greater change in flux, and so on. The current would continue to grow indefinitely, producing power ($= I^2R$) even after the original stimulus ended. This would violate the conservation of energy. Such "perpetual motion" devices do not exist. Thus, Lenz's law as stated above (and not its opposite) is consistent with the law of conservation of energy.

It is important to note that an emf is induced whenever there is a change in flux. Since magnetic flux $\Phi_B = BA \cos \theta$, we see that an emf can be induced in three ways: (1) by a changing magnetic field B; (2) by changing the area of the loop in the field; or (3) by changing the loop's orientation θ with respect to the field. Figures 21–1 and 21–2 illustrated case 1. Examples of cases 2 and 3 are illustrated in Figs. 21–5 and 21–6, respectively.

CONCEPTUAL EXAMPLE 21–1 **Induction stove.** Some modern stove burners are based on induction. That is, an ac current passes around a coil that is the "burner" (a burner that never gets hot). Why will it heat a metal pan but not a glass container?

PHYSICS APPLIED

Induction stove

RESPONSE The ac current sets up a changing magnetic field that passes through the pan bottom. This changing field induces a current through the pan bottom, and since the pan offers resistance, electric energy is transformed to heat, heating the pot and its contents. A glass container offers very high resistance so very little current is induced and very little energy transferred. Recall Eq. 18–6c, $P = V^2/R$.

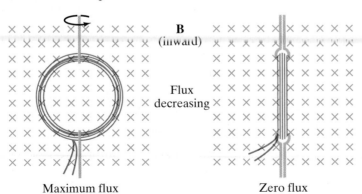

FIGURE 21–6 A current can be induced by rotating a coil in a magnetic field.

(a)	(b)	(c)	(d)	(e)

(a)
N magnetic pole moving toward coil from above the page

(b)
N magnetic pole moving toward the coil in the plane of the page

(c)
Pulling the coil to the right out of a magnetic field that points out of the page

(d)
Shrinking a coil in a magnetic field pointing into the page

(e)
Rotating the coil about the vertical diameter by pulling the left side toward the reader and pushing the right side away from the reader in a magnetic field that points from right to left in the plane of the page

FIGURE 21–7 Example 21–2.

CONCEPTUAL EXAMPLE 21–2 **Practice with Lenz's law.** In which direction is the current induced in the coil for each situation in Fig. 21–7?

RESPONSE (*a*) Magnetic field lines point out from the N pole of a magnet, so as the magnet moves down toward the coil, the field points into the page and is getting stronger. The current will be induced in the counterclockwise direction to produce a field **B** *out* of the page so that its own flux counteracts the externally imposed change.
(*b*) The field is in the plane of the page, so the flux through the coil is zero throughout the process; hence there is no change in magnetic flux with time, and there will be no induced emf or current in the coil.
(*c*) Initially, the magnetic flux pointing out of the page passes through the coil. If you remove the coil, the induced current will be in a direction to make up the deficiency: the current flow will be counterclockwise to produce an outward (toward the reader) magnetic field.
(*d*) The flux is into the page and the coil area shrinks so the flux will decrease; hence the induced current will be clockwise to try to produce its own flux into the page to make up for the flux decrease.
(*e*) Initially there is no flux through the coil (why?). When you start to rotate the coil, the flux begins passing through the coil increasing to the left. To counteract this, the coil will have current induced in a counterclockwise direction so as to produce its own flux to the right.

EXAMPLE 21–3 **Pulling a coil from a magnetic field.** A square coil of side 5.0 cm contains 100 loops and is positioned perpendicular to a uniform 0.60-T magnetic field, as shown in Fig. 21–8. It is quickly and uniformly pulled from the field (moving perpendicular to **B**) to a region where B drops abruptly to zero. It takes 0.10 s for the whole coil to reach the field-free region. Find (*a*) the change in flux through the coil, (*b*) the emf and current induced, and (*c*) how much energy is dissipated in the coil if its resistance is 100 Ω. (*d*) What was the average force required?

SOLUTION (*a*) First we find how the magnetic flux, $\Phi_B = BA$, changes during the time interval $\Delta t = 0.10$ s. The area of the coil is $A = (0.050\,\text{m})^2 = 2.5 \times 10^{-3}\,\text{m}^2$. The flux is initially $\Phi_B = BA = (0.60\,\text{T})(2.5 \times 10^{-3}\,\text{m}^2) = 1.5 \times 10^{-3}\,\text{Wb}$. After 0.10 s, the flux is zero. Hence the change in flux is

$$\Delta\Phi_B = 0 - 1.5 \times 10^{-3}\,\text{Wb} = -1.5 \times 10^{-3}\,\text{Wb}.$$

(*b*) The rate of change of flux is constant during the 0.10 s, so the emf induced (Eq. 21–2) during this period is

$$\mathscr{E} = -(100)\frac{(0 - 1.5 \times 10^{-3}\ \text{Wb})}{(0.10\ \text{s})} = 1.5\ \text{V}.$$

The current is

$$I = \frac{\mathscr{E}}{R} = \frac{1.5\ \text{V}}{100\ \Omega} = 15\ \text{mA}.$$

(c) The total energy dissipated is

$$E = Pt = I^2 Rt = (1.5 \times 10^{-2}\ \text{A})^2(100\ \Omega)(0.10\ \text{s}) = 2.3 \times 10^{-3}\ \text{J}.$$

(d) From the conservation of energy principle, the result in (c) is equal to the work W needed to pull the coil out of the field. Since $W = \bar{F}d$, the average force is

$$\bar{F} = \frac{W}{d} = \frac{2.3 \times 10^{-3}\ \text{J}}{5.0 \times 10^{-2}\ \text{m}} = 0.046\ \text{N},$$

where $d = 5.0\,\text{cm}$ because there is no flux change (hence no force) until one edge of the coil leaves the field.

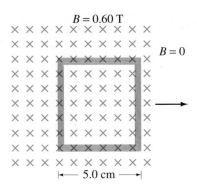

FIGURE 21–8 Example 21–3. The square coil in a magnetic field $B = 0.60\,\text{T}$ is pulled abruptly to the right to a region where $B = 0$.

21–3 EMF Induced in a Moving Conductor

Another way to induce an emf is shown in Fig. 21–9, and this situation helps illuminate the nature of the induced emf. Assume that a uniform magnetic field **B** is perpendicular to the area bounded by the U-shaped conductor and the movable rod resting on it. If the rod is made to move at a speed v, it travels a distance $\Delta x = v\,\Delta t$ in a time Δt. Therefore, the area of the loop increases by an amount $\Delta A = l\,\Delta x = lv\,\Delta t$ in a time Δt. By Faraday's law, there is an induced emf $\mathscr{E}$ whose magnitude is given by

$$\mathscr{E} = \frac{\Delta\Phi_B}{\Delta t} = \frac{B\,\Delta A}{\Delta t} = \frac{Blv\,\Delta t}{\Delta t} = Blv. \qquad (21\text{–}3)$$

This equation is valid as long as B, l, and v are mutually perpendicular. (If they are not, we use only the components of each that are mutually perpendicular.) An emf induced in this way is sometimes called *motional emf*.

Motional emf

We can also obtain Eq. 21–3 without using Faraday's law. We saw in Chapter 20 that a charged particle moving perpendicular to a magnetic field B with speed v experiences a force $F = qvB$. When the rod of Fig. 21–9 moves to the right with speed v, the electrons in the rod move with this same speed. Therefore, each feels a force $F = qvB$, which acts upward in the figure. If the rod were not in contact with the U-shaped conductor, electrons would collect at the upper end of the rod, leaving the lower end positive. There must thus be an induced emf. If the rod does slide on the U-shaped conductor, the electrons will flow into it. There will then be a clockwise (conventional) current flowing in the loop. To calculate the emf, we determine the work W needed to move a charge q from one end of the rod to the other against this potential difference: $W = $ force $\times$ distance $= (qvB)(l)$. The emf equals the work done per unit charge, so $\mathscr{E} = W/q = qvBl/q = Blv$, just as above.[†]

FIGURE 21–9 A conducting rod is moved to the right on a U-shaped conductor in a uniform magnetic field **B** that points out of the paper.

[†]This argument, which is basically the same as for the Hall effect, explains this one way of inducing an emf. It does not explain the general case of electromagnetic induction, however.

FIGURE 21–10 Example 21–4.

Does a moving airplane develop a dangerous emf? An airplane travels 1000 km/h in a region where the Earth's field is 5.0×10^{-5} T and is nearly vertical (Fig. 21–10). What is the potential difference induced between the wing tips that are 70 m apart?

SOLUTION Since $v = 1000 \text{ km/h} = 280 \text{ m/s}$, and $\mathbf{v} \perp \mathbf{B}$, we have

$$\mathcal{E} = Blv = (5.0 \times 10^{-5} \text{ T})(70 \text{ m})(280 \text{ m/s}) = 1.0 \text{ V}.$$

Not much to worry about.

21–4 Changing Magnetic Flux Produces an Electric Field

As we just discussed, the electrons in the moving conductor of Fig. 21–9 feel a force. This implies that there is an electric field in the conductor. Since electric field is defined as the force per unit charge, $E = F/q$, the effective field E in the rod must be (since $F = qvB$)

Electric field is produced by a changing magnetic flux

$$E = \frac{F}{q} = \frac{qvB}{q} = vB. \qquad (21\text{–}4)$$

In the situation where a changing magnetic field (rather than a moving conductor) induces an emf (as, for example, in Fig. 21–2), there also is an induced current. And again this implies that there is an electric field in the wire. Thus we come to the important conclusion that

a changing magnetic flux produces an electric field.

This applies not only to wires and other conductors, but is a general result that applies to any region in space: an electric field will be produced at any point in space where there is a changing magnetic field.

⇒ **PHYSICS APPLIED**

Blood flow measurement

FIGURE 21–11 Measurement of blood velocity from the induced emf.

Electromagnetic blood-flow measurement. The rate of blood flow can be measured using the apparatus shown in Fig. 21–11, since blood contains charged ions. Suppose that the blood vessel is 2.0 mm in diameter, the magnetic field is 0.080 T, and the measured emf is 0.10 mV. What is the flow velocity of the blood?

SOLUTION We solve for v in Eq. 21–3, and we find that

$$v = \frac{\mathcal{E}}{Bl} = \frac{(1.0 \times 10^{-4} \text{ V})}{(0.080 \text{ T})(2.0 \times 10^{-3} \text{ m})} = 0.63 \text{ m/s}.$$

(In actual practice, an alternating current is used to produce an alternating magnetic field. The induced emf is then alternating.)

21–5 Electric Generators

Probably the most important practical result of Faraday's great discovery was the development of the **electric generator** or **dynamo**. A generator transforms mechanical energy into electric energy. This is just the opposite of what a motor does. Indeed, a generator is basically the inverse of a motor.[†] A simplified diagram of an **ac generator** is shown in Fig. 21–12. A generator consists of many coils of wire (only one is shown) wound on an armature that can rotate in a magnetic field. The axle is turned by some mechanical means (falling water, car motor belt), and an emf is induced in the rotating coil. An electric current is thus the *output* of a generator. In Fig. 21–12 the right-hand rule tells us that, with the armature rotating counterclockwise, the (conventional) current in the wire labeled *a* on the armature is outward; therefore it is outward at brush *a*. (Each brush presses against a continuous slip ring.) After one-half revolution, wire *a* will be where wire *b* is now in the drawing, and the current then at brush *a* will be inward. Thus the current produced is alternating. Let us look at this in more detail.

In Fig. 21–13, the loop is being made to rotate clockwise in a uniform magnetic field **B**. The velocity of the two lengths *ab* and *cd* at this instant are shown. Although the sections of wire *bc* and *da* are moving, the force on electrons in these sections is toward the side of the wire, not along its length. The emf generated is thus due only to the force on charges in the sections *ab* and *cd*. From the right-hand rule, we see that the direction of the induced current in *ab* is from *a* toward *b*. And in the lower section, it is from *c* to *d*; so the flow is continuous in the loop. The magnitude of the emf generated in *ab* is given by Eq. 21–3, except that we must take the component of the velocity perpendicular to *B*:

$$\mathscr{E} = Blv_{\perp},$$

where *l* is the length of *ab*. From the diagram we can see that $v_{\perp} = v \sin \theta$, where θ is the angle the face of the loop makes with the vertical. The emf induced in *cd* has the same magnitude and is in the same direction. Therefore they add, and the total emf is

$$\mathscr{E} = 2NBlv \sin \theta,$$

where we have multiplied by *N*, the number of loops in the coil (if there is more than one). If the coil is rotating with constant angular velocity ω, then the angle $\theta = \omega t$. We also have from the angular equations (Chapter 8) that $v = \omega r = \omega(h/2)$, where *h* is the length of *bc* or *ad*. Thus $\mathscr{E} = 2NBl\omega(h/2)\sin \omega t$, or

$$\mathscr{E} = NBA\omega \sin \omega t, \qquad \text{(21–5)}$$

where $A = lh$ is the area of the loop. This equation holds for any shape coil, not just for a rectangle as derived. Thus, the output emf of the generator is sinusoidally alternating (Fig. 21–14 and Section 18–8). Since ω is expressed in radians per second, we can write $\omega = 2\pi f$, where *f* is the frequency.

➡ **PHYSICS APPLIED**

AC generator

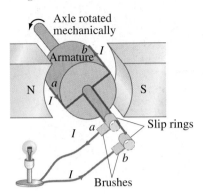

FIGURE 21–12 An ac generator.

FIGURE 21–13 The emf is induced in the segments *ab* and *cd*, whose velocity components perpendicular to the field **B** are $v \sin \theta$.

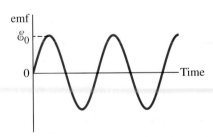

FIGURE 21–14 An ac generator produces an alternating current. The output emf $\mathscr{E} = \mathscr{E}_0 \sin \omega t$, where $\mathscr{E}_0 = NAB\omega$ (Eq. 21–5).

[†]You can, for example, actually run a car generator backward as a motor by connecting its output terminals to a battery.

FIGURE 21-15 Water-driven generators at the base of Boulder Dam, Nevada.

Over 99 percent of the electricity used in the United States is produced from generators (Fig. 21–15). The frequency f is 60 Hz for general use in the United States and Canada, although 50 Hz is used in many countries. In electric power generating plants, the armature is mounted on a heavy axle connected to a turbine, which is the modern equivalent of a waterwheel. Water falling over a dam can turn the turbine at a hydroelectric plant. Most of the power generated at present in the United States, however, is done at steam plants, where the burning of fossil fuels (coal, oil, natural gas) boils water to produce high-pressure steam that turns the turbines. Likewise, at nuclear power plants, the nuclear energy released is used to produce steam to turn turbines. Thus, a heat engine (Chapter 15) connected to a generator is the principal means of generating electric power.

The frequency of 60 Hz is maintained very precisely by power companies, and in doing problems, we will assume it is at least as precise as other numbers given.

EXAMPLE 21–6 **An ac generator.** The armature of a 60-Hz ac generator rotates in a 0.15-T magnetic field. If the area of the coil is $2.0 \times 10^{-2}\,\text{m}^2$, how many loops must the coil contain if the peak output is to be $\mathcal{E}_0 = 170\,\text{V}$?

SOLUTION From Eq. 21–5, we see that the maximum emf is $\mathcal{E}_0 = NBA\omega$. Since $\omega = 2\pi f = (6.28)(60\,\text{s}^{-1}) = 377\,\text{s}^{-1}$, we have

$$N = \frac{\mathcal{E}_0}{BA\omega} = \frac{170\,\text{V}}{(0.15\,\text{T})(2.0 \times 10^{-2}\,\text{m}^2)(377\,\text{s}^{-1})} = 150\,\text{turns}.$$

A **dc generator** is much like an ac generator, except the slip rings are replaced by split-ring commutators, Fig. 21–16a, just as in a dc motor. The output of such a generator is as shown and can be smoothed out by placing a capacitor in parallel with the output (Section 19–7). More common is the use of many armature windings, as in Fig. 21–16b, which produces a smoother output.

In the past, automobiles used dc generators. More common now, however, are ac generators or **alternators**, which avoid the problems of wear and electrical arcing (sparks) across the split-ring commutators of dc generators. Alternators differ from the generators discussed above in the following way. In an alternator, current from the battery produces a magnetic field in an electromagnet, called the *rotor*, which is made to rotate by a belt from the engine. Surrounding the rotating rotor are a set of stationary

FIGURE 21-16 (a) A dc generator with one set of commutators, and (b) a dc generator with many sets of commutators and windings.

(a)

(b)

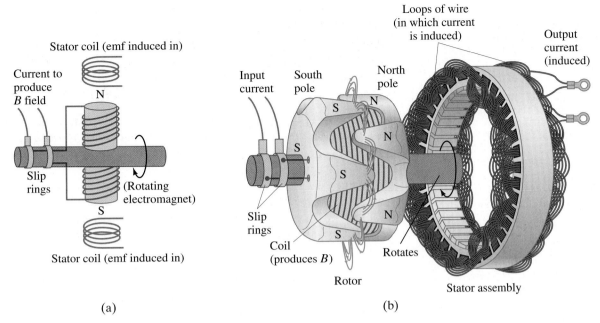

Stator coil (emf induced in)

Current to produce B field

N

Slip rings

S

(Rotating electromagnet)

Stator coil (emf induced in)

(a)

Loops of wire (in which current is induced)

Output current (induced)

Input current

South pole

North pole

S

N

N

S

S

S

N

S

Slip rings

Coil (produces B)

Rotates

Rotor

Stator assembly

(b)

FIGURE 21–17 (a) Schematic (simplified) diagram of an alternator. The input electromagnet current to the rotor is connected through continuous slip rings. Sometimes the rotor electromagnet is replaced by a permanent magnet. (b) Actual shape of an alternator. The rotor is made to turn by a belt from the engine. The current in the wire coil of the rotor produces a magnetic field inside it on its axis that points horizontally from left to right, thus making north and south poles of the plates attached at either end. These end plates are made with triangular fingers that are bent over the coil—hence there are alternating N and S poles quite close to one another, with magnetic field lines between them as shown by the blue lines. As the rotor turns, these field lines pass through the fixed stator coils (shown on the right for clarity, but in operation the rotor rotates within the stator), inducing a current in them, which is the output.

coils called the *stator*, Fig. 21–17. The magnetic field of the rotor passes through the stator coils and, since the rotor is rotating, the field through the fixed stator coils is changing. Hence an alternating current is induced in the stator coils, which is the output. This ac output is changed to dc for charging the battery by the use of semiconductor diodes, which allow current flow in one direction only (see Section 29–8).

Output emf of alternator

*21–6 Counter EMF and Torque; Eddy Currents

A motor turns and produces mechanical energy when a current is made to flow in it. From our description in Section 20–10 of a simple dc motor, you might expect that the armature would accelerate indefinitely due to the torque on it. However, as the armature of the motor turns, the magnetic flux through the coil changes and an emf is generated. This induced emf acts to oppose the motion (Lenz's law) and is called the **back emf** or **counter emf**. The greater the speed of the motor, the greater the counter emf. A motor normally turns and does work on something, but if there were no load, the motor's speed would increase until the counter emf equaled the input voltage. In the normal situation, when there is a mechanical load, the speed of the motor is limited also by the load. The counter emf will then be less than the external voltage. The greater the mechanical load, the slower the motor rotates and the lower is the counter emf ($\mathscr{E} \propto \omega$, Eq. 21–5).

Back emf

Windings of motor 5.0 Ω

Counter emf induced in armature winding

$\mathscr{E}_{induced} = 108$ V

$\mathscr{E} = 120$ V

FIGURE 21–18 Circuit of a motor showing induced counter emf.

Effect of back emf on current

EXAMPLE 21–7 **Counter emf in a motor.** The armature windings of a dc motor have a resistance of 5.0 Ω. The motor is connected to a 120-V line, and when the motor reaches full speed against its normal load, the counter emf is 108 V. Calculate (a) the current into the motor when it is just starting up, and (b) the current when it reaches full speed.

SOLUTION (a) Initially, the motor is not turning (or turning very slowly), so there is no induced counter emf. Hence, from Ohm's law, the current is

$$I = \frac{V}{R} = \frac{120 \text{ V}}{5.0 \ \Omega} = 24 \text{ A}.$$

(b) At full speed, the counter emf is a source of emf that opposes the exterior source. We represent this counter emf as a battery in the equivalent circuit shown in Fig. 21–18. In this case, Ohm's law (or Kirchhoff's rule) gives

$$120 \text{ V} - 108 \text{ V} = I \ (5.0 \ \Omega).$$

Therefore

$$I = 12 \text{ V}/5.0 \ \Omega = 2.4 \text{ A}.$$

This Example illustrates the fact that the current is very high when a motor first starts up. This is why the lights in your house may dim when the motor of the refrigerator (or other large motor) starts up. The large initial current causes the voltage at the outlets to drop (the house wiring has resistance, so there is some voltage drop across it when large currents are drawn).

CONCEPTUAL EXAMPLE 21–8 **Motor overload.** When using an appliance such as a blender, electric drill, or sewing machine, if the appliance is overloaded or jammed so that the motor slows appreciably or stops while the power is still connected, the device can burn out and be ruined. Explain why this happens.

RESPONSE The motors are designed to run at a certain speed for a given applied voltage and the designer must take the expected counter emf into account. If the rotation speed is reduced, the counter emf will not be as high as expected ($\mathscr{E} \propto \omega$, Eq. 21–5), and the current will increase, and may become large enough that the windings of the motor heat up to the point of ruining the motor.

In a generator, the situation is the reverse of that for a motor. As we saw, the mechanical turning of the armature induces an emf in the loops, which is the output. If the generator is not connected to an external circuit, the emf exists at the terminals but no current flows. In this case, it takes little effort to turn the armature. But if the generator *is* connected to a device that draws current, then a current flows in the coils of the armature. Because this current-carrying coil is in a magnetic field, there will be a torque exerted on it (as in a motor), and this torque opposes the motion (use the right-hand rule for the force on a wire, in Fig. 21–12). This is called a **counter torque**. The greater the electrical load—that is, the more current that is drawn—the greater will be the counter torque. Hence the external applied torque will have to be greater to keep the generator turning.

Burning out a motor

Counter torque

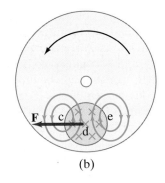

(a)   **B** (inward)   (b)

FIGURE 21–19 Production of eddy currents in a rotating wheel.

This of course makes sense from the conservation-of-energy principle. More mechanical-energy input is needed to produce more electrical-energy output.

Induced currents are not always confined to well-defined paths such as in wires. Consider, for example, the rotating metal wheel in Fig. 21–19a. A magnetic field is applied to a limited area as shown and points into the paper. The section of wheel in the magnetic field has an emf induced in it because the conductor is moving (carrying electrons with it). The flow of (conventional) current is upward in the region of the magnetic field (Fig. 21–19b), and the current follows a downward return path outside that region. Why? According to Lenz's law, the induced currents oppose the change that causes them. Consider the part of the wheel labeled c in Fig. 21–19b, where the magnetic field is zero but is just about to enter a region where **B** points into the page. To oppose this change, the induced current is counterclockwise to produce a field pointing out of the page (right-hand rule). Similarly region d is about to move to e, where **B** is zero; hence the current is clockwise to produce an inward field opposed to this change. These currents are referred to as **eddy currents** and can be present in any conductor that is moving across a magnetic field or through which the magnetic flux is changing. In Fig. 21–19, the magnetic field exerts a force **F** on the induced currents that opposes (use the right-hand rule) the rotational motion. Eddy currents can be used in this way as a smooth braking device on, say, a rapid-transit car. In order to stop the car, an electromagnet can be turned on that applies its field either to the wheels or to the moving steel rail below. Eddy currents can also be used to dampen (reduce) the oscillation of a vibrating system. A common example is in a galvanometer, where induced eddy currents keep the needle from overshooting or oscillating violently. Eddy currents, however, can be a problem. For example, eddy currents induced in the armature of a motor or generator produce heat ($P = I\mathscr{E}$) and waste energy. To reduce the eddy currents, the armatures are *laminated*; that is, they are made of very thin sheets of iron that are well insulated from one another. (See Fig. 21–21 in the next Section.) Thus the total path length of the eddy currents is confined to each slab, which increases the total resistance; hence the current is less and there is less wasted energy.

Eddy currents

➡ PHYSICS APPLIED

Brakes and damping

21–7 Transformers; Transmission of Power

A transformer is a device for increasing or decreasing an ac voltage. Transformers are found everywhere: in TV sets to give the high voltage needed for the picture tube, in converters for plugging in a portable "Walkman,"

➡ PHYSICS APPLIED

Transformers

FIGURE 21–20 Repairing a transformer on a utility pole.

on utility poles (Fig. 21–20) to reduce the high voltage from the electric company to that usable in houses (110 V or 220 V), and in many other applications. A **transformer** consists of two coils of wire known as the **primary** and **secondary** coils. The two coils can be interwoven (with insulated wire); or they can be linked by a soft iron core which is laminated to prevent eddy-current losses (Section 21–6), as shown in Fig. 21–21. Transformers are designed so that (nearly) all the magnetic flux produced by the current in the primary also passes through the secondary coil, and we assume this is true in what follows. We also assume that energy losses in the resistance of the coils and hysteresis in the iron can be ignored—a good approximation for real transformers, which are often better than 99 percent efficient.

When an ac voltage is applied to the primary, the changing magnetic field it produces will induce an ac voltage of the same frequency in the secondary. However, the voltage will be different according to the number of loops in each coil. From Faraday's law, the voltage or emf induced in the secondary is

$$V_S = N_S \frac{\Delta \Phi_B}{\Delta t},$$

where N_S is the number of turns in the secondary coil, and $\Delta \Phi_B / \Delta t$ is the rate at which the magnetic flux changes. The input primary voltage, V_P, is also related to the rate at which the flux changes

$$V_P = N_P \frac{\Delta \Phi_B}{\Delta t},$$

where N_P is the number of turns in the primary coil.[†] We divide these two equations, assuming little or no flux is lost, to find

Transformer equation

$$\frac{V_S}{V_P} = \frac{N_S}{N_P}. \tag{21–6}$$

This *transformer equation* tells how the secondary (output) voltage is related to the primary (input) voltage; V_S and V_P in Eq. 21–6 can be the rms values for both, or peak values for both.

If N_S is greater than N_P, we have a **step-up transformer**. The secondary voltage is greater than the primary voltage. For example, if the secondary has twice as many turns as the primary, then the secondary voltage will be twice that of the primary. If N_S is less than N_P, we have a **step-down transformer**.

Although ac voltage can be increased (or decreased) with a transformer, we don't get something for nothing. Energy conservation tells us that the power output can be no greater than the power input. A well-designed transformer can be greater than 99 percent efficient, so little energy is lost to heat. The power input thus essentially equals the power output. Since power $P = VI$ (Eq. 18–5), we have

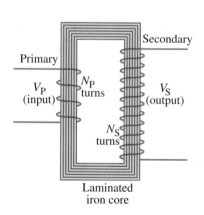

FIGURE 21–21 Step-up transformer ($N_P = 4$, $N_S = 12$).

Transformer equation II

$$V_P I_P = V_S I_S,$$

or

$$\frac{I_S}{I_P} = \frac{N_P}{N_S}. \tag{21–7}$$

[†]This follows because the changing flux produces a counter emf, $N_P \Delta \Phi_B / \Delta t$ in the primary that exactly balances the applied voltage V_P if the resistance of the primary can be ignored (Kirchhoff's rules).

EXAMPLE 21–9 **Portable radio transformer.** A transformer for home use of a portable radio reduces 120-V ac to 9.0-V ac. (Such a device also contains diodes to change the 9.0-V ac to dc. See Chapter 29.) The secondary contains 30 turns and the radio draws 400 mA. Calculate: (*a*) the number of turns in the primary; (*b*) the current in the primary; and (*c*) the power transformed.

SOLUTION (*a*) This is a step-down transformer, and from Eq. 21–6 we have

$$N_P = N_S \frac{V_P}{V_S} = \frac{(30)(120 \text{ V})}{(9.0 \text{ V})} = 400 \text{ turns.}$$

(*b*) From Eq. 21–7:

$$I_P = I_S \frac{N_S}{N_P} = (0.40 \text{ A})\left(\frac{30}{400}\right) = 0.030 \text{ A.}$$

(*c*) The power transformed is

$$P = I_S V_S = (9.0 \text{ V})(0.40 \text{ A}) = 3.6 \text{ W,}$$

which is, assuming 100 percent efficiency, the same as the power in the primary, $P = (120 \text{ V})(0.030 \text{ A}) = 3.6 \text{ W.}$

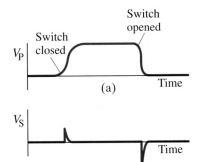

FIGURE 21–22 A dc voltage turned on and off as shown in (a) produces voltage pulses in the secondary (b). Voltage scales in (a) and (b) are not necessarily the same.

It is important to recognize that a transformer operates only on ac. A dc current in the primary does not produce a changing flux and therefore induces no emf in the secondary. However, if a dc voltage is applied to the primary through a switch, at the instant the switch is opened or closed there will be an induced current in the secondary. For example, if the dc is turned on and off as shown in Fig. 21–22a, the voltage induced in the secondary is as shown in Fig. 21–22b. Notice that the secondary voltage drops to zero when the dc voltage is steady.

Transformers play an important role in the transmission of electricity. Power plants are often situated some distance from metropolitan areas. Hydroelectric plants are located at a dam site and nuclear plants need much cooling water. Fossil-fuel plants too are often situated far from a city because of lack of availability of land or to avoid contributing to air pollution. In any case, electricity must often be transmitted over long distances (Fig. 21–23). There is always some power loss in the transmission lines, and this loss can be minimized if the power is transmitted at high voltage, using transformers, as the following Example shows.

➡ **PHYSICS APPLIED**

Transformers help power transmission

FIGURE 21–23 The transmission of electric power from power plants to homes makes use of transformers at various stages.

EXAMPLE 21–10 **Transmission lines.** An average of 120 kW of electric power is sent to a small town from a power plant 10 km away. The transmission lines have a total resistance of 0.40 Ω. Calculate the power loss if the power is transmitted at (a) 240 V and (b) 24,000 V.

SOLUTION For each case we determine the current I in the lines, and then find the power loss from $P = I^2R$. (a) If 120 kW is sent at 240 V, the total current will be

$$I = \frac{P}{V} = \frac{1.2 \times 10^5 \text{ W}}{2.4 \times 10^2 \text{ V}} = 500 \text{ A}.$$

The power loss in the lines, P_L, is then

$$P_L = I^2R = (500 \text{ A})^2(0.40 \text{ } \Omega) = 100 \text{ kW}.$$

Thus, over 80 percent of all the power would be wasted as heat in the power lines!
(b) When V = 24,000 V,

$$I = \frac{P}{V} = \frac{1.2 \times 10^5 \text{ W}}{2.4 \times 10^4 \text{ V}} = 5.0 \text{ A}.$$

The power loss is then

$$P_L = I^2R = (5.0 \text{ A})^2(0.40 \text{ } \Omega) = 10 \text{ W},$$

which is less than $\frac{1}{100}$ of 1 percent.

It should be clear that the greater the voltage, the less the current and thus the less power is wasted in the transmission lines. It is for this reason that power is usually transmitted at very high voltages, as high as 700 kV.

Power is generated at somewhat lower voltages than this, and the voltage in homes and factories is also much lower. The great advantage of ac, and a major reason it is in nearly universal use, is that the voltage can easily be stepped up and down by a transformer. The output voltage of an electric generating plant is stepped up prior to transmission. Upon arrival in a city, it is stepped down in stages at electric substations prior to distribution. The voltage in lines along city streets is typically 2400 V and is stepped down to 240 V or 120 V for home use by transformers (Fig. 21–20).

Direct current transmission has gained in popularity recently. Although changing voltage with dc is more difficult and expensive, it offers some advantages over ac. A few of these are as follows. Alternating current produces alternating magnetic fields which induce current in nearby wires and so reduce transmitted power; this is absent in dc. It is possible to transmit dc at a higher average voltage than ac since for dc, the rms value equals the peak; and breakdown of insulation or of air is determined by the peak voltage.

FIGURE 21–24 Microphone and cassette deck connected to amplifier (or receiver) and to loudspeakers.

* 21–8 Applications of Induction: Sound Systems, Computer Memory, and the Seismograph

Figure 21–24 shows several components of a sound system. Two of these components, the microphone and the tape recorder, make use of the principle of electromagnetic induction. [The output of each goes to an amplifier that amplifies (increases) the signal and sends it to a loudspeaker to be heard; alternately, the microphone signal could go directly to the tape recorder for recording.]

There are various types of *microphones*, and many operate on the principle of induction. In one form, a microphone is just the inverse of a loudspeaker (Section 20–10). A small coil connected to a membrane is suspended close to a small permanent magnet, as shown in Fig. 21–25. The coil moves in the magnetic field when sound waves strike the membrane. The frequency of the induced emf will be just that of the impinging sound waves, and this emf is the "signal" that can be amplified and sent to loud speakers, or sent to a tape recorder to be recorded on tape. In a "ribbon" microphone, a thin metal ribbon is suspended between the poles of a permanent magnet. The ribbon vibrates in response to sound waves, and the emf induced in the ribbon is proportional to its velocity.

Tape recording and tape playback is done by *tape heads* inside a tape recorder (or cassette deck). Recording tape for use in audio and video tape recorders contains a thin layer of magnetic oxide on a thin plastic tape. During recording, the audio or video signal voltage is sent to the recording head, which acts as a tiny electromagnet (Fig. 21–26) that magnetizes the tiny section of tape passing over the narrow gap in the head at each instant. In playback, the changing magnetism of the moving tape at the gap causes corresponding changes in the magnetic field within the soft-iron head, which in turn induces an emf in the coil (Faraday's law). This induced emf is the output signal that can be amplified and sent to a loudspeaker (or, in the case of a video signal, to the picture tube). In audio and video recorders, the signals are usually *analog*—they vary continuously in amplitude over time. The variation in degree of magnetization of the tape at any point reflects the variation in amplitude of the audio or video signal.

➡ **PHYSICS APPLIED**

Microphones

➡ **PHYSICS APPLIED**

Tape recording

FIGURE 21–25 Diagram of microphone that works by induction.

Electric signal
input (or output)

Coil

Recording
or playback
head

Gap

Moving magnetic tape
or disk

FIGURE 21–26 Recording and/or playback head for tape or disk. In recording (or "writing"), the electric input signal to the head, which acts as an electromagnet, magnetizes the passing tape or disk. In playback (or "reading"), the changing magnetic field of the passing tape or disk induces a changing magnetic field in the head, which in turn induces in the coil an emf that is the output signal.

Digital information, such as used on computer disks (floppy disks or hard disks) or on magnetic computer tape and some types of digital tape recorders, is read and written using heads that are basically the same as just described (Fig. 21–26). The essential difference is in the signals, which are not analog, but are digital, and in particular binary, meaning that only two values are possible for each predetermined space on the tape or disk. The two possible values are usually referred to as 1 and 0. The signal voltage does not vary continuously but rather takes on only two values, say +5 volts and 0 volts, corresponding to the 1 or 0. Thus, information is carried as a series of "bits," each of which can have only one of two values, 1 or 0.

In another field, geophysics, an important device based on electromagnetic induction is one type of *seismograph* or *geophone*. A seismograph is placed in direct contact with the Earth and converts the motion of the Earth—whether due to an earthquake or to an explosion (such as for mineral prospecting or for detecting a bomb test)—into an electrical signal. A seismograph contains a magnet and a coil of wire, one of which is fixed rigidly to the case, which moves as the Earth does where it is planted. The other element is inertial and is suspended from the case by a spring. In the type shown in Fig. 21–27, the coil moves with the Earth, and the relative motion of the magnet and coil produces an induced emf in the coil, which is the output of the device. In many geophones, the coil is inertial and the magnet moves with the Earth.

Coil

Suspension
springs

Permanent
magnet

(a)

(b)

FIGURE 21–27 (a) A seismograph or geophone. (b) A seismograph reading (Northridge, California earthquake, January 17, 1994).

* **21–9** **Inductance**

Mutual inductance. If two coils of wire are placed near one another, as in Fig. 21–28, a changing current in one will induce an emf in the other. According to Faraday's law, the emf $\mathscr{E}_2$ induced in coil 2 is proportional to the rate of change of flux passing through it. Since the flux is proportional to the current flowing in coil 1, $\mathscr{E}_2$ must be proportional to the rate of change of the current in coil 1, $\Delta I_1/\Delta t$. Thus we can write

Mutual inductance

$$\mathscr{E}_2 = -M \frac{\Delta I_1}{\Delta t}, \tag{21–8a}$$

where the constant of proportionality, M, is called the **mutual inductance.** (The minus sign is because of Lenz's law.) Mutual inductance has units of $V \cdot s/A = \Omega \cdot s$, which is called the **henry** (H), after Joseph Henry: $1\,H = 1\,\Omega \cdot s$. The value of M depends on whether iron is present or not, and on the geometry of the coil configuration: that is, on the size of the coils, on the number of turns, and on their separation. If we look at the inverse situation—a changing current in coil 2 inducing an emf in coil 1—the proportionality constant, M, turns out to have the same value,

$$\mathscr{E}_1 = -M \frac{\Delta I_2}{\Delta t}. \qquad \text{(21–8b)}$$

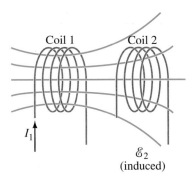

FIGURE 21–28 A changing current in one coil will induce a current in the second coil.

A transformer is an example of mutual inductance in which the coupling is maximized so that nearly all flux lines pass through both coils. However, mutual inductance has other uses as well. For example, some pacemakers, which are used to maintain blood flow in heart patients (Section 19–8), are powered externally. Power in an external coil is transmitted via mutual inductance to a second coil in the pacemaker at the heart. This has the advantage over battery-powered pacemakers in that surgery is not needed to replace a battery when it wears out.

Self-inductance. The concept of inductance applies also to an isolated single coil. When a changing current passes through a coil or solenoid, a changing magnetic flux is produced inside the coil, and this in turn induces an emf. This induced emf opposes the change in flux (Lenz's law); it is much like the back emf generated in a motor. For example, if the current through the coil is increasing, the increasing magnetic flux induces an emf that opposes the original current and tends to retard its increase. If the current is decreasing in the coil, the decreasing flux induces an emf in the same direction as the current, tending to maintain the original current. In either case, the induced emf $\mathscr{E}$ is proportional to the rate of change in current (and is in the direction opposed to the change):

$$\mathscr{E} = -L \frac{\Delta I}{\Delta t}. \qquad \text{(21–9)}$$

Self-inductance
(induced emf for an inductor)

The constant of proportionality L is called the **self-inductance**, or simply the **inductance** of the coil. It, too, is measured in henrys. The magnitude of L depends on the geometry (size and shape) and on whether an iron core is present or not.

An ac circuit always contains some inductance, but often it is quite small unless the circuit contains a coil of many turns. A coil that has significant self-inductance L is called an **inductor** or a **choke coil**. It is shown on circuit diagrams by the symbol

Inductors

$$\text{---}\mathrm{\infty\infty\infty}\text{---}\;.$$ [inductor symbol]

It can serve a useful purpose in certain circuits. Often, inductance is to be avoided in a circuit. Precision resistors are normally wire-wound and thus would have inductance as well as resistance. The inductance can be minimized by winding the insulated wire back on itself so that the current going in the two directions cancels and little magnetic flux is produced; this is called a "noninductive winding."

If an inductor has negligible resistance, it is the inductance, or the back emf, that controls the current. If a source of alternating voltage is applied to the coil, this applied voltage will just be balanced by the induced emf of the coil as given by Eq. 21–9. Thus we can see from Eq. 21–9 that, for a given $\mathscr{E}$, if the inductance L is large, the change in the current—and therefore the current itself—will be small. The greater the inductance, the less the current. An inductance thus acts something like a resistance to impede the flow of alternating current. We use the term **impedance** for this quality of an inductor. We shall discuss impedance more fully in Sections 21–12 to 21–15, and we shall see that it depends not only on L, but also on the frequency. Here we mention one example of its importance. The resistance of the primary in a transformer is usually quite small, perhaps less than 1 Ω. If resistance alone limited the current, tremendous currents would flow when a high voltage was applied. Indeed, a dc voltage applied to a transformer can burn it out. It is the impedance of the coil to an alternating current (or its "back" emf) that limits the current to a reasonable value.

DC can burn out a transformer

EXAMPLE 21–11 **Solenoid inductance.** (*a*) Determine a formula for the self-inductance L of a solenoid (a long coil—see Fig. 20–27b) containing N turns of wire in its length l and whose cross-sectional area is A. (*b*) Calculate the value of L if $N = 100$, $l = 5.0$ cm, $A = 0.30$ cm^2, and the solenoid is air-filled. (*c*) Calculate L if the solenoid has an iron core with $\mu = 4000\, \mu_0$.

Calculating self-inductance of a coil

SOLUTION (*a*) According to Eq. 20–8, the magnetic field inside a solenoid is $B = \mu_0 nI$, where $n = N/l$. From Eqs. 21–2 and 21–9, we have $\mathscr{E} = -N(\Delta\Phi_B/\Delta t) = -L(\Delta I/\Delta t)$. Thus, $L = N(\Delta\Phi_B/\Delta I)$. Since $\Phi_B = BA = \mu_0 NIA/l$, then any change in I causes a change in flux $\Delta\Phi_B = \mu_0 NA\,\Delta I/l$. Thus

$$L = N\frac{\Delta\Phi_B}{\Delta I} = \frac{\mu_0 N^2 A}{l}.$$

(*b*) Since $\mu_0 = 4\pi \times 10^{-7}$ T·m/A,

$$L = \frac{(4\pi \times 10^{-7}\,\text{T·m/A})(100)^2(3.0 \times 10^{-5}\,\text{m}^2)}{(5.0 \times 10^{-2}\,\text{m})} = 7.5\,\mu\text{H}.$$

(*c*) Here we replace μ_0 by $\mu = 4000\mu_0$, so L will be 4000 times larger: $L = 0.030$ H $= 30$ mH.

* 21–10 Energy Stored in a Magnetic Field

In Section 17–9 we saw that the energy stored in a capacitor is equal to $\frac{1}{2}CV^2$. By using a similar argument, we can show that the energy U stored in an inductance L, carrying a current I, is

$$U = \text{energy} = \tfrac{1}{2}LI^2.$$

Just as the energy stored in a capacitor can be considered to reside in the electric field between its plates, so the energy in an inductor can be considered to be stored in its magnetic field.

To write the energy in terms of the magnetic field, let us use the result of Example 21–11 that the inductance of a solenoid is $L = \mu_0 N^2 A/l$. Now the magnetic field B in a solenoid is related to the current I (see Eq. 20–8) by $B = \mu_0 NI/l$. Thus, $I = Bl/\mu_0 N$, and

$$U = \text{energy} = \tfrac{1}{2}LI^2 = \tfrac{1}{2}\left(\frac{\mu_0 N^2 A}{l}\right)\left(\frac{Bl}{\mu_0 N}\right)^2$$

$$= \tfrac{1}{2}\frac{B^2}{\mu_0}Al.$$

We can think of this energy as residing in the volume enclosed by the windings, which is Al. Then the energy per unit volume, or **energy density**, is

$$u = \text{energy density} = \frac{1}{2}\frac{B^2}{\mu_0}.$$ **(21–10)**

Energy density in magnetic field

This formula, which was derived for the special case of a solenoid, can be shown to be valid for any region of space where a magnetic field exists. If a ferromagnetic material is present, μ_0 is replaced by μ. This equation is analogous to that for an electric field, $\frac{1}{2}\epsilon_0 E^2$, Section 17–9.

*21–11 *LR* Circuit

Any inductor will have some resistance. We represent this situation by drawing the inductance L and the resistance R separately, as in Fig. 21–29. The resistance R could also include a separate resistor connected in series. Now we ask, what happens when a dc source is connected in series to such an *LR* circuit? At the instant the switch connecting the battery is closed, the current starts to flow. It is, of course, opposed by the induced emf in the inductor. However, as soon as current starts to flow, there is a voltage drop across the resistance. Hence, the voltage drop across the inductance is reduced and there is then less impedance to the current flow from the inductance. The current thus rises gradually, as shown in Fig. 21–30a, and approaches the steady value $I_{max} = V/R$ when all the voltage drop is across the resistance. The actual shape of the curve for I as a function of time is

$$I = \left(\frac{V}{R}\right)(1 - e^{-t/\tau}),$$

where e is the exponential function (see Section 19–7) and $\tau = L/R$ is called the **time constant** of the circuit. When $t = \tau$, then we have $(1 - e^{-1}) = 0.63$ so we see that τ is the time required for the current to reach $0.63I_{max}$.

If the battery is suddenly removed from the circuit (dashed line in Fig. 21–29), the current drops off as shown in Fig. 21–30b. This is an exponential decay curve given by $I = I_{max}e^{-t/\tau}$. The time constant τ is the time for the current to drop to 37 percent of the original value, and again equals L/R.

These graphs show that there is always some "reaction time" when an electromagnet, for example, is turned on or off. We also see that an *LR* circuit has properties similar to an *RC* circuit (Section 19–7). Unlike the capacitor case, however, the time constant here is *inversely* proportional to R.

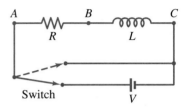

FIGURE 21–29 *LR* circuit.

FIGURE 21–30 (a) Growth of current in an *LR* circuit when connected to a battery; (b) decay of current when the *LR* circuit is shorted out (battery is out of the circuit).

(a)

(b)

> **EXAMPLE 21–12** **Solenoid time constant.** A solenoid has an inductance of 87.5 mH and a resistance of $0.250\,\Omega$. Find (a) the time constant for this circuit, and (b) how long it would take for the current to go from zero to half its final (maximum) value when connected to a battery of voltage V.
>
> **SOLUTION** (a) By definition, $\tau = L/R$, so
>
> $$\tau = \frac{L}{R} = \frac{87.5 \times 10^{-3}\,\text{H}}{0.250\,\Omega} = 0.350\,\text{s}.$$

(b) $I = (V/R)(1 - e^{-t/\tau})$ and we wish to find t such that $I = \frac{1}{2}(V/R)$ where $V/R = I_{max}$. So

$$\frac{1}{2}\frac{V}{R} = \frac{V}{R}(1 - e^{-t/\tau})$$

$$\frac{1}{2} = 1 - e^{-t/\tau}$$

$$e^{-t/\tau} = \frac{1}{2}$$

$$e^{t/\tau} = 2.$$

The inverse operation to the exponential e is the natural log, ln. Thus

$$\ln(e^{t/\tau}) = \frac{t}{\tau} = \ln 2,$$

so

$$t = \tau \ln 2 = (0.350 \text{ s})(0.693) = 0.243 \text{ s}.$$

* 21–12 AC Circuits and Impedance

We have previously discussed circuits that contain combinations of resistor, capacitor, and inductor, but only when they are connected to a dc source of emf or to no source (as in the discharge of a capacitor in an RC circuit). Now we discuss these circuit elements when they are connected to a source of alternating emf.

First we examine, one at a time, how a resistor, a capacitor, and an inductor behave when connected to a source of alternating emf, represented by the symbol •~⊙~•, which produces a sinusoidal voltage of frequency f. We assume in each case that the emf gives rise to a current

$$I = I_0 \cos 2\pi ft,$$

where t is time and I_0 is the peak current. We must remember (Section 18–8) that $V_{rms} = V_0/\sqrt{2}$ and $I_{rms} = I_0/\sqrt{2}$ (Eq. 18–8).

Resistor. When an ac source is connected to a resistor as in Fig. 21–31a, the current increases and decreases with the alternating emf according to Ohm's law, $I = V/R$. Figure 21–31b shows the voltage (red curve) and the current (blue curve). Because the current is zero when the voltage is zero and the current reaches a peak when the voltage does, we say that the current and voltage are **in phase**. Energy is transformed into heat (Section 18–8), at an average rate $\overline{P} = \overline{IV} = I_{rms}^2 R = V_{rms}^2/R$.

Inductor. In Fig. 21–32a an inductor of inductance L, represented by the symbol —ⵔⵔⵔ—, is connected to the ac source. We ignore any resistance it might have (it is usually small). The voltage applied to the inductor will be equal to the "back" emf generated in the inductor by the changing current as given by Eq. 21–9. This is because the sum of the emfs around any closed circuit must be zero, as Kirchhoff's rule tells us. Thus

$$V - L\frac{\Delta I}{\Delta t} = 0 \qquad \text{or} \qquad V = L\frac{\Delta I}{\Delta t},$$

where V is the sinusoidally varying voltage of the source and $L \Delta I/\Delta t$ is the voltage induced in the inductor. According to this equation, I is in-

(a)

(b)

$I = I_0 \cos 2\pi ft$
$V = V_0 \cos 2\pi ft$

FIGURE 21–31 Resistor connected to an ac source. Current is in phase with the voltage.

Resistor: current and voltage are in phase

creasing most rapidly when V has its maximum value, $V = V_0$. And I will be decreasing most rapidly when $V = -V_0$. These two instants correspond to points d and b on the graph of voltage versus time in Fig. 21–32b. At points a and c, $V = 0$. The equation above tells us that $\Delta I / \Delta t = 0$ at these instants, so I is not changing and these points correspond to the maximum and minimum values of the current I (the slope of I versus t is zero), as confirmed by points a and c on the graph. By going point by point in this manner, the curve of I versus t as compared to that for V versus t can be constructed, and they are shown by the blue and red lines, respectively, in Fig. 21–32b. Notice that the current reaches its peaks (and troughs) one quarter of a cycle after the voltage does. We say that

in an inductor, the current lags the voltage by 90°.

Inductor: current lags voltage

Remember that 360° corresponds to a full cycle, so 90° is a quarter cycle. Alternatively, we can say that the voltage leads the current by 90°. Since we originally chose $I = I_0 \cos 2\pi ft$, we see from the graph that V must vary as $V = -V_0 \sin 2\pi ft$.

Because the current and voltage are out of phase by 90°, no energy is transformed in an inductor on the average; and no energy is dissipated as thermal energy. This can be seen as follows from Fig. 21–32b. From point c to d, the voltage is increasing from zero to its maximum. The current, however, is in the opposite direction to the voltage and is approaching zero. The average power over this interval, VI, is negative. From d to e, however, both V and I are positive so VI is positive; this contribution just balances the negative contribution of the previous quarter cycle. Similar considerations apply to the rest of the cycle. Thus, the average power transformed over one or many cycles is zero. We can see that energy from the source passes into the magnetic field of the inductor, where it is stored temporarily. Then the field decreases and the energy is transferred back to the source. None is dissipated in this process. Compare this to a resistor where the current is always in the same direction as the voltage and energy is transferred out of the source and never back into it. (The product VI is never negative.) This energy is not stored in the resistor, but is transformed to thermal energy.

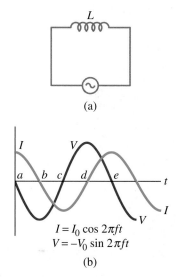

(a)

(b)

$I = I_0 \cos 2\pi ft$
$V = -V_0 \sin 2\pi ft$

FIGURE 21–32 Inductor connected to an ac source. Current lags voltage by a quarter cycle, or 90°.

As mentioned in Section 21–9, the back emf of an inductor impedes the flow of an ac current. Indeed, it is found that the magnitude of the current in an inductor is directly proportional to the applied ac voltage at a given frequency. We can therefore write an equation for a pure inductor (no resistance) that is the equivalent of Ohm's law:

$$V = IX_L, \qquad \textbf{(21–11a)}$$

Reactance and impedance

where X_L is called the **inductive reactance**, or **impedance**, of the inductor. Normally, we use the term "reactance" to refer solely to the inductive properties. We then reserve the term "impedance" to include the total "impeding" qualities of the coil—its inductance plus any resistance it may have (more on this in the next section). In the absence of resistance, the impedance is the same as the reactance.

The quantities V and I in Eq. 21–11a can refer either to rms for both, or to peak values for both. Note, however, that although this equation can relate the peak values, the peak current and voltage are not reached at the same time; so Eq. 21–11a is *not* valid at a particular instant, as is the case for Ohm's law.

From the fact that $V = L\,\Delta I/\Delta t$, we see that the larger L is, the less will be the change in current ΔI during the time Δt. Hence I itself will be smaller at any instant for a given frequency, and we expect $X_L \propto L$. The reactance X_L also depends on the frequency. The greater the frequency, the more rapidly the magnetic flux changes in the inductor. If the emf induced by this field is to remain equal to the source emf, as it must, the magnitude of the current must then be less. Hence the greater the frequency, the greater is the reactance X_L, and we expect $X_L \propto fL$. This is also consistent with the fact that if the frequency f is zero (so the current is dc), there is no back emf and no impedance to the flow of charge. Careful calculation (using calculus), as well as experiment, shows that the constant of proportionality is 2π. Thus

$$X_L = 2\pi fL. \qquad\qquad (21\text{--}11b)$$

EXAMPLE 21–13 **Reactance of a coil.** A coil has a resistance $R = 1.00\,\Omega$ and an inductance of $0.300\,\text{H}$. Determine the current in the coil if: (a) $120\,\text{V}$ dc is applied to it; (b) $120\,\text{V}$ ac (rms) at $60.0\,\text{Hz}$ is applied.

SOLUTION (a) There is no inductive impedance ($X_L = 0$ since $f = 0$), so we apply Ohm's law for the resistance:

$$I = \frac{V}{R} = \frac{120\,\text{V}}{1.00\,\Omega} = 120\,\text{A}.$$

(b) The inductive reactance in this case is

$$X_L = 2\pi fL = (6.28)(60.0\,\text{s}^{-1})(0.300\,\text{H}) = 113\,\Omega.$$

In comparison to this, the resistance can be ignored. Thus,

$$I_{\text{rms}} = \frac{V_{\text{rms}}}{X_L} = \frac{120\,\text{V}}{113\,\Omega} = 1.06\,\text{A}.$$

[It might be tempting to say that the total impedance is $113\,\Omega + 1\,\Omega = 114\,\Omega$. This might imply that about 1 percent of the voltage drop is across the resistor, or about $1\,\text{V}$; and that across the inductance is $119\,\text{V}$. Although the $1\,\text{Vrms}$ across the resistor is correct, the other statements are not true because of the alteration in phase in an inductor. This will be discussed in the next Section.]

Capacitor. When a capacitor is connected to a battery, the capacitor plates quickly acquire equal and opposite charges; but no steady current flows in the circuit. A capacitor prevents the flow of a dc current (as long as the capacitor is not "leaky," in which case a leakage current would exist across the gap). However, if a capacitor is connected to an alternating source of voltage, as in Fig. 21–33a, an alternating current will flow continuously. This can happen because when the ac voltage is first turned on, charge begins to flow so that one plate acquires a negative charge and the other a positive charge. But when the voltage reverses itself, the charges flow in the opposite direction. Thus, for an alternating applied voltage, an ac current is present in the circuit continuously.

Let us look at this process in more detail. First, we recall that the applied voltage must equal the voltage across the capacitor: $V = Q/C$, where C is the capacitance and Q the charge on the plates. Thus the charge Q on the plates follows the voltage; when the voltage is zero, the charge is zero; when the voltage is a maximum, the charge is a maximum. But what about the current I? At point a in Fig. 21–33b, when the voltage starts increasing, the charge on the plates is zero. Thus charge flows readily toward the plates and the current is large. As the voltage approaches its maximum of V_0 (point b), the charge that has accumulated on the plates tends to prevent more charge from flowing, so the current drops to zero at point b. The charge that has accumulated now starts to flow off the plates and the magnitude of current again increases (blue curve), but in the opposite direction; it reaches a maximum (negatively) when the voltage is at point c. Thus the current follows the blue curve in Fig. 21–33b. Like an inductor, the voltage and current are out of phase by 90°. But for a capacitor, the current reaches its peaks $\frac{1}{4}$ cycle before the voltage does, so we say that the

current leads the voltage by 90°.

(Or the voltage lags the current by 90°.) This is the opposite of what happens for an inductor. Again we have chosen $I = I_0 \cos 2\pi ft$ and we see from the graph that $V = V_0 \sin 2\pi ft$.

Because the current and voltage are out of phase, the average power dissipated is zero, just as for an inductor. Energy from the source is fed to the capacitor, where it is stored in the electric field between its plates. As the field decreases, the energy returns to the source. Thus, in an ac circuit, *only a resistance will dissipate energy* to thermal energy.

A relationship between the applied voltage and the current in a capacitor can be written just as for an inductance:

$$V = IX_C, \tag{21–12a}$$

where X_C is the **capacitive reactance** (or **impedance**) of the capacitor. This equation relates the rms or peak values for the voltage and current but is not valid at a particular instant because I and V are out of phase. X_C depends on both the capacitance C and the frequency f. The larger the capacitance, the more charge it can handle, so the less it will retard the flow of an alternating current. Hence, we expect X_C will be inversely proportional to C. It is also inversely proportional to the frequency f, since, when the frequency is higher, there is less time per cycle for the charge to build up on the plates and impede the flow. Again there is a factor of 2π, and

$$X_C = \frac{1}{2\pi fC}. \tag{21–12b}$$

Notice that for dc conditions, $f = 0$ and X_C becomes infinite. This is as it should be, since a capacitor does not pass dc current.

It is interesting to note that the reactance of an inductor increases with frequency, but that of a capacitor decreases with frequency.

(a)

(b)

$I = I_0 \cos 2\pi ft$
$V = V_0 \sin 2\pi ft$

FIGURE 21–33 Capacitor connected to an ac source. Current leads voltage by a quarter cycle, or 90°.

Capacitor: current leads voltage

Only R (not C or L) dissipates energy

Capacitor reactance. What are the peak and rms currents in the circuit of Fig. 21–33a if $C = 1.0 \, \mu\text{F}$ and $V_{\text{rms}} = 120 \, \text{V}$? Calculate for (a) $f = 60 \, \text{Hz}$, and then for (b) $f = 6.0 \times 10^5 \, \text{Hz}$.

SOLUTION (a) $V_0 = \sqrt{2} \, V_{\text{rms}} = 170 \, \text{V}$. Then

$$X_C = \frac{1}{2\pi f C} = \frac{1}{(6.28)(60 \, \text{s}^{-1})(1.0 \times 10^{-6} \, \text{F})} = 2.7 \, \text{k}\Omega.$$

Thus

$$I_0 = \frac{V_0}{X_C} = \frac{170 \, \text{V}}{2.7 \times 10^3 \, \Omega} = 63 \, \text{mA},$$

$$I_{\text{rms}} = \frac{V_{\text{rms}}}{X_C} = \frac{120 \, \text{V}}{2.7 \times 10^3 \, \Omega} = 44 \, \text{mA}.$$

(b) For $f = 6.0 \times 10^5 \, \text{Hz}$, X_C will be $0.27 \, \Omega$, $I_0 = 630 \, \text{A}$, and $I_{\text{rms}} = 440 \, \text{A}$. The dependence on f is dramatic.

Capacitors are used for a variety of purposes, some of which have already been described. Two other applications are illustrated in Fig. 21–34. In Fig. 21–34a, circuit A is said to be capacitively coupled to circuit B. The purpose of the capacitor is to prevent a dc voltage from passing from A to B but allowing an ac signal to pass relatively unimpeded. If C is sufficiently large, the ac signal will not be significantly attenuated, whereas dc is filtered out. The capacitor in Fig. 21–34b also passes ac but not dc. In this case, a dc voltage can be maintained between circuits A and B. If the capacitance C is large enough, the capacitor offers little impedance to an ac signal leaving A. Such a signal then passes to ground instead of into B. Thus the capacitor in Fig. 21–34b acts like a *filter* when a constant dc voltage is required; any sharp variation in voltage will pass to ground instead of into circuit B. Capacitors used in these two ways are very common in circuits.

FIGURE 21–34 Two common uses for a capacitor.

* 21–13 *LRC* Series AC Circuit; Problem Solving

We now examine a circuit containing all three elements in series, a resistor R, an inductor L, and a capacitor C (Fig. 21–35). If a given circuit contains only two of these elements, we can still use the results of this section by setting $R = 0$, $L = 0$, or $C = \infty$ (infinity) as needed. We let V_R, V_L, and V_C represent the voltage across each element at a given instant in time; and V_{R0}, V_{L0}, and V_{C0} represent the *maximum* (peak) values of these voltages. The voltage across each of the elements will follow the phase relations we discussed in the last section. That is, V_R will be in phase with the current, V_L will lead the current by 90°, and V_C will lag behind the current by 90°. Also, at any instant the total voltage V supplied by the source will be $V = V_R + V_L + V_C$. However, because the various voltages are not in phase, the rms voltages (which is what ac voltmeters usually measure) will not simply add up to give the rms voltage of the source. And V_0 will *not* equal $V_{R0} + V_{L0} + V_{C0}$. Let us now examine the circuit in detail. What we would like to find in particular is the impedance of the circuit as a whole, the rms current that flows through the circuit, and the phase difference between the source voltage and the current.

First we note that the current at any instant must be the same at all points in the circuit. Thus, *the currents in each element are in phase with each other, although the voltages are not.* We choose our origin in time ($t = 0$) so that the current I at any time t is $I = I_0 \cos 2\pi f t$.

FIGURE 21–35 An *LRC* circuit.

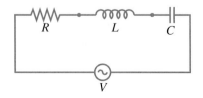

It is convenient to analyze an *LRC* circuit using a **phasor diagram**. Arrows (acting like vectors) are drawn in an *xy* coordinate system to represent each voltage. (These "vectors" are not "real." Phasors are just a useful analytical tool.) *The length of each arrow represents the magnitude of the peak voltage across each element:*

$$V_{R0} = I_0 R, \quad V_{L0} = I_0 X_L, \quad \text{and} \quad V_{C0} = I_0 X_C.$$

V_{R0} is in phase with the current and is initially ($t = 0$) drawn along the positive *x* axis, as is the current. Since V_{L0} leads the current by 90°, it leads V_{R0} by 90°, so is initially drawn along the positive *y* axis. V_{C0} lags the current by 90°, so lags V_{R0} by 90°; hence, V_{C0} is drawn initially along the negative *y* axis. Such a diagram is shown in Fig. 21–36a. If we let the vector diagram rotate counterclockwise at frequency *f*, we get the diagram shown in Fig. 21–36b; after a time, t, each arrow has rotated through an angle $2\pi ft$. Then the *projections of each arrow on the x axis represent the voltages across each element at the instant t* (see Fig. 21–36c). For example, the projection of V_{R0} on the *x* axis is $V_{R0} \cos 2\pi ft$ (as in Fig. 21–31); and the projections of V_{L0} and V_{C0} on the *x* axis are $-V_{L0} \sin 2\pi ft$ and $V_{C0} \sin 2\pi ft$, respectively, as in Figs. 21–32b and 21–33b. Maintaining the 90° angle between each arrow ensures the correct phase relations. Although these results show the validity of a phasor diagram, what we are really interested in is how to add the voltages.

The sum of the projections of the three vectors on the *x* axis is equal to the projection of their sum. But the sum of the projections represents the instantaneous voltage across the whole circuit, *V* (equal to the source voltage). Therefore, the vector sum of these vectors will be the vector that represents the peak source voltage, V_0. This is shown in Fig. 21–37, where it is seen that V_0 makes an angle ϕ with I_0 and V_{R0}. As time passes, V_0 rotates with the other vectors, so the instantaneous voltage *V* (projection of V_0 on the *x* axis) is (see Fig. 21–37)

$$V = V_0 \cos (2\pi ft + \phi).$$

The voltage *V* across the whole circuit must, of course, equal the source voltage (Fig. 21–35). Thus we see that the voltage from the source is out of phase[†] with the current by an angle ϕ.

From this analysis we can now draw some useful conclusions. First, we determine the total **impedance** *Z* of the circuit, which is defined by the relation

$$V_{\text{rms}} = I_{\text{rms}} Z, \quad \text{or} \quad V_0 = I_0 Z. \tag{21–13}$$

From Fig. 21–37 we see, using the Pythagorean theorem (V_0 is the hypotenuse of a right triangle), that

$$V_0 = \sqrt{V_{R0}^2 + (V_{L0} - V_{C0})^2}$$
$$= \sqrt{I_0^2 R^2 + (I_0 X_L - I_0 X_C)^2}$$
$$= I_0 \sqrt{R^2 + (X_L - X_C)^2}.$$

Thus, from Eq. 21–13, and then Eqs. 21–11b and 21–12b,

$$Z = \sqrt{R^2 + (X_L - X_C)^2} \tag{21–14a}$$
$$= \sqrt{R^2 + (2\pi fL - 1/2\pi fC)^2}. \tag{21–14b}$$

This gives the total impedance of the circuit. Also from Fig. 21–37, we can

[†]As a check, note that if $R = X_C = 0$, then $\phi = 90°$, and V_0 would lead the current by 90°, as it must for an inductor alone. Similarly, if $R = L = 0$, $\phi = -90°$ and V_0 would lag the current by 90°, as it must for a capacitor alone.

(a)

(b)

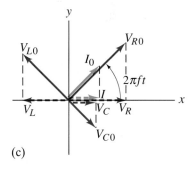

(c)

FIGURE 21–36 Phasor diagram for a series *LRC* circuit

FIGURE 21–37 Phasor diagram for a series *LRC* circuit showing the sum vector, V_0.

Impedance

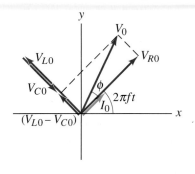

find the phase angle ϕ:

$$\tan \phi = \frac{V_{L0} - V_{C0}}{V_{R0}} = \frac{X_L - X_C}{R} \qquad \text{(21–15a)}$$

or

$$\cos \phi = \frac{V_{R0}}{V_0} = \frac{I_0 R}{I_0 Z} = \frac{R}{Z}. \qquad \text{(21–15b)}$$

Finally, we can determine the power dissipated in the circuit. We saw earlier that power is only dissipated by a resistance; none is dissipated by inductance or capacitance. Therefore, the average power $\overline{P} = I_{rms}^2 R$. But from Eq. 21–15b, $R = Z \cos \phi$. Therefore,

$$\overline{P} = I_{rms}^2 Z \cos \phi$$

$$= I_{rms} V_{rms} \cos \phi. \qquad \text{(21–16)}$$

The factor $\cos \phi$ is referred to as the **power factor** of the circuit. For a pure resistor, $\cos \phi = 1$ and $\overline{P} = I_{rms} V_{rms}$. For a capacitor or inductor, $\phi = -90°$ or $+90°$, respectively, so $\cos \phi = 0$ and no power is dissipated.

The test of this analysis is, of course, in experiment; and experiment is in full agreement with these results.

EXAMPLE 21–15   **LRC Circuit.**   Suppose that $R = 25.0\,\Omega$, $L = 30.0\,\text{mH}$, and $C = 12.0\,\mu\text{F}$ in Fig. 21–35, and that they are connected to a 90.0-V ac (rms) 500-Hz source. Calculate (*a*) the current in the circuit, (*b*) the voltmeter readings (rms) across each element, (*c*) the phase angle ϕ, and (*d*) the power dissipated in the circuit.

SOLUTION   (*a*) First, we find the individual impedances at $f = 500\,\text{s}^{-1}$:

$$X_L = 2\pi f L = 94.2\,\Omega,$$

$$X_C = \frac{1}{2\pi f C} = 26.5\,\Omega.$$

Then

$$Z = \sqrt{R^2 + (X_L - X_C)^2}$$

$$= \sqrt{(25.0\,\Omega)^2 + (94.2\,\Omega - 26.5\,\Omega)^2} = 72.2\,\Omega.$$

From Eq. 21–13,

$$I_{rms} = \frac{V_{rms}}{Z} = \frac{90.0\,\text{V}}{72.2\,\Omega} = 1.25\,\text{A}.$$

(*b*) The rms voltage across each element is

$$(V_R)_{rms} = I_{rms} R = (1.25\,\text{A})(25.0\,\Omega) = 31.2\,\text{V}$$

$$(V_L)_{rms} = I_{rms} X_L = (1.25\,\text{A})(94.2\,\Omega) = 118\,\text{V}$$

$$(V_C)_{rms} = I_{rms} X_C = (1.25\,\text{A})(26.5\,\Omega) = 33.1\,\text{V}.$$

Notice that these do *not* add up to give the source voltage, 90.0 V (rms). Indeed, the rms voltage across the inductance *exceeds* the source voltage. This can happen because the different voltages are out of phase with each other, and at any instant one voltage can be negative, to compensate for a large positive voltage of another. The rms voltages, however, are always positive by definition. Although the rms voltages do not have to add up to the source voltage, the instantaneous voltages at any time do add up, of course.

(c) Since

$$\tan \phi = \frac{X_L - X_C}{R} = \frac{94.2 \ \Omega - 26.5 \ \Omega}{25.0 \ \Omega} = 2.71,$$

then $\phi = 69.7°$.

(d) $\overline{P} = I_{rms} V_{rms} \cos \phi = (1.25 \ \text{A})(90.0 \ \text{V})(25.0 \ \Omega / 72.2 \ \Omega) = 39.0 \ \text{W}.$

*21–14 Resonance in AC Circuits; Oscillators

The rms current in an LRC series circuit is given by (see Eqs. 21–13 and 21–14):

$$I_{rms} = \frac{V_{rms}}{Z} = \frac{V_{rms}}{\sqrt{R^2 + \left(2\pi f L - \dfrac{1}{2\pi f C}\right)^2}}. \qquad \textbf{(21–17)}$$

Because the impedance of inductors and capacitors depends on the frequency f of the source, the current in an LRC circuit will depend on frequency. From Eq. 21–17 we can see that the current will be maximum at a frequency such that

$$2\pi f L - \frac{1}{2\pi f C} = 0.$$

We solve this for f, and call the solution f_0:

$$f_0 = \frac{1}{2\pi} \sqrt{\frac{1}{LC}}. \qquad \textbf{(21–18)}$$

Resonant frequency

This is the **resonant frequency** of the circuit. At this frequency, $X_C = X_L$, so the impedance is purely resistive and $\cos \phi = 1$. A graph of I_{rms} versus f is shown in Fig. 21–38 for particular values of R, L, and C. For smaller R compared to X_L and X_C, the resonance peak will be higher and sharper.

When R is very small, we speak of an **LC circuit**. The energy in an LC circuit oscillates, at frequency f_0, between the inductor and the capacitor, with some being dissipated in R (some resistance is unavoidable). To see this in more detail, consider a perfect LC circuit in which $R = 0$. Assume at $t = 0$ that the capacitor C is charged and the switch is closed (Fig. 21–39). The capacitor immediately begins to discharge. As it does so, the current I through the inductor increases. At every instant, the potential difference across the capacitor, $V = Q/C$ (where Q is the charge on the capacitor at that instant), must equal the potential difference across the inductor, which is equal to its emf, $- L(\Delta I / \Delta t)$. At the instant when the charge on the capacitor reaches zero ($Q = 0$), the current I in the inductor has reached its maximum value, but at this instant I is not changing ($- L \Delta I / \Delta t = Q/C = 0$). At this moment, the magnetic field B in the inductor is also a maximum. The current next begins to decrease as the flowing charge starts to accumulate on the opposite plate of the capacitor. When the current has dropped to zero, the capacitor has again

LC circuit

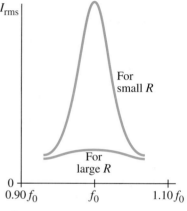

FIGURE 21–38 Current in an LRC circuit as a function of frequency, showing resonance peak at $f = f_0 = (1/2\pi) \sqrt{1/LC}$.

FIGURE 21–39 A pure LC circuit.

attained its maximum charge. The capacitor then begins to discharge again, with the current now flowing in the opposite direction. This process of the charge flowing back and forth from one plate of the capacitor to the other, through the inductor, continues to repeat itself. This is called an **LC oscillation** or an **electromagnetic oscillation**. Not only does the charge oscillate back and forth, but so does the energy, which oscillates between being stored in the electric field of the capacitor and in the magnetic field of the inductor.

EM oscillations

Electric resonance is used in many electronic devices. Radio and television sets, for example, use resonant circuits for tuning in a station. Many frequencies reach the circuit from the antenna, but a significant current flows only for frequencies at or near the resonant frequency. Either L or C is variable so that different stations can be tuned in (more on this in Chapter 22). LC circuits are also used in **oscillators**, which are devices that put out an oscillating signal of particular frequency. Since some resistance is always present, electrical oscillators generally need a periodic input of power to compensate for the energy converted to thermal energy in the resistance.

Electrical resonance is analogous to mechanical resonance, which we discussed in Chapter 11 (see Fig. 11–18). The energy transferred to the system is a maximum at resonance whether it is electrical resonance, the oscillation of a spring, or pushing a child on a swing (Section 11–6). That this is true in the electrical case can be seen from Eqs. 21–14, 21–15, and 21–16. At resonance, $Z = R$, $\cos \phi = 1$, and I_{rms} is a maximum. For constant V_{rms}, the power is then a maximum at resonance. A graph of power versus frequency looks much like that for the current (Fig. 21–38).

An LRC circuit can have the elements arranged in parallel instead of in series. Resonance will occur in this case, too, but the analysis of such circuits is more involved.

---

* **21–15 Impedance Matching**

It is common to connect one electric circuit to a second circuit. For example, a TV antenna is connected to a TV set; an FM tuner is connected to an amplifier; the output of an amplifier is connected to a speaker; electrodes for an ECG or EEG (electrocardiogram and electroencephalogram—electrical traces of heart and brain signals) are connected to an amplifier or a recorder. In many cases it is important that the maximum power be transferred from one to the other, with a minimum of loss. This can be achieved when the output impedance of the one device matches the input impedance of the second.

➡ **PHYSICS APPLIED**

Maximum power transferred when impedances match

FIGURE 21–40 Output of the circuit on the left is input to the circuit on the right.

Circuit 1   Circuit 2

To show why this is true, we consider simple circuits that contain only resistance. In Fig. 21–40, the source in circuit 1 could represent a power supply, the output of an amplifier, or the tiny signal from an antenna, a laboratory probe, or electrodes. R_1 represents the resistance of this device and includes the internal resistance of the source. R_1 is called the output impedance (or resistance) of circuit 1. The output of circuit 1 is across the terminals a and b, which are connected to the input of circuit 2. Circuit 2 may be very complicated. By combining the various resistors, we can find an equivalent resistance. This is represented by R_2, the "input resistance" (or input impedance) of circuit 2.

The power delivered to circuit 2 is $P = I^2 R_2$, where $I = V/(R_1 + R_2)$. Thus

$$P = I^2 R_2 = \frac{V^2 R_2}{(R_1 + R_2)^2}.$$

We divide the top and bottom of the right side by R_1^2 and find

$$P = \frac{V^2}{R_1} \frac{\left(\dfrac{R_2}{R_1}\right)}{\left(1 + \dfrac{R_2}{R_1}\right)^2}. \qquad (21\text{–}19)$$

The question is, if the resistance of the source is R_1, what value should R_2 have so that the maximum power is transferred to circuit 2? To determine this, we plot a graph of P versus (R_2/R_1). This is shown in Fig. 21–41, where representative values are given in the table. For example, for $R_2/R_1 = 1$, Eq. 21–19 gives $P = V^2/4R_1$; for $R_2/R_1 = 3$, $P = 3V^2/16R_1 = 0.19V^2/R_1$; and so on. As can be seen from the graph, P is a maximum when $R_2 = R_1$.

Thus, the maximum power is transmitted when the *output impedance* of one device *equals the input impedance* of the second. This is called **impedance matching**.

In an ac circuit that contains capacitors and inductors, the different phases are important and the analysis is more complicated. However, the same result holds: To maximize power transfer it is important to match impedances ($Z_2 = Z_1$). In addition, one must be aware that it is possible to seriously distort a signal. For example, when a second circuit is connected, it may put the first circuit into resonance, or take it out of resonance for a certain frequency.

Without proper consideration of the impedances involved, one can make measurements that are completely meaningless. These considerations are normally examined by engineers when designing an integrated set of apparatus. It has happened that researchers have connected several components to one another without regard for impedance matching, and made a "new discovery" that later was found, embarrassingly, to be due to impedance mismatch rather than the natural phenomenon they had thought.

In some cases, a transformer is used to alter an impedance, so it can be matched to that of a second circuit. If Z_S is the secondary impedance and Z_P the primary impedance, then $V_S = I_S Z_S$ and $V_P = I_P Z_P$ (I and V are either peak or rms values of current and voltage). Hence,

$$\frac{Z_P}{Z_S} = \frac{V_P I_S}{V_S I_P} = \left(\frac{N_P}{N_S}\right)^2,$$

where we have used Eqs. 21–6 and 21–7 for a transformer. Thus the impedance can be changed with a transformer. A transformer is used for this purpose in some audio amplifiers, which may have several taps corresponding to $4\,\Omega$, $8\,\Omega$, and $16\,\Omega$ so the output can be matched to the impedance of any loudspeaker.

Some instruments, such as oscilloscopes, require only a signal voltage but very little power. Maximum power transfer is then not important and such instruments can have a high input impedance. This has the advantage that the instrument draws very little current and disturbs the original circuit as little as possible. This is often desirable in laboratory experiments.

$\dfrac{R_2}{R_1}$	P
0	0
0.5	$0.22\ V^2/R_1$
1.0	$0.25\ V^2/R_1$
2.0	$0.22\ V^2/R_1$
5.0	$0.14\ V^2/R_1$
100	$0.01\ V^2/R_1$

FIGURE 21–41 Power transferred is at its maximum when $R_2 = R_1$.

Impedance matching

➡ **PHYSICS APPLIED**

Impedance mismatch errors

The **magnetic flux** passing through a loop is equal to the product of the area of the loop times the perpendicular component of the magnetic field strength: $\Phi_B = B_\perp A = BA\cos\theta$.

If the magnetic flux through a coil of wire changes in time, an emf is induced in the coil; the magnitude of the induced emf equals the time rate of change of the magnetic flux through the loop times the number N of loops in the coil:

$$\mathscr{E} = -N\frac{\Delta\phi_B}{\Delta t}.$$

This is **Faraday's law of induction**.

The induced emf produces a current whose magnetic field opposes the original change in flux (**Lenz's law**).

Faraday's law also tells us that a changing magnetic field produces an electric field; and that a straight wire of length l moving with speed v perpendicular to a magnetic field of strength B has an emf induced between its ends equal to:

$$\mathscr{E} = Blv.$$

An electric **generator** changes mechanical energy into electrical energy. Its operation is based on Faraday's law: a coil of wire is made to rotate uniformly by mechanical means in a magnetic field, and the changing flux through the coil induces a sinusoidal current, which is the output of the generator.

A motor, which operates in the reverse of a generator, acts like a generator in that a **counter emf** is induced in its rotating coil; since this counter emf opposes the input voltage, it can act to limit the current in a motor coil.

Similarly, a generator acts somewhat like a motor in that a **counter torque** acts on its rotating coil.

A **transformer**, which is a device to change the magnitude of an ac voltage, consists of a primary and a secondary coil. The changing flux due to an ac voltage in the primary induces an ac voltage in the secondary. In a 100 percent efficient transformer, the ratio of output to input voltages (V_S/V_P) equals the ratio of the number of turns N_S in the secondary to the number N_P in the primary:

$$\frac{V_S}{V_P} = \frac{N_S}{N_P}.$$

The ratio of secondary to primary current is in the inverse ratio of turns:

$$\frac{I_S}{I_P} = \frac{N_P}{N_S}.$$

QUESTIONS

1. What would be the advantage, in Faraday's experiments (Fig. 21–1), of using coils with many turns?

2. What is the difference between magnetic flux and magnetic field? Discuss in detail.

3. Suppose you are holding a circular piece of wire and suddenly thrust a magnet, south pole first, toward the center of the circle. Is a current induced in the wire? Is a current induced when the magnet is held steady within the loop? Is a current induced when you withdraw the magnet? In each case, if your answer is yes, specify the direction.

4. In what direction will the current flow in Fig. 21–9 if the rod moves to the left, which decreases the area of the loop to the left?

5. Two loops of wire are moving in the vicinity of a very long straight wire carrying a steady current as shown in Fig. 21–42. Find the direction of the induced current in each loop.

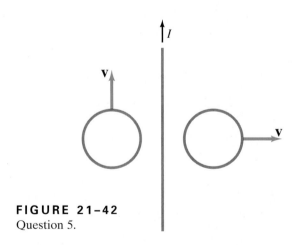

FIGURE 21–42
Question 5.

6. In situations where a small signal must travel over a distance, a "shielded cable" is used in which the signal wire is surrounded by an insulator and then enclosed by a cylindrical conductor. Why is a "shield" necessary?

7. What is the advantage of placing the two insulated electric wires carrying ac close together or even twisted about each other?

8. In some early automobiles, the starter motor doubled as a generator to keep the battery charged once the car was started. Explain how this might work.

9. Explain why, exactly, the lights may dim briefly when a refrigerator motor starts up. When an electric heater is turned on, the lights may stay dimmed as long as it is on. Explain the difference.

10. Explain what is meant by the statement "a motor acts as a motor and generator at the same time." Can the same be said for a generator?

11. Use Fig. 21–12 and the right-hand rules to show why the counter torque in a generator *opposes* the motion.

12. Will an eddy current brake (Fig. 21–19) work on a copper or aluminum wheel, or must it be ferromagnetic?

13. It has been proposed that eddy currents be used to help sort solid waste for recycling. The waste is first ground into tiny pieces and iron removed with a dc magnet. The waste then is allowed to slide down an incline over permanent magnets. How will this aid in the separation of nonferrous metals (Al, Cu, Pb, brass) from nonmetallic materials?

14. The pivoted metal bar with slots in Fig. 21–43 falls much more quickly through a magnetic field than does a solid bar. Explain in detail.

FIGURE 21–43   Question 14.

15. If an aluminum sheet is held between the poles of a large bar magnet, it requires some force to pull it out of the magnetic field even though the sheet is not ferromagnetic and does not touch the pole faces. Explain.

16. A bar magnet falling inside a vertical metal tube reaches a terminal velocity even if the tube is evacuated so that there is no air resistance. Explain.

17. A metal bar, pivoted at one end, oscillates freely in the absence of a magnetic field; but in a magnetic field, its oscillations are quickly damped out. Explain. (This *magnetic damping* is used in a number of practical devices.)

18. An enclosed transformer has four wire leads coming from it. How could you determine the ratio of turns on the two coils without taking the transformer apart? How would you know which wires paired with which?

19. The use of higher-voltage lines in homes, say 600 V or 1200 V, would reduce energy waste. Why are they not used?

* 20. Since a magnetic microphone is basically like a loudspeaker (Section 20–10), could a loudspeaker actually serve as a microphone? That is, could you speak into a loudspeaker and obtain an output signal that could be amplified? Explain. Discuss, in light of your response, how a microphone and loudspeaker differ in construction.

* 21. The primary of a transformer on a telephone pole has a resistance of 0.10 Ω and the input voltage is 2400 V ac. Can you estimate the current that will flow? Will it be 24,000 A? Explain.

* 22. A transformer designed for a 120-V ac input will often "burn out" if connected to a 120-V dc source. Explain. [*Hint*: The resistance of the primary coil is usually very low.]

* 23. How would you arrange two flat circular coils so that their mutual inductance was (*a*) greatest, (*b*) least (without separating them by a great distance)?

* 24. If you are given a fixed length of wire, how would you shape it to obtain the greatest self-inductance? The least?

* 25. Does the emf of the battery in Fig. 21–29a affect the time needed for the *LR* circuit to reach (*a*) a given fraction of its maximum possible current, (*b*) a given value of current?

* 26. Can you tell whether the current in an *LRC* circuit leads or lags the applied voltage from a knowledge of the power factor, cos φ?

* 27. Under what conditions is the impedance in an *LRC* circuit a minimum?

* 28. An *LC* resonance circuit is often called an *oscillator* circuit. What is it that oscillates?

* 29. Compare the oscillations of an *LC* circuit to the vibration of a mass *m* on a spring. What do *L* and *C* correspond to in the mechanical system?

PROBLEMS

SECTIONS 21–1 TO 21–4

1. (I) A 9.2-cm-diameter loop of wire is initially oriented perpendicular to a 1.5-T magnetic field. It is rotated so that its plane is parallel to the field direction in 0.20 s. What is the average induced emf in the loop?

2. (I) A 16-cm-diameter circular loop of wire is in a 1.10-T magnetic field. It is removed from the field in 0.15 s. What is the average induced emf?

FIGURE 21–44
Problem 3.

3. (I) The rectangular loop shown in Fig. 21–44 is pushed into the magnetic field which points inward as shown. In what direction is the induced current?

4. (I) The north pole of the magnet in Fig. 21–45 is being inserted into the coil. In which direction is the induced current flowing through the resistor R?

FIGURE 21–45
Problem 4.

5. (I) The magnetic flux through a coil of wire containing two loops changes from -30 Wb to $+38$ Wb in 0.42 s. What is the emf induced in the coil?

6. (I) A 7.2-cm-diameter wire coil is initially oriented so that its plane is perpendicular to a magnetic field of 0.63 T pointing up. During the course of 0.15 s, the field is changed to one of 0.25 T pointing down. What is the average induced emf in the coil?

7. (II) (a) If the resistance of the resistor in Fig. 21–46 is slowly increased, what is the direction of the current induced in the small circular loop inside the larger loop? (b) What would it be if the small loop were placed outside the larger one, to the left?

FIGURE 21–46
Problem 7.

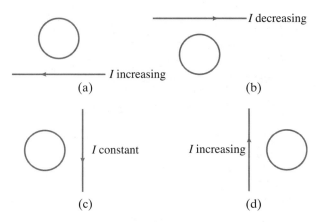

FIGURE 21–47 Problem 8.

8. (II) What is the direction of the induced current in the circular loop due to the current shown in each part of Fig. 21–47?

9. (II) If the solenoid in Fig. 21–48 is being pulled away from the loop shown, in what direction is the induced current in the loop?

FIGURE 21–48
Problem 9.

10. (II) The magnetic field perpendicular to a circular loop of wire 20 cm in diameter is changed from $+0.52$ T to -0.45 T in 180 ms, where $+$ means the field points away from an observer and $-$ toward the observer. (a) Calculate the induced emf. (b) In what direction does the induced current flow?

11. (II) The moving rod in Fig. 21–9 is 12.0 cm long and moves with a speed of 15.0 cm/s. If the magnetic field is 0.800 T, calculate (a) the emf developed, and (b) the electric field in the rod.

12. (II) A circular loop in the plane of the paper lies in a 0.75 T magnetic field pointing into the paper. If the loop's diameter changes from 20.0 cm to 6.0 cm in 0.50 s, (a) what is the direction of the induced current, (b) what is the magnitude of the average induced emf, and (c) if the coil resistance is 2.5 Ω, what is the average induced current?

13. (II) The moving rod in Fig. 21–9 is 13.2 cm long and generates an emf of 100 mV while moving in a 0.90-T magnetic field. (a) What speed is it moving at? (b) What is the electric field in the rod?

14. (II) In Fig. 21–9, the rod moves with a speed of 1.9 m/s, is 30.0 cm long, and has a resistance of 2.5 Ω. The magnetic field is 0.75 T, and the resistance of the U-shaped conductor is 25.0 Ω at a given instant. Calculate (a) the induced emf, (b) the current flowing in the circuit, and (c) the external force necessary to ensure that the rod is moving at constant velocity at that instant.

15. (II) A single rectangular loop of wire with dimensions shown in Fig. 21–49 is situated so that part is inside a region of uniform magnetic field of 0.450 T and part is outside the field. The total resistance of the loop is 0.230 Ω. Calculate the force required to pull the loop from the field (to the right) at a constant velocity of 3.40 m/s. Neglect gravity.

FIGURE 21–49 Problem 15.

16. (II) A 31.0-cm-diameter coil consists of 20 turns of circular copper wire 2.6 mm in diameter. A uniform magnetic field, perpendicular to the plane of the coil, changes at a rate of 8.65×10^{-3} T/s. Determine (a) the current in the loop, and (b) the rate at which thermal energy is produced.

17. (III) If the U-shaped conductor in Fig. 21–9 has resistivity ρ, whereas that of the moving rod is negligible, derive a formula for the current I as a function of time. Assume the rod has length l, starts at the bottom of the U at $t = 0$, and moves with uniform speed v in the magnetic field B. The cross-sectional area of the rod and all parts of the U is A.

18. (III) The magnetic field perpendicular to a single 13.2-cm-diameter circular loop of copper wire decreases uniformly from 0.750 T to zero. If the wire is 2.25 mm in diameter, how much charge moves past a point in the coil during this operation?

SECTION 21–5

19. (I) The generator of a car idling at 1000-rpm produces 12.4 V. What will the output be at a rotation speed of 2500 rpm assuming nothing else changes?

20. (I) A simple generator is used to generate a peak output voltage of 24.0 V. The square armature consists of windings that are 6.0 cm on a side and rotates in a field of 0.420 T at a rate of 60 rev/s. How many loops of wire should be wound on the square armature?

21. (I) Show that the rms output of an ac generator is $V_{rms} = NAB\omega/\sqrt{2}$.

22. (II) A simple generator has a 720-loop square coil 21.0 cm on a side. How fast must it turn in a 0.650-T field to produce a 120-V peak output?

23. (II) A 450-loop circular armature coil with a diameter of 10.0 cm rotates at 60 rev/s in a uniform magnetic field of strength 0.75 T. What is the rms voltage output of the generator? What would you do to the rotation frequency in order to double the rms voltage output?

*SECTION 21–6

* 24. (I) A motor has an armature resistance of 3.75 Ω. If it draws 9.20 A when running at full speed and connected to a 120-V line, how large is the counter emf?

* 25. (I) The counter emf in a motor is 72 V when operating at 1800 rpm. What would be the counter emf at 2500 rpm if the magnetic field is unchanged?

* 26. (II) The counter emf in a motor is 100 V when the motor is operating at 1000 rpm. How would you change the motor's magnetic field if you wanted to reduce the counter emf to 65 V when the motor was running at 2500 rpm?

* 27. (II) What will be the current in the motor of Example 21–7 if the load causes it to run at half speed?

* 28. (II) The magnetic field of a "shunt-wound" dc motor is produced by field coils placed in parallel to the armature coils. Suppose that the field coils have a resistance of 66.0 Ω and the armature coils 5.00 Ω. The back emf at full speed is 105 V when the motor is connected to a 115-V line. (a) Draw the equivalent circuit for the situations when the motor is just starting and when it is running full speed. (b) What is the total current drawn by the motor at start up? (c) What is the total current drawn when the motor runs at full speed?

29. (II) A dc generator is rated at 10 kW, 200 V, and 50 A when it rotates at 1000 rpm. The resistance of the armature windings is 0.40Ω. (a) Calculate the "no-load" voltage at 1000 rpm (when there is no circuit hooked up to the generator). (b) Calculate the full-load voltage (i.e. at 50 A) when the generator is run at 800 rpm. Assume that the magnitude of the magnetic field remains constant.

SECTION 21–7

30. (I) A transformer is designed to change 120 V into 10,000 V, and there are 5000 turns in the primary. How many turns are in the secondary, assuming 100 percent efficiency?

31. (I) A transformer has 420 turns in the primary and 120 in the secondary. What kind of transformer is this and, assuming 100 percent efficiency, by what factor does it change the voltage? By what factor does it change the current?

32. (I) A step-up transformer increases 16 V to 120 V. What is the current in the secondary as compared to the primary? Assume 100 percent efficiency.

33. (I) Neon signs require 12 kV for their operation. To operate from a 220-V line, what must be the ratio of secondary to primary turns of the transformer? What would the voltage output be if the transformer were connected backward?

34. (II) A model-train transformer plugs into 120 V ac, and draws 0.75 A while supplying 15 A to the train. (a) What voltage is present across the tracks? (b) Is the transformer step-up or step-down?

35. (II) The output voltage of a 100-W transformer is 12 V and the input current is 20 A. (a) Is this a step-up or a step-down transformer? (b) By what factor is the voltage multiplied?

36. (II) High-intensity desk lamps are rated at 40 W but require only 12 V. They contain a transformer that converts 120 V household voltage. (a) Is the transformer step-up or step-down? (b) What is the current in the secondary when the lamp is on? (c) What is the current in the primary? (d) What is the resistance of the bulb when on?

37. (II) A transformer has 330 primary turns and 1240 secondary turns. The input voltage is 120 V and the output current is 15.0 A. What is the output voltage and input current assuming 100 percent efficiency?

38. (II) If 30 MW of power at 45 kV (rms) arrives at a town from a generator via 4.0-Ω transmission lines, calculate (a) the emf at the generator end of the lines, and (b) the fraction of the power generated that is lost in the lines.

39. (II) Show that the power loss in transmission lines, P_L, is given by $P_L = (P_T)^2 R_L / V^2$, where P_T is the power transmitted to the user, V is the delivered voltage, and R_L is the resistance of the power lines.

40. (II) If 50 kW is to be transmitted over two 0.100-Ω lines, estimate how much power is saved if the voltage is stepped up from 120 V to 1200 V and then down again, rather than simply transmitting at 120 V. Assume the transformers are each 99 percent efficient.

41. (III) Design a dc transmission line that can transmit 300 MW of electricity 200 km with only a 2 percent loss. The wires are to be made of aluminum and the voltage is 600 kV.

*SECTION 21–9

* **42. (I)** If the current in a 120-mH coil changes steadily from 25.0 A to 10.0 A in 350 ms, what is the direction and magnitude of the induced emf?

* **43. (I)** What is the inductance L of a 0.60-m-long air-filled coil 2.9 cm in diameter containing 10,000 loops?

* **44. (I)** What is the inductance of a coil if it produces an emf of 8.50 V when the current in it changes from -28.0 mA to $+31.0$ mA in 42.0 ms?

* **45. (I)** How many turns of wire would be required to make a 100-mH inductance out of a 30.0-cm-long air-filled coil with a diameter of 5.2 cm?

* **46. (II)** An air-filled cylindrical inductor has 3000 turns, and it is 2.5 cm in diameter and 28.2 cm long. What is its inductance? How many turns would you need to generate the same inductance if the core were iron-filled instead? Assume the magnetic permeability of iron is about 1000 times that of free space.

* **47. (II)** A coil has 2.25-Ω resistance and 440-mH inductance. If the current is 3.00 A and is increasing at a rate of 3.50 A/s, what is the potential difference across the coil at this moment?

* **48. (II)** The wire of a tightly wound solenoid is unwound and used to make another tightly wound solenoid of twice the diameter. By what factor does the inductance change?

* **49. (II)** (a) Show that the self-inductance L of a torus (Figs. 20–43 and 20–57) of radius R containing N loops each of radius r is

$$L \approx \frac{\mu_0 N^2 r^2}{2R}$$

if $R \gg r$. Assume the field is uniform inside the torus; is this actually true? Is this result consistent with L for a solenoid? Should it be? (b) Calculate the inductance L of a large toroid if the diameter of the coils is 1.5 cm and the diameter of the whole ring is 40 cm. Assume the field inside the toroid is uniform. There are a total of 1000 loops of wire.

* **50. (II)** (a) Ignoring any mutual inductance, what is the equivalent inductance of two inductors connected in series? (b) What if they are connected in parallel? (c) How does their mutual inductance (their geometrical relationship to each other) affect the results?

* **51. (II)** A long thin solenoid of length l and cross-sectional area A contains N_1 closely packed turns of wire. Wrapped tightly around it is an insulated coil of N_2 turns, Fig. 21–50. Assume all the flux from coil 1 (the solenoid) passes through coil 2, and calculate the mutual inductance.

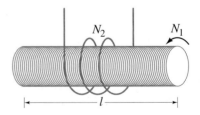

FIGURE 21–50   Problem 51.

* 52. (III) A 30-cm-long coil with 1500 loops is wound on an iron core ($\mu = 3000\,\mu_0$) along with a second coil of 800 loops. The loops of each coil have a radius of 2.0 cm. If the current in the first coil drops uniformly from 3.0 A to zero in 8.0 ms, determine (a) the emf induced in the second coil, and (b) the mutual inductance M.

* 53. (III) The potential difference across a given coil is 22.5 V at an instant when the current is 860 mA and is increasing at a rate of 3.40 A/s. At a later instant, the potential difference is 16.2 V whereas the current is 700 mA and is decreasing at a rate of 1.80 A/s. Determine the inductance and resistance of the coil.

*SECTION 21–10

* 54. (I) The magnetic field inside an air-filled solenoid 36 cm long and 2.0 cm in diameter is 0.80 T. Approximately how much energy is stored in this field?

* 55. (II) At a given instant the current through an inductor is 50.0 mA and is increasing at the rate of 100 mA/s. What is the initial energy stored in the inductor if the inductance is known to be 60.0 mH, and how long does it take for the energy to increase by a factor of 10 from the initial value?

* 56. (II) Assuming the Earth's magnetic field averages about 0.50×10^{-4} T near the surface of the Earth, estimate the total energy stored in this field in the first 10 km above the Earth's surface.

*SECTION 21–11

* 57. (II) It takes 7.20 ms for the current in an LR circuit to reach 80 percent of its maximum value. Determine (a) the time constant of the circuit, and (b) the inductance of the circuit if $R = 250\,\Omega$.

* 58. (II) Determine $\Delta I/\Delta t$ at $t = 0$ (when the battery is connected) for the LR circuit of Fig. 21–29 and show that if I continued to increase at this rate, it would reach its maximum value in one time constant.

* 59. (II) After how many time constants does the current in Fig. 21–29 reach within (a) 10 percent, (b) 1.0 percent, and (c) 0.1 percent of its maximum value?

* 60. (III) Two tightly wound solenoids have the same length and circular cross-sectional area. But solenoid 1 uses wire that is half as thick as solenoid 2. (a) What is the ratio of their inductances? (b) What is the ratio of their inductive time constants (assuming no other resistance in the circuits)?

*SECTION 21–12

* 61. (I) At what frequency will a 160-mH inductor have a reactance of 1.5 kΩ?

* 62. (I) A 9.20-μF capacitor is measured to have a reactance of 250 Ω. At what frequency is it being driven?

* 63. (I) Plot a graph of the impedance of a 1.0-μF capacitor as a function of frequency from 10 to 1000 Hz.

* 64. (I) Plot a graph of the impedance of a 1.0-mH inductor as a function of frequency from 100 to 10,000 Hz.

* 65. (II) Calculate the impedance of, and rms current in, a 160-mH radio coil connected to a 240-V (rms) 10.0-kHz ac line. Ignore resistance.

* 66. (II) An inductance coil operates at 240 V and 60 Hz. It draws 12.8 A. What is the coil's inductance?

* 67. (II) (a) What is the impedance of a well-insulated 0.030-μF capacitor connected to a 2.0-kV (rms) 700-Hz line? (b) What will be the peak value of the current?

* 68. (II) A capacitor is placed in parallel across a load, as in Fig. 21–34b, to filter out stray high-frequency signals, but to allow ordinary 60-Hz ac to pass through with little loss. Suppose that circuit B in the figure is a resistance $R = 300\,\Omega$ connected to ground, and that $C = 0.60\,\mu$F. What percent of the incoming current will pass through C rather than R if (a) it is 60 Hz and (b) it is 60,000 Hz?

* 69. (II) Suppose that circuit B in Fig. 21–34a is a resistance $R = 500\,\Omega$, connected to ground, and the capacitance $C = 2.0\,\mu$F. Will this capacitor act to eliminate 60 Hz ac but pass a high-frequency signal of frequency 60,000 Hz? To check this, determine the voltage drop across R for a 50-mV signal of frequency (a) 60 Hz, and (b) 60,000 Hz.

*SECTION 21–13

* 70. (I) A 30-kΩ resistor is in series with a 0.50-H inductor and an ac source. Calculate the impedance of the circuit if the source frequency is (a) 60 Hz, and (b) 3.0×10^4 Hz.

* 71. (I) A 2.5-kΩ resistor and a 4.0-μF capacitor are connected in series to an ac source. Calculate the impedance of the circuit if the source frequency is (a) 100 Hz, and (b) 10,000 Hz.

* 72. (I) For a 120-V rms 60-Hz voltage, a current of 70 mA passing through the body for 1.0 s could be lethal. What would be the impedance of the body for this to occur?

* 73. (II) A 2.5-kΩ resistor in series with a 420-mH inductor is driven by an ac power supply. At what frequency is the impedance double that of the impedance at 60 Hz?

* 74. (II) (a) What is the rms current in an RC circuit if $R = 28.8\,$kΩ, $C = 0.80\,\mu$F, and the rms applied voltage is 120 V at 60 Hz? (b) What is the phase angle between voltage and current? (c) What is the power dissipated by the circuit? (d) What are the voltmeter readings across R and C?

* 75. (II) (a) What is the rms current in an RL circuit when a 60-Hz 120-V rms ac voltage is applied, where $R = 1.80\,\text{k}\Omega$, and $L = 900\,\text{mH}$? (b) What is the phase angle between voltage and current? (c) How much power is dissipated? (d) What are the rms voltage readings across R and L?

* 76. (II) What is the total impedance, phase angle, and rms current in an LRC circuit connected to a 10.0-kHz, 300-V (rms) source if $L = 22.0\,\text{mH}$, $R = 8.70\,\text{k}\Omega$, and $C = 5000\,\text{pF}$?

* 77. (II) What is the resistance of a coil if its impedance is $35\,\Omega$ and its reactance is $30\,\Omega$?

* 78. (II) A voltage $V = 4.8 \sin 754t$ is applied to an LRC circuit. If $L = 3.0\,\text{mH}$, $R = 1.40\,\text{k}\Omega$, and $C = 3.0\,\mu\text{F}$, how much power is dissipated in the circuit?

* 79. (II) A circuit consists of a 250-Ω resistor in series with a 40.0-mH inductor and a 50.0-V ac generator. The power dissipated by the circuit is 9.50 W. What is the frequency of the generator?

* 80. (II) Show that for the LRC circuit of Fig. 21–35, if we have $I = I_0 \cos \omega t$, then
$$V_R = I_0 R \cos \omega t,$$
$$V_L = I_0 \omega L \cos (\omega t + \pi/2),$$
and
$$V_C = (I_0/\omega C) \cos (\omega t - \pi/2),$$
where $\omega = 2\pi f$.

*SECTION 21–14

* 81. (I) A 3500-pF capacitor is connected to a 50-μH coil of resistance $3.0\,\Omega$. What is the resonant frequency of this circuit?

* 82. (I) The variable capacitor in the tuner of an AM radio has a capacitance of 2800 pF when the radio is tuned to a station at 580 kHz. (a) What must be the capacitance for a station at 1600 kHz? (b) What is the inductance (assumed constant)?

* 83. (II) An LRC circuit has $L = 4.8\,\text{mH}$ and $R = 4.4\,\Omega$. (a) What value must C have to produce resonance at 3600 Hz? (b) What will be the maximum current at resonance if the peak external voltage is 50 V?

* 84. (II) A 3000-pF capacitor is charged to 120 V and then quickly connected to an inductor. The frequency of oscillation is observed to be 20 kHz. Determine (a) the inductance, (b) the peak value of the current, and (c) the maximum energy stored in the magnetic field of the inductor.

*SECTION 21–15

* 85. (I) An audio amplifier has output connections for $4\,\Omega$, $8\,\Omega$, and $16\,\Omega$. If two 8-Ω speakers are to be connected in parallel, to which output terminals should they be connected?

* 86. (I) The output of an amplifier has an impedance of $30\,\text{k}\Omega$. It is to be connected to an 8.0-Ω loudspeaker through a transformer. What should be the turns ratio of the transformer?

GENERAL PROBLEMS

87. Suppose you are looking along a line through the centers of two circular (but separate) wire loops, one behind the other. A battery is suddenly connected to the front loop, establishing a clockwise current. (a) Will a current be induced in the second loop? (b) If so, when does this current start? (c) When does it stop? (d) In what direction is this current? (e) Is there a force between the two loops? (f) If so, in what direction?

88. Suppose you are looking at two current loops in the plane of the page as shown in Fig. 21–51. When the switch is thrown in the left-hand coil, (a) what is the direction of the induced current in the other loop? (b) What is the situation after a "long" time? (c) What is the direction of the induced current in the second loop if the second loop is quickly pulled horizontally to the right?

FIGURE 21–51   Problem 88.

89. A square loop 24.0 cm on a side has a resistance of $6.50\,\Omega$. It is initially in a 0.755-T magnetic field, with its plane perpendicular to **B**, but is removed from the field in 40.0 ms. Calculate the electric energy dissipated in this process.

90. Two conducting rails 30 cm apart rest on a 5.0° ramp. They are joined at the bottom by a $0.60\,\Omega$ resistor, and at the top a copper bar of mass 0.040 kg is laid across the rails. The whole apparatus is immersed in a vertical 0.55 T field. What is the terminal (steady) velocity of the bar as it slides frictionlessly down the rails?

91. A **search coil** for measuring B (also called a *flip coil*) is a small coil with N turns, each of cross-sectional area A. It is connected to a so-called **ballistic galvanometer**, which is a device to measure the total charge Q that passes through it in a short time. The flip coil is placed in the magnetic field to be measured with its face perpendicular to the field. It is then quickly rotated 180°. Show that the total charge Q that flows in the induced current during this short "flip" time is proportional to the magnetic field B; in particular, show that B is given by

$$B = \frac{QR}{2NA}$$

where R is the total resistance of the circuit, including that of the coil and that of the ballistic galvanometer which measures the charge Q.

92. (*a*) Show that the power $P = Fv$ needed to move the conducting rod to the right in Fig. 21–9 is equal to $B^2 l^2 v^2/R$, where R is the total resistance of the circuit. (*b*) Show that this equals the power dissipated in the resistance, I^2R.

93. The primary windings of a transformer, which has an 80% efficiency, are connected to 110 V ac. The secondary windings are connected across a 2.4 Ω, 58 W lightbulb. (*a*) Calculate the current through the primary windings of the transformer. (*b*) Calculate the ratio of the number of primary windings of the transformer to the number of secondary windings of the transformer.

94. A pair of power transmission lines each have a 0.80-Ω resistance and carry 700 A over 9.0 km. If the rms input voltage is 42 kV, calculate (*a*) the voltage at the other end, (*b*) the power input, (*c*) power loss in the lines, and (*d*) the power output.

95. Calculate the peak output voltage of a simple generator whose square armature windings are 6.60 cm on a side if the armature contains 125 loops and rotates in a field of 0.200 T at a rate of 120 rev/s.

96. A small electric car overcomes a 250-N friction force when traveling 30 km/h. The electric motor is powered by ten 12-V batteries connected in series and is coupled directly to the wheels whose diameters are 50 cm. The 300 armature coils are rectangular, 10 cm by 15 cm, and rotate in a 0.60-T magnetic field. (*a*) How much current does the motor draw to produce the required torque? (*b*) What is the back emf? (*c*) How much power is dissipated in the coils? (*d*) What percent of the input power is used to drive the car?

* 97. What is the inductance L of the primary of a transformer whose input is 220 V at 60 Hz and the current drawn is 5.8 A? Assume no current in the secondary.

* 98. A 230-mH coil, whose resistance is 18.5 Ω, is connected to a capacitor C and a 3360-Hz source voltage. If the current and voltage are to be in phase, what value must C have?

* 99. A circuit contains two elements, but it is not known if they are L, R, or C. The current in this circuit when connected to a 120-V 60-Hz source is 5.6 A and lags the voltage by 50°. What are the two elements and what are their values?

*100. A resonant circuit using a 220-pF capacitor is to resonate at 48.0 MHz. The air-core inductor is to be a solenoid with closely packed coils made from 14.0 m of insulated wire 1.1 mm in diameter. How many loops will the inductor contain?

*101. An inductance coil draws 2.5 A dc when connected to a 36-V battery. When connected to a 60-Hz 120-V (rms) source, the current drawn is 3.8 A (rms). Determine the inductance and resistance of the coil.

*102. The **Q factor** of a resonance circuit is defined as the ratio of the voltage across the capacitor (or inductor) to the voltage across the resistor, at resonance. The larger the Q factor, the sharper the resonance curve will be and the sharper the tuning. (*a*) Show that the Q factor is given by the equation $Q = (1/R)\sqrt{L/C}$. (*b*) At a resonant frequency $f_0 = 1.0$ MHz, what must be the value of L and R to produce a Q factor of 1000? Assume that $C = 0.010\ \mu\text{F}$.

*103. In a series LRC circuit, the inductance is 20 mH, the capacitance is 50 nF, and the resistance is 200 Ω. At what frequencies is the power factor equal to 0.17?

Electromagnetic waves, such as radio and TV signals, can be sent, and received, by means of antennas like these. All EM waves, at whatever frequency, travel at the speed of light. And light itself is an EM wave.

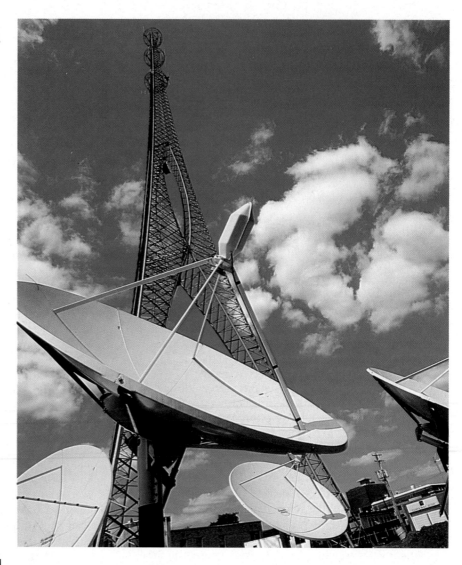

22 ELECTROMAGNETIC WAVES

The culmination of electromagnetic theory in the nineteenth century was the prediction, and the experimental verification, that waves of electromagnetic fields could travel through space. This achievement opened a whole new world of communication—first the wireless telegraph, then radio and television. And it yielded the spectacular prediction that light is an electromagnetic wave.

The theoretical prediction of electromagnetic waves was the work of the Scottish physicist James Clerk Maxwell (1831–1879; Fig. 22–1), who unified, in one magnificent theory, all the phenomena of electricity and magnetism.

22–1 Changing Electric Fields Produce Magnetic Fields; Maxwell's Equations

The development of electromagnetic theory in the early part of the nineteenth century by Oersted, Ampère, and others was not actually done in terms of electric and magnetic fields. The idea of the field was introduced somewhat later by Faraday, and was not generally used until Maxwell showed that all electric and magnetic phenomena could be described using only four equations involving electric and magnetic fields. These equations, known as **Maxwell's equations**, are the basic equations for all electromagnetism. They are fundamental in the same sense that Newton's three laws of motion and the law of universal gravitation are for mechanics. In a sense, they are even more fundamental, since they are consistent with the theory of relativity (Chapter 26), whereas Newton's laws are not. Because all of electromagnetism is contained in this set of four equations, Maxwell's equations are considered one of the great triumphs of the human mind.

Although we will not present Maxwell's equations in mathematical form since they involve calculus, we will summarize them here in words. They are: (1) a generalized form of Coulomb's law known as Gauss's law (discussed in Appendix D) that relates electric field to its sources, electric charge; (2) a similar law for the magnetic field, except that since there seem to be no single magnetic poles (monopoles), there will be no single magnetic charges ($q_B = 0$, let's say), and so magnetic field lines are always continuous—they do not begin or end (as electric field lines do, on charges); (3) an electric field is produced by a changing magnetic field; (4) a magnetic field is produced by an electric current, or by a changing electric field.

Maxwell's equations

Law number (3) is Faraday's law (see Chapter 21, especially Section 21–4). The first part of (4), that a magnetic field is produced by an electric current, was discovered by Oersted, and the mathematical relation is given by Ampère's law (Section 20–8). But the second part of (4) is an entirely new aspect predicted by Maxwell. Maxwell argued that if a changing magnetic field produces an electric field, as given by Faraday's law, then the reverse might be true as well: **a changing electric field will produce a magnetic field**. This was a *hypothesis* by Maxwell, based on the idea of symmetry in nature. Indeed, the size of the effect in most cases is so small that Maxwell recognized it would be difficult to detect it experimentally. We look at this hypothesis in a little detail in the next (optional) Section.

*Changing **E** produces **B***

FIGURE 22–1
James Clerk Maxwell.

22–2 Maxwell's Fourth Equation; Displacement Current

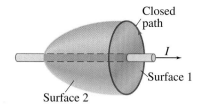

FIGURE 22–2 Ampère's law applied to two different surfaces bounded by the same closed path.

Maxwell used an indirect argument to back up his idea that a changing electric field would produce a magnetic field. In a simplified fashion it goes something like this. According to Ampère's law (Section 20–8), $\Sigma B_{\parallel} \Delta l = \mu_0 I$. That is, you divide any closed path you choose into short segments Δl, multiply each segment by the parallel component of the magnetic field B at that segment, and then sum all these products over the complete closed path; this sum will then equal μ_0 times the net current I that passes through the surface bounded by the path. When we applied Ampère's law to the field around a straight wire, we imagined the current as passing through the circle enclosed by our circular loop. This would be the flat surface 1 in Fig. 22–2. However, we could just as well use the sack-shaped surface 2 in the figure, since the same current I passes through it. This naturally implies that the current passing through any surface bounded by the closed path must be the same. Thus the current flowing into the volume enclosed by surfaces 1 and 2 together equals the current that flows out of this volume. This is basically Kirchhoff's point rule, and tells us that the rate at which charge enters the volume equals the rate at which it leaves.

Now consider the closed path for the situation of Fig. 22–3a, where a capacitor is being discharged. Ampère's law works for surface 1, but it does not work for surface 2, since no current passes through surface 2. There is a magnetic field around the wire, so the left side of Ampère's law is not zero; yet for surface 2, the right side *is* zero, since $I = 0$. We seem to have a contradiction of Ampère's law. Similar things can be said for the closed path indicated in Fig. 22–3b, which surrounds the region of the electric field of the plates; there is a current passing through surface 1, but not through surface 2, and there is a magnetic field around the closed path. There is a magnetic field present in Fig. 22–3, however, only if charge is flowing to or away from the capacitor plates. In this case, the electric field between the plates is changing in time. Maxwell resolved the problem of no current through surface 2 in Figs. 22–3a and b by stating that the

FIGURE 22–3 A capacitor discharging. No conduction current passes through surface 2 in either (a) or (b). An extra term is needed in Ampère's law.

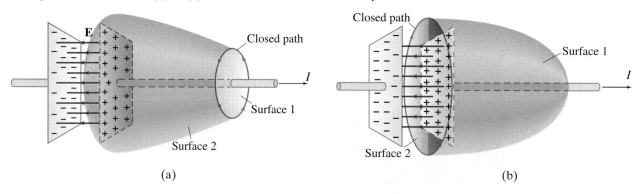

(a)

(b)

changing electric field between the plates is *equivalent to* an electric current. He called this a **displacement current**, I_D. (The name was based on an old discarded theory, and is not especially illuminating today.) An ordinary current is then called a **conduction current**, I_C. Ampère's law, as generalized by Maxwell, becomes

Displacement current defined

$$\Sigma B_{\parallel}\, \Delta l = \mu_0(I_C + I_D).$$

Ampère's law will now apply even for surface 2 in Figs. 22–3a and b, where the displacement current I_D refers to the changing electric field.[†]

We can write I_D in terms of the changing electric field. The charge Q on a capacitor of capacitance C is $Q = CV$, where V is the potential difference between the plates. Also recall that $V = Ed$, where d is the (small) separation of the plates and E is the electric field strength between them; we ignore any fringing of the field. Also, for a parallel-plate capacitor, $C = \epsilon_0 A/d$, where A is the area of each plate (see Chapter 17). We combine these to obtain

$$Q = CV = \left(\epsilon_0 \frac{A}{d}\right)(Ed) = \epsilon_0 AE.$$

Now if the charge Q on the plate changes at a rate $\Delta Q/\Delta t$, the electric field changes at a proportional rate. Then, from the above expression,

$$\frac{\Delta Q}{\Delta t} = \epsilon_0 A \frac{\Delta E}{\Delta t}.$$

But $\Delta Q/\Delta t$ is also the current flowing into or out of the capacitor. If the current flow into the capacitor, $\Delta Q/\Delta t$, is set equal to the displacement current I_D between the plates, then

$$I_D = \frac{\Delta Q}{\Delta t} = \epsilon_0 A \frac{\Delta E}{\Delta t},$$

or

$$I_D = \epsilon_0 \frac{\Delta \Phi_E}{\Delta t},$$

where $\Phi_E = EA$ is the **electric flux**, defined in analogy to magnetic flux (Section 21–2). Then, Ampère's law becomes

$$\Sigma B_{\parallel}\, \Delta l = \mu_0 I_C + \mu_0 \epsilon_0 \frac{\Delta \Phi_E}{\Delta t}. \qquad \textbf{(22–1)}$$

Ampère's law (generalized)

This equation embodies Maxwell's idea that a magnetic field can be caused not only by a normal electric current, but also by a changing electric field or changing electric flux. Eq. 22–1 is essentially Maxwell's fourth equation.[‡]

[†]Notice that this interpretation of the changing electric field as a displacement current fits in well with our discussion in Chapter 21 where we saw that an alternating current passes through a capacitor. It also means that Kirchhoff's point rule will be valid even at a capacitor plate: when a capacitor is being charged, conduction current flows into the capacitor plate, but no conduction current flows out of the plate. Instead, a "displacement current" flows out of one plate toward the other plate.

[‡]Actually, there is also a third term on the right for the case when a magnetic field is produced by magnetized materials; but we assume in what follows that no magnets are present.

Production of Electromagnetic Waves

According to Maxwell (as mentioned at the end of Section 22–1, and expanded upon in Section 22–2), a magnetic field will be produced in empty space if there is a changing electric field. From this, Maxwell derived another startling conclusion. If a changing magnetic field produces an electric field, that electric field is itself changing. This changing electric field will, in turn, produce a magnetic field, which will be changing and so will produce a changing electric field; and so on. When Maxwell manipulated his equations, he found that the net result of these interacting changing fields was to produce a *wave* of electric and magnetic fields that can actually propagate (travel) through space. We now examine, in a simplified way, how such **electromagnetic waves** can be produced.

How EM waves are produced

Consider two conducting rods that will serve as an "antenna" (Fig. 22–4a). Suppose that these two rods are connected by a switch to the opposite terminals of a battery. As soon as the switch is closed, the upper rod quickly becomes positively charged and the lower one negatively charged. Electric field lines are formed as indicated by the lines in Fig. 22–4b. While the charges are flowing, a current exists, whose direction is indicated by the arrows. A magnetic field is therefore produced near the antenna. The magnetic field lines encircle the wires and therefore, in the plane of the page, **B** points into the page ($\otimes$) on the right and out of the page ($\odot$) on the left. Now we ask, how far out do these electric and magnetic fields extend? In the static case, the fields extend outward indefinitely far. However, when the switch in Fig. 22–4 is closed, the fields quickly appear nearby, but it takes time for them to reach distant points. Both electric and magnetic fields store energy, and this energy cannot be transferred to distant points at infinite speed.

Now we look at the situation of Fig. 22–5 where our antenna is connected to an ac generator. In Fig. 22–5a, the connection has just been completed. Charge starts building up and fields form just as in Fig. 22–4. The + and − signs in Fig. 22–5a indicate the net charge on each rod. The arrows indicate the direction of the current. The electric field is represented by lines in the plane of the page; and the magnetic field, according to the right-hand rule, is into ($\otimes$) or out of ($\odot$) the page. In Fig. 22–5b, the emf of the ac generator has reversed in direction; the current is reversed and the new magnetic field is in the opposite direction. Because the new fields have changed direction, the old lines fold back to connect up to some of the new lines and form closed loops as shown.[†] The old fields, however, don't suddenly disappear; they are on their way to distant points. Indeed, because a changing magnetic field produces an electric field, and a changing electric field produces a magnetic field, this combination of changing electric and magnetic fields moving outward is self-supporting, no longer depending on the antenna

(a)

(b)

B is out $\odot$ **B** is in

FIGURE 22–4 Fields produced by charge flowing into conductors. It takes time for the **E** and **B** fields to travel outward to distant points.

[†]We are considering waves traveling through empty space, so there are no charges for lines of **E** to start or stop on, so they form closed loops. Magnetic field lines always form closed loops since there are no single (separate) magnetic poles (as far as we know).

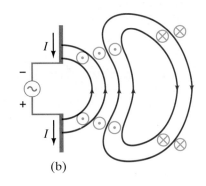

(a)   (b)

FIGURE 22–5 Sequence showing electric and magnetic fields that spread outward from oscillating charges on two conductors connected to an ac source (see the text).

charges. The fields not far from the antenna, referred to as the *near field*, become quite complicated, but we are not so interested in them. We are instead mainly interested in the fields far from the antenna (they are generally what we detect), which we refer to as the **radiation field**. The electric field lines form loops, as shown in Fig. 22–6, and continue moving outward. The magnetic field lines also form closed loops, but are not shown since they are perpendicular to the page. Although the lines are shown only on the right of the source, fields also travel in other directions. (The field strengths are greatest in directions perpendicular to the oscillating charges; and they drop to zero along the direction of oscillation—above and below the antenna in Fig. 22–6.)

The magnitudes of both **E** and **B** in the radiation field are found to decrease with distance as $1/r$. (Compare this to the static electric field given by Coulomb's law where **E** decreases as $1/r^2$.) The energy carried by the electromagnetic wave is proportional (as for any wave, Chapter 11) to the square of the amplitude, E^2 or B^2, as will be discussed further in Section 22–7, so the intensity of the wave decreases as $1/r^2$.

Several things about the radiation field can be noted from Fig. 22–6. First, *the electric and magnetic fields at any point are perpendicular to each other, and to the direction of motion.* Second, we can see that the fields alternate in direction (**B** is into the page at some points and out of the page at others; similarly for **E**). Thus, the field strengths vary from a maximum in one direction, to zero, to a maximum in the other direction. The electric and magnetic fields are "in phase"; that is, they each are zero at the same points and reach their maximum at the same points in space.

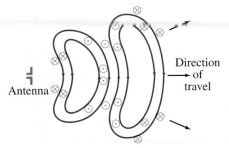

Antenna

Direction of travel

FIGURE 22–6 The radiation fields (far from the antenna) produced by a sinusoidal signal on a dipole antenna. The closed loops represent electric field lines. The magnetic field lines, perpendicular to the page and represented by $\otimes$ and $\odot$, also form closed loops.

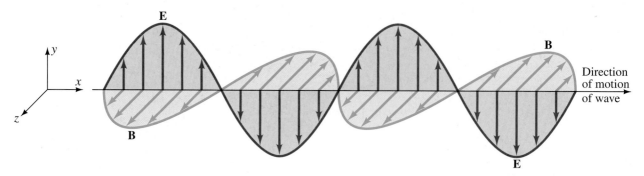

FIGURE 22–7 Electric and magnetic field strengths in an electromagnetic wave. **E** and **B** are at right angles to each other. The entire pattern moves in a direction perpendicular to both **E** and **B**.

If the source emf varies sinusoidally, then the electric and magnetic field strengths in the radiation field will also vary sinusoidally. The sinusoidal character of the waves is diagrammed in Fig. 22–7, which shows the field *strengths* plotted as a function of position. Notice that **B** and **E** are perpendicular to each other and to the direction of travel.

We call these waves electromagnetic (EM) waves. They are *transverse* waves and resemble other types of waves (Chapter 11). However, EM waves are always waves of *fields*, not of matter, as are waves on water or a rope. Because they are fields, EM waves can propagate in empty space.

We have seen in the above analysis that EM waves are produced by electric charges that are oscillating, and hence are undergoing acceleration. In fact, we can say in general that **accelerating electric charges give rise to electromagnetic waves**.

EM waves are produced by accelerating electric charges

In the next (optional) Section, we will derive a formula for the speed of EM waves:

Speed of EM waves

$$v = \frac{1}{\sqrt{\epsilon_0 \mu_0}}. \tag{22–2}$$

which Maxwell himself derived. When Maxwell put in the values for ϵ_0 and μ_0 he found

$$v = \frac{1}{\sqrt{\epsilon_0 \mu_0}}$$

$$= \frac{1}{\sqrt{(8.85 \times 10^{-12} \ \text{C}^2/\text{N·m}^2)(4\pi \times 10^{-7} \ \text{N·s}^2/\text{C}^2)}}$$

$$= 3.00 \times 10^8 \ \text{m/s},$$

which is equal to the measured speed of light.

* 22–4 Calculation of the Speed of Electromagnetic Waves

Maxwell's prediction that EM waves should exist was remarkable. Equally remarkable was the speed at which they were predicted to travel. We now calculate this speed.

We shall consider a region far from the source, so that the wave fronts (the field lines in Fig. 22–6) are essentially flat over a reasonable area. They are then called **plane waves**, meaning that at any instant, **E** and **B** are uniform over a plane perpendicular to the direction of propagation. Let us assume that, in a particular coordinate system, the wave is traveling in the

x direction with speed v; **E** is parallel to the y axis, and **B** is parallel to the z axis, as shown in Fig. 22–7.

Now let us apply Faraday's law to an imaginary rectangle $abcd$ placed in the xy plane as shown with gray shading in Fig. 22–8. We assume that the side ab is in a region where the wave has not yet reached, so $E = B = 0$ there. As the wave moves, its magnetic flux through our rectangular loop changes. Instead of showing the motion of the wave in Fig. 22–8, we show rather the relative position of the rectangle (it's easier to visualize). After a short time Δt, the wave moves to the right a distance $\Delta x = v \Delta t$, where v is the velocity of the wave; or, equivalently, the rectangular loop moves $\Delta x = v \Delta t$ to the left. Rectangle $a'b'c'd'$ represents the new position of the rectangle relative to the wave. The change in magnetic flux $\Delta \Phi_B$ through the loop during the time Δt is just the flux that passes through the small rectangle $dd'c'c$. Thus, $\Delta \Phi_B = B \Delta A = By_0 \Delta x = By_0 v \Delta t$, where y_0 is the width ab or cd, and $\Delta A = y_0 \Delta x = y_0 v \Delta t$ is the area of $dd'c'c$; of course B represents the magnetic field passing through this area (represented by the heavy arrow in Fig. 22–8). We assume Δx is small so that B is essentially constant over this small area. According to Faraday's law, then, the emf induced around the loop $abcd$ equals the rate of change of magnetic flux:

$$\mathscr{E} = \frac{\Delta \Phi_B}{\Delta t} = \frac{B \Delta A}{\Delta t} = \frac{By_0 v \Delta t}{\Delta t} = By_0 v.$$

(We omitted the minus sign for convenience.) The emf around the loop $abcd$ is the sum of the emfs in each straight section:

$$\mathscr{E} = \mathscr{E}_{ab} + \mathscr{E}_{bc} + \mathscr{E}_{cd} + \mathscr{E}_{da}.$$

Now the emf is the work done per unit charge: $\mathscr{E} = W/q = Fd/q = Ed$, where F is the force exerted over the distance d, and the electric field E is the force per unit charge, $E = F/q$. (The result, $\mathscr{E} = Ed$, is the same as Eq. 17–2.) The term $\mathscr{E}_{ab}$ is zero since $E = 0$ in this region; and $\mathscr{E}_{bc} = \mathscr{E}_{da} = 0$, since **E** is perpendicular to the path (Fig. 22–8). Thus

$$\mathscr{E} = \mathscr{E}_{cd} = Ey_0,$$

where E is the magnitude of the electric field along cd (or $c'd'$). From Faraday's law as we wrote it above ($\mathscr{E} = By_0 v$), we thus have

$$Ey_0 = By_0 v$$

$$E = vB. \tag{22–3}$$

Thus we see that at any point in space, E and B are in the ratio $E/B = v$, the velocity of the wave.

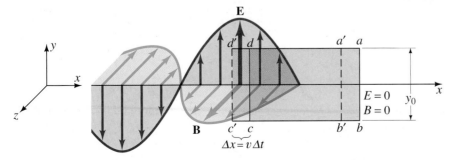

FIGURE 22–8 In a time Δt, the rectangle in the xy plane (actually stationary) moves a distance $\Delta x = v \Delta t$ relative to the wave (the wave moves to the right with speed v).

Now we consider a rectangle in the xz plane, as shown (gray) in Fig. 22–9. Again we show the rectangle moving to the left a distance $\Delta x = v \Delta t$ relative to the wave, although actually the rectangle is stationary and the wave is moving to the right; $abcd$ is the position of the rectangle initially and $a'b'c'd'$ is its position after a time Δt. There is a changing electric flux through this rectangular loop equal to the electric field E (heavy arrow) times the increasing area $\Delta A = z_0 \Delta x = z_0 v \Delta t$ (where z_0 is the width $ab = cd$ of the rectangle). According to Ampère's law, Eq. 22–1 with $I_C = 0$ since there are no conduction currents, we have

$$\Sigma B_\parallel \, \Delta l = \mu_0 \epsilon_0 \frac{\Delta \Phi_E}{\Delta t}$$

$$= \mu_0 \epsilon_0 \frac{(E)(z_0 v \, \Delta t)}{\Delta t} = \mu_0 \epsilon_0 E z_0 v.$$

The sum of $B_\parallel \Delta l$ on the sides ab, bc, and da are all zero, because either $B = 0$ or $\mathbf{B}$ is perpendicular to these sides. But along side cd, the contribution is Bz_0, where B is the magnetic field (heavy arrow) parallel to cd. Thus,

$$Bz_0 = \mu_0 \epsilon_0 E z_0 v$$

so

$$B = \mu_0 \epsilon_0 v E.$$

We combine this with Eq. 22–3 and find that

$$B = \mu_0 \epsilon_0 v (vB) = \mu_0 \epsilon_0 v^2 B.$$

We cancel B on both sides and solve for v:

$$v = \frac{1}{\sqrt{\epsilon_0 \mu_0}}$$

which is just Eq. 22–2 that we quoted before and have now derived.

FIGURE 22–9 Rectangle in the xz plane moves a distance $\Delta x = v \Delta t$ relative to a wave traveling to the right.

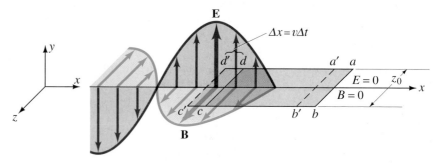

When we put in the values for ϵ_0 and μ_0, we find that

$$v = \frac{1}{\sqrt{\epsilon_0 \mu_0}} = \frac{1}{\sqrt{(8.85 \times 10^{-12}\,\text{C}^2/\text{N·m}^2)(4\pi \times 10^{-7}\,\text{N·s}^2/\text{C}^2)}}$$

$$= 3.00 \times 10^8\,\text{m/s}.$$

Speed of EM waves is

speed of light

This result, that the speed of EM waves is precisely equal to the measured speed of light, is truly remarkable.

22–5 Light as an Electromagnetic Wave and the Electromagnetic Spectrum

The calculation at the end of the last section gives the result that Maxwell himself determined: that the speed of EM waves is 3.00×10^8 m/s, the same as the measured speed of light.

Light had been shown some 60 years previously to behave like a wave (we'll discuss this in Chapter 24). But nobody knew what kind of wave it was—that is, what is it that is oscillating in a light wave? Maxwell, on the basis of the calculated speed of EM waves, argued that light must be an electromagnetic wave. This idea soon came to be generally accepted by scientists, but not fully until after EM waves were experimentally detected. EM waves were first generated and detected experimentally by Heinrich Hertz (1857–1894) in 1887, eight years after Maxwell's death. Hertz used a spark-gap apparatus in which charge was made to rush back and forth for a short time, generating waves whose frequency was about 10^9 Hz. He detected them some distance away using a loop of wire in which an emf was produced when a changing magnetic field passed through. These waves were later shown to travel at the speed of light, 3.00×10^8 m/s, and to exhibit all the characteristics of light such as reflection, refraction, and interference. The only difference was that they were not visible. Hertz's experiment was a strong confirmation of Maxwell's theory.

The wavelengths of visible light were measured in the first decade of the nineteenth century, long before anyone imagined that light was an electromagnetic wave. The wavelengths were found to lie between 4.0×10^{-7} m and 7.5×10^{-7} m; or 400 nm to 750 nm (1 nm $= 10^{-9}$ m). The frequencies of visible light can be found using Eq. 11–12, which we rewrite here:

$$f\lambda = c, \tag{22–4}$$

where f and λ are the frequency and wavelength, respectively, of the wave. Here, c is the speed of light, 3.00×10^8 m/s; it gets the special symbol c because of its universality for all EM waves in free space. Equation 22–4 tells us that the frequencies of visible light are between 4.0×10^{14} Hz and 7.5×10^{14} Hz. (Recall that 1 Hz $= 1$ cycle per second $= 1\,\text{s}^{-1}$.)

c is symbol for speed of light

But visible light is only one kind of EM wave. As we have seen, Hertz produced EM waves of much lower frequency, about 10^9 Hz. These are called **radio waves**, since frequencies in this range are used today to transmit radio and TV signals. Electromagnetic waves, or EM radiation as we sometimes call it, have been produced or detected over a wide range of frequencies. They are usually categorized as shown in Fig. 22–10. This is *EM spectrum* known as the **electromagnetic spectrum**.

Radio waves and microwaves can be produced in the laboratory using electronic equipment, as we saw in Fig. 22–5. Higher-frequency waves are very difficult to produce electronically. These and other types of EM waves are produced in natural processes, as emission from atoms, molecules, and nuclei (more on this in later chapters). Generally, EM waves are produced by the acceleration of electrons or other charged particles, such as electrons accelerating in the antenna of Fig. 22–5. Another example is X-rays, which are produced (see Chapters 25 and 28) when fast-moving electrons are rapidly decelerated upon striking a metal target. Even the visible light emitted by an ordinary incandescent light is due to electrons undergoing acceleration within the hot filament. We will meet various types of EM waves later. However, it is worth mentioning here that infrared (IR) radiation (EM waves whose frequency is just less than that of visible light) is mainly responsible for the heating effect of the Sun. The Sun emits not only visible light but substantial amounts of IR and UV (ultraviolet) as well. The molecules of our skin tend to "resonate" at infrared frequencies, so it is these that are preferentially absorbed and thus warm us up. We humans experience EM waves differently depending on their wavelengths: Our eyes detect wavelengths between 4 and 7×10^{-7} m (visible light), whereas our skin detects longer wavelengths. Many EM wavelengths we don't detect directly at all.

FIGURE 22–10   Electromagnetic spectrum.

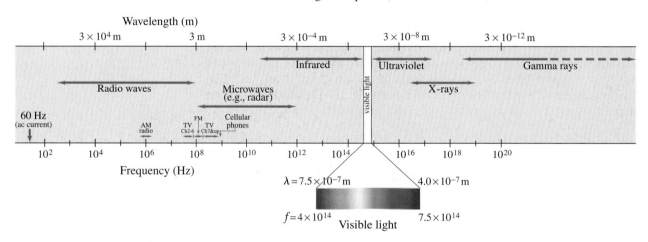

EXAMPLE 22–1 **Wavelengths of EM waves.** Calculate the wavelength: (a) of a 60-Hz EM wave, (b) of a 93.3-MHz FM radio wave, and (c) of a beam of visible red light from a laser at frequency 4.74×10^{14} Hz.

SOLUTION (a) Since $c = \lambda f$,

$$\lambda = \frac{c}{f} = \frac{3.0 \times 10^8 \text{ m/s}}{60 \text{ s}^{-1}} = 5.0 \times 10^6 \text{ m},$$

or 5000 km. 60 Hz is the frequency of ac current in the United States, and, as we see here, one wavelength stretches all the way across the country.

(b)
$$\lambda = \frac{3.00 \times 10^8 \text{ m/s}}{93.3 \times 10^6 \text{ s}^{-1}} = 3.22 \text{ m}.$$

The length of an FM antenna is about half this ($\frac{1}{2}\lambda$).

(c)
$$\lambda = \frac{3.00 \times 10^8 \text{ m/s}}{4.74 \times 10^{14} \text{ s}^{-1}} = 6.33 \times 10^{-7} \text{ m} \ (= 633 \text{ nm}).$$

Electromagnetic waves can travel along transmission lines as well as in empty space. When a source of emf is connected up to a transmission line—be it two parallel wires or a coaxial cable (Fig. 22–11)—the electric field within the wire is not set up immediately at all points along the wires. This is based on the same argument we used in Section 22–3 with reference to Fig. 22–5. Indeed, it can be shown that if the wires are separated by air, the electrical signal travels along the wires at the speed $c = 3.0 \times 10^8$ m/s. For example, when you flip a light switch, the light actually goes on a tiny fraction of a second later. If the wires are in a medium whose electric permittivity is ϵ and magnetic permeability is μ, the speed is not given by Eq. 22–2, but by

$$v = 1/\sqrt{\epsilon\mu}.$$

EXAMPLE 22–2 **Voice speed through the wires.** When you speak on the telephone from Los Angeles to a friend in New York 4000 km away, how long does it take your voice to travel?

SOLUTION Since speed = distance/time, then time = distance/speed = $(4.0 \times 10^6 \text{ m})/(3.0 \times 10^8 \text{ m/s}) = 1.3 \times 10^{-2}$ s, or about $\frac{1}{100}$ s.

FIGURE 22–11
Coaxial cable.

22–6 Measuring the Speed of Light

The first serious attempt to actually measure the speed of light was done by Galileo by trying to measure the time required for light to travel a known distance between two hilltops. He stationed an assistant on one hilltop, and himself on another, and ordered the assistant to lift the cover from a lamp the instant he saw a flash from Galileo's lamp. Galileo measured the time between the flash of his lamp and when he received the light from his assistant's lamp. The time was so short that Galileo concluded it merely represented human reaction time, and that the speed of light must be extremely high.

The first successful determination that the speed of light is finite was made by the Danish astronomer Ole Roemer (1644–1710). Roemer had noted that the carefully measured period of Io, one of Jupiter's moons (an average period of 42.5 h to make one complete orbit around Jupiter), varied slightly, depending on the relative motion of Earth and Jupiter. When Earth was moving away from Jupiter, the period of the moon was slightly longer, and when Earth was moving toward Jupiter, the period was slightly shorter. He attributed this variation to the extra time needed for light to travel the increasing distance to Earth when Earth is receding, or to the shorter travel time for the decreasing distance when the two planets are approaching one another. Roemer concluded that the speed of light—though great—is finite.

Michelson measures c   Since then a number of techniques have been used to measure the speed of light. Among the most important were those carried out by the American Albert A. Michelson (1852–1931). Michelson used the rotating mirror apparatus diagrammed in Fig. 22–12 for a series of high-precision experiments carried out from 1880 to the 1920s. Light from a source was directed at one face of a rotating eight-sided mirror. The reflected light traveled to a stationary mirror a large distance away and back again as shown. If the rotating mirror was turning at just the right rate, the returning beam of light would reflect from one face of the mirror into a small telescope through which the observer looked. If the speed of rotation was only slightly different, the beam would be deflected to one side and would not be seen by the observer. From the required speed of the rotating mirror and the known distance to the stationary mirror, the speed of light could be calculated. In the 1920s, Michelson

FIGURE 22–12
Michelson's speed-of-light apparatus (not to scale).

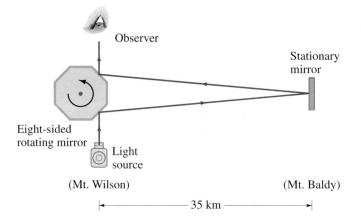

set up the rotating mirror on the top of Mt. Wilson in southern California and the stationary mirror on Mt. Baldy (Mt. San Antonio) 35 km away. He later measured the speed of light in vacuum using a long evacuated tube.

The accepted value today for the speed of light, c, in vacuum is

$$c = 2.99792458 \times 10^8 \text{ m/s}.$$

We usually round this off to

$$c = 3.00 \times 10^8 \text{ m/s}$$

when extremely precise results are not required. In air, the speed is only slightly less.

* 22–7 Energy in EM Waves

Electromagnetic waves carry energy from one region of space to another. This energy is associated with the moving electric and magnetic fields. In Section 17–9, we saw that the energy density (J/m³) stored in an electric field E is $u = \frac{1}{2}\epsilon_0 E^2$, where u is the energy per unit volume. The energy stored in a magnetic field B, as we discussed in Section 21–10 (Eq. 21–10), is given by $u = \frac{1}{2}B^2/\mu_0$. Thus, the total energy stored per unit volume in a region of space where there is an electromagnetic wave is

$$u = \frac{1}{2}\epsilon_0 E^2 + \frac{1}{2}\frac{B^2}{\mu_0}. \tag{22–5}$$

In this equation, E and B represent the electric and magnetic field strengths of the wave at any instant in a small region of space. We can write Eq. 22–5 in terms of the E field only, since from Eq. 22–2 we have $\sqrt{\epsilon_0\mu_0} = 1/c$, and from Eq. 22–3, $B = E/c$. We insert these into Eq. 22–5 to obtain

$$u = \frac{1}{2}\epsilon_0 E^2 + \frac{1}{2}\frac{\epsilon_0\mu_0 E^2}{\mu_0}$$

$$= \epsilon_0 E^2. \tag{22–6a}$$

Notice that the energy density associated with the B field is equal to that associated with the E field, so each contributes half to the total energy. We can also write the energy density in terms of the B field only:

$$u = \epsilon_0 E^2 = \epsilon_0 c^2 B^2 = \frac{\epsilon_0 B^2}{\epsilon_0\mu_0},$$

so

$$u = \frac{B^2}{\mu_0}. \tag{22–6b}$$

Or we can write u in one term containing both E and B:

$$u = \epsilon_0 E^2 = \epsilon_0 EcB = \frac{\epsilon_0 EB}{\sqrt{\epsilon_0\mu_0}},$$

or

$$u = \sqrt{\frac{\epsilon_0}{\mu_0}}\, EB. \tag{22–6c}$$

Equations 22–6 give the energy density in any region of space at any instant.

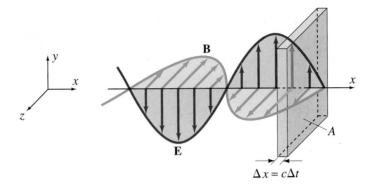

FIGURE 22–13
Electromagnetic wave carrying energy through area A.

Now let us determine the energy that is transported by the wave per unit time per unit area perpendicular to the wave direction. Let us imagine that the wave is passing through an area A perpendicular to the x axis, as shown in Fig. 22–13. In a short time Δt, the wave moves to the right a distance $\Delta x = c\,\Delta t$. The energy that has passed through the area A in the time Δt is the energy that now occupies the volume $\Delta V = A\,\Delta x = Ac\,\Delta t$. The energy density u is $u = \epsilon_0 E^2$, where E is the electric field in this volume at the given instant. So the energy ΔU contained in this volume is the energy density u times the volume: $\Delta U = u\,\Delta V = (\epsilon_0 E^2)(Ac\Delta t)$. Therefore, the energy crossing the area A per time Δt, which we designate[†] S, is

$$S = \frac{\Delta U}{A\,\Delta t} = \epsilon_0 c E^2,$$

and is measured in watts per square meter (W/m²). Since $E = cB$ and $c = 1/\sqrt{\epsilon_0 \mu_0}$, (Eqs. 22–3 and 22–2), this can also be written

Rate energy is transported by EM waves

$$S = \epsilon_0 c E^2 = \frac{cB^2}{\mu_0} = \frac{EB}{\mu_0}. \tag{22–7}$$

Equation 22–7 gives the energy transported per unit area per unit time at any *instant*. We often want to know the *average* over an extended period of time. If E and B are sinusoidal, then $\overline{E^2} = E_0^2/2$, just as for electric currents and voltages (Section 18–8), where E_0 is the *maximum* value of E. Thus we can write

$$\overline{S} = \frac{1}{2}\epsilon_0 c E_0^2 = \frac{1}{2}\frac{c}{\mu_0} B_0^2 = \frac{E_0 B_0}{2\mu_0}, \tag{22–8}$$

where B_0 is the maximum value of B. We can also write

$$\overline{S} = \frac{E_{\text{rms}} B_{\text{rms}}}{\mu_0}$$

where E_{rms} and B_{rms} are the rms values ($E_{\text{rms}} = \sqrt{\overline{E^2}}$, $B_{\text{rms}} = \sqrt{\overline{B_{\text{rms}}^2}}$).

[†]The quantity S is called the *Poynting vector*. Its direction is that in which the energy is being transported, which is the direction the wave is traveling.

EXAMPLE 22–3 E and B from the Sun. Radiation from the Sun reaches the Earth (above the atmosphere) at a rate of about 1350 J/s·m². Assume that this is a single EM wave and calculate the maximum values of E and B.

SOLUTION Since $\bar{S} = 1350 \text{ J/s·m}^2 = \epsilon_0 c E_0^2/2$, then

$$E_0 = \sqrt{\frac{2\bar{S}}{\epsilon_0 c}}$$

$$= \sqrt{\frac{2(1350 \text{ J/s·m}^2)}{(8.85 \times 10^{-12} \text{ C}^2/\text{N·m}^2)(3.0 \times 10^8 \text{ m/s})}}$$

$$= 1.01 \times 10^3 \text{ V/m}.$$

From Eq. 22–3, $B = E/c$, so

$$B_0 = \frac{E_0}{c} = \frac{1.01 \times 10^3 \text{ V/m}}{3.00 \times 10^8 \text{ m/s}} = 3.37 \times 10^{-6} \text{ T}.$$

This Example illustrates that B has a small numerical value compared to E. This is because of the different units for E and B and the way these units are defined. But, as we saw earlier, B contributes the same energy to the wave as E does.

EXAMPLE 22–4 Energy in E and B. (a) Show that Eq. 22–8 connecting the power flux (S has units W/m²) and the electric field works out unit-wise in the metric system. (b) Calculate the energy density due to electric and magnetic contributions in Example 22–3.

SOLUTION (a) The equation reads $\bar{S} = \frac{1}{2}\epsilon_0 c E_0^2$ so the units break down as follows:

$$\epsilon_0 c E_0^2 \text{ units} \Rightarrow \frac{\text{C}^2}{\text{N·m}^2} \times \frac{\text{m}}{\text{s}} \times \frac{\text{N}^2}{\text{C}^2} = \frac{(\text{N·m})}{\text{m}^2 \cdot \text{s}} = \frac{\text{J/s}}{\text{m}^2} = \frac{\text{W}}{\text{m}^2} \Rightarrow \text{OK}.$$

(b) The electric energy density is given by

$$u_E = \frac{1}{2}\epsilon_0 E_0^2$$

$$= \frac{1}{2}(8.85 \times 10^{-12} \text{ C}^2/\text{N·m}^2)(1.01 \times 10^3 \text{ V/m})^2$$

$$= 4.5 \times 10^{-6} \frac{\text{J}}{\text{m}^3}.$$

The magnetic field energy density for the field value calculated in the previous Example is

$$u_B = \frac{B_0^2}{2\mu_0} = \frac{(3.37 \times 10^{-6} \text{ T})^2}{2(4\pi \times 10^{-7} \text{ T·m/A})} = 4.5 \times 10^{-6} \frac{\text{J}}{\text{m}^3}.$$

They are equal as expected. The total energy density in the EM wave is $u = u_E + u_B = 9.0 \times 10^{-6} \text{ J/m}^3$, with equal contributions from electric and magnetic fields.

➡ **PHYSICS APPLIED**

"Wireless" transmission

Electromagnetic waves offer the possibility of transmitting information over long distances. Among the first to realize this and put it in practice was Guglielmo Marconi (1874–1937), who, in the 1890s invented and developed the wireless telegraph. With it, messages could be sent hundreds of kilometers at the speed of light without the use of wires. The first signals were merely long and short pulses that could be translated into words by a code, such as the "dots" and "dashes" of the Morse code. The next decade saw the development of vacuum tubes. Out of this early work radio and television were born. We now discuss briefly (1) how radio and TV signals are transmitted, and (2) how they are received at home.

Transmission of radio waves

The process by which a radio station transmits information (words and music) is outlined in Fig. 22–14. The audio (sound) information is changed into an electrical signal of the same frequencies by, say, a microphone or tape recorder head. This electrical signal is called an audio-frequency (AF) signal, since the frequencies are in the audio range (20 to 20,000 Hz). The signal is amplified[†] electronically and is then mixed with a radio-frequency (RF) signal. The RF frequency is determined by the values of L and C in a resonant LCR circuit (Section 21–14) and is chosen to

Carrier frequency

produce a particular frequency for each station, called its **carrier frequency**. AM radio stations have carrier frequencies from about 530 kHz to 1600 kHz. For example, "710 on your dial" means a station whose carrier frequency is 710 kHz. FM radio stations have much higher carrier frequencies, between 88 MHz and 108 MHz. The carrier frequencies for TV stations in the United States lie between 54 and 88 MHz for channels 2 through 6, and between 174 and 216 MHz for channels 7 through 13; UHF (ultra-high-frequency) stations have even higher carrier frequencies, between 470 and 890 MHz.

FIGURE 22–14 Block diagram of a radio transmitter.

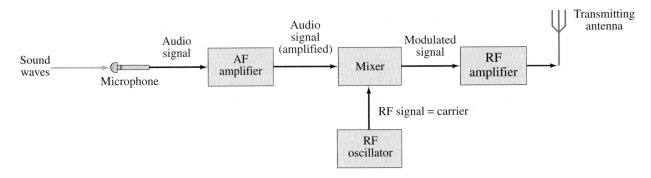

[†]How amplifiers work is discussed in Section 29–9.

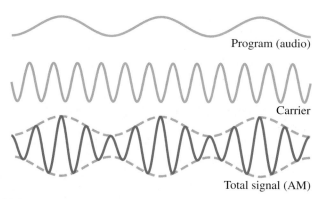

Program (audio)

Carrier

Total signal (AM)

FIGURE 22–15 In amplitude modulation (AM), the amplitude of the carrier signal is made to vary in proportion to the audio signal's amiplitude.

The mixing of the audio and carrier frequencies is done in two ways. In **amplitude modulation** (AM), the amplitude of the higher frequency carrier wave is made to vary so it follows the audio signal as shown in Fig. 22–15. It is called "amplitude modulation" because the amplitude of the carrier is altered ("modulate" means to change or alter). In **frequency modulation** (FM), the *frequency* of the carrier wave is made to change in proportion to the audio signal, as shown in Fig. 22–16.

The mixed signal is amplified further (since the signal contains radio frequencies, this is called an RF amplifier) and is then sent to the antenna, where the complex mixture of frequencies is sent out in the form of EM waves.

A television transmitter works in a similar way, using frequency modulation, except that both audio and video signals are mixed with carrier frequencies.

➡ **PHYSICS APPLIED**

AM and FM

FIGURE 22–16 In frequency modulation (FM), the frequency of the carrier signal is made to change in proportion to the audio signal's amplitude. This method is used by FM radio and television.

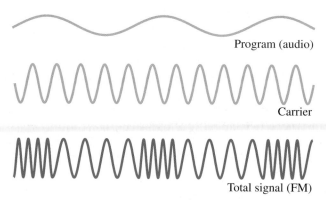

Program (audio)

Carrier

Total signal (FM)

FIGURE 22–17
Block diagram of a
simple radio receiver.

➡ **PHYSICS APPLIED**

Radio and TV receivers

Antennas

FIGURE 22–18 Antennas.
(a) Electric field of EM wave
produces a current in an antenna
consisting of straight wire or rods.
(b) Changing magnetic field induces
an emf and current in a loop antenna.

Now let us look at the other end of the process, the reception of radio and TV programs at home. A simple radio receiver is diagrammed in Fig. 22–17. The EM waves sent out by all stations are received by the antenna. One kind of antenna consists of one or more conducting rods; the electric field in the EM waves exerts a force on the electrons in the conductor, causing them to move back and forth at the frequencies of the waves (Fig. 22–18a). A second type of antenna consists of a tubular coil of wire, often found in AM radios, or the simple loop of a UHF television antenna. These antennas detect the magnetic field of the wave, for the changing B field induces an emf in the coil (Fig. 22–18b). The signal the antenna detects and sends to the receiver is very small and contains frequencies from many different stations. The receiver selects out a particular RF frequency (actually a narrow range of frequencies) corresponding to a particular station using a resonant LC circuit (Section 21–14) with a vari-

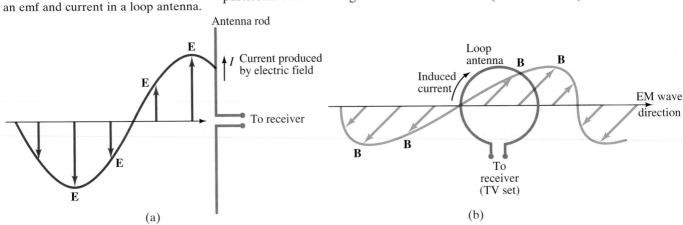

(a)   (b)

able capacitor or inductor. A simple example is shown in Fig. 22–19. A particular station is "tuned-in" by adjusting L and C so that the resonant frequency of the circuit equals that of the station's carrier frequency. The RF signal may be amplified both before and after the tuning is done. The signal, containing both audio and carrier frequencies, next goes to the *detector* (Fig. 22–17) where "demodulation" takes place—that is, the RF carrier frequency is separated from the audio signal. The audio signal is then amplified and sent to a loudspeaker or headphones.

Modern receivers have more stages than those shown. Various means are used to increase the sensitivity and selectivity (ability to detect weak signals and distinguish them from other stations), and to minimize distortion of the original signal.[†]

FIGURE 22–19 Simple tuning
stage of a radio.

[†]For *FM stereo broadcasting*, two signals are carried by the carrier wave. One of these contains frequencies up to about 15 kHz, which includes most audio frequencies. The other signal includes the same range of frequencies, but 19 kHz is added to it. A stereo receiver subtracts this 19,000-Hz signal and distributes the two signals to the left and right channels. The first signal actually consists of the sum of left and right channels ($L + R$), so mono radios detect all the sound. The second signal is the difference between left and right ($L - R$). Hence the receiver must add and subtract the two signals to get pure left and right signals for each channel.

A television receiver does similar things to both the audio and the video signals. The audio signal goes finally to the loudspeaker, and the video signal to the picture tube, a *cathode ray tube* (CRT) whose operation was discussed in Section 17–10.

EXAMPLE 22–5 **Tuning a station.** An FM radio station transmits at 100 MHz. Calculate (a) its wavelength, and (b) the value of the capacitance in the tuning circuit if $L = 0.40\ \mu\text{H}$.

SOLUTION (a) The carrier frequency is $f = 100\ \text{MHz} = 1.0 \times 10^8\ \text{s}^{-1}$. From Eq. 22–4, $\lambda = c/f = (3.0 \times 10^8\ \text{m/s})/(1.0 \times 10^8\ \text{s}^{-1}) = 3.0\ \text{m}$. The wavelengths of other FM signals (88 to 108 MHz) are close to this. FM antennas are typically 1.5 m long, or about a half wavelength. This length is chosen so that the antenna reacts in a resonant fashion and thus is more sensitive.

(b) According to Eq. 21–18, the resonant frequency is $f_0 = 1/(2\pi\sqrt{LC})$. Therefore,

$$C = \frac{1}{4\pi^2 f_0^2 L} = \frac{1}{4(3.14)^2(1.0 \times 10^8\ \text{s}^{-1})^2(4.0 \times 10^{-7}\ \text{H})}$$

$$= 6.3\ \text{pF}.$$

Of course, the capacitor or inductor is variable, so other stations can be selected too.

The various regions of the radio-wave spectrum are assigned by governmental agencies to various purposes. Besides those mentioned above, there are "bands" assigned for use by ships, airplanes, police, military, amateurs, satellites and space, and radar. Cellular telephones, for example, occupy a band from 824 MHz to 894 MHz.

SUMMARY

James Clerk Maxwell synthesized an elegant theory in which all electric and magnetic phenomena could be described using four equations, now called **Maxwell's equations**. They are based on earlier ideas, but Maxwell added one more—that a changing electric field produces a magnetic field.

Maxwell's theory predicted that transverse **electromagnetic** (EM) **waves** would be produced by accelerating electric charges, and these waves would propagate through space at the speed of light, given by the formula

$$c = \frac{1}{\sqrt{\epsilon_0 \mu_0}}.$$

The oscillating electric and magnetic fields in an EM wave are perpendicular to each other and to the direction of propagation.

The wavelength λ and frequency f of EM waves are related to their speed c by

$$c = \lambda f,$$

just as for other waves.

After EM waves were experimentally detected in the late 1800s, the idea that light is an EM wave (of very high frequency) became generally accepted. The **electromagnetic spectrum** includes EM waves of a wide variety of wavelengths, from microwaves and radio waves to visible light to X-rays and γ-rays, all of which travel through space at a speed $c = 3.0 \times 10^8\ \text{m/s}$.

QUESTIONS

1. The electric field in an EM wave traveling north oscillates in an east–west plane. Describe the direction of the magnetic field vector in this wave.

2. Is sound an electromagnetic wave? If not, what kind of wave is it?

3. Can EM waves travel through a perfect vacuum? Can sound waves?

4. How are light and sound alike? How are they different?

5. Are the wavelengths of radio and television signals longer or shorter than those detectable by the human eye?

6. What does the result of Example 22–1 tell you about the phase of a 60-Hz ac current that starts at a power plant as compared to its phase at a house 200 km away?

7. When you connect two loudspeakers to the output of a stereo amplifier, should you be sure the lead wires are equal in length so that there will not be a time lag between speakers? Explain.

8. In the electromagnetic spectrum, what type of EM wave would have a wavelength of 10^3 km? 1 km? 1 m? 1 cm? 1 mm? 1 μm?

* 9. A lost person may signal by flashing a flashlight on and off using Morse code. This is actually a modulated EM wave. Is it AM or FM? What is the frequency of the carrier, approximately?

* 10. Can two radio or TV stations broadcast on the same carrier frequency? Explain.

* 11. If a radio transmitter has a vertical antenna, should a receiver's antenna (rod type) be vertical or horizontal to obtain best reception?

* 12. The carrier frequencies of FM broadcasts are much higher than for AM broadcasts. On the basis of what you learned about diffraction in Chapter 11, explain why AM signals can be detected more readily than FM signals behind low hills or buildings.

13. Discuss how cordless telephones make use of EM waves. What about cellular telephones?

PROBLEMS

*SECTION 22–2

* 1. (I) Calculate the displacement current I_D between the square plates, 2.8 cm on a side, of a capacitor if the electric field is changing at a rate of 2.0×10^6 V/m·s.

* 2. (I) Calculate the displacement current between the round plates of a capacitor, 6.0 cm in diameter, if the plates are spaced 1.3 mm apart and the voltage across them is changing at a rate of 120 V/s.

* 3. (II) At a given instant, a 3.8-A current flows in the wires connected to a parallel-plate capacitor. What is the rate at which the electric field is changing between the plates if the square plates are 1.90 cm on a side?

* 4. (II) A 1200-nF capacitor with circular parallel plates 1.0 cm in diameter is accumulating charge at the rate of 25.0 mC/s at some instant in time. What will be the induced magnetic field strength 10.0 cm radially outward from the center of the plates? What will be the value of the field strength after the capacitor is fully charged?

* 5. (II) Show that the displacement current through a parallel-plate capacitor can be written $I_D = C \Delta V/\Delta t$, where V is the voltage across the capacitor at any instant.

* 6. (III) The electric field between two parallel circular capacitor plates (capacitance C) changes at a rate $\Delta E/\Delta t$. (a) If the radius of the plates is R, show that the magnetic field B a distance r from the center of the plates, if $r \le R$, is $B = \frac{1}{2}\mu_0\epsilon_0 r \, (\Delta E/\Delta t)$; and outside the plates, $r \ge R$, show that $B = (\mu_0\epsilon_0 R^2/2r)(\Delta E/\Delta t)$. (b) Show that the magnetic field beyond the edges of the capacitor plates is given by

$$B = \frac{\mu_0 I_D}{2\pi r},$$

where I_D is the displacement current. (c) This is the same formula for the field outside a straight wire; explain the similarity.

SECTION 22–4

* 7. (I) If the magnetic field in a traveling EM wave has a peak value of 17.5 nT, what is the peak value of the electric field strength?

* 8. (I) If the electric field in an EM wave has a peak of 0.43×10^{-4} V/m, what is the peak value of the magnetic field strength?

* 9. (I) In an EM wave traveling west, the B field oscillates vertically and has a frequency of 80.0 kHz and an rms strength of 9.75×10^{-9} T. What is the frequency and rms strength of the electric field and what is its direction?

10. (I) What is the wavelength of a 27.75×10^9 Hz radar signal?

11. (I) What is the frequency of an X-ray with wavelength 0.10 nm?

12. (I) What is the frequency of a microwave whose wavelength is 1.50 cm?

13. (I) An EM wave has a wavelength of 750 nm. What is its frequency, and what kind of light would we call it?

14. (I) An EM wave has frequency 9.56×10^{14} Hz. What is its wavelength, and how would we classify it?

15. (I) How long does it take light to reach us from the Sun, 1.50×10^8 km away?

16. (I) Our nearest star (other than the Sun) is 4.2 light-years away. That is, it takes 4.2 years for the light it emits to reach Earth. How far away is it in meters?

17. (I) A light-year is a measure of distance (not time). How many meters does light travel in a year?

18. (II) How long would it take a message sent as radio waves from Earth to reach (a) Mars, (b) a spacecraft near Saturn?

19. (II) What is the minimum angular speed at which Michelson's eight-sided mirror would have had to rotate in order that light would be reflected into an observer's eye by succeeding mirror faces (Fig. 22–12)?

20. (II) Who will hear the voice of a singer first—a person in the balcony 50 m away from the stage (Fig. 22–20), or a person 3000 km away at home whose ear is next to the radio? How much sooner? Assume that the microphone is a few centimeters from the singer and the temperature is 20°C.

FIGURE 22–20 Problem 20.

21. (II) Pulsed lasers used for science and medicine produce very short bursts of electromagnetic energy. If the laser light wavelength is 1062 nm (this corresponds to a Neodymium-YAG laser), and the pulse lasts for 30 picoseconds, how many wavelengths are found within the laser pulse? How short would the pulse need to be to fit only one wavelength?

* 22. (I) The **E** field in an EM wave has a peak of 26.5 mV/m. What is the average rate at which this wave carries energy across unit area per unit time?

* 23. (II) The magnetic field in a traveling EM wave has an rms strength of 22.5 nT. How long does it take to deliver 135 J of energy to 1.00 cm² of a wall that it hits perpendicularly?

* 24. (II) How much energy is transported across a 1.00-cm² area per hour by an EM wave whose E field has an rms strength of 18.6 mV/m?

* 25. (II) A spherically spreading EM wave comes from a 1000-W source. At a distance of 10.0 m, what is the average power crossing unit area, and what is the rms value of the electric field?

* 26. (II) What is the energy contained in a 1.00-m³ volume near the Earth's surface due to radiant energy from the Sun? See Example 22–3.

* 27. (II) A 12.8-mW laser puts out a narrow beam 2.00 mm in diameter. What are the average (rms) values of E and B in the beam?

* 28. (II) Estimate the average power output of the Sun, given that about 1350 W/m² reaches the upper atmosphere of the Earth.

* 29. (II) A high-energy pulsed laser emits a 1.0-ns-long pulse of average power 2.5×10^{11} W. The beam is 2.2×10^{-3} m in radius. Determine (a) the energy delivered in each pulse, and (b) the rms value of the electric field.

* 30. (I) The variable capacitor in the tuner of an AM radio has a capacitance of 2400 pF when the radio is tuned to a station at 550 kHz. What must the capacitance be for a station at the other end of the dial, 1550 kHz?

* 31. (I) The oscillator of a 96.1-MHz FM station has an inductance of 1.8 μH. What value must the capacitance be?

* 32. (II) A certain FM radio tuning circuit has a fixed capacitor $C = 840$ pF. Tuning is done by a variable inductance. What range of values must the inductance have to tune stations from 88 MHz to 108 MHz?

* 33. (II) An amateur radio operator wishes to build a receiver that can tune a range from 14.0 MHz to 15.0 MHz. A variable capacitor has a minimum capacitance of 92 pF. (a) What is the required value of the inductance? (b) What is the maximum capacitance used on the variable capacitor?

* 34. (II) A 1.40-m-long FM antenna is oriented parallel to the electric field of an EM wave. How large must the electric field be to produce a 1.00-mV (rms) voltage between the ends of the antenna? What is the rate of energy transport per square meter?

35. What is the wavelength of an AM station at 680 kHz?

36. An FM station broadcasts at 100.7 MHz. What is the wavelength of this wave?

37. Compare 940 on the AM dial to 94 on the FM dial. Which has the longer wavelength, and by what factor is it larger?

38. Television broadcast frequencies range between 54.0 MHz for Channel 2 up to 806 MHz for Channel 69. What are the wavelengths for these channels?

39. If the Sun were to disappear or somehow radically change its output, how long would it take for us on Earth to learn about it?

40. (a) How long did it take for a message sent from Earth to reach the first astronauts on the Moon? (b) How long will it take for a message from Earth to reach the first astronauts who arrive on Mars; assume Mars is at its closest approach to Earth (78×10^6 km)?

41. A radio voice signal from the Apollo crew on the Moon is beamed to a listening crowd from a radio speaker. If you are standing 50 m from the loudspeaker, what is the total time lag between when you hear the sound and when the sound left the Moon?

* 42. A point source emits light energy uniformly in all directions at an average rate P_0 with a single frequency f. Show that the peak electric field in the wave is given by

$$E_0 = \sqrt{\frac{\mu_0 c P_0}{2\pi r^2}}.$$

* 43. What are E_0 and B_0 2.00 m from a 100-W light source? Assume the bulb emits radiation of a single frequency uniformly in all directions.

* 44. Estimate the rms electric field in the sunlight that hits Mars, knowing that the Earth receives about 1350 W/m^2 and that Mars is 1.52 times farther away from the Sun (on average) than is the Earth.

* 45. At a given instant in time, a traveling EM wave is noted to have its maximum magnetic field pointing west and its maximum electric field pointing south. In which direction is the wave traveling? If the rate of energy flow is 500 W/m^2, what are the maximum values for the two fields?

* 46. Light is emitted from an ordinary lightbulb filament in wave-train bursts of about 10^{-8}s in duration. What is the length in space of such wave trains?

* 47. Suppose a 50-kW radio station emits EM waves uniformly in all directions. (a) How much energy per second crosses a 1.0-m^2 area 100 m from the transmitting antenna? (b) What is the rms magnitude of the **E** field at this point, assuming the station is operating at full power? (c) What is the voltage induced in a 1.0-m-long vertical car antenna?

* 48. Repeat Problem 47 for a distance of 100 km from the station.

* 49. Referring to Problem 47, what is the maximum power level of the radio station so as to avoid electrical breakdown of air at a distance of 1.0 m from the antenna? Assume the antenna is a point source. Air breaks down in an electric field of about 3×10^6 V/m. [Hint: See Problem 42.]

* 50. How large an emf (rms) will be generated in an antenna that consists of a 480-loop circular coil of wire 1.7 cm in diameter if the EM wave has a frequency of 810 kHz and is transporting energy at an average rate of 1.0×10^{-4} W/m^2 at the antenna? [Hint: You can use Eq. 21–5, since it could be applied to an observer moving with the coil so that the magnetic field is oscillating with the frequency $f = \omega/2\pi$.]

* 51. The variable capacitance of a radio tuner consists of six plates connected together placed alternately between six other plates, also connected together (Fig. 22–21). Each plate is separated from its neighbor by 1.1 mm of air. One set of plates can move so that the area of overlap varies from 1.0 cm^2 to 9.0 cm^2. (a) Are these capacitors connected in series or in parallel? (b) Determine the range of capacitance values. (c) What value of inductor is needed if the radio is to tune AM stations from 550 kHz to 1600 kHz?

FIGURE 22–21   Problem 51.

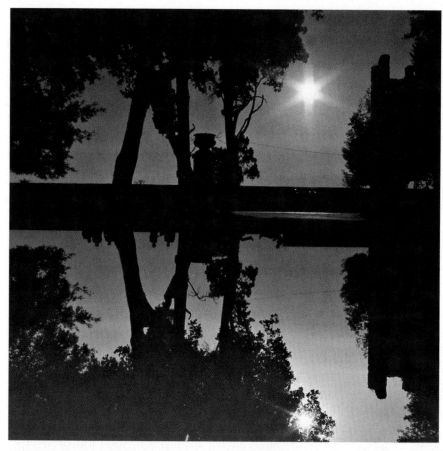

LIGHT: GEOMETRIC OPTICS 23

The sense of sight is extremely important to us, for it provides us with a large part of our information about the world. How do we see? What is the something called *light* that enters our eyes and causes the sensation of sight? How does light behave so that we can see everything that we do? The subject of light will occupy us for the next three chapters, and we will return to it in later chapters.

We see an object in one of two ways: (1) the object may be a source of light, such as a lightbulb, a flame, or a star, in which case we see the light emitted directly from the source; or (2), more commonly, we see an object by light reflected from it. In the latter case, the light may have originated from the sun, artificial lights, or a campfire. An understanding of how bodies *emit* light was not achieved until the 1920s, and this will be discussed in Chapter 27. How light is *reflected* from objects was understood much earlier, and we will discuss this in Section 23–2.

23–1 The Ray Model of Light

A great deal of evidence suggests that *light travels in straight lines* under a wide variety of circumstances. For example, a point source of light like the sun casts distinct shadows, and the beam of a flashlight appears to be a straight line. In fact, we infer the positions of objects in our environment by assuming that light moves from the object to our eyes in straight-line paths. Our whole orientation to the physical world is based on this assumption.

This reasonable assumption has led to the **ray model** of light. This model assumes that light travels in straight-line paths called light **rays**. Actually, a ray is an idealization; it is meant to represent an extremely narrow beam of light. When we see an object, according to the ray model, light reaches our eyes from each point on the object; although light rays leave each point in many different directions, normally only a small bundle of these rays can enter an observer's eye, as shown in Fig. 23–1. If the person's head moves to one side, a different bundle of rays will enter the eye from each point.

We saw in Chapter 22 that light can be considered as an electromagnetic wave. Although the ray model of light does not deal with this aspect of light (we discuss the wave nature of light in Chapter 24), the ray model has been very successful in describing many aspects of light such as reflection, refraction, and the formation of images by mirrors and lenses. Because these explanations involve straight-line rays at various angles, this subject is referred to as **geometric optics**.

Light rays

FIGURE 23–1 Light rays come from each single point on an object. A small bundle of rays leaving one point is shown entering a person's eye.

23–2 Reflection; Image Formation by a Plane Mirror

When light strikes the surface of an object, some of the light is reflected. The rest is either absorbed by the object (and transformed to thermal energy) or, if the object is transparent like glass or water, part of it is transmitted through. For a very shiny object such as a silvered mirror, over 95 percent of the light may be reflected.

When a narrow beam of light strikes a flat surface (Fig. 23–2) we define the **angle of incidence**, θ_i, to be the angle an incident ray makes with the normal to the surface ("normal" means perpendicular) and the **angle of reflection**, θ_r, to be the angle the reflected ray makes with the normal. For flat surfaces, it is found that the *incident and reflected rays lie in the same plane with the normal to the surface*, and that

the angle of incidence equals the angle of reflection.

This is the **law of reflection** and is indicated in Fig. 23–2 (most clearly in (b) which is a side view). It was known to the ancient Greeks, and you can confirm it yourself by shining a narrow flashlight beam at a mirror in a darkened room.

Angle of incidence and Angle of reflection (angle with the normal)

Law of reflection

FIGURE 23–2 Law of reflection. (a) Shows an incident ray being reflected at the top of a flat surface; (b) shows a side or "end-on" view, which we will usually use because of its clarity.

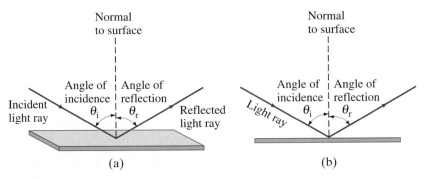

When light is incident upon a rough surface, even microscopically rough such as this page, it is reflected in many directions, Fig. 23–3. This is called **diffuse reflection**. The law of reflection still holds, however, at each small section of the surface. Because of diffuse reflection in all directions, an ordinary object can be seen from many different angles. When you move your head to the side, different reflected rays reach your eye from each point on the object (such as this page), Fig. 23–4a. Let us compare diffuse reflection to reflection from a mirror, which is known as *specular* reflection ("speculum" is Latin for mirror). When a narrow beam of light is shone on a mirror, the light will not reach your eye unless it is placed at just the right place where the law of reflection is satisfied, as shown in Fig. 23–4b. This is what gives rise to the unusual properties of mirrors. (Galileo, using similar arguments, showed that the Moon must have a rough surface rather than a highly polished surface like a mirror, as some people thought.)

FIGURE 23–3 Diffuse reflection from a rough surface.

Mirrors (specular reflection)

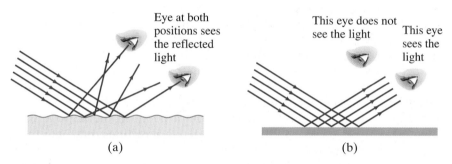

Eye at both positions sees the reflected light

(a)

This eye does not see the light

This eye sees the light

(b)

FIGURE 23–4 A beam of light from a flashlight shines on (a) white paper, and (b) a mirror. In part (a), you can see the white light reflected at various points because of diffuse reflection. But in part (b), you see the reflected light only when your eye is placed correctly ($\theta_r = \theta_i$); this is known as specular reflection.

When you look straight in a mirror, you see what appears to be yourself as well as various objects around and behind you, Fig. 23–5. Your face and the other objects look as if they are in front of you, beyond the mirror; but, of course, they are not. What you see in the mirror is an **image** of the objects.

Image

FIGURE 23–5 When you look in a mirror, you see an image of yourself and objects around you. Note that you don't see yourself as others see you, because left and right are reversed in the image.

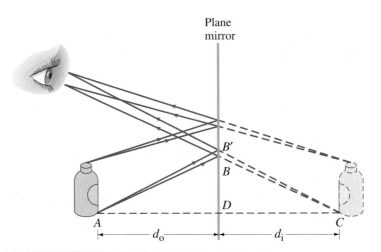

Plane
mirror

B'

B

D

A d_o d_i C

FIGURE 23–6 Formation of a virtual image by a plane mirror.

Ray model of how images are formed

Figure 23–6 shows how an image is formed by a plane mirror (that is, flat), according to the ray model. We are viewing the mirror, on edge, in the diagram of Fig. 23–6, and the rays are shown reflecting from the front surface. (Good mirrors are generally made by putting a highly reflective metallic coating on one surface of a very flat piece of glass.) Rays from two different points on an object are shown in Fig. 23–6: rays leaving from a point on the top of the bottle, and from a point on the bottom. Rays leave each point on the object going in many directions, but only those that enclose the bundle of rays that reach the eye from the two points are shown. The diverging rays that enter the eye *appear* to come from behind the mirror as shown by the dashed lines. That is, our eyes and brain interpret any rays that enter an eye as having traveled a straight-line path. The point from which each bundle of rays seems to come is one point on the image. For each point on the object, there is a corresponding image point. Let us concentrate on the two rays that leave point A on the object and strike the mirror at points B and B'. The angles ADB and CDB are right angles; and angles ABD and CBD are equal because of the law of reflection. Therefore, the two triangles ABD and CBD are congruent, and the length $AD = CD$. Thus the image appears as far behind the mirror as the object is in front: the **image distance**, d_i (distance from mirror to image, Fig. 23–6), equals the **object distance**, d_o. From the geometry, we also see that the height of the image is the same as that of the object.

The light rays do not actually pass through the image location itself. It merely *seems* as though the light is coming from the image because our brains interpret any light entering our eyes as having come in a straight line path from in front of us. Because the rays do not actually pass through the image, the image would not appear on paper or film placed at the location of the image. Therefore, it is called a **virtual image**. This is to distinguish it from a **real image** in which the light does pass through the image and which therefore could appear on paper or film placed at the image position. We will see that curved mirrors and lenses can form real images. A movie projector lens, for example, produces a real image that is visible on the screen.

Real and virtual images

CONCEPTUAL EXAMPLE 23–1 How tall must a full-length mirror be?

A woman 1.60 m tall stands in front of a vertical plane mirror. What is the minimum height of the mirror, and how high must its lower edge be above the floor, if she is to be able to see her whole body? (Assume her eyes are 10 cm below the top of her head.)

RESPONSE The situation is diagrammed in Fig. 23–7. First consider the ray from her foot, AB, which upon reflection becomes BE and enters the eye E. The light from point A (her foot) enters the eye after reflecting at B; so the mirror needs to extend no lower than B. Because the angle of reflection equals the angle of incidence, the height BD is half of the height AE. Because $AE = 1.60\text{ m} - 0.10\text{ m} = 1.50\text{ m}$, then $BD = 0.75\text{ m}$. Similarly, if the woman is to see the top of her head, the top edge of the mirror only needs to reach point F, which is 5 cm below the top of her head (half of $GE = 10$ cm). Thus, $DF = 1.55\text{ m}$, and the mirror need have a vertical height of only $(1.55\text{ m} - 0.75\text{ m}) = 0.80\text{ m}$. And the mirror's bottom edge must be 0.75 m above the floor. In general, a mirror need be only half as tall as a person for that person to see all of himself or herself. Does this result depend on the person's distance from the mirror? (The answer is no—try it and see, it's fun.)

➡ **PHYSICS APPLIED**

How tall a mirror is needed to see your whole reflection?

FIGURE 23–7 Seeing oneself in a mirror. Example 23–1.

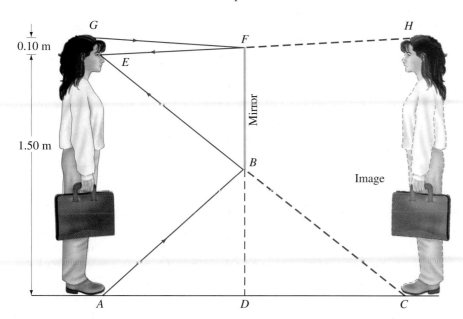

CONCEPTUAL EXAMPLE 23–2 **Is the photo upside down?** Close examination of the photograph on the first page of this chapter reveals that in the top portion, the image of the sun is seen clearly, whereas in the lower portion, the image of the sun is partially blocked by the tree branches. Why isn't the reflection in the water an exact replica of the real scene? Illustrate your answer by drawing a sketch of this situation, showing the sun, the camera, the branch, and two rays going from the sun to the camera (one direct and one reflected). Does your sketch tell you if the photograph is right side up?

RESPONSE The caption on page 683 claims the photo is upside down. Let's try to confirm that, thus assuming that the sun blocked by the tree is actually the direct view, and the full view of the sun is the reflection. Thus, as diagrammed in Fig. 23–8, the ray which reflects off the water and into the camera travels at an angle below the branch, whereas the ray that travels directly to the camera passes through the branches. Try to draw a diagram assuming the photo is right side up (thus assuming that the image of the sun in the reflection is higher above the horizon than it is as viewed directly). It won't work.

Also, what about the people in the photo? Try to draw a diagram showing why they don't appear in the reflection. [*Hint*: assume they are not sitting on the edge of poolside, but back from the edge a bit.] Then try to draw a diagram of the reverse (i.e., assume the photo is right side up so the people are visible in the reflection but not directly).

In general, note that reflected images are not perfect replicas when different planes (distances) are involved.

FIGURE 23–8
Example 23–2

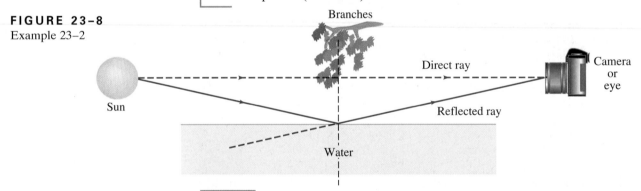

Branches · Direct ray · Camera or eye · Sun · Reflected ray · Water

23–3 Formation of Images by Spherical Mirrors

Reflecting surfaces do not have to be flat. The most common *curved* mirrors are *spherical*, which means they form a section of a sphere. A spherical mirror is called **convex** if the reflection takes place on the outer surface of the spherical shape so that the center of the mirror surface bulges out toward the viewer (Fig. 23–9a). A mirror is called **concave** if the reflecting surface is

Convex and concave mirrors

FIGURE 23–9 Mirrors with convex and concave spherical surfaces.

Convex mirror

(a)

Concave mirror

(b)

(a)

(b)

FIGURE 23–10 (a) A concave makeup mirror gives a magnified image. (b) A convex mirror in a store demagnifies and so includes a wide field of view.

on the inner surface of the sphere so that the center of the mirror sinks away from the viewer (like a "cave") (Fig. 23–9b). Concave mirrors are used as shaving or makeup mirrors (Fig. 23–10a), and convex mirrors are sometimes used on cars and trucks (rearview mirrors) and in shops (to watch for thieves), because they take in a wide field of view (Fig. 23–10b).

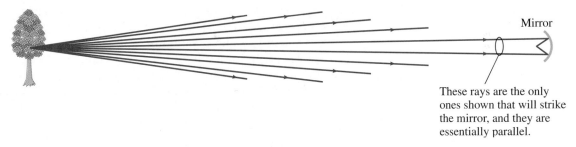

Mirror

These rays are the only ones shown that will strike the mirror, and they are essentially parallel.

FIGURE 23–11 If the object's distance is large compared to the size of the mirror (or lens), the rays are nearly parallel. They are parallel for an object at infinity (the symbol for infinity is ∞).

To see how spherical mirrors form images, we first consider an object that is very far from a concave mirror. For a distant object, as shown in Fig. 23–11, the rays from each point on the object that reach the mirror will be nearly parallel. *For an object infinitely far away* (the Sun and stars approach this), *the rays would be precisely parallel*. Now consider such parallel rays falling on a concave mirror as in Fig. 23–12. The law of reflection holds for each of these rays at the point each strikes the mirror. As can be seen, they are not all brought to a single point. In order to form a sharp image, the rays must come to a point. Thus a spherical mirror will

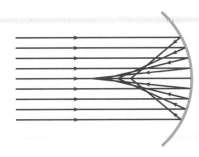

FIGURE 23–12 Parallel rays striking a concave spherical mirror do not focus at precisely a single point.

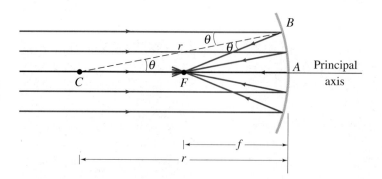

FIGURE 23–13 Rays parallel to the principal axis of a spherical mirror come to a focus at F, called the focal point, as long as the mirror is small in width as compared to its radius of curvature, r.

not make as sharp an image as a plane mirror will. However, if the mirror is small compared to its radius of curvature, so that the reflected rays *Small angle approximation* make only a *small angle* upon reflection, then the rays will cross each other at very nearly a single point, or **focus**, as shown in Fig. 23–13. In the case shown, the rays are parallel to the **principal axis**, which is defined as the straight line perpendicular to the curved surface at its center (line CA in the diagram). The point F, where rays parallel to the principal axis come *Focal point* to a focus, is called the **focal point** of the mirror. The distance between F *Focal length* and the center of the mirror, length FA, is called the **focal length**, f, of the mirror. Another way of defining the focal point is to say that it is the *image point for an object infinitely far away* along the principal axis. The image of the Sun, for example, would be at F.

Now we will show, for a mirror whose reflecting surface is small compared to the radius of curvature, that the rays do indeed meet at a common point, F, and we will also calculate the focal length f. We consider a ray that strikes the mirror at B in Fig. 23–13. The point C is the center of curvature of the mirror (the center of the sphere of which the mirror is a part). So the dashed line CB is equal to r, the radius of curvature, and CB is normal to the mirror's surface at B. The incoming ray that hits the mirror at B makes an angle θ with this normal, and hence the reflected ray, BF, also makes an angle θ with the normal. Note also from the geometry that angle BCF is also θ as shown. The triangle CBF is isosceles because two of its angles are equal. Thus length $CF = BF$. We assume the mirror has a width or diameter that is small compared to its radius of curvature, so the angles are small, and the length FB is nearly equal to length FA. In this approximation, $FA = FC$. But $FA = f$, the focal length, and $CA = 2\,FA = r$. Thus the focal length is half the radius of curvature:

Focal length
of mirror

$$f = \frac{r}{2}.$$ **(23–1)**

This argument assumed only that the angle θ was small, so the same result applies for all the other rays. And, we have also shown that all the rays pass through the same point F in this approximation of a mirror small compared to its radius of curvature.

We saw that, for an object at infinity, the image is located at the focal point of a concave spherical mirror, where $f = r/2$. But where does the image lie for an object not at infinity? First consider the object shown in Fig. 23–14, which is placed on point O between F and C. Let us determine where the image will be for a given point O' on the object. To do this, we can draw several rays and make sure these reflect from the mirror such that the reflection angle equals the incidence angle. This can involve much work and our task is simplified if we deal with three rays that are particularly easy to draw. These are the rays labeled 1, 2, and 3 in Fig. 23–14 and we draw them for object point O' as follows:

Ray 1 is drawn parallel to the axis; therefore it must pass along a line through F after reflection (as we see in Fig. 23–13; see also Fig. 23–14a).

Ray 2 is drawn through F; therefore it must reflect so it is parallel to the axis (Fig. 23–14b).

Ray 3 is chosen to be perpendicular to the mirror, and so is drawn so that it passes through C, the center of curvature; it is along a radius of the spherical surface, and because it is perpendicular to the mirror it will be reflected back on itself (Fig. 23–14c).

The point at which these rays cross is the image point I'. All other rays from the same object point will pass through this image point. To find the image point for any object point, only these three particular rays need to be used. Actually, only two of these rays are needed, but the third serves as a check.

We have shown the image point in Fig. 23–14 only for a single point on the object. Other points on the object are imaged nearby, so a complete image of the object is formed, as shown by the dashed outline. Because the light actually passes through the image itself, this is a real image that will appear on a piece of paper or film placed there. This can be compared to the virtual image formed by a plane mirror (the light does not actually pass through that image, Fig. 23–6).

The image in Fig. 23–14 can be seen by the eye when the eye is placed to the left of the image so that some of the rays diverging from each point on the image (as point I') can enter the eye as shown in part (c). (See also Figs. 23–1 and 23–6.)

➡ **RAY DIAGRAM**

Finding the image position for a curved mirror

Real image

FIGURE 23–14 Rays leave point O' on the object (an arrow). Shown are the three most useful rays for determining where the image I' is formed. [Note that our mirror's height is not small compared to f, so our diagram will not give the precise position of the image.]

(a) Ray 1 goes out from O' parallel to the axis and reflects through F.

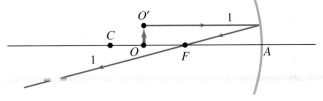

(b) Ray 2 goes through F and then reflects back parallel to the axis.

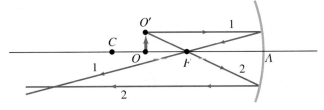

(c) Ray 3 heads out perpendicular to mirror and then reflects back on itself and goes through C (center of curvature).

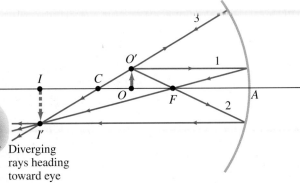

Diverging rays heading toward eye

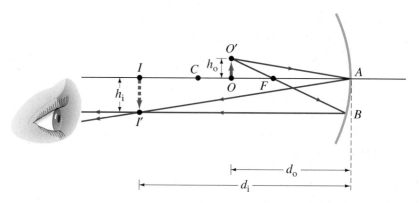

FIGURE 23–15 Diagram for deriving the mirror equation. For the derivation, we assume the mirror size is small compared to its radius of curvature.

Image points can always be determined (at least roughly) by drawing the three rays as described above, although high accuracy is hard to obtain. For one thing it is difficult to draw, maintaining small angles for the rays as we assumed (mirror small compared to f so all rays come to a focus). However, it is possible to derive an equation that gives the image distance if the object distance and radius of curvature of the mirror are known. To do this, we refer to Fig. 23–15. The distance of the object from the center of the mirror, called the **object distance**, is labeled d_o. The **image distance** is labeled d_i. The height of the object OO' is called h_o and the height of the image, $I'I$, is h_i. Two rays are shown, $O'FBI'$ (same as ray 2 in the previous figure) and $O'AI'$. The ray $O'AI'$ obeys the law of reflection, of course, so the two right triangles $O'AO$ and $I'AI$ are similar. Therefore, we have

$$\frac{h_o}{h_i} = \frac{d_o}{d_i}.$$

For the other ray, $O'FBI'$, the triangles $O'FO$ and AFB are also similar since the length $AB = h_i$ (in our approximation of a mirror that is small compared to its radius) and $FA = f$, the focal length of the mirror. Therefore,

$$\frac{h_o}{h_i} = \frac{OF}{FA} = \frac{d_o - f}{f}.$$

The left sides of the two preceding expressions are the same, so we can equate the right sides:

$$\frac{d_o}{d_i} = \frac{d_o - f}{f}.$$

We now divide both sides by d_o and rearrange to obtain

Mirror equation

$$\frac{1}{d_o} + \frac{1}{d_i} = \frac{1}{f}. \tag{23–2}$$

This is the equation we were seeking. It is called the **mirror equation** and relates the object and image distances to the focal length f (where $f = r/2$).

The **lateral magnification**, m, of a mirror is defined as the height of the image divided by the height of the object. From our first set of similar triangles above, we can write:

Lateral magnification for curved mirror

$$m = \frac{h_i}{h_o} = -\frac{d_i}{d_o}, \tag{23–3}$$

where the minus sign is inserted as a convention.

For consistency, we must be careful about the signs of all quantities in Eqs. 23–2 and 23–3. The sign conventions we use are: the image height h_i is positive if the image is upright, and negative if inverted, relative to the object (h_o is always taken as positive); d_i and d_o are positive if image and object are on the reflecting side of the mirror (as in Fig. 23–15), but if either image or object is behind the mirror, the corresponding distance is negative (an example can be seen in Fig. 23–16, Example 23–5). Thus the magnification (Eq. 23–3) is positive for an upright image and negative for an inverted image. We will summarize sign conventions more fully after discussing convex mirrors later in this Section.

EXAMPLE 23–3 **Image in a concave mirror.** A 1.50-cm-high diamond ring is placed 20.0 cm from a concave mirror whose radius of curvature is 30.0 cm. Determine (a) the position of the image, and (b) its size.

SOLUTION The focal length $f = r/2 = 15.0$ cm. The ray diagram is basically like that shown in Fig. 23–14 and Fig. 23–15, since the object is between F and C; a precise ray diagram would have (referring to either Fig. 23–14 or Fig. 23–15): $CA = 30.0$ cm, $FA = 15.0$ cm, and OA (object distance) $= 20.0$ cm. (a) Since $d_o = 20.0$ cm, we have from Eq. 23–2 that

$$\frac{1}{d_i} = \frac{1}{f} - \frac{1}{d_o} = \frac{1}{15.0 \text{ cm}} - \frac{1}{20.0 \text{ cm}} = 0.0167 \text{ cm}^{-1}.$$

So $d_i = 1/0.0167 \text{ cm}^{-1} = 60.0$ cm. The image is 60.0 cm from the mirror on the same side as the object.
(b) From Eq. 23–3, the lateral magnification is

$$m = -\frac{60.0 \text{ cm}}{20.0 \text{ cm}} = -3.00.$$

➡ **PROBLEM SOLVING**

Remember to take the reciprocal

Therefore the image height is

$$h_i = mh_o = (-3.00)(1.5 \text{ cm}) = -4.5 \text{ cm}.$$

The minus sign tells us the image is inverted, as it is in Figs. 23–14 and 23–15.

CONCEPTUAL EXAMPLE 23–4 **Reversible rays.** If the object in Example 23–3 is placed instead where the image is in Fig. 23–15, where will the new image be?

RESPONSE The mirror equation is symmetric in d_o and d_i. Thus the new image will be where the old object was. Indeed, in Fig. 23–15 we need only reverse the direction of the rays to get our new situation.

EXAMPLE 23–5 **Object closer to concave mirror.** A 1.00-cm-high object is placed 10.0 cm from a concave mirror whose radius of curvature is 30.0 cm. (*a*) Draw a ray diagram to locate (approximately) the position of the image. (*b*) Determine the position of the image and the magnification analytically.

SOLUTION (*a*) Since $f = r/2 = 15.0$ cm, the object is between the mirror and the focal point. We draw the three rays as described earlier (Fig. 23–14) and this is shown in Fig. 23–16. Ray 1 leaves the tip of our object heading toward the mirror parallel to the axis, and reflects through *F*. Ray 2 cannot head toward *F* because it would not strike the mirror; so ray 2 must head as if it started at *F* (dashed line) and heads to the mirror and then is reflected parallel to the principal axis. Ray 3 is perpendicular to the mirror, as before. The rays reflected from the mirror diverge and so never meet at a point. They appear, however, to be coming from a point behind the mirror. This point is the image which is thus behind the mirror and *virtual*. (Why?)

(*b*) We use Eq. 23–2 to find d_i when $d_o = 10.0$ cm:

$$\frac{1}{d_i} = \frac{1}{15.0 \text{ cm}} - \frac{1}{10.0 \text{ cm}} = \frac{2 - 3}{30.0 \text{ cm}} = -\frac{1}{30.0 \text{ cm}}.$$

Therefore, $d_i = -30.0$ cm. The minus sign means the image is behind the mirror. The lateral magnification is $m = -d_i/d_o = -(-30.0 \text{ cm})/(10.0 \text{ cm}) = +3.00$. So the image is 3.00 times larger than the object; the plus sign indicates that the image is upright (which is consistent with the ray diagram, Fig. 23–16).

FIGURE 23–16 Object placed within the focal point *F*. The image is *behind* the mirror and is *virtual*, Example 23–5. [Note that the vertical scale (height of object = 1.0 cm) is different from the horizontal (*OA* = 10.0 cm) for ease of drawing, and this will affect the precision of the drawing.]

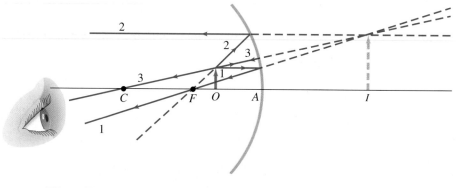

These Examples show that a spherical mirror can produce a magnified image, one that is larger than the object. (There is an old story—maybe fable is a better word—that Julius Caesar spied on the British forces by setting up a curved mirror on the coast of France. Is this reasonable?)

It is useful to compare Figs. 23–14 and 23–16. We can see that if the object is within the focal point, as in Fig. 23–16, the image is virtual, upright, and magnified. This is how a shaving or makeup mirror is used—you must place your head within the focal point if you are to see yourself right side up (Fig. 23–10a). If the object is *beyond* the focal point, as in Fig. 23–14, the image is real and inverted (upside down—and hard to use!). Whether the magnification is greater or less than 1.0 in the latter case depends on the position of the object relative to the center of curvature, point *C*.

The analysis used for concave mirrors can be applied to **convex** mirrors. Even the mirror equation (Eq. 23–2) holds for a convex mirror, although the quantities involved must be carefully defined. Figure 23–17a shows parallel rays falling on a convex mirror. Again spherical aberration

PHYSICS APPLIED

Shaving/makeup mirror

Analysis for convex mirrors

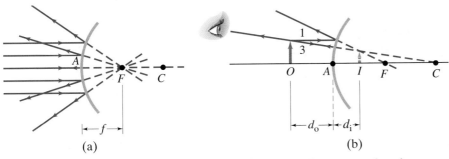

FIGURE 23–17 Convex
mirror: (a) the focal point is
at F, behind the mirror;
(b) the image I of the object
at O is virtual, upright, and
smaller than the object. [Not
to scale for Example 23–6.]

(a)   (b)

will be present, but we assume the mirror's size is small compared to its
radius of curvature. The reflected rays diverge, but seem to come from
point F behind the mirror. This is the **focal point**, and its distance from
the center of the mirror is the **focal length**, f. It is easy to show that again
$f = r/2$. We see that an object at infinity produces a virtual image in a
convex mirror. Indeed, no matter where the object is placed on the reflecting side of a convex mirror, the image will be virtual and erect, as indicated in Fig. 23–17b. To find the image we draw rays 1 and 3 according
to the rules used before on the concave mirror, as shown in Fig. 23–17b.

The mirror equation, Eq. 23–2, holds for convex mirrors but the focal
length f must be considered negative, as must the radius of curvature. The
proof is left as a problem. It is also left as a problem to show that Eq. 23–3
for the magnification is also valid.

EXAMPLE 23–6 **Convex rearview mirror.** A convex rearview car mirror has a radius of curvature of 40.0 cm. Determine the location of the
image and its magnification for an object 10.0 m from the mirror.

➡ PHYSICS APPLIED

Convex rear-view mirror

SOLUTION The ray diagram will be like Fig. 23–17, but the large object distance ($d_o = 10.0$ m) makes a precise drawing difficult; Fig. 23–17 is
not to scale. We have a convex mirror, so r is negative by convention.
Specifically, $r = -40.0$ cm, so $f = -20.0$ cm $= -0.200$ m. The mirror
equation gives

$$\frac{1}{d_i} = \frac{1}{f} - \frac{1}{d_o} = -\frac{1}{0.200 \text{ m}} - \frac{1}{10.0 \text{ m}} = -\frac{51.0}{10.0 \text{ m}}.$$

So $d_i = -10.0 \text{ m}/51.0 = -0.196$ m, or 19.6 cm behind the mirror. The lateral
magnification is $m = -d_i/d_o = -(-0.196 \text{ m})/(10.0 \text{ m}) = 0.0196$ or $1/51$.
So the upright image is smaller than the object by a factor of 51.

➡ **PROBLEM SOLVING** Spherical Mirrors

---

1. Always draw a ray diagram even though you
are going to make an analytic calculation—the
diagram serves as a check, even if not precise.
Draw at least two, and preferably three, of the
easy-to-draw rays as described in Fig. 23–14.
Generally draw the rays starting on a point on
the object to the left of the mirror and moving
to the right.

2. Use Eqs. 23–2 and 23–3; it is crucially important to follow the sign conventions.

3. Sign Conventions
(a) When the object, image, or focal point is on
the reflecting side of the mirror (on the left in
all our drawings), the corresponding distance is
considered positive. If any of these points is behind the mirror (on the right) the corresponding distance is negative.[†]
(b) The image height h_i is positive if the image
is upright, and negative if inverted, relative to
the object (h_o is always taken as positive).

[†]Object distances are positive for material objects, but can be negative in systems with more than one mirror or lens—see
Section 23–10.

23–4 Index of Refraction

We saw in Chapter 22 that the speed of light in vacuum is

$$c = 2.99792458 \times 10^8 \, \text{m/s},$$

which we usually round off to

$$c = 3.00 \times 10^8 \, \text{m/s}.$$

This speed applies to all electromagnetic waves, including visible light.

In air the speed is only very slightly less. In other transparent materials, such as glass and water, the speed is always less than that in vacuum. For example, in water light travels at about $\frac{3}{4}c$. The ratio of the speed of light in vacuum to the speed v in a given material is called the **index of refraction**, n, of that material:

Index of refraction

$$n = \frac{c}{v}. \tag{23–4}$$

The index of refraction is never less than 1 (that is, $n \geq 1$), and its value for various materials is given in Table 23–1. (As we shall see later, n varies somewhat with the wavelength of the light—except in vacuum—so a particular wavelength is specified, that of yellow light with wavelength $\lambda = 589$ nm.)

TABLE 23–1
Indices of Refraction†

Medium	$n = c/v$
Vacuum	1.0000
Air (at STP)	1.0003
Water	1.33
Ethyl alcohol	1.36
Glass	
Fused quartz	1.46
Crown glass	1.52
Light flint	1.58
Lucite or Plexiglas	1.51
Sodium chloride	1.53
Diamond	2.42

†$\lambda = 589$ nm

EXAMPLE 23–7 **Light's speed in diamond.** Calculate the speed of light in diamond.

SOLUTION From Table 23–1, $n = 2.42$ for diamond, so the speed of light traveling inside a diamond is

$$v = \frac{c}{n} = \frac{c}{2.42} = 0.413c$$

or

$$v = \frac{3.00 \times 10^8 \, \text{m/s}}{2.42} = 1.24 \times 10^8 \, \text{m/s}.$$

That light travels more slowly in matter than in vacuum can be explained at the atomic level as being due to the absorption and reemission of light by atoms and molecules of the material.

23–5 Refraction: Snell's Law

When light passes from one medium into another, part of the incident light is reflected at the boundary. The remainder passes into the new medium. If a ray of light is incident at an angle to the surface (other than perpendicular), the ray is bent as it enters the new medium. This bending is called **refraction**. Figure 23–18a shows a ray passing from air into water. The angle θ_1 is the angle of incidence and θ_2 is the **angle of refraction**. Notice that the ray bends toward the normal when entering the water. This is always the case when the ray enters a medium where the speed of light is *less*. If light travels from one

Angle of refraction

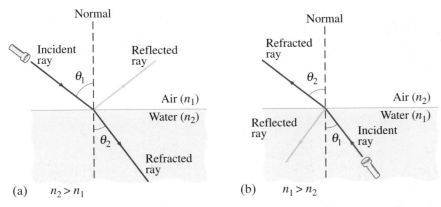

(a) $n_2 > n_1$ (b) $n_1 > n_2$

FIGURE 23–18 Refraction.

medium into a second where its speed is *greater*, the ray bends away from the normal; this is shown in Fig. 23–18b for a ray traveling from water to air.

Refraction is responsible for a number of common optical illusions. For example, a person standing in waist-deep water appears to have shortened legs. As shown in Fig. 23–19, the rays leaving the person's foot are bent at the surface. The observer's eye (and brain) assumes the rays to have traveled a straight-line path, and so the feet appear to be higher than they really are. Similarly, when you put a pencil in water, it appears to be bent (Fig. 23–20).

The angle of refraction depends on the speed of light in the two media and on the incident angle. An analytical relation between θ_1 and θ_2 was arrived at experimentally about 1621 by Willebrord Snell (1591–1626). It is known as **Snell's law** and is written:

$$n_1 \sin \theta_1 = n_2 \sin \theta_2; \tag{23–5}$$

θ_1 is the angle of incidence and θ_2 is the angle of refraction (both measured with respect to a line perpendicular to the surface between the two media, as shown in Fig. 23–18); n_1 and n_2 are the respective indices of refraction in the materials. The incident and refracted rays lie in the same plane, which also includes the perpendicular to the surface. Snell's law is the basic **law of refraction**.

It is clear from Snell's law that if $n_2 > n_1$, then $\theta_2 < \theta_1$; that is, if light enters a medium where n is greater (and its speed less), then the ray is bent toward the normal. And if $n_2 < n_1$, then $\theta_2 > \theta_1$, so the ray bends away from the normal. This is what we saw in Fig. 23–18.

➡ **PHYSICS APPLIED**

Optical illusions

Snell's law
(law of refraction)

FIGURE 23–19 Ray diagram showing why a person's legs look shorter when standing in waist-deep water: the path of light traveling from the bather's foot to the observer's eye bends at the water's surface, and our brain interprets the light as having traveled in a straight line, from higher up (dashed line).

FIGURE 23–20 A pencil in water looks bent even when it isn't.

Foot appears to be here

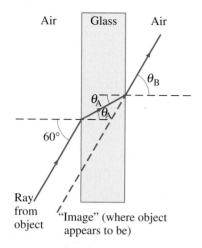

Air   Glass   Air

θ_B

θ_A
θ_A

60°

Ray
from
object

"Image" (where object
appears to be)

FIGURE 23–21 Light
passing through a piece of glass
(Example 23–8).

EXAMPLE 23–8 **Refraction through flat glass.** Light strikes a flat piece
of glass at an incident angle of 60°, as shown in Fig. 23–21. If the index of
refraction of the glass is 1.50, (a) what is the angle of refraction θ_A in the
glass; (b) what is the angle θ_B at which the ray emerges from the glass?

SOLUTION (a) We assume the incident ray is in air, so $n_1 = 1.00$ and
$n_2 = 1.50$. Then, from Eq. 23–5 we have

$$\sin \theta_A = \frac{1.00}{1.50} \sin 60° = 0.577,$$

so $\theta_A = 35.2°$.
(b) Since the faces of the glass are parallel, the incident angle in this case is
just θ_A, so $\sin \theta_A = 0.577$. This time $n_1 = 1.50$ and $n_2 = 1.00$. Thus, $\theta_B (= \theta_2)$ is

$$\sin \theta_B = \frac{1.50}{1.00} \sin \theta_A = 0.866,$$

and $\theta_B = 60.0°$. The direction of the beam is thus unchanged by passing
through a flat piece of glass. It should be clear that this works for any
angle of incidence. The ray is displaced slightly to one side, however. You
can observe this by looking through a piece of glass (near its edge) at
some object and then moving your head to the side so that you see the
object directly.

FIGURE 23–22 Example 23–9.

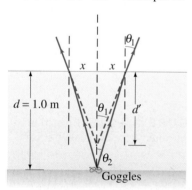

θ_1

x   x

$d = 1.0 \text{ m}$ θ_1 d'

θ_2
Goggles

EXAMPLE 23–9 **Apparent depth of a pool.** A swimmer has dropped
her goggles in the shallow end of a pool, marked as 1.0 m deep. But the
goggles don't look that deep. Why? How deep do the goggles appear to
be when you look straight down into the water?

SOLUTION The ray diagram of Fig. 23–22 shows why the water seems
less deep than it actually is. Rays traveling upward from the goggles on
the bottom are refracted *away* from the normal as they exit the water.
The rays appear to be diverging from a point higher in the water. To calculate the apparent depth d', given a real depth $d = 1.0$ m, we use Snell's
law with $n_1 = 1$ for air and $n_2 = 1.33$ for water:

$$\sin \theta_1 = n_2 \sin \theta_2.$$

We are considering only small angles, so $\sin \theta \approx \tan \theta \approx \theta$, with θ in radians. So Snell's law becomes

$$\theta_1 = n_2 \theta_2.$$

From Fig. 23–22, we see that

$$\theta_1 = \tan \theta_1 = \frac{x}{d'} \qquad \text{and} \qquad \theta_2 = \tan \theta_2 = \frac{x}{d}.$$

Putting these into Snell's law, $\theta_1 = n_2 \theta_2$, we get

$$\frac{x}{d'} = n_2 \frac{x}{d}$$

or

$$d' = \frac{d}{n_2} = \frac{1.0 \text{ m}}{1.33} = 0.75 \text{ m}.$$

23–6 Total Internal Reflection; Fiber Optics

When light passes from one material into a second material where the index of refraction is less (say, from water into air), the light bends away from the normal, as for ray J in Fig. 23–23. At a particular incident angle, the angle of refraction will be 90°, and the refracted ray would skim the surface (ray K) in this case. The incident angle at which this occurs is called the **critical angle**, θ_C. From Snell's law, θ_C is given by

$$\sin \theta_C = \frac{n_2}{n_1} \sin 90° = \frac{n_2}{n_1}. \qquad \text{(23–6)}$$

Critical angle

For any incident angle less than θ_C there will be a refracted ray, although part of the light will also be reflected at the boundary. However, for incident angles greater than θ_C, Snell's law would tell us that $\sin \theta_2$ is greater than 1.00. Yet the sine of an angle can never be greater than 1.00. In this case there is no refracted ray at all, and *all of the light is reflected*, as for ray L in Fig. 23–23. This effect is called **total internal reflection**. But note that total internal reflection can occur only when light strikes a boundary where the medium beyond has a lower index of refraction.

Total internal reflection

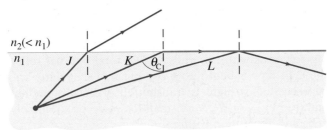

FIGURE 23–23 Since $n_2 < n_1$, light rays are totally internally reflected if $\theta > \theta_C$ as for ray L. If $\theta < \theta_C$, as for ray J, only a part of the light is reflected (this part is not shown), and the rest is refracted.

| **CONCEPTUAL EXAMPLE 23–10** | **View up from under water.** Describe |

what a person would see who looked up at the world from beneath the perfectly smooth surface of a lake or swimming pool.

RESPONSE For an air–water interface, the critical angle is given by

$$\sin \theta_C = \frac{1.00}{1.33} = 0.750.$$

Therefore, $\theta_C = 49°$. Thus the person would see the outside world compressed into a circle whose edge makes a 49° angle with the vertical. Beyond this angle, the person would see reflections from the sides and bottom of the pool or lake (Fig. 23–24).

(a)

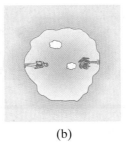

(b)

FIGURE 23–24 View looking upward from beneath the water (the surface of the water must be very smooth).

FIGURE 23–25 Total internal reflection of light by prisms in binoculars.

Many optical instruments, such as binoculars, use total internal reflection within a prism to reflect light. The advantage is that very nearly 100 percent of the light is reflected, whereas even the best mirrors reflect somewhat less than 100 percent. Thus the image is brighter. For glass with $n = 1.50$, $\theta_C = 41.8°$. Therefore, 45° prisms will reflect all the light internally, if oriented as shown in the binoculars of Fig. 23–25.

Fiber optics

Total internal reflection is the principle behind **fiber optics**. Glass and plastic fibers as thin as a few micrometers in diameter can now be made. A bundle of such tiny fibers is called a **light pipe** or cable, and light can be transmitted along it with almost no loss because of total internal reflection. Figure 23–26 shows how light traveling down a thin fiber makes only glancing collisions with the walls so that total internal reflection occurs. Even if the light pipe is bent into a complicated shape, the critical angle still won't be exceeded, so light is transmitted practically undiminished to the other end (see Fig. 23–27). Very small losses do occur, mainly by reflection at the ends and absorption within the fiber.

FIGURE 23–26 Light reflected totally at the interior surface of a glass or transparent plastic fiber.

FIGURE 23–27 Total internal reflection within the tiny fibers of this light pipe makes it possible to transmit light in complex paths with minimal loss.

(a)   (b)

FIGURE 23–28 (a) How a fiber-optic image is made.
(b) Example of a fiber-optic endoscope inserted through
the nose, and the image seen.

Important applications of fiber-optic cables are in telecommunications and medicine. They are being used to transmit telephone calls, video signals, and computer data. The signal is a modulated light beam (a light beam whose intensity can be varied) and is transmitted at a much higher rate and with less loss and less interference than an electrical signal in a copper wire. The sophisticated use of fiber optics to transmit a clear picture, is particularly useful in medicine, Fig. 23–28. For example, a patient's lungs can be examined by inserting a light pipe known as a bronchoscope through the mouth and down the bronchial tube. Light is sent down an outer set of fibers to illuminate the lungs. The reflected light returns up a central core set of fibers. Light directly in front of each fiber travels up that fiber. At the opposite end, a viewer sees a series of bright and dark spots, much like a TV screen—that is, a picture of what lies at the opposite end.[†] The image may be viewed directly or on a TV monitor or film. The fibers must be optically insulated from one another, usually by a thin coating of material whose refractive index is less than that of the fiber. The fibers must be arranged precisely parallel to one another if the picture is to be clear. The more fibers there are, and the smaller they are, the more detailed the picture. Such instruments, including bronchoscopes, colonoscopes (for viewing the colon) and endoscopes (stomach or other organs) are extremely useful for observing hard-to-reach places for surgery or searching for lesions without surgery. The instrument may also be equipped with a mechanism for pinching off tissue samples for biopsies or removal.

➡ **PHYSICS APPLIED**

Fiber optics in telecommunications

➡ **PHYSICS APPLIED**

Medicine—endoscopes

[†]Lenses are used at each end: at the object end to bring the rays in parallel, and at the viewing end as a telescope.

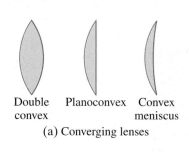

Double convex   Planoconvex   Convex meniscus

(a) Converging lenses

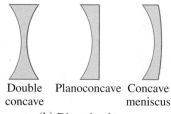

Double concave   Planoconcave   Concave meniscus

(b) Diverging lenses

(c)

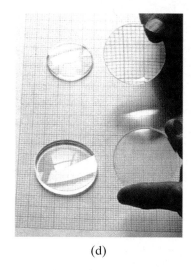

(d)

FIGURE 23–29
(a) Converging and (b) diverging lenses, shown in cross section.
(c) Photo of a converging lens (on the left) and a diverging lens.
(d) Converging lenses (above), and diverging lenses, lying flat, and raised off the paper to form images.

23–7   Thin Lenses; Ray Tracing

The most important simple optical device is no doubt the thin lens. The development of optical devices using lenses dates to the sixteenth and seventeenth centuries, although the earliest record of eyeglasses dates from the late thirteenth century.[†] Today we find lenses in eyeglasses, cameras, magnifying glasses, telescopes, binoculars, microscopes, and medical instruments. A thin lens is usually circular, and its two faces are portions of a sphere. (Although cylindrical surfaces are also possible, we will concentrate on spherical.) The two faces can be concave, convex, or plane; several types are shown in Fig. 23–29, in cross section. The importance of lenses is that they form images of objects, as shown in Fig. 23–30.

[†]Rounded gemstones used as magnifiers probably date from much earlier.

FIGURE 23–30
Converging lens (in holder) forms an image (large "F" on screen at right) of a bright object (illuminated "F" at the left).

Consider the rays parallel to the axis of the double convex lens which is shown in cross section in Fig. 23–31a. We assume the lens is made of glass or transparent plastic, so its index of refraction is greater than that of the air outside. The **axis** of a lens is a straight line passing through the very center of the lens and perpendicular to its two surfaces (Fig. 23–31). From Snell's law, we can see that each ray in Fig. 23–31a is bent toward the axis at both lens surfaces (note the dashed lines indicating the normals to each surface for the top ray). If rays parallel to the axis fall on a thin lens, they will be focused to a point called the **focal point**, F. This will not be precisely true for a lens with spherical surfaces. But it will be very nearly true—that is, parallel rays will be focused to a tiny region that is nearly a point—if the diameter of the lens is small compared to the radii of curvature of the two lens surfaces. This criterion is satisfied by a **thin lens**, one that is very thin compared to its diameter, and we consider only thin lenses here.

The rays from a point on a distant object are essentially parallel—see Fig. 23–11. Therefore we can also say that *the focal point is the image point for an object at infinity on the principal axis*. Thus, the focal point of a lens can be found by locating the point where the Sun's rays (or those of some other distant object) are brought to a sharp image, Fig. 23–32. The distance of the focal point from the center of the lens is called the **focal length**, f. A lens can be turned around so that light can pass through it from the opposite side. The focal length is the same on both sides, as we shall see later, even if the curvatures of the two lens surfaces are different. If parallel rays fall on a lens at an angle, as in Fig. 23–31b, they focus at a point F_a. The plane in which all points such as F and F_a fall is called the **focal plane** of the lens.

Any lens[†] that is thicker in the center than at the edges will make parallel rays converge to a point, and is called a **converging lens** (see Fig. 23–29a). Lenses that are thinner in the center than at the edges (Fig. 23–29b) are called **diverging lenses** because they make parallel light diverge, as shown in Fig. 23–33. The focal point, F, of a diverging lens is defined as that point from which refracted rays, originating from parallel incident rays, seem to emerge as shown in the figure. And the distance from F to the lens is called the **focal length**, just as for a converging lens.

Optometrists and ophthalmologists, instead of using the focal length, use the reciprocal of the focal length to specify the strength of eyeglass (or contact) lenses. This is called the **power**, P, of a lens:

$$P = \frac{1}{f}. \qquad (23–7)$$

The unit for lens power[‡] is the diopter (D), which is an inverse meter: $1\,\text{D} = 1\,\text{m}^{-1}$. For example, a 20-cm-focal-length lens has a power $P = 1/0.20\,\text{m} = 5.0\,\text{D}$. We will mainly use the focal length here, but we will refer again to the power of a lens when we discuss eyeglass lenses in Chapter 25.

[†]We are assuming the lens has an index of refraction greater than that of the surrounding material, such as a glass or plastic lens in air, which is the usual situation.

[‡]Note that lens power has nothing to do with power as the rate of doing work or transforming energy (Section 6–10).

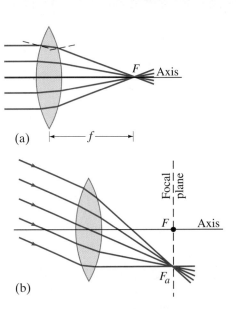

FIGURE 23–31 Parallel rays are brought to a focus by a converging thin lens.

Focal length of lens

FIGURE 23–32 Image of the Sun burning a hole, almost, on a piece of paper.

Power of lens

FIGURE 23–33 Diverging lens.

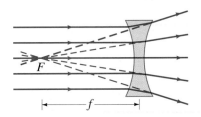

The most important parameter of a lens is its focal length f. For a converging lens, f is easily measured by finding the image point for the Sun or other distant objects. Once f is known, the image position can be found for any object. To find the image point by drawing rays would be difficult if we had to determine all the refractive angles. Instead, we can do it very simply by making use of certain facts we already know, such as that a ray parallel to the axis of the lens passes (after refraction) through the focal point. In fact, to find an image point, we need consider only the three rays indicated in Fig. 23–34, which shows an arrow as the object and a converging lens forming an image to the right. These rays, emanating from a single point on the object, are drawn as if the lens were infinitely thin, and we show only a single sharp bend within the lens instead of the refractions at each surface. These three rays are drawn as follows:

➡ **RAY DIAGRAM**

Finding the image position formed by a thin lens

Ray 1 is drawn parallel to the axis; therefore it is refracted by the lens so that it passes along a line through the focal point F, Fig. 23–34a. (See also Fig. 23–31a.)

Ray 2 is drawn on a line passing through the other focal point F' (front side of lens in Fig. 23–34) and emerges from the lens parallel to the axis, Fig. 23–34b.

Ray 3 is directed toward the very center of the lens, where the two surfaces are essentially parallel to each other; this ray therefore emerges from the lens at the same angle as it entered; as we saw in Example 23–8, the ray would be displaced slightly to one side, but since we assume the lens is thin, we draw ray 3 straight through as shown.

FIGURE 23–34 Finding the image by ray tracing for a converging lens. Rays leave each point on the object. Shown are the three most useful rays, leaving the tip of the object, for determining where the image of that point is formed.

Actually, any two of these rays will suffice to locate the image point, which is the point where they intersect. Drawing the third can serve as a check.

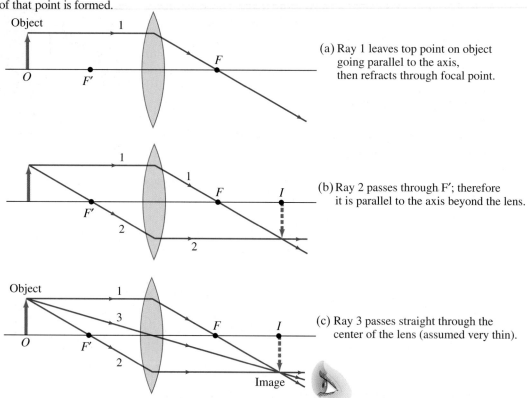

(a) Ray 1 leaves top point on object going parallel to the axis, then refracts through focal point.

(b) Ray 2 passes through F'; therefore it is parallel to the axis beyond the lens.

(c) Ray 3 passes straight through the center of the lens (assumed very thin).

(a)   (b)

FIGURE 23–35   (a) A converging lens can form a real image (here of a distant building) on a screen.
(b) That real image is also directly visible to the eye. Figure 23–29d shows images seen by the eye made
by both diverging and converging lenses.

In this way we can find the image point for one point of the object (the
top of the arrow in Fig. 23–34). The image points for all other points on the object can be found similarly to determine the complete image of the object. Because the rays actually pass through the image for the case shown in Fig. 23–34,
it is a real image (see page 686). The image could be detected by film, or actually seen on a white surface placed at the position of the image (Fig. 23–30).

The image can also be seen directly by the eye when the eye is placed
behind the image, as shown in Fig. 23–34c, so that some of the rays diverging from each point on the image enter the eye.[†] See Fig. 23–35.

By drawing the same three rays we can determine the image position
for a diverging lens, as shown in Fig. 23–36. Note that ray 1 is drawn parallel
to the axis, but does not pass through the focal point F' behind the lens. Instead it seems to come from the focal point F in front of the lens (dashed
line). Ray 2 is directed toward F' and is refracted parallel by the lens. Ray 3
passes directly through the center of the lens. The three refracted rays seem
to emerge from a point on the left of the lens. This is the image, I. Because
the rays do not pass through the image, it is a **virtual image**. Note that the
eye does not distinguish between real and virtual images—both are visible.

➡ **RAY TRACING**

For a diverging lens

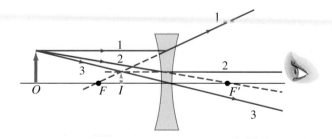

FIGURE 23–36   Finding
the image by ray tracing for a
diverging lens.

[†]Why, in order to see the image, the rays must be diverging from each point on the image will
be discussed in Section 25–2, but is essentially because we see real objects when diverging
rays from each point enter the eye as shown in Fig. 23–1.

23–8 The Lens Equation

We now derive an equation that relates the image distance to the object distance and the focal length of the lens. This will make the determination of image position quicker and more accurate than doing ray tracing. Let d_o be the object distance, the distance of the object from the center of the lens, and d_i be the image distance, the distance of the image from the center of the lens; and let h_o and h_i refer to the heights of the object and image. Consider the two rays shown in Fig. 23–37 for a converging lens (assumed to be very thin). The triangles $FI'I$ and FBA (highlighted in yellow in Fig. 23–37) are similar because angle AFB equals angle IFI'; so

$$\frac{h_i}{h_o} = \frac{d_i - f}{f},$$

since length $AB = h_o$. Triangles OAO' and IAI' are similar. Therefore,

$$\frac{h_i}{h_o} = \frac{d_i}{d_o}.$$

We equate the right sides of these two equations, divide by d_i, and rearrange to obtain

LENS EQUATION

$$\frac{1}{d_o} + \frac{1}{d_i} = \frac{1}{f}. \tag{23–8}$$

This is called the **lens equation**. It relates the image distance d_i to the object distance d_o and the focal length f. It is the most useful equation in geometric optics. (Interestingly, it is exactly the same as the mirror equation, Eq. 23–2). Note that if the object is at infinity, then $1/d_o = 0$, so $d_i = f$. Thus the focal length is the image distance for an object at infinity, as mentioned earlier.

We can derive the lens equation for a diverging lens using Fig. 23–38. Triangles IAI' and OAO' are similar; and triangles IFI' and AFB are similar. Thus (noting that length $AB = h_o$)

$$\frac{h_i}{h_o} = \frac{d_i}{d_o} \qquad \text{and} \qquad \frac{h_i}{h_o} = \frac{f - d_i}{f}.$$

FIGURE 23–37 Deriving the lens equation for a converging lens.

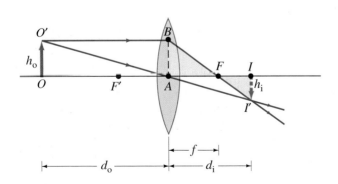

FIGURE 23–38 Deriving the lens equation for a diverging lens.

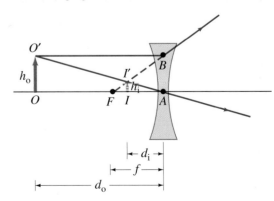

When these are equated and simplified, we obtain

$$\frac{1}{d_o} - \frac{1}{d_i} = -\frac{1}{f}.$$

This equation becomes the same as Eq. 23–8 if we make f and d_i negative. That is, we take f to be *negative for a diverging lens*, and d_i negative when the image is on the same side of the lens as the light comes from. Thus Eq. 23–8 will be valid for both converging and diverging lenses, and for *all* situations, if we use the following **sign conventions**:

1. The focal length is positive for converging lenses and negative for diverging lenses.

2. The object distance is positive if it is on the side of the lens from which the light is coming (this is usually the case, although when lenses are used in combination, it might not be so); otherwise, it is negative.

3. The image distance is positive if it is on the opposite side of the lens from where the light is coming; if it is on the same side, d_i is negative. Equivalently, the image distance is positive for a real image and negative for a virtual image.

4. The height of the image, h_i, is positive if the image is upright, and negative if the image is inverted relative to the object. (h_o is always taken as positive.)

➡ **PROBLEM SOLVING**
Sign conventions for lenses

The **lateral magnification**, m, of a lens is defined as the ratio of the image height to object height, $m = h_i/h_o$. From Figs. 23–37 and 23–38 and the conventions just stated, we have

$$m = \frac{h_i}{h_o} = -\frac{d_i}{d_o}. \tag{23–9}$$

Lateral magnification of a lens

For an upright image the magnification is positive, and for an inverted image m is negative.

From convention 1 above, it follows that the power (Eq. 23–7) of a converging lens, in diopters, is positive, whereas the power of a diverging lens is negative. A converging lens is sometimes referred to as a **positive lens**, and a diverging lens as a **negative lens**.

23–9 Problem Solving for Lenses

In brief outline, solving problems for thin lenses can be approached as follows:

➡ **PROBLEM SOLVING** **Thin Lenses**

1. As always, read and reread the problem,
2. Draw a ray diagram, precise if possible, but even a rough one can serve as confirmation of analytic results. Draw at least two, and preferably three, of the easy-to-draw rays described in Section 23–7 and shown in Figs. 23–34 and 23–36.
3. For analytic solutions, solve for unknowns in the lens equation (Eq. 23–8) and the magnification

(Eq. 23–9). The lens equation (Eq. 23–8) involves reciprocals—avoid the obvious but common error of forgetting to take the reciprocal.
4. Follow the **Sign Conventions** above.
5. Check that your analytic answers are consistent with your ray diagram.

FIGURE 23–39
Example 23–11. (Not to scale.)

EXAMPLE 23–11 **Image formed by converging lens.** What is (*a*) the position, and (*b*) the size, of the image of a large 7.6-cm-high flower placed 1.00 m from a +50.0-mm-focal-length camera lens?

SOLUTION Figure 23–39 is a rough ray diagram, showing only rays 1 and 2 for a single point on the flower. We see that the image ought to be a little behind the focal point, *F*, to the right of the lens. (*a*) We find the image position analytically using the lens equation, Eq. 23–8. The camera lens is converging, with $f = +5.00$ cm, and $d_o = 100$ cm, and so Eq. 23–8 gives

$$\frac{1}{d_i} = \frac{1}{f} - \frac{1}{d_o} = \frac{1}{5.00 \text{ cm}} - \frac{1}{100 \text{ cm}} = \frac{20.0 - 1.0}{100 \text{ cm}}.$$

Then

$$d_i = \frac{100 \text{ cm}}{19.0} = 5.26 \text{ cm},$$

or 52.6 mm behind the lens. Notice that the image is 2.6 mm farther from the lens than would be the image for an object at infinity. This is an example of the fact that when focusing a camera lens, the closer the object is to the camera, the farther the lens must be from the film.
(*b*) The magnification is

$$m = -\frac{d_i}{d_o} = -\frac{5.26 \text{ cm}}{100 \text{ cm}} = -0.0526;$$

so

$$h_i = mh_o = (-0.0526)(7.6 \text{ cm}) = -0.40 \text{ cm}.$$

The image is 4.0 mm high and is inverted ($m < 0$), as in Fig. 23–37, and shown in our sketch, Fig. 23–39.

EXAMPLE 23–12 **Object close to converging lens.** An object is placed 10 cm from a 15-cm-focal-length converging lens. Determine the image position and size (*a*) analytically, and (*b*) using a ray diagram.

SOLUTION (*a*) Given $f = 15$ cm and $d_o = 10$ cm, then

$$\frac{1}{d_i} = \frac{1}{15 \text{ cm}} - \frac{1}{10 \text{ cm}} = -\frac{1}{30 \text{ cm}},$$

➡ **PROBLEM SOLVING**

Don't forget to take the reciprocal

and $d_i = -30$ cm. (Remember to take the reciprocal!) Because d_i is negative, the image must be virtual and on the same side of the lens as the object. The magnification

$$m = -\frac{d_i}{d_o} = -\frac{-30 \text{ cm}}{10 \text{ cm}} = 3.0.$$

The image is three times as large as the object and is upright. (This lens is being used as a simple magnifying glass, which we discuss in more detail in Section 25–3.)

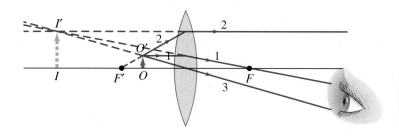

FIGURE 23–40 An object placed within the focal point of a converging lens produces a virtual image. Example 23–12.

(b) The ray diagram is shown in Fig. 23–40, and confirms the result in part (a). For point O' on the top of the object, ray 1 is easy to draw. But ray 2 may take some thought: if we draw it heading toward F', it is going the wrong way—so we have to draw it as if coming from F' (and so dashed), striking the lens, and then going out parallel to the principal axis. We project it backward with a dashed line, as we must do also for ray 1, in order to find where they meet. Ray 3 is easy to draw, through the lens center, and it meets the other two at the image point, I'.

It is a general rule that when an object is placed between a converging lens and its focal point (as in this last Example), the image is virtual.

EXAMPLE 23–13 **Diverging lens.** Where must a small insect be placed if a 25-cm-focal-length diverging lens is to form a virtual image 20 cm in front of the lens?

SOLUTION The ray diagram is basically that of Fig. 23–38 because our lens here is diverging and our image is in front of the lens within the focal distance. (It would be a valuable exercise to draw the ray diagram to scale, precisely, now.) Since $f = -25$ cm and $d_i = -20$ cm, then Eq. 23–8 gives

$$\frac{1}{d_o} = \frac{1}{f} - \frac{1}{d_i} = -\frac{1}{25 \text{ cm}} + \frac{1}{20 \text{ cm}} = \frac{-4 + 5}{100 \text{ cm}} = \frac{1}{100 \text{ cm}}.$$

So the object must be 100 cm in front of the lens. Your diagram should confirm this result.

*23–10 Problem Solving for Combinations of Lenses

We now consider two Examples illustrating how to deal with lenses used in combination. In general, when light passes through more than one lens, we find the image formed by the first lens as if it were alone. This image becomes the *object* for the second lens, and we find the image then formed by this second lens, which is the final image if there are only two lenses. The total magnification will be the product of the separate magnifications of each lens as we shall see.

Multiple lenses:
image formed by first lens
is object for second lens

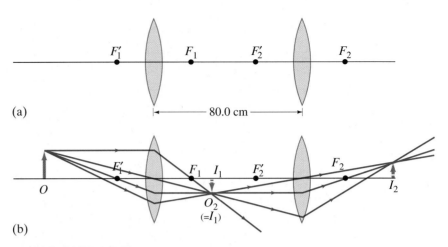

(a)

(b)

FIGURE 23–41
Example 23–14.

EXAMPLE 23–14 **A two-lens system.** Two converging lenses, with focal lengths $f_1 = 20.0$ cm and $f_2 = 25.0$ cm, are placed 80.0 cm apart, as shown in Fig. 23–41a. An object is placed 60.0 cm in front of the first lens as shown. Determine (a) the position, and (b) the magnification, of the final image formed by the combination of the two lenses.

SOLUTION (a) The object is a distance $d_{o1} = +60.0$ cm from the first lens, and this lens forms an image whose position can be calculated using the lens equation:

$$\frac{1}{d_{i1}} = \frac{1}{f_1} - \frac{1}{d_{o1}} = \frac{1}{20.0 \text{ cm}} - \frac{1}{60.0 \text{ cm}} = \frac{3-1}{60.0 \text{ cm}} = \frac{1}{30.0 \text{ cm}}.$$

Careful!
*Note that object distance for second lens is **not** equal to the image distance for first lens*

So the first image is at $d_{i1} = 30.0$ cm behind the first lens. This image becomes the object for the second lens. It is a distance $d_{o2} = 80.0$ cm $-$ 30.0 cm $= 50.0$ cm in front of lens 2, as shown in Fig. 23–41b. The image formed by the second lens, again using the lens equation, is at a distance d_{i2} from the second lens:

$$\frac{1}{d_{i2}} = \frac{1}{f_2} - \frac{1}{d_{o2}} = \frac{1}{25.0 \text{ cm}} - \frac{1}{50.0 \text{ cm}} = \frac{4-2}{100.0 \text{ cm}} = \frac{2}{100.0 \text{ cm}}.$$

Hence $d_{i2} = 50.0$ cm behind lens 2. This is the final image—see Fig. 23–41b.
(b) The first lens has a magnification (Eq. 23–9)

$$m_1 = -\frac{d_{i1}}{d_{o1}} = -\frac{30.0 \text{ cm}}{60.0 \text{ cm}} = -0.500.$$

Thus, the image is inverted and is half as high as the object: again Eq. 23–9,

$$h_{i1} = m_1 h_{o1} = -0.500 \, h_{o1}.$$

The second lens takes this image as object and changes its height by a factor

$$m_2 = \frac{d_{i2}}{d_{o2}} = \frac{50.0 \text{ cm}}{50.0 \text{ cm}} = 1.000.$$

The final image height is (remember h_{o2} is the same as h_{i1}):

$$h_{i2} = m_2 h_{o2} = m_2 h_{i1} = m_2 m_1 h_{o1}.$$

Total magnification is
$m = m_1 m_2$

We see from this equation that the total magnification is the product of m_1 and m_2, which here equals $(-1.000)(-0.500) = +0.500$, or $\frac{1}{2}$ the original height and upright.

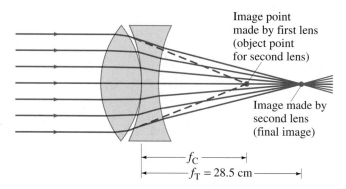

FIGURE 23–42
Determining the focal length of a diverging lens. Example 23–15.

In the figure:
- Image point made by first lens (object point for second lens)
- Image made by second lens (final image)
- f_C
- $f_T = 28.5$ cm

EXAMPLE 23–15 **Measuring *f* for a diverging lens.** To measure the focal length of a diverging lens, a converging lens is placed in contact with it, as shown in Fig. 23–42. The Sun's rays are focused by this combination at a point 28.5 cm behind the lenses as shown. If the converging lens has a focal length f_C of 16.0 cm, what is the focal length f of the diverging lens?

SOLUTION Rays from the Sun are focused 28.5 cm behind the combination, so the focal length of the total combination is $f_T = 28.5$ cm. If the diverging lens were absent, the converging lens would form the image at its focal point—that is, at a distance $f_C = 16.0$ cm behind it (dashed lines in Fig. 23–42). When the diverging lens is placed next to the converging lens (we assume both lenses are thin and the space between them is negligible), we treat the image formed by the first lens as the *object* for the second lens. Since this object lies to the right of the diverging lens, this is a situation where d_o is negative (see the sign conventions, page 707). Thus, for the diverging lens, the object is virtual and $d_o = -16.0$ cm. The diverging lens forms the image of this virtual object at a distance $d_i = 28.5$ cm away (this was given). Thus,

$$\frac{1}{f_D} = \frac{1}{d_o} + \frac{1}{d_i} = \frac{1}{-16.0 \text{ cm}} + \frac{1}{28.5 \text{ cm}} = -0.0274 \text{ cm}^{-1}.$$

We take the reciprocal to find $f_D = -1/(0.0274 \text{ cm}^{-1}) = -36.5$ cm.

Note that the converging lens must be "stronger" than the diverging lens (that is, it must have a focal length whose magnitude is less than that of the diverging lens) if this technique is to work.

*23–11 The Lensmaker's Equation

In this optional Section, we will prove that parallel rays are brought to a focus at a *single* point for a thin lens. At the same time, we will also derive an equation that relates the focal length of a lens to the radii of curvature of its two surfaces, which is known as the lens-maker's equation.

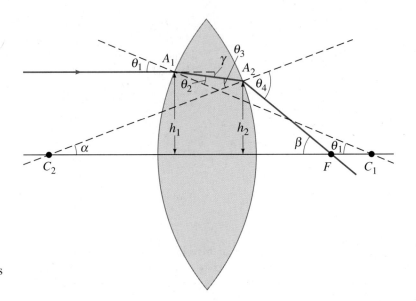

FIGURE 23–43 Diagram of a ray passing through a lens for derivation of the lensmaker's equation.

In Fig. 23–43, a ray parallel to the axis of a lens is refracted at the front surface of the lens at point A_1 and is refracted at the back surface at point A_2. This ray then passes through point F, which we call the focal point for this ray. Point A_1 is a height h_1 above the axis, and point A_2 is height h_2 above the axis. C_1 and C_2 are the centers of curvature of the two lens surfaces; so the length $C_1A_1 = R_1$, the radius of curvature of the front surface, and $C_2A_2 = R_2$ is the radius of the second surface. The thickness of the lens has been grossly exaggerated so the various angles would be clear. But we will assume that the lens is actually very thin and that angles between the rays and the axis are small. In this approximation, $h_1 \approx h_2$, and the sines and tangents of all the angles will be equal to the angles themselves in radians. For example, $\sin \theta_1 \approx \tan \theta_1 \approx \theta_1$ (radians).

To this approximation, then, Snell's law tells us that

$$\theta_1 = n\theta_2$$

$$\theta_4 = n\theta_3$$

where n is the index of refraction of the glass, and we assume that the lens is surrounded by air ($n = 1$). Notice also in Fig. 23–43 that

$$\theta_1 = \sin \theta_1 = \frac{h_1}{R_1}$$

$$\alpha = \frac{h_2}{R_2}$$

$$\beta = \frac{h_2}{f}.$$

The last follows because the distance from F to the lens (assumed very thin) is f. From the diagram, the angle γ was defined as

$$\gamma = \theta_1 - \theta_2.$$

A careful examination of Fig. 23–43 shows also that

$$\alpha = \theta_3 - \gamma.$$

This can be seen by drawing a horizontal line to the left from point A_2, which divides the angle θ_3 into two parts. The upper part equals γ and the lower part equals α. (The opposite angles between an oblique line and two parallel lines are equal.) Thus, $\theta_3 = \gamma + \alpha$. Finally, by drawing a horizontal line to the right from point A_2, we divide θ_4 into two parts. The upper part is α and the lower is β. Thus

$$\theta_4 = \alpha + \beta.$$

We now combine all these equations:

$$\alpha = \theta_3 - \gamma = \frac{\theta_4}{n} - (\theta_1 - \theta_2) = \frac{\alpha}{n} + \frac{\beta}{n} - \theta_1 + \theta_2,$$

or

$$\frac{h_2}{R_2} = \frac{h_2}{nR_2} + \frac{h_2}{nf} - \frac{h_1}{R_1} + \frac{h_1}{nR_1}.$$

Because the lens is thin, $h_1 \approx h_2$ and all h's can be canceled from all the numerators. We then multiply through by n and rearrange to find that

$$\frac{1}{f} = (n - 1)\left(\frac{1}{R_1} + \frac{1}{R_2} \right). \qquad \textbf{(23–10)} \qquad \textit{Lensmaker's equation}$$

This is called the **lensmaker's equation**. It relates the focal length of a lens to the radii of curvature of its two surfaces and its index of refraction. Notice that f does not depend on h_1 or h_2. Thus the position of the point F does not depend on where the ray strikes the lens. Hence, all rays parallel to the axis of a thin lens will pass through the same point F, which we wished to prove.

In our derivation, both surfaces are convex and R_1 and R_2 are considered positive.[†] Equation 23–10 also works for lenses with one or both surfaces concave; but for a concave surface, the radius must be considered *negative*.

Notice in Eq. 23–10 that the equation is symmetrical in R_1 and R_2. Thus, if a lens is turned around so that light impinges on the other surface, the focal length is the same even if the two lens surfaces are different.

[†]Some books use a different convention—for example, R_1 and R_2 are considered positive if their centers of curvature are to the right of the lens, in which case a minus sign appears in their equivalent of Eq. 23–10.

EXAMPLE 23–16 **Calculating f for a converging lens.** A convex meniscus lens (Figs. 23–29a and 23–44) is made from glass with $n = 1.50$. The radius of curvature of the convex surface is 22.4 cm and that of the concave surface is 46.2 cm. (a) What is the focal length? (b) Where will it focus an object 2.00 m away?

SOLUTION (a) $R_1 = 22.4$ cm and $R_2 = -46.2$ cm; the latter is negative because it refers to the concave surface. Then

$$\frac{1}{f} = (1.50 - 1.00)\left(\frac{1}{22.4 \text{ cm}} - \frac{1}{46.2 \text{ cm}}\right)$$

$$= 0.0115 \text{ cm}^{-1}.$$

So

$$f = \frac{1}{0.0115 \text{ cm}^{-1}} = 87 \text{ cm}$$

and the lens is converging. Notice that if we turn the lens around so that $R_1 = -46.2$ cm and $R_2 = +22.4$ cm, we get the same result.
(b) From the lens equation, with $f = 0.87$ m and $d_o = 2.00$ m, we have

$$\frac{1}{d_i} = \frac{1}{f} - \frac{1}{d_o} = \frac{1}{0.87 \text{ m}} - \frac{1}{2.00 \text{ m}}$$

$$= 0.65 \text{ m}^{-1},$$

so $d_i = 1/0.65 \text{ m}^{-1} = 1.53$ m.

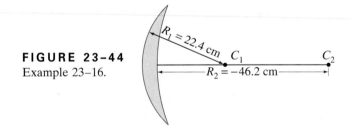

FIGURE 23–44
Example 23–16.

EXAMPLE 23–17 **Calculating f for a diverging lens.** A Lucite planoconcave lens has one flat surface and the other has $R = -18.4$ cm. What is the focal length?

SOLUTION From Table 23–1, n for Lucite is 1.51. A plane surface has infinite radius of curvature; if we call this R_1, then $1/R_1 = 0$. Therefore,

$$\frac{1}{f} = (1.51 - 1.00)\left(-\frac{1}{18.4 \text{ cm}}\right).$$

So $f = (-18.4 \text{ cm})/0.51 = -36$ cm, and the lens is diverging.

S U M M A R Y

Light appears to travel in straight-line paths, called **rays**, at a speed v that depends on the **index of refraction**, n, of the material; that is

$$v = \frac{c}{n},$$

where c is the speed of light in vacuum.

When light reflects from a flat surface, the *angle of reflection equals the angle of incidence.* This **law of reflection** explains why mirrors can form **images**. In a plane mirror, the image is virtual, upright, the same size as the object, and is as far behind the mirror as the object is in front.

A spherical mirror can be concave or convex. A concave spherical mirror focuses parallel rays of light (light from a very distant object) to a point called the **focal point**. The distance of this point from the lens is the **focal length** f of the mirror and

$$f = \frac{r}{2}$$

where r is the radius of curvature of the mirror. Parallel rays falling on a convex mirror reflect from the mirror as if they diverged from a common point behind the mirror. The distance of this point from the mirror is the focal length and is considered negative for a convex mirror. For a given object, the position and size of the image formed by a mirror can be found by ray tracing. Algebraically, the relation between image and object distances, d_i and d_o, and the focal length f, is given by the **mirror equation**:

$$\frac{1}{d_o} + \frac{1}{d_i} = \frac{1}{f}.$$

The ratio of image height to object height, which equals the magnification m, is

$$m = \frac{h_i}{h_o} = -\frac{d_i}{d_o}.$$

If the rays that converge to form an image actually pass through the image, so the image would appear on film or a screen placed there, the image is said to be a **real image**. If the rays do not actually pass through the image, the image is a **virtual image**.

When light passes from one transparent medium into another, the rays bend or refract. The **law of refraction (Snell's law)** states that

$$n_1 \sin \theta_1 = n_2 \sin \theta_2,$$

where n_1 and θ_1 are the index of refraction and angle with the normal to the surface for the incident ray, and n_2 and θ_2 are for the refracted ray.

When light rays reach the boundary of a material where the index of refraction decreases, the rays will be **totally internally reflected** if the incident angle, θ_1, is such that Snell's law would predict $\sin \theta_2 > 1$; this occurs if θ_1 exceeds the critical angle θ_C given by

$$\sin \theta_C = \frac{n_2}{n_1}.$$

A lens uses refraction to produce a real or virtual image. Parallel rays of light are focused to a point, called the **focal point**, by a converging lens. The distance of the focal point from the lens is called the **focal length** f of the lens. After parallel rays pass through a lens, they appear to diverge from a point, its focal point; and the corresponding focal length is considered negative. The **power** P of a lens, which is $P = 1/f$, is given in diopters, which are units of inverse meters (m^{-1}). For a given object, the position and size of the image formed by a lens can be found by ray tracing. Algebraically, the relation between image and object distances, d_i and d_o, and the focal length f, is given by the **lens equation**:

$$\frac{1}{d_o} + \frac{1}{d_i} = \frac{1}{f}.$$

The ratio of image height to object height, which equals the magnification m, is

$$m = \frac{h_i}{h_o} = -\frac{d_i}{d_o}.$$

When using the various equations of geometrical optics, it is important to remember the **sign conventions** for all quantities involved: carefully review them when doing problems.

1. Give arguments to show why the Moon must have a rough surface rather than a polished mirrorlike surface.

2. When you look at the Moon's reflection from a ripply sea, it appears elongated (Fig. 23–45). Explain.

FIGURE 23–45   Question 2.

3. Although a plane mirror reverses left and right, it doesn't reverse up and down. Explain.

4. Archimedes is said to have burned the whole Roman fleet in the harbor of Syracuse by focusing the rays of the Sun with a huge spherical mirror. Is this reasonable?

5. If a concave mirror produces a real image, is the image necessarily inverted?

6. When you use a concave mirror, you cannot see an inverted image of yourself unless you place your head beyond the center of curvature C. Yet you can see an inverted image of another object placed between C and F, as in Fig. 23–14. Explain. [*Hint:* You can see a real image only if your eye is behind the image, so that the image can be formed.]

7. Using the rules for the three rays discussed with reference to Fig. 23–14, draw ray 2 for Fig. 23–17b.

8. What is the focal length of a plane mirror? What is the magnification of a plane mirror?

9. Does the mirror equation, Eq. 23–2, hold for a plane mirror? Explain.

10. How might you determine the speed of light in a solid, rectangular, transparent object?

11. When a wide beam of parallel light enters water at an angle, the beam broadens. Explain.

12. What is the angle of refraction when a light ray meets the boundary between two materials perpendicularly?

13. When you look down into a swimming pool or a lake, are you likely to underestimate or overestimate its depth? Explain. How does the apparent depth vary with the viewing angle? (Use ray diagrams.)

14. Draw a ray diagram to show why a stick looks bent when part of it is under water (Fig. 23–20).

15. Your eye looks into an aquarium and views a fish inside. One ray of light that emerges from the tank from the fish is shown in Fig. 23–46. Also shown is the apparent position of the fish as seen by the eyeball. In the drawing, indicate the approximate position of the actual fish. Briefly justify your answer.

FIGURE 23–46   Question 15.

16. How can you "see" a round drop of water on a table even though the water is transparent and colorless?

17. Can a light ray traveling in air be totally reflected when it strikes a smooth water surface if the incident angle is right?

18. When you look up at an object in air from beneath the surface in a swimming pool, does the object appear to be the same size as when you see it directly in air? Explain.

19. What type of mirror is shown in Fig. 23–47?

FIGURE 23–47
Question 19.

20. Where must the film be placed if a camera lens is to make a sharp image of an object very far away?

21. A photographer moves closer to his subject and then refocuses. Does the camera lens move farther or nearer to the film?

22. Can a diverging lens form a real image under any circumstances? Explain.

23. Show that a real image formed by a thin lens is always inverted, whereas a virtual image is always upright if the object is real.

24. Light rays are said to be "reversible." Is this consistent with the lens equation?

25. Can real images be projected on a screen? Can virtual images? Can either be photographed? Discuss carefully.

26. A thin converging lens is moved closer to a nearby object. Does the real image formed change (*a*) in position, (*b*) in size? If yes, describe how.

27. Compare the mirror equation with the lens equation. Discuss similarities and differences, especially the sign conventions for the quantities involved.

28. A lens is made of a material with an index of refraction *n* = 1.30. In air, it is a converging lens. Will it still be a converging lens if placed in water? Explain, using a ray diagram.

29. A dog stands facing a converging lens with its tail in the air. If the nose and the tail are each focused on a screen in turn, which will have the greater magnification?

30. A cat stands facing a converging lens with its tail in the air. Under what circumstances (if any) would the image of the nose be virtual and the image of the tail be real? Where would the image of the rest of the cat be?

* 31. Why, in Example 23–15, must the converging lens have a shorter focal length than the diverging lens if the latter's focal length is to be determined by combining them?

* 32. An unsymmetrical lens (say, planoconvex) forms an image of a nearby object. Does the image point change if the lens is turned around?

* 33. The thicker a double convex lens is in the center as compared to its edges, the shorter its focal length for a given lens diameter. Explain.

* 34. Does the focal length of a lens depend on the fluid in which it is immersed? What about the focal length of a spherical mirror? Explain.

P R O B L E M S

SECTION 23–2

1. (I) Suppose that you want to take a photograph of yourself as you look at your image in a flat mirror 1.5 m away. For what distance should the camera lens be focused?

2. (I) When you look at yourself in a tall plane mirror, you see the same amount of your body whether you are close to the mirror or far away. (Try it and see.) Use ray diagrams to show why this should be true.

3. (II) Stand up two plane mirrors so they form a right angle as in Fig. 23–48. When you look into this double mirror, you see yourself as others see you, instead of reversed as in a single mirror. Make a ray diagram to show how this occurs.

FIGURE 23–48
Problems 3 and 8.

4. (II) A person whose eyes are 1.62 m above the floor stands 2.10 m in front of a vertical plane mirror whose bottom edge is 43 cm above the floor, Fig. 23–49. What is the horizontal distance *x* to the base of the wall supporting the mirror of the nearest point on the floor that can be seen reflected in the mirror?

FIGURE 23–49   Problem 4.

5. (II) Two mirrors meet at a 135° angle, Fig. 23–50. If light rays strike one mirror at 40° as shown, at what angle ϕ do they leave the second mirror?

FIGURE 23–50   Problem 5.

6. (II) Suppose you are 70 cm from a plane mirror. What area of the mirror is used to reflect the rays entering one eye from a point on the tip of your nose if your pupil diameter is 5.5 mm?

7. (III) Show that if two plane mirrors meet at an angle ϕ, a single ray reflected successively from both mirrors is deflected through an angle of 2ϕ independent of the incident angle. Assume $\phi < 90°$ and that only two reflections, one from each mirror, take place.

8. (III) Suppose a third mirror is placed beneath the two shown in Fig. 23–48, so that all three are perpendicular to each other. (a) Show that for such a "corner reflector," any incident ray will return in its original direction after three reflections. (b) What happens if it makes only two reflections?

SECTION 23–3

9. (I) A solar cooker, really a concave mirror pointed at the Sun, focuses the Sun's rays 17.0 cm in front of the mirror. What is the radius of the spherical surface from which the mirror was made?

10. (I) How far from a concave mirror (radius 27.0 cm) must an object be placed if its image is to be at infinity?

11. (II) Show with ray diagrams that the magnification of a concave mirror is less than 1 if the object is beyond the center of curvature C and is greater than 1 if it is within this point.

12. (II) If you look at yourself in a shiny Christmas tree ball with a diameter of 9.0 cm when your face is 30.0 cm away from it, where is your image? Is it real or virtual? Is it upright or inverted?

13. (II) A mirror at an amusement park shows an upright image of any person who stands 1.3 m in front of it. If the image is three times the person's height, what is the radius of curvature?

14. (II) A dentist wants a small mirror that, when 2.20 cm from a tooth, will produce a $4.5\times$ upright image. What kind of mirror must be used and what must its radius of curvature be?

15. (II) Some rearview mirrors produce images of cars to your rear that are a bit smaller than they would be if the mirror were flat. Are the mirrors concave or convex? What type and height of image would such a mirror produce of a car that was 1.3 m high and was 15.0 m behind you, assuming the mirror's radius of curvature is 3.2 m?

16. (II) A luminous object 3.0 mm high is placed 20 cm from a convex mirror of radius of curvature 20 cm. (a) Show by ray tracing that the image is virtual, and estimate the image distance. (b) Show that to compute this (negative) image distance from Eq. 23–2, it is sufficient to let the focal length be -10 cm. (c) Compute the image size, using Eq. 23–3.

17. (II) (a) Where should an object be placed in front of a concave mirror so that it produces an image at the same location? (b) Is the image real or virtual? (c) Is the image inverted or erect? (d) What is the magnification of the image?

18. (II) The image of a distant tree is virtual and very small when viewed in a curved mirror. The image appears to be 14.0 cm behind the mirror. What kind of mirror is it, and what is its radius of curvature?

19. (II) Use the mirror equation to show that the magnitude of the magnification of a concave mirror is less than 1 if the object is beyond the center of curvature C ($d_o > r$), and is greater than 1 if the object is within C ($d_o < r$).

20. (II) Show, using a ray diagram, that the magnification m of a convex mirror is $m = -d_i/d_o$, just as for a concave mirror. [Hint: Consider a ray from the top of the object that reflects at the center of the mirror.]

21. (II) Use ray diagrams to show that the mirror equation, Eq. 23–2, is valid for a convex mirror as long as f is considered negative.

22. (II) The magnification of a convex mirror is $+0.80\times$ for objects 2.2 m from the mirror. What is the focal length of this mirror?

23. (II) (a) A plane mirror can be considered a limiting case of a spherical mirror. Specify what this limit is. (b) Determine an equation that relates the image and object distances in this limit of a plane mirror. (c) Determine the magnification of a plane mirror in this same limit. (d) Are your results in parts (b) and (c) consistent with the discussion of Section 23–2 on plane mirrors?

24. (III) A 4.5-cm tall object is placed 28 cm in front of a spherical mirror. It is desired to produce a virtual image that is erect and 3.5 cm tall. (a) What type of mirror should be used? (b) Where is the image located? (c) What is the focal length of the mirror? (d) What is the radius of curvature of the mirror?

25. (III) A shaving/makeup mirror is designed to magnify your face by a factor of 1.3 when your face is placed 20.0 cm in front of it. (a) What type of mirror is it? (b) Describe the type of image that it makes of your face. (c) Calculate the required radius of curvature for the mirror.

SECTION 23–4

26. (I) What is the speed of light in (a) crown glass, and (b) Lucite?

27. (I) The speed of light in ice is 2.29×10^8 m/s. What is the index of refraction of ice?

28. (II) The speed of light in a certain substance is 85 percent of its value in water. What is the index of refraction of this substance?

29. (II) If $n = 1.00030 \pm 0.000010$ for air, what uncertainty will this introduce in the value for the speed of light in air?

SECTION 23-5

30. (I) A flashlight beam strikes the surface of a pane of glass ($n = 1.50$) at a 63° angle to the normal. What is the angle of refraction?

31. (I) A diver shines a flashlight upward from beneath the water at a 42.5° angle to the vertical. At what angle does the light leave the water?

32. (I) A light beam coming from an underwater spotlight exits the water at an angle of 60°. At what angle of incidence did it hit the air–water interface from below the surface?

33. (I) Rays of the Sun are seen to make a 21.0° angle to the vertical beneath the water. At what angle above the horizon is the Sun?

34. (II) Light is incident on an equilateral crown glass prism at a 45.0° angle to one face, Fig. 23–51. Calculate the angle at which light emerges from the opposite face. Assume that $n = 1.52$.

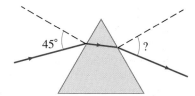

FIGURE 23–51
Problems 34 and 45.

35. (II) In searching the bottom of a pool at night, a watchman shines a narrow beam of light from his flashlight, 1.3 m above the water level, onto the surface of the water at a point 2.7 m from his foot at the edge of the pool (Fig. 23–52). Where does the spot of light hit the bottom of the pool, relative to the edge, if the pool is 2.1 m deep?

FIGURE 23–52
Problem 35.

36. (II) A beam of light in air strikes a slab of crown glass ($n = 1.52$) and is partially reflected and partially refracted. Find the angle of incidence if the angle of reflection is twice the angle of refraction.

FIGURE 23–53   Problem 37.

37. (II) An aquarium filled with water has flat glass sides whose index of refraction is 1.52. A beam of light from outside the aquarium strikes the glass at a 43.5° angle to the perpendicular (Fig. 23–53). What is the angle of this light ray when it enters (*a*) the glass, and then (*b*) the water? (*c*) What would be the refracted angle if the ray entered the water directly?

38. (II) Prove in general that for a light beam incident on a uniform layer of transparent material, as in Fig. 23–21, the direction of the emerging beam is parallel to the incident beam, independent of the incident angle θ.

39. (III) A light ray is incident on a flat piece of glass as in Fig. 23–21. Show that if the incident angle θ is small, the ray is displaced a distance $d = t\theta(n - 1)/n$, where t is the thickness of the glass and θ is in radians.

SECTION 23-6

40. (I) What is the critical angle for the interface between water and Lucite? To be internally reflected, the light must start in which material?

41. (I) The critical angle for a certain liquid-air surface is 44.7°. What is the index of refraction of the liquid?

42. (II) A beam of light is emitted in a pool of water from a depth of 62.0 cm. Where must it strike the air–water interface, relative to the spot directly above it, in order that the light does *not* exit the water?

43. (II) A ray of light enters a light fiber at an angle of 15° with the long axis of the fiber, as in Fig. 23–54. Calculate the distance the light ray travels between successive reflections off the sides of the fiber. Assume that the fiber has an index of refraction of 1.6 and is 10^{-4} m in diameter.

FIGURE 23–54   Problem 43.

44. (II) A beam of light is emitted 8.0 cm beneath the surface of a liquid and strikes the surface 7.0 cm from the point directly above the source. If total internal reflection occurs, what can you say about the index of refraction of the liquid?

45. (III) Suppose a ray strikes the left face of the prism in Fig. 23–51 at 45° as shown, but is totally internally reflected at the opposite side. If the apex angle (at the top) is $\theta = 75°$, what can you say about the index of refraction of the prism?

46. (III) A beam of light enters the end of an optic fiber as shown in Fig. 23–55. Show that we can guarantee total internal reflection at the side surface of the material (at point a), if the index of refraction is greater than about 1.42. In other words, regardless of the angle α, the light beam reflects back into the material at point a.

FIGURE 23–55
Problem 46.

Air   Transparent
             material

47. (III) (a) What is the minimum index of refraction for a glass or plastic prism to be used in binoculars (Fig. 23–25) so that total internal reflection occurs at 45°? (b) Will binoculars work if its prisms (assume $n = 1.50$) are immersed in water? (c) What minimum n is needed if the prisms are immersed in water?

SECTIONS 23–7 TO 23–9

48. (I) A sharp image is located 78.0 mm behind a 65.0-mm-focal-length converging lens. Find the object distance (a) using a ray diagram, (b) by calculation.

49. (I) Sunlight is observed to focus at a point 18.5 cm behind a lens. (a) What kind of lens is it? (b) What is its power in diopters?

50. (I) A certain lens focuses an object 2.25 m away as an image 48.3 cm on the other side of the lens. What type of lens is it and what is its focal length? Is the image real or virtual?

51. (I) (a) What is the power of a 29.5-cm-focal-length lens? (b) What is the focal length of a -6.25-diopter lens? Are these lenses converging or diverging?

52. (II) A stamp collector uses a converging lens with focal length 24 cm to view a stamp 18 cm in front of the lens. (a) Where is the image located? (b) What is the magnification?

53. (II) How large is the image of the Sun on the film used in a camera with (a) a 28-mm-focal-length lens, (b) a 50-mm-focal-length lens, and (c) a 200-mm-focal-length lens? (d) If the 50-mm lens is considered normal for this camera, how would you characterize the other two lenses? The Sun's diameter is 1.4×10^6 km and it is 1.5×10^8 km away.

54. (II) An 80-mm-focal-length lens is used to focus an image on the film of a camera. The maximum distance allowed between the lens and the film plane is 120 mm. (a) How far ahead of the film should the lens be if the object to be photographed is 10.0 m away? (b) 3.0 m away? (c) 1.0 m away? (d) What is the closest object this lens could photograph sharply?

55. (II) A -6.0-diopter lens is held 14.0 cm from an ant 1.0 mm high. What is the position, type, and height of the image?

56. (II) It is desired to magnify reading material by a factor of $2.5 \times$ when a book is placed 8.0 cm behind a lens. (a) Draw a ray diagram and describe the type of image this would be. (b) What type of lens is needed for this? (c) What is the power of the lens in diopters?

57. (II) (a) How far from a 50.0-mm-focal-length lens must an object be placed if its image is to be magnified $2.00 \times$ and be real? (b) What if the image is to be virtual and magnified $2.00 \times$?

58. (II) Repeat Problem 57 for a -50.0-mm-focal-length lens.

59. (II) (a) An object 31.5 cm in front of a certain lens is imaged 8.20 cm in front of that lens (on the same side as the object). What type of lens is this and what is its focal length? Is the image real or virtual? (b) If the image were located, instead, 38.0 cm in front of the lens, what type of lens would it be and what focal length would it have?

60. (II) (a) A 2.20-cm-high insect is 1.20 m from a 135-mm-focal-length lens. Where is the image, how high is it, and what type is it? (b) What if $f = -135$ mm?

61. (II) A bright object and a viewing screen are separated by a distance of 60 cm. At what location(s) between the object and the screen should a lens of focal length 15 cm be placed in order to produce a crisp image on the screen? [*Hint*: First draw a diagram.]

62. (III) How far apart are an object and an image formed by a 75-cm-focal-length converging lens if the image is $2.75 \times$ larger than the object and is real?

63. (III) (a) Show that the lens equation can be written in the *Newtonian form*:
$$xx' = f^2,$$
where x is the distance of the object from the focal point on the front side of the lens, and x' is the distance of the image to the focal point on the other side of the lens. Calculate the location of an image if the object is placed 45.0 cm in front of a convex lens with a focal length of 32.0 cm using (b) the standard form of the thin lens formula, and (c) the Newtonian form, derived above.

*64. (II) Two 27.0-cm-focal-length converging lenses are placed 16.5 cm apart. An object is placed 35.0 cm in front of one. Where will the final image formed by the second lens be located? What is the total magnification?

*65. (II) A diverging lens with $f = -31.5$ cm is placed 14.0 cm behind a converging lens with $f = 20.0$ cm. Where will an object at infinity be focused?

*66. (II) A 31.0-cm-focal-length converging lens is 21.0 cm behind a diverging lens. Parallel light strikes the diverging lens. After passing through the converging lens, the light is again parallel. What is the focal length of the diverging lens? [*Hint*: First draw a ray diagram.]

*67. (III) A diverging lens is placed next to a converging lens of focal length f_C, as in Fig. 23–42. If f_T represents the focal length of the combination, show that the focal length of the diverging lens, f_D, is given by

$$\frac{1}{f_D} = \frac{1}{f_T} - \frac{1}{f_C}.$$

*68. (I) A double concave lens has surface radii of 31.2 cm and 23.8 cm. What is the focal length if $n = 1.52$?

*69. (I) Both surfaces of a double convex lens have radii of 31.0 cm. If the focal length is 28.9 cm, what is the index of refraction of the lens material?

*70. (I) Show that if the lens of Example 23–17 is reversed so the light enters the curved face, the focal length is unchanged.

*71. (I) A planoconvex lens (Fig. 23–29) is to have a focal length of 28.5 cm. If made from fused quartz, what must be the radius of curvature of the convex surface?

*72. (I) A glass ($n = 1.50$) planoconcave lens has a focal length of -25.4 cm. What is the radius of the concave surface?

*73. (II) A prescription for a corrective lens calls for $+1.50$ diopters. The lensmaker grinds the lens from a "blank" with $n = 1.56$ and a preformed convex front surface of radius of curvature of 20.0 cm. What should be the radius of curvature of the other surface?

*74. (II) An object is placed 100 cm from a glass lens ($n = 1.56$) with one concave surface of radius 21.0 cm and one convex surface of radius 18.5 cm. Where is the final image? What is the magnification?

*75. (III) A glass lens ($n = 1.50$) in air has a power of $+5.2$ diopters. What would its power be if it were submerged in water?

GENERAL PROBLEMS

76. (II) Two plane mirrors are facing each other at a distance of 2.0 m apart as in Fig. 23–56. You stand 1.5 m away from one of these mirrors and look into it. You will see multiple images of yourself. (*a*) How far away from you are the first three images in the mirror in front of you? (*b*) Which way are these first three images facing, toward you or away from you?

FIGURE 23–56
Problem 76.

77. We wish to determine the depth of a swimming pool filled with water by measuring the width ($x = 5.50$ m) and then noting that the bottom edge of the pool is just visible at an angle of 14.0° above the horizontal as shown in Fig. 23–57. Calculate the depth of the pool.

FIGURE 23–57   Problem 77.

78. The critical angle of a certain piece of plastic in air is $\theta_C = 37.3°$. What is the critical angle of the same plastic if it is immersed in water?

79. In a slide or movie projector, the film acts as the object whose image is projected on a screen (Fig. 23–58). If a 100 mm-focal-length lens is to project an image on a screen 7.50 m away, how far from the lens should the slide be? If the slide is 36 mm wide, how wide will the picture be on the screen?

FIGURE 23–58   Problem 79.

FIGURE 23–59
Problem 80.

80. (II) A kaleidoscope makes symmetric patterns with two plane mirrors with a 60° angle between them as shown in Fig. 23–59. Draw the location of the images (some of them images of images) of the object placed between the mirrors.

81. A 35-mm slide (picture size is actually 24 by 36 mm) is to be projected on a screen 1.80 by 2.70 m placed 9.00 m from the projector. What focal-length lens should be used if the image is to cover the screen?

82. Show analytically that the image formed by a converging lens is real and inverted if the object is beyond the focal point ($d_o > f$), and is virtual and upright if the object is within the focal point ($d_o < f$). Describe the image if the object is an image, formed by another lens, so its position is beyond the lens, for which $-d_o > f$, and for which $0 < -d_o < f$.

83. Show analytically that a diverging lens can never form a real image of a real object. Can you describe a situation in which a diverging lens can form a real image?

FIGURE 23–60   Problem 84.

84. If the apex angle of a prism is $\phi = 72°$ (see Fig. 23–60), what is the minimum incident angle for a ray if it is to emerge from the opposite side (i.e., not be totally internally reflected), given $n = 1.52$?

* 85. A lighted candle is placed 30 cm in front of a converging lens of focal length $f_1 = 15$ cm, which in turn is 50 cm in front of another converging lens of focal length $f_2 = 10$ cm (see Fig. 23–61). (a) Draw a ray diagram for this problem. Estimate from your diagram, and then calculate (b) the location and the size of the final image.

$f_1 = 15$ cm $f_2 = 10$ cm

|← 30 cm →|← 50 cm →|

FIGURE 23–61   Problem 85.

* 86. (a) Show that if two lenses of focal lengths f_1 and f_2, respectively, are placed next to each other, the focal length of the combination is given by $f_T = f_1 f_2 / (f_1 + f_2)$. (b) Show that the power P of the combination of two lenses is the sum of their separate powers, $P = P_1 + P_2$.

87. A bright object is placed on one side of a converging lens of focal length f, and a white screen for viewing the image is on the opposite side. The distance $d_T = d_i + d_o$ between the object and the screen is kept fixed, but the lens can be moved. Show that (a) if $d_T > 4f$, there will be two positions where the lens can be placed and a sharp image will be produced on the screen, and (b) if $d_T < 4f$, there will be no lens position where a sharp image is formed. (c) Determine the distance between the two lens positions in part (a), and the ratio of the image sizes.

* 88. A converging lens with focal length of 10.0 cm is placed in contact with a diverging lens with a focal length of 20.0 cm. What is the focal length of the combination and is the combination converging or diverging?

89. A 1.65-m-tall person stands 3.25 m from a convex mirror and notices that he looks precisely half as tall as he does in a plane mirror placed at the same distance. What is the radius of curvature of the convex mirror? (Assume that $\sin \theta \approx \theta$.)

90. Each student in a physics lab is assigned to find the location where a bright object may be placed in order that a concave mirror with radius of curvature $r = 40$ cm will produce an image three times the size of the object. Two students complete the assignment at different times using identical equipment, but when they compare notes later, they discover that their answers for the object distance are not the same. Explain why they do not necessarily need to repeat the lab, and justify your response with a calculation.

* 91. A lens whose index of refraction is n is submerged in a material whose index of refraction is n' ($n' \neq 1$). Derive the equivalent of Eqs. 23–8, 23–9, and 23–10 for this lens.

The wave nature of light nicely explains how light reflected from the front and back surfaces of this very thin film of soapy water interferes constructively to produce the bright colors. Which color we see depends on the thickness of the film at that point—the film is very thin at the top and gets thicker going down toward the bottom (because of gravity). Can you use the clues (colors of light) to determine the thickness of the film at any point? You should be able to after reading this chapter.

THE WAVE NATURE OF LIGHT

That light carries energy is obvious to anyone who has focused the Sun's rays with a magnifying glass on a piece of paper and burned a hole in it. But how does light travel, and in what form is this energy carried? In our discussion of waves in Chapter 11, we noted that energy can be carried from place to place in basically two ways: by particles or by waves. In the first case, material bodies or particles can carry energy, such as an avalanche or rushing water. In the second case, water waves and sound waves, for example, can carry energy over long distances even though mass itself does not travel these distances. In view of this, what can we say about the nature of light: does light travel as a stream of particles away from its source, or does it travel in the form of waves that spread outward from the source?

Historically, this question has turned out to be a difficult one. For one thing, light does not reveal itself in any obvious way as being made up of tiny particles nor do we see tiny light waves passing by as we do water waves. The evidence seemed to favor first one side and then the other until about 1830, when most physicists had accepted the wave theory. By the end of the nineteenth century, light was considered to be an *electromagnetic wave* (Chapter 22). In the early twentieth century, light was shown to have a particle nature as well, as we shall discuss in Chapter 27. Nonetheless, the wave theory of light remains valid and has proved very successful. We now investigate the evidence for the wave theory and how it has explained a wide range of phenomena.

24–1 Waves Versus Particles; Huygens' Principle and Diffraction

Huygens' principle

The Dutch scientist Christiaan Huygens (1629–1695), a contemporary of Newton, proposed a wave theory of light that had much merit. Still useful today is a technique he developed for predicting the future position of a wave front when an earlier position is known. This is known as **Huygens' principle** and can be stated as follows: *Every point on a wave front can be considered as a source of tiny wavelets that spread out in the forward direction at the speed of the wave itself. The new wave front is the envelope of all the wavelets*—that is, *the tangent to all of them.*

As a simple example of the use of Huygens' principle, consider the wave front *AB* in Fig. 24–1, which is traveling away from a source *S*. We assume the medium is *isotropic*—that is, the speed *v* of the waves is the same in all directions. To find the wave front a short time *t* after it is at *AB*, tiny circles are drawn with radius $r = vt$. The centers of these tiny circles are on the original wave front *AB* and the circles represent Huygens' (imaginary) wavelets. The tangent to all these wavelets, the line *CD*, is the new position of the wave front.

FIGURE 24–1 Huygens' principle used to determine wave front *CD* when *AB* is given.

Huygens' principle is particularly useful when waves impinge on an obstacle and the wave fronts are partially interrupted. Huygens' principle predicts that waves bend in behind an obstacle, as shown in Fig. 24–2. This is just what water waves do, as we saw in Chapter 11 (Figs. 11–43 and 11–44). The bending of waves behind obstacles into the "shadow region" is known as **diffraction**. Since diffraction occurs for waves, but not for particles, it can serve as one means for distinguishing the nature of light.

Does light exhibit diffraction? In the mid-seventeenth century, a Jesuit priest, Francesco Grimaldi (1618–1663), had observed that when sunlight entered a darkened room through a tiny hole in a screen, the spot on the opposite wall was larger than would be expected from geometric rays. He also observed that the border of the image was not clear but was surrounded by colored fringes. Grimaldi attributed this to the diffraction of light.

Note that the ray model (Chapter 23) cannot account for diffraction, and it is important to be aware of the limitations of the ray model. Geometric op-

FIGURE 24–2 Huygens' principle is consistent with diffraction (a) around the edge of an obstacle, (b) through a large hole, (c) through a small hole whose size is on the order of the wavelength of the wave.

(a)   (b)   (c)

tics using rays is so successful in its limited sphere because normal openings and obstacles are much larger than the wavelength of the light, and so relatively little diffraction or bending occurs.

24–2 Huygens' Principle and the Law of Refraction

The laws of reflection and refraction were well known in Newton's time. The law of reflection could not distinguish between the two theories: waves versus particles. For when waves reflect from an obstacle, the angle of incidence equals the angle of reflection (Fig. 11–34). The same is true of particles—think of a tennis ball without spin striking a flat surface.

The law of refraction is another matter. Consider light entering a medium where it is bent toward the normal, as when it travels from air into water. As shown in Fig. 24–3, this effect can be constructed using Huygens' principle if we assume the speed of light is less in the second medium ($v_2 < v_1$). That is, in time t, the point B on wave front AB goes a distance $v_1 t$ to reach point D. Point A, on the other hand, travels a distance $v_2 t$ to reach point C. Huygens' principle is applied to points A and B to obtain the curved wavelets shown at C and D. The wave front is tangent to these two wavelets, so the new wave front is the line CD. Hence the rays, which are perpendicular to the wave fronts, bend toward the normal if $v_2 < v_1$, as drawn.[†] Newton favored a particle theory of light which predicted the opposite result that the speed of light would be greater in the second medium ($v_2 > v_1$). Thus the wave theory predicts that the speed of light in water, for example, is less than in air; and Newton's particle theory predicts the reverse. An experiment to actually measure the speed of light in water was performed in 1850 by the French physicist Jean Foucault, and it confirmed the wave-theory prediction. By then, however, the wave theory was already fully accepted, as we shall see in the next Section.

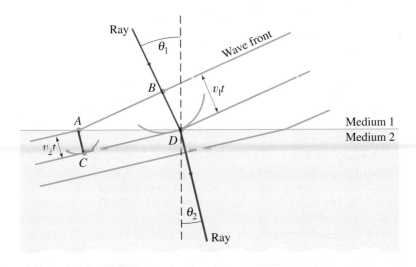

FIGURE 24–3 Refraction explained, using Huygens' principle.

[†]This is basically the same as the discussion around Fig. 11–42.

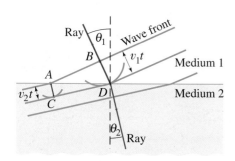

FIGURE 24–3 (Repeated.) Refraction explained, using Huygens' principle.

It is easy to show that Snell's law of refraction follows directly from Huygens' principle, given that the speed of light v in any medium is related to the speed in a vacuum, c, and the index of refraction, n, by Eq. 23–4: that is, $v = c/n$. From the Huygens' construction of Fig. 24–3, angle ADC is equal to θ_2 and angle BAD is equal to θ_1. Then for the two triangles that have the common side AD, we have

$$\sin \theta_1 = \frac{v_1 t}{AD}, \qquad \sin \theta_2 = \frac{v_2 t}{AD}.$$

We divide these two equations and obtain:

$$\frac{\sin \theta_1}{\sin \theta_2} = \frac{v_1}{v_2}.$$

Then, since $v_1 = c/n_1$ and $v_2 = c/n_2$,

$$n_1 \sin \theta_1 = n_2 \sin \theta_2,$$

which is Snell's law of refraction, Eq. 23–5. (The law of reflection can be derived from Huygens' principle in a similar way, and this is given as Problem 1 at the end of the chapter.)

When a light wave travels from one medium to another, its frequency does not change, but its wavelength does. This can be seen from Fig. 24–3, where we assume each of the blue lines representing a wave front corresponds to a crest (peak) of the wave. Then

Wavelength depends on n

$$\frac{\lambda_2}{\lambda_1} = \frac{v_2 t}{v_1 t} = \frac{v_2}{v_1} = \frac{n_1}{n_2},$$

where, in the last step, we used Eq. 23–4, $v = c/n$. If medium 1 is a vacuum (or air), so $n_1 = 1$, $v_1 = c$, and we call λ_1 simply λ, then the wavelength in another medium of index of refraction n $(=n_2)$ will be

$$\lambda_n = \frac{\lambda}{n}. \tag{24–1}$$

This result is consistent with the frequency f being unchanged since $c = f\lambda$. Combining this with $v = f\lambda_n$ in a medium where $v = c/n$ gives $\lambda_n = v/f = c/nf = f\lambda/nf = \lambda/n$, which checks.

Wave fronts can be used to explain how mirages are produced by refraction of light. Let us explain why, for example, on a hot day motorists sometimes see a mirage of water on the highway ahead of them, with distant vehicles seemingly reflected in it (Fig. 24–4a). On a hot day, there can be a layer of very hot air next to the roadway (made hot by the sun beating on the road). Hot air is less dense than cooler air, so the index of refraction is slightly lower in the hot air. In Fig. 24–4b, we see a diagram of light coming from one point on a distant car (on the right) heading left toward the observer. Wave fronts and two rays are shown. Ray A heads directly at the observer and follows a straight-line path, and represents the normal view of the car. Ray B is a ray initially directed slightly downward. But it does not hit the ground. Instead, ray B is bent slightly as it moves through layers of air of different index of refraction. The wave fronts, shown in blue in Fig. 24–4b, move slightly

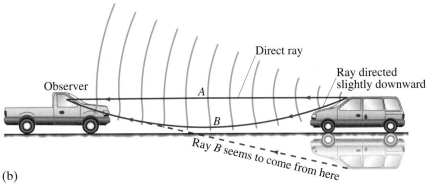

(a)   (b)

faster in the layers of air nearer the ground (see Fig. 24–3, and also the soldier analogy in Fig. 11–42). Thus ray B is bent as shown, and seems to the observer to be coming from below (dashed line) as if reflected off the road. Hence the mirage.

FIGURE 24–4 (a) A highway mirage. (b) Drawing (greatly exaggerated) showing wave fronts and rays to explain highway mirages. Note how the sections of the wavefronts near the ground move faster and so are further apart.

24–3 Interference—Young's Double-Slit Experiment

In 1801, the Englishman Thomas Young (1773–1829) obtained convincing evidence for the wave nature of light and was even able to measure the wavelengths for visible light. Figure 24–5a shows a schematic diagram of Young's famous double-slit experiment. Light from a single source (Young used the Sun) falls on a screen containing two closely spaced slits S_1 and S_2. If light consists of tiny particles, we might expect to see two bright lines on a screen placed behind the slits as in (b). But Young observed instead a series of bright lines as in (c). Young was able to explain this result as a **wave-interference** phenomenon. To see this, imagine plane waves of light of a single wavelength—called **monochromatic**, meaning "one color"—falling on the two slits as shown in Fig. 24–6. Because of diffraction, the waves leaving the two small slits spread out as shown. This is equivalent to the interference pattern produced when two rocks are thrown into a lake (Fig. 11–36), or when sound from two loudspeakers interferes (Fig. 12–17).

FIGURE 24–5 (a) Young's double-slit experiment. (b) If light consists of particles, we would expect to see two bright lines on the screen behind the slits. (c) Young observed many lines.

Viewing screen
(a)

(b)

Viewing screen
(particle theory prediction)

(c)

Viewing screen
(actual)

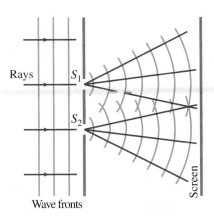

FIGURE 24–6 If light is a wave, light passing through one of two slits should interfere with light passing through the other slit.

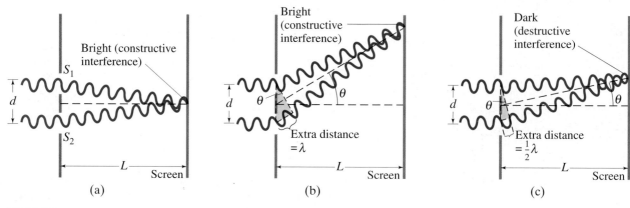

FIGURE 24–7 How the wave theory explains the pattern of lines seen in the double-slit experiment. (a) At the center of the screen the waves from each slit travel the same distance and are in phase. (b) At this angle θ, the lower wave travels an extra distance of one whole wavelength, and the waves are in phase; note from the shaded triangle that the extra distance equals $d \sin \theta$. (c) For this angle θ, the lower wave travels an extra distance equal to one-half wavelength, so the two waves arrive at the screen fully out of phase.

FIGURE 24–8
(a) Constructive interference.
(b) Destructive interference.
(See also Section 11–11.)

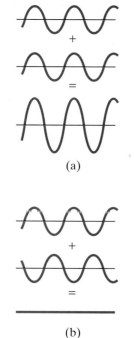

To see how an interference pattern is produced on the screen, we make use of Fig. 24–7. Waves of wavelength λ are shown entering the slits S_1 and S_2, which are a distance d apart. The waves spread out in all directions after passing through the slits, but they are shown only for three different angles θ. In Fig. 24–7a, the waves reaching the center of the screen are shown ($\theta = 0$). The waves from the two slits travel the same distance, so they are in phase: a crest of one wave arrives at the same time as a crest of the other wave. Hence the amplitudes of the two waves add to form a larger amplitude as shown in Fig. 24–8a. This is **constructive interference**, and there is a bright spot at the center of the screen. Constructive interference also occurs when the paths of the two rays differ by one wavelength (or any whole number of wavelengths), as shown in Fig. 24–7b. But if one ray travels an extra distance of one-half wavelength (or $\frac{3}{2}\lambda$, $\frac{5}{2}\lambda$, and so on), the two waves are exactly out of phase when they reach the screen: the crests of one wave arrive at the same time as the troughs of the other wave, and so they add to produce zero amplitude (Fig. 24–8b). This is **destructive interference**, and the screen is dark, Fig. 24–7c. Thus, there will be a series of bright and dark lines (or **fringes**) on the viewing screen.

To determine exactly where the bright lines fall, first note that Fig. 24–7 is somewhat exaggerated; in real situations, the distance d between the slits is very small compared to the distance L to the screen. The rays from each slit for each case will therefore be essentially parallel and θ is the angle they make with the horizontal. From the shaded right triangle in Fig. 24–7b, we can see that the extra distance traveled by the lower ray is $d \sin \theta$. **Constructive interference** will occur, and a bright fringe will appear on the screen, when $d \sin \theta$ equals a whole number of wavelengths:

$$d \sin \theta = m\lambda, \qquad m = 0, 1, 2, \cdots. \qquad \begin{bmatrix} \text{constructive} \\ \text{interference} \end{bmatrix} \quad \textbf{(24–2a)}$$

The value of m is called the **order** of the interference fringe. The first order ($m = 1$), for example, is the first fringe on each side of the central

(a)

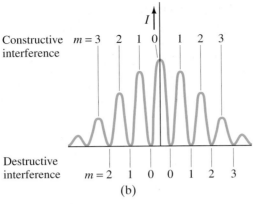

Constructive interference $m = 3$ 2   1   0   1   2   3

Destructive interference $m = 2$ 1   0   0   1   2   3

(b)

FIGURE 24–9
(a) Interference fringes produced by double-slit experiment and detected by photographic film placed on the viewing screen. The arrow marks the central fringe. (b) Intensity of light in the interference pattern. Also shown are values of m for Eq. 24–2a (constructive interference) and Eq. 24–2b (destructive interference).

fringe (at $\theta = 0$). Destructive interference occurs when the extra distance $d \sin \theta$ is $\frac{1}{2}, \frac{3}{2}$, and so on, wavelengths:

$$d \sin \theta = (m + \tfrac{1}{2})\lambda, \qquad m = 0, 1, 2, \cdots. \qquad \begin{bmatrix} \text{destructive} \\ \text{interference} \end{bmatrix} \quad \textbf{(24–2b)}$$

The intensity of the bright fringes is greatest for the central fringe ($m = 0$) and decreases for higher orders, as shown in Fig. 24–9.

EXAMPLE 24–1 **Line spacing for double-slit interference.** A screen containing two slits 0.100 mm apart is 1.20 m from the viewing screen. Light of wavelength $\lambda = 500$ nm falls on the slits from a distant source. Approximately how far apart will the bright interference fringes be on the screen?

SOLUTION Given $d = 0.100$ mm $= 1.00 \times 10^{-4}$ m, $\lambda = 500 \times 10^{-9}$ m, and $L = 1.20$ m, the first-order fringe ($m = 1$) occurs at an angle θ given by

$$\sin \theta_1 = \frac{m\lambda}{d} = \frac{(1)(500 \times 10^{-9}\,\text{m})}{1.00 \times 10^{-4}\,\text{m}} = 5.00 \times 10^{-3}.$$

This is a very small angle, so we can take $\sin \theta = \theta$, with θ in radians. The first-order fringe will occur a distance x_1 above the center of the screen (see Fig. 24–10) given by $x_1/L = \tan \theta_1 = \theta_1$, so

$$x_1 = L\theta_1 = (1.20\,\text{m})(5.00 \times 10^{-3}) = 6.00\,\text{mm}.$$

The second fringe ($m = 2$) will occur at

$$x_2 = L\theta_2 = L\frac{2\lambda}{d} = 12.0\,\text{mm}$$

above the center, and so on. Thus the fringes are 6.00 mm apart.

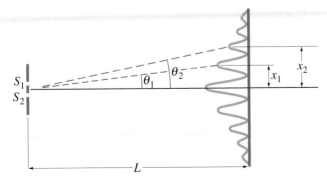

FIGURE 24–10 For small angles, the interference fringes occur at distance $x = \theta L$ above the center fringe ($m = 0$); θ_1 and x_1 are for the first order fringe ($m = 1$), θ_2 and x_2 are for $m = 2$.

CONCEPTUAL EXAMPLE 24–2 **Changing the wavelength.** (*a*) What happens to the interference pattern shown in Fig. 24–10, Example 24–1, if the incident light (500 nm) is replaced by light of wavelength 700 nm? (*b*) What happens instead if the slits are moved farther apart?

RESPONSE (*a*) When λ increases in Eq. 24–2a but d stays the same, then θ increases and the interference pattern spreads out. (*b*) Increasing the slit spacing d reduces θ for each order, so the lines are closer together.

Wavelength (or frequency) determines color

From Eqs. 24–2 we can see that, except for the zeroth-order fringe at the center, the position of the fringes depends on wavelength. Consequently, when white light falls on the two slits, as Young found in his experiments, the central fringe is white, but the first- (and higher-) order fringes contain a spectrum of colors like a rainbow; θ was found to be smallest for violet light and largest for red. By measuring the position of these fringes, Young was the first to determine the wavelengths of visible light (using Eqs. 24–2). In doing so, he showed that what distinguishes different colors physically is their wavelength, an idea put forward earlier by Grimaldi in 1665.

White

|◄2.0 mm►|

|◄———3.5 mm———►|

FIGURE 24–11 Example 24–3.

EXAMPLE 24–3 **Wavelengths from double-slit interference.** White light passes through two slits 0.50 mm apart and an interference pattern is observed on a screen 2.5 m away. The first-order fringe resembles a rainbow with violet and red light at either end. The violet light falls about 2.0 mm and the red 3.5 mm from the center of the central white fringe (Fig. 24–11). Estimate the wavelengths of the violet and red lights.

SOLUTION We use Eq. 24–2a with $m = 1$ and $\sin \theta = \theta$. Then for violet light, $x = 2.0$ mm, so (see also Fig. 24–10)

$$\lambda = \frac{d\theta}{m} = \frac{d}{m}\frac{x}{L} = \left(\frac{5.0 \times 10^{-4}\text{ m}}{1}\right)\left(\frac{2.0 \times 10^{-3}\text{ m}}{2.5\text{ m}}\right) = 4.0 \times 10^{-7}\text{ m},$$

or 400 nm. For red light, $x = 3.5$ mm, so

$$\lambda = \frac{d}{m}\frac{x}{L} = \left(\frac{5.0 \times 10^{-4}\text{ m}}{1}\right)\left(\frac{3.5 \times 10^{-3}\text{ m}}{2.5\text{ m}}\right) = 7.0 \times 10^{-7}\text{ m} = 700\text{ nm}.$$

Coherent and incoherent sources

The two slits in Fig. 24–7 act as if they were two sources of radiation. They are called **coherent sources** because the waves leaving them bear the same phase relationship to each other at all times (because ultimately the waves come from a single source to the left of the two slits in Fig. 24–7). An interference pattern is observed only when the sources are coherent. If two tiny lightbulbs replaced the two slits, an interference pattern would not be seen. The light emitted by one lightbulb would have a random phase with respect to the second bulb, and the screen would be more or less uniformly illuminated. Two such sources, whose output waves bear no fixed relationship to each other, are called **incoherent sources**.

24-4 The Visible Spectrum and Dispersion

The two most obvious properties of light are readily describable in terms of the wave theory of light: intensity (or brightness) and color. The **intensity** of light is the energy it carries per unit time, and is related to the square of the amplitude of the wave, just as for any wave (see Section 11–9, or Eqs. 22–7 and 22–8). The **color** of the light is related to the wavelength or frequency of the light. Visible light—that to which our eyes are sensitive—falls in the wavelength range of about 400 nm to 750 nm.[†] This is known as the **visible spectrum**, and within it lie the different colors from violet to red, as shown in Fig. 24–12. Light with wavelength shorter than

FIGURE 24–12 The spectrum of visible light, showing the range of wavelengths for the various colors.

400 nm is called **ultraviolet** (UV), and light with wavelength greater than 750 nm is called **infrared** (IR).[‡] Although human eyes are not sensitive to UV or IR, some types of photographic film do respond to them.

A prism separates white light into a rainbow of colors, as shown in Fig. 24–13. This happens because the index of refraction of a material depends on the wavelength, as shown for several materials in Fig. 24–14. White light is a mixture of all visible wavelengths, and when incident on a prism, as in Fig. 24–15, the different wavelengths are bent to varying degrees. Because the index of refraction is greater for the shorter wavelengths, violet light is bent the most and red the least as indicated. This spreading of white light into the full spectrum is called **dispersion**.

FIGURE 24–13 White light passing through a prism is broken down into its constituent colors.

Dispersion

FIGURE 24–14 Index of refraction as a function of wavelength for various transparent solids.

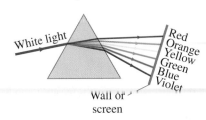

FIGURE 24–15 White light dispersed by a prism into the visible spectrum.

[†]Sometimes the angstrom (Å) unit is used when referring to light: $1 \text{ Å} = 1 \times 10^{-10}$ m. Then visible light falls in the wavelength range of 4000 Å to 7500 Å.

[‡]The complete electromagnetic spectrum is illustrated in Fig. 22–10.

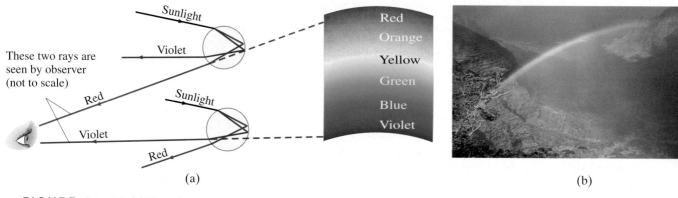

FIGURE 24–16 (a) Ray diagram explaining how a rainbow (b) is formed.

➡ PHYSICS APPLIED

➡ PHYSICS APPLIED

Rainbows

Rainbows are a spectacular example of dispersion—by drops of water. You can see rainbows when you look at falling water droplets with the Sun at your back. Figure 24–16 shows how red and violet rays are bent by spherical water droplets and are reflected off the back surface. Red is bent the least and so reaches the observer's eyes from droplets higher in the sky, as shown in the diagram. Thus the top of the rainbow is red.

Diamonds

Diamonds achieve their brilliance from a combination of dispersion and total internal reflection. Because diamonds have a very high index of refraction of about 2.4, the critical angle for total internal reflection is only 25°. Incident light therefore strikes many of the internal surfaces before it strikes one at less than 25° and emerges. After many such reflections, the light has traveled far enough that the colors have become sufficiently separated to be seen individually and brilliantly by the eye after leaving the crystal.

24–5 Diffraction by a Single Slit or Disk

Young's double-slit experiment put the wave theory of light on a firm footing. But full acceptance came only with studies on diffraction more than a decade later.

We have already discussed diffraction briefly with regard to water waves (Section 11–13) as well as for light (Section 24–1), and we have seen that it refers to the spreading or bending of waves around edges. Let's look in more detail.

A part of the history of the wave theory of light belongs to Augustin Fresnel (1788–1827), who in 1819 presented to the French Academy a wave theory of light that predicted and explained interference and diffraction effects. Almost immediately Siméon Poisson (1781–1840) pointed out a counterintuitive inference: according to Fresnel's wave theory, if light from a point source were to fall on a solid disk, then light diffracted around the edges should interfere constructively at the center of the shadow (Fig. 24–17). That prediction seemed very unlikely. But when the experiment was actually carried out (by Francois Arago), the bright spot was seen at the very center of the shadow! This was strong evidence for the wave theory.

Figure 24–18a is a photograph of the shadow cast by a coin using a (nearly) point source of light (a laser in this case). The bright spot is clearly present at the center. Note also that there are bright and dark fringes beyond the shadow. These resemble the interference fringes of a double slit. Indeed, they are due to interference of waves diffracted around different parts of the disk, and the whole is referred to as a **diffraction pattern**. A diffraction pattern exists around any sharp object illuminated by a point source, as shown in Figs. 24–18b and c. We are not always aware of them because most sources of light in everyday life are not points, so light from different parts of the source washes out the pattern.

To see how a diffraction pattern arises, we will analyze the important case of monochromatic light passing through a narrow slit. We will assume that parallel rays (or plane waves) of light fall on the slit of width D, and that the viewing screen is far away. As we know from studying water waves and from Huygens' principle, the waves passing through the slit spread out in all directions. We will now examine how the waves passing through different parts of the slit interfere with each other.

The (un)expected diffraction spot

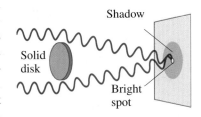

FIGURE 24–17 If light is a wave, a bright spot will appear at the center of the shadow of a solid disk illuminated by a point source of monochromatic light.

FIGURE 24–18 Diffraction pattern of (a) a penny, (b) a razor blade, (c) a single slit, each illuminated by a (nearly) point source of monochromatic light.

(a) (b) (c)

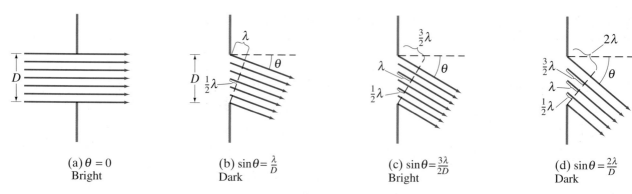

(a) $\theta = 0$
Bright

(b) $\sin\theta = \frac{\lambda}{D}$
Dark

(c) $\sin\theta = \frac{3\lambda}{2D}$
Bright

(d) $\sin\theta = \frac{2\lambda}{D}$
Dark

FIGURE 24–19 Analysis of diffraction pattern formed by light passing through a narrow slit.

Parallel rays of monochromatic light pass through the narrow slit as shown in Fig. 24–19a. The light falls on a screen which is assumed to be very far away, so the rays heading for any point are essentially parallel. First we consider rays that pass straight through as in Fig. 24–19a. They are all in phase, so there will be a central bright spot on the screen. In Fig. 24–19b, we consider rays moving at an angle θ such that the ray from the top of the slit travels exactly one wavelength farther than the ray from the bottom edge. The ray passing through the very center of the slit will travel one-half wavelength farther than the ray at the bottom of the slit. These two rays will be exactly out of phase with one another and so will destructively interfere. Similarly, a ray slightly above the bottom one will cancel a ray that is the same distance above the central one. Indeed, each ray passing through the lower half of the slit will cancel with a corresponding ray passing through the upper half. Thus, all the rays destructively interfere in pairs, and so no light will reach the viewing screen at this angle. The angle θ at which this takes place can be seen from the diagram to occur when $\lambda = D\sin\theta$, so

Diffraction equation (angular width of central spot)

$$\sin\theta = \frac{\lambda}{D}. \qquad \text{[first minimum]} \quad \textbf{(24–3a)}$$

The light intensity is a maximum at $\theta = 0°$ and decreases to a minimum (intensity = zero) at the angle θ given by Eq. 24–3a.

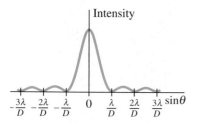

FIGURE 24–20 Intensity in the diffraction pattern of a single slit as a function of $\sin\theta$. Note that the central maximum is not only much higher than the maxima to each side, but it is also twice as wide ($2\lambda/D$ wide) as any of the others (only λ/D wide each).

Single slit diffraction minima

Now consider a larger angle θ such that the top ray travels $\frac{3}{2}\lambda$ farther than the bottom ray, as in Fig. 24–19c. In this case, the rays from the bottom third of the slit will cancel in pairs with those in the middle third because they will be $\lambda/2$ out of phase. However, light from the top third of the slit will still reach the screen, so there will be a bright spot, but not nearly as bright as the central spot at $\theta = 0°$. For an even larger angle θ such that the top ray travels 2λ farther than the bottom ray, Fig. 24–19d, rays from the bottom quarter of the slit will cancel with those in the quarter just above it because the path lengths differ by $\lambda/2$. And the rays through the quarter of the slit just above center will cancel with those through the top quarter. At this angle there will again be a minimum of zero intensity in the diffraction pattern. A plot of the intensity as a function of angle is shown in Fig. 24–20. This corresponds well with the photo of Fig. 24–18c. Notice that minima (zero intensity) occur at

$$D\sin\theta = m\lambda, \qquad m = 1, 2, 3, \cdots, \qquad \text{[minima]} \quad \textbf{(24–3b)}$$

but *not* at $m = 0$ where there is the strongest maximum. Between the minima, smaller intensity maxima occur. [Note that the *minima* for a diffraction pattern, Eq. 24–3b, satisfy a criterion very similar to that for the *maxima* (bright spots) for double-slit interference, Eq. 24–2a.]

EXAMPLE 24–4 **Single-slit diffraction maximum.** Light of wavelength 750 nm passes through a slit 1.0×10^{-3} mm wide. How wide is the central maximum (*a*) in degrees, and (*b*) in centimeters, on a screen 20 cm away?

SOLUTION (*a*) The first minimum occurs at

$$\sin \theta = \frac{\lambda}{D} = \frac{7.5 \times 10^{-7} \text{ m}}{1 \times 10^{-6} \text{ m}} = 0.75.$$

So $\theta = 49°$. This is the angle between the center and the first minimum, Fig. 24–21. The angle subtended by the whole central maximum, between the minima above and below the center, is twice this, or 98°.
(*b*) The width of the central maximum is $2x$, where $\tan \theta = x/20$ cm. So $2x = 2(20 \text{ cm})(\tan 49°) = 46$ cm. A large width of the screen will be illuminated, but it will not normally be very bright since the amount of light that passes through such a small slit will be small and it is spread over a large area.

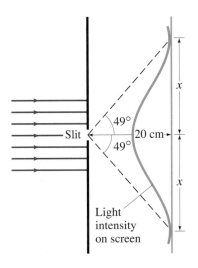

FIGURE 24–21
Example 24–4.

CONCEPTUAL EXAMPLE 24–5 **Diffraction spreads.** Light shines through a rectangular hole that is narrower in the vertical direction than the horizontal, Fig. 24–22. (*a*) Would you expect the light to diffract more in the vertical direction or in the horizontal direction? (*b*) Should a rectangular loudspeaker horn at a stadium be high and narrow, or wide and flat?

RESPONSE (*a*) From Eq. 24–3a we can see that if we make the slit (width D) narrower, the pattern spreads out more. (This is consistent with our earlier study of waves in Chapter 11.) Hence the diffraction through the rectangular hole will be wider vertically, since the aperture in that direction is smaller.
(*b*) For the loudspeaker, the sound pattern desired is one spread out horizontally, so the horn should be tall and narrow.

FIGURE 24–22
Example 24–5.

24–6 Diffraction Grating

A large number of equally spaced parallel slits is called a **diffraction grating**, although the term "interference grating" might be as appropriate. Gratings can be made by precision machining of very fine parallel lines on a glass plate. The untouched spaces between the lines serve as the slits. Photographic transparencies of an original grating serve as inexpensive gratings. Gratings containing 10,000 lines per centimeter are common today, and are very useful for precise measurements of wavelengths. A diffraction grating containing slits is called a **transmission grating**. **Reflection gratings** are also possible; they can be made by ruling fine lines on a metallic or glass surface from which light is reflected and analyzed. The analysis is basically the same as for a transmission grating, which we now discuss.

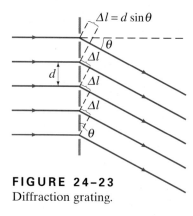

FIGURE 24–23
Diffraction grating.

Diffraction grating maxima (m = order)

Why more slits yield sharper peaks

FIGURE 24–24 Intensity as a function of viewing angle θ (or position on screen) for (a) two slits, (b) six slits. For a diffraction grating, the number of slits is very large ($\sim 10^4$) and the peaks are narrower still.

(a)

(b)

The analysis of a diffraction grating is much like that of Young's double-slit experiment. We assume parallel rays of light are incident on the grating as shown in Fig. 24–23. We also assume that the slits are narrow enough so that diffraction by each of them spreads light over a very wide angle on a distant screen behind the grating, and interference can occur with light from all the other slits. Light rays that pass through each slit without deviation ($\theta = 0°$) interfere constructively to produce a bright line at the center of the screen. Constructive interference also occurs at an angle θ such that rays from adjacent slits travel an extra distance of $\Delta l = m\lambda$, where m is an integer. Thus, if d is the distance between slits, then we see from Fig. 24–23 that $\Delta l = d \sin \theta$, and

$$\sin \theta = \frac{m\lambda}{d}, \qquad m = 0, 1, 2, \cdots \qquad \text{[principal maxima]} \quad \textbf{(24–4)}$$

is the criterion to have a brightness maximum. This is the same equation as for the double-slit situation, and again m is called the order of the pattern.

There is an important difference between a double-slit and a multiple-slit pattern, however. The bright maxima are much *sharper* and *narrower* for a grating. Why this happens can be seen as follows. Suppose that the angle θ is increased just slightly beyond that required for a maximum. In the case of only two slits, the two waves will be only slightly out of phase, so nearly full constructive interference occurs. This means the maxima are wide (see Fig. 24–10). For a grating, the waves from two adjacent slits will also not be significantly out of phase. But waves from one slit and those from a second one a few hundred slits away may be exactly out of phase; all or nearly all the light will cancel in pairs in this way. For example, suppose the angle θ is different from its first-order maximum so that the extra path length for a pair of adjacent slits is not exactly λ but rather 1.0010λ. The wave through one slit and another one 500 slits below will be out of phase by 1.5000λ, or exactly $1\frac{1}{2}$ wavelengths, so the two will cancel. A pair of slits, one below each of these, will also cancel. That is, the light from slit 1 cancels with that from slit 501; light from slit 2 cancels with that from slit 502, and so on. Thus even for a tiny angle[†] corresponding to an extra path length of $\frac{1}{1000} \lambda$, there is much destructive interference, and so the maxima are very narrow. The more lines there are in a grating, the sharper will be the peaks (see Fig. 24–24). Because a grating produces much sharper (and brighter) lines than two slits alone, it is a far more precise device for measuring wavelengths.

Suppose the light striking a diffraction grating is not monochromatic, but rather consists of two or more distinct wavelengths. Then for all orders other than $m = 0$, each wavelength will produce a maximum at a different angle (Fig. 24–25a), just as for a double slit. If white light strikes a grating, the central ($m = 0$) maximum will be a sharp white peak. But for all other orders, there will be a distinct spectrum of colors spread out over a certain angular width, Fig. 24–25b. Because a diffraction grating spreads out light into its component wavelengths, the resulting pattern is called a **spectrum**.

[†]Depending on the total number of slits, there may or may not be complete cancellation for such an angle, so there will be very tiny peaks between the main maxima (see Fig. 24–24b), but they are usually much too small to be seen.

| $m=2$ | | $m=2$ | $m=1$ | $m=1$ | | $m=0$ | | $m=1$ | $m=1$ | $m=2$ | | $m=2$ |

| 700 | | 400 | 700 | 400 | | Both λ | | 400 | 700 | 400 | | 700 |
| nm | | nm | nm | nm | | | | nm | nm | nm | | nm |

(a)

| $m=2$ | $m=1$ | $m=0$ | $m=1$ | $m=2$ |

| Rainbow | Rainbow | White | Rainbow | Rainbow |
| (fainter) | | | | (fainter) |

(b)

FIGURE 24–25 Spectra produced by a grating: (a) two wavelengths, 400 nm and 700 nm; (b) white light. The second order will normally be dimmer than the first order. (Higher orders are not shown.) If grating spacing is small enough, the second and higher orders will be missing.

EXAMPLE 24–6 **Diffraction grating: lines.** Calculate the first- and second-order angles for light of wavelength 400 nm and 700 nm if the grating contains 10,000 lines/cm.

SOLUTION The grating contains 10^4 lines/cm $= 10^6$ lines/m, which means the separation between slits is $d = (1/10^6)$ m $= 1.0 \times 10^{-6}$ m. In first order ($m = 1$), the angles are

$$\sin \theta_{400} = \frac{m\lambda}{d} = \frac{(1)(4.0 \times 10^{-7}\,\text{m})}{1.0 \times 10^{-6}\,\text{m}} = 0.400$$

$$\sin \theta_{700} = 0.700$$

so $\theta_{400} = 23.6°$ and $\theta_{700} = 44.4°$. In second order,

$$\sin \theta_{400} = \frac{(2)(4.0 \times 10^{-7}\,\text{m})}{1.0 \times 10^{-6}\,\text{m}} = 0.800$$

$$\sin \theta_{700} = 1.40$$

so $\theta_{400} = 53.1°$, but the second order does not exist for $\lambda = 700$ nm because $\sin \theta$ cannot exceed 1. No higher orders will appear.

EXAMPLE 24–7 **Spectra overlap.** White light containing wavelengths from 400 nm to 750 nm strikes a grating containing 4000 lines/cm. Show that the blue at $\lambda = 450$ nm of the third-order spectrum overlaps the red at 700 nm of the second order.

SOLUTION The grating spacing is $d = (1/4000)$ cm $= 2.50 \times 10^{-6}$ m. The blue of the third order occurs at an angle θ given by

$$\sin \theta = \frac{m\lambda}{d} = \frac{(3)(4.50 \times 10^{-7}\,\text{m})}{(2.50 \times 10^{-6}\,\text{m})} = 0.540.$$

Red in second order occurs at

$$\sin \theta = \frac{(2)(7.00 \times 10^{-7}\,\text{m})}{(2.50 \times 10^{-6}\,\text{m})} = 0.560$$

which is a greater angle; so the second order overlaps into the beginning of the third-order spectrum.

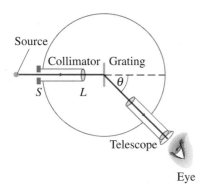

FIGURE 24–26 Spectrometer or spectroscope.

The diffraction grating is the essential component of a spectrometer, a device for precise measurement of wavelengths, and we discuss it next.

24–7 The Spectrometer and Spectroscopy

A **spectrometer** or **spectroscope**, Fig. 24–26, is a device[†] to measure wavelengths accurately using a diffraction grating (discussed in Section 24–6), or a prism, to separate different wavelengths of light. Light from a source passes through a narrow slit S in the collimator. The slit is at the focal point of the lens L, so parallel light falls on the grating. The movable telescope can bring the rays to a focus. Nothing will be seen in the viewing telescope unless it is positioned at an angle θ that corresponds to a diffraction peak (first order is usually used) of a wavelength emitted by the source. The angle θ can be measured[‡] to very high accuracy, so the wavelength of a line can be determined to high accuracy using Eq. 24–4:

$$\lambda = \frac{d}{m} \sin \theta,$$

where m is an integer representing the order, and d is the distance between grating lines. (The line you see in a spectrometer corresponding to each wavelength is actually an image of the slit S. The narrower the slit, the narrower— but dimmer—the line is, and the more precisely we can measure its angular position. If the light contains a continuous range of wavelengths, then a continuous spectrum is seen in the spectroscope.)

In some spectrometers, a reflection grating is used, and sometimes a prism. A prism works because of dispersion (Section 24–4), bending light of different wavelengths into different angles.

An important use of a spectrometer is for the identification of atoms or molecules. When a gas is heated or a large electric current is passed through *Line spectra* it, the gas emits a characteristic **line spectrum**. That is, only certain discrete wavelengths of light are emitted, and these are different for different elements and compounds.[§] Figure 24–27 shows the line spectra for a number of elements in the gas state. Line spectra occur only for gases at high temperatures and low pressure. The light from heated solids, such as a lightbulb filament, and even from a dense gaseous object such as the Sun, produces a **continuous spectrum** including a wide range of wavelengths.

Figure 24–27 also shows the Sun's "continuous spectrum," which contains a number of *dark* lines (only the most prominent are shown), called **absorption lines**. Atoms and molecules can absorb light at the same wavelengths at which they emit light. The Sun's absorption lines are due to absorption by atoms and molecules in the cooler outer atmosphere of the Sun, as well as by atoms and molecules in the Earth's atmosphere. A careful analysis of all these thousands of lines reveals that at least two thirds of all elements are present in

[†]If the spectrum of a source is recorded (say, on film) rather than viewed by the eye, the device is called a **spectrometer** or spectrograph, as compared to a **spectroscope**, which is for viewing only; but these terms are often used interchangeably. Devices that can also measure the intensity of light of a given wavelength are called **spectrophotometers**.

[‡]The angle θ for a given wavelength is usually measured on both sides of center because the grating cannot always be aligned precisely; the average of the two values is then taken.

[§]Why atoms and molecules emit line spectra was a great mystery for many years and played a central role in the development of modern quantum theory, as we shall see in Chapter 27.

Atomic hydrogen

Mercury

Sodium

Solar absorption spectrum

FIGURE 24–27 Line spectra for the gases indicated, and spectrum from the Sun showing absorption lines.

the Sun's atmosphere. The presence of elements in the atmosphere of other planets, in interstellar space, and in stars is also determined by spectroscopy.

Spectroscopy is useful for determining the presence of certain types of molecules in laboratory specimens where chemical analysis would be difficult. For example, biological DNA and different types of protein absorb light in particular regions of the spectrum (such as in the UV). The material to be examined, which is often in solution, is placed in a monochromatic light beam whose wavelength is chosen by placement of a prism or diffraction grating. The amount of absorption, as compared to a standard solution without the specimen, can reveal not only the presence of a particular type of molecule, but also its concentration.

Light emission and absorption also occur outside the visible part of the spectrum, such as in the UV and IR regions. Glass absorbs light in these regions, so reflection gratings and mirrors (in place of lenses) are used. Special types of film or detectors are used for detection.

FIGURE 24–27 Line spectra for the gases indicated, and spectrum from the Sun showing absorption lines.

➡ **PHYSICS APPLIED**

Chemical and biochemical analysis

24–8 Interference by Thin Films

Interference of light gives rise to many everyday phenomena such as the bright colors reflected from soap bubbles and from thin oil films on water (Fig. 24–28). In these and other cases, the colors are a result of constructive interference between light reflected from the two surfaces of the thin film.

FIGURE 24–28 Thin film interference patterns seen in (a) soap bubbles, (b) thin films of soapy water, and (c) a thin layer of oil on the water of a street puddle.

(a)

(b)

(c)

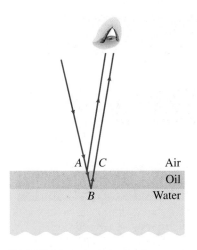

FIGURE 24–29 Light reflected from the upper and lower surfaces of a thin film of oil lying on water. Analysis assumes the light strikes the surface perpendicularly, but is shown here at a slight angle for clarity.

Newton's rings

To see how this happens, consider a smooth surface of water on top of which is a thin uniform layer of another substance, say an oil whose index of refraction is less than that of water (we'll see why we assume this in a moment); see Fig. 24–29. Part of the incident light (say, from the Sun or street lights) is reflected at A on the top surface, and part of that transmitted is reflected at B on the lower surface. The part reflected at the lower surface must travel the extra distance ABC. If the distance ABC is equal to one or a whole number of wavelengths, the two waves will interfere constructively and the light will be bright. But if ABC equals $\frac{1}{2}\lambda, \frac{3}{2}\lambda$, and so on, the two waves will be exactly out of phase and destructive interference occurs. The wavelength λ is *the wavelength in the film* (see Eq. 24–1).

When white light falls on such a film, the path ABC will equal λ (or $m\lambda$, with $m =$ an integer) for only one wavelength at a given viewing angle. This color will be seen as very bright. For light viewed at a slightly different angle, the path ABC will be longer or shorter and a different color will undergo constructive interference. Thus, for an extended (nonpoint) source emitting white light, a series of bright colors will be seen next to one another. Variations in thickness of the film will also alter the path length ABC and therefore affect the color of light that is most strongly reflected.

When a curved glass surface is placed in contact with a flat glass surface, Fig. 24–30, a series of concentric rings is seen when illuminated from above by monochromatic light. These are called **Newton's rings**[†] and they are due to interference between rays reflected by the top and bottom surfaces of the very thin *air gap* between the two pieces of glass. Because this gap (which is equivalent to a thin film) increases in width from the central contact point out to the edges, the extra path length for the lower ray (equal to BCD) varies; where it equals $0, \frac{1}{2}\lambda, \lambda, \frac{3}{2}\lambda, 2\lambda$, and so on, it corresponds to constructive and destructive interference; and this gives rise to the series of bright and dark lines seen in Fig. 24–30b.

The point of contact of the two glass surfaces (A in Fig. 24–30a) is dark in Fig. 24–30b. Since the path difference is zero here, we expect the rays reflected from each surface to be in phase and so this central point ought to be bright. But it is dark, which tells us the two rays must be completely out of phase. This can happen only because one of the waves undergoes a change in phase of 180° upon reflection, corresponding to $\frac{1}{2}$ cycle or $\frac{1}{2}\lambda$. Indeed, this and other experiments reveal that *a beam of light reflected by a material whose*

FIGURE 24–30 Newton's rings.

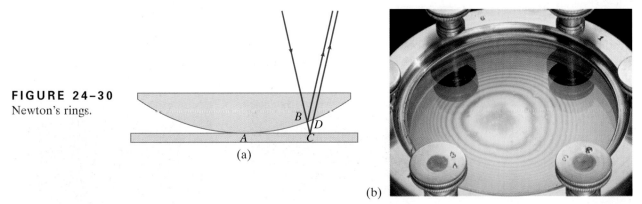

(a)

(b)

[†]Although Newton gave an elaborate description of them, they had been first observed and described by his contemporary, Robert Hooke. Newton did not realize their significance in support of a wave theory of light.

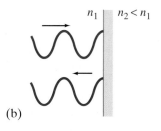

FIGURE 24–31 (a) Reflected ray changes phase by $\frac{1}{2}\lambda$ if $n_2 > n_1$, but does not (b) if $n_2 < n_1$.

index of refraction is greater than that of the material in which it is traveling, changes phase by $\frac{1}{2}$ cycle. See Fig. 24–31. If the refractive index is less than that of the material in which the light is traveling, no phase change occurs.[†] (This corresponds to the reflection of a wave traveling along a rope when it reaches the end; as we saw in Fig. 11–35, if the end is tied down, the wave changes phase and the pulse flips over, but if the end is free, no phase change occurs.) Thus the ray reflected by the curved surface above the air gap in Fig. 24–30a undergoes no change in phase. That reflected at the lower surface, where the beam in air strikes the glass, undergoes a $\frac{1}{2}\lambda$ phase change. Thus the two rays reflected at the point of contact A of the two glass surfaces (where the air gap approaches zero thickness) will be $\frac{1}{2}\lambda$ out of phase, and a dark spot occurs. Other dark bands will occur when the path difference BCD in Fig. 24–30a is equal to an integral number of wavelengths. Bright bands will occur when the path difference is $\frac{1}{2}\lambda, \frac{3}{2}\lambda$, and so on, because the phase change at one surface effectively adds another $\frac{1}{2}\lambda$.

EXAMPLE 24–8 **Thin film of air, wedge-shaped.** A very fine wire 7.35×10^{-3} mm in diameter is placed between two flat glass plates as in Fig. 24–32a. Light whose wavelength in air is 600 nm falls (and is viewed) perpendicular to the plates, and a series of bright and dark bands is seen, Fig. 24–32b. How many light and dark bands will there be in this case? Will the area next to the wire be bright or dark?

SOLUTION The thin film is the wedge of air between the two glass plates. Because of the phase change at the lower surface, there will be a dark band when the path difference is $0, \lambda, 2\lambda, 3\lambda$, and so on. Since the light rays are perpendicular to the plates, the extra path length equals $2t$, where t is the thickness of the air gap at any point; so dark bands occur where

$$2t = m\lambda, \qquad m = 0, 1, 2, \cdots.$$

Bright bands occur when $2t = (m + \frac{1}{2})\lambda$, where m is an integer. At the position of the wire, $t = 7.35 \times 10^{-6}$ m. At this point there will be $(2)(7.35 \times 10^{-6}\,\text{m})/(6.00 \times 10^{-7}\,\text{m}) = 24.5$ wavelengths; this is a "half integer," so the area next to the wire will be bright. There will be a total of 25 dark lines along the plates, corresponding to path lengths of $0\lambda, 1\lambda, 2\lambda, 3\lambda, \cdots, 24\lambda$, including the one at the point of contact A ($m = 0$). Between them, there will be 24 bright lines plus the one at the end, or 25. The bright and dark bands will be straight only if the glass plates are extremely flat. If they are not, the pattern is uneven, as in Fig. 24–32c. This is a very precise way of testing a glass surface for flatness. Curved lens surfaces can be tested for precision similarly by placing the lens on a flat glass surface and observing Newton's rings (Fig. 24–30b) for perfect circularity.

[†]Note that in Fig. 24–29, the light reflecting at both interfaces, air–oil and oil–water, underwent a phase change of $\frac{1}{2}\lambda$, since $n_{\text{water}} > n_{\text{oil}} > n_{\text{air}}$. Since the phase changes were equal, they don't affect our analysis.

FIGURE 24–32 (a) Light rays reflected from the upper and lower surfaces of a thin wedge of air interfere to produce bright and dark bands. (b) Pattern observed when glass plates are optically flat; (c) pattern when plates are not so flat. See Example 24–8.

(a)

(b)

(c)

If the wedge between the two glass plates of Example 24–8 is filled with some transparent substance other than air—say, water—the pattern shifts because the wavelength of the light changes. In a material where the index of refraction is n, the wavelength is $\lambda_n = \lambda/n$ where λ is the wavelength in vacuum (see Eq. 24–1). For instance, if the thin wedge of Example 24–8 were filled with water, $\lambda_n = 600\,\text{nm}/1.33 = 450\,\text{nm}$; and instead of 25 dark lines, there would be 33.

When white light (rather than monochromatic light) is incident on the thin wedge of Figs. 24–30a or 24–32a, a colorful series of fringes is seen. This is because constructive interference occurs in the reflected light at different locations along the wedge for different wavelengths. Such a difference in thickness is part of the reason bright colors appear when light is reflected from a soap bubble or a thin layer of oil on a puddle or lake (Fig. 24–28). Which wavelengths appear brightest also depends on the viewing angle, as we saw earlier.

EXAMPLE 24–9 **Thickness of soap bubble skin.** A soap bubble appears green ($\lambda = 540\,\text{nm}$) at the point on its front surface nearest the viewer. What is its minimum thickness? Assume $n = 1.35$.

SOLUTION The light is reflected perpendicularly from the point on a spherical surface nearest the viewer. Therefore the path difference is $2t$, where t is the thickness of the soap film. Light reflected from the first (outer) surface undergoes a $\frac{1}{2}\lambda$ phase change (index of refraction of soap is greater than that of air), whereas that at the second (inner) surface does not. Therefore, green light is bright when the minimum path difference equals $\frac{1}{2}\lambda_n$. Thus, $2t = \lambda/2n$, so

$$t = \frac{\lambda}{4n} = \frac{(540\,\text{nm})}{(4)(1.35)} = 100\,\text{nm}.$$

This is the minimum thickness. The front surface would also appear green if $2t = 3\lambda/2n$, and, in general, if $2t = (2m + 1)\,\lambda/2n$ where m is an integer.

An important application of thin-film interference is in the coating of glass to make it "nonreflecting," particularly for lenses. A glass surface reflects about 4 percent of the light incident upon it. Good-quality cameras, microscopes, and other optical devices may contain six to ten thin lenses. Reflection from all these surfaces can reduce the light level considerably and multiple reflections produce a background haze that reduces the quality of the image. By reducing reflection, transmission is increased. A very thin coating on the lens surfaces can reduce reflections considerably: the thickness of the film is chosen so that light (at least for one wavelength) reflecting from the front and rear surfaces of the film destructively interferes. The amount of reflection at a boundary depends on the difference in index of refraction between the two materials. Ideally, the coating material should have an index of refraction which is the geometric mean of those for air and glass, so that the amount of reflection at each surface is about equal. Then destructive interference can occur nearly completely for one particular wavelength depending on the thickness of the coating. Nearby wavelengths will at least partially destructively interfere, but it is clear that a single coating cannot eliminate reflections for all wavelengths. Nonetheless, a single coating can reduce total reflection from 4 percent to 1 percent of the incident light. Often the coating is designed to eliminate the center of the reflected spectrum (around 550 nm). The extremes of the spectrum—red and violet—will not be reduced as much. Since a mixture of red and violet produces purple, the light seen reflected from such coated lenses is purple (Fig. 24–33). Lenses containing two or three separate coatings can more effectively reduce a wider range of reflecting wavelengths.

FIGURE 24–33 A coated lens. Note color of light reflected from the front lens surface.

EXAMPLE 24–10 **Nonreflective coating.** What is the thickness of an optical coating of MgF_2 whose index of refraction is $n = 1.38$ and is designed to eliminate reflected light at wavelengths centered at 550 nm when incident normally on glass for which $n = 1.50$?

SOLUTION Figure 24–34 shows an incoming ray and two rays reflected from the front and rear surfaces of the coating on the lens. The rays are drawn not quite perpendicular to the lens so we can see each of them. To eliminate reflection, we want the reflected rays 1 and 2 to be $\frac{1}{2}$ wavelength out of phase with each other so they destructively interfere. Rays 1 and 2 *both* undergo a change of phase by $\frac{1}{2}\lambda$ when they reflect, respectively, from the front and rear surfaces of the coating. Therefore, we want the extra distance traveled by ray 2 $(=2t)$ to be a half integral number of wavelengths. That is, $2t = (m + \frac{1}{2})\lambda_n$, where m is an integer and λ_n is the wavelength inside the MgF_2 coating. The minimum thickness $(m = 0)$ is usually chosen because destructive interference will then occur over the widest angle. Then

$$t = \frac{\lambda_n}{4} = \frac{\lambda}{4n} = \frac{(550 \text{ nm})}{(4)(1.38)} = 99.6 \text{ nm}.$$

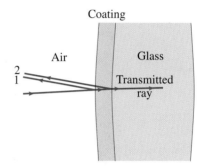

FIGURE 24–34 Incident ray of light is partially reflected at front surface of a lens coating (ray 1) and again partially reflected at the rear surface of the coating (ray 2), with most of the energy passing as the transmitted ray into the glass.

*24–9 Michelson Interferometer

Interference by thin films is the basis of the **Michelson interferometer** (Fig. 24–35),[†] invented by the American Albert A. Michelson (Section 22–6). Monochromatic light from a single point on an extended source is shown striking a half-silvered mirror M_S. This **beam splitter** mirror M_S has a thin layer of silver that reflects only half the light that hits it, so that half of the beam passes through to a fixed mirror M_2, where it is reflected back. The other half is reflected by M_S up to a mirror M_1 that is movable (by a fine-thread screw), where it is also reflected back. Upon its return, part of beam 1 passes through M_S and reaches the eye; and part of beam 2, on its return, is reflected by M_S into the eye. If the two path lengths are identical, the two coherent beams entering the eye constructively interfere and brightness will be seen. If the movable mirror is moved a distance $\lambda/4$, one beam will travel an extra distance equal to $\lambda/2$ (because it travels back and forth over the distance $\lambda/4$). In this case, the two beams will destructively interfere and darkness will be seen. As M_1 is moved farther, brightness will recur (when the path difference is λ), then darkness, and so on.

FIGURE 24–35 Michelson interferometer.

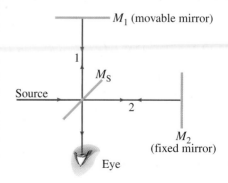

[†]There are other types of interferometer, but Michelson's is the best known.

Very precise length measurements can be made with an interferometer. The motion of mirror M_1 by only $\frac{1}{4}\lambda$ produces a clear difference between brightness and darkness. For $\lambda = 400$ nm, this means a precision of 100 nm or 10^{-4} mm! If mirror M_1 is tilted very slightly, the sequence of bright and dark spots is seen instead as a series of fringes. By counting the number of fringes, or fractions thereof, extremely precise length measurements can be made.

Michelson saw that the interferometer could be used to determine the length of the standard meter in terms of the wavelength of a particular light. In 1960, that standard was chosen to be a particular orange line in the spectrum of krypton-86 (krypton atoms with atomic mass 86). Careful repeated measurements of the old standard meter (the distance between two marks on a platinum–iridium bar kept in Paris) were made to establish 1 meter as being 1,650,763.73 wavelengths of this light, which was *defined* to be the meter. In 1983, the meter was redefined in terms of the speed of light (Section 1–5).

24–10 Polarization

An important and useful property of light is that it can be *polarized*. To see what this means, let us examine waves traveling on a rope. A rope can be set into vibration in a vertical plane as in Fig. 24–36a, or in a horizontal plane as in Fig. 24–36b. In either case, the wave is said to be **plane-polarized**—that is, the oscillations are in a plane.

If we now place an obstacle containing a vertical slit in the path of the wave, Fig. 24–37, a vertically polarized wave passes through, but a horizontal-

FIGURE 24–36 Transverse waves on a rope polarized (a) in a vertical plane and (b) in a horizontal plane.

(a)

(b)

FIGURE 24–37 Vertically polarized wave passes through a vertical slit, but a horizontally polarized wave will not. [*Note*: This diagram would apply to the magnetic field **B** in an EM wave—see footnote on next page.]

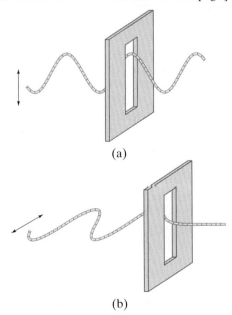

(a)

(b)

ly polarized wave will not. If a horizontal slit were used, the vertically polarized wave would be stopped. If both types of slit were used, both types of wave would be stopped. Note that polarization can exist *only* for *transverse waves*, and not for longitudinal waves such as sound. The latter vibrate only along the direction of motion, and neither orientation of slit would stop them.

Maxwell's theory of light as electromagnetic (EM) waves predicted that light can be polarized since an EM wave is a transverse wave. The direction of polarization in a plane-polarized EM wave is taken as the direction of the electric field vector (the y direction in Fig. 22–7).

Light is not necessarily polarized. It can be **unpolarized**, which means that the source has vibrations in many planes at once, as shown in Fig. 24–38. An ordinary incandescent lightbulb emits unpolarized light, as does the Sun.

Plane-polarized light can be obtained from unpolarized light using certain crystals such as tourmaline. Or, more commonly today, we can use a **Polaroid sheet**. (Polaroid materials were invented in 1929 by Edwin Land.) A Polaroid sheet consists of complicated long molecules arranged parallel to one another. Such a Polaroid acts like a series of parallel slits to allow one orientation of polarization to pass through nearly undiminished (this direction is called the *axis* of the Polaroid), whereas a perpendicular polarization is absorbed almost completely.[†] If a beam of plane-polarized light strikes a Polaroid whose axis is at an angle θ to the incident polarization direction, the beam will emerge plane-polarized parallel to the Polaroid axis and its amplitude will be reduced by $\cos \theta$, Fig. 24–39. Thus, a Polaroid passes only that component of polarization (the electric field vector, **E**) that is parallel to its axis. Because the intensity of a light beam is proportional to the square of the amplitude (Sections 11–9 and 22–7), we see that the intensity of a plane-polarized beam transmitted by a polarizer is

$$I = I_0 \cos^2 \theta, \qquad\qquad (24\text{–}5)$$

where θ is the angle between the polarizer axis and the plane of polarization of the incoming wave, and I_0 is the incoming intensity.[‡]

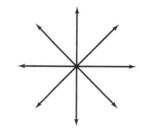

FIGURE 24–38 Vibration of the electric field vectors in unpolarized light. The light is traveling into or out of the page.

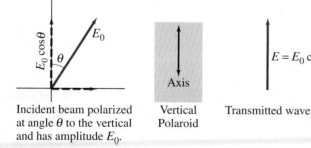

Incident beam polarized at angle θ to the vertical and has amplitude E_0.

Vertical Polaroid

Transmitted wave

FIGURE 24–39 Vertical Polaroid transmits only the vertical component of a wave incident upon it.

[†]How this occurs can be explained at the molecular level. An electric field **E** that oscillates parallel to the long molecules can set electrons into motion along the molecules, thus doing work on them and transferring energy. Hence, if **E** is parallel to the molecules, it gets absorbed. An electric field **E** perpendicular to the long molecules does not have this possibility of doing work and transferring its energy, and so passes through freely. When we speak of the *axis* of a Polaroid, we mean the direction for which **E** is passed, so a Polaroid axis is perpendicular to the long molecules. (If we want to think of there being slits between the parallel molecules in the sense of Fig. 24–37, then Fig. 24–37 would apply for the **B** field in the EM wave, not the **E** field.)

[‡]Equation 24–5 is often referred to as **Malus' law**, after Etienne Malus, a contemporary of Fresnel.

A Polaroid can be used as a **polarizer** to produce plane-polarized light from unpolarized light, since only the component of light parallel to the axis is transmitted. A Polaroid can also be used as an **analyzer** to determine (1) if light is polarized and (2) what is the plane of polarization. A Polaroid acting as an analyzer will pass the same amount of light independent of the orientation of its axis if the light is unpolarized; try rotating one lens of a pair of Polaroid sunglasses while looking through it at a lightbulb. If the light is polarized, however, when you rotate the Polaroid the transmitted light will be a maximum when the plane of polarization is parallel to the Polaroid's axis, and a minimum when perpendicular to it. If you do this while looking at the sky, preferably at right angles to the Sun's direction, you will see that skylight is polarized. (Direct sunlight is unpolarized, but don't look directly at the Sun, even through a polarizer, for damage to the eye may occur.)

Unpolarized light consists of light with random directions of polarization (electric field vector). Each of these polarization directions can be resolved into components along two mutually perpendicular directions. Thus, an unpolarized beam can be thought of as two plane-polarized beams of equal magnitude perpendicular to one another. When unpolarized light passes through a polarizer, one of the components is eliminated. So the intensity of the light passing through is reduced by half since half the light is eliminated, $I = \frac{1}{2}I_0$ (Fig. 24–40).

When two Polaroids are *crossed*—that is, their axes are perpendicular to one another—unpolarized light can be entirely stopped. As shown in Fig. 24–41, unpolarized light is made plane-polarized by the first Polaroid (the polarizer). The second Polaroid, the analyzer, then eliminates this component since its axis is perpendicular to the first. You can try this with Polaroid sunglasses (Fig. 24–42). Note that Polaroid sunglasses eliminate 50 percent of unpolarized light because of their polarizing property; they absorb even more because they are colored.

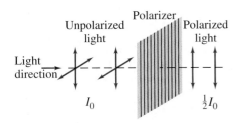

FIGURE 24–40 Unpolarized light has both vertical and horizontal components. After passing through a polarizer, one of these components is eliminated. The intensity of the light is reduced to half.

Crossed Polaroids

FIGURE 24–41 Crossed Polaroids completely eliminate light.

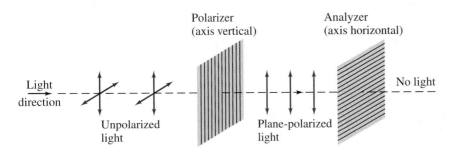

FIGURE 24–42 Crossed Polaroids. When the two polarized sunglass lenses overlap, with axes perpendicular, almost no light passes through.

EXAMPLE 24–11   **Two Polaroids at 60°.**   Unpolarized light passes through two Polaroids; the axis of one is vertical and that of the other is at 60° to the vertical. What is the orientation and intensity of the transmitted light?

SOLUTION   The first Polaroid eliminates half the light so the intensity is reduced by half: $I_1 = \frac{1}{2}I_0$. The light reaching the second polarizer is vertically polarized and so is reduced in intensity (Eq. 24–5) to

$$I_2 = I_1(\cos 60°)^2 = \frac{1}{4}I_1.$$

Thus, $I_2 = \frac{1}{8}I_0$. The transmitted light has an intensity one-eighth that of the original and is plane-polarized at a 60° angle to the vertical.

CONCEPTUAL EXAMPLE 24–12   **Three Polaroids.**   We saw in Fig. 24–41 that when unpolarized light falls on two crossed Polaroids (axes at 90°), no light passes through. What happens if a third Polaroid, with axis at 45° to each of the other two, is placed between them?

RESPONSE   We start just as in Example 24–11. The first Polaroid changes the unpolarized light to plane-polarized and reduces the intensity from I_0 to $I = \frac{1}{2}I_0$. The second polarizer further reduces the intensity by $(\cos 45°)^2$, Eq. 24–5:

$$I_2 = I_1(\cos 45°)^2 = \frac{1}{2}I_1 = \frac{1}{4}I_0.$$

The light leaving the second polarizer is plane polarized at 45° (Fig. 24–43) relative to the third polarizer, so the latter reduces the intensity to

$$I_3 = I_2(\cos 45°)^2 = \frac{1}{2}I_2$$

or $I_3 = \frac{1}{8}I_0$. Thus $\frac{1}{8}$ of the original intensity gets transmitted. But if we don't put the 45° Polaroid in at all, zero intensity results (Fig. 24–41). The 45° Polaroid must be inserted *between* the other two if transmission is to occur. Placing it before or after the other two results in zero intensity.

FIGURE 24–43
Example 24–12.

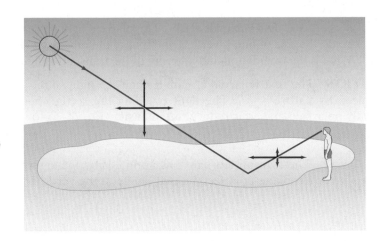

FIGURE 24–44 Light reflected from a nonmetallic surface, such as the smooth surface of water in a lake, is partially polarized parallel to the surface.

FIGURE 24–45 Photographs of a river, (a) allowing all light into the camera lens, and (b) using a polarizer which is adjusted to absorb most of the (polarized) light reflected from the water's surface, allowing the dimmer light from the bottom of the river, and any fish swimming there, to be seen more readily.

(a)

(b)

Another means of producing polarized light from unpolarized light is by reflection. When light strikes a nonmetallic surface at any angle other than perpendicular, the reflected beam is polarized preferentially in the plane parallel to the surface, Fig. 24–44. In other words, the component with polarization in the plane perpendicular to the surface is preferentially transmitted or absorbed. You can check this by rotating Polaroid sunglasses while looking through them at a flat surface of a lake or road. Since most outdoor surfaces are horizontal, Polaroid sunglasses are made with their axes vertical to eliminate the more strongly reflected horizontal component, and thus reduce glare. Fishermen wear Polaroids to eliminate reflected glare from the surface of a lake or stream and thus see beneath the water more clearly (Fig. 24–45).

The amount of polarization in the reflected beam depends on the angle, varying from no polarization at normal incidence to 100 percent polarization at an angle known as the **polarizing angle**, θ_p.[†] This angle is related to the index of refraction of the two materials on either side of the boundary by the equation

$$\tan \theta_p = \frac{n_2}{n_1}, \qquad\qquad \textbf{(24–6a)}$$

where n_1 is the index of refraction of the material in which the beam is traveling, and n_2 is that of the medium beyond the reflecting boundary. If the beam is traveling in air, $n_1 = 1$, and Eq. 24–6a becomes

$$\tan \theta_p = n. \qquad\qquad \textbf{(24–6b)}$$

The polarizing angle θ_p is also called **Brewster's angle**, and Eqs. 24–6 called *Brewster's law*, after the Scottish physicist David Brewster (1781–1868),

---

[†]Only a fraction of the incident light is reflected at the surface of the transparent medium. Although this reflected light is 100 percent polarized (if $\theta = \theta_p$), the remainder of the light, which is transmitted into the new medium, is only partially polarized.

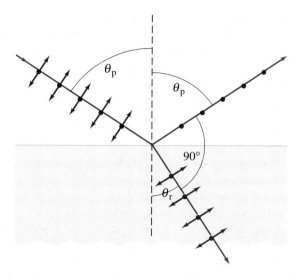

FIGURE 24-46 At θ_p the reflected light is plane-polarized parallel to the surface, and $\theta_p + \theta_r = 90°$, where θ_r is the refraction angle. (The large dots represent vibrations perpendicular to the page.)

who worked it out experimentally in 1812. Equations 24–6 can be derived from the electromagnetic wave theory of light. It is interesting that at Brewster's angle, the reflected and transmitted rays make a 90° angle to each other; that is, $\theta_p + \theta_r = 90°$, Fig. 24–46. This can be seen as follows: we substitute Eq. 24–6a, $n_2 = n_1 \tan \theta_p = n_1 \sin \theta_p / \cos \theta_p$, into Snell's law, $n_1 \sin \theta_p = n_2 \sin \theta_r$, and get $\cos \theta_p = \sin \theta_r$, which can only hold if $\theta_p = 90° - \theta_r$.

EXAMPLE 24–13 **Polarizing angle.** (a) At what incident angle is sunlight reflected fully plane-polarized from a lake? (b) What is the refraction angle?

SOLUTION (a) We use Eq. 24–6b with $n = 1.33$, so $\tan \theta_p = 1.33$ and $\theta_p = 53.1°$.
(b) $\theta_r = 90.0° - \theta_p = 36.9°$.

* 24–11 Scattering of Light by the Atmosphere

Sunsets are red, the sky is blue, and skylight is polarized (at least partially). These phenomena can be explained on the basis of the *scattering* of light by the molecules of the atmosphere. In Fig. 24–47 we see unpolarized light from the Sun impinging on a molecule of the Earth's atmosphere. The electric field of the EM wave sets the electric charges within the molecule into motion, and the molecule absorbs some of the incident radiation. But it quickly reemits this light because the charges are oscillating. As discussed in Section 22–3, oscillating electric charges produce EM waves. The electric field of these waves is in the plane that includes the line of oscillation. The intensity is strongest along a line perpendicular to the oscillation, and drops to zero along the line of oscillation (Fig. 22–6). In Fig. 24–47, the motion of the charges is resolved into two components. An observer viewing at right angles to the direction of the sunlight, as shown, will see plane-polarized light because no light is emitted along the line of the other component of the oscillation. (Another way to understand this is to note that when viewing along the line of oscillation, one doesn't see the oscillation, and hence sees no waves made by it.) At other viewing angles, both

FIGURE 24-47 Unpolarized sunlight scattered by molecules of the air. An observer at right angles sees plane-polarized light, because the component of vibration along the line of sight emits no light along that line.

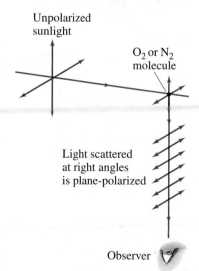

Unpolarized sunlight

O_2 or N_2 molecule

Light scattered at right angles is plane-polarized

Observer

components will be present; one will be stronger, however, so the light appears partially polarized. Thus, the process of scattering explains the polarization of skylight. [It can also explain complete polarization by reflection, Fig. 24–46. At Brewster's angle, as we saw, the angle between the reflected and refracted ray is 90°. If we think of the molecules of the medium oscillating perpendicular to the direction of the refracted ray, we can see that at 90° to this direction (the direction of the reflected ray) there will be only one component of polarization (Fig. 24–46) just as for scattering (Fig. 24–47).]

Scattering of light by the Earth's atmosphere depends on λ. For particles much smaller than the wavelength of light (such as molecules of air), the particles will be less of an obstruction to long wavelengths than to short ones. The scattering decreases, in fact, as $1/\lambda^4$. Red and orange light is thus scattered much less than blue and violet, which is why the sky looks blue. At sunset, on the other hand, the Sun's rays pass through a maximum length of atmosphere. Much of the blue has been taken out by scattering. The light that reaches the surface of the Earth is thus lacking in blue, which is why sunsets appear reddish.

The dependence of scattering on $1/\lambda^4$ is valid only if the scattering objects are much smaller than the wavelength of the light. This is valid for oxygen and nitrogen molecules whose diameters are about 0.2 nm. Clouds, however, contain water droplets or crystals that are much larger than λ; they scatter all frequencies of light nearly uniformly. Hence clouds appear white (or gray, if shadowed).

➥ **P R O B L E M   S O L V I N G**   **Interference**

1. Interference effects depend on the simultaneous arrival of two or more waves at the same point in space.

2. Constructive interference occurs when waves arrive in phase with each other: a crest of one wave arrives at the same time as a crest of the other wave. The amplitudes of the waves then add to form a larger amplitude. Constructive interference also occurs when the phase difference is exactly one full wavelength or any integer multiple of a full wavelength: $1\lambda, 2\lambda, 3\lambda, \cdots$.

3. Destructive interference occurs when a crest of one wave arrives at the same time as a trough of the other wave. The amplitudes add, but they are of opposite sign, so the total amplitude is reduced to zero if the two amplitudes are equal. Destructive interference occurs whenever the phase difference is a half-integral number of wavelengths. Thus, the total amplitude will be zero if two identical waves arrive one-half wavelength out of phase, or $(m + \frac{1}{2})\lambda$ out of phase where m is an integer.

4. For thin-film interference, don't forget an extra one-half wavelength phase shift that occurs when light reflects from an optically more dense medium (going from a medium of lesser toward greater index of refraction).

The wave theory of light is strongly supported by the observations that light exhibits **interference** and **diffraction**. Wave theory also explains the refraction of light and the fact that light travels more slowly in transparent solids and liquids than it does in air. The wavelength of light in a medium with index of refraction n is

$$\lambda_n = \lambda/n,$$

where λ is the wavelength in vacuum; the frequency is not changed.

Young's double-slit experiment clearly demonstrated the interference of light. The observed bright spots of the interference pattern were explained as constructive interference between the beams coming through the two slits, where the beams differ in path length by an integral number of wavelengths. The dark areas in between are due to destructive interference when the path lengths differ by $\frac{1}{2}\lambda, \frac{3}{2}\lambda$, and so on. The angles θ at which **constructive interference** occurs are given by

$$\sin \theta = m\frac{\lambda}{d},$$

where λ is the wavelength of the light, d the separation of the slits, and m an integer $(0, 1, 2, \cdots)$. **Destructive interference** occurs at angles θ given by

$$\sin \theta = (m + \tfrac{1}{2})\frac{\lambda}{d}$$

where m is an integer $(0, 1, 2, \cdots)$.

The wavelength of light determines its color. The **visible spectrum** extends from about 400 nm (violet) to about 750 nm (red). Glass prisms break down white light into its constituent colors because the index of refraction varies with wavelength, a phenomenon known as **dispersion**.

The formula $\sin \theta = m\lambda/d$ for constructive interference also holds for a **diffraction grating**, which consists of many parallel slits or lines, separated from each other by a distance d. The peaks of constructive interference are much brighter and sharper for a diffraction grating than for the simple two-slit apparatus. A diffraction grating (or a prism) is used in a **spectroscope** to separate different colors or to observe **line spectra**, since for a given order m, θ depends on λ. Precise determination of wavelength can thus be done with a spectroscope by careful measurement of θ.

Diffraction refers to the fact that light, like other waves, bends around objects it passes, and spreads out after passing through narrow slits. This bending gives rise to a **diffraction pattern** due to interference between rays of light that travel different distances. Light passing through a very narrow slit of width D will produce a pattern with a bright central maximum of half-width θ given by

$$\sin \theta = \lambda/D,$$

flanked by fainter lines to either side.

Light reflected from the front and rear surfaces of a thin film of transparent material can interfere. A phase change of $180°$ ($\frac{1}{2}\lambda$) occurs when the light reflects at a surface where the index of refraction increases. Such **thin-film interference** has many practical applications, such as lens coatings and Newton's rings.

In **unpolarized light**, the electric field vectors vibrate at all angles. If the electric vector vibrates only in one plane the light is said to be **plane-polarized**. Light can also be partially polarized. When an unpolarized light beam passes through a Polaroid sheet, the emerging beam is plane-polarized. When a light beam is polarized and passes through a Polaroid, the intensity varies as the Polaroid is rotated. Thus a Polaroid can act as a polarizer or as an analyzer.

Light can also be partially or fully **polarized by reflection**. If light traveling in air is reflected from a medium of index of refraction n, the reflected beam will be *completely* plane-polarized if the incident angle θ_p is given by $\tan \theta_p = n$. The fact that light can be polarized shows that it must be a transverse wave.

1. Does Huygens' principle apply to sound waves? To water waves?

2. What is the evidence that light is energy?

3. Why is light sometimes described as rays and sometimes as waves?

4. Two rays of light from the same source destructively interfere if their path lengths differ by how much?

5. If Young's double-slit experiment were submerged in water, how would the fringe pattern be changed?

6. Monochromatic red light is incident on a double slit and the interference pattern is viewed on a screen some distance away. Explain how the fringe pattern would change if the red light source is replaced by a blue light source.

7. Why was the observation of the double-slit interference pattern more convincing evidence for the wave theory of light than the observation of diffraction?

8. Compare a double-slit experiment for sound waves to that for light waves. Discuss the similarities and differences.

9. Why doesn't the light from the two headlights of a distant car produce an interference pattern?

10. When white light passes through a flat piece of window glass, it is not broken down into colors as it is by a prism. Explain.

11. For both converging and diverging lenses, discuss how the focal length for red light differs from that for violet light.

12. We can hear sounds around corners, but we cannot see around corners; yet both sound and light are waves. Explain the difference.

13. For diffraction by a single slit, what is the effect of increasing (a) the slit width, and (b) the wavelength?

14. What is the difference in the interference patterns formed by two slits 10^{-4} cm apart and by a diffraction grating containing 10^4 lines/cm?

15. Explain why there are tiny peaks between the main peaks produced by a diffraction grating illuminated with monochromatic light. Why are the peaks so tiny?

16. For a diffraction grating, what is the advantage of (a) many slits, (b) closely spaced slits.

17. White light strikes (a) a diffraction grating, and (b) a prism. A rainbow appears on a wall just below the direction of the horizontal incident beam in each case. What is the color of the top of the rainbow in each case?

18. Why are interference fringes noticeable only for a *thin* film like a soap bubble and not for a thick piece of glass, say?

19. Why are Newton's rings (Fig. 24–30) closer together farther from the center?

20. Some coated lenses appear greenish yellow when seen by reflected light. What wavelengths do you suppose they are designed to eliminate completely?

* 21. Describe how a Michelson interferometer could be used to measure the index of refraction of air.

22. What does polarization tell us about the nature of light?

23. Explain the advantage of polarized sunglasses over normal tinted sunglasses.

24. How can you tell if a pair of sunglasses is polarizing or not?

* 25. What would be the color of the sky if the Earth had no atmosphere?

* 26. If the Earth's atmosphere were 50 times denser than it is, would sunlight still be white, or would it be some other color?

■ PROBLEMS

SECTION 24–2

1. (II) Derive the law of reflection—namely, that the angle of incidence equals the angle of reflection from a flat surface—using Huygens' principle for waves.

SECTION 24–3

2. (I) Monochromatic light falling on two slits 0.042 mm apart produces the fifth-order fringe at a 7.8° angle. What is the wavelength of the light used?

3. (I) The third-order fringe of 650 nm light is observed at an angle of 15° when the light falls on two narrow slits. How far apart are the slits?

4. (II) Monochromatic light falls on two very narrow slits 0.040 mm apart. Successive fringes on a screen 5.00 m away are 5.5 cm apart near the center of the pattern. What is the wavelength and frequency of the light?

5. (II) A parallel beam of light from a He-Ne laser, with a wavelength 656 nm, falls on two very narrow slits 0.050 mm apart. How far apart are the fringes in the center of the pattern if the screen is 2.6 m away?

6. (II) In a water tank experiment, water waves are generated with their crests 2.5 cm apart and parallel. They pass through two openings 5.0 cm apart in a long wooden board. If the end of the tank is 2.0 m beyond the boards, where would you stand, relative to the "straight-through" direction, so that you received little or no wave action?

7. (II) Light of wavelength 680 nm falls on two slits and produces an interference pattern in which the fourth-order fringe is 48 mm from the central fringe on a screen 1.5 m away. What is the separation of the two slits?

8. (II) If 480-nm and 620-nm light passes through two slits 0.54 mm apart, how far apart are the second-order fringes for these two wavelengths on a screen 1.6 m away?

9. (II) Suppose a thin piece of glass were placed in front of the lower slit in Fig. 24–7 so that the two waves enter the slits 180° out of phase (Fig. 24–48). Describe in detail the interference pattern on the screen.

FIGURE 24–48   Problem 9.

10. (II) In a double-slit experiment it is found that blue light of wavelength 460 nm gives a second-order maximum at a certain location on the screen. What wavelength of visible light would have a minimum at the same location?

11. (III) Light of wavelength 400 nm in air falls on two slits 5.00×10^{-2} mm apart. The slits are immersed in water, as is a viewing screen 40.0 cm away. How far apart are the fringes on the screen?

12. (III) A very thin sheet of plastic ($n = 1.60$) covers one slit of a double-slit apparatus illuminated by 540-nm light. The center point on the screen, instead of being a maximum, is dark. What is the (minimum) thickness of the plastic?

SECTION 24–4

13. (I) By what percent, approximately, does the speed of red light (700 nm) exceed that of violet light (400 nm) in silicate flint glass? (See Fig. 24–14).

14. (I) How much less is the speed of light in silicate flint glass (see Fig. 24–14) for blue light of wavelength 450 nm than red light of 700 nm? Express as a percent.

15. (II) A light beam strikes a piece of glass at a 60.00° incident angle. The beam contains two wavelengths, 450.0 nm and 700.0 nm, for which the index of refraction of the glass is 1.4820 and 1.4742, respectively. What is the angle between the two refracted beams?

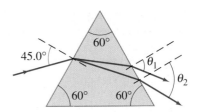

FIGURE 24–49   Problem 16.

16. (II) A parallel beam of light containing two wavelengths, $\lambda_1 = 400$ nm and $\lambda_2 = 650$ nm, enters the silicate flint glass of an equilateral prism as shown in Fig. 24–49. At what angle does each beam leave the prism (give angle with normal to the face)?

*17. (III) A double convex lens whose radii of curvature are both 17.5 cm is made of crown glass. Find the distance between the focal points for 400-nm and 700-nm light. [*Hint*: Use the lensmaker's equation, Eq. 23–10, and Fig. 24–14.]

SECTION 24–5

18. (I) If 520-nm light falls on a slit 0.0440 mm wide, what is the angular width of the central diffraction peak?

19. (I) Monochromatic light falls on a slit that is 3.00×10^{-3} mm wide. If the angle between the first dark fringes on either side of the central maximum is 37.0°, what is the wavelength of the light used?

20. (I) Light of wavelength 550 nm falls on a slit that is 3.50×10^{-3} mm wide. How far from the central maximum will the first diffraction maximum fringe be if the screen is 10.0 m away?

21. (II) Monochromatic light of wavelength 633 nm falls on a slit. If the angle between the first bright fringes on either side of the central maximum is 19.5°, what is the slit width?

22. (II) How wide is the central diffraction peak on a screen 2.50 m behind a 0.0348-mm-wide slit illuminated by 589-nm light?

23. (II) When violet light of wavelength 415 nm falls on a single slit, it creates a central diffraction peak that is 9.20 cm wide on a screen that is 2.55 m away. How wide is the slit?

24. (II) If a slit diffracts 550-nm light so that the diffraction maximum is 3.0 cm wide on a screen 1.50 m away, what will be the width of the diffraction maximum for light with a wavelength of 400 nm?

25. (II) (a) For a given wavelength λ, what is the maximum slit width for which there will be no diffraction minima? (b) What is the maximum slit width so that no visible light exhibits a diffraction minimum?

SECTIONS 24–6 AND 24–7

26. (I) At what angle will 650-nm light produce a second-order maximum when falling on a grating whose slits are 1.15×10^{-3} cm apart?

27. (I) A 3500-line/cm grating produces a third-order fringe at a 22.0° angle. What wavelength of light is being used?

28. (II) The first-order line of 589-nm light falling on a diffraction grating is observed at a 15.5° angle. How far apart are the slits? At what angle will the third order be observed?

29. (II) Light falling normally on a 10,000-line/cm grating is revealed to contain three lines in the first-order spectrum at angles of 31.2°, 36.4°, and 47.5°. What wavelengths are these?

30. (II) How many lines per centimeter does a grating have if the third-order occurs at a 23.0° angle for 630-nm light?

31. (II) A grating has 7000 lines/cm. How many spectral orders can be seen when it is illuminated by white light?

32. (II) What is the highest spectral order that can be seen if a grating with 6000 lines per cm is illuminated with 633-nm laser light? Assume normal incidence.

33. (II) Two and only two full spectral orders can be seen on either side of the central maximum when white light is sent through a diffraction grating. What is the maximum number of lines per cm for the grating?

34. (II) White light containing wavelengths from 400 nm to 750 nm falls on a grating with 7500 lines/cm. How wide is the first-order spectrum on a screen 2.30 m away?

35. (II) The α and δ lines of the atomic hydrogen spectrum have wavelengths of 656 nm and 410 nm. If these fall at normal incidence on a grating with 6600 lines per cm, what will be the angular separation of the two wavelengths in the first-order spectrum?

36. (II) Two first-order spectrum lines are measured by an 8500-line/cm spectroscope at angles, on each side of center, of $+26°38'$, $+41°08'$ and $-26°48'$, $-41°19'$. What are the wavelengths?

37. (III) Suppose the angles measured in Problem 36 were produced when the spectrometer (but not the source) was submerged in water. What then would be the wavelengths?

SECTION 24–8

38. (I) If a soap bubble is 120 nm thick, what color will appear at the center when illuminated normally by white light? Assume that $n = 1.34$.

39. (I) How far apart are the dark fringes in Example 24–8 if the glass plates are each 26.5 cm long?

40. (II) What is the minimum thickness of a soap film ($n = 1.42$) that would appear black if illuminated with 480-nm light? Assume there is air on both sides of the soap film.

41. (II) A lens appears greenish yellow ($\lambda = 570$ nm is strongest) when white light reflects from it. What minimum thickness of coating ($n = 1.25$) do you think is used on such a (glass) lens, and why?

42. (II) A total of 31 bright and 31 dark Newton's rings (not counting the dark spot at the center) are observed when 550-nm light falls normally on a planoconvex lens resting on a flat glass surface (Fig. 24–30). How much thicker is the center than the edges?

43. (II) A fine metal foil separates one end of two pieces of optically flat glass, as in Fig. 24–32. When light of wavelength 670 nm is incident normally, 28 dark lines are observed (with one at each end). How thick is the foil?

44. (II) How thick (minimum) should the air layer be between two flat glass surfaces if the glass is to appear bright when 450-nm light is incident normally? What if the glass is to appear dark?

45. (II) A thin film of alcohol ($n = 1.36$) lies on a flat glass plate ($n = 1.51$). When monochromatic light, whose wavelength can be changed, is incident normally, the reflected light is a minimum for $\lambda = 512$ nm and a maximum for $\lambda = 640$ nm. What is the thickness of the film?

46. (III) When a Newton's ring apparatus (Fig. 24–30) is immersed in a liquid, the diameter of the eighth dark ring decreases from 2.92 cm to 2.48 cm. What is the refractive index of the liquid?

47. (III) Show that the radius r of the mth dark Newton's ring, as viewed from directly above (Fig. 24–30), is given by $r = \sqrt{m\lambda R}$ where R is the radius of curvature of the curved glass surface and λ is the wavelength of light used. Assume that the thickness of the air gap is much less than R at all points and that $r \ll R$.

48. (III) A planoconvex Lucite lens 6.1 cm in diameter is placed on a flat piece of glass as in Fig. 24–30. When 610-nm light is incident normally, 48 bright rings are observed, the last one right at the edge. What is the radius of curvature of the lens surface?

*SECTION 24–9

*49. (II) What is the wavelength of the light entering an interferometer if 644 bright fringes are counted when the movable mirror moves 0.225 mm?

*50. (II) A micrometer is connected to the movable mirror of an interferometer. When the micrometer bears on a thin metal foil, the net number of bright fringes that move, compared to the empty micrometer, is 272. What is the thickness of the foil? The wavelength of light used is 589 nm.

*51. (II) How far must the mirror M_1 in a Michelson interferometer be moved if 850 fringes of 589-nm light are to pass by a reference line?

to mirror M_1

Glass container

Source

M_S

1.30 cm

M_2

FIGURE 24–50   Problem 52.

* 52. (III) One of the beams of an interferometer (Fig. 24–50) passes through a small glass container containing a cavity 1.30 cm deep. When a gas is allowed to slowly fill the container, a total of 236 dark fringes are counted to move past a reference line. The light used has a wavelength of 610 nm. Calculate the index of refraction of the gas, assuming that the interferometer is in vacuum.

* 53. (III) The yellow sodium D lines have wavelengths of 589.0 and 589.6 nm. When they are used to illuminate a Michelson interferometer, it is noted that the interference fringes disappear and reappear periodically as the mirror M_1 is moved. Why does this happen? How far must the mirror move between one disappearance and the next?

SECTION 24–10

54. (I) The axes of two polarizers make a 70° angle to one another. Unpolarized light falls on them. What fraction of the light intensity is transmitted?

55. (I) What is Brewster's angle for an air–glass ($n = 1.52$) surface?

56. (I) What is Brewster's angle for a diamond submerged in water if the light is hitting the diamond while traveling in the water?

57. (II) The critical angle for total internal reflection at a boundary between two materials is 52°. What is Brewster's angle at this boundary?

58. (II) Two Polaroids are aligned so that the light passing through them is a maximum. At what angle should one of them be placed so that the intensity is subsequently reduced by half?

59. (II) At what angle should the axes of two Polaroids be placed so as to reduce the intensity of the incident unpolarized light by an additional factor (after the first Polaroid cuts it in half) of (a) 25 percent, (b) 10 percent, (c) 1 percent?

60. (II) Two polarizers are oriented at 40° to each other and plane-polarized light is incident on them. If only 15 percent of the light gets through both of them, what was the initial polarization direction of the incident light?

61. (II) Two polarizers are oriented at 58.0° to one another. Light polarized at a 29.0° angle to each polarizer passes through both. What reduction in intensity takes place?

62. (II) What would Brewster's angle be for reflections off the surface of water for light coming from beneath the surface? Compare to the angle for total internal reflection, and to Brewster's angle from above the surface.

63. (III) Describe how to rotate the plane of polarization of a plane-polarized beam of light by 90° and produce only a 10 percent loss in intensity using "perfect" polarizers.

GENERAL PROBLEMS

64. Stealth aircraft are designed to not reflect radar, whose wavelength is typically 2 cm, by using an anti-reflecting coating. Ignoring any change in wavelength in the coating, estimate its thickness.

65. A teacher stands well back from an outside doorway 0.88 m wide, and blows a whistle of frequency 750 Hz. Ignoring reflections, estimate at what angle(s) it is *not* possible to hear the whistle clearly on the playground outside the doorway.

66. The wings of a certain beetle have a series of parallel lines across them. When normally incident 460-nm light is reflected from the wing, the wing appears bright when viewed at an angle of 50°. How far apart are the lines?

67. How many lines per centimeter must a grating have if there is to be no second-order spectrum for any visible wavelength?

68. Show that the second- and third-order spectra of white light produced by a diffraction grating always overlap. What wavelengths overlap exactly?

69. Television and radio waves can reflect from nearby mountains or from airplanes, and the reflections can interfere with the direct signal from the station. (a) Determine what kind of interference will occur when 75-MHz television signals arrive at a receiver directly from a distant station, and are reflected from an airplane 118 m directly above the receiver. (Assume $\frac{1}{2}\lambda$ change in phase of the signal upon reflection.) (b) What kind of interference will occur if the plane is 22 m closer to the receiver?

FIGURE 24–51 Problem 70.

70. A radio station operating at 102.1 MHz broadcasts from two identical antennae at the same elevation but separated by a 9.0-m horizontal distance d, Fig. 24–51. A maximum signal is found along the midline, perpendicular to d at its midpoint and extending horizontally in both directions. If the midline is taken as 0°, at what other angle(s) θ is a maximum signal detected? A minimum signal? Assume all measurements are made much farther than 9 m from the antenna towers.

71. Light of wavelength 590 nm passes through two narrow slits 0.60 mm apart. The screen is 1.70 m away. A second source of unknown wavelength produces its second-order fringe 1.33 mm closer to the central maximum than the 590-nm light. What is the wavelength of the unknown light?

72. At what angle above the horizon is the Sun when light reflecting off a smooth lake is polarized most strongly?

73. Unpolarized light falls on two polarizer sheets whose axes are at right angles. (a) What fraction of the incident light intensity is transmitted? (b) What fraction is transmitted if a third polarizer is placed between the first two so that its axis makes a 60° angle with the axis of the first polarizer? (c) What if the third polarizer is in front of the other two?

74. What is the index of refraction of a clear material if a minimum of 150 nm thickness of it, when laid on glass, is needed to reduce reflection to nearly zero when light of 600 nm is incident normally upon it? Do you have a choice for an answer?

75. Monochromatic light of variable wavelength is incident normally on a thin sheet of plastic film in air. The reflected light is a minimum only for $\lambda = 512$ nm and $\lambda = 640$ nm in the visible spectrum. What is the thickness of the film ($n = 1.58$)?

76. If parallel light falls on a single slit of width D at a 30° angle to the normal, describe the diffraction pattern.

77. When yellow sodium light, $\lambda = 589$ nm, falls on a diffraction grating, its first-order peak on a screen 60.0 cm away falls 3.32 cm from the central peak. Another source produces a line 3.71 cm from the central peak. What is its wavelength? How many lines/cm are on the grating?

78. Compare the minimum thickness needed for an antireflective coating ($n = 1.38$) applied to a glass lens in order to eliminate (a) blue (450 nm), or (b) red (700 nm) reflections for light at normal incidence.

79. Suppose you viewed the light *transmitted* through a thin film on a flat piece of glass. Draw a diagram, similar to Fig. 24–29 or 24–34, and describe the conditions required for maxima and minima; consider all possible values of index of refraction. Discuss the relative size of the minima compared to the maxima and to zero.

80. Light of wavelength λ strikes a screen containing two slits a distance d apart at an angle θ_i to the normal. Determine the angle θ_m at which the mth-order maximum occurs.

81. Four ideal polarizers are placed in succession with their axes vertical, at 30° to the vertical, at 60° to the vertical, and at 90° to the vertical respectively. (a) Calculate what fraction of the incident unpolarized light is transmitted by the four polarizers. (b) Can the transmitted light be *decreased* by removing one of the polarizers? If so, which one? (c) Can the transmitted light intensity be extinguished by removing polarizers? If so which one(s)?

82. Very highly reflective mirrors for a particular wavelength can be made by alternating many layers of *transparent* materials of indices of refraction n_1 and $n_2(1 < n_1 < n_2)$. What should the minimum thicknesses d_1 and d_2 of Fig. 24–52 be, in terms of the incident wavelength λ to maximize reflection?

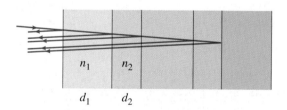

FIGURE 24–52 Problem 82.

83. Light is incident on a diffraction grating with 7500 lines per centimeter and the pattern is viewed on a screen located 2.5 m from the grating. The incident light beam consists of two wavelengths, $\lambda_1 = 4.4 \times 10^{-7}$ m and $\lambda_2 = 6.3 \times 10^{-7}$ m. Calculate the linear distance between the first-order bright fringes of these two wavelengths on the screen.

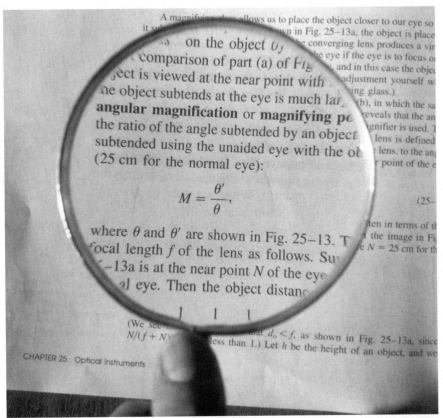

Of the many optical devices we discuss in this chapter, the magnifying glass is the simplest. Here it is magnifying a page that describes how it works according to the ray model.

OPTICAL INSTRUMENTS 25

In our discussion of the behavior of light in the two previous chapters, we also described a few instruments such as the spectrometer and the Michelson interferometer. In this chapter, we will discuss some other, more common, instruments, most of which use lenses, such as the camera, telescope, microscope, and the human eye. To describe their operation, we will use ray diagrams. However, we will see that understanding some aspects of their operation will require the wave nature of light.

25–1 The Camera

The basic elements of a **camera** are a lens, a light-tight box, a shutter to let light pass through the lens only briefly, and a sensitized plate or piece of film (Fig. 25–1). When the shutter is opened, light from external objects in the field of view is focused by the lens as an image on the film. The film contains light-sensitive chemicals that undergo change when light strikes them. In the development process, chemical reactions cause the changed areas to turn opaque so that the image is recorded on the film.[†] You can

➡ PHYSICS APPLIED

The camera

[†]This is called a *negative*, because the black areas correspond to bright objects and vice versa. The same process occurs during printing to produce a black-and-white "positive" picture from the negative. Color film makes use of three dyes corresponding to the primary colors.

see an image yourself by removing the camera back and viewing through a piece of tissue or wax paper (on which the image can form) placed at the position of the film with the shutter open.

There are three main adjustments on good-quality cameras: shutter speed, *f*-stop, and focusing, and we now discuss them. Although many cameras today make these adjustments automatically, it is valuable to understand these adjustments to use any camera effectively. For special or top-quality work, a camera that allows manual adjustments is indispensable (see Fig. 25–2).

Shutter speed **Shutter speed.** This refers to how long the shutter is open and the film exposed. It may vary from a second or more ("time exposures") to $\frac{1}{1000}$ s or less. To avoid blurring from camera movement, speeds faster than $\frac{1}{100}$ s are normally used. If the object is moving, faster shutter speeds are needed to "stop" the action. Figure 25–1 shows the common type of shutter just behind the lens. A second type used in SLR (single-lens reflex) cameras (discussed later in this Section) is the "focal-plane" shutter, which is an opaque curtain just in front of the film whose opening can move quickly across the film to expose it.

f-stop. The amount of light reaching the film must be carefully controlled to avoid **underexposure** (too little light for any but the brightest objects to show up) or **overexposure** (too much light, so that all bright objects look the same, with a consequent lack of contrast and a "washed-out" appearance). To control the exposure, a "stop" or iris diaphragm, whose opening is of variable diameter, is placed behind the lens (Fig. 25–1). The size of the opening is varied to compensate for bright or dark days, the sensitivity of the film[†] used, and for different shutter speeds. The size of the opening is specified by the **f-stop**, defined as

f-stop
$$f\text{-stop} = \frac{f}{D},$$

where f is the focal length of the lens and D is the diameter of the opening. For example, when a 50-mm-focal-length lens has an opening $D = 25$ mm, we say it is set at $f/2$. When the lens is set at $f/8$, the opening is only $6\frac{1}{4}$ mm $(50/6\frac{1}{4} = 8)$. The faster the shutter speed, or the darker the day, the greater the opening must be to get a proper exposure. This corresponds to a smaller f-stop number. The smaller the f-stop number, the more light passes through the lens to the film. The smallest f-number of a lens (largest opening) is referred to as the *speed* of the lens. It is common to find $f/2.0$ lenses today, and even some as fast as $f/1.0$. Fast lenses are expensive to make and require many elements in order to reduce the defects present in simple thin lenses (Section 25–6). The advantage of a fast lens is that it allows pictures to be taken under poor lighting conditions.

Lenses normally stop down to $f/16$, $f/22$, or $f/32$. Although the lens opening can usually be varied continuously, there are nearly always markings for specific lens openings: the standard f-stop markings are 1.0, 1.4, 2.0, 2.8, 4.0, 5.6, 8, 11, 16, 22, and 32. Notice that each of these stops corresponds to a diameter reduction by a factor of about $\sqrt{2} = 1.4$. Because the amount of light reaching the film is proportional to the *area* of the opening,

[†]Different films have different sensitivities to light, referred to as the "film speed," and specified as an "ASA" number; a "faster" film is more sensitive and needs less light to produce a good image.

Lens
Viewfinder
Iris diaphragm or "stop" **Shutter** **Film**

FIGURE 25–1 A simple camera.

FIGURE 25–2 On this camera, the *f*-stops and the focusing ring are on the camera lens. Shutter speeds are selected on the small wheel on top of the camera body.

(a)

(b)

FIGURE 25–3 Photos with camera focused (a) on nearby object with distant object blurry, and (b) on more distant object with nearby object blurry.

and therefore proportional to the diameter squared, we see that each standard f-stop corresponds to a factor of 2 in light intensity reaching the film.

Focusing. Focusing is the operation of placing the lens at the correct position relative to the film for the sharpest image. The image distance is a minimum for objects at infinity (the symbol ∞ is used for infinity) and is equal to the focal length. For closer objects, the image distance is greater than the focal length, as can be seen from the lens equation, $1/f = 1/d_o + 1/d_i$. To focus on nearby objects, the lens must therefore be moved away from the film, and this is usually done by turning a ring on the lens.

Focusing

If the lens is focused on a nearby object, a sharp image of it will be formed, but distant objects may be blurry (Fig. 25–3). The rays from a point on the distant object will be out of focus—they will form a circle on the film as shown (exaggerated) in Fig. 25–4. The distant object will thus produce an image consisting of overlapping circles and will be blurred. These circles are called **circles of confusion**. If you want to have near and distant objects sharp at the same time, you can try setting the lens focus at an intermediate position. Neither near nor distant objects will then be perfectly sharp, but the circles of confusion may be small enough that the blurriness is not too noticeable. For a given distance setting, there is a range of distances over which the circles of confusion will be small enough that the images will be reasonably sharp. This is called the **depth of field.** For a particular choice of circle of confusion diameter as upper limit (typically taken to be 0.03 mm for 35-mm cameras), the depth of field varies

Depth of field

Rays from nearby object

Rays from distant object

"Circle of confusion" for distant object (greatly exaggerated)

FIGURE 25–4 When the lens is positioned to focus on a nearby object, points on a distant object produce circles and are therefore blurred. (The effect is shown greatly exaggerated.)

with the lens opening. If the lens opening is smaller, only rays through the central part of the lens are accepted, and these form smaller circles of confusion for a given object distance. Hence, at smaller lens openings, a greater range of object distances will fit within the circle of confusion criterion, so that the depth of field is greater.

Other factors also affect the sharpness of the image, such as the graininess of the film, diffraction, and lens aberrations relating to the quality of the lens itself. (Lens quality and diffraction effects will be discussed in Sections 25–6 and 25–7.)

Camera lenses are categorized into normal, telephoto, and wide angle, according to focal length and film size. A **normal lens** is one that covers the film with a field of view that corresponds approximately to that of normal vision. A normal lens for 35-mm film has a focal length in the vicinity of 50 mm.[†] A **telephoto lens**, as its name implies, acts like a telescope to magnify images. They have longer focal lengths than a normal lens: as we saw in Chapter 23 (Eq. 23–9), the height of the image for a given object distance is proportional to the image distance, and the image distance will be greater for a lens with longer focal length. For distant objects, the image height is very nearly proportional to the focal length (can you prove this?). Thus a 200-mm telephoto lens for use with a 35-mm camera gives a $4\times$ magnification over the normal 50-mm lens. A **wide-angle lens** has a shorter focal length than normal: a wider field of view is included and objects appear smaller. A **zoom lens** is one whose focal length can be changed so that you seem to zoom up to, or away from, the subject as you change the focal length.

Two types of viewing systems are common in cameras today. In many cameras, you view through a small window just above the lens as in Fig. 25–1. In a **single-lens reflex** camera (SLR), you actually view through the lens with the use of prisms and mirrors (Fig. 25–5). A mirror hangs at a 45° angle behind the lens and flips up out of the way just before the

Telephoto and wide angle lenses

FIGURE 25–5 Single-lens reflex (SLR) camera, showing how the image is viewed through the lens with the help of a movable mirror and prism.

[†]Note that a "35-mm camera" uses film that is 35 mm wide; that 35 mm is not to be confused with a focal length.

shutter opens. SLRs have the great advantage that you can see almost exactly what you will get on film.

EXAMPLE 25–1 **Camera focus.** How far must a 50.0-mm-focal-length camera lens be moved from its infinity setting in order to sharply focus an object 3.00 m away?

SOLUTION When focused at infinity, the lens is 50.0 mm from the film. When focused at $d_o = 3.00$ m, the image distance is given by the lens equation,

$$\frac{1}{d_i} = \frac{1}{f} - \frac{1}{d_o} = \frac{1}{50.0 \text{ mm}} - \frac{1}{3000 \text{ mm}}.$$

We solve for d_i and find $d_i = 50.8$ mm, so the lens moves 0.8 mm.

CONCEPTUAL EXAMPLE 25–2 **Shutter speed.** To improve the depth of field, you stop down your camera lens by two f-stops (say, from $f/4$ to $f/8$). What should you do to the shutter speed to maintain the same exposure?

RESPONSE The amount of light admitted by the lens is proportional to the area of the lens opening. Reducing the lens opening by one f-stop reduces the diameter by a factor of $\sqrt{2}$, and the area by a factor of 2. Stopping down by two f-stops reduces the area of the lens opening by a factor of 4. To maintain the same exposure, the shutter must be open 4 times as long. So if the shutter speed had been $\frac{1}{250}$ s, you have to increase it to $\frac{1}{60}$ s.

25–2 The Human Eye; Corrective Lenses

The human eye resembles a camera in its basic structure (Fig. 25–6). The eye is an enclosed volume into which light passes through a lens. A diaphragm, called the **iris** (the colored part of your eye), adjusts automatically to control the amount of light entering the eye. The hole in the iris through which light passes (the **pupil**) is black because no light is reflected from it (it's a hole), and very little light is reflected back out from the interior of the eye. The **retina**, which plays the role of the film in a camera, is on the curved rear surface. It consists of a complex array of nerves and receptors known as *rods* and *cones* which act to change light energy into electrical signals that travel along the nerves. The reconstruction of the image from all these tiny receptors is done mainly in the brain, although some analysis is apparently done in the complex interconnected nerve network at the retina itself. At the center of the retina is a small area called the **fovea**, about 0.25 mm in diameter, where the cones are very closely packed and the sharpest image and best color discrimination are found.

Unlike a camera, the eye contains no shutter. The equivalent operation is carried out by the nervous system, which analyzes the signals to form images at the rate of about 30 per second. This can be compared to motion picture or television cameras, which operate by taking a series of still pictures at a rate of 24 (movies) or 30 (U.S. television) per second. The rapid projection of these on the screen gives the appearance of motion.

⇒ **PHYSICS APPLIED**

The eye

Anatomy of the eye

FIGURE 25–6 Diagram of a human eye.

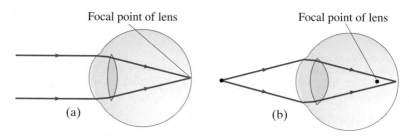

FIGURE 25-7 Accommodation by a normal eye: (a) lens relaxed, focused at infinity; (b) lens thickened, focused on a nearby object.

Focusing the eye

The lens of the eye does little of the bending of the light rays. Most of the refraction is done at the front surface of the **cornea** (index of refraction = 1.376), which also acts as a protective covering. The lens acts as a fine adjustment for focusing at different distances. This is accomplished by the ciliary muscles (Fig. 25–6), which change the curvature of the lens so that its focal length is changed. To focus on a distant object, the muscles are relaxed and the lens is thin, Fig. 25–7a, and parallel rays focus at the focal point (on the retina). To focus on a nearby object, the muscles contract, causing the center of the lens to be thicker, Fig. 25–7b, thus shortening the focal length so that images of nearby objects can be focused on the retina, behind the focal point. This focusing adjustment is called **accommodation**.

The closest distance at which the eye can focus clearly is called the **near point** of the eye. For young adults it is typically 25 cm, although younger children can often focus on objects as close as 10 cm. As people grow older, the ability to accommodate is reduced and the near point increases. A given person's **far point** is the farthest distance at which an object can be seen clearly. For some purposes it is useful to speak of a **normal eye** (a sort of average over the population), which is defined as one having a near point of 25 cm and a far point of infinity. To check your own near point, place this book close to your eye and slowly move it away until the type is sharp.

"Normal eye"

The "normal" eye is more of an ideal than a commonplace. A large part of the population have eyes that do not accommodate within the normal range of 25 cm to infinity, or have some other defect. Two common defects are nearsightedness and farsightedness. Both can be corrected to a large extent with lenses—either eyeglasses or contact lenses.

Corrective lenses

Nearsightedness

Nearsightedness, or *myopia*, refers to an eye that can focus only on nearby objects. The far point is not infinity but some shorter distance, so distant objects are not seen clearly. It is usually caused by an eyeball that is too long, although sometimes it is the curvature of the cornea that is too great. In either case, images of distant objects are focused in front of the retina. A diverging lens, because it causes parallel rays to diverge, allows the rays to be focused at the retina (Fig. 25–8a) and thus corrects this defect.

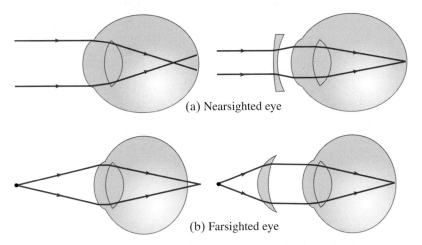

(a) Nearsighted eye

(b) Farsighted eye

FIGURE 25–8 Correcting eye defects with lenses: (a) a nearsighted eye, which cannot focus clearly on distant objects, can be corrected by use of a diverging lens; (b) a farsighted eye, which cannot focus clearly on nearby objects, can be corrected by use of a converging lens.

Farsightedness

Astigmatism

Farsightedness, or *hyperopia*, refers to an eye that cannot focus on nearby objects. Although distant objects are usually seen clearly, the near point is somewhat greater than the "normal" 25 cm, which makes reading difficult. This defect is caused by an eyeball that is too short or (less often) by a cornea that is not sufficiently curved. It is corrected by a converging lens, Fig. 25–8b. Similar to hyperopia is *presbyopia*, which refers to the lessening ability of the eye to accommodate as one ages, and the near point moves out. Converging lenses also compensate for this.

Astigmatism is usually caused by an out-of-round cornea or lens so that point objects are focused as short lines, which blurs the image. It is as if the cornea were spherical with a cylindrical section superimposed. As shown in Fig. 25–9, a cylindrical lens focuses a point into a line parallel to its axis. An astigmatic eye focuses rays in a vertical plane, say, at a shorter distance than it does for rays in a horizontal plane. Astigmatism is corrected with the use of a compensating cylindrical lens. Lenses for eyes that are nearsighted or farsighted as well as astigmatic are ground with superimposed spherical and cylindrical surfaces, so that the radius of curvature of the correcting lens is different in different planes. Astigmatism is tested for by looking with one eye at a pattern like that in Fig. 25–10. Sharply focused lines appear dark, whereas those that are not in focus appear dimmer or gray.

FIGURE 25–9 A cylindrical lens forms a line image of a point object because it is converging in one plane only.

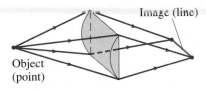

Image (line)

Object (point)

FIGURE 25–10 Test for astigmatism.

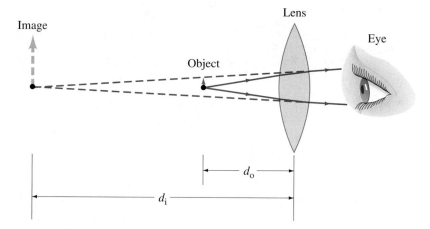

FIGURE 25–11 Lens of reading glasses (Example 25–3).

EXAMPLE 25–3 **Farsighted eye.** A particular farsighted person has a near point of 100 cm. Reading glasses must have what lens power so that this person can read a newspaper at a distance of 25 cm? Assume the lens is very close to the eye.

SOLUTION When the object is placed 25 cm from the lens, we want the image to be 100 cm away on the *same* side of the lens, and so it will be virtual, Fig. 25–11. Thus, $d_o = 25$ cm, $d_i = -100$ cm, and the lens equation gives

$$\frac{1}{f} = \frac{1}{25 \text{ cm}} + \frac{1}{-100 \text{ cm}} = \frac{4-1}{100 \text{ cm}} = \frac{1}{33 \text{ cm}}.$$

So $f = 33$ cm $= 0.33$ m. The power P of the lens is $P = 1/f = +3.0$ D. The plus sign indicates that it is a converging lens.

EXAMPLE 25–4 **Nearsighted eye.** A nearsighted eye has near and far points of 12 cm and 17 cm, respectively. What lens power is needed for this person to see distant objects clearly, and what then will be the near point? Assume that the lens is 2.0 cm from the eye (typical for eyeglasses).

SOLUTION (*a*) First we determine the power of the lens needed to focus objects at infinity, when the eye is relaxed. For a distant object ($d_o = \infty$), as shown in Fig. 25–12a, the lens must put the image 17 cm from the eye (its far point), which is 15 cm in front of the lens; hence $d_i = -15$ cm. We use the lens equation to solve for the focal length of the needed lens:

$$\frac{1}{f} = -\frac{1}{15 \text{ cm}} + \frac{1}{\infty} = -\frac{1}{15 \text{ cm}}.$$

So $f = -15$ cm $= -0.15$ m or $P = 1/f = -6.7$ D. The minus sign indicates that it must be a diverging lens.

FIGURE 25–12
Example 25–4.

17 cm
(far point)

(a)

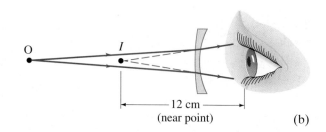

12 cm
(near point)

(b)

(b) To determine the near point when wearing the glasses, we note that a sharp image will be 12 cm from the eye (its near point, see Fig. 25–12b), which is 10 cm from the lens; so $d_i = -0.10$ m and the lens equation gives

$$\frac{1}{d_o} = \frac{1}{f} - \frac{1}{d_i} = -\frac{1}{0.15 \text{ m}} + \frac{1}{0.10 \text{ m}} = \frac{1}{0.30 \text{ m}}.$$

So $d_o = 30$ cm, which means the near point when the person is wearing glasses is 30 cm in front of the lens.

Contact lenses could be used to correct the eye in Example 25–4. Since contacts are placed directly on the cornea, we would not subtract out the 2.0 cm for the image distances. That is, for distant objects $d_i = -17$ cm, so $P = 1/f = -5.9 D$ (diopters). The new near point would be 41 cm. Thus we see that a contact lens and an eyeglass lens will require slightly different powers, or focal lengths, for the same eye because of their different placements relative to the eye.

Contact lenses

25–3 The Magnifying Glass

Much of the remainder of this chapter will deal with optical devices that are used to produce magnified images of objects. We first discuss the **simple magnifier**, or **magnifying glass**, which is simply a converging lens, Fig. 25–13.

How large an object appears, and how much detail we can see on it, depends on the size of the image it makes on the retina. This, in turn, depends on the angle subtended by the object at the eye. For example, a penny held 30 cm from the eye looks twice as high as one held 60 cm away because the angle it subtends is twice as great (Fig. 25–14). When we want to examine detail on an object, we bring it up close to our eyes so that it subtends a greater angle. However, our eyes can accommodate only up to a point (the near point), and we will assume a standard distance of 25 cm as the near point in what follows.

FIGURE 25–13 Photo of a magnifying glass and the image it makes.

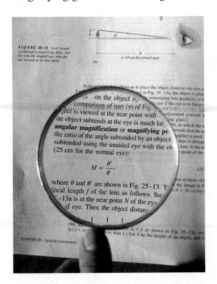

FIGURE 25–14 When the same object is viewed at a shorter distance, the image on the retina is greater, so the object appears larger and more detail can be seen. The angle θ that the object subtends in (a) is greater than in (b).

(a)

(b)

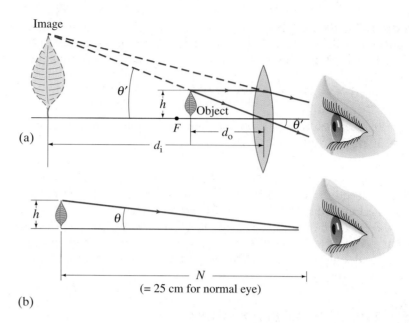

(a)

(b)

N
(= 25 cm for normal eye)

FIGURE 25–15 Leaf viewed (a) through a magnifying glass, and (b) with the unaided eye, with the eye focused at its near point.

Magnification of a magnifying glass

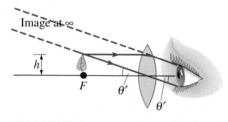

Image at ∞

FIGURE 25–16 With the eye relaxed, the object is placed at the focal point, and the image is at infinity. Compare to Fig. 25–15a where the image is at the eye's near point.

Magnification of a simple magnifier

A magnifying glass allows us to place the object closer to our eye so that it subtends a greater angle. As shown in Fig. 25–15a, the object is placed at the focal point or just within it. Then the converging lens produces a virtual image, which must be at least 25 cm from the eye if the eye is to focus on it. If the eye is relaxed, the image will be at infinity, and in this case the object is exactly at the focal point. (You make this slight adjustment yourself when you "focus" on the object by moving the magnifying glass.)

A comparison of part (a) of Fig. 25–15 with part (b), in which the same object is viewed at the near point with the unaided eye, reveals that the angle the object subtends at the eye is much larger when the magnifier is used. The **angular magnification** or **magnifying power**, M, of the lens is defined as the ratio of the angle subtended by an object when using the lens, to the angle subtended using the unaided eye with the object at the near point N of the eye ($N = 25$ cm for the normal eye):

$$M = \frac{\theta'}{\theta}, \qquad \textbf{(25–1)}$$

where θ and θ' are shown in Fig. 25–15. We can write M in terms of the focal length by noting that $\theta = h/N$ (Fig. 25–15b) and $\theta' = h/d_o$ (Fig. 25–15a), where h is the height of the object and we assume the angles are small so θ and θ' equal their sines and tangents. If the eye is relaxed (for least eye strain), the image will be at infinity and the object will be precisely at the focal point; see Fig. 25–16. Then $d_o = f$ and $\theta' = h/f$. Thus

$$M = \frac{\theta'}{\theta} = \frac{h/f}{h/N} = \frac{N}{f}. \qquad \begin{bmatrix} \text{eye focused at } \infty\,; \\ N = 25 \text{ cm for normal eye} \end{bmatrix} \quad \textbf{(25–2a)}$$

We see that the shorter the focal length of the lens, the greater the magnification. [However, simple single-lens magnifiers are limited to about 2 or $3\times$ because of distortion due to spherical aberration (Section 25–6).] The magnification of a given lens can be increased a bit by moving the lens and adjusting your eye so it focuses on the image at the eye's near point. In this case, $d_i = -N$ (see Fig. 25–15a) if your eye is very near the magnifier.

Then the object distance d_o is given by

$$\frac{1}{d_o} = \frac{1}{f} - \frac{1}{d_i} = \frac{1}{f} + \frac{1}{N}.$$

(We see from this equation that $d_o < f$, as shown in Fig. 25–15a, since $N/(f + N)$ must be less than 1.) Now $\theta' = h/d_o$ so

$$M = \frac{\theta'}{\theta} = \frac{h/d_o}{h/N} = \frac{N}{d_o} = N\left(\frac{1}{f} + \frac{1}{N}\right)$$

or

$$M = \frac{N}{f} + 1. \qquad \begin{bmatrix} \text{eye focused at near point, } N; \\ N = 25 \text{ cm for normal eye} \end{bmatrix} \quad \textbf{(25–2b)}$$

We see that the magnification is slightly greater when the eye is focused at its near point, rather than relaxed.

EXAMPLE 25–5 **ESTIMATE** **A jeweler's loupe.** An 8-cm-focal-length converging lens is used as a "jeweler's loupe," which is a magnifying glass. Estimate (a) the magnification when the eye is relaxed, and (b) the magnification if the eye is focused at its near point $N = 25$ cm.

SOLUTION (a) With the relaxed eye focused at infinity, $M = N/f = 25 \text{ cm}/8 \text{ cm} \approx 3\times$. (b) The magnification when the eye is focused at its near point ($N = 25$ cm), and the lens is near the eye, is:

$$M = 1 + \frac{N}{f} = 1 + \frac{25}{8} \approx 4\times.$$

25–4 Telescopes

A telescope is used to magnify objects that are very far away. In most cases, the object can be considered to be at infinity.

Galileo, although he did not invent it,[†] developed the telescope into a usable and important instrument. He was the first to examine the heavens with the telescope (Fig. 25–17), and he made world-shaking discoveries

FIGURE 25–17 (a) Objective lens (mounted now in an ivory frame) from the telescope with which Galileo made his world-shaking discoveries, including the moons of Jupiter. (b) Later telescopes made by Galileo.

(a)

(b)

[†]Galileo built his first telescope in 1609 after having heard of such an instrument existing in Holland. The first telescopes magnified only 3 to 4 times, but Galileo soon made a 30-power instrument. The first Dutch telescope seems to date from about 1604, but there is a reference suggesting it may have been copied from an Italian telescope built as early as 1590. Kepler (see Chapter 5) gave a ray description (1611) of the Keplerian telescope, which is named for him because he first described it, although he did not build it.

(the moons of Jupiter, the phases of Venus, sunspots, the structure of the Moon's surface, that the Milky Way is made up of a huge number of individual stars, among others).

Refracting telescope

Several types of **astronomical telescope** exist. The common **refracting** type, sometimes called **Keplerian**, contains two converging lenses located at opposite ends of a long tube, as diagrammed in Fig. 25–18. The lens closest to the object is called the **objective lens** and forms a real image I_1 of the distant object in the plane of its focal point F_o (or near it if the object is not at infinity). Although this image, I_1, is smaller than the original object, it subtends a greater angle and is very close to the second lens, called the **eyepiece**, which acts as a magnifier. That is, the eyepiece magnifies the image produced by the objective to produce a second, greatly magnified image, I_2, which is virtual, and inverted. If the viewing eye is relaxed, the eyepiece is adjusted so the image I_2 is at infinity. Then the real image I_1 is at the focal point F_e' of the eyepiece, and the distance between the lenses is $f_o + f_e$ for an object at infinity.

To find the total magnification of this telescope, we note that the angle an object subtends as viewed by the unaided eye is just the angle θ subtended at the telescope objective. From Fig. 25–18 we can see that $\theta \approx h/f_o$, where h is the height of the image I_1 and we assume θ is small so that $\tan\theta \approx \theta$. Note, too, that the thickest of the three rays drawn in the figure is parallel to the axis before it strikes the eyepiece and therefore passes through the eyepiece focal point F_e. Thus, $\theta' \approx h/f_e$ and the total magnifying power (angular magnification) of this telescope is

Telescope magnification

$$M = \frac{\theta'}{\theta} = -\frac{f_o}{f_e}, \qquad (25\text{–}3)$$

where we have inserted a minus sign to indicate that the image is inverted. To achieve a large magnification, the objective lens should have a long focal length and the eyepiece a short focal length.

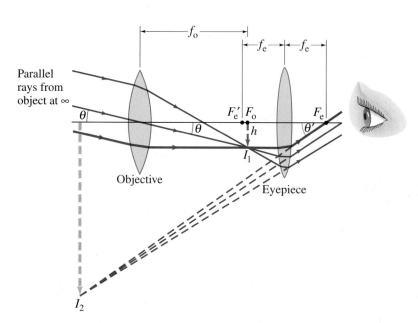

FIGURE 25–18
Astronomical telescope (refracting). Parallel light from one point on a distant object ($d_o = \infty$) is brought to a focus by the objective lens in its focal plane. This image (I_1) is magnified by the eyepiece to form the final image I_2. Only two of the rays shown are standard rays described in Fig. 23–34.

EXAMPLE 25–6 **Telescope magnification.** The largest refracting telescope in the world is located at the Yerkes Observatory in Wisconsin, Fig. 25–19. It is referred to as a "40-inch" telescope, meaning that the diameter of the objective is 40 inches, or 102 cm. The objective has a focal length of 19 m, and the eyepiece has a focal length of 10 cm. (*a*) Calculate the total magnifying power of this telescope. (*b*) Estimate the length of the telescope.

SOLUTION (*a*) From Eq. 25–3 we find,

$$M = -\frac{f_o}{f_e} = -\frac{19 \text{ m}}{0.10 \text{ m}} = -190\times.$$

(*b*) For a relaxed eye, the image I_1 is at the focal point of both the eyepiece and the objective lenses. The distance between the two lenses is thus $f_o + f_e \approx 19 \text{ m}$, which is essentially the length of the telescope.

FIGURE 25–19 This large refracting telescope was built in 1897 and is housed at Yerkes Observatory in Wisconsin. The objective lens is 102 cm (40 inches) in diameter, and the telescope tube is about 19 m long.

Reflecting telescopes

For an astronomical telescope to produce bright images of distant stars, the objective lens must be large to allow in as much light as possible. Indeed, the diameter of the objective (and hence its "light-gathering power") is the most important parameter for an astronomical telescope, which is why the largest ones are specified by giving the objective diameter (such as the 200-inch Hale telescope on Palomar Mountain). The construction and grinding of large lenses is very difficult. Therefore, the largest telescopes are **reflecting telescopes** that use a curved mirror as the objective, Fig. 25–20, since a mirror has only one surface to be ground and can be supported along its entire surface[†] (a large lens, supported at its edges, would sag under its own weight). Normally, the eyepiece lens or mirror (see Fig. 25–20) is removed so that the real image formed by the objective mirror can be recorded directly on film.

(a)   (b)   (c)   (d)

FIGURE 25–20 A concave mirror can be used as the objective of an astronomical telescope. Either a lens (a) or a mirror (b) can be used as the eyepiece. Arrangement (a) is called the Newtonian focus and (b) the Cassegrainian focus. Other arrangements are also possible. (c) The 200-inch (mirror diameter) Hale telescope on Palomar Mountain in California. (d) The 10-meter Keck telescope on Mauna Kea, Hawaii. The Keck combines thirty-six 1.8 meter six-sided mirrors into the equivalent of a very large (10-m diameter) single reflector.

[†]Another advantage of mirrors is that they exhibit no chromatic aberration (Section 25–6) because the light doesn't pass through them. Also, they can be ground in a parabolic shape to correct for spherical aberration (Section 25–6). The reflecting telescope was first proposed by Newton.

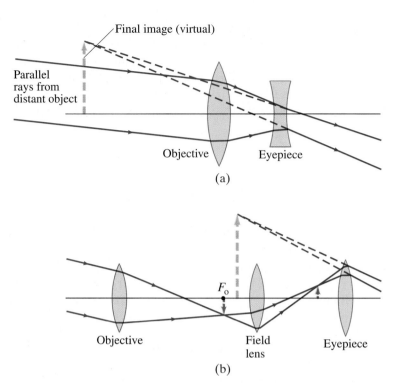

Final image (virtual)

Parallel rays from distant object

Objective   Eyepiece

(a)

F_o

Objective   Field lens   Eyepiece

(b)

FIGURE 25–21 Terrestrial telescopes that produce an upright image: (a) Galilean; (b) spyglass, or field-lens, type.

A **terrestrial telescope** (for use in viewing objects on Earth), unlike its astronomical counterpart, must provide an upright image. Two designs are shown in Fig. 25–21. The **Galilean** type shown in part (a), which Galileo used for his great astronomical discoveries, has a diverging lens as eyepiece which intercepts the converging rays from the objective lens before they reach a focus, and acts to form a virtual upright image. This design is often used in opera glasses. The tube is reasonably short, but the field of view is small. The second type, shown in Fig. 25–21b, is often called a **spyglass** and makes use of a third lens ("field lens") that acts to make the image upright as shown. A spyglass must be quite long. The most practical design today is the **prism binocular** which was shown in Fig. 23–25. The objective and eyepiece are converging lenses. The prisms reflect the rays by total internal reflection and shorten the physical size of the device, and they also act to produce an upright image. One prism reinverts the image in the vertical plane, the other in the horizontal plane.

25–5 Compound Microscope

The compound **microscope**, like the telescope, has both objective and eyepiece (or ocular) lenses, Fig. 25–22. The design is different from that for a telescope because a microscope is used to view objects that are very close, so the object distance is very small. The object is placed just beyond the objective's focal point as shown in Fig. 25–22a. The image I_1 formed by the objective lens is real, quite far from the lens, and much enlarged. This image is magnified by the eyepiece into a very large virtual image, I_2, which is seen by the eye and is inverted.

The overall magnification of a microscope is the product of the magnifications produced by the two lenses. The image I_1 formed by the objec-

FIGURE 25–22
Compound microscope:
(a) ray diagram, (b) photograph
(illumination comes from the
lower right, then up through
the slide holding the object).

(a)   (b)

tive is a factor m_o greater than the object itself. From Fig. 25–22a and
Eq. 23–9 for the lateral magnification of a simple lens, we have

$$m_o = \frac{h_i}{h_o} = \frac{d_i}{d_o} = \frac{l - f_e}{d_o}, \tag{25–4}$$

where d_o and d_i are the object and image distances for the objective lens, l
is the distance between the lenses (equal to the length of the barrel), and
we ignored the minus sign in Eq. 23–9 which only tells us that the image is
inverted. The eyepiece acts like a simple magnifier. If we assume that the
eye is relaxed, its angular magnification M_e is (from Eq. 25–2a)

$$M_e = \frac{N}{f_e}, \tag{25–5}$$

where the near point $N = 25$ cm for the normal eye. Since the eyepiece enlarges the image formed by the objective, the overall angular magnification
M is the product of the lateral magnification of the objective lens, m_o, times
the angular magnification, M_e, of the eyepiece lens (Eqs. 25–4 and 25–5):

$$M = M_e m_o = \left(\frac{N}{f_e}\right)\left(\frac{l - f_e}{d_o}\right) \tag{25–6a}$$

$$\approx \frac{Nl}{f_e f_o}. \tag{25–6b}$$

Magnification

of

microscope

The approximation, Eq. 25–6b, is accurate when f_e and f_o are small compared to l, so $l - f_e \approx l$ and $d_o \approx f_o$ (Fig. 25–22a). This is a good approximation for large magnifications, since these are obtained when f_o and f_e
are very small (they are in the denominator of Eq. 25–6b). In order to
make lenses of very short focal length, which can be done best for the objective, compound lenses involving several elements must be used to avoid
serious aberrations, as discussed in the next Section.

EXAMPLE 25-7 **Microscope.** A compound microscope consists of a 10× eyepiece and a 50× objective 17.0 cm apart. Determine (a) the overall magnification, (b) the focal length of each lens, and (c) the position of the object when the final image is in focus with the eye relaxed. Assume a normal eye, so $N = 25$ cm.

SOLUTION (a) The overall magnification is $10 \times 50 = 500\times$. (b) The eyepiece focal length is (Eq. 25-5) $f_e = N/M_e = 25$ cm/10 = 2.5 cm. For the objective lens, it is easier to next find d_o (part (c)) before we find f_o because we can use Eq. 25-4. Solving for d_o, we find

$$d_o = (l - f_e)/m_o = (17.0 \text{ cm} - 2.5 \text{ cm})/50 = 0.29 \text{ cm}.$$

Then, from the lens equation with $d_i = l - f_e = 14.5$ cm (see Fig. 25-22a),

$$\frac{1}{f_o} = \frac{1}{d_o} + \frac{1}{d_i} = \frac{1}{0.29 \text{ cm}} + \frac{1}{14.5 \text{ cm}} = 3.52;$$

so $f_o = 0.28$ cm.
(c) We just calculated $d_o = 0.29$ cm, which is very close to f_o.

Opaque objects are generally illuminated by a source placed above them. If the objects to be viewed are transparent, such as cells or tissue, light is normally passed through the object from a source beneath the microscope stage (see Fig. 25-22b). The illumination system must be carefully designed if maximum sharpness and contrast are to be achieved. Usually, a *condenser* is employed, which is a set of two or three lenses, although inexpensive condensers may be a single lens or curved mirror. The purpose of the condenser is to gather light over a wide angle from the source, and to "condense" it down to a narrow beam that will illuminate the object strongly and uniformly. A number of different designs are employed. The source is often placed in the focal plane of the condenser so that light from each point on the source is parallel when it passes through the object.

Some specialized microscope types are described in Section 25-10.

25-6 Aberrations of Lenses and Mirrors

In Chapter 23, we developed a theory of image formation by a thin lens. We found, for example, that all rays from each point on an object are brought to a single point as the image point. This, and other results, were based on approximations such as that all rays make small angles with one another and we can use $\sin\theta \approx \theta$. Because of these approximations, we expect deviations from the simple theory and these are referred to as **lens aberrations**. There are several types of aberration; we will briefly discuss each of them separately but all may be present at one time.

Consider an object at any point (even at infinity) on the axis of a lens. Rays from this point that pass through the outer regions of the lens are brought to a focus at a different point from those that pass through the center of the lens. This is called **spherical aberration**, and is shown exaggerated in Fig. 25-23. Consequently, the image seen on a piece of film (for

Spherical aberration

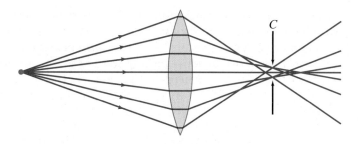

example) will not be a point but a tiny circular patch of light. If the film is placed at the point C, as indicated, the circle will have its smallest diameter, which is referred to as the **circle of least confusion**. Spherical aberration is present whenever spherical surfaces are used. It can be corrected by using nonspherical lens surfaces, but to grind such lenses is difficult and expensive. It can be minimized with spherical surfaces by choosing the curvatures so that equal amounts of bending occur at each lens surface; a lens can be designed like this for only one particular object distance. Spherical aberration is usually corrected (by which we mean reduced greatly) by the use of several lenses in combination, and by using only the central part of lenses.

Off-axis aberrations

For object points off the lens axis, additional aberrations occur. Rays passing through the different parts of the lens cause spreading of the image that is noncircular. We won't go into the details but merely point out that there are two effects: **coma** (because the image of a point is comet-shaped rather than a circle) and **off-axis astigmatism**.[†] Furthermore, the image points for objects off the axis but at the same distance from the lens do not fall on a flat plane but on a curved surface—that is, the focal plane is not flat. (We expect this because the points on a flat plane, such as the film in a camera, are not equidistant from the lens.) This aberration is known as **curvature of field** and is obviously a problem in cameras and other devices where the film is placed in a flat plane. In the eye, however, the retina is curved, which compensates for this effect. Another aberration, known as **distortion**, is a result of variation of magnification at different distances from the lens axis. Thus a straight line object some distance from the axis may form a curved image. A square grid of lines may be distorted to produce "barrel distortion," or "pincushion distortion," Fig. 25–24. The latter is common in extreme wide-angle lenses.

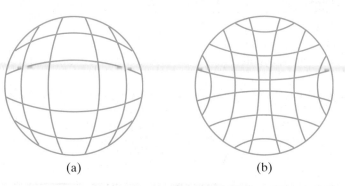

(a)   (b)

FIGURE 25–24 Distortion. Lenses may image a square grid of perpendicular lines to produce (a) barrel distortion or (b) pincushion distortion. These distortions can be seen in the photographs of Fig. 23-29d.

[†] Although the effect is the same as for astigmatism in the eye (Section 25–2), the cause is different. Off-axis astigmatism is no problem in the eye because objects are clearly seen only at the fovea, which is on the lens axis.

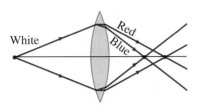

FIGURE 25–25 Chromatic aberration. Different colors are focused at different points.

FIGURE 25–26 Achromatic doublet.

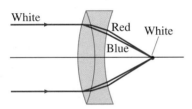

All the above aberrations occur for monochromatic light and hence are referred to as *monochromatic aberrations*. Normal light is not monochromatic, and there will also be **chromatic aberration**. This aberration arises because of dispersion—the variation of index of refraction of transparent materials with wavelength (Section 24–4). For example, blue light is bent more than red light by glass. So if white light is incident on a lens, the different colors are focused at different points, Fig. 25–25, and there will be colored fringes in the image. Chromatic aberration can be eliminated for any two colors (and reduced greatly for all others) by the use of two lenses made of different materials with different indices of refraction and dispersion. Normally one lens is converging and the other diverging, and they are often cemented together (Fig. 25–26). Such a lens combination is called an **achromatic doublet** (or "color-corrected" lens).

It is not possible to fully correct all aberrations. Combining two or more lenses together can reduce them. High-quality lenses used in cameras, microscopes, and other devices are **compound lenses** consisting of many simple lenses (referred to as **elements**). A typical high-quality camera lens may contain six to eight (or more) elements.

For simplicity we will normally indicate lenses in diagrams as if they were simple lenses. But it must be remembered that good-quality lenses are compound.

The human eye is also subject to aberrations, but they are minimal. Spherical aberration, for example, is minimized because (1) the cornea is less curved at the edges than at the center, and (2) the lens is less dense at the edges than at the center. Both effects cause rays at the outer edges to be bent less strongly, and thus help to reduce spherical aberration. Chromatic aberration is partially compensated for because the lens absorbs the shorter wavelengths appreciably and the retina is less sensitive to the blue and violet wavelengths. This is just the region of the spectrum where dispersion—and thus chromatic aberration—is greatest (Fig. 24–14).

Spherical mirrors (Section 23–3) also suffer aberrations including spherical aberration (see Fig. 23–12). Mirrors can be ground in a parabolic shape to correct for spherical aberration, but they are much harder to make and so very expensive. Spherical mirrors do not, however, exhibit chromatic aberration because the light does not pass through them.

25–7 Limits of Resolution; the Rayleigh Criterion

The ability of a lens to produce distinct images of two point objects very close together is called the **resolution** of the lens. The closer the two images can be and still be seen as distinct (rather than overlapping blobs), the higher the resolution. The resolution of a camera lens, for example, is often specified as so many lines per millimeter,[†] and can be determined by photographing a standard set of parallel lines on fine-grain film. The minimum spacing of lines distinguishable on film using the lens gives the resolution.

Two principal factors limit the resolution of a lens. The first is lens aberrations. As we saw, because of spherical and other aberrations, a point object is not a point on the image but a tiny blob. Careful design of compound lenses can reduce aberrations significantly, but they cannot be elim-

[†] This may be specified at the center of the field of view as well as at the edges, where it is usually less because of off-axis aberrations.

inated entirely. The second factor that limits resolution is *diffraction*, which cannot be corrected for because it is a natural result of the wave nature of light. We discuss it now.

In Section 24–5 we saw that because light travels as a wave, light from a point source passing through a slit is spread out into a diffraction pattern (Figs. 24–18 and 24–20). A lens, because it has edges, acts like a slit. When a lens forms the image of a point object, the image of that point is actually a tiny diffraction pattern. Thus, *an image would be blurred even if aberrations were absent.*

Image point is a diffraction pattern (i.e. blurred)

In the analysis that follows we assume that the lens is free of aberrations, so that we can focus our attention on diffraction effects and how much they limit the resolution of a lens. In Fig. 24–20 we saw that the diffraction pattern produced by light passing through a rectangular slit has a central maximum in which most of the light falls. This central peak falls to a minimum on either side of its center at an angle $\theta \approx \sin\theta = \lambda/D$ (this is Eq. 24–3), where D is the width of the slit, λ is the wavelength of light used, and we assume θ is small. There are also low-intensity fringes beyond. For a lens, or any circular hole, the image of a point object will consist of a *circular* central peak (called the *diffraction spot* or *Airy disk*) surrounded by faint circular fringes, as shown in Fig. 25–27a. The central maximum has an angular half width given by

$$\theta = \frac{1.22\lambda}{D},$$

where D is the diameter.

This differs from the formula for a slit (Eq. 24–3) by the factor 1.22. This factor comes from the fact that the width of a circular hole is not uniform (like a rectangular slit) but varies from its diameter D to zero. A careful analysis shows that the "average" width is $D/1.22$. Hence we get the equation above rather than Eq. 24–3. The intensity of light in the diffraction pattern of light from a point source passing through a circular opening is shown in Fig. 25–28. (The image for a non-point source is a superposition of such patterns.) For most purposes, we need consider only the central spot since the concentric rings are much dimmer.

If two point objects are very close, the diffraction patterns of their images will overlap as shown in Fig. 25–27b. As the objects are moved closer,

FIGURE 25–27 Photographs of images (greatly magnified) formed by a lens, showing diffraction pattern of image for: (a) a single point object; (b) two point objects whose images are barely resolved.

FIGURE 25–28 Intensity of light across the diffraction pattern of a circular hole.

(a)

(b)

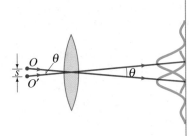

a separation is reached where you can't tell if there are two overlapping images or a single image. The separation at which this happens may be judged differently by different observers. However, a generally accepted criterion is one proposed by Lord Rayleigh (1842–1919). This **Rayleigh criterion** states that *two images are just resolvable when the center of the diffraction disk of one is directly over the first minimum in the diffraction pattern of the other.* This is shown in Fig. 25–29. Since the first minimum is at an angle $\theta = 1.22\lambda/D$ from the central maximum, Fig. 25–29 shows us that two objects can be considered just resolvable if they are separated by this angle θ:

Rayleigh criterion (resolution limit)

$$\theta = \frac{1.22\lambda}{D}. \qquad (25\text{–}7)$$

This is the limit on resolution set by the wave nature of light due to diffraction.

EXAMPLE 25–8 **Hubble space telescope.** The Hubble Space Telescope (HST) is a reflecting telescope that has been placed in orbit above the Earth's atmosphere, so its resolution is not limited by turbulence in the atmosphere (Fig. 25–30). Its objective diameter is 2.4 m. [The Hubble can also observe radiation in the near ultraviolet (wavelengths as small as 115 nm) and infrared (wavelengths as long as 1 mm) which are ranges of the spectrum blocked by the atmosphere.] For visible light (take λ = 550 nm), estimate the improvement in resolution the Hubble offers over Earth-bound telescopes, which are limited in resolution by movement of the Earth's atmosphere to about half an arc second (each degree is divided into 60 minutes each containing 60 seconds, so 1° = 3600 arc seconds).

SOLUTION Earth-bound telescopes are limited to an angular resolution of

$$\theta = \frac{1}{2}\left(\frac{1}{3600}\right)^{\circ}\left(\frac{2\pi \, \text{rad}}{360°}\right) = 2.4 \times 10^{-6} \, \text{rad}.$$

The Hubble, on the other hand, is limited by diffraction (Eq. 25–7) which for λ = 550 nm is

$$\theta = \frac{1.22(550 \times 10^{-9} \, \text{m})}{2.4 \, \text{m}} = 2.8 \times 10^{-7} \, \text{rad},$$

which is almost ten times better.

FIGURE 25–30 Hubble Space Telescope, with Earth in the background. The flat orange panels are solar cells that collect energy from the Sun.

25-8 Resolution of Telescopes and Microscopes

You might think that a microscope or telescope could be designed to produce any desired magnification, depending on the choice of focal lengths and quality of the lenses. But this is not possible, because of diffraction. An increase in magnification above a certain point merely results in magnification of the diffraction patterns. This would be highly misleading since we might think we are seeing details of an object when we are really seeing details of the diffraction pattern. To examine this, we apply the Rayleigh criterion: two objects (or two nearby points on one object) are just resolvable if they are separated by an angle θ (Fig. 25–29) given by Eq. 25–7:

$$\theta = \frac{1.22\lambda}{D}.$$

This is valid for either a microscope or a telescope, where D is the diameter of the objective lens. For a telescope, the resolution is specified by stating θ as given by this equation.[†]

For a microscope, it is more convenient to specify the actual distance, s, between two points that are just barely resolvable, Fig. 25–29. Since objects are normally placed near the focal point of the microscope objective, $\theta = s/f$, or $s = f\theta$. If we combine this with Eq. 25–7, we obtain for the **resolving power (RP)**:

$$\text{RP} = s = f\theta = \frac{1.22\lambda f}{D}. \qquad \textbf{(25–8)}$$

Resolving power

This distance s is called the resolving power of the lens because it is the minimum separation of two object points that can just be resolved, assuming the highest quality lens since this limit is imposed by the wave nature of light.

Equation 25–8 is often written in terms of the *angle of acceptance*, α, of the objective lens as defined in Fig. 25–31. The derivation is long and we only quote the result:

$$\text{RP} = s = \frac{1.22\lambda}{2\sin\alpha} = \frac{0.61\lambda}{\sin\alpha}.$$

The resolving power can be increased by placing a drop of oil that encloses the object and the front surface of the objective. This is called an **oil-immersion objective**. In the oil, the wavelength of the light is reduced to λ/n (Eq. 24–1), where n is the oil's index of refraction. Thus the resolving power becomes

$$\text{RP} = \frac{0.61\lambda}{n\sin\alpha}. \qquad \textbf{(25–9)}$$

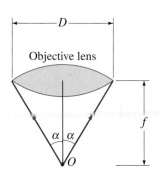

FIGURE 25–31 Objective lens of a microscope, showing the angle of acceptance, α.

The oil typically has $n \approx 1.5$, although n may be as great as 1.8. Thus oil immersion increases the resolution by 50 percent or more.

The quantity $(n\sin\alpha)$ is called the **numerical aperture** (NA) of the lens:

$$\text{NA} = n\sin\alpha. \qquad \textbf{(25–10)}$$

It is usually specified on the objective lens housing along with the magnification. The larger the value of the NA, the finer the resolving power.

[†] Telescopes with large-diameter objectives are usually limited not by diffraction but by other effects such as turbulence in the atmosphere. The resolution of a high-quality microscope, on the other hand, normally *is* limited by diffraction because microscope objectives are complex compound lenses containing many elements of small diameter (since f is small).

FIGURE 25–32 The 300-meter radiotelescope in Arecibo, Puerto Rico, uses radio waves (Fig. 22–10) instead of visible light.

EXAMPLE 25–9 **Telescope resolution (radio wave vs. visible light).** What is the theoretical minimum angular separation of two stars that can just be resolved by: (*a*) the 200-inch telescope on Palomar Mountain (Fig. 25–20c); and (*b*) the Arecibo radiotelescope (Fig. 25–32), whose diameter is 300 m and whose radius of curvature is also 300 m. Assume $\lambda = 550$ nm for the visible-light telescope in part (*a*), and $\lambda = 4$ cm (the shortest wavelength at which the radiotelescope has been operated) in part (*b*).

SOLUTION (*a*) Since $D = 200$ inch $= 5.1$ m, we have from Eq. 25–7 that

$$\theta = \frac{1.22\lambda}{D} = \frac{(1.22)(5.50 \times 10^{-7}\,\text{m})}{(5.1\,\text{m})} = 1.3 \times 10^{-7}\,\text{rad},$$

or 0.75×10^{-5} deg. (Note that this is equivalent to resolving two points less than 1 cm apart from a distance of 100 km!) This is the limit set by diffraction. The resolution is not this good because of aberrations and, more importantly, turbulence in the atmosphere. In fact, large-diameter objectives are not justified by increased resolution, but by their greater light-gathering ability—they allow more light in, so fainter objects can be seen. (*b*) Radiotelescopes are not hindered by atmospheric turbulence, and for radio waves with $\lambda = 0.04$ m the resolution is

$$\theta = \frac{(1.22)(0.04\,\text{m})}{(300\,\text{m})} = 1.6 \times 10^{-4}\,\text{rad}.$$

EXAMPLE 25–10 **Microscope resolution.** Determine the NA and RP of the best oil-immersion microscopes, where the index of refraction of the oil is $n = 1.8$ and $\sin \alpha \approx 0.90$. Assume that $\lambda = 550$ nm.

SOLUTION The NA $= n \sin \alpha = 1.6$. The resolving power is

$$\text{RP} = \frac{0.61\lambda}{\text{NA}} = \frac{(0.61)(5.50 \times 10^{-7}\,\text{m})}{(1.6)} \approx 2 \times 10^{-7}\,\text{m} = 200\,\text{nm}.$$

This is the best resolution that a visible-light microscope can attain.

Diffraction sets an ultimate limit on the detail that can be seen on any object. In Eq. 25–8 we note that the focal length of a lens cannot be made less than (approximately) the radius of the lens, and even that is very difficult—see the lensmaker's equation (Eq. 23–10). In this best case,[†] Eq. 25–8 gives, with $f \approx D/2$,

$$\text{RP} \approx \frac{\lambda}{2}. \tag{25–11}$$

Thus we can say, to within a factor of 2 or so, that

Resolution limited to λ

it is not possible to resolve detail of objects smaller than the wavelength of the radiation being used.

This is an important and useful rule of thumb.

[†]The same result can be obtained from Eq. 25–9 since $\sin \alpha$ can never exceed 1 and typically is 0.6 to 0.9 at most. With oil immersion, Eq. 25–9 gives, *at best*, RP $\approx \lambda/3$, which corresponds to the result of Example 25–10.

Compound lenses are now designed so well that the actual limit on resolution is often set by diffraction—that is, by the wavelength of the light used. To obtain greater detail, one must use radiation of shorter wavelength. The use of UV radiation can increase the resolution by a factor of perhaps 2. Far more important, however, was the discovery in the early twentieth century that electrons have wave properties (Chapter 27) and that their wavelengths can be very small. The wave nature of electrons is utilized in the electron microscope (Section 27–7), which can magnify 100 to 1000 times more than a visible-light microscope because of the much shorter wavelengths. X-rays, too, have very short wavelengths and are often used to study objects in great detail (Section 25–11).

25–9 Resolution of the Human Eye and Useful Magnification

The resolution of the human eye is limited by several factors, all of roughly the same order of magnitude. The resolution is best at the fovea, where the cone spacing is smallest, about $3 \, \mu\text{m}$ ($= 3000 \, \text{nm}$). The diameter of the pupil varies from about 0.1 cm to about 0.8 cm. So for $\lambda = 550 \, \text{nm}$ (where the eye's sensitivity is greatest), the diffraction limit is about $\theta \approx 1.22\lambda/D \approx 8 \times 10^{-5} \, \text{rad}$ to $6 \times 10^{-4} \, \text{rad}$. Since the eye is about 2 cm long, this corresponds to a resolving power of $s \approx (8 \times 10^{-5} \, \text{rad})(2 \times 10^{-2} \, \text{m}) \approx 2 \, \mu\text{m}$ at best, to about $15 \, \mu\text{m}$ at worst (pupil small). Spherical and chromatic aberration also limit the resolution to about $10 \, \mu\text{m}$. The net result is that the eye can resolve objects whose angular separation is about $5 \times 10^{-4} \, \text{rad}$ at best. This corresponds to objects separated by 1 cm at a distance of about 20 m.

The typical near point of a human eye is about 25 cm. At this distance, the eye can just resolve objects that are $(25 \, \text{cm})(5 \times 10^{-4} \, \text{rad}) \approx 10^{-4} \, \text{m} = \frac{1}{10} \, \text{mm}$ apart. Since the best light microscopes can resolve objects no smaller than about 200 nm (see Example 25–10), the useful magnification $[= (\text{resolution by naked eye})/(\text{resolution by microscope})]$ is limited to about

$$\frac{10^{-4} \, \text{m}}{200 \times 10^{-9} \, \text{m}} = 500\times.$$

In practice, magnifications of about $1000\times$ are often used to minimize eyestrain. Any greater magnification would simply make visible the diffraction pattern produced by the microscope objective.

*25–10 Specialty Microscopes and Contrast

All the resolving power a microscope can attain will be useless if the object to be seen cannot be distinguished from the background. The difference in brightness between the image of an object and the image of the surroundings is called **contrast**. Achieving high contrast is an important problem in microscopy and other forms of imaging. The problem arises in biology, for example, because cells consist largely of water and are almost uniformly transparent to light. We now discuss two special types of microscope that can increase contrast: the interference and phase-contrast microscopes.

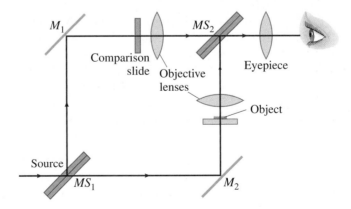

FIGURE 25–33 Object—say, a bacterium—in a water solution.

Interference microscope. The interference microscope makes use of the wave properties of light in a direct way. It is one of the most effective means to increase contrast in a transparent object. To see how it works, let us consider a transparent object—say, a bacterium—in water solution (Fig. 25–33). Light enters uniformly from the left and is coherent (meaning in phase) at all points such as a and b. If the object is as transparent as the water solution, the beam leaving at d will be as bright as that at c. There will be no contrast and the object will not be seen. However, if the object's refractive index is slightly different from that of the surrounding medium, the wavelength within the object will be altered as shown. Hence the waves at points c and d will differ in phase, if not in amplitude. This appears at first to be of no help, since the eye responds only to differences in amplitude or brightness, and does not detect this difference in phase. What the interference microscope does is to change this difference in phase into a difference of amplitude. It does so by superimposing the light that passes through the sample onto a reference beam that does not pass through the object, so that they interfere. One way of doing this is shown in Fig. 25–34. Light from a source is split into two equal beams by a half-silvered mirror, MS_1. One beam passes through the object and the second (comparison beam) passes through an identical system without the object. The two meet again and are superposed by the half-silvered mirror MS_2 before entering the eyepiece and the eye. The path length (and amplitude) of the comparison beam is adjustable. It can be adjusted, for example, so that the background is dark; that is, full destructive interference occurs. Light passing through the object (beam bd in Fig. 25–33) will also interfere with the comparison beam. But because of its different phase, the interference will not be completely destructive. Thus it will appear brighter than the background. Where the object varies in thickness, the phase difference between beams ac and bd in Fig. 25–33 will be different; and this will affect the amount of interference. Hence *variation in the thickness of the object will appear as variations in brightness in the image.* As an example, suppose that the object is a bacterium 1.0 μm thick and has a refractive index of 1.35. Then if light of wavelength $\lambda = 550$ nm (in air) is used, there will be $(1.0 \, \mu\text{m})/(550 \, \text{nm}/1.35) = 2.46$ wavelengths in the bacterium and $(1.0 \mu\text{m})/(550 \, \text{nm}/1.33) = 2.42$ wavelengths in the water. Thus the two waves will be out of phase by 0.04 wavelengths, or 14° $(=0.04 \times 360°)$.

FIGURE 25–34 Diagram of an interference microscope.

Phase-contrast microscope. The phase-contrast microscope also makes use of interference and differences in phase to produce a high-contrast image. Although it has certain limitations, it is far simpler to construct and operate than an interference microscope. To describe the operation of a phase-contrast microscope in detail, we would have to discuss the diffraction theory of image formation—how the diffraction pattern produced by each point on the object contributes to the final image. This is quite complicated, so we give only a simplified description. Figure 25–35 is a simplified diagram of a phase-contrast microscope. The object to be viewed is illuminated from below as usual. To be specific, we assume that rays from each point on the source are made parallel by a set of condensing lenses (not shown). However, a plate with a ring-shaped hole is placed above the source, so the light can pass only through this annular ring. Light that is not deviated by the object (the beam is shown shaded) is brought into focus by the objective lens in the *source image plane*. If the source is effectively at infinity (because of the condensing lenses), as is assumed here, the source image plane is at the focal point of the lens. Light that strikes the object, on the other hand, is diffracted or scattered by the object. Each point on the object then serves as a source for rays diverging from that point (dashed lines in the figure). These rays are brought to a focus in the *object image plane*, which is behind the source image plane (because the object is so close to the lens). The undeviated light from the source diverges, meanwhile, from its image plane and provides a broad bright background at the object image plane. The object is transparent, however, and the image will not be seen clearly since there will be little contrast. Contrast is achieved by inserting a circular glass *phase plate* at the source image plane. The phase plate has a groove as shown (or a raised portion) in the shape of a ring. This ring is positioned so that all the undeviated rays pass through it. Most of the rays deviated by the object, on the other hand, do not pass through this ring (see Fig. 25–35). Because the rays deviated by the object travel through a different thickness of glass than the undeviated source rays, the two can be out of phase and can interfere destructively at the object image plane. Thus the image of the object will contrast sharply with the background. Actually, because only a small fraction of the light is deviated by the object, the background light will be much stronger, and so the contrast will not be great. To compensate for this, the grooved ring on the phase plate is darkened to absorb a good part of the undeviated light so that its intensity is more nearly equal to that of the deviated light. Then nearly complete destructive interference can occur at particular points, and the contrast will be very high. The chief limitation of the phase-contrast microscope is that images tend to have "halos" around them as a result of diffraction from the phase-plate opening. Because of this artifact, care must be taken in the interpretation of images.

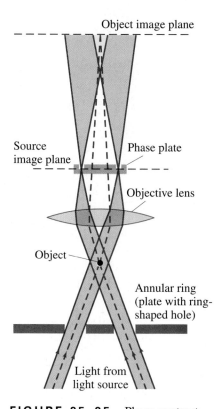

FIGURE 25–35 Phase-contrast microscope. Light beam from the source that is undeviated is shown shaded (pinkish) for clarity. Rays deviated by the object, and which form the image of the object, are shown dashed.

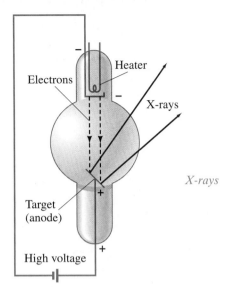

FIGURE 25-36 X-ray tube. Electrons emitted by a heated filament in a vacuum tube are accelerated by high voltage. When they strike the surface of the anode, the "target," X-rays are emitted.

FIGURE 25-37 This X-ray diffraction pattern is one of the first observed by Max von Laue in 1912 when he aimed a beam of X-rays at a zinc sulfide crystal. The diffraction pattern was detected directly on a photographic plate.

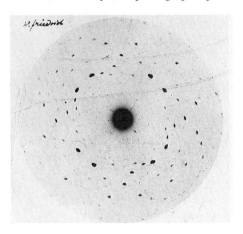

25-11 X-Rays and X-Ray Diffraction

In 1895, W. C. Roentgen (1845–1923) discovered that when electrons were accelerated by a high voltage in a vacuum tube and allowed to strike a glass (or metal) surface inside the tube, fluorescent minerals some distance away would glow, and photographic film would become exposed. Roentgen attributed these effects to a new type of radiation (different from cathode rays). They were given the name **X-rays** after the algebraic symbol x, meaning an unknown quantity. He soon found that X-rays penetrated through some materials better than through others, and within a few weeks he presented the first X-ray photograph (of his wife's hand). The production of X-rays today is usually done in a tube (Fig. 25–36) similar to Roentgen's, using voltages of typically 30 kV to 150 kV.

Investigations into the nature of X-rays indicated they were not charged particles (such as electrons) since they could not be deflected by electric or magnetic fields. It was suggested that they might be a form of invisible light. However, they showed no diffraction or interference effects using ordinary gratings. Of course, if their wavelengths were much smaller than the typical grating spacing of 10^{-6} m ($= 10^3$ nm), no effects would be expected. Around 1912, it was suggested by Max von Laue (1879–1960) that if the atoms in a crystal were arranged in a regular array (see Fig. 13–2a), such a crystal might serve as a diffraction grating for very short wavelengths on the order of the spacing between atoms, estimated to be about 10^{-10} m ($= 10^{-1}$ nm). Experiments soon showed that X-rays scattered from a crystal did indeed show the peaks and valleys of a diffraction pattern (Fig. 25–37). Thus it was shown, in a single blow, that X-rays have a wave nature and that atoms are arranged in a regular way in crystals. Today, X-rays are recognized as electromagnetic radiation with wavelengths in the range of about 10^{-2} nm to 10 nm, the range readily produced in an X-ray tube.

We saw in Sections 25–7 and 25–8 that light of shorter wavelength provides greater resolution when we are examining an object microscopically. Since X-rays have much shorter wavelengths than visible light, in principle, they should offer much greater resolution. However, there seems to be no effective material to use as lenses for the very short wavelengths of X-rays. Instead, the clever but complicated technique of **X-ray diffraction** (or **crystallography**) has proved very effective for examining the microscopic world of atoms and molecules. In a simple crystal such as NaCl, the atoms are arranged in an orderly cubical fashion, Fig. 25–38, with atoms spaced a distance d apart. Suppose that a beam of X-rays is incident on the crystal at an angle ϕ to the surface, and that the two rays shown are reflected from two subsequent planes of atoms as shown. The two rays will constructively interfere if the extra distance ray I travels is a whole number of wavelengths farther than what ray II travels. This extra distance is $2d \sin \phi$. Therefore, constructive interference will occur when

Bragg equation

$$m\lambda = 2d \sin \phi, \qquad m = 1, 2, 3, \cdots, \qquad \textbf{(25–12)}$$

where m can be any integer. (Notice that ϕ is *not* the angle with respect to the normal to the surface.) This is called the **Bragg equation** after W. L. Bragg (1890–1971), who derived it and who, together with his father W. H. Bragg (1862–1942), developed the theory and technique of X-ray diffraction by crystals in 1912–13. Thus, if the X-ray wavelength is known and the angle ϕ at which constructive interference occurs is measured, d can be obtained. This is the basis for X-ray crystallography.

Actual X-ray diffraction patterns are quite complicated. First of all, a crystal is a three-dimensional object, and X-rays can be diffracted from different planes at different angles within the crystal, as shown in Fig. 25–39. Although the analysis is complex, a great deal can be learned about any substance that can be put in crystalline form. If the substance is not a single crystal but a mixture of many tiny crystals—as in a metal or a powder—then instead of a series of spots, as in Fig. 25–37, a series of circles is obtained, Fig. 25–40; each circle corresponds to diffraction of a certain order m (Eq. 25–12), from a particular set of parallel planes.

X-ray diffraction has been very useful in determining the structure of biologically important molecules. Often it is possible to make a crystal of such molecules. The analysis is complex, and it is usually necessary to make various guesses of the structure of the molecule. Predictions of the diffraction pattern for each guessed structure can then be compared to that actually obtained. For larger molecules, such as proteins and nucleic acids, an important innovation has been the "heavy-atom technique." Since very large atoms scatter X-rays much more strongly than the ordinary C, N, O, and H atoms of biological molecules, heavy atoms can be used as "markers." The heavy atoms are chemically added to particular spots on the molecule (say, a protein)—hopefully without disturbing its structure significantly. Analysis of the *changes* in the resulting diffraction pattern gives helpful information.

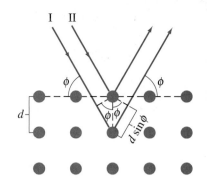

FIGURE 25–38 X-ray diffraction by a crystal.

FIGURE 25–39 There are many possible planes existing within a crystal from which X-rays can be diffracted.

(a)   (b)

FIGURE 25–40 (a) Diffraction of X-rays from a polycrystalline substance produces a set of circular rings as in (b), which is for polycrystalline sodium acetoacetate.

Even when a good crystal cannot be obtained, if the molecule under study has a regularly repeating shape (such as many proteins and DNA have), X-ray diffraction can reveal it. In a sense, each molecule is then like a single crystal and a sample is a collection of such tiny crystals. Indeed, it was with the help of X-ray diffraction that, in 1953, J. D. Watson and F. H. C. Crick worked out the double-helix structure of DNA. See Fig. 25–41, and for models of the double helix, Figs. 16–34a and 16–35.

Around 1960, the first detailed structure of a protein molecule was elucidated with the aid of X-ray diffraction; this was for myoglobin, a relative of the important constituent of blood, hemoglobin. Soon the structure of hemoglobin itself was worked out, and since then the structures of a great many molecules have been determined with the help of X-rays.

FIGURE 25–41 X-ray diffraction photo of DNA molecules taken by Rosalind Franklin in the early 1950's. The cross of spots suggested that DNA is a helix.

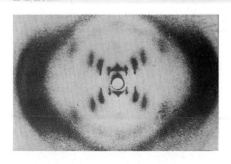

25–12 X-Ray Imaging and Computerized Axial Tomography (CAT Scan)

For a conventional medical (or dental) X-ray photograph, the X-rays emerging from the tube (Fig. 25–36, Section 25–11) pass through the body and are detected on photographic film or a fluorescent screen, Fig. 25–42a. The rays travel in very nearly straight lines through the body with minimal deviation since at X-ray wavelengths there is little diffraction or refraction. There is absorption (and scattering), however; and the difference in absorption by different structures in the body is what gives rise to the image produced by the transmitted rays. The less the absorption, the greater the transmission and the darker the film. The image is, in a sense, a "shadow" of what the rays have passed through. (The X-ray image is *not* produced by focusing rays with lenses as is the case for the instruments discussed earlier in this chapter.)

Within months of Roentgen's 1895 discovery, X-rays had already become a powerful tool for medical diagnosis, and they have remained so to this day. Although many technical advances have been made over the years, the basic principles for normal X-rays have not changed significantly. However, in the 1970s, a revolutionary new technique called **computerized tomography** (CT) using X-rays was developed.

In conventional X-ray images, the entire thickness of the body is projected onto the film; structures overlap and in many cases are difficult to distinguish. A tomographic image, on the other hand, is an image of a *slice* through the body. (The word **tomography** comes from the Greek: *tomos* = slice, *graph* = picture.) Structures and lesions previously impossible to visualize can now be seen with remarkable clarity. The principle behind CT is shown in Fig. 25–42b: a thin collimated beam of X-rays (to "collimate" means to "make straight") passes through the body to a detector that measures the transmitted intensity. Measurements are made at a large number of points as the source and detector are moved past the body together. The apparatus is then rotated slightly about the body axis and again scanned; this is repeated at (perhaps) 1° intervals for 180°. The intensity of the transmitted beam for the many points of each scan, and for each angle, are sent to a computer that

FIGURE 25–42 (a) Conventional X-ray imaging, which is essentially shadowing. (b) Tomographic imaging: the X-ray source and detector move together across the body, the transmitted intensity being measured at a large number of points. Then the source-detector assembly is rotated slightly (say, 1°) and another scan is made. This process is repeated for perhaps 180°. The computer reconstructs the image of the slice and it is presented on a TV monitor (cathode-ray tube).

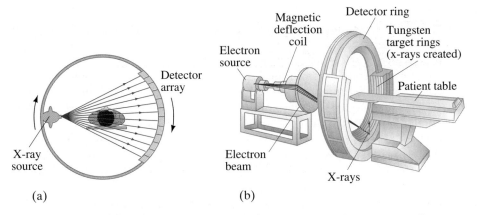

reconstructs the image of the slice. Note that the imaged slice is perpendicular to the long axis of the body. For this reason, CT is sometimes called **computerized axial tomography** (CAT), although the abbreviation CAT, as in CAT scan, can also be read as **computer-assisted tomography**.

CAT scans

The use of a single detector as in Fig. 25–42b requires a few minutes for the many scans needed to form a complete image. Much faster scanners use a fan beam, Fig. 25–43a, in which beams passing through the entire cross section of the body are detected simultaneously by many detectors. The source and detectors are then rotated about the patient. At each of the hundreds of angular positions of the apparatus, several hundred detectors can measure the intensity of transmitted rays simultaneously, so an image requires only a few seconds. Even faster, and therefore useful for heart scans, are fixed source machines wherein an electron beam is directed (by magnetic fields) to tungsten targets surrounding the patient, creating the X-rays. See Fig. 25–43b.

But how is the image formed? We can think of the slice to be imaged as being divided into many tiny picture elements (or **pixels**), which could be squares, as in Fig. 25–44. For CT, the width of each pixel is chosen according to the width of the detectors and/or the width of the X-ray beams, and this determines the resolution of the image, which is typically about 2 mm. An X-ray detector measures the intensity of the transmitted beam after it has passed through the body. Subtracting this value from the intensity of the beam at the source, we get the total absorption. Note that only the total absorption (called "a projection") along each beam line can be measured (the absorption by all the pixels in a line). To form an image, we need to determine how much radiation is absorbed at *each* pixel. (How that can be done will be discussed in a moment.) We can then assign a "grayness value" to each pixel according to how much radiation was absorbed. The image, then, is made up of tiny spots (pixels) of varying shades of gray, as is a black-and-white television picture. Often the amount of absorption is color-coded. The colors in the resulting ("false color") image have nothing to do, however, with the actual color of the object.

Finally, we must discuss how the "grayness" of each pixel can be determined even though all we can measure is the total absorption along each beam line in the slice. It can be done only by using the many beam scans made at a great many different angles. Suppose the image is to be an array of 100 × 100 elements for a total of 10^4 pixels. If we have 100 detectors and measure the absorption projections at 100 different angles, then we get 10^4 pieces of information. From this information, an image can be reconstructed, but not precisely. If more angles are measured, the reconstruction of the image can be done more accurately.

FIGURE 25-44 Example of an image made up of many small squares called *pixels* (picture elements). This one has rather poor resolution.

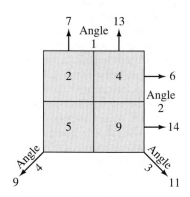

FIGURE 25–45 A simple 2 × 2 image showing true absorption values and measured projections.

There are a number of mathematical reconstruction techniques, all of which are complicated and require the use of a computer. To suggest how it is done, we consider a very simple case using the so-called "iterative" technique ("to iterate" is from the Latin "to repeat"). Although this technique is less used now than the more direct "Fourier transform" and "back projection" techniques, it is the simplest to explain. Suppose our sample slice is divided into the simple 2 × 2 pixels as shown in Fig. 25–45. The number in each pixel represents the amount of absorption by the material in that area (say, in tenths of a percent): that is, 4 represents twice as much absorption as 2. But we cannot directly measure these values—they are the unknowns we want to solve for. All we can measure are the projections—the total absorption along each beam line—and these are shown in the diagram as the sum of the absorptions for the pixels along each line at four different angles. These projections (given at the tip of each arrow) are what we can measure, and we now want to work back from them to see how close we can get to the true absorption value for each pixel. We start our analysis with each pixel being assigned a zero value, Fig. 25–46a. In the iterative technique, we use the projections to estimate the absorption value in each square, and repeat for each angle. The angle 1 projections are 7 and 13. We divide each of these equally between their two squares: each square in the left column gets $3\frac{1}{2}$ (half of 7), and each square in the right column gets $6\frac{1}{2}$ (half of 13); see Fig. 25–46b. Next we use the projections at angle 2. We calculate the difference between the measured projections at angle 2 (6 and 14) and the projections based on the previous estimate (top row: $3\frac{1}{2} + 6\frac{1}{2} = 10$; same for bottom row). Then we distribute this difference equally to the squares in that row. For the top row, we have

$$3\frac{1}{2} + \frac{6 - 10}{2} = 1\frac{1}{2} \qquad \text{and} \qquad 6\frac{1}{2} + \frac{6 - 10}{2} = 4\frac{1}{2};$$

and for the bottom row,

$$3\frac{1}{2} + \frac{14 - 10}{2} = 5\frac{1}{2} \qquad \text{and} \qquad 6\frac{1}{2} + \frac{14 - 10}{2} = 8\frac{1}{2}.$$

These values are inserted as shown in Fig. 25–46c. Next, the projection at angle 3 gives

(upper left) $\quad 1\frac{1}{2} + \dfrac{11 - 10}{2} = 2 \quad$ and $\quad$ (lower right) $\quad 8\frac{1}{2} + \dfrac{11 - 10}{2} = 9;$

and that for angle 4 gives

FIGURE 25–46 Reconstructing the image using projections in an iterative procedure.

(lower left) $\quad 5\frac{1}{2} + \dfrac{9 - 10}{2} = 5 \quad$ and $\quad$ (upper right) $\quad 4\frac{1}{2} + \dfrac{9 - 10}{2} = 4.$

The result, shown in Fig. 25–46d, corresponds exactly to the true values.

(a)

(b)

(c)

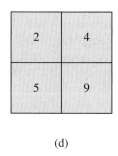

(d)

(Note that in real situations, the true values are not known, which is why these computer techniques are required.) To obtain these numbers exactly, we used six pieces of information (two each at angles 1 and 2, one each at angles 3 and 4). For the much larger number of pixels used for actual images, exact values are generally not attained. Many iterations may be needed, and the calculation is considered sufficiently precise when the difference between calculated and measured projections is sufficiently small. The above example illustrates the "convergence" of the process: the first iteration (b to c in Fig. 25–46) changed the values by 2, the last iteration (c to d) by only $\frac{1}{2}$.

Figure 25–47 illustrates what actual CT images look like. It is generally agreed that CT scanning has revolutionized some areas of medicine by providing much less invasive, and/or more accurate, diagnosis.

Computerized tomography can also be applied to ultrasound imaging (Section 12–10) and to emissions from radioisotopes and nuclear magnetic resonance (Sections 31–8 and 31–9).

(a)   (b)

FIGURE 25–47 Two CT images, with different resolutions, each showing a cross section of a brain. The left photo is of low resolution; the right photo, of higher resolution, shows a brain tumor (dark area on the right).

SUMMARY

A **camera** lens forms an image on film by allowing light in through a shutter. The lens is focused by moving it relative to the film, and its *f*-stop (or lens opening) must be adjusted for the brightness of the scene and the chosen shutter speed. The *f*-stop is defined as the ratio of the focal length to the diameter of the lens opening.

The human **eye** also adjusts for the available light—by opening and closing the iris. It focuses not by moving the lens, but by adjusting the shape of the lens to vary its focal length. The image is formed on the retina, which contains an array of receptors known as rods and cones. Diverging eyeglass or contact lenses are used to correct the defect of a nearsighted eye, which cannot focus well on distant objects. Converging lenses are used to correct for defects in which the eye cannot focus on close objects.

A **simple magnifier** is a converging lens that forms a virtual image of an object placed at (or within) the focal point. The **angular magnification**, when viewed by a relaxed normal eye, is

$$M = \frac{N}{f},$$

where f is the focal length of the lens and N is the near point of the eye (25 cm for a "normal" eye).

An **astronomical telescope** consists of an **objective** lens or mirror and an **eyepiece** that magnifies the real image formed by the objective. The **magnification** is equal to the ratio of the objective and eyepiece focal lengths, and the image is inverted:

$$M = -\frac{f_o}{f_e}.$$

A compound **microscope** also uses objective and eyepiece lenses, and the final image is inverted. The total magnification is the product of the magnifications of the two lenses and is approximately

$$M \approx \left(\frac{N}{f_e}\right)\left(\frac{l}{f_o}\right),$$

where l is the distance between the lenses, N is the near point of the eye, and f_o and f_e are the focal lengths of objective and eyepiece, respectively.

Microscopes, telescopes, and other optical instruments are limited in the formation of sharp images by **lens aberrations**. These include **spherical aberration**, in which rays passing through the edge of a lens are not focused at the same point as those that pass near the center; and **chromatic aberration**, in which different colors are focused at different points. Compound lenses, consisting of several elements, can largely correct for aberrations.

The wave nature of light also limits the sharpness, or **resolution**, of images. Because of diffraction, it is *not possible to discern details smaller than the wavelength* of the radiation being used. This limits the useful magnification of a light microscope to about 1000×.

X-rays are a form of electromagnetic radiation of very short wavelength. They are produced when high-speed electrons, accelerated by high voltage in an evacuated tube, strike a glass or metal target.

QUESTIONS

1. Why is the depth of field greater, and the image sharper, when a camera lens is "stopped down" to a larger *f*-number? Ignore diffraction.

2. Describe how diffraction affects the statement of Question 1. [*Hint*: See Eq. 24–3.]

3. Why must a camera lens be moved farther from the film to focus on a closer object?

4. Why are bifocals needed mainly by older persons and not generally by younger people?

5. Explain why swimmers with good eyes see distant objects as blurry when they are underwater. Use a diagram and also show why goggles correct this problem.

6. Will a nearsighted person who wears corrective lenses be able to see clearly underwater when wearing glasses? Use a diagram to show why or why not.

7. You can tell whether a person is nearsighted or farsighted by looking at the width of the face through their glasses. If the person's face appears shrunk through the glasses, is the person farsighted or nearsighted?

8. The human eye is much like a camera—yet, when a camera shutter is left open and the camera moved, the image will be blurred; but when you move your head with your eyes open, you still see clearly. Explain.

9. In attempting to discern distant details, people will sometimes squint. Why does this help?

10. Is the image formed on the retina of the human eye upright or inverted? Discuss the implications of this for our perception of objects.

11. Reading glasses use converging lenses. A simple magnifier is also a converging lens. Are reading glasses therefore magnifiers? Discuss the similarities and differences between converging lenses as used for these two different purposes.

12. Inexpensive microscopes for children's use usually produce images that are colored at the edges. Why?

13. Spherical aberration in a thin lens is minimized if rays are bent equally by the two surfaces. If a planoconvex lens is used to form a real image of an object at infinity, which surface should face the object? Use ray diagrams to show why.

14. Which aberrations present in a simple lens are not present (or are greatly reduced) in the human eye?

15. Explain why chromatic aberration occurs for thin lenses but not for mirrors.

16. What are the advantages (give at least two) for the use of large reflecting mirrors in astronomical telescopes?

17. Which color of visible light would give the best resolution in a microscope?

18. Atoms have diameters of about 10^{-8} cm. Can visible light be used to "see" an atom? Why or why not?

PROBLEMS

SECTION 25–1

1. (I) A 55-mm-focal-length lens has *f*-stops ranging from *f*/1.4 to *f*/22. What is the corresponding range of lens diaphragm diameters?

2. (I) A television camera lens has a 14-cm focal length and a lens diameter of 6.0 cm. What is its *f*-number?

3. (I) A light meter reports that a camera setting of $\frac{1}{250}$ s at *f*/5.6 will give a correct exposure. But the photographer wishes to use *f*/11 to increase the depth of field. What should the shutter speed be?

4. (I) A properly exposed photograph is taken at *f*/16 and $\frac{1}{60}$ s. What lens opening would be required if the shutter speed were $\frac{1}{1000}$ s?

5. (II) A "pinhole" camera uses a tiny pinhole instead of a lens. Show, using ray diagrams, how reasonably sharp images can be formed using such a pinhole camera. In particular, consider two point objects 2.0 cm apart that are 1.0 m from a 1.0-mm-diameter pinhole. Show that on a piece of film 7.0 cm behind the pinhole, each object produces a tiny, easily resolvable spot.

6. (II) Suppose that a correct exposure is $\frac{1}{250}$s at *f*/11. Under the same conditions, what exposure time would be needed for a pinhole camera if the pinhole diameter is 1.0 mm and the film is 7.0 cm from the hole?

7. (II) If a 135-mm telephoto lens is designed to cover object distances from 1.2 m to ∞, over what distance must the lens move relative to the plane of the film?

8. (II) A 200-mm-focal-length lens can be adjusted so that it is 200.0 mm to 206.0 mm from the film. For what range of object distances can it be adjusted?

9. (II) A nature photographer wishes to photograph a 22-m tall tree from a distance of 50 m. What focal-length lens should be used if the image is to fill the 24-mm height of the film?

SECTION 25–2

10. (I) A human eyeball is about 2.0 cm long and the pupil has a maximum diameter of about 5.0 mm. What is the "speed" of this lens?

11. (II) A person struggles to read by holding a book at arm's length, a distance of 50 cm away. What power of reading glasses should be prescribed for him, assuming they will be placed 2.0 cm from the eye and he wants to read at the normal near point of 25 cm?

12. (II) Reading glasses of what power are needed for a person whose near point is 120 cm, so that he can read a computer screen at 50 cm? Assume a lens–eye distance of 1.8 cm.

13. (II) Show that if the nearsighted person in Example 25–4 wore contact lenses corrected for the far point $(=\infty)$, that the near point would be 41 cm. (Would glasses be better in this case?)

14. (II) A person's left eye is corrected by a -5.0 diopter lens, 2.0 cm from the eye. (a) Is this person near- or farsighted? (b) What is this person's far point without glasses?

15. (II) A person's right eye can see objects clearly only if they are between 20 cm and 70 cm away. (a) What power of contact lens is required so that objects far away are sharp? (b) What will be the near point with the lens in place?

16. (II) About how much longer is the nearsighted eye in Example 25–4 than the 2.0 cm of a normal eye?

17. (II) One lens of a nearsighted person's eyeglasses has a focal length of -25.0 cm and the lens is 1.8 cm from the eye. If the person switches to contact lenses that are placed directly on the eye, what should be the focal length of the corresponding contact lens?

18. (II) What is the focal length of the eye lens system when viewing an object (a) at infinity, and (b) 30 cm from the eye? Assume that the lens–retina distance is 2.0 cm.

SECTION 25–3

19. (I) What is the magnification of a lens used with a relaxed eye if its focal length is 12 cm?

20. (I) What is the focal length of a magnifying glass of $3.0\times$ magnification for a relaxed normal eye?

21. (I) A magnifier is rated at $2.5\times$ for a normal eye focusing on an image at the near point. (a) What is its focal length? (b) What is its focal length if the $2.5\times$ refers to a relaxed eye?

22. (II) Sherlock Holmes is using a 10.0-cm-focal-length lens as his magnifying glass. To obtain maximum magnification, where must the object be placed (assume a normal eye), and what will be the magnification?

23. (II) A 3.10-mm-wide beetle is viewed with a 9.00-cm-focal-length lens. A normal eye views the image at its near point. Calculate (a) the angular magnification, (b) the width of the image, and (c) the object distance from the lens.

24. (II) A small insect is placed 5.35 cm from a $+6.00$-cm-focal-length lens. Calculate (a) the position of the image, and (b) the angular magnification.

25. (II) A magnifying glass with a focal length of 9.5 cm is used to read print placed at a distance of 8.5 cm. Calculate: (a) the position of the image; (b) the linear magnification; and (c) the angular magnification.

26. (II) A magnifying glass is rated at $3.0\times$ for a normal eye that is relaxed. What would be the magnification for a relaxed eye whose near point is (a) 50 cm, and (b) 16 cm? Explain the differences.

27. (II) A nearsighted person has near and far points of 10.0 and 20.0 cm respectively. If she puts on a pair of glasses with power $P = -4.0D$, what are her new near and far points? Neglect the distance between her eye and the lens.

SECTION 25–4

28. (I) What is the magnification of an astronomical telescope whose objective lens has a focal length of 80 cm and whose eyepiece has a focal length of 2.8 cm? What is the overall length of the telescope when adjusted for a relaxed eye?

29. (I) The overall magnification of an astronomical telescope is desired to be $25\times$. If an objective of 80 cm focal length is used, what must be the focal length of the eyepiece? What is the overall length of the telescope when adjusted for use by the relaxed eye?

30. (I) An $8.0\times$ binocular has 3.0-cm-focal-length eyepieces. What is the focal length of the objective lenses?

31. (II) An astronomical telescope has an objective with focal length 85 cm and a $+40$-D eyepiece. What is the total magnification?

32. (II) An astronomical telescope has its two lenses spaced 76 cm apart. If the objective lens has a focal length of 74.5 cm, what is the magnification of this telescope? Assume a relaxed eye.

33. (II) A Galilean telescope adjusted for a relaxed eye is 33 cm long. If the objective lens has a focal length of 36 cm, what is the magnification?

34. (II) What is the magnifying power of an astronomical telescope using a reflecting mirror whose radius of curvature is 5.0 m and an eyepiece whose focal length is 3.2 cm?

35. (II) The Moon's image appears to be magnified $120\times$ by a reflecting astronomical telescope with an eyepiece having a focal length of 3.5 cm. What are the focal length and radius of curvature of the main mirror?

36. (III) A $180\times$ astronomical telescope is adjusted for a relaxed eye when the two lenses are 1.25 m apart. What is the focal length of each lens?

37. (III) A $6.0\times$ pair of binoculars has an objective focal length of 26 cm. If the binoculars are focused on an object 4.0 m away (from the objective), what is the magnification? (The $6.0\times$ refers to objects at infinity; Eq. 25–3 holds only for objects at infinity and not for nearby ones.)

SECTION 25–5

38. (I) A microscope uses an eyepiece with a focal length of 1.5 cm. Using a normal eye with a final image at infinity, the tube length is 17.5 cm and the focal length of the objective lens is 0.65 cm. What is the magnification of the microscope?

39. (I) A $720\times$ microscope uses a 0.40-cm-focal-length objective lens. If the tube length is 17.5 cm, what is the focal length of the eyepiece? Assume a normal eye and that the final image is at infinity.

40. (II) A microscope has a 12.0× eyepiece and a 62.0× objective 20.0 cm apart. Calculate (a) the total magnification, (b) the focal length of each lens, and (c) where the object must be for a normal relaxed eye to see it in focus.

41. (II) A microscope has a 1.8-cm-focal-length eyepiece and 0.80-cm objective. Calculate (a) the position of the object if the distance between the lenses is 16.0 cm, and (b) the total magnification assuming a relaxed normal eye.

42. (II) Repeat Problem 41 assuming that the final image is located 25 cm from the eyepiece (near point of a normal eye).

43. (II) The eyepiece of a compound microscope has a focal length of 2.70 cm and the objective has $f = 0.740$ cm. If an object is placed 0.790 cm from the objective lens, calculate (a) the distance between the lenses when the microscope is adjusted for a relaxed eye, and (b) the total magnification.

SECTION 25–6

44. (II) An achromatic lens is made of two very thin lenses placed in contact that have focal lengths of $f_1 = -28$ cm and $f_2 = +23$ cm. (a) Is the combination converging or diverging? (b) What is the net focal length?

*45. (III) Let's examine spherical aberration in a particular situation. A planoconvex lens of index of refraction 1.50 and a radius of curvature $R = 12.0$ cm is shown in Fig. 25–48. Consider an incoming ray parallel to the principle axis and a height h above it as shown. Determine the distance d, from the flat face of the lens, to where this ray crosses the principle axis if (a) $h = 1.0$ cm, and (b) $h = 6.0$ cm. (c) How far apart are these "focal points"? (d) How large is the "circle of confusion" at the "focal point" for $h = 1.0$ cm?

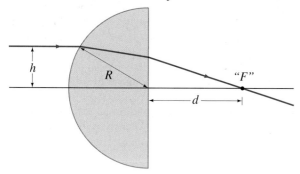

FIGURE 25–48   Problem 45.

SECTIONS 25–7 TO 25–9

46. (I) What is the angular resolution limit set by diffraction for the Mt. Wilson 100-in. (mirror diameter) telescope ($\lambda = 550$ nm)?

47. (I) What is the angle of acceptance α of a microscope oil-immersion objective, and its resolving power, if $n = 1.80$ and the NA = 1.41? Use $\lambda = 500$ nm.

48. (II) A microscope objective is immersed in oil ($n = 1.60$) and accepts light scattered from the object up to 60° on either side of vertical. (a) What is the numerical aperture? (b) What is the approximate resolution of the microscope for 550-nm light?

49. (II) Two stars 20 light-years away are barely resolved by a 12-in. (mirror diameter) amateur telescope. How far apart are the stars? Assume $\lambda = 500$ nm.

50. (II) The numerical aperture of an instrument is 0.95. How far apart are the lines on a certain shell if they are just barely resolved when viewed with green light (540 nm)? Will they be resolved if viewed by violet (400 nm) or red (700 nm) light?

51. (II) A certain sea organism has a pattern of dots on its surface with an average spacing of 0.63 μm. If the specimen is viewed using 480-nm light, what minimum value must the numerical aperture be in order that the dots be resolved?

52. (II) What minimum magnification would be required to see the dots on the organism of Problem 51?

53. (II) (a) How far away can a human eye distinguish two car headlights 2.0 m apart? Consider only diffraction effects and assume an eye diameter of 5.0 mm and a wavelength of 500 nm. (b) What is the minimum angular separation an eye could resolve when viewing two stars, considering only diffraction effects? In reality, it is about 1′ of arc. Why is it not equal to your answer in (b)?

*SECTION 25–10

*54. (II) Show that the phase difference (in radians) between the two waves ac and bd in Fig. 25–33 is $\delta = (2\pi/\lambda)(n_2 - n_1)t$, where n_2 and n_1 are the refractive indices of the object and the medium, t the thickness of the object, and λ the wavelength of light used.

*SECTION 25–11

*55. (II) X-rays of wavelength 0.135 nm fall on a crystal whose atoms, lying in planes, are spaced 0.280 nm apart. At what angle must the X-rays be directed if the first diffraction maximum is to be observed?

*56. (II) X-rays of wavelength 0.0973 nm are directed at an unknown crystal. The second diffraction maximum is recorded at 23.4°. What is the spacing between crystal planes?

*57. (II) First-order Bragg diffraction is observed at 23.4° from a crystal with spacing between atoms of 0.24 nm. (a) At what angle will second order be observed? (b) What is the wavelength of the X-rays?

*58. (III) If X-ray diffraction peaks corresponding to the first three orders ($m = 1, 2,$ and 3) are measured, can both the X-ray wavelength λ and lattice spacing d be determined? Prove your answer.

*59. (II) (a) Suppose for a conventional X-ray image that the X-ray beam consists of parallel rays. What would be the magnification of the image? (b) Suppose, instead, the X-rays come from a point source (as in Fig. 25–42a) that is 15 cm in front of a human body 25 cm thick, and the film is pressed against the person's back. Determine and discuss the range of magnifications that results.

GENERAL PROBLEMS

60. Sam purchases +3.2 diopter eyeglasses which correct his faulty vision to put his near point at 25 cm. (Assume he wears the lenses 2.0 cm from his eyes.) (a) Is Sam nearsighted or farsighted? (b) Calculate the focal length of Sam's glasses. (c) Calculate Sam's near point without glasses. (d) Pam, who has normal eyes with near point at 25 cm, puts on Sam's glasses. Calculate Pam's near point with Sam's glasses on.

61. As early morning passed toward midday, and the sunlight got more intense, a photographer who was taking repeated shots of the same subject noted that, if she kept her shutter speed constant, she had to change the f-number from $f/5.6$ to $f/22$. By how much had the sunlight intensity increased during that time?

62. Show that for objects very far away (assume infinity), the magnification of a camera lens is proportional to its focal length.

63. For a camera equipped with a 50-mm-focal-length lens, what is the object distance if the image height equals the object height? How far is the object from the film?

64. A woman can see clearly with her right eye only when objects are between 40 cm and 180 cm away. Prescription bifocals should have what powers so that she can see distant objects clearly (upper part) and be able to read a book 25 cm away (lower part)? Assume that the glasses will be 2.0 cm from the eye.

65. A child has a near point of 15 cm. What is the maximum magnification the child can obtain using an 8.0-cm focal-length magnifier? Compare to that for a normal eye.

66. What is the magnifying power of a +4.0-D lens used as a magnifier? Assume a relaxed normal eye.

67. A physicist lost in the mountains tries to make a telescope using the lenses from his reading glasses. They have powers of +2.0 D and +4.5 D, respectively. (a) What maximum magnification telescope is possible? (b) Which lens should be used as the eyepiece?

68. The normal lens on a 35-mm camera has a focal length of 50 mm. Its aperture diameter varies from a maximum of 25 mm ($f/2$) to a minimum of 3.0 mm ($f/16$). Determine the resolution limit set by diffraction for $f/2$ and $f/16$. Specify as the number of lines per millimeter resolved on the film. Take $\lambda = 500$ nm.

69. A 50-year-old man uses +2.5 diopter lenses to be able to read a newspaper 25 cm away. Ten years later, he finds that he must hold the paper 35 cm away to see clearly with the same lenses. What power lenses does he need now? (Distances are measured from the lens.)

70. Spy planes fly at extremely high altitudes (25 km) to avoid interception. Their cameras are reportedly able to discern features as small as 5 cm. What must be the minimum aperture of the camera lens to afford this resolution? (Use $\lambda = 550$ nm.)

71. Suppose that you wish to construct a telescope that can resolve features 10 km across on the Moon, 384,000 km away. You have a 2.2-m-focal-length objective lens whose diameter is 12 cm. What focal-length eyepiece is needed if your eye can resolve objects 0.10 mm apart at a distance of 25 cm? What is the resolution limit set by the size of the objective lens (that is, by diffraction)? Use $\lambda = 500$ nm.

72. Exposure times must be increased for pictures taken at very short distances, because of the increased distance of the lens from the film for a focused image. (a) Show that when the object is so close to the camera that the image height equals the object height, the exposure time must be four times longer than when the object is a long distance away (say, ∞), given the same illumination and f-stop. (b) Show that if d_o is at least four or five times the focal length f of the lens, the exposure time is increased negligibly relative to the same object being a great distance away.

73. The objective lens and the eyepiece of a telescope are spaced 85 cm apart. If the eyepiece is 20 diopters, what is the total magnification of the telescope?

*74. A Lucite planoconvex lens (Fig. 23–29a) has one flat surface and the other has $R = 18.4$ cm. It is used to view an object, located 66.0 cm away from the lens, which is a mixture of red and yellow. The index of refraction of the glass is 1.5106 for red light and 1.5226 for yellow light. What are the locations of the red and yellow images formed by the lens?

75. A movie star catches a reporter shooting pictures of her at home. She claims the reporter was trespassing and to prove her point, she gives as evidence the film she seized. Her 1.75 m height is 8.25 mm high on the film, and the focal length of the camera lens was 200 mm. How far away from the subject was the reporter standing?

CHAPTER

26 SPECIAL THEORY OF RELATIVITY

Physics at the end of the nineteenth century looked back on a period of great progress. The theories developed over the preceding three centuries had been very successful in explaining a wide range of natural phenomena. Newtonian mechanics beautifully explained the motion of objects on Earth and in the heavens. Furthermore, it formed the basis for successful treatments of fluids, wave motion, and sound. Kinetic theory explained the behavior of gases and other materials. Maxwell's theory of electromagnetism not only brought together and explained electric and magnetic phenomena, but it predicted the existence of electromagnetic (EM) waves that would behave in every way just like light—so light came to be thought of as an electromagnetic wave. Indeed, it seemed that the natural world, as seen through the eyes of physicists, was very well explained. A few puzzles remained, but it was felt that these would soon be explained using already known principles.

But it did not turn out so simply. Instead, these few puzzles were to be solved only by the introduction, in the early part of the twentieth century, of two revolutionary new theories that changed our whole conception of nature: the *theory of relativity* and *quantum theory*.

Physics as it was known at the end of the nineteenth century (what we've covered up to now in this book) is referred to as **classical physics**. The new physics that grew out of the great revolution at the turn of the twentieth century is now called **modern physics**. In this chapter, we present the special theory of relativity, which was first proposed by Albert Einstein (1879–1955; Fig. 26–1) in 1905. In the following chapter, we introduce the equally momentous quantum theory.

Classical vs. modern physics

26-1 Galilean–Newtonian Relativity

Einstein's special theory of relativity deals with how we observe events, particularly how objects and events are observed from different frames of reference.[†] This subject had, of course, already been explored by Galileo and Newton. We first briefly discuss these earlier ideas, before seeing (starting in Section 26–3) how the theory of relativity changed them.

The special theory of relativity deals with events that are observed and measured from so-called **inertial reference frames**, which (as mentioned in Chapter 4) are reference frames in which Newton's first law, the law of inertia, is valid. (Newton's first law states that, if an object experiences no net force, the object either remains at rest or continues in motion with constant velocity in a straight line.) It is easiest to analyze events when they are observed and measured from inertial frames, and the Earth, though not quite an inertial frame (it rotates), is close enough that for most purposes we can consider it an inertial frame. Rotating or otherwise accelerating frames of reference are noninertial frames,[‡] and Einstein dealt with such complicated frames of reference in his general theory of relativity (Chapter 33).

A reference frame that moves with constant velocity with respect to an inertial frame is itself also an inertial frame, since Newton's laws hold in it as well. When we say that we observe or make measurements from a certain reference frame, it means that we are at rest in that reference frame.

Both Galileo and Newton were aware of what we now call the **relativity principle** applied to mechanics: that *the basic laws of physics are the same in all inertial reference frames.* You may have recognized its validity in everyday life. For example, objects move in the same way in a smoothly moving (constant-velocity) train or airplane as they do on Earth. (This assumes no vibrations or rocking—for they would make the reference frame noninertial.) When you walk, drink a cup of soup, play Ping-Pong, or drop a pencil on the floor while traveling in a train, airplane, or ship moving at constant velocity, the bodies move just as they do when you are at rest on Earth. Suppose you are in a car traveling rapidly along at constant velocity. If you release a coin from above your head inside the car, how will it fall? It falls straight downward with respect to the car, and hits the floor

Relativity principle: the laws of physics are the same in all inertial reference frames

---

[†]A reference frame is a set of coordinate axes fixed to some body such as the Earth, a train, the Moon, and so on. See Section 2–1.

[‡]On a rotating platform (say a merry-go-round), for example, an object at rest starts moving outward even though no body exerts a force on it. This is therefore not an inertial frame. See Appendix C, Fig. C–1.

(a)
Reference frame = car

(b)
Reference frame = Earth

FIGURE 26–2 A coin is dropped by a person in a moving car. (a) In the reference frame of the car, the coin falls straight down. (b) In a reference frame fixed on the Earth, the coin follows a curved (parabolic) path. The upper views show the moment of the coin's release, and the lower views are a short time later.

directly below the point of release, Fig. 26–2a. (If you drop the coin out the car's window, this won't happen because the moving air drags the coin backward relative to the car.) This is just how objects fall on the Earth—straight down—and thus our experiment in the moving car is in accord with the relativity principle.

Note in this example, however, that to an observer on the Earth, the coin follows a curved path, Fig. 26–2b. The actual path followed by the coin is different as viewed from different frames of reference. This does not violate the relativity principle because this principle states that the *laws* of physics are the same in all inertial frames. The same law of gravity, and the same laws of motion, apply in both reference frames. And the acceleration of the coin is the same in both reference frames. The difference in Figs. 26–2a and b is that in the Earth's frame of reference, the coin has an initial velocity (equal to that of the car). The laws of physics therefore predict it will follow a parabolic path like any projectile. In the car's reference frame, there is no initial velocity, and the laws of physics predict that the coin will fall straight down. The laws are the same in both reference frames, although the specific paths are different.[†]

Galilean–Newtonian relativity involves certain unprovable assumptions that make sense from everyday experience. It is assumed that the lengths of objects are the same in one reference frame as in another, and that time passes at the same rate in different reference frames. In classical mechanics, then, space and time are considered to be **absolute**: their measurement doesn't change from one reference frame to another. The mass of an object, as well as all forces, are assumed to be unchanged by a change in inertial reference frame.

The position of an object is, of course, different when specified in different reference frames, and so is velocity. For example, a person may walk inside a bus toward the front with a speed of 5 km/h. But if the bus moves 40 km/h with respect to the Earth, the person is then moving with a speed of 45 km/h with respect to the Earth. The acceleration of a body, however, is the same in any inertial reference frame according to classical mechanics. This is because the change in velocity, and the time interval, will be the same.

[†]Galileo, in his great book *Dialogues on the Two Chief Systems of the World*, described a similar experiment and predicted the same results. Galileo's example involved a sailor dropping a knife from the top of the mast of a sailing vessel. If the vessel moves at constant speed, where will the knife hit the deck (ignoring Earth's rotation and air resistance)?

For example, the person in the bus may accelerate from 0 to 5 km/h in 1.0 seconds, so $a = 5$ km/h/s in the reference frame of the bus. With respect to the Earth, the acceleration is $(45$ km/h $- 40$ km/h$)/(1.0$ s$) = 5$ km/h/s, which is the same.

Since neither F, m, nor a changes from one inertial frame to another, then Newton's second law, $F = ma$, does not change. Thus Newton's second law satisfies the relativity principle. It is easily shown that the other laws of mechanics also satisfy the relativity principle.

That the laws of mechanics are the same in all inertial reference frames implies that no one inertial frame is special in any sense. We express this important conclusion by saying that **all inertial reference frames are equivalent** for the description of mechanical phenomena. No one inertial reference frame is any better than another. A reference frame fixed to a car or an aircraft traveling at constant velocity is as good as one fixed on the Earth. When you travel smoothly at constant velocity in a car or airplane, it is just as valid to say you are at rest and the Earth is moving as it is to say the reverse. There is no experiment you can do to tell which frame is "really" at rest and which is moving. Thus, there is no way to single out one particular reference frame as being at absolute rest.

All inertial reference frames are equally valid

A complication arose, however, in the last half of the nineteenth century. When Maxwell presented his comprehensive and very successful theory of electromagnetism (Chapter 22), he showed that light can be considered an electromagnetic wave. Maxwell's equations predicted that the velocity of light c would be 3.00×10^8 m/s; and this is just what is measured, within experimental error. The question then arose: in what reference frame does light have precisely the value predicted by Maxwell's theory? For it was assumed that light would have a different speed in different frames of reference. For example, if observers were traveling on a rocket ship at a speed of 1.0×10^8 m/s away from a source of light, we might expect them to measure the speed of the light reaching them to be 3.0×10^8 m/s $- 1.0 \times 10^8$ m/s $= 2.0 \times 10^8$ m/s. But Maxwell's equations have no provision for relative velocity. They predicted the speed of light to be $c = 3.0 \times 10^8$ m/s. This seemed to imply there must be some special reference frame where c would have this value.

We discussed in Chapters 11 and 12 that waves travel on water and along ropes or strings, and sound waves travel in air and other materials. Nineteenth-century physicists viewed the material world in terms of the laws of mechanics, so it was natural for them to assume that light too must travel in some *medium*. They called this transparent medium the **ether** and assumed it permeated all space.[†] It was therefore assumed that the velocity of light given by Maxwell's equations must be with respect to the ether.

The "ether"

However, it appeared that Maxwell's equations did *not* satisfy the relativity principle. They were not the same in all inertial reference frames. They were simplest in the frame where $c = 3.00 \times 10^8$ m/s; that is, in a reference frame at rest in the ether. In any other reference frame, extra terms would have to be added to take into account the relative velocity. Thus, although most of the laws of physics obeyed the relativity principle,

---

[†]The medium for light waves could not be air, since light travels from the Sun to Earth through nearly empty space. Therefore, another medium was postulated, the ether. The ether was not only transparent, but, because of difficulty in detecting it, was assumed to have zero density.

the laws of electricity and magnetism apparently did not. Instead, they seemed to single out one reference frame that was better than any other—a reference frame that could be considered absolutely at rest.

Scientists soon set out to determine the speed of the Earth relative to this absolute frame, whatever it might be. A number of clever experiments were designed. The most direct were performed by A. A. Michelson and E. W. Morley in the 1880s. The details of their experiment are discussed in the next Section. Briefly, what they did was measure the difference in the speed of light in different directions. They expected to find a difference depending on the orientation of their apparatus with respect to the ether. For just as a boat has different speeds relative to the land when it moves upstream, downstream, or across the stream, so too light would be expected to have different speeds depending on the velocity of the ether past the Earth.

Strange as it may seem, they detected no difference at all. This was a great puzzle. A number of explanations were put forth over a period of years, but they led to contradictions or were otherwise not generally accepted.

Then in 1905, Albert Einstein proposed a radical new theory that reconciled these many problems in a simple way. But at the same time, as we shall soon see, it completely changed our ideas of space and time.

* 26–2 The Michelson–Morley Experiment

The Michelson–Morley experiment was designed to measure the speed of the *ether*—the medium in which light was assumed to travel—with respect to the Earth. The experimenters thus hoped to find an absolute reference frame, one that could be considered to be at rest.

One of the possibilities nineteenth-century scientists considered was that the ether is fixed relative to the Sun, for even Newton had taken the Sun as the center of the universe. If this were the case (there was no guarantee, of course), the Earth's speed of about 3×10^4 m/s in its orbit around the Sun would produce a change of 1 part in 10^4 in the speed of light $(3.0 \times 10^8$ m/s). Direct measurement of the speed of light to this accuracy was not possible. But A. A. Michelson, later with the help of E. W. Morley, was able to use his interferometer (Section 24–9) to measure the difference in the speed of light in different directions to this accuracy. This famous experiment is based on the principle shown in Fig. 26–3. Part (a) is a diagram of the Michelson interferometer, and it is assumed that the "ether wind" is moving with speed v to the right. (Alternatively, the Earth is assumed to move to the left with respect to the ether at speed v.) The light from the source is split into two beams by the half-silvered mirror M_S. One beam travels to mirror M_1 and the other to mirror M_2. The beams are reflected by M_1 and M_2 and are joined again after passing through M_S. The now superposed beams interfere with each other and the resultant is viewed by the observer's eye as an interference pattern (discussed in Section 24–9).

Whether constructive or destructive interference occurs at the center of the interference pattern depends on the relative phases of the two beams after they have traveled their separate paths. To examine this let us consider an analogy of a boat traveling up and down, and across, a river whose current moves with speed v, as shown in Fig. 26–3b. In still water, the boat can travel with speed c (not the speed of light in this case).

First we consider beam 2 in Fig. 26–3a, which travels parallel to the "ether wind." In its journey from M_S to M_2, we expect the light to travel

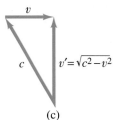

(a)

(b)

(c)

FIGURE 26–3 The Michelson–Morley experiment. (a) Michelson interferometer. (b) Boat analogy: boat 1 goes across the stream and back; boat 2 goes downstream and back upstream. (c) Calculation of the velocity of boat (or light beam) traveling perpendicular to the current (or ether wind).

with speed $c + v$, just as a boat traveling downstream (see Fig. 26–3b) acquires the speed of the river current. Since the beam travels a distance l_2, the time it takes to go from M_S to M_2 is $t = l_2/(c + v)$. To make the return trip from M_2 to M_S, the light must move against the ether wind (like the boat going upstream), so its relative speed is expected to be $c - v$. The time for the return trip is $l_2/(c - v)$. The total time required for beam 2 to travel from M_S to M_2 and back to M_S is

$$t_2 = \frac{l_2}{c + v} + \frac{l_2}{c - v}$$

$$= \frac{2l_2}{c(1 - v^2/c^2)}.$$

The second line was obtained from the first by finding the common denominator and factoring out c^2 in the denominator.

Now let us consider beam 1, which travels crosswise to the ether wind. Here the boat analogy (part b) is especially helpful. The boat is to go from wharf A to wharf B directly across the stream. If it heads directly across, the stream's current will drag it downstream. To reach wharf B, the boat must head at an angle upstream. The precise angle depends on the magnitudes of c and v, but is of no interest to us in itself. Part (c) of Fig. 26–3 shows how to calculate the velocity v' of the boat relative to Earth as it crosses the stream. Since c, v, and v' form a right triangle, we have that $v' = \sqrt{c^2 - v^2}$. The boat has the same velocity when it returns. If we now apply these principles to light beam 1 in Fig. 26–3a, we see that the beam travels with a speed $\sqrt{c^2 - v^2}$ in going from M_S to M_1 and back again. The total distance traveled is $2l_1$, so the time required for beam 1 to make the round trip is $2l_1/\sqrt{c^2 - v^2}$, or

$$t_1 = \frac{2l_1}{c\sqrt{1 - v^2/c^2}}.$$

Notice that the denominator in this equation for t_1 involves a square root, whereas that for t_2 does not.

If $l_1 = l_2 = l$, we see that beam 2 will lag behind beam 1 by an amount

$$\Delta t = t_2 - t_1 = \frac{2l}{c}\left(\frac{1}{1 - v^2/c^2} - \frac{1}{\sqrt{1 - v^2/c^2}}\right).$$

If $v = 0$, then $\Delta t = 0$, and the two beams will return in phase since they were initially in phase. But if $v \neq 0$, then $\Delta t \neq 0$, and the two beams will return out of phase. If this change of phase from the condition $v = 0$ to that for $v = v$ could be measured, then v could be determined. But the Earth cannot be stopped. Furthermore, it is not possible to independently assume $l_1 = l_2$.

Michelson and Morley realized that they could detect the difference in phase (assuming that $v \neq 0$) if they rotated their apparatus by 90°, for then the interference pattern between the two beams should change. In the rotated position, beam 1 would now move parallel to the ether and beam 2 perpendicular to it. Thus the roles could be reversed, and in the rotated position the times (designated by primes) would be

$$t_1' = \frac{2l_1}{c(1 - v^2/c^2)} \quad \text{and} \quad t_2' = \frac{2l_2}{c\sqrt{1 - v^2/c^2}}.$$

The time lag between the two beams in the nonrotated position (unprimed) would be

$$\Delta t = t_2 - t_1 = \frac{2l_2}{c(1 - v^2/c^2)} - \frac{2l_1}{c\sqrt{1 - v^2/c^2}}.$$

In the rotated position, the time difference would be

$$\Delta t' = t_2' - t_1' = \frac{2l_2}{c\sqrt{1 - v^2/c^2}} - \frac{2l_1}{c(1 - v^2/c^2)}.$$

When the rotation is made, the fringes of the interference pattern (Section 24–9) will shift an amount determined by the difference:

$$\Delta t - \Delta t' = \frac{2}{c}(l_1 + l_2)\left(\frac{1}{1 - v^2/c^2} - \frac{1}{\sqrt{1 - v^2/c^2}}\right).$$

This expression can be considerably simplified if we assume that $v/c \ll 1$. For in this case we can use the binomial expansion,[†] so

$$\frac{1}{1 - v^2/c^2} \approx 1 + \frac{v^2}{c^2} \quad \text{and} \quad \frac{1}{\sqrt{1 - v^2/c^2}} \approx 1 + \frac{1}{2}\frac{v^2}{c^2}.$$

[†]The binomial expansion (see Appendix A) states that $(1 \pm x)^n = 1 \pm nx + [n(n-1)/2]x^2 + \cdots$. In our case we have, therefore, $(1 - x)^{-1} \approx 1 + x$, and $(1 - x)^{-1/2} \approx 1 + \frac{1}{2}x$, where only the first term is kept, since $x = v^2/c^2$ is assumed to be small.

Then

$$\Delta t - \Delta t' \approx \frac{2}{c}(l_1 + l_2)\left(1 + \frac{v^2}{c^2} - 1 - \frac{1}{2}\frac{v^2}{c^2}\right)$$

$$\approx (l_1 + l_2)\frac{v^2}{c^3}.$$

Now we take $v = 3.0 \times 10^4$ m/s, the speed of the Earth in its orbit around the Sun. In Michelson and Morley's experiments, the arms l_1 and l_2 were about 11 m long. The time difference would then be about

$$(22 \text{ m})(3.0 \times 10^4 \text{ m/s})^2/(3.0 \times 10^8 \text{ m/s})^3 \approx 7.0 \times 10^{-16} \text{ s}.$$

For visible light of wavelength $\lambda = 5.5 \times 10^{-7}$ m, say, the frequency would be $f = c/\lambda = (3.0 \times 10^8 \text{ m/s})/(5.5 \times 10^{-7} \text{ m}) = 5.5 \times 10^{14}$ Hz, which means that wave crests pass by a point every $1/(5.5 \times 10^{14} \text{ Hz}) = 1.8 \times 10^{-15}$ s. Thus, with a time difference of 7.0×10^{-16} s, Michelson and Morley should have noted a movement in the interference pattern of $(7.0 \times 10^{-16} \text{ s})/(1.8 \times 10^{-15} \text{ s}) = 0.4$ fringe. They could easily have detected this, since their apparatus was capable of observing a fringe shift as small as 0.01 fringe.

But they found *no significant fringe shift whatever*! They set their apparatus at various orientations. They made observations day and night so that they would be at various orientations with respect to the Sun (due to the Earth's rotation). They tried at different seasons of the year (the Earth at different locations due to its orbit around the Sun). Never did they observe a significant fringe shift.

The null result

This "null" result was one of the great puzzles of physics at the end of the nineteenth century. To explain it was a difficult challenge. One possibility to explain the null result was to apply an idea put forth independently by G. F. Fitzgerald and H. A. Lorentz (in the 1890s) in which they proposed that any length (including the arm of an interferometer) contracts by a factor $\sqrt{1 - v^2/c^2}$ in the direction of motion through the ether. According to Lorentz, this could be due to the ether affecting the forces between the molecules of a substance, which were assumed to be electrical in nature. This theory was eventually replaced by the far more comprehensive theory proposed by Albert Einstein in 1905—the special theory of relativity.

26–3 Postulates of the Special Theory of Relativity

The problems that existed at the turn of the century with regard to electromagnetic theory and Newtonian mechanics were beautifully resolved by Einstein's introduction of the theory of relativity in 1905. Einstein, however, was apparently not influenced directly by the null result of the Michelson–Morley experiment. What motivated Einstein were certain questions regarding electromagnetic theory and light waves. For example, he asked himself: "What would I see if I rode a light beam?" The answer was that instead of a traveling electromagnetic wave, he would see alternating electric and magnetic fields at rest whose magnitude changed in space, but did not change in time. Such fields, he realized, had never been detected and indeed were not consistent with Maxwell's electromagnetic theory. He argued, therefore, that it was unreasonable to think that the speed of light relative to any observer could be reduced to zero, or in fact reduced at all. This idea became the second postulate of his theory of relativity.

Einstein concluded that the inconsistencies he found in electromagnetic theory were due to the assumption that an absolute space exists. In his famous 1905 paper, he proposed doing away completely with the idea of the ether and the accompanying assumption of an absolute reference frame at rest. This proposal was embodied in two postulates. The first postulate was an extension of the Newtonian relativity principle to include not only the laws of mechanics but also those of the rest of physics, including electricity and magnetism:

First postulate (*the relativity principle*): **The laws of physics have the same form in all inertial reference frames.**

The second postulate is consistent with the first:

Second postulate (*constancy of the speed of light*): **Light propagates through empty space with a definite speed c independent of the speed of the source or observer.**

These two postulates form the foundation of Einstein's **special theory of relativity**. It is called "special" to distinguish it from his later "general theory of relativity," which deals with noninertial (accelerating) reference frames (discussed in Chapter 33). The special theory, which is what we discuss here, deals only with inertial frames.

The second postulate may seem hard to accept, for it violates commonsense notions. First of all, we have to think of light traveling through empty space. Giving up the ether is not too hard, however, for after all, it had never been detected. But the second postulate also tells us that the speed of light in vacuum is always the same, 3.00×10^8 m/s, no matter what the speed of the observer or the source. Thus, a person traveling toward or away from a source of light will measure the same speed for that light as someone at rest with respect to the source. This conflicts with our everyday notions, for we would expect to have to add in the velocity of the observer. Part of the problem is that in our everyday experience, we do not measure velocities anywhere near as large as the speed of light. Thus we can't expect our everyday experience to be helpful when dealing with such a high velocity. On the other hand, the Michelson–Morley experiment is fully consistent with the second postulate.[†]

Einstein's proposal has a certain beauty. For by doing away with the idea of an absolute reference frame, it was possible to reconcile classical mechanics with Maxwell's electromagnetic theory. The speed of light predicted by Maxwell's equations *is* the speed of light in vacuum in *any* reference frame.

Einstein's theory required giving up commonsense notions of space and time, and in the following sections we will examine some strange but interesting consequences of Einstein's theory. Our arguments for the most part will be simple ones. We will use a technique that Einstein himself did:

[†]The Michelson–Morley experiment can also be considered as evidence for the first postulate, for it was intended to measure the motion of the Earth relative to an absolute reference frame. Its failure to do so implies the absence of any such preferred frame.

we will imagine very simple experimental situations in which little mathematics is needed. In this way, we can see many of the consequences of relativity theory without getting involved in detailed calculations. Einstein called these "gedanken" experiments, which is German for "thought" experiments. Some of the more mathematical details of special relativity are treated in Appendix E.

26–4 Simultaneity

One of the important consequences of the theory of relativity is that we can no longer regard time as an absolute quantity. No one doubts that time flows onward and never turns back. But, as we shall see in this section and the next, the time interval between two events, and even whether two events are simultaneous, depends on the observer's reference frame.

Two events are said to occur simultaneously if they occur at exactly the same time. But how do we know if two events occur precisely at the same time? If they occur at the same point in space—such as two apples falling on your head at the same time—it is easy. But if the two events occur at widely separated places, it is more difficult to know whether the events are simultaneous since we have to take into account the time it takes for the light from them to reach us. Because light travels at finite speed, a person who sees two events must calculate back to find out when they actually occurred. For example, if two events are *observed* to occur at the same time, but one actually took place farther from the observer than the other, then the former must have occurred earlier, and the two events were not simultaneous.

We will now make use of a simple thought experiment. We assume an observer, called O, is located exactly halfway between points A and B where two events occur, Fig. 26–4. The two events may be lightning that strikes the points A and B, as shown, or any other type of events. For brief events like lightning, only short pulses of light will travel outward from A and B and reach O. O "sees" the events when the pulses of light reach point O. If the two pulses reach O at the same time, then the two events had to be simultaneous. This is because the two light pulses travel at the same speed (postulate 2), and since the distance OA equals OB, the time for the light to travel from A to O and B to O must be the same. Observer O can then definitely state that the two events occurred simultaneously. On the other hand, if O sees the light from one event before that from the other, then it is certain the former event occurred first.

A "thought" experiment

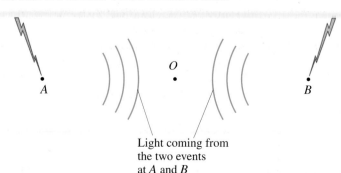
Light coming from the two events at A and B

FIGURE 26–4 A moment after lightning strikes points A and B, the pulses of light are traveling toward the observer O, but O "sees" the lightning only when the light reaches O.

(a)

(b)

FIGURE 26–5 Observers O_1 and O_2, on two different trains (two different reference frames), are moving with relative velocity v. O_2 says that O_1 is moving to the right (a); O_1 says that O_2 is moving to the left (b). Both viewpoints are legitimate—it all depends on your reference frame.

The question we really want to examine is this: if two events are simultaneous to an observer in one reference frame, are they also simultaneous to another observer moving with respect to the first? Let us call the observers O_1 and O_2 and assume they are fixed in reference frames 1 and 2 that move with speed v relative to one another. These two reference frames can be thought of as trains (Fig. 26–5). O_2 says that O_1 is moving to the right with speed v, as in (a); and O_1 says O_2 is moving to the left with speed v, as in (b). Both viewpoints are legitimate according to the relativity principle. (There is, of course, no third point of view which will tell us which one is "really" moving.)

Now suppose two events occur that are observed and measured by both observers. Let us assume again that the two events are the striking of lightning and that the lightning marks both trains where it struck: at A_1 and B_1 on O_1's train, and at A_2 and B_2 on O_2's train. For simplicity, we assume that O_1 happens to be exactly halfway between A_1 and B_1, and that O_2 is halfway between A_2 and B_2. We now put ourselves in one reference frame or the other, from which we make our observations and measurements. Let us put ourselves in O_2's reference frame, so we observe O_1 moving to the right with speed v. Let us also assume that the two events occur *simultaneously* in O_2's frame, and just at the instant when O_1 and O_2 are opposite each other, Fig. 26–6a. A short time later, Fig. 26–6b, the light from A_2 and B_2 reaches O_2 at the same time (we assumed this). Since O_2 knows (or measures) the distances O_2A_2 and O_2B_2 as equal, O_2 knows the two events are simultaneous in the O_2 reference frame.

FIGURE 26–6 Thought experiment on simultaneity. To observer O_2, the reference frame of O_1 is moving to the right. In (a), one lightning bolt strikes the two reference frames at A_1 and A_2, and a second lightning bolt strikes at B_1 and B_2. (b) A moment later, the light from the two events reaches O_2 at the same time, so according to observer O_2, the two bolts of lightning strike simultaneously. But in O_1's reference frame, the light from B_1 has already reached O_1, whereas the light from A_1 has not yet reached O_1. So in O_1's reference frame, the event at B_1 must have preceded the event at A_1. Time is not absolute.

(a)

(b)

But what does observer O_1 observe and measure? From our (O_2) reference frame, we can predict what O_1 will observe. We see that O_1 moves to the right during the time the light is traveling to O_1 from A_1 and B_1. As shown in Fig. 26–6b, we can see from our O_2 reference frame that the light from B_1 has already passed O_1, whereas the light from A_1 has not yet reached O_1. Therefore, it is clear that O_1 will observe the light coming from B_1 before he observes the light coming from A_1. Now O_1's frame is as good as O_2's. Light travels at the same speed c for O_1 as for O_2 (the second postulate)[†]; and in the O_1 reference frame, this speed c is of course the same for light traveling from A_1 to O_1 as it is for light traveling from B_1 to O_1. Furthermore the distance O_1A_1 equals O_1B_1. Hence, since O_1 observes the light from B_1 before he observes the light from A_1 (we established this above, looking from the O_2 reference frame, Fig. 26–6b), then observer O_1 can only conclude that the event at B_1 occurred before the event at A_1. The two events are not simultaneous for O_1, even though they are for O_2.

We thus find that two events which are simultaneous to one observer are not necessarily simultaneous to a second observer.

Simultaneity is relative

It may be tempting to ask: "Which observer is right, O_1 or O_2?" The answer, according to relativity, is that they are *both* right. There is no "best" reference frame we can choose to determine which observer is right. Both frames are equally good. We can only conclude that simultaneity is not an absolute concept, but is relative. We are not aware of it in everyday life, however, because the effect is noticeable only when the relative speed of the two reference frames is very large (near c), or the distances involved are very large.

Because of the principle of relativity, the argument we gave for the thought experiment of Fig. 26–6 can be done from O_1's reference frame as well. In this case, O_1 will be at rest and will see event B_1 occur before A_1. But O_1 will recognize (by drawing a diagram equivalent to Fig. 26–6—try it and see!) that O_2, who is moving with speed v to the left, will see the two events as simultaneous.

26–5 | Time Dilation and the Twin Paradox

The fact that two events simultaneous to one observer may not be simultaneous to a second observer suggests that time itself is not absolute. Could it be that time passes differently in one reference frame than in another? This is, indeed, just what Einstein's theory of relativity predicts, as the following thought experiment shows.

[†]Note that O_1 does not see himself catching up with one light beam and running away from the other (that is O_2's viewpoint of what happens for O_1). O_1 sees both light beams traveling at the same speed, c.

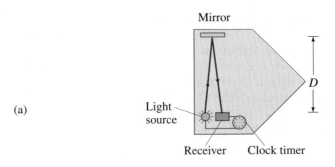

(a)

Mirror

Light source

Receiver   Clock timer

D

(b)

$\sqrt{D^2 + L^2}$

$\sqrt{D^2 + L^2}$

L L

Earth

FIGURE 26–7 Time dilation can be shown by a thought experiment: the time it takes for light to travel over and back on a spaceship is longer for the observer on Earth (b) than for the observer on the spaceship (a).

Figure 26–7 shows a spaceship traveling past Earth at high speed. The point of view of an observer on the spaceship is shown in part (a), and that of an observer on Earth in part (b). Both observers have accurate clocks. The person on the spaceship (a) flashes a light and measures the time it takes the light to travel across the spaceship and return after reflecting from a mirror. The light travels a distance $2D$ at speed c, so the time required, which we call Δt_0, is

$$\Delta t_0 = \frac{2D}{c}.$$

This is the time as measured by the observer on the spaceship.

The observer on Earth, Fig. 26–7b, observes the same process. But to this observer, the spaceship is moving. So the light travels the diagonal path shown going across the spaceship, reflecting off the mirror, and returning to the sender. Although the light travels at the same speed to this observer (the second postulate), it travels a greater distance. Hence the time required, as measured by the observer on Earth, will be *greater* than that measured by the observer on the spaceship. The time interval, Δt, as observed by the observer on Earth can be calculated as follows. In the time Δt, the spaceship travels a distance $2L = v\,\Delta t$ where v is the speed of the spaceship (Fig. 26–7b). Thus, the light travels a total distance on its diagonal path of $2\sqrt{D^2 + L^2}$, and therefore

$$c = \frac{2\sqrt{D^2 + L^2}}{\Delta t} = \frac{2\sqrt{D^2 + v^2(\Delta t)^2/4}}{\Delta t}.$$

We square both sides, and then solve for Δt, to find

$$c^2 = \frac{4D^2}{(\Delta t)^2} + v^2,$$

$$\Delta t = \frac{2D}{c\sqrt{1 - v^2/c^2}}.$$

We combine this with the formula above for Δt_0 ($\Delta t_0 = 2D/c$) and find:

$$\Delta t = \frac{\Delta t_0}{\sqrt{1 - v^2/c^2}}. \qquad \textbf{(26–1)} \qquad \textit{Time-dilation formula}$$

Since $\sqrt{1 - v^2/c^2}$ is always less than 1, we see that $\Delta t > \Delta t_0$. That is, the time interval between the two events (the sending of the light, and its reception on the spaceship) is *greater* for the observer on Earth than for the observer on the spaceship. This is a general result of the theory of relativity, and is known as **time dilation**. Stated simply, the time-dilation effect says that

> **clocks moving relative to an observer are measured by that observer to run more slowly (as compared to clocks at rest).**

Time dilation: moving clocks run slowly

However, we should not think that the clocks are somehow at fault. Time is actually measured to pass more slowly in any moving reference frame as compared to your own. This remarkable result is an inevitable outcome of the two postulates of the theory of relativity.

The concept of time dilation may be hard to accept, for it violates our commonsense understanding. We can see from Eq. 26–1 that the time dilation effect is negligible unless v is reasonably close to c. If v is much less than c, then the term v^2/c^2 is much smaller than the 1 in the denominator of Eq. 26–1, and then $\Delta t \approx \Delta t_0$ (see Example 26–2). The speeds we experience in everyday life are much smaller than c, so it is little wonder we don't ordinarily notice time dilation. Experiments have tested the time-dilation effect, and have confirmed Einstein's predictions. In 1971, for example, extremely precise atomic clocks were flown around the world in jet planes. The speed of the planes (10^3 km/h) was much less than c, so the clocks had to be accurate to nanoseconds (10^{-9} s) in order to detect any time dilation. They were this accurate, and they confirmed Eq. 26–1 to within experimental error. Time dilation had been confirmed decades earlier, however, by observation on "elementary particles" (see Chapter 32) which have very small masses (typically 10^{-30} to 10^{-27} kg) and so require little energy to be accelerated to speeds close to c. Many of these elementary particles are not stable and decay after a time into smaller particles. One example is the muon, whose mean lifetime is 2.2 μs when at rest. Careful experiments showed that when a muon is traveling at high speeds, its lifetime is measured to be longer than when it is at rest, just as predicted by the time-dilation formula.

Why we don't usually notice time dilation

EXAMPLE 26–1 **Lifetime of a moving muon.** (*a*) What will be the mean lifetime of a muon as measured in the laboratory if it is traveling at $v = 0.60c = 1.8 \times 10^8$ m/s with respect to the laboratory? Its mean life at rest is 2.2×10^{-6} s. (*b*) How far does a muon travel in the laboratory, on average, before decaying?

SOLUTION (*a*) If an observer were to move along with the muon (the muon would be at rest to this observer), the muon would have a mean life of 2.2×10^{-6} s. To an observer in the lab, the muon lives longer because of time dilation. From Eq. 26–1 with $v = 0.60c$, we have

$$\Delta t = \frac{\Delta t_0}{\sqrt{1 - \frac{v^2}{c^2}}} = \frac{2.2 \times 10^{-6} \text{ s}}{\sqrt{1 - \frac{0.36c^2}{c^2}}} = \frac{2.2 \times 10^{-6} \text{ s}}{\sqrt{0.64}} = 2.8 \times 10^{-6} \text{ s}.$$

(*b*) At a speed of 1.8×10^8 m/s, classical physics would tell us that with a mean life of $2.2 \, \mu$s, an average muon would travel $d = vt = (1.8 \times 10^8 \text{ m/s})(2.2 \times 10^{-6} \text{ s}) = 400$ m. But relativity predicts an average distance of $(1.8 \times 10^8 \text{ m/s})(2.8 \times 10^{-6} \text{ s}) = 500$ m, and it is this longer distance that is measured experimentally.

We need to make a comment about the use of Eq. 26–1 and the meaning of Δt and Δt_0. The equation is true only when Δt_0 represents the time interval between the two events in a reference frame where the two events occur at *the same point in space* (as in Fig. 26–7a where the two events are the light flash being sent and being received). This time interval, Δt_0, is called the **proper time**. Then Δt in Eq. 26–1 represents the time interval between the two events as measured in a reference frame moving with speed v with respect to the first. In Example 26–1 above, Δt_0 (and not Δt) was set equal to 2.2×10^{-6} s because it is only in the rest frame of the muon that the two events ("birth" and "decay") occur at the same point in space.

Proper time

EXAMPLE 26–2 **Time dilation at 100 km/h.** Let's check time dilation for everyday speeds. A car traveling 100 km/h covers a certain distance in 10.00 s according to the driver's watch. What does an observer on Earth measure for the time interval?

SOLUTION The car's speed relative to Earth is 100 km/h = $(1.00 \times 10^5 \text{ m})/(3600 \text{ s}) = 27.8$ m/s. We set $\Delta t_0 = 10.00$ s in the time-dilation formula (the driver is at rest in the reference frame of the car), and then Δt is

$$\Delta t = \frac{\Delta t_0}{\sqrt{1 - \frac{v^2}{c^2}}} = \frac{10.00 \text{ s}}{\sqrt{1 - \left(\frac{27.8 \text{ m/s}}{3.00 \times 10^8 \text{ m/s}}\right)^2}} = \frac{10.00 \text{ s}}{\sqrt{1 - 8.59 \times 10^{-15}}}.$$

If you put these numbers into a calculator, you will obtain $\Delta t = 10.00$ s, since the denominator differs from 1 by such a tiny amount. Indeed, the time measured by an observer on Earth would be no different from that measured by the driver, even with the best of today's instruments. A computer that could calculate to a large number of decimal places could reveal a difference between Δt and Δt_0. But we can estimate the difference quite easily using the binomial expansion (Appendix A), which says that in a formula of the form $(1 \pm x)^n$, if $x \ll 1$, then to a good approximation,

$$(1 \pm x)^n \approx 1 \pm nx.$$

➡ PROBLEM SOLVING

Use of the binomial expansion

In our time-dilation formula, we have the factor $1/\sqrt{1 - v^2/c^2} = (1 - v^2/c^2)^{-1/2}$. Thus (setting $x = v^2/c^2$ and $n = -\frac{1}{2}$ in the binomial expansion):

$$\Delta t = \Delta t_0 \left(1 - \frac{v^2}{c^2}\right)^{-1/2} \approx \Delta t_0 \left(1 + \frac{1}{2}\frac{v^2}{c^2}\right)$$

$$\approx 10.00 \text{ s} \left[1 + \frac{1}{2}\left(\frac{27.8 \text{ m/s}}{3.00 \times 10^8 \text{ m/s}}\right)^2\right] \approx 10.00 \text{ s} + 4 \times 10^{-15} \text{ s}.$$

So the difference between Δt and Δt_0 is predicted to be 4×10^{-15} s, an unmeasurably small amount.

Time dilation has aroused interesting speculation about space travel. According to classical (Newtonian) physics, to reach a star 100 light-years away would not be possible for ordinary mortals (1 light-year is the distance light can travel in 1 year $= 3.0 \times 10^8$ m/s $\times 3.15 \times 10^7$ s $= 9.5 \times 10^{15}$ m). Even if a spaceship could travel at close to the speed of light, it would take over 100 years to reach such a star. But time dilation tells us that the time involved would be less for an astronaut. In a spaceship traveling at $v = 0.999c$, the time for such a trip would be only about $\Delta t_0 = \Delta t \sqrt{1 - v^2/c^2} = (100 \text{ yr})\sqrt{1 - (0.999)^2} = 4.5 \text{ yr}$. Thus time dilation allows such a trip, but the enormous practical problems of achieving such speeds will not be overcome in the near future.

Notice, in this example, that whereas 100 years would pass on Earth, only 4.5 years would pass for the astronaut on the trip. Is it just the clocks that would slow down for the astronaut? The answer is no. All processes, including life processes, run more slowly for the astronaut according to the Earth observer. But to the astronaut, time would pass in a normal way. The astronaut would experience 4.5 years of normal sleeping, eating, reading, and so on. And people on Earth would experience 100 years of ordinary activity.

Not long after Einstein proposed the special theory of relativity, an apparent paradox was pointed out. According to this **twin paradox**, suppose one of a pair of 20-year-old twins takes off in a spaceship traveling at very high speed to a distant star and back again, while the other twin remains on Earth. According to the Earth twin, the traveling twin will age less. Whereas 20 years might pass for the Earth twin, perhaps only 1 year (depending on the spacecraft's speed) would pass for the traveler. Thus, when the traveler returns, the earthbound twin could expect to be 40 years old whereas the traveling twin would be only 21.

Twin paradox

This is the viewpoint of the twin on the Earth. But what about the traveling twin? If all inertial reference frames are equally good, won't the traveling twin make all the claims the Earth twin does, only in reverse? Can't the astronaut twin claim that since the Earth is moving away at high speed, time passes more slowly on Earth and the twin on Earth will age less? This is the opposite of what the Earth twin predicts. They cannot both be right, for after all the spacecraft returns to Earth and a direct comparison of ages and clocks can be made.

There is, however, not a paradox at all. The consequences of the special theory of relativity—in this case, time dilation—can be applied only by observers in inertial reference frames. The Earth is such a frame (or nearly so), whereas the spacecraft is not. The spacecraft accelerates at the start and end of its trip and, more importantly, when it turns around at the far point of its journey. During these acceleration periods, the spacecraft's predictions based on special relativity are not valid. The twin on Earth is in an inertial frame and can make valid predictions. Thus, there is no paradox. The traveling twin's point of view expressed above is not correct. The predictions of the Earth twin *are* valid, and the prediction that the traveling twin returns having aged less is the proper one.[†]

26–6 Length Contraction

Not only time intervals are different in different reference frames. Space intervals—lengths and distances—are different as well, according to the special theory of relativity, and we illustrate this with a thought experiment.

Observers on Earth watch a spacecraft traveling at speed v from Earth to, say, Neptune, Fig. 26–8a. The distance between the planets, as measured by the Earth observers, is L_0. The time required for the trip, measured from Earth, is $\Delta t = L_0/v$. In Fig. 26–8b we see the point of view of observers on the spacecraft. In this frame of reference, the spaceship is at rest; Earth and Neptune move with speed v. (We assume v is much greater than the relative speed of Neptune and Earth, so the latter can be ignored.) The time between the departure of Earth and arrival of Neptune (as observed from the spacecraft) is the "proper time" (since the two events occur at the same point in space—i.e., on the spacecraft). Therefore the time interval is less for the spacecraft observers than for the Earth observers, because of time dilation. From Eq. 26–1, the time for the trip as viewed by the spacecraft is $\Delta t_0 = \Delta t \sqrt{1 - v^2/c^2}$. Since the spacecraft observers measure the same speed but less time between these two events, they must also measure the

FIGURE 26–8 (a) A spaceship traveling at very high speed from Earth to Neptune, as seen from Earth's frame of reference. (b) As viewed by an observer on the spaceship, Earth and Neptune are moving at the very high velocity v: Earth leaves the spaceship, and a time Δt_0 later planet Neptune arrives at the spaceship. [Note in (b) that each planet does not look shortened because at high speeds we see the trailing edge (as in Fig. 26–10), and the net effect is to leave its appearance as a circle.]

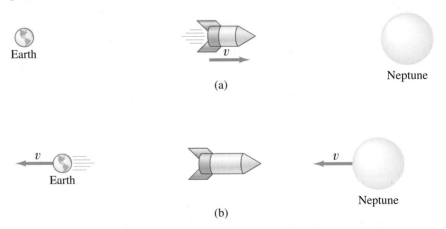

[†]Einstein's general theory of relativity, which deals with accelerating reference frames, confirms this result.

distance as less. If we let L be the distance between the planets as viewed by the spacecraft observers, then $L = v \Delta t_0$. We have already seen that $\Delta t_0 = \Delta t \sqrt{1 - v^2/c^2}$ and $\Delta t = L_0/v$, so we have $L = v \Delta t_0 = v \Delta t \sqrt{1 - v^2/c^2} = L_0 \sqrt{1 - v^2/c^2}$. That is,

$$L = L_0 \sqrt{1 - v^2/c^2}. \qquad (26\text{--}2)$$

Length-contraction formula

This is a general result of the special theory of relativity and applies to lengths of objects as well as to distance. The result can be stated most simply in words as:

the length of an object is measured to be shorter when it is moving relative to the observer than when it is at rest.

Length contraction: moving objects are shorter (in the direction of motion)

This is called **length contraction**. The length L_0 in Eq. 26–2 is called the **proper length**. It is the length of the object—or distance between two points whose *positions are measured at the same time*—as measured by observers at rest with respect to it. Equation 26–2 gives the length L that will be measured by observers when the object travels past them at speed v. It is important to note, however, that length contraction occurs *only along the direction of motion*. For example, the moving spaceship in Fig. 26–8a is shortened in length, but its height is the same as when it is at rest.

Length contraction, like time dilation, is not noticeable in everyday life because the factor $\sqrt{1 - v^2/c^2}$ in Eq. 26–2 differs from 1.00 significantly only when v is very large.

EXAMPLE 26–3 **Painting's contraction.** A rectangular painting measures 1.00 m tall and 1.50 m wide. It is hung on the side wall of a spaceship which is moving past the Earth at a speed of 0.90c. See Fig. 26–9a. (*a*) What are the dimensions of the picture according to the captain of the spaceship? (*b*) What are the dimensions as seen by an observer on the Earth?

SOLUTION (*a*) The painting (as well as everything else in the spaceship) looks perfectly normal to everyone on the spaceship, so the captain sees a 1.00 m by 1.50 m painting.
(*b*) Only the dimension in the direction of motion is shortened, so the height is unchanged at 1.00 m, Fig. 26–9b. The length, however, is contracted to

$$L = L_0 \sqrt{1 - \frac{v^2}{c^2}}$$
$$= (1.50 \text{ m}) \sqrt{1 - (0.90)^2} = 0.65 \text{ m}.$$

So the picture has dimensions 1.00 m × 0.65 m.

(a)

(b)

FIGURE 26–9 Example 26–3.

Equation 26–2 tells us what the length of an object will be *measured* to be when traveling at speed v. The *appearance* of the object is another matter. Suppose, for example, you are traveling to the left past a small building at speed $v = 0.85c$. This is equivalent to the building moving past

(a)

(b)

(c)

FIGURE 26–10 Building seen (a) at rest, and (b) moving at high speed. (c) Diagram explains why the side of the building is seen (see the text).

you to the right at speed v. The building will look narrower (and the same height), but you will also be able to see the side of the building even if you are directly in front of it. This is shown in Fig. 26–10b—part (a) shows the building at rest. The fact that you see the side is not really a relativistic effect, but is due to the finite speed of light. To see how this occurs, we look at Fig. 26–10c which is a top view of the building, looking down. At the instant shown, the observer O is directly in front of the building. Light from points A and B reach O at the same time. If the building were at rest, light from point C could never reach O. But the building is moving at very high speed and does "get out of the way" so that light from C can reach O. Indeed, at the instant shown, light from point C when it was at an earlier location (C' on the diagram) can reach O because the building has moved. In order to reach the observer at the same time as light from A and B, light from C had to leave at an earlier time since it must travel a greater distance. Thus it is light from C' that reaches the observer at the same time as light from A and B. This, then, is how an observer might see both the front and side of an object at the same time even when directly in front of it.[†] It can be shown, by the same reasoning, that spherical objects will actually still have a circular outline even at high speeds. That is why the planets in Fig. 26–8b are drawn round rather than contracted.

26–7 Four-Dimensional Space–Time

Let us imagine a person is on a train moving at a very high speed, say $0.65c$, Fig. 26–11. This person begins a meal at 7:00 and finishes at 7:15, according to a clock on the train. The two events, beginning and ending the meal, take place at the same point on the train. So the proper time between these two events is 15 min. To observers on Earth, the meal will take longer—20 min according to Eq. 26–1. Let us assume that the meal was served on a 20-cm-diameter plate. To observers on the Earth, the plate

FIGURE 26–11 According to an accurate clock on a fast-moving train, a person (a) begins dinner at 7:00 and (b) finishes at 7:15. At the beginning of the meal, observers on Earth set their watches to correspond with the clock on the train. These observers measure the eating time as 20 minutes.

(a)

(b)

[†]It would be an error to think that the building in Fig. 26–10b would look rotated. This is not correct since in that case side A would look shorter than side B. In fact, if the observer is directly in front, these sides appear equal in height. Thus the building looks contracted in its front face, but we also see the side, as described above. Also, though not shown in Fig. 26–10b, the walls of the building would appear curved, because of differing distances from the observer's eye of the various points from top to bottom along a vertical wall.

is only 15 cm wide (length contraction). Thus, to observers on the Earth, the meal looks smaller but lasts longer.

In a sense these two effects, time dilation and length contraction, balance each other. When viewed from the Earth, what the meal seems to lose in size it gains in length of time it lasts. Space, or length, is exchanged for time.

Considerations like this led to the idea of **four-dimensional space–time**: space takes up three dimensions and time is a fourth dimension. Space and time are intimately connected. Just as when we squeeze a balloon we make one dimension larger and another smaller, so when we examine objects and events from different reference frames, a certain amount of space is exchanged for time, or vice versa.

Although the idea of four dimensions may seem strange, it refers to the idea that any object or event is specified by four quantities—three to describe where in space, and one to describe when in time. The really unusual aspect of four-dimensional space–time is that space and time can intermix: a little of one can be exchanged for a little of the other when the reference frame is changed.

It is difficult for most of us to understand the idea of four-dimensional space–time. Somehow we feel, just as physicists did before the advent of relativity, that space and time are completely separate entities. Yet we have found in our thought experiments that they are not completely separate. Our difficulty in accepting this is reminiscent of the situation in the seventeenth century at the time of Galileo and Newton. Before Galileo, the vertical direction, that in which objects fall, was considered to be distinctly different from the two horizontal dimensions. Galileo showed that the vertical dimension differs only in that it happens to be the direction in which gravity acts. Otherwise, all three dimensions are equivalent, a viewpoint we all accept today. Now we are asked to accept one more dimension, time, which we had previously thought of as being somehow different. This is not to say that there is no distinction between space and time. What relativity has shown is that space and time determinations are not independent of one another.

26–8 Momentum and Mass

The three basic mechanical quantities are length, time intervals, and mass. The first two have been shown to be relative—their value depends on the reference frame from which they are measured. We might ask if mass, too, is a relative quantity.

Analysis of collision processes between two particles shows that if we want to preserve conservation of momentum as a principle also in relativity, we must redefine momentum as

$$p = \frac{m_0 v}{\sqrt{1 - \dfrac{v^2}{c^2}}}.$$

(26–3)   *Relativistic momentum*

For speeds much less than the speed of light, Eq. 26–3 gives the classical momentum, $p = m_0 v$. We have written m_0 rather than m because Eq. 26–3 suggests a relativistic interpretation of mass. Namely, that *the mass of an object is measured to increase as its speed increases* according to the formula

Mass increase formula

$$m = \frac{m_0}{\sqrt{1 - v^2/c^2}}.$$ (26–4)

In this **mass-increase** formula, m_0 is the **rest mass** of the object—the mass it has as measured in a reference frame in which it is at rest; and m is the mass it will be measured to have in a reference frame in which it moves at speed v.

Relativistic momentum and mass increase have been tested countless times on tiny elementary particles (such as muons), and they have been found to increase in accord with Eqs. 26–3 and 26–4.

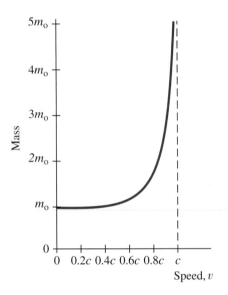

FIGURE 26–12 Mass of a particle (rest mass m_0) as a function of speed v (given as a fraction of c, the speed of light).

EXAMPLE 26–4 **Mass of moving electron.** Calculate the mass of an electron when it has a speed of (*a*) 4.00×10^7 m/s in the CRT of a television set, and (*b*) $0.98c$ in an accelerator used for cancer therapy.

SOLUTION The rest mass of an electron is $m_0 = 9.11 \times 10^{-31}$ kg. (*a*) At $v = 4.00 \times 10^7$ m/s, the electron's mass will be

$$m = \frac{m_0}{\sqrt{1 - \dfrac{v^2}{c^2}}} = \frac{9.11 \times 10^{-31}\text{ kg}}{\sqrt{1 - \dfrac{(4.00 \times 10^7\text{ m/s})^2}{(3.00 \times 10^8\text{ m/s})^2}}} = 9.19 \times 10^{-31}\text{ kg}.$$

Even at such a high speed ($v \approx 0.1c$), the electron's mass is only about 1 percent higher than its rest mass. But in (*b*), we have

$$m = \frac{m_0}{\sqrt{1 - \dfrac{v^2}{c^2}}} = \frac{m_0}{\sqrt{1 - \dfrac{(0.98c)^2}{c^2}}} = \frac{m_0}{\sqrt{1 - (0.98)^2}} = 5.0m_0.$$

An electron traveling at 98 percent the speed of light has a mass five times its rest mass!

Figure 26–12 is a graph of mass versus speed for any particle.

26–9 The Ultimate Speed

A basic result of the special theory of relativity is that the speed of an object cannot equal or exceed the speed of light. That the speed of light is a natural upper speed limit in the universe can be seen from any one of Eqs. 26–1 through 26–4. It is perhaps easiest to see it from Eq. 26–4, the mass-increase formula, $m = m_0/\sqrt{1 - v^2/c^2}$. As an object is accelerated to greater and greater speeds, its mass becomes larger and larger. Indeed, if v were to equal c, the denominator in this equation would be zero and the mass m would become infinite. To accelerate an object up to $v = c$ would thus require infinite energy, and so is not possible. Similarly, Eqs. 26–1 and 26–2 tell us that length would disappear and time become infinite as v approaches c.

26-10 $E = mc^2$; Mass and Energy

When a steady net force is applied to an object of rest mass m_0, the object increases in speed. Since the force is acting over a distance, work is done on the object and its kinetic energy increases. As the speed of the object approaches c, the speed cannot increase indefinitely because it cannot exceed c. On the other hand, the mass of the object increases with increasing speed. That is, the work done on an object not only increases its speed but also contributes to increasing its *mass*. Since the work done on an object increases its energy, this new twist from the theory of relativity leads to the idea that mass is a form of energy, a crucial part of Einstein's theory.

To find the mathematical relationship between mass and energy, Einstein assumed that the work–energy theorem (Chapter 6) is still valid in relativity. That is, the net work done on a particle is equal to its change in kinetic energy (KE). Using this theorem, Einstein showed that at high speeds the formula $\text{KE} = \frac{1}{2}mv^2$ is not correct. You might think that using Eq. 26–4 for m would give $\text{KE} = \frac{1}{2}m_0v^2/\sqrt{1 - v^2/c^2}$, but this formula, too, is wrong. Instead, Einstein showed that the kinetic energy of a particle is given by

$$\text{KE} = mc^2 - m_0c^2, \tag{26–5}$$

Relativistic kinetic energy

where m is the mass of the particle traveling at speed v and m_0 is its rest mass.

But what does the second term in Eq. 26–5—the m_0c^2—mean? Consistent with the idea that mass is a form of energy, Einstein called m_0c^2 the **rest energy** of the object. We can rearrange Eq. 26–5 to get $mc^2 = m_0c^2 + \text{KE}$. We call mc^2 the *total energy* E of the particle (assuming no potential energy), and we see that the total energy equals the rest energy plus the kinetic energy:

$$E = mc^2, \tag{26–6a}$$

$E = mc^2$, mass related to energy

or

$$E = m_0c^2 + \text{KE}. \tag{26–6b}$$

Here we have Einstein's famous formula $E = mc^2$.

For a particle at rest in a given reference frame, its total energy is $E_0 = m_0c^2$, which we have called its rest energy. This formula mathematically relates the concepts of energy and mass. But if this idea is to have any meaning from a practical point of view, then mass ought to be convertible to energy and vice versa. That is, if mass is just one form of energy, then it should be convertible to other forms of energy just as other types of energy are interconvertible. Einstein suggested that this might be possible, and indeed changes of mass to other forms of energy, and vice versa, have been experimentally confirmed countless times. The interconversion of mass and energy is most easily detected in nuclear and elementary particle physics. For example, the neutral pion (π^0) of rest mass 2.4×10^{-28} kg is observed to decay into pure electromagnetic radiation (photons). The π^0 completely disappears in the process. The amount of electromagnetic energy produced is found to be exactly equal to that predicted by Einstein's formula, $E = m_0c^2$. The reverse process is also commonly observed in the laboratory: electromagnetic radiation under certain conditions can be converted into material particles such as electrons. On a larger scale, the energy produced in nuclear power plants is a result of the loss in mass of the uranium fuel as it undergoes the process called fission

Mass and energy interchangeable

(Chapter 31). Even the radiant energy we receive from the Sun is an example of $E = mc^2$; the Sun's mass is continually decreasing as it radiates electromagnetic energy outward.

The relation $E = mc^2$ is now believed to apply to all processes, although the changes are often too small to measure. That is, when the energy of a system changes by an amount ΔE, the mass of the system changes by an amount Δm given by

$$\Delta E = (\Delta m)(c^2).$$

In a chemical reaction where heat is gained or lost, the masses of the reactants and the products will be different. Even when water is heated on a stove, the mass of the water increases very slightly. This example is also easy to understand from the point of view of kinetic theory (Chapter 13), because as heat is added, the temperature and therefore the average speed of the molecules increases; and Eq. 26–4 tells us that the mass also increases.

EXAMPLE 26–5 **Pion's KE.** A π^0 meson ($m_0 = 2.4 \times 10^{-28}$ kg) travels at a speed $v = 0.80c = 2.4 \times 10^8$ m/s. What is its kinetic energy? Compare to a classical calculation.

SOLUTION The mass of the π^0 moving with a speed of $v = 0.80c$ is

$$m = \frac{m_0}{\sqrt{1 - v^2/c^2}} = \frac{2.4 \times 10^{-28}\text{ kg}}{\sqrt{1 - (0.80)^2}} = 4.0 \times 10^{-28}\text{ kg}.$$

Thus its KE is

➡ **PROBLEM SOLVING**

Relativistic KE

$$\text{KE} = (m - m_0)c^2 = (4.0 \times 10^{-28}\text{ kg} - 2.4 \times 10^{-28}\text{ kg})(3.0 \times 10^8\text{ m/s})^2$$
$$= 1.4 \times 10^{-11}\text{ J}.$$

Notice that the units of mc^2 are kg·m²/s², which is the joule. A classical calculation would give $\text{KE} = \frac{1}{2}m_0v^2 = \frac{1}{2}(2.4 \times 10^{-28}\text{ kg})(2.4 \times 10^8\text{ m/s})^2 = 6.9 \times 10^{-12}$ J, about half as much, but this is not a correct result.

EXAMPLE 26–6 **Energy from pion mass.** How much energy would be released if the π^0 meson in the last example is transformed completely into electromagnetic radiation?

SOLUTION The rest energy of the π^0 is

Rest energy

$$E_0 = m_0c^2 = (2.40 \times 10^{-28}\text{ kg})(3.00 \times 10^8\text{ m/s})^2 = 2.16 \times 10^{-11}\text{ J}.$$

This is how much energy would be released if the π^0 decayed at rest. We saw in Chapter 17, Section 17–4, that the energies of atomic particles are often expressed in terms of the electron volt (eV) unit:

$$1\text{ eV} = 1.60 \times 10^{-19}\text{ J}, \qquad \text{and} \qquad 1\text{ MeV} = 10^6\text{ eV} = 1.60 \times 10^{-13}\text{ J}.$$

Thus the rest mass of the π^0 is equivalent to

$$\frac{2.16 \times 10^{-11}\text{ J}}{1.60 \times 10^{-13}\text{ J/MeV}} = 135\text{ MeV}$$

of energy. If the π^0 had $\text{KE} = 1.4 \times 10^{-11}$ J, the total energy released would be $(2.16 + 1.4) \times 10^{-11}$ J $= 3.6 \times 10^{-11}$ J, or 230 MeV.

EXAMPLE 26–7 **Energy from nuclear decay.** The energy required or released in nuclear reactions and decays comes from a change in mass between the initial and final particles. In one type of radioactive decay (Chapter 30), an atom of uranium ($m = 232.03714$ u) decays to an atom of thorium ($m = 228.02873$ u) plus an atom of helium ($m = 4.00260$ u) where the masses given are in atomic mass units (1 u $= 1.6605 \times 10^{-27}$ kg). Calculate the energy released in this decay.

SOLUTION The initial mass is 232.03714 u, and after the decay it is 228.02873 u $+ 4.00260$ u $= 232.03133$ u, so there is a decrease in mass of 0.00581 u. This mass, which equals $(0.00581 \text{ u})(1.66 \times 10^{-27} \text{ kg}) = 9.64 \times 10^{-30}$ kg, is changed into energy. By $E = mc^2$, we have

Energy released in nuclear process

$$E = (9.64 \times 10^{-30} \text{ kg})(3.0 \times 10^8 \text{ m/s})^2 = 8.68 \times 10^{-13} \text{ J}.$$

Since 1 MeV $= 1.60 \times 10^{-13}$ J, the energy released is 5.4 MeV.

Equation 26–5 for the kinetic energy can be written in terms of the speed v of the object with the help of Eq. 26–4:

$$\text{KE} = m_0 c^2 \left(\frac{1}{\sqrt{1 - v^2/c^2}} - 1 \right). \tag{26–7}$$

At low speeds, $v \ll c$, we can expand the square root in Eq. 26–7 using the binomial expansion (see Appendix A or Example 26–2). Then we get

$$\text{KE} \approx m_0 c^2 \left(1 + \frac{1}{2} \frac{v^2}{c^2} + \cdots - 1 \right)$$

$$\approx \tfrac{1}{2} m_0 v^2,$$

where the dots in the first expression represent very small terms in the expansion which we have neglected since we assumed that $v \ll c$. Thus at low speeds, the relativistic form for kinetic energy reduces to the classical form, $\text{KE} = \tfrac{1}{2} m_0 v^2$. This is, of course, what we would like. It makes relativity a more valuable theory in that it can predict accurate results at low speed as well as at high. Indeed, the other equations of special relativity also reduce to their classical equivalents at ordinary speeds: length contraction, time dilation, and mass increase all disappear for $v \ll c$ since $\sqrt{1 - v^2/c^2} \approx 1$.

A useful relation between the total energy E of a particle and its momentum p can also be derived. The relativistic momentum of a particle of mass m and speed v is given by Eq. 26–3:

$$p = mv = \frac{m_0 v}{\sqrt{1 - v^2/c^2}}.$$

Relativistic momentum

Then, since $E = mc^2$, we can write (in the first line we insert "$v^2 - v^2$" which is zero, but will help us):

$$E^2 = m^2 c^4 = m^2 c^2 (c^2 + v^2 - v^2)$$

$$= m^2 c^2 v^2 + m^2 c^2 (c^2 - v^2)$$

$$= p^2 c^2 + \frac{m_0^2 c^4 (1 - v^2/c^2)}{1 - v^2/c^2},$$

or

$$E^2 = p^2 c^2 + m_0^2 c^4, \tag{26–8}$$

Energy–momentum relation

where we have assumed there is no potential energy. Thus, the total energy can be written in terms of the momentum p, or in terms of the kinetic energy (Eq. 26–6).

815

Units:
eV/c for p
eV/c² for m

In the tiny world of atoms and nuclei, it is common to quote energies in eV (electron volts) or multiples such as MeV (10^6 eV). Momentum (see Eq. 26–8) can be quoted in units of eV/c (or MeV/c). And mass can be quoted (from $E = mc^2$) in units of eV/c^2 (or MeV/c^2).

26–11 Relativistic Addition of Velocities

Consider a rocket ship that travels away from the Earth with speed v, and assume that this rocket has fired off a second rocket that travels at speed u' with respect to the first (Fig. 26–13). We might expect that the speed u of rocket 2 with respect to Earth is $u = v + u'$, which in the case shown in the figure is $u = 0.60c + 0.60c = 1.20c$. But, as discussed in Section 26–9, no object can travel faster than the speed of light in any reference frame. Indeed, Einstein showed that since length and time are different in different reference frames, the old addition-of-velocities formula is no longer valid. Instead, the correct formula is

*Relativistic addition of velocities formula (**u** and **v** along same line)*

$$u = \frac{v + u'}{1 + vu'/c^2}$$ (26–9)

for motion along a straight line. We derive this formula in Appendix E. If u' is in the opposite direction from v, then u' must have a minus sign and $u = (v - u')/(1 - vu'/c^2)$.

$u' = 0.60c$ with respect to rocket 1

Earth

$v = 0.60c$ with respect to Earth

FIGURE 26–13 Rocket 2 is fired from rocket 1 with speed $u' = 0.60c$. What is the speed of rocket 2 with respect to the Earth?

EXAMPLE 26–8 **Relative velocity, relativistically.** Calculate the speed of rocket 2 in Fig. 26–13 with respect to Earth.

SOLUTION Rocket 2 moves with speed $u' = 0.60c$ with respect to rocket 1. Rocket 1 has speed $v = 0.60c$ with respect to Earth. The speed of rocket 2 with respect to Earth is therefore

$$u = \frac{0.60c + 0.60c}{1 + \dfrac{(0.60c)(0.60c)}{c^2}} = \frac{1.20c}{1.36} = 0.88c.$$

Notice that Eq. 26–9 reduces to the classical form for velocities small compared to the speed of light since $1 + vu'/c^2 \approx 1$ for v and $u' \ll c$. Thus, $u \approx v + u'$.

Let us test our formula in one more case, that of the speed of light. Suppose that rocket 1 in Fig. 26–13 sends out a beam of light so that

$u' = c$. Equation 26–9 tells us that the speed of this light with respect to Earth is

$$u = \frac{0.60c + c}{1 + \frac{(0.60c)(c)}{c^2}} = \frac{1.60c}{1.60} = c,$$

which is fully consistent with the second postulate of relativity.

26–12 The Impact of Special Relativity

A great many experiments have been performed to test the predictions of the special theory of relativity. Within experimental error, no contradictions have been found. Scientists have therefore accepted relativity as an accurate description of nature.

At speeds much less than the speed of light, the relativistic formulas reduce to the old classical ones, as we have discussed. We would, of course, hope—or rather, insist—that this be true since Newtonian mechanics works so well for objects moving with speeds $v \ll c$. This insistence that a more general theory (such as relativity) give the same results as a more restricted theory (such as classical mechanics which works for $v \ll c$) is called the **correspondence principle**. The two theories must correspond where their realms of validity overlap. Relativity thus does not contradict classical mechanics. Rather, it is a more general theory, of which classical mechanics is now considered to be a limiting case.

Correspondence principle

The importance of relativity is not simply that it gives more accurate results, especially at very high speeds. Much more than that, it has changed the way we view the world. The concepts of space and time are now seen to be relative, and intertwined with one another, whereas before they were considered absolute and separate. Even our concepts of matter and energy have changed: either can be converted to the other. The impact of relativity extends far beyond physics. It has influenced the other sciences, and even the world of art and literature; it has, indeed, entered the general culture.

From a practical point of view, we do not have much opportunity in our daily lives to use the mathematics of relativity. For example, the factor $\sqrt{1 - v^2/c^2}$, which appears in many relativistic formulas, has a value of 0.995 when $v = 0.10c$. Thus, for speeds even as high as $0.10c = 3.0 \times 10^7$ m/s, the factor $\sqrt{1 - v^2/c^2}$ in relativistic formulas gives a numerical correction of less than 1 percent. For speeds less than $0.10c$, or unless mass and energy are interchanged, we thus don't usually need to use the more complicated relativistic formulas, and can use the simpler classical formulas.

The special theory of relativity we have studied in this chapter deals with inertial (nonaccelerating) reference frames. In Chapter 33 we will discuss briefly the more complicated "general theory of relativity" which can deal with noninertial reference frames.

SUMMARY

An **inertial reference frame** is one in which Newton's law of inertia holds. Inertial reference frames can move at constant velocity relative to one another; accelerating reference frames are noninertial.

The **special theory of relativity** is based on two principles: the **relativity principle**, which states that the laws of physics are the same in all inertial reference frames, and the principle of the **constancy of the speed of light**, which states that the speed of light in empty space has the same value in all inertial reference frames.

One consequence of relativity theory is that two events that are simultaneous in one reference frame may not be simultaneous in another. Other effects are **time dilation**: moving clocks are measured to run slowly; **length contraction**: the length of a moving object is measured to be shorter (in its direction of motion) than when it is at rest; *mass increase*: the mass of a body is measured to increase with speed. Quantitatively,

$$L = L_0\sqrt{1 - v^2/c^2}$$

$$\Delta t = \frac{\Delta t_0}{\sqrt{1 - v^2/c^2}}$$

$$m = \frac{m_0}{\sqrt{1 - v^2/c^2}}$$

where L, Δt, and m are the length, time interval, and mass of objects (or events) that are observed as they move by at speed v; L_0, Δt_0, and m_0 are the **proper length**, **proper time**, and **rest mass**—that is, the same quantities as measured in the rest frame of the objects or events. Velocity addition also must be done in a special way. All these effects are significant only at high speeds, close to the speed of light, which itself is the ultimate speed in the universe.

The theory of relativity has changed our notions of space and time, and of mass and energy. Space and time are seen to be intimately connected, with time being the fourth dimension in addition to the three dimensions of space. Mass and energy are interconvertible. The equation

$$E = mc^2$$

tells how much energy E is needed to create a mass m, or vice versa. Said another way, $E = mc^2$ is the amount of energy an object has because of its mass m. The law of conservation of energy must include mass as a form of energy. The kinetic energy of an object moving at speed v is given by

$$\text{KE} = mc^2 - m_0c^2 = \left(\frac{1}{\sqrt{1 - v^2/c^2}} - 1\right)m_0c^2,$$

where m_0 is the rest mass of the object. The momentum p of an object is related to its total energy E (assuming no potential energy) by

$$E^2 = p^2c^2 + m_0^2c^4.$$

QUESTIONS

1. You are in a windowless car in an exceptionally smooth train. Is there any physical experiment you can do in the train car to determine whether you are moving?

2. You might have had the experience of being at a red light when, out of the corner of your eye, you see the car beside you creep forward. Instinctively you stomp on the brake pedal, thinking that you are rolling backward. What does this say about absolute and relative motion?

3. A worker stands on top of a moving railroad car, and throws a heavy ball straight up (from his point of view). Ignoring air resistance, will the ball land on the car or behind it?

4. Does the Earth really go around the Sun? Or is it also valid to say that the Sun goes around the Earth? Discuss in view of the first principle of relativity (that there is no best reference frame).

5. If you were on a spaceship traveling at 0.5c away from a star, at what speed would the starlight pass you?

6. Will two events that occur at the same place and same time for one observer be simultaneous to a second observer moving with respect to the first?

7. Analyze the thought experiment of Section 26–4 from O_1's point of view. (Make a diagram analogous to Fig. 26–6.)

8. The time-dilation effect is sometimes expressed as "moving clocks run slowly." Actually, this effect has nothing to do with motion affecting the functioning of clocks. What then does it deal with?

9. Does time dilation mean that time actually passes more slowly in moving reference frames or that it only *seems* to pass more slowly?

10. A young-looking woman astronaut has just arrived home from a long trip. She rushes up to an old gray-haired man and in the ensuing conversation refers to him as her son. How might this be possible?

11. If you were traveling away from Earth at speed $0.5c$, would you notice a change in your heartbeat? Would your mass, height, or waistline change? What would observers on Earth using telescopes say about you?

12. Discuss how our everyday lives would be different if the speed of light were only 25 m/s.

13. Do mass increase, time dilation, and length contraction occur at ordinary speeds, say 90 km/h?

14. Suppose the speed of light were infinite. What would happen to the relativistic predictions of length contraction, time dilation, and mass increase?

15. Explain how the length-contraction and time-dilation formulas might be used to indicate that c is the limiting speed in the universe.

16. Consider an object of mass m to which is applied a constant force for an indefinite period of time. Discuss how its velocity and mass change with time.

17. A white-hot iron bar is cooled to room temperature. Does its mass change?

18. Does the equation $E = mc^2$ conflict with the conservation of energy principle? Explain.

19. Does $E = mc^2$ apply to particles that travel at the speed of light? Does it apply only to them?

20. An electron is limited to travel at speeds less than c. Does this put an upper limit on the momentum of an electron? If so, what is this upper limit?

21. If mass is a form of energy, does this mean that a spring has more mass when compressed than when relaxed?

22. It is not correct to say that "matter can neither be created nor destroyed." What must we say instead?

23. Is our intuitive notion that velocities simply add, as we did in Section 3–8, completely wrong?

PROBLEMS

SECTIONS 26–5 AND 26–6

1. (I) Lengths and time intervals (as well as mass) depend on the factor

$$\sqrt{1 - v^2/c^2}$$

according to the theory of relativity (Eqs. 26–1, 26–2, 26–4). Evaluate this correction factor for speeds of: (a) $v = 20,000$ m/s (typical speed of a satellite); (b) $v = 0.0100c$; (c) $v = 0.100c$; (d) $v = 0.900c$; (e) $v = 0.990c$; (f) $v = 0.999c$.

2. (I) A spaceship passes you at a speed of $0.850c$. You measure its length to be 48.2 m. How long would it be when at rest?

3. (I) A beam of a certain type of elementary particle travels at a speed of 2.70×10^8 m/s. At this speed, the average lifetime is measured to be 4.76×10^{-6} s. What is the particle's lifetime at rest?

4. (I) If you were to travel to a star 100 light-years from Earth at a speed of 2.60×10^8 m/s, what would you measure this distance to be?

5. (II) You are sitting in your car when a very fast sports car passes you at a speed of $0.37c$. A person in that car says his car is 6.00 m long and yours is 6.21 m long. What do you measure for these two lengths?

6. (II) What is the speed of a beam of pions if their average lifetime is measured to be 4.10×10^{-8} s? At rest, their lifetime is 2.60×10^{-8} s.

7. (II) Suppose you decide to travel to a star 90 light-years away. How fast would you have to travel so the distance would be only 25 light-years?

8. (II) At what speed do the relativistic formulas for length and time intervals differ from classical values by 1.00 percent? (This is a reasonable way to estimate when to do relativistic calculations rather than classical.)

9. (II) Suppose a news report stated that starship *Enterprise* had just returned from a 5-year voyage while traveling at $0.89c$. (a) If the report meant 5.0 years of *Earth time*, how much time elapsed on the ship? (b) If the report meant 5.0 years of *ship time*, how much time passed on Earth?

10. (II) A certain star is 75.0 light-years away. How long would it take a spacecraft traveling $0.950c$ to reach that star from Earth, as measured by observers: (a) on Earth, (b) on the spacecraft? (c) What is the distance traveled according to observers on the spacecraft? (d) What will the spacecraft occupants compute their speed to be from the results of (b) and (c)?

11. (II) A friend of yours travels by you in her fast sports vehicle at a speed of 0.580c. You measure it to be 5.80 m long and 1.20 m high. (a) What will be its length and height at rest? (b) How many seconds would you say elapsed on your friend's watch when 20.0 s passed on yours? (c) How fast did you appear to be traveling to your friend? (d) How many seconds would she say elapsed on your watch when she saw 20.0 s pass on hers?

12. (III) How fast must a pion be moving, on average, to travel 10.0 m before it decays? The average lifetime, at rest, is 2.60×10^{-8} s.

SECTION 26–8

13. (I) What is the mass of a proton traveling at $v = 0.90c$?

14. (I) At what speed will an object's mass be twice its rest mass?

15. (II) At what speed v will the mass of an object be 10 percent greater than its rest mass?

16. (II) Escape velocity from the Earth is 40,000 km/h. What would be the percent increase in mass of a 7.2×10^5-kg spacecraft traveling at that speed?

17. (II) (a) What is the speed of an electron whose mass is 10,000 times its rest mass? Such speeds are reached in the Stanford Linear Accelerator, SLAC. (b) If the electrons travel in the lab through a tube 3.0 km long (as at SLAC), how long is this tube in the electron's reference frame?

SECTION 26–10

18. (I) What is the kinetic energy of an electron whose mass is 3.0 times its rest mass?

19. (I) A certain chemical reaction requires 4.82×10^4 J of energy input for it to go. What is the increase in rest mass of the products over the reactants?

20. (I) When a uranium nucleus at rest breaks apart in the process known as fission in a nuclear reactor, the resulting fragments have a total kinetic energy of about 200 MeV. How much mass was lost in the process?

21. (I) Calculate the rest energy of an electron in joules and in MeV ($1 \text{ MeV} = 1.60 \times 10^{-13}$ J).

22. (I) Calculate the rest mass of a proton in MeV/c^2.

23. (I) The total annual energy consumption in the United States is about 8×10^{19} J. How much mass would have to be converted to energy to fuel this need?

24. (II) How much energy can be obtained from conversion of 1.0 gram of mass? How much mass could this energy raise to a height of 100 m?

25. (II) Show that when the kinetic energy of a particle equals its rest energy, the speed of the particle is about 0.866c.

26. (II) (a) How much work is required to accelerate a proton from rest up to a speed of 0.998c? (b) What would be the momentum of this proton?

27. (II) (a) By how much does the mass of the Earth increase each year as a result solely of the sunlight reaching it? (b) How much mass does the Sun lose per year? (Radiation from the Sun reaches the Earth at a rate of about 1400 W/m^2 of area perpendicular to the energy flow.)

28. (II) Calculate the kinetic energy and momentum of a proton traveling 2.50×10^8 m/s.

29. (II) What is the momentum of a 750-MeV proton (that is, one with KE = 750 MeV)?

30. (II) What is the speed of a proton accelerated by a potential difference of 75 MV?

31. (II) What is the speed of an electron whose KE is 1.00 MeV?

32. (II) What is the speed and apparent rest mass of an electron when it hits a television screen after being accelerated by the 25,000 V of the picture tube?

33. (II) Two identical particles of rest mass m_0 approach each other at equal and opposite speeds, v. The collision is completely inelastic and results in a single particle at rest due to momentum conservation. What is the rest mass of the new particle? How much energy was lost in the collision? How much kinetic energy is lost in this collision?

34. (II) Calculate the mass of a proton ($m_0 = 1.67 \times 10^{-27}$ kg) whose kinetic energy is half its total energy. How fast is it traveling?

35. (II) What is the speed and momentum of an electron ($m_0 = 9.11 \times 10^{-31}$ kg) whose kinetic energy equals its rest energy?

36. (II) Suppose a spacecraft of rest mass 37,000 kg is accelerated to 0.21c. (a) How much kinetic energy would it have? (b) If you used the classical formula for KE, by what percentage would you be in error?

37. (II) Calculate the kinetic energy and momentum of a proton ($m_0 = 1.67 \times 10^{-27}$ kg) traveling 9.8×10^7 m/s. By what percentages would your calculations have been in error if you had used classical formulas?

38. (II) The americium nucleus, $^{241}_{95}$Am, decays to a neptunium nucleus, $^{237}_{93}$Np, by emitting an alpha particle of mass 4.00260 u and kinetic energy 5.5 MeV. Estimate the mass of the neptunium nucleus, ignoring its recoil, given that the americium mass is 241.05682 u.

39. (II) An electron ($m_0 = 9.11 \times 10^{-31}$ kg) is accelerated from rest to speed v by a conservative force. In this process, its potential energy decreases by 7.60×10^{-14} J. Determine the electron's speed, v.

40. (II) Make a graph of the kinetic energy versus momentum for (a) a particle of nonzero rest mass, and (b) a particle with zero rest mass.

41. (II) What magnetic field intensity is needed to keep 900-GeV protons revolving in a circle of radius 1.0 km (at, say, the Fermilab synchrotron)? Use the relativistic mass. The proton's rest mass is 0.938 GeV/c^2. (1 GeV $= 10^9$ eV.)

42. (II) A negative muon traveling at 33 percent the speed of light collides head on with a positive muon traveling at 50 percent the speed of light. The two muons (each of mass 105.7 MeV/c^2) annihilate, and produce electromagnetic energy of what total amount?

43. (II) Show that the energy of a particle of charge e revolving in a circle of radius r in a magnetic field B is given by E (in eV) $= Brc$ in the relativistic limit $(v \approx c)$.

44. (III) Show that the kinetic energy (KE) of a particle of rest mass m_0 is related to its momentum p by the equation $p = \sqrt{(\text{KE})^2 + 2(\text{KE})(m_0c^2)}/c$.

SECTION 26-11

45. (I) A person on a rocket traveling at 0.50c (with respect to the Earth) observes a meteor come from behind and pass her at a speed she measures as 0.50c. How fast is the meteor moving with respect to the Earth?

46. (II) Two spaceships leave the Earth in opposite directions, each with a speed of 0.50c with respect to the Earth. (a) What is the velocity of spaceship 1 relative to spaceship 2? (b) What is the velocity of spaceship 2 relative to spaceship 1?

47. (II) An observer on Earth sees an alien vessel approach at a speed of 0.60c. The *Enterprise* comes to the rescue (Fig. 26–14), overtaking the aliens while moving directly toward Earth at a speed of 0.90c relative to Earth. What is the relative speed of one vessel as seen by the other?

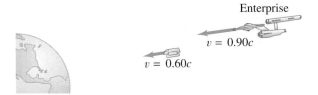

Enterprise

$v = 0.90c$

$v = 0.60c$

FIGURE 26-14   Problem 47.

48. (II) A spaceship leaves Earth traveling 0.65c. A second spaceship leaves the first at a speed of 0.91c with respect to the first. Calculate the speed of the second ship with respect to Earth if it is fired (a) in the same direction the first spaceship is already moving, (b) directly backward toward Earth.

GENERAL PROBLEMS

49. As a rule of thumb, anything traveling faster than about 0.1c is called *relativistic*—i.e., for which the correction using special relativity is a significant effect. Is the electron in a hydrogen atom (radius 0.5×10^{-10} m) relativistic? (Treat the electron as though it were in a circular orbit around the proton.)

50. An atomic clock is taken to the North Pole, while another stays at the Equator. How far will they be out of synchronization after a year has elapsed?

51. The nearest star to Earth is Proxima Centauri, 4.3 light-years away. (a) At what constant velocity must a spacecraft travel from Earth if it is to reach the star in 4.0 years, as measured by travelers on the spacecraft? (b) How long does the trip take according to Earth observers?

52. Derive a formula showing how the density of an object changes with speed v relative to an observer.

53. An airplane travels 1500 km/h around the world, returning to the same place, in a circle of radius essentially equal to that of the Earth. Estimate the difference in time to make the trip as seen by Earth and airplane observers. [*Hint*: Use the binomial expansion, Appendix A.]

54. How many grams of matter would have to be totally destroyed to run a 100-W lightbulb for 1 year?

55. What minimum amount of electromagnetic energy is needed to produce an electron and a positron together? A positron is a particle with the same rest mass as an electron, but has the opposite charge. (Note that electric charge is conserved in this process. See Section 27–4.)

56. A 1.68-kg mass oscillates on the end of a spring whose spring constant is $k = 48.7$ N/m. If this system is in a spaceship moving past Earth at 0.900c, what is its period of oscillation according to (a) observers on the ship, and (b) observers on Earth?

57. An electron ($m_0 = 9.11 \times 10^{-31}$ kg) enters a uniform magnetic field $B = 1.8$ T, and moves perpendicular to the field lines with a speed $v = 0.92c$. What is the radius of curvature of its path?

58. A free neutron can decay into a proton, an electron, and a neutrino. The neutrino's rest mass is zero, and the other masses can be found in the table inside the front cover. Determine the total kinetic energy shared among the three particles when a neutron decays at rest.

59. The Sun radiates energy at a rate of about 4×10^{26} W. (a) At what rate is the Sun's mass decreasing? (b) How long does it take for the Sun to lose a mass equal to that of Earth? (c) Estimate how long the Sun could last if it radiated constantly at this rate.

60. An unknown particle is measured to have a negative charge and a speed of 2.24×10^8 m/s. Its momentum is determined to be 3.07×10^{-22} kg·m/s. Identify the particle by finding its rest mass.

61. How much energy would be required to break a helium nucleus into its constituents, two protons and two neutrons? The rest masses of a proton (including an electron), a neutron, and helium are, respectively, 1.00783 u, 1.00867 u, and 4.00260 u. (This is called the *total binding energy* of the $^4_2$He nucleus.)

62. What is the percentage increase in the mass of a car traveling 110 km/h as compared to at rest?

63. Two protons, each having a speed of $0.935c$ in the laboratory, are moving toward each other. Determine (a) the momentum of each proton in the laboratory, (b) the total momentum of the two protons in the laboratory, and (c) the momentum of one proton as seen by the other proton.

64. A pi meson of rest mass m_π decays at rest into a muon (rest mass m_μ) and a neutrino of zero rest mass. Show that the kinetic energy of the muon is $\text{KE}_\mu = (m_\pi - m_\mu)^2 c^2 / 2m_\pi$.

65. A farm boy studying physics believes that he can fit a 15.0-m-long pole into a 12.0-m-long barn if he runs fast enough (carrying the pole). Can he do it? Explain in detail. How does this fit with the idea that when he is running the barn looks even shorter than 12.0 m?

66. Show analytically that a particle with momentum p and energy E has a speed given by

$$v = \frac{pc^2}{E} = \frac{pc}{\sqrt{m_0^2 c^2 + p^2}}.$$

Electron microscopes produce images based on wave properties of electrons. Since the wavelength of electrons can be much smaller than that of visible light, much greater resolution and magnification can be obtained. The scanning electron microscope can produce images with a three-dimensional quality, as for these brain cells. The large grey objects are neurons. Magnification is about 2000×.

EARLY QUANTUM THEORY AND MODELS OF THE ATOM

27

The second aspect of the revolution that shook the world of physics in the early part of the twentieth century (the first half was Einstein's theory of relativity) was the quantum theory. Unlike the special theory of relativity, the revolution of quantum theory required almost three decades to unfold, and many scientists contributed to its development. It began in 1900 with Planck's quantum hypothesis, and culminated in the mid-1920s with the theory of quantum mechanics of Schrödinger and Heisenberg which has been so effective in explaining the structure of matter. The discovery of the electron in the 1890s, with which we begin this chapter, might be said to mark the beginning of modern physics, and is a sort of precursor to the quantum theory.

27–1 Discovery and Properties of the Electron

Toward the end of the nineteenth century, studies were being done on the discharge of electricity through rarefied gases. One apparatus, diagrammed

FIGURE 27–1 Discharge tube. In some models, one of the screens is the anode (positive plate).

in Fig. 27–1, was a glass tube fitted with electrodes and evacuated so only a small amount of gas remained inside. The negative electrode is called the *cathode*, and the positive one the *anode* (Section 17–10). When a very high voltage was applied to the electrodes, a dark space seemed to extend outward from the cathode toward the opposite end of the tube; and that far end of the tube would glow. If one or more screens containing a small hole were inserted as shown, the glow was restricted to a tiny spot on the end of the tube. It seemed as though something being emitted by the cathode traveled to the opposite end of the tube. These "somethings" were given the name **cathode rays**.

There was much discussion at the time about what these rays might be. Some scientists thought they might resemble light. But the observation that the bright spot at the end of the tube could be deflected to one side by an electric or magnetic field suggested that cathode rays could be charged particles; and the direction of the deflection was consistent with a negative charge. Furthermore, if the tube contained certain types of rarefied gas, the path of the cathode rays was made visible by a slight glow.

e/m measured

Estimates of the charge e of the (assumed) cathode-ray particles, as well as of their charge-to-mass ratio, e/m, had been made by 1897. But in that year, J. J. Thomson (1856–1940) was able to measure e/m directly, using the apparatus shown in Fig. 27–2. Cathode rays are accelerated by a high voltage and then pass between a pair of parallel plates built into the tube. The voltage applied to the plates produces an electric field, and a pair of coils produces a magnetic field. When only the electric field is present, say with the upper plate positive, the cathode rays are deflected upward as in path a in the figure. If only a magnetic field exists, say inward in the figure, the rays are deflected downward along path c. These observations are just what is expected for a negatively charged particle. The force on the rays due to the magnetic field B is (Eq. 20–4)

$$F = evB,$$

where e is the charge and v is the velocity of the cathode rays. In the absence

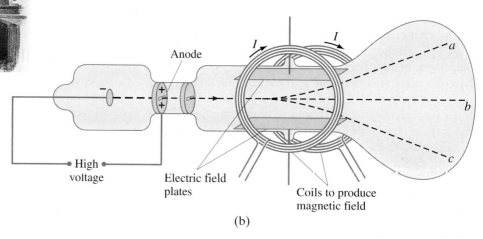

(a)

(b)

FIGURE 27–2 (a) J. J. Thomson and his cathode ray tube. (b) Cathode rays deflected by electric and magnetic fields.

of an electric field, the rays are bent into a curved path, so from $F = ma$, we have

$$evB = \frac{mv^2}{r},$$

and thus

$$\frac{e}{m} = \frac{v}{Br}.$$

The radius of curvature r can be measured, and so can B. The velocity v is found by applying an electric field in addition to the magnetic field. The electric field E is adjusted so that the cathode rays are undeflected and follow path b in Fig. 27–2. In this situation, the force due to the electric field, $F = eE$, is just balanced by the force due to the magnetic field, $F = evB$. Thus we have $eE = evB$ and

$$v = \frac{E}{B}.$$

Combining this with the above equation, we have

$$\frac{e}{m} = \frac{E}{B^2 r}. \qquad\qquad \textbf{(27–1)}$$

The quantities on the right side can all be measured, so that although e and m could not be determined separately, the ratio e/m could be determined. The accepted value today is $e/m = 1.76 \times 10^{11}$ C/kg. Cathode rays soon came to be recognized as beams of particles, which we now call **electrons**.

It is worth noting that the "discovery" of the electron, like many others *"Discovery"* in science, is not quite so obvious as discovering gold or oil. Should the *of the electron* discovery of the electron be credited to the person who first saw a glow in the tube? Or to the person who first called them cathode rays? Perhaps neither one, for they had no conception of the electron as we know it today. In fact, the credit for the discovery is generally given to Thomson, but not because he was the first to see the glow in the tube. Rather it is because he believed that this phenomenon was due to tiny negatively charged particles and made careful measurements on them; furthermore he argued that these particles were constituents of atoms, and not ions or atoms themselves as many thought, and he developed an electron theory of matter. His view is close to what we accept today, and this is why Thomson is credited with the "discovery." Note, however, that neither he nor anyone else ever actually saw an electron itself. We discussed this, briefly, for it illustrates that discovery in science is not always a clear-cut matter. In fact, some philosophers of science think the word "discovery" is not always appropriate, such as in this case.

Thomson believed that an electron was not an atom, but rather a constituent, or part, of an atom. Convincing evidence for this came soon with the determination of the charge and the mass of the cathode ray particles. Thomson's student, J. S. Townsend, made the first direct (but rough) measurements of e in 1897. But it was the more refined **oil-drop experiment** of *Millikan oil-drop experiment* Robert A. Millikan (1868–1953) that yielded a precise value for the charge *to determine e* on the electron and showed that electric charge comes in discrete amounts.

FIGURE 27–3 Millikan's oil-drop experiment.

In this experiment, tiny droplets of mineral oil carrying an electric charge were allowed to fall under gravity between two parallel plates, Fig. 27–3. The electric field E between the plates was adjusted until the oil drop was suspended in midair. The downward pull of gravity, mg, was then just balanced by the upward force due to the electric field. Thus, $qE = mg$, so the charge $q = mg/E$. The mass of the droplet was determined by measuring its terminal velocity in the absence of the electric field (see Problem 4). Sometimes the drop was charged negatively and sometimes positively, suggesting that the drop had acquired or lost electrons (by friction, leaving the atomizer). Millikan's painstaking observations and analysis presented convincing evidence that any electric charge was an integral multiple of a smallest charge, e, that was ascribed to the electron, and that the value of e was 1.6×10^{-19} C. (Today's value of e, as mentioned in Chapter 16, is $e = 1.602 \times 10^{-19}$ C.)

This value for e, combined with the measurement of e/m (see above), gives the mass of the electron to be $(1.6 \times 10^{-19}$ C$)/(1.76 \times 10^{11}$ C/kg$) = 9.1 \times 10^{-31}$ kg. This mass is less than a thousandth the mass of the smallest atom, and thus confirmed the idea that the electron is only a part of an atom. The accepted value today for the mass of the electron is $m_e = 9.11 \times 10^{-31}$ kg. The experimental results that any charge seems to be an integral multiple of e means that electric charge is *quantized* (exists only in discrete amounts), as we discussed in Chapter 16.

27–2 Planck's Quantum Hypothesis

One of the observations that was unexplained at the end of the nineteenth century was the spectrum of light emitted by hot objects. We saw in Section 14–9 that all objects emit radiation whose total intensity is proportional to the fourth power of the Kelvin temperature (T^4). At normal temperatures, we are not aware of this electromagnetic radiation because of its low intensity. At higher temperatures, there is sufficient infrared radiation that we can feel heat if we are close to the object. At still higher temperatures (on the order of 1000 K), objects actually glow, such as a red-hot electric stove burner or the element in a toaster. At temperatures above 2000 K, objects glow with a yellow or whitish color, such as white-hot iron and the filament of a lightbulb. As the temperature increases, the electromagnetic radiation emitted by bodies is strongest at higher and higher frequencies.

Blackbody radiation
The spectrum of light emitted by a hot dense object is shown in Fig. 27–4 for an idealized **blackbody**. A blackbody is a body that would absorb all the radiation falling on it (and so would appear black under reflection when illuminated from outside). The radiation such a blackbody would emit when hot and luminous, called **blackbody radiation** (though not necessarily black in color), is the easiest to deal with, and the radiation approximates that of many real objects. As can be seen, the spectrum contains a continuous range of frequencies. Such a continuous spectrum is emitted by any heated solid or liquid, and even by dense gases. The 6000-K curve in Fig. 27–4, corresponding to the temperature of the surface of the Sun, peaks in the visible part of the spectrum. For lower temperatures, the total

FIGURE 27-4 Spectrum of frequencies emitted by a blackbody at two different temperatures.

radiation drops considerably and the peak occurs at longer wavelengths. Hence the blue end of the visible spectrum (and the UV) is relatively weaker. (This is why objects glow with a red color at around 1000 K.) It is found that the wavelength at the peak of the spectrum, λ_P, is related to the Kelvin temperature T by

$$\lambda_P T = 2.90 \times 10^{-3} \text{ m} \cdot \text{K}. \tag{27-2}$$

This is known as **Wien's law**.

EXAMPLE 27-1 **The Sun's surface temperature.** Estimate the temperature of the surface of our Sun, given that the Sun emits light whose peak is in the visible spectrum at around 500 nm.

SOLUTION Wien's law gives

$$T = \frac{2.90 \times 10^{-3} \text{ m} \cdot \text{K}}{\lambda_P} = \frac{2.90 \times 10^{-3} \text{ m} \cdot \text{K}}{500 \times 10^{-9} \text{ m}} \approx 6000 \text{ K}.$$

EXAMPLE 27-2 **Star color.** Suppose a star has a surface temperature of 32,500 K. What color would this star appear?

SOLUTION From Wien's law we have,

$$\lambda_P = \frac{2.90 \times 10^{-3} \text{ m} \cdot \text{K}}{T} = \frac{2.90 \times 10^{-3} \text{ m} \cdot \text{K}}{3.25 \times 10^4 \text{ K}} = 89.2 \text{ nm}.$$

The peak is in the UV range of the spectrum. In the visible region, the curve will be descending (see Fig. 27–4), so the shortest visible wavelengths will be strongest. Hence the star will appear bluish (or blue-white).

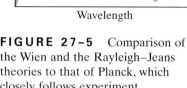

FIGURE 27–5 Comparison of the Wien and the Rayleigh–Jeans theories to that of Planck, which closely follows experiment.

A major problem facing scientists in the 1890s was to explain black-body radiation. Maxwell's electromagnetic theory had predicted that oscillating electric charges produce electromagnetic waves, and the radiation emitted by a hot object could be due to the oscillations of electric charges in the molecules of the material. Although this would explain where the radiation came from, it did not correctly predict the observed spectrum of emitted light. Two important theoretical curves based on classical ideas were those proposed by W. Wien (in 1896) and by Lord Rayleigh (in 1900). The latter was slightly modified later by J. Jeans and since then has been known as the Rayleigh–Jeans theory. As experimental data came in, it became clear that neither Wien's nor the Rayleigh–Jeans formulations were in accord with experiment. Wien's was accurate at short wavelengths but deviated from experiment at longer wavelengths, whereas the reverse was true for the Rayleigh–Jeans theory (see Fig. 27–5).

The break came in late 1900 when Max Planck (1858–1947; Fig. 27–6) proposed an empirical formula that nicely fit the data. He then sought a theoretical basis for the formula and within two months found that he could obtain the formula by making a new and radical (though not so recognized at the time) assumption: that the energy distributed among the oscillating electric charges of the molecules is not continuous, but instead consists of a finite number of very small discrete amounts, each related to the frequency of oscillation by

$$E_{\min} = hf.$$

Here h is a constant, now called **Planck's constant**, whose value was estimated by Planck by fitting his formula for the blackbody radiation curve to experiment. The value accepted today is

$$h = 6.626 \times 10^{-34} \, \text{J·s}.$$

Planck's assumption suggests that the energy of any molecular vibration could be only some whole number multiple of hf:

Planck's quantum hypothesis

$$E = nhf, \qquad n = 1, 2, 3, \cdots. \qquad \textbf{(27–3)}$$

This idea is often called **Planck's quantum hypothesis** ("quantum" means "fixed amount"), although little attention was brought to this point at the time. In fact, it appears that Planck considered it more as a mathematical device to get the "right answer" rather than as a discovery comparable to those of Newton. Planck himself continued to seek a classical explanation for the introduction of h. The recognition that this was an important and radical innovation did not come until later, after about 1905 when others, particularly Einstein, entered the field.

The quantum hypothesis, Eq. 27–3, states that the energy of an oscillator can be $E = hf$, or $2hf$, or $3hf$, and so on, but there cannot be vibrations whose energy lies between these values. That is, energy would not be a continuous quantity as had been believed for centuries; rather it is **quantized**—it exists only in discrete amounts. The smallest amount of energy possible (hf) is called the **quantum of energy**. Another way of expressing the quantum hypothesis is that not just any amplitude of vibration is possible. The possible values for the amplitude are related to the frequency f.

A simple analogy may help. A stringed instrument such as a violin or guitar can be played over a continuous range of frequencies by moving

FIGURE 27–6 Max Planck.

FIGURE 27–7 Ramp versus stair analogy. (a) On a ramp, a box can have continuous values of energy. (b) But on stairs, the box can have only discrete (quantized) values of energy.

(a)　　　　　(b)

your finger along the string. A flute or piano, on the other hand, is "quantized" in the sense that only certain frequencies (notes) can be played. Or compare a ramp, on which a box can be placed at any height, to a flight of stairs on which the box can have only certain discrete amounts of potential energy, as shown in Fig. 27–7.

27–3 Photon Theory of Light and the Photoelectric Effect

In 1905, the same year that he introduced the special theory of relativity, Einstein made a bold extension of the quantum idea by proposing a new theory of light. Planck's work had suggested that the vibrational energy of molecules in a radiating object is quantized with energy $E = nhf$, where n is an integer. Einstein argued that therefore, when light is emitted by a molecular oscillator, its energy of nhf must decrease by an amount hf (or by $2hf$, etc.) to another integer times hf, namely $(n - 1)hf$. Then to conserve energy, the light ought to be emitted in packets or quanta, each with an energy

$$E = hf, \qquad (27\text{–}4)$$

Again h is Planck's constant. Since all light ultimately comes from a radiating source, this suggests that perhaps *light is transmitted as tiny particles*, or **photons**, as they are now called, rather than as waves. This, too, was a radical departure from classical ideas. Einstein proposed a test of the quantum theory of light: quantitative measurements on the photoelectric effect.

The **photoelectric effect** is the phenomenon that when light shines on a metal surface, electrons are emitted from the surface. (The photoelectric effect occurs in other materials, but is most easily observed with metals.) It can be observed using the apparatus shown in Fig. 27–8. A metal plate P and a smaller electrode C are placed inside an evacuated glass tube, called a **photocell**. The two electrodes are connected to an ammeter and a source of emf, as shown. When the photocell is in the dark, the ammeter reads zero. But when light of sufficiently high frequency is shone on the plate, the ammeter indicates a current flowing in the circuit. To explain completion of the circuit, we can imagine electrons, ejected by the impinging radiation, flowing across the tube from the plate to the "collector" C as shown in the diagram.

That electrons should be emitted when light shines on a metal is consistent with the electromagnetic (EM) wave theory of light: the electric

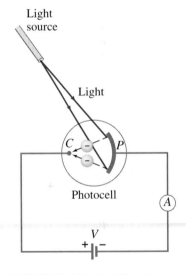

FIGURE 27–8 The photoelectric effect.

Photoelectric effect

field of an EM wave could exert a force on electrons in the metal and eject some of them. Einstein pointed out, however, that the wave theory and the photon theory of light give very different predictions on the details of the photoelectric effect. For example, one thing that can be measured with the apparatus of Fig. 27–8 is the maximum kinetic energy (KE_{max}) of the emitted electrons. This can be done by using a variable voltage source and reversing the terminals so that electrode C is negative and P is positive. The electrons emitted from P will be repelled by the negative electrode, but if this reverse voltage is small enough, the fastest electrons will still reach C and there will be a current in the circuit. If the reversed voltage is increased, a point is reached where the current reaches zero—no electrons have sufficient kinetic energy to reach C. This is called the *stopping potential*, or *stopping voltage*, V_0, and from its measurement, KE_{max} can be determined using conservation of energy (loss of KE = gain in PE):

$$KE_{max} = eV_0.$$

Now let us examine the details of the photoelectric effect from the point of view of the wave theory versus Einstein's particle theory. First the wave theory, assuming monochromatic light. The two important properties of a light wave are its intensity and its frequency (or wavelength). When these two quantities are varied, the wave theory makes the following predictions:

Wave

theory

predictions

1. If the light intensity is increased, the number of electrons ejected and their maximum KE should be increased because the higher intensity means a greater electric field amplitude, and the greater electric field should eject electrons with higher speed.

2. The frequency of the light should not affect the KE of the ejected electrons. Only the intensity should affect KE_{max}.

The photon theory makes completely different predictions. First we note that in a monochromatic beam, all photons have the same energy ($= hf$). Increasing the intensity of the light beam means increasing the number of photons in the beam, but does not affect the energy of each photon as long as the frequency is not changed. According to Einstein's theory, an electron is ejected from the metal by a collision with a single photon. In the process, all the photon energy is transferred to the electron and the photon ceases to exist. Since electrons are held in the metal by attractive forces, some minimum energy W_0 (called the **work function**, which is on the order of a few electron volts for most metals) is required just to get an electron out through the surface. If the frequency f of the incoming light is so low that hf is less than W_0, then the photons will not have enough energy to eject any electrons at all. If $hf > W_0$, then electrons will be ejected and energy will be conserved in the process. That is, the input energy (of the photon), hf, will equal the outgoing KE of the electron plus the energy required to get it out of the metal, W:

$$hf = KE + W. \qquad \textbf{(27–5a)}$$

For the least tightly held electrons, W is the work function W_0, and KE in this equation becomes KE_{max}:

$$hf = \text{KE}_{max} + W_0. \qquad \textbf{(27–5b)}$$

Many electrons will require more energy than the bare minimum (W_0) to get out of the metal, and thus the KE of such electrons will be less than the maximum.

From these considerations, the photon theory makes the following predictions:

1. An increase in intensity of the light beam means more photons are incident, so more electrons will be ejected; but since the energy of each photon is not changed, the maximum KE of electrons is not changed.

2. If the frequency of the light is increased, the maximum KE of the electrons increases linearly, according to Eq. 27–5b. That is,

$$\text{KE}_{max} = hf - W_0.$$

This relationship is plotted in Fig. 27–9.

3. If the frequency f is less than the "cutoff" frequency f_0, where $hf_0 = W_0$, no electrons will be ejected at all, no matter how great the intensity.

These predictions of the photon theory are clearly very different from the predictions of the wave theory. In 1913–1914, careful experiments were carried out by R. A. Millikan. The results were fully in agreement with Einstein's photon theory.

Photon

theory

predictions

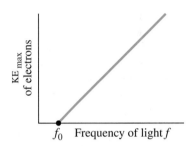

FIGURE 27–9 Photoelectric effect: maximum kinetic energy of ejected electrons increases linearly with frequency of incident light. No electrons are emitted if $f < f_0$.

EXAMPLE 27–3 **Photon energy.** Calculate the energy of a photon of blue light, $\lambda = 450\,\text{nm}$.

SOLUTION Since $f = c/\lambda$, we have

$$E = hf = \frac{hc}{\lambda} = \frac{(6.63 \times 10^{-34}\,\text{J·s})(3.0 \times 10^8\,\text{m/s})}{(4.5 \times 10^{-7}\,\text{m})} = 4.4 \times 10^{-19}\,\text{J},$$

or $(4.4 \times 10^{-19}\,\text{J})/(1.6 \times 10^{-19}\,\text{J/eV}) = 2.7\,\text{eV}$.

EXAMPLE 27–4 **ESTIMATE** **Photons from a lightbulb.** Estimate how many visible light photons a 100-W lightbulb emits per second.

SOLUTION Let's assume an average wavelength in the middle of the visible spectrum, $\lambda \approx 500\,\text{nm}$. The energy emitted in one second ($= 100\,\text{J}$) is $E = nhf$ when n is the number of photons emitted per second and $f = c/\lambda$. Hence

$$n = \frac{E}{hf} = \frac{E\lambda}{hc} = \frac{(100\,\text{J})(500 \times 10^{-9}\,\text{m})}{(6.63 \times 10^{-34}\,\text{J·s})(3.0 \times 10^8\,\text{m/s})} = 2.5 \times 10^{20}.$$

This is an overestimate since much of the 100 J of electric energy input is transformed into heat rather than light. If the efficiency is between 1 percent and 10 percent, then the number of photons emitted is on the order of 10^{19}, still an enormous number.

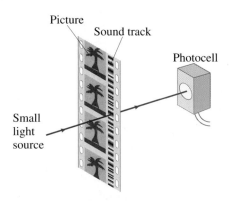

FIGURE 27–10 Optical sound track on movie film. In the projector, light from a small source (different from that for the picture) passes through the sound track on the moving film. The light and dark areas on the sound track vary the intensity of the transmitted light which reaches the photocell, whose current output is then a replica of the original sound. This output is amplified and sent to the loudspeakers. High-quality projectors can show movies containing several parallel sound tracks to go to different speakers around the theater.

➡ **PHYSICS APPLIED**

Alarms,
door openers,
smoke detectors,
light meters,
film soundtrack

➡ **PHYSICS APPLIED**

Photosynthesis

EXAMPLE 27–5 **Photoelectron speed and energy.** What is the maximum kinetic energy and speed of an electron ejected from a sodium surface whose work function is $W_0 = 2.28\,\text{eV}$ when illuminated by light of wavelength: (a) 410 nm; (b) 550 nm?

SOLUTION (a) For $\lambda = 410\,\text{nm}$,

$$hf = \frac{hc}{\lambda} = 4.85 \times 10^{-19}\,\text{J} \quad \text{or} \quad 3.03\,\text{eV}.$$

From Eq. 27–5b, $KE_{max} = 3.03\,\text{eV} - 2.28\,\text{eV} = 0.75\,\text{eV}$, or $1.2 \times 10^{-19}\,\text{J}$. Since $KE = \frac{1}{2}mv^2$ where $m = 9.1 \times 10^{-31}\,\text{kg}$,

$$v = \sqrt{2\,KE/m} = 5.1 \times 10^5\,\text{m/s}.$$

Notice that we used the nonrelativistic equation for KE. If v had turned out to be more than about $0.1c$, our calculation would have been inaccurate by more than a percent or so, and we would probably prefer to redo it using the relativistic form (Eq. 26–7).
(b) For $\lambda = 550\,\text{nm}$, $hf = 3.60 \times 10^{-19}\,\text{J} = 2.25\,\text{eV}$. Since this photon energy is less than the work function, no electrons are ejected.

The photoelectric effect, besides playing an important historical role in confirming the photon theory of light, also has many practical applications. Burglar alarms and automatic door openers often make use of the photocell circuit of Fig. 27–8. When a person interrupts the beam of light, the sudden drop in current in the circuit activates a switch—often a solenoid—which operates a bell or opens the door. UV or IR light is sometimes used in burglar alarms because of its invisibility. Many smoke detectors use the photoelectric effect to detect tiny amounts of smoke that interrupt the flow of light and so alter the electric current. Photographic light meters use this circuit as well. Photocells are used in many other devices, such as absorption spectrophotometers, to measure light intensity. One type of film sound track is a variably shaded narrow section at the side of the film. Light passing through the film is thus "modulated," and the output electrical signal of the photocell detector follows the frequencies on the sound track. See Fig. 27–10. For many applications today, the vacuum-tube photocell of Fig. 27–8 has been replaced by a semiconductor device known as a **photodiode**. In these semiconductors, the absorption of a photon liberates a bound electron, which changes the conductivity of the material, so the current through a photodiode is altered.

The photon theory of light is useful in biology and medicine. Here is an Example.

EXAMPLE 27–6 **Photosynthesis.** In **photosynthesis**, which is the process by which pigments such as chlorophyll in plants capture the energy of sunlight to change CO_2 to useful carbohydrate, about nine photons are needed to transform one molecule of CO_2 to carbohydrate and O_2. Assuming light of wavelength $\lambda = 670\,\text{nm}$ (chlorophyll absorbs most strongly in the range 650 nm to 700 nm), how efficient is the photosynthetic process? The reverse chemical reaction has a heat of combustion of $4.9\,\text{eV/molecule}$ of CO_2.

SOLUTION The energy of nine photons, each of energy $hf = hc/\lambda$, is $(9)(6.6 \times 10^{-34}\,\text{J·s})(3.0 \times 10^8\,\text{m/s})/(6.7 \times 10^{-7}\,\text{m}) = 2.7 \times 10^{-18}\,\text{J}$, or $17\,\text{eV}$. Thus the process is $(4.9\,\text{eV}/17\,\text{eV}) = 29$ percent efficient.

27–4 Photon Interactions; Compton Effect and Pair Production

A number of other experiments were carried out in the early twentieth century that also supported the photon theory. One of these was the **Compton effect** (1923), named after its discoverer, A. H. Compton (1892–1962). Compton scattered short-wavelength light (actually X-rays) from various materials. He found the scattered light had a slightly lower frequency than did the incident light, indicating a loss of energy. This finding, he showed, could be explained on the basis of the photon theory, as incident photons colliding with electrons of the material, Fig. 27–11. He applied the laws of conservation of energy and momentum to such collisions and found that the predicted energies of scattered photons was in accord with experimental results.

To analyze the Compton effect (and other photon interactions), we must recognize that the photon is truly a relativistic particle—it travels at the speed of light. Thus we must use relativistic formulas for dealing with its mass, energy, and momentum. The mass m of any particle is given by $m = m_0/\sqrt{1 - v^2/c^2}$. Since $v = c$ for a photon, the denominator is zero. So the rest mass, m_0, of a photon must also be zero, or its energy $E = mc^2$ would be infinite. Of course, a photon is never at rest. The momentum of a photon, from Eq. 26–8 with $m_0 = 0$, is $E^2 = p^2c^2$, or

$$p = \frac{E}{c}.$$

Since $E = hf$, the momentum of a photon is related to its wavelength by

$$p = \frac{hf}{c} = \frac{h}{\lambda}. \qquad \textbf{(27–6)}$$

Using Eq. 27–6 for momentum of a photon, Compton applied the laws of conservation of momentum and energy to the collision of Fig. 27–11 and derived the following equation for the wavelength of the scattered photons:

$$\lambda' = \lambda + \frac{h}{m_0 c}(1 - \cos \phi), \qquad \textbf{(27–7)}$$

where m_0 is the rest mass of the electron. (The quantity h/m_0c, which has the dimensions of length, is called the **Compton wavelength** of the electron.) We see that the wavelength of scattered photons depends on the angle ϕ at which they are detected. Compton's measurements of 1923 were consistent with this formula. The wave theory of light predicts no such shift: an incoming EM wave of frequency f should set electrons into oscillation at frequency f; and such oscillating electrons would reemit EM waves of this same frequency f (Section 22–3), which would not change with angle (ϕ). Hence the Compton effect adds to the firm experimental foundation for the photon theory of light.

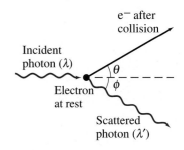

FIGURE 27–11 The Compton effect. A single photon of wavelength λ strikes an electron in some material, knocking it out of its atom. The scattered photon has less energy (since some is given to the electron) and hence has a longer wavelength λ'. Experiments found scattered X-rays of just the wavelengths predicted by conservation of energy and momentum using the photon model.

EXAMPLE 27–7 **X-ray scattering.** X-rays of wavelength 0.140 nm are scattered from a block of carbon. What will be the wavelengths of X-rays scattered at (a) 0°, (b) 90°, (c) 180°?

SOLUTION (a) For $\phi = 0°$, $\cos \phi = 1$, and Eq. 27–7 gives $\lambda' = \lambda = 0.140$ nm. This makes sense since for $\phi = 0°$, there really isn't any collision as the photon goes straight through without interacting.
(b) For $\phi = 90°$, $\cos \phi = 0$, so

$$\lambda' = \lambda + \frac{h}{m_0 c} = 0.140 \text{ nm} + \frac{6.63 \times 10^{-34} \text{ J·s}}{(9.11 \times 10^{-31} \text{ kg})(3.00 \times 10^8 \text{ m/s})}$$

$$= 0.140 \text{ nm} + 2.4 \times 10^{-12} \text{ m} = 0.142 \text{ nm};$$

that is, the wavelength is longer by one Compton wavelength ($= 0.0024$ nm for an electron).
(c) For $\phi = 180°$, which means the photon is scattered backward, returning in the direction from which it came (a direct "head-on" collision), $\cos \phi = -1$, so

$$\lambda' = \lambda + 2\frac{h}{m_0 c} = 0.140 \text{ nm} + 2(0.0024 \text{ nm}) = 0.145 \text{ nm}.$$

When a photon passes through matter, it interacts with the atoms and electrons. There are four important types of interactions that a photon can undergo:

Photon

interactions

1. The photon can be scattered off an electron (or a nucleus) and in the process lose some energy; this is the *Compton effect* (Fig. 27–11). But notice that the photon is not slowed down. It still travels with speed c, but its frequency will be lower.

2. The *photoelectric effect*: a photon may knock an electron out of an atom and in the process itself disappear.

3. The photon may knock an atomic electron to a higher energy state in the atom if its energy is not sufficient to knock the electron out altogether. In this process the photon also disappears, and all its energy is given to the atom. Such an atom is then said to be in an *excited state*, and we shall discuss this more later.

4. *Pair production*: A photon can actually create matter, such as the production of an electron and a positron, Fig. 27–12. (A positron has the same mass as an electron, but the opposite charge, $+e$.)

Pair production

FIGURE 27–12 Pair production: a photon disappears and produces an electron and a positron.

This last process is called **pair production**, and the photon disappears in the process of creating the electron-positron pair. This is an example of rest mass being created from pure energy, and it occurs in accord with Einstein's equation $E = mc^2$. Notice that a photon cannot create an electron alone since electric charge would not then be conserved. The inverse of pair production also occurs: if an electron collides with a positron, the two **annihilate** each other and their energy, including their mass, appears as electromagnetic energy of photons. Because of this process, positrons usually do not last long in nature.

EXAMPLE 27–8 **Pair production.** What is (a) the minimum energy of a photon that can produce an electron–positron pair? (b) What is this photon's wavelength?

SOLUTION (a) Because $E = mc^2$, and the mass created is equal to two electron masses, the photon must have energy

$$E = 2(9.11 \times 10^{-31} \text{ kg})(3.0 \times 10^8 \text{ m/s})^2 = 1.64 \times 10^{-13} \text{ J}$$
$$= 1.02 \text{ MeV}.$$

A photon with less energy cannot undergo pair production.

(b) Since $E = hf = hc/\lambda$, the wavelength of a 1.02-MeV photon is

$$\lambda = \frac{hc}{E} = \frac{(6.6 \times 10^{-34} \text{ J·s})(3.0 \times 10^8 \text{ m/s})}{(1.64 \times 10^{-13} \text{ J})} = 1.2 \times 10^{-12} \text{ m},$$

which is 0.0012 nm. Such photons are in the gamma-ray (or very short X-ray) region of the electromagnetic spectrum.

Pair production cannot occur in empty space, for energy and momentum could not simultaneously be conserved. In Example 27–8, for instance, energy is conserved, but the electron–positron pair would be created with no momentum to carry away the initial momentum of the photon. Indeed, it can be shown that at any energy, an additional massive object, such as an atomic nucleus, must take part in the interaction to carry off some of the momentum.

The Compton effect has been used to diagnose bone disease such as osteoporosis. Gamma rays, which are photons of even shorter wavelength than X-rays, coming from a radioactive source are scattered off bone material. The total intensity of the scattered radiation is proportional to the density of electrons which is in turn proportional to the bone density. Changes in the density of bone material can indicate the onset of osteoporosis.

27–5 Wave–Particle Duality; the Principle of Complementarity

The photoelectric effect, the Compton effect, and other experiments (see, for example, Section 28–9 on X-rays) have placed the particle theory of light on a firm experimental basis. But what about the classic experiments of Young and others (Chapter 24) on interference and diffraction which showed that the wave theory of light also rests on a firm experimental basis?

We seem to be in a dilemma. Some experiments indicate that light behaves like a wave; others indicate that it behaves like a stream of particles. These two theories seem to be incompatible, but both have been shown to have validity. Physicists have finally come to the conclusion that this duality of light must be accepted as a fact of life. It is referred to as the **wave–particle duality**. Apparently, light is a more complex phenomenon than just a simple wave or a simple beam of particles.

To clarify the situation, the great Danish physicist Niels Bohr (1885–1962, Fig. 27–13) proposed his famous **principle of complementarity**. It states that to understand any given experiment, we must use either the wave or the photon theory, but not both. Yet we must be aware of both the wave and particle aspects of light if we are to have a full understanding of light. Therefore these two aspects of light complement one another.

It is not possible to "visualize" this duality. We cannot picture a combination of wave and particle. Instead, we must recognize that the two aspects of light are different "faces" that light shows to experimenters.

FIGURE 27–13 Niels Bohr, walking with Enrico Fermi along the Appian Way outside Rome.

Wave–particle duality

Principle of complementarity

Part of the difficulty stems from how we think. Visual pictures (or models) in our minds are based on what we see in the everyday world. We apply the concepts of waves and particles to light because in the macroscopic world we see that energy is transferred from place to place by these two methods. We cannot see directly whether light is a wave or particle—so we do indirect experiments. To explain the experiments, we apply the models of waves or of particles to the nature of light. But these are abstractions of the human mind. When we try to conceive of what light really "is," we insist on a visual picture. Yet there is no reason why light should conform to these models (or visual images) taken from the macroscopic world. The "true" nature of light—if that means anything—is not possible to visualize. The best we can do is recognize that our knowledge is limited to the indirect experiments, and that in terms of everyday language and images, light reveals both wave and particle properties.

It is worth noting that Einstein's equation $E = hf$ itself links the particle and wave properties of a light beam. In this equation, E refers to the energy of a particle; and on the other side of the equation, we have the frequency f of the corresponding wave.

27–6 Wave Nature of Matter

In 1923, Louis de Broglie (1892–1987; Fig. 27–14) extended the idea of the wave–particle duality. He sensed deeply the symmetry in nature and argued that if light sometimes behaves like a wave and sometimes like a particle, then perhaps those things in nature thought to be particles—such as electrons and other material objects—might also have wave properties. De Broglie proposed that the wavelength of a material particle would be related to its momentum in the same way as for a photon,[†] Eq. 27–6, $p = h/\lambda$. That is, for a particle of mass m traveling with speed v, the wavelength λ is given by

de Broglie wavelength

$$\lambda = \frac{h}{mv}. \qquad (27-8)$$

This is sometimes called the **de Broglie wavelength** of a particle.

EXAMPLE 27–9 **Wavelength of a ball.** Calculate the de Broglie wavelength of a 0.20-kg ball moving with a speed of 15 m/s.

SOLUTION $\lambda = \dfrac{h}{mv} = \dfrac{(6.6 \times 10^{-34}\,\text{J·s})}{(0.20\,\text{kg})(15\,\text{m/s})} = 2.2 \times 10^{-34}\,\text{m}.$

This is an incredibly small wavelength. Even if the speed were extremely small, say 10^{-4} m/s, the wavelength would be about 10^{-29} m. Indeed, the wavelength of any ordinary object is much too small to be measured and detected. The problem is that the properties of waves, such as interference and diffraction, are significant only when the size of objects or slits is not much larger than the wavelength. And there are no known objects or slits to diffract waves only 10^{-30} m long, so the wave properties of ordinary objects go undetected.

But tiny elementary particles, such as electrons, are another matter. Since the mass m appears in the denominator in Eq. 27–8, a very small mass should have a much larger wavelength.

FIGURE 27–14
Louis de Broglie.

[†]De Broglie chose this formula (rather than, say, $E = hf$, which is not consistent with $p = h/\lambda$ for a particle with non-zero rest mass), because it allowed him to explain, or give a reason for, Bohr's model of the atom. This will be discussed in Section 27–11.

EXAMPLE 27–10 **Wavelength of an electron.** Determine the wavelength of an electron that has been accelerated through a potential difference of 100 V.

SOLUTION We assume that the speed of the electron will be much less than c, so we use nonrelativistic mechanics. (If this assumption were to come out wrong, we would have to recalculate using relativistic formulas—see Section 26–10.) The gain in KE will equal the loss in PE, so $\frac{1}{2}mv^2 = eV$ and

$$v = \sqrt{2eV/m} = \sqrt{(2)(1.6 \times 10^{-19} \text{ C})(100 \text{ V})/(9.1 \times 10^{-31} \text{ kg})}$$
$$= 5.9 \times 10^6 \text{ m/s}.$$

Then

$$\lambda = \frac{h}{mv} = \frac{(6.6 \times 10^{-34} \text{ J·s})}{(9.1 \times 10^{-31} \text{ kg})(5.9 \times 10^6 \text{ m/s})} = 1.2 \times 10^{-10} \text{ m},$$

or 0.12 nm.

From this Example, we see that electrons can have wavelengths on the order of 10^{-10} m. Although small, this wavelength can be detected: the spacing of atoms in a crystal is on the order of 10^{-10} m and the orderly array of atoms in a crystal could be used as a type of diffraction grating, as was done earlier for X-rays (see Section 25–11). C. J. Davisson and L. H. Germer performed the crucial experiment; they scattered electrons from the surface of a metal crystal and, in early 1927, observed that the electrons were scattered into a pattern of regular peaks. When they interpreted these peaks as a diffraction pattern, the wavelength of the diffracted electron wave was found to be just that predicted by de Broglie, Eq. 27–8. In the same year, G. P. Thomson (son of J. J. Thomson, who is credited with the discovery of the particle nature of electrons as we saw in Section 27–1), using a different experimental arrangement, also detected diffraction of electrons. (See Fig. 27–15.) Later experiments showed that protons, neutrons, and other particles also have wave properties.

Thus the wave–particle duality applies to material objects as well as to light. The principle of complementarity applies to matter as well. That is, we must be aware of both the particle and wave aspects in order to have an understanding of matter, including electrons. But again we must recognize that a visual picture of a "wave–particle" is not possible.

We might ask ourselves: "What is an electron?" The early experiments of J. J. Thomson (see Section 27–1) indicated a glow in a tube that moved when a magnetic field was applied. The results of these and other experiments were best interpreted as being caused by tiny negatively charged particles which we now call electrons. No one, however, has actually seen an electron directly. The drawings we sometimes make of electrons as tiny spheres with a negative charge on them are merely convenient pictures (now recognized to be inaccurate). Again we must rely on experimental results, some of which are best interpreted using the particle model and others using the wave model. These models are mere pictures that we use to extrapolate from the macroscopic world to the tiny microscopic world of the atom. And there is no reason to expect that these models somehow reflect the reality of an electron. We thus use a wave or a particle model (whichever works best in a situation) so that we can talk about what is happening. But we shouldn't be led to believe that an electron *is* a wave or

FIGURE 27–15 Diffraction pattern of electrons scattered from aluminum foil, as recorded on film.

Wave-particle duality and the complementarity principle apply also to matter

What is an electron?

a particle. Instead, we could say that an electron is the set of its properties that we can measure. Bertrand Russell said it well when he wrote that an electron is "a logical construction".

*27–7 Electron Microscopes

The idea that electrons have wave properties led to the development of the **electron microscope**, which can produce images of much greater magnification than a light microscope. Figures 27–16 and 27–17 are diagrams of two types, the **transmission electron microscope**, which produces a two-dimensional image, and the **scanning electron microscope**, which produces images with a three-dimensional quality. In each design, the objective and eyepiece lenses are actually magnetic fields that exert forces on the electrons to bring them to a focus. The fields are produced by carefully designed current-carrying coils of wire. Photographs using each type are shown in Fig. 27–18.

As discussed in Sections 25–7 and 25–8, the best resolution of details on an object is on the order of the wavelength of the radiation used to view it. Electrons accelerated by voltages on the order of 10^5 V have wavelengths on the order of 0.004 nm. The maximum resolution obtainable would be on this order, but in practice, aberrations in the magnetic lenses limit the resolution in transmission electron microscopes to at best about 0.2 to 0.5 nm. This is still 10^3 times finer than that attainable

FIGURE 27–17 Scanning electron microscope. Scanning coils move an electron beam back and forth across the specimen. Secondary electrons produced when the beam strikes the specimen are collected and modulate the intensity of the beam in the CRT to produce a picture.

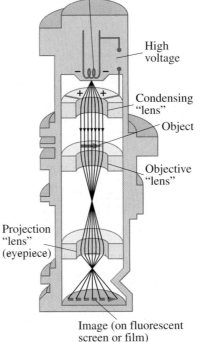

Hot filament (source of electrons)

High voltage

Condensing "lens"

Object

Objective "lens"

Projection "lens" (eyepiece)

Image (on fluorescent screen or film)

FIGURE 27–16 Transmission electron microscope. The magnetic-field coils are designed to be "magnetic lenses," which bend the electron paths and bring them to a focus, as shown.

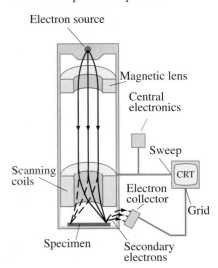

Electron source

Magnetic lens

Central electronics

Sweep

CRT

Scanning coils

Electron collector

Grid

Specimen

Secondary electrons

(a)

(b)

FIGURE 27–18 Electron micrographs (in false color) of neurons of the human cerebral cortex: (a) transmission electron micrograph of synapse (junction) between two neurons ($\approx 40,000\times$); (b) scanning electron micrograph of a single neuron ($\approx 4000\times$).

FIGURE 27–19 Probe tip of scanning tunneling electron microscope moves up and down to maintain constant tunneling current, producing an image of the surface.

➡ PHYSICS APPLIED
STM and AFM

FIGURE 27–20 Image of cellular DNA, magnified about 2 million times, taken with a scanning tunneling microscope. Three turns of the DNA double helix can be seen in this false-color image.

with a light microscope, and corresponds to a useful magnification of about a million. Such magnifications are difficult to attain, and more common magnifications are 10^4 to 10^5. The maximum resolution attainable with a scanning electron microscope is somewhat less, about 5 to 10 nm at best.

The **scanning tunneling electron microscope** (STM), developed in the 1980s, contains a tiny probe, whose tip may be only one (or a few) atoms wide, that is moved across the specimen to be examined in a series of linear passes, like those made by the electron beam in a TV tube (CRT, Section 17–10). The tip, as it scans, remains very close to the surface of the specimen, about 1 nm above it, Fig. 27–19. A small voltage applied between the probe and the surface causes electrons to leave the surface and pass through the vacuum to the probe, by a process known as *tunneling* (discussed in Section 30–12). This "tunneling" current is very sensitive to the gap width, so that a feedback mechanism can be used to raise and lower the probe to maintain a constant electron current. The probe's vertical motion, following the surface of the specimen, is then plotted as a function of position, producing a three-dimensional image of the surface. (See, for example, Fig. 27–20.) Surface features as fine as the size of an atom can be resolved: a resolution better than 0.1 nm laterally and 10^{-2} to 10^{-3} nm vertically. This kind of resolution was not available previously and has given a great impetus to the study of the surface structure of materials. The "topographic" image of a surface actually represents the distribution of electron charge.

The new **atomic force microscope** (AFM) is in many ways similar to an STM, but can be used on a wider range of sample materials. Instead of detecting an electric current, the AFM measures the force between a cantilevered tip and the sample, a force which depends strongly on the tip-sample separation at each point. The tip is moved as for the STM.

≈10^{-10}m

Positively
charged
material

FIGURE 27–21
Plum-pudding model of the atom.

FIGURE 27–22
Ernest Rutherford (left), with
J. J. Thomson.

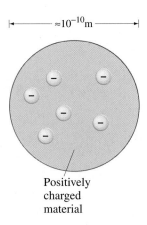

27–8 Early Models of the Atom

The idea that matter is made up of atoms was accepted by most scientists by 1900. With the discovery of the electron in the 1890s, scientists began to think of the atom itself as having a structure and electrons as part of that structure. We now trace, in the remainder of this chapter and the next, the development of our modern understanding of the atom, and of the quantum theory with which it is intertwined.[†]

A typical model of the atom in the 1890s visualized the atom as a homogeneous sphere of positive charge inside of which there were tiny negatively charged electrons, a little like plums in a pudding, Fig. 27–21. J. J. Thomson, soon after his discovery of the electron in 1897, argued that the electrons in this model should be moving.

Around 1911, Ernest Rutherford (1871–1937; Fig. 27–22) and his colleagues performed experiments whose results contradicted Thomson's model of the atom. In these experiments a beam of positively charged "alpha (α) particles" was directed at a thin sheet of metal foil such as gold, Fig. 27–23a. (These newly discovered α particles were emitted by certain radioactive materials and were soon shown to be doubly ionized helium atoms—that is, having a charge of $+2e$; more on alpha particles in Chapter 30.) It was expected from Thomson's model that the alpha particles would not be deflected significantly since electrons are so much lighter than alpha particles, and the alpha particles should not have encountered any massive concentration of positive charge in that model to strongly repel them. The experimental results completely contradicted these predictions. It was found that most of the alpha particles passed through the foil unaffected, as if the foil were mostly empty space. And of those deflected, a few were deflected at very large angles—some even nearly back in the direction from

FIGURE 27–23
(a) Experimental setup for Rutherford's experiment: α particles emitted by radium strike metallic foil and some rebound backward; (b) backward rebound of α particles explained as repulsion from heavy positively charged nucleus.

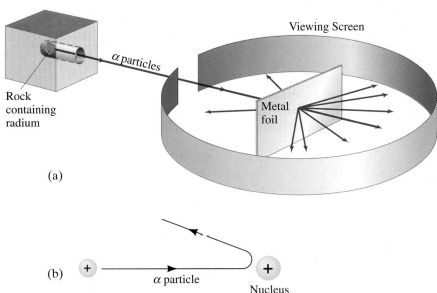

Viewing Screen

α particles

Rock containing radium

Metal foil

(a)

(b)

+

α particle

+

Nucleus

[†]Some readers may say: "Tell us the facts as we know them today, and don't bother us with the historical background and its outmoded theories." Such an approach would ignore the creative aspect of science and thus give a false impression of how science develops. Moreover, it is not really possible to understand today's view of the atom without insight into the concepts that led to it.

which they had come. This could happen, Rutherford reasoned, only if the positively charged alpha particles were being repelled by a massive positive charge concentrated in a very small region of space (see Fig. 27–23b). He theorized that the atom must consist of a tiny but massive positively charged nucleus, containing over 99.9 percent of the mass of the atom, surrounded by electrons some distance away. The electrons would be moving in orbits about the nucleus—much as the planets move around the Sun—because if they were at rest, they would fall into the nucleus due to electrical attraction, Fig. 27–24. Rutherford's experiments suggested that the nucleus must have a radius of about 10^{-15} to 10^{-14} m. From kinetic theory, and especially Einstein's analysis of Brownian movement (see Section 13–1), the radius of atoms was estimated to be about 10^{-10} m. Thus the electrons would seem to be at a distance from the nucleus of about 10,000 to 100,000 times the radius of the nucleus itself (if the nucleus were the size of a baseball, the atom would have the diameter of a big city several kilometers across). So an atom would be mostly empty space.

Rutherford's "planetary" model of the atom (also called the "nuclear model of the atom") was a major step toward how we view the atom today. It was not, however, a complete model and presented some major problems, as we shall see.

Rutherford's planetary model

FIGURE 27–24 Rutherford's model of the atom, in which electrons orbit a tiny positive nucleus (not to scale). The atom is visualized as mostly empty space.

27–9 Atomic Spectra: Key to the Structure of the Atom

Earlier in this chapter we saw that heated solids (as well as liquids and dense gases) emit light with a continuous spectrum of wavelengths. This radiation is assumed to be due to oscillations of atoms and molecules, which are largely governed by the interaction of each atom or molecule with its neighbors.

Rarefied gases can also be excited to emit light. This is done by intense heating, or more commonly by applying a high voltage to a "discharge tube" containing the gas at low pressure, Fig. 27–25. The radiation from excited gases had been observed early in the nineteenth century, and it was found that the spectrum was not continuous, but *discrete*. Since excited gases emit light of only certain wavelengths, when

(a)

(b)

FIGURE 27–25 Gas-discharge tube: (a) is a diagram; (b) photo of actual discharge tube for hydrogen.

(a)

(b)

(c)

FIGURE 27–26 Emission spectra of the gases (a) atomic hydrogen, (b) helium, and (c) the *solar absorption* spectrum.

FIGURE 27–27 Balmer series of lines for hydrogen.

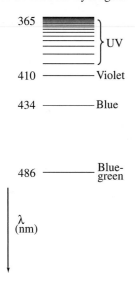

this light is analyzed through the slit of a spectroscope or spectrometer, a **line spectrum** is seen rather than a continuous spectrum. The line spectra emitted by a number of elements in the visible region are shown here in Fig. 27–26, and also in Chapter 24, Fig. 24–27. The **emission spectrum** is characteristic of the material and can serve as a type of "fingerprint" for identification of the gas.

We also saw (Chapter 24) that if a continuous spectrum passes through a gas, dark lines are observed in the emerging spectrum, at wavelengths corresponding to lines normally emitted by the gas. This is called an **absorption spectrum** (Fig. 27–26c), and it became clear that gases can absorb light at the same frequencies at which they emit. Using film sensitive to ultraviolet and to infrared light, it was found that gases emit and absorb discrete frequencies in these regions as well as in the visible.

For our purposes here, the importance of the line spectra is that they are emitted (or absorbed) by gases with low density. In such thin gases, the atoms are far apart on the average and hence the light emitted or absorbed is assumed to be by *individual atoms* rather than through interactions between atoms, as in a solid, liquid, or dense gas. Thus the line spectra serve as a key to the structure of the atom: any theory of atomic structure must be able to explain why atoms emit light only of discrete wavelengths, and it should be able to predict what these wavelengths are.

Hydrogen is the simplest atom—it has only one electron orbiting its nucleus. It also has the simplest spectrum. The spectrum of most atoms shows little apparent regularity. But the spacing between lines in the hydrogen spectrum decreases in a regular way (Fig. 27–27). Indeed, in 1885, J. J. Balmer (1825–1898) showed that the four visible lines in the hydrogen spectrum (with measured wavelengths 656 nm, 486 nm, 434 nm, and 410 nm) fit the following formula

$$\frac{1}{\lambda} = R\left(\frac{1}{2^2} - \frac{1}{n^2}\right), \qquad n = 3, 4, \cdots, \qquad \textbf{(27–9)}$$

where n takes on the values 3, 4, 5, 6 for the four lines, and R, called the **Rydberg constant**, has the value $R = 1.097 \times 10^7 \, \mathrm{m}^{-1}$. Later it was found

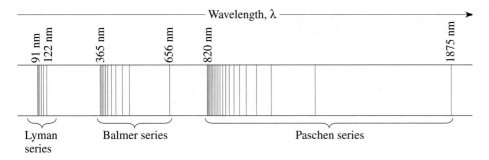

Wavelength, λ

91 nm
122 nm
365 nm
656 nm
820 nm
1875 nm

Lyman series

Balmer series

Paschen series

FIGURE 27–28 Line spectrum of atomic hydrogen. Each series fits the formula $\frac{1}{\lambda} = R\left(\frac{1}{n'^2} - \frac{1}{n^2}\right)$ where $n' = 1$ for the Lyman series, $n' = 2$ for the Balmer series, $n' = 3$ for the Paschen series, and so on; n can take on all integer values from $n = n' + 1$ up to infinity. The only lines in the visible region of the electromagnetic spectrum are part of the Balmer series.

that this **Balmer series** of lines extended into the UV region, ending at $\lambda = 365$ nm, as shown in Fig. 27–27. Balmer's formula, Eq. 27–9, also worked for these lines with higher integer values of n. The lines near 365 nm became too close together to distinguish, but the limit of the series at 365 nm corresponds to $n = \infty$ (so $1/n^2 = 0$ in the formula).

Later experiments on hydrogen showed that there were similar series of lines in the UV and IR regions, and each series had a pattern just like the Balmer series, but at different wavelengths, Fig. 27–28. Each of these series was found to fit a formula with the same form as Eq. 27–9 but with the $1/2^2$ replaced by $1/1^2$, $1/3^2$, $1/4^2$, and so on. For example, the so-called **Lyman series** contains lines with wavelengths from 91 nm to 122 nm (in the UV region) and fits the formula

$$\frac{1}{\lambda} = R\left(\frac{1}{1^2} - \frac{1}{n^2}\right), \qquad n = 2, 3, \cdots.$$

And the wavelengths of the **Paschen series** (in the IR region) fit

$$\frac{1}{\lambda} = R\left(\frac{1}{3^2} - \frac{1}{n^2}\right), \qquad n = 4, 5, \cdots.$$

The Rutherford model, as it stood, was unable to explain why atoms emit line spectra. It had other difficulties as well. According to the Rutherford model, electrons orbit the nucleus, and since their paths are curved the electrons are accelerating. Hence they should give off light like any other accelerating electric charge (Chapter 22). Then, since energy is conserved, the electron's own energy must decrease to compensate. Hence electrons would be expected to spiral into the nucleus. As they spiraled inward, their frequency would increase in a short time and so too would the frequency of the light emitted. Thus the two main difficulties of the Rutherford model are these: (1) it predicts that light of a continuous range of frequencies will be emitted, whereas experiment shows line spectra; (2) it predicts that atoms are unstable—electrons should quickly spiral into the nucleus—but we know that atoms in general are stable, since the matter around us is stable.

Clearly Rutherford's model was not sufficient. Some sort of modification was needed, and it was Niels Bohr who provided it by adding an essential idea—the quantum hypothesis.

27–10 The Bohr Model

Bohr had studied in Rutherford's laboratory for several months in 1912 and was convinced that Rutherford's planetary model of the atom had validity. But in order to make it work, he felt that the newly developing quantum theory would somehow have to be incorporated in it. The work of Planck and Einstein had shown that in heated solids, the energy of oscillating electric charges must change discontinuously—from one discrete energy state to another, with the emission of a quantum of light. Perhaps, Bohr argued, the electrons in an atom also cannot lose energy continuously, but must do so in quantum "jumps." In working out his theory during the next year, Bohr postulated that electrons move about the nucleus in circular orbits, but that only certain orbits are allowed. He further postulated that an electron in each orbit would have a definite energy and would move in the orbit *without radiating energy* (even though this violated classical ideas since accelerating electric charges are supposed to emit EM waves; see Chapter 22). He thus called the possible orbits **stationary states**. Light is emitted, he hypothesized, only when an electron jumps from one stationary state to another of lower energy. When such a jump occurs, a single photon of light is emitted whose energy, since energy is conserved, is given by

Stationary state

$$hf = E_u - E_l, \qquad (27\text{–}10)$$

where E_u refers to the energy of the upper state and E_l the energy of the lower state. See Fig. 27–29.

FIGURE 27–29 An atom emits a photon (energy $= hf$) when its energy changes from E_u to a lower energy E_l.

Bohr set out to determine what energies these orbits would have, since the spectrum of light emitted could then be predicted from Eq. 27–10. In the Balmer formula he had the key he was looking for. Bohr quickly found that his theory would be in accord with the Balmer formula if he assumed that the electron's angular momentum L is quantized and equal to an integer n times $h/2\pi$. As we saw in Chapter 8, angular momentum is given by $L = I\omega$, where I is the moment of inertia and ω is the angular velocity. For a single particle of mass m moving in a circle of radius r with speed v, $I = mr^2$ and $\omega = v/r$; hence, $L = I\omega = (mr^2)(v/r) = mvr$. Bohr's **quantum condition** is

Angular momentum

$$L = mvr_n = n\frac{h}{2\pi}, \qquad n = 1, 2, 3, \cdots, \qquad (27\text{–}11)$$

where n is an integer and r_n is the radius of the nth possible orbit. The allowed orbits are numbered $1, 2, 3, \cdots$, according to the value of n, which is called the **quantum number** of the orbit.

Quantum number, n

Equation 27–11 did not have a firm theoretical foundation. Bohr had searched for some "quantum condition," and such tries as $E = hf$ (where E represents the energy of the electron in an orbit) did not give results in accord with experiment. Bohr's reason for using Eq. 27–11 was simply that it worked; and we now look at how.

An electron in a circular orbit of radius r_n (Fig. 27–30) would have a centripetal acceleration v^2/r_n produced by the electrical force of attraction between the negative electron and the positive nucleus. This force is given by Coulomb's law,

$$F = k\frac{(Ze)(e)}{r^2},$$

where $k = 1/4\pi\epsilon_0 = 9.00 \times 10^9 \text{ N·m}^2/\text{C}^2$. The charge on the electron is $q_1 = -e$, and that on the nucleus is $q_2 = +Ze$, where Z is the number of positive charges[†] (i.e., protons). For the hydrogen atom, $Z = +1$.

In Newton's second law, $F = ma$, we substitute $a = v^2/r_n$ and Coulomb's law for F, and obtain

$$F = ma$$

$$k\frac{Ze^2}{r_n^2} = \frac{mv^2}{r_n}.$$

We solve this for r_n, and then substitute for v from Eq. 27–11 (which says $v = nh/2\pi mr_n$):

$$r_n = \frac{kZe^2}{mv^2} = \frac{kZe^2 4\pi^2 mr_n^2}{n^2h^2}.$$

We solve for r_n (it appears on both sides, so we cancel one of them) and find

$$r_n = \frac{n^2h^2}{4\pi^2 mkZe^2} = \frac{n^2}{Z}r_1 \qquad \textbf{(27–12)}$$

where

$$r_1 = \frac{h^2}{4\pi^2 mke^2}. \qquad \textbf{(27–13a)}$$

Equation 27–12 gives the radii of all possible orbits. The smallest orbit is for $n = 1$, and for hydrogen ($Z = 1$) has the value

$$r_1 = \frac{(1)^2(6.626 \times 10^{-34} \text{ J·s})^2}{4(3.14)^2(9.11 \times 10^{-31} \text{ kg})(9.00 \times 10^9 \text{ N·m}^2/\text{C}^2)(1)(1.602 \times 10^{-19} \text{ C})^2}$$

$$r_1 = 0.529 \times 10^{-10} \text{ m}. \qquad \textbf{(27–13b)}$$

The radius of the smallest orbit in hydrogen, r_1, is sometimes called the **Bohr radius**. From Eq. 27–12, we see that the radii of the larger orbits[‡] increase as n^2, so

$$r_2 = 4r_1 = 2.12 \times 10^{-10} \text{ m},$$

$$r_3 = 9r_1 = 4.76 \times 10^{-10} \text{ m},$$

$$\vdots$$

$$r_n = n^2 r_1.$$

FIGURE 27–30 Electric force (Coulomb's law) keeps the negative electron in orbit around the positively charged nucleus.

Bohr radius

[†]We include Z in our derivation so that we can treat other single-electron ("hydrogenlike") atoms such as the ions He^+ ($Z = 2$) and Li^{2+} ($Z = 3$). Helium in the neutral state has two electrons; if one electron is missing, the remaining He^+ ion consists of one electron revolving around a nucleus of charge $+2e$. Similarly, doubly ionized lithium, Li^{2+}, also has a single electron, and in this case $Z = 3$.

[‡]Be careful not to believe that these well-defined orbits actually exist. The Bohr model is only a model, not reality. The idea of electron orbits was rejected a few years later, and today electrons are better thought of as forming "clouds," as discussed in Chapter 28.

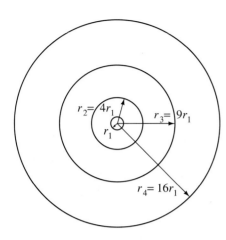

FIGURE 27–31 Possible orbits in the Bohr model of hydrogen; $r_1 = 0.529 \times 10^{-10}$ m.

The first four orbits are shown in Fig. 27–31. Notice that, according to Bohr's model, an electron can exist only in the orbits given by Eq. 27–12. There are no allowable orbits in between.

In each of its possible orbits, the electron would have a definite energy, as the following calculation shows. The total energy equals the sum of the kinetic and potential energies. The potential energy of the electron is given by PE $= qV = -eV$, where V is the potential due to a point charge $+Ze$ as given by Eq. 17–3: $V = kQ/r = kZe/r$. So

$$\text{PE} = -eV = -k\frac{Ze^2}{r}.$$

The total energy E_n for an electron in the nth orbit of radius r_n is the sum of the kinetic and potential energies:

$$E_n = \frac{1}{2}mv^2 - \frac{kZe^2}{r_n}.$$

When we substitute v from Eq. 27–11 and r_n from Eq. 27–12 into this equation, we obtain

Energy levels

$$E_n = -\frac{2\pi^2 Z^2 e^4 m k^2}{h^2}\frac{1}{n^2} \qquad n = 1, 2, 3, \cdots \qquad \textbf{(27–14)}$$

or

$$E_n = \frac{Z^2}{n^2} E_1 \qquad n = 1, 2, 3, \cdots$$

where E_1 is the lowest energy level ($n = 1$) for hydrogen ($Z = 1$). The value of E_1 is

Ground state of hydrogen

$$E_1 = -\frac{2\pi^2 e^4 m k^2}{h^2} = -2.17 \times 10^{-18}\,\text{J} = -13.6\,\text{eV},$$

where we have converted joules to electron volts, as is customary in atomic physics. Since n^2 appears in the denominator of Eq. 27–14, the energies of the larger orbits in hydrogen ($Z = 1$) are given by

$$E_n = \frac{-13.6\,\text{eV}}{n^2}.$$

For example,

Excited states (first two)

$$E_2 = \frac{-13.6\,\text{eV}}{4} = -3.40\,\text{eV},$$

$$E_3 = \frac{-13.6\,\text{eV}}{9} = -1.51\,\text{eV}.$$

We see that not only are the orbit radii quantized, but from Eq. 27–14 so is the energy. The quantum number n that labels the orbit radii also labels the energy levels. The lowest **energy level** or **energy state** has energy E_1, and is called the **ground state**. The higher states, E_2, E_3, and so on, are called **excited states**.

Notice that although the energy for the larger orbits has a smaller numerical value, all the energies are less than zero. Thus, $-3.4\,\text{eV}$ is a greater energy than $-13.6\,\text{eV}$. Hence the orbit closest to the nucleus (r_1) has the lowest energy. The reason the energies have negative values has to do with the way we defined the zero for potential energy. For two point charges, $PE = kq_1q_2/r$ corresponds to zero PE when the two charges are infinitely far apart. Thus, an electron that can just barely be free from the atom by reaching $r = \infty$ (or, at least, far from the nucleus) with zero KE will have $E = 0$, corresponding to $n = \infty$ in Eq. 27–14. If an electron is free and has kinetic energy, then $E > 0$. To remove an electron that is part of an atom requires an energy input (otherwise atoms would not be stable). Since $E \geq 0$ for a free electron, then an electron bound to an atom must have $E < 0$. That is, energy must be added to bring its energy up, from a negative value, to at least zero in order to free it.

The minimum energy required to remove an electron from the ground state of an atom is called the **binding energy** or **ionization energy**. The ionization energy for hydrogen has been measured to be 13.6 eV, and this corresponds precisely to removing an electron from the lowest state, $E_1 = -13.6$ eV, up to $E = 0$ where it can be free.

Binding energy
Ionization energy

It is useful to show the various possible energy values as horizontal lines on an energy-level diagram.[†] This is shown for hydrogen in Fig. 27–32. The electron in a hydrogen atom can be in any one of these levels according to Bohr theory. But it could never be in between, say at -9.0 eV. At room temperature, nearly all H atoms will be in the ground state ($n = 1$). At higher temperatures, or during an electric discharge when there are many collisions between free electrons and atoms, many atoms can be in excited states ($n > 1$). Once in an excited state, an atom's electron can jump down to a lower state, and give off a photon in the process. This is, according to the Bohr model, the origin of the emission spectra of excited gases. The vertical arrows in Fig. 27–32 represent the

Line spectra
emission
explained

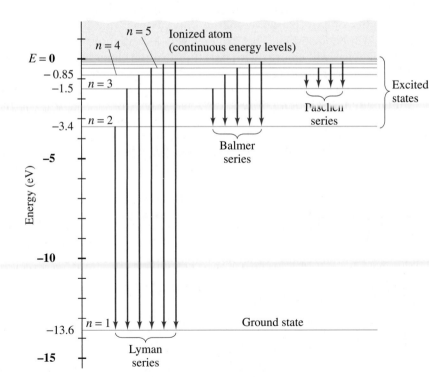

FIGURE 27–32 Energy-level diagram for the hydrogen atom, showing origin of spectral lines for the Lyman, Balmer, and Paschen series.

[†]Note that above $E = 0$, an electron is free and can have any energy (E is not quantized). Thus there is a continuum of energy states above $E = 0$, as indicated in the energy-level diagram of Fig. 27–32.

transitions or jumps that correspond to the various observed spectral lines. For example, an electron jumping from the level $n = 3$ to $n = 2$ would give rise to the 656-nm line in the Balmer series, and the jump from $n = 4$ to $n = 2$ would give rise to the 486-nm line (see Fig. 27–27). We can predict wavelengths of the spectral lines emitted by combining Eq. 27–10 with Eq. 27–14. Since $hf = hc/\lambda$, we have from Eq. 27–10

$$\frac{1}{\lambda} = \frac{hf}{hc} = \frac{1}{hc}(E_n - E_{n'}),$$

and then using Eq. 27–14,

$$\frac{1}{\lambda} = \frac{2\pi^2 Z^2 e^4 m k^2}{h^3 c}\left(\frac{1}{n'^2} - \frac{1}{n^2}\right), \qquad \textbf{(27–15)}$$

where n refers to the upper state and n' to the lower state. This theoretical formula has the same form as the experimental Balmer formula, Eq. 27–9, with $n' = 2$. Thus we see that the Balmer series of lines corresponds to transitions or "jumps" that bring the electron down to the second energy level. Similarly, $n' = 1$ corresponds to the Lyman series and $n' = 3$ to the Paschen series (see Fig. 27–32). When the constant $(2\pi^2 Z^2 e^4 m k^2/h^3 c)$ in Eq. 27–15 is evaluated with $Z = 1$, it is found to have the measured value of the Rydberg constant, $R = 1.0974 \times 10^7 \, \text{m}^{-1}$ in Eq. 27–9, in accord with experiment (see Problem 52).

The great success of Bohr's theory is that it explains why atoms emit line spectra and accurately predicts, for hydrogen, the wavelengths of emitted light. The Bohr theory also explains absorption spectra: photons of just the right wavelength can knock an electron from one energy level to a higher one. To conserve energy, only photons that have just the right energy will be absorbed. This explains why a continuous spectrum of light entering a gas will emerge with dark (absorption) lines at frequencies that correspond to emission lines.

Absorption lines explained

The Bohr theory also ensures the stability of atoms. It establishes stability by fiat: the ground state is the lowest state for an electron and there is no lower energy level to which it can go and emit more energy. Finally, as we saw above, the Bohr theory accurately predicts the ionization energy of 13.6 eV for hydrogen. However, the Bohr theory was not so successful for other atoms, and has been superseded as we shall discuss in the next chapter.

EXAMPLE 27–11 **Wavelength of a Lyman line.** Use Fig. 27–32 to determine the wavelength of the first Lyman line, the transition from $n = 2$ to $n = 1$. In what region of the electromagnetic spectrum does this lie?

SOLUTION In this case, $hf = E_2 - E_1 = 13.6 \, \text{eV} - 3.4 \, \text{eV} = 10.2 \, \text{eV} = 1.63 \times 10^{-18}$ J. Since $\lambda = c/f$, we have

$$\lambda = \frac{c}{f} = \frac{hc}{E_2 - E_1} = \frac{(6.63 \times 10^{-34}\,\text{J}\cdot\text{s})(3.00 \times 10^8\,\text{m/s})}{1.63 \times 10^{-18}\,\text{J}}$$

$$= 1.22 \times 10^{-7} \, \text{m},$$

or 122 nm, which is in the UV region. See Fig. 27–28.

EXAMPLE 27–12 **Wavelength of a Balmer line.** Determine the wavelength of light emitted when a hydrogen atom makes a transition from the $n = 6$ to the $n = 2$ energy level according to the Bohr model.

SOLUTION We can use Eq. 27–15 or its equivalent, Eq. 27–9, with $R = 1.097 \times 10^7 \, \text{m}^{-1}$. Thus

$$\frac{1}{\lambda} = (1.097 \times 10^7 \, \text{m}^{-1})\left(\frac{1}{4} - \frac{1}{36}\right) = 2.44 \times 10^6 \, \text{m}^{-1}.$$

So $\lambda = 1/(2.44 \times 10^6 \, \text{m}^{-1}) = 4.10 \times 10^{-7} \, \text{m}$ or 410 nm. This is the fourth line in the Balmer series, Fig. 27–27, and is violet in color.

EXAMPLE 27–13 **Absorption wavelength.** Use Figure 27–32 to determine the maximum wavelength that hydrogen in its ground state can absorb. What would be the next smaller wavelength that would work?

SOLUTION Maximum λ corresponds to minimal energy, and thus the jump from the ground state to the first excited state (Fig. 27–32) for which the energy is $13.6 \, \text{eV} - 3.4 \, \text{eV} = 10.2 \, \text{eV}$; the required wavelength, as we saw in Example 27–11, is 122 nm. The next possibility is to jump from the ground state to the second excited state, which requires $13.6 \, \text{eV} - 1.5 \, \text{eV} = 12.1 \, \text{eV}$ and corresponds to a wavelength

$$\lambda = \frac{c}{f} = \frac{hc}{E_3 - E_1}$$

$$= \frac{(6.63 \times 10^{-34} \, \text{J·s})(3.00 \times 10^8 \, \text{m/s})}{(12.1 \, \text{eV})(1.60 \times 10^{-19} \, \text{J/eV})} = 103 \, \text{nm}.$$

EXAMPLE 27–14 **Ionization energy.** Use the Bohr model to determine the ionization energy of the He^+ ion, which has a single electron. Also calculate the minimum wavelength a photon must have to cause ionization.

SOLUTION We want to determine the minimum energy required to lift the electron from its ground state and to barely reach the free state at $E = 0$. The ground state energy of He^+ is given by Eq. 27–14 with $n = 1$ and $Z = 2$. Since all the symbols in Eq. 27–14 are the same as for the calculation for hydrogen, except that Z is 2 instead of 1, we see that E_1 will be $Z^2 = 2^2 = 4$ times the E_1 for hydrogen. That is,

$$E_1 = 4(-13.6 \, \text{eV}) = -54.4 \, \text{eV}.$$

Thus, to ionize the He^+ ion should require 54.4 eV, and this value agrees with experiment. The minimum wavelength photon that can cause ionization will have energy $hf = 54.4 \, \text{eV}$ and wavelength $\lambda = c/f = hc/hf = (6.63 \times 10^{-34} \, \text{J·s})(3.00 \times 10^8 \, \text{m/s})/(54.4 \, \text{eV})(1.60 \times 10^{-19} \, \text{J/eV}) = 22.8 \, \text{nm}$. If the atom absorbed a photon of greater energy (wavelength shorter than 22.8 nm), the atom could still be ionized and the freed electron would have kinetic energy of its own.

In this last Example, we saw that E_1 for the He$^+$ ion is four times more negative than that for hydrogen. Indeed, the energy-level diagram for He$^+$ looks just like that for hydrogen, Fig. 27–32, except that the numerical values for each energy level are four times larger. It is important to note, however, that we are talking here about the He$^+$ *ion*. Normal (neutral) helium has two electrons and its energy level diagram is entirely different.

CONCEPTUAL EXAMPLE 27–15 **Hydrogen at 20° C.** Estimate the average kinetic energy of hydrogen atoms (or molecules) at room temperature, and use the result to explain why nearly all H atoms are in the ground state at room temperature, and hence emit no light.

RESPONSE According to kinetic theory (Chapter 13), the average KE of atoms or molecules in a gas is given by Eq. 13–8:

$$\overline{KE} = \tfrac{3}{2} kT,$$

where $k = 1.38 \times 10^{-23}$ J/K is Boltzmann's constant, and T is the kelvin (absolute) temperature. Room temperature is about $T = 300$ K, so

$$\overline{KE} = \tfrac{3}{2}(1.38 \times 10^{-23}\,\text{J/K})(300\,\text{K}) = 6.2 \times 10^{-21}\,\text{J},$$

or, in electron volts:

$$\overline{KE} = \frac{6.2 \times 10^{-21}\,\text{J}}{1.6 \times 10^{-19}\,\text{J/eV}} = 0.04\,\text{eV}.$$

The average KE is thus very small compared to the energy between the ground state and the next higher energy state (13.6 eV − 3.4 eV = 10.2 eV). Any atoms in excited states emit light and eventually fall to the ground state. Once in the ground state, collisions with other atoms can transfer energy of only 0.04 eV on the average. A small fraction of atoms can have much more energy (see Section 13–12 on the distribution of molecular speeds), but even a KE that is 10 times the average is not nearly enough to excite atoms above the ground state. Thus, at room temperature, nearly all atoms are in the ground state. Atoms can be excited to upper states by very high temperatures, or by passing a current of high energy electrons through the gas, as in a discharge tube (Fig. 27–25).

We should note that Bohr made some radical assumptions that were at variance with classical ideas. He assumed that electrons in fixed orbits do not radiate light even though they are accelerating (moving in a circle), and he assumed that angular momentum is quantized. Furthermore, he was not able to say how an electron moved when it made a transition from one energy level to another. On the other hand, there is no real reason to expect that in the tiny world of the atom electrons would behave as ordinary-sized objects do. Nonetheless, he felt that where quantum theory overlaps with the macroscopic world, it should predict classical results. This is the **correspondence principle**, already mentioned in regard to relativity (Section 26–12). This principle does work for Bohr's theory of the hydrogen atom. The orbit sizes and energies are quite different for $n = 1$ and $n = 2$, say. But orbits with $n = 100{,}000{,}000$ and $100{,}000{,}001$ would be very close in size and energy (see Fig. 27–32). Indeed, jumps between such large orbits, which would approach macroscopic sizes, would be imperceptible. Such orbits would thus appear to be continuously spaced, which is what we expect in the everyday world.

Correspondence principle

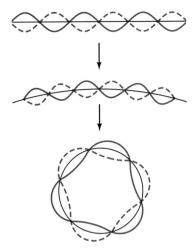

27–11 de Broglie's Hypothesis Applied to Atoms

Bohr's theory was largely of an *ad hoc* nature. Assumptions were made so that theory would agree with experiment. But Bohr could give no reason why the orbits were quantized, nor why there should be a stable ground state. Finally, ten years later, a reason was proposed by de Broglie.

We saw in Section 27–6 that in 1923, Louis de Broglie proposed that material particles, such as electrons, have a wave nature; and that this hypothesis was confirmed by experiment several years later.

One of de Broglie's original arguments in favor of the wave nature of electrons was that it provided an explanation for Bohr's theory of the hydrogen atom. According to de Broglie, a particle of mass m moving with speed v would have a wavelength (Eq. 27–8) of

$$\lambda = \frac{h}{mv}.$$

Each electron orbit in an atom, he proposed, is actually a standing wave. As we saw in Chapter 11, when a violin or guitar string is plucked, a vast number of wavelengths are excited. But only certain ones—those that have nodes at the ends—are sustained. These are the *resonant* modes of the string. Waves with other wavelengths interfere with themselves upon reflection and their amplitudes quickly drop to zero. With electrons moving in circles, according to Bohr's theory, de Broglie argued that the electron wave was a *circular* standing wave that closes on itself, Fig. 27–33. If the wavelength of a wave does not close on itself, as in Fig. 27–34, destructive interference takes place as the wave travels around the loop, and the wave quickly dies out. Thus, the only waves that persist are those for which the circumference of the circular orbit contains a whole number of wavelengths, Fig. 27–35. The circumference of a Bohr orbit of radius r_n is $2\pi r_n$, so we have

$$2\pi r_n = n\lambda, \qquad n = 1, 2, 3, \cdots.$$

When we substitute $\lambda = h/mv$, we get

$$2\pi r_n = \frac{nh}{mv},$$

or

$$mvr_n = \frac{nh}{2\pi}$$

This is just the *quantum condition* proposed by Bohr on an *ad hoc* basis, Eq. 27–11. And it is from this equation that the discrete orbits and energy

FIGURE 27–33 An ordinary standing wave compared to a circular standing wave.

FIGURE 27–34 When a wave does not close (and hence interferes destructively with itself), it rapidly dies out.

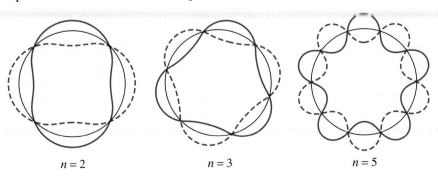

$n = 2$ $n = 3$ $n = 5$

FIGURE 27–35 Standing circular waves for two, three, and five wavelengths on the circumference; n, the number of wavelengths, is also the quantum number.

Quantized orbits
a result of wave nature

levels were derived. Thus we have an explanation for the quantized orbits and energy states in the Bohr model: they are due to the wave nature of the electron, and that only resonant "standing" waves can persist. This implies that the *wave–particle duality* is at the root of atomic structure.

It should be noted in viewing the circular electron waves of Fig. 27–35 that the electron is not to be thought of as following the oscillating wave pattern. In the Bohr model of hydrogen, the electron, considered as a particle, moves in a circle. The circular wave, on the other hand, represents the *amplitude* of the electron "matter wave," and in Fig. 27–35 the wave amplitude is shown superimposed on the circular path of the particle orbit for convenience.

Bohr's theory worked well for hydrogen and for one-electron ions. It did not prove as successful for multielectron atoms. We will discuss this and other problems with the Bohr theory in the next chapter, and we will see how a new and radical theory, quantum mechanics, finally solved the problem of atomic structure and gave us a very different view of the atom: the idea of electrons in well-defined orbits was replaced with the idea of electron "clouds." And this new theory of quantum mechanics has given us a wholly different view of the basic mechanisms underlying physical processes.

SUMMARY

Quantum theory has its origins in **Planck's quantum hypothesis** that molecular oscillations are quantized: their energy E can only be integer (n) multiples of hf, where h is Planck's constant and f is the natural frequency of oscillation: $E = nhf$. This hypothesis explained the spectrum of radiation emitted by (black) bodies at high temperature.

Einstein proposed that for some experiments, light could be pictured as being emitted and absorbed as quanta (particles), which we now call **photons,** each with energy

$$E = hf.$$

He proposed the photoelectric effect as a test for the photon theory of light. In the **photoelectric effect**, the photon theory says that each incident photon can strike an electron in a material and eject it if it has sufficient energy. The maximum energy of ejected electrons is then linearly related to the frequency of the incident light. The photon theory is also supported by the Compton effect and the observation of electron–positron **pair production**.

The wave–particle duality refers to the idea that light and matter (such as electrons) have both wave and particle properties. The wavelength of a material object is given by

$$\lambda = \frac{h}{mv},$$

where mv is the momentum of the object. The **principle of complementarity** states that we must be aware of both the particle and wave properties of light and of matter for a complete understanding of them.

Early models of the atom include the plum-pudding model and Rutherford's planetary (or nuclear) model. Rutherford's model, which was created to explain the backscattering of alpha particles from thin metal foils, assumes that an atom consists of a tiny but massive positively charged nucleus surrounded (at a relatively great distance) by electrons.

To explain the line spectra emitted by atoms, as well as the stability of atoms, **Bohr theory** postulated (1) that electrons bound in an atom can only occupy orbits for which the angular momentum is quantized, which results in discrete values for the radius and energy; (2) that an electron in such a **stationary state** emits no radiation; (3) that, if an electron jumps to a lower state, it emits a photon whose energy equals the difference in energy between the two states; (4) that the angular momentum L of atomic electrons is quantized by the rule

$$L = \frac{nh}{2\pi},$$

where n is an integer called a **quantum number**. The $n = 1$ state in hydrogen is the **ground state**, which has an energy $E_1 = -13.6 \text{ eV}$; higher values of n correspond to **excited states** and their energies are

$$E_n = -\frac{13.6 \text{ eV}}{n^2}.$$

Atoms are excited to these higher states by collisions with other atoms or electrons or by absorption of a photon of just the right frequency.

de Broglie's hypothesis that electrons (and other matter) have a wavelength $\lambda = h/mv$ gave an explanation for Bohr's quantized orbits by bringing in the wave–particle duality: the orbits correspond to circular standing waves in which the circumference of the orbit equals a whole number of wavelengths.

QUESTIONS

1. What can be said about the relative temperature of whitish-yellow, reddish, and bluish stars?

2. If energy is radiated by all objects, why can't we see them in the dark? (See also Section 14–9.)

3. Does a lightbulb at a temperature of 2500 K produce as white a light as the Sun at 6000 K? Explain.

4. An ideal blackbody can be approximated by a small hole in an otherwise enclosed cavity. Explain. [*Hint*: The pupil of your eye is an approximate case.]

5. Why do jewelers often examine diamonds in daylight rather than with indoor light?

6. Darkrooms for developing black and white film are sometimes lit by a red bulb. Why red? Would such a bulb work in a darkroom for developing color photographs?

7. If the threshold wavelength in the photoelectric effect increases when the emitting metal is changed to a different metal, what can you say about the work functions of the two metals?

8. Explain why the existence of a cutoff frequency in the photoelectric effect more strongly favors a particle theory rather than a wave theory of light.

9. UV light causes sunburn, whereas visible light does not. Explain.

10. If an X-ray photon is scattered by an electron, does its wavelength change? If so, does it increase or decrease?

11. In both the photoelectric effect and in the Compton effect we have a photon colliding with an electron causing the electron to fly off. What then, is the difference between the two processes?

12. Explain how the photoelectric circuit of Fig. 27–8 could be used in (*a*) a burglar alarm, (*b*) a smoke detector, (*c*) a photographic light meter, and (*d*) a spectrophotometer (see Section 24–7).

13. Why do we say that light has wave properties? Why do we say that light has particle properties?

14. Why do we say that electrons have wave properties? Why do we say that electrons have particle properties?

15. What is the difference between a photon and an electron? Be specific: make a list.

16. If an electron and a proton travel at the same speed, which has the shorter wavelength?

17. In Rutherford's planetary model of the atom, what keeps the electrons from flying off into space?

18. Which of the following can emit a line spectrum: (*a*) gases, (*b*) liquids, (*c*) solids? Which can emit a continuous spectrum?

19. Why doesn't the O_2 gas in the air around us give off light?

20. How can you tell if there is oxygen on the Sun?

21. When a wide spectrum of light passes through hydrogen gas at room temperature, absorption lines are observed that correspond only to the Lyman series. Why don't we observe the other series?

22. Explain how the closely spaced energy levels for hydrogen near the top of Fig. 27–32 correspond to the closely spaced spectral lines at the top of Fig. 27–27.

23. Discuss the differences between Rutherford's and Bohr's theory of the atom.

24. In a helium atom, which contains two electrons, do you think that on the average the electrons are closer to the nucleus or farther away than in a hydrogen atom? Why?

25. How can the spectrum of hydrogen contain so many lines when hydrogen contains only one electron?

26. The Lyman series is brighter than the Balmer series, because this series of transitions ends up in the most common state for hydrogen, the ground state. Why then was the Balmer series discovered first?

27. The shortest wavelength light in the Lyman *absorption* spectrum is 91 nm. Could light of 89 nm (slightly higher energy) also be absorbed by hydrogen? What would be the final state of the electron? Why isn't the light from the Sun absorbed at this frequency?

28. Use conservation of momentum to explain why photons emitted by hydrogen atoms have slightly less energy than that predicted by Eq. 27–10.

PROBLEMS

SECTION 27–1

1. (I) What is the value of e/m for a particle that moves in a circle of radius 7.0 mm in a 0.86-T magnetic field if a crossed 300-V/m electric field will make the path straight?

2. (II) What is the velocity of a beam of electrons that go undeflected when passing through crossed electric and magnetic fields of magnitude 1.38×10^4 V/m and 2.90×10^{-3} T, respectively? What is the radius of the electron orbit if the electric field is turned off?

3. (II) An oil drop whose mass is determined to be 2.8×10^{-15} kg is held at rest between two large plates separated by 1.0 cm when the potential difference between them is 340 V. How many excess electrons does this drop have?

4. (III) In Millikan's oil-drop experiment, the mass of the oil drop is obtained by observing the terminal speed v_T of the freely falling drop in the absence of an electric field. Under these circumstances, the "effective" weight equals the viscous force given by Stokes's law, $F = 6\pi\eta r v_T$, where η is the viscosity of air and r the radius of the drop. Also, the actual weight, $mg = \frac{4}{3}\pi r^3 \rho g$, must be corrected for the buoyant force of the air. This is done by replacing ρ with $\rho - \rho_A$, where ρ is the density of the oil and ρ_A the density of air. With these preliminaries, show that the charge on the drop is given by

$$q = 18\pi \frac{d}{V} \sqrt{\frac{\eta^3 v_T^3}{2(\rho - \rho_A)g}},$$

where d is the separation of the plates (Fig. 27–3) and V is the voltage across them that just keeps the drop stationary. All of the quantities on the right side of this equation are known or can be measured. The terminal velocity v_T is determined by measuring the time it takes the drop to fall a measured distance, which is observed through a small telescope.

SECTION 27–2

5. (I) How hot is a metal being welded if it radiates most strongly at 410 nm?

6. (I) (*a*) At what temperature will the peak of a blackbody spectrum be at 15.0 nm? (*b*) What is the wavelength at the peak of a blackbody spectrum if the body is at a temperature of 2000 K?

7. (I) Estimate the peak wavelength for radiation from (*a*) ice at 0° C, (*b*) a floodlamp at 3000 K, (*c*) helium at 4 K, assuming blackbody emission. In what region of the EM spectrum is each?

8. (I) An HCl molecule vibrates with a natural frequency of 8.1×10^{13} Hz. What is the difference in energy (in joules and electron volts) between possible values of the oscillation energy?

9. (II) A child's swing has a natural frequency of 0.90 Hz. (*a*) What is the separation between possible energy values (in joules)? (*b*) If the swing reaches a vertical height of 45 cm above its lowest point and has a mass of 20 kg (including the child), what is the value of the quantum number n? (*c*) What is the fractional change in energy between levels whose quantum numbers are n (as just calculated) and $n + 1$? Would quantization be measurable in this case?

10. (II) The steps of a flight of stairs are 20.0 cm high (vertically). If a 58.0-kg person stands with both feet on the same step, what is the gravitational potential energy of this person, relative to the ground, on (*a*) the first step, (*b*) the second step, (*c*) the third step, (*d*) the nth step? (*e*) What is the change in energy as the person descends from step 6 to step 2?

11. (II) Estimate the peak wavelength of light issuing from the pupil of the human eye (which approximates a blackbody) assuming normal body temperature.

SECTION 27–3

12. (I) What is the energy range in eV, of photons in the visible spectrum, of wavelength 400 nm to 700 nm?

13. (I) What is the energy of photons (in eV) emitted by a 102.1-MHz FM radio station?

14. (I) A typical gamma ray emitted from a nucleus during radioactive decay may have an energy of 200 keV. What is its wavelength? Would we expect significant diffraction of this type of light when it passes through an everyday opening, like a door?

15. (I) About 0.1 eV is required to break a "hydrogen bond" in a protein molecule. What are the minimum frequency and maximum wavelength of a photon that can accomplish this?

16. (I) What minimum frequency of light is needed to eject electrons from a metal whose work function is 4.3×10^{-19} J?

17. (II) What is the longest wavelength of light that will emit electrons from a metal whose work function is 3.10 eV?

18. (II) Barium has a work function of 2.48 eV. What is the maximum kinetic energy of electrons if the metal is illuminated by light of wavelength 390 nm? What is their speed?

19. (II) The work functions for sodium, cesium, copper and iron are 2.3, 2.1, 4.7 and 4.5 eV respectively. Which of these metals will not emit electrons when visible light shines on it?

20. (II) In a photoelectric effect experiment it is observed that no current flows unless the wavelength is less than 570 nm. (a) What is the work function of this material? (b) What is the stopping voltage required if light of wavelength 400 nm is used?

21. (II) What is the maximum kinetic energy of electrons ejected from barium ($W_0 = 2.48$ eV) when illuminated by white light, $\lambda = 400$ to 700 nm?

22. (II) When UV light of wavelength 225 nm falls on a metal surface, the maximum kinetic energy of emitted electrons is 1.40 eV. What is the work function of the metal?

23. (II) The threshold wavelength for emission of electrons from a given surface is 320 nm. What will be the maximum kinetic energy of ejected electrons when the wavelength is changed to (a) 250 nm, (b) 350 nm?

24. (II) A certain type of film is sensitive only to light whose wavelength is less than 660 nm. What is the energy (eV and kcal/mol) needed for the chemical reaction to occur which causes the film to change?

25. (II) When 230-nm light falls on a metal, the current through a photoelectric circuit (Fig. 27–8) is brought to zero at a reverse voltage of 1.64 V. What is the work function of the metal?

SECTION 27–4

26. (I) How much total kinetic energy will an electron–positron pair have if produced by a 2.54-MeV photon?

27. (II) What is the momentum and the effective mass of a 0.35-nm X-ray photon?

28. (II) What is the longest wavelength photon that could produce a proton–antiproton pair? (Each has a mass of 1.67×10^{-27} kg.)

29. (II) What is the minimum photon energy needed to produce a $\mu^+ - \mu^-$ pair? The mass of each μ (muon) is 207 times the mass of the electron. What is the wavelength of such a photon?

30. (II) A gamma-ray photon produces an electron–positron pair, each with a kinetic energy of 345 keV. What was the energy and wavelength of the photon?

31. (II) The quantity $h/m_0 c$, which has the dimensions of length, is called the *Compton wavelength*. Determine the Compton wavelength for (a) an electron, (b) a proton. (c) Show that if a photon has wavelength equal to the Compton wavelength of a particle, the photon's energy is equal to the rest energy of the particle.

32. (II) X-rays of wavelength $\lambda = 0.120$ nm are scattered from a carbon block. What is the Compton wavelength shift for photons detected at angles (relative to the incident beam) of (a) 45°, (b) 90°, (c) 180°?

33. (III) For each of the scattering angles in the previous Problem, determine (a) the fractional energy loss of the photon, and (b) the energy given to the scattered electron.

34. (III) Derive Eq. 27–7 using conservation of energy and momentum.

35. (III) In the Compton effect, a 0.100-nm photon strikes a free electron in a head-on collision and knocks it into the forward direction. The rebounding photon recoils directly backward. Use conservation of (relativistic) energy and momentum to determine (a) the kinetic energy of the electron, and (b) the wavelength of the recoiling photon. Assume the electron's kinetic energy is given by the nonrelativistic formula. (Note: use Eq. 27–6, but not Eq. 27–7.)

SECTION 27–6

36. (I) Calculate the wavelength of a 0.21 kg ball traveling at 0.10 m/s.

37. (I) What is the wavelength of a neutron ($m_0 = 1.67 \times 10^{-27}$ kg) traveling 5.5×10^4 m/s?

38. (I) Through how many volts of potential difference must an electron be accelerated to achieve a wavelength of 0.23 nm?

39. (II) Calculate the de Broglie wavelength of an electron in your TV picture tube if it is accelerated by 20,000 V. Is it relativistic? How does its wavelength compare to the size of the "neck" of the tube, typically 5 cm? Do we have to worry about diffraction problems blurring our picture on the screen?

40. (II) What is the wavelength of an electron of energy (a) 10 eV, (b) 100 eV, (c) 1.0 keV?

41. (II) Show that if an electron and a proton have the same nonrelativistic kinetic energy, the proton has the shorter wavelength.

42. (II) Calculate the ratio of the KE of an electron to that of a proton if their wavelengths are equal. Assume that the speeds are non-relativistic.

43. (II) What is the wavelength of an O_2 molecule in the air at room temperature? [*Hint:* See Chapter 13.]

44. (III) A Cadillac with a mass of 2000 kg approaches a freeway underpass that is 10 m across. At what speed must the car be moving, in order for it to have a wavelength such that it might somehow "diffract" after passing through this single "slit"? How do these conditions compare to normal freeway speeds of 30 m/s?

45. (III) Show that for a particle of rest mass m_0, if $\lambda = h/mv$ then it cannot be true that $E = hf$ where E is (a) kinetic energy, or (b) KE plus rest mass energy, and $v = f\lambda$ is the speed of the particle.

*SECTION 27–7

* 46. (II) What voltage is needed to produce electron wavelengths of 0.10 nm? (Assume that the electrons are nonrelativistic.)

* 47. (III) Electrons are accelerated by 2250 V in an electron microscope. To achieve a resolution of 5.0 nm, what numerical aperture (Section 25–8) is required?

SECTION 27–10

48. (I) For the three hydrogen transitions indicated below, with n being the initial state and n' being the final state, is the transition an absorption or an emission? Which is higher, the initial state energy or the final state energy of the atom? Finally, which of these transitions involves the largest energy photon? (a) $n = 1, n' = 3$ (b) $n = 6, n' = 2$ (c) $n = 4, n' = 5$.

49. (I) How much energy is needed to ionize a hydrogen atom in the $n = 2$ state?

50. (I) The third longest wavelength in the Paschen series in hydrogen (Fig. 27–32) corresponds to what transition?

51. (I) Calculate the ionization energy of doubly ionized lithium, Li^{2+}, which has $Z = 3$.

52. (I) Evaluate the Rydberg constant R using Bohr theory (compare Eqs. 27–9 and 27–15) and show that its value is $R = 1.0974 \times 10^7 \, m^{-1}$.

53. (II) What wavelength photon would be required to ionize a hydrogen atom in the ground state and give the ejected electron a kinetic energy of 10.0 eV?

54. (II) (a) Determine the wavelength of the second Balmer line ($n = 4$ to $n = 2$ transition) using Fig. 27–32. Determine likewise (b) the wavelength of the second Lyman line and (c) the wavelength of the third Balmer line.

55. (II) In the Sun, an ionized helium (He^+) atom makes a transition from the $n = 6$ state to the $n = 2$ state, emitting a photon. Can that photon be absorbed by hydrogen atoms present in the Sun? If so, between what energy states will the hydrogen atom jump?

56. (II) What is the longest wavelength light capable of ionizing a hydrogen atom in the ground state?

57. (II) Construct the energy-level diagram for the He^+ ion (see Fig. 27–32).

58. (II) Construct the energy-level diagram for doubly ionized lithium, Li^{2+}.

59. (II) What is the potential energy and the kinetic energy of an electron in the ground state of the hydrogen atom?

60. (II) An excited hydrogen atom could, in principle, have a radius of 1.00 mm. What would be the value of n for a Bohr orbit of this size? What would its energy be?

61. (II) Is the use of nonrelativistic formulas justified in the Bohr atom? To check, calculate the electron's velocity, v, in terms of c for the ground state of hydrogen, and then calculate $\sqrt{1 - v^2/c^2}$.

62. (II) Suppose an electron was bound to a proton, as in the hydrogen atom, by the gravitational force rather than by the electric force. What would be the radius, and energy, of the first Bohr orbit?

63. (III) Show that the magnitude of the PE of an electron in any Bohr orbit of a hydrogen atom is twice the magnitude of its KE in that orbit.

64. (III) *Correspondence principle*: Show that for large values of n, the difference in radius Δr between two adjacent orbits (with quantum numbers n and $n - 1$) is given by

$$\Delta r = r_n - r_{n-1} \approx \frac{2r_n}{n},$$

so $\Delta r/r_n \to 0$ as $n \to \infty$, in accordance with the correspondence principle. [Note that we can check the correspondence principle by either considering large values of $n(n \to \infty)$ or by letting $h \to 0$. Are these equivalent?]

65. (III) Suppose a particle of mass m is confined to a one-dimensional box of width L. According to quantum theory, the particle's wave (with $\lambda = h/mv$) is a standing wave with nodes at the edges of the box. (a) Show the possible modes of vibration on a diagram. (b) Show that the kinetic energy of the particle has quantized energies given by $\text{KE} = n^2 h^2 / 8mL^2$, where n is an integer.

(c) Calculate the ground-state energy ($n = 1$) for an electron confined to a box of width 0.50×10^{-10} m. (d) What is the ground-state energy, and speed, of a baseball ($m = 140$ g) in a box 0.50 m wide? (e) An electron confined to a box has a ground-state energy of 20 eV. What is the width of the box?

GENERAL PROBLEMS

66. The Big Bang theory states that the beginning of the Universe was accompanied by a huge burst of photons. Those photons are still present today and make up the so called Cosmic Microwave Background Radiation. The Universe radiates like a blackbody with a temperature of about 2.7 K. Calculate the peak wavelength of this radiation.

67. At low temperatures, nearly all the atoms in hydrogen gas will be in the ground state. What minimum frequency photon is needed if the photoelectric effect is to be observed?

68. A beam of 85-eV electrons is scattered from a crystal, as in X-ray diffraction, and a first-order peak is observed at $\theta = 38°$. What is the spacing between planes in the diffracting crystal? (See Section 25–11.)

69. Show that the energy E (in electron volts) of a photon whose wavelength is λ (meters) is given by E (in eV) $= 1.24 \times 10^{-6}/\lambda$.

70. A microwave oven produces electromagnetic radiation at $\lambda = 12.2$ cm and produces a power of 760 W. Calculate the number of microwave photons produced by the microwave oven each second.

71. Sunlight reaching the Earth has an intensity of about 1000 W/m². Estimate how many photons per square meter per second this represents. Take the average wavelength to be 550 nm.

72. A beam of red laser light ($\lambda = 633$ nm) hits a black wall and is fully absorbed. If this exerts a total force $F = 5.5$ nN on the wall, how many photons per second are hitting the wall?

73. If a 100-W light bulb emits 3.0 percent of the input energy as visible light (average wavelength 550 nm) uniformly in all directions, estimate how many photons per second of visible light will strike the pupil (4.0 mm diameter) of the eye of an observer 1.0 km away.

74. By what potential difference must (a) a proton ($m_0 = 1.67 \times 10^{-27}$ kg), and (b) an electron ($m_0 = 9.11 \times 10^{-31}$ kg), be accelerated to have a wavelength $\lambda = 5.0 \times 10^{-12}$ m?

75. In certain of Rutherford's experiments (Fig. 27–23), the α particles (mass $= 6.64 \times 10^{-27}$ kg) had a kinetic energy of 4.8 MeV. How close could they get to a gold nucleus (charge $= +79e$)? Ignore the recoil motion of the nucleus.

76. By what fraction does the mass of an H atom decrease when it makes an $n = 4$ to $n = 1$ transition?

77. For what maximum kinetic energy is a collision between an electron and a hydrogen atom in its ground state definitely elastic?

78. Using Bohr theory, derive an equation for the angular velocity ω, and frequency f, of an electron in a hydrogen atom. Determine (a) for the ground state and (b) for the first excited state ($n = 2$).

79. Calculate the ratio of the gravitational to electric force for the electron in a hydrogen atom. Can the gravitational force be safely ignored?

80. Electrons accelerated by a potential difference of 12.3 V pass through a gas of hydrogen atoms at room temperature. What wavelengths of light will be emitted?

81. Atoms can be formed in which a muon (mass $= 207$ times the mass of an electron) replaces one of the electrons in an atom. Calculate, using Bohr theory, the energy of the photon emitted when a muon makes a transition from $n = 2$ to $n = 1$ in a muonic $^{208}_{82}$Pb atom (lead whose nucleus has a mass 208 times the proton mass and charge $+82e$).

82. In an X-ray tube (see Fig. 25–36 and discussion in Section 25–11), the high voltage between filament and target is V. After being accelerated through this voltage, an electron strikes the target where it is decelerated (by positively charged nuclei) and in the process one or more X-ray photons are emitted. (a) Show that the photon of shortest wavelength will have

$$\lambda_0 = \frac{hc}{eV}.$$

(b) What is the shortest wavelength of X-ray emitted when accelerated electrons strike the face of a 27-kV television picture tube?

83. Show that the wavelength of a particle of mass m_0 with kinetic energy KE is given by the relativistic formula $\lambda = hc/\sqrt{(\text{KE})^2 + 2m_0c^2(\text{KE})}$.

84. What is the kinetic energy and wavelength of a "thermal" neutron (one that is in equilibrium at room temperature—see Chapter 13)?

85. What is the theoretical limit of resolution for an electron microscope whose electrons are accelerated through 60 kV? (Relativistic formulas should be used.)

86. The intensity of the Sun's light in the vicinity of the Earth is about 1000 W/m². Imagine a spacecraft with a mirrored square sail of dimension 1.0 km. Estimate how much thrust (in newtons) this craft will experience due to collisions with the Sun's photons. [*Hint*: Assume the photons bounce off the sail with no change in magnitude of their momentum.]

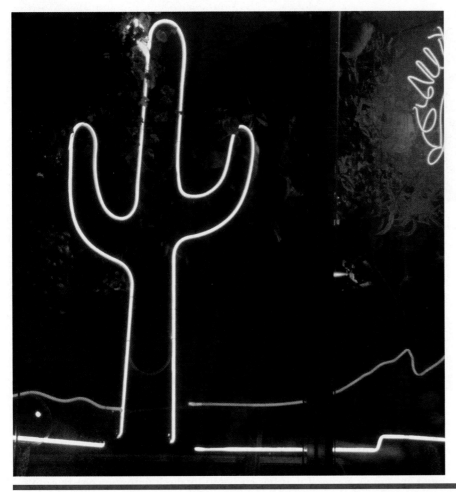

Neon tubes are thin glass tubes (moldable into various shapes) filled with neon (or other) gas that glow with a particular color when high voltage current passes through them. The atoms of the gas are excited to upper energy levels, and when their electrons jump down to lower energy levels they emit light (photons) whose wavelengths (color) are characteristic of the type of gas. A tube filled with neon produces a red-orange color, helium produces pink, and a mixture of Ne, Ar, and Hg produces blue, for example. See Fig. 24–27 or 27–26. The color of the glass, if not clear, also affects the color seen.

28

QUANTUM MECHANICS OF ATOMS

Bohr's model of the atom gave us a first (though rough) picture of what an atom is like. It proposed an explanation for why there should be emission and absorption of light by atoms at discrete wavelengths, as well as for the stability of atoms. The wavelengths of the line spectra and the ionization energy for hydrogen (and one-electron ions) that is predicted are in excellent agreement with experiment. But the Bohr theory had important limitations. It was not able to predict the line spectra for more complex atoms—not even for the neutral helium atom, which has only two electrons. Nor could it explain why emission lines, when viewed with great precision, consist of two or more very closely spaced lines (referred to as *fine structure*). The Bohr theory also didn't explain why some spectral lines were brighter than others. And it couldn't explain the bonding of atoms in molecules or in solids and liquids.

From a theoretical point of view, too, the Bohr theory was not satisfactory. For it was a strange mixture of classical and quantum ideas. And the wave–particle duality was still not really resolved.

FIGURE 28–1 Erwin Schrödinger with Lise Meitner (see Chapter 31).

FIGURE 28–2 Werner Heisenberg (center) on Lake Como with Wolfgang Pauli (right) and Enrico Fermi (left).

We mention these limitations of the Bohr theory not to disparage it—for it was a landmark in the history of science. Rather, we mention them to show why, in the early 1920s, it became increasingly evident that a new, more comprehensive theory was needed. It was not long in coming. Less than two years after de Broglie gave us his matter–wave hypothesis, Erwin Schrödinger (1887–1961; Fig. 28–1) and Werner Heisenberg (1901–1976; Fig. 28–2) independently developed a new comprehensive theory. Their separate approaches were quite different but were soon shown to be fully compatible.

28–1 Quantum Mechanics—A New Theory

The new theory, called **quantum mechanics**, unifies the wave–particle duality into a single consistent theory. As a theory, quantum mechanics has been extremely successful. It has successfully dealt with the spectra emitted by complex atoms, even the fine details. It explains the relative brightness of spectral lines and how atoms form molecules. It is also a much more general theory that covers all quantum phenomena from blackbody radiation to atoms and molecules. It has explained a wide range of natural phenomena and from its predictions many new practical devices have become possible. Indeed, it has been so successful that it is accepted today by nearly all physicists as the fundamental theory underlying physical processes.

Quantum mechanics deals mainly with the microscopic world of atoms and light. But in our everyday macroscopic world, we do perceive light and we accept that ordinary objects are made up of atoms. This new theory must therefore also account for the verified results of classical physics. That is, when it is applied to macroscopic phenomena, quantum mechanics must be able to produce the old classical laws. This, the **correspondence principle** (already mentioned in Section 27–10), is met fully by quantum mechanics. This doesn't mean we throw away classical theories such as Newton's laws. In the everyday world, the latter are far easier to apply and they give an accurate description. But when we deal with high speeds, close to the speed of light, we must use the theory of relativity; and when we deal with the tiny world of the atom, we use quantum mechanics.

Although we won't go into the detailed mathematics of quantum mechanics, we will discuss the main ideas and how they involve the wave and particle properties of matter to explain atomic structure and other applications.

28–2 | The Wave Function and Its Interpretation; the Double-Slit Experiment

The important properties of any wave are its wavelength, frequency, and amplitude. For an electromagnetic wave, the wavelength determines whether the light is visible or not, and if so, what color it is. We also have seen that the wavelength (or frequency) is a measure of the energy of the corresponding photon ($E = hf$). The amplitude of an electromagnetic wave at any point is the strength of the electric (or magnetic) field at that point, and is related to the intensity of the wave (the brightness of the light).

For material particles such as electrons, quantum mechanics relates the wavelength to momentum according to de Broglie's formula, $\lambda = h/mv$. But what about the amplitude of a matter wave? In quantum mechanics the amplitude of, say, an electron wave is called the **wave function** and is given the symbol Ψ (the Greek letter psi). Thus Ψ represents the amplitude, as a function of time and position, of a new kind of field which we might call a "matter" field or a matter wave.

To calculate the wave function Ψ in a given situation (say, for an electron in an atom) is one of the basic tasks of quantum mechanics. Indeed, the development of an equation to do so was Schrödinger's great contribution. The *Schrödinger wave equation*, as it is called, is a differential equation (so we will not deal with it here) and is considered to be the basic equation for the description of nonrelativistic material particles. Since our treatment of quantum mechanics is necessarily brief (whole books are needed for the subject), the actual form of the Schrödinger equation, and how it is solved, will not concern us. What will be useful are the solutions obtained and the meaning of the wave function, Ψ, itself. One way to interpret Ψ is simply as the amplitude, at any point in space and time, of a "matter wave," so it plays the role that E (the electric field) does for an electromagnetic wave. Another interpretation is possible, however, based on the wave–particle duality. To understand this, we make an analogy with light.

We saw in Chapter 11 that the intensity I of any wave is proportional to the square of the amplitude. This holds true for light waves as well, as we saw in Chapter 22; that is,

$$I \propto E^2,$$

where E is the electric field strength. From the *particle* point of view, the intensity of a light beam (of given frequency) is proportional to the number of photons, N, that pass through a given area per unit time. The more photons there are, the greater the intensity. Thus

$$I \propto E^2 \propto N.$$

This proportion can be turned around so that we have

$$N \propto E^2.$$

That is, the number of photons (striking a page of this book, say) is proportional to the square of the electric-field strength.

If the light beam is very weak, only a few photons will be involved. Indeed, it is possible to "build up" a photograph in a camera using very weak

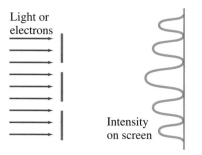

FIGURE 28-3 Parallel beam, of light or electrons, falls on two slits whose sizes are comparable to the wavelength. An interference pattern is observed.

Probability $\propto \Psi^2$

FIGURE 28-4 Young's double-slit experiment done with electrons—note that the pattern is not evident with only a few electrons (top photo), but with more and more electrons (second and third photos), the familiar double-slit interference pattern (Chapter 24) is seen.

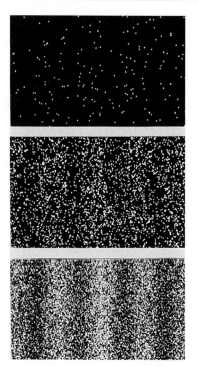

light so the effect of individual photons can be seen. If we are dealing with only one photon, the relationship above ($N \propto E^2$) can be interpreted in a slightly different way. At any point the square of the electric field strength, E^2, is a measure of the *probability* that a photon will be at that location. At points where E^2 is large, there is a high probability the photon will be there; where E^2 is small, the probability is low.

We can interpret matter waves in the same way, as was first suggested by Max Born (1882–1970) in 1927. The wave function Ψ may vary in magnitude from point to point in space and time. If Ψ describes a collection of many electrons, then Ψ^2 at any point will be proportional to the number of electrons expected to be found at that point. When dealing with small numbers of electrons we can't make very exact predictions, so Ψ^2 takes on the character of a probability. If Ψ, which depends on time and position, represents a single electron (say, in an atom), then Ψ^2 is interpreted as follows: Ψ^2 *at a certain point in space and time represents the probability of finding the electron at the given position and time.*

To understand this better, we take as a thought experiment the familiar double-slit experiment, and consider it both for light and for electrons.

Consider two slits whose size and separation are on the order of the wavelength of whatever we direct at them, either light or electrons, Fig. 28–3. We know very well what would happen in this case for light, since this is just Young's double-slit experiment (Section 24–3): an interference pattern would be seen on the screen behind. If light were replaced by electrons with wavelength comparable to the slit size, they too would produce an interference pattern (recall Fig. 27–15). In the case of light, the pattern would be visible to the eye or could be recorded on film. For electrons, a fluorescent screen could be used (it glows where an electron strikes).

Now, if we reduced the flow of electrons (or photons) so that they passed through the slits one at a time, we would see a flash each time one struck the screen. At first, the flashes would seem random. Indeed, there is no way to predict just where any one electron would hit the screen. If we let the experiment run for a long time, however, and kept track of where each electron hit the screen, we would soon see a pattern emerging—namely the interference pattern predicted by the wave theory; see Fig. 28–4. Thus, although we could not predict where a given electron would strike the screen, we could predict probabilities. (The same can be said for photons.) The probability, as mentioned before, is proportional to Ψ^2. Where Ψ^2 is zero, we would get a minimum in the interference pattern. And where Ψ^2 is a maximum, we would get a peak in the interference pattern.

Since the interference pattern would occur even when electrons (or photons) passed through the slits one at a time, it is clear that the interference pattern would not arise from the interaction of one electron with another. It is as if the electron passed through both slits at the same time, interfering with itself. This is possible because, remember, an electron is not precisely a particle. It is as much a wave as it is a particle, and a wave could certainly travel through both slits at once. But what would happen if we covered one of the slits so we knew that the electron passed through the other one, and after a time we covered the second slit so the electron had to have passed through the first? The result would be that no interference pattern would be seen. We would see, instead, two bright areas (or diffraction patterns) on the screen behind the slits. This confirms our idea that if both slits were open, each electron would pass through both, as if it were a wave, forming an interference pattern. Yet each electron would make a tiny spot on the screen as if it were a particle.

The main point of this discussion is this: if we treat electrons (and other matter) as if they were waves, then Ψ represents the wave amplitude. If we treat them as particles, then we must treat them on a *probabilistic* basis. The square of the wave function, Ψ^2, gives the probability of finding a given electron at a given point. We cannot predict—or even follow—the path of a single electron precisely through space and time.

28–3 | The Heisenberg Uncertainty Principle

Whenever a measurement is made, some uncertainty or error is always involved. For example, you cannot make an absolutely exact measurement of the length of a table. Even with a measuring stick that has markings 1 mm apart, there will be an inaccuracy of about $\frac{1}{2}$ mm or so. More precise instruments will produce more precise measurements. But there is always some uncertainty involved in a measurement no matter how good the measuring device. We expect that by using more precise instruments, the uncertainty in a measurement can be made indefinitely small.

But according to quantum mechanics, there is actually a limit to the accuracy of certain measurements. This limit is not a restriction on how well instruments can be made; rather, it is inherent in nature. It is the result of two factors: the wave–particle duality, and the unavoidable interaction between the thing observed and the observing instrument. Let us look at this in more detail.

To make a measurement of an object without somehow disturbing it, at least a little, is not possible. Consider trying to locate a Ping-pong ball in a completely dark room. You grope about trying to find its position; and just when you touch it with your finger, it bounces away. Whenever we measure the position of an object, whether it's a Ping-pong ball or an electron, we always touch it with something else that gives us the information about its position. To locate a lost Ping-pong ball in a dark room, you could probe about with your hand or a stick; or you could shine a light and detect the light reflecting off the ball. When you search with your hand or a stick, you find the ball's position when you touch it. But when you touch the ball you unavoidably bump it, and give it some momentum. Thus you won't know its *future* position. The same would be true, but to a much lesser extent, if you observe the Ping-pong ball using light. In order to "see" the ball, at least one photon must scatter from it, and the reflected photon must enter your eye or some other detector. When a photon strikes an ordinary-sized object, it does not appreciably alter the motion or position of the object. But when a photon strikes a very tiny object like an electron, it can transfer momentum to the object and thus greatly change the object's motion and position in an unpredictable way. The mere act of measuring the position of an object at one time makes our knowledge of its future position imprecise.

FIGURE 28-5 Thought experiment for observing an electron with a powerful light microscope. At least one photon must scatter from the electron (transferring some momentum to it) and enter the microscope.

Now let us see where the wave–particle duality comes in. Imagine a thought experiment in which we are trying to measure the position of an object, say an electron, with photons, Fig. 28–5. (The arguments would be similar if we were using, instead, an electron microscope.) As we saw in Chapter 25, objects can be seen to an accuracy at best of about the wavelength of the radiation used. If we want an accurate position measurement, we must use a short wavelength. But a short wavelength corresponds to high frequency and high energy (since $E = hf$); and the more energy the photons have, the more momentum they can give the object when they strike it. If photons of longer wavelength, and correspondingly lower energy are used, the object's motion when struck by the photons will not be affected as much. But the longer wavelength means lower resolution, so the object's position will be less accurately known. Thus the act of observing produces a significant uncertainty in either the *position* or the *momentum* of the electron. This is the essence of the *uncertainty principle* first enunciated by Heisenberg in 1927.

Quantitatively, we can make an approximate calculation of the magnitude of this effect. If we use light of wavelength λ, the position can be measured at best to an accuracy of about λ. That is, the uncertainty in the position measurement, Δx, is approximately

$$\Delta x \approx \lambda.$$

Suppose that the object can be detected by a single photon. The photon has a momentum $p = h/\lambda$. When the photon strikes our object, it will give some or all of this momentum to the object, Fig. 28–5. Therefore, the final momentum of our object will be uncertain in the amount

$$\Delta p \approx \frac{h}{\lambda}$$

since we can't tell beforehand how much momentum will be transferred. The product of these uncertainties is

$$(\Delta x)(\Delta p) \approx h.$$

Of course, the uncertainties could be worse than this, depending on the apparatus and the number of photons needed for detection. In Heisenberg's more careful calculation, he found that at the very best

$$(\Delta x)(\Delta p) \gtrsim \frac{h}{2\pi}. \tag{28-1}$$

This is a mathematical statement of the **Heisenberg uncertainty principle**, or, as it is sometimes called, the **indeterminancy principle**. It tells us that we cannot measure both the position and momentum of an object precisely at the same time. The more accurately we try to measure the position, so that Δx is small, the greater will be the uncertainty in momentum, Δp. If we try to measure the momentum very precisely, then the uncertainty in the position becomes large. The uncertainty principle does not forbid single exact measurements, however. For example, in principle we could measure the position of an object exactly. But then its momentum would be completely unknown. Thus, although we might know the position of the object exactly at one instant, we could have no idea at all where it would be a moment later.

Another useful form of the uncertainty principle relates energy and time, and we examine this as follows. The object to be detected has an uncertainty in position $\Delta x \approx \lambda$. Now the photon used to detect it travels with speed c, and it takes a time $\Delta t \approx \Delta x/c \approx \lambda/c$ to pass through the distance of uncertainty. Hence, the measured time when our object is at a given position is uncertain by about

$$\Delta t \approx \frac{\lambda}{c}.$$

Since the photon can transfer some or all of its energy ($=hf = hc/\lambda$) to our object, the uncertainty in energy of our object as a result is

$$\Delta E \approx \frac{hc}{\lambda}.$$

The product of these two uncertainties is

$$(\Delta E)(\Delta t) \approx h.$$

Heisenberg's more careful calculation gives

$$(\Delta E)(\Delta t) \gtrsim \frac{h}{2\pi}. \qquad \textbf{(28–2)}$$

UNCERTAINTY PRINCIPLE
(ΔE and Δt)

This form of the uncertainty principle tells us that the energy of an object can be uncertain, or may even be nonconserved, by an amount ΔE for a time $\Delta t \approx h/(2\pi \Delta E)$.

The quantity ($h/2\pi$) appears so often in quantum mechanics that for convenience it is given the symbol $\hbar$ ("h-bar"). That is

$$\hbar = \frac{h}{2\pi} = \frac{6.626 \times 10^{-34} \text{ J} \cdot \text{s}}{2\pi} = 1.055 \times 10^{-34} \text{ J} \cdot \text{s}.$$

By using this notation, Eqs. 28–1 and 28–2 for the uncertainty principle can be written

$$(\Delta x)(\Delta p) \gtrsim \hbar \qquad \text{and} \qquad (\Delta E)(\Delta t) \gtrsim \hbar.$$

We have been discussing the position and velocity of an electron as if it were a particle. But it isn't a particle. Indeed, we have the uncertainty principle because an electron—and matter in general—has wave as well as particle properties. What the uncertainty principle really tells us is that if we insist on thinking of the electron as a particle, then there are certain limitations on this simplified view—namely, that the position and velocity cannot both be known precisely at the same time; and that the energy can be uncertain (or nonconserved) in the amount ΔE for a time $\Delta t \approx \hbar/\Delta E$.

Because Planck's constant, h, is so small, the uncertainties expressed in the uncertainty principle are usually negligible on the macroscopic level. But at the level of the atom, the uncertainties are significant. Because we consider ordinary objects to be made up of atoms containing nuclei and electrons, the uncertainty principle is relevant to our understanding of all of nature. The uncertainty principle expresses, perhaps most clearly, the probabilistic nature of quantum mechanics. It thus is often used as a basis for philosophic discussion.

EXAMPLE 28–1 **Position uncertainty of electron.** An electron moves in a straight line with a constant speed $v = 1.10 \times 10^6$ m/s which has been measured to a precision of 0.10 percent. What is the maximum precision with which its position could be simultaneously measured?

SOLUTION The momentum of the electron is $p = mv = (9.11 \times 10^{-31}$ kg$) \cdot (1.10 \times 10^6$ m/s$) = 1.00 \times 10^{-24}$ kg·m/s. The uncertainty in the momentum is 0.10 percent of this, or

$$\Delta p = 1.0 \times 10^{-27} \text{ kg·m/s}.$$

From the uncertainty principle, the best simultaneous position measurement will have an uncertainty of

$$\Delta x = \frac{\hbar}{\Delta p} = \frac{1.06 \times 10^{-34} \text{ J·s}}{(1.0 \times 10^{-27} \text{ kg·m/s})} = 1.1 \times 10^{-7} \text{ m},$$

or 110 nm. This is about 1000 times the diameter of an atom.

EXAMPLE 28–2 **Position uncertainty of a baseball.** What is the uncertainty in position, imposed by the uncertainty principle, on a 150-g baseball thrown at (93 ± 2) mph $= (42 \pm 1)$ m/s? (Should the umpire be concerned? Can he use Heisenberg as an excuse?)

SOLUTION The uncertainty in the momentum is

$$\Delta p = m \, \Delta v = (0.150 \text{ kg})(1 \text{ m/s}) = 0.15 \text{ kg·m/s}.$$

Hence the uncertainty in a position measurement could be as small as

$$\Delta x = \frac{\hbar}{\Delta p} = \frac{1.06 \times 10^{-34} \text{ J·s}}{(0.15 \text{ kg·m/s})} = 7 \times 10^{-34} \text{ m},$$

which is a distance incredibly smaller than any we could imagine observing or measuring. Indeed, the uncertainty principle sets no relevant limit on measurement for macroscopic objects. (The ump is on his own for excuses.)

EXAMPLE 28–3 **ESTIMATE** **J/ψ lifetime calculated.** The J/ψ meson, discovered in 1974, was measured to have an average mass of 3100 MeV/c^2 (note the use of energy units since $E_0 = m_0 c^2$) and an intrinsic width of 63 keV/c^2. By this we mean that the masses of different J/ψ mesons were actually measured to be slightly different from one another. This mass "width" is related to the very short lifetime of the J/ψ before it decays into other particles. Estimate its lifetime using the uncertainty principle.

SOLUTION The uncertainty of 63 keV/c^2 in the J/ψ's mass is an uncertainty in its rest energy, which in joules is

$$\Delta E = (63 \times 10^3 \text{ eV})(1.60 \times 10^{-19} \text{ J/eV}) = 1.01 \times 10^{-14} \text{ J}.$$

Then we expect its lifetime $\tau \, (= \Delta t$ here) to be

$$\tau \approx \frac{\hbar}{\Delta E} = \frac{1.06 \times 10^{-34} \text{ J·s}}{1.01 \times 10^{-14} \text{ J}} \approx 1.01 \times 10^{-20} \text{ s}.$$

Lifetimes this short are difficult to measure directly, and the assignment of very short lifetimes depends on this use of the uncertainty principle. (See Chapter 32.)

28–4 Philosophic Implications; Probability versus Determinism

The classical Newtonian view of the world is a deterministic one (see Section 5–9). One of its basic ideas is that once the position and velocity of an object are known at a particular time, its future position can be predicted if the forces on it are known. For example, if a stone is thrown a number of times with the same initial velocity and angle, and the forces on it remain the same, the path of the projectile will always be the same. If the forces are known (gravity and air resistance, if any), the stone's path can be precisely predicted. This mechanistic view implies that the future unfolding of the universe, assumed to be made up of particulate bodies, is completely determined.

This classical deterministic view of the physical world has been radically altered by quantum mechanics. As we saw in the analysis of the double-slit experiment (Section 28–2), electrons all prepared in the same way will not all end up in the same place. According to quantum mechanics, certain probabilities exist that an electron will arrive at different points. This is very different from the classical view, in which the path of a particle is precisely predictable from the initial position and velocity and the forces exerted on it. According to quantum mechanics, the position and velocity of an object cannot even be known accurately at the same time. This is expressed in the uncertainty principle, and arises because basic entities, such as electrons, are not considered simply as particles: they have wave properties as well. Quantum mechanics allows us to calculate only the probability[†] that, say, an electron (when thought of as a particle) will be observed at various places. Quantum mechanics says there is some inherent unpredictability in nature.

Since matter is considered to be made up of atoms, even ordinary-sized objects are expected to be governed by probability, rather than by strict determinism. For example, there is a finite (but very small) probability that when you throw a stone, its path will suddenly curve upward instead of following the downward-curved parabola of normal projectile motion. Quantum mechanics predicts with very high probability that ordinary objects will behave just as the classical laws of physics predict. But these predictions are considered probabilities, not certainties. The reason that macroscopic objects behave in accordance with classical laws with very high probability is due to the large number of molecules involved: when large numbers of objects are present in a statistical situation, deviations from the average (or most probable) are negligible. It is the average configuration of vast numbers of molecules that follows the so-called fixed laws of classical physics with such high probability, and gives rise to an apparent "determinism." Deviations from classical laws are observed when small numbers of molecules are dealt with. We can say, then, that although there are no precise deterministic laws in quantum mechanics, there are statistical laws based on probability.

It is important to note that there is a difference between the probability imposed by quantum mechanics and that used in the nineteenth century to understand thermodynamics and the behavior of gases in terms of molecules (Chapters 13 and 15). In thermodynamics, probability is used because there

[†]Note that these probabilities can be calculated precisely, just like exact predictions of probabilities at dice or playing cards, but unlike predictions of probabilities at sporting events or for natural or man-made disasters, which are only estimates.

are far too many molecules to be kept track of. But the molecules were still assumed to move and interact in a deterministic way according to Newton's laws. Probability in quantum mechanics is quite different. It is seen as *inherent* in nature, and not as a limitation on our abilities to calculate.

Although a few physicists have not given up the deterministic view of nature and have been reluctant to accept quantum mechanics as a complete theory—one was Einstein—nonetheless, the vast majority of physicists do accept quantum mechanics and the probabilistic view of nature. This view, which as presented here is the generally accepted one, is called the **Copenhagen interpretation** of quantum mechanics in honor of Niels Bohr's home, since it was largely developed there through discussions between Bohr and other prominent physicists.

Because electrons are not simply particles, they cannot be thought of as following particular paths in space and time. This suggests that a description of matter in space and time may not be completely correct. This deep and far-reaching conclusion has been a lively topic of discussion among philosophers. Perhaps the most important and influential philosopher of quantum mechanics was Bohr. He argued that a space–time description of actual atoms and electrons is not possible. Yet a description of experiments on atoms or electrons must be given in terms of space and time and other concepts familiar to ordinary experience, such as waves and particles. We must not let our *descriptions* of experiments lead us into believing that atoms or electrons themselves actually exist in space and time as particles. This distinction between our interpretations of experiments and what is "really" happening in nature is crucial.

28–5 Quantum-Mechanical View of Atoms

At the beginning of this chapter, we discussed the limitations of the Bohr theory of atomic structure. Now we examine the quantum-mechanical theory of atoms, which is far more complete than the old Bohr theory. Although the Bohr model has been discarded as an accurate description of nature, nonetheless, quantum mechanics reaffirms certain aspects of the older theory, such as that electrons in an atom exist only in discrete states of definite energy, and that a photon of light is emitted (or absorbed) when an electron makes a transition from one state to another. But quantum mechanics is a much deeper theory, and has provided us with a very different view of the atom. According to quantum mechanics, electrons do not exist in well-defined circular orbits as in the Bohr theory. Rather, the electron (because of its wave nature) can be thought of as spread out in space as a "cloud." The size and shape of the electron cloud can be calculated for a given state of an atom. For the ground state in the hydrogen atom, the electron cloud is spherically symmetric, as shown in Fig. 28–6. The electron cloud roughly indicates the "size" of an atom; but just as a cloud may not have a distinct border, atoms do not have a precise boundary or a well-defined size. Not all electron clouds have a spherical shape, as we shall see later in this chapter.

The electron cloud can be interpreted using either the particle or the wave viewpoint. Remember that by a particle we mean something that is localized in space—it has a definite position at any given instant. By contrast, a wave is spread out in space. The electron cloud, spread out in space

FIGURE 28–6
Electron cloud or "probability distribution" for the ground state of the hydrogen atom. The circle represents the Bohr radius.

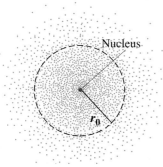

as in Fig. 28–6, is a result of the wave nature of electrons. Electron clouds can also be interpreted as **probability distributions** for a particle. If you were to make 500 different measurements of the position of an electron (considering it as a particle), the majority of the results would show the electron at points where the probability is high (darker areas with more dots in Fig. 28–6). Only occasionally would the electron be found where the probability is low. We cannot predict the path an electron will follow. As we saw in Section 28–3, after one measurement of its position we cannot predict exactly where it will be at a later time. We can only calculate the probability that it will be found at different points. This is clearly different from classical Newtonian physics. Indeed, as Bohr later pointed out, it is meaningless even to ask how an electron gets from one state to another when the atom emits a photon of light.

Probability distributions

28–6 Quantum Mechanics of the Hydrogen Atom; Quantum Numbers

We now look more closely at what quantum mechanics tells us about the hydrogen atom. Much of what we say here also applies to more complex atoms, which are discussed in the next Section.

Quantum mechanics predicts basically the same energy levels (Fig. 27–32) for the hydrogen atom as does the Bohr theory. That is,

$$E_n = -\frac{13.6 \text{ eV}}{n^2}, \qquad n = 1, 2, 3, \cdots,$$

where n is an integer. In the simple Bohr theory, there was only one quantum number, n. In quantum mechanics, it turns out that four different quantum numbers are needed to specify each state in the atom.

Quantum numbers

The *quantum number*, n, from the Bohr theory is found also in quantum mechanics and is called the **principal quantum number**. It can have any integer value from 1 to ∞. The total energy of a state in the hydrogen atom depends on n, as we saw above.

The **orbital quantum number**, l, is related to the magnitude of the angular momentum of the electron; l can take on integer values from 0 to $(n - 1)$. For the ground state ($n = 1$), l can only be zero.[†] But for $n = 3$, l can be 0, 1, or 2. The actual magnitude of the angular momentum L is related to the quantum number l by

$$L = \sqrt{l(l + 1)}\hbar \qquad \textbf{(28–3)}$$

(where again $\hbar = h/2\pi$). The value of l has almost no effect on the total energy in the hydrogen atom; only n does to any appreciable extent.[‡] But in atoms with two or more electrons, the energy does depend on l as well as n, as we shall see in the next Section.

The **magnetic quantum number**, m_l, is related to the direction of the electron's angular momentum, and it can take on integer values ranging from $-l$ to $+l$. For example, if $l = 2$, then m_l can be $-2, -1, 0, +1,$ or $+2$. Since angular momentum is a vector, it is not surprising that both its mag-

[†] Contrast this with the Bohr theory, which assigned $l = 1$ to the ground state (Eq. 27–11).

[‡] See discussion of *fine structure* on the next page.

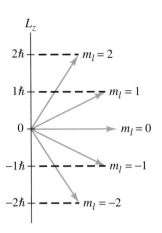

FIGURE 28–7 Quantization of angular momentum direction for $l = 2$.

FIGURE 28–8 When a magnetic field is applied, an $n = 3$, $l = 2$ energy level is split into five separate levels, corresponding to the five values of m_l $(2, 1, 0, -1, -2)$. An $n = 2, l = 1$ level is split into three levels $(m_l = 1, 0, -1)$. Transitions can occur between levels (not all transitions are shown), with photons of several slightly different frequencies being given off (the Zeeman effect).

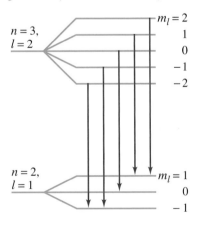

nitude and its direction would be quantized. For $l = 2$, the five different directions allowed can be represented by the diagram of Fig. 28–7. This limitation on the direction of **L** is often called **space quantization**. In quantum mechanics, the direction of the angular momentum is usually specified by giving its component along the z axis (this choice is arbitrary). Then L_z is related to m_l by the equation

$$L_z = m_l \hbar.$$

The name for m_l derives not from theory (which relates it to L_z), but from experiment. It was found that when a gas-discharge tube was placed in a magnetic field, the spectral lines were split into several very closely spaced lines. This splitting, known as the **Zeeman effect**, implies that the energy levels must be split (Fig. 28–8), and thus that the energy of a state depends not only on n but also on m_l when a magnetic field is applied—hence the name "magnetic quantum number." (Why the energy should depend on the direction of **L** can be seen from a semiclassical view of a moving electron as an electric current that interacts with the magnetic field.)

Finally, there is the **spin quantum number**, m_s, which for an electron can have only two values, $m_s = +\frac{1}{2}$ and $m_s = -\frac{1}{2}$. The existence of this quantum number did not come out of Schrödinger's original theory, as did n, l, and m_l. Instead, a subsequent modification by P. A. M. Dirac (1902–1984) explained its presence as a relativistic effect. The first hint that m_s was needed, however, came from experiment. A careful study of the spectral lines of hydrogen showed that each actually consisted of two (or more) very closely spaced lines even in the absence of an external magnetic field. It was at first hypothesized that this tiny splitting of energy levels, called **fine structure**, was due to angular momentum associated with a spinning of the electron. That is, the electron might spin on its axis as well as orbit the nucleus, just as the Earth spins on its axis as it orbits the Sun. The interaction between the tiny current of the spinning electron could then interact with the magnetic field due to the orbiting charge and cause the small observed splitting of energy levels. (So the energy depends slightly on m_l and m_s.) Today we consider this picture of a spinning electron as not legitimate. We cannot even view an electron as a localized object, much less a spinning one. What is important is that the electron can have two different states due to some intrinsic property that behaves as an angular momentum, and we still call this property "spin." The two possible values of m_s $(+\frac{1}{2}$ and $-\frac{1}{2})$ are often said to be "spin up" and "spin down," referring to the two possible directions of the spin angular momentum.

The possible values of the four quantum numbers for an electron in the hydrogen atom are summarized in Table 28–1.

TABLE 28–1 Quantum Numbers for an Electron

Name	Symbol	Possible Values
Principal	n	$1, 2, 3, \cdots, \infty.$
Orbital	l	For a given n: l can be $0, 1, 2, \cdots, n - 1.$
Magnetic	m_l	For given n and l: m_l can be $l, l - 1, \cdots, 0, \cdots, -l.$
Spin	m_s	For each set of n, l, and m_l: m_s can be $+\frac{1}{2}$ or $-\frac{1}{2}.$

CONCEPTUAL EXAMPLE 28–4 | **Possible states for n = 3.** How many different states are possible for an electron whose principal quantum number is $n = 3$?

RESPONSE For $n = 3$, l can have the values $l = 2, 1, 0$. For $l = 2$, m_l can be $2, 1, 0, -1, -2$, which is five different possibilities. For each of these, m_s can be either up or down ($+\frac{1}{2}$ or $-\frac{1}{2}$); so for $l = 2$, there are $2 \times 5 = 10$ states. For $l = 1$, m_l can be $1, 0, -1$, and since m_s can be $+\frac{1}{2}$ or $-\frac{1}{2}$ for each of these, we have six more possible states. Finally, for $l = 0$, m_l can only be 0, and there are only two states corresponding to $m_s = +\frac{1}{2}$ and $-\frac{1}{2}$. The total number of states is $10 + 6 + 2 = 18$, as detailed in the following table:

n	l	m_l	m_s	n	l	m_l	m_s
3	2	2	$\frac{1}{2}$	3	2	-2	$-\frac{1}{2}$
3	2	2	$-\frac{1}{2}$	3	1	1	$\frac{1}{2}$
3	2	1	$\frac{1}{2}$	3	1	1	$-\frac{1}{2}$
3	2	1	$-\frac{1}{2}$	3	1	0	$\frac{1}{2}$
3	2	0	$\frac{1}{2}$	3	1	0	$-\frac{1}{2}$
3	2	0	$-\frac{1}{2}$	3	1	-1	$\frac{1}{2}$
3	2	-1	$\frac{1}{2}$	3	1	-1	$-\frac{1}{2}$
3	2	-1	$-\frac{1}{2}$	3	0	0	$\frac{1}{2}$
3	2	-2	$\frac{1}{2}$	3	0	0	$-\frac{1}{2}$

EXAMPLE 28–5 | **E and L for n = 3.** Determine (a) the energy and (b) the orbital angular momentum for each of the states in Example 28–4.

SOLUTION (a) The energy of a state depends only on n, except for the very small corrections mentioned above, which we will ignore. Since $n = 3$ for all these states, they all have the same energy,

$$E_3 = -\frac{13.6 \text{ eV}}{(3)^2} = -1.51 \text{ eV.}$$

(b) For $l = 0$,

$$L = \sqrt{l(l + 1)}\hbar = 0.$$

For $l = 1$,

$$L = \sqrt{1(1 + 1)}\hbar = \sqrt{2}\hbar$$

$$= 1.49 \times 10^{-34} \text{ J·s.}$$

Atomic angular momenta are generally given as a multiple of $\hbar$ (e.g., $\sqrt{2}\hbar$ in this case), rather than in SI units. But note that for $l = 1$, L is on the order of 10^{-34} J·s. This means that macroscopic angular momenta will have such extremely high quantum numbers that the quantization of angular momentum will not be detectable: L will appear continuous, in accordance with the correspondence principle. Finally, for $l = 2$,

$$L = \sqrt{2(2 + 1)}\hbar = \sqrt{6}\hbar.$$

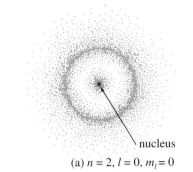

(a) $n = 2, l = 0, m_l = 0$

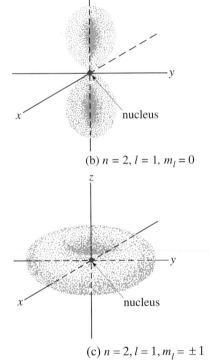

(b) $n = 2, l = 1, m_l = 0$

(c) $n = 2, l = 1, m_l = \pm 1$

FIGURE 28–9 Electron cloud, or probability distribution, for $n = 2$ states in hydrogen.

Another prediction of quantum mechanics is that when a photon is emitted or absorbed, transitions can occur only between states with values of l that differ by one unit:

$$\Delta l = \pm 1.$$

According to this **selection rule**, an electron in an $l = 2$ state can jump only to a state with $l = 1$ or $l = 3$. It cannot jump to a state with $l = 2$ or $l = 0$. A transition such as $l = 2$ to $l = 0$ is called a **forbidden transition**. Actually, such a transition is not absolutely forbidden and can occur, but only with very low probability compared to **allowed transitions**—those that satisfy the selection rule $\Delta l = \pm 1$. Since the orbital angular momentum of an H atom must change by one unit when it emits a photon, conservation of angular momentum tells us that the photon must carry off angular momentum. Indeed, experimental evidence of many sorts shows that the photon can be assigned a spin of 1.

Although l and m_l do not significantly affect the energy levels in hydrogen, they do affect the electron probability distribution in space. For $n = 1$, l and m_l can only be zero and the electron distribution is as shown in Fig. 28–6. For $n = 2$, l can be 0 or 1. The distribution for $n = 2$, $l = 0$ is shown in Fig. 28–9a, and it is seen to differ from that for the ground state, although it is still spherically symmetric. For $n = 2$, $l = 1$, the distributions are not spherically symmetric as shown in Figs. 28–9b (for $m_l = 0$) and 28–9c (for $m_l = +1$ or -1).

Although the spatial distributions of the electron can be calculated for the various states, it is difficult to measure them experimentally. Most of the experimental information about the atom has come from a careful examination of the emission spectra under various conditions.

28–7 Complex Atoms; the Exclusion Principle

We have discussed the hydrogen atom in detail because it is the simplest to deal with. Now we briefly discuss more complex atoms, those that contain more than one electron, and whose energy levels can be determined experimentally from an analysis of the emission spectra. The energy levels are *not* the same as in the H atom, since the electrons interact with each other as well as with the nucleus. For atoms with more than one electron, the energy levels depend on both n and l.

The number of electrons in a neutral atom is called its **atomic number**, Z; Z is also the number of positive charges (protons) in the nucleus, and determines what kind of atom it is. That is, Z determines most of the properties that distinguish one atom from another.

Although modifications of the Bohr theory had been attempted in order to deal with complex atoms, the development of quantum mechanics in the years after 1925 proved far more successful. The mathematics becomes very difficult, however, since in multielectron atoms, each electron is not only attracted to the nucleus but is repelled by the other electrons.

The simplest approach has been to treat each electron in an atom as occupying a particular state characterized by the quantum numbers n, l, m_l, and m_s. But to understand the possible arrangements of electrons in an atom, a new principle was needed. It was introduced by Wolfgang Pauli (1900–1958; Fig. 28–2) and is called the **Pauli exclusion principle**. It states:

No two electrons in an atom can occupy the same quantum state.

Pauli exclusion principle

Thus, no two electrons in an atom can have exactly the same set of the quantum numbers n, l, m_l, and m_s. The Pauli exclusion principle[†] forms the basis not only for understanding complex atoms, but also for understanding molecules and bonding, and other phenomena as well.

Let us now look at the structure of some of the simpler atoms when they are in the ground state. After hydrogen, the next simplest atom is *helium* with two electrons. Both electrons can have $n = 1$, since one can have spin up ($m_s = +\frac{1}{2}$) and the other spin down ($m_s = -\frac{1}{2}$), thus satisfying the exclusion principle. Of course, since $n = 1$, l and m_l must be zero (Table 28–1). Thus the two electrons have the quantum numbers indicated in the table in the margin.

Lithium has three electrons, two of which can have $n = 1$. But the third cannot have $n = 1$ without violating the exclusion principle. Hence the third electron must have $n = 2$. Since it happens that the $n = 2$, $l = 0$ level has a lower energy than $n = 2$, $l = 1$, the electrons in the ground state have the quantum numbers indicated in the table in the margin. Of course, the quantum numbers of the third electron could also be, say, $(3, 1, -1, \frac{1}{2})$. But the atom in this case would be in an excited state since it would have greater energy. It would not be long before it jumped to the ground state with the emission of a photon. At room temperature, unless extra energy is supplied (as in a discharge tube), the vast majority of atoms are in the ground state.

We can continue in this way to describe the quantum numbers of each electron in the ground state of larger and larger atoms. That for sodium, with its eleven electrons, is shown in the table in the margin.

Figure 28–10 shows a simple energy level diagram where occupied states are shown as up or down arrows ($m_s = +\frac{1}{2}$ or $-\frac{1}{2}$), and possible empty states are shown as a small circle.

The ground-state configuration for all atoms is given in the **periodic table**, which is displayed inside the back cover of this book, and discussed in the next Section.

Helium, Z = 2

n	l	m_l	m_s
1	0	0	$\frac{1}{2}$
1	0	0	$-\frac{1}{2}$

Lithium, Z = 3

n	l	m_l	m_s
1	0	0	$\frac{1}{2}$
1	0	0	$-\frac{1}{2}$
2	0	0	$\frac{1}{2}$

Sodium, Z = 11

n	l	m_l	m_s
1	0	0	$\frac{1}{2}$
1	0	0	$-\frac{1}{2}$
2	0	0	$\frac{1}{2}$
2	0	0	$-\frac{1}{2}$
2	1	1	$\frac{1}{2}$
2	1	1	$-\frac{1}{2}$
2	1	0	$\frac{1}{2}$
2	1	0	$-\frac{1}{2}$
2	1	-1	$\frac{1}{2}$
2	1	-1	$-\frac{1}{2}$
3	0	0	$\frac{1}{2}$

FIGURE 28–10 Energy level diagram showing occupied states (arrows) and unoccupied states (○) for He, Li, and Na. Note that we have shown the $n = 2$, $l = 1$ level of Li even though it is empty.

[†]The exclusion principle applies to identical particles whose spin quantum number is a half-integer ($\frac{1}{2}$, $\frac{3}{2}$, and so on), including electrons, protons, and neutrons; such particles are called **fermions** (after E. Fermi who derived a statistical theory describing them). The exclusion principle does not apply to particles with integer spin (0, 1, 2, and so on), such as the photon and π meson, all of which are referred to as **bosons** (after S. N. Bose, who derived a statistical theory for them).

TABLE 28–2 Values of l

Value of l	Letter Symbol	Maximum Number of Electrons in Subshell
0	s	2
1	p	6
2	d	10
3	f	14
4	g	18
5	h	22

Shells and subshells

TABLE 28–3 Electron Configuration of Some Elements

Z (No. of Electrons)	Element[†]	Ground State Configuration (outer electrons)
1	H	$1s^1$
2	He	$1s^2$
3	Li	$2s^1$
4	Be	$2s^2$
5	B	$2s^2 2p^1$
6	C	$2s^2 2p^2$
7	N	$2s^2 2p^3$
8	O	$2s^2 2p^4$
9	F	$2s^2 2p^5$
10	Ne	$2s^2 2p^6$
11	Na	$3s^1$
12	Mg	$3s^2$
13	Al	$3s^2 3p^1$
14	Si	$3s^2 3p^2$
15	P	$3s^2 3p^3$
16	S	$3s^2 3p^4$
17	Cl	$3s^2 3p^5$
18	Ar	$3s^2 3p^6$
19	K	$4s^1$
20	Ca	$4s^2$
21	Sc	$3d^1 4s^2$
22	Ti	$3d^2 4s^2$
23	V	$3d^3 4s^2$
24	Cr	$3d^5 4s^1$
25	Mn	$3d^5 4s^2$
26	Fe	$3d^6 4s^2$

[†]Names of elements can be found in Appendix F.

28–8 The Periodic Table of Elements

A century ago, Dmitri Mendeleev (1834–1907) arranged the then known elements into what we now call the **periodic table** of the elements. The atoms were arranged according to increasing mass, but also so that elements with similar chemical properties would fall in the same column. Today's version is shown inside the back cover. Each square contains the atomic number Z, the symbol for the element, and the atomic mass (in atomic mass units). Finally, in the lower left corner the configuration of the ground state of the atom is given. This requires some explanation. Electrons with the same value of n are referred to as being in the same **shell**. Electrons with $n = 1$ are in one shell (the K shell), those with $n = 2$ are in a second shell (the L shell), those with $n = 3$ are in the third (M) shell, and so on. Electrons with the same values of n and l are referred to as being in the same **subshell**. Letters are often used to specify the value of l as shown in Table 28–2. That is, $l = 0$ is the s subshell; $l = 1$ is the p subshell; $l = 2$ is the d subshell; beginning with $l = 3$, the letters follow the alphabet, f, g, h, i, and so on. (The first letters s, p, d, and f were originally abbreviations of "sharp," "principal," "diffuse," and "fundamental," experimental terms referring to the spectra.)

The Pauli exclusion principle limits the number of electrons possible in each shell and subshell. For any value of l, there are $2l + 1$ different m_l values (m_l can be any integer from 1 to l, from -1 to $-l$, or zero), and two different m_s values. There can be, therefore, at most $2(2l + 1)$ electrons in any l subshell. For example, for $l = 2$, five m_l values are possible $(2, 1, 0, -1, -2)$, and for each of these, m_s can be $+\frac{1}{2}$ or $-\frac{1}{2}$, for a total of $2(5) = 10$ states. Table 28–2 lists the maximum number of electrons that can occupy each subshell.

Since the energy levels depend almost entirely on the values of n and l, it is customary to specify the electron configuration simply by giving the n value and the appropriate letter for l, with the number of electrons in each subshell given as a superscript. The ground-state configuration of sodium, for example, is written as $1s^2 2s^2 2p^6 3s^1$. This is simplified in the periodic table by specifying the configuration only of the outermost electrons and any other nonfilled subshells (see Table 28–3 here, and the periodic table inside the back cover).

CONCEPTUAL EXAMPLE 28–6 **Electron configurations.** Which of the following electron configurations are possible, and which forbidden? (a) $1s^2 2s^2 2p^6 3s^3$; (b) $1s^2 2s^2 2p^6 3s^2 3p^5 4s^2$; (c) $1s^2 2s^2 2p^6 2d^1$.

RESPONSE (a) This is not allowed, because there are too many electrons in the s subshell in the M ($n = 3$) shell. The s subshell has $m_l = 0$, with two slots only: for "spin up" and "spin down" electrons. (b) This is allowed, but it is an excited state. One of the electrons in the $3p$ subshell has jumped up to the $4s$ subshell. Since there are 19 electrons, the element is potassium. (c) This is not allowed, because there is no $d(l = 2)$ subshell in the $n = 2$ shell (Table 28–1). The outermost electron will have to be (at least) in the $n = 3$ shell.

The grouping of atoms in the periodic table is according to increasing mass. There is also a regularity according to chemical properties. And although this is treated in chemistry textbooks, we discuss it here briefly because it is a result of quantum mechanics. See the periodic table on the inside back cover.

All the noble gases (in the last column of the periodic table) have completely filled shells or subshells. That is, their outermost subshell is completely full, and the electron distribution is spherically symmetric. With such full spherical symmetry, other electrons are not attracted nor are electrons readily lost (ionization energy is high). This is why the noble gases are nonreactive (more on this when we discuss molecules and bonding in Chapter 29). Column seven contains the **halogens**, which lack one electron from a filled shell. Because of the shapes of the orbits (see Section 29–1), an additional electron can be accepted from another atom, and hence these elements are quite reactive. They have a valence of -1, meaning that when an extra electron is acquired, the resulting ion has a net charge of $-1e$. At the left of the periodic table, column I contains the **alkali metals**, all of which have a single outer s electron. This electron spends most of its time outside the inner closed shells and subshells which shield it from most of the nuclear charge. Indeed, it is relatively far from the nucleus and is attracted to it by a net charge of only about $+1e$, because of the shielding effect of the other electrons. Hence this outer electron is easily removed and can spend much of its time around another atom, forming a molecule. This is why the alkali metals are highly reactive and have a valence of $+1$. The other columns of the table can be treated similarly.

The presence of the transition elements in the center of the table, as well as the lanthanides (rare earths) and actinides below, is a result of incomplete inner shells. For the lowest Z elements, the subshells are filled in a simple order: first $1s$, then $2s$, followed by $2p$, $3s$, and $3p$. You might expect that $3d$ ($n = 3, l = 2$) would be filled next, but it isn't. Instead, the $4s$ level actually has a slightly lower energy than the $3d$ (due to electrons interacting with each other), so it fills first (K and Ca). Only then does the $3d$ shell start to fill up, beginning with Sc. (The $4s$ and $3d$ levels are close, so some elements have only one $4s$ electron, such as Cr.) Most of the chemical properties of these **transition elements** are governed by the relatively loosely held $4s$ electrons and hence they usually have valences of $+1$ or $+2$. See also Table 28–3. A similar effect is responsible for the rare earths, which are shown at the bottom of the periodic table for convenience. All have very similar chemical properties, which are determined by their two outer $6s$ or $7s$ electrons, whereas the different numbers of electrons in the unfilled inner shells have little effect.

28–9 X-Ray Spectra and Atomic Number

The line spectra of atoms in the visible, UV, and IR regions of the EM spectrum are mainly due to transitions between states of the outer electrons. Much of the charge of the nucleus is shielded from these electrons by the negative charge on the inner electrons. But the innermost electrons in the $n = 1$ shell "see" the full charge of the nucleus. Since the energy of a level is proportional to Z^2 (see Eq. 27–14), for an atom with $Z = 50$, we would expect wavelengths about $50^2 = 2500$ times shorter than those found in the Lyman series of hydrogen (around 100 nm), or 10^{-2} to 10^{-1} nm. Such short wavelengths lie in the X-ray region of the spectrum.

X-rays are produced when electrons accelerated by a high voltage strike the metal target inside the X-ray tube (Section 25–11). If we look at the spectrum of wavelengths emitted by an X-ray tube, we see that the spectrum consists of two parts: a continuous spectrum with a cutoff at some λ_0

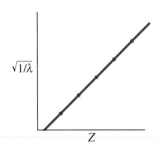

FIGURE 28-11 Spectrum of X-rays emitted from a molybdenum target in an X-ray tube operated at 50 kV.

FIGURE 28-12 Plot of $\sqrt{1/\lambda}$ vs. Z for K_α X-ray lines.

which depends only on the voltage across the tube, and a series of peaks superimposed. A typical example is shown in Fig. 28–11. The smooth curve and the cutoff wavelength λ_0 move to the left as the voltage across the tube increases. The peaks (labeled K_α and K_β in Fig. 28–11), however, remain at the same wavelength when the voltage is changed, although they are located at different wavelengths when different target materials are used. This observation suggests that the peaks are characteristic of the material used. Indeed, we can explain them by imagining that the electrons accelerated by the high voltage of the tube can reach sufficient energies that when they collide with the atoms of the target, they can knock out one of the very tightly held inner electrons. Then we explain these **characteristic X-rays** (the peaks in Fig. 28–11) as photons emitted when an electron in an upper state drops down to fill the vacated lower state.

The K lines result from transitions *into* the K shell ($n = 1$). The K_α line is a transition that originates from the $n = 2$ (L) shell, the K_β line from the $n = 3$ (M) shell. An L line is due to a transition into the L shell, and so on.

Measurement of the characteristic X-ray spectra has allowed a determination of the inner energy levels of atoms. It has also allowed the determination of Z values for many atoms, since (as we have seen) the wavelength of the shortest X-rays emitted will be inversely proportional to Z^2. Actually, for an electron jumping from, say, the $n = 2$ to the $n = 1$ level, the wavelength is inversely proportional to $(Z - 1)^2$ because the nucleus is shielded by the one electron that still remains in the $1s$ level. In 1914, H. G. J. Moseley (1887–1915) found that a plot of $\sqrt{1/\lambda}$ versus Z produced a straight line, Fig. 28–12. The Z values of a number of elements were determined by fitting them to such a **Moseley plot**. The work of Moseley put the concept of atomic number on a firm experimental basis.

EXAMPLE 28–7 **X-ray wavelength.** Estimate the wavelength for an $n = 2$ to $n = 1$ transition in molybdenum ($Z = 42$). What is the energy of such a photon?

SOLUTION We use the Bohr formula, Eq. 27–15, with Z^2 replaced by $(Z - 1)^2 = (41)^2$. Or, more simply, we can use the result of Example 27–11 for the $n = 2$ to $n = 1$ transition in hydrogen ($Z = 1$). Since

$$\lambda \propto \frac{1}{(Z - 1)^2},$$

we will have

$$\lambda = (1.22 \times 10^{-7} \text{ m})/(41)^2 = 0.073 \text{ nm}.$$

This is close to the measured value (Fig. 28–11) of 0.071 nm. Each of these photons would have energy (in eV) of:

$$E = hf = \frac{hc}{\lambda} = \frac{(6.63 \times 10^{-34} \text{ J·s})(3.00 \times 10^8 \text{ m/s})}{(7.3 \times 10^{-11} \text{ m})(1.60 \times 10^{-19} \text{ J/eV})} = 17 \text{ keV}.$$

EXAMPLE 28–8 **Determining atomic number.** High-energy photons are used to bombard an unknown material. The strongest peak is found for X-rays with an energy of 66 keV. Guess what the material is.

SOLUTION The strongest X-rays are generally for the K_α line (see Fig. 28–11) which occurs when photons knock out K shell electrons (the innermost orbit) and their place is taken by electrons from the L shell. We use the Bohr model, and assume the electrons "see" a nuclear charge of $Z - 1$ (screened by one electron). The hydrogen transition $n = 2$ to $n = 1$ would yield about 10.2 eV (see Fig. 27–32 or Example 27–11). Then since energy E is proportional to Z^2 (Eq. 27–14), or rather $(Z - 1)^2$ as we've just discussed, we can write

$$\frac{(Z - 1)^2}{1^2} = \frac{66 \times 10^3 \text{ eV}}{10.2 \text{ eV}} = 6.5 \times 10^3,$$

so $Z - 1 = \sqrt{6500} = 81$, and $Z = 82$, which makes it lead.

Now we briefly analyze the continuous part of an X-ray spectrum (Fig. 28–11) based on the photon theory of light. When electrons strike the target, they collide with atoms of the material and give up most of their energy as heat (about 99 percent, so X-ray tubes must be cooled). However, electrons can also give up energy by emitting a photon of light. An electron can be decelerated by interaction with atoms of the target (Fig. 28–13). But an accelerating charge can emit radiation (Chapter 22), and in this case it is called **bremsstrahlung** (German for "braking radiation"). Because energy is conserved, the energy of the emitted photon, hf, must equal the loss of kinetic energy of the electron, $\Delta\text{KE} = \text{KE} - \text{KE}'$, so

$$hf = \Delta\text{KE}.$$

An electron may lose all or a part of its energy in such a collision. The continuous X-ray spectrum (Fig. 28–11) is explained as being due to such bremsstrahlung collisions in which varying amounts of energy are lost by the electrons. The shortest-wavelength X-ray (the highest frequency) must be due to an electron that gives up *all* its kinetic energy to produce one photon in a single collision. Since the initial kinetic energy of an electron is equal to the energy given it by the accelerating voltage, V, then $\text{KE} = eV$. In a single collision in which the electron is brought to rest, we have

$$hf_0 = eV$$

or

$$\lambda_0 = \frac{hc}{eV}, \qquad \text{(28–4)}$$

where $\lambda_0 = c/f_0$ is the cutoff wavelength, Fig. 28–11. This prediction for λ_0 corresponds precisely with that observed experimentally. This result is further evidence that X-rays are a form of electromagnetic radiation (light)[†] and that the photon theory of light is valid.

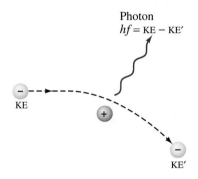

FIGURE 28–13
Bremsstrahlung photon produced by an electron decelerated by interaction with a target atom.

EXAMPLE 28–9 **Cutoff wavelength.** What is the shortest-wavelength X-ray photon emitted in an X-ray tube subjected to 50 kV?

SOLUTION From Eq. 28–4,

$$\lambda_0 = \frac{(6.6 \times 10^{-34} \text{ J·s})(3.0 \times 10^8 \text{ m/s})}{(1.6 \times 10^{-19} \text{ C})(5.0 \times 10^4 \text{ V})} = 2.5 \times 10^{-11} \text{ m},$$

or 0.025 nm. This agrees well with experiment, Fig. 28–11.

[†]If X-rays were not photons but rather neutral particles with rest mass m_0, Eq. 28–4 would not hold.

X-rays have many applications, some of which we discussed in Chapter 25, including imaging techniques.

* 28–10 Fluorescence and Phosphorescence

One photon absorbed — Two photons emitted

FIGURE 28–14 Fluorescence.

FIGURE 28–15 When UV light illuminates these various "fluorescent" rocks, they fluoresce in the visible region of the spectrum.

When an atom is excited from one energy state to a higher one by the absorption of a photon, it may return to the lower level in a series of two (or more) jumps if there is an energy level in between (Fig. 28–14). The photons emitted will consequently have lower energy and frequency than the absorbed photon. When the absorbed photon is in the UV and the emitted photons are in the visible region of the spectrum, this phenomenon is called **fluorescence** (Fig. 28–15).

The wavelength for which fluorescence will occur depends on the energy levels of the particular atoms. Because the frequencies are different for different substances, and because many substances fluoresce readily, fluorescence is a powerful tool for identification of compounds. It is also used for assaying—determining how much of a substance is present—and for following substances along a natural pathway as in plants and animals. For detection of a given compound, the stimulating light must be monochromatic, and solvents or other materials present must not fluoresce in the same region of the spectrum. Sometimes the observation of fluorescent light being emitted is sufficient to detect a compound. In other cases, spectrometers are used to measure the wavelengths and intensities of the emitted light.

Fluorescent lightbulbs work in a two-step process. The applied voltage accelerates electrons that strike atoms of the gas in the tube and cause them to be excited. When the excited atoms jump down to their normal levels, they emit UV photons which strike a fluorescent coating on the inside of the tube. The light we see is a result of this material fluorescing in response to the UV light striking it.

Materials such as those used for luminous watch dials are said to be **phosphorescent**. When an atom is raised to a normal excited state, it drops back down within about 10^{-8} s. In phosphorescent substances, atoms can be excited by photon absorption to energy levels, said to be **metastable**, which are states that last much longer—even a few seconds or longer. In a collection of such atoms, many of the atoms will descend to the lower state fairly soon, but many will remain in the excited state for over an hour. Hence light will be emitted even after long periods. When you put your watch dial close to a bright lamp, it excites many atoms to metastable states, and you can see the glow a long time after.

* 28–11 Lasers

A laser is a device that can produce a very narrow intense beam of monochromatic coherent light. (By *coherent*, we mean that across any cross section of the beam, all parts have the same phase.) The emitted beam is a nearly perfect plane wave. An ordinary light source, on the other hand, emits light in all directions (so the intensity decreases rapidly with distance), and the emitted light is incoherent (the different parts of the beam are not in phase with each other). The excited atoms that emit the light in an ordinary lightbulb act independently, so each photon emitted can be considered as a short wave train, typically 30 cm long and lasting 10^{-8} s. These wave trains bear no phase relation to one another. Just the opposite is true of lasers.

The action of a laser is based on quantum theory. We have seen that a photon can be absorbed by an atom if (and only if) its energy hf corresponds to the energy difference between an occupied energy level of the atom and an available excited state, Fig. 28–16a. This is, in a sense, a resonant situation. If the atom is already in the excited state, it may of course jump spontaneously (i.e. no apparent stimulus) to the lower state with the emission of a photon. However, if a photon with this same energy strikes the excited atom, it can stimulate the atom to make the transition sooner to the lower state, Fig. 28–16b. This phenomenon is called **stimulated emission**, and it can be seen that not only do we still have the original photon, but also a second one of the same frequency as a result of the atom's transition. And these two photons are exactly *in phase*, and they are moving in the same direction. This is how coherent light is produced in a laser. Hence the name "laser," which is an acronym for **l**ight **a**mplification by **s**timulated **e**mission of **r**adiation.

Normally, most atoms are in the lower state, so incident photons will mostly be absorbed. In order to obtain the coherent light from stimulated emission, two conditions must be satisfied. First, the atoms must be excited to the higher state. That is, an **inverted population** is needed, one in which more atoms are in the upper state than in the lower one (Fig. 28–17), so that *emission* of photons will dominate over absorption. And second, the higher state must be a **metastable state**—a state in which the electrons remain longer than usual[†] so that the transition to the lower state occurs by stimulated emission rather than spontaneously. How these conditions are achieved for different lasers will be discussed shortly. For now, we assume that the atoms have been excited to an upper state. Figure 28–18 is a schematic diagram of a laser: the "lasing" material is placed in a long narrow tube at the ends of which are two mirrors, one of which is partially transparent (perhaps 1 or 2 percent). Some of the excited atoms drop down fairly soon after being excited. One of these is the atom shown on the far left in Fig. 28–18. If the emitted photon strikes another atom in the excited state, it stimulates this atom to emit a photon of the *same* frequency, moving in the *same* direction, and *in phase* with it. These two photons then move on to strike other atoms

Stimulated emission

FIGURE 28–16 (a) Absorption of a photon. (b) Stimulated emission.

FIGURE 28–17 Two energy levels for a collection of atoms. Each dot represents the energy state of one atom. (a) A normal situation; (b) an inverted population.

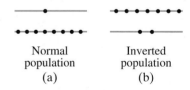

| Normal population (a) | Inverted population (b) |

FIGURE 28–18 Laser diagram, showing excited atoms stimulated to emit light.

[†]An excited atom may land in such a state and can jump to a lower state only by a so-called forbidden transition (discussed in Section 28–6), which is why its lifetime is longer than normal

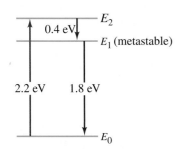

FIGURE 28–19 Energy levels of chromium in a ruby crystal. Photons of energy 2.2 eV "pump" atoms from E_0 to E_2, which then decay to metastable state E_1. Lasing action occurs by stimulated emission of photons in transition from E_1 to E_0.

Ruby laser

He–Ne laser

causing more stimulated emission. As the process continues, the number of photons multiplies. When the photons strike the end mirrors, most are reflected back, and as they move in the opposite direction, they continue to stimulate more atoms to emit photons. As the photons move back and forth between the mirrors, a small percentage passes through the partially transparent mirror at one end. These photons make up the narrow coherent external laser beam.

Inside the tube, some spontaneously emitted photons will be emitted at an angle to the axis, and these will merely go out the side of the tube and not affect the narrowness of the main beam. In a well-designed laser, the spreading of the beam is limited only by diffraction, so the angular spread is $\approx \lambda/D$ (see Eq. 24–3) where D is the diameter of the end mirror. The diffraction spreading can be incredibly small. The light energy, instead of spreading out in space as it does for an ordinary light source, is directed in a pencil-thin beam.

The excitation of the atoms in a laser can be done in several ways to produce the necessary inverted population. In a ruby laser, the lasing material is a ruby rod consisting of Al_2O_3 with a small percentage of aluminum (Al) atoms replaced by chromium (Cr) atoms. The Cr atoms are the ones involved in lasing. The atoms are excited by strong flashes of light of wavelength 550 nm, which corresponds to a photon energy of 2.2 eV. As shown in Fig. 28–19, the atoms are excited from state E_0 to state E_2. This process is called **optical pumping**. The atoms quickly decay either back to E_0 or to the intermediate state E_1, which is metastable with a lifetime of about 3×10^{-3} s (compared to 10^{-8} s for ordinary levels). With strong pumping action, more atoms can be forced into the E_1 state than are in the E_0 state. Thus we have the inverted population needed for lasing. As soon as a few atoms in the E_1 state jump down to E_0, they emit photons that produce stimulated emission of the other atoms and the lasing action begins. A ruby laser thus emits a beam whose photons have energy 1.8 eV and a wavelength of 694.3 nm (or "ruby-red" light).

In a helium–neon (He–Ne) laser, the lasing material is a gas, a mixture of about 15 percent He and 85 percent Ne. The atoms are excited by applying a high voltage to the tube so that an electric discharge takes place within the gas. In the process, some of the He atoms are raised to the metastable state E_1 shown in Fig. 28–20, which corresponds to a jump of 20.61 eV, almost exactly equal to an excited state in neon, 20.66 eV. The

FIGURE 28–20 Energy levels for He and Ne. He is excited in the electric discharge to the E_1 state. This energy is transferred to the E_3' level of the Ne by collision. E_3' is metastable and decays to E_2' by stimulated emission.

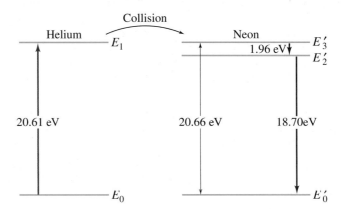

He atoms do not quickly return to the ground state by spontaneous emission, but instead often give their excess energy to a Ne atom when they collide—see Fig. 28–20. In such a collision, the He drops to the ground state and the Ne atom is excited to the state E_3' (the prime refers to neon states). The slight difference in energy (0.05 eV) is supplied by the kinetic energy of the moving molecules. In this manner, the E_3' state in Ne—which is metastable—becomes more populated than the E_2' level. This inverted population between E_3' and E_2' is what is needed for lasing.

Other types of laser include: chemical lasers, in which the energy input comes from the chemical reaction of highly reactive gases; dye lasers, whose frequency is tunable; CO_2 gas lasers, capable of high power output in the infrared; rare-earth solid-state lasers such as the high-power Nd:Yag laser; and the *pn* junction laser in which the transitions occur between the bottom of the conduction band and the upper part of the valence band (Section 29–6).

Other lasers

The excitation of the atoms in a laser can be done continuously or in pulses. In a **pulsed laser**, the atoms are excited by periodic inputs of energy. The multiplication of photons continues until all the atoms have been stimulated to jump down to the lower state, and the process is repeated with each input pulse. In a **continuous laser**, the energy input is continuous so that as atoms are stimulated to jump down to the lower level, they are soon excited back up to the upper level so that the output is a continuous laser beam. Any laser, of course, is not a source of energy. Energy must be put in, and the laser converts a part of this input energy into an intense narrow beam output.

The unique feature of light from a laser is, as mentioned before, that it is a coherent narrow beam of a single frequency (or several distinct frequencies). Because of this feature, the laser has found many applications. Lasers are a useful surgical tool. The narrow intense beam can be used to destroy tissue in a localized area, or to break up gallstones and kidney stones. Because of the heat produced, a laser beam can be used to "weld" broken tissue, such as a detached retina. For some types of internal surgery, the laser beam can be carried by an optical fiber (Section 23–6) to the surgical point, sometimes as an additional fiber-optic path on an endoscope, Fig. 28–21 (also discussed in Section 23–6). An example is the removal of plaque clogging human arteries. Tiny organelles within a living cell have been destroyed using lasers by researchers studying how the absence of that organelle affects the behavior of the cell. Laser beams have been used to destroy cancerous and precancerous cells; at the same time, the heat seals off capillaries and lymph vessels, thus "cauterizing" the wound in the process to prevent spread of the disease. The intense heat produced in a small area by a laser beam is also used for welding and machining metals and for drilling tiny holes in hard materials. The beam of a laser is narrow in itself (typically, a few mm). But because the beam is coherent, monochromatic, and essentially parallel and narrow, lenses can be used to focus the light into incredibly small areas without the usual aberration problems. The limiting factor thus becomes diffraction, and the energy crossing unit area per unit time can be very large. The precise straightness of a laser beam is also useful to surveyors for lining up equipment precisely, especially in inaccessible places.

➡ **PHYSICS APPLIED**

Medical and other uses of lasers

FIGURE 28–21 Laser being used in eye surgery.

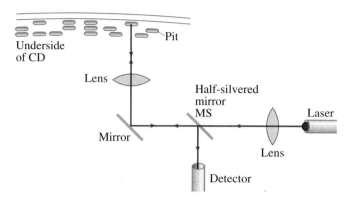

FIGURE 28–22 Reading a CD. The fine beam of a laser, focused even more finely with lenses, is directed at the undersurface of a rotating compact disk. The beam is reflected back from the areas between pits but reflects much less from pits. The reflected light is detected as shown, reflected by a half-reflecting mirror MS. The strong and weak reflections correspond to the 0s and 1s of the binary code representing the audio or video signal.

➡ **PHYSICS APPLIED**

*CD players
and bar codes*

In everyday life, lasers are used as bar-code readers (at store checkout stands) and in compact disc (CD) players. The laser beam reflects off the stripes and spaces of a bar code, and off the tiny pits of a CD as shown in Fig. 28–22. The recorded information on a CD is a series of pits and spaces representing 0s and 1s (or "off" and "on") of a digitized code that must be decoded electronically before being sent to the audio or video system. A bar-code reader is similar.

* 28–12 Holography

➡ **PHYSICS APPLIED**

Holography

One of the most interesting applications of laser light is the production of three-dimensional images called **holograms** (see Fig. 28–23). In an ordinary photograph, the film simply records the intensity of light reaching it at each point. When the photograph or transparency is viewed, light reflecting from it or passing through it gives us a two-dimensional picture. In holography, the images are formed by interference, without lenses. When a laser hologram is made on film, a broadened laser beam is split into two parts by a half-silvered mirror, Fig. 28–23. One part goes directly to the film; the rest passes to the object to be photographed, from which it is reflected to the film. Light from every point on the object reaches each point on the film, and the interference of the two beams allows the film to record both the intensity and relative phase of the light at each point. It is crucial that the light be coherent—that is, in phase at all points—which is why a laser is used. After the film is developed, it is placed again in a laser beam and a three-dimensional image of the object is created. You can walk around such an

FIGURE 28–23 Making a hologram. Light reflected from various points on the object interferes (at the film) with light from the direct beam.

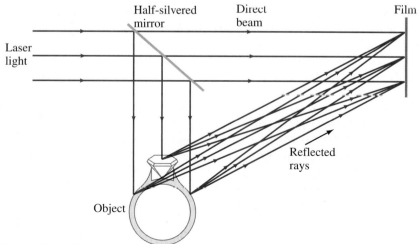

image and see it from different sides as if it were the original object. Yet, if you try to touch it with your hand, there will be nothing material there.

So-called **volume** or **white-light holograms** do not require a laser to see the image, but can be viewed with ordinary white light (preferably a nearby point source, such as the Sun or a clear bulb with a small bright filament). Such holograms must be made, however, with a laser. They are made not on thin film, but on a *thick* emulsion. The interference pattern, instead of being two-dimensional as for an ordinary hologram, is actually three-dimensional (hence the name "volume hologram"). The interference pattern in the film emulsion can be thought of as an array of bands or ribbons (consisting of the silver grains from the development process) where constructive interference occurred. This array, and the reconstruction of the image, can be compared to Bragg scattering of X-rays from the atoms in a crystal (see Section 25–11). White light can reconstruct the image because the Bragg condition ($m\lambda = 2d \sin \theta$) selects out the appropriate single wavelength. If the hologram is originally produced by lasers emitting the three additive primary colors (red, green, and blue), the three-dimensional image can be seen in full color when viewed with white light.

White-light holograms

SUMMARY

In 1925, Schrödinger and Heisenberg separately worked out a new theory, **quantum mechanics**, which is now considered to be the basic theory at the atomic level. It is a statistical theory rather than a deterministic one.

An important aspect of quantum mechanics is the **Heisenberg uncertainty principle**. It results from the wave–particle duality and the unavoidable interaction between the observed object and the observer. One form of the uncertainty principle states that both the position x and momentum p of an object cannot be measured precisely at the same time. The products of the uncertainties, $(\Delta x)(\Delta p)$, can be no less than $\hbar \ (= h/2\pi)$:

$$(\Delta p)(\Delta x) \gtrsim \hbar.$$

Another form states that the energy can be uncertain or nonconserved, by an amount ΔE for a time Δt where

$$(\Delta E)(\Delta t) \gtrsim \hbar.$$

In the quantum mechanical view of the atom, the electrons do not have well-defined orbits, but instead exist as a "cloud." Electron clouds can be interpreted as an electron wave spread out in space, or as a **probability distribution** for electrons considered as particles.

According to quantum mechanics, the state of an electron in an atom is specified by four **quantum numbers**: n, l, m_l, and m_s. The principal quantum number, n, can take on any integer value $(1, 2, 3, \cdots)$ and corresponds to the quantum number of the old Bohr theory; l can take on values from 0 up to $n - 1$; m_l can take on integer values from $-l$ to $+l$; and m_s can be $+\frac{1}{2}$ or $-\frac{1}{2}$. The energy levels in the hydrogen atom depend on n, whereas in other atoms they depend on n and l. When an external magnetic field is applied, the spectral lines are split (the **Zeeman effect**), indicating that the energy depends also on m_l in this case. Even in the absence of a magnetic field, precise measurements of spectral lines show a tiny splitting of the lines called **fine structure**, whose explanation is that the energy depends very slightly on m_l and m_s.

The arrangement of electrons in multielectron atoms is governed by the **Pauli exclusion principle**, which states that no two electrons can occupy the same quantum state—that is, they cannot have the same set of quantum numbers n, l, m_l, and m_s. Electrons, as a result, are grouped into **shells** (according to the value of n) and **subshells** (according to l).

Electron configurations are specified using the numerical values of n, but using letters for l: s, p, d, f, etc., for $l = 0, 1, 2, 3$, and so on, plus a superscript for the number of electrons in that subshell. Thus, the ground state of hydrogen is $1s^1$, whereas that for oxygen is $1s^2 2s^2 2p^4$. The **periodic table** arranges the elements in horizontal rows according to increasing atomic number (number of electrons in the neutral atom); each vertical column contains elements with similar chemical properties.

X-rays, which are a form of electromagnetic radiation of very short wavelength, are produced when high-speed electrons strike a target. The spectrum of X-rays so produced consists of two parts, a continuous spectrum produced when the electrons are decelerated by atoms of the target, and peaks representing photons emitted by atoms of the target after being excited by collision with the high-speed electrons.

QUESTIONS

1. Compare a matter wave Ψ to (a) a wave on a string, (b) an EM wave. Discuss similarities and differences.

2. Explain why Bohr's theory of the atom is not compatible with quantum mechanics, particularly the uncertainty principle.

3. Explain why it is that the more massive an object is, the easier it becomes to predict its future position.

4. In view of the uncertainty principle, why does a baseball seem to have a well-defined position and speed whereas an electron does not?

5. Would it ever be possible to balance a very sharp needle precisely on its point? Explain.

6. A cold thermometer is placed in a hot bowl of soup. Will the temperature reading of the thermometer be the same as the temperature of the hot soup before the measurement was made?

7. When you check the pressure in a tire, doesn't some air inevitably escape? Is it possible to avoid this escape of air altogether? What is the relation to the uncertainty principle?

8. It has been said that the ground-state energy in the hydrogen atom can be precisely known but the excited states have some uncertainty in their values (an "energy width"). Is this consistent with the uncertainty principle in its energy form? Explain.

9. Which model of the hydrogen atom, the Bohr model or the quantum-mechanical model, predicts that the electron spends more time near the nucleus?

10. The size of atoms varies by only a factor of three or so from largest to smallest, yet the number of electrons varies from one to over 100. Why?

11. Excited hydrogen and excited helium atoms both radiate light as they jump down to the $n = 1$, $l = 0$, $m_l = 0$ state. Yet the two elements have very different emission spectra. Why?

12. The 589-nm yellow line in sodium is actually two very closely spaced lines. This splitting is due to an "internal" Zeeman effect. Can you explain this? [Hint: Put yourself in the reference frame of the electron.]

13. Which of the following electron configurations are forbidden? (a) $1s^2 2s^2 2p^4 3s^2 4p^2$; (b) $1s^2 2s^2 2p^8 2s^1$, (c) $1s^2 2s^2 2p^6 3s^2 3p^5 4s^2 4d^5 4f^1$.

14. Give the complete electron configuration for a uranium atom (careful scrutiny across the periodic table on the inside back cover will provide useful hints).

15. In what column of the periodic table would you expect to find the atom with each of the following configurations? (a) $1s^2 2s^2 2p^6 3s^2$; (b) $1s^2 2s^2 2p^6 3s^2 3p^6$; (c) $1s^2 2s^2 2p^6 3s^2 3p^6 4s^1$; (d) $1s^2 2s^2 2p^5$.

16. The ionization energy for neon ($Z = 10$) is 21.6 eV and that for sodium ($Z = 11$) is 5.1 eV. Explain the large difference.

17. Why do chlorine and iodine exhibit similar properties?

18. Explain why potassium and sodium exhibit similar properties.

19. Why are the chemical properties of the rare earths so similar?

20. Why do we not expect perfect agreement between measured values of X-ray line wavelengths and those calculated using Bohr theory, as in Example 28–7?

21. How would you figure out which lines in an X-ray spectrum correspond to K_α, K_β, L, etc., transitions?

22. Why do we expect electron transitions deep within an atom to produce shorter wavelengths than transitions by outer electrons?

* 23. Compare spontaneous emission to stimulated emission.

* 24. How does laser light differ from ordinary light? How is it the same?

* 25. Explain how a 0.0005-W laser beam, photographed at a distance, can seem much stronger than a 1000-W street lamp.

PROBLEMS

SECTION 28–2

1. (II) The neutrons in a parallel beam, each having kinetic energy $\frac{1}{40}$ eV, are directed through two slits 0.50 mm apart. How far apart will the interference peaks be on a screen 1.0 m away?

2. (II) Bullets of mass 3.0 g are fired in parallel paths with speeds of 200 m/s through a hole 3.0 mm in diameter. How far from the hole must you be to detect a 1.0-cm diameter spread in the beam?

3. (I) A proton is traveling with a speed of $(6.560 \pm 0.012) \times 10^5$ m/s. With what maximum accuracy can its position be ascertained?

4. (I) If an electron's position can be measured to an accuracy of 2.0×10^{-8} m, how accurately can its velocity be known?

5. (I) An electron remains in an excited state of an atom for typically 10^{-8} s. What is the minimum uncertainty in the energy of the state (in eV)?

6. (I) The Z^0 boson, discovered in 1985, is the mediator of the weak nuclear force, and it typically decays very quickly. Its average rest energy is 91.19 GeV, but its short lifetime shows up as an intrinsic width of 2.5 GeV. What is the lifetime of this particle?

7. (II) What is the uncertainty in the mass of a muon $(m = 105.7 \text{ MeV}/c^2)$, specified in eV/c^2, given its lifetime of 2.20 μs?

8. (II) A free neutron $(m = 1.67 \times 10^{-27}$ kg) has a mean life of 900 s. What is the uncertainty in its mass (in kg)?

9. (II) An electron and a 140-g baseball are each traveling 150 m/s measured to an accuracy of 0.055 percent. Calculate and compare the uncertainty in position of each.

10. (II) Estimate the lowest possible energy of a neutron contained in a typical nucleus of radius 1.0×10^{-15} m. [*Hint*: A particle can have an energy at least as large as its uncertainty.]

11. (II) Use the uncertainty principle to show that if an electron were present in the nucleus $(r \approx 10^{-15}$ m), its kinetic energy (use relativity) would be hundreds of MeV. (Since such electron energies are not observed, we conclude that electrons are not present in the nucleus.) [*Hint*: See hint for Problem 10.]

12. (III) How accurately can the position of a 3.00-keV electron be measured assuming its energy is known to 1.00 percent?

13. (III) The uncertainty principle can be stated in terms of angular quantities as follows:

$$\Delta L \Delta \phi \gtrsim \hbar.$$

Here, L stands for angular momentum along a given axis, and ϕ for the angular position measured in a plane perpendicular to that axis. (a) Make a plausibility argument for this relation. (b) Electrons in atoms have well-defined quantized values of angular momentum, with no uncertainty. What does this say about the uncertainty in angular position and the concept of electron orbits?

SECTIONS 28-6 TO 28-8

14. (I) For $n = 6$, what values can l have?

15. (I) For $n = 5$, $l = 4$, what are the possible values of m_l and m_s?

16. (I) How many electrons can be in the $n = 6$, $l = 3$ subshell?

17. (I) List the quantum numbers for each electron in the ground state of nitrogen $(Z = 7)$.

18. (I) List the quantum numbers for each electron in the ground state of magnesium $(Z = 12)$.

19. (I) How many different states are possible for an electron whose principal quantum number is $n = 4$? Write down the quantum numbers for each state.

20. (I) A hydrogen atom is known to have $l = 4$. What are the possible values for n, m_l, and m_s?

21. (I) If a hydrogen atom has $m_l = -3$, what are the possible values of n, l, and m_s?

22. (I) Calculate the magnitude of the angular momentum of an electron in the $n = 4$, $l = 3$ state of hydrogen.

23. (II) Show that there can be 18 electrons in a "g" subshell.

24. (II) What is the full electron configuration in the ground state for elements with Z equal to (a) 27, (b) 36, (c) 38? [*Hint*: See the periodic table inside the back cover.]

25. (II) What is the full electron configuration for (a) selenium (Se), (b) gold (Au), (c) uranium (U)? [*Hint*: See the periodic table inside the back cover.]

26. (II) For each of the following atomic transitions, state whether the transition is *allowed* or *forbidden*, and if forbidden, what rule is being violated: (a) $4p \rightarrow 3p$; (b) $2p \rightarrow 1s$; (c) $3d \rightarrow 2d$; (d) $4d \rightarrow 3s$; (e) $4s \rightarrow 2p$.

27. (II) A hydrogen atom is in the $6h$ state. Determine (a) the principal quantum number, (b) the energy of the state, (c) the orbital angular momentum and its quantum number l, and (d) the possible values for the magnetic quantum number.

28. (II) Estimate the binding energy of the third electron in lithium using the Bohr theory. [*Hint*: This electron has $n = 2$ and "sees" a net charge of approximately $+1e$.] The measured value is 5.36 eV.

29. (II) The ionization (binding) energy of the outermost electron in boron is 8.26 eV. (a) Use the Bohr model to estimate the "effective charge," Z_{eff}, seen by this electron. (b) Estimate the average orbital radius.

30. (II) Show that the total angular momentum is zero for a filled subshell.

31. (II) Show that the maximum number of electrons allowed in any subshell is equal to $2(2l + 1)$ where l is the angular momentum quantum number of the subshell.

SECTION 28-9

32. (I) What are the shortest-wavelength X-rays emitted by electrons striking the face of a 30-kV TV picture tube? What are the longest wavelengths?

33. (I) If the shortest-wavelength bremsstrahlung X-rays emitted from an X-ray tube have $\lambda = 0.030$ nm, what is the voltage across the tube?

34. (I) Show that the cutoff wavelength λ_0 is given by

$$\lambda_0 = 1240 \text{ nm}/V,$$

where V is the X-ray tube voltage in volts.

35. (II) Use the result of Example 28–7 to estimate the X-ray wavelength emitted when a Co ($Z = 27$) atom jumps from $n = 2$ to $n = 1$.

36. (II) Estimate the wavelength for an $n = 2$ to $n = 1$ transition in chromium ($Z = 24$).

37. (II) Use the Bohr theory to estimate the wavelength for an $n = 3$ to $n = 1$ transition in molybdenum. The measured value is 0.063 nm. Why do we not expect perfect agreement?

38. (II) A mixture of iron and an unknown material is bombarded with electrons. The wavelength of the K_α lines are 194 pm for iron and 229 pm for the unknown. What is the unknown material?

*39. (II) A laser used to weld detached retinas puts out 30-ms-long pulses of 640-nm light which average 0.60-W output during a pulse. How much energy can be deposited per pulse and how many photons does each pulse contain?

*40. (II) Estimate the angular spread of a laser beam due to diffraction if the beam emerges through a 3.0-mm-diameter mirror. Assume that $\lambda = 694$ nm. What would be the diameter of this beam if it struck a satellite 300 km above the Earth?

GENERAL PROBLEMS

41. Use the uncertainty principle to estimate the position uncertainty for the electron in the ground state of the hydrogen atom. [Hint: Determine the momentum using the Bohr model of Section 27–10 and assume the momentum can be anywhere between this value and zero.] How does this result compare to the Bohr radius?

42. An electron in the $n = 2$ state of hydrogen remains there on the average about 10^{-8} s before jumping to the $n = 1$ state. (a) Estimate the uncertainty in the energy of the $n = 2$ state. (b) What fraction of the transition energy is this? (c) What is the wavelength, and width (in nm), of this line in the spectrum of hydrogen?

43. What are the largest and smallest possible values for the angular momentum L of an electron in the $n = 5$ shell?

44. Estimate (a) the quantum number l for the orbital angular momentum of the Earth about the Sun, and (b) the number of possible orientations for the plane of Earth's orbit.

45. A 12-g bullet leaves a rifle at a speed of 180 m/s. (a) What is the wavelength of this bullet? (b) If the position of the bullet is known to an accuracy of 0.60 cm (radius of the barrel), what is the minimum uncertainty in its momentum? (c) If the accuracy of the bullet were determined only by the uncertainty principle (an unreasonable assumption), by how much might the bullet miss a pinpoint target 200 m away?

46. Using the Bohr formula for the radius of an electron orbit, estimate the average distance from the nucleus for an electron in the innermost ($n = 1$) orbit of a lead atom ($Z = 82$). Approximately how much energy would be required to remove this innermost electron?

47. An X-ray tube operates at 100 kV with a current of 25 mA and nearly all the electron energy goes into heat. If the specific heat capacity of the 0.085-kg plate is 0.11 kcal/kg·C°, what will be the temperature rise per minute if no cooling water is used?

48. Use the Bohr theory (especially Eq. 27–15) to show that the Moseley plot (Fig. 28–12) can be written

$$\sqrt{\frac{1}{\lambda}} = a(Z - b),$$

where $b \approx 1$, and evaluate a.

49. In the so-called *vector model* of the atom, space quantization of angular momentum (Fig. 28–7) is illustrated as shown in Fig. 28–24. The angular momentum vector of magnitude $L = \sqrt{l(l + 1)}$ is thought of as precessing around the z axis (like a spinning top or gyroscope) in such a way that the z component of angular momentum, $L_z = m_l$, also stays constant. Calculate the possible values for the angle θ between **L** and the z axis (a) for $l = 1$, (b) $l = 2$, and (c) $l = 3$. (d) Determine the minimum value of θ for $l = 100$ and $l = 10^6$. Is this consistent with the correspondence principle?

FIGURE 28–24
The vector model for orbital angular momentum. The orbital angular momentum vector **L** is imagined to precess about the z axis; L and L_z remain constant, but L_x and L_y continually change. Problem 49.

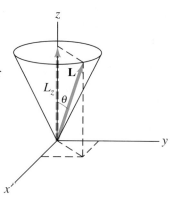

50. Show that the diffractive spread of a laser beam, $\approx \lambda/D$ as described in Section 28–11, is precisely what you might expect from the uncertainty principle. [Hint: Since the beam's width is constrained by the dimension of the aperture D, the component of the light's momentum perpendicular to the laser axis is uncertain.]

A silicon "chip" containing a complex of electronic elements.

MOLECULES AND SOLIDS 29

Since its development in the 1920s, quantum mechanics has had a profound influence on our lives, both intellectually and technologically. Even the way we view the world has changed, as we saw in Chapter 20. In the present chapter, we will discuss how quantum mechanics has given us an understanding of the structure of molecules and matter in bulk, as well as a number of important applications including semiconductor devices and applications to biology. Our discussions will necessarily be qualitative for the most part.

* 29-1 Bonding in Molecules

One of the great successes of quantum mechanics was to give scientists, at last, an understanding of the nature of chemical bonds. Since it is based in physics, and because this understanding is so important in many fields, we discuss it here.

By a molecule, we mean a group of two or more atoms that are strongly held together so as to function as a single unit. When atoms make such an attachment, we say that a chemical **bond** has been formed. There are two main types of strong chemical bond: covalent and ionic. Many bonds are actually intermediate between these two types.

Covalent Bonds. To understand how covalent bonds are formed, we take the simplest case, the bond that holds two hydrogen atoms together to form the hydrogen molecule, H_2. The mechanism is basically the same for

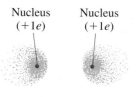

FIGURE 29–1 Electron probability distribution (electron cloud) for two H atoms when their spins are the same ($S = 1$).

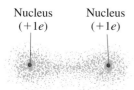

FIGURE 29–2 Electron probability distribution (cloud) around two H atoms when their spins are opposite ($S = 0$). In this case, a bond is formed because the positive nuclei are attracted to the concentration of negative charge between them. This is a hydrogen molecule, H_2.

Covalent bond

other covalent bonds. As two H atoms approach each other, the electron clouds begin to overlap, and the electrons from each atom can "orbit" both nuclei. (This is sometimes called "sharing" electrons.) If both electrons are in the ground state ($n = 1$) of their respective atoms, there are two possibilities: their spins can be parallel (both up or both down), in which case the total spin is $S = \frac{1}{2} + \frac{1}{2} = 1$; or their spins can be opposite ($m_s = +\frac{1}{2}$ for one, $m_s = -\frac{1}{2}$ for the other), so that the total spin $S = 0$. We shall now see that a bond is formed only for the $S = 0$ state, when the spins are opposite. First we consider the $S = 1$ state, for which the spins are the same. The two electrons cannot both be in the lowest energy state and be attached to the same atom, for then they would have identical quantum numbers in violation of the exclusion principle. The exclusion principle tells us that since no two electrons can occupy the same quantum state, if two electrons have the same quantum numbers, they must be different in some other way—namely, by being in different places in space (for example, attached to different atoms). When the two atoms approach, the electrons will stay away from each other as shown by the probability distribution of Fig. 29–1. The positively charged nuclei then repel each other, and no bond is formed.

For the $S = 0$ state, on the other hand, the spins are opposite and the two electrons are consequently in different quantum states. Hence they can come close together spatially. In this case, the probability distribution looks like Fig. 29–2. As can be seen, the electrons spend much of their time between the two positively charged nuclei, and the latter are attracted to the negatively charged electron cloud between them. This attraction, which holds the two atoms together to form a molecule, constitutes a *covalent bond*.

The probability distributions of Figs. 29–1 and 29–2 can perhaps be better understood on the basis of waves. What the exclusion principle requires is that when the spins are the same, there is destructive interference of the electron wave functions in the region between the two atoms. But when the spins are opposite, constructive interference occurs in the region between the two atoms, resulting in a large amount of negative charge there. Thus a covalent bond can be said to be the result of constructive interference of the electron wave functions in the space between the two atoms, and of the electrostatic attraction of the two positive nuclei for the negative charge concentration between them.

Why a bond is formed can also be understood from the energy point of view. When the two H atoms approach close to one another, if the spins of their electrons are opposite, the electrons can occupy the same space, as discussed above. This means that each electron can now move about in the space of two atoms instead of in the volume of only one. Because each electron now occupies more space, it is less well localized. From the uncertainty principle, with Δx increased, the momentum, and hence the energy, can be less. Another way of understanding this is from the point of view of de Broglie waves (Section 27–11). Because each electron has a larger "orbit," its wavelength λ can be longer, so its momentum $p = h/\lambda$ (Eq. 27–8) can be less. With less momentum, each electron has less energy when the two atoms combine than when they are separate. That is, the molecule has less energy than the two separate atoms, and so is more stable. An energy input is required to break the H_2 molecule into two separate H atoms, so the H_2 molecule is a stable entity. This is what we mean by a *bond*. The energy required to break a bond is called the **bond energy**, the **binding energy**, or the **dissociation energy**.

Bond energy

Ionic Bonds. An ionic bond is, in a sense, a special case of the covalent bond. Instead of the electrons being shared equally, they are shared unequally. For example, in sodium chloride (NaCl), the outer electron of the sodium spends nearly all its time around the chlorine (Fig. 29–3). The chlorine atom acquires a net negative charge as a result of the extra electron, whereas the sodium atom is left with a net positive charge. The electrostatic attraction between these two charged atoms holds them together. The resulting bond is called an *ionic bond* because it is created by the attraction between the two ions (Na^+ and Cl^-). But to understand the ionic bond, we must understand why the extra electron from the sodium spends so much of its time around the chlorine. After all, the chlorine is neutral; why should it attract another electron?

The answer lies in the probability distributions of the two neutral atoms. Sodium contains 11 electrons, 10 of which are in spherically symmetric closed shells (Fig. 29–4). The last electron spends most of its time beyond these closed shells. Because the closed shells have a total charge of $-10e$ and the nucleus has charge $+11e$, the outermost electron in sodium "feels" a net attraction due to $+1e$. It is not held very strongly. On the other hand, 12 of chlorine's 17 electrons form closed shells, or subshells (corresponding to $1s^2 2s^2 2p^6 3s^2$). These 12 form a spherically symmetric shield around the nucleus. The other five electrons are in $3p$ states whose probability distributions are not spherically symmetric and have a form similar to those for the $2p$ states in hydrogen shown in Fig. 28–9b and c. Four of these $3p$ electrons can have "doughnut-shaped" distributions symmetric about the z axis, as shown in Fig. 29–5. The fifth can have a "barbell-shaped" distribution (as for $m_l = 0$ in Fig. 28–9b), which in Fig. 29–5 is shown only in faint outline because it is half empty. That is, the exclusion principle allows one more electron to be in this state (it will have spin opposite to that of the electron already there). If an extra electron—say from a Na atom—happens to be in the vicinity, it can be in this state, say at point x in Fig. 29–5. It could experience an attraction due to as much as $+5e$ because the $+17e$ of the nucleus is partly shielded at this point by the 12 inner electrons. Thus, the outer electron of a Na atom will be more strongly attracted by the $+5e$ of the chlorine atom than by the $+1e$ of its own atom. This, combined with the strong attraction between the two ions when the extra electron stays with the Cl^-, produces the charge distribution of Fig. 29–3, and hence the ionic bond.

Ionic bond

FIGURE 29–3 Probability distribution for last electron of Na in NaCl.

FIGURE 29–4 In a neutral sodium atom, the 10 inner electrons shield the nucleus, so the single outer electron is attracted by a net charge of $+1e$.

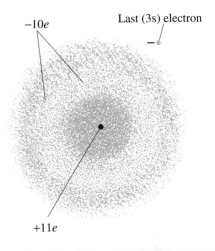

$-10e$

Last (3s) electron

$+11e$

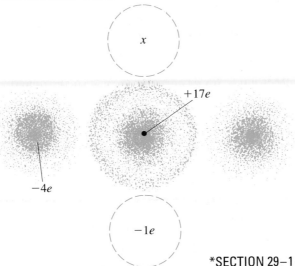

x

$+17e$

$-4e$

$-1e$

FIGURE 29–5 Neutral chlorine atom. The $+17e$ of the nucleus is shielded by the 12 electrons in the inner shells and subshells. Four of the five $3p$ electrons are shown in doughnut-shaped clouds, and the fifth is in the (dashed-line) cloud concentrated about the z axis (vertical). An extra electron at x will be attracted by a net charge that can be as much as $+5e$.

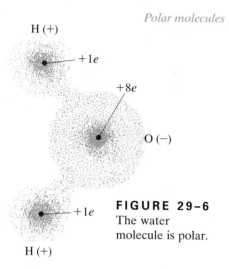

Polar molecules

H (+)

+1e

+8e

O (−)

+1e

H (+)

FIGURE 29–6
The water
molecule is polar.

Partial Ionic Character of Covalent Bonds. A pure covalent bond in which the electrons are shared equally occurs mainly in symmetrical molecules such as H_2, O_2, and Cl_2. When the atoms involved are different from each other, it is usual to find that the shared electrons are more likely to be in the vicinity of one atom than the other. The extreme case is an ionic bond; in intermediate cases the covalent bond is said to have a *partial ionic character*. The molecules themselves are **polar**—that is, one part (or parts) of the molecule has a net positive charge and other parts a net negative charge. An example is the water molecule, H_2O (Fig. 29–6). The shared electrons are more likely to be found around the oxygen atom than around the two hydrogens. The reason is similar to that discussed above in connection with ionic bonds. Oxygen has eight electrons ($1s^2 2s^2 2p^4$), of which four form a spherically symmetric core and the other four could have, for example, a doughnut-shaped distribution. The barbell-shaped distribution on the z axis (like that shown dashed in Fig. 29–5) could be empty, so electrons from hydrogen atoms can be attracted by a net charge of $+4e$. They are also attracted by the H nuclei, so they partly orbit the H atoms as well as the O atom. The net effect is that there is a net positive charge on each H atom (less than $+1e$), because the electrons spend only part of their time there. And, there is a net negative charge on the O atom.

FIGURE 29–7 Potential energy as a function of separation for two point charges of (a) like sign and (b) opposite sign.

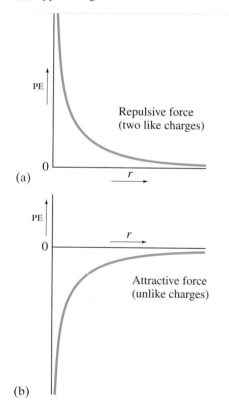

PE

Repulsive force
(two like charges)

0

r

(a)

PE

r

0

Attractive force
(unlike charges)

(b)

* **29–2 Potential-Energy Diagrams for Molecules**

It is useful to analyze the interaction between two objects—say, between two atoms or molecules—with the use of a potential-energy diagram, a plot of the potential energy versus the separation distance.

For the simple case of two point charges, q_1 and q_2, the PE is given by (see Chapter 17):

$$PE = \frac{1}{4\pi\epsilon_0} \frac{q_1 q_2}{r},$$

where r is the distance between the charges, and the constant $(1/4\pi\epsilon_0)$ is equal to $9.0 \times 10^9 \, N\cdot m^2/C^2$. If the two charges have the same sign, the PE is positive for all values of r, and a graph of PE versus r in this case is shown in Fig. 29–7a. The force is repulsive (the charges have the same sign) and the curve rises as r decreases; this makes sense since work is done to bring the charges together, thereby increasing their potential energy. If, on the other hand, the two charges are of opposite sign, the PE is negative because the product $q_1 q_2$ is negative. The force is attractive in this case and the graph of PE versus r looks like Fig. 29–7b. The PE becomes more *negative* as r decreases.

Now let us look at the potential-energy diagram for the formation of a covalent bond, such as for the hydrogen molecule. The potential energy of one H atom in the presence of the other is plotted in Fig. 29–8. Starting at large r, the PE decreases as the atoms approach, because the electrons concen-

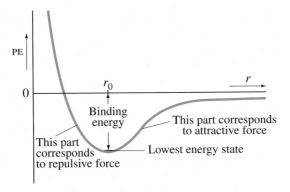

FIGURE 29–8 Potential-energy diagram for H_2 molecule; r is the separation of the two H atoms. The binding energy (the energy difference between PE = 0 and the lowest energy state near the bottom of the well) is 4.5 eV, and $r_0 = 0.074$ nm.

trate between the two nuclei (Fig. 29–2), so attraction occurs. However, at very short distances, the electrons would be "squeezed out"—there is no room for them between the two nuclei. Without the electrons between them, each nucleus would feel a repulsive force due to the other, so the curve rises as r decreases further. There is an optimum separation of the atoms, r_0 in Fig. 29–8, at which the energy is lowest. This is the point of greatest stability for the hydrogen molecule, and r_0 is the average separation of atoms in the H_2 molecule. The depth of this "well" is the *binding energy*,[†] as shown. This is how much energy must be put into the system to separate the two atoms to infinity, where the PE = 0.

Binding energy

For many bonds, the potential-energy curve has the shape shown in Fig. 29–9. There is still an optimum distance r_0 at which the molecule is stable. But when the atoms approach from a large distance, the force is initially repulsive rather than attractive. The atoms thus do not interact spontaneously. Instead, some additional energy must be injected into the system to get it over the "hump" (or barrier) in the potential-energy diagram. This required energy is called the **activation energy**.

Activation energy

The curve of Fig. 29–9 is much more common than that of Fig. 29–8. The activation energy often reflects a need to break other bonds, before

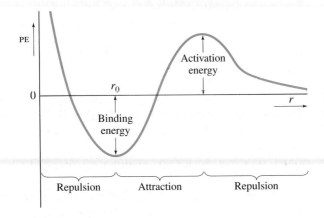

FIGURE 29–9 Potential-energy diagram for a bond requiring an activation energy.

[†]The binding energy corresponds not quite to the bottom of the PE curve, but to the lowest energy state, slightly above it, as shown in Fig. 29–8.

the one under discussion can be made. For example, to make water from O_2 and H_2, the H_2 and O_2 molecules must first be broken into H and O atoms by an input of energy; this is what the activation energy represents. Then the H and O atoms can combine to form H_2O with the release of a great deal more energy than was put in initially. The initial activation energy can be provided by applying an electric spark to a mixture of H_2 and O_2, breaking a few of these molecules into H and O atoms. The resulting explosive release of energy when these atoms combine to form H_2O quickly provides the activation energy needed for further reactions, so additional H_2 and O_2 molecules are broken up and recombined to form H_2O.

The potential-energy diagrams for ionic bonds can have similar shapes. In NaCl, for example, the Na^+ and Cl^- ions attract each other at distances a bit larger than some r_0, but at shorter distances the overlapping of inner electron shells gives rise to repulsion. The two atoms thus are most stable at some intermediate separation, r_0, and there often is an activation energy.

Sometimes the potential energy of a bond looks like that of Fig. 29–10. In this case, the energy of the bonded molecule, at a separation r_0, is greater than when there is no bond ($r = \infty$). That is, an energy *input* is required to make the bond (hence the binding energy is negative), and there is energy release when the bond is broken. Such a bond is stable only because there is the barrier of the activation energy. This type of bond is important in living cells, for it is in such bonds that energy can be stored efficiently in certain molecules, particularly ATP (adenosine triphosphate). The bond that connects the last phosphate group (designated Ⓟ in Fig. 29–10) to the rest of the molecule (ADP, meaning adenosine diphosphate, since it contains only two phosphates) is of the form shown in Fig. 29–10. Energy is actually stored in this bond. When the bond is broken (ATP → ADP + Ⓟ), energy is released and this energy can be used to make other chemical reactions "go."

In living cells, many chemical reactions have activation energies that are often on the order of several eV. Such energy barriers are not easy to overcome in the cell. This is where enzymes come in. They act as catalysts, which means that they act to lower the activation energy so that reactions can occur that otherwise would not. Enzymes act by distorting the bonding electrons so that the initial bonds are easily broken.

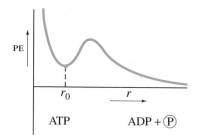

FIGURE 29–10 Potential-energy diagram for the formation of ATP from ADP and phosphate (Ⓟ).

* 29–3 Weak (van der Waals) Bonds

Once a bond between two atoms or ions is made, energy must normally be supplied to break the bond and separate the atoms. As mentioned in Section 29–1, this energy is called the *bond energy* or *binding energy*. The binding energy for covalent and ionic bonds is typically 2 to 5 eV. These bonds, which hold atoms together to form molecules, are often called **strong bonds** to distinguish them from so-called "weak bonds." The term **weak bond**, as we use it here, refers to an attachment between molecules due to simple electrostatic attraction—such as *between* polar molecules (and not *within* a polar molecule, which is a strong bond). The strength of the attachment is much less than for the strong bonds. Binding energies are typically in the range 0.04 to 0.3 eV—hence their name "weak bonds."

Weak bonds are generally the result of attraction between dipoles. For example, Fig. 29–11 shows two molecules, which have permanent dipole

FIGURE 29–11 The C^+—O^- and H^+—N^- dipoles attract each other. (These dipoles may be part of, for example, cytosine and guanine molecules. See Fig. 29–12.) The + and − charges typically have magnitudes of a fraction of e.

moments, attracting one another. Besides such **dipole–dipole bonds**, there can also be **dipole–induced dipole bonds**, in which a polar molecule with a permanent dipole moment can induce a dipole moment in an otherwise electrically balanced (nonpolar) molecule, just as a single charge can induce a separation of charge in a nearby object (see Fig. 16–7). There can even be an attraction between two nonpolar molecules. Even though a molecule may not have a permanent dipole moment on the average, we can think of its electrons as moving about so that at any instant there may be a separation of charge. Such transient dipoles can induce a dipole moment in a nearby molecule, creating a weak attraction. All these weak bonds are referred to as **van der Waals bonds**, and the forces involved **van der Waals forces**. The potential energy has the general shape shown in Fig, 29–8, with the attractive van der Waals PE varying as $1/r^6$.

When one of the atoms in a dipole–dipole bond is hydrogen, as in Fig. 29–11, it is called a **hydrogen bond**. A hydrogen bond is generally the strongest of the weak bonds. This is because the hydrogen atom is the smallest atom and thus can be approached more closely. Hydrogen bonds also have a partial "covalent" character. That is, electrons between the two dipoles may be shared to a small extent.

Weak bonds are important in liquids and solids when strong bonds are absent (see Section 29–5). They are also very important for understanding the activities of cells, such as the double helix shape of DNA (Fig. 29–12), and DNA replication (see Section 16–10). The average kinetic energy of molecules in a cell is around $\frac{3}{2}kT \approx 0.04\,\text{eV}$, about the magnitude of weak bonds. This means that a weak bond can readily be broken just by a molecular collision. Hence weak bonds are not very permanent—they are, instead, brief attachments. But because of this, they play an important role in the cell. On the other hand, strong bonds—those that hold molecules together—are almost never broken simply by molecular collision. Thus they are relatively permanent. They can be broken by chemical action (the making of even stronger bonds), and this usually happens in the cell with the aid of an enzyme, which is a protein molecule.

FIGURE 29–12 (a) Section of a DNA double helix. The red dots represent hydrogen bonds between the two chains. (b) "Close-up" view of helix: cytosine (C) and guanine (G) molecules on separate chains of a DNA double helix are held together by the hydrogen bonds (red dots) involving an H^+ on one molecule attracted to an N^- or C^+—O^- on the other molecule. See also Section 16–10 and Figs. 16–34 and 16–35.

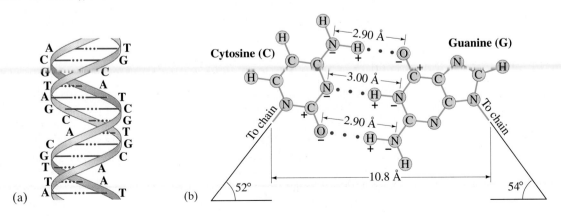

EXAMPLE 29–1 **Nucleotide energy.** Calculate the interaction energy between the C   O dipole of thymine and the H   N dipole of adenine, assuming that the two dipoles are lined up as shown in Fig. 29–11. Dipole moment measurements (see Table 17–2) give $q_H = -q_N = 0.19e = 3.0 \times 10^{-20}$ C, and $q_C = -q_O = 0.41e = 6.7 \times 10^{-20}$ C.

SOLUTION The interaction energy, for which we use the symbol U, will be equal to the potential energy of one dipole in the presence of the other, since this will be equal to the work needed to pull them infinitely far apart. U will consist of four terms:

$$U = U_{CH} + U_{CN} + U_{OH} + U_{ON},$$

where U_{CH} means the potential energy of C in the presence of H, and similarly for the other terms. We do not have terms corresponding to C and O, or N and H, because the two dipoles are assumed to be stable entities. Since the potential energy $U = qV$, where V is the electric potential (Eq. 17–3), then for two point charges $U_{12} = kq_1q_2/r$, where r is the distance between them. So

$$U = k\frac{q_C q_H}{r_{CH}} + k\frac{q_C q_N}{r_{CN}} + k\frac{q_O q_H}{r_{OH}} + k\frac{q_O q_N}{r_{ON}}.$$

Using the distances shown in Fig. 29–11, we get:

$$U = (9.0 \times 10^9 \text{ N·m}^2/\text{C}^2)\left(\frac{(6.7)(3.0)}{0.31} + \frac{(6.7)(-3.0)}{0.41} + \frac{(-6.7)(3.0)}{0.19} + \right.$$

$$\left. \frac{(-6.7)(-3.0)}{0.29}\right)\frac{(10^{-20} \text{ C})^2}{(10^{-9} \text{ m})}$$

$$= -1.83 \times 10^{-20} \text{ J} = -0.11 \text{ eV}.$$

The PE is negative, meaning 0.11 eV of work (or energy input) is required to separate the molecules. That is, the binding energy of this "weak" or hydrogen bond is 0.11 eV. This is only an estimate, of course, since other charges in the vicinity would have an influence too.

➡ **PHYSICS APPLIED**

Protein synthesis

Weak bonds, especially hydrogen bonds, are crucial to the process of protein synthesis. Proteins serve as structural parts of the cell and as enzymes to catalyze chemical reactions needed for the growth and survival of the organism. A protein molecule consists of one or more chains of small molecules known as *amino acids*. Each gene of the DNA contains the information needed to produce a particular type of protein molecule. The four types of bases (A, T, C, G) of the DNA are arranged according to a code, the "genetic code," which is translated into the amino acids that form the protein molecule, as discussed in Fig. 29–13. This process of protein synthesis is often presented as if it occurred in clockwork fashion—as if each molecule knew its role and went to its assigned place. But this is not the case. The forces of attraction between the electric charges of the molecules are rather weak and become significant only when the molecules can come close together and several weak bonds can be made. Indeed, if the shapes are not just right, there is almost no electrostatic attraction, which is why there are few mistakes. The fact that weak bonds are weak is very important. If they were strong, collisions with other molecules

FIGURE 29–13 Model of protein synthesis: messenger-RNA (m-RNA), a molecule much like DNA except that uracil (U) replaces T, is synthesized (on the right), one base (A, C, G, U) after the other, in just the order they appear on the DNA; the bases are selected via weak bonds and "shapes," much as in DNA replication (Section 16–10). The completed m-RNA (for one gene) then is buffeted about in the cell until it is attracted, and becomes attached to, a "ribosome" by electrostatic attraction. Transfer RNA (t-RNA) molecules, each carrying a particular amino acid and a corresponding "anticodon" of three bases, "line up" via electrostatic attraction at certain positions on the ribosome. Only the proper anticodon will line up with an m-RNA codon, as shown. The t-RNA molecules act as "translators" of the genetic code; the aligned amino acids (on the other end of the t-RNA) are attached together by enzymes (at which point the t-RNA, held tenuously by weak bonds, drifts away and the ribosome can move to the next codon on the m-RNA), thus making a protein chain molecule.

would not allow a t-RNA molecule to be released from the ribosome, or the m-RNA to be released from the DNA. If they were not temporary encounters, metabolism would grind to a halt.

* 29–4 Molecular Spectra

When atoms combine to form molecules, the probability distributions of the outer electrons overlap and this interaction alters the energy levels. Nonetheless, molecules can undergo transitions between electron energy levels just as atoms do. For example, the H_2 molecule can absorb a photon of just the right frequency to excite one of its $1s$ electrons to a $2p$ state. The excited electron can then return to the ground state, emitting a photon. The energy of photons emitted by molecules is of the same order of magnitude as for atoms, typically 1 to 10 eV.

Additional energy levels become possible for molecules (but not for atoms) because the molecule as a whole can rotate, and the atoms of the molecule can vibrate relative to each other. The energy levels for both rotational and vibrational levels are quantized, and are generally spaced much more closely (10^{-3} to 10^{-1} eV) than the electronic levels. Each atomic energy level thus becomes a set of closely spaced levels corresponding to the vibrational and rotational motions, Fig. 29–14. Transitions from one level to another appear as many very closely spaced lines. In fact, the lines are not always distinguishable, and these spectra are called **band spectra**. Each type of molecule has its own characteristic spectrum, which can be used for identification and for determination of structure.

Molecular Rotation. Let us now look in more detail at rotational and vibrational states in molecules. We begin with rotation, considering here only diatomic molecules, although the analysis can be extended to polyatomic molecules. When a diatomic molecule rotates about its center of mass, as shown in Fig. 29–15, its kinetic energy of rotation (see Section 8–7) is

$$E_{rot} = \frac{1}{2}I\omega^2 = \frac{(I\omega)^2}{2I},$$

where ($I\omega$) is the angular momentum (Section 8–8). Quantum mechanics predicts quantization of angular momentum just as in atoms (see Eq. 28–3):

$$I\omega = \sqrt{L(L+1)}\hbar, \qquad L = 0, 1, 2, \cdots,$$

where L is an integer called the **rotational angular momentum quantum number**. Thus the rotational energy is quantized:

$$E_{rot} = \frac{(I\omega)^2}{2I} = L(L+1)\frac{\hbar^2}{2I}, \qquad L = 0, 1, 2, \cdots. \qquad \textbf{(29–1)}$$

Transitions between rotational energy levels are subject to the *selection rule*:

$$\Delta L = \pm 1.$$

The energy of a photon emitted or absorbed for a transition between rotational states with angular-momentum quantum number L and $L - 1$ will be

$$\Delta E_{rot} = E_L - E_{L-1} = \frac{\hbar^2}{2I}L(L+1) - \frac{\hbar^2}{2I}(L-1)(L)$$

$$= \frac{\hbar^2}{I}L. \qquad \textbf{(29–2)}$$

We see that the transition energy increases directly with L. Figure 29–16 shows some of the allowed rotational energy levels and transitions. Measured

Rotational levels

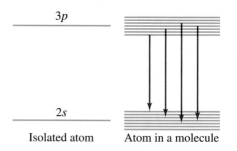

3p

2s

Isolated atom   Atom in a molecule

FIGURE 29–14 The individual energy levels of an isolated atom become bands of closely spaced levels in molecules, as well as in solids and liquids.

Selection rule

FIGURE 29–15 Diatomic molecule rotating about a vertical axis.

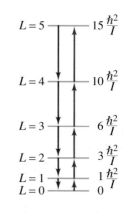

$L = 5$ ———— $15\frac{\hbar^2}{I}$

$L = 4$ ———— $10\frac{\hbar^2}{I}$

$L = 3$ ———— $6\frac{\hbar^2}{I}$

$L = 2$ ———— $3\frac{\hbar^2}{I}$

$L = 1$ ———— $1\frac{\hbar^2}{I}$
$L = 0$ ———— 0

FIGURE 29–16 Rotational energy levels and allowed transitions (emission and absorption) for a diatomic molecule. Upward-pointing arrows represent absorption of a photon, and downward arrows represent emission.

absorption lines fall in the microwave or far-infrared regions of the spectrum, and their frequencies are generally 2, 3, 4, $\cdots$ times higher than the lowest one, as predicted by Eq. 29–2.

EXAMPLE 29–2 **Rotational transition.** A rotational transition $L = 1$ to $L = 0$ for the molecule CO has a measured absorption wavelength $\lambda_1 = 2.60$ mm (microwave region). Use this to calculate (a) the moment of inertia of the CO molecule, and (b) the CO bond length, r. (c) Calculate the wavelengths of the next three rotational transitions, and the energies of the photon emitted for each of these four transitions

SOLUTION (a) From Eq. 29–2, we can write

$$\frac{\hbar^2}{I}L = \Delta E = hf = \frac{hc}{\lambda_1}.$$

With $L = 1$ (the upper state) in this case, we solve for I:

$$I = \frac{\hbar^2 L}{hc}\lambda_1 = \frac{h\lambda_1}{4\pi^2 c} = \frac{(6.63 \times 10^{-34}\,\text{J·s})(2.60 \times 10^{-3}\,\text{m})}{4\pi^2(3.00 \times 10^8\,\text{m/s})}$$

$$= 1.46 \times 10^{-46}\,\text{kg·m}^2.$$

(b) We write the moment of inertia of the rotating molecule, which rotates about its center of mass (CM), in terms of the CO separation r. In Fig. 29–15, let m_1 be the mass of the C, which is $m_1 = 12$ u and let m_2 be the mass of the O ($m_2 = 16$ u). The distance of the CM from the C atom, which is r_1 in Fig. 29–16, is given by the CM formula, Eq. 7–9:

$$r_1 = \frac{0 + m_2 r}{m_1 + m_2} = \frac{16}{12 + 16}r = 0.57r.$$

The O atom is a distance $r_2 = r - r_1 = 0.43r$ from the CM. The moment of inertia of the CO molecule about its CM is then (see Example 8–11)

$$I = m_1 r_1^2 + m_2 r_2^2$$

$$= [(12\,\text{u})(0.57r)^2 + (16\,\text{u})(0.43r)^2][1.66 \times 10^{-27}\,\text{kg/u}]$$

$$= (1.14 \times 10^{-26}\,\text{kg})r^2.$$

We solve for r and use the result of part (a) for I:

$$r = \sqrt{\frac{1.46 \times 10^{-46}\,\text{kg·m}^2}{1.14 \times 10^{-26}\,\text{kg}}} = 1.13 \times 10^{-10}\,\text{m},$$

or 0.113 nm.

(c) From Eq. 29–2, $\Delta E \propto L$. Hence $\lambda = c/f = hc/\Delta E$ is proportional to $1/L$. Thus, for $L = 2$ to $L = 1$ transitions, $\lambda_2 = \frac{1}{2}\lambda_1 = 1.30$ mm. For $L = 3$ to $L = 2$, $\lambda_3 = \frac{1}{3}\lambda_1 = 0.87$ mm. And for $L = 4$ to $L = 3$, $\lambda_4 = 0.65$ mm. All are close to measured values. The energies of the photons, $hf = hc/\lambda$, are respectively 4.8×10^{-4} eV, 9.5×10^{-4} eV, 1.4×10^{-3} eV, and 1.9×10^{-3} eV.

The spacing of rotational energy levels is thus of the order 10^{-3} eV for the CO molecule. For the H_2 molecule, the wavelengths are on the order of 10^{-4} m, which is in the far-infrared region of the spectrum, and the transition energies are on the order of 10^{-2} eV (see Problem 10).

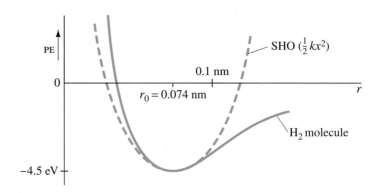

FIGURE 29–17 Potential energy for the H_2 molecule and for a simple harmonic oscillator (PE $= \frac{1}{2}kx^2$, with $|x| = |r - r_0|$).

Vibrational levels **Molecular Vibration.** The potential energy of the two atoms in a typical diatomic molecule has the shape shown in Fig. 29–8 or 29–9, and Fig. 29–17 again shows the PE for the H_2 molecule. We note that the PE, at least in the vicinity of the equilibrium separation r_0, closely resembles the potential energy of a harmonic oscillator, PE $= \frac{1}{2}kx^2$, which is shown superimposed in dashed lines. Thus, for small displacements from r_0, each atom experiences a restoring force proportional to the displacement, and the molecule vibrates as a simple harmonic oscillator (SHO)—see Chapter 11. According to quantum mechanics, the possible energy levels are quantized according to

$$E_{\text{vib}} = (\nu + \tfrac{1}{2})hf, \qquad \nu = 0, 1, 2, \cdots, \qquad \textbf{(29–3)}$$

where f is the classical frequency (see Chapter 11—f depends on the mass of the atoms and on the bond strength or "stiffness") and ν is an integer called the **vibrational quantum number**. The lowest energy state ($\nu = 0$) is not zero (as for rotation), but has $E = \frac{1}{2}hf$. This is called the **zero-point energy**. Higher states have energy $\frac{3}{2}hf, \frac{5}{2}hf$, and so on, as shown in Fig. 29–18. Transitions are subject to the *selection rule*:

$$\Delta\nu = \pm 1,$$

so allowed transitions occur only between adjacent states and all give off

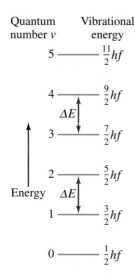

Quantum
number v

Vibrational
energy

5 ——— $\frac{11}{2}hf$

4 ——— $\frac{9}{2}hf$

ΔE

3 ——— $\frac{7}{2}hf$

2 ——— $\frac{5}{2}hf$

Energy $\quad \Delta E$

1 ——— $\frac{3}{2}hf$

0 ——— $\frac{1}{2}hf$

FIGURE 29–18 Allowed vibrational energies for a diatomic molecule, where f is the fundamental frequency of vibration (see Chapter 11). The energy levels are equally spaced. Transitions are allowed only between adjacent levels ($\Delta v = \pm 1$).

photons of energy

$$\Delta E_{vib} = hf. \tag{29-4}$$

This is very close to experimental values for small v, but for higher energies, the PE curve (Fig. 29–17) begins to deviate from a perfect SHO curve, and this then affects the wavelengths and frequencies of the transitions. Typical transition energies are on the order of $10^{-1}\,\text{eV}$, about 10 times larger than for rotational transitions, with wavelengths in the infrared region of the spectrum ($\approx 10^{-5}\,\text{m}$).

EXAMPLE 29–3 **Vibrational energy levels in hydrogen.** The hydrogen molecule emits infrared radiation of wavelength around 2300 nm. (*a*) What is the separation in energy between different vibrational levels? (*b*) What is the lowest vibrational energy state?

SOLUTION

(*a*) $\Delta E_{vib} = hf = \dfrac{hc}{\lambda}$

$$= \frac{(6.63 \times 10^{-34}\,\text{J·s})(3.00 \times 10^{8}\,\text{m/s})}{(2300 \times 10^{-9}\,\text{m})(1.60 \times 10^{-19}\,\text{J/eV})} = 0.54\ \text{eV}.$$

(*b*) The lowest vibrational energy has $v = 0$ in Eq. 29–3: $E = \frac{1}{2}hf = 0.27\ \text{eV}$.

When energy is imparted to a molecule, both the rotational and vibrational modes can be excited. Because rotational energies are an order of magnitude or so smaller than vibrational energies, which in turn are smaller than the electronic energy levels, we can represent the grouping of levels as shown in Fig. 29–19. Transitions between energy levels, with emission of a photon, are subject to the *selection rules*:

$$\Delta \nu = \pm 1$$

and

$$\Delta L = \pm 1.$$

Some allowed and forbidden (marked ×) transitions are indicated in Fig. 29–19. Not all transitions and levels are shown, and the separation between vibrational levels, and (even more) between rotational levels, has been exaggerated. But we can clearly see the origin of the very closely spaced lines that give rise to the band spectra, as mentioned with reference to Fig. 29–14 earlier in this Section.

The spectra are quite complicated, so we consider briefly only transitions within the same electronic level, such as those at the top of Fig. 29–19. A transition from a state with quantum numbers ν and L, to one with quantum numbers $\nu + 1$ and $L \pm 1$ (see the selection rules above), will absorb[†] a photon of energy:

$$\Delta E = \Delta E_{vib} + \Delta E_{rot}$$

$$= hf + (L + 1)\frac{\hbar^2}{I} \qquad [L \to L + 1], \qquad L = 0, 1, 2, \cdots$$

(29–5)

$$= hf - L\frac{\hbar^2}{I} \qquad [L \to L - 1], \qquad L = 1, 2, 3, \cdots,$$

where we have used Eqs. 29–2 and 29–4. Note that for $L \to L - 1$ transitions, L cannot be zero since there is then no state with $L = -1$. Equations 29–5 predict an absorption spectrum like that shown schematically in Fig. 29–20, with transitions $L \to L - 1$ on the left and $L \to L + 1$ on the

FIGURE 29–19 Combined electronic, vibrational, and rotational energy levels. Transitions marked with an × are not allowed by the selection rules.

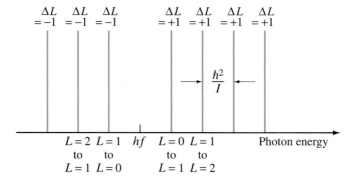

FIGURE 29–20 Expected spectrum for transitions between combined rotational and vibrational states.

[†]This is for absorption; for emission of a photon, the transition would be $\nu \to \nu - 1, L \to L \pm 1$.

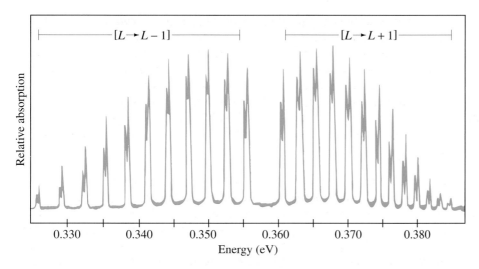

FIGURE 29–21
Absorption spectrum for HCl molecules. Lines on the left correspond to transitions where $L \to L - 1$; those on the right are for $L \to L + 1$. Each line has a double peak because chlorine has two isotopes of different mass and different moments of inertia.

right. Figure 29–21 shows the molecular absorption spectrum of HCl, which follows this pattern very well. (Each line in that spectrum is split into two because Cl consists of two isotopes of different mass; hence there are two kinds of HCl molecule with different moments of inertia I.)

* 29–5 Bonding in Solids

Quantum mechanics has been a great tool for understanding the structure of solids. This active field of research today is called **solid-state physics**, or **condensed-matter physics** so as to include liquids as well. The rest of this chapter is devoted to this subject, and we begin with a brief look at the structure of solids and the bonds that hold them together.

Although some solid materials are *amorphous* in structure (such as glass), in that the atoms and molecules show no long-range order, we will be interested here in the large class of *crystalline* substances whose atoms, ions, or molecules are generally accepted to form an orderly array in a geometric arrangement known as a **lattice**. Figure 29–22 shows three of the possible arrangements of atoms in a crystal: simple cubic, face-centered cubic, and body-centered cubic. The NaCl crystal lattice is face-centered

FIGURE 29–22   Arrangement of atoms in (a) a simple cubic crystal, (b) face-centered cubic crystal (note the atom at the center of each face), and (c) body-centered cubic crystal. Each diagram shows the relationship of the bonds. Each of these "cells" is repeated in three dimensions to the edges of the macroscopic crystal.

(a)

(b)

(c)

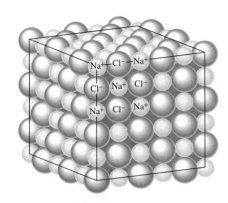

FIGURE 29–23 Diagram of NaCl crystal.

cubic, with one Na^+ ion or one Cl^- ion at each lattice point (see Fig. 29–23).

The molecules of a solid are held together in a number of ways. The most common are by *covalent* bonding (as between the carbon atoms of the diamond crystal) or *ionic* bonding (as in a NaCl crystal). Often the bonds are partially covalent and partially ionic. Our discussion of these bonds earlier in this chapter for molecules applies equally well here to solids.

Let us look for a moment at the NaCl crystal of Fig. 29–23. Each Na^+ ion feels an attractive Coulomb potential due to each of the six "nearest neighbor" Cl^- ions surrounding it. Note that one Na^+ does not "belong" exclusively to one Cl^-, so we must not think of ionic solids as consisting of individual molecules. Each Na^+ also feels a repulsive Coulomb potential due to other Na^+ ions, although this is weaker since the Na^+ ions are farther away.

A different type of bond occurs in metals. Metal atoms have relatively loosely held outer electrons. Present-day **metallic bond** theories propose that in a metallic solid, these outer electrons roam rather freely among all the metal atoms which, without their outer electrons, act like positive ions. The electrostatic attraction between the metal ions and this negative electron "gas" is at least in part responsible for holding the solid together. The binding energy of metal bonds is typically 1 to 3 eV, somewhat weaker than ionic or covalent bonds (5 to 10 eV in solids). The "free electrons" are responsible for the high electrical and thermal conductivity of metals. This theory also nicely accounts for the shininess of smooth metal surfaces: the free electrons can vibrate at any frequency, so when light of a range of frequencies falls on a metal, the electrons can vibrate in response and reemit light of those same frequencies. Hence, the reflected light will consist largely of the same frequencies as the incident light. Compare this to nonmetallic materials that have a distinct color—the atomic electrons exist only in certain energy states, and when white light falls on them, the atoms absorb at certain frequencies, and reflect other frequencies which make up the color we see.

Here is a brief comparison of important strong bonds:

- ionic: an electron is stolen from one atom by another
- covalent: electrons are shared by atoms within a single molecule
- metallic: electrons are shared by all atoms in the metal

The atoms or molecules of some materials, such as the noble gases, can form only **weak bonds** with each other. As we saw in Section 29–3, weak bonds have very low binding energies and would not be expected to hold atoms together as a liquid or solid at room temperature. Indeed, the noble gases condense only at very low temperatures, where the atomic (thermal) kinetic energy is small and the weak attraction can then hold the atoms together.

* 29–6 Band Theory of Solids

We saw in Section 29–1 that when two hydrogen atoms approach each other, the wave functions overlap, and the two $1s$ states (one for each atom) divide into two states of different energy. (As we saw, only one of these states, $S = 0$, has low enough energy to give a bound H_2 molecule.)

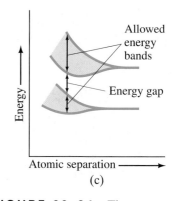

(a)   (b)   (c)

FIGURE 29–24 The splitting of 1s and 2s atomic energy levels as (a) two atoms approach each other (the atomic separation decreases, moving toward the left); (b) the same for six atoms, and (c) for many atoms when they come together to form a solid.

Figure 29–24a shows this situation for 1s and 2s states for two atoms. If six atoms come together, as in Fig. 29–24b, each of the states splits into six levels. If a large number of atoms come together to form a solid, then each of the original atomic levels becomes a **band** as shown in Fig. 29–24c. The energy levels are so close together in each band that they seem essentially continuous. This is why the spectrum of heated solids (Section 27–2) appears continuous.

The crucial aspect of a good **conductor** is that the highest energy band containing electrons is only partially filled. Consider sodium, for example, whose energy bands are shown in Fig. 29–25. The 1s, 2s, and 2p bands are full (just as in a Na atom) and don't concern us. The 3s band, however, is only half full. To see why, recall that the exclusion principle stipulates that in an atom, only two electrons can be in the 3s state, one with spin up and one with spin down. These two states have slightly different energy. For a solid consisting of N atoms, the 3s band will contain 2N possible energy states. Now a sodium atom has a single 3s electron, so in a sample of sodium metal containing N atoms, there are N electrons in the 3s band, and N unoccupied states. When a potential difference is applied across the metal, electrons can respond by accelerating and increasing their energy, since there are plenty of unoccupied states of slightly higher energy available. Hence, a current flows readily and sodium is a good conductor. The characteristic of all good conductors is that the highest energy band is only partially filled, or two bands overlap so that unoccupied states are available. An example of the latter is magnesium, which has two 3s electrons, so its 3s band is filled. But the unfilled 3p band overlaps the 3s band in energy, so there are lots of available states for the electrons to move into. Thus magnesium, too, is a good conductor.

In a material that is a good **insulator**, on the other hand, the highest band containing electrons, called the **valence band**, is completely filled. The next higher energy band, called the **conduction band**, is separated from the valence band by a "forbidden" **energy gap**, (or **band gap**), E_g, of typically 5 to 10 eV. So at room temperature (300 K), where thermal energies (that is, average kinetic energy—see Chapter 13) are on the order of $\frac{3}{2}kT \approx 0.04$ eV, almost no electrons can acquire the 5 eV needed to reach the conduction band. When a potential difference is applied across the material, no available states are accessible to the electrons, and no current flows. Hence, the material is a good insulator.

Conductors

FIGURE 29–25 Energy bands for sodium.

3s

2p

2s

1s

Insulators

Valence and conduction bands

Energy gap

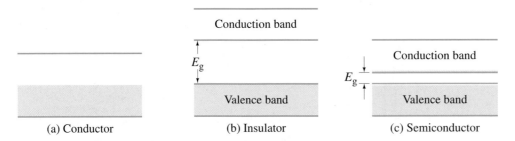

(a) Conductor   (b) Insulator   (c) Semiconductor

FIGURE 29–26 Energy bands for (a) a conductor, (b) an insulator, and (c) a semiconductor. Shading represents occupied states. Pale shading in (c) represents electrons that can pass from the top of the valence band to the bottom of the conduction band due to thermal agitation at room temperature (exaggerated).

Semiconductors (pure)

Figure 29–26 compares the relevant energy bands (a) for conductors, (b) for insulators, and also (c) for the important class of materials known as **semiconductors**. The bands for a pure (or **intrinsic**) semiconductor, such as silicon or germanium, are like those for an insulator, except that the unfilled conduction band is separated from the filled valence band by a much smaller energy gap, E_g, typically on the order of 1 eV. At room temperature, a few electrons can acquire enough thermal energy to reach the conduction band, and so a very small current may flow when a voltage is applied. At higher temperatures, more electrons have enough energy to jump the gap. This effect can often more than offset the effects of more frequent collisions due to increased disorder at higher temperature, so that the resistivity of semiconductors can *decrease* with increasing temperature (see Table 18–1). But this is not the whole story of semiconductor conduction. When a potential difference is applied to a semiconductor, the few electrons in the conduction band move toward the positive electrode. Electrons in the valence band try to do the same thing, and a few can because there are a small number of unoccupied states which were left empty by the electrons reaching the conduction band. Such unfilled electron states are called **holes**. Each electron in the valence band that fills a hole in this way as it moves toward the positive electrode leaves behind a hole, so that the holes migrate toward the negative electrode. As the electrons tend to accumulate at one side of the material, the holes tend to accumulate on the opposite side. We will look at this phenomenon in more detail in the next Section.

Holes (in semiconductor)

EXAMPLE 29–4 **Calculating the energy gap.** It is found that the conductivity of a certain semiconductor increases when light of wavelength 345 nm or shorter strikes it, suggesting that electrons are being promoted from the valence band to the conduction band. What is the energy gap, E_g, for this semiconductor?

SOLUTION The longest wavelength, or lowest energy, photon to cause an increase in conductivity has $\lambda = 345$ nm, and it can transfer to an electron an energy

$$E_g = hf = \frac{hc}{\lambda} = \frac{(6.63 \times 10^{-34}\,\text{J·s})(3.00 \times 10^8\,\text{m/s})}{(345 \times 10^{-9}\,\text{m})(1.60 \times 10^{-19}\,\text{J/eV})} = 3.6\,\text{eV}.$$

> **CONCEPTUAL EXAMPLE 29–5** **Which is transparent?** The energy gap for silicon is 1.14 eV at room temperature while that of zinc sulfide (ZnS) is 3.6 eV. Which one of these is opaque while the other is transparent to visible light?
>
> **RESPONSE** Visible light photons span energies from roughly 1.8 eV to 3.2 eV ($E = hf = hc/\lambda$ where λ = 400 nm to 700 nm and 1 eV = $1.6. \times 10^{-19}$ J). Light is absorbed by the electrons in a material. Zinc sulfide's energy gap is too wide to absorb visible photons, and so the light can pass through the material; it can be transparent. On the other hand, silicon's gap is small enough to absorb these photons (thus bumping electrons well up into the conduction band), and this is why silicon is opaque.

* 29–7 Semiconductors and Doping

The most commonly used semiconductors in modern electronics are silicon (Si) and germanium (Ge). An atom of silicon or germanium has four outer electrons that act to hold the atoms in the regular lattice structure of the crystal, shown schematically in Fig. 29–27a. Germanium and silicon acquire useful properties for use in electronics only when a tiny amount of impurity is introduced into the crystal structure (perhaps 1 part in 10^6 or 10^7). This is called **doping** the semiconductor. Two kinds of doped semiconductor can be made, depending on the type of impurity used. If the impurity is an element whose atoms have five outer electrons, such as arsenic, we have the situation shown in Fig. 29–27b. Only four of arsenic's electrons fit into the bonding structure. The fifth does not fit in and can move relatively freely, somewhat like the electrons in a conductor. Because of this small number of extra electrons, a doped semiconductor becomes slightly conducting. The density of conduction electrons in an intrinsic (pure) semiconductor is about 1 per 10^{10} atoms. With an impurity concentration of 1 in 10^6 or 10^7 when doped, the conductivity will be much higher and it can be controlled with great precision. An arsenic-doped silicon crystal is called an **n-type semiconductor** because negative charges (electrons) carry the electric current.

Doped semiconductors

n-type

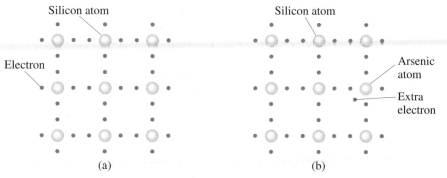

FIGURE 29–27 Two-dimensional representation of a silicon crystal. (a) Four (outer) electrons surround each silicon atom. (b) Silicon crystal doped with a few arsenic atoms: the extra electron doesn't fit into the crystal lattice and so is free to move about. This is an *n*-type semiconductor.

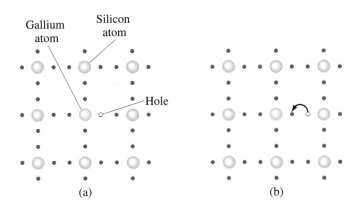

(a)　　　　　　　(b)

p-type

In a *p*-type semiconductor, a small amount of an element with three outer electrons—such as gallium—is added to the semiconductor. As shown in Fig. 29–28a, there is a "hole" in the lattice structure near a gallium atom since it has only three outer electrons. Electrons from nearby silicon atoms can jump into this hole and fill it. But this leaves a hole where that electron had previously been, Fig. 29–28b. The vast majority of atoms are silicon, so holes are almost always next to a silicon atom. Since silicon atoms require four outer electrons to be neutral, this means that there is a net positive charge at the hole. Whenever an electron moves to fill a hole, the positive hole is then at the previous position of that electron. Another electron can then fill this hole, and the hole thus moves to a new location; and so on. This type of semiconductor is called *p-type* because it is the positive holes that seem to carry the electric current. Note, however, that both *p*-type and *n*-type semiconductors have *no net charge* on them.

Holes are positive

According to the band theory (Section 29–6), in a doped semiconductor, the impurity provides additional energy states between the bands, as shown in Fig. 29–29. In an *n*-type semiconductor, the impurity energy level lies just below the conduction band. Electrons in this energy level need only about 0.05 eV (in Si; even less in Ge) of energy to reach the conduction band; this is on the order of $\frac{3}{2}kT \, (= \overline{KE} = 0.04 \text{ eV})$ at 300 K, so transitions occur readily at room temperature. Since this energy level supplies electrons to the conduction band, it is called a **donor** level. In *p*-type semiconductors, the impurity energy level is just above the valence band (Fig. 29–29). It is called an **acceptor** level because electrons from the valence band can easily jump into it. Positive holes are left behind in the valence band, and as other electrons move into these holes, the holes move about as discussed earlier.

FIGURE 29-29
Impurity energy levels in doped semiconductors.

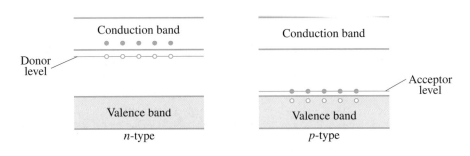

29–8 Semiconductor Diodes

Semiconductor diodes and transistors are essential components of modern electronic devices. The miniaturization achieved today allows many thousands of diodes, transistors, resistors, and so on, to be placed on a single *chip* only a millimeter on a side. We now discuss, briefly and qualitatively, the operation of diodes and transistors.

When an *n*-type semiconductor is joined to a *p*-type, a **p-n junction diode** is formed. Separately, the two semiconductors are electrically neutral. When joined, a few electrons near the junction diffuse from the *n*-type into the *p*-type semiconductor, where they fill a few of the holes. The *n*-type is left with a positive charge, and the *p*-type acquires a net negative charge. Thus a potential difference is established, with the *n* side positive relative to the *p* side, and this prevents further diffusion of electrons.

If a battery is connected to a diode with the positive terminal to the *p* side and the negative terminal to the *n* side as in Fig. 29–30a, the externally applied voltage opposes the internal potential difference and the diode is said to be **forward biased**. If the voltage is great enough (about 0.3 V for Ge, 0.6 V for Si at room temperature), a current will flow. The positive holes in the *p*-type semiconductor are repelled by the positive terminal of the battery and the electrons in the *n*-type are repelled by the negative terminal of the battery. The holes and electrons meet at the junction, and the electrons cross over and fill the holes. A current is flowing. Meanwhile, the positive terminal of the battery is continually pulling electrons off the *p* end, forming new holes, and electrons are being supplied by the negative terminal at the *n* end. Consequently, a large current flows through the diode.

When the diode is **reverse biased**, as in Fig. 29–30b, the holes in the *p* end are attracted to the battery's negative terminal and the electrons in the *n* end are attracted to the positive terminal. The current carriers do not meet near the junction and, ideally, no current flows.

A graph of current versus voltage for a typical diode is shown in Fig. 29–31. As can be seen, a real diode does allow a small amount of reverse current to flow. For most practical purposes, this is negligible.[†]

(a)

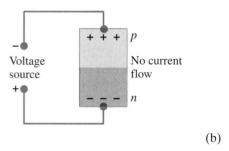

(b)

FIGURE 29–30 Schematic diagram showing how a semiconductor diode operates. Current flows when the voltage is connected in forward bias, as in (a), but not when connected in reverse bias, as in (b).

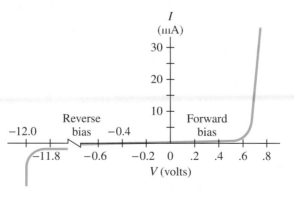

FIGURE 29–31 Current through a diode as a function of applied voltage.

[†]Reverse current in Ge at room temperature is typically a few μA; for Si, a few pA. The reverse current increases rapidly with temperature, however, and may render a diode ineffective above 200°C.

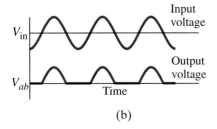

Input voltage

V_{in}

Output voltage

V_{ab}

Time

(b)

FIGURE 29–32 (a) A simple (half-wave) rectifier circuit using a semiconductor diode. (b) AC source input voltage, and output voltage across R, as functions of time.

Rectifier

EXAMPLE 29–6 **A diode.** The diode whose current–voltage characteristics are shown in Fig. 29–31 is connected in series with a 4.0-V battery and a resistor. If a current of 10 mA is to pass through the diode, what resistance must the resistor have?

SOLUTION In Fig. 29–31, we see that the voltage drop across the diode is about 0.7 V when the current is 10 mA. Therefore, the voltage drop across the resistor is $4.0\,V - 0.7\,V = 3.3\,V$, so $R = V/I = (3.3\,V)/(1.0 \times 10^{-2}\,A) = 330\,\Omega$.

If the voltage across a diode connected in reverse bias is increased greatly, a point is reached where breakdown occurs. The electric field across the junction becomes so large that ionization of atoms results. The electrons thus pulled off their atoms contribute to a larger and larger current as breakdown continues. The voltage remains constant over a wide range of currents. This is shown on the far left in Fig. 29–31. This property of diodes can be used to accurately regulate a voltage supply. A diode designed for this purpose is called a **zener diode**. When placed across the output of an unregulated power supply, a zener diode can maintain the voltage at its own breakdown voltage as long as the supply voltage is always above this point. Zener diodes can be obtained corresponding to voltages of a few volts to hundreds of volts.

Since a *p-n* junction diode allows current to flow only in one direction (as long as the voltage is not too high), it can serve as a **rectifier**—to change ac into dc. A simple rectifier circuit is shown in Fig. 29–32a where the arrow inside the symbol for a diode indicates the direction in which a diode conducts conventional (+) current. The ac source applies a voltage across the diode alternately positive and negative. Only during half of each cycle will a current pass through the diode; so only then is there a current through the resistor R. Hence, a graph of the voltage V_{ab} across R as a function of time looks like Fig. 29–32b. This **half-wave rectification** is not exactly dc, but it is unidirectional. More useful is a **full-wave rectifier** circuit, which uses two diodes (or sometimes four) as shown in Fig. 29–33a. At any given instant, either one diode or the other will conduct current to the right. Therefore, the output across the load resistor R will be as shown in Fig. 29–33b. Actually this is the voltage if the capacitor C were not in the circuit. The capacitor tends to store charge, and thus helps to smooth out the current as shown in Fig. 29–33c.

FIGURE 29–33 (a) Full-wave-rectifier circuit (including a transformer so the magnitude of the voltage can be changed). (b) Output voltage in the absence of capacitor C. (c) Output voltage with the capacitor in the circuit.

(a)

V output

(b) Without capacitor

V output

(c) With capacitor

Rectifier circuits are important because most line voltage is ac, and most electronic devices require a dc voltage for their operation. Hence, diodes are found in nearly all electronic devices including radio and TV sets, calculators, and computers.

Another useful device is a **light-emitting diode** (LED). When a *p-n* junction is forward biased, a current begins to flow. Electrons cross from the *n* region into the *p* region and combine with holes, and a photon can be emitted with an energy approximately equal to the band gap, E_g (see Figs. 29–26c and 29–29). Often the energy, and hence the wavelength, is in the red region of the visible spectrum, producing the familiar LED displays on VCRs, CD players, car instrument panels, clocks, and so on. Infrared (i.e., nonvisible) LEDs are used in remote controls for TV, VCRs, and stereos.

LED

Photodiodes and **solar cells** are *p-n* junctions used in the reverse way. Photons are absorbed, creating electron–hole pairs if the photon energy is greater than the band gap energy, E_g. The created electrons and holes produce a current that, when connected to an external circuit, becomes a source of emf and power. Applications were discussed in Section 15–12, and you may find solar cells as a source of emf and power in a handheld calculator.

Solar cells

A diode is called a **nonlinear device** because the current is not proportional to the voltage; that is, a graph of current versus voltage (Fig. 29–31) is not a straight line as it is for a resistor (which ideally *is* linear). Transistors are also *nonlinear* devices.

*29–9 Transistors and Integrated Circuits

A simple **junction transistor** consists of a crystal of one type of doped semiconductor sandwiched between two crystals of the opposite type. Both *pnp* and *npn* transistors are made, and they are shown schematically in Fig. 29–34a. The three semiconductors are given the names *collector*, *base*, and *emitter*. The symbols for *npn* and *pnp* transistors are shown in Fig. 29–34b. The arrow is always placed on the emitter and indicates the direction of (conventional) current flow in normal operation.

Transistors

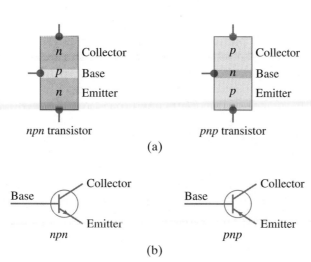

FIGURE 29–34
(a) Schematic diagram of *npn* and *pnp* transistors. (b) Symbols for *npn* and *pnp* transistors.

FIGURE 29–35 An *npn* transistor used as an amplifier.

The operation of a transistor can be analyzed qualitatively—very briefly—as follows. Consider an *npn* transistor connected as shown in Fig. 29–35. A voltage V_{CE} is maintained between the collector and emitter by the battery $\mathscr{E}_C$. The voltage applied to the base is called the *base bias voltage*, V_{BE}. If V_{BE} is positive, conduction electrons in the emitter are attracted into the base. Since the base region is very thin (perhaps 1 μm), most of these electrons flow right across into the collector, which is maintained at a positive voltage. A large current, I_C, flows between collector and emitter and a much smaller current, I_B, through the base. A small variation in the base voltage due to an input signal causes a large change in the collector current and therefore a large change in the voltage drop across the output resistor R_C. Hence a transistor can *amplify* a small signal into a larger one. In fact, transistors are the basic elements in modern electronic **amplifiers** of all sorts.

Amplifiers

A *pnp* transistor operates in the same fashion, except that holes move instead of electrons. The collector voltage is negative, and so is the base voltage in normal operation.

In Fig. 29–35 two batteries were shown: $\mathscr{E}_C$ supplied the collector voltage and $\mathscr{E}_B$ the base bias voltage. In practice, only one source is often used, and the base bias voltage can be obtained using a resistance voltage divider as in Fig. 29–36. Transistors can be connected in many other ways as well, and many new and innovative uses have been found for them.

Transistors were a great advance in miniaturization of electronic circuits. Although individual transistors are very small compared to the once used vacuum tubes, they are huge compared to **integrated circuits** or **chips** (see photo at start of this chapter). Tiny amounts of impurities can be placed at particular locations within a single silicon crystal. These can be arranged to form diodes, transistors, and resistors (undoped semiconductors). Capacitors and inductors can also be formed, although they are often connected separately. A tiny chip, only 1 mm on a side, may contain thousands of transistors and other circuit elements. Integrated circuits are now the heart of computers, television, calculators, cameras, and the electronic instruments that control aircraft, space vehicles, and automobiles. The "miniaturization" produced by integrated circuits not only allows extremely complicated circuits to be placed in a small space, but also has allowed a great increase in the speed of operation of, say, computers, because the distances the electronic signals travel are so tiny.

FIGURE 29–36 Typical transistor circuit involving an *npn* transistor. $\mathscr{E}_C = +12$ V and $\mathscr{E}_B$ is determined by the resistors.

SUMMARY

Quantum mechanics explains the bonding together of atoms to form **molecules**. In a **covalent bond**, the electron clouds of two or more atoms overlap because of constructive interference between the electron waves. The positive nuclei are attracted to this concentration of negative charge between them, forming the bond. An **ionic bond** is an extreme case of a covalent bond in which one or more electrons from one atom spend much more time around the other atom than around their own. The atoms then act as oppositely charged ions that attract each other, forming the bond. These **strong bonds** hold molecules together, and also hold atoms and molecules together in solids. Also important are **weak bonds** (or **van der Waals bonds**), which are generally dipole attractions between molecules.

When atoms combine to form molecules, the energy levels of the outer electrons are altered because they now interact with each other. Additional energy levels also become possible because the atoms can vibrate with respect to each other, and the molecule as a whole can rotate. The energy levels for both vibrational and rotational motion are quantized, and are very close together (typically, 10^{-1} eV to 10^{-3} eV apart). Each atomic energy level thus becomes a set of closely spaced levels corresponding to the vibrational and rotational motions. Transitions from one level to another appear as many very closely spaced lines. The resulting spectra are called **band spectra**. The quantized rotational energy levels are given by

$$E_{\text{rot}} = L(L + 1)\frac{\hbar^2}{2I}, \qquad L = 0, 1, 2, \cdots,$$

where I is the moment of inertia of the molecule. The energy levels for vibrational motion are given by

$$E_{\text{vib}} = (\nu + \tfrac{1}{2})hf, \qquad \nu = 0, 1, 2, \cdots,$$

where f is the classical natural frequency of vibration for the molecule. Transitions between energy levels are subject to the selection rules $\Delta L = \pm 1$ and $\Delta \nu = \pm 1$.

Some **solids** are bound together by covalent and ionic bonds, just as molecules are. In metals, the electrostatic force between free electrons and positive ions helps form the **metallic bond**.

In a crystalline solid, the possible energy states for electrons are arranged in **bands**. Within each band the levels are very close together, but between the bands there may be forbidden **energy gaps**. Good conductors are characterized by the highest occupied band (the **conduction band**) being only partially full, so there are many accessible states available to electrons to move about and accelerate when a voltage is applied. In a good insulator, the highest occupied energy band (the **valence band**) is completely full, and there is a large energy gap (5 to 10 eV) to the next highest band, the *conduction band*. At room temperature, molecular kinetic energy (thermal energy) available due to collisions is only about 0.04 eV, so almost no electrons can jump from the valence to the conduction band. In a **semiconductor**, the gap between valence and conduction bands is much smaller, on the order of 1 eV, so a few electrons can make the transition from the essentially full valence band to the nearly empty conduction band.

In a **doped** semiconductor, a small percentage of impurity atoms with five or three valence electrons replace a few of the normal silicon atoms with their four valence electrons. A five-electron impurity produces an **n-type** semiconductor with negative electrons as carriers of current. A three-electron impurity produces a **p-type** semiconductor in which positive **holes** carry the current. The energy level of impurity atoms lies slightly below the conduction band in an *n*-type semiconductor, and acts as a **donor** from which electrons readily pass into the conduction band. The energy level of impurity atoms in a *p*-type semiconductor lies slightly above the valence band and acts as an **acceptor** level, since electrons from the valence band easily reach it, leaving holes behind to act as charge carriers.

A semiconductor **diode** consists of a **p-n-junction** and allows current to flow in one direction only; it can be used as a **rectifier** to change ac to dc. Common **transistors** consist of three semiconductor sections, either as **pnp** or **npn**. Transistors can amplify electrical signals and find many other uses. An integrated circuit consists of a tiny semiconductor crystal or "chip" on which many transistors, diodes, resistors, and other circuit elements have been constructed using careful placement of impurities.

QUESTIONS

* **1.** What type of bond would you expect for (*a*) the N_2 molecule, (*b*) the HCl molecule, (*c*) Fe atoms in a solid?

* **2.** Describe how the molecule $CaCl_2$ could be formed.

* **3.** Does the H_2 molecule have a permanent dipole moment? Does O_2? Does H_2O? Explain.

* **4.** Although the molecule H_3 is not stable, the ion H_3^+ is. Explain, using the Pauli exclusion principle.

* **5.** The energy of a molecule can be divided into four categories. What are they?

* **6.** If conduction electrons are free to roam about in a metal, why don't they leave the metal entirely?

* **7.** A silicon semiconductor is doped with phosphorus. Will these atoms be donors or acceptors? What type of semiconductor will this be?

* **8.** Explain why the resistivity of metals increases with temperature whereas the resistivity of semiconductors decreases with increasing temperature.

* **9.** Can a diode be used to amplify a signal?

* **10.** Figure 29–37 shows a "bridge-type" full-wave rectifier. Explain how the current is rectified and how current flows during each half cycle.

* **11.** Compare the resistance of a *p-n* junction diode connected in forward bias to its resistance when connected in reverse bias.

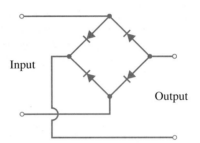

FIGURE 29–37   Question 10.

* **12.** Explain how a transistor could be used as a switch.

* **13.** If $\mathscr{E}_C$ were reversed in Fig. 29–35, how would the amplification be altered?

* **14.** Describe how a *pnp* transistor can operate as an amplifier.

* **15.** In a transistor, the base–emitter junction and the base–collector junction are essentially diodes. Are these junctions reverse-biased or forward-biased in the application shown in Fig. 29–35?

* **16.** What purpose does the capacitor in Fig. 29–36 serve?

* **17.** Do diodes and transistors obey Ohm's law?

PROBLEMS

*SECTIONS 29–1 TO 29–3

* **1.** (I) Estimate the binding energy of a KCl molecule by calculating the electrostatic potential energy when the K^+ and Cl^- ions are at their stable separation of 0.28 nm. Assume each has a charge of magnitude $1.0e$.

* **2.** (II) The measured binding energy of KCl is 4.43 eV. From the result of Problem 1, estimate the contribution to the binding energy of the repelling electron clouds at the equilibrium distance $r_0 = 0.28$ nm.

* **3.** (II) Estimate the binding energy of the H_2 molecule, assuming the two H nuclei are 0.074 nm apart and the two electrons spend 33 percent of their time midway between them.

* **4.** (II) Binding energies are often measured experimentally in kcal per mole, and then the binding energy in eV per molecule is calculated from that result. What is the conversion factor in going from kcal per mole to eV per molecule? What is the binding energy of KCl ($= 4.43$ eV) in kcal per mole?

* **5.** (II) Estimate the total binding energy for the three hydrogen bonds between cytosine and guanine shown in Fig. 29–12b. Assume each of the bonds is between the H—N dipole ($q_H = -q_N = 0.19e$) and an O or N atom of charge $-1.0e$.

* **6.** (III) Apply reasoning similar to that in the text for the $S = 0$ and $S = 1$ states in the formation of the H_2 molecule to show why the molecule He_2 is *not* formed. Show also why the He_2^+ molecular ion *is* formed (with a binding energy of 3.1 eV at $r_0 = 0.11$ nm).

*SECTION 29–4

* **7.** (I) Show that the quantity $\hbar^2/I$ has units of energy.

* **8.** (II) The so-called "characteristic rotational energy," $\hbar^2/2I$, for N_2 is 2.48×10^{-4} eV. Calculate the N_2 bond length.

* 9. (II) (a) Calculate the characteristic rotational energy, $\hbar^2/2I$, for the O_2 molecule whose bond length is 0.121 nm. (b) What are the energy and wavelength of photons emitted in a $L = 2$ to $L = 1$ transition?

* 10. (II) The equilibrium separation of H atoms in the H_2 molecule is 0.074 nm (Fig. 29–8). Calculate the energies and wavelengths of photons for the rotational transitions (a) $L = 1$ to $L = 0$, (b) $L = 2$ to $L = 1$, and (c) $L = 3$ to $L = 2$.

* 11. (II) Calculate the bond length for the NaCl molecule given that three successive wavelengths for rotational transitions are 23.1 mm, 11.6 mm, and 7.71 mm.

* 12. (II) In the absorption spectrum of the HCl molecule (Fig. 29–21), the lines are separated by 2.6×10^{-3} eV. Determine (a) the moment of inertia of the HCl molecule, and (b) the bond length. Assume that Cl has mass of 35 u.

* 13. (II) Derive Eqs. 29–5.

* 14. (II) (a) Show that the moment of inertia of a diatomic molecule can be written as $I = \mu r_0^2$ where r_0 is the distance between atoms and $\mu = m_1 m_2/(m_1 + m_2)$ is called the reduced mass. Determine I for the following molecules: (b) H_2 ($r_0 = 0.074$ nm); (c) O_2 ($r_0 = 0.121$ nm); (d) NaCl ($r_0 = 0.24$ nm); (e) CO ($r_0 = 0.113$ nm).

* 15. (III) (a) Use the curve of Fig. 29–17 to estimate the stiffness constant k for the H_2 molecule. (Recall that $\text{PE} = \frac{1}{2}kx^2$.) (b) Then estimate the fundamental wavelength for vibrational transitions using the classical formula (Chapter 11).

*SECTION 29–5

* 16. (II) The spacing between "nearest neighbor" Na and Cl ions in a NaCl crystal is 0.24 nm. What is the spacing between two nearest neighbor Na ions?

* 17. (II) Common salt, NaCl, has a density of 2.165 g/cm^3. The molecular weight of NaCl is 58.44. Estimate the distance between nearest neighbor Na and Cl ions. [Hint: Each ion can be considered to have one "cube" or "cell" of side s (our unknown) extending out from it.]

* 18. (II) Repeat the previous problem for KCl whose density is 1.99 g/cm^3.

*SECTION 29–6

* 19. (I) Explain on the basis of energy bands why the sodium chloride crystal is a good insulator. [Hint: Consider the shells of Na$^+$ and Cl$^-$ ions.]

* 20. (I) A semiconductor, bombarded with light of slowly increased frequency, begins to conduct when the wavelength of light is 640 nm; estimate the size of the energy gap E_g.

* 21. (I) Calculate the longest-wavelength photon that can cause an electron in silicon ($E_g = 1.1$ eV) to jump from the valence band to the conduction band.

* 22. (II) The energy gap between valence and conduction bands in germanium is 0.72 eV. What range of wavelengths can a photon have to excite an electron from the top of the valence band into the conduction band?

* 23. (II) A TV remote control emits IR light. If the detector on the TV set is not to react to visible light, could it make use of silicon as a "window" with its energy gap $E_g = 1.14$ eV? What is the shortest-wavelength light that can strike silicon without causing electrons to jump from the valence band to the conduction band?

* 24. (II) We saw that there are $2N$ possible electron states in the $3s$ band of Na, where N is the total number of atoms. How many possible electron states are there in the (a) $2s$ band, (b) $2p$ band, and (c) $3p$ band? (d) State a general formula for the total number of possible states in any given electron band.

* 25. (II) The energy gap E_g in germanium is 0.72 eV. When used as a photon detector, roughly how many electrons can be made to jump from the valence to the conduction band by the passage of a 660-keV photon that loses all its energy in this fashion?

*SECTION 29–7

* 26. (II) Suppose that a silicon semiconductor is doped with phosphorus so that one silicon atom in 10^6 is replaced by a phosphorus atom. Assuming that the "extra" electron in every phosphorus atom is donated to the conduction band, by what factor is the density of conduction electrons increased? The density of silicon is 2330 kg/m^3, and the density of conduction electrons in pure silicon is about 10^{16} m^{-3} at room temperature.

*SECTION 29–8

* 27. (I) At what wavelength will an LED radiate if made from a material with an energy gap $E_g = 1.4$ eV?

* 28. (I) If an LED emits light of wavelength $\lambda = 650$ nm, what is the energy gap (in eV) between valence and conduction bands?

* 29. (II) A silicon diode, whose current–voltage characteristics are given in Fig. 29–31, is connected in series with a battery and a 660-Ω resistor. What battery voltage is needed to produce a 12-mA current?

* 30. (II) Suppose that the diode of Fig. 29–31 is connected in series to a 100-Ω resistor and a 2.0-V battery. What current flows in the circuit? [Hint: Draw a line on Fig. 29–31 representing the current in the resistor as a function of the voltage across the diode; the intersection of this line with the characteristic curve will give the answer.]

31. (II) Sketch the resistance as a function of current, for $V > 0$, for the diode shown in Fig. 29–31.

* 32. (II) An ac voltage of 120 V rms is to be rectified. Estimate very roughly the average current in the output resistor R (18 kΩ) for (a) a half-wave rectifier (Fig. 29–32), and (b) a full-wave rectifier (Fig. 29–33) without capacitor.

* 33. (III) A silicon diode passes significant current only if the forward-bias voltage exceeds about 0.6 V. Make a rough estimate of the average current in the output resistor R of (a) a half-wave rectifier (Fig. 29–32), and (b) a full-wave rectifier (Fig. 29–33) without a capacitor. Assume that $R = 150\,\Omega$ in each case and that the ac voltage is 12.0 V rms in each case.

* 34. (III) A 120-V rms 60-Hz voltage is to be rectified with a full-wave rectifier as in Fig. 29–33, where $R = 18$ kΩ, and $C = 25\,\mu$F. (a) Make a rough estimate of the average current. (b) What happens if $C = 0.10\,\mu$F? [Hint: See Section 19–7.]

*SECTION 29–9

* 35. (II) Suppose that the *current gain* of the transistor in Fig. 29–35 is $\beta = I_C/I_B = 80$. If $R_C = 3.3$ kΩ, calculate the output voltage for a time-varying input current of $2.0\,\mu$A.

* 36. (II) If the current gain of the transistor amplifier in Fig. 29–35 is $\beta = I_C/I_B = 100$, what value must R_C have if a 1.0-μA base current is to produce an output voltage of 0.40 V?

* 37. (II) A transistor, whose current gain $\beta = I_C/I_B = 90$, is connected as in Fig. 29–35 with $R_B = 2.2$ kΩ and $R_C = 6.2$ kΩ. Calculate (a) the voltage gain, and (b) the power amplification.

* 38. (II) An amplifier has a *voltage gain* of 70 and a 14-kΩ load (output) resistance. What is the peak output current through the load resistor if the input voltage is an ac signal with a peak of 0.080 V?

GENERAL PROBLEMS

* 39. Estimate the binding energy of the H_2 molecule by calculating the difference in KE of the electrons between when they are in separate atoms and when they are in the molecule, using the uncertainty principle. Take Δx for the electrons in the separated atoms to be the radius of the first Bohr orbit, 0.053 nm, and for the molecule take Δx to be the separation of the nuclei, 0.074 nm.

* 40. The average translational kinetic energy of an atom or molecule is about $\overline{KE} = \frac{3}{2}kT$ (see Chapter 13), where $k = 1.38 \times 10^{-23}$ J/K is Boltzmann's constant. At what temperature T will $\overline{KE}$ be on the order of the bond energy (and hence the bond able to be broken by thermal motion) for (a) a covalent bond (say H_2) of binding energy 4.5 eV, and (b) a "weak" hydrogen bond of binding energy 0.15 eV?

* 41. In the ionic salt KF, the separation distance between ions is about 0.27 nm. (a) Estimate the electrostatic potential energy between the ions assuming them to be point charges (magnitude $1e$). (b) It is known that F releases 4.07 eV of energy when it "grabs" an electron, and 4.34 eV is required to ionize K. Find the binding energy of KF relative to free K and F atoms, neglecting the energy of repulsion.

* 42. Most of the Sun's radiation has wavelengths shorter than 1000 nm. For a solar cell to absorb all this, what energy gap ought the material have?

* 43. For an arsenic donor atom in a doped silicon semiconductor, assume that the "extra" electron moves in a Bohr orbit about the arsenic ion. For this electron in the ground state, take into account the dielectric constant $K = 12$ of the Si lattice (which represents the weakening of the Coulomb force due to all the other atoms or ions in the lattice), and estimate (a) the binding energy, and (b) the orbit radius for this extra electron. [Hint: Substitute $\epsilon = K\epsilon_0$ in Coulomb's law; see Section 17–8.]

* 44. A full-wave rectifier (Fig. 29–33) uses two diodes to rectify a 95-V rms 60 Hz ac voltage. If $R = 10$ kΩ and $C = 30\,\mu$F, what will be the approximate percent variation in the output voltage? The variation in output voltage (Fig. 29–33c) is called *ripple voltage*. [Hint: See Section 19–7 and assume the discharge of the capacitor is approximately linear.]

* 45. A zener diode voltage regulator is shown in Fig. 29–38. Suppose that $R = 1.80$ kΩ and that the diode breakdown voltage is 130 V; the diode is rated at a maximum current of 100 mA. (a) If $R_{load} = 15.0$ kΩ, over what range of supply voltages will the circuit maintain the voltage at 130 V? (b) If the supply voltage is 200 V, over what range of load resistance will the voltage be regulated?

FIGURE 29–38 Problem 45.

*46. Do we need to consider quantum effects for everyday rotating objects? Estimate the differences between rotational energy levels for a spinning baton compared to the energy of the baton. Assume the baton consists of a 30-cm bar with a mass of 200 g and two small end masses, each of mass 300 g, and that the baton rotates uniformly at 2.0 rev/s about the bar's center.

*47. A strip of silicon 1.5 cm wide and 1.0 mm thick is immersed in a magnetic field of strength 1.5 T perpendicular to the strip (Fig. 29–39). When a current of 0.20 mA is run through the strip, there is a resulting Hall effect voltage of 18 mV across the strip (Section 20–11). How many electrons per silicon atom are in the conduction band? The density of silicon is 2330 kg/m³.

FIGURE 29–39   Problem 47.

*48. When solid argon melts at $-189°C$, its latent heat of fusion goes directly into breaking the bonds between the atoms. Solid argon is a weakly bound cubic lattice, with each atom connected to six neighbors, each bond having a binding energy of 3.9×10^{-3} eV. What is the latent heat of fusion for argon, in J/kg? [*Hint*: Show that in a simple cubic lattice (Fig. 29–40), there are *three* times as many bonds as there are atoms, when the number of atoms is large.]

FIGURE 29–40   Problem 48.

A geologist can search for radioactive elements (perhaps uranium) using a portable Geiger counter or scintillation counter.

30 NUCLEAR PHYSICS AND RADIOACTIVITY

In the early part of the twentieth century, Rutherford's experiments led to the idea that at the center of an atom there is a tiny but massive nucleus. At the same time that the quantum theory was being developed and scientists were attempting to understand the structure of the atom and its electrons, investigations into the nucleus itself had also begun. In this chapter and the next, we take a brief look at *nuclear physics*.

30–1 Structure and Properties of the Nucleus

An important question to physicists in the early part of this century was whether the nucleus had a structure, and what that structure might be. It turns out that the nucleus is a complicated entity and is not fully understood even today. However, by the early 1930s, a model of the nucleus had been developed that is still useful. According to this model, a nucleus is considered as an aggregate of two types of particles: protons and neutrons. (Of course, we must remember that these "particles" also have wave properties, but for ease of visualization and language, we often refer to them simply as "particles.") A **proton** is the nucleus of the simplest atom, hydro-

Proton

gen. It has a positive charge ($= +e = +1.60 \times 10^{-19}$ C) and a mass

$$m_p = 1.6726 \times 10^{-27} \text{ kg.}$$

Neutron

The **neutron**, whose existence was ascertained only in 1932 by the Englishman James Chadwick (1891–1974), is electrically neutral ($q = 0$), as its name implies. Its mass, which is almost identical to that of the proton, is

$$m_n = 1.6749 \times 10^{-27} \text{ kg}.$$

These two constituents of a nucleus, neutrons and protons, are referred to collectively as **nucleons**.

Nucleons

Although the hydrogen nucleus consists of a single proton alone, the nuclei of all other elements consist of both neutrons and protons. The different types of nuclei are often referred to as **nuclides**. The number of protons in a nucleus (or nuclide) is called the **atomic number** and is designated by the symbol Z. The total number of nucleons, neutrons plus protons, is designated by the symbol A and is called the **atomic mass number**. This name is used since the mass of a nucleus is very closely A times the mass of one nucleon. A nuclide with 7 protons and 8 neutrons thus has $Z = 7$ and $A = 15$. The **neutron number** N is $N = A - Z$.

Z and A

To specify a given nuclide, we need give only A and Z. A special symbol is commonly used which takes the form

$$^A_Z X,$$

where X is the chemical symbol for the element (see Appendix F, and the periodic table inside the back cover), A is the atomic mass number, and Z is the atomic number. For example, $^{15}_7\text{N}$ means a nitrogen nucleus containing 7 protons and 8 neutrons for a total of 15 nucleons. In a neutral atom, the number of electrons orbiting the nucleus is equal to the atomic number Z (since the charge on an electron has the same magnitude but opposite sign to that of a proton). The main properties of an atom, and how it interacts with other atoms, are largely determined by the number of electrons. Hence Z determines what kind of atom it is: carbon, oxygen, gold, or whatever. It is redundant to specify both the symbol of a nucleus and its atomic number Z as described above. If the nucleus is nitrogen, for example, we know immediately that $Z = 7$. The subscript Z is thus sometimes dropped and $^{15}_7\text{N}$ is then written simply ^{15}N; in words we say "nitrogen fifteen."

For a particular type of atom (say, carbon), nuclei are found to contain different numbers of neutrons, although they all have the same number of protons. For example, carbon nuclei always have 6 protons, but they may have 5, 6, 7, 8, 9, or 10 neutrons. Nuclei that contain the same number of protons but different numbers of neutrons are called **isotopes**. Thus, $^{11}_6\text{C}$, $^{12}_6\text{C}$, $^{13}_6\text{C}$, $^{14}_6\text{C}$, $^{15}_6\text{C}$, and $^{16}_6\text{C}$ are all isotopes of carbon. Of course, the isotopes of a given element are not all equally common. For example, 98.9 percent of naturally occurring carbon (on Earth) is the isotope $^{12}_6\text{C}$ and about 1.1 percent is $^{13}_6\text{C}$. These percentages are referred to as the **natural abundances**.[†] Many isotopes that do not occur naturally can be produced in the laboratory by means of nuclear reactions (more on this later). Indeed, all elements beyond uranium ($Z > 92$) do not occur naturally and are only produced artificially, as are many nuclides with $Z \leq 92$.

Isotopes

The approximate size of nuclei was determined originally by Rutherford from the scattering of charged particles. Of course, we cannot speak about a definite size for nuclei because of the wave–particle duality: their spatial extent must remain somewhat fuzzy. Nonetheless a rough "size" can be measured by scattering high-speed electrons off nuclei. It is found

---

[†]The mass value for each element as given in the periodic table (inside back cover) is an average weighted according to the natural abundances of its isotopes.

that nuclei have a roughly spherical shape with a radius that increases with A according to the approximate formula

Nuclear radii

$$r \approx (1.2 \times 10^{-15} \text{ m})(A^{\frac{1}{3}}). \tag{30-1}$$

Since the volume of a sphere is $V = \frac{4}{3}\pi r^3$, we see that the volume of a nucleus is proportional to the number of nucleons, $V \propto A$. This is what we would expect if nucleons were like impenetrable billiard balls: if you double the number of balls, you double the total volume. Hence, all nuclei have nearly the same density.

EXAMPLE 30–1 **ESTIMATE** **Nuclear sizes.** Estimate the diameter of the following nuclei: (*a*) $^{1}_{1}\text{H}$, (*b*) $^{40}_{20}\text{Ca}$, (*c*) $^{208}_{82}\text{Pb}$, (*d*) $^{235}_{92}\text{U}$.

SOLUTION (*a*) For hydrogen, $A = 1$, Eq. 30–1 gives

$$d = \text{diameter} = 2r \approx 2.4 \times 10^{-15} \text{ m}$$

since $A^{\frac{1}{3}} = 1^{\frac{1}{3}} = 1$.
(*b*) For calcium

$$d = 2r \approx (2.4 \times 10^{-15} \text{ m})(40)^{\frac{1}{3}} = 8.2 \times 10^{-15} \text{ m}.$$

(*c*) For lead

$$d \approx (2.4 \times 10^{-15} \text{ m})(208)^{\frac{1}{3}} = 14 \times 10^{-15} \text{ m}.$$

(*d*) For uranium

$$d \approx (2.4 \times 10^{-15} \text{ m})(235)^{\frac{1}{3}} = 15 \times 10^{-15} \text{ m}.$$

The masses of nuclei can be determined, in one method, by measuring the radius of curvature of fast-moving nuclei in a magnetic field using a mass spectrometer, as discussed in Section 20–12. Indeed, as mentioned there, the existence of different isotopes of the same element was discovered using this device. Nuclear masses are specified in **unified atomic mass units** (u). On this scale, a neutral $^{12}_{6}\text{C}$ atom is given the precise value 12.000000 u. A neutron then has a measured mass of 1.008665 u, a proton 1.007276 u, and a neutral hydrogen atom, $^{1}_{1}\text{H}$ (proton plus electron) 1.007825 u. The masses of many nuclides are given in Appendix F. It should be noted that the masses in this table, as is customary, are for the *neutral atom*, and not for a bare nucleus.

➥ **PROBLEM SOLVING**

Masses are for neutral atom

Masses are often specified using the electron-volt energy unit. This can be done because mass and energy are related, and the precise relationship is given by Einstein's equation $E = mc^2$ (Chapter 26). Since the mass of a neutral $^{1}_{1}\text{H}$ atom is 1.67353×10^{-27} kg, or 1.007825 u then $1.0000 \text{ u} = (1.00000/1.007825) \cdot (1.67353 \times 10^{-27} \text{ kg}) = 1.66054 \times 10^{-27}$ kg; this is equivalent to an energy $E = mc^2 = (1.66054 \times 10^{-27} \text{ kg}) \cdot (2.9979 \times 10^{8} \text{ m/s})^2/(1.6022 \times 10^{-19} \text{ J/eV}) = 931.5 \text{ MeV}$. Thus

Atomic mass unit

$$\begin{aligned} 1 \text{ u} &= 1.6605 \times 10^{-27} \text{ kg} \\ &= 931.5 \text{ MeV}/c^2. \end{aligned}$$

The rest masses of some of the basic particles are given in Table 30–1.

Just as an electron has an intrinsic spin and angular momentum, so too do nuclei and their constituents, the proton and neutron. Both the proton and the neutron are spin-$\frac{1}{2}$ particles. A nucleus, made up of protons and neutrons, has a **nuclear spin,** I, that can be either integer or half integer, depending on whether it is made up of an even or an odd number of nucleons. The *nuclear angular momentum* of a nucleus is given, as might be expected (see Section 28–6), by $\sqrt{I(I + 1)}\hbar$.

TABLE 30–1
Rest Masses in Kilograms, Unified Atomic Mass Units, and MeV/c^2

Object	Mass		
	kg	u	MeV/c^2
Electron	9.1094×10^{-31}	0.00054858	0.51100
Proton	1.67262×10^{-27}	1.007276	938.27
$^1_1$H atom	1.67353×10^{-27}	1.007825	938.78
Neutron	1.67493×10^{-27}	1.008665	939.57

30–2 Binding Energy and Nuclear Forces

The total mass of a stable nucleus is always less than the sum of the masses of its constituent protons and neutrons, as the following Example shows.

EXAMPLE 30–2 $^4_2$**He mass compared to its constituents.** Compare the mass of a $^4_2$He nucleus to that of its constituent nucleons.

SOLUTION The mass of a neutral $^4_2$He atom, from Appendix F, is 4.002602 u. The mass of two neutrons and two protons (including the two electrons) is

$$2m_n = 2.017330 \text{ u}$$
$$2m(^1_1\text{H}) = \underline{2.015650 \text{ u}}$$
$$4.032980 \text{ u}.$$

(We almost always deal with masses of neutral atoms—that is, nuclei with Z electrons—since this is how masses are measured. We must therefore be sure to balance out the electrons when we compare masses, which is why we used the mass of $^1_1$H in this Example rather than that of the proton alone.)

➡ **PROBLEM SOLVING**

Keep track of electron masses

Thus the mass of $^4_2$He is measured to be 4.032980 u − 4.002602 u = 0.030378 u less than the masses of its constituents. How can this be? If four nucleons come together to form $^4_2$He, where can this lost mass have gone?

It has, in fact, gone into energy of another kind (such as radiation, or kinetic energy, for example). The mass (or energy) difference in the case of $^4_2$He, given in energy units, is (0.030378 u)(931.5 MeV/u) = 28.30 MeV. This difference is referred to as the **total binding energy** of the nucleus. The total binding energy represents the amount of energy that must be put into a nucleus in order to break it apart into its constituent protons and neutrons. If the mass of, say, a $^4_2$He nucleus were exactly equal to the mass of two neutrons plus two protons, the nucleus could fall apart without any input of energy. To be stable, the mass of a nucleus *must* be less than that of its constituent nucleons, so that energy input is needed to break it apart. Note that the binding energy is not something a nucleus has—it is energy it "lacks" relative to the total mass of its separate constituents.

[Nuclear binding energy can be compared to the binding energy of electrons in an atom. We saw in Chapter 27 that the binding energy of the one electron in the hydrogen atom, for example, is 13.6 eV. The mass of a $^1_1$H atom is less than that of a single proton plus a single electron by

Binding energy

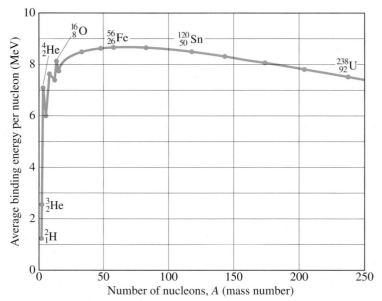

FIGURE 30-1 Average binding energy per nucleon as a function of mass number A for stable nuclei.

Binding energy per nucleon

13.6 eV. Compared to the total mass of the atom (938 MeV), this is incredibly small (1 part in 10^8), and for practical purposes the mass difference can be ignored. The binding energies of nuclei are on the order of 10^6 times greater than the binding energies of electrons in atoms.]

The **average binding energy per nucleon** is defined as the total binding energy of a nucleus divided by A, the total number of nucleons. For $^4_2$He, it is 28.3 MeV/4 = 7.1 MeV. Figure 30–1 shows the average binding energy per nucleon as a function of A for stable nuclei. The curve rises as A increases and reaches a plateau at about 8.7 MeV per nucleon above about $A \approx 40$. Beyond about $A \approx 80$, the curve decreases slowly, indicating that larger nuclei are held together a little less tightly than those in the middle of the periodic table. (We will see later that these figures allow the release of nuclear energy in the processes of fission and fusion.)

EXAMPLE 30–3 Binding energy for iron. Calculate the total binding energy and the average binding energy per nucleon for $^{56}_{26}$Fe, the most common stable isotope of iron.

SOLUTION $^{56}_{26}$Fe has 26 protons and 30 neutrons whose separate masses are

$$(26)(1.007825 \text{ u}) = 26.2035 \text{ u (includes electrons)}$$
$$(30)(1.008665 \text{ u}) = \underline{30.2600 \text{ u}}$$
$$\text{Total} = 56.4635 \text{ u.}$$

$$\text{Subtract mass of } ^{56}_{26}\text{Fe:} \quad \underline{-55.9349 \text{ u}} \text{ (Appendix F)}$$
$$\Delta m = 0.5286 \text{ u.}$$

The total binding energy is thus

$$(0.5286 \text{ u})(931.5 \text{ MeV/u}) = 492.4 \text{ MeV}$$

and the average binding energy per nucleon is

$$\frac{492.4 \text{ MeV}}{56 \text{ nucleons}} = 8.8 \text{ MeV.}$$

| EXAMPLE 30–4 | **Binding energy of last neutron.** What is the binding energy of the last neutron in $^{13}_{6}\text{C}$?

SOLUTION We compare the mass of $^{13}_{6}\text{C}$ to that of the atom with one less neutron, $^{12}_{6}\text{C}$, plus a free neutron (Appendix F):

$$\text{Mass } ^{12}_{6}\text{C} = \quad 12.000000 \text{ u}$$

$$\text{Mass } ^{1}_{0}\text{n} = \quad \underline{1.008665 \text{ u}}$$

$$13.008665 \text{ u.}$$

$$\text{Subtract mass of } ^{13}_{6}\text{C:} \quad \underline{-13.003355 \text{ u}}$$

$$\Delta m = \quad 0.005310 \text{ u}$$

which in energy is $(931.5 \text{ MeV/u})(0.005310 \text{ u}) = 4.95 \text{ MeV}$. That is, it would require 4.95 MeV input of energy to remove one neutron from $^{13}_{6}\text{C}$.

We can analyze nuclei not only from the point of view of energy, but also from the point of view of the forces that hold them together. We would not expect a collection of protons and neutrons to come together spontaneously, since protons are all positively charged and thus exert repulsive forces on each other. Indeed, the question arises as to how a nucleus stays together at all in view of the fact that the electric force between protons would tend to break it apart. Since stable nuclei *do* stay together, it is clear that another force must be acting. Because this new force is stronger than the electric force (which, in turn, is much stronger than gravity at the nuclear level) it is called the **strong nuclear force**. The strong nuclear force is an attractive force that acts between all nucleons—protons and neutrons alike. Thus protons attract each other via the nuclear force at the same time they repel each other via the electric force. Neutrons, since they are electrically neutral, only attract other neutrons or protons via the nuclear force.

Strong nuclear force

The nuclear force turns out to be far more complicated than the gravitational and electromagnetic forces. A precise mathematical description is not yet possible. Nonetheless, a great deal of work has been done to try to understand the nuclear force. One important aspect of the strong nuclear force is that it is a **short-range** force: it acts only over a very short distance. It is very strong between two nucleons if they are less than about 10^{-15} m apart, but it is essentially zero if they are separated by a distance greater than this. Compare this to electric and gravitational forces, which can act over great distances and are therefore called **long-range** forces. The strong nuclear force has some strange quirks. For example, if a nuclide contains too many or too few neutrons relative to the number of protons, the binding of the nucleons is reduced; nuclides that are too unbalanced in this regard are unstable. As shown in Fig. 30–2, stable nuclei tend to have the same number of protons as neutrons ($N = Z$) up to about $A \approx 30$ or 40. Beyond this, stable nuclei contain more neutrons than protons. This makes sense since, as Z increases, the electrical repulsion increases, so a greater number of neutrons—which exert only the attractive nuclear force—are required to maintain stability. For very large Z, no number of neutrons can overcome the greatly increased electric repulsion. Indeed, there are no completely stable nuclides above $Z = 82$.

Long- and short-range forces

FIGURE 30–2 Number of neutrons versus number of protons for stable nuclides, which are represented by dots.

What we mean by a stable nucleus is one that stays together indefinitely. What then is an unstable nucleus? It is one that comes apart; and this results in radioactive decay. Before we discuss the important subject of radioactivity (the next Section), we note that there is a second type of nuclear force that is much weaker than the strong nuclear force. It is called the **weak nuclear force**, and we are aware of its existence only because it shows itself in certain types of radioactive decay. These two nuclear forces, the strong and the weak, together with the gravitational and electromagnetic forces, comprise the four known types of force in nature (more on this in Chapter 32).

Weak force

30–3 Radioactivity

Discovery of radioactivity

Nuclear physics had its beginnings in 1896. In that year, Henri Becquerel (1852–1908) made an important discovery: in his studies of phosphorescence, he found that a certain mineral (which happened to contain uranium) would darken a photographic plate even when the plate was wrapped to exclude light. It was clear that the mineral emitted some new kind of radiation that, unlike X-rays, occurred without any external stimulus. This new phenomenon eventually came to be called **radioactivity**.

Soon after Becquerel's discovery, Marie Curie (1867–1934) and her husband, Pierre Curie (1859–1906), isolated two previously unknown elements that were very highly radioactive (Fig. 30–3). These were named polonium and radium. Other radioactive elements were soon discovered as well. The radioactivity was found in every case to be unaffected by the strongest physical and chemical treatments, including strong heating or cooling and the action of strong chemical reagents. It was clear that the source of radioactivity must be deep within the atom, that it must emanate from the nucleus. And it became apparent that radioactivity is the result of the *disintegration* or *decay* of an unstable nucleus. Certain isotopes are not stable under the action of the nuclear force, and they decay with the emission of some type of radiation or "rays."

Many unstable isotopes occur in nature, and such radioactivity is called "natural radioactivity." Other unstable isotopes can be produced in the laboratory by nuclear reactions (Section 31–1); these are said to be produced "artificially" and to have "artificial radioactivity."

FIGURE 30–3 Marie and Pierre Curie in their laboratory (about 1906) where radium was discovered.

Rutherford and others began studying the nature of the rays emitted in radioactivity about 1898. They found that the rays could be classified into three distinct types according to their penetrating power. One type of radiation could barely penetrate a piece of paper. The second type could pass through as much as 3 mm of aluminum. The third was extremely penetrating: it could pass through several centimeters of lead and still be detected on the other side. They named these three types of radiation alpha (α), beta (β), and gamma (γ), respectively, after the first three letters of the Greek alphabet.

Each type of ray was found to have a different charge and hence is bent differently in a magnetic field, Fig. 30–4; α rays are positively charged, β rays are negatively charged, and γ rays are neutral. It was soon found that all three types of radiation consisted of familiar kinds of particles. Gamma rays are very high-energy photons whose energy is even higher than that of X-rays. Beta rays are electrons, identical to those that orbit the nucleus (but they are created within the nucleus itself). Alpha rays (or α particles) are simply the nuclei of helium atoms, $_2^4He$; that is, an α ray consists of two protons and two neutrons bound together.

We now discuss each of these three types of radioactivity, or decay, in more detail.

30–4 Alpha Decay

When a nucleus emits an α particle ($_2^4He$), it is clear that the remaining nucleus will be different from the original: for it has lost two protons and two neutrons. Radium 226 ($_{88}^{226}Ra$), for example, is an α emitter. It decays to a nucleus with $Z = 88 - 2 = 86$ and $A = 226 - 4 = 222$. The nucleus with $Z = 86$ is radon (Rn)—see Appendix F or the periodic table. Thus the radium decays to radon with the emission of an α particle. This is written

$$_{88}^{226}Ra \rightarrow {}_{86}^{222}Rn + {}_2^4He.$$

See Fig. 30–5.

It is clear that when α decay occurs, a new element is formed. The **daughter** nucleus ($_{86}^{222}Rn$ in this case) is different from the **parent** nucleus ($_{88}^{226}Ra$ in this case). This changing of one element into another is called **transmutation**.

Alpha decay can be written

$$_Z^A N \rightarrow {}_{Z-2}^{A-4}N' + {}_2^4He \qquad \text{[}\alpha\text{ decay]}$$

where N is the parent, N' the daughter, and Z and A are the atomic number and atomic mass number, respectively, of the parent.

Alpha decay occurs because the strong nuclear force is unable to hold very large nuclei together. Because the nuclear force is a short-range force, it acts only between neighboring nucleons. But the electric force can act all the way across a large nucleus. For very large nuclei, the large Z means the repulsive electric force becomes very large (Coulomb's law); and it acts between all protons. The strong nuclear force, since it acts only between neighboring nucleons, is overpowered and is unable to hold the nucleus together.

We can express the instability in terms of energy (or mass): the mass of the parent nucleus is greater than the mass of the daughter nucleus plus the mass of the α particle. The mass difference appears as kinetic energy, which is carried away by the α particle and the recoiling daughter nucleus.

α, β, γ radiation

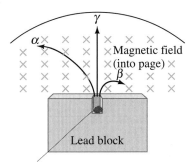

FIGURE 30–4 Alpha and beta rays are bent in opposite directions by a magnetic field, whereas gamma rays are not bent at all.

FIGURE 30–5 Radioactive decay of radium to radon with emission of an alpha particle.

α decay

Daughter nucleus
Parent nucleus
Transmutation

Why the strong nuclear force cannot hold a nucleus together

The total energy released is called the **disintegration energy**, Q, or the **Q-value** of the decay. Q is defined as

Q-value

$$Q = (M_P - M_D - m_\alpha)c^2, \qquad (30\text{-}2)$$

where M_P, M_D, and m_α are the masses of the parent, daughter, and α particle, respectively. If the parent had *less* mass than the daughter plus the α particle (so $Q < 0$), the decay could not occur, for the conservation of energy law would be violated.

EXAMPLE 30–5 **Uranium decay energy release.** Calculate the disintegration energy when $^{232}_{92}$U (mass = 232.037131 u) decays to $^{228}_{90}$Th (228.028716 u) with the emission of an α particle. (As always, masses are for neutral atoms.)

SOLUTION Since the mass of the $^4_2$He is 4.002602 u (Appendix F), the total mass in the final state is

$$228.028716 \text{ u} + 4.002602 \text{ u} = 232.031318 \text{ u}.$$

The mass lost when the $^{232}_{92}$U decays is

$$232.037131 \text{ u} - 232.031318 \text{ u} = 0.005813 \text{ u}.$$

Since 1 u = 931.5 MeV, the energy Q released is

$$Q = (0.005813 \text{ u})(931.5 \text{ MeV/u}) \approx 5.4 \text{ MeV},$$

and this energy appears as KE of the α particle and the daughter nucleus. (Using conservation of momentum, it can be shown that the α particle in this decay has a KE of about 5.3 MeV. Thus, the daughter nucleus—which recoils in the opposite direction from the emitted α particle—has about 0.1 MeV of kinetic energy. See Problem 34.)

Why α particles?

Why, you may wonder, do nuclei emit this combination of four nucleons called an α particle? Why not just four nucleons, or even one? The answer is that the α particle is very strongly bound, so that its mass is significantly less than that of four separate nucleons. As we saw in Example 30–2, two protons and two neutrons separately have a total mass of about 4.032980 u. The total mass of a $^{228}_{90}$Th nucleus plus four separate nucleons is 232.061696 u, which is greater than the mass of the parent nucleus. Such a decay could not occur because it would violate the conservation of energy. Similarly, it is almost always true that the emission of a single nucleon is energetically not possible. Or, to put it another way, for any nuclide for which it could be possible (mass of parent > mass of daughter + mass of nucleon), the decay happens so fast after formation of the parent that we don't see the parent in nature.

➥ PHYSICS APPLIED

Smoke detector

One widespread application of nuclear physics is present in nearly every home in the form of the ordinary **smoke detector**. There are two technologies on which these devices are based. One involves an infrared light source and detector to measure the attenuation of the light if smoke particles are present. This type does not work nearly as well as the more common approach, which is based upon ionization of air by radioactive decay. This type of detector contains about 0.2 mg of the americium isotope $^{241}_{95}$Am, in the form of AmO_2, which is radioactive. The radiation ionizes the nitrogen and oxygen molecules in the air space between two oppositely charged plates. The resulting conductivity allows a small steady current. If smoke enters, the radiation is absorbed by the smoke particles rather than by the air molecules, thus reducing the current. The current drop is detected by the device's electronics and sets off the alarm. The radiation dose that escapes from an intact americium smoke detector is much less than the natural radioactive background, and so can be considered harmless. There is no question that smoke detectors save lives and reduce property damage.

30–5 Beta Decay

Transmutation of elements also occurs when a nucleus decays by β decay—that is, with the emission of an electron or β^- particle. The nucleus $^{14}_{6}C$, for example, emits an electron when it decays:

$$^{14}_{6}C \rightarrow {}^{14}_{7}N + e^- + \text{a neutrino},$$

where e^- is the symbol for the electron. (The symbol $_{-1}^{0}e$ is sometimes used for the electron whose charge corresponds to $Z = -1$ and, since it is not a nucleon and has very small mass, has $A = 0$.) The particle known as the neutrino, with rest mass $m_0 = 0$ and charge $q = 0$, was not initially detected and was only later hypothesized to exist, as we shall discuss later in this section. No nucleons are lost when an electron is emitted, and the total number of nucleons, A, is the same in the daughter nucleus as in the parent. But because an electron has been emitted, the charge on the daughter nucleus is different from that on the parent. The parent nucleus had $Z = +6$. In the decay, the nucleus loses a charge of -1; so, from charge conservation, the nucleus remaining behind must have an extra positive charge for a total of 7. So the daughter nucleus has $Z = 7$, which is nitrogen.

It must be carefully noted that the electron emitted in β decay is *not* an orbital electron. Instead, the electron is created *within the nucleus itself*. What happens is that one of the neutrons changes to a proton and in the process (to conserve charge) throws off an electron. Indeed, free neutrons actually do decay in this fashion: $n \rightarrow p + e^-$ (plus a neutrino). Because of their origin in the nucleus, the electrons emitted in β decay are often referred to as "β particles," rather than as electrons, to remind us of their origin. They are, nonetheless, indistinguishable from orbital electrons.

Emitted e^- comes from nucleus

EXAMPLE 30–6 **Energy release in $^{14}_{6}C$ decay.** How much energy is released when $^{14}_{6}C$ decays to $^{14}_{7}N$ by β emission? Use Appendix F.

SOLUTION The masses given in Appendix F are those of the neutral atom, and we have to keep track of the electrons involved. Assume the parent nucleus has six orbiting electrons so it is neutral, and its mass is 14.003242 u. The daughter, in this decay $^{14}_{7}N$, is not neutral since it has the same six electrons circling it but the nucleus has a charge of $+7e$. However, the mass of this daughter with its six electrons, plus the mass of the emitted electron (which makes a total of seven electrons), is just the mass of a neutral nitrogen atom. That is, the total mass in the final state is

$$\text{(mass of } {}^{14}_{7}N \text{ nucleus} + 6 \text{ electrons}) + (\text{mass of 1 electron}),$$

and this is equal to

$$\text{mass of neutral } {}^{14}_{7}N \text{ (includes 7 electrons)},$$

which, from Appendix F is a mass of 14.003074 u. (Note that the neutrino doesn't contribute to either the mass or charge balance since it has $m_0 = 0$ and $q = 0$.) Hence the mass after decay is 14.003074 u, whereas before decay, it was 14.003242 u. So the mass difference is 0.000168 u, which corresponds to 0.156 MeV or 156 keV.

FIGURE 30-6 Enrico Fermi. Fermi contributed significantly to both theoretical and experimental physics, a feat almost unique in this century.

According to this Example, we would expect the emitted electron to have a kinetic energy of 156 keV. (The daughter nucleus, because its mass is very much larger than that of the electron, recoils with very low velocity and hence gets very little of the kinetic energy.) Indeed, very careful measurements indicate that a few emitted β particles do have kinetic energy close to this calculated value. But the vast majority of emitted electrons have somewhat less energy. In fact, the energy of the emitted electron can be anywhere from zero up to the maximum value as calculated above. This range of electron KE was found for any β decay. It was as if the law of conservation of energy was being violated, and indeed Bohr actually considered this possibility. Careful experiments indicated that linear momentum and angular momentum also did not seem to be conserved. Physicists were troubled at the prospect of having to give up these laws, which had worked so well in all previous situations. In 1930, Wolfgang Pauli proposed an alternate solution: perhaps a new particle that was very difficult to detect was emitted during β decay in addition to the electron. This hypothesized particle could be carrying off the energy, momentum, and angular momentum required to maintain the conservation laws. This new particle was named the **neutrino**—meaning "little neutral one"—by the great Italian physicist Enrico Fermi (1901–1954; Fig. 30–6), who in 1934 worked out a detailed theory of β decay. (It was Fermi who, in this theory, postulated the existence of the fourth force in nature, which we call the weak nuclear force.) The neutrino has zero charge and seems to have zero rest mass, although we cannot yet rule out the possibility that it might have a very tiny rest mass. If its rest mass is zero, it is much like a photon in that it is neutral and travels at the speed of light. But the neutrino is far more difficult to detect. In 1956, complex experiments produced further evidence for the existence of the neutrino; but by then, most physicists had already accepted its existence.

The symbol for the neutrino is the Greek letter nu (ν). The correct way of writing the decay of $^{14}_{6}\text{C}$ is then

β^- decay

$$^{14}_{6}\text{C} \rightarrow {}^{14}_{7}\text{N} + \text{e}^- + \bar{\nu}.$$

The bar ($^-$) over the neutrino symbol is to indicate that it is an "antineutrino." (Why this is called an antineutrino rather than simply a neutrino need not concern us now; it is discussed in Chapter 32.)

Many isotopes decay by electron emission. They are always isotopes that have too many neutrons compared to the number of protons. That is, they are isotopes that lie above the stable isotopes plotted in Fig. 30–2. But what about unstable isotopes that have too few neutrons compared to their number of protons—those that fall below the stable isotopes of Fig. 30–2? These, it turns out, decay by emitting a **positron** instead of an electron. A positron (sometimes called an e$^+$ or β^+ particle) has the same mass as the electron, but it has a positive charge of $+1e$. Because it is so like an electron, except for its charge, the positron is called the **antiparticle**[†] to the electron. An example of a β^+ decay is that of $^{19}_{10}\text{Ne}$:

Positron (β^+) decay

$$^{19}_{10}\text{Ne} \rightarrow {}^{19}_{9}\text{F} + \text{e}^+ + \nu,$$

where e$^+$ (or $^{0}_{1}\text{e}$) stands for a positron. (Note that the ν emitted here is a neutrino, whereas that emitted in β^- decay is called an antineutrino. Thus

[†]Discussed in Chapter 32. Briefly, an antiparticle has the same mass as its corresponding particle, but opposite charge.

an antielectron (= positron) is emitted with a neutrino, whereas an anti-neutrino is emitted with an electron; this is discussed in Chapter 32.)

We can write β^- and β^+ decay, in general, as follows:

$$^A_Z N \to ^{\ \ A}_{Z+1} N' + e^- + \bar{\nu} \qquad\qquad [\beta^- \text{ decay}]$$

$$^A_Z N \to ^{\ \ A}_{Z-1} N' + e^+ + \nu, \qquad\qquad [\beta^+ \text{ decay}]$$

where N is the parent nucleus and N' the daughter.

Besides β^- and β^+ emission, there is a third related process. This is **electron capture** (abbreviated EC in Appendix F) and occurs when a nucleus absorbs one of its orbiting electrons. An example is $^7_4$Be, which as a result becomes $^7_3$Li. The process is written

$$^7_4\text{Be} + e^- \to ^7_3\text{Li} + \nu,$$

Electron capture

or, in general,

$$^A_Z N + e^- \to ^{\ \ A}_{Z-1} N' + \nu \qquad\qquad [\text{electron capture}]$$

Usually it is an electron in the innermost (K) shell that is captured, in which case it is called "K-capture." The electron disappears in the process and a proton in the nucleus becomes a neutron; a neutrino is emitted as a result. This process is inferred experimentally by detection of emitted X-rays (due to electrons jumping down to fill the empty state) of just the proper energy.

K-capture

In β decay, it is the weak nuclear force that plays the crucial role. The neutrino is unique in that it interacts with matter only via the weak force, which is why it is so hard to detect.

30–6 Gamma Decay

Gamma rays are photons having very high energy. The decay of a nucleus by emission of a γ ray is much like emission of photons by excited atoms. Like an atom, a nucleus itself can be in an excited state. When it jumps down to a lower energy state, or to the ground state, it emits a photon. The possible energy levels of a nucleus are much farther apart than those of an atom: on the order of keV or MeV, as compared to a few eV for electrons in an atom. Hence, the emitted photons have energies that can range from a few keV to several MeV. For a given decay, the γ ray always has the same energy. Since a γ ray carries no charge, there is no change in the element as a result of a γ decay.

How does a nucleus get into an excited state? It may occur because of a violent collision with another particle. More commonly, the nucleus remaining after a previous radioactive decay may be in an excited state. A typical example is shown in the energy-level diagram of Fig. 30–7. $^{12}_5$B can decay by β decay directly to the ground state of $^{12}_6$C; or it can go by β decay to an excited state of $^{12}_6$C, which then decays by emission of a 4.4 MeV γ ray to the ground state.

We can write γ decay as

$$^A_Z N^* \to ^A_Z N + \gamma. \qquad\qquad [\gamma \text{ decay}]$$

In some cases, a nucleus may remain in an excited state for some time before it emits a γ ray. The nucleus is then said to be in a **metastable state** and is called an **isomer**.

Isomer

FIGURE 30–7 Energy-level diagram showing how $^{12}_5$B can decay to the ground state of $^{12}_6$C by β decay (total energy released = 13.4 MeV), or can instead β-decay to an excited state of $^{12}_6$C (indicated by *), which subsequently decays to its ground state by emitting a 4.4-MeV γ ray.

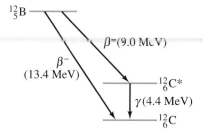

An excited nucleus can sometimes return to the ground state by another process known as **internal conversion** with no γ ray emitted. In this process, the excited nucleus interacts with one of the orbital electrons and ejects this electron from the atom with the same KE (minus the binding energy of the electron) that an emitted γ ray would have had.

What, you may wonder, is the difference between a γ ray and an X-ray? They both are electromagnetic radiation (photons) and, though γ rays usually have higher energy than X-rays, their range of energies overlap to some extent. The difference is not intrinsic. We use the term X-ray if the photon is produced by an electron–atom interaction, and γ ray if the photon is produced in a nuclear process.

TABLE 30–2 The Three Types of Radioactive Decay

α *decay:*
$$\ce{_Z^A N} \rightarrow \ce{_{Z-2}^{A-4} N'} + \ce{_2^4He}$$

β *decay:*
$$\ce{_Z^A N} \rightarrow \ce{_{Z+1}^A N'} + \text{e}^- + \bar{\nu}$$

$$\ce{_Z^A N} \rightarrow \ce{_{Z-1}^A N'} + \text{e}^+ + \nu$$

$$\ce{_Z^A N} + \text{e}^- \rightarrow \ce{_{Z-1}^A N'} + \nu \ [\text{EC}]^\dagger$$

γ *decay:*
$$\ce{_Z^A N}^* \rightarrow \ce{_Z^A N} + \gamma$$

*Indicates the excited state of a nucleus.
†Electron capture.

30–7 Conservation of Nucleon Number and Other Conservation Laws

In all three types of radioactive decay, the classical conservation laws hold. Energy, linear momentum, angular momentum, and electric charge are all conserved. These quantities are the same before the decay as after. But a new conservation law is also revealed, the **law of conservation of nucleon number**. According to this law, the total number of nucleons (A) remains constant in any process, although one type can change into the other type (protons into neutrons or vice versa). This law holds in all three types of decay. Table 30–2 gives a summary of α, β, and γ decay.

30–8 Half-Life and Rate of Decay

A macroscopic sample of any radioactive isotope consists of a vast number of radioactive nuclei. These nuclei do not all decay at one time. Rather, they decay one by one over a period of time. This is a random process: we can't predict exactly when a given nucleus will decay. But we can determine, on a probabilistic basis, approximately how many nuclei in a sample will decay over a given time period, by assuming that each nucleus has the same probability of decaying in each second that it exists.

The number of decays ΔN that occur in a very short time interval Δt is then proportional to Δt and to the total number N of radioactive nuclei present:

$$\Delta N = -\lambda N \, \Delta t. \tag{30–3}$$

In this equation, λ is a constant of proportionality called the **decay constant**, which is different for different isotopes. The greater λ is, the greater the rate of decay and the more radioactive that isotope is said to be. The number of decays that occur in the short time interval Δt is designated ΔN because each decay that occurs corresponds to a decrease by one in the number N of nuclei present. That is, radioactive decay is a "one-shot" process, Fig. 30–8. Once a particular parent nucleus decays into its daughter, it cannot do it again. The minus sign in Eq. 30–3 is needed to indicate that N is decreasing.

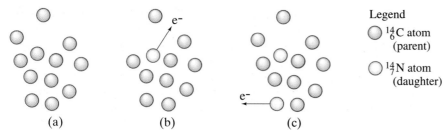

Legend

○ $^{14}_{6}$C atom (parent)

○ $^{14}_{7}$N atom (daughter)

FIGURE 30–8 Radioactive nuclei decay one by one. Hence, the number of parent nuclei in a sample is continually decreasing. When a $^{14}_{6}$C nucleus emits the electron, it becomes a $^{14}_{7}$N nucleus.

Equation 30–3 can be solved for N (using calculus) and the result is

$$N = N_0 e^{-\lambda t}, \qquad (30\text{–}4) \qquad \textit{Radioactive decay law}$$

where N_0 is the number of nuclei present at time $t = 0$, and N the number remaining after a time t. The symbol e is the natural exponential (encountered earlier in Sections 19–7 and 21–11) whose value is $e = 2.718\cdots$. Thus the number of parent nuclei in a sample decreases exponentially in time, as shown in Fig. 30–9a for $^{14}_{6}$C decay. Equation 30–4 is called the **radioactive decay law**.

The number of decays per second, $\Delta N / \Delta t$, is called the **activity** of the sample. Since $\Delta N / \Delta t$ is proportional to N (see Eq. 30–3), it, too, decreases exponentially in time with the same time constant (Fig. 30–9b). The activity at time t is given by

Activity

$$\frac{\Delta N}{\Delta t} = \left(\frac{\Delta N}{\Delta t}\right)_0 e^{-\lambda t}, \qquad (30\text{–}5)$$

where $(\Delta N / \Delta t)_0$ is the activity at $t = 0$.

The rate of decay of any isotope is often specified by giving its "half-life" rather than the decay constant. The **half-life** of an isotope is defined as the time it takes for half the original amount of isotope in a given sample to decay. For example, the half-life of $^{14}_{6}$C is 5730 years. If at some time a piece of petrified wood contains, say, 1.00×10^{22} nuclei of $^{14}_{6}$C, then 5730 years later it will contain only 0.50×10^{22} of these nuclei. After another 5730 years it will contain 0.25×10^{22} nuclei, and so on. This is shown in Fig. 30–9a. Since the rate of decay $\Delta N / \Delta t$ is proportional to N, it, too, decreases by a factor of 2 every half-life (Fig. 30–9b).

Half-life

FIGURE 30–9 (a) The number N of parent nuclei in a given sample of $^{14}_{6}$C decreases exponentially. (b) The number of decays per second also decreases exponentially. The half-life of $^{14}_{6}$C is 5730 yr, which means that the number of parent nuclei, N, and the rate of decay, $\Delta N / \Delta t$, decrease by half every 5730 yr.

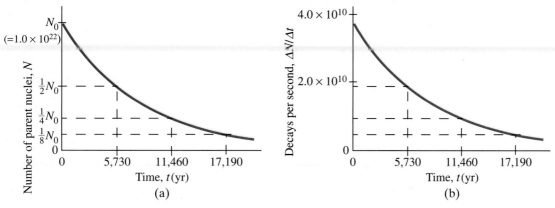

The half-lives of known radioactive isotopes vary from about 10^{-22} s to 10^{28} s (about 10^{21} yr). The half-lives of many isotopes are given in Appendix F. It should be clear that the half-life (which we designate $T_{\frac{1}{2}}$) bears an inverse relationship to the decay constant. The longer the half-life of an isotope, the more slowly it decays, and hence λ is smaller. Conversely, very active isotopes (large λ) have very short half-lives. The precise relationship between half-life and decay constant is

$$T_{\frac{1}{2}} = \frac{0.693}{\lambda}. \qquad \qquad (30\text{–}6)$$

We can derive this starting with Eq. 30–4. At $t = 0$, $N = N_0 e^0 = N_0$. At $t = T_{\frac{1}{2}}$, $N = N_0/2$ by definition of $T_{\frac{1}{2}}$ (half the parent nuclei remain). Then, from Eq. 30–4 evaluated at time $T_{\frac{1}{2}}$,

$$\frac{N_0}{2} = N_0 e^{-\lambda T_{\frac{1}{2}}} \qquad \text{or} \qquad \frac{1}{2} = e^{-\lambda T_{\frac{1}{2}}},$$

so

$$e^{\lambda T_{\frac{1}{2}}} = 2.$$

We take natural logs of both sides (remember that "ln" and "e" are inverse operations, meaning $\ln(e^x) = x$) and find

$$\ln\left(e^{\lambda T_{\frac{1}{2}}}\right) = \ln 2,$$

so

$$\lambda T_{\frac{1}{2}} = 0.693 \qquad \text{and} \qquad T_{\frac{1}{2}} = \frac{0.693}{\lambda},$$

which is Eq. 30–6. In the next Section we will see how to make calculations involving $T_{\frac{1}{2}}$.

30–9 Calculations Involving Decay Rates and Half-Life

Let us now consider Examples of what we can determine about a sample of radioactive material if we know the half-life.

EXAMPLE 30–7 **Sample activity.** The isotope $_{6}^{14}C$ has a half-life of 5730 yr. If at some time a sample contains 1.00×10^{22} carbon-14 nuclei, what is the activity of the sample?

SOLUTION First we calculate the decay constant λ from Eq. 30–6, and obtain

$$\lambda = \frac{0.693}{T_{\frac{1}{2}}} = \frac{0.693}{(5730 \text{ yr})(3.156 \times 10^7 \text{ s/yr})} = 3.83 \times 10^{-12} \text{ s}^{-1},$$

since the number of seconds in a year is $(60)(60)(24)(365\frac{1}{4}) = 3.156 \times 10^7$ s. From Eq. 30–3, the activity or rate of decay (we ignore the minus sign) is

$$\frac{\Delta N}{\Delta t} = \lambda N = (3.83 \times 10^{-12} \text{ s}^{-1})(1.00 \times 10^{22})$$

$$= 3.83 \times 10^{10} \text{ decays/s}.$$

(The unit "decays/s" is often written simply as s^{-1} since "decays" is not a unit but refers only to the number.) Note that the graph of Fig. 30–9b starts at this value, corresponding to the original value of $N = 1.0 \times 10^{22}$ nuclei in Fig. 30–9a.

EXAMPLE 30-8 **A sample of radioactive $^{13}_7N$.** A laboratory has $1.49 \, \mu g$ of pure $^{13}_7N$, which has a half-life of 10.0 min (600 s). (a) How many nuclei are present initially? (b) What is the activity initially? (c) What is the activity after 1.00 h? (d) After approximately how long will the activity drop to less than one per second?

SOLUTION (a) Since the atomic mass is 13.0, then 13.0 g will contain 6.02×10^{23} nuclei (Avogadro's number). Since we have only 1.49×10^{-6} g, the number of nuclei, N_0, that we have initially is given by the ratio

$$\frac{N_0}{1.49 \times 10^{-6} \, \text{g}} = \frac{6.02 \times 10^{23}}{13.0 \, \text{g}},$$

so $N_0 = 6.90 \times 10^{16}$ nuclei.
(b) From Eq. 30–6, $\lambda = (0.693)/(600 \, \text{s}) = 1.16 \times 10^{-3} \, \text{s}^{-1}$. Then, at $t = 0$ (Eq. 30–3),

$$\left(\frac{\Delta N}{\Delta t}\right)_0 = \lambda N_0 = (1.16 \times 10^{-3} \, \text{s}^{-1})(6.90 \times 10^{16}) = 8.00 \times 10^{13} \, \text{decays/s}.$$

(c) Since the half-life is 10.0 min, the decay rate decreases by half every 10.0 min. We can make the following table of activity after given periods of time:

Time (min)	Activity (decays/s)
0	8.00×10^{13}
10	4.00×10^{13}
20	2.00×10^{13}
30	1.00×10^{13}
40	0.500×10^{13}
50	0.250×10^{13}
60	0.125×10^{13}

Thus, after 1.0 h, the activity is 1.25×10^{12} decays/s. A simpler way of seeing this is to note that 60 min is 6 half-lives, so the activity will decrease to $(\frac{1}{2})(\frac{1}{2})(\frac{1}{2})(\frac{1}{2})(\frac{1}{2})(\frac{1}{2}) = (\frac{1}{2})^6 = \frac{1}{64}$ of its original value, or $(8.00 \times 10^{13})/(64) = 1.25 \times 10^{12}$ per second. A third way to arrive at this answer is to use Eq. 30–5 (setting $t = 60.0 \, \text{min} = 3600 \, \text{s}$):

$$\frac{\Delta N}{\Delta t} = \left(\frac{\Delta N}{\Delta t}\right)_0 e^{-\lambda t}$$

$$= (8.00 \times 10^{13} \, \text{s}^{-1})e^{-(1.16 \times 10^{-3} \, \text{s}^{-1})(3600 \, \text{s})} = 1.23 \times 10^{12} \, \text{s}^{-1}.$$

(The slight discrepancy in results arises because we kept only three significant figures.)

(d) After each hour, the activity drops by a factor of 64, so our table is extended to read as follows:

Time (h)	Activity (s^{-1})
0	8.00×10^{13}
1	1.25×10^{12}
2	1.95×10^{10}
3	3.05×10^{8}
4	4.77×10^{6}
5	7.45×10^{4}
6	1.16×10^{3}
7	$1.8 \ \ \times 10^{1} = 18$
8	< 1

So the activity drops to less than one per second within 8 h. We can obtain a more precise result using the properties of the exponential and its inverse, the natural logarithm (ln). We want to determine the time t when $\Delta N / \Delta t = 1.00 \, \text{s}^{-1}$. From Eq. 30–5, we have

$$e^{-\lambda t} = \frac{(\Delta N / \Delta t)}{(\Delta N / \Delta t)_0} = \frac{1.00 \, \text{s}^{-1}}{8.00 \times 10^{13} \, \text{s}^{-1}} = 1.25 \times 10^{-14}.$$

We take the natural log (ln) of both sides (remember $\ln e^{-\lambda t} = -\lambda t$) and divide by λ to find

$$t = -\frac{\ln (1.25 \times 10^{-14})}{\lambda} = 2.76 \times 10^4 \, \text{s} = 7.66 \, \text{h}.$$

30–10 Decay Series

It is often the case that one radioactive isotope decays to another isotope that is also radioactive. Sometimes this daughter decays to yet a third isotope which also is radioactive. Such successive decays are said to form a **decay series**. An important example is illustrated in Fig. 30–10. As can be seen, $^{238}_{92}\text{U}$ decays by α emission to $^{234}_{90}\text{Th}$, which in turn decays by β decay to $^{234}_{91}\text{Pa}$.. The series continues as shown, with several possible branches near the bottom. For example, $^{218}_{84}\text{Po}$ can decay either by α decay to $^{214}_{82}\text{Pb}$ or by β decay to $^{218}_{85}\text{At}$. The series ends at the stable lead isotope $^{206}_{82}\text{Pb}$. Other radioactive series also exist.

Because of such decay series, certain radioactive elements are found in nature that otherwise would not be. For when the solar system acquired its present form about 5 billion years ago, it is believed that nearly all nuclides were formed (by the fusion process, Sections 31–3 and 33–2). Many isotopes with short half-lives decayed quickly and no longer exist in nature today. But long-lived isotopes, such as $^{238}_{92}\text{U}$ with a half-life of 4.5×10^9 yr, still do exist in nature today. Indeed, about half of the original $^{238}_{92}\text{U}$ still remains (assuming that the origin of the solar system was about 5×10^9 yr ago). We might expect, however, that radium ($^{226}_{88}\text{Ra}$), with a half-life of 1600 yr, would long since have disappeared from the Earth. Indeed, the original $^{226}_{88}\text{Ra}$ nuclei must by now have all decayed. However, because $^{238}_{92}\text{U}$ decays (in several steps) to

FIGURE 30–10 Decay series beginning with $^{238}_{92}U$. Nuclei in the series are specified by a dot representing A and Z values. Half-lives are given in seconds (s), minutes (min), hours (h), days (d), or years (yr). Note that a horizontal arrow represents β decay (A does not change), whereas a diagonal line represents α decay (A changes by 4, Z changes by 2).

$^{226}_{00}Ra$, the supply of $^{226}_{88}Ra$ is continually replenished, which is why it is still found on Earth today. The same can be said for many other radioactive nuclides.

| **CONCEPTUAL EXAMPLE 30–9** | **Decay chain.** The decay chain start- |

ing with ^{234}U in Fig. 30–10 has nuclides with half-lives of 250,000 yr, 75,000 yr, 1600 yr, and a little under 4 days, respectively. Each decay in the chain has an alpha particle of a characteristic energy, and so we can monitor the radioactive decay rate of each nuclide. Given a sample that was pure ^{234}U a million years ago, which alpha decay would you expect to have the highest activity rate in the sample?

RESPONSE The first instinct is to say that the process with the shortest half-life would show the highest activity. Surprisingly, however, the activity rates in this sample are all the same! The reason is that in each case the decay of the parent acts as a bottleneck to the decay of the daughter. Compared to the 1600-yr half-life of ^{226}Ra, for example, its daughter ^{222}Rn decays almost immediately, but it cannot decay until it is made. (This is like an automobile assembly line: If worker A takes 20 minutes to do a task and then worker B takes only 1 minute to do the next task, worker B still does only one car every 20 minutes).

30–11 Radioactive Dating

Radioactive decay has many interesting applications. One is the technique of *radioactive dating* by which the age of ancient materials can be determined.

➡ PHYSICS APPLIED

Carbon–14 dating

The age of any object made from once-living matter, such as wood, can be determined using the natural radioactivity of $^{14}_{6}C$. All living plants absorb carbon dioxide (CO_2) from the air and use it to synthesize organic molecules. The vast majority of these carbon atoms are $^{12}_{6}C$, but a small fraction, about 1.3×10^{-12}, is the radioactive isotope $^{14}_{6}C$. The ratio of $^{14}_{6}C$ to $^{12}_{6}C$ in the atmosphere has remained roughly constant over many thousands of years, in spite of the fact that $^{14}_{6}C$ decays with a half-life of about 5730 yr. This is because neutrons in the cosmic radiation that impinges on the Earth from outer-space collide with atoms of the atmosphere. In particular, collisions with nitrogen nuclei produce the following nuclear transformation: $n + {}^{14}_{7}N \rightarrow {}^{14}_{6}C + p$. That is, a neutron strikes and is absorbed by a $^{14}_{7}N$ nucleus, and a proton is knocked out in the process. The remaining nucleus is $^{14}_{6}C$. This continual production of $^{14}_{6}C$ in the atmosphere roughly balances the loss of $^{14}_{6}C$ by radioactive decay. As long as a plant or tree is alive, it continually uses the carbon from carbon dioxide in the air to build new tissue and to replace old. Animals eat plants, so they too are continually receiving a fresh supply of carbon for their tissues. Organisms cannot distinguish[†] $^{14}_{6}C$ from $^{12}_{6}C$, and since the ratio of $^{14}_{6}C$ to $^{12}_{6}C$ in the atmosphere remains nearly constant, the ratio of the two isotopes within the living organism remains nearly constant as well. But when an organism dies, carbon dioxide is no longer absorbed and utilized. Because the $^{14}_{6}C$ decays radioactively, the ratio of $^{14}_{6}C$ to $^{12}_{6}C$ in a dead organism decreases in time. Since the half-life of $^{14}_{6}C$ is about 5730 yr, the $^{14}_{6}C/^{12}_{6}C$ ratio decreases by half every 5730 yr. If, for example, the $^{14}_{6}C/^{12}_{6}C$ ratio of an ancient wooden tool is half of what it is in living trees, then the object must have been made from a tree that was felled about 5700 years ago. Actually, corrections must be made for the fact that the $^{14}_{6}C/^{12}_{6}C$ ratio in the atmosphere has not remained precisely constant over time. The determination of what this ratio has been over the centuries has required using techniques such as comparing the expected ratio to the actual ratio for objects whose age is known, such as very old trees whose annual rings can be counted.

➡ PHYSICS APPLIED

Archeological dating

EXAMPLE 30–10 **An ancient animal.** An animal bone fragment found in an archeological site has a carbon mass of 200 g. It registers an activity of 15 decays/s. What is the age of the bone?

SOLUTION When the animal was alive, the ratio of $^{14}_{6}C$ to $^{12}_{6}C$ in the 200-g piece of bone was 1.3×10^{-12}. The number of $^{14}_{6}C$ nuclei at that time was

$$N_0 = \left(\frac{6.02 \times 10^{23} \text{ atoms}}{12 \text{ g}} \right)(200 \text{ g})(1.3 \times 10^{-12}) = 1.3 \times 10^{13}.$$

From Eq. 30–3 we can see that

$$\left(\frac{\Delta N}{\Delta t} \right)_0 = \lambda N_0,$$

[†] Organisms operate almost exclusively via chemical reactions—which involve only the outer orbital electrons of the atom; extra neutrons in the nucleus have essentially no effect.

where $\lambda = 3.83 \times 10^{-12}\,\text{s}^{-1}$ (Example 30–7). So the original activity was

$$\left(\frac{\Delta N}{\Delta t}\right)_0 = (3.83 \times 10^{-12}\,\text{s}^{-1})(1.3 \times 10^{13}) = 50\,\text{s}^{-1}.$$

Combining Eq. 30–3 with Eq. 30–4, and ignoring the minus sign, gives

$$\frac{\Delta N}{\Delta t} = \lambda N = \lambda N_0 e^{-\lambda t},$$

and we rewrite this as

$$e^{\lambda t} = \frac{\lambda N_0}{\Delta N / \Delta t}.$$

Now we take the natural log (ln) of both sides to get

$$t = \frac{1}{\lambda}\ \ln \frac{\lambda N_0}{\Delta N / \Delta t}$$

$$= \frac{1}{3.83 \times 10^{-12}\,\text{s}^{-1}}\ \ln \left[\frac{(3.83 \times 10^{-12}\,\text{s}^{-1})(1.3 \times 10^{13})}{15\,\text{s}^{-1}}\right]$$

$$= 3.13 \times 10^{11}\,\text{s} = 9900\,\text{yr},$$

which is the time elapsed since the death of the animal.

Carbon dating is useful only for determining the age of objects less than about 60,000 years old. The amount of $^{14}_{6}\text{C}$ remaining in older objects is usually too small to measure accurately, although new techniques are allowing detection of even smaller amounts of $^{14}_{6}\text{C}$, pushing the time frame further back. On the other hand, radioactive isotopes with longer half-lives can be used in certain circumstances to obtain the age of older objects. For example, the decay of $^{238}_{92}\text{U}$, because of its long half-life of 4.5×10^9 years, is useful in determining the ages of rocks on a geologic time scale. When molten material solidified into rock as the temperature dropped, different compounds solidified according to the melting points, and thus different compounds separated to some extent. Uranium present in a material became fixed in position and the daughter nuclei that result from the decay of uranium were also fixed in that position. Thus, by measuring the amount of $^{238}_{92}\text{U}$ remaining in the material relative to the amount of daughter nuclei, the time when the rock solidified can be determined.

➡ **PHYSICS APPLIED**

Geological dating

Radioactive dating methods using $^{238}_{92}\text{U}$ and other isotopes have shown the age of the oldest Earth rocks to be about 4×10^9 yr. The age of rocks in which the oldest fossilized organisms are embedded indicates that life appeared at least 3 billion years ago. The earliest fossilized remains of mammals are found in rocks 200 million years old, and the first humanlike creatures seem to have appeared about 2 million years ago. Radioactive dating has been indispensable for the reconstruction of Earth's history and the evolution of its biological organisms.

➡ **PHYSICS APPLIED**

Oldest Earth rocks and earliest life

FIGURE 30–11 Potential energy for alpha particle and nucleus, showing the Coulomb barrier through which the α particle must tunnel to escape. The Q-value of the reaction is also shown.

We have seen that radioactive decay occurs only when the mass of the parent nucleus is greater than the sum of the masses of the daughter nucleus and all particles emitted. For example, $^{238}_{92}\text{U}$ can decay to $^{234}_{90}\text{Th}$ because the mass of $^{238}_{92}\text{U}$ is greater than the mass of the $^{234}_{90}\text{Th}$ plus the mass of the α particle. Since systems tend to go in the direction that reduces their internal or potential energy (a ball rolls downhill, a positive charge moves toward a negative charge), you may wonder why an unstable nucleus doesn't fall apart immediately. In other words, why do $^{238}_{92}\text{U}$ nuclei ($T_{\frac{1}{2}} = 4.5 \times 10^9 \text{ yr}$) and other isotopes have such long half-lives? Why don't parent nuclei all decay at once?

The answer has to do with quantum theory and the nature of the forces involved. One way to view the situation is with the aid of a potential-energy diagram, as in Fig. 30–11. Let us consider the particular case of the decay $^{238}_{92}\text{U} \rightarrow {}^{234}_{90}\text{Th} + {}^4_2\text{He}$. The blue line represents the potential energy, including rest mass, where we imagine the α particle as a separate entity within the $^{238}_{92}\text{U}$ nucleus. The region labeled A in Fig. 30–11 represents the PE of the α particle when it is held within the uranium nucleus by the nuclear force (R_0 is the nuclear radius). Region C represents the PE when the α particle is free of the nucleus. The downward-curving PE (proportional to $1/r$) represents the electrical (Coulomb's law) repulsion between the positively charged α and the $^{234}_{90}\text{Th}$ nucleus. In order to get to region C, the α particle has to get by the barrier shown. Since the PE just beyond $r = R_0$ (region B) is greater than the energy of the alpha particle (dashed line), the α particle could not escape the nucleus if it were governed by classical physics. It could escape only if there were an input of energy equal to the height of the barrier. Nuclei decay spontaneously, however, without any input of energy. How, then, does the α particle get from region A to region C?

Tunneling It actually passes through the barrier in a process known as **tunneling**. Classically, this could not happen, because an α particle in region B (within the barrier) would be violating the conservation-of-energy principle.[†] The uncertainty principle, however, tells us that energy conservation can be violated by an amount ΔE for a length of time Δt given by

$$(\Delta E)(\Delta t) \approx \frac{h}{2\pi}.$$

We saw in Section 28–3 that this is a result of the wave–particle duality. Thus quantum mechanics allows conservation of energy to be violated for brief periods that may be long enough for an α particle to "tunnel" through the barrier. ΔE would represent the energy difference between the average barrier height and the particle's energy, and Δt the time to pass through the barrier. The higher and wider the barrier, the less time the α particle has to escape and the less likely it is to do so. It is therefore the height and width of this barrier that controls the rate of decay and half-life of an isotope.

[†]Since $\text{KE} = \frac{1}{2}mv^2 \geq 0$, then $E = \text{KE} + \text{PE} \geq \text{PE}$. But in Fig. 30–11, the total E (dashed line) is less than the PE in region B, so a particle reaching region B would violate conservation of energy.

*30–13 Detection of Radiation

Individual particles such as electrons, protons, α particles, neutrons, and γ rays are not detected directly by our senses. Consequently, a variety of instruments have been developed to detect them.

One of the most common is the **Geiger counter**. As shown in Fig. 30–12, it consists of a cylindrical metal tube filled with a certain type of gas. A long wire runs down the center and is kept at a high positive voltage ($\approx 10^3$ V) with respect to the outer cylinder. The voltage is just slightly less than that required to ionize the gas atoms. When a charged particle enters through the thin "window" at one end of the tube, it ionizes a few atoms of the gas. The freed electrons are attracted toward the positive wire and as they are accelerated they strike and ionize additional atoms. An "avalanche" of electrons is quickly produced, and when it reaches the wire anode, it produces a voltage pulse. The pulse, after being amplified, can be sent to an electronic counter, which counts how many particles have been detected. Or the pulses can be sent to a loudspeaker and each detection of a particle is heard as a "click."

A **scintillation counter** makes use of a solid, liquid, or gas known as a **scintillator** or **phosphor**. The atoms of a scintillator are easily excited when struck by an incoming particle and emit visible light when they return to their ground states. Typical scintillators are crystals of NaI and certain plastics. One face of a solid scintillator is cemented to a photomultiplier tube, and the whole is wrapped with opaque material to keep it light-tight or is placed within a light-tight container. The **photomultiplier** (PM) **tube** converts the energy of the scintillator-emitted photon(s) into an electric signal. A PM tube is a vacuum tube containing several electrodes (typically 8 to 14), called *dynodes*, which are maintained at successively higher voltages as shown in Fig. 30–13. At its top surface is a photoelectric surface, called the *photocathode*, whose work function (Section 27–3) is low enough that an electron is easily released when struck by a photon from the scintillator. Such an electron is accelerated toward the first dynode. When it strikes the first dynode, the electron has acquired sufficient kinetic energy so that it can eject two to five more electrons. These, in turn, are accelerated to the second dynode, and a multiplication process begins. The number of electrons striking the last dynode may be 10^6 or more. Thus the passage of a particle through the scintillator results in an electric signal at the output of the PM tube that can be sent to an electronic counter just as for a Geiger tube. Because a scintillator crystal is a solid and therefore much more dense than the gas of a Geiger counter, it is a more efficient detector—especially for γ rays, which interact less with matter than do β rays.

In tracer work and other biological experiments (Section 31–7), **liquid scintillators** are often used. Radioactive samples taken at different times or from different parts of an organism are placed directly in small bottles containing the liquid scintillator. This is particularly convenient for detection of β rays from $^3_1$H and $^{14}_6$C, which have very low energies and have difficulty passing through the outer covering of a crystal scintillator or Geiger tube. A PM tube is still used to produce the electric signal.

A **semiconductor detector** consists of a reverse-biased *p-n* junction diode (Section 29–8). A particle passing through the junction can excite electrons into the conduction band, leaving holes in the valence band. The freed charges produce a short electrical pulse that can be counted just as for Geiger and scintillation counters.

FIGURE 30–12 Diagram of a Geiger counter.

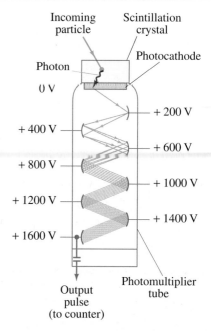

FIGURE 30–13 Scintillation counter with a photomultiplier tube.

The three devices discussed so far are used for counting the number of particles (or decays of a radioactive isotope). Other devices have been developed that allow the track of a charged particle to be *seen*. The simplest is the **photographic emulsion** (which, being small and simple and therefore portable, is now used particularly for cosmic-ray studies from balloons). A particle passing through a layer of photographic emulsion ionizes the atoms along its path. This results in a chemical change at these points, and when the emulsion is developed, the particle's path is revealed.

In a **cloud chamber**, a gas is cooled to a temperature slightly below its usual condensation point. (It is said to be "supercooled.") The gas molecules begin to condense on any ionized molecules present. Thus the ions produced when a charged particle passes through serve as centers on which tiny droplets form (Fig. 30–14). Light scatters more from these droplets than from the gas background, so a photo of the cloud chamber at the right moment shows the track of the particle. An important instrument in the early days of nuclear physics, it is little used today.

The **bubble chamber**, invented in 1952 by D. A. Glaser (1926–   ), makes use of a superheated liquid. The liquid is kept close to its normal boiling point and the bubbles characteristic of boiling form around ions produced by the passage of a charged particle (Fig. 30–15). A photograph of the interior of the chamber thus reveals the paths of particles that recently passed through. Because the bubble chamber uses a liquid—often liquid hydrogen—the density of atoms is much greater than in a cloud chamber. Hence it is a much more efficient device for observing the tracks of charged particles and their interactions with the nuclei of the liquid. Usually, a magnetic field is applied across the chamber and the momentum of the moving particles can be determined from the radius of curvature of their paths.

A **wire chamber** consists of a set of closely spaced fine parallel wires. Alternate planes of wires are grounded and the ones in between are kept at very high voltage. When a charged particle passes through, the ions produced in the gas between the wires become an avalanche, and a large current results which produces a visible spark. Thus the path of the particle is made visible and can be photographed. Alternatively, the sparks produce electric pulses in the wires and the positions of the sparks can be determined by the time it takes the pulses to reach detectors placed at the ends of the wires. Then the particle's path is reconstructed electronically with computers which can then "draw" a picture of the tracks.

FIGURE 30–14 In a cloud or bubble chamber, droplets or bubbles are formed around ions produced by the passage of a charged particle.

Path of particle

FIGURE 30–15 (a) Bubble chamber; (b) particle tracks in a bubble chamber.

(a)

(b)

SUMMARY

Nuclear physics is the study of atomic nuclei. Nuclei contain **protons** and **neutrons**, which are collectively known as **nucleons**. The total number of nucleons, A, is the **atomic mass number**. The number of protons, Z, is the **atomic number**. The number of neutrons equals $A - Z$. **Isotopes** are nuclei with the same Z, but with different numbers of neutrons. For an element X, an isotope of given Z and A is represented by

$$_Z^A X.$$

The nuclear radius is proportional to $A^{\frac{1}{3}}$, indicating that all nuclei have about the same density. Nuclear masses are specified in **unified atomic mass units** (u), where the mass of $_6^{12}C$ (including its 6 electrons) is defined as exactly 12.000000 u, or in terms of their energy equivalent (because $E = mc^2$), where

$$1\,u = 931.5\,MeV/c^2 = 1.66 \times 10^{-27}\,kg.$$

The mass of a stable nucleus is less than the sum of the masses of its constituent nucleons. The difference in mass (times c^2) is the **total binding energy**. It represents the energy needed to break the nucleus into its constituent nucleons. The **binding energy per nucleon** averages about 8 MeV per nucleon, and is lowest for low mass and high mass nuclei.

Unstable nuclei undergo **radioactive decay**; they change into other nuclei with the emission of an α, β, or γ particle. An α particle is a $_2^4He$ nucleus; a β particle is an electron or positron; and a γ ray is a high-energy photon. In β decay, a **neutrino** is also emitted. The transformation of the parent into the daughter nucleus is called **transmutation** of the elements. Radioactive decay occurs spontaneously only when the rest mass of the products is less than the mass of the parent nucleus. The loss in mass appears as kinetic energy of the products.

Nuclei are held together by the **strong nuclear force**. The **weak nuclear force** makes itself apparent in β decay. These two forces, plus the gravitational and electromagnetic forces, are the four known types of force. Electric charge, linear and angular momentum, mass–energy, and **nucleon number** are **conserved** in all decays.

Radioactive decay is a statistical process. For a given type of radioactive nucleus, the number of nuclei that decay (ΔN) in a time Δt is proportional to the number N of parent nuclei present:

$$\Delta N = -\lambda N \,\Delta t.$$

The proportionality constant, λ, is called the **decay constant** and is characteristic of the given nucleus. The number N of nuclei remaining after a time t decreases exponentially

$$N \propto e^{-\lambda t}$$

as does the **activity**, $\Delta N/\Delta t$. The **half-life**, $T_{\frac{1}{2}}$, is the time required for half the nuclei of a radioactive sample to decay. It is related to the decay constant by $T_{\frac{1}{2}} = 0.693/\lambda$.

QUESTIONS

1. What do different isotopes of a given element have in common? How are they different?

2. What are the elements represented by the X in the following: (a) $_{92}^{232}X$; (b) $_7^{18}X$; (c) $_1^1 X$; (d) $_{38}^{82}X$; (e) $_{97}^{247}X$?

3. How many protons and how many neutrons do each of the isotopes in Question 2 have?

4. Why are the atomic masses of many elements (see the periodic table) not close to whole numbers?

5. How do we know there is such a thing as the strong nuclear force?

6. What are the similarities and the differences between the strong nuclear force and the electric force?

7. What is the experimental evidence in favor of radioactivity being a nuclear process?

8. The isotope $_{29}^{64}Cu$ is unusual in that it can decay by γ, β^-, and β^+ emission. What is the resulting nuclide for each case?

9. A $_{92}^{238}U$ nucleus decays to a nucleus containing how many neutrons?

10. Describe, in as many ways as possible, the difference between α, β, and γ rays.

11. What element is formed by the radioactive decay of (a) $_{11}^{24}Na$ (β^-); (b) $_{11}^{22}Na$ (β^+); (c) $_{84}^{210}Po$ (α)?

12. What element is formed by the decay of (a) $_{15}^{32}P$ (β^-); (b) $_{16}^{35}S$ (β^-); (c) $_{83}^{211}Bi$ (α)?

13. Fill in the missing particle or nucleus:
   (a) $_{20}^{45}Ca \rightarrow ? + e^- + \bar{\nu}$
   (b) $_{29}^{58}Cu \rightarrow ? + \gamma$
   (c) $_{24}^{46}Cr \rightarrow _{23}^{46}V + ?$
   (d) $_{94}^{234}Pu \rightarrow ? + \alpha$
   (e) $_{93}^{239}Np \rightarrow _{94}^{239}Pu + ?$

14. Immediately after a $^{238}_{92}U$ nucleus decays to $^{234}_{90}Th$ + $^{4}_{2}He$, the daughter thorium nucleus still has 92 electrons circling it. Since thorium normally holds only 90 electrons, what do you suppose happens to the two extra ones?

15. When a nucleus undergoes beta decay, what happens to the energy levels of the atomic electrons? What is likely to happen to these electrons following the decay?

16. The alpha particles from a given alpha-emitting nuclide are monoenergetic; that is, they all have the same kinetic energy. But the beta particles from a beta-emitting nuclide have a spectrum of energies. Explain the difference between these two cases.

17. Do isotopes that undergo electron capture generally lie above or below the line of stability in Fig. 30–2?

18. Can hydrogen or deuterium emit an α particle?

19. Why are many artificially produced radioactive isotopes rare in nature?

20. An isotope has a half-life of one month. After two months, will a given sample of this isotope have completely decayed? If not, how much remains?

21. Explain the absence of β^{+} emitters in the radioactive decay series of Fig. 30–10.

22. Can $^{14}_{6}C$ dating be used to measure the age of stone walls and tablets of ancient civilizations?

23. What assumptions are made in carbon-dating? What do you think could affect these assumptions?

*24. Describe how the potential energy curve for an α particle in an α-emitting nucleus differs from that for a stable nucleus.

PROBLEMS

SECTION 30–1

1. (I) What is the rest mass of an α particle in MeV/c^2?

2. (I) A pi meson has a mass of $139\ MeV/c^2$. What is this in atomic mass units?

3. (I) What is the approximate radius of an alpha particle ($^{4}_{2}He$)?

4. (I) By what percentage is the radius of the isotope $^{14}_{6}C$ greater than that of its sister $^{12}_{6}C$?

5. (II) (a) What is the fraction of the hydrogen atom's mass that is in the nucleus? (b) What is the fraction of the hydrogen atom's volume that is occupied by the nucleus? (c) What is the density of nuclear matter? Compare this with water.

6. (II) (a) What is the approximate radius of a $^{64}_{29}Cu$ nucleus? (b) Approximately what is the value of A for a nucleus whose radius is 3.7×10^{-15} m?

7. (II) How much energy must an α particle have to just "touch" the surface of a $^{238}_{92}U$ nucleus?

8. (II) If an alpha particle were released from rest near the surface of a $^{243}_{95}Am$ nucleus, what would its kinetic energy be when far away?

9. (II) What stable nucleus has approximately half the radius of a uranium nucleus? [*Hint*: Find A and use Appendix F to get Z.]

SECTION 30–2

10. (I) Estimate the total binding energy for $^{40}_{20}Ca$, using Fig. 30–1.

11. (I) Use Fig. 30–1 to estimate the total binding energy of (a) $^{238}_{92}U$, and (b) $^{84}_{36}Kr$.

12. (II) Use Appendix F to calculate the binding energy of $^{2}_{1}H$ (deuterium).

13. (II) Calculate the binding energy per nucleon for a $^{14}_{7}N$ nucleus.

14. (II) Calculate the total binding energy and the binding energy per nucleon for $^{6}_{3}Li$. Use Appendix F.

15. (II) Calculate the binding energy of the last neutron in a $^{12}_{6}C$ nucleus. [*Hint*: Compare the mass of $^{12}_{6}C$ with that of $^{11}_{6}C$ + $^{1}_{0}n$; use Appendix F.]

16. (II) Compare the binding energy of a neutron in $^{23}_{11}Na$ to that in $^{24}_{11}Na$.

17. (II) How much energy is required to remove (a) a proton, (b) a neutron, from $^{16}_{8}O$? Explain the difference in your answers.

18. (II) (a) Show that the nucleus $^{8}_{4}Be$ (mass = 8.005305 u) is unstable to decay into two α particles. (b) Is $^{12}_{6}C$ stable against decay into three α particles? Show why or why not.

SECTIONS 30–3 TO 30–7

19. (I) $^{60}_{27}Co$ in an excited state emits a 1.33-MeV γ ray as it jumps to the ground state. What is the mass (in u) of the excited cobalt atom?

20. (I) How much energy is released when tritium, $^{3}_{1}H$, decays by β^{-} emission?

21. (I) What is the maximum KE of an electron emitted in the β decay of a free neutron?

22. (I) Show that the decay $^{11}_{6}C \rightarrow {}^{10}_{5}B + p$ is not possible because energy would not be conserved.

23. (II) $^{22}_{11}Na$ is radioactive. Is it a β^{-} or β^{+} emitter? Write down the decay reaction, and estimate the maximum KE of the emitted β.

24. (II) Give the result of a calculation that shows whether or not the following decays are possible:
(a) $^{236}_{92}U \rightarrow {}^{235}_{92}U + n$; (b) $^{16}_{8}O \rightarrow {}^{15}_{8}O + n$;
(c) $^{23}_{11}Na \rightarrow {}^{22}_{11}Na + n$.

25. (II) A $^{232}_{92}U$ nucleus emits an α particle with KE = 5.32 MeV. What is the final nucleus and what is the approximate atomic mass (in u) of the final atom?

26. (II) When $^{23}_{10}$Ne (mass = 22.9945 u) decays to $^{23}_{11}$Na (mass = 22.9898 u), what is the maximum kinetic energy of the emitted electron? What is its minimum energy? What is the energy of the neutrino in each case?

27. (II) The nuclide $^{32}_{15}$P decays by emitting an electron whose maximum kinetic energy can be 1.71 MeV. (a) What is the daughter nucleus? (b) What is its atomic mass (in u)?

28. (II) The isotope $^{218}_{84}$Po can decay by either α or β^- emission. What is the energy release in each case? The mass of $^{218}_{84}$Po is 218.008965 u.

29. (II) How much energy is released in electron capture by beryllium: $^7_4$Be $+ \, ^0_{-1}$e $\rightarrow \, ^7_3$Li $+ \, \nu$?

30. (II) What is the energy of the α particle emitted in the decay $^{210}_{84}$Po $\rightarrow \, ^{206}_{82}$Pb $+ \alpha$?

31. (III) The α particle emitted when $^{238}_{92}$U decays has KE = 4.20 MeV. Calculate the recoil KE of the daughter nucleus and the Q-value of the decay.

32. (III) Show that when a nucleus decays by β^+ decay, the total energy released is equal to

$$(M_P - M_D - 2m_e)c^2,$$

where M_P and M_D are the masses of the parent and daughter atoms (neutral), and m_e is the mass of an electron or positron.

33. (III) Use the result of Problem 32 to determine the maximum kinetic energy of β^+ particles released when $^{11}_6$C decays to $^{11}_5$B. What is the maximum energy the neutrino can have? What is its minimum energy?

34. (III) In α decay of, say, a $^{226}_{88}$Ra nucleus, show that the nucleus carries away a fraction $1/(1 + A_D/4)$ of the total energy available, where A_D is the mass number of the daughter nucleus. [*Hint*: Use conservation of momentum as well as conservation of energy.] Approximately what percentage of the energy available is thus carried off by the α particle in the case cited?

SECTIONS 30–8 TO 30–11

35. (I) A radioactive material produces 1280 decays per minute at one time, and 6 h later produces 320 decays per minute. What is its half-life?

36. (I) Look up the half life of $^{238}_{92}$U in Appendix F, and then determine its decay constant.

37. (I) The decay constant of a given nucleus is 5.4×10^{-3} s^{-1}. What is its half-life?

38. (I) What is the activity of a sample of $^{14}_6$C that contains 4.1×10^{20} nuclei?

39. (I) What fraction of a sample of $^{68}_{32}$Ge, whose half-life is about 9 months, will remain after 3.0 yr?

40. (I) How many nuclei of $^{238}_{92}$U remain in a rock if the activity registers 275 decays per second?

41. (II) What fraction of a sample is left after (a) exactly 4 half-lives, (b) exactly $4\frac{1}{2}$ half-lives?

42. (II) In a series of decays, the nuclide $^{235}_{92}$U becomes $^{207}_{82}$Pb. How many α and β^- particles are emitted in this series?

43. (II) The iodine isotope $^{131}_{53}$I is used in hospitals for diagnosis of thyroid function. If 532 μg are ingested by a patient, determine the activity (a) immediately, (b) 1.0 h later when the thyroid is being tested, and (c) 6 months later. Use Appendix F.

44. (II) $^{124}_{55}$Cs has a half-life of 30.8 s. (a) If we have 7.8 μg initially, how many nuclei are present? (b) How many are present 2.0 min later? (c) What is the activity at this time? (d) After how much time will the activity drop to less than about 1 per second?

45. (II) Calculate the activity of a pure 4.7-μg sample of $^{32}_{15}$P ($T_{\frac{1}{2}} = 1.23 \times 10^6$ s).

46. (II) The activity of a sample of $^{35}_{16}$S ($T_{\frac{1}{2}} = 7.56 \times 10^6$ s) is 3.55×10^5 decays per second. What is the mass of sample present?

47. (II) A sample of $^{233}_{92}$U ($T_{\frac{1}{2}} = 1.59 \times 10^5$ yr) contains 6.50×10^{19} nuclei. (a) What is the decay constant? (b) Approximately how many disintegrations will occur per minute?

48. (II) The activity of a sample drops by a factor of 10 in 9.6 minutes. What is its half-life?

49. (II) A 135-g sample of pure carbon contains 1.3 parts in 10^{12} (atoms) of $^{14}_6$C. How many disintegrations occur per second?

50. (II) A radioactive nuclide produces 2880 decays per minute at one time, and 1.6 h later produces 820 decays per minute. What is the half-life of the nuclide?

51. (II) A sample of $^{40}_{19}$K is decaying at a rate of 8.70×10^2 decays/s. What is the mass of the sample?

52. (II) The rubidium isotope $^{87}_{37}$Rb, a β emitter with a half-life of 4.75×10^{10} yr, is used to determine the age of rocks and fossils. Rocks containing fossils of early animals contain a ratio of $^{87}_{38}$Sr to $^{87}_{37}$Rb of 0.0160. Assuming that there was no $^{87}_{38}$Sr present when the rocks were formed, calculate the age of these fossils. [*Hint*: Use Eq. 30–3.]

53. (II) Use Fig. 30–10 and calculate the relative decay rates for α decay of $^{218}_{84}$Po and $^{214}_{84}$Po.

54. (II) $^7_4$Be decays with a half-life of about 53 d. It is produced in the upper atmosphere, and filters down onto the Earth's surface. If a plant leaf is detected to have 350 decays/s of $^7_4$Be, how long do we have to wait for the decay rate to drop to 10/s? Estimate the initial mass of $^7_4$Be on the leaf.

55. (II) Which radioactive isotope of lead is being produced in a reaction where the measured activity of a sample drops to 1.050 percent of its original activity in 4.00 h?

56. (III) An ancient club is found that contains 170 g of carbon and has an activity of 5.0 decays per second. Determine its age assuming that in living trees the ratio of ^{14}C/^{12}C atoms is about 1.3×10^{-12}.

57. (III) At $t = 0$, a pure sample of radioactive nuclei contains N_0 nuclei whose decay constant is λ. Determine a formula for the number of daughter nuclei, N_D, as a function of time; assume that $N_D = 0$ at $t = 0$.

58. Show that the radius of the largest nuclei (say, $^{238}_{92}U$) is only about 6 times greater than that of the smallest (^1_1H).

59. (a) Determine the density of nuclear matter in kg/m^3, and show that it is essentially the same for all nuclei. (b) What would be the radius of the Earth if it had its actual mass but had the density of nuclei? (c) What would be the radius of a $^{238}_{92}U$ nucleus if it had the density of the Earth?

60. Using the uncertainty principle, use the size of the nucleus to estimate the kinetic energy of a nucleon in, say, iron. Ignore relativistic corrections. [*Hint*: A particle can have an energy at least as large as its uncertainty.]

61. How much recoil energy does a $^{40}_{19}K$ nucleus get when it emits a 1.46 MeV gamma ray?

62. An old wooden tool is found to contain only 10 percent of $^{14}_6C$ that a sample of fresh wood would. How old is the tool?

63. The ^3_1H isotope of hydrogen, which is called *tritium* (because it contains three nucleons), has a half-life of 12.33 yr. It can be used to measure the age of objects up to about 100 yr. It is produced in the upper atmosphere by cosmic rays and brought to Earth by rain. As an application, determine approximately the age of a bottle of wine whose ^3_1H radiation is about $\frac{1}{10}$ that present in new wine.

64. A neutron star consists of neutrons at approximately nuclear density. Estimate, for a 10-km-diameter neutron star, (a) its mass number, (b) its mass (kg), and (c) the acceleration of gravity at its surface.

65. Recent elementary particle theories (Section 32–11) suggest that the proton may be unstable, with a half-life $\geq 10^{32}$ yr. How long would you expect to wait for one proton in your body to decay (consider that your body is all water)?

66. When water is placed near an intense neutron source, the neutrons can be slowed down by collisions with the water molecules and eventually captured by a hydrogen nucleus to form the stable isotope called deuterium, ^2_1H, giving off a gamma ray. What is the energy of the gamma ray?

67. How long must you wait (in half-lives) for a radioactive sample to drop to 1.00 percent of its original activity?

68. If the potassium isotope $^{40}_{19}K$ gives 50 decays/s in a liter of milk, estimate how much $^{40}_{19}K$ and regular $^{39}_{19}K$ are in a liter of milk. Use Appendix F.

69. Decay series, such as that shown in Fig. 30–10, can be classified into four families, depending on whether the mass numbers have the form $4n$, $4n + 1$, $4n + 2$, $4n + 3$, where n is an integer. Justify this statement and show that for a nuclide in any family, all its daughters will be in the same family.

70. Strontium-90 is produced as a nuclear fission product of uranium in both reactors and atomic bombs. Look at its location in the periodic table to see what other elements it might be similar to chemically, and tell why you think it might be dangerous to ingest. It has too many neutrons, and it decays with a half-life of about 29 yr. How long will we have to wait for the amount of $^{90}_{38}Sr$ on the Earth's surface to reach 1 percent of its current level, assuming no new material is scattered about? Write down the decay reaction, including the daughter nucleus. Is the daughter radioactive? If so, write down its decay. Finish the decay scheme until you reach a stable nucleus.

71. The nuclide $^{191}_{76}Os$ decays with β^- energy of 0.14 MeV accompanied by γ rays of energy 0.042 MeV and 0.129 MeV. (a) What is the daughter nucleus? (b) Draw an energy-level diagram showing the ground states of the parent and daughter and excited states of the daughter. To which of the daughter states does β^- decay of $^{191}_{76}Os$ occur?

72. Use the uncertainty principle to argue why electrons are unlikely to be found in the nucleus. Use relativity. [See hint for Problem 60.]

73. Estimate the total binding energy for copper and then estimate the energy, in joules, needed to break a 3-g copper penny into its constituent nucleons.

74. Instead of giving atomic masses for nuclides as in Appendix F, some tables give the *mass excess*, Δ, defined as $\Delta = M - A$, where A is the atomic number and M is the mass in u. Determine the mass excess, in u and in MeV/c^2, for: (a) ^4_2He; (b) $^{12}_6C$; (c) $^{107}_{47}Ag$; (d) $^{235}_{92}U$. (e) From a glance at Appendix F, can you make a generalization about the sign of Δ as a function of Z or A?

75. (a) A 100-gram sample of natural carbon contains the usual fraction of $^{14}_6C$. How long will it take on average before there is only one $^{14}_6C$ nucleus left? (b) How does the answer in (a) change if the sample is 200 grams? What does this tell you about the limits of carbon dating?

76. If the mass of the proton were just a little closer to the mass of the neutron, the following reaction would be possible even at low collision energies:

$$e^- + p \rightarrow n + \nu.$$

Why would this situation be catastrophic? By what percentage would the proton's mass have to be increased to make this reaction possible?

Technician is looking at a positron emission tomography (PET) image of a patient's brain. PET is one of several powerful types of medical imaging based on physics used by doctors to diagnose illnesses.

NUCLEAR ENERGY; EFFECTS AND USES OF RADIATION

We continue our study of nuclear physics in this chapter. We begin with a discussion of nuclear reactions, after which we examine the important large energy-releasing processes of fission and fusion. The remainder of the chapter deals with the effects of nuclear radiation when it passes through matter, particularly biological matter, and how radiation is used medically for therapy and diagnosis, including recently developed imaging techniques.

31–1 Nuclear Reactions and the Transmutation of Elements

When a nucleus undergoes α or β decay, the daughter nucleus is that of a different element from the parent. The transformation of one element into another, called *transmutation*, also occurs by means of nuclear reactions. A **nuclear reaction** is said to occur when a given nucleus is struck by another nucleus, or by a simpler particle such as a γ ray or neutron, so that an interaction takes place. Ernest Rutherford was the first to report seeing a nuclear reaction. In 1919 he observed that some of the α particles passing through nitrogen gas were absorbed and protons emitted. He concluded that nitrogen nuclei had been transformed into oxygen nuclei via the reaction

$$^{4}_{2}\text{He} + {}^{14}_{7}\text{N} \rightarrow {}^{17}_{8}\text{O} + {}^{1}_{1}\text{H},$$

where $^{4}_{2}\text{He}$ is an α particle, and $^{1}_{1}\text{H}$ is a proton.

Since then, a great many nuclear reactions have been observed. Indeed, many of the radioactive isotopes used in the laboratory are made by means of nuclear reactions. Nuclear reactions can be made to occur in the laboratory, but they also occur regularly in nature. In Chapter 30 we saw an example of this: $^{14}_{6}\text{C}$ is continually being made in the atmosphere via the reaction $\text{n} + {}^{14}_{7}\text{N} \rightarrow {}^{14}_{6}\text{C} + \text{p}$.

Nuclear reactions are sometimes written in a shortened form: for example, the reaction

$$\text{n} + {}^{14}_{7}\text{N} \rightarrow {}^{14}_{6}\text{C} + \text{p}$$

is written

$$^{14}_{7}\text{N} (\text{n}, \text{p}){}^{14}_{6}\text{C}.$$

The symbols outside the parentheses on the left and right represent the initial and final nuclei, respectively. The symbols inside the parentheses represent the bombarding particle (first) and the emitted small particle (second).

In any nuclear reaction, both electric charge and nucleon number are conserved. These conservation laws are often useful, as the following Example shows.

CONCEPTUAL EXAMPLE 31–1 | **Deuterium reaction.** A neutron is observed to strike an $^{16}_{8}\text{O}$ nucleus and a deuteron is given off. (A **deuteron**, or **deuterium**, is the isotope of hydrogen containing one proton and one neutron, $^{2}_{1}\text{H}$; it is sometimes given the symbol d or D.) What is the nucleus that results?

Deuterium

RESPONSE We have the reaction $\text{n} + {}^{16}_{8}\text{O} \rightarrow ? + {}^{2}_{1}\text{H}$. The total number of nucleons initially is $1 + 16 = 17$, and the total charge is $0 + 8 = 8$. The same totals apply to the right side of the reaction. Hence the product nucleus must have $Z = 7$ and $A = 15$. From the periodic table, we find that it is nitrogen that has $Z = 7$, so the nucleus produced is $^{15}_{7}\text{N}$. The reaction can be written $^{16}_{8}\text{O} (\text{n}, \text{d}){}^{15}_{7}\text{N}$, where d represents deuterium, $^{2}_{1}\text{H}$.

Energy (as well as momentum) is conserved in nuclear reactions, and we can use this to determine whether a given reaction can occur or not. For example, if the total mass of the products is less than the total mass of the initial particles, this decrease in mass (recall $E = mc^2$) appears as kinetic energy of the outgoing particles. But if the total mass of the products is greater than the total mass of the initial reactants, the reaction requires energy. The reaction will then not occur unless the bombarding particle has sufficient kinetic energy. Consider a nuclear reaction of the general form

$$\text{a} + \text{X} \rightarrow \text{Y} + \text{b},$$

where a is a projectile particle (or small nucleus) that strikes nucleus X, producing nucleus Y and particle b (typically, p, n, α, γ). We define the **reaction energy**, or **Q-value**, in terms of the masses involved, as

$$Q = (M_a + M_X - M_b - M_Y)c^2. \qquad \textbf{(31–1a)} \qquad \textit{Q-value}$$

Since energy is conserved, Q is equal to the change in kinetic energy (final minus initial):

$$Q = \mathrm{KE_b} + \mathrm{KE_Y} - \mathrm{KE_a} - \mathrm{KE_X}. \qquad \textbf{(31–1b)}$$

Normally, $\mathrm{KE_X} = 0$ since X is the target nucleus at rest (or nearly so) struck by an incoming particle a. For $Q > 0$, the reaction is said to be *exothermic* or *exoergic*; energy is released in the reaction, so the total KE is greater after the reaction than before. If Q is negative ($Q < 0$), the reaction is said to be *endothermic* or *endoergic*. In this case the final total KE is less than the initial KE, and an energy input is required to make the reaction happen; the energy input comes from the kinetic energy of the initial colliding particles (a and X).

EXAMPLE 31–2 **A slow-neutron reaction.** The nuclear reaction

$$\mathrm{n} + {}^{10}_{5}\mathrm{B} \rightarrow {}^{7}_{3}\mathrm{Li} + {}^{4}_{2}\mathrm{He}$$

is observed to occur even when very slow-moving neutrons ($M_n = 1.0087\,\mathrm{u}$) strike a boron atom at rest. For a particular reaction in which $\mathrm{KE_n} \approx 0$, the helium ($M_{He} = 4.0026\,\mathrm{u}$) is observed to have a speed of $9.30 \times 10^6\,\mathrm{m/s}$. Determine (a) the KE of the lithium ($M_{Li} = 7.0160\,\mathrm{u}$), and (b) the Q-value of the reaction.

SOLUTION (a) Since the neutron and boron are both essentially at rest, the total momentum before the reaction is zero, and afterward is also zero. Therefore,

$$M_{Li}v_{Li} = M_{He}v_{He}.$$

We solve this for v_{Li} and substitute it into the equation for kinetic energy. We can use classical KE with little error, rather than relativistic formulas, because $v_{He} = 9.30 \times 10^6\,\mathrm{m/s}$ is not close to the speed of light c, and v_{Li} will be even less since $M_{Li} > M_{He}$. Thus we can write:

$$\mathrm{KE_{Li}} = \frac{1}{2}M_{Li}v_{Li}^2 = \frac{1}{2}M_{Li}\left(\frac{M_{He}v_{He}}{M_{Li}}\right)^2 = \frac{M_{He}^2 v_{He}^2}{2M_{Li}}.$$

We put in numbers, changing the mass in u to kg and recalling that $1.60 \times 10^{-13}\,\mathrm{J} = 1\,\mathrm{MeV}$:

$$\mathrm{KE_{Li}} = \frac{(4.0026\,\mathrm{u})^2(1.66 \times 10^{-27}\,\mathrm{kg/u})^2(9.30 \times 10^6\,\mathrm{m/s})^2}{2(7.0160\,\mathrm{u})(1.66 \times 10^{-27}\,\mathrm{kg/u})}$$

$$= 1.64 \times 10^{-13}\,\mathrm{J} = 1.02\,\mathrm{MeV}.$$

(b) We are given the data $\mathrm{KE_a} = \mathrm{KE_X} = 0$ in Eq. 31–1b, so $Q = \mathrm{KE_{Li}} + \mathrm{KE_{He}}$, where

$$\mathrm{KE_{He}} = \tfrac{1}{2}M_{He}v_{He}^2 = \tfrac{1}{2}(4.0026\,\mathrm{u})(1.66 \times 10^{-27}\,\mathrm{kg/u})(9.30 \times 10^6\,\mathrm{m/s})^2$$

$$= 2.84 \times 10^{-13}\,\mathrm{J} = 1.78\,\mathrm{MeV}.$$

Hence, $Q = 1.02\,\mathrm{MeV} + 1.78\,\mathrm{MeV} = 2.80\,\mathrm{MeV}.$

Will the reaction "go"? Can the reaction $^{13}_{6}C\,(p,n)^{13}_{7}N$ occur when $^{13}_{6}C$ is bombarded by 2.0-MeV protons?

SOLUTION We use Eq. 31–1a, looking up the masses of the nuclei in Appendix F. The total masses before and after the reaction are:

Before	After
$M(^{13}_{6}C) = 13.003355$	$M(^{13}_{7}N) = 13.005738$
$M(^{1}_{1}H) = \underline{\ 1.007825}$	$M(n) = \underline{\ 1.008665}$
14.011180	14.014403

(We must use the mass of the $^{1}_{1}H$ atom rather than that of the bare proton because the masses of $^{13}_{6}C$ and $^{13}_{7}N$ include the electrons, and we must include an equal number of electrons on each side of the equation since none are created or destroyed.) The products have an excess mass of $(14.014403 - 14.011180)u = 0.003223\,u \times 931.5\,MeV/u = 3.00\,MeV$. Thus $Q = -3.00\,MeV$, and the reaction is endoergic. This reaction requires energy, and the 2.0 MeV protons do not have enough to make it go.

The proton in this last Example would have to have somewhat more than 3.00 MeV of KE to make this reaction go; 3.00 MeV would be enough to conserve energy, but a proton of this energy would produce the $^{13}_{7}N$ and n with no KE and hence no momentum. Since an incident 3.0-MeV proton has momentum, conservation of momentum would be violated. A more complicated calculation shows that to conserve both energy and momentum, the minimum proton energy, called the **threshold energy**, is 3.23 MeV in this case (see Problem 15).

The artificial transmutation of elements took a great leap forward in the 1930s when Enrico Fermi realized that neutrons would be the most effective projectiles for causing nuclear reactions and in particular for producing new elements. Because neutrons have no net electric charge, they are not repelled by positively charged nuclei as are protons or alpha particles. Hence the probability of a neutron reaching the nucleus and causing a reaction is much greater than for charged projectiles,[†] particularly at low energies. Between 1934 and 1936, Fermi and his co-workers in Rome produced many previously unknown isotopes by bombarding different elements with neutrons. Fermi realized that if the heaviest known element, uranium, were bombarded with neutrons, it might be possible to produce new elements whose atomic numbers were greater than that of uranium. After several years of hard work, it was suspected that two new elements had been produced, neptunium ($Z = 93$) and plutonium ($Z = 94$). The full confirmation that such "transuranic" elements could be produced came several years later at the University of California, Berkeley. The reactions are shown in Fig. 31–1.

It was soon shown that what Fermi actually had observed when he bombarded uranium was an even stranger process—one that was destined to play an extraordinary role in the world at large.

FIGURE 31–1 Neptunium and plutonium are produced in this series of reactions, after bombardment of $^{238}_{92}U$ by neutrons.

(a) $n + ^{238}_{92}U \rightarrow ^{239}_{92}U$

Neutron captured by $^{238}_{92}U$.

(b) $^{239}_{92}U \rightarrow ^{239}_{93}Np + e^{-} + \bar{\nu}$

$^{239}_{92}U$ decays by β decay to neptunium-239.

(c) $^{239}_{93}Np \rightarrow ^{239}_{94}Pu + e^{-} + \bar{\nu}$

$^{239}_{93}Np$ itself decays by β decay to produce plutonium-239.

[†]That is, positively charged particles. Electrons rarely cause nuclear reactions because they do not interact via the strong nuclear force.

31–2 Nuclear Fission; Nuclear Reactors

In 1938, the German scientists Otto Hahn and Fritz Strassmann made an amazing discovery. Following up on Fermi's work, they found that uranium bombarded by neutrons sometimes produced smaller nuclei that were roughly half the size of the original uranium nucleus. Lise Meitner and Otto Frisch, two refugees from Nazi Germany working in Scandinavia, quickly realized what had happened: the uranium nucleus, after absorbing a neutron, actually had split into two roughly equal pieces. This was startling, for until then the known nuclear reactions involved knocking out only a tiny fragment (for example, n, p, or α) from a nucleus.

This new phenomenon was named **nuclear fission** because of its resemblance to biological fission (cell division). It occurs much more readily for $^{235}_{92}U$ than for the more common $^{238}_{92}U$. The process can be visualized by imagining the uranium nucleus to be like a liquid drop. According to this **liquid-drop model**, the neutron absorbed by the $^{235}_{92}U$ nucleus gives the nucleus extra internal energy (like heating a drop of water). This intermediate state, or **compound nucleus**, is $^{236}_{92}U$ (because of the absorbed neutron). The extra energy of this nucleus—it is in an excited state—appears as increased motion of the individual nucleons, which causes the nucleus to take on abnormal elongated shapes, Figure 31–2. When the nucleus elongates (in this model) into the shape shown in Fig. 31–2c, the attraction of the two ends via the short-range nuclear force is greatly weakened by the increased separation distance, and the electric repulsive force becomes dominant. So the nucleus splits in two. The two resulting nuclei, N_1 and N_2, are called **fission fragments**, and in the process a number of neutrons (typically two or three) are also given off. The reaction can be written

$$n + {}^{235}_{92}U \rightarrow {}^{236}_{92}U \rightarrow N_1 + N_2 + \text{neutrons.} \qquad \textbf{(31–2a)}$$

The compound nucleus, $^{236}_{92}U$, exists for less than 10^{-12} s, so the process occurs very quickly. The two fission fragments more often split the original uranium mass as about 40%–60% rather than precisely half and half. A typical fission reaction is

$$n + {}^{235}_{92}U \rightarrow {}^{141}_{56}Ba + {}^{92}_{36}Kr + 3n, \qquad \textbf{(31–2b)}$$

although many others also occur.

A tremendous amount of energy is released in a fission reaction because the mass of $^{235}_{92}U$ is considerably greater than the total mass of the fission fragments plus neutrons. This can be seen from the binding-energy-per-nucleon curve of Fig. 30–1; the binding energy per nucleon for uranium is about 7.6 MeV/nucleon, but for fission fragments that have intermediate mass (in the center portion of the graph, $A \approx 100$), the average binding energy per nucleon is about 8.5 MeV/nucleon. Since the fission fragments are more tightly bound, they have less mass. The difference in mass (or energy) between the original uranium nucleus and the fission fragments is about $8.5 - 7.6 = 0.9$ MeV per nucleon. Since there are 236 nucleons involved in each fission, the total energy released per fission is

$$(0.9 \text{ MeV/nucleon})(236 \text{ nucleons}) \approx 200 \text{ MeV.}$$

This is an enormous amount of energy for one single nuclear event. At a practical level, the energy from one fission is, of course, tiny. But if many

(a)

(b) ^{236}U (excited)

(c)

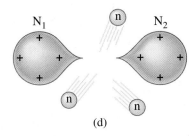

(d)

FIGURE 31–2 Fission of a $^{235}_{92}U$ nucleus after capture of a neutron, according to the liquid drop model.

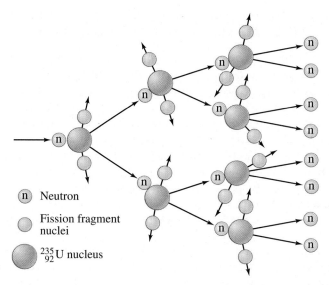

FIGURE 31–3 Chain reaction.

n Neutron

Fission fragment nuclei

$^{235}_{92}$U nucleus

such fissions could occur in a short time, an enormous amount of energy at the macroscopic level would be available. A number of physicists, including Fermi, recognized that the neutrons released in each fission (Eqs. 31–2) could be used to create a **chain reaction**. That is, one neutron initially causes one fission of a uranium nucleus; the two or three neutrons released can go on to cause additional fissions, so the process multiplies as shown schematically in Fig. 31 3. If a **self-sustaining chain reaction** was actually possible in practice, the enormous energy available in fission could be released on a larger scale. Fermi and his co-workers (at the University of Chicago) showed it was possible by constructing the first **nuclear reactor** in 1942 (Fig. 31–4).

Chain reaction

Several problems have to be overcome to make any nuclear reactor function. First, the probability that a $^{235}_{92}$U nucleus will absorb a neutron is large only for slow neutrons, but the neutrons emitted during a fission, and which are needed to sustain a chain reaction, are moving very fast. A sub-stance known as a **moderator** must be used to slow down the neutrons. The most effective moderator will consist of atoms whose mass is as close

Moderator

FIGURE 31–4 Color painting of the first nuclear reactor, built by Fermi under the grandstand of Stagg Field at the University of Chicago. (There are no photographs of the original reactor because of military secrecy.) Natural uranium was used with graphite as moderator. On December 2, 1942, Fermi slowly withdrew the cadmium control rods and the reactor went critical. This first self-sustaining chain reaction was announced to Washington, by telephone, by Arthur Compton who witnessed the event and reported: "The Italian navigator has just landed in the new world."

as possible to that of the neutrons. (To see why this is true, recall from Chapter 7 that a billiard ball striking an equal mass ball at rest can itself be stopped in one collision; but a billiard ball striking a heavy object bounces off with nearly the same speed it had.) The best moderator would thus contain $_1^1\text{H}$ atoms. Unfortunately, $_1^1\text{H}$ tends to absorb neutrons. But the isotope of hydrogen called *deuterium*, $_1^2\text{H}$, does not absorb many neutrons and is thus an almost ideal moderator. Either $_1^1\text{H}$ or $_1^2\text{H}$ can be used in the form of water. In the latter case, it is **heavy water**, in which the hydrogen atoms have been replaced by deuterium. Another common moderator is *graphite,* which consists of $_6^{12}\text{C}$ atoms.

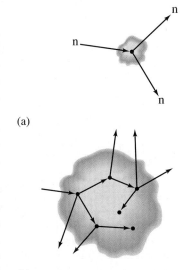

A second problem is that the neutrons produced in one fission may be absorbed and produce other nuclear reactions with other nuclei in the reactor, rather than produce further fissions. In a "light-water" reactor, the $_1^1\text{H}$ nuclei absorb neutrons, as does $_{92}^{238}\text{U}$ to form $_{92}^{239}\text{U}$ in the reaction $n + _{92}^{238}\text{U} \rightarrow _{92}^{239}\text{U} + \gamma$. Naturally occurring uranium† contains 99.3 percent $_{92}^{238}\text{U}$ and only 0.7 percent fissionable $_{92}^{235}\text{U}$. To increase the probability of fission of $_{92}^{235}\text{U}$ nuclei, natural uranium can be **enriched** to increase the percentage of $_{92}^{235}\text{U}$ using processes such as diffusion or centrifugation. (Enrichment is not usually necessary for reactors using heavy water as moderator since heavy water doesn't absorb neutrons.)

The third problem is that some neutrons will escape through the surface of the reactor core before they cause further fissions (Fig. 31–5). Thus the mass of fuel must be sufficiently large for a self-sustaining chain reaction to take place. The minimum mass of uranium needed is called the **critical mass**. The value of the critical mass depends on the moderator, the fuel ($_{94}^{239}\text{Pu}$ may be used instead of $_{92}^{235}\text{U}$), and how much the fuel is enriched, if at all. Typical values are on the order of a few kilograms (that is, not grams nor thousands of kilograms).

FIGURE 31–5 If the amount of uranium exceeds the critical mass, as in (b), a sustained chain reaction is possible. If the mass is less than critical, as in (a), most neutrons escape before additional fissions occur, and the chain reaction is not sustained.

Critical mass

To have a self-sustaining chain reaction, it is clear that on the average at least one neutron produced in each fission must go on to produce another fission. The average number of neutrons per each fission that do go on to produce further fissions is called the **multiplication factor**, f. For a self-sustaining chain reaction, we must have $f \geq 1$. If $f < 1$, the reactor is "subcritical." If $f > 1$, it is "supercritical." Reactors are equipped with movable **control rods** (usually of cadmium or boron), whose function is to absorb neutrons and maintain the reactor at just barely "critical," $f = 1$. The release of neutrons and subsequent fissions caused by them occurs so quickly that manipulation of the control rods to maintain $f = 1$ would not be possible if it weren't for the small percentage (~1%) of so-called **delayed neutrons**. They come from the decay of neutron-rich fission fragments (or their daughters) having lifetimes on the order of seconds—sufficient to allow enough reaction time to operate the control rods and maintain $f = 1$.

Control rods

Delayed neutrons

Nuclear reactors have been built for use in research and to produce electric power. Fission produces many neutrons and a "research reactor" is basically an intense source of neutrons. These neutrons can be used as projectiles in nuclear reactions to produce nuclides not found in nature, including isotopes used as tracers and for medical therapy. A "power reactor" is used to produce electric power. The energy released in the fission

Types of nuclear reactor

---

†$_{92}^{238}\text{U}$ will fission, but only with fast neutrons ($_{92}^{238}\text{U}$ is more stable than $_{92}^{235}\text{U}$). The probability of absorbing a fast neutron and producing a fission is too low to produce a self-sustaining chain reaction.

FIGURE 31–6 A nuclear reactor. The heat generated by the fission process in the fuel rods is carried off by hot water or liquid sodium and is used to boil water to steam in the heat exchanger. The steam drives a turbine to generate electricity and is then cooled in the condenser.

Primary system

Secondary system

Core (fuel and moderator)

Hot water or liquid sodium

Heat exchanger

Steam turbine

Electric generator

Steam

Water

Control rods

Containment vessel (shielding)

Condenser

Shielding

Pump

Cooling water

FIGURE 31–7 Devastation around Chernobyl in Russia, after the nuclear power plant disaster in 1986.

process appears as heat, which is used to boil water and produce steam to drive a turbine connected to an electric generator (Fig. 31–6). The **core** of a nuclear reactor consists of the fuel and a moderator (water in most U.S. commercial reactors). The fuel is usually uranium enriched so that it contains 2 to 4 percent $^{235}_{92}U$. Water or other liquid (such as liquid sodium) is allowed to flow through the core. The thermal energy it absorbs is used to produce steam in the heat exchanger, so the fissionable fuel acts as the heat input for a heat engine (see Sections 15–5 and 15–12).

Many problems are associated with nuclear power plants. Besides the usual thermal pollution associated with any heat engine (Section 15–12), there is the problem of disposal of the radioactive fission fragments produced in the reactor, plus radioactive nuclides produced by neutrons interacting with the structural parts of the reactor. Fission fragments, like their uranium or plutonium parents, have about 50 percent more neutrons than protons. Nuclei with atomic number in the typical range for fission fragments ($Z \approx 30$ to 60) are stable only if they have more nearly equal numbers of protons and neutrons (see Fig. 30–2). Hence the highly neutron-rich fission fragments are very unstable and decay radioactively. The accidental release of highly radioactive fission fragments into the atmosphere poses a serious threat to human health (Section 31–4), as does possible leakage of the radioactive wastes when they are disposed of. The accidents at Three Mile Island (1979) and at Chernobyl (1986) have illustrated some of the dangers and have shown that nuclear plants must be constructed, maintained, and operated with great care and precision (Fig. 31–7). Finally, the lifetime of nuclear power plants is limited to 30-some years, due to buildup of radioactivity and the fact that the structural materials themselves are weakened by the intense conditions inside. Decommissioning of a power plant could take a number of forms, but the cost of any method of decommissioning a large plant will be very great.

So-called **breeder reactors** were proposed as a solution to the problem of limited supplies of fissionable uranium. A breeder reactor is one in which some of the neutrons produced in the fission of $^{235}_{92}U$ are absorbed by $^{238}_{92}U$, and $^{239}_{94}Pu$ is produced via the set of reactions shown in Fig. 31–1. $^{239}_{94}Pu$ is fissionable with slow neutrons, so after separation it can be used as a fuel in a nuclear reactor. Thus a breeder reactor "breeds" new fuel† ($^{239}_{94}Pu$) from otherwise useless $^{238}_{92}U$. Since natural uranium is 99.3 percent $^{238}_{92}U$, this means that the supply of fissionable fuel could be increased by more than a factor of 100. But breeder reactors not only have the same problems as other reactors, but in addition present other serious problems. Not only is plutonium considered by many to be a serious health hazard in itself (radioactive with a half-life of 24,000 years), but plutonium produced in a reactor can readily be used in a bomb. Thus the use of a breeder reactor, even more than a conventional uranium reactor, presents the danger of nuclear proliferation, and the possibility of theft of fuel by terrorists who could produce a bomb.

It is clear that nuclear power presents many risks. Other large-scale energy-conversion methods, such as conventional coal-burning steam plants, also present health and environmental hazards, some of which were discussed in Chapter 15, including air pollution, oil spills, and the release of CO_2 gas which may be trapping heat as in a greenhouse and raising the Earth's temperature. The solution to the world's needs for energy is not only technological, but economic and political as well. A major factor surely is to "conserve"—to not waste energy and use as little as possible.

EXAMPLE 31–4 **ESTIMATE** | **Uranium fuel amount.** Estimate the minimum amount of $^{235}_{92}U$ that needs to undergo fission in order to run a 1000-MW power reactor per year of continuous operation. Assume an efficiency (Chapter 15) of about 33%.

SOLUTION For 1000 MW output, the total power generation is 3000 MW, of which 2000 MW is dumped as "waste" heat. Thus the total energy release in 1 yr (3×10^7 s) from fission is about

$$(3 \times 10^9 \text{ J/s})(3 \times 10^7 \text{ s}) \approx 10^{17} \text{ J}.$$

If each fission releases 200 MeV of energy, the number of fissions required is

$$\frac{(10^{17} \text{ J})}{(2 \times 10^8 \text{ eV/fission})(1.6 \times 10^{-19} \text{ J/eV})} \approx 3 \times 10^{27} \text{ fissions}.$$

The mass of a single uranium atom is about $(235 \text{ u})(1.66 \times 10^{-27} \text{ kg/u}) \approx 4 \times 10^{-25}$ kg, so the total mass needed is $(4 \times 10^{-25} \text{ kg})(3 \times 10^{27} \text{ fissions}) \approx$ 1000 kg, or about a ton. Since $^{235}_{92}U$ is only a fraction of normal uranium, and even when enriched it is never more than 10 percent of the total, the yearly requirement for uranium is on the order of tens of tons. This is orders of magnitude less than coal, both in mass and volume (see Problem 26 in Chapter 15).

The first use of fission, however, was not to produce electric power. Instead, it was first used as a fission bomb (the "atomic bomb"). In early 1940, with Europe already at war, Hitler banned the sale of uranium from

Atom bomb

---

†A breeder reactor does *not* produce more fuel than it uses.

FIGURE 31–8 J. Robert Oppenheimer, on the left, with General Leslie Groves, who was the administrative head of Los Alamos during the war. The photograph was taken at the Trinity site in the New Mexico desert, where the first atomic bomb was exploded.

FIGURE 31–9 Photo taken a month after the bomb was dropped on Nagasaki. The shacks were constructed afterwards from debris in the ruins.

the Czech mines he had recently taken over. Research into the fission process suddenly was enshrouded in secrecy. Physicists in the United States were alarmed. A group of them approached Einstein—a man whose name was a household word—to send a letter to President Roosevelt about the possibilities of using nuclear fission for a bomb far more powerful than any previously known, and inform him that Germany might already have begun development of such a bomb. Roosevelt responded by authorizing the program known as the Manhattan Project, to see if a bomb could be built. Work began in earnest after Fermi's demonstration in 1942 that a sustained chain reaction was possible. A new secret laboratory was developed on an isolated mesa in New Mexico known as Los Alamos. Under the direction of J. Robert Oppenheimer (1904–1967; Fig. 31–8), it became the home of famous scientists from all over Europe and the United States.

To build a bomb that was subcritical during transport but that could be made supercritical (to produce a chain reaction) at just the right moment, two pieces of uranium were used, each less than the critical mass but together greater than the critical mass. The two masses would be kept separate until the moment of detonation arrived. Then a kind of gun would force the two pieces together very quickly, a chain reaction of explosive proportions would occur, and a tremendous amount of energy would be released very suddenly. The first fission bomb was tested in the New Mexico desert in July 1945. It was successful. In early August, a fission bomb using uranium was dropped on Hiroshima and a second, using plutonium, was dropped on Nagasaki (Fig. 31–9). World War II ended shortly thereafter, but the dropping of these bombs aroused controversy.

Scientists were later criticized for working on the bomb, and many regretted having done so. Critical judgment, however, should perhaps be placed in the context of the times. In the early 1940s, Hitler's armies were overrunning Europe, and it was believed that German scientists were trying to build a bomb (they never did). Japan had entered the war and the world seemed in danger of being overwhelmed by oppressive powerful regimes. The scientists at Los Alamos were aware they were developing a dangerous weapon. But they had to make a choice. Stopping Hitler was uppermost in their minds.

Besides its great destructive power, a fission bomb produces many highly radioactive fission fragments, as does a nuclear reactor. When a fission bomb explodes, these radioactive isotopes are released into the atmosphere and are known as *radioactive fallout.*

Testing of nuclear bombs in the atmosphere after World War II was a cause of concern, for the movement of air masses spread the fallout all over the globe. Radioactive fallout eventually settles to the Earth, particularly in rainfall, and is absorbed by plants and grasses and enters the food chain. This is a far more serious problem than the same radioactivity on the exterior of our bodies, since α and β particles are largely absorbed by clothing and the outer (dead) layer of skin. But once inside our bodies via food, the isotopes are in direct contact with living cells. One particularly dangerous radioactive isotope is $^{90}_{38}\text{Sr}$, which is chemically much like calcium and becomes concentrated in bone, where it causes bone cancer and destruction of bone marrow. The 1963 treaty signed by over 100 nations that bans nuclear weapons testing in the atmosphere was motivated because of the hazards of fallout.

31–3 Fusion

The mass of every stable nucleus is less than the sum of the masses of its constituent protons and neutrons. For example, the mass of the helium isotope ^4_2He is less than the mass of two protons plus the mass of two neutrons, as we saw in Example 30–2. Thus, if two protons and two neutrons were to come together to form a helium nucleus there would be a loss of mass. This mass loss is manifested in the release of a large amount of energy. The process of building up nuclei by bringing together individual protons and neutrons, or building larger nuclei by combining small nuclei, is called **nuclear fusion**. A glance at Fig. 31–10 (same as Fig. 30–1) shows why small nuclei can combine to form larger ones with the release of energy: it is because the binding energy per nucleon is smaller for light nuclei than it is for those of increasing mass (up to about $A \approx 60$). It is believed that many of the elements in the universe were originally formed through the process of fusion (see Chapter 33), and that today, fusion is continually taking place within the stars, including our Sun, producing the prodigious amounts of radiant energy they emit.

EXAMPLE 31–5 **Fusion energy release.** One of the simplest fusion reactions involves the production of deuterium, ^2_1H, from a neutron and a proton: $^1_1\text{H} + \text{n} \rightarrow {}^2_1\text{H} + \gamma$. How much energy is released in this reaction?

SOLUTION From Appendix F, the initial rest mass is $1.007825\,\text{u} + 1.008665\,\text{u} = 2.016490\,\text{u}$, and after the reaction the mass is that of the ^2_1H, namely $2.014102\,\text{u}$. The mass difference is $0.002388\,\text{u}$, so the energy released is $(0.002388\,\text{u})(931.5\,\text{MeV/u}) = 2.22\,\text{MeV}$, and it is carried off by the ^2_1H nucleus and the γ ray.

FIGURE 31–10 Average binding energy per nucleon as a function of mass number A for stable nuclei. Same as Fig. 30–1.

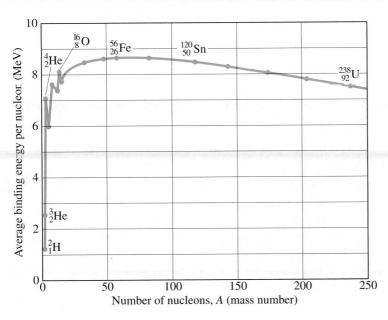

The energy output of our Sun is believed to be due principally to the following sequence of fusion reactions:

Fusion reactions

$$_1^1\text{H} + {}_1^1\text{H} \rightarrow {}_1^2\text{H} + e^+ + \nu \qquad (0.42\,\text{MeV}) \quad \textbf{(31–3a)}$$

in the Sun

$$_1^1\text{H} + {}_1^2\text{H} \rightarrow {}_2^3\text{He} + \gamma \qquad (5.49\,\text{MeV}) \quad \textbf{(31–3b)}$$

(proton–proton cycle)

$$_2^3\text{He} + {}_2^3\text{He} \rightarrow {}_2^4\text{He} + {}_1^1\text{H} + {}_1^1\text{H} \qquad (12.86\,\text{MeV}) \quad \textbf{(31–3c)}$$

where the energy released (Q-value) for each reaction is given in parentheses. The net effect of this sequence, which is called the **proton–proton cycle**, is for four protons to combine to form one ${}_2^4\text{He}$ nucleus plus two positrons, two neutrinos, and two gamma rays:

$$4\,_1^1\text{H} \rightarrow {}_2^4\text{He} + 2e^+ + 2\nu + 2\gamma. \qquad \textbf{(31–4)}$$

Note that it takes two of each of the first two reactions (Eqs. 31–3a and b) to produce the two ${}_2^3\text{He}$ for the third reaction, so the total energy release for the net reaction, Eq. 31–4, is $(2 \times 0.42\,\text{MeV} + 2 \times 5.49\,\text{MeV} + 12.86\,\text{MeV}) = 24.7\,\text{MeV}$. However, each of the two e^+ (formed in reaction Eq. 31–3a) quickly annihilates with an electron to produce $2m_e c^2 = 1.02\,\text{MeV}$, so the total energy released is $(24.7\,\text{MeV} + 2 \times 1.02\,\text{MeV}) = 26.7\,\text{MeV}$. The first reaction, the formation of deuterium from two protons (Eq. 31–3a), has a very low probability, and the infrequency of that reaction serves to limit the rate at which the Sun produces energy.

In stars hotter than the Sun, it is more likely that the energy output comes principally from the **carbon cycle**, which comprises the following sequence of reactions:

$$_6^{12}\text{C} + {}_1^1\text{H} \rightarrow {}_7^{13}\text{N} + \gamma$$

Carbon

$$_7^{13}\text{N} \rightarrow {}_6^{13}\text{C} + e^+ + \nu$$

$$_6^{13}\text{C} + {}_1^1\text{H} \rightarrow {}_7^{14}\text{N} + \gamma$$

cycle

$$_7^{14}\text{N} + {}_1^1\text{H} \rightarrow {}_8^{15}\text{O} + \gamma$$

(some stars)

$$_8^{15}\text{O} \rightarrow {}_7^{15}\text{N} + e^+ + \nu$$

$$_7^{15}\text{N} + {}_1^1\text{H} \rightarrow {}_6^{12}\text{C} + {}_2^4\text{He}.$$

(See Problem 34.) The theory of the proton–proton cycle and of the carbon cycle as the source of energy for the Sun and stars was first worked out by Hans Bethe (1906–   ) in 1939.

> **CONCEPTUAL EXAMPLE 31–6** **Stellar fusion.** What is the heaviest element likely to be produced in fusion processes in stars?
>
> **RESPONSE** Fusion is possible as long as the final product has more binding energy (less mass) than the reactants, for then there is net release of energy. Since the binding energy curve in Fig. 31–10 (or Fig. 30–1) peaks near $A \approx 56$, which corresponds to iron, it would not be energetically favorable to produce elements heavier than that. Nevertheless, in the center of massive stars or in supernova explosions, there is enough energy available to drive endothermic reactions that produce heavier elements, as well.

The possibility of utilizing the energy released in fusion to make a power reactor is very attractive. The fusion reactions most likely to succeed in a reactor involve the isotopes of hydrogen, $^2_1$H (deuterium) and $^3_1$H (tritium), and are as follows, with the energy released given in parentheses:

$$^2_1\text{H} + {^2_1\text{H}} \rightarrow {^3_1\text{H}} + {^1_1\text{H}} \qquad\qquad (4.03\,\text{MeV}) \quad \textbf{(31–5a)}$$

$$^2_1\text{H} + {^2_1\text{H}} \rightarrow {^3_2\text{He}} + \text{n} \qquad\qquad (3.27\,\text{MeV}) \quad \textbf{(31–5b)}$$

$$^2_1\text{H} + {^3_1\text{H}} \rightarrow {^4_2\text{He}} + \text{n}. \qquad\qquad (17.59\,\text{MeV}) \quad \textbf{(31–5c)}$$

Comparing these energy yields with that for the fission of $^{235}_{92}$U, we can see that the energy released in fusion reactions can be greater for a given mass of fuel than in fission. Furthermore, as fuel, a fusion reactor could use deuterium, which is very plentiful in the water of the oceans (the natural abundance of $^2_1$H is 0.015 percent, or about 1 g of deuterium per 60 L of water). The simple proton–proton reaction of Eq. 31–3a, which could use a much more plentiful source of fuel, $^1_1$H, has such a small probability of occurring that it cannot be considered a possibility on Earth.

Although a successful fusion reactor has not yet been achieved, considerable progress has been made in overcoming the inherent difficulties. The problems are associated with the fact that all nuclei have a positive charge and thus repel each other. However, if they can be brought close enough together so that the short-range attractive nuclear force can come into play, the latter can pull the nuclei together and fusion will occur. Thus, in order for the nuclei to get close enough together, they must have rather large kinetic energy to overcome the electric repulsion. High kinetic energies are easily attainable with particle accelerators such as the cyclotron, but the number of particles involved is too small. To produce realistic amounts of energy, we must deal with matter in bulk, for which high KE means higher temperatures. Indeed, very high temperatures are required for fusion to occur, and fusion devices are often referred to as **thermonuclear devices**. The Sun and other stars are very hot, many millions of degrees, so the nuclei are moving fast enough for fusion to take place, and the energy released keeps the temperature high so that further fusion reactions can occur. The Sun and the stars represent huge self-sustaining thermonuclear reactors, but on Earth the high temperatures and densities required are not easily attained in a controlled manner.

It was realized after World War II that the temperature produced within a fission (or "atomic") bomb was close to 10^8 K. This suggested that a fission bomb could be used to ignite a fusion bomb (popularly known as a thermonuclear or hydrogen bomb) to release the vast energy of fusion. The uncontrollable release of fusion energy in an H-bomb was relatively easy to obtain. But to realize usable energy from fusion at a slow and controlled rate turned out to be difficult. The temperature required for a usable fusion reactor has been estimated to be in the range $T = 2$ to 4×10^8 K.

EXAMPLE 31–7 **ESTIMATE** **Temperature needed for d–t fusion.** Estimate the temperature required for deuterium–tritium fusion (d–t) to occur.

SOLUTION We assume the nuclei approach head-on, each with kinetic energy KE, and that their "surfaces" must just "touch"—that is, the distance between them is equal to the sum of their radii. From Eq. 30–1, $r_d \approx 1.5$ fm and $r_t \approx 1.7$ fm. The electrostatic potential energy (Chapter 17) of the 2 particles at this distance must equal the total KE of the 2 particles when far apart:

$$2 \, \text{KE} \approx k \frac{e^2}{(r_d + r_t)}$$

$$\approx \left(9.0 \times 10^9 \frac{\text{N} \cdot \text{m}^2}{\text{C}^2} \right) \frac{(1.6 \times 10^{-19} \text{ C})^2}{(3.2 \times 10^{-15} \text{ m})(1.6 \times 10^{-19} \text{ J/eV})}$$

$$\approx 0.45 \text{ MeV}.$$

Thus, KE ≈ 0.22 MeV, and if we ask that the average KE be this high, then from Eq. 13–8, $\frac{3}{2}kT = \overline{\text{KE}}$, we have

$$T = \frac{2 \, \overline{\text{KE}}}{3 \, k} = \frac{2(0.22 \text{ MeV})(1.6 \times 10^{-13} \text{ J/MeV})}{3(1.38 \times 10^{-23} \text{ J/K})} \approx 2 \times 10^9 \text{ K}.$$

The required temperature is actually about an order of magnitude less, partly because it is not necessary that the *average* KE be 0.22 MeV—a small percentage with this much energy would be sufficient.

It is not only a high temperature that is required for a fusion reactor. There must also be a high density of nuclei to ensure a sufficiently high collision rate. A real difficulty with controlled fusion is to contain nuclei long enough and at a high enough density for sufficient reactions to occur that a usable amount of energy is obtained. At the temperatures needed for fusion, the atoms are ionized, and the resulting collection of nuclei and electrons is referred to as a **plasma**. Ordinary materials vaporize at a few thousand degrees at best, and hence cannot be used to contain a high-temperature plasma. Two major containment techniques are being investigated at present: *magnetic confinement* and *inertial confinement*.

Plasma

In **magnetic confinement**, magnetic fields are used to try to contain the hot plasma. One possibility is the "magnetic bottle" shown in Fig. 31–11. The paths of the charged particles in the plasma are bent by the magnetic field, and where the lines are close together, the force on the particles is such that they are reflected back toward the center by this "magnetic mirror." Unfortunately, magnetic bottles develop "leaks" and the charged particles slip out before sufficient fusion takes place. A more complex device called the **tokamak**, first developed in the USSR, shows considerable promise. A tokamak (Fig. 31–12) is toroid-shaped and utilizes a combination of two magnetic fields: one directed along the axis of the toroid produced by current-carrying conductors ("toroidal" field); and a second produced by a current that passes through the plasma

Magnetic confinement

FIGURE 31–11 "Magnetic bottle" used to confine a plasma.

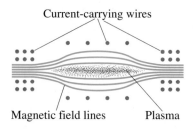

Current-carrying wires

Magnetic field lines   Plasma

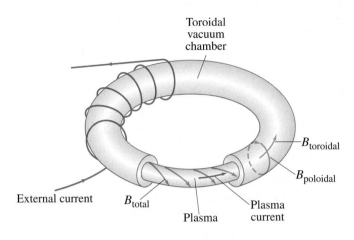

Toroidal
vacuum
chamber

B_{toroidal}

B_{poloidal}

External current

B_{total}

Plasma

Plasma
current

FIGURE 31–12 Tokamak configuration showing total **B** field due to external current plus current in the plasma.

("poloidal" field), which not only helps confinement, but heats the plasma as well. The combination of these two fields produces a helical field as shown in Fig. 31–12.

In 1957, J. D. Lawson showed that the product of ion density n and confinement time τ must satisfy, at a minimum, approximately

$$n\tau \gtrsim 3 \times 10^{20} \text{ s/m}^3.$$

Lawson criterion

Ignition

Break-even

This **Lawson criterion** must be reached to produce **ignition**, by which we mean a self-sustaining thermonuclear "burn" that continues after all external heating is turned off. To reach **break-even**, the point at which the energy output due to fusion is equal to the energy input to heat the plasma, requires an $n\tau$ about an order of magnitude less. The break-even point was very closely approached at the Tokamak Fusion Test Reactor (TFTR) at Princeton. The very high temperature needed for ignition (4×10^8 K) was exceeded, and the Lawson criterion closely approached—although not both of these at the same time. Nonetheless, the great progress made in recent years, after more than 40 years of research, gives hope that a working fusion reactor might be a reality by the early part of the next century.

The second method for containing the fuel for fusion is **inertial confinement**: a small pellet of deuterium and tritium is struck simultaneously from several directions by very intense laser beams. The intense influx of energy in such a laser fusion device heats and ionizes the pellet into a plasma. The outer layers evaporate, but the collisions they make with ions in the core of the pellet drive the latter inward. This implosion raises the density to about 10^3 times normal and further heats the core to temperatures at which fusion occurs. The confinement time is very short, on the order of

(a)

(b)

FIGURE 31–13 (a) Target chamber (5 m in diameter) of NOVA laser at Lawrence Livermore Laboratory, into which 10 laser beams converge on a target. (b) A 1-mm-diameter DT (deuterium–tritium) target, on its support, at the center of the target chamber.

Biological damage

10^{-11} to 10^{-9} s, during which time the ions don't move appreciably because of their own inertia. Very soon thereafter fusion takes place and the pellet explodes. The NOVA laser used for fusion research at the Lawrence Livermore Laboratory (Fig. 31–13) can deposit 10^{5} J into the pellet over a 10^{-9} s pulse. This is a power input of 10^{14} W, more than the total electric-generator capacity of the United States! Of course, it is sustained only over an extremely short interval of time. A future reactor might implode 100 such pellets per second, thus requiring an input of 10^{5} J $\times 100\,\text{s}^{-1} = 10^{7}$ W on the average. Inertial confinement depends on high particle density, n, over very short time intervals, τ. Recently, the Lawson criterion has been achieved, but at a temperature less than that needed for ignition.

Besides the problems of confinement, the building of a practical fusion reactor will require the development of materials used in its construction that can withstand high temperatures and high levels of radiation.

31–4 Passage of Radiation Through Matter; Radiation Damage

When we speak of *radiation,* we include α, β, γ, and X-rays, as well as protons, neutrons, and other particles such as pions (see Chapter 32). Because charged particles can ionize the atoms or molecules of any material they pass through, they are referred to as **ionizing radiation**. And because radiation produces ionization, it can cause considerable damage to materials, particularly to biological tissue.

Charged particles, such as α and β rays and protons, cause ionization because of the electric force. That is, when they pass through a material, they can attract or repel electrons strongly enough to remove them from the atoms of the material. Since the α and β rays emitted by radioactive substances have energies on the order of 1 MeV (10^{4} to 10^{7} eV), whereas ionization of atoms and molecules requires on the order of 10 eV, it is clear that a single α or β particle can cause thousands of ionizations.

Neutral particles also give rise to ionization when they pass through materials. For example, X-ray and γ-ray photons can ionize atoms by knocking out electrons by means of the photoelectric and Compton effects (Chapter 27). Furthermore, if a γ ray has sufficient energy (greater than 1.02 MeV), it can undergo pair production: an electron and a positron are produced (Section 27–4). The charged particles produced in all three of these processes can themselves go on to produce further ionization. Neutrons, on the other hand, interact with matter mainly by collisions with nuclei, with which they interact strongly. Often the nucleus is broken apart by such a collision, altering the molecule of which it was a part. And the fragments produced can in turn cause ionization.

Radiation passing through matter can do considerable damage. Metals and other structural materials become brittle and their strength can be weakened if the radiation is very intense, as in nuclear reactor power plants and for space vehicles that must pass through areas of intense cosmic radiation.

The radiation damage produced in biological organisms is due primarily to ionization produced in cells. Several related processes can occur. Ions or radicals are produced that are highly reactive and take part in chemical reactions that interfere with the normal operation of the cell. All forms of radiation can ionize atoms by knocking out electrons. If these are bonding electrons, the molecule may break apart, or its structure may be altered so that it does not perform its normal function or performs a harmful function. In the case of proteins, the loss of one molecule is not serious if there are other copies of that particular one in the cell, and additional ones can be made from its gene. However, large doses of radiation may damage so many molecules that new copies cannot be made quickly enough, and the cell dies. Damage to the DNA is more serious, since a cell may have only one copy. Each alteration in the DNA affects a gene and can alter the molecule it codes for, so that needed proteins or other materials may not be made at all. Again the cell may die. The death of a single cell is not normally a problem, since the body can replace it with a new one. (There are exceptions, such as neurons, which are *not* replaceable, so their loss is serious.) But if many cells die, the organism may not be able to recover. On the other hand, a cell may survive but be defective. It may go on dividing and produce many more defective cells, to the detriment of the whole organism. Thus radiation can cause cancer—the rapid production of defective cells.

Radiation damage to biological organisms is often separated into categories according to its location in the body: "somatic" and "genetic." *Somatic damage* refers to any part of the body except the reproductive organs. Somatic damage affects that particular organism, causing cancer and, at high doses, radiation sickness (characterized by nausea, fatigue, loss of body hair, and other symptoms) or even death. *Genetic damage* refers to damage to reproductive cells and so affects an individual's offspring. Damage to the genes results in mutations, the majority of which are harmful; and mutations, if they occur in the reproductive organs, are transmitted to future generations. Radiation, including that from diagnostic use of X-rays, is commonplace, and in the latter case, the possible damage done by the radiation must be balanced against the medical benefits and prolongation of life as a result of their diagnostic use.

31–5 Measurement of Radiation—Dosimetry

Although the passage of ionizing radiation through the human body can cause considerable damage, radiation can also be used to treat certain diseases, particularly cancer, often by using very narrow beams directed at the cancerous tumor to destroy it (Section 31–6). It is therefore important to be able to quantify the amount, or **dose**, of radiation. This is the subject of **dosimetry**.

The strength of a source can be specified at a given time by stating the **source activity**, or how many disintegrations occur per second. The traditional unit is the **curie** (Ci), defined as *Source activity*

$$1 \text{ Ci} = 3.70 \times 10^{10} \text{ disintegrations per second.}$$

(This figure comes from the original definition as the activity of exactly 1 gram of radium.) Although the curie is still in common use, the proper SI unit for source activity is the **becquerel** (Bq), defined as

$$1 \text{ Bq} = 1 \text{ disintegration/s.}$$

Commercial suppliers of **radionuclides** (radioactive nuclides) specify the activity at a given time. Since the activity decreases over time, particularly for short-lived isotopes, it is important to take this into account.

The source activity ($\Delta N/\Delta t$) is related to the half-life, $T_{\frac{1}{2}}$, by (see Section 30–8):

$$\frac{\Delta N}{\Delta t} = \lambda N = \frac{0.693}{T_{\frac{1}{2}}} N.$$

EXAMPLE 31–8 **Radioactivity taken up by cells.** In a certain experiment, 0.016 μCi of $^{32}_{15}$P is injected into a medium containing a culture of bacteria. After 1 h, the cells are washed and a detector that is 70 percent efficient (counts 70 percent of emitted β rays) records 720 counts per minute from all the cells. What percentage of the original $^{32}_{15}$P was taken up by the cells?

SOLUTION The total number of disintegrations per second originally was $(0.016 \times 10^{-6})(3.7 \times 10^{10}) = 590$. The counter could be expected to count 70 percent of this, or 410 per second. Since it counted $720/60 = 12$ per second, then $12/410 = 0.029$ or 2.9 percent was incorporated into the cells.

Absorbed dose

Another type of measurement is the exposure or **absorbed dose**—that is, the effect the radiation has on the absorbing material. The earliest unit of dosage was the **roentgen** (R), which was defined in terms of the amount of ionization produced by the radiation (1.6×10^{12} ion pairs per gram of dry air at standard conditions). Today, 1 R is defined as the amount of X or γ radiation that deposits 0.878×10^{-2} J of energy per kilogram of air. The roentgen was largely superseded by another unit of absorbed dose applicable to any type of radiation, the **rad**: *1 rad is that amount of radiation which deposits energy at a rate of 1.00×10^{-2} J/kg in any absorbing material.* (This is quite close to the roentgen for X- and γ rays.) The proper SI unit for absorbed dose is the **gray** (Gy):

The rad

$$1 \text{ Gy } = 1 \text{ J/kg} = 100 \text{ rad}, \tag{31–6}$$

and is slowly coming into use. The absorbed dose depends not only on the strength of a given radiation beam (number of particles per second) and the energy per particle, but also on the type of material that is absorbing the radiation. Since bone, for example, is denser than flesh and absorbs more of the radiation normally used, the same beam passing through a human body deposits a greater dose (in rads or grays) in bone than in flesh.

The gray and the rad are physical units of dose—the energy deposited per unit mass of material. They are, however, not the most meaningful units for measuring the biological damage produced by radiation. This is because equal doses of different types of radiation cause differing amounts of damage. For example, 1 rad of α radiation does 10 to 20 times the amount of damage as 1 rad of β or γ rays. This difference arises largely because α rays (and other heavy particles such as protons and neutrons) move much more slowly than β and γ rays of equal energy due to their greater mass. Hence, ionizing collisions occur closer together, so more irreparable damage can be done. The **relative biological effectiveness** (RBE) or **quality factor** (QF) of a given type of radiation is defined as the number of rads of X or γ radiation that produces the same biological damage as 1 rad of the given radiation. Table 31–1 gives the QF for several types of radiation. The numbers are approximate since they depend somewhat on the energy of the particles and on the type of damage that is used as the criterion.

RBE or QF

The **effective dose** can be given as the product of the dose in rads and the QF, and this unit is known as the **rem** (which stands for *rad equivalent man*):

Effective dose

$$\text{effective dose (in rem)} = \text{dose (in rad)} \times \text{QF.} \qquad \textbf{(31–7a)}$$

The rem

This unit is being replaced by the SI unit for "effective dose," the **sievert** (Sv):

$$\text{effective dose (Sv)} = \text{dose (Gy)} \times \text{QF.} \qquad \textbf{(31–7b)}$$

By this definition, 1 rem of any type of radiation does approximately the same amount of biological damage. For example, 50 rem of fast neutrons does the same damage as 50 rem of γ rays. But note that 50 rem of fast neutrons is only 5 rads, whereas 50 rem of γ rays is 50 rads.

We are constantly exposed to low-level radiation from natural sources: cosmic rays, natural radioactivity in rocks and soil, and naturally occurring radioactive isotopes in our food, such as $^{40}_{19}\text{K}$. Radon, $^{222}_{86}\text{Rn}$, is of considerable concern today. It is the product of radium decay and is an intermediate in the decay series from uranium (see Fig. 30–10). Most intermediates remain in the rocks where formed, but radon is a gas that can escape from rock (and from building material like concrete) to enter the atmosphere we breathe. Although radon is inert chemically (it is a noble gas), it is not inert physically—it decays by alpha emission, and its products, also radioactive, are *not* chemically inert and can attach to the interior of the lung.

Natural radioactivity

Radon

The natural radioactive background averages about 0.36 rem (360 mrem) per year per person, although there are large variations. From medical X-rays, the average person receives about 40 mrem per year. The U.S. government specifies the recommended upper limit of allowed radiation for an individual in the general populace at about 0.5 rem (500 mrem) per year, exclusive of natural sources. It is not known if low doses of radiation increase the chances of cancer or genetic defects, so the attitude today is to play safe and keep the radiation dose as low as possible.

People who work around radiation—in hospitals, in power plants, in research—often are subjected to much higher doses than 0.5 rem/yr. The upper limit for such occupational exposures has been set somewhat higher, on the order of 5 rem/yr whole-body dose (presumably because such people know what they are getting into). To protect them from excess exposure, those people who work around radiation generally carry some type of dosimeter, typically a **radiation film badge**, which is a piece of film wrapped in light-tight material. The passage of ionizing radiation through the film (see Section 30–13) changes it so that the film is darkened upon development, and so indicates the received dose.

Large doses of radiation can cause reddening of the skin, drop in white-blood-cell count, and a large number of unpleasant symptoms such as nausea, fatigue, and loss of body hair. Such effects are sometimes referred to as **radiation sickness**. Large doses can also be fatal, although the

Radiation sickness

**TABLE 31–1
Quality Factor (QF) of
Different Kinds of Radiation**

Type	QF
X and γ rays	≈ 1
β (electrons)	≈ 1
Fast protons	1
Slow neutrons	≈ 3
Fast neutrons	Up to 10
α particles and heavy ions	Up to 20

time span of the dose is important. A short dose of 1000 rem is nearly always fatal. A 400-rem dose in a short period of time is fatal in 50 percent of the cases. However, the body possesses remarkable repair processes, so that a 400-rem dose spread over several weeks is not usually fatal. It will, nonetheless, cause considerable damage to the body.

CONCEPTUAL EXAMPLE 31–9 | **Limiting the dose.** A worker in an environment with a radioactive source is warned that she is accumulating a dose too quickly and will have to lower her exposure by a factor of ten to continue working for the rest of the year. If the worker is able to work farther away from the source, how much farther away is necessary?

RESPONSE If the energy is radiated uniformly in all directions, then the intensity (dose/area) should fall off proportionately to the distance squared, just as it does for sound and light. Thus, moving four times farther away lowers the exposure by a factor of sixteen, enough to do the trick.

FIGURE 31–14 Radiation spreads out in all directions. A person 4.0 m away intercepts only a fraction: his cross-sectional area divided by the area of a sphere of radius 4.0 m.

EXAMPLE 31–10 | **Whole-body dose.** What whole-body dose is received by a 70-kg laboratory worker exposed to a 40 mCi $^{60}_{27}$Co source, assuming the person's body has cross-sectional area 1.5 m^2 and is, on average, 4.0 m from the source for 4.0 h per day? $^{60}_{27}$Co emits γ rays of energy 1.33 MeV and 1.17 MeV in quick succession. Approximately 50 percent of the γ rays interact in the body and deposit all their energy. (The rest pass through.)

SOLUTION The total γ-ray energy per decay is $(1.33 + 1.17)$ MeV = 2.50 MeV, so the total energy emitted by the source is

$$(0.040 \text{ Ci})(3.7 \times 10^{10} \text{ decays/Ci·s})(2.50 \text{ MeV}) = 3.7 \times 10^{9} \text{MeV/s}.$$

The proportion of this intercepted by the body is its 1.5 m^2 area divided by the area of a sphere of radius 4.0 m (Fig. 31–14)

$$\frac{1.5 \text{ m}^2}{4\pi r^2} = \frac{1.5 \text{ m}^2}{4\pi(4.0 \text{ m})^2} = 7.5 \times 10^{-3}.$$

So the rate energy is deposited in the body (remembering that only $\frac{1}{2}$ of the γ rays interact in the body) is

$$E = (\tfrac{1}{2})(7.5 \times 10^{-3})(3.7 \times 10^{9} \text{ MeV/s})(1.6 \times 10^{-13} \text{ J/MeV})$$

$$= 2.2 \times 10^{-6} \text{ J/s}.$$

Since 1 Gy = 1 J/kg, the whole-body dose rate for this 70-kg person is $(2.2 \times 10^{-6} \text{ J/s})/(70 \text{ kg}) = 3.1 \times 10^{-8}$ Gy/s. In the space of 4.0 h, this amounts to a dose of $(4.0 \text{ h})(3600 \text{ s/h})(3.1 \times 10^{-8} \text{ Gy/s}) = 4.5 \times 10^{-4}$ Gy. Since QF $\approx$ 1 for gammas, the effective dose (Eq. 31–7) is 450 μSv or (see Eq. 31–6):

$$(100 \text{ rad/Gy})(4.5 \times 10^{-4} \text{ Gy})(1) = 45 \text{ mrem}.$$

This 45-mrem effective dose is nearly 10 percent of the normal allowed dose for a whole year (500 mrem/yr), or 1 percent of the yearly allowance for radiation workers. This worker should not receive such a dose every day and should seek ways to reduce it (shield the source, vary the work, work farther away, etc.).

*31–6 Radiation Therapy

The applications of radioactivity and radiation to human beings and other organisms is a vast field that has filled many books. In the medical field there are two basic aspects: (1) **radiation therapy**—the treatment of disease (mainly cancer)—which we discuss in this Section; and (2) the *diagnosis* of disease, which we discuss in the following Sections of this chapter.

Radiation can cause cancer. It can also be used to treat it. Rapidly growing cancer cells are especially susceptible to destruction by radiation. Nonetheless, large doses are needed to kill the cancer cells, and some of the surrounding normal cells are inevitably killed as well. It is for this reason that cancer patients receiving radiation therapy often suffer side effects characteristic of radiation sickness. To minimize the destruction of normal cells, a narrow beam of γ or X-rays is often used when the cancerous tumor is well localized. The beam is directed at the tumor, and the source (or body) is rotated so that the beam passes through various parts of the body to keep the dose at any one place as low as possible—except at the tumor and its immediate surroundings, where the beam passes at all times (Fig. 31–15). The radiation may be from a radioactive source such as $^{60}_{27}$Co, or it may be from an X-ray machine that produces photons in the range 200 keV to 5 MeV. Protons, neutrons, electrons, and pions, which are produced in particle accelerators (Section 32–2), are also being used in cancer therapy.

In some cases, a tiny radioactive source may be inserted directly inside a tumor, which will eventually kill the majority of the cells. A similar technique is used to treat cancer of the thyroid with the radioactive isotope $^{131}_{53}$I. The thyroid gland tends to concentrate any iodine present in the bloodstream; so when $^{131}_{53}$I is injected into the blood, it becomes concentrated in the thyroid, particularly in any area where abnormal growth is taking place. The intense radioactivity emitted can then destroy the defective cells.

Although radiation can increase the lifespan of many patients, it is not always completely effective. It may not be possible to kill all the diseased cancer cells, so a recurrence of the disease is possible. Many cases, especially when the cancerous cells are not well localized in one area, are difficult to treat at all without damaging the rest of the organism.

Another application of radiation is for sterilizing bandages, surgical equipment, and even packaged foods, since bacteria and viruses can be killed or deactivated by large doses of radiation.

➡ **PHYSICS APPLIED**

Radiation therapy

FIGURE 31–15 Radiation source rotates so that the beam always passes through the diseased tissue, but minimizing the dose in the rest of the body.

*31–7 Tracers and Imaging in Research and Medicine

Radioactive isotopes are commonly used in biological and medical research as **tracers**. A given compound is artificially synthesized using a radioactive isotope such as $^{14}_{6}$C or $^{3}_{1}$H. Such "tagged" molecules can then be traced as they move through an organism or as they undergo chemical reaction. The presence of these tagged molecules (or parts of them, if they undergo chemical change) can be detected by a Geiger or scintillation counter, which detects emitted radiation (see Section 30–13). The details of how food molecules are digested, and to what parts of the body they are diverted, can be traced in this way. Radioactive tracers have been used to determine how amino acids and other essential compounds are synthe-

➡ **PHYSICS APPLIED**

Tracers in medicine and biology

FIGURE 31–16

(a) Autoradiograph of a mature leaf of the squash plant *Cucurbita melopepo* exposed for 30 s to $^{14}CO_2$. The photosynthetic (green) tissue has become radioactive; the nonphotosynthetic tissue of the veins is free of ^{14}C and therefore does not blacken the X-ray sheet. This technique is very useful in following patterns of nutrient transport in plants. (b) An autoradiograph of a fiber of chromosomal DNA isolated from the higher plant *Arabidopsis thaliana*. The dashed arrays of silver grains show the Y-shaped growing point of replicating DNA.

(a)

(b)

sized by organisms. The permeability of cell walls to various molecules and ions can be determined using radioactive isotopes: the tagged molecule or ion is injected into the extracellular fluid and the radioactivity present inside and outside the cells is measured as a function of time.

In many cases, the cells or chemicals involved are removed from the organism and placed in the Geiger or scintillation counter. To follow a process in time, different sets of organisms are treated for different periods of time before the cells or chemicals are extracted. (See Example 31–8.)

In a technique known as **autoradiography**, the position of the radioactive isotopes is detected on film. For example, the distribution of carbohydrates produced in the leaves of plants from absorbed CO_2 can be observed by keeping the plant in an atmosphere where the carbon atom in the CO_2 is $^{14}_6C$. After a time, a leaf is placed firmly on a photographic plate and the emitted radiation darkens the film most strongly where the isotope is most strongly concentrated (Fig. 31–16a). Autoradiography using labeled nucleotides (components of DNA) has revealed much about the details of DNA replication (Fig. 31–16b).

For medical diagnosis, the radionuclide commonly used today is $^{99m}_{43}Tc$, a long-lived excited state of technetium-99 (the "m" in the symbol stands for "metastable" state). It is formed when $^{99}_{42}Mo$ decays. The great usefulness of $^{99m}_{43}Tc$ derives from its convenient half-life of 6 h (short, but not too short) and the fact that it can combine with a large variety of compounds. The compound to be labeled with the radionuclide is so chosen because it concentrates in the organ or region of the anatomy to be studied. Detectors outside the body then record, or image, the distribution of the radioactively labeled compound. The detection can be done by a single detector (Fig. 31–17) which is moved across the body, measuring the intensity of radioactivity at a large number of points. The image represents the relative intensity of radioactivity at each point. The relative radioactivity is a diagnostic tool. For example, high or low radioactivity may represent overactivity or underactivity of an organ or part of an organ, or in another case may represent a lesion or tumor. More complex *gamma cameras* make use of many detectors which simultaneously record the radioactivity at many points. The measured intensities can be displayed on a CRT (TV monitor or oscilloscope screen), and allow "dynamic" studies (that is, images that change in time) to be performed.

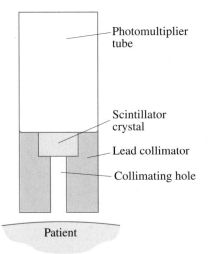

Photomultiplier tube

Scintillator crystal

Lead collimator

Collimating hole

Patient

FIGURE 31–17 Collimated gamma-ray detector for scanning (moving) over patient. The collimator is necessary to select γ rays that come in a straight line from the patient. Without the collimator, γ rays from all parts of the body could strike the scintillator, producing a very poor image.

*31–8 Emission Tomography

The images formed using the standard techniques of nuclear medicine, as briefly discussed in the previous section, are produced from radioactive tracer sources within the *volume* of the body. It is also possible to image the radioactive emissions in a single plane or slice through the body using the computed tomography techniques discussed in Section 25–12. A basic gamma camera is moved around the patient to measure the radioactive intensity from the tracer at many points and angles; the data are processed in much the same way as for X-ray CT scans (Section 25–12). This technique is referred to as **single photon emission tomography** (SPET).[†]

Another important technique is **positron emission tomography** (PET), which makes use of positron emitters such as $^{11}_{6}C$, $^{13}_{7}N$, $^{15}_{8}O$, and $^{18}_{9}F$. These isotopes are incorporated into molecules that, when inhaled or injected, accumulate in the organ or region of the body to be studied. When such a nuclide β decays, the emitted positron travels at most a few millimeters before it collides with a normal electron. In this collision, the positron and electron are annihilated, producing two γ rays ($e^+ + e^- \rightarrow 2\gamma$), each having energy of 510 keV. The two γ rays fly off in opposite directions ($180° \pm 0.25°$) since they must have almost exactly equal and opposite momenta to conserve momentum (the momenta of the e^+ and e^- are essentially zero compared to the momenta of the γ rays). Because the photons travel along the same line in opposite directions, their detection in coincidence by rings of detectors surrounding the patient (Fig. 31–18) readily establishes the line along which the emission took place. If the difference in time of arrival of the two photons could be determined accurately, the actual position of the emitting nuclide along that line could be calculated. Present-day electronics can measure times to at best ± 300 ps, so at the γ ray's speed ($c = 3 \times 10^8$ m/s), the actual position could be determined to an accuracy on the order of about $vt \approx (3 \times 10^8 \text{ m/s}) \cdot (300 \times 10^{-2} \text{ s}) \approx 10$ cm, which is not very useful. Although there may be future potential for time-of-flight measurements to determine position, today computed tomography techniques are used instead, similar to those for X-ray CT, which can reconstruct PET images with a resolution on the order of 3–5 mm. One big advantage of PET is that no collimators are needed (as for detection of a single photon—see Fig. 31–17). Thus, fewer photons are "wasted" and lower doses can be administered to the patient with PET.

► **PHYSICS APPLIED**

Medical imaging: PET and SPET

SPET

PET

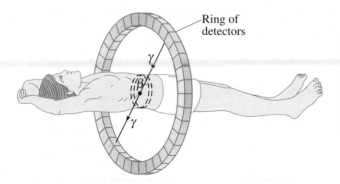

Ring of detectors

FIGURE 31–18 Positron emission tomography (PET) system showing a ring of detectors to detect the two annihilation γ rays ($e^+ + e^- \rightarrow 2\gamma$) emitted at 180° to each other. See also photo at start of this chapter.

[†]Also known as SPECT, "single photon emission computed tomography."

Both PET and SPET systems can give images related to biochemistry, metabolism, and function. This is to be compared to X-ray CT scans (Section 25–12), whose images reflect shape and structure—that is, the anatomy of the imaged region.

*31–9 Nuclear Magnetic Resonance (NMR) and Magnetic Resonance Imaging (MRI)

NMR **Nuclear magnetic resonance** (NMR) is a phenomenon that soon after its discovery in 1946 became a powerful research tool in a variety of fields from physics to chemistry and biochemistry. In recent years, it has been developed into an important and widely used medical imaging technique. We first briefly discuss the phenomenon, and then we will look at its applications.

We saw in Chapter 20 (Section 20–9) that a loop of wire carrying an electric current has a torque on it when placed in a magnetic field. The loop is said to have a magnetic moment, and the torque on the loop tends to align it so that the direction of its magnetic moment (taken to be perpendicular to the plane of the coil) is parallel to the field, **B**. (The coil's plane is perpendicular to **B**.) When the coil is so aligned, it has its lowest potential energy. Then in Chapter 28 (Section 28–6), we saw that atomic electrons in their orbital motions act as tiny current loops, and furthermore that they have an intrinsic angular momentum called *spin*, which also acts like the magnetic moment of a current loop. When atoms are placed in a magnetic field, atomic energy levels split into several closely spaced levels (see Fig. 28–8).

Nuclei, too, exhibit these magnetic properties. Many nuclides have magnetic moments, but we will examine only the simplest, the hydrogen (H) nucleus, since it is the one most used even for medical imaging. The $^1_1$H nucleus consists of a single proton. Its spin angular momentum (and its magnetic moment), like that of the electron, can take on only two values when placed in a magnetic field: we call these "spin up" (parallel to the field) and "spin down" (antiparallel to the field), as suggested in Fig. 31–19. When a magnetic field is present, the energy of the nucleus splits into two levels as shown in Fig. 31–20, with the spin up (parallel to field) having the lower energy. (This is like the Zeeman effect for atomic levels—see Fig. 28–8.) The difference in energy, ΔE, between these two levels is proportional to the total magnetic field, B_T, at the nucleus:

$$\Delta E = kB_T,$$

where k is a proportionality constant that is different for different nuclides.

FIGURE 31–19 Schematic picture of a proton represented in a magnetic field **B** (pointing upward) with its two possible states of spin, up and down.

FIGURE 31–20 Energy E_0 in the absence of a magnetic field splits into two levels in the presence of a magnetic field.

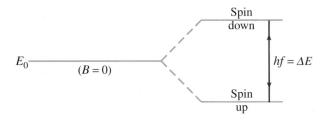

In a standard nuclear magnetic resonance (NMR) setup, the sample to be examined is placed in a static magnetic field. A radiofrequency (RF) pulse of electromagnetic radiation (that is, photons) is applied to the sample. If the frequency, f, of this pulse corresponds precisely to the energy difference between the two energy levels (Fig. 31–20), so that

$$hf = \Delta E = kB_T, \tag{31–8}$$

then the photons of the RF beam will be absorbed, exciting many of the nuclei from the lower state to the upper state. This is a resonance phenomenon since there is significant absorption only if f is very near $f = kB_T/h$. Hence the name "nuclear magnetic resonance." For free $_1^1H$ nuclei, the frequency is 42.58 MHz for a field $B_T = 1.0$ T. If the H atoms are bound in a molecule, on the other hand, the total magnetic field B_T at the H nuclei will be the sum of the external applied field plus the magnetic field due to electrons and nuclei of neighboring atoms. Since f is proportional to B_T, the value of f will be slightly different for the bound H atoms than for free atoms in the same external field. This change in frequency, which can be measured, is called the "chemical shift." A great deal has been learned about the structure of molecules and bonds using such NMR measurements.

For producing medically useful NMR images—now commonly called MRI, or **magnetic resonance imaging**—the element most used is hydrogen since it is the commonest element in the human body and gives the strongest NMR signals. The experimental apparatus is shown in Fig. 31–21. The large coils set up the static magnetic field, and the RF coils produce the RF pulse of electromagnetic waves (photons) that cause the nuclei to jump from the lower state to the upper one (Fig. 31–20). These same coils (or another coil) can detect the absorption of energy or the emitted radiation (also of frequency $f = \Delta E/h$, Eq. 31–8) when the nuclei jump back down to the lower state.

The formation of a two-dimensional or three-dimensional image can be done using techniques similar to those for computed tomography (Section 25–12). The simplest thing to measure for creating an image is the intensity of absorbed and/or reemitted radiation from many different points of the

➡ **PHYSICS APPLIED**

NMR imaging (MRI)

FIGURE 31–21 Typical NMR imaging setup: (a) diagram; (b) photograph.

Magnets

RF coils

(a)   (b)

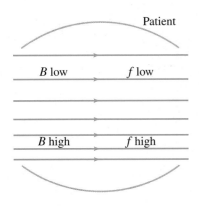

Patient

B low   *f* low

B high   *f* high

FIGURE 31–22 A static field that is stronger at the bottom than at the top. The frequency of absorbed or emitted radiation is proportional to *B* in NMR.

FIGURE 31–23 False-color NMR image (MRI) of a vertical section through the head showing structures in the normal brain.

body, and this would be a measure of the density of H atoms at each point. But how do we determine from what part of the body a given photon comes? One technique is to give the static magnetic field a gradient; that is, instead of applying a uniform magnetic field, B_T, the field is made to vary with position across the width of the sample (or patient). Since the frequency absorbed by the H nuclei is proportional to B_T (Eq. 31–8), only one plane within the body will have the proper value of B_T to absorb photons of a particular frequency *f*. By varying *f*, absorption by different planes can be measured. Alternately, if the field gradient is applied *after* the RF pulse, the frequency of the emitted photons will be a measure of where they were emitted. See Fig. 31–22. If a magnetic field gradient in one direction is applied during excitation (absorption of photons) and photons of a single frequency are transmitted, only H nuclei in one thin slice will be excited. By applying a gradient in a different direction, perpendicular to the first, during reemission, the frequency *f* of the reemitted radiation will represent depth in that slice. Other ways of varying the magnetic field throughout the volume of the body can be used in order to correlate NMR frequency with position.

A reconstructed image based on the density of H atoms (that is, the intensity of absorbed or emitted radiation) is not very interesting. More useful are images based on the rate at which the nuclei decay back to the ground state, and such images can produce resolution of 1 mm or better. This NMR technique (sometimes called **spin-echo**) is producing images of great diagnostic value, both in the delineation of structure (anatomy) and in the study of metabolic processes. An NMR image is shown in Fig. 31–23.

Table 31–2 lists the recently developed techniques we have discussed for imaging the interior of the body, along with the optimum resolution attainable today. Of course, resolution is only one factor that must be considered; it must be remembered that the different imaging techniques provide different types of information, useful for different types of diagnosis.

TABLE 31–2 Medical Imaging Techniques

Technique	Where Discussed In This Book	Resolution
Conventional X-ray	Section 25–12	$\frac{1}{2}$ mm
CT scan, X-ray	Section 25–12	$\frac{1}{2}$ mm
Nuclear medicine (tracers)	Section 31–7	1 cm
SPET (single photon emission)	Section 31–8	1 cm
PET (positron emission)	Section 31–8	3–5 mm
NMR	Section 31–9	$\frac{1}{2}$–1 mm
Ultrasound	Section 12–10	2 mm

SUMMARY

A **nuclear reaction** occurs when two nuclei collide and two or more other nuclei (or particles) are produced. In this process, as in radioactivity, **transmutation** (change) of elements occurs.

In **fission**, a heavy nucleus such as uranium splits into two intermediate-sized nuclei after being struck by a neutron. $^{235}_{92}$U is fissionable by slow neutrons, whereas some fissionable nuclei require fast neutrons. Much energy is released in fission because the binding energy per nucleon is lower for heavy nuclei than it is for intermediate-sized nuclei, so the mass of a heavy nucleus is greater than the total mass of its fission products. The fission process releases neutrons, so that a **chain reaction** is possible. The **critical mass** is the minimum mass of fuel needed to sustain a chain reaction. In a **nuclear reactor** or nuclear bomb, a moderator is needed to slow down the released neutrons.

The **fusion** process, in which small nuclei combine to form larger ones, also releases energy. The energy from our Sun is believed to originate in the fusion reactions known as the **proton–proton cycle** in which four protons fuse to form a $^4_2$He nucleus producing over 25 MeV. A working fusion reactor for power generation has not yet proved possible because of the difficulty in containing the fuel (e.g., deuterium) long enough at the high temperature required. Nonetheless, great progress has been made in confining the collection of charged ions known as a **plasma**. The two main methods are **magnetic confinement**, using a magnetic field in a device such as the toroidal-shaped **tokamak**, and **inertial confinement**, in which intense laser beams compress a fuel pellet of deuterium and tritium.

Radiation can cause damage to materials, including biological tissue. Quantifying amounts of radiation is the subject of **dosimetry**. The **curie** (Ci) and the **becquerel** (Bq) are units that measure the **source activity** or rate of decay of a sample: $1 \text{ Ci} = 3.70 \times 10^{10}$ disintegrations per second, whereas $1 \text{ Bq} = 1$ disintegration/s. The **absorbed dose**, often specified in **rads**, measures the amount of energy deposited per unit mass of absorbing material: 1 rad is the amount of radiation that deposits energy at the rate of 10^{-2} J/kg of material. The SI unit of absorbed dose is the **gray**: $1 \text{ Gy} = 1 \text{ J/kg} = 100 \text{ rad}$. The **effective dose** is often specified by the **rem** $= \text{rad} \times \text{QF}$, where QF is the "quality factor" of a given type of radiation; 1 rem of any type of radiation does approximately the same amount of biological damage. The average dose received per person per year in the United States is about 0.36 rem. The SI unit for effective dose is the **sievert**: $1 \text{ Sv} = 10^2 \text{ rem}$.

QUESTIONS

1. Fill in the missing particles or nuclei: (a) $^{137}_{56}$Ba(n, γ)?; (b) $^{137}_{56}$Ba$(n, ?)^{137}_{55}$Cs; (c) $^2_1$H$(d, ?)^4_2$He; (d) $^{197}_{79}$Au(α, d)? where d stands for deuterium.

2. The isotope $^{32}_{15}$P is produced by an (n, p) reaction. What must be the target nucleus?

3. When $^{22}_{11}$Na is bombarded by deuterons ($^2_1$H), an α particle is emitted. What is the resulting nuclide?

4. Why are neutrons such good projectiles for producing nuclear reactions?

5. A proton strikes a $^{20}_{10}$Ne nucleus, and an α particle is observed to emerge. What is the residual nucleus? Write down the reaction equation.

6. Are fission fragments β^+ or β^- emitters?

7. If $^{235}_{92}$U released only 1.5 neutrons per fission on the average, would a chain reaction be possible? If so, what would be different?

8. $^{235}_{92}$U releases an average of 2.5 neutrons per fission compared to 2.7 for $^{239}_{94}$Pu. Pure samples of which of these two nuclei do you think would have the smaller critical mass?

9. The energy from nuclear fission appears in the form of thermal energy—but the thermal energy of what?

10. Why can't uranium be enriched by chemical means?

11. How can a neutron, with practically no KE, excite a nucleus to the extent shown in Fig. 31–2?

12. Why would a porous block of uranium be more likely to explode if kept under water rather than in air?

13. A reactor that uses highly enriched uranium can use ordinary water (instead of heavy water) as a moderator and still have a self-sustaining chain reaction. Explain.

14. Why must the fission process release neutrons if it is to be useful?

15. Discuss the relative merits and disadvantages, including pollution and safety, of power generation by fossil fuels, nuclear fission, and nuclear fusion.

16. What is the reason for the "secondary system" in a nuclear reactor, Fig. 31–6? That is, why is the water heated by the fuel in a nuclear reactor not used directly to drive the turbines?

17. Discuss how the course of history might have been changed if, during World War II, scientists had refused to work on developing a nuclear bomb. Do you think it would have been possible to delay the building of a bomb indefinitely?

18. Research in molecular biology is moving toward the ability to perform genetic manipulations on human beings. The moral implications of future discoveries along these lines has led to a warning that this may be the molecular biologists' "Hiroshima." Discuss.

19. Does $E = mc^2$ apply in (a) fission, (b) fusion, (c) nuclear reactions?

20. Light energy emitted by the Sun and stars comes from the fusion process. What conditions in the interior of stars makes this possible?

21. How do stars, and our Sun, maintain confinement of the plasma for fusion?

22. What is the basic difference between fission and fusion?

23. People who work around metals that emit alpha particles are trained that there is little danger from proximity or even touching the material, but that they must take extreme precautions against ingesting it. Hence,

there are strong rules against eating and drinking while working, and against machining the metal. Why?

24. Why is the recommended maximum radiation dose higher for women beyond the child-bearing age than for younger women?

* 25. How might radioactive tracers be used to find a leak in a pipe?

PROBLEMS

SECTION 31–1

1. (I) Natural aluminum is all $^{27}_{13}$Al. If it absorbs a neutron, what does it become? Does it decay by β^+ or β^-? What will be the product nucleus?

2. (I) Determine whether the reaction $^2_1$H(d, n)$^3_2$He requires a threshold energy. (d stands for deuterium, $^2_1$H.)

3. (I) Is the reaction $^{238}_{92}$U (n, γ)$^{239}_{92}$U possible with slow neutrons? Explain.

4. (II) Does the reaction $^7_3$Li(p, α)$^4_2$He require energy or does it release energy? How much energy?

5. (II) The reaction $^{18}_8$O(p, n)$^{18}_9$F requires an input of energy equal to 2.453 MeV. What is the mass of $^{18}_9$F?

6. (II) Calculate the energy released (or energy input required) for the reaction $^9_4$Be(α, n)$^{12}_6$C.

7. (II) (*a*) Can the reaction $^{24}_{12}$Mg(n, d)$^{23}_{11}$Na occur if the bombarding particles have 10.00 MeV of KE? (*b*) If so, how much energy is released?

8. (II) (*a*) Can the reaction $^7_3$Li(p, α)$^4_2$He occur if the incident proton has KE = 2500 keV? (*b*) If so, what is the total kinetic energy of the products?

9. (II) In the reaction $^{14}_7$N(α, p)$^{17}_8$O, the incident α particles have 7.68 MeV of kinetic energy. (*a*) Can this reaction occur? (*b*) If so, what is the total kinetic energy of the products? The mass of $^{17}_8$O is 16.999131 u.

10. (II) Calculate the Q-value for the "capture" reaction $^{16}_8$O(α, γ)$^{20}_{10}$Ne.

11. (II) Calculate the total KE of the products of the reaction $^{13}_6$C(d, n)$^{14}_7$N if the incoming deuteron (d) has KE = 36.3 MeV.

12. (II) An example of a "stripping" nuclear reaction is $^6_3$Li(d, p)X. (*a*) What is X, the resulting nucleus? (*b*) Why is it called a "stripping" reaction? (*c*) What is the Q-value of this reaction? Is the reaction endothermic or exothermic?

13. (II) An example of a "pick-up" nuclear reaction is $^{12}_6$C($^3_2$He, α)X. (*a*) Why is it called a "pickup" reaction? (*b*) What is the resulting nucleus? (*c*) What is the Q-value of this reaction? Is the reaction endothermic or exothermic?

14. (II) (*a*) Complete the following nuclear reaction, ?(p, γ)$^{32}_{16}$S. (*b*) What is the Q-value?

15. (III) Use conservation of energy and momentum to show that a bombarding proton must have an energy of 3.23 MeV in order to make the reaction $^{13}_6$C(p, n)$^{13}_7$N occur. (See Example 31–3.)

16. (III) Show, using the laws of conservation of energy and momentum, that for a nuclear reaction requiring energy, the minimum kinetic energy of the bombarding particle (the *threshold energy*) is equal to $[Qm_{pr}/(m_{pr} - m_b)]$, where $-Q$ is the energy required (difference in total mass between products and reactants), m_b is the rest mass of the bombarding particle, and m_{pr} the total rest mass of the products. Assume the target nucleus is at rest before an interaction takes place, and that all particles are nonrelativistic.

SECTION 31–2

17. (I) Calculate the energy released in the fission reaction n + $^{235}_{92}$U $\rightarrow$ $^{88}_{38}$Sr + $^{136}_{54}$Xe + 12n. Use Appendix F; assume the initial KE of the neutron is very small.

18. (I) What is the energy released in the fission reaction of Eq. 31–2b? (The masses of $^{141}_{56}$Ba and $^{92}_{36}$Kr are 140.91440 u and 91.92630 u, respectively.)

19. (I) How many fissions take place per second in a 200-MW reactor? Assume 200 MeV is released per fission.

20. (II) Suppose that the average power consumption, day and night, in a typical house is 300 W. What initial mass of $^{235}_{92}$U would have to undergo fission to supply the electrical needs of such a house for a year? (Assume 200 MeV is released per fission.)

21. (II) What initial mass of $^{235}_{92}$U is required to operate a 500-MW reactor for 1 yr? Assume 40% efficiency.

22. (II) If a 1.0-MeV neutron emitted in a fission reaction loses one-half of its KE in each collision with moderator nuclei, how many collisions must it make to reach thermal energy ($\frac{3}{2}kT$ = 0.040 eV)?

23. (III) Suppose that the neutron multiplication factor is 1.0004. If the average time between successive fissions in a chain of reactions is 1.0 ms, by what factor will the reaction rate increase in 1.0 s?

SECTION 31–3

24. (I) What is the average kinetic energy of protons at the center of a star where the temperature is 10^7 K?

25. (II) Show that the energy released in the fusion reaction $^2_1$H + $^3_1$H $\rightarrow$ $^4_2$He + n is 17.59 MeV.

26. (II) Show that the energy released when two deuterium nuclei fuse to form $^3_2$He with the release of a neutron is 3.27 MeV.

27. (II) Verify the energy output stated for each of the reactions of Eq. 31–3. [*Hint*: Be careful with electrons.]

28. (II) Calculate the energy release per gram of fuel for the reactions of Eqs. 31–5a, b, and c. Compare to the energy release per gram of uranium in fission.

29. (II) If a typical house requires 300 W of electric power on average, how much deuterium fuel would have to be used in a year to supply these electrical needs? Assume the reaction of Eq. 31–5b.

30. (II) Show that the energies carried off by the $^4_2$He nucleus and the neutron for the reaction of Eq. 31–5c are about 3.5 MeV and 14 MeV, respectively. Are these fixed values, independent of the plasma temperature?

31. (II) Some stars, in a later stage of evolution, may begin to fuse two $^{12}_6$C nuclei into one $^{24}_{12}$Mg nucleus. (*a*) How much energy would be released in such a reaction? (*b*) What KE must two carbon nuclei each have when far apart, if they can then approach each other to within 6.0 fm, center-to-center? (*c*) Approximately what temperature would this require?

32. (III) Suppose a fusion reactor ran on "d–d" reactions, Eqs. 31–5a and b. Estimate how much water, for fuel, would be needed per hour to run a 1000-MW reactor, assuming 30 percent efficiency.

33. (III) How much energy (J) is contained in 1.00 kg of water if its natural deuterium is used in the fusion reaction of Eq. 31–5a? Compare to the energy obtained from the burning of 1.0 kg of gasoline, about 5×10^7 J.

34. (III) The energy output of massive stars is believed to be due to the *carbon cycle* (see text). (*a*) Show that no carbon is consumed in this cycle and that the net effect is the same as for the proton–proton cycle. (*b*) What is the total energy release? (*c*) Determine the energy output for each reaction and decay. (*d*) Why does the carbon cycle require a higher temperature ($\approx 2 \times 10^7$ K) than the proton–proton cycle ($\approx 1.5 \times 10^7$ K)?

35. (III) The deuterium–tritium pellet in a laser fusion device contains equal numbers of $^2_1$H and $^3_1$H atoms raised to a density of 200×10^3 kg/m^3 by the laser pulses. Estimate (*a*) the density of particles in this compressed state, and (*b*) the length of time τ the pellet must be confined to meet the Lawson criterion for ignition.

SECTION 31–5

36. (I) A dose of 500 rem of γ rays in a short period would be lethal to about half the people subjected to it. How many rads is this?

37. (I) Fifty rads of α-particle radiation is equivalent to how many rads of X-rays in terms of biological damage?

38. (I) How many rads of slow neutrons will do as much biological damage as 50 rads of fast neutrons?

39. (I) How much energy is deposited in the body of a 70-kg adult exposed to a 50-rad dose?

40. (II) A 0.018-μCi sample of $^{32}_{15}$P is injected into an animal for tracer studies. If a Geiger counter intercepts 20 percent of the emitted β particles, what will be the counting rate, assumed 90% efficient?

41. (II) A 1.0-mCi source of $^{32}_{15}$P (in NaHPO$_4$), a β^- emitter, is implanted in an organ where it is to administer 5000 rad. The half-life of $^{32}_{15}$P is 14.3 days and 1 mCi delivers about 1 rad/min. Approximately how long should the source remain implanted?

42. (II) About 35 eV is required to produce one ion pair in air. Show that this is consistent with the two definitions of the roentgen given in the text.

43. (II) $^{57}_{27}$Co emits 122-keV γ rays. If a 70-kg person swallowed 2.0 μCi of $^{57}_{27}$Co, what would be the dose rate (rad/day) averaged over the whole body? Assume that 50 percent of the γ-ray energy is deposited in the body. [*Hint*: Determine the rate of energy deposited in the body and use the definition of the rad.]

44. (II) What is the mass of a 1.00-μCi $^{14}_6$C source?

45. (II) Huge amounts of radioactive $^{131}_{53}$I were released in the accident at Chernobyl in 1986. Chemically, iodine goes to the human thyroid. (Doctors can use it for diagnosis and treatment of thyroid problems.) In a normal thyroid, $^{131}_{53}$I absorption can cause damage to the thyroid. (*a*) Write down the decay scheme for $^{131}_{53}$I. (*b*) Its half-life is 8.0 d; how long would it take for ingested $^{131}_{53}$I to become 10 percent of the initial value? (*c*) Absorbing 1 mCi of $^{131}_{53}$I can be harmful; what mass of iodine is this?

46. (II) Assume a liter of milk typically has an activity of 2000 pCi due to $^{40}_{19}$K. If a person drinks 2 glasses (0.5 L) per day, estimate the total dose (rem) received in a year. As a crude model, assume the milk stays in the stomach 12 hr and is then released. Assume also that very roughly 10 percent of the 1.5 MeV released per decay is absorbed by the body. Compare your result to the normal allowed dose per year. Make your estimate for (*a*) a 50-kg adult, and (*b*) a 5-kg baby.

47. (II) Radon gas, $^{222}_{86}$Rn, is considered a serious health hazard (see discussion in text). It decays by α-emission. (*a*) What is the daughter nucleus? (*b*) Is the daughter nucleus stable or radioactive? If the latter, how does it decay, and what is its lifetime? (*c*) Is the daughter nucleus also a noble gas, or is it chemically reacting? (*d*) Suppose 1.0 ng of $^{222}_{86}$Rn seeps into a basement. What will be its activity? If the basement is then sealed, what will be the activity 1 month later? [*Hint*: See the periodic table as well as Section 30–10 and Fig. 30–10.]

*SECTION 31–9

48. (II) Calculate the wavelength of photons needed to produce NMR transitions in free protons in a 1.000-T field. In what region of the spectrum does it lie?

49. J. Chadwick discovered the neutron by bombarding $^9_4$Be with the popular projectile of the day, alpha particles. (a) If one of the reaction products was the then unknown neutron, what was the other product? (b) What is the Q of this reaction?

50. Fusion temperatures are often given in keV. Determine the conversion factor from kelvins to keV using, as is usual in this field, $\text{KE} = kT$ without the factor $\frac{3}{2}$.

51. One means of enriching uranium is by diffusion of the gas UF_6. Calculate the ratio of the speeds of molecules of this gas containing $^{235}_{92}$U and $^{238}_{92}$U, on which this process depends.

52. (a) What mass of $^{235}_{92}$U was actually fissioned in the first atomic bomb, whose energy was the equivalent of about 20 kilotons of TNT (1 kiloton of TNT releases 5×10^{12} J)? (b) What was the actual mass transformed to energy?

53. In a certain town the average yearly background radiation consists of 25 mrad of X-rays and γ rays plus 3.0 mrad of particles having a QF of 10. How many rem will a person receive per year on the average?

54. Deuterium makes up 0.015 percent of natural hydrogen. Make a rough estimate of the total deuterium in the Earth's oceans and estimate the total energy released if all of it were used in fusion reactors.

55. A shielded γ-ray source yields a dose rate of 0.050 rad/h at a distance of 1.0 m for an average-sized person. If workers are allowed a maximum dose rate of 5.0 rem/yr, how close to the source may they operate, assuming a 40-h work week? Assume that the intensity of radiation falls off as the square of the distance. (It actually falls off more rapidly than $1/r^2$ because of absorption in the air, so the answer above will give a better-than-permissible value.)

56. Radon gas, $^{222}_{86}$Rn, is formed by α decay. (a) Write the decay equation. (b) Ignoring the KE of the daughter nucleus (it's so massive), estimate the KE of the α particle produced. (c) Estimate the momentum of the alpha and of the daughter nucleus. (d) Estimate the KE of the daughter, and show that your approximation in (b) was valid.

57. Consider a system of nuclear power plants that produce 4000 MW. (a) What total mass of $^{235}_{92}$U fuel would be required to operate these plants for 1 yr, assuming that 200 MeV is released per fission? (b) Typically 6 percent of the $^{235}_{92}$U nuclei that fission produce $^{90}_{38}$Sr, a β^- emitter with a half-life of 29 yr. What is the total radioactivity of the $^{90}_{38}$Sr, in curies, produced in 1 yr? (Neglect the fact that some of it decays during the 1-yr period.)

58. In the net reaction, Eq. 31–4, for the proton–proton cycle in the Sun, the neutrinos escape from the Sun with energy of about 0.5 MeV. The remaining energy, 26.2 MeV, is available within the Sun. Use this value to calculate the "heat of combustion" per kilogram of hydrogen fuel and compare it to the heat of combustion of coal, about 3×10^7 J/kg.

59. Energy reaches the Earth from the Sun at a rate of about 1400 W/m². Calculate (a) the total energy output of the Sun, and (b) the number of protons consumed per second in the reaction of Eq. 31–4, assuming that this is the source of all the Sun's energy. (c) Assuming that the Sun's mass of 2.0×10^{30} kg was originally all protons and that all could be involved in nuclear reactions in the Sun's core, how long would you expect the Sun to "glow" at its present rate?

60. Estimate how many solar neutrinos pass through a 100 m² room in a year. [Hint: See Problems 58 and 59.]

61. An average adult body contains about 0.10 μCi of $^{40}_{19}$K, which comes from food. (a) How many decays occur per second? (b) The potassium decays produce beta particles with energies of around 1.4 MeV. Calculate the dose per year in rem for a 50-kg adult. Is this a significant fraction of the 0.36 rem/year background rate?

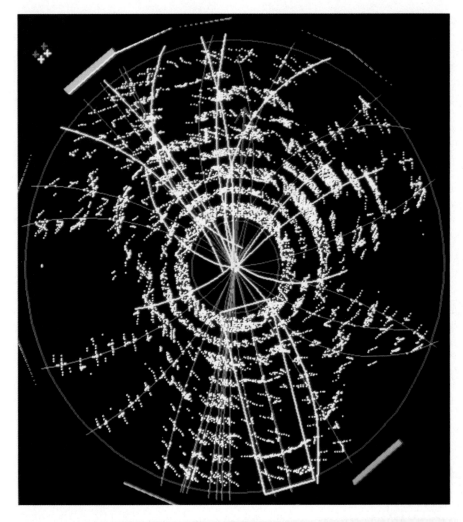

This computer-generated reconstruction of a proton-antiproton collision at Fermilab (Fig. 32–4) occurred at a combined energy of nearly 2 TeV. It is one of the events that provide evidence for the top quark, and that led to the announcement of its discovery in 1995. Because the detector is in a magnetic field, the particle tracks are curved; the radius of curvature is a measure of each particle's momentum (Chapter 20). The top quark (t) has too brief a lifetime ($\approx 10^{-23}$ s) to be detected itself, so experimenters look for its possible decay products. This photo is believed to show the following interaction and subsequent decays:

$$p + \bar{p} \longrightarrow t + \bar{t}$$

$$\begin{array}{l} \quad\quad\quad\quad \hookrightarrow W^- + \bar{b} \\ \quad\quad\quad\quad\quad\quad \hookrightarrow \text{jet} \\ \quad\quad\quad\quad\quad\quad\quad\quad \rightarrow \bar{\mu} + \bar{\nu}_\mu \\ \quad\quad \hookrightarrow W^+ + b \\ \quad\quad\quad\quad\quad \hookrightarrow \text{jet} \\ \quad\quad\quad \hookrightarrow u + \bar{d} \\ \quad\quad\quad\quad\quad\quad \hookrightarrow \text{jet} \\ \quad\quad\quad\quad\quad\quad \rightarrow \text{jet} \end{array}$$

The tracks visible in the photo include the jets (each a group of particles moving in roughly the same direction), and a muon (μ^-) whose track is the pink one enclosed by a yellow rectangle to make it stand out. After reading this chapter, try to give the name for each symbol above and comment on whether all conservation laws hold.

ELEMENTARY PARTICLES 32

I n the final two chapters of this book we discuss two of the most exciting areas of contemporary physics: (1) elementary particles, and (2) cosmology and astrophysics. These are subjects at the forefront of knowledge—one treats the smallest objects in the universe, the other the largest (and oldest) aspects of the universe. Both of these chapters could be considered optional. Nonetheless, the reader who wants an understanding of the great beauties of present-day science—and/or wants to be a good citizen—will want to read these chapters, even if there is not time to cover them in a physics course.

In this penultimate chapter we discuss *elementary particle* physics, which represents the human endeavor to understand the basic building blocks of all matter. In the years after World War II, it was found that if the incoming particle in a nuclear reaction has sufficient energy, new types of particles can be produced. The earliest experiments used **cosmic rays**—particles that impinge on the Earth from space. But in order to produce high-energy particles in the laboratory, various types of particle accelerators have been constructed. Most commonly they accelerate protons or electrons, although heavy ions can also be accelerated, depending on the design. These high-energy accelerators have been used to probe the nucleus more deeply, to produce and study new particles, and to give us information about the basic forces and constituents of nature. Because the projectile particles are at high-energy, this field is sometimes called **high-energy physics**.

32–1 High-Energy Particles

Particles accelerated to high energy are projectiles which can probe the interior of nuclei and nucleons that they strike. An important factor is that faster-moving projectiles can reveal more detail. The wavelength of projectile particles is given by de Broglie's wavelength formula (Eq. 27–8),

de Broglie wavelength

$$\lambda = \frac{h}{mv},$$

showing that the greater the momentum of the bombarding particle, the shorter its wavelength. As discussed in Chapter 25 on optical instruments, resolution of details in images is limited by the wavelength: the shorter the wavelength, the finer the detail that can be obtained. This is one reason why particle accelerators of higher and higher energy have been built in recent years.

EXAMPLE 32–1 **High resolution with electrons.** What is the wavelength, and hence the expected resolution, for a beam of 1.3-GeV electrons?

SOLUTION 1.3 GeV = 1300 MeV, which is about 2500 times the mass of the electron ($0.51 \text{ MeV}/c^2$). We are clearly dealing with relativistic speeds here, and it is easily shown that the speed of the electron is nearly $c = 3.0 \times 10^8$ m/s. [From Eq. 26–5, KE $= mc^2 - m_0 c^2 \approx mc^2$; and from Eq. 26–8, $E^2 = p^2 c^2 + m_0^2 c^4 \approx p^2 c^2$ since $m_0 c^2$ is small compared to pc. Hence, $p^2 c^2 = m^2 v^2 c^2 \approx m^2 c^4$, so $v \approx c$.] Therefore

$$\lambda = \frac{h}{mv} \approx \frac{h}{mc} = \frac{hc}{mc^2}$$

where $mc^2 = 1.3$ GeV. Hence

$$\lambda = \frac{(6.6 \times 10^{-34} \text{ J·s})(3.0 \times 10^8 \text{ m/s})}{(1.3 \times 10^9 \text{ eV})(1.6 \times 10^{-19} \text{ J/eV})} = 0.96 \times 10^{-15} \text{ m},$$

or 0.96 fm. The maximum possible resolution of this beam of electrons is far greater than for a light beam in a light microscope ($\lambda \approx 500$ nm). Indeed, this resolution of about 1 fm is on the order of the size of nuclei (see Eq. 30–1).

Another major reason for building high-energy accelerators is that new particles of greater mass can be produced at higher energies, as we will discuss shortly. Now we look briefly at several types of particle accelerator.

GENERATOR

Conductor

ACCELERATOR

B Pulley

Ion source

Target

Belt

Insulator

Motor-driven pulley

A

50 kV

(a)

(b)

FIGURE 32–1 Van de Graaff generator and accelerator: (a) diagram; (b) in use at Brookhaven National Laboratory.

32–2 Particle Accelerators

Van de Graaff accelerator. The key component of a Van de Graaff accelerator (Fig. 32–1) is a Van de Graaff generator, invented in 1931. A large, hollow, spherical conductor is supported by an insulating column above the base. The sphere is charged to a high potential by a nonconducting moving belt in the following way. A high voltage of typically 50,000 V is applied to a pointed conductor *A*, which "sprays" positive charge onto the moving belt. (Actually, electrons are pulled off the belt onto the electrode *A*.) The belt carries the positive charge into the interior of the sphere, where it is "wiped off" the belt at *B* and races to the outer surface of the spherical conductor. (Remember that charge collects on the outer surface of any conductor, since the charges repel each other and try to get as far from each other as possible.) As more and more charge is brought upward, the sphere becomes more highly charged and reaches greater voltage. The process requires energy, of course, since the upward-moving charged belt is repelled by the charged sphere. The energy is supplied by the motor driving the belt. Very high potential differences can be obtained in this way, often giving the accelerated particles as much as 30 MeV of KE when two or more generators are used in tandem. Connected to the Van de Graaff generator is an evacuated tube which serves as the particle accelerator. A source of H or He ions (p or α) is located inside the tube, and the large positive voltage repels them so they are accelerated toward the grounded target at the far end of the tube (Fig. 32–1).

Cyclotron. The cyclotron was developed in 1930 by E. O. Lawrence (1901–1958; Fig. 32–2) at the University of California, Berkeley. It uses a magnetic field to maintain charged ions—usually protons—in nearly circular paths (Chapter 20). The protons move within two **D**-shaped cavities, as

Van de Graaff generator and accelerator

FIGURE 32–2 Ernest O. Lawrence holding the first cyclotron, around 1930.

Cyclotron

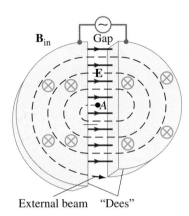

FIGURE 32–3 Diagram of a cyclotron. The magnetic field, applied by a large electromagnet, points into the page. *A* is the ion source. The field lines shown are for the electric field in the gap.

shown in Fig. 32–3. Each time they pass into the gap between the "dees," a voltage is applied that accelerates them via the electric force. This increases their speed and also increases the radius of curvature of their path. After many revolutions, the protons acquire high kinetic energy and reach the outer edge of the cyclotron. They then either strike a target placed inside the cyclotron or leave the cyclotron with the help of a carefully placed "bending magnet" and are directed to an external target. Acceleration only occurs when the protons are in the gap *between* the dees. (The magnetic field merely keeps the protons revolving so they can be repeatedly accelerated when they reach the gap.) The voltage applied to the dees to produce the acceleration must be alternating. When the protons are moving to the right across the gap in Fig. 32–3, the right dee must be electrically negative and the left one positive. A half-cycle later, the protons are moving to the left, so the left dee must be negative in order to accelerate them. The frequency, *f*, of the applied voltage must be equal to that of the circulating protons, and this can be determined as follows. When ions of charge *q* (for a proton, $q = +e$) are circulating within the dees, the net force *F* on each is simply that due to the magnetic field *B*, so $F = qvB$, where *v* is the speed of the ion at a given moment (Eq. 20–4). Since the ions move in circles, the acceleration is centripetal and equals v^2/r, where *r* is the radius of the ion's path at a given moment. We use Newton's second law, $F = ma$, and find that

$$F = ma$$

$$qvB = \frac{mv^2}{r}$$

or

$$v = \frac{qBr}{m}.$$

The time required for a complete revolution is the period *T* and is equal to

$$T = \frac{\text{distance}}{\text{speed}} = \frac{2\pi r}{qBr/m} = \frac{2\pi m}{qB}.$$

Hence the frequency of revolution *f* is

Cyclotron frequency

$$f = \frac{1}{T} = \frac{qB}{2\pi m}. \tag{32–1}$$

This is known as the **cyclotron frequency**.

EXAMPLE 32–2 Cyclotron. A small cyclotron of maximum radius $R = 0.25$ m accelerates protons in a 1.7-T magnetic field. Calculate (*a*) what frequency is needed for the applied alternating voltage, and (*b*) the kinetic energy of protons when they leave the cyclotron.

SOLUTION (*a*) From Eq. 32–1,

$$f = \frac{qB}{2\pi m}$$

$$= \frac{(1.6 \times 10^{-19}\ \text{C})(1.7\ \text{T})}{(6.28)(1.67 \times 10^{-27}\ \text{kg})} = 2.6 \times 10^7\ \text{Hz} = 26\ \text{MHz},$$

which is in the radio wave region of the EM spectrum (Fig. 22–10).

(*b*) The protons leave the cyclotron at $r = R = 0.25\,\text{m}$. Then, since $v = qBr/m$,

$$\text{KE} = \frac{1}{2}mv^2 = \frac{1}{2}m\frac{q^2B^2R^2}{m^2} = \frac{q^2B^2R^2}{2m}$$

$$= \frac{(1.6 \times 10^{-19}\,\text{C})^2(1.7\,\text{T})^2(0.25\,\text{m})^2}{(2)(1.67 \times 10^{-27}\,\text{kg})} = 1.4 \times 10^{-12}\,\text{J} = 8.7\,\text{MeV}.$$

Note that the magnitude of the voltage applied to the dees does not affect the final energy. But the higher this voltage, the fewer revolutions are required to bring the protons to full energy.

An interesting aspect of the cyclotron is that the frequency of the applied voltage, as given by Eq. 32–1, does not depend on the radius r. That is, the frequency does not have to be changed as the ions start from the source and are accelerated to paths of larger and larger radii. Unfortunately, this is only true at nonrelativistic energies. For at higher speeds, the mass of the ions will increase according to Einstein's formula, $m = m_0/\sqrt{1 - v^2/c^2}$, where m_0 is the rest mass. As can be seen from Eq. 32–1, as the mass increases, the frequency of the applied voltage must be reduced. To achieve large energies, complex electronics is needed to decrease the frequency as a packet of protons increases in speed and reaches larger orbits. Such a modified cyclotron is called a **synchrocyclotron**.

Synchrocyclotron

Synchrotron

Synchrotron.   Another way to deal with the relativistic increase in mass with speed is to increase the magnetic field B as the particles speed up. Such devices are called *synchrotrons*, and today they can be enormous. The Fermi National Accelerator Laboratory (Fermilab) at Batavia, Illinois, has a radius of 1.0 km, and that at CERN (European Center for Nuclear Research) in Geneva, Switzerland, is 1.1 km in radius. These machines can accelerate protons to energies of 500 GeV. The new *Tevatron* at Fermilab uses superconducting magnets to accelerate protons to about $1000\,\text{GeV} = 1\,\text{TeV}$ (hence its name; $1\,\text{TeV} = 10^{12}\,\text{eV}$). These large synchrotrons do not use enormous magnets 1 km in radius. Instead, a narrow ring of magnets is used (see Fig. 32–4) with each magnet placed at the same radius from the center of the circle. The magnets are interrupted by gaps where high voltage accelerates the particles. Thus, once the particles are injected, they must move in a

FIGURE 32–4 (a) Aerial view of Fermilab at Batavia, Illinois; the accelerator is a circular ring 1.0 km in radius. (b) The interior of the tunnel of the main accelerator at Fermilab. The upper (rectangular-shaped) ring of magnets is for the 500-GeV accelerator. Below it is the ring of superconducting magnets for the Tevatron.

(a)

(b)

circle of constant radius. This is accomplished by giving them considerable energy initially in a much smaller accelerator, and then slowly increasing the magnetic field as they speed up in the large synchrotron.

One problem of any accelerator is that accelerating electric charges radiate electromagnetic energy (see Chapter 22). Since ions or electrons are accelerated in an accelerator, we can expect considerable energy to be lost by radiation. The effect increases with speed and is especially important in circular machines, where centripetal acceleration is present, particularly in synchrotrons, and hence is called **synchrotron radiation**. Synchrotron radiation can actually be useful, however. Intense beams of photons are sometimes needed, and they are usually obtained from an electron synchrotron.

Linac **Linear accelerators.**   A Van de Graaff accelerator is essentially a linear accelerator since the ions move in a linear path. But the name *linear accelerator* is usually reserved for a more complex arrangement in which particles are accelerated many, many times along a straight-line path. Figure 32–5 is a diagram of a simple "linac." The ions pass through a series of tubular conductors. The voltage applied to the tubes must be alternating so that when positive ions (say) reach a gap, the tube in front of them is negative and the one they just left is positive. This assures that they are accelerated at each gap. As the ions increase in speed, they cover more distance in the same amount of time. Consequently, the tubes must be longer the farther they are from the source. Linear accelerators are of particular importance for accelerating electrons. Because of their small mass, electrons reach high speeds very quickly; an electron linac such as the one shown in Fig. 32–5a would have tubes nearly equal in length, since the electrons would be traveling close to $c = 3.0 \times 10^8$ m/s for almost the entire distance. The amount of energy radiated away by electrons in a linear machine is much less than for a circular machine (see mention of synchrotron radiation above). The largest electron linear accelerator is that at Stanford (Stanford Linear Accelerator Center, or SLAC), Fig. 32–5b. It is over 3 km (2 miles) long and can accelerate electrons to 50 GeV (recall 1 GeV = 10^9 eV). Many hospitals have 10-MeV electron linacs that produce photons to irradiate tumors.

Colliders **Colliding beams.**   The typical way in which high-energy-physics experiments are carried out is to allow the beam of particles from an accelerator to strike a stationary target. An alternative method, which increases the energy of a collision, is to use **colliding beams**—that is, the target particles as well as the projectile particles are moving. The two

FIGURE 32–5   (a) Diagram of a simple linear accelerator. (b) Photo of the Stanford Linear Accelerator (SLAC) in California.

(a)

(b)

FIGURE 32–6 The 8.5 km diameter LEP collider, about 100 m below ground, straddles the French–Swiss border near Geneva.

beams of particles are made to collide head-on. One way to accomplish this with only one accelerator is through the use of **storage rings**. The accelerator accelerates one type of particle (say electrons or protons) to a maximum energy and then magnets are used to steer these particles into a circular storage ring where the particles can continue to circulate for many hours. It then accelerates a second type of particle (say positrons or antiprotons), or another group of the first type (such as protons). The two beams are directed so they intersect and collide head-on with each other. Colliding-beam experiments are in use at a number of facilities around the world, including CERN, Fermilab, and the Stanford Linear Collider (SLC), and they have played an important role in recent advances in elementary particle physics. For example, in the recent experiments that provided strong evidence for the top quark (see chapter opening photo and Section 32–9), the Fermilab Tevatron accelerated protons and antiprotons each to 900 GeV, so that the combined energy of head-on collisions was 1.8 TeV. The largest collider accelerator today is the truly enormous Large Electron-Positron (LEP) Collider at CERN, which has a circumference of 26.7 km (Fig. 32–6). Inaugurated in 1989, it produces oppositely revolving beams of e^+ and e^-, each of energy 93 GeV, for a total interaction energy of 186 GeV.

32–3 Beginnings of Elementary Particle Physics—the Yukawa Particle

By the mid-1930s, it was recognized that all atoms can be considered to be made up of neutrons, protons, and electrons. The basic constituents of the universe were no longer considered to be atoms but rather the proton, neutron, and electron. Besides these three *elementary particles*, several others were also known: the positron (a positive electron), the neutrino, and the γ particle (or photon), for a total of six elementary particles.

In the decades that followed, hundreds of other subnuclear particles were discovered. The properties and interactions of these particles, and which ones should be considered as fundamental or "elementary," became the substance of research in **elementary particle physics**. Today, many of these particles are considered to be made up of two or three more fundamental entities called *quarks*. But some of the early-discovered particles are also considered elementary, as we shall see.

(a) Repulsive force (children throwing pillows)

(b) Attractive force (children grabbing pillows from each other's hands)

FIGURE 32–7 Forces equivalent to particle exchange. (a) Repulsive force (children throwing pillows at each other). (b) Attractive force (children grabbing pillows from each other's hands).

Feynman diagram

QED

Particles that mediate or "carry" forces

Elementary particle physics, as it exists today, can be said to have begun in 1935 when the Japanese physicist Hideki Yukawa (1907–1981) predicted the existence of a new particle that would in some way mediate the strong nuclear force. To understand Yukawa's idea, we first look at the electromagnetic force. When we first discussed electricity, we saw that the electric force acts over a distance, without contact. To better perceive how a force can act over a distance, we saw that Faraday introduced the idea of a *field*. The force that one charged particle exerts on a second can be said to be due to the electric field set up by the first. Similarly, the magnetic field can be said to carry the magnetic force. Later (Chapter 22), we saw that electromagnetic fields can travel through space as waves. Finally, in Chapter 27, we saw that electromagnetic radiation (light) can be considered as either a wave or as a collection of particles called *photons*. Because of this wave–particle duality, it is possible to imagine that the electromagnetic force between charged particles is due (1) to the EM field set up by one charged particle and felt by the other, or (2) to an exchange of photons or γ particles between them. It is (2) that we want to concentrate on here, and a crude analogy for how an exchange of particles could give rise to a force is suggested in Fig. 32–7. In part (a), two children start throwing heavy pillows at each other; each catch results in the child being pushed backward by the impulse. This is the equivalent of a repulsive force. On the other hand, if the two children exchange pillows by grabbing them out of the other person's hand, they will be pulled toward each other, as when an attractive force acts.

For the electromagnetic force between two charged particles, it is photons that are exchanged between the two particles that give rise to the force. A simple diagram describing this photon exchange is shown in Fig. 32–8. Such a diagram, called a **Feynman diagram** (after its inventor, the American physicist, Richard Feynman (1918–1988)), is based on the theory called **quantum electrodynamics** (QED). Figure 32–8 represents the simplest case, in which a single photon is exchanged. One of the charged particles emits the photon and recoils somewhat as a result; and the second particle absorbs the photon. In any collision or *interaction*, energy and momentum are transferred from one particle to the other, carried by the photon. Because the photon is absorbed by the second particle very shortly after it is emitted by the first, it is not observable, and is referred to as a *virtual* photon, in contrast to one that is free and can be detected by instruments. The photon is said to *mediate*, or *carry*, the electromagnetic force.

Now to Yukawa's prediction. By analogy with photon exchange that mediates the electromagnetic force, Yukawa argued that there ought to be a particle that mediates the strong nuclear force—the force that holds nucleons together in the nucleus. Just as the photon is called the quantum of the electromagnetic field or force, so the Yukawa particle would represent the quantum of the strong nuclear force.

FIGURE 32–8 Feynman diagram showing a photon acting as the carrier of the electromagnetic force between two electrons. This is sort of an *x* vs. *t* graph, with *t* increasing upward. Starting at the bottom, two electrons approach each other (the distance between them decreases in time). As they get close, momentum and energy get transferred from one to the other, carried by a photon (or, perhaps, by more than one), and the two electrons bounce apart.

Yukawa predicted that this new particle would have a mass intermediate between that of the electron and the proton. Hence it was called a **meson**, meaning "in the middle," and Fig. 32–9 is a Feynman diagram of meson exchange. We can make a rough estimate of the mass of the meson as follows. Suppose the proton on the left in Fig. 32–9 is at rest. For it to emit a meson would require energy (to make the mass) that would have to come from nowhere; such a process would violate conservation of energy. But the uncertainty principle allows nonconservation of energy by an amount ΔE if it occurs only for a time Δt given by $(\Delta E)(\Delta t) \approx h/2\pi$. We set ΔE equal to the energy needed to create the mass m of the meson: $\Delta E = mc^2$. Now conservation of energy is violated only as long as the meson exists, which is the time Δt required for the meson to pass from one nucleon to the other. If we assume the meson travels at relativistic speed, close to the speed of light c, then Δt need be at most about $\Delta t = d/c$, where d is the maximum distance that can separate the interacting nucleons. Thus we have

FIGURE 32–9 Meson exchange when a proton and neutron interact via the strong nuclear force.

$$\Delta E \, \Delta t \approx \frac{h}{2\pi}$$

$$mc^2\left(\frac{d}{c}\right) \approx \frac{h}{2\pi}$$

or

$$mc^2 \approx \frac{hc}{2\pi d}. \tag{32–2}$$

The range of the strong nuclear force (the maximum distance away it can be felt), is small—not much more than the size of a nucleon or small nucleus (see Eq. 30–1); so let us take $d \approx 1.5 \times 10^{-15}$ m, and then from Eq. 32–2,

$$mc^2 \approx \frac{hc}{2\pi d} = \frac{(6.6 \times 10^{-34} \, \text{J·s})(3.0 \times 10^8 \, \text{m/s})}{(6.28)(1.5 \times 10^{-15} \, \text{m})}$$

$$\approx 2.2 \times 10^{-11} \, \text{J} = 130 \, \text{MeV}.$$

The mass of the predicted meson is thus very roughly $130 \, \text{MeV}/c^2$ or about 250 times the electron mass of $0.51 \, \text{MeV}/c^2$. [Note, incidentally, that since the electromagnetic force has infinite range ($d = \infty$), Eq. 32–2 tells us that the exchanged particle for the electromagnetic force, the photon, will have zero rest mass.]

Just as photons can be observed as free particles, as well as acting in an exchange, so it was expected that mesons might be observed directly. Such a meson was searched for in the cosmic radiation that enters the Earth's atmosphere from the Sun and other sources in the universe. In 1937 a new particle was discovered whose mass was $106 \, \text{MeV}/c^2$ (207 times the electron mass). This is close to the mass predicted. But it turned out that this new particle, called the **muon**, did not interact strongly with matter. It could hardly mediate the strong nuclear force if it didn't interact via the strong nuclear force. Thus the muon, which can have either a positive or a negative charge and seems to be nothing more than a very massive electron, is not the Yukawa particle.

Muon

The particle predicted by Yukawa was finally discovered in cosmic rays by C. F. Powell and G. Occhialini in 1947. It is called the "π" or pi meson, or simply the **pion**. It comes in three charge states: $+$, $-$, or 0. The π^+ and π^- have mass of $139.6 \, \text{MeV}/c^2$ and the π^0 a mass of $135.0 \, \text{MeV}/c^2$. All three

Pion

(a)

(b)

FIGURE 32–10
(a) Computer reconstruction of a Z-particle decay into an electron and a positron ($Z^0 \rightarrow e^+e^-$) whose tracks are shown in white, which took place in the UA1 detector at CERN. (b) Photo of the UA1 detector at CERN as it was being built.

interact strongly with matter. Reactions observed in the laboratory, using a particle accelerator, include

$$p + p \rightarrow p + p + \pi^0,$$
$$p + p \rightarrow p + n + \pi^+. \quad \textbf{(32–3)}$$

The incident proton from the accelerator must have sufficient energy to produce the additional mass of the pion. A number of other mesons were discovered in subsequent years which were also considered to mediate the strong nuclear force. The recent theory of quantum chromodynamics, however, which involves quarks, has replaced mesons with *gluons* as the basic carriers of the strong force, as we shall discuss in Section 32–10.

So far we have discussed the particles that mediate the electromagnetic and strong nuclear forces. But there are four known types of force—or inter-action—in nature. What about the other two: the weak nuclear force, and gravity? Theorists believe that these are also mediated by particles. The par-ticles presumed to transmit the weak force are referred to as the W^+, W^-, and Z^0, and were only detected in 1983 (see Fig. 32–10). The quantum (or carrier) of the gravitational force, called the **graviton**, has not yet been found. A comparison of the four forces is given in Table 32–1, where they are listed according to their (approximate) relative strengths. Notice that although gravity may be the most obvious force in daily life (because of the huge mass of the Earth), on a nuclear scale, it is much the weakest of the four forces and its effect at the nuclear or atomic level can nearly always be ignored.

TABLE 32–1   The Four Forces in Nature

Type	Relative Strength (approx., for 2 protons in nucleus)	Field Particle
Strong nuclear	1	Gluons[†] (mesons)
Electromagnetic	10^{-2}	Photon
Weak nuclear	10^{-6}	$W^\pm$ and Z^0
Gravitational	10^{-38}	Graviton (?)

[†]Until the 1970s, thought to be mesons, now gluons (see Section 32–10).

32–4 Particles and Antiparticles

The positron, as we have seen, is basically a positive electron. That is, many of its properties are the same as for the electron, such as mass, but it has the opposite charge. The positron is said to be the **antiparticle** to the electron. After the positron was discovered in 1932, it was predicted that other particles also ought to have antiparticles. In 1955 the antiparticle to the proton was found, the **antiproton** ($\overline{p}$), which carries a negative charge; see Fig. 32–11. (The bar over the p is used to indicate antiparticle.) Soon after, the antineutron ($\overline{n}$) was found. Most other particles also have antiparticles. But the photon, the π^0, and a few other particles do not have distinct antiparticles—or we say that they are their own antiparticles.

Antiparticles are produced in nuclear reactions when there is sufficient energy available, and they do not live very long in the presence of matter. For example, a positron is stable when by itself, but if it encounters an electron, the two annihilate each other. The energy of their vanished mass, plus any kinetic energy they possessed, is converted to the energy of γ rays or of other particles. Annihilation also occurs for all other particle–antiparticle pairs.

32–5 Particle Interactions and Conservation Laws

One of the important uses of high-energy accelerators is to study the interactions of elementary particles with each other. As a means of ordering this subnuclear world, the conservation laws are indispensable. The laws of conservation of energy, of momentum, of angular momentum, and of electric charge are found to hold precisely in all particle interactions.

A study of particle interactions has revealed a number of new conservation laws which (just like the old ones) are ordering principles: they help to explain why some reactions occur and others do not. For example, the following reaction has never been found to occur:

$$p + n \not\rightarrow p + p + \overline{p}$$

even though charge, energy, and so on, are conserved ($\overline{p}$ means an antiproton and $\not\rightarrow$ means the reaction does not occur). To understand why such a reaction doesn't occur, physicists hypothesized a new conservation law, the conservation of **baryon number**. (Baryon number is a generalization of nucleon number, which we saw earlier is conserved in nuclear reactions and decays.) An important addition to this law is the proposal that whereas all nucleons have baryon number $B = +1$, all antinucleons (antiprotons, antineutrons) have $B = -1$. The reaction above does not conserve baryon number since on the left side we have $B = (+1) + (+1) = +2$, and on the right $B = (+1) + (+1) + (-1) = +1$. On the other hand, the following reaction does conserve B and *does* occur if the incoming proton has sufficient energy:

$$p + n \rightarrow p + n + \overline{p} + p,$$

$$B = +1 + 1 = +1 + 1 - 1 + 1.$$

As indicated, $B = +2$ on both sides of this equation. From these and other reactions, the conservation of baryon number has been established as a basic law of physics.

FIGURE 32–11 Liquid-hydrogen bubble-chamber photograph of an antiproton ($\overline{p}$) colliding with a proton, producing a Xi–anti-Xi pair ($\overline{p} + p \rightarrow \Xi^- + \overline{\Xi}^+$) that subsequently decay into other particles. The drawing indicates the assignment of particles to each track, which is based on how or if that particle decays, and on mass values estimated from measurement of momentum (curvature of track in magnetic field) and energy (heaviness of track, for example). Neutral particle paths are shown by dashed lines since neutral particles produce no bubbles and hence no tracks.

Also useful are the conservation laws for the three **lepton numbers**, associated with weak interactions including decays. In ordinary β decay, an electron or positron is emitted along with a neutrino or antineutrino. In a similar type of decay, a muon is emitted instead of an electron. The neutrino (ν_e) that accompanies an emitted electron is found to be different from the neutrino (ν_μ) that accompanies an emitted muon. Each of these neutrinos has an antiparticle: $\bar{\nu}_e$ and $\bar{\nu}_\mu$. In ordinary β decay we have, for example,

$$n \rightarrow p + e^- + \bar{\nu}_e$$

but never $n \rightarrow p + e^- + \bar{\nu}_\mu$ or $n \rightarrow p + e^- + \bar{\nu}_e + \nu_e$. To explain why these do not occur, the concept of electron lepton number, L_e, was invented. If the electron (e^-) and the electron neutrino (ν_e) are assigned $L_e = +1$, and e^+ and $\bar{\nu}_e$ are assigned $L_e = -1$, whereas all other particles have $L_e = 0$, then all observed decays conserve L_e. For example, in $n \rightarrow p + e^- + \bar{\nu}_e$, $L_e = 0$ initially, and $L_e = 0 + (+1) + (-1) = 0$ after the decay. Decays that do not conserve L_e but would obey the other conservation laws are not observed to occur. Hence it is believed that L_e is conserved in all interactions.

In a decay involving muons, such as

$$\pi^+ \rightarrow \mu^+ + \nu_\mu,$$

a second quantum number, muon lepton number (L_μ), is conserved. The μ^- and ν_μ are assigned $L_\mu = +1$, and μ^+ and $\bar{\nu}_\mu$ have $L_\mu = -1$, whereas other particles have $L_\mu = 0$. It is believed that L_μ is also conserved in all interactions or decays. Similar assignments can be made for a third lepton number, L_τ, associated with the recently discovered τ lepton and its neutrino, ν_τ.

Keep in mind that antiparticles have not only opposite electric charge from their particles, but also opposite B, L_e, L_μ, and L_τ.

CONCEPTUAL EXAMPLE 32–3 | **Lepton number in muon decay.** Which of the following decay schemes is possible for muon decay: (*a*) $\mu^- \rightarrow e^- + \bar{\nu}_e$; (*b*) $\mu^- \rightarrow e^- + \bar{\nu}_e + \nu_\mu$; (*c*) $\mu^- \rightarrow e^- + \nu_e$? All of these particles have $L_\tau = 0$.

RESPONSE A μ^- has $L_\mu = +1$ and $L_e = 0$. This is the initial state, and the final state (after decay) must also have $L_\mu = +1$, $L_e = 0$. In (*a*), the final state has $L_\mu = 0 + 0 = 0$, and $L_e = +1 - 1 = 0$; L_μ would not be conserved and indeed this decay is not observed to occur. The final state of (*b*) has $L_\mu = 0 + 0 + 1 = +1$ and $L_e = +1 - 1 + 0 = 0$, so both L_μ and L_e are conserved. This is in fact the most common decay mode of the μ^-. Finally, (*c*) does not occur because L_e ($= +2$ in final state) is not conserved, nor is L_μ.

32–6 Particle Classification

In the decades following the discovery of the π meson in the late 1940s, a great many other subnuclear particles were discovered. Today they number in the hundreds. Much theoretical and experimental work has been done to try to understand this multitude of particles. An important aid to understanding is to arrange the particles in categories according to their

TABLE 32–2 Particles (stable under strong decay)[†]

Category	Particle Name	Symbol	Anti-particle	Spin	Rest Mass (MeV/c^2)	B	L_e	L_μ	L_τ	S	Lifetime (s)	Principal Decay Modes
Gauge bosons	Photon	γ	Self	1	0	0	0	0	0	0	Stable	
	W	W^+	W^-	1	80.33×10^3	0	0	0	0	0	3×10^{-25}	$e\nu_e, \mu\nu_\mu, \tau\nu_\tau$, hadrons
	Z	Z^0	Self	1	91.19×10^3	0	0	0	0	0	3×10^{-25}	$e^+e^-, \mu^+\mu^-, \tau^+\tau^-$, hadrons
Leptons	Electron	e^-	e^+	$\frac{1}{2}$	0.511	0	+1	0	0	0	Stable	
	Neutrino (e)	ν_e	$\bar{\nu}_e$	$\frac{1}{2}$	$0(<7.0 \times 10^{-6})$[‡]	0	+1	0	0	0	Stable	
	Muon	μ^-	μ^+	$\frac{1}{2}$	105.7	0	0	+1	0	0	2.20×10^{-6}	$e^- \bar{\nu}_e \nu_\mu$
	Neutrino (μ)	ν_μ	$\bar{\nu}_\mu$	$\frac{1}{2}$	$0(<0.17)$[‡]	0	0	+1	0	0	Stable	
	Tau	τ^-	τ^+	$\frac{1}{2}$	1777	0	0	0	+1	0	2.91×10^{-13}	$\mu^- \bar{\nu}_\mu \nu_\tau, e^- \bar{\nu}_e \nu_\tau$, hadrons $+ \nu_\tau$
	Neutrino (τ)	ν_τ	$\bar{\nu}_\tau$	$\frac{1}{2}$	$0(<24)$[‡]	0	0	0	+1	0	Stable	
Hadrons (selected)												
Mesons	Pion	π^+	π^-	0	139.6	0	0	0	0	0	2.60×10^{-8}	$\mu^+ \nu_\mu$
		π^0	Self	0	135.0	0	0	0	0	0	0.84×10^{-16}	2γ
	Kaon	K^+	K^-	0	493.7	0	0	0	0	+1	1.24×10^{-8}	$\mu^+ \nu_\mu, \pi^+ \pi^0$
		K^0_S	$\bar{K}^0_S$	0	497.7	0	0	0	0	+1	0.89×10^{-10}	$\pi^+ \pi^-, 2\pi^0$
		K^0_L	$\bar{K}^0_L$	0	497.7	0	0	0	0	+1	5.17×10^{-8}	$\pi^\pm e^\mp \overset{(-)}{\nu_e}, \pi^\pm \mu^\mp \overset{(-)}{\nu_\mu}, 3\pi$
	Eta	η^0	Self	0	547.5	0	0	0	0	0	5×10^{-19}	$2\gamma, 3\pi^0, \pi^+ \pi^- \pi^0$
	and others											
Baryons	Proton	p	$\bar{p}$	$\frac{1}{2}$	938.3	+1	0	0	0	0	Stable	
	Neutron	n	$\bar{n}$	$\frac{1}{2}$	939.6	+1	0	0	0	0	887	$pe^- \bar{\nu}_e$
	Lambda	Λ^0	$\bar{\Lambda}^0$	$\frac{1}{2}$	1115.7	+1	0	0	0	-1	2.63×10^{-10}	$p\pi^-, n\pi^0$
	Sigma	Σ^+	$\bar{\Sigma}^-$	$\frac{1}{2}$	1189.4	+1	0	0	0	-1	0.80×10^{-10}	$p\pi^0, n\pi^+$
		Σ^0	$\bar{\Sigma}^0$	$\frac{1}{2}$	1192.6	+1	0	0	0	-1	7.4×10^{-20}	$\Lambda^0 \gamma$
		Σ^-	$\bar{\Sigma}^+$	$\frac{1}{2}$	1197.4	+1	0	0	0	-1	1.48×10^{-10}	$n\pi^-$
	Xi	Ξ^0	$\bar{\Xi}^0$	$\frac{1}{2}$	1314.9	+1	0	0	0	-2	2.90×10^{-10}	$\Lambda^0 \pi^0$
		Ξ^-	Ξ^+	$\frac{1}{2}$	1321.3	+1	0	0	0	-2	1.64×10^{-10}	$\Lambda^0 \pi^-$
	Omega	Ω^-	Ω^+	$\frac{3}{2}$	1672.5	+1	0	0	0	-3	0.82×10^{-10}	$\Xi^0 \pi^-, \Lambda^0 K^-, \Xi^- \pi^0$
	and others											

[†]See also Table 32–4 for particles with charm and bottomness.

[‡]Experimental upper limits on neutrino masses are given in parentheses.

properties. One way of doing this is according to their interactions. Since not all particles interact by means of all four of the forces known in nature (though all interact via gravity), this is used as a classification scheme. Table 32–2 lists some of the more common particles classified in this way along with many of their properties. The particles listed are those that are stable, and many that are unstable. At the top of the table are the **gauge bosons** (so-named[†] after the theory that describes them, "gauge theory"), which include the *photon*, and the W and Z particles, that carry the electromagnetic and weak interactions, respectively.

Gauge bosons

Next in Table 32–2 are the **leptons**, which are particles that do not interact via the strong force but do interact via the weak nuclear force (as well as the much weaker gravitational force); those that carry electric charge also interact via the electromagnetic force. The leptons include the electron, the muon, and the tau (or τ, discovered in 1976 and more than 3000 times heavier than the electron), and three types of neutrino: the electron neutrino (ν_e), the muon neutrino (ν_μ), and the tau neutrino (ν_τ). They each have antiparticles, as indicated in Table 32–2.

Leptons

[†]Bosons are particles that are not governed by the Pauli exclusion principle (see Sections 28–7 and 32–10).

The third category of particle in Table 32–2 is the **hadron**. Hadrons are those particles that interact via the strong nuclear force. Hence they are said to be **strongly interacting particles**. They also interact via the other forces, but the strong force predominates at short distances. The hadrons include nucleons, pions, and a large number of other particles.

They are divided into two subgroups:[†] **baryons**, which are those particles that have baryon number $+1$ (or -1 in the case of their antiparticles); and

mesons, which have baryon number $= 0$.

Notice that the baryons Λ, Σ, Ξ, and Ω all decay to lighter-mass baryons, and eventually to a proton or neutron. All these processes conserve baryon number. Since there is no lighter particle than the proton with $B = +1$, if baryon number is strictly conserved, the proton itself cannot decay and is stable.

32–7 Particle Stability and Resonances

Many of the particles listed in Table 32–2 are unstable. The lifetime of an unstable particle depends on which force is most active in causing the decay. When we say the strong nuclear force is stronger than the electromagnetic, we mean that two particles will interact more strongly and more quickly if this force is acting. When a stronger force influences a decay, that decay occurs more quickly. Decays caused by the weak force typically have lifetimes of 10^{-13} s or longer. Particles that decay via the electromagnetic force have much shorter lifetimes, typically about 10^{-16} to 10^{-19} s. (Exceptions to this scheme are the W and Z particles, which decay via the weak interaction but have very short lifetimes due to their special nature as exchange particles.) The unstable particles listed in Table 32–2 decay either via the weak or the electromagnetic interaction. Decays that involve a γ (photon) are electromagnetic (such as $\pi^0 \rightarrow 2\gamma$). The other decays shown take place via the weak interaction, often accompanied by a neutrino which interacts only via the weak interaction: examples are $\pi^+ \rightarrow \mu^+ \nu_\mu$ and $\Sigma^- \rightarrow n\pi^-$.

A great many particles have been found that decay via the strong interaction, and these are not listed in Table 32–2. Such particles decay into other strongly interacting particles (say, n, p, π, but not involving γ, e, ν, and so on) and their lifetimes are very short, typically about 10^{-23} s. In fact their

lifetimes are so short that they do not travel far enough to be detected before decaying. Their decay products can be detected, however, and it is from them that the existence of such short-lived particles is inferred. To see how this is done, we consider the first such particle discovered (by Fermi). Fermi used a beam of π^+ directed through a hydrogen target (protons) with varying amounts of energy. A graph of the number of interactions (π^+ scattered) versus the pion's kinetic energy is shown in Fig. 32–12. The large

---

[†]Originally, particles were divided according to their mass into leptons (meaning light particles), baryons (meaning "heavy"), and those of intermediate mass, the mesons (meaning "middle"). The newer classification according to their interactions is not always consistent with this. For example, the J/ψ particles are very heavy but have $B = 0$ and so are classified as mesons (Section 32–9), and the τ is a lepton although it is heavier than some baryons.

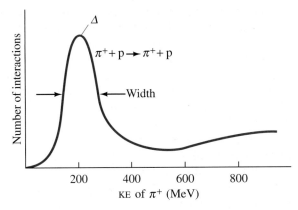

FIGURE 32–12 Number of π^+ particles scattered by a proton target as a function of the incident π^+ kinetic energy.

peak around 200 MeV was much higher than expected and certainly much higher than the number of interactions at neighboring energies. This led Fermi to conclude that the π^+ and proton combined momentarily to form a short-lived particle before coming apart again, or at least that they resonated together for a short time. Indeed, the large peak in Fig. 32–12 resembles a resonance curve (see Figs. 11–18 and 21–38), and this new "particle"—now called the Δ—is referred to as a **resonance**. Hundreds of other resonances have been found in a similar way. Many resonances are regarded as excited states of other particles such as of the nucleon (proton or neutron).

Resonance

The width of a resonance—in Fig. 32–12 the width of the Δ peak is on the order of 100 MeV—is an interesting application of the uncertainty principle. If a particle lives only 10^{-23} s, then its mass (i.e., its rest energy) will be uncertain by an amount $\Delta E \approx h/2\pi\Delta t \approx (6.6 \times 10^{-34}\,\text{J·s})/(6)(10^{-23}\,\text{s}) \approx 10^{-11}\,\text{J} \approx 100$ MeV, which is what is observed. Actually, the lifetimes of $\approx 10^{-23}$ s for such resonances are inferred by the reverse process: from the measured width being ≈ 100 MeV.

32–8 Strange Particles

In the early 1950s, certain of the newly found particles, namely, the K, Λ, and Σ, were found to behave rather strangely in two ways. First, they were always produced in pairs. For example, the reaction

$$\pi^- + p \rightarrow K^0 + \Lambda^0$$

occurred with high probability, but the reaction $\pi^- + p \rightarrow K^0 + n$ was never observed to occur. This seemed strange because the unobserved reaction would not have violated any known conservation law, and plenty of energy was available. The second feature of these **strange particles** (as they came to be called) was that, although they were clearly produced via the strong interaction (that is, at a high rate), they did not decay at a rate characteristic of the strong interaction even though they decayed into strongly interacting particles (for example, $K \rightarrow 2\pi$, $\Sigma^+ \rightarrow p + \pi^0$). Instead of lifetimes of 10^{-23} s as expected for strongly interacting particles, strange particles have lifetimes of 10^{-10} to 10^{-8} s, which are characteristic of the weak interaction.

To explain these observations, a new quantum number, **strangeness**, and a new conservation law, conservation of strangeness, were introduced. By assigning the strangeness numbers (S) indicated in Table 32–2, the production of strange particles in pairs was readily explained. Antiparticles were assigned opposite strangeness from their particles: one of each pair was assigned $S = +1$ and the other $S = -1$ (see Table 32–2). For example, in the reaction $\pi^- + \mathrm{p} \rightarrow \mathrm{K}^0 + \Lambda^0$, the initial state has strangeness $S = 0 + 0 = 0$, and the final state has $S = +1 - 1 = 0$, so strangeness is conserved. But for $\pi^- + \mathrm{p} \rightarrow \mathrm{K}^0 + \mathrm{n}$, the initial state has $S = 0$ and the final state has $S = +1 + 0 = +1$, so strangeness would not be conserved; and this reaction isn't observed.

Strangeness and its conservation

To explain the decay of strange particles, it is assumed that strangeness is conserved in the strong interaction but is *not* conserved in the weak interaction. Thus, although strange particles were forbidden by strangeness conservation to decay to lower-mass nonstrange particles via the strong interaction, they could undergo such decay by means of the weak interaction. This would occur much more slowly, of course, which accounts for their longer lifetimes of 10^{-10} to 10^{-8} s.

Strangeness conserved in strong interactions but not in weak

The conservation of strangeness was the first example of a "partially conserved" quantity. In this case, the quantity strangeness is conserved by strong interactions but not by weak.

32–9 Quarks

We saw in our discussion of Table 32–2 that all particles, except the gauge bosons, fall into either of two categories: leptons and hadrons. The principal difference between these two groups is that the hadrons interact via the strong interaction, whereas the leptons do not. Another important difference that physicists had to deal with in the 1960s was that there were

TABLE 32–3 Properties of Quarks and Antiquarks

				Quarks				
Name	**Symbol**	**Spin**	**Charge**	**Baryon Number**	**Strangeness**	**Charm**	**Bottomness**	**Topness**
Up	u	$\frac{1}{2}$	$+\frac{2}{3}e$	$\frac{1}{3}$	0	0	0	0
Down	d	$\frac{1}{2}$	$-\frac{1}{3}e$	$\frac{1}{3}$	0	0	0	0
Strange	s	$\frac{1}{2}$	$-\frac{1}{3}e$	$\frac{1}{3}$	-1	0	0	0
Charmed	c	$\frac{1}{2}$	$+\frac{2}{3}e$	$\frac{1}{3}$	0	$+1$	0	0
Bottom	b	$\frac{1}{2}$	$-\frac{1}{3}e$	$\frac{1}{3}$	0	0	-1	0
Top	t	$\frac{1}{2}$	$+\frac{2}{3}e$	$\frac{1}{3}$	0	0	0	$+1$
				Antiquarks				
Name	**Symbol**	**Spin**	**Charge**	**Baryon Number**	**Strangeness**	**Charm**	**Bottomness**	**Topness**
Up	$\bar{u}$	$\frac{1}{2}$	$-\frac{2}{3}e$	$-\frac{1}{3}$	0	0	0	0
Down	$\bar{d}$	$\frac{1}{2}$	$+\frac{1}{3}e$	$-\frac{1}{3}$	0	0	0	0
Strange	$\bar{s}$	$\frac{1}{2}$	$+\frac{1}{3}e$	$-\frac{1}{3}$	$+1$	0	0	0
Charmed	$\bar{c}$	$\frac{1}{2}$	$-\frac{2}{3}e$	$-\frac{1}{3}$	0	-1	0	0
Bottom	$\bar{b}$	$\frac{1}{2}$	$+\frac{1}{3}e$	$-\frac{1}{3}$	0	0	$+1$	0
Top	$\bar{t}$	$\frac{1}{2}$	$-\frac{2}{3}e$	$-\frac{1}{3}$	0	0	0	-1

only four known leptons (e^-, μ^-, ν_e, ν_μ; the τ and ν_τ were not yet discovered), but there were well over a hundred hadrons.

The leptons are considered to be truly elementary particles since they do not seem to break down into smaller entities, do not show any internal structure, and have no measurable size. (Attempts to determine the size of leptons have put an upper limit of about 10^{-18} m.)

The hadrons, on the other hand, are more complex. Experiments indicate they do have an internal structure. And the fact that there are so many of them suggests that they can't all be elementary. To deal with this problem, M. Gell-Mann and G. Zweig in 1963 independently proposed that none of the hadrons so far observed, not even the proton and neutron, was elementary. Instead, they proposed that the hadrons are made up of combinations of three, more fundamental, pointlike entities called **quarks**.[†] Quarks, then, would be considered truly elementary particles, like leptons. The three quarks were labeled u, d, s, and given the names *up*, *down*, and *sideways* (or, more commonly now, *strange*). They were assumed to have fractional charge ($\frac{1}{3}$ or $\frac{2}{3}$ the charge on the electron—that is, less than the previously thought smallest charge). Other properties of quarks and antiquarks are indicated in Table 32–3. All hadrons known at the time could be constructed in theory from these three types of quark. Mesons would consist of a quark–antiquark pair. For example, a π^+ meson is considered a u$\overline{d}$ pair (note that for the u$\overline{d}$ pair, $Q = \frac{2}{3}e + \frac{1}{3}e = +1e$, $B = \frac{1}{3} - \frac{1}{3} = 0$, $S = 0 + 0 = 0$, as they must for a π^+); and a $K^+ = u\overline{s}$, with $Q = +1$, $B = 0$, $S = +1$. Baryons, on the other hand, would consist of three quarks. For example, a neutron is n = ddu, whereas an antiproton is $\overline{p} = \overline{u}\overline{u}\overline{d}$. See Fig. 32–13.

Soon after the quark theory was proposed, physicists began looking for these fractionally charged particles, but direct detection has not been successful. Indeed, it may be that quarks are so tightly bound together that they may not ever exist singly in the free state.

In 1964 several physicists proposed that there ought to be a fourth quark. Their argument was based on the expectation that there exists a deep symmetry in nature, including a connection between quarks and leptons. If there are four leptons (as was thought in the 1960s), then symmetry in nature would suggest there should also be four quarks. The fourth quark was said to be *charmed*. Its charge would be $+\frac{2}{3}e$ and it would have another property to distinguish it from the other three quarks. This new property, or quantum number, was called **charm** (see Table 32–3). Charm was assumed to be like strangeness: it would be conserved in strong and electromagnetic interactions, but would not be conserved by the weak. The new charmed quark would have charm $C = +1$ and its antiquark $C = -1$. The first charmed particle, the J/ψ meson, was discovered in 1974.

Also in the 1970s strong evidence appeared for the tau (τ) lepton, with a mass of 1777 MeV/c^2. This lepton, like the electron and muon, presumably has a neutrino associated with it. Thus, the family of leptons is at present believed to have six members. This would upset the balance between leptons and quarks, the presumed basic building blocks of matter,

Hadrons can't all be elementary

Quarks

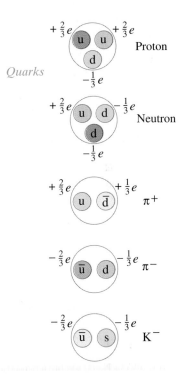

Charm

Charmed quark

Two more leptons

FIGURE 32–13 Quark compositions for several particles.

[†]Gell-Mann chose the word from a phrase in James Joyce's *Finnegans Wake*.

TABLE 32–4 Partial List of Hadrons Associated with Charm and Bottomness ($L_e = L_\mu = L_\tau = 0$)

Category	Particle	Anti-particle	Spin	Rest Mass (MeV/c^2)	Baryon Number	Strange-ness	Charm	Bottom-ness	Lifetime (s)	Principal Decay Modes
Mesons	D^+	D^-	0	1869.4	0	0	+1	0	10.6×10^{-13}	K + others, e + others
	D^0	$\overline{D}^0$	0	1864.6	0	0	+1	0	4.2×10^{-13}	K + others, μ or e + others
	D_S^+	$D_{\overline{S}}^-$	0	1969	0	+1	+1	0	4.7×10^{-13}	K + others
	J/ψ (3097)	Self	1	3096.9	0	0	0	0	0.8×10^{-20}	Hadrons, e^+e^-, $\mu^+\mu^-$
	Υ (9460)	Self	1	9460.4	0	0	0	0	1.3×10^{-20}	Hadrons, $\mu^+\mu^-$, e^+e^-, $\tau^+\tau^-$
	B^-	B^+	0	5279	0	0	0	-1	1.5×10^{-12}	D^0 + others
	B^0	$\overline{B}^0$	0	5279	0	0	0	-1	1.5×10^{-12}	D^0 + others
Baryons	Λ_c^+	Λ_c^-	$\tfrac{1}{2}$	2285	+1	0	+1	0	2.0×10^{-13}	Hadrons (e.g., Λ + others)
	Σ_c^{++}	Σ_c^{--}	$\tfrac{1}{2}$	2453	+1	0	+1	0	?	$\Lambda_c^+ \pi^+$
	Σ_c^+	Σ_c^-	$\tfrac{1}{2}$	2454	+1	0	+1	0	?	$\Lambda_c^+ \pi^0$
	Σ_c^0	$\overline{\Sigma}_c^0$	$\tfrac{1}{2}$	2452	+1	0	+1	0	?	$\Lambda_c^+ \pi^-$
	Λ_b^0	Λ_b^0	$\tfrac{1}{2}$	5640	+1	0	0	-1	1.1×10^{-12}	$J/\psi\Lambda^0$, $pD^0\pi^-$, $\Lambda_c^+\pi^+\pi^-\pi^-$

t and b quarks

unless two additional quarks also exist. Indeed, theoretical physicists postulated the existence of a fifth and sixth quark, named **top** and **bottom**. (Some physicists prefer the names *truth* and *beauty* for these t and b quarks.) The names apply also to the new properties (quantum numbers) that distinguish the new quarks from the old quarks (see Table 32–3), and which (like strangeness) are conserved in strong, but not weak, interactions. New mesons involving b quarks were soon detected (Table 32–4). Strong evidence for the top quark was not forthcoming until 1995 (see photo at start of this chapter) after years of searching. Its extremely high mass of almost 200 GeV/c^2 contributed to its elusiveness because of the huge energy required to produce it.

Today, the "truly" elementary particles are considered to be the six quarks, the six leptons, and the gauge bosons that carry the fundamental forces. See Table 32–5, where the quarks and leptons are arranged in three groups (generations).

TABLE 32–5 The Elementary Particles[†] as Seen Today

	First generation	Second generation	Third generation
Quarks	u, d	s, c	b, t
Leptons	e, ν_e	μ, ν_μ	τ, ν_τ
Gauge bosons	γ(photon)	$W^\pm$, Z^0	gluons

[†]Note that the quarks and leptons are arranged into three generations each, and the gauge particles are arranged in groups for the forces they mediate.

EXAMPLE 32–4 Quark combinations. Find the baryon number, charge, and strangeness for the following quark combinations and identify the hadron particle that is made up of these quark combinations: (*a*) udd, (*b*) u$\overline{u}$, (*c*) uss, (*d*) sdd, and (*e*) b$\overline{u}$.

SOLUTION We use Table 32–3 to get the properties of the quarks, then Table 32–2 or 32–4 to find the particle that has these properties. (a) udd has $Q = +\frac{2}{3}e - \frac{1}{3}e - \frac{1}{3}e = 0$, $B = \frac{1}{3} + \frac{1}{3} + \frac{1}{3} = 1$, $S = 0 + 0 + 0 = 0$, $C = 0$, bottomness $= 0$, topness $= 0$; the only baryon ($B = +1$) that has $Q = 0, S = 0$, etc., is the neutron (Table 32–2).

(b) u$\bar{u}$ has $Q = \frac{2}{3}e - \frac{2}{3}e = 0$, $B = 0$, and all other quantum numbers $= 0$. Sounds like a π^0.

(c) uss has $Q = 0, B = +1, S = -2$, others $= 0$. This is a Ξ^0.

(d) sdd has $Q = -1, B = +1, S = -1$, so must be a Σ^-.

(e) b$\bar{u}$ has $Q = -1$, $B = 0$, $S = 0$, $C = 0$, bottomness $= -1$, topness $= 0$. This must be a B$^-$ meson (Table 32–4).

32–10 The "Standard Model": Quantum Chromodynamics (QCD) and the Electroweak Theory

Not long after the quark theory was proposed, it was suggested that quarks have another property (or quality) called **color**. The distinction between the six quarks (u, d, s, c, b, t) was referred to as **flavor**. According to theory, each of the flavors of quark can have three colors, usually designated red, green, and blue. (These are the three primary colors which, when added together in equal amounts, as on a TV screen, produce white.) Note that the names "color" and "flavor" have nothing to do with our senses, but are purely whimsical—as are other names, such as charm, in this new field. (We did, however, "color" the quarks in Fig. 32–13.) The antiquarks are colored antired, antigreen, and antiblue. Baryons are made up of three quarks, one of each color. Mesons consist of a quark–antiquark pair of a particular color and its anticolor. Thus baryons can be said to be white, and mesons colorless.

Originally, the idea of quark color was proposed to preserve the Pauli exclusion principle (Section 28–7). Not all particles obey the exclusion principle. Those that do, such as electrons, protons, and neutrons, are called **fermions**. Those that don't are called **bosons**. These two categories are distinguished also in their spin (Section 28–6): bosons have integer spin (0, 1, 2, etc.) whereas fermions have half-integer spin, usually $\frac{1}{2}$, as for electrons and nucleons, but can also be $\frac{3}{2}, \frac{5}{2}$, etc. Matter is made up mainly of fermions, but the carriers of the forces (γ, W, Z, and gluons, as we'll see) are all bosons. Quarks are fermions (they have spin $\frac{1}{2}$) and therefore should obey the exclusion principle. Yet for three particular baryons (uuu, ddd, and sss), all three quarks would have the same quantum numbers, and at least two of them have their spin in the same direction (since there are only two choices, spin up [$m_s = +\frac{1}{2}$] or spin down [$m_s = -\frac{1}{2}$]). This would seem to violate the exclusion principle, but if quarks have an additional quantum number (color), which could be different for each quark, it would serve to distinguish them and the exclusion principle would hold. Although quark color, and the resulting threefold increase in the number of quarks, was thus originally an *ad hoc* idea, it also served to bring the theory into better agreement with experiment, such as predicting the correct lifetime of the π^0 meson. The idea of color soon became, in addition, a central feature of the theory as determining the force binding quarks together in a hadron. Each quark is assumed to carry a *color charge*, analogous to electric charge, and the strong force

FIGURE 32–14 (a) The force between two quarks holding them together as part of a proton, for example, is carried by a gluon, which in this case involves a change in color. (b) Strong interaction np → np with the exchange of a charged π meson (+ or −, depending on whether it is considered moving to the left or to the right). (c) Quark representation of the same interaction np → np. The wavy lines between quarks represent gluon exchanges holding the hadrons together.

(a) (b) (c)

QCD

between quarks is often referred to as the **color force**. This new theory of the strong force is called **quantum chromodynamics** (*chrome* = color in Greek), or **QCD**, to indicate that the force acts between color charges (and not between, say, electric charges). The strong force between two hadrons[†] is considered to be a force between the quarks that make them up, as suggested in Fig. 32–14. The particles that transmit the color force (analogous to photons for the EM force) are called **gluons** (a play on "glue"). They are included in Table 32–5. There are eight gluons, according to the theory, all massless, and six of them have color charge.[‡] Thus gluons have replaced mesons (Table 32–1) as the particles carrying the strong (color) force.

Gluons

The color force has the interesting property that its strength increases with increasing distance (as does the force exerted by a coiled spring, Chapter 11); as two quarks approach each other very closely (equivalently, have high energy), the force between them becomes very small. This aspect is referred to as **asymptotic freedom**.

The weak force, as we have seen, is thought to be mediated by the W^+, W^-, and Z^0 particles. It acts between the "weak charges" that each particle has. Each elementary particle can thus have electric charge, weak charge, color charge, and gravitational mass, although one or more of these could be zero. For example, all leptons have color charge of zero, so they do not interact via the strong force.

FIGURE 32–15 Quark representation of the Feynman diagram for beta decay of a neutron into a proton.

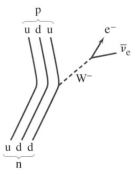

CONCEPTUAL EXAMPLE 32–5 **Beta decay.** Draw a Feynman diagram, showing what happens in beta decay using quarks.

RESPONSE Beta decay is a result of the weak interaction, and the mediator is either a $W^\pm$ or Z^0 particle. What happens, in part, is that a neutron (udd quarks) decays into a proton (uud). Apparently a d quark (charge $-\frac{1}{3}e$) has turned into a u quark (charge $+\frac{2}{3}e$). Charge conservation means that a negatively charged particle, namely a W^-, was emitted by the d quark. Since an electron and an antineutrino appear in the final state, they must have come from the decay of the virtual W^-, as shown in Fig. 32–15.

[†]The strong force between hadrons appears feeble, however, in comparison to the force directly between quarks within hadrons.

[‡]Compare to the EM interaction, where the photon has no electric charge. Because gluons have color charge, they could attract each other and form composite particles (photons cannot). Such "glueballs" are being searched for.

To summarize, the latest theories propose that the truly elementary particles (Table 32–5) are the leptons, the quarks, and the gauge bosons (photon, W and Z, and the gluons). Some theories suggest there may be other bosons as well. The photon and the leptons are observed in experiments, and finally so too were the W^+, W^-, and Z^0. But so far only combinations of quarks (baryons and mesons) have been observed, and it seems likely that free quarks and gluons are unobservable.

One important aspect of new theoretical work is the attempt to find a unified basis for the different forces in nature. This was a long-held hope of Einstein, which he was never able to fulfill. A so-called **gauge theory** that unifies the weak and electromagnetic interactions was put forward in the 1960s by S. Weinberg, S. Glashow, and A. Salam. In this **electroweak theory**, the weak and electromagnetic forces are seen as two different manifestations of a single, more fundamental, *electroweak* interaction. The electroweak theory has had many successes, including the prediction of the $W^\pm$ particles as carriers of the weak force, with masses of $81 \pm 2\,\text{GeV}/c^2$ in excellent agreement with the measured values of $80.33 \pm 0.15\,\text{GeV}/c^2$ (and similar accuracy for the Z^0). The electroweak theory plus QCD for the strong interaction are often referred to today as the **standard model**.

Electroweak theory

Standard model

Theoreticians have wondered why the W and Z have large masses rather than being massless like the photon. Electroweak theory suggests an explanation by means of a new **Higgs field** and its particle, the **Higgs boson**, which interact with the W and Z to "slow them down." In being forced to go slower than the speed of light, they must acquire mass.

32–11 Grand Unified Theories

With the success of the unified electroweak theory, attempts have recently been made to incorporate it and QCD for the strong (color) force into a so-called **grand unified theory** (GUT). One type of such a grand unified theory of the electromagnetic, weak, and strong forces has been worked out in which there is only one class of particle—leptons and quarks belong to the same family and are able to change freely from one type to the other—and the three forces are different aspects of a single underlying force. The unity is predicted to occur, however, only on a scale of less than about $10^{-30}\,\text{m}$. If two elementary particles (leptons or quarks) approach each other to within this **unification scale**, the apparently fundamental distinction between them would not exist at this level, and a quark could readily change to a lepton, or vice versa. Baryon and lepton numbers would not be conserved. The weak, electromagnetic, and strong (color) force would blend to a force of a single strength.

GUT

Unification of forces

How could a lepton become a quark, or vice versa? The theory predicts the existence of particles, called X bosons, that can be exchanged between a quark and a lepton allowing one to change into the other, somewhat like the charged pion exchanged between the p and n in Fig. 32–14b allows the proton on the right to become a neutron. The mass of the proposed X boson, consistent with the uncertainty principle as applied earlier in this chapter (see Eq. 32–2 with $d \approx 10^{-30}\,\text{m}$), would be about $10^{14}\,\text{GeV}/c^2$, or 10^{14} times the proton mass. With such an incredibly large mass, there is little hope of seeing them in the laboratory. It is also this huge mass that would keep baryon and lepton numbers conserved in observed reactions, since the likelihood of producing such a massive particle, even as a virtual exchange particle, is extremely small at even the highest laboratory energies.

What happens between the unification distance of 10^{-30} m and more normal (larger) distances is referred to as **symmetry breaking**. As an analogy, consider an atom in a crystal. Deep within the atom, there is much symmetry—in the innermost regions the electron cloud is spherically symmetric (Chapter 28). Farther out, this symmetry breaks down—the electron clouds are distributed preferentially along the lines (bonds) joining the atoms in the crystal. In a similar way, at 10^{-30} m the force between elementary particles is theorized to be a single force—it is symmetrical and does not single out one type of "charge" over another. But at larger distances, that symmetry is broken and we see three distinct forces. (In the "standard model" of electroweak interactions, Section 32–10, the symmetry breaking between the electromagnetic and the weak interactions occurs at about 10^{-18} m.)

Symmetry breaking

FIGURE 32–16 Symmetry around a table. Example 32–6.

CONCEPTUAL EXAMPLE 32–6 **Symmetry.** The table in Fig. 32–16 has four identical place settings. Four people sit down to eat. Describe the symmetry of this table and what happens to it when someone starts the meal.

RESPONSE The table has several kinds of symmetry. It is symmetric to rotations of 90°: that is, the table will look the same if everyone moved one chair to the left or to the right. It is also north–south symmetric and east–west symmetric, so that swaps across the table don't affect the way the table looks. It also doesn't matter whether any person picks up the fork to the left of the plate or the fork to the right. But once that first person picks up either fork, the choice is set for all the rest at the table as well. The symmetry has been *broken*. The underlying symmetry is still there—the water glasses could still be chosen either way—but some choice must get made and at that moment the symmetry of the diners is broken.

Connection with cosmology

An interesting prediction of unified theories relates to cosmology (see Chapter 33). It is thought that during the first 10^{-35} s after the theorized big bang that created the universe, the temperature was so extremely high that particles had energies corresponding to the unification scale. Baryon number would not have been conserved then, thus allowing an imbalance that might account for the observed predominance of matter ($B > 0$) over antimatter ($B < 0$) in the universe.

This last example is interesting, for it illustrates a deep connection between investigations at either end of the size scale: theories about the tiniest objects (elementary particles) have a strong bearing on the understanding of the universe as a whole. We will look at this more in the next chapter.

Even more ambitious than grand unified theories are attempts to also incorporate gravity, and thus unify all four forces in nature into a single theory. (Such theories are sometimes referred to misleadingly as **theories of everything**.) The only consistent theory so far that attempts to unify all four forces is called **string theory**, in which the elementary particles (Table 32–5) are imagined not as points but as one-dimensional strings about 10^{-35} m long.

String theory

A related idea is **supersymmetry**, which applied to strings is known as **superstring theory**. Supersymmetry predicts that interactions exist that

Supersymmetry

would change fermions into bosons and vice versa, and that all known fermions have supersymmetric boson partners. Thus, for every quark there would be a *squark*, and for every lepton there would be a *slepton*. Likewise, for every known boson (photons and gluons, for example), there would be a supersymmetric fermion (*photinos* and *gluinos*). But why hasn't this "missing half" of the universe ever been detected? The best guess is that supersymmetric particles might be heavier than their conventional counterparts, perhaps too heavy to have been produced in today's accelerators. Until a supersymmetric particle is found, however, supersymmetry is just an elegant guess.

The world of elementary particles is opening new vistas. What happens in the near future is bound to be exciting.

■ SUMMARY

Particle accelerators are used to accelerate charged particles, such as electrons and protons, to very high energy. High-energy particles have short wavelength and so can be used to probe the structure of matter at very small distances in great detail. High kinetic energy also allows the creation of new particles through collision (via $E = mc^2$). Van De Graaff and linear accelerators use high voltage to accelerate particles along a line. Cyclotrons and synchrotrons use a magnetic field to keep the particles in a circular path and accelerate them at intervals by high voltage.

An **antiparticle** has the same mass as a particle but opposite charge. Certain other properties may also be opposite: for example, the antiproton has **baryon number** (nucleon number) opposite to that for the proton. In all nuclear and particle reactions, the following conservation laws hold: momentum, angular momentum, mass–energy, electric charge, baryon number, and the three **lepton numbers**. Certain particles have a property, called **strangeness**, which is conserved by the strong force but not by the weak force. The more recently noted properties, **charm**, **bottomness**, and **topness**, also are believed to be conserved by the strong force but not by the weak.

Just as the electromagnetic force can be said to be due to an exchange of photons, the strong nuclear force is thought to be carried by *mesons* that have rest mass, or, according to more recent theory, by massless **gluons**. The W and Z particles carry the weak force. These fundamental force carriers (photon, W and Z, gluons) are called **gauge bosons**.

Other particles can be classified as either *leptons* or *hadrons*. **Leptons** participate in the weak and electromagnetic interactions. **Hadrons** participate in the strong interaction as well. The hadrons can be classified as **mesons**, with baryon number zero, and **baryons**, with nonzero baryon number.

All particles, except for the photon, electron, neutrinos, and proton, decay with measurable half-lives varying from 10^{-25} s to 10^3 s. The half-life depends on which force is predominant in the decay. Weak decays usually have half-lives greater than about 10^{-13} s. Electromagnetic decays have half-lives on the order of 10^{-16} to 10^{-19} s. The shortest lived particles, called **resonances**, decay via the strong interaction and live typically for only about 10^{-23} s.

The latest theories of elementary particle physics postulate the existence of **quarks** as the basic building blocks of the hadrons. Initially, three quarks were proposed (called **up**, **down**, and **strange**). Recent evidence suggests that three additional quarks are needed, called the **charmed**, **bottom**, and **top** quarks. It is expected that there are the same number of quarks as leptons (six of each), and that quarks and leptons are the truly elementary particles along with the gauge bosons (γ, W, Z, gluons). Quarks are said to have **color**, and, according to **quantum chromodynamics** (QCD), the strong color force acts between their color charges and is transmitted by **gluons**. **Electroweak theory** views the weak and electromagnetic forces as two aspects of a single underlying interaction. QCD plus the electroweak theory are referred to as the **standard model**.

Grand unified theories of forces suggest that at very short distances (10^{-30} m) and very high energy, the weak, electromagnetic, and strong forces appear as a single force, and the fundamental difference between quarks and leptons disappears.

QUESTIONS

1. What limits the maximum energy attainable for protons in an ordinary cyclotron? How is this limitation overcome in a synchrotron?

2. Give a reaction between two nucleons, similar to Eq. 32–3, that could produce a π^-.

3. If a proton is moving at very high speed, so that its KE is much greater than its rest energy (m_0c^2), can it then decay via $p \rightarrow n + \pi^+$?

4. What would an "antiatom," made up of the antiparticles to the constituents of normal atoms, consist of? What might happen if *antimatter*, made of such antiatoms, came in contact with our normal world of matter?

5. What particle in a decay signals the electromagnetic interaction?

6. Does the presence of a neutrino among the decay products of a particle necessarily mean that the decay occurs via the weak interaction? Do all decays via the weak interaction produce a neutrino?

7. Why is it that a neutron decays via the weak interaction even though the neutron and one of its decay products (proton) are strongly interacting?

8. Which of the four interactions (strong, electromagnetic, weak, gravitational) does an electron take part in? A neutrino? A proton?

9. Check that charge and baryon number are conserved in each of the decays in Table 32–2.

10. Which of the particle decays in Table 32–2 occur via the electromagnetic interaction?

11. Which of the particle decays in Table 32–2 occur by the weak interaction?

12. By what interaction, and why, does $\Sigma^{\pm}$ decay to Λ^0? What about Σ^0 decaying to Λ^0?

13. The Δ baryon has spin $\frac{3}{2}$, baryon number 1, and charge $Q = +2, +1, 0$, or -1. Why is there no charge state $Q = -2$?

14. Which of the particle decays in Table 32–4 occur via the electromagnetic interaction?

15. Which of the particle decays in Table 32–4 occur by the weak interaction?

16. Quarks have spin $\frac{1}{2}$. How do you account for the fact that baryons have spin $\frac{1}{2}$ or $\frac{3}{2}$, and mesons have spin 0 or 1?

17. Suppose there were a kind of "neutrinolet" that was massless, had no color charge or electrical charge, and did not feel the weak force. Could you say that this particle even exists?

PROBLEMS

SECTIONS 32–1 AND 32–2

1. (I) What is the total energy of a proton whose kinetic energy is 8.50 GeV?

2. (I) Calculate the wavelength of 40-GeV electrons.

3. (I) What strength of magnetic field is used in a cyclotron in which protons make 2.4×10^7 revolutions per second?

4. (I) What is the time for one complete revolution for a very high-energy proton in the 1.0-km-radius Fermilab accelerator?

5. (I) If α particles are accelerated by the cyclotron of Example 32–2, what must be the frequency of the voltage applied to the dees?

6. (II) Is a 30-MeV alpha particle or 30-MeV proton better for picking out details of the nucleus? Compare each of their wavelengths with the size of a nucleon in a nucleus.

7. (II) (a) If the cyclotron of Example 32–2 accelerated α particles, what maximum energy could they attain? What would their speed be? (b) Repeat for deuterons ($^2_1$H). (c) In each case, what frequency of voltage is required?

8. (II) The voltage across the dees of a cyclotron is 45 kV. How many revolutions do protons make to reach a kinetic energy of 25 MeV?

9. (II) A cyclotron with a radius of 1.0 m is to accelerate deuterons ($^2_1$H) to an energy of 10 MeV. (a) What is the required magnetic field? (b) What frequency is needed for the voltage between the dees? (c) If the potential difference between the dees averages 25 kV, how many revolutions will the particle make before exiting? (d) How much time does it take for one deuteron to go from start to exit, and (e) how far does it travel during this time?

10. (II) What is the wavelength, and maximum resolving power attainable, using 900-GeV protons at Fermilab?

11. (II) Protons are injected into the 1.0-km-radius Fermilab Tevatron with an energy of 8.0 GeV. If they are accelerated by 2.5 MV each revolution, how far do they travel and approximately how long does it take for them to reach 900 GeV?

12. (II) The Fermilab Tevatron takes about 20 seconds to bring the energies of the stored protons from 150 GeV to 900 GeV. The acceleration is done once per turn. Estimate the energy given to the protons on each turn. (You can assume that the speed of the protons is essentially c the whole time.)

13. (III) What magnetic field intensity is needed at the 1.0-km-radius Fermilab synchrotron for 900-GeV protons? Use the relativistic mass.

14. (III) Show that the energy of a particle (charge e) in a synchrotron, in the relativistic limit ($v \approx c$), is given by E (in eV) $= Brc$, where B is magnetic field strength and r the radius of the orbit (SI units).

SECTIONS 32–3 TO 32–6

15. (I) Draw a Feynman diagram for $n + p \rightarrow n + p + \pi^0$.

16. (I) Draw a Feynman diagram for $p\bar{p}$ annihilation with the production of two pions.

17. (I) Draw a Feynman diagram for the photoelectric effect.

18. (I) Two protons are heading toward each other with equal speeds. What minimum kinetic energy must each have if a π^0 meson is to be created in the process? (See Table 32–2.)

19. (I) How much energy is released in the decay $\pi^+ \rightarrow \mu^+ + \nu_\mu$?

20. (I) About how much energy is released when a Λ^0 decays to $n\pi^0$? (See Table 32–2.)

21. (I) How much energy is required to produce a neutron–antineutron pair?

22. (I) Estimate the range of the strong force if the mediating particle were the kaon instead of the pion.

23. (II) Which of the following decays are possible? For those that are forbidden, explain which laws are violated.
    (a) $\Xi^0 \rightarrow \Sigma^+ + \pi^-$ (b) $\Omega^- \rightarrow \Sigma^0 + \pi^- + \nu$
    (c) $\Sigma^0 \rightarrow \Lambda^0 + \gamma + \gamma$.

24. (II) Estimate the range of the weak force using Eq. 32–2, given the masses of the W and Z particles as about 80 to 90 GeV/c^2.

25. (II) What are the wavelengths of the two photons produced when a proton and antiproton at rest annihilate?

26. (II) (a) Show, by conserving momentum and energy, that it is impossible for an isolated electron to radiate only a single photon. (b) With this result in mind, how can you defend the photon exchange diagram in Fig. 32–8?

27. (II) What would be the wavelengths of the two photons produced when an electron–positron pair, each with 300 keV of KE, annihilate?

28. (II) In the rare decay $\pi^+ \rightarrow e^+ + \nu_e$, what is the kinetic energy of the positron? Assume the π^+ decays from rest.

29. (II) What minimum kinetic energy must a neutron and proton each have if they are traveling at the same speed toward each other, collide, and produce a K^+K^- pair in addition to themselves? (See Table 32–2.)

30. (II) Calculate the KE of each of the two products in the decay $\Xi^- \rightarrow \Lambda^0 \pi^-$. Assume the Ξ^- decays from rest.

31. (III) Could a π^+ meson be produced if a 100-MeV proton struck a proton at rest? What minimum KE must the incoming proton have?

32. (III) For the reaction $p + p \rightarrow 3p + \bar{p}$, where one of the initial protons is at rest, use relativistic formulas to show that the threshold energy is $6m_p c^2$, equal to three times the Q-value of the reaction, where m_p is the proton rest mass.

SECTIONS 32–7 TO 32–11

33. (I) The measured width of the J/ψ meson is 88 keV. Estimate its lifetime.

34. (I) The measured width of the $\psi(3685)$ meson is 277 keV. Estimate its lifetime.

35. (I) What is the energy width (or uncertainty) of (a) η^0, and (b) Σ^0?

36. (I) Use Fig. 32–12 to estimate the energy width and then the lifetime of the Δ resonance.

37. (I) The B^- meson is presumed to be a $b\bar{u}$ quark combination. (a) Show that this is consistent for all quantum numbers. (b) What are the quark combinations for B^+, B^0, $\bar{B}^0$?

38. (II) What are the quark combinations that can form (a) a neutron, (b) an antineutron, (c) a Λ^0, (d) a $\bar{\Sigma}^0$?

39. (II) What particles do the following quark combinations produce? (a) uud, (b) $\bar{u}\bar{u}\bar{s}$, (c) $\bar{u}s$, (d) $d\bar{u}$, (e) $\bar{c}s$.

40. (II) What is the quark combination needed to produce a D^0 meson ($Q = B = S = 0$, $C = +1$)?

41. (II) The D_S^+ meson has $S = C = +1$, $B = 0$. What quark combination would produce it?

42. (II) (a) Show that the so-called unification distance of 10^{-30} m in grand unified theory is equivalent to an energy of about 10^{14} GeV. Use either the uncertainty principle or de Broglie's wavelength formula, and explain how they apply. (b) Calculate what temperature this corresponds to.

43. (II) Draw possible Feynman diagrams using quarks (as in Fig. 32–14c) for the reactions (a) $\pi^- p \rightarrow \pi^0 n$, (b) $\bar{p}p \rightarrow \pi^+ \pi^-$.

44. (II) Draw a possible quark Feynman diagram (see Fig. 32–14c) for the reaction $K^- p \rightarrow K^- p$.

45. There are typically 5×10^{13} protons at 900 GeV stored in the 1.0-km-radius ring of the Tevatron. (a) How much current (amperes) is carried by this beam? (b) How fast would a 1500 kg car have to move to carry the same kinetic energy as this beam?

46. The 4.25-km-radius LEP tunnel will be reused to house the magnets for the Large Hadron Collider (LHC), scheduled for completion around 2005. If the design calls for proton beams of energy 7.0 TeV, what magnetic field will be required?

47. (a) How much energy is released when an electron and a positron annihilate each other? (b) How much energy is released when a proton and an antiproton annihilate each other?

48. Which of the following reactions are possible, and by what interaction could they occur? For those forbidden, explain why.
   (a) $\pi^- p \rightarrow K^+ \Sigma^-$ (f) $\pi^- p \rightarrow K^0 p \pi^0$
   (b) $\pi^+ p \rightarrow K^+ \Sigma^+$ (g) $K^- p \rightarrow \Lambda^0 \pi^0$
   (c) $\pi^- p \rightarrow \Lambda^0 K^0 \pi^0$ (h) $K^+ n \rightarrow \Sigma^+ \pi^0 \gamma$
   (d) $\pi^+ p \rightarrow \Sigma^0 \pi^0$ (i) $K^+ \rightarrow \pi^0 \pi^0 \pi^+$
   (e) $\pi^- p \rightarrow p e^- \bar{\nu}_e$ (j) $\pi^+ \rightarrow e^+ \nu_e$

49. For the decay $\Lambda^0 \rightarrow p \pi^-$, calculate (a) the Q-value (energy released), and (b) the KE of the p and π^-, assuming the Λ^0 decays from rest. (Use relativistic formulas.)

50. Symmetry breaking occurs in the electroweak theory at about 10^{-18} m. Show that this corresponds to an energy that is on the order of the mass of the $W^\pm$.

51. The mass of a π^0 can be measured by observing the reaction $\pi^- + p \rightarrow \pi^0 + n$ at very low incident π^- kinetic energy (assume it is zero). The neutron is observed to be emitted with a KE of 0.60 MeV. Use conservation of energy and momentum to determine the π^0 mass.

52. Calculate the Q value for each of the reactions, Eq. 32–3, for producing a pion.

53. Calculate the maximum KE of the electron in the decay $\mu^- \rightarrow e^- + \bar{\nu}_e + \nu_\mu$. [Hint: In what direction do the two neutrinos move relative to the electron in order to give the latter the maximum KE? Both energy and momentum are conserved; use relativistic formulas.]

54. Calculate the Q-value for the reaction $\pi^- p \rightarrow \Lambda^0 K^0$, when negative pions strike stationary protons. Estimate the minimum pion KE needed to produce this reaction.

55. Draw the normal Feynman diagrams for (a) Compton scattering, (b) bremsstrahlung by an electron, and (c) electron–positron pair creation.

56. How many fundamental fermions are there in a water molecule?

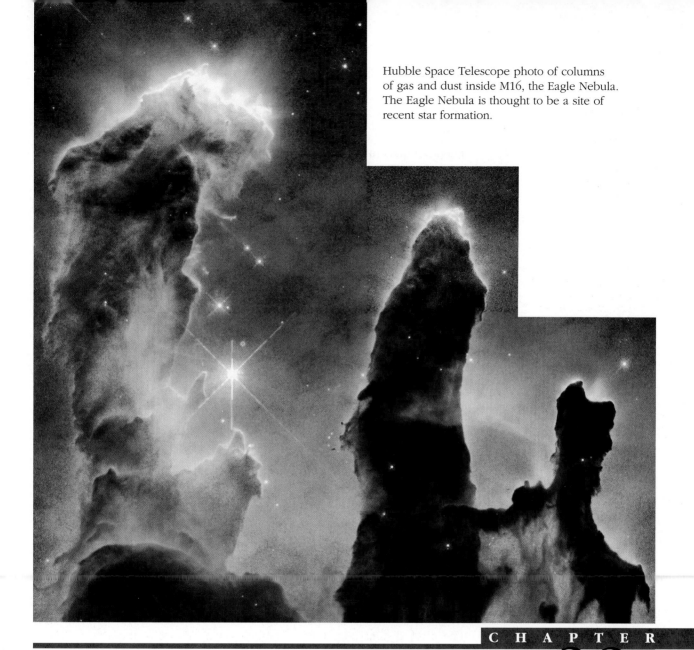

Hubble Space Telescope photo of columns of gas and dust inside M16, the Eagle Nebula. The Eagle Nebula is thought to be a site of recent star formation.

ASTROPHYSICS AND COSMOLOGY

I n the previous chapter, we studied the tiniest objects in the universe—the elementary particles. Now we leap to the largest—stars and galaxies. These two extreme realms, elementary particles and the cosmos, are among the most intriguing and exciting subjects in science. And, surprising though it may seem, these distant realms are related in a fundamental way, as already hinted in Chapter 32.

Use of the techniques and ideas of physics to study the heavens is often referred to as **astrophysics**. At the base of our present theoretical understanding of the universe is Einstein's *general theory of relativity* and its theory of gravitation—for in the large-scale structure of the universe, gravity is the dominant force. General relativity serves also as the foundation for modern **cosmology**, which is the study of the universe as a whole. Cosmology deals especially with the search for a theoretical framework to understand the observed universe, its origin, and its future. The questions posed by cosmology are complex and difficult; the possible answers are often unimaginable. They are questions like "Has the universe always existed, or did it have a beginning in time?" Either alternative is difficult to imagine: time going back indefinitely into the past, or an actual moment when the universe began (but, then, what was there before?). And what about the size of the universe? Is it infinite in size? (It is hard to imagine infinity.) Or is it finite in size? (This is also hard to imagine, for if the universe is finite, it does not make sense to ask what is beyond it, because the universe is all there is.)

Our survey of astrophysics and cosmology will be necessarily brief and qualitative, but we will nonetheless touch on the major ideas. We begin with a look at what can be seen beyond the Earth.

33-1 Stars and Galaxies

According to the ancients, the stars, except for the few that seemed to move (the planets), were fixed on a sphere beyond the last planet. The universe was neatly self-contained, and we on Earth were at or near its center. But in the centuries following Galileo's first telescopic observations of the heavens in 1610, our view of the universe has changed dramatically. We no longer place ourselves at the center, and we view the universe as vastly larger. The distances involved are so great that we specify them in terms of the time it takes light to travel the given distance: for example, 1 light-second = $(3.0 \times 10^8 \text{ m/s})(1.0 \text{ s}) = 3.0 \times 10^8 \text{ m} = 3.0 \times 10^5 \text{ km}$; 1 light-minute = $18 \times 10^6 \text{ km}$; and 1 **light-year** (ly) is

Light-year

$$1 \text{ ly} = (2.998 \times 10^8 \text{ m/s})(3.156 \times 10^7 \text{ s/yr}) = 9.46 \times 10^{15} \text{ m} \approx 10^{13} \text{ km}.$$

FIGURE 33-1 A section of the Milky Way. The thin line is the trail of an artificial Earth satellite.

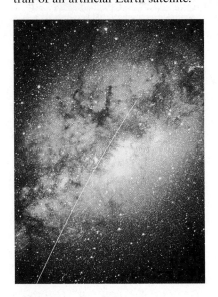

For specifying distances to the Sun and Moon, we usually use meters or kilometers, but we could specify them in terms of light. The Earth–Moon distance is 384,000 km, which is 1.28 light-seconds. The Earth–Sun distance is 1.50×10^{11} m, or 150,000,000 km; this is equal to 8.3 light-minutes. The most distant planet in the solar system, Pluto, is about 6×10^9 km from the Sun, or 6×10^{-4} ly. The nearest star to us, other than the Sun, is Proxima Centauri, about 4.3 ly away. (Note that the nearest star is 10,000 times farther from us than the farthest planet.)

On a clear moonless night, thousands of stars of varying degrees of brightness can be seen, as well as the elongated cloudy stripe known as the Milky Way (Fig. 33–1). It was Galileo who first observed, about 1610, that the Milky Way is comprised of countless individual stars. A century and a half later (about 1750), Thomas Wright suggested that the Milky Way was a flat disc of stars extending to great distances in a plane, which we call the **Galaxy** (Greek for "milky way").

Our Galaxy has a diameter of almost 100,000 light-years and a thickness of very roughly 2000 light-years. It has a bulging central nucleus and spiral arms (Fig. 33–2). Our Sun, which seems to be just another star, is located more than halfway from the center to the edge, about 28,000 ly from the center. Our Galaxy contains about 10^{11} stars. The Sun orbits the galactic center approximately once every 200 million years, so its speed is

FIGURE 33–2 Our Galaxy, as it would appear from the outside: (a) "end view," in the plane of the disc; (b) "top view," looking down on the disc. (If only we could see it like this—from the outside!) (c) Infrared photograph of the inner reaches of the Milky Way, showing the central bulge of our Galaxy. This very wide angle photo extends over 180° of sky, and to be viewed properly it should be wrapped in a semicircle with your eyes at the center. The white dots are nearby stars.

100,000 ly

Our Sun

Our Sun

2000 ly

(a)

(b)

(c)

about 250 km/s relative to the center of the Galaxy. The total mass of all the stars in our Galaxy is about 3×10^{41} kg.

EXAMPLE 33–1 **ESTIMATE** **Our Galaxy's mass.** Estimate the total mass of our Galaxy using the orbital data of the Sun (including our solar system) about the center of the Galaxy. Assume that most of the mass of the Galaxy can be approximated as a uniform sphere of mass (the central bulge, Fig. 33–2a).

SOLUTION Our Sun and solar system orbit the center of the Galaxy, according to the best measurements as mentioned above, with a speed of about $v = 250$ km/s at a distance from the Galaxy center of about $r = 28,000$ ly. We use Newton's second law, $F = ma$, with a being the centripetal acceleration, $a = v^2/r$, and F being the universal law of gravitation (Chapter 5)

$$F = ma$$

$$G\frac{Mm}{r^2} = m\frac{v^2}{r}$$

where M is the mass of the Galaxy and m is the mass of our Sun and solar system. Solving this, we find

$$M = \frac{rv^2}{G} \approx \frac{(28{,}000 \text{ ly})(10^{16} \text{ m/ly})(2.5 \times 10^5 \text{ m/s})^2}{6.67 \times 10^{-11} \text{ N·m}^2/\text{kg}^2} \approx 3 \times 10^{41} \text{ kg}.$$

In terms of *numbers* of stars, if they are like our Sun ($m = 2.0 \times 10^{30}$ kg), there would be about $(3 \times 10^{41} \text{ kg})/(2 \times 10^{30} \text{ kg}) \approx 10^{11}$ or about 100 billion stars.

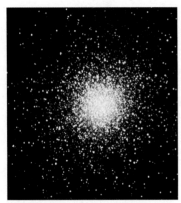

FIGURE 33-3 This globular star cluster is located in the constellation Hercules.

FIGURE 33-4 This gaseous nebula, found in the constellation Carina, is about 9000 light-years from us.

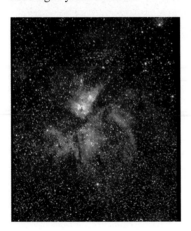

We can see by telescope, in addition to stars both within and outside the Milky Way, many faint cloudy patches in the sky which were all referred to once as "nebulae" (Latin for "clouds"). A few of these, such as those in the constellations Andromeda and Orion, can actually be discerned with the naked eye on a clear night. Some are **star clusters** (Fig. 33–3), groups of stars that are so numerous they appear to be a cloud. Others are glowing clouds of gas or dust (Fig. 33–4), and it is for these that we now mainly reserve the word **nebula**. Most fascinating are those that belong to a third category: they often have fairly regular elliptical shapes and seem to be a great distance beyond our Galaxy. Immanuel Kant (about 1755) seems to have been the first to suggest that these latter might be circular discs like our Galaxy: they appear elliptical because we see them at an angle, and are faint because they are so distant. At first it was not universally accepted that these objects were **extragalactic**—that is, outside our Galaxy. The very large telescopes constructed in this century revealed that individual stars could be resolved within these extragalactic objects and that many contained spiral arms. Edwin Hubble (1889–1953), who did much of this observational work in the 1920s using the 2.5-m (100-inch) telescope[†] on Mt. Wilson near Los Angeles, California, was also able to demonstrate that these objects were indeed extragalactic because of their great distances. The distance to the Andromeda nebula (a galaxy), for example, is over 2 million light-years, a distance 20 times greater than the diameter of our Galaxy. Thus it was determined that these nebulae are **galaxies** similar to ours. Today, the largest telescopes can see about 10^{11} galaxies. See Fig. 33–5. (Note that it is usual to capitalize the word galaxy only when it refers to our own.)

Galaxies tend to be grouped in **galaxy clusters**, with anywhere from a few to many thousands of galaxies in each cluster. Furthermore, clusters themselves seem to be organized into even larger aggregates: clusters of clusters of galaxies, or **superclusters**. The galaxies nearest us are about 2 million light-years away. The farthest detectable galaxies are thousands of times farther away, on the order of 10^{10} ly. (See Table 33–1.)

TABLE 33-1 Heavenly Distances

Object	Approx. Distance from Earth (ly)
Moon	4×10^{-8}
Sun	1.6×10^{-5}
Farthest planet in our solar system (Pluto)	6×10^{-4}
Nearest star (Proxima Centauri)	4.3
Center of our Galaxy	3×10^4
Nearest galaxy	2×10^6
Farthest galaxies	10^{10}

[†]2.5 m (=100 inches) refers to the diameter of the curved objective mirror. The bigger the mirror, the more light it collects and the less diffraction there is, so more and fainter stars can be seen. See Chapter 25.

(a)   (b)   (c)

CONCEPTUAL EXAMPLE 33–2 | **Looking back in time.** Astronomers often think of their telescopes as time machines, looking back toward the origin of the universe. How far back do they look?

RESPONSE The distance in light-years measures exactly how long in years the light has been traveling to reach us, so Table 33–1 tells us also how far back in time we are looking. For example, if we saw Proxima Centauri explode into a supernova today, then the event would have really occurred 4.3 years ago. The most distant objects, galaxies 10^{10} ly away, emitted the light we see now 10^{10} years ago, so what we see was how they were then, close to the beginning of the universe.

FIGURE 33–5 Photographs of galaxies. (a) Spiral galaxy in the constellation Hydra. (b) Two galaxies: the larger and more dramatic one is known as the Whirlpool galaxy. (c) The same galaxies as in (b), but this is a false-color infrared image which shows the arms of the spiral as being more regular than in the visible light photo (b); the different colors correspond to different light intensities.

From this perspective of a vast universe, we living creatures on Earth may feel very small and anonymous—a far cry from the view held only a few centuries ago that placed us at the very center of the universe. If this modern view of the universe seems too stark, however, we can take a philosophical approach: all of these observations and interpretations have been made here on Earth, and the ideas were created here on Earth. Who is to say that we are in any way unimportant in the universe, even though we may not hold a geometrically central position?

Besides the usual stars, clusters of stars, galaxies, and clusters and super-clusters of galaxies, the universe contains a number of other interesting objects. Among these are stars known as *red giants*, *white dwarfs*, *neutron stars*, *black holes* (at least theoretically), and exploding stars called *novae* and *supernovae*. In addition there are *quasars* ("quasistellar radio sources"), which, if we judge their distance correctly, are galaxies thousands of times brighter than ordinary galaxies. Furthermore, there is radiation that reaches the Earth but does not emanate from the bright pointlike objects we call stars: it is a background radiation that seems to arrive uniformly from all directions in the universe. We discuss all these phenomena in due course.

We have talked about the vast distances of objects in the universe. But how do we measure these distances? One basic technique employs simple geometry to measure the **parallax** of a star. By parallax we mean the apparent motion of a star, against the background of more distant stars,

How astronomical distances are measured

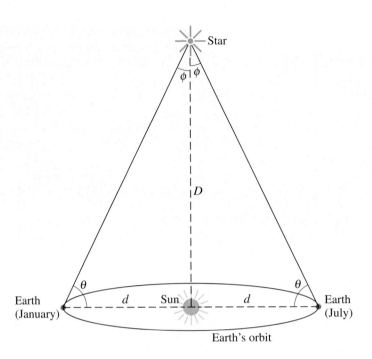

FIGURE 33–6 Distance to a star determined by parallax. (Not to scale.)

due to the Earth's motion about the Sun. As shown in Fig. 33–6, the sighting angle of a star relative to the plane of Earth's orbit (angle θ) is measured at different times of the year. Since we know the distance d from Earth to Sun, we can reconstruct the right triangles shown in the figure and can determine[†] the distance D to the star.

EXAMPLE 33–3 **ESTIMATE** **Distance to a star using parallax.** Estimate the distance D to a star if the angle θ in Fig. 33–6 is measured to be 89.99994°.

SOLUTION The angle ϕ in Figure 33–6 is 0.00006°, or about 1.0×10^{-6} radians. Since $\tan \phi = d/D$, then the distance D to the star is

$$D = \frac{d}{\tan \phi} = \frac{d}{\phi} = \frac{1.5 \times 10^8 \text{ km}}{1.0 \times 10^{-6} \text{ rad}} = 1.5 \times 10^{14} \text{ km},$$

or about 15 ly. (We can use $\tan \phi \approx \phi$ since ϕ is very small.)

The distances to stars are often specified in terms of parallax angle given in seconds of arc: 1 second (1″) is $\frac{1}{60}$ of a minute (′) of arc, which is $\frac{1}{60}$ of a degree, so $1″ = \frac{1}{3600}$ of a degree. The distance, then, is specified in parsecs (meaning *par*allax angle in *sec*onds of arc), where the **parsec** (pc) is defined as $1/\phi$ and ϕ is given in seconds. In the Example we just did, $\psi - (6 \times 10^{-5})° (3600) = 0.22″$ of arc, so we would say the star is at a distance of $1/0.22″ = 4.5$ pc. It is easy to show that the parsec is given by

Parsec
$$1 \text{ pc} = 3.26 \text{ ly}.$$

Parallax can be used to determine the distance to stars as far away as 100 light-years (≈ 30 parsecs). Beyond that distance, parallax angles are too small to measure. For greater distances, more subtle techniques must

---

[†]This is essentially the way the heights of mountains are determined, by "triangulation."

be employed. We can, for example, compare the apparent brightnesses of galaxies, and—using the inverse square law (intensity drops off as the square of the distance)—estimate their relative distances. This method cannot be very precise since we cannot expect all galaxies to have the same intrinsic brightness. A better estimate compares the brightest stars in galaxies or the brightest galaxies in galaxy clusters. It is reasonable to assume, for example, that the brightest stars in all galaxies are similar and have about the same intrinsic luminosity. Consequently, their *apparent* brightness would be a measure of how far away they are.

Another important technique for estimating distance is via the "red-shift" in the line spectra of elements, which is related to the expansion of the universe, as will be discussed in Section 33–4.

As we look farther and farther away, the measurement techniques are less and less reliable, so there is more and more uncertainty in the measurements of large distances.

33–2 Stellar Evolution: The Birth and Death of Stars

The stars appear unchanging. Night after night the heavens reveal no significant variations. Indeed, on a human time scale, the vast majority of stars (except novae and supernovae) change very little. Although stars *seem* fixed in relation to each other, many move sufficiently for the motion to be detected. Indeed, the speeds of stars relative to neighboring stars can be hundreds of km/s, but at their great distance from us, this motion is detectable only by careful measurement. Furthermore, there is a great range of brightness among stars. The differences in brightness are due both to differences in the amount of light stars emit and to their distances from us.

A useful parameter for a star or galaxy is its **absolute luminosity**, L, by which we mean the total power radiated in watts. Also important is the **apparent brightness**, l, defined as the power crossing unit area perpendicular to the path of the light at the Earth. Given that energy is conserved, and ignoring any absorption in space, the total emitted power L at a distance d from the star will be spread over a sphere of surface area $4\pi d^2$. If we let d be the distance from the star to the Earth, then L must be equal to $4\pi d^2$ times l (power per unit area at Earth). That is,

Luminosity and brightness of stars

$$l = \frac{L}{4\pi d^2}. \tag{33–1}$$

EXAMPLE 33–4 **Apparent brightness.** Suppose a particular star has absolute luminosity equal to that of the Sun but is 10 pc away from Earth. By what factor will it appear dimmer than the Sun?

SOLUTION The luminosity L is the same for both stars, so the apparent brightness depends only on their relative distances. We use the inverse square law as stated above: the star appears dimmer by a factor

$$\frac{l_1}{l_2} = \frac{d_2^2}{d_1^2} = \frac{(1.5 \times 10^8 \text{ km})^2}{(10 \text{ pc})^2 (3.26 \text{ ly/pc})^2 (10^{13} \text{ km/ly})^2} \approx 2 \times 10^{-13},$$

where d_1 and d_2 are the distances from Earth to the star and to the Sun, respectively.

Careful study of nearby stars has shown that for most stars, the absolute luminosity depends on the mass: *the more massive the star, the greater its luminosity*. Another important parameter of a star is its surface temperature, which can be determined from the spectrum of electromagnetic frequencies it emits, just as for a blackbody (Section 27–2). As we saw in Chapter 27, the spectrum of hotter and hotter bodies shifts from predominantly lower frequencies (and longer wavelengths, such as red) to higher frequencies (and shorter wavelengths such as blue). Quantitatively, the relation is given by Wien's law (Eq. 27–2): the peak wavelength λ_P in the spectrum of light emitted by a blackbody (and stars are fairly good approximations to blackbodies) is inversely proportional to its kelvin temperature T; that is, $\lambda_P T = 2.90 \times 10^{-3}\,\text{m·K}$. The surface temperatures of stars typically range from about 3500 K (reddish) to perhaps 50,000 K (bluish).

An important astronomical discovery, made near the turn of the century, was that for most stars, the color is related to the absolute luminosity and therefore to the mass. A useful way to present this relationship is by the so-called Hertzsprung–Russell (H–R) diagram. On the H–R diagram, the horizontal axis shows the temperature T whereas the vertical axis is the luminosity L, and each star is represented by a point on the diagram, Fig. 33–7. Most stars fall along the diagonal band termed the **main sequence**. Starting at the lower right we find the coolest stars, reddish in color; they are the least luminous and therefore of low mass. Farther up toward the left we find hotter and more luminous stars that are yellowish-white, like our Sun. Still farther up we find still more massive and more luminous stars, bluish in color. Stars that fall on this diagonal band are called *main-sequence stars*. There are also stars that fall outside the main sequence. Above and to the right we find extremely large stars, with high luminosities but with low (reddish) color temperature: these are called *red giants*. At the lower left, there are a few stars of low luminosity but with high temperature: these are the *white dwarfs*.

Before considering the different types of stars, let us first look at a couple of Examples that show how physical principles can be used to gain information about stars.

Luminosity increases with star's mass

H–R diagram

Main-sequence stars

Red giants

White dwarfs

FIGURE 33–7
Hertzsprung–Russell (H–R) diagram.

EXAMPLE 33–5 **Determining star temperatures and size.** Suppose that the distances to two nearby stars can be reasonably estimated and this data, together with their measured apparent brightnesses, suggests that the two stars have about the same absolute luminosity, L. The spectrum of one of the stars peaks at about 700 nm (so it is reddish). The spectrum of the other peaks at about 350 nm (bluish). Use Wien's law (Eq. 27–2) and the Stefan–Boltzmann law (Section 14–9) to determine (a) the surface temperature of each star, and (b) how much larger one star is than the other.

SOLUTION (a) Wien's law (Eq. 27–2) states that $\lambda_P T = 2.90 \times 10^{-3}\,\text{m·K}$. So the temperature of the reddish star is

$$T_r = \frac{2.90 \times 10^{-3}\,\text{m·K}}{700 \times 10^{-9}\,\text{m}} = 4140\,\text{K}.$$

The temperature of the bluish star will be double this since its peak wavelength is half (350 nm vs 700 nm); just to check:

$$T_b = \frac{2.90 \times 10^{-3}\,\text{m·K}}{350 \times 10^{-9}\,\text{m}} = 8280\,\text{K}.$$

(b) The Stefan-Boltzmann law, which we discussed in Chapter 14 (see Eq. 14–4), states that the power radiated *per unit area* of surface from a body is proportional to the fourth power of the kelvin temperature, T^4. Now the temperature of the bluish star is double that of the reddish star, so the bluish one must radiate $(2)^4 = 16$ times as much energy per unit area. But we are given that they have the same luminosity (the same total power output), so the surface area of the blue star must be $\frac{1}{16}$ that of the red one. Since the surface area of a sphere is $4\pi r^2$, we conclude that the radius of the reddish star is $\sqrt{16} = 4$ times larger than the radius of the bluish star (or $4^3 = 64$ times the volume).

EXAMPLE 33–6 **ESTIMATE** **Distance to a star using H–R and color.** Suppose that detailed study of a certain star suggests that it most likely fits on the main sequence of an H–R diagram. Its measured apparent brightness is $l = 1.0 \times 10^{-12}\,\text{W/m}^2$, and the peak wavelength of its spectrum is $\lambda_P \approx 600\,\text{nm}$. Estimate its distance from us.

SOLUTION The star's temperature, from Wien's law (Eq. 27–2), is

$$T \approx \frac{2.90 \times 10^{-3}\,\text{m·K}}{600 \times 10^{-9}\,\text{m}} \approx 4800\,\text{K}.$$

A star on the main sequence of an H–R diagram at this temperature has absolute luminosity of about $L \approx 1 \times 10^{26}\,\text{W}$, read off of Fig. 33–7. Then, from Eq. 33–1,

$$d = \sqrt{\frac{L}{4\pi l}} \approx \sqrt{\frac{1 \times 10^{26}\,\text{W}}{4(3.14)(1.0 \times 10^{-12}\,\text{W/m}^2)}} \approx 3 \times 10^{18}\,\text{m}.$$

Its distance from us in light-years and parsecs is

$$d = \frac{3 \times 10^{18}\,\text{m}}{10^{16}\,\text{m/ly}} \approx 300\,\text{ly} \approx \frac{300\,\text{ly}}{3.26\,\text{ly/pc}} \approx 90\,\text{pc}.$$

Why are there different types of stars, such as red giants and white dwarfs, as well as main-sequence stars? Were they all born this way, in the beginning? Or might each different type represent a different age in the life cycle of a star? Astronomers and astrophysicists today believe the latter is most likely the case. Note, however, that we cannot actually follow any but the tiniest part of the life cycle of any given star since they live for ages vastly greater than ours, on the order of millions or billions of years. Nonetheless, let us follow the process of **stellar evolution** from the birth to the death of a star, as astrophysicists have theoretically reconstructed it today.

Birth of a star

Contraction due to gravity

Fusion begins when T (and $\overline{\text{KE}}$) is large enough

Reaching the main sequence

Stars are born, it is believed, when gaseous clouds (mostly hydrogen) contract due to the pull of gravity. A huge gas cloud might fragment into numerous contracting masses, each mass centered in an area where the density was only slightly greater than that at nearby points. Once such "globules" formed, gravity would cause each to contract in toward its center of mass. As the particles of such a *protostar* accelerate inward, their kinetic energy increases. When the kinetic energy is sufficiently high, the Coulomb repulsion that keeps the hydrogen nuclei apart can be overcome and nuclear fusion can take place. In a star like our Sun, the "burning" of hydrogen[†] occurs via the proton–proton cycle (Section 31–3, Eqs. 31–3), in which four protons fuse to form a $^4_2$He nucleus with the release of γ rays and neutrinos. These reactions require a temperature of about 10^7 K, corresponding to an average KE (kT) of about 1 keV. In more massive stars, the carbon cycle produces the same net effect: Four $^1_1$H produce a $^4_2$He— see Section 31–3. The fusion reactions take place primarily in the core of a star, where T is sufficiently high. (The surface temperature is, of course, much lower—on the order of a few thousand kelvins.) The tremendous release of energy in these fusion reactions produces a pressure sufficient to halt the gravitational contraction, and our protostar, now really a young *star*, stabilizes on the main sequence. Exactly where the star falls along the main sequence depends on its mass. The more massive the star, the farther up (and to the left) it falls on the diagram. To reach the main sequence requires perhaps 30 million years, if it is a star like our Sun, and (assuming our theory is right) it will remain there[‡] about 10 billion years (10^{10} yr). Although most stars are billions of years old, there is evidence that stars are actually being born at this moment.

As hydrogen "burns"—that is, fuses to form helium—the helium that is formed is denser and tends to accumulate in the central core where it was formed. As the core of helium grows, hydrogen continues to "burn" in a shell around it, Fig. 33–8. When much of the hydrogen within the core has been consumed, the production of energy decreases and is no longer sufficient to prevent the huge gravitational forces from once again causing the core to contract and heat up. The hydrogen in the shell around the core then "burns" even more fiercely because of this rise in temperature,

FIGURE 33–8 A shell of "burning" hydrogen (fusing to become helium) surrounds the core where the newly formed helium gravitates.

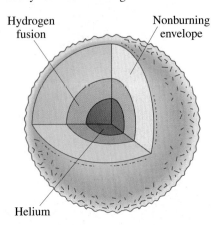

Hydrogen fusion

Nonburning envelope

Helium

[†]The word "burn" is put in quotation marks because these high-temperature fusion reactions occur via a *nuclear* process, and must not be confused with ordinary burning (of, say, paper, wood, or coal) in air, which is a chemical reaction at the *atomic* level (and at a much lower temperature).

[‡]More massive stars, since they are hotter and the Coulomb repulsion is more easily overcome, "burn" much more quickly, and so use up their fuel faster, resulting in shorter lives. A star 10 times more massive than our Sun, for example, will remain on the main sequence only for about 10^7 years. Stars less massive than our Sun live much longer than our Sun's 10^{10} yr.

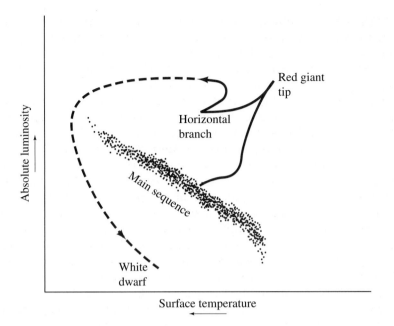

FIGURE 33-9 Evolution of a star like our Sun represented on an H–R diagram.

causing the outer envelope of the star to expand and to cool. The surface temperature, thus reduced, produces a spectrum of light that peaks at longer wavelength. By this time the star has left the main sequence. It has become redder, and as it has grown in size, it has become more luminous. So it will have moved to the right and upward on the H–R diagram, as shown in Fig. 33–9. As it moves upward, it enters the **red giant** stage. Thus, theory explains the origin of red giants as a natural step in a star's evolution. Our Sun, for example, has been on the main sequence for about $4\frac{1}{2}$ billion years. It will probably remain there another 4 or 5 billion years. (We can take comfort in that!) When our Sun leaves the main sequence, it is expected to grow in size (as a red giant) until it occupies all the volume out to approximately the present orbit of Earth. *Red giants*

As a star's envelope expands, the core is shrinking and heating up. When the temperature reaches about 10^8 K, helium nuclei, with their greater charge and hence greater electrical repulsion, can then undergo fusion. The reactions are

$$\begin{aligned} {}^4_2\text{He} + {}^4_2\text{He} &\rightarrow {}^8_4\text{Be} + \gamma, \\ {}^4_2\text{He} + {}^8_4\text{Be} &\rightarrow {}^{12}_6\text{C} + \gamma. \end{aligned} \qquad \textbf{(33–2)}$$ *Nucleosynthesis*

These two reactions occur in quick succession, and the net effect is

$$3 \, {}^4_2\text{He} \rightarrow {}^{12}_6\text{C}.$$

The burning of helium causes a rapid and major change in the star, and the star moves rapidly to the "horizontal branch" on the H–R diagram (see Fig. 33–9). Further reactions are possible, creating elements of higher and higher Z, up to $Z = 10$ or 12. If the mass of the star is great enough, higher Z elements can be formed. The star can get even hotter and in the range $T = 2.5$–5×10^9 K, nuclei as heavy as ${}^{56}_{26}\text{Fe}$ and ${}^{56}_{28}\text{Ni}$ can be made. But here the process of **nucleosynthesis**, the formation of heavy nuclei from lighter ones by fusion, ends. As we saw in Fig. 30–1, the average binding energy per nucleon begins to decrease for A greater than about *Production of heavy elements*

60. Further fusions would *require* energy, rather than release it. (Elements heavier than Ni are probably formed mainly by neutron capture, in supernova explosions, as we shall discuss shortly.)

What happens next depends on the mass of the star. If the star has a residual mass less than about 1.4 solar masses[†] (known as the *Chandrasekhar limit*), no further fusion energy can be obtained and the star collapses under the action of gravity. As it shrinks, the star cools and typically follows the route shown in Fig. 33–9, according to theory, descending from the upper left downward, becoming a **white dwarf**. A white dwarf with a mass equal to that of the Sun would be about the size of the Earth. What determines its size is that it collapses to the point at which the electron clouds of the atoms start to overlap: at this point it collapses no further because, as the Pauli exclusion principle claims, no two electrons can be in the same quantum state. A white dwarf continues to lose internal energy, decreasing in temperature and becoming dimmer and dimmer until its light goes out. It has then become a *black dwarf*, a dark cold chunk of ash.

More massive stars (residual mass greater than 1.4 solar masses) are thought to follow a quite different scenario. A star with this great a mass can contract under gravity and heat up even further, reaching extremely high temperatures. The KE of the nuclei is then so high that fusion of elements heavier than iron is possible even though the reactions are energy-requiring. Furthermore the high-energy collisions can also cause the breaking apart of iron and nickel nuclei into He nuclei, and eventually into protons and neutrons:

$$^{56}_{26}\text{Fe} \rightarrow 13\ ^{4}_{2}\text{He} + 4\text{n}.$$

$$^{4}_{2}\text{He} \rightarrow 2\text{p} + 2\text{n}.$$

These are energy-requiring (endoergic) reactions, but at such extremely high temperature and pressure there is plenty of energy available, enough even to force electrons and protons together to form neutrons in inverse β decay:

$$e^- + p \rightarrow n + \nu.$$

As the core contracts under the huge gravitational forces, the tremendous mass becomes essentially an enormous nucleus made up almost exclusively of neutrons. The size of the star is no longer limited by the exclusion principle applied to electrons, but rather applied to neutrons, and the star begins to contract rapidly toward forming an enormously dense **neutron star**. The contraction of the core would mean a great reduction in gravitational potential energy. Somehow this energy would have to be released. Indeed, it was suggested in the 1930s that the final core collapse to a neutron star may be accompanied by a catastrophic explosion whose tremendous energy could form virtually all elements of the periodic table and blow away the entire outer envelope of the star (Fig. 33–10), spreading its contents into interstellar space. Such explosions are believed to produce one type of observed **supernova** (there may be other mechanisms as well). Indeed, the presence of heavy elements on Earth and in our solar system suggests that our solar system may have formed from the debris from at least one supernova.

White dwarf

Neutron stars

Supernovae

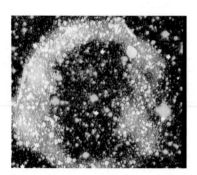

FIGURE 33–10 Computer-enhanced false-color image of the shell of material expanding outward from "Tycho's supernova" (seen by Tycho Brahe in 1572).

[†]This cutoff refers to the mass of the star remaining *after* mass loss from the red giant stage. A star whose residual mass is 1.4 solar masses had an original mass of perhaps 6 to 8 solar masses.

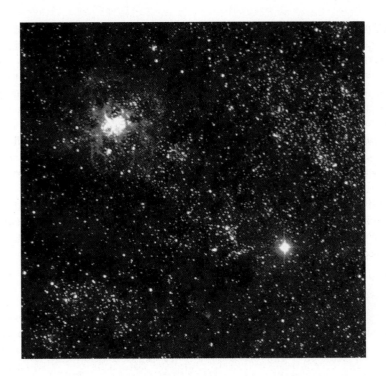

FIGURE 33–11 SN1987a, the supernova that appeared in February 1987, is the very bright star on the right. The bright nebulous object is the Tarantula Nebula.

In a supernova explosion, a star's brightness is observed to suddenly increase billions of times in a period of just a few days and then fade away over the next few months. In February 1987 such a supernova occurred in an appendage to our Galaxy known as the Large Magellanic Cloud, only 170,000 ly away, and was visible to the naked eye in the southern hemisphere (Fig. 33–11). Supernovae are seen relatively often in distant galaxies, but this recent one, called SN1987a, is the first to have occurred close enough to be visible to the naked eye since the famous one of 1604 observed by Kepler and Galileo (and thus SN1987a is the first since the invention of the telescope). A supernova visible to the naked eye was observed by Chinese astronomers in A.D. 1054, and its remains are still visible in the sky (in the Crab Nebula), in the midst of which there is now a **pulsar**. Pulsars are astronomical objects that emit sharp pulses of radiation at regular intervals, on the order of a second. They are now believed to be neutron stars which, because of conservation of angular momentum, increase greatly in rotational speed as their moment of inertia decreases during their contraction. The intense magnetic field of such a rapidly rotating star (perhaps as high as 10^6 to 10^{10} T) could trap and accelerate charged particles, which then give off radiation in a beam that rotates with the star. The discovery of pulsars in 1967 has lent support to the theory that neutron stars are the result of supernova explosions.

Pulsars

The core of a neutron star contracts to the point at which all neutrons are as close together as they are in a nucleus. That is, the density of a neutron star is on the order of 10^{14} times that of normal solids and liquids on Earth. A neutron star that has a mass 1.5 times that of our Sun would have a diameter of only about 10 km.

If the mass of a neutron star is less than about two or three solar masses, its subsequent evolution is thought to be similar to that of a white dwarf. If the mass is greater than this, the gravitational force may become so strong that the star contracts to an even smaller diameter and an even greater density, overcoming, in a sense, even the neutron exclusion principle. Gravity would then be so strong that even light emitted from it could not escape—it would be pulled back in by the force of gravity. In other words, the escape velocity from such a star is greater than c, the speed of light. Since no radiation could escape from such a star, we could not see it—it would be black. A body may pass by it and be deflected by its gravitational field, but if it came too close, it would be swallowed up, never to escape. This is a **black hole**. Black holes are predicted by theory to exist. Experimentally, evidence for their existence is strong but not everyone agrees it is sufficient to fully confirm the existence of black holes. One possibility is there may be a giant black hole at the center of our Galaxy—its mass is estimated at several million times that of the Sun—and it seems possible that many or all galaxies have black holes at their centers.

Black holes

33–3 General Relativity: Gravity and the Curvature of Space

We have seen that the force of gravity plays a dominant role in the processes that occur in stars and, in general, in the evolution of the universe as a whole. The reasons gravity, and not one of the other of the four forces in nature, plays the dominant role in the universe are (1) it is long-range and (2) it is always attractive. The strong and weak nuclear forces act over very short distances only, on the order of the size of a nucleus; hence they do not act over astronomical distances (although of course they act between nuclei and nucleons in stars to produce nuclear reactions). The electromagnetic force, like gravity, acts over great distances; but it can be either attractive or repulsive. And since the universe does not seem to contain large areas of net electric charge, a large net force does not occur. But gravity acts as an attractive force between *all* masses, and there are large accumulations in the universe of only the one "sign" of mass (not + and − as with electric charge).

However, the force of gravity as Newton described it in his law of universal gravitation shows discrepancies on a cosmological scale. Einstein, in his general theory of relativity, developed a theory of gravity that resolves these problems and forms the basis of cosmological dynamics.

In the *special theory of relativity* (Chapter 26), Einstein concluded that there is no way for an observer to determine whether a given frame of reference is at rest or is moving at constant velocity in a straight line. Thus, Einstein said, the laws of physics must be the same in different inertial reference frames. (As we saw, the laws can be the same even though the specific *paths* of objects are different in different reference frames.) To consider only uniformly moving reference frames is somewhat restricting. What about the general case of motion, where reference frames can be *accelerating*?

It is in the **general theory of relativity** that Einstein tackled the problem of accelerating reference frames and developed a theory of gravity. The mathematics of general relativity is very complex, so our discussion will be mainly qualitative.

We begin with Einstein's famous **principle of equivalence**, which states that

> **No observer can determine by experiment whether he or she is accelerating or is rather in a gravitational field.**

If some observers sensed that they were accelerating (as in a vehicle speeding around a sharp curve), they could not prove by any experiment that in fact they weren't simply experiencing the pull of a gravitational field. Conversely, we might think we are being pulled by gravity when in fact we are undergoing an "inertial" acceleration having nothing to do with gravity. For example, pilots making a steeply banked turn in fog often have this experience, and cannot tell in which direction the Earth lies.

As a thought experiment, consider a person in a freely falling elevator near the Earth's surface. If our observer held out a book and let go of it, what would happen? Gravity would pull it downward toward the Earth, but at the same rate ($g = 9.8 \, \text{m/s}^2$) at which the person and elevator were falling. So the book would hover right next to the person's hand (Fig. 33–12). The effect is exactly the same as if this reference frame were at rest and *no* forces were acting. On the other hand, suppose the elevator were far out in space where there is no gravitational field. If the person released the book, it would float, just as it does in Fig. 33–12. If, instead, the elevator (out in space) were accelerating upward at an acceleration of $9.8 \, \text{m/s}^2$, the book as seen by our observer would fall to the floor with an acceleration of $9.8 \, \text{m/s}^2$, just as if it were falling because of gravity. According to the principle of equivalence, the observer could do no experiment to determine whether the book fell because the elevator was accelerating upward at $a = 9.8 \, \text{m/s}^2$ in the absence of gravity, or because a gravitational field with $g = 9.8 \, \text{m/s}^2$ was acting downward and he was at rest (say on the Earth). The two descriptions are equivalent.

The principle of equivalence is related to the concept of mass and to the idea that there are two types of mass. For any force, Newton's second law says that $F = ma$, where m is the mass — or more precisely, the **inertial mass**. The more inertial mass a body has, the less it is affected by a given force (the less acceleration it undergoes). You might say that inertial mass represents resistance to any type of force whatever. The second type of mass is **gravitational mass**. When one body attracts another by the gravitational force (Newton's law of universal gravitation, $F \propto mm'/d^2$, Chapter 5), the strength of the force is proportional to the product of the gravitational masses of the two bodies. This is much like the electric force between two bodies that is proportional to the product of their electric charges. The electric charge of a body is clearly not related to its inertial mass; so why should we expect that a body's gravitational mass (call it gravitational charge if you like) be related to its inertial mass? We have, up to now, assumed they were the same. Why? Because no experiment—not even high-precision experiments—has been able to discern any measurable difference between inertial and gravitational mass. This, then, is another way to state the equivalence principle: *gravitational mass is equivalent to inertial mass.*

The principle of equivalence can be used to show that light ought to be deflected due to the gravitational force of a massive body. Let us consider a thought experiment to get the idea. Consider an elevator in free space where no gravity acts. If there is a hole in the side of the elevator

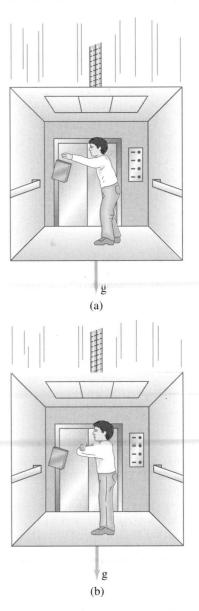

FIGURE 33–12 A freely falling elevator. The released book hovers next to the owner's hand.

(a)

(b)

FIGURE 33–13 (a) Light beam goes straight across an elevator not accelerating. (b) The light beam bends (exaggerated) in an elevator accelerating in an upward direction.

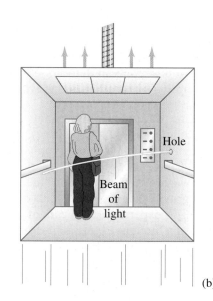

(a)

(b)

and a beam of light enters from outside, the beam travels straight across the elevator and makes a spot on the opposite side if the elevator is at rest (Fig. 33–13a). If the elevator is accelerating upward as in Fig. 33–13b, the light beam still travels straight across in a reference frame at rest. In the upwardly accelerating elevator, however, the beam is observed to curve downward. Why? Because during the time the light travels from one side of the elevator to the other, the elevator is moving upward at ever-increasing speed. Now, according to the equivalence principle, an upwardly accelerating reference frame is equivalent to a downward gravitational field. Hence, we can picture the curved light path in Fig. 33–13b as being the effect of a gravitational field. Thus we expect gravity to exert a force on a beam of light and to bend it out of a straight-line path!

That light is affected by gravity is an important prediction of Einstein's general theory of relativity. And it can be tested. The amount a light beam would be deflected from a straight-line path must be small even when passing a massive body. (For example, light near the Earth's surface after traveling 1 km is predicted to drop only about 10^{-10} m, which is equal to the diameter of a small atom and not detectable.) The most massive body near us is the Sun, and it was calculated that light from a distant star would be deflected by 1.75″ of arc (tiny but detectable) as it passed near the Sun (Fig. 33–14). However, such a measurement could be made only

FIGURE 33–14 (a) Three stars in the sky. (b) If the light from one of these stars passes very near the Sun, whose gravity bends the rays, the star will appear higher than it actually is.

(a)

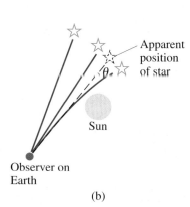

(b)

during a total eclipse of the Sun, so the Sun's tremendous brightness would not overwhelm the starlight passing near its edge. An opportune eclipse occurred in 1919 and scientists journeyed to the South Atlantic to observe it. Their photos of stars around the Sun revealed shifts in accordance with Einstein's prediction. One newspaper flaunted the headline "Light Caught Bending." Newton's universal law of gravitation, coupled with the equivalence principle, predicts a gravitational deflection of light even when light is considered as photons of mass $m = E/c^2$, but the predicted deflection is half that predicted by Einstein's general relativity; and it is the latter that is observed. (See Fig. 33–15.)

Light bending

Now a light beam must travel by the shortest, most direct, path between two points. If it didn't, some other object could travel between the two points in a shorter time and thus have a greater speed than the speed of light—a clear contradiction of the special theory of relativity. If a light beam can follow a curved path (as discussed above), then this curved path must be the shortest distance between the two points—which suggests that *space itself is curved* and that it is the gravitational field that causes the curvature. Indeed, the curvature of space—or rather, of four-dimensional space-time—is a basic aspect of general relativity.

What is meant by *curved space*? To understand, let us recall that our normal method of viewing the world is via Euclidean plane geometry. In Euclidean geometry, there are many axioms and theorems we take for granted, such as that the sum of the angles of any triangle is 180°. Other geometries, non-Euclidean which involve curved space, have also been imagined by mathematicians. Now it is hard enough to imagine three-dimensional curved space, much less curved four-dimensional space–time. So let us explain the idea of curved space by using two-dimensional surfaces.

Consider, for example, the two-dimensional surface of a sphere. It is clearly curved, Fig. 33–16, at least to us who view it from the outside—from our three-dimensional world. But how would hypothetical two-dimensional creatures determine whether their two-dimensional space were flat (a plane) or curved? One way would be to measure the sum of the angles of a triangle. If the surface is a plane, the sum of the angles is 180°, as we learn in plane geometry. But if the space is curved, and a sufficiently large triangle is constructed, the sum of the angles will *not* be 180°. To construct a triangle on a curved surface, say a sphere, we must use the equivalent of a straight line: that is, the shortest distance between two points, which is called a **geodesic**. On a sphere, a geodesic is an arc of a great circle (an arc contained in a plane passing through the center of the sphere) such as the Earth's equator and the Earth's longitude lines. Consider, for example, the large triangle of Fig. 33–16 whose sides are two "longitude" lines passing from the "north pole" to the equator, a part of which forms the third side. The two longitude lines make 90° angles with the equator (look at a world globe to see this more clearly). If they make, say, a 90° angle with each other at the north pole, the sum of these angles is 90° + 90° + 90° = 270°. This is clearly *not* a Euclidean space. Note, however, that if the triangle is small in comparison to the radius of the sphere, the angles will add up to nearly 180°, and the triangle (and space) will seem flat.

Another way to test the curvature of space is to measure the radius r and circumference C of a large circle. On a plane surface, $C = 2\pi r$. But on a two-dimensional spherical surface, C is *less* than $2\pi r$, as can be seen in

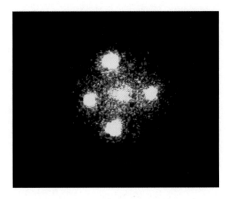

FIGURE 33–15 Photograph, taken by the Hubble Space Telescope, of the so-called "Einstein cross" (or Huchra's cross after its discoverer, John Huchra). It is thought to represent "gravitational lensing": the central spot is a relatively nearby galaxy, whereas the four other spots are thought to be images of a single quasar *behind* the galaxy, and the galaxy bends the light coming from the quasar to produce the four images. See also Fig. 33–14. [If the shape of the nearby galaxy were perfectly symmetric, we would expect the "image" of the distant quasar to be a circular ring or halo, instead of the four spots seen here.]

FIGURE 33–16 On a two-dimensional curved surface, the sum of the angles of a triangle may not be 180°.

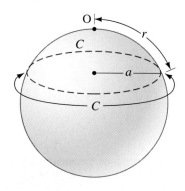

FIGURE 33–17 On a spherical surface, a circle of circumference C is drawn about point O as the center. The radius of the circle (not the sphere) is the distance r along the surface. (Note that in our three-dimensional view, we can tell that $2\pi a = C$; since $r > a$, then $2\pi r > C$.)

The universe: open or closed?

FIGURE 33–18 Example of a two-dimensional surface with negative curvature.

Fig. 33–17. The proportionality constant between C and r is *less* than 2π. Such a surface is said to have *positive curvature*. On the saddlelike surface of Fig. 33–18, the circumference of a circle is greater than $2\pi r$, and the sum of the angles of a triangle is less than 180°. Such a surface is said to have a *negative curvature*.

Now, what about our universe? On a large scale (not just near a large mass), what is the overall curvature of the universe? Does it have positive curvature, negative curvature, or is it flat (zero curvature)? In the nineteenth century, Carl Friedrich Gauss (1777–1855) tried to determine whether our natural three-dimensional space deviated from Euclidean space by measuring the angles of a triangle formed by three mountain peaks using light rays as sides of the triangle. He was unable to detect any deviation from 180°, nor have experiments today detected any deviation.

Nonetheless, the question of the curvature of space in the real world is an important one in cosmology. And the answer is still not known. If the universe has a positive curvature, then the universe is *finite*, or *closed*. This does not mean that in such a universe the stars and galaxies would extend out to a certain boundary, beyond which there is empty space. Rather, galaxies would be spread throughout the space, and the space would fold back and "close on itself." There is no boundary or edge in such a universe. If a particle were to move in a straight line in a particular direction, it would eventually return to the starting point—albeit eons of time later. To ask "What is beyond such a closed universe?" is futile, since the space of the universe is all there is. If the curvature of space is zero or negative, on the other hand, the universe would be *open*. It would just go on and on and never fold back on itself. An open universe would be *infinite*. Whether the universe is open or closed depends, in part, on how much total mass there is in the universe, as we will discuss in Section 33–7. If the mass is great enough, it bends space into a positively curved, closed, and finite space.

According to Einstein's theory, space–time is curved, especially near massive bodies. To visualize this, we might think of space as being like a thin rubber sheet; if a heavy weight is hung from it, it curves as shown in Fig. 33–19. The weight corresponds to a huge mass that causes space (space itself!) to curve. Thus, in Einstein's theory[†] we do not speak of the

FIGURE 33–19 Rubber-sheet analogy for space–time curved by matter.

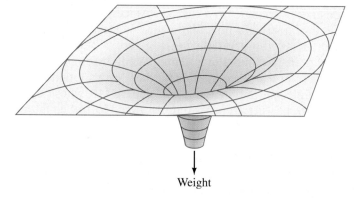

Weight

[†]Alexander Pope (1688–1744) wrote an epitaph for Newton:
"Nature, and Nature's laws lay hid in night:
God said, *Let Newton be!* and all was light."
Sir John Squire (1884–1958), perhaps uncomfortable with Einstein's profound thoughts, added:
"It did not last: the Devil howling '*Ho!*
Let Einstein be!' restored the status quo."

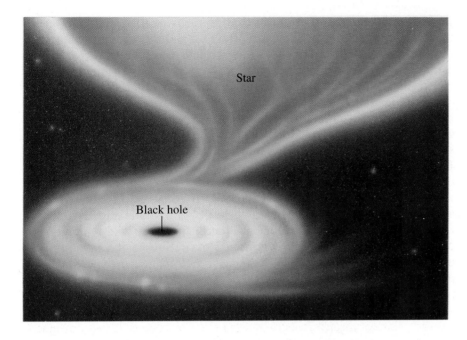

Black holes

FIGURE 33–20 Artist's conception of how a black hole, which is one star of a binary pair, might pull matter from the other (more normal) star.

"force" of gravity acting on bodies. Instead we say that bodies and light rays move along geodesics (the equivalent of straight lines in plane geometry) in curved space–time. Thus, a body at rest or moving slowly near the great mass of Fig. 33–19 would follow a geodesic toward that body.

The extreme curvature of space–time shown in Fig. 33–19 could be produced by a **black hole**. A black hole, as we saw in the previous Section, is so dense that even light cannot escape from it. To become a black hole, a body of mass M must undergo **gravitational collapse**, contracting by gravitational self-attraction to within a radius called the **Schwarzschild radius**:

$$R = \frac{2GM}{c^2},$$

where G is the gravitational constant and c is the speed of light.

How might we observe black holes? We cannot see them because no light can escape from them. They would be black objects against a black sky. But they do exert a gravitational force on nearby bodies. The suspected black hole at the center of our Galaxy was inferred by examining the motion of matter in its vicinity. Another technique is to examine stars which appear to be rotating as if they were members of a *binary system* (two stars rotating about their common center of mass), although the companion is invisible. If the unseen star is a black hole, it might be expected to pull off gaseous material from its visible companion (Fig. 33–20). As this matter approached the black hole, it would be highly accelerated and should emit X-rays of a characteristic type. One of the candidates for a black hole is in the binary star Cygnus X-1.

33–4 The Expanding Universe

We discussed in Section 33–2 how individual stars evolve from their birth to their death as white dwarfs, neutron stars, and black holes. But what about the universe as a whole: is it static or does it evolve? The evolution of stars suggests that the universe as a whole evolves.

One of the most important scientific ideas of this century proposes that distant galaxies are racing away from us, and that the farther they are from us, the faster they are moving away. How astronomers arrived at this astonishing idea and what it means for the past history of the universe as well as its future, will occupy us for the remainder of the book.

Hubble and the redshift

The idea that the universe is expanding was first put forth by Hubble in 1929. It was based on observations of the Doppler shift of light emitted by stars. In Chapter 12 we discussed how the frequency and wavelength of sound are altered if the source is moving toward or away from an observer. If the source moves toward us, the frequency is higher and the wavelength is shorter. If the source moves away from us, the frequency is lower and the wavelength is longer. The **Doppler effect** occurs also for light, but the shifted wavelength or frequency is given by a formula slightly different from that for sound because for light (according to special relativity), we can make no distinction between motion of the source and motion of the observer. (Recall that sound travels in a medium, such as air, but light does not; according to relativity, there is no ether.) According to special relativity, the formula is (see Problem 31)

$$\lambda' = \lambda \sqrt{\frac{1 + v/c}{1 - v/c}}, \tag{33-3}$$

where λ is the emitted wavelength as seen in the source's reference frame, and λ' is the wavelength measured in a frame moving with velocity v away from the source along the line of sight. Note that this relation depends only on the relative velocity v. (For relative motion toward each other, $v < 0$ in this formula.) Thus, when a source emits light of a particular wavelength and the source is moving away from us, the wavelength appears longer to us: the color of the light (if it is visible) is shifted toward the red end of the visible spectrum, an effect known as a **redshift**. If the source moves toward us, the color shifts toward the blue (shorter wavelength) end of the spectrum. The amount of shift depends on the velocity of the source (Eq. 33-3). For speeds not too close to the speed of light, it is easy to show (Problem 30) that the fractional change in wavelength is proportional to the speed of the source to or away from us (as was the case for sound):

Redshift

$$\frac{\Delta\lambda}{\lambda} = \frac{\lambda' - \lambda}{\lambda} \approx \frac{v}{c}. \qquad [v \ll c] \quad \textbf{(33-4)}$$

In the spectra of stars and galaxies, lines are observed that correspond to lines in the known spectra of particular atoms (see Section 27-9). What Hubble found was that the lines seen in the spectra of galaxies were generally *redshifted*, and that the amount of shift seemed to be approximately proportional to the distance of the galaxy from us. That is, the velocity, v, of a galaxy moving away from us is proportional to its distance, d, from us:

Hubble's law

$$v = Hd. \tag{33-5}$$

This is **Hubble's law**, one of the most important astronomical ideas of this century. The constant H is called the **Hubble parameter**. Hubble's law does not work well for nearby galaxies—in fact some are actually moving toward us ("blueshifted"); but this is believed to merely represent random motion of the galaxies. For more distant galaxies, the velocity of recession (Hubble's law) is much greater than that of random motion, and so is dominant.

The value of H is not known very precisely. It is generally taken to be about

$$H \approx 80 \text{ km/s/Mpc}$$

(that is, 80 km/s per megaparsec of distance). If we use light-years for distance, then $H \approx 25$ km/s per million light-years of distance. However, H could be as low as 50 km/s/Mpc or as high as about 100 km/s/Mpc (15 to 30 km/s/Mly). This rather large uncertainty arises mainly from the uncertainty in distance measurement for remote galaxies.

What does it mean that distant galaxies are all moving away from us, and with ever greater speed the farther they are from us? It is as if there had been a great explosion at some distant time in the past. And at first sight we seem to be in the middle of it all. But we aren't, necessarily. The expansion appears the same from any other point in the universe. To understand why, see Fig. 33–21. In Fig. 33–21a we have the view from Earth (or from our Galaxy). The velocities of surrounding galaxies are indicated by arrows, pointing away from us, and greater for galaxies more distant from us. Now what if we were on the galaxy labeled A in Fig. 33–21a. From Earth, galaxy A appears to be moving to the right at a velocity, call it $\mathbf{v}_A$, represented by the arrow pointing to the right. If we were *on* galaxy A, Earth would appear to be moving to the left at velocity $-\mathbf{v}_A$. To determine the velocities of other galaxies relative to A, we vectorially add the velocity vector, $-\mathbf{v}_A$, to all the velocity arrows shown in Fig. 33–21a. This yields Fig. 33–21b, where we see clearly that the universe is expanding away from galaxy A as well; and the velocities of galaxies receding from A are proportional to their distance from A.

Thus the expansion of the universe can be stated as follows: all galaxies are racing away from *each other* at an average rate of about 80 km/s per megaparsec of distance between them (or 25 km/s per million light-years of separation). The ramifications of this idea are profound, and we discuss them in a moment.

There is, however, a class of objects called **quasars**, or "quasistellar radio sources," that do not seem to conform to Hubble's law. Quasars are as bright as nearby stars but display very large redshifts. If quasars are normal participants in the general expansion of the universe according to Hubble's law, their large redshifts suggest they are very distant. If they are so far away, they must be incredibly bright—sometimes thousands of times brighter than normal galaxies. On the other hand, it has been suggested that some quasars, at least, are not of abnormal brightness because

FIGURE 33-21 Expansion of the universe looks the same from any point in the universe.

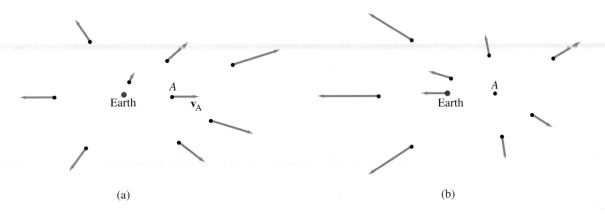

(a)

(b)

they are nearer than their redshifts suggest. In the first case we would have an unresolved brightness problem, in the second case an unresolved redshift problem. An interesting piece of evidence is that the population density of quasars seems to increase with distance from us. If they are at great distances from us—as their redshifts suggest—this would merely mean they were more common in the early universe than they are now. But if instead they are much closer to us—as their brightness suggests—it would seem that *we* are in a special place in the universe, the place where quasars are least populous. Most astronomers are unwilling to accept this, for it would violate a basic assumption of uniformity, the so-called *cosmological principle*, which we discuss in the next paragraph. A recent suggestion that could resolve the quasar question is that these mysterious objects may be cores of galaxies that emit enormous amounts of energy when they draw in nearby material, for example when galaxies "collide." If the center of a galaxy is a black hole, a large amount of matter accelerated toward it would give off a prodigious amount of energy. It seems possible, then, that quasars are powered by black holes.

Cosmological principle   A basic assumption in cosmology has been that on a large scale, the universe looks the same to observers at different places at the same time. In other words, the universe is both isotropic (looks the same in all directions) and homogeneous (would look the same if we were located elsewhere, say in another galaxy). This assumption is called the **cosmological principle**. On a local scale, say in our solar system or within our Galaxy, it clearly does not apply (the sky looks different in different directions). But it has long been thought to be valid if we look on a large enough scale, so that the average population density of galaxies and clusters of galaxies ought to be the same in different areas of the sky. The expansion of the universe (Fig. 33–21) is consistent with the cosmological principle, and the uniformity of the cosmic microwave background radiation (discussed in the next Section) supports it. But matter (i.e., galaxies), even on the largest scale, seems not to be homogeneous but tends to clump, as already discussed. Although astronomers are reluctant to give up the cosmological principle, there is now some doubt about its validity. One possible resolution might be that over 90 percent of the universe may be nonluminous dark matter, which might be uniformly distributed. In any case, the cosmological principle has aided us in treating the universe as a single evolving entity and not as a random collection of material bodies. Another way of stating the cosmological principle is this: There is nothing special about the Earth on a cosmological scale; our large-scale observations are no different from those that might be made elsewhere in the universe. What a change has occurred since that time, only a few hundred years ago, when we saw ourselves at the center of the universe!

The expansion of the universe, as described by Hubble's law, strongly suggests that galaxies must have been closer together in the past than they are now. Further, Hubble's law is consistent with all galaxies having been quite close together at some time in the past. This is, in fact, the basis of the *Big Bang* theory of the origin of the universe (discussed in the next section) which pictures the beginning of the universe as a great explosion. Let us see what can then be said about the age of the universe.

Age of the universe   We can estimate the age of the universe using the Hubble parameter, and let us choose the midpoint between the accepted extremes of $H = 15$ to 30 km/s/Mly, namely $H \approx 22$ km/s per million light-years. Then the time required for the galaxies to arrive at their present separations would be approximately (starting with $v = d/t$ and using Eq. 33–5):

$$t = \frac{d}{v} = \frac{d}{Hd} = \frac{1}{H} \approx \frac{(10^6 \, \text{ly})(10^{13} \, \text{km/ly})}{(22 \, \text{km/s})(3 \times 10^7 \, \text{s/yr})} \approx 15 \times 10^9 \, \text{yr},$$

or roughly 15 billion years. The age of the universe calculated in this way is called the *characteristic expansion time* or "Hubble age." It is not very precise since we don't know the value of H precisely, and may be an overestimate since the galaxies have not been moving with fixed velocities but have been slowing down under the action of their mutual gravitational attraction.

There are two other independent checks on the age of the universe. The first is determination of the age of the Earth (and solar system) from radioactivity, primarily using uranium, which places the age of the solar system at about $4\frac{1}{2}$ billion years. Second, by using the theory of stellar evolution, the ages of stars have been estimated to be about 10 to 15 billion years. These independent and unrelated determinations are consistent with a Big Bang occurring, according to our best estimates today, 10 to 15 billion years ago. The lower value determined from radioactivity is consistent since we would expect the origin of the Earth (and the solidification of rocks) to have occurred somewhat after the origin of the universe as a whole.

Before discussing the Big Bang theory in more detail, we briefly discuss one of the alternatives to the Big Bang—the **steady-state model**— which assumes that the universe is infinitely old and on the average looks the same now as it always has.[†] Thus, according to the steady-state model, no large-scale changes have taken place in the universe as a whole, particularly no Big Bang. To maintain this view in the face of the recession of galaxies away from each other, mass–energy conservation must be violated. That is, matter is assumed to be created continuously, keeping the density of the universe constant. The rate of mass creation required is very small— about one nucleon per cubic meter every 10^9 years. This is much too small to be detected, and thus cannot be tested.

Steady-state model

The steady-state model provided the Big Bang model with healthy competition in the 1940s and 50s. However, the discovery of the cosmic microwave background radiation (next Section) pushed the Big Bang model to the forefront.

33–5 The Big Bang and the Cosmic Microwave Background

The expansion of the universe seems to suggest, as we have seen, that the matter of the universe was once much closer together than it is now. This is the basis for the idea that the universe began 10 to 15 billion years ago with a huge explosion known affectionately as the **Big Bang**.

The Big Bang

If there was a Big Bang, it must have occurred simultaneously at all points in the universe. If the universe is *finite*, the explosion would have taken place in a tiny volume approaching a point. However, this point of extremely dense matter is not to be thought of as a concentrated mass in the midst of a much larger space around it. Rather, the initial dense point

[†]That the universe should be uniform in time as well as in space is essentially an extension of the cosmological principle and is sometimes referred to as the *perfect cosmological principle*.

was the universe—the entire universe. There wouldn't have been anything else. If, on the other hand, the universe is *infinite*, then the explosion would have occurred at *all* points in the universe since an infinite universe, even if smaller at an earlier time, would still have been infinite. In either case, when we say the universe was once smaller than it is now, we mean that the average separation between galaxies was less. Thus, it is the *size of the universe itself* that has increased since the Big Bang.

What is the evidence supporting the Big Bang? First, the age of the universe as calculated from the Hubble expansion, from stellar evolution, and from radioactivity, all point to a consistent time of origin for the universe, as we saw in the last Section. Another, and crucial, piece of evidence was the discovery in the 1960s of the **cosmic microwave background** radiation (or CMB), which came about as follows.

In 1964, Arno Penzias and Robert Wilson were experiencing difficulty with what they assumed to be background noise, or "static," in their radio telescope (a large antenna device for detecting radio waves from the heavens, Fig. 33–22). Eventually, they became convinced that it was real and that it was coming from outside our Galaxy. They made precise measurements at a wavelength $\lambda = 7.35$ cm, which is in the microwave region of the electromagnetic spectrum. (See Fig. 22–10; this radiation is called "microwave" because the wavelength, though much greater than that of visible light, is somewhat smaller than wavelengths for ordinary radio waves, which are typically meters or hundred of meters.) The intensity of this radiation was found initially not to vary by day or night or time of year, nor to depend on direction. It came from all directions in the universe with equal intensity (within less than one part per thousand). It could only be concluded that this radiation came from beyond our Galaxy, from the universe as a whole. The remarkable uniformity of the cosmic microwave background radiation was in accordance with the cosmological principle. But theorists also felt that there needed to be some small inhomogeneities in the CMB that would have provided "seeds" around which galaxy formation could have started. Such small inhomogeneities, on the order of parts per million, were finally detected in 1992.

The intensity of this radiation as measured at $\lambda = 7.35$ cm corresponds to blackbody radiation (see Section 27–2) at a temperature of about 3 K, now measured more precisely to be 2.7 ± 0.1 K. When radiation at other wavelengths was measured, the intensities were found to fall on a blackbody curve (see Fig. 27–4), with the peak of the curve a little above 0.1 cm corresponding to a temperature of 2.7 K.

The discovery of the CMB at a temperature of 2.7 K ranks as one of the two most significant cosmological discoveries of this century. (The other was Hubble's expanding universe.) It is highly significant because it provides strong evidence in support of the Big Bang, and it gives us some idea of conditions in the very early universe. In fact, in the late 1940s, George Gamow and his collaborators calculated that a Big Bang origin of the universe should have generated just such a microwave background radiation.

To understand this, let us look at what a Big Bang might have been like. There must have been a tremendous release of concentrated energy. The temperature must have been extremely high, so high that there could not have been any atoms in the very early stages of the universe. Instead,

The 2.7 K cosmic microwave background radiation

FIGURE 33–22 Robert Wilson (left) and Arno Penzias, and behind them their "horn antenna."

Importance of CMB: the Big Bang

the universe must have consisted solely of radiation (photons) and elementary particles. The universe would have been opaque—the photons in a sense "trapped," since as soon as they were emitted they would have been scattered or absorbed, primarily by electrons. Indeed, the microwave background radiation is strong evidence that matter and radiation were once in equilibrium at a very high temperature. As the universe expanded, the energy would have spread out over an increasingly larger volume and the temperature would have dropped. Only when the temperature had reached about 3000 K some 300,000 years later, could nuclei and electrons have stayed together as atoms. With the disappearance of free electrons, as they combined with nuclei to form atoms, the radiation would have been freed—"decoupled" from matter, we could say—to spread throughout the universe. As the universe expanded, so too the wavelengths of the radiation expanded, redshifting to longer wavelengths that correspond to lower temperature (recall Wien's law, $\lambda_{max} T = $ constant, Section 27–2), until they would have reached the 2.7 K background radiation we observe today.

Although the total energy associated with the cosmic microwave background is much larger than that from other radiation sources (such as the light emitted by stars), it is small compared to the energy, and mass (remember $E = mc^2$), associated with matter. In fact today, radiation is believed to make up less than $\frac{1}{1000}$ of the energy of the universe: today the universe is *matter-dominated*. But it was not always so. The cosmic microwave background radiation strongly suggests that early in its history the universe was *radiation-dominated*.[†] But, as we shall see in the next Section, that period lasted less than $\frac{1}{10,000}$ of the history of the universe (thus far).

33–6 The Standard Cosmological Model: The Early History of the Universe

It is now almost generally agreed that the evolution of the universe must have been determined in the first few moments of the Big Bang. In the last decade or two, a convincing theory of the origin and evolution of the universe has developed, now known as the **standard model**. Although a few cosmologists hold other views, most favor the standard model. Much of this theory is based on recent theoretical and experimental advances in elementary particle physics. Indeed, in the last few years cosmology and elementary particle physics have cross-fertilized to a surprising extent.

Let us go back now to the earliest of times—as close as possible to the Big Bang—and follow a standard model scenario of events as the universe expanded and cooled after the Big Bang. Initially, we will be talking of extremely small time intervals, as well as extremely high temperatures, far

---

[†]If there had not been such intense radiation in the first few minutes of the universe, nuclear reactions might have produced a much larger percentage of heavy nuclei than we see. Instead, nearly 75 percent of visible matter is hydrogen, presumably because the intense radiation immediately blasted apart any heavy nuclei that formed into their constituent protons and neutrons.

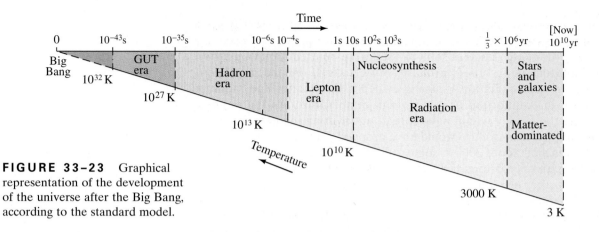

FIGURE 33–23 Graphical representation of the development of the universe after the Big Bang, according to the standard model.

beyond anything in the universe today. Figure 33–23 is a graphical representation of the events, and it may be helpful to consult it as we go along.

The standard-model "scenario" of the history of the universe after the Big Bang

We begin at a time only a minuscule fraction of a second after the Big Bang, 10^{-43} s. Although this is an unimaginably short time, predictions as early as this can be made based on present theory, albeit somewhat speculatively. Earlier than this instant, we can say nothing since we do not yet have a theory of quantum effects on gravity which would be needed for the incredibly high densities and temperatures then. It is imagined, however,

All four forces unified

that prior to 10^{-43} s, the four forces in nature were unified—there was only one force. The temperature would have been about 10^{32} K, corresponding to particles moving about every which way with an average kinetic energy of 10^{19} GeV (see Eq. 13–8):

$$\text{KE} \approx kT \approx \frac{(1.4 \times 10^{-23}\ \text{J/K})(10^{32}\ \text{K})}{1.6 \times 10^{-19}\ \text{J/eV}} \approx 10^{28}\ \text{eV} = 10^{19}\ \text{GeV}.$$

(Note that the factor $\frac{3}{2}$ in Eq. 13–8 is usually ignored in such calculations.)

Symmetry broken (gravity condensed out)

At $t = 10^{-43}$ s, a kind of "phase transition" is believed to have occurred during which the gravitational force, in effect, "condensed out" as a separate force. This, and subsequent phase transitions, are somewhat analogous to the phase transitions water undergoes as it cools from a gas, condenses into a liquid, and with further cooling freezes into ice. The symmetry of the four forces was broken, but the strong, weak, and electromagnetic forces were still unified. At this point, the universe entered the so-called **grand unified era** (after grand unified theory—see Chapter 32). There was no distinction between quarks and leptons; baryon and lepton numbers were not conserved. Very shortly thereafter, as the universe expanded considerably and the temperature had dropped to about 10^{27} K, there was another phase transition during which the strong force condensed out. This probably occurred about 10^{-35} s after the Big Bang. Now the universe was filled with a soup of leptons and quarks. The leptons included electrons, muons, taus, neutrinos, and all their antiparticles. The quarks were initially free (something we have not seen in our present universe), but soon they began to "condense" into more normal particles: nucleons and the other hadrons and their antiparti-

Quark confinement (hadron era)

cles. With this **confinement of quarks**, the universe entered the **hadron era**.

We can think of this "soup" as a grand mixture of particles and antiparticles, as well as photons—all in roughly equal numbers—colliding with one another frequently and exchanging energy.

By the time the universe was only about a microsecond (10^{-6} s) old, it had cooled to about 10^{13} K, corresponding to an average KE of 1 GeV, and the vast majority of hadrons disappeared. To see why, let us focus on the most familiar hadrons: nucleons and their antiparticles. When the average kinetic energy of particles was somewhat higher than 1 GeV, protons, neutrons, and their antiparticles were continually being created out of the energies of collisions involving photons and other particles, such as

$$\text{photons} \rightarrow \text{p} + \bar{\text{p}}, \qquad \text{n} + \bar{\text{n}}.$$

But just as quickly, particles and antiparticles would annihilate: for example

$$\text{p} + \bar{\text{p}} \rightarrow \text{photons or leptons}.$$

So the processes of creation and annihilation of nucleons were in equilibrium. The numbers of nucleons and antinucleons were high—roughly as many as there were electrons, positrons, or photons. But as the universe cooled and the average kinetic energy of particles dropped below about 1 GeV, which is the minimum energy needed to create nucleons and antinucleons (about 940 MeV each) in a typical collision, the process of nucleon creation could not continue. The process of annihilation could continue, however, with antinucleons annihilating nucleons, until there were almost no nucleons left. But not quite zero. To explain our present world, which consists mainly of matter (nucleons and electrons) with very little antimatter in sight, we must suppose that earlier in the universe, perhaps around 10^{-35} s after the Big Bang, a slight excess of quarks over antiquarks was formed.[†] This would have resulted in a slight excess of nucleons over antinucleons. And it is these "leftover" nucleons that we are made of today. The excess of nucleons over antinucleons was about one part in 10^9. Earlier, during the hadron era, there should have been about as many nucleons as photons. After it ended, the "leftover" nucleons thus numbered only about one nucleon per 10^9 photons, and this ratio has persisted to this day. Protons, neutrons, and all other heavier particles were thus tremendously reduced in number by about 10^{-6} s after the Big Bang. The lightest hadrons, the pions, disappeared as the nucleons had; because they are the lightest mass hadrons (140 MeV), they were the last hadrons to go, around 10^{-4} s after the Big Bang. Lighter particles, including electrons, positrons, neutrinos, photons—in roughly equal numbers—dominated, and the universe entered the **lepton era**.

By the time the first full second had passed (clearly the most eventful second in history!), the universe had cooled to about 10 billion degrees, 10^{10} K. The average KE was about 1 MeV. This was still sufficient energy to create electrons and positrons and balance their annihilation reactions, since their masses correspond to about 0.5 MeV. So there were about as many e^+ and e^- as there were photons. But within a few more seconds, the temperature had dropped sufficiently so that e^+ and e^- could no longer be formed. Annihilation ($e^+ + e^- \rightarrow$ photons) continued. And, like nucleons before them, electrons and positrons all but disappeared from the universe—except for a slight excess of electrons over positrons (later to join with nuclei to form atoms). Thus, about $t = 10$ s after the Big Bang, the universe entered the **radiation era**. Its major constituents were now pho-

---

[†]An alternative possibility is that there was perfect symmetry between quarks and antiquarks, matter and antimatter, but that the universe somehow separated into domains, some containing only matter, others only antimatter. If this were true, we would expect antiparticles from such distant domains to reach us, at least occasionally, in cosmic rays; but none has ever been detected.

tons and neutrinos. But the neutrinos, partaking only in the weak force, rarely interacted. So the universe, until then experiencing an energy balance between matter and radiation, became **radiation-dominated**: much more energy was contained in radiation than in matter, a situation that would last hundreds of thousands of years (Fig. 33–23).

Radiation-dominated universe

Meanwhile, during the next few minutes, crucial events were taking place. Beginning about 2 or 3 minutes after the Big Bang, nuclear fusion began to occur. The temperature had dropped to about 10^9 K, corresponding to $\overline{\text{KE}} \approx 100$ keV, where nucleons could strike each other and be able to fuse (Section 31–3), but now cool enough that newly formed nuclei would not be immediately broken apart by subsequent collisions. Deuterium, helium, and very tiny amounts of lithium nuclei were probably made. But the universe was cooling too quickly, and larger nuclei were not made. After only a few minutes, probably not even a quarter of an hour after the Big Bang, the temperature dropped far enough that nucleosynthesis stopped, not to start again for millions of years (in stars). Thus, after the first hour or so of the universe, matter consisted mainly of bare nuclei of hydrogen (about 75 percent) and helium (about 25 percent)[†] and electrons. But radiation (photons) continued to dominate.

Making He

Our story is almost complete. The next important event is presumed to have occurred about 300,000 years later. The universe had expanded to about $\frac{1}{1000}$ of its present size, and the temperature had cooled to about 3000 K. The average KE of nuclei, electrons, and photons was less than an electron volt. Since ionization energies of atoms are on the order of eV, then as the temperature dropped below this point, electrons could orbit the bare nuclei and remain there (without being ejected by collisions), thus forming atoms. With the birth of atoms, the photons which were continually scattering from the free electrons became much freer to spread throughout the universe. The total energy contained in radiation had been decreasing (redshifting as the universe expanded), until at this point it was about equal to the total energy contained in matter. As the universe continued to expand, the radiation cooled further (to 2.7 K today, forming the cosmic microwave background radiation we detect from everywhere in the universe), and lost energy. But the mass of material particles did not decrease, so beginning at about this point the energy of the universe became increasingly concentrated in matter rather than in radiation: the universe became **matter-dominated**, as it remains today.[‡]

Birth of stable atoms

Matter-dominated universe

Shortly after the birth of atoms, stars and galaxies formed—probably by self-gravitation around mass concentrations (inhomogeneities). This transpired about a million years after the Big Bang. The universe continued to evolve (see Section 33–2) until today, 10 to 15 billion years later.

*   *   *

[†]This standard model prediction of a 25 percent primordial production of helium is fully in accord with what we observe today—the universe *does* contain about 25 percent He—and it is strong evidence in support of the standard Big Bang model. Furthermore, the theory says that 25 percent He abundance is fully consistent with there being three neutrino types, which is the number we observe so far. And it sets an upper limit of four to the maximum number of possible neutrino types. Actually, the fourth could be another type of low-mass particle, a *photino* or a *gravitino*, for example. Here we have a situation where cosmology actually makes a specific prediction about fundamental physics.

[‡]Although today matter contains more of the energy of the universe than does radiation, there are many more photons (perhaps 10^9 times more) than atoms, nuclei, and electrons. But each photon (at $T \approx 2.7$ K) has very little energy.

This scenario is by no means "proven" in any sense. Nor does it answer all questions. But it does provide a tentative picture, for the first time, of how the universe may have begun and evolved. It does have problems, however, some of which have been resolved by a modification first proposed in the early 1980s known as the **inflationary scenario**. According to the inflationary scenario, at the earliest stages of cosmological evolution, around 10^{-35} s after the Big Bang, the universe underwent a very rapid exponential expansion associated with the phase transition (symmetry breaking) that separated the strong force from the electroweak (as discussed earlier in this Section). After the brief inflationary period, the scenario settles back to the standard expansion already discussed. The appeal of the inflationary scenario is that it provides natural explanations for a number of problems not resolved by the standard model, such as why the universe is as close to being as flat as it seems to be, and why the cosmic microwave background radiation is so uniform.

Inflation

There are, however, some questions we haven't yet treated such as: was there a stage before the Big Bang, or did time just begin with the Big Bang? And what of the future of the universe? We'll look at these questions next, in the final Section of this book.

33–7 The Future of the Universe?

According to the standard Big Bang model, the universe is evolving and changing. Individual stars are evolving and dying as white dwarfs, neutron stars, black holes. At the same time, the universe as a whole is expanding. One important question is whether the universe will continue to expand forever. This question is connected to the curvature of space–time (Section 33–3) and to whether the universe is open (and infinite) or closed (and finite). There are three possibilities as shown in Fig. 33–24. If the curvature of the universe is *negative*, the expansion of the universe will never stop, although it should decrease due to the gravitational attraction of its parts. Such a universe would be *open* and infinite. If the universe is *flat* (no curvature), it will still be open and infinite but its expansion will slowly approach a zero rate. Finally, if the universe has *positive* curvature, it will be *closed* and finite. The effect of gravity in this case would be strong enough so that the expansion would eventually stop and the universe would then begin to contract. All matter eventually would collapse back onto itself in a **big crunch**. If, in this last case, the maximum expansion of the universe corresponded to, say, an intergalactic separa-

An open or closed universe?

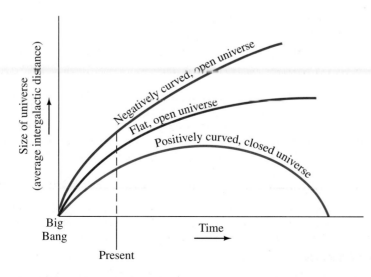

FIGURE 33–24 Three future possibilities for the universe.

1027

tion twice what it is now, the maximum expansion would occur about 30 or 40 billion years from now. Then, as the universe began to contract, the big crunch would occur about 100 billion years after the Big Bang.

Whether we live in an open and continually expanding universe, or a closed one that eventually will contract, is a basic question in cosmology. But we don't know the answer. How might we find out? One way is to determine the average mass density in the universe. If the average mass density is above a critical value known as the **critical density**, estimated to be about

$$\rho_c \approx 10^{-26} \text{ kg/m}^3$$

Critical density of the universe

(i.e., a few nucleons/m^3 on average throughout the universe), then gravity will prevent expansion from continuing forever, and will eventually pull the universe back into a big crunch. To say it another way, if $\rho > \rho_c$, there would be sufficient mass that gravity would give space–time a positive curvature. If instead the actual density is equal to the critical density, $\rho = \rho_c$, the universe will be flat and open. If the actual density is less than the critical density, $\rho < \rho_c$, the universe will have negative curvature and be open, expanding forever.

Great efforts have gone into measuring the actual density of the universe. Estimates of the amount of visible matter in the universe put the actual density at between one and two orders of magnitude less than the critical density, thereby suggesting an open universe. However, there is evidence for a significant amount of nonluminous matter in the universe, often referred to as the **missing mass** or **dark matter**, enough to bring the density to almost exactly ρ_c.

"Missing" mass or dark matter

For example, observations of the rotation of galaxies suggest that they rotate as if they had considerably more mass than we can see. And observations of the motion of galaxies within clusters also suggest that they have considerably more mass than can be seen. Furthermore, the highly regarded inflationary theory offers a strong argument in favor of ρ being precisely equal to ρ_c, and thus that space on a grand scale is Euclidean. If there is nonluminous matter in the universe, what might it be? One suggestion is that the missing mass consists of previously unknown weakly interacting massive particles (WIMPS), or perhaps small primordial black holes made in the early stages of the universe. Another possibility for the missing mass (or part of it) are MACHOS (massive compact halo objects), which would be chunks of matter in the form of large planets (like Jupiter) or stars that are too small to sustain fusion; they might glow faintly due to energy released from gravitational contraction (thus they are sometimes referred to as *brown dwarfs*), but too faintly to be seen. Evidence for MACHOS was found in 1993, inferred from their (assumed) gravitational effect on light passing by them from a distant star.

Another proposal for the missing mass is that neutrinos, once believed to be massless, may actually have nonzero rest mass. Since the universe probably contains roughly as many neutrinos of each type as it does photons (that is, about 10^9 times the number of nucleons, although this neutrino background has yet to be detected), neutrino masses of only a few eV could help to bring the actual density of the universe up to the critical density. The supernova of 1987 offered a fine opportunity to measure the electron neutrino mass. If neutrinos have nonzero rest mass, then their velocities are less than the speed of light ($v < c$). High-energy neutrinos emitted from the supernova should in this case have arrived at Earth earlier than lower-energy (and therefore slower) neutrinos emitted at the same instant. Since they traveled a distance of about 170,000 ly from SN1987a, the time difference ought to be measurable. To get an idea of the size of the effect, let us consider the following Example.

EXAMPLE 33–7 | ESTIMATE | **Neutrino mass estimate from supernova.**
[A challenging Example.] Suppose two neutrinos from SN1987a were detected on Earth (via the reaction $\bar{\nu}_e + p \rightarrow n + e^+$) 10 seconds apart, with measured kinetic energies of about 20 MeV and 10 MeV. (a) Estimate the rest mass of the neutrino, m_0, using this data, assuming both neutrinos were emitted at the same time. (b) Theoretical models of supernova explosions suggest that the neutrinos are emitted in a burst that lasts from a second or two up to perhaps 10 s. If we assume the neutrinos are not emitted simultaneously but rather at any time over a 10-s interval, what then can we say about the neutrino mass based on the data given above?

SOLUTION (a) We expect the neutrino mass to be less than 100 eV (from laboratory measurements). Since our neutrinos have KE of 20 MeV and 10 MeV, we can make the approximation $m_0 c^2 \ll E$, where E (the total energy) is essentially equal to the kinetic energy. From Eqs. 26–4 and 26–6, we have

$$E = mc^2 = \frac{m_0 c^2}{\sqrt{1 - v^2/c^2}}.$$

We solve this for v, the velocity of a neutrino with energy E:

$$v = c\left(1 - \frac{m_0^2 c^4}{E^2}\right)^{\frac{1}{2}} = c\left(1 - \frac{m_0^2 c^4}{2E^2} + \cdots\right),$$

where we have used the binomial expansion $(1 + x)^{\frac{1}{2}} = 1 + \frac{1}{2}x + \cdots$ (see Appendix A), and we ignore higher-order terms since $m_0^2 c^4 \ll E^2$. The time t for a neutrino to travel a distance $d (= 170{,}000 \text{ ly})$ is

$$t = \frac{d}{v} = \frac{d}{c\left(1 - \frac{m_0^2 c^4}{2E^2}\right)} \approx \frac{d}{c}\left(1 + \frac{m_0^2 c^4}{2E^2}\right),$$

where again we used the binomial expansion $[(1 + x)^{-1} = 1 - x + \cdots]$. The difference in arrival times for our two neutrinos of energies $E_1 = 20$ MeV and $E_2 = 10$ MeV is

$$t_2 - t_1 = \frac{d}{c}\frac{m_0^2 c^4}{2}\left(\frac{1}{E_2^2} - \frac{1}{E_1^2}\right).$$

We solve this for $m_0 c^2$ and set $t_2 - t_1 = 10$ s:

$$m_0 c^2 = \left[\frac{2c(t_2 - t_1)}{d}\frac{E_1^2 E_2^2}{E_1^2 - E_2^2}\right]^{\frac{1}{2}}$$

$$= \left[\frac{2(3 \times 10^8 \text{ m/s})(10 \text{ s})}{(1.7 \times 10^5 \text{ ly})(10^{16} \text{ m/ly})}\frac{(400 \text{ MeV}^2)(100 \text{ MeV}^2)}{(400 \text{ MeV}^2 - 100 \text{ MeV}^2)}\right]^{\frac{1}{2}}$$

$$= 22 \times 10^{-6} \text{ MeV} = 22 \text{ eV}.$$

We thus estimate the mass of the neutrino to be 22 eV, but there would of course be experimental uncertainties.
(b) If the two neutrinos of energy $E_1 = 20$ MeV and $E_2 = 10$ MeV were emitted at unknown times over a 10-s interval, then the 10-s difference in their arrival times could be due to a 10-s difference in their emission time. In this case our data are consistent with zero rest mass and it puts an approximate *upper limit* on the neutrino mass of 22 eV.

In the actual experiments, the most sensitive detector consisted of several thousand tons of water in an underground chamber. It detected 11 events in 12 seconds, probably via the reaction $\bar{\nu}_e + p \rightarrow n + e^+$. There was not a clear correlation between energy and time of arrival. Nonetheless, a careful analysis of this experiment and others, has set a rough upper limit on the electron anti-neutrino mass of about 7 eV. The upper limits for the masses of the other neutrinos are much higher (see Table 32–2), so we cannot yet rule out neutrinos as being able to "close" the universe.

Deceleration parameter

Another factor that could provide the answer to the question of an open or closed universe, if we could measure it accurately enough, is the so-called **deceleration parameter**. It is a measure of the rate at which the expansion of the universe is slowing. But to measure this rate requires looking far back in time, to the galaxies farthest away, whose light we now receive was emitted at a time closer to the beginning of the universe. At that time, the rate of expansion was much faster than today. Unfortunately, we would have to know the distance to these galaxies more precisely than is now possible. So this method does not at present yield an answer to whether the universe is open or closed. A recent related analysis of the number of galaxies as a function of redshift (i.e., distance) gives a result consistent with $\rho = \rho_c$, although the experimental uncertainty allows fairly wide limits on ρ. Nonetheless, the suggestion of the simplest universe of all, with no overall curvature, is appealing.

If the universe is open, how would it evolve in the future? According to the latest theories, which rely to a large extent on elementary particle theory, after about 10^{18} years, galaxies would have much of their matter knocked away and scattered throughout the universe by collisions with other stars. The remaining matter would eventually condense into massive "galactic black holes." Clusters of these would then coalesce into extremely massive "supergalactic black holes." Finally, the black holes themselves would "evaporate"—the matter within them, through the slow quantum-mechanical process of tunneling (see Section 30–12), would "leak out." This process is so slow that it would take on the order of 10^{100} years. The universe would then be mainly a thin gas of electrons, positrons, neutrinos, and photons.

On the other hand, if the universe is closed, it might turn around and begin to contract even before all the stars have burnt out. As the universe contracts, the background radiation would increase in energy and temperature. The universe might simply retrace its steps, if it weren't for black holes. As density increases and the universe rushes toward its inevitable end in the "big crunch," black holes might gobble up more and more matter until the entire universe coalesced into a single supermassive black hole—which would then be the universe.

If the universe is closed, what happens after the big crunch? We don't know, of course. What is possible, though, is a "bounce." That is, the dense fiery nucleus of the big crunch might explode again, resulting once more in an expanding universe. Thus the universe might be cyclic as shown in Fig. 33–25. Such a **cyclic** or **pulsating** universe proposes a possible answer to one of our favorite "unanswerable" questions: what happened before the Big Bang? In this model there was simply a previous cycle. But left unanswered would be a number of other questions such as "When did it all begin?"

The questions raised by cosmology can seem absurd at times, they are so removed from everyday "reality." We can always say: the Sun is shining, it's going to burn on for an unimaginably long time, all is well. Nonethe-

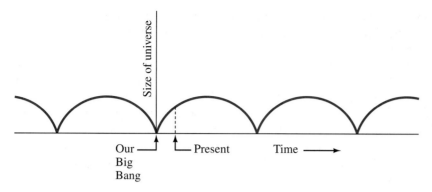

FIGURE 33-25 Cyclic model of the universe. Although the cycles shown here are the same, they could have different periods and different expansion rates.

less, the questions of cosmology are deep ones that fascinate the human intellect. One aspect that is especially intriguing is this: calculations on the formation and evolution of the universe have been performed that deliberately varied the values—just slightly—of certain fundamental physical constants. The result? A universe in which life as we know it could not exist. [For example, if the difference in mass between proton and neutron were zero, or small (less than $0.5 \, \mathrm{MeV}/c^2$), there would be no atoms: electrons would be captured by protons never to be freed again.] Such results have given rise to the so-called **Anthropic principle**, which says that if the universe were even a little different than it is, we couldn't be here. It's as if the universe were exquisitely tuned, almost as if to accommodate us.

SUMMARY

The night sky contains myriads of stars including those in the Milky Way, which is a "side view" of our **Galaxy** looking along the plane of the disc. Our Galaxy includes about 10^{11} stars. Beyond our Galaxy are billions of other galaxies. Astronomical distances are measured in **light-years** ($1 \, \mathrm{ly} \approx 10^{13} \, \mathrm{km}$). The nearest star is about 4 ly away and the nearest other galaxy is 2 million ly away. Our galactic disc has a diameter of about 100,000 ly. Distances are often specified in **parsecs**, where 1 parsec = 3.26 ly.

Stars are believed to begin life as collapsing masses of hydrogen gas (protostars). As they contract, they heat up (PE is transformed to KE). When the temperature reaches 10 million degrees, nuclear fusion begins and forms heavier elements (**nucleosynthesis**), mainly helium at first. The energy released during these reactions balances the gravitational force, and the young star stabilizes as a **main-sequence** star. The tremendous luminosity of stars comes from the energy released during these thermonuclear reactions. After billions of years, as helium is collected in the core and hydrogen is used up, the core contracts and heats further. The envelope expands and cools, and the star becomes a **red giant** (larger diameter, redder color). The next stage of

stellar evolution depends on the mass of the star. Stars of residual mass less than about 1.4 solar masses cool further and become **white dwarfs**, eventually fading and going out altogether. Heavier stars contract further due to their greater gravity: the density approaches nuclear density, the huge pressure forces electrons to combine with protons to form neutrons, and the star becomes essentially a huge nucleus of neutrons. This is a **neutron star**, and the energy released from its final core contraction is believed to produce **supernovae**. If the star's residual mass is greater than two or three solar masses, it may contract even further and form a **black hole**, which is so dense that no matter or light can escape from it.

In the **general theory of relativity**, the **equivalence principle** states that an observer cannot distinguish acceleration from a gravitational field. Said another way, gravitational and inertial mass are the same. The theory predicts gravitational bending of light rays to a degree consistent with experiment. Gravity is treated as a curvature in space and time, the curvature being greater near massive bodies. The universe as a whole may be curved. If there is sufficient mass, the curvature of the universe is positive, and the universe is **closed** and **finite**; otherwise, it is **open** and **infinite**.

Distant galaxies display a **redshift** in their spectral lines, interpreted as a Doppler shift. The universe seems to be expanding, its galaxies racing away from each other at speeds (v) proportional to the distance (d) between them:

$$v = Hd,$$

which is known as **Hubble's law** (H is the **Hubble parameter**). This expansion of the universe suggests an explosive origin, the **Big Bang**, which probably occurred 10 to 15 billion years ago. **Quasars** are objects with a large redshift (suggesting great distance) and high luminosity (suggesting closeness or, more likely, extraordinary energy output). The **cosmological principle** assumes that the universe, on a large scale, is homogeneous and isotropic.

Important evidence for the Big Bang model of the universe was the discovery of the **cosmic microwave background** radiation (CMB), which conforms to a blackbody radiation curve at a temperature of 2.7 K. The **standard model** of the Big Bang provides a possible scenario as to how the universe developed as it expanded and cooled after the Big Bang. Starting at 10^{-43} seconds after the Big Bang, according to this model, there was a series of **phase transitions** during which previously unified forces of nature "condensed out" one by one. The **inflationary scenario** assumes that during one of these phase transitions, the universe underwent a brief but rapid exponential expansion. Until about 10^{-35} s, there was no distinction between quarks and leptons. Shortly thereafter, quarks were **confined** into hadrons (the **hadron era**). About 10^{-6} s after the Big Bang, the majority of hadrons disappeared, introducing the **lepton era**. By the time the universe was about 10 s old, the electrons too had mostly disappeared and the universe became **radiation-dominated**. A couple of minutes later, nucleosynthesis began, but lasted only a few minutes. It was then several hundred thousand years before the universe was cool enough for electrons to combine with nuclei to form atoms. Also about this time, the background radiation had expanded and cooled so much that its total energy equaled the energy in matter. As the radiation cooled further, losing energy, the universe became **matter-dominated** (not in numbers, but in energy). Then stars and galaxies formed, producing a universe not much different than it is today—10 or 15 billion years later.

If the universe is **open**, it will continue to expand indefinitely. If it is **closed**, gravity is sufficiently strong to halt expansion and the universe will eventually begin to collapse back on itself, ending in a "big crunch." Whether the universe is open or closed depends on whether its average mass density is above or below a critical density. If the universe is closed, it may rebound from the big crunch and perhaps reexpand in a **cyclic** manner.

QUESTIONS

1. The Milky Way was once thought to be "cloudy," but no longer is considered so. Explain.

2. Give an explanation for why some galaxies have arms.

3. If you were measuring star parallaxes from the Moon instead of Earth, what corrections would you have to make? What changes would occur if you were measuring parallaxes from Mars?

4. A star is in equilibrium when it radiates at its surface all the energy generated at its core. What happens when it begins to generate more energy than it radiates? Less energy? Explain.

5. Describe a red giant star. List some of its properties.

6. Select a point on the H–R diagram. Mark several directions away from this point. Now describe the changes that would take place in a star moving in each of these directions.

7. Does the H–R diagram reveal anything about the core of a star?

8. Why do some stars end up as white dwarfs, and others as neutron stars or black holes?

9. What is a geodesic? What is its role in general relativity?

10. If it were discovered that the redshift of spectral lines of galaxies was due to something other than expansion, how might our view of the universe change? Would there be conflicting evidence? Discuss.

11. All galaxies appear to be moving away from us. Are we therefore at the center of the universe? Explain.

12. If you were located in a galaxy near the boundary of our observable universe, would galaxies in the direction of the Milky Way appear to be approaching you or receding from you? Explain.

13. What is the difference between the Hubble age of the universe and the actual age? Which is greater?

14. Compare an explosion on Earth to the Big Bang. Consider such questions as: would the debris spread at a higher speed for more distant particles, as in the Big Bang? Would the debris come to rest? What type of universe would this correspond to, open or closed?

15. When the primordial nucleus exploded, thus creating the universe, into what did it expand? Discuss.

16. If nothing, not even light, escapes from a black hole, then how can we tell if one is there?

17. Explain what the 2.7 K cosmic microwave background radiation is. Where does it come from? Why is its temperature now so low?

18. The birth of atoms—that is, the combination of electrons with nuclei about which they orbit—occurred when the universe had cooled to about 3000 K and is generally called *recombination*. Why is this term misleading?

19. Why were atoms unable to exist until hundreds of thousands of years after the Big Bang?

20. Explain why today the universe is said to be matter-dominated, yet there are probably 10^9 times as many photons as there are massive particles.

21. If the universe is open, what will eventually happen to the cosmic microwave background radiation? If it is closed, what will happen to it?

22. Under what circumstances would the universe eventually collapse in on itself?

PROBLEMS

SECTIONS 33–1 AND 33–2

1. (I) The parallax angle of a star is 0.00017°. How far away is the star?

2. (I) A star exhibits a parallax of 0.28 seconds of arc. How far away is it?

3. (I) Using the definitions of the parsec and the light-year, show that 1 pc = 3.26 ly.

4. (I) A star is 24 parsecs away. What is its parallax angle? State (a) in seconds of arc, and (b) in degrees.

5. (I) What is the parallax angle for a star that is 42 light-years away? How many parsecs is this?

6. (I) A star is 25 parsecs away. How long does it take for its light to reach us?

7. (I) If one star is twice as far away from us as a second star, will the parallax angle of the first star be greater or less than that of the second star? By what factor?

8. (II) We saw earlier (Chapter 14) that the rate energy reaches the Earth from the Sun (the "solar constant") is about 1.3×10^3 W/m². What is (a) the apparent brightness l of the Sun, and (b) the absolute luminosity L of the Sun?

9. (II) What is the apparent brightness of the Sun as seen on Jupiter? (Jupiter is 5.2 times farther from the Sun than the Earth.)

10. (II) Estimate the angular width that our Galaxy would subtend if observed from the nearest galaxy to us. Compare to the angular width of the Moon from Earth.

11. (II) When our Sun becomes a red giant, what will be its average density if it expands out to the orbit of Earth (1.5×10^{11} m from the Sun)?

12. (II) When our Sun becomes a white dwarf, it is expected to be about the size of the Moon. What angular width will it subtend from Earth?

13. (II) Calculate the density of a white dwarf whose mass is equal to the Sun's and whose radius is equal to the Earth's. How many times larger than Earth's density is this?

14. (II) A neutron star whose mass is 1.5 solar masses has a radius of about 11 km. Calculate its average density and compare to that for a white dwarf (Problem 13) and to that of nuclear matter.

15. (II) Calculate the Q-values for the He burning reactions of Eqs. 33–2. (The mass of $^{8}_{4}$Be is 8.005305 u.)

16. (II) In the later stages of stellar evolution, a star (if massive enough) will begin fusing carbon nuclei to form, for example, magnesium:

$$^{12}_{6}C + {}^{12}_{6}C \rightarrow {}^{24}_{12}Mg + \gamma.$$

(a) How much energy is released in this reaction (see Appendix F). (b) How much kinetic energy must each carbon nucleus have (assume equal) in a head-on collision if they are just to touch (use Eq. 30–1) so that the strong force can come into play? (c) What temperature does this KE correspond to?

17. (II) Repeat Problem 16 for the reaction

$$^{16}_{8}O + {}^{16}_{8}O \rightarrow {}^{28}_{14}Si + {}^{4}_{2}He.$$

18. (II) Suppose two stars of the same apparent brightness l are also believed to be the same size. The spectrum of one star peaks at 800 nm whereas that of the other peaks at 400 nm. Use Wien's law (Section 27–2) and the Stefan-Boltzmann law (Eq. 14–4) to estimate their relative distances from us. [*Hint*: See Examples 33–5 and 33–6.]

19. (III) Stars located in a certain cluster are assumed to be about the same distance from us. Two such stars have spectra that peak at $\lambda_1 = 500$ nm and $\lambda_2 = 700$ nm, and the ratio of their apparent brightness is $l_1/l_2 = 0.091$. Estimate their relative sizes (give ratio of their diameters). [*Hint*: Use the Stefan-Boltzmann law, Eq. 14–4.]

SECTION 33–3

20. (I) Describe a triangle, drawn on the surface of a sphere, for which the sum of the angles is (*a*) 360°, and (*b*) 180°.

21. (I) Show that the Schwarzschild radius for a star with mass equal to that (*a*) of our Sun is 2.95 km, and (*b*) of Earth is 8.9 mm.

22. (II) What is the Schwarzschild radius for a typical galaxy?

23. (II) What is the maximum sum-of-the-angles for a triangle on a sphere?

SECTION 33–4

24. (I) If a galaxy is traveling away from us at 1.0 percent of the speed of light, roughly how far away is it?

25. (I) The redshift of a galaxy indicates a velocity of 2500 km/s. How far away is it?

26. (I) Estimate the speed of a galaxy (relative to us) that is near the "edge" of the universe, say 10 billion light-years away.

27. (I) Make an approximate calculation for the age of the universe using Hubble's constant, assuming (*a*) $H = 50$ km/s per Mpc and (*b*) 100 km/s per Mpc.

28. (II) Estimate the observed wavelength for the 656-nm line in the Balmer series of hydrogen in the spectrum of a galaxy whose distance from us is (*a*) 1.0×10^6 ly, (*b*) 1.0×10^8 ly, (*c*) 1.0×10^{10} ly.

29. (II) Estimate the speed of a galaxy, and its distance from us, if the wavelength for the hydrogen line at 434 nm is measured on Earth as being 610 nm.

30. (II) Starting from Eq. 33–3, show that the Doppler shift in wavelength is $\Delta\lambda/\lambda \approx v/c$ (Eq. 33–4) for $v \ll c$. [*Hint*: Use the binomial expansion—see Appendix A.]

31. (III) Derive the Doppler shift formula for light (Eq. 33–3) using the special theory of relativity. [*Hint*: Proceed as in Section 12–8 for sound, but invoke special relativity including time dilation.]

SECTIONS 33–5 TO 33–7

32. (I) Calculate the wavelength at the peak of the blackbody radiation distribution at 2.7 K using the Wien law.

33. (II) The critical density for closure of the universe is $\rho_c \approx 10^{-26}$ kg/m³. State ρ_c in terms of the average number of nucleons per cubic meter.

34. (II) The size of the universe (the average distance between galaxies) at any one moment is believed to have been inversely proportional to the absolute temperature. Estimate the size of the universe, compared to today, at (*a*) $t = 10^6$ yr, (*b*) $t = 1$ s, (*c*) $t = 10^{-6}$ s, and (*d*) $t = 10^{-35}$ s.

35. (II) At approximately what time had the universe cooled below the threshold temperature for producing (*a*) kaons ($M \approx 500$ MeV/c^2), (*b*) Y ($M \approx 9500$ MeV/c^2), and (*c*) muons ($M \approx 100$ MeV/c^2)?

GENERAL PROBLEMS

36. Suppose that three main-sequence stars could undergo the three changes represented by the three arrows, *A*, *B*, and *C*, in the H–R diagram of Fig. 33–26. For each case, describe the changes in temperature, luminosity, and size.

37. Assume that the nearest stars to us have an intrinsic luminosity about the same as the Sun's. Their apparent brightness, however, is about 10^{11} times fainter than the Sun. From this, estimate the distance to the nearest stars. (Newton did this calculation, although he made a numerical error of a factor of 100.)

38. Use conservation of angular momentum to estimate the angular velocity of a neutron star which has collapsed to a diameter of 10 km, from a star whose radius was equal to that of our Sun (7×10^8 m), of mass 1.5 times that of the Sun, and which rotated (like our Sun) about once a month.

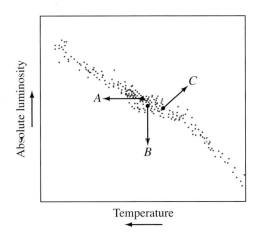

FIGURE 33–26 Problem 36.

39. By what factor does the rotational KE change when the star in Problem 38 collapses to a neutron star?

40. A certain pulsar, believed to be a neutron star of mass 1.5 times that of the Sun, with diameter 10 km, is observed to have a rotation speed of 1.0 rev/s. If it loses rotational KE at the rate of 1 part in 10^9 per day, which is all transformed into radiation, what is the power output of the star?

41. Estimate the rate at which hydrogen atoms would have to be created, according to the steady-state model, to maintain the present density of the universe of about 10^{-27} kg/m^3, assuming the universe is expanding with Hubble constant $H = 80$ km/s/Mpc.

42. Estimate what neutrino rest mass (in eV) would provide the critical density to close the universe. Assume the neutrino density is, like photons, about 10^9 times that of nucleons, and that nucleons make up only (a) 2 percent of the mass needed, or (b) 5 percent of the mass needed.

43. Two stars, whose spectra peak at 600 nm and 400 nm, respectively, both lie on the main sequence. Use Wien's law, the Stefan-Boltzmann law, and the H–R diagram (Fig. 33–7) to estimate the ratio of their diameters. [*Hint*: See Examples 33–5 and 33–6.]

44. The farthest we can measure with parallax is about 30 parsecs. What is our minimum angular resolution (in degrees), based on this information?

45. Through some coincidence, the Balmer lines from singly ionized helium in a distant star happen to overlap with the Balmer lines from hydrogen in the Sun. How fast is the star receding from us?

46. What is the temperature that corresponds to the 1.8 TeV collisions at the Fermilab collider? To what era in cosmological history does this correspond?

47. Astronomers have recently measured the rotation of gas around what might be a supermassive black hole of about 2 billion solar masses at the center of a galaxy. If the radius from the galactic center to the gas clouds is 60 light-years, what Doppler shift $\Delta\lambda/\lambda$ do you estimate they saw?

48. Astronomers use an *apparent magnitude* (*m*) scale to describe apparent brightness. It uses a logarithmic scale, where a higher number corresponds to a less bright star. (For example, the Sun has magnitude -27, whereas most stars have positive magnitudes.) On this scale, a change in apparent magnitude by $+5$ corresponds to a decrease in apparent brightness by a factor of 100. If Venus has an apparent magnitude of -4.4, whereas Sirius has an apparent magnitude of -1.4, which is brighter? What is the ratio of the apparent brightness of these two objects?

A MATHEMATICAL REVIEW

A–1 Relationships, Proportionality, and Equations

One of the important aspects of physics is the search for relationships between different quantities—that is, determining how one quantity affects another. For example, how does temperature affect the air pressure in a tire? Or how does the net force on an object affect its acceleration? Sometimes a given quantity is affected by two or more quantities; for instance, the acceleration of an object is related to both its mass and the applied force. If it is suspected that a relationship exists between two or more quantities, one can try to determine the precise nature of this relationship. This is done by varying one of the quantities and measuring how the other varies as a result. If it is likely that a particular quantity will be affected by more than one factor or quantity, only one of the latter is varied at a time, while the others are held constant.[†]

Direct proportion

As a simple example, the ancients found that if one circle has twice the diameter of a second circle, the first also has twice the circumference. If the diameter is three times as large, the circumference is also three times as large. In other words, an increase in the diameter results in a proportional increase in the circumference. We say that the circumference is *directly proportional to* the diameter. This can be written in symbols as $C \propto D$, where "$\propto$" means "is proportional to," and C and D refer to the circumference and diameter of a circle, respectively. The next step is to change this proportionality to an equation, which will make it possible to link the two quantities numerically. This merely entails inserting a proportionality constant, which in many cases is determined by measurement. (In some cases it can be chosen arbitrarily, if it involves only the definition of a new unit.) The ancients found that the ratio of the circumference to the diameter of any circle was 3.1416 (to keep only the first few decimal places). This number is designated by the Greek letter π. It is the constant of proportionality for the relationship $C \propto D$, and to obtain an equation we insert it into the proportion and change the $\propto$ to $=$. Thus, $C = \pi D$.

Other kinds of proportionality occur as well. For example, the area of a circle is proportional to the *square* of its radius. That is, if the radius is doubled, the area becomes four times as large; and so on. In this case we can write $A \propto r^2$, where A stands for the area and r for the radius of the circle.

Inverse proportion

Sometimes two quantities are related in such a way that an increase in one leads to a proportional *decrease* in the other. This is called *inverse proportion*. For example, the time required to travel a given distance is in-

[†]When one quantity affects another, we often use the expression "is a function of" to indicate this dependence; for example, we say that the pressure in a tire is a function of the temperature.

versely proportional to the speed of travel. The greater the speed, the less time it takes. We can write this inverse proportion as: time $\propto$ 1/speed. The larger the denominator of a fraction, the smaller the fraction is as a whole. For example, $\frac{1}{4}$ is smaller than $\frac{1}{2}$. Thus, if the speed is doubled, the time is halved, which is what we want to express by this inverse proportionality relationship.

Whatever kind of proportion is found to hold, it can be changed to an equality by insertion of the proper proportionality constant. Quantitative statements or predictions about the physical world can then be made with the equation.

A–2 Exponents

When we write 10^4 we mean that you multiply 10 times itself four times: $10^4 = 10 \times 10 \times 10 \times 10 = 10,000$. The superscript 4 is called an *exponent*, and 10 is said to be raised to the fourth power. Any number or symbol can be raised to a power; special names are used when the exponent is 2 (a^2 is "a squared") or 3 (a^3 is "a cubed"). For any other power, we say a^n is "a to the nth power." If the exponent is 1, it is usually dropped: $a^1 = a$, since no multiplication is involved.

The rules for multiplying numbers expressed as powers are as follows:

$$(a^n)(a^m) = a^{n+m}. \tag{A–1}$$

That is, the exponents are added. To see why, consider the result of the multiplication of 3^3 by 3^4:

$$(3^3)(3^4) = (3)(3)(3) \times (3)(3)(3)(3) = (3)^7.$$

Here the sum of the exponents is $3 + 4 = 7$, so rule A–1 works. Notice that this rule works only if the base numbers (a in Eq. A–1) are the same. Thus we *cannot* use the rule of summing exponents for $(6^3)(5^2)$; these numbers would have to be written out. However, if the base numbers are different but the exponents are the same, we can write a second rule:

$$(a^n)(b^n) = (ab)^n. \tag{A–2}$$

For example, $(5^3)(6^3) = (30)^3$ since

$$(5)(5)(5)(6)(6)(6) = (30)(30)(30).$$

The third rule involves a power raised to another power: $(a^3)^2$ means $(a^3)(a^3)$, which is equal to $a^{3+3} = a^6$. The general rule is then

$$(a^n)^m = a^{nm}. \tag{A–3}$$

In this case, the exponents are multiplied.

Negative exponents are used for reciprocals. Thus,

$$\frac{1}{a} = a^{-1}, \qquad \frac{1}{a^3} = a^{-3},$$

and so on. The reason for using negative exponents is that it allows us to use the multiplication rules given above. For example, $(a^5)(a^{-3})$ means

$$\frac{(a)(a)(a)(a)(a)}{(a)(a)(a)} = a^2.$$

Rule A-1 gives us the same result:

$$(a^5)(a^{-3}) = a^{5-3} = a^2.$$

What does an exponent of zero mean? That is, what is a^0? Any number raised to the zeroth power is defined as being equal to 1:

$$a^0 = 1.$$

This definition is used because it follows from the rules for adding exponents. For example,

$$a^3 a^{-3} = a^{3-3} = a^0 = 1.$$

But *does* $a^3 a^{-3}$ actually equal 1? Yes, because

$$a^3 a^{-3} = \frac{a^3}{a^3} = 1.$$

Fractional exponents are used to represent *roots*. For example, $a^{\frac{1}{2}}$ means the square root of a; that is $a^{\frac{1}{2}} = \sqrt{a}$. Similarly, $a^{\frac{1}{3}}$ means the cube root of a, and so on. The fourth root of a means that if you multiply the fourth root of a by itself four times, you again get a:

$$(a^{\frac{1}{4}})^4 = a.$$

This is consistent with rule A-3 since $(a^{\frac{1}{4}})^4 = a^{\frac{4}{4}} = a^1 = a$.

A-3 Powers of 10, or Exponential Notation

Writing out very large and very small numbers such as the distance of Neptune from the Sun, 4,500,000,000 km, or the diameter of a typical atom, 0.00000001 cm, is inconvenient and prone to error. It also leaves in question (see Section 1–4) the number of significant figures. (How many of the zeros are significant in the number 4,500,000,000 km?) We therefore make use of the "powers of 10," or exponential notation. The distance from Neptune to the Sun is then expressed as 4.50×10^9 km (assuming that the value is significant to three digits) and the diameter of an atom 1.0×10^{-8} cm. This way of writing numbers is based on the use of exponents, where a^n signifies a multiplied by itself n times. For example, $10^4 = 10 \times 10 \times 10 \times 10 = 10,000$. Thus, $4.50 \times 10^9 = 4.50 \times 1,000,000,000 = 4,500,000,000$. Notice that the exponent (9 in this case) is just the number of places the decimal point is moved to the right to obtain the fully written out number (4.500,000,000.)

When two numbers are multiplied (or divided), you first multiply (divide) the simple parts and then the powers of 10. Thus, 2.0×10^3 multiplied by 5.5×10^4 equals $(2.0 \times 5.5) \times (10^3 \times 10^4) = 11 \times 10^7$, where we have used the rule for adding exponents (Appendix A–2). Similarly, 8.2×10^5 divided by 2.0×10^2 equals

$$\frac{8.2 \times 10^5}{2.0 \times 10^2} = \frac{8.2}{2.0} \times \frac{10^5}{10^2} = 4.1 \times 10^3.$$

For numbers less than 1, say 0.01, the exponent power of 10 is written with a negative sign: $0.01 = 1/100 = 1/10^2 = 1 \times 10^{-2}$. Similarly, $0.002 = 2 \times 10^{-3}$. The decimal point has again been moved the number of places expressed in the exponent. Thus, $0.020 \times 3600 = 72$; in exponential notation $(2.0 \times 10^{-2}) \times (3.6 \times 10^3) = 7.2 \times 10^1 = 72$.

Notice also that $10^1 \times 10^{-1} = 10 \times 0.1 = 1$, and by the law of exponents, $10^1 \times 10^{-1} = 10^0$. Therefore, $10^0 = 1$.

When writing a number in exponential notation, it is usual to make the simple number be between 1 and 10. Thus it is conventional to write 4.5×10^9 rather than 45×10^8, although they are the same number.[†] This notation also allows the number of *significant figures* to be clearly expressed. We write 4.50×10^9 if this value is accurate to three significant figures, but 4.5×10^9 if it is accurate to only two.

A–4 Algebra

Physical relationships between quantities can be represented as equations involving symbols (usually letters of the alphabet) that represent the quantities. The manipulation of such equations is the field of algebra, and is used a great deal in physics. An equation involves an equals sign, which tells us that the quantities on either side of the equals sign have the same value. Examples of equations are

$$3 + 8 = 11$$
$$2x + 7 = 15$$
$$a^2b + c = 6.$$

The first equation involves only numbers, so is called an arithmetic equation. The other two equations are algebraic since they involve symbols. In the third equation, the quantity a^2b means the product of a times a times b: $a^2b = a \times a \times b$.

Solving for an Unknown
Often we wish to solve for one (or more) symbols, and we treat it as an *unknown*. For example, in the equation $2x + 7 = 15$, x is the unknown; this equation is true, however, only when $x = 4$. Determining what value (or values) the unknown(s) can have to satisfy the equation(s) is called *solving the equation*. To solve an equation, the following rule can be used:

> *An equation will remain true if any operation performed on one side is also performed on the other side:* for example, (*a*) addition or subtraction of a number or symbol; (*b*) multiplication or division by a number or symbol; (*c*) raising each side of the equation to the same power, or taking the same root (such as square root).

[†]Another convention used, particularly with computers, is that the simple number be between 0.1 and 1. Thus we could write 4,500,000,000 as 0.450×10^{10}.

EXAMPLE A–1 Solve for x in the equation

$$2x + 7 = 15.$$

SOLUTION We first subtract 7 from both sides:

$$2x + 7 - 7 = 15 - 7$$

$$2x = 8.$$

Then we divide both sides by 2 to get

$$\frac{2x}{2} = \frac{8}{2}$$

$$x = 4,$$

and this solves the equation.

EXAMPLE A–2 (*a*) Solve the equation

$$a^2b + c = 24$$

for the unknown a in terms of b and c. (*b*) Solve for a assuming that $b = 2$ and $c = 6$.

SOLUTION (*a*) We are trying to solve for a, so we first subtract c from both sides:

$$a^2b = 24 - c,$$

then divide by b:

$$a^2 = \frac{24 - c}{b},$$

and finally take square roots:

$$a = \sqrt{\frac{24 - c}{b}}.$$

(*b*) If we are given that $b = 2$ and $c = 6$, then

$$a = \sqrt{\frac{24 - 6}{2}} = 3.$$

To check a solution, we put it back into the original equation (this is really a check that we did all the manipulations correctly). In the equation

$$a^2b + c = 24,$$

we put in $a = 3, b = 2, c = 6$ and find

$$(3)^2(2) + (6) \overset{?}{=} 24$$

$$24 = 24,$$

which checks.

Two or More Unknowns

If we have two or more unknowns, one equation is not sufficient to find them. In general, if there are n unknowns, n independent equations are needed. For example, if there are two unknowns, we need two equations. If the unknowns are called x and y, a typical procedure is to solve one equation for x in terms of y, and substitute this into the second equation.

EXAMPLE A–3 Solve the following pair of equations for x and y.

$$3x - 2y = 19$$
$$x + 4y = -3.$$

SOLUTION We solve the second equation for x in terms of y by subtracting $4y$ from both sides:

$$x = -3 - 4y.$$

We substitute this expression for x into the first equation, and simplify:

$$3(-3 - 4y) - 2y = 19$$
$$-9 - 12y - 2y = 19 \qquad \text{(carried out the multiplication by 3)}$$
$$-14y = 28 \qquad \text{(added 9 to both sides)}$$
$$y = -2 \qquad \text{(divided both sides by } -14\text{).}$$

Now that we know $y = -2$, we substitute this into the expression for x:

$$x = -3 - 4y$$
$$= -3 - 4(-2) = -3 + 8 = 5.$$

Our solution is $x = 5$, $y = -2$. We check this solution by putting these values back into the original equations:

$$3x - 2y \stackrel{?}{=} 19$$
$$3(5) - 2(-2) \stackrel{?}{=} 19$$
$$15 + 4 \stackrel{?}{=} 19$$
$$19 = 19 \qquad \text{(it checks)}$$

and

$$x + 4y \stackrel{?}{=} -3$$
$$5 + 4(-2) \stackrel{?}{=} -3$$
$$-3 = -3 \qquad \text{(it checks).}$$

Other methods for solving two or more equations, such as the method of determinants, can be found in an algebra textbook.

The Quadratic Formula

We sometimes encounter equations that involve an unknown, say x, that appears not only to the first power, but squared as well. Such a *quadratic equation* can be written in the form

$$ax^2 + bx + c = 0.$$

The quantities a, b, and c are typically numbers or constants that are given.[†] The general solutions to such an equation are given by the *quadratic formula*:

Quadratic formula

$$x = \frac{-b \pm \sqrt{b^2 - 4ac}}{2a}.$$

(A–4)

The $\pm$ sign indicates that there are two solutions for x: one where the plus sign is used, the other where the minus sign is used.[‡]

EXAMPLE A–4 Find the solutions for x in the equation

$$3x^2 - 5x = 2.$$

SOLUTION First we write this equation in the standard form

$$ax^2 + bx + c = 0$$

by subtracting 2 from both sides:

$$3x^2 - 5x - 2 = 0.$$

In this case, a, b, and c in the standard formula take the values $a = 3$, $b = -5$, and $c = -2$. The two solutions for x are

$$x = \frac{+5 + \sqrt{25 - (4)(3)(-2)}}{(2)(3)} = \frac{5 + 7}{6} = 2$$

and

$$x = \frac{+5 - \sqrt{25 - (4)(3)(-2)}}{(2)(3)} = \frac{5 - 7}{6} = -\frac{1}{3}.$$

In this Example, the two solutions are $x = 2$ and $x = -\frac{1}{3}$. In physics problems, it sometimes happens that only one of the solutions corresponds to a real-life situation; in this case, the other solution is discarded. In other cases, both solutions may correspond to physical reality.

Notice, incidentally, that b^2 must be greater than $4ac$, so that $\sqrt{b^2 - 4ac}$ yields a real number. If $(b^2 - 4ac)$ is less than zero (negative), there is no real solution. The square root of a negative number is called *imaginary*.

[†]Or one or more of them could be variables, in which case additional equations are needed.
[‡]A second-order equation—one in which the highest power of x is 2—has two solutions: a third-order equation—involving x^3—has three solutions, and so on.

A–5 The Binomial Expansion

Sometimes we end up with a quantity of the form $(1 + x)^n$. That is, the quantity $(1 + x)$ is raised to the nth power. This can be written as an infinite sum of terms, known as a *series expansion*, as follows:

$$(1 + x)^n = 1 + nx + \frac{n(n - 1)}{2!}x^2 + \cdots. \qquad \textbf{(A–5)}$$

This formula is useful for us mainly when x is very small compared to one ($x \ll 1$). In this case, each successive term is much smaller than the preceding term. For example, if $x = 0.01$, and $n = 2$, say, then whereas the first term equals 1, the second term is $nx = (2)(0.01) = 0.02$, and the third term is $[(2)(1)/2](0.01)^2 = 0.0001$, and so on. Thus, when x is small, we can ignore all but the first two (or three) terms and can write

$$(1 + x)^n \approx 1 + nx. \qquad \textbf{(A–6)}$$

This approximation often allows us to easily solve an equation that otherwise might be very difficult. Some examples are

$$(1 + x)^2 \approx 1 + 2x$$

$$\frac{1}{1 + x} = (1 + x)^{-1} \approx 1 - x$$

$$\sqrt{1 + x} = (1 + x)^{\frac{1}{2}} \approx 1 + \tfrac{1}{2}x$$

$$\frac{1}{\sqrt{1 + x}} = (1 + x)^{-\frac{1}{2}} \approx 1 - \tfrac{1}{2}x$$

where $x \ll 1$.

As a numerical example, let us evaluate $\sqrt{1.02}$ using the binomial theorem since $x = 0.02$ is much smaller than 1:

$$\sqrt{1.02} = (1.02)^{\frac{1}{2}} = (1 + 0.02)^{\frac{1}{2}} \approx 1 + \tfrac{1}{2}(0.02) = 1.01.$$

You can check with a calculator (and maybe not even more quickly) that $\sqrt{1.02} \approx 1.01$.

A–6 Plane Geometry

We review here a number of theorems involving angles and triangles that are useful in physics.

FIGURE A–2

1. *Equal angles*. Two angles are equal if any of the following conditions are true:
   (*a*) They are vertical angles (Fig. A–1); *or*
   (*b*) the left side of one is parallel to the left side of the other, and the right side of one is parallel to the right side of the other (the left and right sides are as seen from the vertex, where the two sides meet; Fig. A–2); *or*
   (*c*) the left side of one is perpendicular to the left side of the other, and the right sides are likewise perpendicular (Fig. A–3).

2. *The sum of the angles* in any plane triangle is 180°.

FIGURE A–3

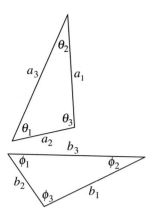

FIGURE A–4

3. *Similar triangles.* Two triangles are said to be similar if all three of their angles are equal (in Fig. A–4, $\theta_1 = \phi_1$, $\theta_2 = \phi_2$, $\theta_3 = \phi_3$). Similar triangles thus have the same basic shape but may be different sizes and have different orientations. Two useful theorems about similar triangles are:

(*a*) Two triangles are similar if any two of their angles are equal. (This follows because the third angles must also be equal since the sum of the angles of a triangle is 180°.)

(*b*) The ratio of corresponding sides of two similar triangles are equal. That is (Fig. A–4),

$$\frac{a_1}{b_1} = \frac{a_2}{b_2} = \frac{a_3}{b_3}.$$

4. *Congruent triangles.* Two triangles are congruent if one can be placed precisely on top of the other. That is, they are similar triangles and they have the same size. Two triangles are congruent if any of the following holds:

(*a*) The three corresponding sides are equal.

(*b*) Two sides and the enclosed angle are equal ("side-angle-side").

(*c*) Two angles and the enclosed side are equal ("angle-side-angle").

5. *Right triangles.* A right triangle has one angle that is 90° (a *right angle*); that is, the two sides that meet at the right angle are perpendicular. The two other (acute) angles in the right triangle add up to 90°.

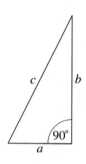

FIGURE A–5

6. *Pythagorean theorem.* In any right triangle, the square of the length of the hypotenuse (the side opposite the right angle) is equal to the sum of the squares of the lengths of the other two sides. In Fig. A–5,

$$c^2 = a^2 + b^2.$$

A–7 Areas and Volumes

Object	Surface Area	Volume
Circle, radius r	πr^2	—
Sphere, radius r	$4\pi r^2$	$\frac{4}{3}\pi r^3$
Right circular cylinder, radius r, height h	$2\pi r^2 + 2\pi rh$	$\pi r^2 h$
Right circular cone, radius r, height h	$\pi r^2 + \pi r\sqrt{r^2 + h^2}$	$\dfrac{\pi r^2 h}{3}$

(See Fig. A–6.)

$$\sin \theta = \frac{o}{h} \qquad\qquad \csc \theta = \frac{1}{\sin \theta} = \frac{h}{o}$$

$$\cos \theta = \frac{a}{h} \qquad\qquad \sec \theta = \frac{1}{\cos \theta} = \frac{h}{a}$$

$$\tan \theta = \frac{o}{a} = \frac{\sin \theta}{\cos \theta} \qquad\qquad \cot \theta = \frac{1}{\tan \theta} = \frac{a}{o}$$

$$a^2 + o^2 = h^2 \qquad\qquad \text{(Pythagorean theorem)}.$$

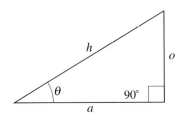

FIGURE A–6

Figure A–7 shows the signs ($+$ or $-$) that cosine, sine, and tangent take on for angles θ in the four quadrants (0° to 360°). Note that angles are measured counterclockwise from the x axis as shown; negative angles are measured from *below* the x axis, clockwise: for example, $-30° = +330°$, and so on. The following are some useful identities among the trigonometric functions:

$$\sin^2 \theta + \cos^2 \theta = 1, \quad \sec^2 \theta - \tan^2 \theta = 1, \quad \csc^2 \theta - \cot^2 \theta = 1$$

$$\sin 2\theta = 2 \sin \theta \cos \theta$$

$$\cos 2\theta = \cos^2\theta - \sin^2 \theta = 2 \cos^2 \theta - 1 = 1 - 2 \sin^2 \theta$$

$$\tan 2\theta = \frac{2 \tan \theta}{1 - \tan^2 \theta}$$

$$\sin (A \pm B) = \sin A \cos B \pm \cos A \sin B$$

$$\cos (A \pm B) = \cos A \cos B \mp \sin A \sin B$$

$$\tan (A \pm B) = \frac{\tan A \pm \tan B}{1 \mp \tan A \tan B}$$

$$\sin (180° - \theta) = \sin \theta$$

$$\cos (180° - \theta) = -\cos \theta$$

$$\sin (90° - \theta) = \cos \theta$$

$$\cos (90° - \theta) = \sin \theta$$

$$\sin \tfrac{1}{2}\theta = \sqrt{\frac{1 - \cos \theta}{2}}, \; \cos \tfrac{1}{2}\theta = \sqrt{\frac{1 + \cos \theta}{2}}, \; \tan \tfrac{1}{2}\theta = \sqrt{\frac{1 - \cos \theta}{1 + \cos \theta}}$$

$$\sin A \pm \sin B = 2 \sin \left(\frac{A \pm B}{2}\right) \cos \left(\frac{A \mp B}{2}\right).$$

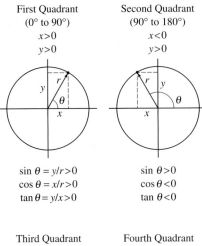

First Quadrant
(0° to 90°)
$x > 0$
$y > 0$

$\sin \theta = y/r > 0$
$\cos \theta = x/r > 0$
$\tan \theta = y/x > 0$

Second Quadrant
(90° to 180°)
$x < 0$
$y > 0$

$\sin \theta > 0$
$\cos \theta < 0$
$\tan \theta < 0$

Third Quadrant
(180° to 270°)
$x < 0$
$y < 0$

$\sin \theta < 0$
$\cos \theta < 0$
$\tan \theta > 0$

Fourth Quadrant
(270° to 360°)
$x > 0$
$y < 0$

$\sin \theta < 0$
$\cos \theta > 0$
$\tan \theta < 0$

FIGURE A–7

For any triangle (see Fig. A–8):

$$\frac{\sin \alpha}{a} = \frac{\sin \beta}{b} = \frac{\sin \gamma}{c} \qquad \text{(Law of sines)}$$

$$c^2 = a^2 + b^2 - 2ab \cos \gamma. \qquad \text{(Law of cosines)}$$

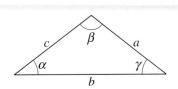

FIGURE A–8

A–9 Logarithms

Logarithms are defined in the following way:

$$\text{if } y = A^x, \qquad \text{then } x = \log_A y.$$

That is, the logarithm of a number y to the base A is that number which, as the exponent of A, gives back the number y. For *common logarithms*, the base is 10, so

$$\text{if } y = 10^x, \qquad \text{then } x = \log y.$$

The subscript 10 on $\log_{10}$ is usually omitted when dealing with common logs. Another base sometimes used is the natural number $e = 2.718 \cdots.^{\dagger}$ Such logarithms are called *natural logarithms* and are written ln. Thus,

$$\text{if } y = e^x, \qquad \text{then } x = \ln y.$$

For any number y, the two types of logarithm are related by

$$\ln y = 2.3026 \log y.$$

Some simple rules for logarithms are as follows:

$$\log (ab) = \log a + \log b. \qquad\qquad \text{(A–7)}$$

This is true because if $a = 10^n$ and $b = 10^m$, then $ab = 10^{n + m}$. From the definition of logarithm, $\log a = n$, $\log b = m$, and $\log (ab) = n + m$; hence, $\log (ab) = n + m = \log a + \log b$. In a similar way, we can show that

$$\log \left(\frac{a}{b}\right) = \log a - \log b \qquad\qquad \text{(A–8)}$$

and

$$\log a^n = n \log a. \qquad\qquad \text{(A–9)}$$

These three rules apply not only to common logs but to natural or any other kind of logarithm.

Logs were once used as a technique for simplifying certain types of calculation. Because of the advent of electronic calculators and computers, they are not often used any more for this purpose. However, logs do appear in certain physical equations, so it is helpful to know how to deal with them. If you do not have a calculator that calculates logs, you can easily use a *log table*, such as the small one shown here (Table A–1). The number N is given to two digits (some tables give N to three or more digits); the first digit is in the vertical column to the left, the second digit is in the horizontal row across the top. For example, the table tells us that $\log 1.0 = 0.000$, $\log 1.1 = 0.041$, and $\log 4.1 = 0.613$. Note that the table does not include the decimal point—it is understood. The table gives logs for numbers between 1.0 and 9.9; for larger or smaller numbers we use rule A–7:

$$\log (ab) = \log a + \log b.$$

For example,

$$\log (380) = \log (3.8 \times 10^2) = \log (3.8) + \log (10^2).$$

From the table, $\log 3.8 = 0.580$; and from rule A–9, $\log (10^2) = 2 \log (10) = 2$, since $\log (10) = 1$. [This follows from the definition of the

---

†The natural number e can be written as an infinite series:

$$e = 1 + \frac{1}{1} + \frac{1}{1 \cdot 2} + \frac{1}{1 \cdot 2 \cdot 3} + \frac{1}{1 \cdot 2 \cdot 3 \cdot 4} + \cdots.$$

TABLE A–1   Short Table of Common Logarithms

N	0.0	0.1	0.2	0.3	0.4	0.5	0.6	0.7	0.8	0.9
1	000	041	079	114	146	176	204	230	255	279
2	301	322	342	362	380	398	415	431	447	462
3	477	491	505	519	531	544	556	568	580	591
4	602	613	623	633	643	653	663	672	681	690
5	699	708	716	724	732	740	748	756	763	771
6	778	785	792	799	806	813	820	826	833	839
7	845	851	857	863	869	875	881	886	892	898
8	903	908	914	919	924	929	935	940	944	949
9	954	959	964	968	973	978	982	987	991	996

logarithm: if $10 = 10^1$, then $1 = \log (10)$.] Thus,

$$\log (380) = \log (3.8) + \log (10^2)$$
$$= 0.580 + 2$$
$$= 2.580.$$

Similarly,

$$\log (0.081) = \log (8.1) + \log (10^{-2})$$
$$= 0.908 - 2 = -1.092.$$

Sometimes we need to do the reverse process: find the number N whose log is, say, 2.670. This is called "taking the antilogarithm." To do so, *Antilog* we separate our number 2.670 into two parts, making the separation at the decimal point:

$$\log N = 2.670 = 2 + 0.670$$
$$= \log 10^2 + 0.670.$$

We now look at Table A–1 to see what number has its log equal to 0.670; none does, so we must *interpolate*: we see that $\log 4.6 = 0.663$ and $\log 4.7 = 0.672$. So the number we want is between 4.6 and 4.7, and closer to the latter by 7/9. Approximately we can say that $\log 4.68 = 0.670$. Thus

$$\log N = 2 + 0.670$$
$$= \log (10^2) + \log (4.68) = \log (4.68 \times 10^2),$$

so $N = 4.68 \times 10^2 = 468$. If the given logarithm is negative, say, -2.180, we proceed as follows:

$$\log N = -2.180 = -3 + 0.820$$
$$= \log 10^{-3} + \log 6.6 = \log 6.6 \times 10^{-3},$$

so $N = 6.6 \times 10^{-3}$. Notice that what we did was to add to our given logarithm the next largest integer (3 in this case) so that we have an integer, plus a decimal number between 0 and 1.0 whose antilogarithm can be looked up in the table.

$\mathbf{B}$ DIMENSIONAL ANALYSIS

When we speak of the *dimensions* of a quantity, we are referring to the type of units that must be used. The dimensions of area, for example, are always length squared (abbreviated $[L^2]$, using square brackets) and the units can be square meters, square feet, and so on. Velocity, on the other hand, can be measured in units of km/h, m/s, and mi/h, but the dimensions are always a length $[L]$ divided by a time $[T]$, that is, $[L/T]$. The formula for a quantity may be different in different cases, but the dimensions remain the same. For example, the area of a triangle of base b and height h is $A = \frac{1}{2}bh$, whereas the area of a circle of radius r is $A = \pi r^2$. The formulas are different in the two cases, but the dimensions in both cases are the same: $[L^2]$.

When we specify the dimensions of a quantity, we usually do so in terms of basic quantities such as length $[L]$, time $[T]$, mass $[M]$, and electric current $[I]$. Thus, force, which by Newton's second law has the same units as mass $[M]$ times acceleration $[L/T^2]$, has dimensions of $[ML/T^2]$.

Dimensions can be used as a help in working out relationships, and such a procedure is referred to as *dimensional analysis*. One useful technique is the use of dimensions to check a relationship for correctness. Two simple rules apply here. First, we can add or subtract quantities only if they have the same dimensions (we do not add centimeters and pounds); second, the quantities on each side of an equals sign must have the same dimensions.

For example, suppose that you derived the equation $v = v_0 + \frac{1}{2}at^2$, where v is the speed of an object after a time t, when it starts with an initial speed v_0 and undergoes an acceleration a. Let us do a dimensional check to see if this equation can be correct. We write a dimensional equation as follows, remembering that the dimensions of speed are $[L/T]$ and of acceleration are $[L/T^2]$:

$$\left[\frac{L}{T}\right] \stackrel{?}{=} \left[\frac{L}{T}\right] + \left[\frac{L}{T^2}\right][T^2]$$

$$\stackrel{?}{=} \left[\frac{L}{T}\right] + [L].$$

The dimensions are incorrect: on the right side, we have the sum of two quantities whose dimensions are not the same. Thus, we conclude that an error was made in the derivation of the original equation.

If such a dimensional check does come out correct, it does not prove that the equation is correct; for example, a dimensionless numerical factor (such as $\frac{1}{2}$ or 2π) could be wrong. Thus a dimensional check can only tell you when a relationship is wrong; it cannot tell you if it is completely right.

Another use of dimensional analysis is for a quick check on an equation you are not sure about. For example, suppose that you cannot remember whether the equation for the period T of an oscillating mass m on the end of a spring with constant k is $T = 2\pi\sqrt{k/m}$ or is $T = 2\pi\sqrt{m/k}$. A dimensional check can tell you. The dimensions of k, since from Hooke's law $k = \text{force/distance}$, are $[ML/T^2/L] = [M/T^2]$. Thus the formula $T = 2\pi\sqrt{m/k}$ is correct:

$$[T] = \sqrt{\frac{[M]}{[M/T^2]}},$$

whereas the formula $T = 2\pi\sqrt{k/m}$ is not:

$$[T] \neq \sqrt{\frac{[M/T^2]}{[M]}}.$$

Finally, an important use of dimensional analysis, but one with which much care must be taken, is to obtain the *form* of an equation. That is, we may want to determine how one quantity depends on others. To take a concrete example, let us try to find an expression for the period T of a simple pendulum. First, we try to figure out what T could depend on, and make a list of these variables. It might depend on its length l, on the mass m of the bob, on the angle of swing θ, and on the acceleration due to gravity, g. It might also depend on air resistance (we would use the viscosity of air), the gravitational pull of the Moon, and so on; but everyday experience suggests that the Earth's gravity is the major force involved, so we ignore the other possible forces. So let us assume that T is a function of l, m, θ, and g, and that each of these factors is present to some power:

$$T = Cl^w m^x \theta^y g^z.$$

C is a dimensionless constant, and w, x, y, and z are exponents we want to solve for. We now write down the dimensional equation for this relationship:

$$[T] = [L]^w[M]^x[L/T^2]^z;$$

because θ has no dimensions (a radian is a length over a length—see Eq. 8–1), it does not appear. We simplify and obtain

$$[T] = [L]^{w+z}[M]^x[T]^{-2z}.$$

To have dimensional consistency, we must have

$$1 = -2z$$
$$0 = w + z$$
$$0 = x.$$

We solve these equations and find that $z = -\frac{1}{2}$, $w = \frac{1}{2}$, and $x = 0$. Thus our desired equation must be

$$T = C\sqrt{l/g}\, f(\theta), \tag{B–1}$$

where $f(\theta)$ is some function of θ that we cannot determine using this technique. Nor can we determine in this way the dimensionless constant C. (Of course, to obtain C and f, we would have to do an analysis such as that in Chapter 11 using Newton's laws, which reveals that $C = 2\pi$ and $f \approx 1$ for small θ). But look what we *have* found, using only dimensional consistency. We obtained the form of the expression that relates the period of a simple pendulum to the major variables of the situation, l and g (see Eq. 11–11a) and saw that it indeed does not depend on the mass m.

How did we do it? And how useful is this technique? Basically, we had to use our intuition as to which variables were important and which were not. This is not always easy, and often requires a lot of insight. As to usefulness, the final result in our example could have been obtained from Newton's laws, as in Chapter 11. But in many physical situations, such a derivation from other laws cannot be done. In those situations, dimensional analysis can be a powerful tool.

In the end, any expression derived by the use of dimensional analysis (or by any other means, for that matter) must be checked against experiment. For example, in our derivation of Eq. B–1, we can compare the periods of two pendula of different lengths, l_1 and l_2, whose amplitudes (θ) are the same. For, using Eq. B–1, we would have

$$\frac{T_1}{T_2} = \frac{C\sqrt{l_1/g}\, f(\theta)}{C\sqrt{l_2/g}\, f(\theta)} = \sqrt{\frac{l_1}{l_2}}.$$

Because C and $f(\theta)$ are the same for both pendula, they cancel out, so we can experimentally determine if the ratio of the periods varies as the ratio of the square roots of the lengths. This comparison to experiment checks our derivation, at least in part. C and $f(\theta)$ could be determined by further experiments.

ROTATING FRAMES OF REFERENCE; INERTIAL FORCES; CORIOLIS EFFECT

Inertial and Noninertial Reference Frames

In Chapters 5 and 8 we examined the motion of bodies, including circular and rotational motion, from the outside, as observers fixed on the ground. Sometimes it is convenient to place ourselves (in theory, if not physically) into a rotating system. Let us, for example, examine the motion of objects from the point of view, or frame of reference, of persons seated on a rotating platform such as a merry-go-round. It looks to them as if the rest of the world is going around *them*. But let us focus attention on what they observe when they place a tennis ball on the floor of the rotating platform, which we assume is frictionless. If they put the ball down gently, without giving it any push, they will observe that it accelerates from rest and moves outward as shown in Fig. C–1a. According to Newton's first law, an object initially at rest should stay at rest if no force acts on it. But, according to the observers on the rotating platform, the ball starts moving even though there is no force applied to it. To observers on the ground, this is all very clear: the ball has an initial velocity when it is released (because the platform is moving), and it simply continues moving in a straight-line path as shown in Fig. C–1b, in accordance with Newton's first law.

But what shall we do about the frame of reference of the observers on the rotating platform? Clearly, Newton's first law, the law of inertia, does not hold in this rotating frame of reference. For this reason, such a frame is called a **noninertial reference frame**. An **inertial reference frame** is one in which the law of inertia—Newton's first law—does hold, and so do Newton's second and third laws. In a noninertial reference frame, such as our rotating platform, Newton's second law also does not hold. For instance in the situation described above, there is no net force on the ball; yet, with respect to the rotating platform, the ball accelerates.

Fictitious (Inertial) Forces

Because Newton's laws do not hold when observations are made with respect to a rotating frame of reference, calculation of motion can be complicated. However, we can still apply Newton's laws in such a reference frame if we make use of a trick. The ball on the rotating platform of Fig. C–1a flies outward when released (as if a force were acting on it—though as we saw above, no force actually does act on it); so the trick we use is to write down the equation $\Sigma F = ma$ as if a force equal to mv^2/r (or $m\omega^2 r$) were acting radially outward on the object in addition to any other forces that may be acting. This extra force, which might be designated as "centrifugal force" since it *seems* to act outward, is called a **fictitious force** or **pseudoforce**. It is a pseudoforce ("pseudo" means "false") because there is no object that exerts

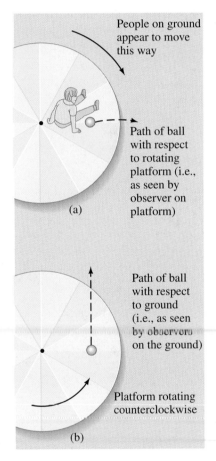

FIGURE C–1 Path of a ball released on a rotating merry-go-round (a) in the reference frame of the merry-go-round, and (b) in a reference frame on the ground.

People on ground appear to move this way

Path of ball with respect to rotating platform (i.e., as seen by observer on platform)

(a)

Path of ball with respect to ground (i.e., as seen by observers on the ground)

Platform rotating counterclockwise

(b)

this force. Furthermore, when viewed from an inertial reference frame, the effect doesn't exist at all. We have made up this pseudoforce so that we can make calculations in a noninertial frame using Newton's second law, $\Sigma F = ma$. Thus the observer in Fig. C–1a can determine the motion of the ball by assuming that a force equal to mv^2/r acts on it. Such pseudoforces are also called **inertial forces** since they arise only because the reference frame is not an inertial one.

We can examine the motion of a particle in a centrifuge (Section 5–5) from the frame of reference of the rotating rotor. In this frame of reference, the particles move in a more-or-less straight path down the tube. (From the reference frame of the Earth, the particles go round and round.) The acceleration of a particle with respect to the rotor can then be calculated using $F = ma$ if we include a pseudoforce, "F," equal to $m\omega^2 r = m(v^2/r)$ acting down the tube, in addition to the drag force F_D exerted by the fluid on the particle (Fig. C–2) up the tube.

In Section 5–3 we discussed the forces on a person in a car going around a curve (Fig. 5–11) from the point of view of an inertial frame. The car, on the other hand, is not an inertial frame. Passengers in such a car could interpret this being pressed outward as the effect of a "centrifugal" force. But they need to recognize that it is a pseudoforce because there is no identifiable object exerting it. It is an effect of being in a noninertial frame of reference.

The Earth itself is rotating on its axis. Thus, strictly speaking, Newton's laws are not valid on the Earth. However, the effect of the Earth's rotation is usually so small that it can be ignored, although it does influence the movement of large air masses and ocean currents. Because of the Earth's rotation, the material of the Earth is concentrated slightly at the equator. The Earth is thus not a perfect sphere but is slightly fatter at the equator than it is at the poles.

Coriolis Effect

In a reference frame that rotates at a constant angular speed ω (relative to an inertial system), there exists another pseudoforce known as the *Coriolis force*. It appears to act on a body in a rotating reference frame only if the body is moving relative to that reference frame, and it acts to deflect the body sideways. It, too, is an effect of the reference frame being noninertial and hence is referred to as an *inertial force*. To see how the Coriolis force arises, consider two people, A and B, at rest on a platform rotating with angular speed ω, as shown in Fig. C–3a. They are situated at distances r_A and r_B from the axis of rotation (at O). The woman at A throws a ball with a horizontal velocity $\mathbf{v}$ (in her reference frame) radially outward toward the man at B on the outer edge of the platform. In Fig. C–3a, we view the situation from an inertial reference frame. The ball initially has not only the velocity $\mathbf{v}$ radially outward, but also a tangential velocity $\mathbf{v}_A$ due to the rotation of the platform. Now Eq. 8–4 tells us that $v_A = r_A\omega$, where r_A is the woman's radial distance from the axis of rotation at O. If the man at B had this same velocity v_A, the ball would reach him perfectly. But his speed is greater than v_A (Fig. C–3a) since he is farther from the axis of rotation. His speed is $v_B = r_B\omega$, which is greater than v_A because $r_B > r_A$. Thus, when the ball reaches the outer edge of the platform, it passes a point that the man at B has already passed because his speed in that direction is greater than the ball's. So the ball passes behind him.

FIGURE C–2 The forces on a particle in a rotating centrifuge, in the reference frame of the centrifuge.

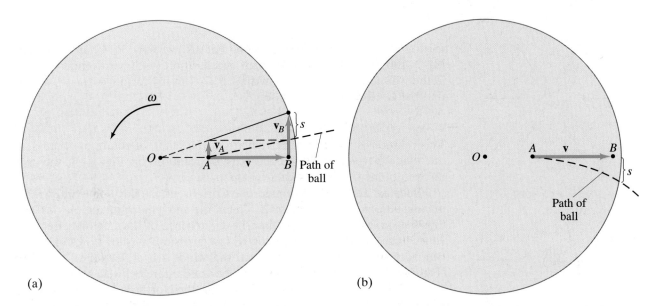

(a)

(b)

FIGURE C–3 The origin of the Coriolis effect. Looking down on a rotating platform, (a) as seen from a nonrotating inertial system, and (b) as seen from the rotating platform as frame of reference.

Figure C–3b shows the situation as seen from the rotating platform as frame of reference. Both A and B are at rest, and the ball is thrown with velocity $\mathbf{v}$ toward B, but the ball deflects to the right as shown and passes behind B as previously described. This is not a centrifugal-force effect, for the latter acts radially outward. Instead, this effect acts sideways, perpendicular to $\mathbf{v}$, and is called a **Coriolis acceleration**; it is said to be due to the Coriolis force, which is a fictitious, inertial force. Its explanation as seen from an inertial system was given above: it is an effect of being in a rotating system, wherein points that are farther from the rotation axis have higher linear speeds. On the other hand, when viewed from the rotating system, we can describe the motion using Newton's second law, $\Sigma \mathbf{F} = m\mathbf{a}$, if we add a "pseudoforce" term corresponding to this Coriolis effect.

Let us determine the magnitude of the Coriolis acceleration for the simple case described above. (We assume v is large and distances short, so we can ignore gravity.) We do the calculation from the inertial reference frame (Fig. C–3a). The ball moves radially outward a distance $r_B - r_A$ at speed v in a time t given by

$$r_B - r_A = vt.$$

During this time, the ball moves to the side a distance s_A given by

$$s_A = v_A t.$$

The man at B, in this time t, moves a distance

$$s_B = v_B t.$$

The ball therefore passes behind him a distance s (Fig. C–3a) given by

$$s = s_B - s_A = (v_B - v_A)t.$$

We saw earlier that $v_A = r_A \omega$ and $v_B = r_B \omega$, so

$$s = (r_B - r_A)\omega t.$$

We substitute $r_B - r_A = vt$ (see above) and get

$$s = \omega v t^2. \qquad \textbf{(C–1)}$$

This equals the sideways displacement as seen from the noninertial rotating system (Fig. C–3b). We see immediately that Eq. C–1 corresponds to

(a)

(b)

(c)

FIGURE C–4 (a) Winds (moving air masses) would flow directly toward a low-pressure area if the Earth did not rotate; (b) and (c): because of the Earth's rotation, the winds are deflected to the right in the Northern Hemisphere (as in Fig. C–3) as if a fictitious (Coriolis) force were acting.

motion at a constant acceleration. For as we saw in Chapter 2 (see Eq. 2–10b), $y = \frac{1}{2}at^2$ for a constant acceleration (with zero initial velocity in the y direction). Thus, if we write Eq. C–1 in the form $s = \frac{1}{2}a_{\text{cor}}t^2$, we see that the Coriolis acceleration a_{cor} is

$$a_{\text{cor}} = 2\omega v. \tag{C–2}$$

This relation is valid for any velocity in the plane of rotation—that is, in the plane perpendicular to the axis of rotation (in Fig. C–3, the axis through point O perpendicular to the page).

Because the Earth rotates, the Coriolis effect has some interesting manifestations on the Earth. It affects the movement of air masses and thus has an influence on weather. In the absence of the Coriolis effect, air would rush directly into a region of low pressure, as shown in Fig. C–4a. But because of the Coriolis effect, the winds are deflected to the right (Fig. C–4b), so that there tends to be a counterclockwise wind pattern around a low-pressure area. This is true for the Northern Hemisphere—for which the situation of Fig. C–3b applies since the Earth rotates from west to east. The reverse is true in the Southern Hemisphere. Thus cyclones rotate counterclockwise in the Northern Hemisphere and clockwise in the Southern Hemisphere. The same effect explains the easterly trade winds near the equator: any winds heading south toward the equator will be deflected toward the west (that is, as if coming from the east).

The Coriolis effect also acts on a falling body. A body released from the top of a high tower will not hit the ground directly below the release point, but will be deflected slightly to the east. Viewed from an inertial frame, this is because the top of the tower revolves with a slightly higher speed than does the bottom of the tower.

GAUSS'S LAW D

An important relation in electricity is Gauss's law, developed by the great mathematician Karl Friedrich Gauss (1777–1855). It relates electric charge and electric field, and is a more general and elegant version of Coulomb's law.

Gauss's law involves the concept of *electric flux*, which refers to the electric field passing through a given area. For a uniform electric field **E** passing through an area A, as shown in Fig. D–1a, the electric flux Φ_E is defined as

$$\Phi_E = EA \cos \theta, \qquad\qquad (\textbf{D–1})$$

where θ is the angle between the electric-field direction and a line drawn perpendicular to the area, as shown. The flux can be written equivalently as

$$\Phi_E = E_\perp A = EA_\perp,$$

where $E_\perp = E \cos \theta$ is the component of **E** perpendicular to the area (Fig. D–1b), and, similarly, $A_\perp = A \cos \theta$ is the projection of the area A perpendicular to the field **E** (Fig. D–1c).

Electric flux has a simple intuitive interpretation in terms of field lines. We saw in Section 16–8 that field lines can always be drawn so that the number (N) passing through unit area perpendicular to the field ($A_\perp$) is proportional to the magnitude of the field (E): that is, $E \propto N/A_\perp$. Hence,

$$N \propto EA_\perp = \Phi_E, \qquad\qquad (\textbf{D–2})$$

so the flux through an area is proportional to the number of lines passing through that area.

Gauss's law involves the *total* flux through a closed surface. For any such surface, such as that shown in Fig. D–2, we divide the surface up into many tiny areas, $\Delta A_1, \Delta A_2, \Delta A_3, \cdots$, and so on. We make the division so that each ΔA is small enough that it can be considered flat and so that the electric field can be considered constant within each ΔA. Then the *total* flux through the entire surface is the sum over all the individual fluxes through each of the tiny areas:

$$\Phi_E = E_1 \Delta A_1 \cos \theta_1 + E_2 \Delta A_2 \cos \theta_2 + \cdots$$
$$= \sum E \, \Delta A \cos \theta = \sum E_\perp \Delta A,$$

where the symbol Σ means "sum of." We saw in Section 16–8 that the number of field lines starting on a positive charge or ending on a negative charge is proportional to the magnitude of the charge. Hence, the *net* number of lines N pointing out of any closed surface (number of lines pointing out minus the number pointing in) must be proportional to the *net* enclosed

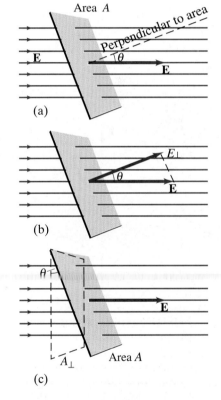

FIGURE D–1 (a) A uniform electric field **E** passing through a flat area A. (b) $E_\perp = E \cos \theta$ is the component of **E** perpendicular to the plane of the area A. (c) $A_\perp = A \cos \theta$ is the projection (shown dashed) of the area A perpendicular to the field **E**.

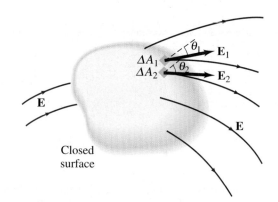

Electric field lines passing through a closed surface. The surface is divided up into many tiny areas, ΔA_1, $\Delta A_2, \cdots$, and so on, of which only two are shown.

Closed surface

charge, Q. But from Eq. D–2, we have that the net number of lines N is proportional to the total flux Φ_E. Therefore,

$$\Phi_E = \sum_{\substack{\text{closed} \\ \text{surface}}} E_\perp \, \Delta A \propto Q.$$

The constant of proportionality is $1/\epsilon_0$, as we shall see, so we have

Gauss's law

$$\sum_{\substack{\text{closed} \\ \text{surface}}} E_\perp \Delta A = \frac{Q}{\epsilon_0}, \tag{D–3}$$

where the sum (Σ) is over any closed surface, and Q is the net charge enclosed within that surface. This is **Gauss's law**.

This argument to justify Gauss's law may seem rather abstract. So let us show, for a particular example—that of a single point charge—that Gauss's law is equivalent to Coulomb's law, and at the same time show that the proportionality constant is indeed $1/\epsilon_0$. In Fig. D–3, we have a single charge Q, and for our "gaussian surface," we choose an imaginary sphere of radius r centered on the charge. Because Gauss's law is supposed to be valid for any surface, we have chosen one that will make our calculation easy, namely a sphere centered on the charge. Since the charge is at the center of the sphere, the magnitude of the electric field E at all points on the sphere is the same, namely

$$E = \frac{1}{4\pi\epsilon_0} \frac{Q}{r^2},$$

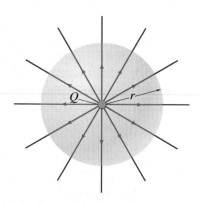

FIGURE D–3 A single point charge Q at the center of an imaginary sphere of radius r (our "gaussian surface"—that is, the closed surface we choose to use for applying Gauss's law in this case).

from Coulomb's law (Eq. 16–4b). The direction of $\mathbf{E}$ at all points is perpendicular to the sphere, so $E_\perp = E$. Hence, the total flux is

$$\sum_{\substack{\text{closed} \\ \text{surface}}} E_\perp \, \Delta A = E \sum \Delta A = \frac{1}{4\pi\epsilon_0} \frac{Q}{r^2} \sum \Delta A.$$

Because E is the same through each of the areas ΔA, we were able to factor it out of the sum. Now the sum over the entire area of the spherical surface, $\Sigma \Delta A$, is simply the surface area of a sphere, $4\pi r^2$. Hence,

$$\sum E_\perp \, \Delta A = \frac{1}{4\pi\epsilon_0} \frac{Q}{r^2} (4\pi r^2) = \frac{Q}{\epsilon_0},$$

which is just Gauss's law, as we set out to show.

We can argue that Gauss's law will hold for any closed surface other than a sphere by noting that the number of lines N passing through it would be the same as through our sphere (Fig. D–4). Since the flux through the surface is proportional to N, then $\Sigma E_\perp \Delta A = Q/\epsilon_0$ is valid for any surface surrounding a point charge.

If we have many charges, $Q_1, Q_2, \cdots$, then for each charge individually,

$$\sum E_{i\perp} \Delta A = \frac{Q_i}{\epsilon_0}$$

where the subscript i refers to any one of the charges. The sum of all the charges is the total charge $Q \,(= \Sigma Q_i)$. And, by the superposition principle, the total electric field E is equal to the sum of the fields due to each separate charge: $E = \Sigma E_i$. Hence,

$$\sum E_\perp \Delta A = \frac{\sum Q_i}{\epsilon_0} = \frac{Q}{\epsilon_0}.$$

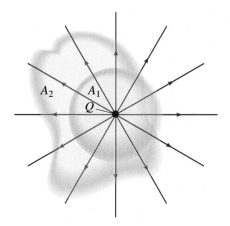

FIGURE D–4 A single point charge Q, surrounded by a spherical surface A_1 and an irregular surface A_2.

So Gauss's law is consistent with Coulomb's law for any number of charges enclosed within a closed surface of any shape.

Coulomb's law and Gauss's law are equivalent for electric fields produced by static charges. However, electric fields can also be produced by (changing) magnetic fields, as was discussed in Chapter 21. Coulomb's law does not apply for such electric fields, but Gauss's law does apply. Hence, Gauss's law is considered a more general law than Coulomb's law.

Coulomb's law and Gauss's law can be used to determine the electric field due to a given (static) charge distribution. Gauss's law is useful when the charge distribution is simple and symmetrical. However, we must choose the "gaussian" surface very carefully so we can determine $\mathbf{E}$. We normally choose a surface that has just the symmetry needed so E will be constant on all or on parts of its surface.

EXAMPLE D–1 **Spherical shell.** A thin spherical shell of radius r_0 possesses a total net charge Q that is uniformly distributed on it, Fig. D–5. Determine the electric field at points (a) outside the shell, and (b) inside the shell.

SOLUTION (a) Because the charge is distributed symmetrically, the electric field must also be symmetric. Thus the field must be directed radially outward (inward if $Q < 0$) and must depend only on r. The electric field will thus have the same magnitude at all points on an imaginary gaussian surface, which we draw as a sphere of radius r, concentric with the shell, and shown in Fig. D–5 as a dashed circle labeled A_1. Because $\mathbf{E}$ is perpendicular to this surface, Gauss's law gives

$$\sum E_\perp \Delta A = E \sum \Delta A = E(4\pi r^2) = \frac{Q}{\epsilon_0},$$

or

$$E = \frac{1}{4\pi\epsilon_0} \frac{Q}{r^2}. \qquad\qquad [r > r_0]$$

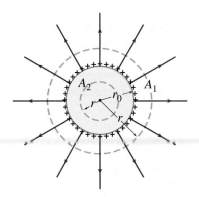

FIGURE D–5 Uniform distribution of charge on a thin spherical shell of radius r_0. Example D–1.

Thus the field outside a uniformly charged spherical shell is the same as if all the charge were concentrated at the center as a point charge.

(*b*) Inside the shell, the field must also be symmetric. So E must again have the same value at all points on a spherical gaussian surface (A_2 in Fig. D–5) concentric with the shell. Thus, E can be factored out of the sum and we have

$$\sum E_\perp \, \Delta A = E \sum \Delta A = E(4\pi r^2) = \frac{Q}{\epsilon_0} = 0$$

since the charge Q inside the shell is zero. Hence

$$E = 0 \qquad\qquad\qquad [r < r_0]$$

inside a uniform spherical shell of charge.

The useful results of Example D–1 also apply to a uniform *solid* spherical conductor that is charged, since all the charge would lie in a thin layer at the surface (Section 16–9).

FIGURE D–6 Calculation of **E** due to a very long line of charge. Example D–2.

FIGURE D–7 Electric field near the surface of a conductor. Two small cylindrical boxes are shown dashed. Either one can serve as our gaussian surface. See Example D–3.

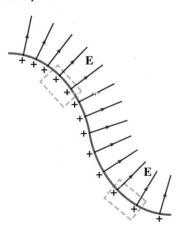

EXAMPLE D–2 **Long uniform line of charge.** A very long straight wire possesses a uniform charge per unit length, λ. Calculate the electric field at points near (but outside) the wire, far from the ends.

SOLUTION Because of the symmetry, we expect the field to be directed radially outward (if $Q > 0$) and to depend only on the perpendicular distance, r, from the wire. Because of this cylindrical symmetry, the field will be the same at all points on a gaussian surface that is a cylinder with the wire along its axis, Fig. D–6. **E** is perpendicular to this surface at all points. But for Gauss's law, we need a closed surface, so we must include the flat ends of the cylinder. Since **E** is parallel to the ends, there is no flux through the ends (the $\cos\theta$ in Gauss's law is $\cos 90° = 0$). So Gauss's law tells us

$$\sum E_\perp \, \Delta A = E \sum \Delta A = E(2\pi r l) = \frac{Q}{\epsilon_0} = \frac{\lambda l}{\epsilon_0},$$

where l is the length of our chosen gaussian surface ($l \ll$ length of wire), and $2\pi r$ is its circumference. Hence

$$E = \frac{1}{2\pi\epsilon_0} \frac{\lambda}{r}.$$

EXAMPLE D–3 ***E* at surface of conductor.** Show that the electric field just outside the surface of any good conductor of arbitrary shape is given by

$$E = \frac{\sigma}{\epsilon_0},$$

where σ is the surface charge density (Q/A) on the conductor at that point.

SOLUTION We choose as our gaussian surface a small cylindrical box, very small in height so that one of its circular ends is just above the conductor (Fig. D–7); the other end is just below the conductor's surface, and the sides are perpendicular to it. The electric field is zero inside a conductor and is perpendicular to the surface just outside it (Section 16–9), so electric flux passes only through the outside end of our cylindrical box. We choose the area A (of the flat cylinder end) small enough so that E is essentially uniform over it. Then Gauss's law gives

$$\sum E_\perp \, \Delta A = EA = \frac{Q}{\epsilon_0} = \frac{\sigma A}{\epsilon_0},$$

so that

$$E = \frac{\sigma}{\epsilon_0}. \qquad\qquad \text{[at surface of conductor]}$$

This useful result applies for any shape conductor, including a large uniformly charged flat sheet: the electric field will be constant and equal to σ/ϵ_0.

This last Example also gives us the field between the two parallel plates we discussed in Fig. 16–29d, and also relative to Eq. 17–6. If the plates are large compared to their separation, then the field lines are perpendicular to the plates and, except near the edges, they are parallel to each other. Therefore the electric field (see Fig. D–8, which shows the same gaussian surface as Fig. D–7) is also

$$E = \frac{\sigma}{\epsilon_0} = \frac{Q/A}{\epsilon_0} \qquad\qquad \textbf{(D–4)}$$

where $Q = \sigma A$ is the charge on one of the plates.

We saw in Section 16–9 that in the static situation, the electric field inside any conductor must be zero. (Otherwise, the free charges in the conductor would move—until the net force on each, and hence $\mathbf{E}$, was zero.) We also saw there that any net electric charge on a conductor must all reside on its outer surface. This is readily seen using Gauss's law. Consider any charged conductor of any shape, such as that shown in Fig. D–9, which carries a net charge Q. We choose the gaussian surface, shown dashed in the diagram, so that it lies below the surface of the conductor. Our gaussian surface can be arbitrarily close to the surface, but still *inside* the conductor. The electric field is zero at all points on this gaussian surface since it is inside the conductor. Hence, from Gauss's law, Eq. D–3, the net charge within the surface must be zero. Hence, there can be no net charge within the conductor. Any net charge must lie on the surface of the conductor.

FIGURE D–8 The electric field between two parallel plates is uniform and equal to $E = \sigma/\epsilon_0$.

Surface of conductor

Gaussian surface

Q

FIGURE D–9 An insulated charged conductor of arbitrary shape, showing (dashed) the gaussian surface just below the surface of the conductor.

APPENDIX

E GALILEAN AND LORENTZ TRANSFORMATIONS

W
e now examine in detail the mathematics of relating quantities in one inertial reference frame to the equivalent quantities in another. In particular, we will see how positions and velocities *transform* (that is, change) as we go from one frame of reference to another.

We begin with the classical or Galilean viewpoint. Consider two reference frames S and S' which are each characterized by a set of coordinate axes, Fig. E–1. The axes x and y (z is not shown) refer to S, and x' and y' refer to S'. The x' and x axes overlap one another, and we assume that frame S' moves to the right (in the x direction) at speed v with respect to S. And for simplicity let us assume the origins O and O' of the two reference frames are superimposed at time $t = 0$.

Now consider an event that occurs at some point P (Fig. E–1) represented by the coordinates x', y', z' in reference frame S' at the time t'. What will be the coordinates of P in S? Since S and S' overlap precisely initially, after a time t, S' will have moved a distance vt'. Therefore, at time t', $x = x' + vt'$. The y and z coordinates, on the other hand, are not altered by motion along the x axis; thus $y = y'$ and $z = z'$. Finally, since time is assumed to be absolute in Galilean–Newtonian physics, clocks in the two frames will agree with each other; so $t = t'$. We summarize these in the following **Galilean transformation equations**:

Galilean

transformations

$$x = x' + vt'$$
$$y = y'$$
$$z = z'$$
$$t = t'.$$

(E–1)

These equations give the coordinates of an event in the S frame when those in the S' frame are known. If those in the S system are known, then

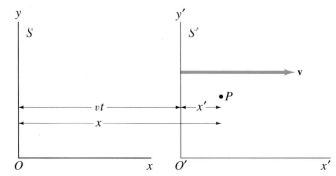

FIGURE E–1 Inertial reference frame S' moves to the right at speed v with respect to frame S.

the S' coordinates are obtained from

$$x' = x - vt, \qquad y' = y, \qquad z' = z, \qquad t' = t.$$

These four equations are the "inverse" transformation and are very easily obtained from Eqs. E–1. Notice that the effect is merely to exchange primed and unprimed quantities and replace v by $-v$. This makes sense because from the S' frame, S moves to the left (negative x direction) with speed v.

Now suppose that the point P in Fig. E–1 represents an object that is moving. Let the components of its velocity vector in S' be u'_x, u'_y, and u'_z (we use u to distinguish it from the relative velocity of the two frames, v). Now $u'_x = \Delta x'/\Delta t'$, $u'_y = \Delta y'/\Delta t'$, and $u'_z = \Delta z'/\Delta t'$, where all quantities are as measured in the S' frame. For example, if at time t'_1 the particle is at x'_1 and a short time later, t'_2, it is at x'_2, then

$$u'_x = \frac{x'_2 - x'_1}{t'_2 - t'_1} = \frac{\Delta x'}{\Delta t'}.$$

Now the velocity of P as seen from S will have components u_x, u_y, and u_z. We can show how these are related to the velocity components in S' by using Eqs. E–1. For example,

$$u_x = \frac{\Delta x}{\Delta t} = \frac{x_2 - x_1}{t_2 - t_1} = \frac{(x'_2 + vt'_2) - (x'_1 + vt'_1)}{t'_2 - t'_1}$$

$$= \frac{(x'_2 - x'_1) + v(t'_2 - t'_1)}{t'_2 - t'_1}$$

$$= \frac{\Delta x'}{\Delta t'} + v = u'_x + v.$$

For the other components, $u'_y = u_y$ and $u'_z = u_z$, so we have

$$u_x = u'_x + v,$$

$$u_y = u'_y, \qquad\qquad\qquad \textbf{(E–2)}$$

$$u_z = u'_z.$$

Galilean

velocity

transformations

These are known as the **Galilean velocity transformation** equations. We see that the y and z components of velocity are unchanged, but the x components differ by v. This is just what we have used before when dealing with relative velocity. For example, if S' is a train and S the Earth, and the train moves with speed v with respect to Earth, a person walking toward the front of the train with speed u'_x will have a speed with respect to the Earth of $u_x = u'_x + v$.

The Galilean transformations, Eqs. E–1 and E–2, are valid only when the velocities involved are not relativistic (Chapter 26)—that is, much less than the speed of light, c. We can see, for example, that the first of Eqs. E–2 will not work for the speed of light; for light traveling in S' with speed $u'_x = c$ will have speed $c + v$ in S, whereas the theory of relativity insists it must be c in S. Clearly, then, a new set of transformation equations is needed to deal with relativistic velocities.

Relativity theory

We will derive the required equations in a simple way, again looking at Fig. E–1. We assume the transformation is linear and of the form

$$x = \gamma(x' + vt'), \qquad y = y', \quad z = z'.$$

That is, we modify the first of Eqs. E–1 by multiplying by a factor γ which is yet to be determined. But we assume the y and z equations are unchanged because we expect no length contraction in these directions. We won't assume a form for t, but will derive it. The inverse equations must have the same form with v replaced by $-v$. (The principle of relativity demands it, since S' moving to the right with respect to S is equivalent to S moving to the left with respect to S'.) Therefore

$$x' = \gamma(x - vt).$$

Now if a light pulse leaves the common origin of S and S' at time $t = t' = 0$, after a time t it will have traveled along the x axis a distance $x = ct$ (in S), or $x' = ct'$ (in S'). Therefore, from the equations for x and x' above,

$$ct = \gamma(ct' + vt') = \gamma(c + v)t',$$

$$ct' = \gamma(ct - vt) = \gamma(c - v)t.$$

We substitute t' from the second equation into the first and find $ct = \gamma(c + v)\gamma(c - v)(t/c) = \gamma^2(c^2 - v^2)t/c$. We cancel out the t on each side and solve for γ to find

$$\gamma = \frac{1}{\sqrt{1 - v^2/c^2}}.$$

Now that we have found γ, we need only find the relation between t and t'. To do so, we combine $x' = \gamma(x - vt)$ with $x = \gamma(x' + vt')$:

$$x' = \gamma(x - vt) - \gamma[\gamma(x' + vt') - vt].$$

We solve for t and find $t = \gamma(t' + vx'/c^2)$. In summary,

$$
\begin{aligned}
x &= \frac{1}{\sqrt{1 - v^2/c^2}}(x' + vt') \\
y &= y' \\
z &= z' \\
t &= \frac{1}{\sqrt{1 - v^2/c^2}}\left(t' + \frac{vx'}{c^2}\right).
\end{aligned}
$$

Lorentz

transformations

(E–3)

These are called the **Lorentz transformation** equations. They were first proposed, in a slightly different form, by Lorentz in 1904 to explain the null result of the Michelson–Morley experiment and to make Maxwell's equations take the same form in all inertial systems. A year later, Einstein derived them independently based on his theory of relativity. Notice that not only is the x equation modified as compared to the Galilean transformation, but so is the t equation. Indeed, we see directly in this last equation, as well as in the first, how the space and time coordinates mix.

The relativistically correct velocity equations are readily obtained. For example, using Eqs. E–3 (we let $\gamma = 1/\sqrt{1 - v^2/c^2}$),

$$
u_x = \frac{\Delta x}{\Delta t} = \frac{\gamma(\Delta x' + v\,\Delta t')}{\gamma(\Delta t' + v\,\Delta x'/c^2)} = \frac{(\Delta x'/\Delta t') + v}{1 + (v/c^2)(\Delta x'/\Delta t')}
$$

$$
= \frac{u_x' + v}{1 + vu_x'/c^2}.
$$

The others are obtained in the same way and we collect them here:

$$u_x = \frac{u'_x + v}{1 + vu'_x/c^2}$$

$$u_y = \frac{u'_y\sqrt{1 - v^2/c^2}}{1 + vu'_x/c^2}$$ **(E–4)** *Relativistic velocity transformations*

$$u_z = \frac{u'_z\sqrt{1 - v^2/c^2}}{1 + vu'_x/c^2}.$$

The first of these equations is Eq. 26–9, which we used in Section 26–11 where we discussed how velocities do not add in our commonsense (Galilean) way, because of the denominator $(1 + vu'_x/c^2)$. We can now also see that the y and z components of velocity are also altered and that they depend on the x' component of velocity.

EXAMPLE E–1 **Length contraction.** Derive the length-contraction formula, Eq. 26–2, from the Lorentz transformation equations.

SOLUTION Let an object of length L_0 be at rest on the x axis in S. The coordinates of its two end points are x_1 and x_2, so that $x_2 - x_1 = L_0$. At any instant in S', the end points will be at x'_1 and x'_2 as given by the Lorentz transformation equations. The length measured in S' is $L = x'_2 - x'_1$. An observer in S' measures this length by measuring x'_2 and x'_1 at the same time (in the S' frame), so $t'_2 = t'_1$. Then, from the first of Eqs. E–3,

$$L_0 = x_2 - x_1 = \frac{1}{\sqrt{1 - v^2/c^2}}(x'_2 + vt'_2 - x'_1 - vt'_1).$$

Since $t'_2 = t'_1$, we have

$$L_0 = \frac{1}{\sqrt{1 - v^2/c^2}}(x'_2 - x'_1) = \frac{L}{\sqrt{1 - v^2/c^2}},$$

or

$$L = L_0\sqrt{1 - v^2/c^2},$$

which is Eq. 26–2.

EXAMPLE E–2 **Time dilation.** Derive the time-dilation formula, Eq. 26–1, from the Lorentz transformation equations.

SOLUTION The time Δt_0 between two events that occur at the same place $(x'_2 = x'_1)$ in S' is measured to be $\Delta t_0 = t'_2 - t'_1$. Since $x'_2 = x'_1$, then from the last of Eqs. E–3, the time Δt between the events as measured in S is

$$\Delta t = t_2 - t_1 = \frac{1}{\sqrt{1 - v^2/c^2}}\left(t'_2 + \frac{vx'_2}{c^2} - t'_1 - \frac{vx'_1}{c^2}\right)$$

$$= \frac{1}{\sqrt{1 - v^2/c^2}}(t'_2 - t'_1) = \frac{\Delta t_0}{\sqrt{1 - v^2/c^2}},$$

which is Eq. 26–1. Notice that we chose S' to be the frame in which the two events occur at the same place, so that $x'_1 = x'_2$, and the terms containing x'_1 and x'_2 cancel out.

F SELECTED ISOTOPES

(1) Atomic Number Z	(2) Element	(3) Symbol	(4) Mass Number A	(5) Atomic Mass[†]	(6) % Abundance (or Radioactive Decay Mode)	(7) Half-life (if radioactive)
0	(Neutron)	n	1	1.008665	β^-	10.4 min
1	Hydrogen	H	1	1.007825	99.985%	
	Deuterium	D	2	2.014102	0.015%	
	Tritium	T	3	3.016049	β^-	12.33 yr
2	Helium	He	3	3.016029	0.000137%	
			4	4.002602	99.999863%	
3	Lithium	Li	6	6.015121	7.5%	
			7	7.016003	92.5%	
4	Beryllium	Be	7	7.016928	EC, γ	53.29 days
			9	9.012182	100%	
5	Boron	B	10	10.012936	19.9%	
			11	11.009305	80.1%	
6	Carbon	C	11	11.011433	β^+, EC	20.385 min
			12	12.000000	98.90%	
			13	13.003355	1.10%	
			14	14.003242	β^-	5730 yr
7	Nitrogen	N	13	13.005738	β^+	9.965 min
			14	14.003074	99.63%	
			15	15.000108	0.37%	
8	Oxygen	O	15	15.003065	β^+, EC	122.24 s
			16	15.994915	99.76%	
			18	17.999160	0.20%	
9	Fluorine	F	19	18.998404	100%	
10	Neon	Ne	20	19.992435	90.48%	
			22	21.991383	9.25%	
11	Sodium	Na	22	21.994434	β^+, EC, γ	2.6088 yr
			23	22.989767	100%	
			24	23.990961	β^-, γ	14.9590 h
12	Magnesium	Mg	24	23.985042	78.99%	
13	Aluminum	Al	27	26.981538	100%	
14	Silicon	Si	28	27.976927	92.23%	
			31	30.975362	β^-, γ	157.3 min

[†]The masses given in column (5) are those for the neutral atom, including the Z electrons.

(1) Atomic Number Z	(2) Element	(3) Symbol	(4) Mass Number A	(5) Atomic Mass	(6) % Abundance (or Radioactive Decay Mode)	(7) Half-life (if radioactive)
15	Phosphorus	P	31	30.973762	100%	
			32	31.973908	β^-	14.262 days
16	Sulfur	S	32	31.972071	95.02%	
			35	34.969033	β^-	87.51 days
17	Chlorine	Cl	35	34.968853	75.77%	
			37	36.965903	24.23%	
18	Argon	Ar	40	39.962384	99.600%	
19	Potassium	K	39	38.963708	93.2581%	
			40	39.964000	0.01117%	
					$\beta^-, EC, \gamma, \beta^+$	1.28×10^9 yr
20	Calcium	Ca	40	39.962591	96.941%	
21	Scandium	Sc	45	44.955911	100%	
22	Titanium	Ti	48	47.947947	73.8%	
23	Vanadium	V	51	50.943962	99.750%	
24	Chromium	Cr	52	51.940511	83.79%	
25	Manganese	Mn	55	54.938048	100%	
26	Iron	Fe	56	55.934940	91.72%	
27	Cobalt	Co	59	58.933198	100%	
			60	59.933820	β^-, γ	5.2714 yr
28	Nickel	Ni	58	57.935346	68.077%	
			60	59.930789	26.233%	
29	Copper	Cu	63	62.929599	69.17%	
			65	64.927791	30.83%	
30	Zinc	Zn	64	63.929144	48.6%	
			66	65.926035	27.9%	
31	Gallium	Ga	69	68.925580	60.108%	
32	Germanium	Ge	72	71.922079	27.66%	
			74	73.921177	35.94%	
33	Arsenic	As	75	74.921594	100%	
34	Selenium	Se	80	79.916519	49.61%	
35	Bromine	Br	79	78.918336	50.69%	
36	Krypton	Kr	84	83.911508	57.0%	
37	Rubidium	Rb	85	84.911793	72.17%	
38	Strontium	Sr	86	85.909266	9.86%	
			88	87.905618	82.58%	
			90	89.907737	β^-	29.1 yr
39	Yttrium	Y	89	88.905847	100%	
40	Zirconium	Zr	90	89.904702	51.45%	
41	Niobium	Nb	93	92.906376	100%	
42	Molybdenum	Mo	98	97.905407	24.13%	
43	Technetium	Tc	98	97.907215	β^-, γ	4.2×10^6 yr
44	Ruthenium	Ru	102	101.904348	31.6%	
45	Rhodium	Rh	103	102.905502	100%	
46	Palladium	Pd	106	105.903481	27.33%	

(1) Atomic Number Z	(2) Element	(3) Symbol	(4) Mass Number A	(5) Atomic Mass	(6) % Abundance (or Radioactive Decay Mode)	(7) Half-life (if radioactive)
47	Silver	Ag	107	106.905091	51.839%	
			109	108.904754	48.161%	
48	Cadmium	Cd	114	113.903359	28.73%	
49	Indium	In	115	114.903876	95.7%; β^-, γ	4.41×10^{14} yr
50	Tin	Sn	120	119.902197	32.59%	
51	Antimony	Sb	121	120.903820	57.36%	
52	Tellurium	Te	130	129.906228	33.87%	2.5×10^{21} yr
53	Iodine	I	127	126.904474	100%	
			131	130.906111	β^-, γ	8.04 days
54	Xenon	Xe	132	131.904141	26.9%	
			136	135.90721	8.9%	
55	Cesium	Cs	133	132.905436	100%	
56	Barium	Ba	137	136.905816	11.23%	
			138	137.905236	71.70%	
57	Lanthanum	La	139	138.906346	99.9098%	
58	Cerium	Ce	140	139.905434	88.43%	
59	Praseodymium	Pr	141	140.907647	100%	
60	Neodymium	Nd	142	141.907718	27.13%	
61	Promethium	Pm	145	144.912745	EC, γ, α	17.7 yr
62	Samarium	Sm	152	151.919728	26.7%	
63	Europium	Eu	153	152.921226	52.2%	
64	Gadolinium	Gd	158	157.924099	24.84%	
65	Terbium	Tb	159	158.925344	100%	
66	Dysprosium	Dy	164	163.929172	28.2%	
67	Holmium	Ho	165	164.930320	100%	
68	Erbium	Er	166	165.930292	33.6%	
69	Thulium	Tm	169	168.934213	100%	
70	Ytterbium	Yb	174	173.938861	31.8%	
71	Lutecium	Lu	175	174.940772	97.4%	
72	Hafnium	Hf	180	179.946547	35.100%	
73	Tantalum	Ta	181	180.947993	99.988%	
74	Tungsten (wolfram)	W	184	183.950929	30.7%	
75	Rhenium	Re	187	186.955746	62.60%; β^-	4.35×10^{10} yr
76	Osmium	Os	191	190.960922	β^-, γ	15.4 days
			192	191.961468	41.0%	
77	Iridium	Ir	191	190.960585	37.3%	
			193	192.962916	62.7%	
78	Platinum	Pt	195	194.964765	33.8%	
79	Gold	Au	197	196.966543	100%	
80	Mercury	Hg	199	198.968253	16.87%	
			202	201.970617	29.86%	
81	Thallium	Tl	205	204.974401	70.476%	

(1) Atomic Number Z	(2) Element	(3) Symbol	(4) Mass Number A	(5) Atomic Mass	(6) % Abundance (or Radioactive Decay Mode)	(7) Half-life (if radioactive)
82	Lead	Pb	206	205.974440	24.1%	
			207	206.975871	22.1%	
			208	207.976627	52.4%	
			210	209.984163	β^-, γ, α	22.3 yr
			211	210.988734	β^-, γ	36.1 min
			212	211.991872	β^-, γ	10.64 h
			214	213.999798	β^-, γ	26.8 min
83	Bismuth	Bi	209	208.980374	100%	
			211	210.987254	α, γ, β^-	2.14 min
84	Polonium	Po	210	209.982848	α, γ	138.376 days
			214	213.995177	α, γ	164.3 μs
85	Astatine	At	218	218.00868	α, β^-	1.6 s
86	Radon	Rn	222	222.017571	α, γ	3.8235 days
87	Francium	Fr	223	223.019733	β^-, γ, α	21.8 min
88	Radium	Ra	226	226.025402	α, γ	1600 yr
89	Actinium	Ac	227	227.027749	β^-, γ, α	21.773 yr
90	Thorium	Th	228	228.028716	α, γ	1.9131 yr
			232	232.038051	100%; α, γ	1.405×10^{10} yr
91	Protactinium	Pa	231	231.035880	α, γ	3.276×10^4 yr
92	Uranium	U	232	232.037131	α, γ	68.9 yr
			233	233.039630	α, γ	1.592×10^5 yr
			235	235.043924	0.720%; α, γ	7.038×10^8 yr
			236	236.045562	α, γ	2.3415×10^7 yr
			238	238.050784	99.2745%; α, γ	4.468×10^9 yr
			239	239.054289	β^-, γ	23.50 min
93	Neptunium	Np	239	239.052932	β^-, γ	2.355 days
94	Plutonium	Pu	239	239.052157	α, γ	24,119 yr
95	Americium	Am	243	243.061373	α, γ	7380 yr
96	Curium	Cm	245	245.065484	α, γ	8500 yr
97	Berkelium	Bk	247	247.07030	α, γ	1380 yr
98	Californium	Cf	249	249.074844	α, γ	351 yr
99	Einsteinium	Es	254	254.08802	α, γ, β^-	275.7 days
100	Fermium	Fm	253	253.085174	EC, α, γ	3.00 days
101	Mendelevium	Md	255	255.09107	EC, α, γ	27 min
102	Nobelium	No	255	255.09324	α, γ, EC	3.1 min
103	Lawrencium	Lr	257	257.0995	α, EC	0.646 s
104	Rutherfordium[†]	Rf	261	261.10869	α	65 s
105	Dubnium[†]	Db	262	262.11376	α, EC	34 s
106	Seaborgium[†]	Sg	263	263.1182	α, fission	0.9 s
107	Bohrium[†]	Bh	262	262.1231	α	0.10 s
108	Hassium[†]	Hs	264	264.1285	α, fission	0.08 ms
109	Meitnerium[†]	Mt	266	266.1378	α	3.4 ms
110			269	269	α	0.27 ms
111			272	272	α	1.5 ms
112			277	277	α	0.24 ms

[†]Provisional names for elements 104–109.

ANSWERS TO ODD-NUMBERED PROBLEMS

CHAPTER 1

1. (*a*) 1×10^{10} yr; (*b*) 3×10^{17} s.
3. (*a*) 1.156×10^6; (*b*) 2.18×10^2;
   (*c*) 6.8×10^{-3};
   (*d*) 2.7635×10^1;
   (*e*) 2.1×10^{-1}; (*f*) 2.2×10^1.
5. 11%.
7. (*a*) 1×10^1%; (*b*) 1%; (*c*) 0.2%.
9. 1.8×10^5 s.
11. 6%.
13. (*a*) 0.086 6 m; (*b*) 0.000 035 V;
   (*c*) 0.860 g;
   (*d*) 0.000 000 000 600 s;
   (*e*) 0.000 000 000 000 012 5 m;
   (*f*) 250,000,000,000 volts.
15. 1.8 m.
17. (*a*) 3.9×10^{-9} in;
   (*b*) 1.0×10^8 atoms.
19. (*a*) 0.621 mi/h; (*b*) 3.28 ft/s;
   (*c*) 0.278 m/s.
21. (*a*) 9.46×10^{15} m;
   (*b*) 6.31×10^4 AU;
   (*c*) 7.20 AU/h.
23. (*a*) 10^4; (*b*) 10^3; (*c*) 10^{-3};
   (*d*) 10^{10}.
25. $\approx 20\%$.
27. $\approx 1 \times 10^5$ cm^3.
29. (*a*) 500 dentists.
31. $\approx 3 \times 10^8$ kg/yr.
33. (*a*) 3.16×10^7 s;
   (*b*) 3.16×10^{16} ns;
   (*c*) 3.17×10^{-8} yr.
35. 50,000 chips.
37. 1.4×10^3 gumballs.
39. ≈ 3 ft ≈ 1 m.
41. $\approx 4 \times 10^3$ km.
43. ≈ 4 yr.

CHAPTER 2

1. 70.8 km/h.
3. 61 m.
5. (*a*) 203 mi; (*b*) 61 mi/h.
7. (*a*) 4.29 m/s; (*b*) 0.
9. 2.7 min.
11. 55 km/h, 0.
13. 4.3 m/s^2.
15. (*a*) 7.41 m/s^2;
   (*b*) 9.60×10^4 km/h^2.
17.

t(s)	v(m/s)	t(s)	a(m/s^2)
0.0	0.0		
0.125	0.44	0.25	3.8
0.375	1.4	0.50	4.0
0.625	2.4	0.75	4.5
0.875	3.5	1.06	4.9
1.25	5.36	1.50	5.0
1.75	7.85	2.00	5.2
2.25	10.5	2.50	5.3
2.75	13.1	3.00	5.5
3.25	15.9	3.50	5.6
3.75	18.7	4.00	5.5
4.25	21.4	4.50	4.8
4.75	23.9	5.00	4.1
5.25	25.9	5.50	3.8
5.75	27.8		

19. $a = 2.2$ m/s^2, 1.1×10^2 m.
21. 1.5×10^2 m.

23. 62.5 m.
25. (*a*) 1.6×10^2 m; (*b*) 25 s;
   (*c*) 12 m, 10 m.
27. (*a*) 103 m; (*b*) 64 m.
29. 24.0 s, 240 km/h.
31. 3.1 s.
33. 2.6 s.
35. (*a*) 8.81 s; (*b*) 86.3 m/s.
37. 1.5 s.
39.
(*a*)

(*b*)

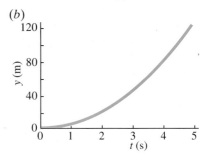

41. 5.22 s.
45. 0.038 s.
47. 52 m.
49. (*a*) 5.33 s; (*b*) 40.2 m/s;
   (*c*) 89.7 m.
51. (*a*) 0.28 m/s; (*b*) 1.2 m/s;
   (*c*) 0.28 m/s; (*d*) 1.6 m/s;
   (*e*) -1.0 m/s.

53. (*a*) 50 s; (*b*) 90 s to 107 s; (*c*) 0 to 20 s, and 90 s to 107 s; (*d*) 75 s.

55. (*a*) 4.7 m/s^2; (*b*) 2.2 m/s^2; (*c*) 0.3 m/s^2; (*d*) 1.6 m/s^2.

57.

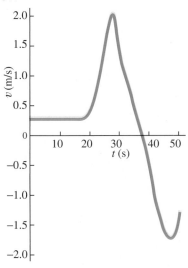

59. (*a*) negative; (*b*) speeding up; (*c*) negative; (*d*) positive; (*e*) speeding up; (*f*) positive; (*g*) not moving, zero.

61. $6h_{\text{Earth}}$.

63. 209.5 km/h.

65. (*a*) 3.88 s; (*b*) 73.9 m; (*c*) 38.0 m/s, 48.4 m/s;

67.
(*a*)

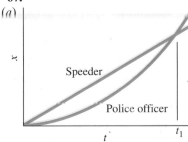

(*b*) 22.9 s; (*c*) 2.67 m/s^2; (*d*) 61.1 m/s = 220 km/h.

69. 3.3 m.

71. The passing attempt should not be made.

73. 2 poles.

CHAPTER 3

1. 177 km, 15° S of W.

5. 31 m, 44° N of E.

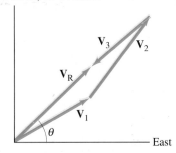

7. (*a*) 2.76 in the $+x$-direction; (*b*) 13.86 in the $+x$-direction; (*c*) 13.86 in the $-x$-direction.

9. (*a*) 614 km/h, 489 km/h; (*b*) 1.84×10^3 km, 1.47×10^3 km.

11. (*a*) 35.9, 17.3; (*b*) 39.9, 25.8° above $+x$-axis.

13. (*a*) 80.7, 1.56° above $-x$-axis; (*b*) 80.7, 1.56° below $+x$-axis.

15. (*a*) 117, 72.1° below $-x$-axis; (*b*) 226, 35.5° below $+x$-axis.

17. 5032 m.

19. 5.6 m.

21. 14° and 76°.

23. Unsuccessful, 34.7 m.

27. 12.9 m.

29. Six times as far.

31. 5.71 s.

35. (*a*) 92.6 m; (*b*) 8.69 s; (*c*) 539 m; (*d*) 68.0 m/s, 24.2° above the horizontal.

39. $\theta = \tan^{-1}(gt/v_0)$ below the horizontal.

41. 2.9 m/s, 20° from the river bank.

43. (*a*) 2.59 m/s, 62.4° from the shore; (*b*) 7.77 m at 62.4° to the shore.

45. (*a*) 435 km/h, 9.36° east of south; (*b*) 12 km.

47. 1.41 m/s.

49. (*a*) 1.66 m/s; (*b*) 3.19 m/s.

51. (*a*) 120 m; (*b*) 150 s (2.5 min).

53. 42.9° N of E.

55. 58 km/h, 31°, 58 km/h opposite to $\mathbf{v}_{12}$.

57. 109 km/h.

59. (*a*) Parallel; (*b*) perpendicular; (*c*) $\mathbf{V}_2 = 0$.

61. -2.4 m/s^2 (opposite to the truck's motion); -1.4 m/s^2 (down).

63. $v_{\text{r}} = v_{\text{T}}/\tan \theta$.

65. 1.8×10^2 s, 4.8 km, 21 s, 0.56 km.

67. 1.6 m/s^2.

69. 2.7 s, 1.9 m/s.

71. 61.8° below the horizontal.

73. 63 m/s, 66° above the horizontal.

CHAPTER 4

1. 69.0 N.

3. 8.8×10^2 N.

5. (*a*) 6.5×10^2 N; (*b*) 1.1×10^2 N; (*c*) 2.4×10^2 N; (*d*) 0.

7. 3.4×10^3 N, opposite to the velocity.

9. $m > 1.5$ kg.

11. 2.1×10^2 N.

13. 3.5 m/s^2 (down).

15. a (downward) ≥ 2.2 m/s^2.

17. 0.557 m/s^2.

19. (*a*) 2.2 m/s^2; (*b*) 18 m/s; (*c*) 93 s.

21. (*a*) 9.4 m/s; (*b*) 3.3×10^3 N up.

23. (*a*) 40 N; (*b*) 10 N; (*c*) 0.

25. (*a*) (*b*)

27. (*a*) 19.0 N, 0.702 m/s^2, both 57.5° below $-x$-axis; (*b*) 14.0 N, 0.519 m/s^2, both 51.0° above $+x$-axis.

29. (*a*) 11 m/s^2; (*b*) 4.3 m/s.

31. (*a*) 8.01×10^4 N; (*b*) 1.25×10^4 N.

35. (a)

(b) $a = F/(m_1 + m_2 + m_3)$;
(c) $F_{net1} = m_1F/(m_1 + m_2 + m_3)$,
$F_{net2} = m_2F/(m_1 + m_2 + m_3)$,
$F_{net3} = m_3F/(m_1 + m_2 + m_3)$;
(d) $F_{12} = F_{21} = F(m_2 + m_3)/(m_1 + m_2 + m_3)$,
$F_{23} = F_{32} = Fm_3/(m_1 + m_2 + m_3)$;
(e) 2.67 m/s^2, $F_{net} = 32$ N,
$F_{21} = F_{12} = 64$ N,
$F_{32} = F_{23} = 32$ N.

37. 1.74 m/s^2, $F_{T1} = 22.6$ N,
$F_{T2} = 20.9$ N.

39. (a) $\mu_s = 0.82$; (b) $\mu_k = 0.74$.

41. $\mu_s = 0.41$.

43. 6.7 kg.

45. (a) 2.5 m/s^2; (b) 4.4×10^2 N.

47. -7.4 m/s^2.

49. $\mu_k = 0.40$.

51. 1.8 s, no change.

53. (a) 1.2 m up the plane; (b) 1.6 s.

55. 23 m/s (85 km/h).

57. Too steep.

59. $\theta_{max} = 31°$.

61. (a) $a = (m_2 - m_1 \sin\theta)g/(m_1 + m_2)$; (b) a down requires $m_2 < m_1 \sin\theta$, a up requires $m_2 > m_1 \sin\theta$.

63. $\mu_k = 0.64$.

65. 2.0×10^{-2} N.

67. (b) Will slide.

69. 35 m/s (130 km/h).

71. (a) $m_{max} = 4.6$ kg;
(b) $m_{max} = 3.8$ kg.

73. (a) 4.1 m/s^2, 3.2 m/s^2;
(b) $a_1 = 4.1 \text{ m/s}^2$,
$a_2 = 3.2 \text{ m/s}^2$; (c) 3.5 m/s^2.

75. (a) $\mu_k = (v_0^2/2gd \cos\theta) - \tan\theta$; (b) $\mu_s \geq \tan\theta$.

77. 8.8°.

79. 82 m/s (300 km/h).

81. (a) 11.6 kg; (b) 0.879 m/s^2.

CHAPTER 5

1. $4.25g$ up.

3. $5.93 \times 10^{-3} \text{ m/s}^2$ toward the Sun, 3.55×10^{22} N toward the Sun, Sun.

7. 23 m/s, independent of the mass.

9. $\mu_s \geq 0.84$.

11. 0.16.

13. (a) 5.8×10^3 N;
(b) 4.1×10^2 N; (c) 31 m/s.

17. 8.0×10^3 N down the slope.

19. 5.91°, 14.3 N.

21. 1.1 km.

23. $a_{tan} = 6.29 \text{ m/s}^2$,
$a_R = 19.7 \text{ m/s}^2$, $\mu_s \geq 2.11$.

25. 1.52×10^3 N.

27. 1.6 m/s^2.

29. (a) 2.10 kg on both;
(b) $w_{Earth} = 20.6$ N,
$w_{planet} = 25.2$ N.

31. $6.7 \times 10^{12} \text{ m/s}^2$.

33. $4.4 \times 10^7 \text{ m/s}^2$.

35. 2.0×10^{-8} N toward center of the square.

37. 6.4×10^{23} kg.

39. 6.3×10^3 m/s.

41. 11 s.

43. 2.0 h.

45. 5 h 35 min, 19 h 50 min.

47. (a) 20 N (toward the Moon);
(b) 1.8×10^2 N (away from the Moon).

49. 5.5×10^{29} kg.

51. 0.0587 days (1.41 h).

53. 1.6×10^2 yr.

55. 5.4×10^{12} m, in the Solar System, Pluto.

57. (a) 1.90×10^{27} kg;
(b) 1.90×10^{27} kg,
1.89×10^{27} kg, 1.89×10^{27} kg.

59. (a) 5.2 yr; (b) No.

61. 9.0 Earth-days.

63. 6.1 m/s.

65. 2.1×10^2 N.

67. 3.46×10^8 m from Earth's center.

69. 3.3 m/s^2 upward.

71. $M_P = g_P r^2/G$.

73. 74 km/h $< v <$ 140 km/h.

75. (a) Attractive gravitational force; (b) 9.6×10^{26} kg.

77. $2.3g_{Earth}$.

CHAPTER 6

1. 7.35×10^3 J.

3. 1.3×10^4 J.

5. 36.1 m.

7. (a) 8.8×10^5 J; (b) 1.6×10^6 J.

9. 23 J.

11. (a) $1.10Mg$; (b) $1.10Mgh$.

13. (a) 2.8×10^3 J; (b) 2.1×10^3 J.

15. 1.9×10^2 J.

17. 484 m/s.

19. -1.64×10^{-18} J.

21. 23%.

23. 3.4×10^2 N in the direction of the motion of the ball.

25. 27 m/s (97 km/h or 60 mi/h), mass cancels.

27. 7.1 m/s, 3.5 m/s.

29. 0.34 m.

31. 2.5×10^3 J.

33. (a) 8.1×10^5 J; (b) 8.1×10^5 J;
(c) yes.

35. $h = 1.6$ m, no.

37. 5.14 m/s.

39. (a) 9.2 m/s; (b) 0.35 m.

41. 199 m/s.

43. (a) 8.03 m/s; (b) 3.44 m.

45. $E = \frac{1}{2}kx_0^2 = \frac{1}{2}mv^2 + \frac{1}{2}kx^2$.

47. 3.7×10^3 N/m.

49. 4.5×10^6 J.

51. (a) 22 m/s; (b) 2.9×10^2 m.

53. 20 m/s.

55. 0.48.

57. (a) 1.07×10^3 km/h;
(b) 2.1×10^3 N.

59. 5.4×10^2 N.

61. (b) 360 kWh; (c) 1.3×10^9 J;
(d) $43.20, independent of rate.

63. 8.1×10^6 J.

65. 4.57 W.

67. 2.6 m/s.

69. 1.8×10^6 W.

71. (a) -3.6×10^4 J;
(b) -3.3×10^3 N;
(c) -2.5×10^5 J.

73. 1.5×10^3 J.

75. (a) 19 m/s; (b) -1.4 N.

77. (a) 1.1×10^6 J;
(b) 60 W = 0.081 hp;
(c) 4.0×10^2 W = 0.54 hp.

79. (a) $h = 2.5r$; (b) $11mg$; (c) $5mg$;
(d) mg.

81. (a) 40 m/s; (b) 2.6×10^5 W.

85. (a) 29°; (b) 6.4×10^2 N;
(c) 9.2×10^2 N.

CHAPTER 7

1. 0.18 kg·m/s.

3. 5.2×10^7 N, up.

5. 12.3 m/s.

7. 0.491 m.

9. 6.7×10^3 m/s.

11. 0.39 m/s.

13. 140 kg.

15. 1.32×10^4 N.

17. 2.1 N·s.

19. (a) ≈ 5.0 N·s; (b) 83 m/s.

21. $v_1' = 1.23$ m/s, $v_2' = 4.93$ m/s.

23. $v_1' = -3.00$ m/s (rebound),
$v_2' = 2.00$ m/s.

25. (a) $v_2' = 1.8$ m/s;
(b) $m_2 = 1.1$ kg.

27. (a) $m_1 = 0.840$ kg; (b) 0.750.

29. (b) $v_1' \approx -v_1$, $v_2' \approx 0$;
(c) $v_1' \approx v_1$, $v_2' \approx 2v_1$;
(d) $v_1' \approx 0$, $v_2' \approx v_1$

31. (a) $+M/(m + M)$; (b) 0.964.

33. 15 m/s.

35. 9.

37. 1.08×10^{-22} kg·m/s, 30.1° from the direction opposite to the electron's.

39. (a) $m_A v_A + 0 = m_A v_A' \cos\theta_A' + m_B v_B' \cos\theta_B'$, $0 + 0 = m_A v_A' \sin\theta_A' - m_B v_B' \sin\theta_B'$;
(b) 0.808 m/s, 33.0°.

41. 141°.

43. $\theta_1' = 76°$, $v_n' = 5.1 \times 10^5$ m/s,
$v_{He}' = 1.8 \times 10^5$ m/s.

47. 2.74 m from the front of the car.

49. 1.10 m (east), -1.10 m (south).

51. Along the line joining the centers $0.07R$ outside the hole.

53. 19% of the height.

55. 9.4% of the body height, yes.

57. (a) 6.2 m; (b) 3.4 m; (c) 3.8 m.

59. (a) $7D/3$, or $2D/3$ closer to the launch site; (b) $5D$, or $2D$ farther from the launch site.

61. 8.3×10^2 N, about the same.

63. 1.4×10^4 N, 43.3°.

65. A "scratch shot".

67. 4.00 m.

69. 59 mi/h.

71. (a) 2.5×10^{-13} m/s;
(b) 1.7×10^{-17}; (c) 0.19 J.

73. (b) 0.93 N·s; (c) 4.2 g.

75. $m \leq M/3$.

77. 29.6 km/s.

CHAPTER 8

1. (a) $\pi/6$ rad = 0.524 rad;
(b) $19\pi/60$ = 0.995 rad;
(c) $\pi/2$ = 1.571 rad;
(d) 2π = 6.283 rad;
(e) $7\pi/3$ = 7.330 rad.

3. $\theta_{Sun} = 9.30 \times 10^{-3}$ rad (0.53°),
$\theta_{Moon} = 9.06 \times 10^{-3}$ rad (0.52°).

5. D_{spot} = 6.8 km.

7. 33 m/s, 6.2×10^3 m/s^2.

9. (a) 0.105 rad/s;
(b) 1.75×10^{-3} rad/s;
(c) 1.45×10^{-4} rad/s; (d) zero.

11. 9.5 cm.

13. $D_{Moon} \approx 3.8 \times 10^3$ km.

15. (a) 464 m/s;
(b) 185 m/s; (c) 328 m/s.

17. (a) 3.1 rad/s^2;
(b) $a_R = 1.9 \times 10^2$ m/s^2,
$a_{tan} = 1.1$ m/s^2.

19. (a) 1.8×10^{-4} rad/s^2;
(b) $a_R = 1.2 \times 10^{-2}$ m/s^2,
$a_{tan} = 7.7 \times 10^{-4}$ m/s^2.

21. 2.75×10^4 rev.

23. (a) $\alpha = 0.070$ rad/s^2;
(b) 40 rpm.

27. (a) -4.4 rad/s^2; (b) 7.9 s.

29. 92 m·N.

31. 1.1 m·N (clockwise).

33. 2.7×10^2 N, 1.8×10^3 N.

35. 0.139 kg·m^2, distance of its mass from the axis is so small.

37. 1.2×10^{-10} m.

39. 10 N.

41. 62 m·N.

43. 8.80×10^2 rev, 10.6 s.

45. (a) 95.2 rad/s^2; (b) 9.4×10^2 N.

47. (a) 2.9 m; (b) 3.6 m.

49. (a) 10.8 rad/s^2; (b) 13.0 m/s^2;
(c) 653 m/s^2; (d) 4.77×10^3 N;
(e) 1.14°.

51. 1.11×10^4 J.

53. 1.70×10^4 J.

55. 3.22 m/s.

57. 2.64 kg·m^2/s.

59. 0.38 rev/s.

61. (a) 2×10^{17} J;
(b) 5×10^{20} kg·m^2/s.

63. (a) 7.1×10^{33} kg·m^2/s;
(b) 2.7×10^{40} kg·m^2/s.

65. 4.2 rev/s.

67. 3×10^{-18}.

69. 4.8×10^{-2} rad/s,
$KE_2 = 2.0 \times 10^4$ KE_1.

71. (a) $\omega_P = -(I_W/I_P)\omega_W$ (down);
(b) $\omega_P = -(I_W/2I_P)\omega_W$ (down);
(c) $\omega_P = (I_W/I_P)\omega_W$ (up);
(d) $\omega_P = 0$.

73. 8.21×10^{-6}.

75. (a) 0.84 m/s; (b) 0.964.

77. 2.0×10^9 rev/day.

79. (a) 7.3 m; (b) 6.8 s.

81. $F_{min} = Mg[h(2R - h)]^{1/2}/(R - h)$.

83. $h_{min} = 2.7R_0$.

85. $v_{CM} = (3gL/4)^{1/2}$.

CHAPTER 9

1. 370 N, 116°.

3. 1.8×10^3 m·N.

5. 9.2×10^2 N.

7. 2.94×10^3 N down,
1.52×10^3 N down.

9. 2.3 N.

11. 3.3 m from the adult.

13. $F_{T1} = 1.8 \times 10^2$ N,
$F_{T2} = 2.4 \times 10^2$ N.

15. $F_1 = -2.94 \times 10^3$ N (down),
$F_2 = 1.47 \times 10^4$ N.

17. Top hinge: $F_{Ax} = 55.2$ N, $F_{Ay} = 63.7$ N; bottom hinge: $F_{Ax} = -55.2$ N, $F_{Ay} = 63.7$ N.

19. $F_1 = -1.8 \times 10^3$ N (down), $F_2 = 2.4 \times 10^3$ N (up).

21. 4.38 kg.

23. $F_T = 1.9 \times 10^2$ N, $\mathbf{F}_W = 1.9 \times 10^2$ N, 50° above the horizontal.

25. 89.5 cm from the feet.

27. $F_T = 2.5 \times 10^2$ N, $F_{AH} = 2.5 \times 10^2$ N, $F_{AV} = 2.0 \times 10^2$ N.

29. 6.5×10^2 N.

31. 0.50.

33. 1.0×10^2 N.

35. 9.9×10^2 N.

37. 2.7×10^3 N.

39. 1.4×10^3 N (up), 2.1×10^3 N (down).

41. 2.7×10^3 N.

43. (b) beyond the table; (c) $D = L \sum\limits_{i=1}^{n} \dfrac{1}{2i}$; (d) 32 bricks.

45. (a) 1.2×10^5 N/m^2; (b) 2.4×10^{-6}.

47. (a) 1.3×10^5 N/m^2; (b) 6.6×10^{-7}; (c) 0.0062 mm.

49. 997 cm^3.

51. 9.0×10^7 N/m^2, 9.0×10^2 atm.

53. 0.017 J.

55. 5.1×10^4 N.

57. (a) Not break; (b) 1.3 mm.

59. $A_{min2} = 7.7 \times 10^{-3}$ m^2, $A_{min1} = 1.4 \times 10^{-3}$ m^2.

61. 2.5 cm.

63. 2.5×10^6 N.

65. $+2.3 \times 10^9$ m·N, will not topple.

67. (a) $Mg[h/(2R - h)]^{1/2}$; (b) $Mg[h(2R - h)]^{1/2}/(R - h)$.

69. (a) 3.5×10^8 N/m^2; (b) will break; (c) 8.2×10^6 N/m^2, will not break.

71. $F_{T1} = 4.54mg$, $F_{T2} = 4.96mg$, $h = 158$ m.

73. 36 kg.

75. 3.8.

77. 6.3 m.

79. 1.3 m.

81. (a) 2.6×10^2 N; (b) 88 N.

CHAPTER 10

1. 2.7×10^{11} kg.

3. 4.3×10^2 kg.

5. 0.8477.

7. 1.65×10^4 N/m^2.

9. (a) 3.1×10^5 N (down); (b) 3.1×10^5 N.

11. 1.96×10^3 kg.

13. 13 m.

15. 8.0 km.

17. 9.0×10^5 N/m^2 (gauge); 92 m.

19. (a) 0.34 kg; (b) 1.5×10^4 N (up).

21. 983 kg/m^3.

23. 0.199.

25. 12 kg.

27. (a) 1.03×10^3 kg/m^3; (b) $\rho_{liquid} = [(m - m_{apparent})/m]\rho_{object}$.

29. 0.85.

31. 7.9×10^2 kg.

35. 3.8 m/s.

37. 1.2×10^5 N/m^2 = 1.2 atm.

39. 1.4×10^5 N.

41. 9.7×10^4 N/m^2.

43. 2.22 m/s, 1.8 atm (gauge).

45. 0.072 Pa·s.

47. 9.9×10^2 N/m^2.

49. 0.12 m.

51. 0.96 N/m^2/cm.

53. 1×10^{-20}.

55. 0.036 N/m.

57. Would not remain on the surface.

59. 3.

61. (a) 0.88 m; (b) 0.55 m; (c) 0.24 m.

63. 1.5×10^2 N $\leq F \leq 2.2 \times 10^2$ N.

65. 4×10^2 atm.

67. 0.63 N.

69. 1.4×10^2 N.

71. 1.1 m.

73. 2.44×10^7 kg.

75. 1.1 W.

77. 1.2×10^4 N/m^2.

79. (a) 2.8×10^2 m/s; (b) 1.03×10^5 N/m^2 (1.01 atm).

CHAPTER 11

1. 0.74 Hz.

3. 1.00 m.

5. Cosine wave.

7. (a) 12 N/m; (b) 1.1 Hz.

9. (a) 2.8 m/s; (b) 2.1 m/s; (c) 2.0 J; (d) (0.15 m) cos $[2\pi(3.0$ Hz$)t]$.

11. $(2k/m)^{1/2}/2\pi$.

13. 10.3 m/s.

15. (a) 4.2×10^2 N/m; (b) 3.3 kg.

17. (a) 0.45 m; (b) 1.34 Hz; (c) 3.6 J; (d) 2.0 J, 1.6 J.

19. (a) $x = (0.280$ m$)$ sin $[(29.0 s^{-1})t]$; (b) maximum at 0.0542 s, 0.271 s, 0.488 s, $\cdots$, minimum at 0.163 s, 0.379 s, 0.596 s, $\cdots$.

21. 557 m/s.

23. (a) $0.707A$; (b) $0.866A$.

29. (a) 0.993 m; (b) slow.

31. (a) 0.61 Hz; (b) 0.53 m/s; (c) 0.044 J.

33. (a) $-15°$, (b) $+4.6°$, (c) $+15°$.

35. 1.26 m.

37. (a) 1.4×10^3 m/s; (b) 4.1×10^3 m/s; (c) 5.1×10^3 m/s.

39. 0.33 s.

41. 2.1 km.

43. 1.0 m.

45. (a) 5.0×10^9 J/m^2·s; (b) 5.0×10^{10} W.

47. 1.41.

49. 0.56.

51. 440 Hz, 880 Hz, 1320 Hz, 1760 Hz.

53. 70 Hz, 140 Hz, 210 Hz.

55. 0.097 m.

57. 300 Hz, 600 Hz, 900 Hz.

61. 4.

63. 5.4 km/s.

65. 41°.

67. 0.16 m/s.

69. 0.38 s.

71. 4.0 Hz.

73. (a) 1.4 Hz; (b) 15 J.

75. (a) 784 Hz, 1176 Hz; 880 Hz, 1320 Hz; (b) 1.26; (c) 1.12; (d) 0.794.

77. (a) $v_2/v_1 = (\mu_1/\mu_2)^{1/2}$;
(b) $\lambda_2/\lambda_1 = (\mu_1/\mu_2)^{1/2}$;
(c) lighter cord.

79. $k = \rho_{water}gA$.

81. Slow, -12.3 s.

CHAPTER 12

1. 2.6×10^2 m.

3. (a) 1.7 cm $\leq \lambda \leq 17$ m;
(b) 3.4×10^{-5} m.

5. 5.4×10^2 m.

7. 1200 m, 300 m.

9. 1.0 W/m^2, 1.0×10^{-10} W/m^2.

11. 1.3.

13. 6.3×10^5.

15. (a) 5.0×10^{-12} W;
(b) 6.3×10^3 yr.

17. (a) 7.9×10^2 W;
(b) 2.5×10^2 m.

19. (a) 136 dB; (b) 99 dB; (c) 57 dB.

21. 150 dB.

23. 25 dB.

25. (a) 2×10^8; (b) 2×10^{11}.

27. 0.17 m.

29. (a) 570 Hz; (b) 860 Hz.

31. (a) 76.6 Hz, 230 Hz, 383 Hz,
536 Hz; (b) 153 Hz, 306 Hz,
459 Hz, 612 Hz.

33. 8.0 Hz, Yes.

35. (a) 0.289 m; (b) 869 Hz.

37. 198 m/s.

39. (a) 283 overtones;
(b) 284 overtones.

41. 0.64, 0.20, -2 dB, -7 dB.

43. 15 Hz, not audible, 3.8 Hz,
audible.

45. 346 Hz.

47. 2.2 Hz.

49. (a) 5.6 Hz; (b) 61 m.

51. (a) 1950 Hz; (b) 1640 Hz.

53. 2 Hz.

55. 9 Hz.

57. 13 Hz.

59. 90 beats/min.

61. (a) 1.1×10^2 m/s;
(b) 2.6×10^2 m/s.

63. (a) 26°; (b) 21 s.

65. (a) 2.46°; (b) 11.1°.

67. 10.

69. 30 dB.

71. 333 Hz.

73. (a) 2.8×10^2 m/s, 44 N;
(b) 19.5 cm; (c) 880 Hz,
1320 Hz.

75. 635 Hz.

77. (a) 0; (b) 12 Hz; (c) 0.

79. 6.4×10^2 N.

81. $\Delta d = (\lambda/4d)[(\ell/2)^2 + d^2]^{1/2}$.

83. 780 Hz.

85. 11.5 m.

CHAPTER 13

1. 0.548.

3. (a) 20°C; (b) 3272°F.

5. 58°C, -89°C.

7. 65°.

9. 0.50 cm.

11. 5/9.

15. 5.1 mL.

17. 0.27 cm^3.

23. -2.8×10^{-3} (0.28%).

25. 3.5×10^7 N/m^2.

27. (a) 27°C; (b) 5.2×10^3 N.

29. -459.7°F.

31. 0.854 m^3.

33. 2.32 atm.

35. (a) 0.439 m^3; (b) -63°C.

37. -0.074 (7.4%).

39. 0.588 kg/m^3, water vapor is not
an ideal gas.

41. 5.50 cm^3.

43. 55.6 mol, 3.34×10^{25}
molecules.

45. 6.4×10^3 N.

47. 6.1×10^3 m/s.

49. 5.2.

51. 31.7°C.

57. (a) Solid or vapor;
(b) 5.11 atm $< P < 73$ atm,
-56.6°C $< T < 31$°C.

59. (a) Vapor; (b) solid.

61. 0.69 atm.

63. 11°C.

65. 3.1 kg.

67. 120°C.

69. 29%.

71. 0.28 s, 5.4×10^{-5} m/s,
3.1×10^2 m/s.

73. (b) Nitrogen, 6.9%.

75. -3.8 mL (1.3%).

77. $P = \rho RT/M$.

79. 3×10^2 molecules/cm^3.

81. 11 L, not advisable.

83. 1.4×10^5 K.

85. (a) 9.2×10^2 kg;
(b) 1.1×10^2 kg.

87. 1.1×10^{44} molecules.

89. (a) 6.09 cm; (b) $L =$
$4V_0(\beta_{mercury} - \beta_{glass})\Delta T/\pi d^2$.

91. 3.0×10^{-10} J.

93. 7.6×10^2 m.

CHAPTER 14

1. 6.7×10^6 J.

3. 10.6°C.

7. 2.1×10^2 kg/h.

9. 2.6×10^3 J/kg·C°.

11. 10.7 : 4.65 : 1.

13. 0.124 kg.

15. 61 C°.

17. 87°C.

19. 2.3×10^3 J/kg·C°.

21. 5.8 min.

23. 0.334 kg (0.334 L).

25. 9.6 g.

27. 4.7×10^3 kcal.

29. 1.12×10^4 J/kg.

31. 360 m/s.

33. (a) 95 W; (b) 33 W.

35. 4.2×10^{30} W, $1.2 \times 10^4 P_{Sun}$.

37. 22 bulbs.

39. 164°C.

41. 10 C°.

43. 36 W.

45. 14 h.

47. 9.8×10^3 kg.

49. (a) 3.2×10^{26} W;
(b) 1.1×10^3 W/m^2.

51. 0.80 C°.

53. 1.96 cm, rod vaporizes.

55. 4.3 kg.

59. (a) 44 C°; (b) no melting.

61. 1.1×10^{-6} kg/s (4.1 g/h).

CHAPTER 15

1. -1.8×10^3 J.

3. P (atm)

5.

7. (a) 0; (b) $+1350$ J; (c) rise.

9. (a) 3.5×10^2 J; (b) $\Delta U = 0$;
   (c) $+3.5 \times 10^2$ J (into the gas).

11. (a) -76 J; (b) $+24$ J; (c) $+52$ J;
   (d) $+28$ J; (e) $+23$ J.

13. (a) $+57$ J; (b) $+57$ J; (c) 0;
   (d) 48%.

15. 1.6×10^2 W.

17. 28%.

19. 30.5%.

21. 1.7×10^{13} J/h.

23. (a) 28%; (b) 7.46×10^4 W,
   1.52×10^9 J/h
   $(3.64 \times 10^5$ kcal/h).

25. 238°C.

27. 4.3×10^8 kg/h.

29. 7.0.

31. (a) 2.1×10^2 J; (b) 3.5×10^2 J.

33. 3.2 min.

35. $+0.31$ kcal/K.

37. $+11$ kcal/K.

39. $+2.2$ cal/K.

41. (a) 75.8% of ideal;
   (b) $+0.270$ J/K; (c) 0.

43. (b), (a), (c), (d).

45. (a) 1/379; (b) $1/1.59 \times 10^{11}$.

47. 15 MW.

49. 3.3×10^{13} J.

51. -401 C°.

53. (a) 2.2×10^5 J; (b) -1.7×10^5 J;
   (c) P (atm)

55. 69%.

57. 2.9×10^3 J/K.

59. (a) 17; (b) 6.1×10^7 J/h.

61. 0.15 J/K.

63. 17 km³/day, 83 km².

CHAPTER 16

1. 1.88×10^{14} electrons.

3. 11.5 cm.

5. 2.7×10^{-3} N.

7. 2.5×10^3 N.

9. 6.8×10^3 C.

11. -1.4×10^2 N (left),
   $+5.3 \times 10^2$ N (right),
   -3.9×10^2 N (left).

13. 6.20×10^5 N away from the
   center of the square.

15. $F_{\text{electric}} = 8.2 \times 10^{-8}$ N,
   $F_{\text{gravitational}} = 3.6 \times 10^{-47}$ N,
   2.3×10^{39}.

17. (a) $Q_1 = \frac{1}{2} Q_T$;
   (b) Q_1 (or Q_2) = 0.

19. 0.91 m beyond the negative
   charge.

21. 1.05×10^{14} m/s², opposite to
   the direction of the electric
   field, independent of the
   velocity.

23. $+2.0 \times 10^5$ N/C (south).

25. 3.30×10^6 N/C (up).

27. 7.12×10^{-10} N/C (south).

29. $2kQa/(a^2 + x^2)^{3/2}$ parallel to
   the line of the charges.

31. 3.80×10^6 N/C away from the
   positive charge.

33. (a) $1.73kQ/L^2$ 60° below the
   $-x$-axis; (b) kQ/L^2 30° below
   the $+x$-axis.

35. 0.10 N/C.

37. $Q_2 = 4Q_1$.

39. (a) In the direction of the
   velocity, to the right;
   (b) 5.1×10^2 N/C.

41. 5.08 m.

43. 1.02×10^{-7} N/C (up).

45. 1.0×10^7 electrons.

47. 2.4×10^{-10} N (attraction).

49. Positive, 2.2×10^{-7} C.

51. (a) 0.83 mm; (b) 2.2 ns.

53. 5.4 μC.

55. $+0.19$ N (right).

CHAPTER 17

1. -6.5×10^{-4} J (done by the
   field).

3. 3.4×10^{-15} J, 21 keV.

5. 4.2×10^4 V/m.

7. -32.5 kV.

9. $V_a - V_b = 269$ V.

11. 7.32×10^7 m/s.

13. 2.40×10^5 V.

15. $+1.08$ J.

17. 4.2 MV.

19. (a) $+6.1 \times 10^3$ V;
   (b) 8.8×10^4 N/C, 53° N of E.

21. 3.0×10^3 m/s.

23. $2kq(2b - d)/b(d - b)$.

25. (a) 0.036 V; (b) 0.025 V;
   (c) -0.025 V.

27. 5.2×10^{-20} C.

29. Zero, resultant force.

31. 0.79 pF.

33. 108 μC.

35. 1.5×10^{-10} F.

37. 26.3 nC.

39. 4.5×10^4 V/m.

41. 56 μF.

43. 1.09×10^{-3} J.

45. (a) 3.6 pF; (b) 32 pC;
   (c) 90 V/m; (d) 1.5×10^{-10} J;
   (e) capacitance, charge, and
   work done by the battery

47. (a) 4×; (b) 4×; (c) $\frac{1}{2}$×.

49. (a) 2.73×10^{-3} J;
   (b) 1.10×10^{-3} J;
   (c) -1.63×10^{-3} J; (d) the
   stored potential energy is not
   conserved.

51. 5.4×10^5 V/m.

53. 9.96×10^{-8} J/m^3.

55. (a) 0.039 eV; (b) 0.039 eV; (c) 0.3 keV; (d) 0.029 eV.

57. 24 mC.

59. 0.14 μC.

61. 10.9 V.

63. (a) 4.9 cm from the negative charge, and 7.4 cm from the positive charge; (b) 0.56 cm from the negative charge toward the positive charge, 2.1 cm from the negative charge away from the positive charge.

65. $Q_1 = Q_0 C_1/(C_1 + C_2)$, $Q_2 = Q_0 C_2/(C_1 + C_2)$, $V = Q_0/(C_1 + C_2)$.

67. 1.03×10^6 m/s.

69. 0.10 m/s.

71. (a) 0.649 μF; (b) 23 C; (c) 4.0×10^8 J.

CHAPTER 18

1. 1.4×10^5 C.

3. 2.5×10^{-11} A.

5. 35 Ω.

7. 2.1×10^{21} electron/min.

9. 24 Ω, 3.6×10^2 J.

11. 0.029 Ω.

13. $R_{Al} = 1.6 R_{Cu}$.

15. 1/8 the length, 8.75 Ω, 1.25 Ω.

17. 58.3°C.

19. 1.8×10^{3}°C.

21. (a) 0.28 Ω; (b) 2.9×10^2 A; (c) 31 V, 49 V.

23. 17 Ω.

25. 26 V.

27. (a) 240 Ω, 0.50 A; (b) 32.7 Ω, 3.67 A.

29. 0.092 kWh, 18¢/month.

31. (a) 8.6 Ω, 1.1 W; (b) 4×.

33. (a) 0.23 W; (b) 32 mA.

35. 71%.

37. (a) 26.2 Ω; (b) 19 s; (c) 0.12¢.

39. 26 A, 0.46 Ω.

41. 0.545 A, 0.386 A.

43. 5.3×10^2 V.

45. 6.0 hp, 13 A.

47. 7.8×10^{-10} m/s.

49. 6.25 mol/m^3.

51. 7.0×10^6 V/m.

53. 5×10^{-11} J, 5×10^{-5} W.

55. 3.60×10^3 C.

57. 12 h.

59. 0.23 S.

61. $R_2 = \frac{1}{4}R_1$.

63. 4.0 mm.

65. 8.4%.

67. 9.00 Ω.

69. 1.5 kW.

71. 0.258 mm, 38.8 m.

73. 2.1×10^{-3} F.

75. 5.5×10^{-7} m.

CHAPTER 19

1. 560 Ω, 35 Ω.

3. 30 Ω, 80 Ω, 19 Ω.

5. Connect three 1.0-Ω resistors in series.

7. 4.6 kΩ.

9. 450 Ω, 2.6%.

11. 960 Ω in parallel.

13. 105 Ω.

15. (a) V_1 and V_2 increase; V_3 and V_4 decrease; (b) I_1 and I_2 increase; I_3 and I_4 decrease; (c) increases; (d) 0, 0.300 A, 0.150 A, 0.338 A, 0.113 A.

17. 50 V.

19. 0.23 Ω.

21. 0.000 Ω.

23. 0.16 A.

25. 17.4 V, 13.3 V.

27. 43 V, 77 V.

29. $I_1 = 0.60$ A, $I_2 = -0.33$ A, $I_3 = 0.93$ A.

31. $I_1 = 0.15$ A right, $I_2 = 0.33$ A left, $I_3 = 0.18$ A up.

33. 0.46 A up.

35. 1.30 A.

37. 22 μF, 0.62 μF.

39. 10,560 pF, yes.

41. 300 pF in parallel.

43. 0.053 J.

47. $2C_0 K_1 K_2/(K_1 + K_2)$.

49. (a) 6.1 μF; (b) 7.2 V, 16.8 V, 24 V.

51. (a) 41 μC; (b) 0.63 A, 0.21 A, 0.42 A.

53. 93 ms.

55. 7.50 MΩ.

57. (a) 50×10^{-6} Ω in parallel; (b) 20 MΩ in series.

59. 1.0 kΩ in series, 100 Ω/V.

61. 5.54 mA.

63. (a) 2.00 Ω, 3.00 Ω, circuit 1 is better; (b) 99 Ω, 101 Ω, both circuits give about the same inaccuracy; (c) 3.3 kΩ, 5.0 kΩ, circuit 2 is better.

65. $R_V \geq 240$ kΩ.

67. $R_1 = 9.4$ kΩ, $R_2 = 6.8$ kΩ.

69. 288 Ω, 144 Ω.

71. 7.3 mA.

73. 0.11 MΩ.

75. (a) 0.12 A; (b) 0.12 A; (c) 64 mA.

77. 26.4 m.

79. (a) Parallel; (b) 6.2 pF $\leq C \leq$ 37 pF.

81. 10 0.22-kΩ resistors in series, 10 22-kΩ resistors in parallel.

83. 13 Ω in series, 30 kΩ in parallel.

85. (a) 25 kΩ; (b) between b and c.

CHAPTER 20

1. (a) 7.8 N/m; (b) 5.5 N/m.

3. 2.68 A.

5. 7.45×10^{-13} N south.

7. (a) left; (b) left; (c) up; (d) in; (e) zero; (f) down.

9. 1.3 T.

11.

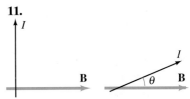

$\theta = 27°$.

13. 4.47 keV.

17. 2.0 A, down.

19. 2.0×10^{-5} T, 36%.

21. 0.18 N attraction.

23. 3 A.

25. $\theta = 34°$ W of N.

27. 1.6×10^{-17} T.

29. 1.0×10^{-6} T up, 2% of the Earth's field.

31. 5.8×10^{-5} up, 3.4×10^{-5} N/m 60° below the line toward *C*, 3.4×10^{-5} N/m 60° below the line toward *B*.

33. 4.1×10^{-5} T.

35. 5.18×10^{-5} T 10.9° below the plane parallel to the two wires.

37. Between long, thin and short, fat.

41. 73.3 μA.

43. decrease by 28%.

47. (*a*) 2.2×10^{-4} V/m; (*b*) 2.7×10^{-4} m/s; (*c*) 4.6×10^{28} electrons/m^3.

49. (*a*) determine the polarity of the emf; (*b*) 0.43 m/s.

51. 70 u, 72 u, 73 u, 74 u.

53. (*a*) 2.08×10^{-3} T; (*b*) out of the page; (*c*) 5.8×10^{7} Hz.

57. 3.0 T up.

59. 42.8.

61. 0.93 N.

63. (*a*) 4.1 cm; (*b*) 5.4×10^{-7} s.

65. 1.50×10^{4} A.

67. (*a*) 6.6 mm; (*b*) 1.7 mm.

69. (*c*) 48 MeV; (*d*) resonance

71. (*a*) 0.512 mm; (*b*) 0.103 Ω.

CHAPTER 21

1. 0.050 V.

3. Counterclockwise.

5. -3.2×10^{2} V.

7. (*a*) Counterclockwise; (*b*) clockwise.

9. Counterclockwise.

11. (*a*) 14.4 mV; (*b*) 0.120 V/m down.

13. (*a*) 0.84 m/s; (*b*) 0.758 V/m down.

15. 0.367 N.

17. $B\ell vA/\rho(2vt + \ell)$.

19. 31.0 V.

23. 0.71 kV, 120 rev/s.

25. 100 V.

27. 13 A.

29. (*a*) 220 V; (*b*) 156 V.

31. Step-down, $V_S/V_P = 0.285$, $I_S/I_P = 3.50$.

33. 55, 4.0 V.

35. (*a*) Step-up; (*b*) 2.4.

37. 451 V, 56.4 A.

41. 2.35 cm.

43. 0.14 H.

45. 3.4×10^{3} turns.

47. 8.29 V.

49. (*a*) Consistent; (*b*) 0.18 mH.

51. $M = \mu_0 N_1 N_2 A/\ell$.

53. 24.3 Ω, 0.463 H.

55. 7.5×10^{-5} J, 1.08 s.

57. (*a*) 4.47 ms. (*b*) 1.12 H.

59. (*a*) 2.3; (*b*) 4.6; (*c*) 6.9.

61. 1.5 kHz.

63.

65. 10.1 kΩ, 23.9 mA.

67. (*a*) 7.6 kΩ; (*b*) 0.37 A.

69. (*a*) 18 mV; (*b*) 50 mV.

71. (*a*) 1.6 kΩ; (*b*) 1.5 kΩ.

73. 1.6 kHz.

75. (*a*) 65.5 mA; (*b*) +10.4°; (*c*) 7.73 W; (*d*) $V_R = 118$ V, $V_L = 22$ V.

77. 18 Ω.

79. 228 Hz.

81. 3.8×10^{5} Hz.

83. (*a*) 0.41 μF; (*b*) 11A.

85. 4 Ω.

87. (*a*) Yes; (*b*) immediately; (*c*) when the current in the first loop reaches the steady state; (*d*) counterclockwise; (*e*) yes; (*f*) away from the other loop.

89. 7.33×10^{-3} J.

93. (*a*) 0.66 A; (*b*) 9.3.

95. 82 V.

97. 0.10 H.

99. 13.8 Ω, 43.5 mH.

101. 14 Ω, 75 mH.

103. 2.2×10^{3} Hz, 1.1×10^{4} Hz.

CHAPTER 22

1. 1.4×10^{-8} A.

3. 1.2×10^{15} V/m·s.

7. 5.25 V/m.

9. 80.0 kHz, 2.33 V/m, horizontal north-south line.

11. 3.0×10^{18} Hz.

13. 3.33×10^{14} Hz, infrared.

15. 8.33 min.

17. 3.4×10^{3} rad/s.

19. 9.47×10^{15} m.

21. 8.5×10^{3} wavelengths, 3.5 fs.

23. 4.24 months.

25. 0.796 W/m^2, 17.3 V/m.

27. 1.24×10^{3} V/m, 4.13×10^{-6} T.

29. (*a*) 2.5×10^{2} J; (*b*) 2.5×10^{9} V/m.

31. 1.5 pF.

33. (*a*) 1.22 μH; (*b*) 106 pF.

35. 441 m.

37. AM station, 100×.

39. 8.33 min.

41. 1.43 s.

43. 38.7 V/m, 1.29×10^{-7} T.

45. Downward, 614 V/m, 2.05×10^{-6} T.

47. (*a*) 0.40 W; (*b*) 12 V/m; (*c*) 12 V.

49. 1.5×10^{11} W.

51. (*a*) Parallel; (*b*) 8.9 pF $\leq C \leq$ 80 pF; (*c*) 1.05 mH $\leq L \leq$ 1.22 mH.

CHAPTER 23

1. 3.0 m.

5. 5°.

9. 34.0 cm.

13. 3.9 m.

15. Convex, virtual, 0.13 m.

17. (*a*) Center of curvature; (*b*) real; (*c*) inverted; (*d*) -1.

23. (*a*) $f = \infty$; (*b*) $d_i = -d_o$; (*c*) $+1$; (*d*) yes.

25. (*a*) Concave mirror; (*b*) erect and virtual; (*c*) 173 cm.

27. 1.31.

29. $(2.997026 \pm 0.000030) \times 10^{8}$ m/s.

31. 64.0°.

33. 61.5°.

35. 4.6 m.

37. (a) 26.9°; (b) 31.2°; (c) 31.2°.

41. 1.42.

43. 3.9×10^{-4} m.

45. $n \geq 1.414$.

47. (a) 1.414; (b) no; (c) 1.88.

49. (a) Converging lens; (b) 5.41 D.

51. (a) 3.39 D, converging; (b) −16.0 cm, diverging.

53. (a) −0.26 mm; (b) −0.47 mm; (c) −1.9 mm; (d) 28 mm: wide-angle lens, 200 mm: telephoto lens.

55. −7.6 cm (virtual image behind the lens), 0.54 mm (upright).

57. (a) 75.0 mm; (b) 25.0 mm.

59. (a) −11.1 cm (diverging), virtual; (b) +184 cm (converging).

61. Midway between the object and screen.

63. (b) 110.8 cm; (c) 110.8 cm.

65. Real, 7.4 cm beyond second lens.

69. 1.54.

71. 13.1 cm.

73. −43.1 cm (concave).

75. +1.3 D.

77. 5.2 m.

79. 0.101 m, −2.7 m.

81. +12 cm.

83. $0 < -d_o < -f$.

85. (b) Real, 20 cm beyond second lens, upright, same size as object.

87. (c) $\Delta d = (d_T^2 - 4d_T f)^{1/2}$, $\{[d_T + (d_T^2 - 4d_T f)^{1/2}]/[d_T - (d_T^2 - 4d_T f)^{1/2}]\}^2$.

89. −3.25 m.

91. $1/f' = [(n/n')-1][(1/R_1)+(1/R_2)]$; $(1/d_o) + (1/d_{i2}) = 1/f'$, where $1/f' = [(n/n')-1]/f(n-1)$; $m = h_i/h_o = -d_i/d_o$.

CHAPTER 24

3. 7.5 μm.

5. 3.4 cm.

7. 0.085 mm.

9. Reverse of the usual double-slit pattern.

11. 2.40 mm.

13. 3.0%.

15. 0.22°.

17. 0.61 cm.

19. 952 nm.

21. 5.61 μm.

23. 0.0230 mm.

25. (a) $D_{max} = \lambda$; (b) 400 nm.

27. 357 nm.

29. 518 nm, 593 nm, 737 nm.

31. One full order.

33. 6.67×10^3 lines/cm.

35. 10.0°.

37. 704 nm, 1030 nm.

39. 1.08 cm.

41. 228 nm.

43. 9.05 μm.

45. 471 nm.

49. 699 nm.

51. 0.250 mm.

53. 0.289 mm.

55. 56.7°.

57. 52°, 38°.

59. (a) 30°; (b) 18°; (c) 5.7°.

61. 78.5%.

63. 24 polarizers, with each at an angle of 3.75° with the next.

65. 31° on either side of the normal.

67. 12,500 lines/cm.

69. (a) Constructive; (b) destructive.

71. (a) 355 nm.

73. (a) 0; (b) $0.094I_0$; (c) no light gets transmitted.

75. 810 nm.

77. 658 nm, 938 lines/cm.

79. $\frac{1}{2}(m - \frac{1}{2})\lambda/n_{film}, m = 1,2,3,\cdots$; $\frac{1}{2}m\lambda/n_{film}, m = 0,1,2,3,\cdots$; $\frac{1}{2}m\lambda/n_{film}, m = 0,1,2,3,\cdots$; $\frac{1}{2}(m - \frac{1}{2})\lambda/n_{film}, m = 1,2,3,\cdots$.

81. (a) $0.21I_0$; (b) second or third polarizer; (c) second and third polarizers.

83. 0.47 m.

CHAPTER 25

1. 2.5 mm $\leq D \leq$ 39 mm.

3. (1/60) s.

7. 17 mm.

9. 54 mm.

11. +2.3 D.

13. Glasses would be better.

15. (a) −1.43 D; (b) 28 cm.

17. −26.8 cm.

19. 2.1×.

21. (a) 17 cm; (b) 10 cm.

23. (a) 3.78×; (b) 11.7 mm; (c) 6.62 cm from the lens.

25. (a) −81 cm; (b) 9.5×; (c) 2.9×.

27. 17 cm, 100 cm.

29. 3.2 cm, 83 cm.

31. −34×.

33. 12×.

35. 4.2 m, 8.4 m.

37. 6.4×.

39. 1.5 cm.

41. (a) 0.85 cm; (b) 230×.

43. (a) 14.4 cm; (b) 137×.

45. (a) 15.9 cm; (b) 14.8 cm; (c) 1.1 cm; (d) 0.19 cm.

47. 51.6°, 216 nm.

49. 3.8×10^{11} m.

51. 0.46.

53. (a) 16 km; (b) 0.42′.

55. 14.0°.

57. (a) 52.6°; (b) 0.19 nm.

59. (a) 1; (b) 2.7, 1.

61. $I_2 = 16I_1$.

63. 100 mm, 200 mm.

65. 2.9×, 4.1×.

67. (a) −2.25×; (b) 4.5 D.

69. +3.6 D.

71. 14 cm, 5.1×10^{-6} rad.

73. −16×.

75. 42.7 m.

CHAPTER 26

1. (a) 1.000, (b) 0.99995, (c) 0.995, (d) 0.436; (e) 0.141; (f) 0.0447.

3. 2.07×10^{-6} s.

5. 5.57 m, 6.68 m.

7. 0.96c.

9. (a) 2.3 yr; (b) 11 yr.

11. (a) 7.12 m, 1.20 m; (b) 16.3 s; (c) 0.580c; (d) 16.3 s.

13. 3.8×10^{-27} kg.

15. $0.417c$.

17. (a) 1.5 m/s less than c; (b) 30 cm.

19. 5.36×10^{-13} kg.

21. 8.20×10^{-14} J, 0.511 MeV.

23. 9×10^2 kg.

27. (a) 6.3×10^7 kg;
(b) 1.4×10^{17} kg.

29. 7.49×10^{-19} kg·m/s.

31. $0.941c$.

33. $2m_0/[1 - (v^2/c^2)]^{1/2}$, no loss,
$2m_0c^2(\{1/[1 - (v^2/c^2)]^{1/2}\} - 1)$.

35. $0.866c$, 4.73×10^{-22} kg·m/s.

37. 55 MeV (8.7×10^{-12} J),
1.7×10^{-19} kg·m/s, -8%, -6%.

39. $0.855c$.

41. 3.0 T.

45. $0.80c$.

47. $0.65c$.

49. Not relativistic.

51. (a) $0.73c$; (b) 5.9 yr.

53. 9.3×10^{-8} s.

55. 1.02 MeV (1.64×10^{13}) J).

57. 2.2 mm.

59. (a) 4.4×10^9 kg/s;
(b) 4.3×10^7 yr;
(c) 1.4×10^{13} yr.

61. 28.3 MeV.

63. (a) 1.32×10^{-18} kg·m/s; (b) 0;
(c) 7.45×10^{-18} kg·m/s.

65. Not possible.

CHAPTER 27

1. 5.8×10^4 C/kg.

3. 5.

5. 7.07×10^3 K.

7. (a) 10.6 μm, near infrared;
(b) 967 nm, infrared;
(c) 0.73 mm, far infrared.

9. (a) 6.0×10^{-34} J;
(b) 1.5×10^{35}; (c) 6.8×10^{-36},
not measurable.

11. 9.4 μm.

13. 4.23×10^{-7} eV.

15. 2.4×10^{13} Hz, 1.2×10^{-5} m.

17. 400 nm.

19. Copper and iron.

21. 0.62 eV.

23. (a) 1.08 eV; (b) no ejected
electrons.

25. 3.75 eV.

27. 1.9×10^{-24} kg·m/s,
6.3×10^{-33} kg.

29. 212 MeV, 5.85×10^{-15} m.

31. (a) 2.43×10^{-12} m;
(b) 1.32×10^{-15} m.

33. (a) 5.90×10^{-3}, 1.98×10^{-2},
3.89×10^{-2}; (b) 60.8 eV, 204 eV,
401 eV.

35. (a) 9.21×10^{-17} J (576 eV);
(b) 0.105 nm.

37. 7.2×10^{-12} m.

39. 8.6×10^{-12} m, diffraction
effects are negligible.

43. 2.6×10^{-11} m.

47. 3.2×10^{-3}.

49. 3.4 eV.

51. 122 eV.

53. 52.5 nm.

55. $n = 1$ to $n = 3$.

57.

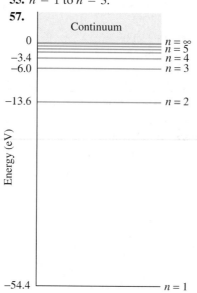

59. -27.2 eV, $+13.6$ eV.

61. Justified.

65. (a)

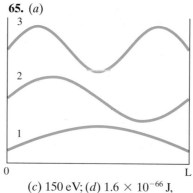

(c) 150 eV; (d) 1.6×10^{-66} J,
4.8×10^{-33} m/s; (e) 0.14 nm.

67. 3.28×10^{15} Hz.

71. 2.78×10^{21} photons/s·m^2.

73. 8.3×10^6 photons/s.

75. 4.7×10^{-14} m.

77. 10.2 eV.

79. 4.4×10^{-40}, yes.

81. 14.2 MeV.

85. 5×10^{-12} m.

CHAPTER 28

1. 3.6×10^{-7} m.

3. $\pm 1.3 \times 10^{-11}$ m.

5. 6.6×10^{-8} eV.

7. 3.0×10^{-10} eV/c^2.

9. 1.4×10^{-3} m, 9.1×10^{-33} m.

13. (b) Infinite uncertainty.

15. $m_\ell = -4, -3, -2, -1, 0, 1, 2, 3,$
$4; m_s = -\frac{1}{2}, +\frac{1}{2}$.

17. $(1, 0, 0, -\frac{1}{2}), (1, 0, 0, +\frac{1}{2}), (2, 0, 0,$
$-\frac{1}{2}), (2, 0, 0, +\frac{1}{2}), (2, 1, -1, -\frac{1}{2}),$
$(2, 1, -1, +\frac{1}{2}), (2, 1, 0, -\frac{1}{2})$.

19. 32 states.

21. $\ell \geq 3, n \geq \ell + 1$ (minimum 4),
$m_s = -\frac{1}{2}, +\frac{1}{2}$.

25. (a) $1s^2 2s^2 2p^6 3s^2 3p^6 3d^{10} 4s^2 4p^4$;
(b) $1s^2 2s^2 2p^6 3s^2 3p^6 3d^{10} 4s^2 4p^6$
$4d^{10} 4f^{14} 5s^2 5p^6 5d^{10} 6s^1$;
(c) $1s^2 2s^2 2p^6 3s^2 3p^6 3d^{10} 4s^2 4p^6 4d^{10}$
$4f^{14} 5s^2 5p^6 5d^{10} 6s^2 6p^6 5f^3 6d^1 7s^2$.

27. (a) 6; (b) -0.378 eV;
(c) 5.78×10^{-34} kg·m^2/s;
(d) $m_\ell = -5, -4, -3, -2, -1,$
$0, 1, 2, 3, 4, 5$.

29. (a) 1.56; (b) 1.4×10^{-10} m.

33. 41 kV.

35. 0.18 nm.

37. 0.061 nm, partial shielding
provided by the $n = 2$ shell.

39. 0.018 J, 5.8×10^{16} photons.

41. r_1, Bohr radius.

43. 0, 4.72×10^{-34} kg·m^2/s.

45. (a) 3.1×10^{-34} m;
(b) 1.8×10^{-32} kg·m/s;
(c) 1.6×10^{-30} m.

47. 3.8×10^3 C°/min.

49. (a) $45°, 90°, 135°$; (b) $35.3°,$
$65.9°, 90°, 114.1°, 144.7°$;
(c) $30.0°, 54.7°, 73.2°, 90°,$
$106.8°, 125.3°, 150.0°$; (d) $5.71°,$
$0.0573°$.

1. 5.13 eV.

3. 4.6 eV.

5. 1.4 eV.

9. (a) 1.79×10^{-4} eV;
   (b) 7.16×10^{-4} eV, 1.73 mm.

11. 2.35×10^{-10} m.

15. (a) 680 N/m; (b) 2.1 μm.

17. 0.283 nm.

21. 1.1 μm.

23. Could be used, 1.09 μm.

25. 9.2×10^5.

27. 0.89 μm.

29. 8.6 V.

33. (a) 38 mA; (b) 76 mA.

35. 0.53 V.

37. (a) 2.5×10^2; (b) 2.3×10^4.

39. 13 eV.

41. (a) -5.3 eV; (b) 5.0 eV.

43. (a) 0.094 eV; (b) 0.63 nm.

45. (a) 146 V $\leq$ V $\leq$ 326 V;
    (b) 3.34 kΩ $\leq$ R_{load} $< \infty$.

47. 2.0×10^{-9}.

1. 3727 MeV/c^2.

3. 1.9×10^{-15} m.

5. (a) 0.99945; (b) 1.2×10^{-14};
   (c) 2.3×10^{17} kg/m^3, $10^{14}\times$.

7. 28 MeV.

9. $^{31}_{15}$P.

11. (a) 1.8×10^3 MeV; (b) 730 MeV.

13. 7.48 MeV.

15. 18.7 MeV.

17. (a) 12.1 MeV; (b) 15.7 MeV,
    repulsive electric force.

19. 59.93525 u.

21. 0.783 MeV.

23. β^+, $^{22}_{11}$Na $\rightarrow$ $^{22}_{10}$Ne $+ \beta^+ + \nu$,
    1.82 MeV.

25. $^{228}_{90}$Th, 228.02883 u.

27. (a) $^{32}_{16}$S; (b) 31.97207 u.

29. 0.861 MeV.

31. 0.0718 MeV, 4.27 MeV.

33. 0.960 MeV, 0.960 MeV, 0.

35. 3.0 h.

37. 2.1 min.

39. 0.0625.

41. (a) 0.0625; (b) 0.0442.

43. (a) 2.44×10^{12} decays/s;
    (b) 2.43×10^{12} decays/s;
    (c) 4.48×10^5 decays/s.

45. 5.0×10^{10} decays/s.

47. (a) 1.38×10^{-13} s^{-1};
    (b) 5.38×10^8 decays/min.

49. 34 decays/s.

51. 3.4 mg.

53. 8.6×10^{-7}.

55. $^{211}_{82}$Pb.

57. $N_0(1 - e^{-\lambda t})$.

59. (a) 2.29×10^{17} kg/m^3;
    (b) 180 m; (c) 2.58×10^{-10} m.

61. 28.6 eV.

63. 41 yr.

65. 6×10^3 yr.

67. $6.65 T_{1/2}$.

71. (a) $^{191}_{77}$Ir;
    (b)

73. (a) 550 MeV; (b) 2.5×10^{12} J.

75. (a) 2.5×10^5 yr; (b) no change.

1. $^{28}_{13}$Al, β^-, $^{28}_{14}$Si.

3. Yes.

5. 18.000953 u.

7. (a) Yes; (b) 0.53 MeV.

9. (a) Yes; (b) 6.49 MeV.

11. 41.6 MeV.

13. (a) Incident ^{3}He picks up a
    neutron; (b) $^{11}_6$C; (c) 1.857 MeV,
    exothermic.

17. 126.5 MeV.

19. 6.3×10^{18} fissions/s.

21. 482 kg.

23. 1.49.

29. 0.120 g.

31. (a) 13.93 MeV; (b) 4.3 MeV;
    (c) 3.3×10^{10} K.

33. 3.23×10^9 J, 65$\times$.

35. (a) 4.8×10^{31} particles/m^3;
    (b) 6×10^{-12} s.

37. 1000 rad.

39. 35 J.

41. 4 days.

43. 8.9×10^{-5} rad/day.

45. (a) $^{131}_{53}$I $\rightarrow$ $^{131}_{54}$Xe $+ {}^{0}_{-1}$e $+ \bar{\nu}$;
    (b) 27 days; (c) 8.0×10^{-9} g.

47. (a) $^{218}_{84}$Po; (b) radioactive,
    α-emission, β^- emission,
    3.1 min; (c) chemically reacting;
    (d) 150 μCi, 0.63 μCi.

49. (a) $^{12}_6$C; (b) 5.70 MeV.

51. 1.0043.

53. 55 mrem/yr.

55. 4.5 m.

57. (a) 1.54×10^3 kg/yr;
    (b) 4.8×10^6 Ci.

59. (a) 4.0×10^{26} J/s;
    (b) 3.8×10^{38} protons/s;
    (c) 10^{11} yr.

61. (a) 3.7×10^3 decays/s;
    (b) 5.2×10^{-2} rem, ≈ 0.15
    background.

1. 9.44 GeV.

3. 1.6 T.

5. 13 MHz.

7. (a) 8.7 MeV, 2.0×10^7 m/s;
   (b) 4.3 MeV, 2.0×10^7 m/s;
   (c) 13 MHz (both).

9. (a) 0.65 T; (b) 4.9 MHz;
   (c) 200 rev; (d) 41 μs;
   (e) 0.63 km.

11. 2.2×10^6 km, 7.5 s.

13. 3.0 T.

15.

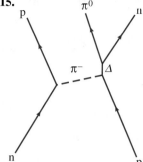

19. 33.9 MeV.

21. 1.879 GeV.

23. (a) Forbidden; (b) forbidden; (c) possible.

25. 1.32×10^{-15} m.

27. 1.53×10^{-3} nm.

29. $KE_n = 494.0$ MeV, $KE_p = 493.4$ MeV.

31. No, 292.4 MeV.

33. 7.5×10^{-21} s.

35. (a) 1.3 keV; (b) 8.9 keV.

37. (b) $B^+ = \bar{b}u$, $B^0 = b\bar{d}$, $\bar{B}^0 = \bar{b}d$.

39. (a) p; (b) $\bar{\Sigma}^+$; (c) K^-; (d) π^-; (e) D_S^-.

41. $c\bar{s}$.

43.
(a)

(b)

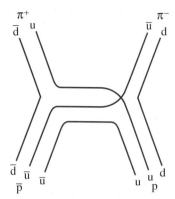

45. (a) 0.38 A; (b) 98 m/s (≈ 350 km/h).

47. (a) 1.002 MeV; (b) 1876.6 MeV.

49. (a) 37.8 MeV; (b) $KE_p = 5.37$ MeV, $KE_\pi = 32.4$ MeV.

51. 133.5 MeV/c^2.

53. 52.4 MeV.

55. (a)

CHAPTER 33

1. 5.3 ly.

5. $0.078''$, 13 pc.

7. $\phi_1 = \frac{1}{2}\phi_2$.

9. 48 W/m².

11. 1.4×10^{-4} kg/m³.

13. 1.8×10^9 kg/m³, $3 \times 10^5 \times$.

15. -0.0941 MeV, 7.365 MeV.

17. (a) 9.595 MeV; (b) 7.6 MeV; (c) 6×10^{10} K.

19. $d_2 = 6.5d_1$.

23. 540°.

25. 1.0×10^8 ly.

27. (a) 2×10^{10} yr; (b) 1×10^{10} yr.

29. $0.328c$, 4.0×10^9 ly.

33. 6 nucleons/m³.

35. (a) 10^{-5} s; (b) 10^{-7} s; (c) 10^{-4} s.

37. 5.0 ly.

39. 2×10^{10}.

41. 1 H atom/km³ every 7 yr.

43. 2.

45. $0.88c$.

47. $\pm 2.3 \times 10^{-3}$.

Classical physics (*defn*), 793
Clausius, R. J. E., 451, 458
Clausius statement of second law, 451, 456
Clock paradox (*see* Twin paradox)
Closed tube, 358–60
Cloud chamber, 938
CO molecule, 897–98
Coating of lenses, optical, 742–43
Coaxial cable, 602, 671
Coefficient (*see* name of)
Coefficient of performance, 456–57
Coherent source of light, 730, 878
Cohesion, 299
Coil (*see* Inductor)
Collector (of transistor), 910
Colliding beams, 978–79
Collisions, 185–94
    elastic, 187–91
    inelastic, 188, 190–92
Color:
    of quarks, 991–93
    related to wavelength, 730–32
    of sky, 749–50
Color charge, 991
Color force, 991–93
Coma, 773
Commutators, 605, 630
Compass, magnetic, 590–91, 597
Complementary, principle of, 835–36
Completely inelastic collision, 190–92
Complex atoms, 872–73
Components of vector, 52–57
Compound lenses, 774
Compound microscope, 770–72
Compound nucleus, 947
Compounds (*defn*), 382 *fn*
Compression (longitudinal wave), 326
Compressive stress, 256
Compton, A. H., 833
Compton effect, 833–35
Compton wavelength, 833
Computerized axial tomography, 784–87
Computers, 637–38
    digital information and, 638
    hard drive, 214
    keyboards, 516
    monitors, 518–19
Concave mirror, 688 *ff*
Concentration gradient, 409
Concrete, prestressed and reinforced, 259
Condensation, 406
Condensed-matter physics (*see* Solid-state physics)
Condenser (*see* Capacitor)
Condenser lens, 772
Conductance, 553 *pr*
Conduction:
    electric, 479, 530–49, 903
    heat, 428–31
    nerve, 546–49
Conduction band, 903, 906
Conduction current (*defn*), 663
Conductivity, thermal, 429
Conductors:
    electrical, 479, 492–93, 508, 528–49, 627–28, 903
    heat, 428–31
Confinement (fusion), 956
Conservation laws:
    of angular momentum, 229–31
    apparent violation of in beta decay, 926
    of baryon number, 983

of electric charge, 478–79, 564
    in elementary particle interactions, 983–84
    of energy, 158–65, 444 *ff*, 564
    of lepton numbers, 984
    of linear momentum, 182–85 and *ff*
    in nuclear processes, 928
    of nucleon number, 928, 983
    of strangeness, 988
Conservative force, 157–58
Conserved quantity, 159
Constant-volume gas thermometer, 387–88
Constructive interference, 333–34, 363, 728
Contact force, 87, 486
Contact lenses, 765
Continental drift, 286
Continuity, equation of, 287–89
Continuous laser, 881
Continuous spectrum, 738, 826
Contrast in microscope images, 779–81
Control rods, 949
Convection, 431–33
    on hiking trail, 432–33
Conventional current (*defn*), 532
Conventions, sign, 695, 707
Converging lens, 703
Conversion factors, 11–12
Convex mirror, 688, 694–95
Coordinate systems, 20
Copenhagen interpretation, of quantum mechanics, 868
Copernicus, 3
Copier, electrostatic, 487
Core, of reactor, 950
Coriolis acceleration, 1052–54
Coriolis force, 1052–54
Cornea, 762
Corrective lenses, 761–65
Correspondence principle, 817, 850, 856 *pr*, 860
Cosmic microwave background radiation, 1021–23
Cosmic rays, 911, 974
Cosmological principle, 1020
Cosmology, 994, 999–1032
Coulomb (unit), 481
    operational definition of, 599
Coulomb, Charles, 481
Coulomb force (*defn*), 483
Coulomb's law, 481–85, 487
Counter emf, 631–33
Counter torque, 631–33
Covalent bond, 887–88, 890–92, 901
Creativity, in science, 2–4
Crick, F. H. C., 783
Critical angle, 699
Critical damping, 321
Critical density of universe, 1028 *ff*
Critical mass, 949
Critical point, 104
Critical temperature, 404
Crossed Polaroids, 746
CRT, 518–19, 679
Crystal lattice, 901–2
Crystallography, 782
CT scan, 784–87
Curie (unit), 959
Curie, Marie, 922
Curie, Pierre, 922
Curie temperature, 609
Current (*see* Electric current)
Current gain, 914 *pr*
Current sensitivity of meters, 575–78

Curvature of field, 773
Curvature of space, 1012–17, 1027–31
Curvature of universe, 1012–17, 1027–31
Cycle (*defn*), 310
Cyclotron, 621 *pr*, 975–77
Cyclotron frequency, 976

Dalton, John, 382
Damped harmonic motion, 320–21
Dark matter, 1028
Dart gun, 163
Dating, radioactive, 934–35
Daughter nucleus (*defn*), 923
da Vinci, Leonardo, 4
Davisson, C. J., 837
dB (unit), 350
DC (*defn*), 541
Dc circuits, 555–78
Dc generator, 630
Dc motor, 605
de Broglie, Louis, 836, 851–52, 860
de Broglie's hypothesis, 851–52
de Broglie's wavelength, 836, 851–52, 974
Debye (unit), 512
Decay:
    alpha, 923–25, 928, 936
    beta, 923, 925–28
    of elementary particles, 985 *ff*
    gamma, 923, 927
    rate of, 928–32
    types of radioactive, 923, 928 (table)
Decay constant, 928
Decay series, 932–33
Deceleration (*defn*), 26
Deceleration parameter, 1030
Decibel (dB) (unit), 350
Declination, magnetic, 590
Decommissioning nuclear plant, 950
Defibrillator, 573 *fn*
Definite proportions, law of, 381–82
Degradation of energy, 462
Dehumidifier, 415 *pr*
Delayed neutrons, 949
Delbrück, Max, 4
Demagnetization, 613
Democritus, 381
Dendrite, 546–49
Density, 275–76
    and floating, 285
    of universe, 1028
Depth of field, 759–60
Descartes, R., 136
Destructive interference, 333–34, 363, 728
Detectors, of particles and radiation, 937–38
Determinism, 136–37, 867–68
Deuterium, 944, 949, 953–54
Deuteron, 944
Dew point, 408
Diamagnetism, 611 *fn*
Diastolic pressure, 301
Dielectric constant, 514
Dielectrics, 514–16
Diffraction, 339–40, 724–25, 732–35, 774–76
    by circular opening, 775
    of electrons, 837
    of light, 724–25, 732–35
    of matter, 837
    as limit to resolution, 775
    by single slit, 732–35
    of water waves, 339–40
    X-ray, 782–83
Diffraction grating, 735–38

PHOTO CREDITS

Periodic Table of the Elements §

Transition Elements

Legend (example cell):
- Symbol — Cl
- Atomic Number — 17
- Atomic Mass § — 35.4527
- Electron Configuration (outer shells only) — $3p^5$

Main Table

Group I	Group II											Group III	Group IV	Group V	Group VI	Group VII	Group VIII
H 1 1.00794 $1s^1$																	He 2 4.002602 $1s^2$
Li 3 6.941 $2s^1$	Be 4 9.012182 $2s^2$											B 5 10.811 $2p^1$	C 6 12.011 $2p^2$	N 7 14.00674 $2p^3$	O 8 15.9994 $2p^4$	F 9 18.9984032 $2p^5$	Ne 10 20.1797 $2p^6$
Na 11 22.989768 $3s^1$	Mg 12 24.3050 $3s^2$											Al 13 26.98154 $3p^1$	Si 14 28.0855 $3p^2$	P 15 30.97376 $3p^3$	S 16 32.066 $3p^4$	Cl 17 35.4527 $3p^5$	Ar 18 39.948 $3p^6$
K 19 39.0983 $4s^1$	Ca 20 40.078 $4s^2$	Sc 21 44.955910 $3d^1 4s^2$	Ti 22 47.88 $3d^2 4s^2$	V 23 50.9415 $3d^3 4s^2$	Cr 24 51.9961 $3d^5 4s^1$	Mn 25 54.93805 $3d^5 4s^2$	Fe 26 55.847 $3d^6 4s^2$	Co 27 58.93320 $3d^7 4s^2$	Ni 28 58.6934 $3d^8 4s^2$	Cu 29 63.546 $3d^{10} 4s^1$	Zn 30 65.39 $3d^{10} 4s^2$	Ga 31 69.723 $4p^1$	Ge 32 72.61 $4p^2$	As 33 74.92159 $4p^3$	Se 34 78.96 $4p^4$	Br 35 79.904 $4p^5$	Kr 36 83.80 $4p^6$
Rb 37 85.4678 $5s^1$	Sr 38 87.62 $5s^2$	Y 39 88.90585 $4d^1 5s^2$	Zr 40 91.224 $4d^2 5s^2$	Nb 41 92.90638 $4d^4 5s^1$	Mo 42 95.94 $4d^5 5s^1$	Tc 43 (97.9072) $4d^5 5s^2$	Ru 44 101.07 $4d^7 5s^1$	Rh 45 102.90550 $4d^8 5s^1$	Pd 46 106.42 $4d^{10} 5s^0$	Ag 47 107.8682 $4d^{10} 5s^1$	Cd 48 112.411 $4d^{10} 5s^2$	In 49 114.818 $5p^1$	Sn 50 118.710 $5p^2$	Sb 51 121.757 $5p^3$	Te 52 127.60 $5p^4$	I 53 126.90447 $5p^5$	Xe 54 131.29 $5p^6$
Cs 55 132.90543 $6s^1$	Ba 56 137.327 $6s^2$	57–71 †	Hf 72 178.49 $5d^2 6s^2$	Ta 73 180.9479 $5d^3 6s^2$	W 74 183.84 $5d^4 6s^2$	Re 75 186.207 $5d^5 6s^2$	Os 76 190.23 $5d^6 6s^2$	Ir 77 192.22 $5d^7 6s^2$	Pt 78 195.08 $5d^9 6s^1$	Au 79 196.96654 $5d^{10} 6s^1$	Hg 80 200.59 $5d^{10} 6s^2$	Tl 81 204.3833 $6p^1$	Pb 82 207.2 $6p^2$	Bi 83 208.98037 $6p^3$	Po 84 (208.9824) $6p^4$	At 85 (209.9871) $6p^5$	Rn 86 (222.0176) $6p^6$
Fr 87 (223.0197) $7s^1$	Ra 88 (226.0254) $7s^2$	89–103 ‡	Rf* 104 (261.10869) $6d^2 7s^2$	Db* 105 (262.11376) $6d^3 7s^2$	Sg* 106 (263.1182) $6d^4 7s^2$	Bh* 107 (262.1231) $6d^5 7s^2$	Hs* 108 (264.1285) $6d^6 7s^2$	Mt* 109 (266.1378)	110 (269)	111 (272)	112 (277)						

† Lanthanide Series

La 57	Ce 58	Pr 59	Nd 60	Pm 61	Sm 62	Eu 63	Gd 64	Tb 65	Dy 66	Ho 67	Er 68	Tm 69	Yb 70	Lu 71
138.9055 $5d^1 6s^2$	140.115 $4f^1 5d^1 6s^2$	140.90765 $4f^3 5d^0 6s^2$	144.24 $4f^4 5d^0 6s^2$	(144.9127) $4f^5 5d^0 6s^2$	150.36 $4f^6 5d^0 6s^2$	151.965 $4f^7 5d^0 6s^2$	157.25 $4f^7 5d^1 6s^2$	158.92534 $4f^9 5d^0 6s^2$	162.50 $4f^{10} 5d^0 6s^2$	164.93032 $4f^{11} 5d^0 6s^2$	167.26 $4f^{12} 5d^0 6s^2$	168.93421 $4f^{13} 5d^0 6s^2$	173.04 $4f^{14} 5d^0 6s^2$	174.967 $4f^{14} 5d^1 6s^2$

‡ Actinide Series

Ac 89	Th 90	Pa 91	U 92	Np 93	Pu 94	Am 95	Cm 96	Bk 97	Cf 98	Es 99	Fm 100	Md 101	No 102	Lr 103
(227.0278) $6d^1 7s^2$	232.0381 $6d^2 7s^2$	231.03588 $5f^2 6d^1 7s^2$	238.0289 $5f^3 6d^1 7s^2$	(237.0482) $5f^4 6d^1 7s^2$	(244.0642) $5f^6 6d^0 7s^2$	(243.0614) $5f^7 6d^0 7s^2$	(247.07035) $5f^7 6d^1 7s^2$	(247.07030) $5f^9 6d^0 7s^2$	(251.0796) $5f^{10} 6d^0 7s^2$	(252.0803) $5f^{11} 6d^0 7s^2$	(257.0951) $5f^{12} 6d^0 7s^2$	(258.0984) $5f^{13} 6d^0 7s^2$	(259.10011) $5f^{14} 6d^0 7s^2$	(262.1098) $5f^{14} 6d^1 7s^2$

§ Atomic mass values averaged over isotopes in percentages they occur on Earth's surface. For many unstable elements, mass of the most stable known isotope is given in parentheses. 1993 revisions.

* Recommended symbols and names (see Appendix F) as of 1997.